Environmental Science
Toward a Sustainable Future

ELEVENTH EDITION

Richard T. Wright
Dorothy F. Boorse
Gordon College

Benjamin Cummings

Boston Columbus Indianapolis New York San Francisco Upper Saddle River
Amsterdam Cape Town Dubai London Madrid Milan Munich Paris Montréal Toronto
Delhi Mexico City São Paulo Sydney Hong Kong Seoul Singapore Taipei Tokyo

Executive Editor: Chalon Bridges
Project Editor/Development Editor: Nora Lally-Graves
Executive Director of Development: Deborah Gale
Editor-in-Chief: Beth Wilbur
Vice President and Editorial Director: Frank Ruggirello
Editorial Assistant: Joshua Taylor
Assistant Media Producer: Lee Ann Doctor
Senior Media Producer: Deb Greco
Director of Media Strategy: Stacy Treco
Vice President and Director of Media Development: Lauren Fogel
Executive Marketing Manager: Lauren Harp
Managing Editor, Production: Gina M. Cheselka
Project Manager, Production: Ed Thomas

Art Project Manager: Connie Long
Senior Media Production Supervisor: Liz Winer
Art Director: Maureen Eide
Interior Design: Kenny Beck
Cover Design: Riezebos Holzbaur
Photo Research Manager: Elaine Soares
Photo Researcher: Clare Maxwell
Senior Operations Supervisor: Nick Sklitsis
Operations Supervisor: Maura Zaldivar
Full Service/Composition: GGS Higher Education Resources, PMG
Production Editor, Full Service: Suganya Karuppasamy
Illustrations: Precision Graphics

Cover Photograph: Gardener working on an organic herb and vegetable garden on a high-rise rooftop in Bangkok, Thailand. (Photo © Barbar Walton/epa/Corbis)

Photo credits continue following the appendices.

Library of Congress Cataloging-in-Publication Data

Wright, Richard T.
  Environmental science : toward a sustainable future / Richard T. Wright, Dorothy F. Boorse. — 11th ed.
    p. cm.
  Includes bibliographical references and index.
  ISBN-13: 978-0-321-59870-7
  ISBN-10: 0-321-59870-9
  1. Environmental sciences—Textbooks.   I. Boorse, Dorothy.   II. Title.
GE105.N42   2011
628—dc22

                              2009041839

ISBN 0-321-59870-9 / 978-0-321-59870-7 [Student Edition]
ISBN 0-321-62370-3 / 978-0-321-62370-6 [Books à là Carte]

Printed in the United States
10 9 8 7 6 5 4 3 2 1

Benjamin Cummings
is an imprint of

www.pearsonhighered.com

Register Now to Improve Your Grade!

Environmental Science

ABOUT THE AUTHORS

Richard T. Wright is Professor Emeritus of Biology at Gordon College in Massachusetts, where he taught environmental science for 28 years. He earned a B.A. from Rutgers University and a M.A. and Ph.D. in biology from Harvard University. For many years Wright received grant support from the National Science Foundation for his work in marine microbiology, and, in 1981, he was a founding faculty member of Au Sable Institute of Environmental Studies in Michigan, where he also served as Academic Chairman for 11 years. He is a Fellow of the American Association for the Advancement of Science, Au Sable Institute, and the American Scientific Affiliation. In 1996, Wright was appointed a Fulbright Scholar to Daystar University in Kenya, where he taught for two months. He is a member of many environmental organizations, including The Nature Conservancy, Habitat for Humanity, the Union of Concerned Scientists, and the Audubon Society, and is a supporting member of the Trustees of Reservations. He volunteers his services at the Parker River National Wildlife Refuge in Newbury, MA, is a board member for Kupenda for the Children, and is an elder in First Presbyterian Church of the North Shore. Wright and his wife Ann live in Byfield, Massachusetts, and they drive a Toyota Camry hybrid vehicle as a means of reducing their environmental impact. Wright spends his spare time birding, fishing, hiking, and enjoying his three children and seven grandchildren.

Dorothy F. Boorse is an associate professor of biology at Gordon College, in Wenham Massachusetts. Her research interest is in drying wetlands such as vernal pools and prairie pot-holes, and in salt marshes. Her research with undergraduates has included wetland and invasive species projects. She earned a BS in biology from Gordon College, a MS in entomology from Cornell University, and a PhD in oceanography and limnology from the University of Wisconsin-Madison. Boorse teaches, writes, and speaks about biology, the environment, and care of creation. An essay she wrote on climate change was recently published in a book by the Union of Concerned Scientists and Penguin Press. In 2005 Boorse provided expert testimony on wildlife corridors and environmental ethics for a congressional House subcommittee hearing.

Boorse is a member of a number of ecological and environmental societies including The Ecological Society of America, the Society of Wetland Scientists, the Nature Conservancy, The Audubon Society, The New England Wildflower Society, and the Trustees of Reservations (the oldest land conservancy group in the United States). She and her family live in Beverly Massachusetts. She and her family belong to Appleton farms, a CSA community supported agriculture farm. They are also members of a newly established community supported fishery, Cape Ann Fresh Catch. At home Boorse has a native plant garden and has recently planted two disease resistant elm trees.

ABOUT OUR SUSTAINABILITY INITIATIVES

This book is carefully crafted to minimize environmental impact. The materials used to manufacture this book originated from sources committed to responsible forestry practices. The paper is FSC certified. The binding, cover, and paper come from facilities that minimize waste, energy consumption, and the use of harmful chemicals.

Pearson closes the loop by recycling every out-of-date text returned to our warehouse. We pulp the books, and the pulp is used to produce items such as paper coffee cups and shopping bags. In addition, Pearson aims to become the first climate neutral educational publishing company.

The future holds great promise for reducing our impact on Earth's environment, and Pearson is proud to be leading the way. We strive to publish the best books with the most up-to-date and accurate content, and to do so in ways that minimize our impact on Earth.

Mixed Sources
Product group from well-managed forests, controlled sources and recycled wood or fibre
www.fsc.org   Cert no. SW-COC-002985
© 1996 Forest Stewardship Council

Benjamin Cummings
is an imprint of

CONTENTS

PART FOUR

Harnessing Energy
for Human Societies   347

ESSAYS

Sound Science

Sustainability

Stewardship

PREFACE

If there is one word that can be summoned to characterize the first decade of the 21st century, it is the word *change*. Globally, change is happening to the atmosphere and climate, the human population and its well-being, and the Earth's natural resources. We have a major economic recession, pointing to the reality of globalization; public health alerts as avian and swine influenzas spread everywhere; accelerating climate change, brought on by increased fossil fuel burning; the emergence of India and China as major global players in trade and resource use; terrorism and conflict that are gripping the Middle East and spreading; and crises in food and agricultural production, leading to greater numbers of hungry and impoverished people.

In contrast to these and other disturbing changes, there are some changes that point to a brighter future. Renewable energy is ramping up swiftly in its share of the world's energy portfolio; many of the UN Millennium Development Goals (MDGs) are on track to be achieved by their target date of 2015; major climate agreements are on the table that will bring real reductions in greenhouse gas emissions; death by tobacco use is being addressed in a global campaign; AIDS, tuberculosis, and malaria are on the defensive as public-health agencies expand treatment options and research; and population growth in the developing countries is continuing to decline.

The most profound change that must happen, and soon, is the transition to a sustainable civilization—one in which a stable human population recognizes the finite limits of Earth's systems to produce resources and absorb wastes and acts accordingly. This is hard to picture at present, but it is the only future that makes any sense. If we fail to achieve it by our deliberate actions, the natural world will impose it on us in highly undesirable ways.

New Coauthor

Like the global scene, the field of environmental science evolves and continues to change, and so has this text. The most significant change has been the addition of a coauthor, Dr. Dorothy Boorse. For me (Richard Wright), this has been an especially gratifying development, as Dr. Boorse was one of my students at Gordon College and is now back at the college as Associate Professor of Biology, teaching courses I once taught. You can check out her qualifications in About the Authors, but you will see her work in the five chapters she has contributed to this edition: Chapters 3, 4, 5, 6, and 8. I wish you could meet her personally because Dr. Boorse is a remarkably gifted woman and teacher. It has been a delight to work together on this edition.

Core Values

Environmental science stands at the interface between humans and Earth and explores the interactions and relations between them. This relationship will need to be considered in virtually all future decision making. This text considers a full spectrum of views and information in an effort to establish a solid base of understanding and a sustainable formula for the future. What you have in your hands is a readable guide and up-to-date source of information that will help you to explore the issues in more depth. It will also help you to connect them to a framework of ideas and values that will equip you to become part of the solution to many of the environmental problems confronting us.

In this new edition, we hope to continue to reflect accurately the field of environmental science; in so doing, we have constantly attempted to accomplish each of the following objectives:

- To write in a style that makes learning about environmental science both interesting to read and easy to understand, without overwhelming students with details;
- To present well-established scientific principles and concepts that form the knowledge base for an understanding of our interactions with the natural environment;
- To organize the text in a way that promotes sequential learning, yet allows individual chapters to stand on their own;
- To address all of the major environmental issues that confront our society and help to define the subject matter of environmental science;
- To present the latest information available by making full use of resources such as the Internet, books, and journals; every possible statistic has been brought up to date.
- To assess options or progress made in solving environmental problems; and
- To support the text with excellent supplements for instructors and students that strongly enhance the teaching and learning processes.

Because we believe that learning how to live in the environment is one of the most important subjects in every student's educational experience, we have made every effort to put in their hands a book that will help the study of environmental science come alive.

New to This Edition

To help implement our core values, we have made several changes to the eleventh edition. The major changes are listed here, followed by a list of updates for each part and chapter. All of these changes were made in the hope of creating a book with a solid, rigorous framework that is also easily comprehensible.

- The table of contents has been reorganized. The policy chapter has been moved up (from Chapter 22 to Chapter 2) to provide this information early in the text, and the ecology content has been reorganized so topics flow from species to ecosystems to humans (Unit Two, Chapters 3-7).

- We have both simplified and strengthened the thematic approach to environmental science that has been a hallmark of this text for several editions. We now focus on only three themes to help readers focus on the significance of the issues and conceptualize the task of forging a sustainable future:

  Sound Science—the basis for our understanding of how the world works and how human systems interact with it;

  Sustainability—the practical goal that our interactions with the natural world should be working toward; and

  Stewardship—the actions and programs that manage natural resources and human well-being for the common good.

- The design, the graphics, and the photos have been substantially updated. Over 50% of the photos are new to this edition.

- Over 64% of the research cited is from 2008 or later.

- The sections titled **Making a Difference,** offering suggestions to students for what they can do to help, have been expanded and are now found in each chapter rather than at the end of each part.

- In this new edition, we are using Forest Stewardship Council certified paper, made from trees managed sustainably, and we are switching from being a hardcover book to being a paperback.

- New Video Field Trips give students a tour of coal-fired power plants, waste water treatment facilities, farms, recycling facilities and more.

- New MasteringEnvironmentalScience online assessment and tutorial system helps instructors maximize class time and students to learn outside of class and arrive prepared for lecture.

Content Updates to the Eleventh Edition of *Environmental Science*

Part One—Framework for a Sustainable Future

Chapter 1 (Science and the Environment) contains a new part opener and new material on environmental history and the scientific method. A new section introduces and distinguishes between environmental science and environmentalism. The chapter then introduces the three unifying themes for this edition: sound science, sustainability, and stewardship. Each theme is thoroughly defined and explored; the science theme in particular is expanded in this edition. A new essay on the decline of vultures in south Asia illustrates the scientific method at work, as researchers traced the anti-inflammatory drug diclofenac from its use on domestic animals to the vultures. Experimental tests showed how the diclofenac caused kidney failure in vultures that consumed the dead domestic animals. Chapter 2 (Economics, Politics, and Public Policy) is moved here from the back of the book in order to highlight the crucial environmental roles played by these elements of human societies. The chapter opens with a new essay on China's remarkable economic growth and its environmental consequences. The chapter has also been reorganized, to make a more clear distinction between economics and public policy. A new section compares GDP with the Genuine Progress Indicator. A new case study illustrates the Policy Life Cycle: the discovery of ozone depletion in the stratosphere by Sherwood Rowland and Mario Molina, a Nobel Prize-winning research team. The chapter ends with a look at politics, culminating with a section on changes brought about by the Obama administration.

Part Two—Ecology: The Science of Organisms and Their Environment

Chapters 3–5 have been thoroughly reorganized, with topics flowing from basic to more complex, small to large, and species to ecosystems and humans. Chapter 3 (Ecology: Basic Needs of Living Things) opens with the plight of the giant panda, and moves into a description of ecology, environmental factors, matter and energy laws, and the cycling of matter in ecosystems. A new essay explores the controversial proposal to seed the oceans with nutrients in order to stimulate phytoplankton growth that would pull CO_2 from the atmosphere. Chapter 4 (Populations and Communities) has a new opening story on the golden frog of Panama and substantial new material on the mathematics of population growth. There are also new sections on r and K strategies, survivorship curves, and territoriality, and a new essay on spotted knapweed. Chapter 5 (Ecosystems: Energy, Patterns and Disturbance) moves right into biomes and aquatic ecosystems, which are now illustrated with photos and simple tables. Energy flow is then explored, and the section on restoration ecology is rewritten. Chapter 6 (Wild Species and Biodiversity) is reorganized to combine wild species and biodiversity issues together. The section on Valuing now includes Aldo Leopold and the land ethic. There are several updates to include the latest legislation and statistics. Chapter 7 (The Use and Restoration of Ecosystems) has a new essay, The Turtle Island Commons, which presents a homegrown stewardship conservation effort that was trumped by a government edict. The Everglades restoration effort is brought up to date—a needed restoration that has been stifled by politics and the recession.

Part Three—The Human Population and Essential Resources

Chapter 8 (The Human Population) has a new opener essay that looks at three college-aged people representing three very different groups, and then moves on to a new section on ways in which human populations are like and unlike other organisms, using ecological concepts to support the discussion. Dynamics of the human population are brought up to date with the 2008 World Population Data Sheet from the Population Reference Bureau. Chapter 9 (Population and Development) highlights the Millennium Development Goals and their accompanying targets for the year 2015, together with MDG Support, the new action plan for achieving the targets. The social modernization section contains a discussion of the discrepancy in mortality data between developed and developing countries, with a special focus on the role of AIDS and its prevention. Chapter 10 (Water: Hydrologic Cycle and Human Use) has an updated chapter opener, a new major heading on water use in food production, updates for the 2008 Iowa floods, and new material on wetlands destruction. Chapter 11 (Soil: Foundation for Land Ecosystems) has new content on Land Degradation Assessment in Drylands (LADA) and the latest farm bill in the U.S, as well as new stewardship content on the Keita Project. Confined animal feeding operations (CAFOs), the controversy over biofuels and food production, and the global food crisis are newly introduced in Chapter 12 (The Production and Distribution of Food). Chapter 13 (Pests and Pest Control) discusses new information on glyphosate-resistant weeds, the cancelled pesticide carbofuran, and the Asian Longhorned beetle invasion. It should be mentioned that for all of these chapters, photos, graphics and statistical data have been brought up-to-date.

Part Four—Harnessing Energy for Human Societies

Chapter 14 (Energy from Fossil Fuels) has a new part opener on oil and coal, and a new chapter opener on the issue of mountaintop-removal coal mining. New perspectives on energy are provided by a new section on power blackouts and a strong emphasis on "peak oil." Information on the 2007 Energy Independence and Security Act is added in the policy analysis of supply-side and demand-side policies. Extensive data and content updates were made due to new reports from the Department of Energy.

Chapter 15 (Nuclear Power) opens with a new story on the July 2007 earthquake off Japan that shut down a large nuclear power plant. A new section, Making a Bomb, explores the possibility of terrorist theft of nuclear materials. The chapter includes updated content about a renewed interest in nuclear power, and updated data on nuclear plants around the world. New legislation and administrative decisions are included (such as Yucca Mountain). Chapter 16 (Renewable Energy) begins with a new opener on the story of Cape Wind, a proposed and controversial offshore wind farm. New MIT water-splitting research and fuel-cell development is included, as well as a discussion of Miscanthus grass as a second-generation biofuel. Overall energy policy is updated at the chapter's end.

Part Five—Pollution and Prevention

Chapter 17 (Environmental Hazards and Human Health) opens with an essay examining the threats of the new swine flu (H1N1) and avian flu, two of many emerging diseases. All disease statistics are brought up to date, including new outbreaks, and more data is presented on natural hazards and the "Hurricane Coast." The Sound Science essay has been updated with MSRA and West Nile data, and a new vaccine for malaria is discussed. The section on Tobacco Use is expanded with the World Health Organization's report on smoking and the global MPOWER campaign. There is a thoroughly rewritten presentation of Global Climate Change (Chapter 18), featuring the work of the Intergovernmental Panel on Climate Change's 2007 physical science assessment and the studies of the U.S. Global Change Research Program. A new Sound Science essay explores recent corrections of ocean temperature figures and the importance of checking data with independent measurements. Chapter 19 (Atmospheric Pollution) now contains the ozone layer depletion material formerly coupled with global climate change, and the new CAFE standards for vehicle mileage. New content on atmospheric brown clouds has been added. Chapter 20 (Water Pollution and Its Prevention) has new content on the history of U.S. water pollution and the latest water-saving techniques (such as composting toilets). There is a new graphic on the oxygen sag curve that results from organic pollution of streams. A Sound Science essay on harmful algal blooms includes the impact of the algal toxin domoic acid. Chapter 21 (Municipal Solid Waste) has a new essay on the extent of e-waste and how it is being (mis)handled, as well as new content on bioplastics and plastic bag pollution. Chapter 22 (Hazardous Chemicals: Pollution and Prevention) has updated content on superfund and underground tank storage cleanup, and a discussion of several synthetic organic chemicals (like bisphenol A) that have been in the news. A new discussion of the administration of the Toxic Substances Control Act is included.

Part Six—Stewardship for a Sustainable Future

Chapter 23 (Sustainable Communities and Lifestyles) contains new material on mass transit ridership and the direction of stimulus funds towards transit systems, the UN's Localizing Agenda 21, sustainability rankings by SustainLane, and recent changes where U.S. governments and industries put resources into sustainability.

Reviewers

Clark Adams, *Texas A & M University*

Anthony Akubue, *St. Cloud State University*

Mary Allen, *Hartwick College*

Walter Arenstein, *San Jose State University*

Abbed Babaei, *Cleveland State University*

Kenneth Banks, *University of North Texas*

Raymond Beiersdorfer, *Youngstown State University*

Lisa Bonneau, *Metropolitan Community College*

Geralyn Caplan, *Owensboro Community and Technical College*

Kelly Cartwright, *College of Lake County*

Jason Cashmore, *College of Lake County*

Anne Ehrlich, *Stanford University*

Cory Etchberger, *Johnson County Community College*

Marcia Gillette, *Indiana University—Kokomo*

Sue Habeck, *Tacoma Community College*

Crista Haney, *Mississippi State University*

Les Kanat, *Johnson State College*

Barry King, *College of Santa Fe*

Cindy Klevickis, *James Madison University*

Steven Kolmes, *University of Portland*

Peter Kyem, *Central Connecticut State University*

David Lawrence, *Brown Cal State*

Marcie Lehman, *Shippensburg University*

Kurt Leuschner, *College of the Desert*

Gary Li, *Cal State—East Bay Campus*

Kenneth Mantai, *SUNY Fredonia*

Heidi Marcum, *Baylor University*

Allan Matthias, *University of Arizona*

Dan McNally, *Bryant University*

Kiran Misra, *Edinboro University of Pennsylvania*

Robert Andrew Nichols, *City College—Gainesville Campus*

Tim Nuttle, *Indiana University of Pennsylvania*

Bruce Olszewski, *San Jose State University*

John Pleasants, *Iowa State University*

John Reuter, *Portland State University*

Kim Schulte, *Georgia Perimeter College*

Brian Shmaefsky, *Lone Star College—Kingwood*

Annelle Soponis, *Reading Area Community College*

Shamili Stanford, *College of DuPage*

Keith Summerville, *Drake University*

Todd Tarrant, *Michigan State University*

Dave Wartell, *Harrisburg Area Community College*

John Weishampel, *University of Central Florida*

Arlene Westhoven, *Ferris State University*

Danielle Wirth, *Des Moines Area Community College*

Todd Yetter, *University of the Cumberlands*

ACKNOWLEDGMENTS

Although the content and accuracy of this text are the responsibility of the authors, it would never have seen the light of day without the dedicated work of many other people. We want to express our heartfelt thanks to all those at Pearson who have contributed to the book in so many ways.

We salute our editor, Chalon Bridges, for her encouragement and persistent prodding in helping us reorganize and write this new edition. Our developmental editor, Nora Lally-Graves, worked closely with us on every aspect of the book; thank you, Nora, for your good contributions. Sherry Goldbecker took care of the copy editing and made many helpful suggestions. Suganya Karuppasamy was our production editor, keeping us focused on the details of transforming manuscripts into a published product—and doing it so cheerfully. We are very grateful to Gina Cheselka for her management and guidance, to Ed Thomas for leading our production team, and to Lee Ann Doctor for her work on the media. Thanks also go to Clare Maxwell, photo researcher, who scoured the Web and other media sources for the pictures we were looking for. Benjamin Wright contributed the end-of-chapter features Making a Difference and Thinking Environmentally, and we thank him for his skillful handling of them. In addition, we thank Ed Zalisko for doing a fine job with the test bank and for his work on the Mastering quizzes. We would like to thank Lauren Harp for her help in marketing this text. Finally, we would like to thank Deborah Gale, Beth Wilbur, and Frank Ruggirello for their leadership.

Almost 30 years ago Bernard Nebel published the first edition of this text. He did it because he was frustrated with existing environmental science texts and was convinced he could produce a more readable and effective book—and he did! By the fourth edition, Richard Wright joined Nebel and collaborated for two more editions. From the seventh to tenth editions, Wright was responsible for the text, and now with this eleventh edition, there are once again two authors. This is a good development, one that is already adding richness and a new voice to many of the chapters. We are deeply indebted to Nebel for his diligent work in developing the text and producing successive editions. We offer this new edition of the book as our contribution to the students who are now well into this new century, in the hope that they will join us in helping to bring about the environmental revolution—the transition to a sustainable society—that must come, hopefully sooner rather than later.

I (Richard Wright) wish to offer some very personal thanks to my wife, Ann, who has been my companion since the beginning of my work in biology and has provided the emotional base and care without which I would be far less of a person and a biologist. Her love and patience have sustained me in immeasurable ways.

I (Dorothy Boorse) would like to thank my husband, Gary, my biggest supporter. I am also very thankful to my mentors—particularly Richard Wright, who has known me for years and helped me inestimably, and Calvin DeWitt of the University of Wisconsin–Madison, who has been one of the foremost figures in motivating young people to care about the environment and ethics.

Finally, it is our hope that this book can inspire a new generation to work toward bringing healing to a Creation suffering from human misuse.

Richard T. Wright
Dorothy F. Boorse

DEDICATION

We take pleasure in dedicating this edition of *Environmental Science* to Cynthia Bauer, founder and director of Kupenda for the Children (www.kupenda.org). Cynthia graduated from the University of Maine, majoring in Wildlife Management/Conservation Biology. She then earned a master's degree in Applied Ecology at Eastern Kentucky University, with thesis work on the coast of Kenya studying the impact of illegal logging on the golden-rumped elephant shrew. While in Kenya, Cynthia, who was born without a left hand, saw many children with disabilities who were in need of education and special equipment. On her return to the U.S. she raised funds to support school fees for 26 handicapped children to attend a special school and receive hearing aids and wheelchairs. This led to further visits to Kenya and more involvement from friends and contacts in the developed countries, and the birth of Kupenda for the Children (Kupenda is the Kiswahili word for "to love"). Now, five years later, Kupenda employs six Kenyan staff members (the only paid employees) who work with special needs children in the Malindi district of Kenya's coast. Kupenda enlists sponsors who pay for school needs and medical prostheses for

some 500 children, and has a budget in excess of $140,000. The need is great (there is never enough money to meet the number of handicapped children), and Cynthia Bauer still works as an unpaid volunteer, supporting herself with environmental consulting and teaching. We salute you, Cynthia, and may Kupenda grow and flourish as you help handicapped children in Kenya.

Mastering ENVIRONMENTAL SCIENCE ™

The Mastering system is an online assessment and tutorial system designed to help instructors teach more efficiently, and pedagogically proven to help students learn. It helps instructors maximize class time with customizable, easy-to-assign, and automatically graded assessments that motivate students to learn outside of class and arrive prepared for lecture. The powerful gradebook provides unique insight into student and class performance even before the first test. As a result, instructors can spend class time where students need it most. The Mastering system empowers students to take charge of their learning through activities aimed at different learning styles, and engages them in learning science through practice and step-by-step guidance—at their convenience, 24/7.

Student Resources

1. Get personalized coaching and feedback

Automatically-graded assignments offer a wide variety of activities aimed at different learning styles:

- Pre-lecture Quizzes help students keep on track with reading assignments and check their understanding of the content.
- Current Events quizzes prompt students to read recent *New York Times* articles.
- Video Field Trips take students to coal-fired power plants, waste water treatment facilities, community-supported agriculture farms, recycling facilities and more. Each Video Field Trip includes assessment questions to assign as homework.

- BioFlix activities are highly interactive and use 3D animations. Topics include photosynthesis, cellular respiration, the carbon cycle, mechanisms of evolution, and population ecology.

- GraphIt activities help students better understand how to work with and interpret graphs and real data.

2. Pearson eText

The Pearson eText gives students access to the text whenever and wherever they can access the Internet. The eText pages look exactly like those in the printed text, plus they offer powerful interactive and customization functions.

ANNOTATION FUNCTION allows students to take notes.

HIGHLIGHT FUNCTION lets students highlight what they want to remember.

GOOGLE-BASED search function.

HYPERLINKS to quizzes, tests, activities, and animations.

ZOOM in and out for better viewing.

INTERACTIVE GLOSSARY provides pop-up definitions and terms.

INSTRUCTORS CAN SHARE their notes and highlights with students and can also hide chapters.

3. Self-Study Area

The MasteringEnvironmentalScience **Study Area** offers a self-study resource that students can access 24/7.

Mastering**ENVIRONMENTALSCIENCE**™

Instructor Resources

1. Motivate students to come to class prepared

PRE-LECTURE QUIZZES
Assign reading quizzes before lecture to help students arrive prepared.

COLOR BAR DATA
Green Indicates correct answers, Red shows the percentage of students who requested an answer. Orange Indicates the average number of wrong answers per student.

CUSTOMIZE CONTENT
Questions and answers can be easily edited.

AT-A-GLANCE STATISTICS
Reveal data for your class as well as national results.

WRONG ASNWERS SUMMARY
Gives unique insight into your student's misunderstandings and allows for just-in-time teaching adjustments.

Multiple Choice 1

Part A

If a toxin bioaccumulates in a terrestrial ecosystem, which of the following will most likely have the highest concentration of the toxin in its cells?

ANSWER:
- producer
- first-level consumer
- second-level consumer
- ● third-level consumer

Answer Stats:	Students	% Correct	% Unfinished	% Req'd Solution	Wrong/student	Hints/student
Overall	157	93%	5.1%	1.9%	0.3	0
Envi Sci 101	94	80%	0%	20%	2.8	0

Wrong Answers for ENVISCIGRADES

% Wrong	Answer	Response
15.2%	producer	
11.0%	first-level consumer	
7.4%	second-level consumer	

2. Gain insight into students' misunderstandings with at-a-glance data

Gradebook Display Mode: Spreadsheet

Gradebook for Course: Course ENVISCIGRADES

Filter Showing score on All Activities for All Students

Score Time Difficulty Students Per Page: 25 D

Name	In..gy	Ch..02	Ch..03	Ch..04	Ch..05	Ch..06	Ch..07	Ch..08	Ch..09	Ch..10	Ch..11	Ch.	Total
Class Average	--	66.6	68.7	67.5	69.1	68.1	66.2	47.9	84.0	71.7	70.6		0.0
Last02, First02	--	74.6	70.0	74.2	75.0	83.0	79.1	90.5	166	90.0	89.0		0.0
Last03, First03	--	22.9	62.0	50.4	45.3	52.5	48.9	28.7	40.9	56.0	30.0		0.0
Last04, First04	--	59.5	85.0	82.9	90.0	89.0	80.0	64.2	144	85.0	81.0		0.0
Last06, First06	--	66.0	75.0	67.0	88.2	57.5	77.0	28.8	50.0	70.0	72.5		0.0
Last07, First07	--	72.0	77.5	72.9	77.2	62.5	71.9	28.7	50.0	70.0	82.5		0.0
Last11, First11	--	91.6	72.5	72.9	65.9	72.5	62.5	69.1	90.9	82.5	80.0		0.0
Last13, First13	--	74.2	70.0	78.0	77.1	85.0	73.6	62.1	131	76.0	80.0		0.0
Last15, First15	--	72.2	37.5	41.6	34.6	42.5	36.4	10.7	0.0	44.4	50.0		0.0

Every assignment is **automatically graded.**
At a glance, **shades of red highlights** a student in trouble or a difficult assignment for your class as a whole.

3. Additional tools for lectures

Instructor Resource DVD with TestGen, Instructor's Manual, and course management system cartridges including Blackboard and WebCT. Go to www.pearsonhighered.com for more information on any of these resources and a complete list of all supplements.

Food riot in Haiti. Protesters angered by high food prices flooded the streets of Port-au-Prince, forcing businesses and schools to close.

PART ONE

Framework for a Sustainable Future

CHAPTER 1
Science and the Environment

CHAPTER 2
Economics, Politics, and Public Policy

THE YEAR 2008 brought two markets into sharp focus: food and energy. The demand for oil pushed the price for a barrel of crude oil to record highs as the emerging economies of India and China competed with the developed countries for what never seemed to be enough oil. In turn, countries everywhere reached out for fuel substitutes, and corn and other plant oils were drawn into refineries to generate alcohol and biodiesel fuel. High oil prices and the drive for biofuel production combined to bring on an unprecedented rise in global food prices: Wheat, rice, and corn prices doubled or tripled within a year. This surge in food prices created havoc in the food-importing countries, especially those that were economically on the edge. Even in the United States, considered to be well off, the food banks were brought to new lows, and food stamp use surged. Food riots occurred in numerous countries as people reacted to escalating prices for their basic food needs. The high food prices were believed to be responsible for pushing an additional 100 million people more deeply into poverty and hunger.

EVEN THOUGH THE 2009 global recession brought some relief to food and oil prices, these markets will rise again when economic growth resumes. Such trends are symptomatic of a world system that is careening toward greater chaos and social instability. The drive for more and more energy has other consequences: global climate change that is leading to a warmer earth and rising seas, wars and conflicts over oil that consume billions of dollars, and increased pressure to explore for oil and gas in fragile ecosystems. The food and energy clash is symptomatic of the unsustainable direction of our societies. Other symptoms will be discussed in Chapter 1, as we begin the text with a look at what happened on Easter Island in the Pacific Ocean. We will end on a more positive note, however, because there are encouraging trends, too, and we will consider these as we close the chapter.

CHAPTER 1

Science and the Environment

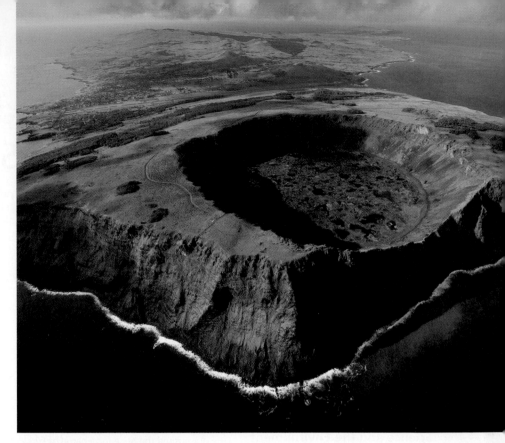

Aerial view of Easter Island.

O N EASTER SUNDAY, 1722, three Dutch ships commanded by Admiral Jacob Roggeveen sighted a previously uncharted island in the South Pacific. They named it Easter Island. The island, some 166 km^2 (65 mi^2), is perhaps the most remote spot on the planet—1,200 miles from the nearest island and 2,300 miles from the coast of South America. The Dutch found the island to be inhabited by Polynesians who were living primitively. Roggeveen wrote, "We originally, from a further distance, have considered the said Easter Island as sandy; the reason for that is this, that we counted as sand the withered grass, hay, or other scorched and burnt vegetation, because its wasted appearance could give no other impression than of a singular poverty and barrenness." The sailors were amazed to find many large stone statues scattered about the island (Fig. 1–1), evidence of a once sophisticated civilization. The past culture and civilization of the island had vanished.

THE PAST. Working from the legends that islanders told, conducting excavations for evidence, and using radiocarbon dating methods, archaeologists have pieced together the following possible chronology of events: The original inhabitants of Easter Island were Polynesians who arrived on the island as part of a deliberate colonization mission sometime around 1200 A.D. They had landed on an environmentally fragile island: small, isolated, dry, relatively cold, nutrient poor. The evidence from pollen grains and plant remains found in corings from the bottom of volcanic lakes shows that these early arrivals encountered an island abundantly forested with a wide

Figure 1–1 Easter Island. The great stone heads and other artifacts found on Easter Island show that a prosperous culture once existed there. The present barren, eroded landscape indicates that the civilization collapsed from the overexploitation of forest and soil resources. Is the story of Easter Island a parable for modern civilization?

variety of trees, including palms, conifers, and sandalwood. As their population grew and flourished, they cut trees to make room for agriculture, to provide structural materials, and to move the huge stone heads from the quarries where they were shaped to the sites at which they would be erected. By 1650, all the trees were gone. Without a protecting tree canopy, the cleared land failed to hold water, and the soil washed into the sea. The eroded soil baked hard and dry after rains, offering poor support for agriculture.

As the forest was depleted and the soil and water resources were degraded, the work necessary for existence became harder and the rewards fewer. The gap between the ruling religious elites and secular warrior-workers widened, apparently becoming intolerable. Sometime during the 17th century, there was a revolt of the workers. In the great war that ensued, virtually the entire ruling religious class was killed. Still, the situation worsened. Anarchy broke out among the workers, who splintered into groups and continued to fight among themselves. Starvation and disease became epidemic. Without any trees, no one could escape the island by boat. A population that had once numbered 10 to 20 thousand was down to a few thousand at the time of its "discovery" by the Dutch. Many have puzzled over the reasons why the residents of Easter Island, who could walk around their island in a day, failed to foresee the consequences of their practices.

The residents of Easter Island, who call themselves the Rapa Nui, suffered terribly from their subsequent contacts with the "civilized" world. In the 19th century, they were visited repeatedly by whalers who were seeking food and water—and who also infected the Rapa Nui with venereal diseases. Then, in midcentury, Peruvian slavers raided the island and captured over 2,000 Rapa Nui for the South American slave trade. Eventually, smallpox came to the island, and by 1877, only 111 Rapa Nui remained. Easter Island was then annexed by Chile, which fostered sheep farming, while enclosing all of the Rapa Nui in one village. Visits by archaeologists in the mid–20th century brought the island to the world's attention, and things began to improve for the Rapa Nui. Today many of the stone statues have been restored, the Rapa Nui have regained a measure of control over their own destiny, and a significant tourist trade has brought hotels, inns, restaurants, and curio shops to Hanga Roa, the only town on the island (Fig. 1–2). The Rapa Nui now depend on imported food, however, and unemployment and alcoholism have become serious problems.

LESSONS. This is a sad story. It is not over, however, and the Rapa Nui are working to preserve Easter Island's cultural heritage and restore the soil to allow the cultivation of some agricultural specialties that could diversify the island's economy. The story also carries some powerful lessons, as we look back at what happened. When a society fails to care for the environment that sustains it, when its population increases beyond the capacity of the land and water to provide adequate food for all, and when the disparity between haves and have-nots widens into a gulf of

Figure 1–2 Hanga Roa. The tiny harbor in Hanga Roa, where nearly all of Easter Island's 3,000 inhabitants live. It is the only area with running water and electricity.

social injustice, the civilization collapses. History is replete with the ruins of other civilizations, such as the Mayans, Greeks, Incas, and Romans, that failed to recognize the constraints of their environment.

The more developed world had a devastating impact on Easter Island, too. Foreign contact brought disease, slavery, and subjugation that lasted for two centuries. More recently, outsiders have brought economic and social help to the Rapa Nui, but their language and culture are in danger of being lost. The future is uncertain for Easter Island. Much depends on the efforts of the Rapa Nui themselves, for they have the most to gain and the most to lose as they undergo continued development. We will revisit this story at the end of the chapter and draw some additional lessons from it.

MOVING ON.   In the meantime, this chapter briefly explores the current condition of our planet and then introduces three themes that provide structure to the primary goal of this text: *to use the insights of environmental science to promote a sustainable future.*

1.1 The State of the Planet

Picture our planet as an important patient whose good health, of great concern to all of us, may be in question. A host of organizations is involved in diagnosing the patient's vital signs and what needs to be done to maintain the patient's health. Four global trends are particularly unhealthy: (1) increasing population growth and its detrimental effects on human well-being, (2) a decline of vital ecosystem services, (3) the negative impacts of global climate change, and (4) a loss of biodiversity. Each of these is explored in greater depth in later chapters.

Increasing Population Growth and Declining Human Well-Being

The world's human population, over 6.8 billion persons in 2010, has grown by 2 billion in just the last 25 years. It is continuing to grow, adding about 75 million persons per year.

Even though the growth rate is gradually slowing, the world population in 2050 could be 9.1 billion, according to the most recent projections from the United Nations (UN) Population Division (Fig. 1–3). The 2.2 billion persons added to the human population by 2050 will all have to be fed, clothed, housed, and, hopefully, supported by gainful employment. Virtually all of the increase will be in the developing countries.

Yet in these same countries, an estimated 1.2 billion already experience the extreme poverty that accompanies an income of $1.25 a day, while some 2.5 billion exist on $2 a day. Over 1 billion—one out of every five in developing countries—remain malnourished. More than 6 million preschool children die each year of hunger and malnutrition. Addressing these tragic outcomes of severe poverty has been a major concern of the United Nations Human Development Program, and in 2000, all UN member countries adopted a set of goals—the **Millennium Development Goals (MDGs)**—to

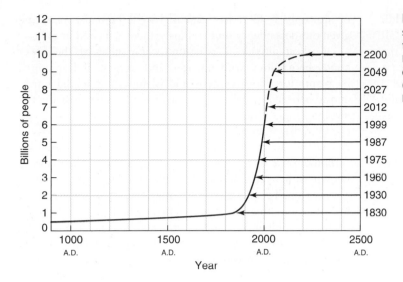

Figure 1–3 World population explosion. World population started a rapid growth phase in the early 1800s and has increased sixfold in the last 200 years. It is growing by 75 million people per year. (See Chapter 8.) Future projections are based on assumptions that birthrates will continue to decline. Go to **GRAPHit!** at www.masteringenvironmentalscience.com. (Data from UN Population Division, 2004 revision, and from Population Reference Bureau reports.)

reduce extreme poverty and its effects on human well-being (see Table 9–2 for a list of the eight goals). If the goals are achieved by the target date of 2015, more than 400 million people will be lifted out of extreme poverty, and many millions of lives will be saved. Indeed, much progress has been made toward the goals in many of the developing countries. For example, measles deaths in Africa have been reduced by 91%, and the number of people falling victim to AIDS has declined for several straight years. However, the neglect of family planning in many of the least developed countries is likely to undermine the efforts to achieve the MDGs in those countries.

At the same time, global economic production continues to rise, having doubled since 1986. And average gross domestic product (GDP) growth in low-income countries has improved; it was 4.5% per year in the 1990s, but since 2000, it has increased to 6.5% per year. However, real income in most developing countries is falling further behind that of the developed countries because of the large inequalities in wealth that exist between them. Stabilizing population growth in the developing countries, important for achieving the MDGs, is also essential for closing the economic gap between those nations and the industrialized countries.

The Decline of Ecosystems

Natural and managed ecosystems support human life and economies with a range of goods and services. As crucial as they are, these vital resources are not being managed well. Around the world, human societies are depleting groundwater supplies, degrading agricultural soils, overfishing the oceans, and cutting forests faster than they can regrow. The world economy depends heavily on many renewable resources, as we exploit these systems for **goods**—fresh water (water with less than 0.1% salt), all of our food, much of our fuel, wood for lumber and paper, leather, furs, raw materials for fabrics, oils and alcohols, and much more. Just three sectors—agriculture, forestry, and fishing—are responsible for 50% of all jobs worldwide and 70% of all jobs in sub-Saharan Africa, eastern Asia, and the Pacific islands.

These same ecosystems also provide a flow of **services** that support human life and economic well-being, such as the breakdown of waste, regulation of the climate, erosion control, pest management, maintenance of crucial nutrient cycles, and so forth. In a very real sense, these *goods and services* can be thought of as capital—*ecosystem capital*. Human well-being and economic development are absolutely dependent on the products of this capital—its income, so to speak. As a result, the stock of ecosystem capital in a nation and its income-generating capacity represents a major form of the wealth of the nation. These goods and services are provided year after year, as long as the ecosystems producing them are protected.

In 2007, the World Resources Institute, an environmental think tank, published a report,[1] that focused on the services that ecosystems provide to human societies and produced an agenda for ecosystem restoration—in essence, examining the vital signs of Earth's ecosystems and prescribing the medicine that would help to heal those whose health is failing. This work by the World Resources Institute depends heavily on a recent, comprehensive effort to understand the status of global ecosystems and human impacts on them, the **Millennium Ecosystem Assessment.**

Millennium Ecosystem Assessment. Launched on World Environment Day, 2001, the Millennium Ecosystem Assessment took four years to gather available information on the state of ecosystems across the globe. This monumental effort involved some 1,360 scientists from 95 countries gathering, analyzing, and synthesizing information from published, peer-reviewed research. The project focused especially on the linkages between ecosystem goods and services and human well-being on global, regional, and local scales. The reports are now published and are also available on the Internet.

In a summary report,[2] the most prominent finding of the scientists was the widespread abuse and overexploitation

---

[1]Irwin, *Restoring Nature's Capital: An Action Agenda to Sustain Ecosystem Services*, 2007.
[2]*Ecosystems and Human Well-Being: Synthesis*, 2005.

of ecosystem resources. Humans have altered the world's ecosystems more rapidly and profoundly over the past 50 years than at any time in human history. Over 60% of the classes of ecosystem goods and services assessed by the team are being degraded or used unsustainably (Table 1–1), and if this is not reversed, the next half-century could see deadly consequences for humans as the ecosystem services that sustain life are further degraded. The overall intent of the project is to build a knowledge base for sound policy decisions and management interventions; it remains for policy makers and managers to act on that knowledge.

Global Climate Change

Historically, pollution has been a relatively local problem, affecting a given river, lake, or bay or a slightly larger area such as the air in a city. Today scientists are analyzing pollution on a global scale. For example, concern about depletion of the stratospheric ozone layer has already led to international action—the Montreal Protocol in 1987—aimed at curbing pollution from the release of chlorofluorocarbon refrigerants into the atmosphere.

A more serious problem today is the danger of global climate change due to the accumulation of greenhouse gases—especially carbon dioxide (CO_2)—in the atmosphere. Carbon dioxide is an unavoidable by-product of burning fossil fuels—crude oil, coal, and natural gas. Because of the large amount of fossil fuels currently being burned, CO_2 levels in the atmosphere have grown from about 280 parts per million (ppm) in 1900 to over 390 ppm in 2010. For the last few years, the level of atmospheric CO_2 has increased by 2 ppm per year, and given our dependency on fossil fuels, there is no end in sight.

Carbon dioxide is a natural component of the lower atmosphere, along with nitrogen and oxygen. It is required by plants for photosynthesis and is important to the Earth-atmosphere energy system. Carbon dioxide gas is transparent to incoming light from the Sun, but it absorbs infrared (heat) energy radiated from Earth's surface, thus slowing the loss of this energy to space. The absorption of infrared energy by carbon dioxide warms the lower atmosphere in a phenomenon known as the greenhouse effect. Although the concentration of CO_2 is a small percentage of the atmospheric gases, increases in the volume of this gas affect temperatures. Figure 1–4 graphs changes in global air temperatures and carbon dioxide from 1880 to the present and illustrates the fact that both of these parameters are increasing—and rather steeply. This is not proof that the CO_2 caused the increase, but the argument based on the known greenhouse effect is quite convincing.

The Intergovernmental Panel on Climate Change (IPCC) was established by the UN in 1988 and given the responsibility to report its assessment of climate change at five-year intervals. The latest of these assessments, the Fourth Assessment Report (FAR), was released during 2007. The work of thousands of scientific expert reviewers and hundreds of authors, this assessment produced convincing evidence of human-induced climate change that is already severely impacting global climate and sea level. The IPCC concluded that future climate change could be catastrophic if something is not done to bring the rapidly rising emission of CO_2 and other greenhouse gases under control.

TABLE 1–1 The Global Status of Ecosystem Services. Human use has degraded almost two-thirds of these identified services; 20% are mixed, meaning they are degraded in some areas and enhanced in others; and 17% have been enhanced by human use.

Ecosystem Services	Degraded	Mixed	Enhanced
Provisioning (goods obtained from ecosystems)	Capture fisheries Wild foods Wood fuel Genetic resources Biochemicals Fresh water	Timber Fiber	Crops Livestock Aquaculture
Regulating (services obtained from the regulation of ecosystem processes)	Air quality regulation Regional and local climate regulation Erosion regulation Water purification Pest regulation Pollination Natural hazard regulation	Water regulation (flood protection, aquifer recharge) Disease regulation	Carbon sequestration (trapping atmospheric carbon in trees, etc.)
Cultural (nonmaterial benefits from ecosystems)	Spiritual and religious values Aesthetic values	Recreation and ecotourism	

Source: Millennium Ecosystem Assessment, *Ecosystems and Human Well-Being: Synthesis* (Washington, D.C.: Island Press, 2005).

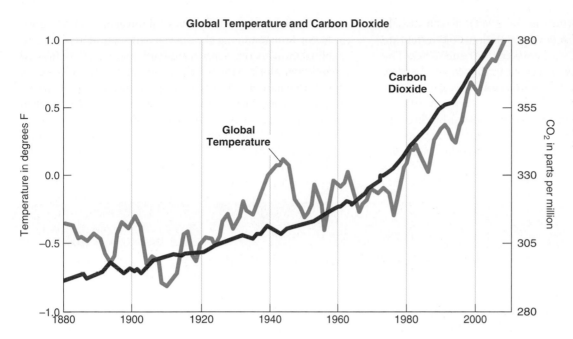

Global Temperature and Carbon Dioxide

Figure 1–4 Global temperature and carbon dioxide, 1880 to present. Half of the increase in CO_2 from human activities has been added since 1975. The sharpest rise in temperature accompanies this 30-year increase in CO_2. Is this a coincidence? (Adapted from zFacts.com.)

The Kyoto Protocol. Concern about global climate change as a result of earlier IPCC assessments led representatives of 166 nations to meet in Kyoto, Japan, in December of 1997 to negotiate a treaty to reduce emissions of CO_2 and other greenhouse gases. At that meeting, most of the industrialized nations (including the United States) agreed to reduce emissions to below 1990 levels, a goal that is to be achieved by the year 2012. The United States, at the time the world's biggest emitter, withdrew from the Kyoto agreement in 2001, but the treaty was ratified in 2005 and is now in force in most industrialized nations.

Kyoto, however, is only a first step; even if all parties adhere to the treaty, the levels of greenhouse gases will continue to rise indefinitely. At issue for many countries are the concerns surrounding the conflict between the short-term economic impacts of reducing the use of fossil fuels and the long-term consequences of climate change for the planet and all its inhabitants. Future climate changes are likely to disrupt the ecosystem goods and services essential to human well-being, and since the extreme poor depend especially on natural ecosystems, they will suffer disproportionately. New international agreements are being negotiated, with the hope that the United States will join the Kyoto nations and the larger developing countries in reducing greenhouse gas emissions to levels below what is considered dangerous. Without doubt, climate change is one of the defining environmental issues of the 21st century.

Loss of Biodiversity

The Millennium Ecosystem Assessment defines biodiversity as "the variability among living organisms and the ecological complexes of which they are part." The rapidly growing human population, with its growing appetite for food, water, timber, fiber, and fuel, is accelerating the conversion of forests, grasslands, and wetlands to agriculture and urban development (Fig. 1–5). The inevitable result is the loss of most of the wild plants and animals that occupy those natu-

ral habitats. Pollution also degrades habitats—particularly aquatic and marine habitats—destroying the species they support. Further, hundreds of species of mammals, reptiles, amphibians, fish, birds, and butterflies, as well as innumerable plants, are exploited for their commercial value. Even when species are protected by law, many are hunted, killed, and marketed illegally. According to the assessment, the majority of wild plant and animal species are declining in their range and/or population size.

As a result, Earth is rapidly losing many of its species, although no one knows exactly how many. About 2 million species have been described and classified, but scientists estimate that 5 to 30 million species may exist on Earth. Because so many species remain unidentified, the exact number of species becoming extinct can only be estimated. However,

Figure 1–5 Natural ecosystems giving way to development. Continuing growth requires a massive reorganization and exploitation of natural resources, bringing record levels of degradation of natural ecosystems. Here we see a mature forest in the western United States being stripped away.

the Living Planet Index (Lob et al. 2005), which closely follows trends in nearly 4,000 populations of vertebrate species, has declined by 27% between 1970 and 2005. The most dramatic declines have occurred in the tropics, in both ground-dwelling species and ocean species.

Risks of Losing Biodiversity. Why is losing biodiversity so critical? Biodiversity is the mainstay of agricultural crops and of many medicines. The loss of biodiversity can only

curtail development in these areas. Biodiversity is also a critical factor in maintaining the stability of natural systems and enabling them to recover after disturbances such as fires or volcanic eruptions. Most of the essential goods and services provided by natural systems are derived directly from various living organisms, and we threaten our own well-being when we diminish the biodiversity within those natural systems. (An outstanding case in point is presented in "Sound Science: Keeping the Vultures Flying in India and Pakistan.") These

Sound Science

Keeping the Vultures Flying in India and Pakistan

In countries of the Indian subcontinent, several species of vultures play a key role by rapidly disposing of carcasses of cows and other large animals that die in the cities, villages, and countryside (by far the most common of these species is the Oriental white-backed vulture—see the photo). This practice is broadly approved by the populace—by doing this, the vultures reduce the risk of disease and other unpleasant consequences of the presence of dead animals. Starting in the 1990s, vultures began a population decline that reduced numbers to less than 5% of their normal levels within 10 years. Tens of millions of vultures died, with major ecological consequences, such as an increase in numbers of feral dogs and rats and an associated risk of a rise in diseases (for example, rabies) carried by these animals.

The vulture decline was the most rapid ever seen in populations of raptors and quickly led to the designation of the vultures as critically endangered—meaning that they were at serious risk of becoming extinct. In 2000, the Peregrine Fund established the Asian Vulture Crisis Project (AVCP) and asked Dr. Lindsay Oaks, assistant professor of Microbiology and Pathology at Washington State University, to assemble an international team of workers to investigate the decline as quickly as possible. The AVCP team gathered in Pakistan and began with postmortem examinations of hundreds of vultures. Over 85% of the dead vultures showed evidence of visceral gout, which is commonly the result of kidney failure leading to the deposit of uric acid on and within the visceral organs. Further investigations of freshly dead vultures, including a number without visceral gout, failed to reveal the presence of any known causes of avian disease or toxic concentrations of any known pesticides or heavy metals.

Since the primary food source of the vultures is dead domestic livestock, the team hypothesized that some veterinary substance being given to the farm animals might be responsible for the deaths. Investigations led to diclofenac, a

Dogs and vultures compete for a carcass on the Ganges River, India.

nonsteroidal anti-inflammatory drug often administered to animals to treat inflammation, pain, and fever. It is known to be toxic to kidneys. This drug is inexpensive, is sold over the counter and had become available in the mid-1990s. A survey of druggists and veterinarians revealed that all sold diclofenac, usually on a daily basis.

The team then established the kidney toxicity of the drug by feeding oral doses in food to captive, nonreleasable vultures. The vultures quickly developed renal failure and died within a few days of administration of the drug. Untreated control birds fed a similar diet, but no diclofenac, showed none of the effects of the experimental birds. To show that the carcasses of livestock treated with diclofenac could be the source of the scourge, the team administered various levels of diclofenac to several domestic animals and fed them to captive vultures. The higher the exposure, the greater the tendency to develop visceral gout and kidney failure. Six control vultures were fed

meat that did not contain the drug, and they all survived. Out of 20 exposed vultures, 13 died, with the 7 survivors having received the lowest levels of diclofenac. The team concluded that the widespread veterinary use of diclofenac was indeed responsible for the death of millions of vultures across the Indian subcontinent. Other researchers in India and Nepal confirmed the statistically high association of dead vultures and significant diclofenac residues in their carcasses.

Reacting to this information, the Indian government phased out the veterinary use of diclofenac, and there are signs that vulture numbers are slowly rising since the ban on the drug. A safe substitute—meloxicam—is replacing the use of diclofenac for treating animals. It looks like the vultures will once again be flying over India and Pakistan, carrying out their crucial disposal work.

Source: J. Lindsay Oaks et al., "Diclofenac residues as the cause of vulture population decline in Pakistan," *Nature* 427 (2004): 630–633.

goods and services are especially important in sustaining the poor in developing countries. There are also aesthetic and moral arguments for maintaining biodiversity. Shall we continue to erase living species from the planet, or do we have a moral responsibility to protect and preserve the amazing diversity of life on Earth? Once a species is gone, it is gone forever. Finding ways to protect the planet's biodiversity is one of the many challenges of environmental science.

1.2 Environmental Science and the Environmental Movement

As you read the chapters in this book, you will encounter descriptions of the natural world and how it works. You will also encounter a great diversity of issues and problems that have arisen because human societies live in this natural world, and as we do, we use materials from it (the goods we take from natural ecosystems); convert parts of it into the built environment of our towns, cities, factories, and highways; and transform many natural ecosystems into food-producing agricultural systems. We also use the environment as a place to dump our wastes—solids, liquids, gases—all proceeding from human activities, most of them legitimate, but all of them impacting the environment. The **environment**, then, includes the natural world, human societies, and the human-built world; it is an extremely inclusive concept.

Environmental Science

Things go wrong in the environment, sometimes badly. We have already considered four trends (human population growth, ecosystem decline, global climate change, loss of biodiversity) that signal to us that we are creating problems for ourselves that we ignore to our peril. Lest you think that all we do is create problems, however, consider some of the great successes human societies have achieved. We have learned how to domesticate landscapes and ecosystems, converting them into highly productive food-producing systems that provide enough sustenance for almost 7 billion people. We have largely brought under control the major infectious diseases that for millennia had decimated human populations. We have learned how to convert natural materials into an endless number of manufactured goods and structures, all useful for the successful building of cities, roadways, vehicles, and all that makes up a 21st-century human world. We have successfully tapped into the abundant pool of carbon-based fuels laid down in the geological past, using it for energy to produce electricity, heat our homes, power our vehicles, and manufacture our products. All of these successes, however, carry with them dangers. Burning fossil fuels creates air pollutants that can make us sick and even raise the temperature of our planet; using chemicals to control diseases and pests leaves a residue of contaminants that can poison whole food chains—and ourselves; overharvesting forests and fisheries can lead to their collapse, erasing vital services that natural ecosystems provide.

This is where environmental science comes in. Simply put, **environmental science** is the study of how the world works. It looks at the cause-and-effect relationships underlying the issues and problems that arise from our use of the natural world and provides answers that enable human societies to make changes consistent with a sustainable future. All sorts of disciplines contribute to environmental science: history, engineering of many sorts, geology, physics, medicine, biology, and sociology, to name a few. It is perhaps the most multidisciplinary of all sciences. As you study environmental science, you have an opportunity to engage in something that can change your life and that will certainly equip you to better understand the world you live in now and into the future.

The Environmental Movement

Although what we now term the environmental movement began less than 60 years ago, it had its roots in the late 19th century, when some people realized that the unique, wild aspects of the United States were disappearing. The 1890 U.S. Census highlighted the "closing of the frontier," an event that was noted with some sadness. No longer could any area of the country be classified as totally uninhabited. Around that time, several groups devoted to conservation formed: the National Audubon Society, the National Wildlife Federation, and the Sierra Club, founded in California by naturalist John Muir, who helped popularize the idea of wilderness. To counter the wide-scale misuse of natural resources, President Theodore Roosevelt promoted the conservation of public lands, placing some 230 million acres of wild lands under public protection. A national environmental consciousness was stirring.

The first half of the 20th century was marked by increasing awareness of the environment. Agricultural practices following World War I eventually helped create an environmental crisis—the Dust Bowl of the 1930s, when wheat failed to hold soil in place and the topsoil eroded (Fig. 1–6). During the Great Depression (1930–1936), conservation provided a means of both restoring the land and providing work for the unemployed. Many trails, erosion-control projects, and other improvements in national parks and forests were originally installed by the Civilian Conservation Corps (CCC), which played a major role in pulling the country out of the depression. The two decades following World War II (1945–1965) were full of technological optimism. The United States and its allies had won the "greatest of all wars," in part because of new technologies. The tremendous productive capacity built up during the war, together with new developments ranging from rocket science to computers and from pesticides to antibiotics, could be redirected to peacetime applications. Except for the tensions of the Cold War during those years and some apprehension about atomic energy, it seemed as if nothing but opportunity and prosperity lay ahead.

Although economic expansion enabled most families to have a home, a car, and other possessions, certain problems became obvious. The air in and around cities was becoming murky and irritating to people's eyes and respiratory systems. Streetlights from St. Louis to Pittsburgh were left on during the day because of the smoke from coal-burning industries.

Figure 1–6 The Dust Bowl. In the 1930s, modern farming methods combined with an extended drought led to the loss of enormous amounts of topsoil due to wind erosion.

In winter, even freshly fallen snow soon turned gray as the soot fell from city chimneys. Rivers and beaches were increasingly fouled with raw sewage, garbage, and chemical wastes. Conspicuous declines occurred in many bird populations (including our national symbol, the bald eagle), aquatic species, and other animals.

It was easy to identify some culprits: belching industrial smokestacks, open burning dumps, and municipal and industrial sewers, discharging raw sewage and chemical wastes into waterways. The decline of the bald eagle and other bird populations was traced to the accumulation in their bodies of DDT, the long-lasting pesticide that had been used in large amounts since the 1940s. In short, it was clear that we were seriously contaminating our environment.

The Modern Environmental Movement.   In 1962, biologist Rachel Carson wrote *Silent Spring*, presenting her scenario of a future with no songbirds and with other dire consequences if pollution of the environment with DDT and other pesticides continued (see Chapter 13). Carson's voice was soon joined by others, many of whom formed organizations to focus and amplify the voices of thousands more in demanding a cleaner environment. This was the beginning of the modern **environmental movement**, in which a newly militant citizenry demanded the curtailment of pollution, the cleanup of polluted environments, and the protection of pristine areas. It is significant that the environmental movement began as a grassroots initiative, and the movement maintains its momentum and force today only by continuing to command public interest and support.

Members of the environmental movement of the 1960s joined forces with older organizations, such as the National Audubon Society and the Sierra Club, that had a considerable history of dedication to preserving wildlife and natural habitats. There is an obvious overlap between the goal of reducing pollution and that of protecting wildlife. Wildlife cannot survive in polluted surroundings, and preserving an uncontaminated environment presupposes that the plant and animal species within that environment will be protected. Therefore, established wildlife-preservation organizations and their members became significant players in the blossoming environmental movement, along with newly formed organizations such as the Environmental Defense Fund, the Natural Resources Defense Council, Greenpeace, and the Union of Concerned Scientists. Persons and organizations with this strong focus on environmental concerns became known as **environmentalists**, and the widespread development of this movement is often called **environmentalism**.

By almost any measure, the environmental movement has been successful. Pressured by a concerned citizenry, Congress created the Environmental Protection Agency (EPA) in 1970 and passed numerous laws promoting pollution control and wildlife protection, including the National Environmental Policy Act of 1969, the Clean Air Act of 1970, the Clean Water Act of 1972, the Marine Mammals Protection Act of 1972, the Endangered Species Act of 1973, and the Safe Drinking Water Act of 1974. You will read more about these and more recent laws and their effects in subsequent chapters. In the area of pollution abatement alone, our society has spent hundreds of billions of dollars, both on installing pollution control devices and on redesigning procedures and products so that less pollution is created. The pollution control devices on our cars are an example. Governments have spent additional billions on upgrading sewage treatment, refuse disposal, and other measures to reduce pollution. As a result of these expenditures, the air in our cities and the water in our lakes and rivers are far cleaner than they were in the late 1960s. Without a doubt, absent the environmental movement, our air and water would now be a toxic brew.

Environmentalism Acquires Its Critics. In the early stages of the environmental movement, the sources of the problems were specific and visible. That made it easy to identify certain polluters as the "bad guys" and take the side of the environmental movement. Likewise, the solutions seemed relatively straightforward: Install waste treatment and pollution control equipment and ban the use of DDT, substituting safer pesticides. At first, many industries conformed to the reforms, but as regulatory demands increased, a number of the more polluting industries, such as mining and utilities, pressed for a halt in regulation, and some began demanding a significant *deregulation.* They believed that environmental regulation was harmful to the economy, and they were often able to find political allies in the Congress and even at the presidential level.

Nowadays, there are political battles surrounding almost every environmental issue. Bitter conflicts have emerged over issues involving access to publicly owned resources such as water, grazing land, and timber. The Endangered Species Act, which protects wild plants and animals whose existence is threatened, has been a lightning rod for controversy. When the EPA develops regulations to address air and water pollution, for example, business special interests oppose the regulations, and environmental special interests favor them. Sadly, the Congress and certain administrations have shown a tendency to side with the deregulation forces, and the EPA is forced to go along with them. As environmental public policies are developed and enforced, the controversies that accompany those policies remind us that *politics always accompanies policy.*

1.3 Three Unifying Themes

What will it take to move our civilization in the direction of a long-term sustainable relationship with the natural world? The answer to this question is not simple, but we would like to present three themes that provide coherence to the issues and topics covered in this text. In reality, these themes deal with *how we should conceptualize our task of forging a sustainable future* (Fig. 1–7). Each theme is a concept that we

Figure 1–7 Three unifying themes. Sustainability, stewardship, and sound science are three vital concepts or ideals that can move societies toward a sustainable future if they are applied to public policies and private environmental actions.

believe is essential to environmental science. The first is **sound science**—the basis for our understanding of how the world works and how human systems interact with it. The second theme is **sustainability**—the practical goal toward which our interactions with the natural world should be working. The third is **stewardship**—the actions and programs that manage natural resources and human well-being for the common good. These themes will be applied to public policy and individual responsibility throughout the text and will also be explored in brief essays within each chapter. As each chapter of this text develops, the concepts behind these themes will come into play at different points. At the end of each chapter, we will revisit the themes, summarizing their relevance to the chapter topics and often adding our own perspective. We explore each of these themes here, beginning with sound science.

Sound Science

Many environmental issues are embroiled in controversies so polarized that no middle ground seems possible. On the one side are persons who argue from apparently sound facts and proven theories. On the other are persons who disagree and present opposing explanations of the same facts. Both groups may have motives for arguing their case that are not all apparent to the public. In the face of such controversy, many people are understandably left confused. It is our objective to give a brief overview of the nature of science and the scientific method so that you can evaluate for yourself the two sides of such controversies.

The Scientific Method.   In its essence, science is simply a way of gaining knowledge; that way is called the **scientific method**. The term *science* further refers to all the knowledge gained through that method. We employ the term *sound science* to distinguish legitimate science from what has been called *junk science*, information that is presented as valid science, but that does not conform to the rigors of the methods and practice of legitimate science. Sound science involves an approach to understanding how the natural world works, an approach that has become known as the scientific method.

Assumptions.   To begin, the scientific method rests on four basic assumptions that most of us accept without argument. The first assumption is that *what we perceive with our basic five senses represents an objective reality*; that is, our perceptions are not some kind of mirage or dream. The second assumption is that this "objective reality" functions according to *certain basic principles and natural laws that remain consistent through time and space*.

The third assumption, which follows directly from the second, is that *every result has a cause and every event in turn will cause other events*. In other words, we assume that events do not occur without reason and that there is an explainable cause behind every happening. The fourth and final assumption is that *through our powers of observation, manipulation, and reason, we can discover and understand the basic principles and natural laws by which the universe functions*.

Although assumptions, by definition, are premises that usually cannot be proved, the fact is that the assumptions underlying science have served us well and are borne out by everyday experience. For example, we suffer severe consequences if we do not accept our perception of fire as real. Similarly, our experience confirms that gravity is a predictable force acting throughout the universe and that it is not subject to unpredictable change. And the same can be said for any number of other phenomena that we observe. If our car fails to start, we know that there is a logical reason, and we call a mechanic to fix it.

Thus, whether we are conscious of the fact or not, we all accept the basic assumptions of science in the conduct and understanding of our everyday lives. Scientists and scientific investigations focused on the natural world (the **natural sciences**) extend the boundaries of everyday experience, deepen our understanding of cause-and-effect relationships, and enable us to have a greater appreciation of the principles and natural laws that seem to determine the behavior of all things, from the outcome of a chemical reaction to the functioning of the biosphere.

Observation.   In previous schooling, you may have learned that the scientific method consists of the following sequence: observation, hypothesis, test (experiment), and explanation or conclusion. This sequence is an oversimplification in that it fails to describe the unique thought processes often involved in the four steps it comprises.

The foundation of all science and scientific discovery is **observation** (seeing, hearing, smelling, tasting, feeling). Indeed, many branches of science, such as natural history (the study of where and how various plants and animals live, mate, reproduce, etc.), astronomy, anthropology, and evolutionary biology, are based almost entirely on observation because experimentation is either inappropriate or impossible. For example, experimentation is obviously counterproductive if you want to discover what plants and animals do under completely natural conditions, and it simply is impossible to conduct experiments on stars or past events. Figure 1–8 shows how observations alone can lead to answers or explanations of some natural phenomena.

Likewise, much of the data and many of the conclusions of other sciences, such as zoology, botany, geology, comparative anatomy, and taxonomy (the classification of organisms), are based on nothing more (or less) than the careful observing and chronicling of things and events by persons taking the pains and time to do so. Even experimentation, as we will discuss shortly, is conducted to gain another window of observation. Simply put, in all science, *careful observation is the keystone.* So how can we be sure that observations are accurate? Well, as a matter of fact, not every reported observation is accurate, for reasons ranging from honest misperceptions to calculated mischief. Therefore, an important aspect of science, and a trait of scientists, is to be skeptical of any new report until it is confirmed or verified. Such confirmation usually involves other investigators' repeating and checking out the observations of the first investigator and validating (or invalidating) their accuracy. As observations are confirmed by more and more investigators, they gain the status of factual data. In other

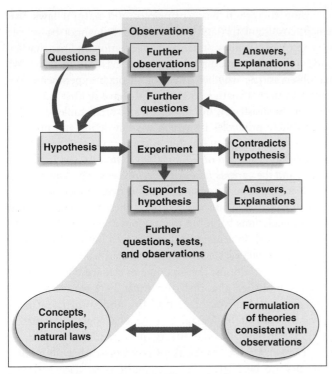

Figure 1–8 Steps of the scientific method. The process leading to the formation of theories and the postulation of natural laws and concepts is a continual interplay among observations, hypotheses, tests (experiments), explanations or theories, and further refinement.

words, **facts** are things or events that have been confirmed by more than one observer and remain open to be reconfirmed by additional people. Things or events that do not allow such confirmation—UFOs, for example—remain in the realm of speculation from a scientific standpoint.

Various observations by themselves, like the pieces of a puzzle, may be put together into a larger picture—a model of how a system works. To give a simple example, we observe that water evaporates and leads to moist air and that water from moist air condenses on a cool surface. We also observe clouds and precipitation. Putting these observations together logically, we derive the concept of the hydrologic cycle, described in Chapter 10. Water evaporates and then condenses as air is cooled, condensation forms clouds, and precipitation follows. Water thus makes a cycle from the surface of Earth into the atmosphere and back to Earth (Fig. 10–3). Note how this simple example incorporates the four assumptions described previously: There is an objective reality, it operates according to principles, every result has a cause, and we can discover the principles according to which reality operates. Note also how the example broadens our everyday experiences of the evaporation of water and the falling of rain into an understanding of a cycle involving both.

Thus, the essence of science and the scientific method may be seen as a process of making observations and logically integrating those observations into a model of how the world works. To be sure, many areas of science get more complex and difficult to comprehend. Still, the basic

process—constructing a logically coherent picture of causes and effects from basic observations—is the same.

Where, then, does experimentation, that additional hallmark of science, fit in?

Experimentation.   **Experimentation** is basically setting up situations to make more systematic observations regarding causes and effects. For example, the number of chemical reactions one can readily observe in nature is limited. However, in the laboratory, it is possible to purify elements or compounds, mix them together in desired combinations and proportions, and carefully observe and measure how they react (or fail to react). From the way chemicals react, chemists have constructed a coherent cause-and-effect picture, the **atomic theory**, and they have determined the attributes of each element. Similarly, biologists put plants or animals into specific situations in which they can carefully observe and measure their responses to particular conditions or treatments (Fig. 1–9). (Again, note that experimentation is necessarily limited to things that lend themselves to artificial manipulation. In many cases, such as stars, geological events, atmospheric events, and events that have occurred in the past, one has only observations to work with in the construction of the broader picture.)

Some experimentation may be more or less spontaneous and random—the childhood inclination to "do this and see what happens." Sometimes, valuable information may be obtained in this way if careful accounts are kept so that one has an accurate record of causes and effects. However, to solve a particular problem (i.e., What is the cause of *this* event?), a more systematic line of experimentation is used. This is where the sequence consisting of observation, hypothesis, test (experiment), and explanation comes in.

Figure 1–9 Experiment demonstrating the importance of mycorrhizae. Mycorrhizae is a root-fungus symbiosis that greatly enhances the ability of a plant to absorb nutrients. The pine seedlings shown were grown under identical conditions, except that the seedling on the left was grown with mycorrhizal fungi, and the one on the right was grown without the fungi.

Let's consider the problem of the decline of vultures in the Indian subcontinent, as described in "Keeping the Vultures Flying in India and Pakistan" (p. 8). The question is, What has caused the unusual decline of vultures? The Asian Vulture Crisis Project team collected crucial observations on dead vultures and found that most had died of kidney failure. The next step was to make educated guesses as to the cause. Each such educated guess is a **hypothesis**. Each hypothesis is then tested by making further field observations or conducting experiments to determine whether the hypothesis really accounts for the observed effect. The team hypothesized that the vultures were being poisoned by something in their food—mainly dead domestic animals. Further sharpening of this hypothesis led the team to the veterinary drug diclofenac. Other hypotheses, such as some avian disease or toxic concentrations of pesticides or heavy metals, were ruled out by more testing and observations. The team then conducted several crucial experiments, each time testing the effects of the suspected drug on living vultures and maintaining a control set of birds that did not receive the drug. The experimental results brought what started out as a hypothesis to the status of a satisfactory explanation, which we might express as a *theory* or *concept*. Figure 1–8 shows how such a process moves from observations to hypotheses, experiments, and a satisfactory answer.

Theories. We have already noted how various specific observations (e.g., evaporation, precipitation) may fit together to give a logically coherent cause-and-effect picture of a broader phenomenon (i.e., the hydrologic cycle). At first, the broader picture is termed a hypothesis because it is really a tentative explanation as to how various observations are related. It becomes a **theory** only after much testing and confirmation showing that it is logically consistent with all observations (Fig. 1–8). (Of course, a hypothesis may be *dis*confirmed, too, if it is inconsistent with an observation.)

Through logical reasoning, theories will generally suggest or predict certain events. If an event predicted by a theory is indeed observed, the observation provides strong evidence for the truth of the theory. Predictions require experiments, testing, further data gathering, and more observation. When theories reach a state of providing a logically consistent framework for all relevant observations (facts) and when they can be used to reliably predict outcomes, they represent a valid interpretation of reality. For example, we have never seen atoms directly (or, until recently, even indirectly), but innumerable observations and experiments are coherently explainable by the theory that all gases, liquids, and solids consist of various combinations of slightly over a hundred kinds of atoms. Hence, we fully accept the atomic theory of matter.

Sometimes people will argue that because a theory is not proven fact, one theory is as good as another. That notion is false! One theory may have overwhelming supporting evidence, whereas much evidence may contradict another theory. In evaluating a theory, we must ask, What is the supporting evidence? Is there more evidence, or less, supporting an alternative theory?

Natural Laws and Concepts. Like a theory, the second assumption underlying science—that the universe functions according to certain basic principles and natural laws that remain consistent through time and space—cannot be proved with absolute certainty. Still, every observation and test that scientists have performed up to the present have borne it out. All our observations, whether direct or through experimentation, demonstrate that matter and energy behave neither randomly nor even inconsistently, but in precise and predictable ways. We refer to these principles by which we can define and precisely predict the behavior of matter and energy as **natural laws.** Examples are the *Law of Gravity*, the *Law of Conservation of Matter*, and the various *Laws of Thermodynamics*. Our technological success in space exploration and many other fields is in no small part due to our recognition of these principles and our precise calculations based on them. Conversely, trying to make something work in a manner contrary to a natural law invariably results in failure.

In many situations, the outcome of scientific work results in well-established theories that must be expressed in the mathematical language of probability and statistics. This is especially true of biological phenomena such as predator-prey relationships and the effects of pesticides. Accordingly, here it is appropriate to speak of *concepts* rather than laws. **Concepts** are perfectly valid explanations of data gathered from the natural world, and they can also be predictive, but they never reach the status of laws. They model the way we believe the natural world works and enable us to make *qualified* predictions of future outcomes. Thus, we might say that, on the basis of our understanding of the effects of DDT (a potent pesticide) on mosquitoes and on the basis of our observation that mosquitos can, and do, develop resistance to DDT, continued spraying of the local salt marshes with DDT to eradicate unwelcome mosquitoes is *likely* to result in the development of a more resistant mosquito population in the future. Here the *concept* of pest resistance is used to modify public policy. In general, explanations in biology do not lead to the development of statements with such high predictability that we are justified in calling them laws.

The Role of Instruments in Science. Complex instrumentation is another hallmark of science, one that often gives it an aura of mystery. Yet, regardless of complexity, all scientific **instruments** perform one of three basic functions. First, they may extend our powers of observation, as do telescopes, microscopes, X-ray machines, and CAT scans, for example (Fig. 1–10). Second, instruments are used to quantify observations—that is, they enable us to measure exact quantities. For example, we may feel cold, but a thermometer enables us to measure and quantify exactly how cold it is. Comparisons, communication, and verification of different observations and events would be impossible if it were not for such measurement and quantification. Third, instruments such as growth chambers and robots help us achieve conditions and perform manipulations required to make certain observations or perform certain experiments.

All instruments used in science need to be subjected to testing and verification to be sure that they are giving as accurate a representation of reality as we can attain with their use. The need for this can be seen in the case of satellite instruments measuring atmospheric temperature (see Chapter 18).

Figure 1–10 Mass spectrometer. This machine, the Quadrupole Mass Spectrometer, greatly extends our ability to make chemical analyses. The mass spectrometer produces charged particles (ions) from a chemical substance to be analyzed and then determines the masses and relative abundances of the ions present. For example, it is used to measure the different carbon isotopes (C-12, C-13, C-14) in organic compounds and thus trace their origins in the natural world.

Failure to calibrate the instruments properly led to a serious underestimation of atmospheric temperature changes over time. It is safe to say that modern scientific instruments have made possible many important discoveries that were out of reach of the human senses.

The Scientific Community. Science has successfully brought us to a high degree of understanding of the natural world and has also brought a host of technologies based on that understanding. Doesn't this success mean that our theories about how the world works must be true? A philosopher of science would answer, "No, not necessarily." The best we can do is to establish a theory beyond a reasonable doubt. Theories are always less than absolutely certain because we can never know whether some other theory exists that will do a better job of explaining the data. Generally speaking, *our confidence in scientific knowledge should be proportional to the evidence supporting it.*

It is important to understand that science and its outcomes take place in the context of a scientific community and a larger society. There is no single authoritative source that makes judgments on the validity of scientific theories. Instead, it is the collective body of scientists working in a given field who, because of their competence and experience, establish what is authentic science and what is not. They do so by communicating their findings to each other and to the public as they publish their work in peer-reviewed journals. The process of peer review is crucial—it requires experts in a given field to analyze the results of their colleagues' work. Careful scrutiny is given to research papers, with the objective of rooting out poor or sloppy science and affirming work that is clearly meritorious.

Even with these checks and balances that work to establish scientific knowledge and with all the procedures of the scientific method, controversies arise that pit one group against another, with both groups claiming that they have science on their side. For example, we will make the claim in Chapter 18 that Earth is warming because of human activities. Not all scientists agree with this claim, although for most the evidence is convincing. There are at least four reasons why such controversies occur.

1. **New information.** We are continually confronted by new observations—the hole in the ozone layer, for instance, or the dieback of certain forests. It takes some time before all the hypotheses regarding the cause of what we have observed can be adequately tested. During this time, there may be honest disagreement as to which hypothesis

is most likely. Such controversies are gradually settled by further observations and testing, but the process leads into the second reason for continuing controversy.

2. **Complex phenomena.** Certain phenomena, such as the hole in the ozone layer or the loss of forests, do not lend themselves to simple tests or experiments. Therefore, it is difficult and time consuming to prove the causative role of one factor or to rule out the involvement of another. Gradually, different lines of evidence come to support one hypothesis and exclude another, enabling the issue to be resolved. When is there enough evidence to say unequivocally that one hypothesis is right and another wrong? At some point, the scientists working in the appropriate field will reach a consensus in making that judgment.

3. **Bias.** The third reason for controversy is that there are many vested interests that wish to maintain and promote disagreement because they stand to profit by doing so. For example, tobacco interests argued for years that the connection between smoking and illness had not been proved and that more studies were necessary. By harping on the absence of absolute proof (a scientific impossibility anyway) and downplaying the overwhelming body of evidence supporting the connection between smoking and illness, the tobacco lobby succeeded in keeping the issue controversial and thereby delayed regulatory restrictions on smoking. In fact, businesses often hire "product-defense firms," whose strategy is to do everything possible to challenge scientific evidence damaging to their products or interests.

4. **Subjective values.** The fourth reason for controversy is that subjective value judgments may be involved. This is particularly true in environmental science because the discipline deals with the human response to environmental issues. For example, there is virtually no controversy regarding nuclear power, as long as it is considered at the purely scientific level of physics. However, when it comes to deciding whether to promote the further use of nuclear energy to generate electrical power, controversy arises because different people have different subjective feelings about the relative risks and benefits involved.

Some controversy is the inevitable outcome of the scientific process itself, but much of it is attributable to less noble causes. Unfortunately, the media and the public may be unaware of the true nature of the information and will often give equal credibility to opposing views on an issue. This can be seen, for example, in the media's coverage of global warming, where the efforts of a few industry-sponsored skeptics have often been given equal coverage with the overwhelming scientific consensus. This leads us to the topic of junk science.

Junk Science.   **Junk science** traditionally refers to information that is presented as scientifically valid, but that does not conform to the rigors imposed by the scientific community. In recent years, another use of this term has appeared that has created some confusion. In essence, special interests—usually antienvironmental interests—have used the term *junk science* to refer to anything that threatens their preferred viewpoint; conversely, *sound science* becomes anything that supports their viewpoint. In spite of this distortion, we will hold to the original meanings for sound and junk science, but be aware of this perverted usage.

True junk science can take many forms: the presentation of selective results (picking and choosing only those results that agree with one's preconceived ideas), politically motivated distortions of scientifically sound information, and the publication of poor work in quasi-scientific (unreviewed) journals and books. Quite often, junk science is generated by special interest groups trying to influence the public debate about science-related concerns. The result is that the junk science may be given equal respectability with the sound science in the public arena. Information presented by the Tobacco Institute, for example, is clearly suspect, due to its corporate-related funding biases. In particular, junk science has been heavily employed in the debate surrounding climate change and what to do about it in the United States.

Evaluating Science.   Whether scientist or layperson, we can use facets of the scientific method to judge the relative validity of alternative viewpoints and to develop our own capacity for logical reasoning. Here are some basic questions to ask:

- What are the observations underlying the conclusion or explanation? Can they be satisfactorily confirmed?

- Do the explanations and theories follow logically from the data?

- Does the explanation account for all of the observations? (If the conclusion is logically inconsistent with any observations, it must be judged as questionable at best.)

- Are there reasons that a particular explanation is favored? Who profits from having that explanation accepted broadly?

- Is the conclusion supported by the *community of scientists* with the greatest competence to judge the work? If not, it is highly suspect.

In sum, sound science is absolutely essential to forging a sustainable relationship with the natural world. Our planet is dominated by human beings—our activities have reached such an intensity and such a scale that we are now one of the major forces affecting nature. Our influence on the natural ecosystems that support most of the world economy and process our wastes is strong and widespread, and we need to know how to manage the planet so as to maintain a sustainable relationship with it. As a result, the information gathered by scientists needs to be accurate, credible, and communicated clearly to policy makers and the public, and the policy makers need to handle that information responsibly. The interaction between science and politics will be explored in Chapter 2.

Sustainability

A system or process is *sustainable* if it can be continued indefinitely, without depleting any of the material or energy resources required to keep it running. The term was first applied to the idea of **sustainable yields** in human endeavors such as forestry and fisheries. Trees, fish, and other biological species normally grow and reproduce at rates faster than that required just to keep their populations stable. This built-in capacity allows every species to increase or replace a population following some natural disaster.

Thus, it is possible to harvest a certain percentage of trees or fish every year without depleting the forest or reducing the fish population below a certain base number. As long as the number harvested stays within the capacity of the population to grow and replace itself, the practice can be continued indefinitely. The harvest then represents a sustainable yield. It becomes unsustainable only when trees are cut or fish are caught at a rate that exceeds the capacity of their present population to reproduce and grow. The concept of a sustainable yield can also be applied to freshwater supplies, soils, and the ability of natural systems to absorb pollutants without being damaged.

The notion of sustainability can be extended to include ecosystems. *Sustainable ecosystems* are entire natural systems that persist and thrive over time by recycling nutrients, maintaining a diversity of species in balance, and using the Sun as a source of sustainable energy. As Chapters 3, 4, and 5 show, ecosystems are enormously successful at being sustainable.

Sustainable Societies.   Applying the concept of sustainability to human systems, we say that a *sustainable society* is a society in balance with the natural world, continuing generation after generation, neither depleting its resource base by exceeding sustainable yields nor producing pollutants in excess of nature's capacity to absorb them. Many primitive societies were sustainable in this sense for thousands of years.

When the concept of sustainability is applied to modern societies, we generally picture certain desirable, healthy characteristics of the people, their communities, and the ecosystems on which they depend. Many of our interactions with the environment are *not* sustainable, however, demonstrated by such global trends as the decline of biodiversity and essential ecosystems and the increased emissions of greenhouse gases. Although population growth in the industrialized countries has almost halted, these countries are using energy and other resources at unsustainable rates, producing pollutants that are accumulating in the atmosphere, water, and land. In contrast, developing countries are experiencing continued population growth, yet are often unable to meet the needs of many of their people in spite of heavy exploitation of their natural resources. As our modern societies pursue continued economic growth and progress, we are apparently locked into an unsustainable mode. How do we resolve this dilemma? One answer is the concept of sustainable development.

Sustainable Development. *Sustainable development* is a term that was first brought into common use by the World Commission on Environment and Development, a group appointed by the United Nations. The commission made sustainable development the theme of its final report, *Our Common Future*, published in 1987. The report defined sustainable development as a form of development or progress that "meets the needs of the present without compromising the ability of future generations to meet their own needs." The concept arose in the context of a debate between the respective *environmental* and *developmental* concerns of groups of developed and developing countries. **Development** refers to the continued improvement of human well-being, usually in the developing countries. Both developed and developing countries have embraced the concept of sustainable development, although the industrialized countries are usually more concerned about environmental sustainability, while the developing countries are more concerned about economic development. The basic idea, however, is to maintain and improve the well-being of both humans and ecosystems.

The concept of sustainable development is now so well entrenched in international circles that it has become almost an article of faith. It sounds comforting, so people want to believe that it is possible, and it appears to incorporate some ideals that are sorely needed, such as **equity**—whereby the needs of the present are actually met and future generations are seen as equally deserving as those living now. Sustainable development means different things to different people, however, as illustrated by the viewpoints of three important disciplines traditionally concerned with the processes involved. *Economists* are concerned mainly with growth, efficiency, and the optimum use of resources. *Sociologists* mainly focus on human needs and on concepts like equity, empowerment, social cohesion, and cultural identity. *Ecologists* show their greatest concern for preserving the integrity of natural systems, for living within the carrying capacity of the environment, and for dealing effectively with pollution. It can be argued, however, that sustainable solutions will be found mainly where the concerns of these three groups intersect, as illustrated in Figure 1–11.

An Ideal. There are many dimensions to sustainable development—environmental, social, economic, political—and no societies today have achieved anything resembling it. Nevertheless, as with justice, equality, and freedom, it is important to uphold sustainable development as an ideal—a goal toward which all human societies need to be moving, even if we have not achieved it completely anywhere. For example, policies and actions in a society that reduce infant mortality, increase the availability of family planning, improve the air quality, provide more abundant and pure water, preserve and protect natural ecosystems, reduce soil erosion, reduce the release of toxic chemicals to the environment, restore healthy coastal fisheries, and so on are all moving that society in the right direction—toward a sustainable future. At the same time, these actions are all quite measurable, so progress in achieving sustainable development can be assessed. Communities and organizations are forging sustainable development indicators and goals to track their progress, such as the Environmental Sustainability Index

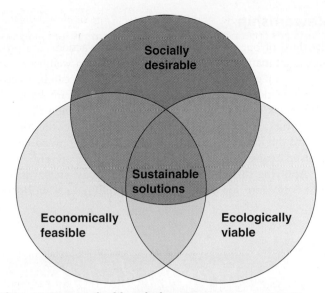

Figure 1–11 Sustainable solutions. If the concerns of sociologists, economists, and ecologists can be brought into intersection, sustainable solutions in a society are much more achievable.

(Yale University, 2005), which evaluates the ability of nations to protect the environment, and the Millennium Development Goals referred to earlier.

An Essential Transition. The transition to a truly sustainable civilization is hard to picture at present. It requires achieving a stable human population that recognizes the finite limits of the ability of Earth's systems to produce resources and absorb wastes and that acts accordingly. However, if we fail to achieve sustainability by our deliberate actions, the natural world will impose it on us in highly undesirable ways, such as through famine, disease, and deprivation. To achieve sustainability will require a special level of dedication and commitment to care for the natural world and to act with justice and equity toward one another. What will it take to make the transition to a sustainable future? There is broad agreement on the following major points:

- A demographic transition from a continually increasing human population to one that is stable.

- A resource transition to an economy that is not obsessed with growth, but instead relies on nature's income and protects ecosystem capital from depletion.

- A technology transition from pollution-intensive economic production to environmentally friendly processes.

- A political/sociological transition to societies that embrace a stewardly and just approach to people's needs and that eliminate large-scale poverty.

- A community transition from the present car-dominated urban sprawl of developed countries to the "smart growth" concepts of smaller, functional settlements and more livable cities.

This is not an exhaustive list, but it does give a glimpse of what a sustainable future should look like. Note that these transitions incorporate the characteristics of our third unifying theme, stewardship.

Stewardship

The third of our themes is *stewardship*—the actions and programs that manage natural resources and human well-being for the common good. Stewardship is a concept that can be traced back to ancient civilizations. A steward was put in charge of his master's household, responsible for maintaining the welfare of the people and the property of the owner. Because a steward did not own the property himself, the steward's ethic involved caring for something on behalf of someone else.

Applying this concept to the world today, stewards are those who care for something—from the natural world or from human culture—that they do not own and that they will pass on to the next generation. Modern-day environmental stewardship, therefore, incorporates an ethic that guides actions taken to benefit the natural world and other people. It recognizes that even our ownership of land is temporary; the land will be there after we die, and others will own it in turn. Stewardly care is compatible with the goal of sustainability, but it is different from it, too, because stewardship deals more directly with how sustainability is to be achieved—what actions are taken, and what values and ethical considerations are behind those actions. (See Stewardship, "Exploring a Stewardship Ethic.")

Stewardship

Exploring a Stewardship Ethic

One of the major concerns of philosophers is the development of codes of practice and thought that define right and wrong, known as *ethics*. Ethics is about the good—those values and virtues we should encourage—and about the right—our moral duties as we face practical problems. Ethics, therefore, is a "normative" discipline—it tells us what we ought to do. Some things are right, and some things are wrong. How do we know what we ought to do? In other words, how does an ethic work? A fully developed ethic has four ingredients:

1. **Cases.** These refer to specific acts and ask whether a particular act is morally justified. The answer must be based on moral rules.

2. **Moral rules.** These are general guidelines that can be applied to various areas of concern, such as the rules that govern how we should treat endangered species.

3. **Moral principles.** Moral rules are based on more general principles, which are the broadest ethical concepts. They are considered to be valid in all cases. An example is the principle of distributive justice, which states that all persons should be treated equitably.

4. **Bases.** Ethical principles are justified by reference to some philosophical or theological basis. This is the foundation for an ethical system.

A stewardship ethic is concerned with right and wrong as they apply to actions taken to care for the natural world and the people in it. Because a steward is someone who cares for the natural world on behalf of others, we might ask, "To whom is the steward responsible?" Many would answer, "The steward is responsible to present and future generations of people who depend on the natural world as their life-support system." For people with religious convictions, stewardship stems from a belief that the world and everything in it belongs to a higher being; thus, they may be stewards on behalf of a higher power. For others, stewardship becomes a matter of concern that stems from a deep understanding and love of the natural world and the necessary limitations on our use of that world. For yet others, stewardship provides an opportunity for greed and self-serving; such stewardship is reprehensible.

Is there a well-established stewardship ethic? Insofar as human interests are concerned, ethical principles and rules are fairly well established, even if they are often violated by public and private acts. However, there is no firmly established ethic that deals with care for natural lands and creatures for their own sake. Most of our ethic concerning natural things really deals with how those things serve human purposes; that is, our current ethic is highly anthropocentric. For example, a basic ethical principle from the UN Declaration on Human Rights and the Environment is "All persons have the right to a secure, healthy, and ecologically sound environment." The environment, in other words, exists to serve humanity. So, just as there are good stewardship and poor stewardship in the actions taken to care for the natural world and humankind, there are also good and poor stewardship ethics.

Many organizations have developed stewardship principles for their own guidance, and these principles can sometimes serve as broader guidelines for society, although they usually address specific areas of concern and are generally anthropocentric. For example, the U.S. Forest Service has published a "Land and Service Ethic," which is supposed to guide Forest Service personnel as they carry out their activities. The Forest Service's Land Ethic principle is to "promote the sustainability of ecosystems by ensuring their health, diversity, and productivity." The agency goes on to explain: "Through ecosystem sustainability, present and future generations will reap the benefits that healthy, diverse, and productive ecosystems provide."*

Here are some of the issues that a stewardship ethic must grapple with in order to be truly helpful in accomplishing our stated goal of providing the "actions and programs that manage natural resources and human well-being for the common good":

1. How to define the common good in cases where conflicting needs emerge, such as the economic need to extract resources from natural systems versus the need to maintain those systems in a healthy state.

2. How to balance the needs of present generations versus those of future generations.

3. How to preserve species when doing so clearly means limiting some of the property rights commonly enjoyed by people and organizations.

4. How to encourage people and governments to exercise compassion and care for others who suffer profoundly from a lack of access to the basic human needs of food, shelter, health, and gainful work.

5. How to promote virtues that are conducive to stewardly care of the natural world, such as benevolence, accountability, frugality, and responsibility. (There are many others.)

6. How to limit consumption while at the same time allowing people the freedom to choose their lifestyles. In other words, how do you balance individual environmental rights and responsibilities with those of the community?

*Forest Service, U.S. Department of Agriculture, *The Forest Service: Ethics and Course to the Future* (Washington, DC.: USDA Forest Service, 1994).

Who Are the Stewards? How is stewardship achieved? Sometimes stewardship leads people to try to stop the destruction of the environment or to stop the pollution that is degrading human neighborhoods and health. For example, Rachel Carson, in her 1962 book *Silent Spring*, alerted the public about the dangers of pesticides and their role in rapidly declining bird populations. Pablo Fajardo Mendoza and Luis Yanza, of Ecuador (Fig. 1–12), battled oil giant Chevron by bringing it to court and demanding that it pay for massive petroleum contamination. Their work has led the government of Ecuador to pass stronger environmental protection laws; they received the 2008 Goldman Prize for their work. Stewardship is not without risks, however—the two men and their families have received frequent death threats and intimidation. Dr. Wangari Maathai (the first Kenyan woman to earn a Ph.D.) founded the Green Belt Movement, which has planted 30 million trees in that country, but she was also beaten and jailed when she went on to protest government corruption. Eventually, she was appointed deputy environment minister under a new Kenyan president and received the Nobel Peace Prize in 2004 (Fig. 1–13)—the first environmental activist to receive that honor.

More often, stewardship is a matter of everyday people caring enough for each other and for the natural world that they do the things that are compatible with that care. These include participating fully in recycling efforts, purchasing cars that pollute less and use less energy, turning off the lights in an empty room, refusing to engage in the conspicuous consumption constantly being urged on them by commercial advertising, supporting organizations that promote sustainable practices, staying informed on environmentally sensitive issues, and expressing their citizenship by voting for candidates who are sympathetic to environmental concerns and the need for sustainable development.

Justice and Equity. The stewardship ethic is concerned not only with caring for the natural world, but also with establishing just relationships among humans. This concern

Figure 1–13 Stewardship at work. Wangari Maathai receiving the 2004 Nobel Peace Prize from Ole Danbolt Mjøs, chairman of the Norwegian Nobel Committee. She is an example of unusual environmental stewardship at work.

for justice is seen in the United States in the *environmental justice movement*. The major problem addressed by the movement is **environmental racism**—the placement of waste sites and other hazardous industries in towns and neighborhoods where most of the residents are nonwhite. The flip side of this problem is seen when wealthier, more politically active, and often predominantly white communities receive a disproportionately greater share of facility improvements, such as new roads, public buildings, and water and sewer projects.

People of color are seizing the initiative to correct these wrongs, creating citizen groups and watchdog agencies to bring effective action and to monitor progress. For example, Piney Woods and Alton Park are suburbs of Chattanooga, Tennessee, and are populated largely by African Americans. In those neighborhoods, there are 42 known hazardous waste sites, 12 of which have been listed as Superfund sites (polluted sites deemed serious enough to require government intervention and remediation). Cancer and asthma rates there were unusually high, so people organized the group Stop TOxic Pollution (STOP) to draw attention to the toxic waste and health problems and, eventually, to acquire a technical assistance grant from the Environmental Protection Agency (EPA) to help fence off the worst sites. Sociologist Robert Bullard of Clark Atlanta University has established the Environmental Justice Resource Center as a clearinghouse of resources to fight environmental racism.

Justice for the Developing World. Justice is especially crucial for the developing world, where unjust relationships often leave people without land, with inadequate food, and in poor health. Extreme poverty is the condition of a billion people, whose poverty is often brought on by injustices within

Figure 1–12 Goldman Prize winners. Luis Yanza (left) and Pablo Fajardo Mendoza received the 2008 Goldman Prize for their work in bringing Chevron to court in a lawsuit demanding that the company clean up the oil sludge and waste pits left behind when they were drilling oil in Ecuador.

societies where wealthy elites maintain political power and, through corruption and nepotism, steal money and create corporations that receive preferential treatment. It will take significant government reform to turn this situation around so that the poor can achieve justice in securing property rights and gain greater access to ecosystem goods and services.

Some of the poverty of the developing countries can be attributed to unjust economic practices of the wealthy industrialized countries. The current pattern of international trade is a prime example. The industrialized countries have maintained inequities that discriminate against the developing countries by taxing and restricting imports from the developing countries and by flooding the world markets with agricultural products that are subsidized (and, therefore, priced below real cost). This is the number one issue at every meeting of the World Trade Organization, where global trade agreements are negotiated. Such barriers deprive people in developing countries of jobs and money that would go far to improve their living conditions. Although international justice is a difficult issue, it is part of the mission of stewardly action to address it.

1.4 Moving Toward a Sustainable Future

Without doubt one of the emerging trends of our times is *globalization*—the accelerating interconnectedness of human activities, ideas, and cultures. For centuries, global connections have brought people together through migration, trade, and the exchange of ideas. What is new about the current scene is the increasing intensity, speed, extent, and impact of connectedness, to the point where its consequences are profoundly changing many human enterprises. These changes are most evident in the globalization of economies, cultural patterns, political arrangements, environmental resources, and pollution. For many in the world, globalization has brought undeniable improvements in well-being and health, as economic exchanges have connected people to global markets, information exchanges have improved public-health practices, and agricultural research has improved crop yields. For some, however, globalization has brought the dilution and even destruction of cultural and religious ideals and norms and has done little to improve economic well-being. Whether helpful or harmful, globalization is taking place, and it is important not only to recognize its impacts, but also to point out where its impacts are harmful or undesirable. Moving toward a sustainable future will occur in the context of globalization.

Economic Changes

The major element of globalization may be the economic reorganization of the world. This transformation has been facilitated by the globalization of communication, whereby people are instantly linked through the Internet, satellites, and cables. Other components of the global economic reorganization are the relative ease of transportation and financial transactions and the dominance of transnational corpora-

tions (e.g., Nokia, Roche, Philips Electronics, Nestlé) with unprecedented wealth and power. Also, there are special trade arrangements between countries that give preferential treatment to some and deny trade access to others. Moreover, economic changes bring cultural, environmental, and technological changes (to name a few) in their wake. Western diets, styles, and culture are marketed throughout the world, to the detriment of local customs and diversity.

Environmental Changes

On the one hand, the increased dissemination of information has made it possible for environmental organizations and government agencies to connect with the public and has enabled people to become politically involved with current issues. It has also enabled consumers to find more environmentally friendly consumer goods and services, such as shade-grown coffee and sustainably harvested timber. On the other hand, globalization has contributed to some notoriously harmful outcomes, such as the worldwide spread of emerging diseases like the SARS virus, H1N1 swine influenza, and AIDS; the global dispersion of exotic species, the trade in hazardous wastes; the spread of persistent organic pollutants; the radioactive fallout from nuclear accidents; the exploitation of oceanic fisheries; the destruction of the ozone layer; and global climate change. These outcomes must be addressed in the move toward a sustainable future.

A New Commitment

People in all walks of life—scientists, sociologists, workers and executives, economists, government leaders, and clergy, as well as traditional environmentalists—are recognizing that "business as usual" is not sustainable. Many global trends are on a collision course with the fundamental systems that maintain our planet as a tolerable place to live. A finite planet cannot continue to grow by 75 million persons annually without significant detrimental effects. The current degradation of ecosystems, atmospheric changes, losses of species, and depletion of water resources inevitably lead to a point where resources are no longer adequate to support the human population and where, consequently, civil order will break down. As one observer put it, "If we don't change direction, we will end up where we are heading."

The news is not all bad, however. Food production has improved the nutrition of millions in the developing world, life expectancy continues to rise, and the percentage of individuals who are desperately poor continues to decline. Population growth rates continue to fall in many of the developing countries. A rising tide of environmental awareness in the industrialized countries has led to the establishment of policies, laws, and treaties that have improved the protection of natural resources and significantly reduced the pollution load.

Numerous caring people are beginning to play an important role in changing society's treatment of Earth. For example, many are now adding their voices to the literally hundreds of environmental and professional organizations devoted to controlling pollution and protecting wildlife. People in business have formed the Business Council for Sustainable Development, economists have formed the International Society for Ecological Economics, religious leaders have formed the National Religious Partnership for the Environment, and environmental philosophers are calling for a new ethic of "caring for creation."

The movement toward sustainability is growing as people, businesses, and governments slowly but surely begin to change direction. Some examples: Wind energy is the fastest-growing source of electricity; cap-and-trade markets are being established to curb carbon emissions; car-sharing companies are springing up in our cities; engineers are designing manufacturing on principles of no waste and carbon neutrality; an estimated $100 billion a year are directly addressed to environmental concerns; microfinance is opening up people's small-scale business opportunities; "sustainability investments" are attracting increasing capital; and as seen with the Kyoto Protocol, the nations of the world are working on agreements to bring down greenhouse gas emissions.

These developments show that environmental degradation *can* be slowed down and reversed, people *can* be freed from hunger and poverty, and people's behavior toward the environment *can* be transformed from exploitative to conserving.

Revisiting the Themes

It will be our practice in each chapter to revisit the three themes that we have introduced in this chapter, summarizing the ways in which the chapter topics have connected to the themes and sometimes adding editorial comments.

Sound Science

Our brief look at the global environment was based largely on the information uncovered by sound science. The Millennium Ecosystem Assessment and the Intergovernmental Panel on Climate Change demonstrate the crucial value of scientific study of the status of and trends in global ecosystems and the ways in which human systems affect them. Environmental science itself is a multidisciplinary enterprise, one that can readily lead to disputes and misuse. To understand these, it is essential to understand the scientific method as a disciplined way to investigate the natural world. The essay about the sudden decline of vultures in the Indian subcontinent illustrates how rigorous application of the scientific method led to discovery of the cause of the decline. The issue of "junk science" points to the importance of a functioning scientific community and the need for an educated public able to evaluate opposing claims of scientific support for a given issue.

Sustainability

The Easter Island story starkly illustrates the consequences for a society that was unable to adapt sustainably to its limited resources. The result was not only the loss of natural goods and services that the island's ecosystems could provide, but also the loss of an entire culture. As we examine the global environment, ask yourself whether there are any parallels between what happened on Easter Island and what is happening now across the entire planet. The human population continues to expand. With expansion comes greater consumption, so natural and managed ecosystems are being pressed to provide increasing goods and services. We may have reached the limit in some resources, and most of the ways we use the other resources are currently unsustainable. In addition, we extract and burn fossil fuels much faster than they could ever be replenished, bringing on global climate change, and we tolerate the continued loss of biodiversity. Like Easter Island, this is still a world of haves and have-nots, and the haves still seem unwilling to take the necessary steps to help those suffering from hunger and the other effects of poverty. The Millennium Development Goals are moving us in a better direction, but it remains to be seen if those goals will be met. A movement toward sustainability is under way, and it will occur within the often disturbing context of globalization.

Stewardship

The religion and culture of the Rapa Nui of Easter Island did not include an effective stewardship ethic. If they had, their story might have had a happier ending. Although current cultures and religions place some value on stewardship, it is not a fundamental part of the way we think about caring for the natural world and for our fellow humans. If it were, there would not be the widespread degradation of essential ecosystems and the devastating impacts of poverty and injustice in the developing countries. As the essay about the stewardship ethic shows, it is no easy matter to establish a valid stewardship ethic to guide the actions and programs that manage natural resources and human well-being for the common good.

REVIEW QUESTIONS

1. What factors brought about the collapse of the Easter Island civilization? How did later contact with the rest of the world affect the islanders?

2. Cite four global trends that indicate that the health of planet Earth is suffering.

3. Define *environment*, and describe the general development and successes of modern environmentalism.

4. What are the concepts behind each of the three unifying themes—sound science, sustainability, and stewardship?

5. Explain the role of assumptions, observation, experimentation, and theory formation in the operation of scientific research and thinking.

6. What led to the decline of vultures in India and Pakistan? How did scientific work lead to the explanation for the decline?

7. Cite some reasons for the existence of controversies within science, and carefully distinguish between sound science and junk science.

8. Define *sustainability* and *sustainable development*. What is a sustainable society?

9. List five transitions that are necessary for a future sustainable civilization.

10. Describe the origins of stewardship and its modern applications.

11. What is an ethic, and what are some of the difficult questions confronting the stewardship ethic?

12. What are the concerns of the environmental justice movement?

13. How does justice become an issue between the industrialized countries and the developing countries?

14. What is globalization? What are its most significant elements?

15. Cite some of the recent developments in the movement toward sustainability.

THINKING ENVIRONMENTALLY

1. Imagine a class debate between people representing the developing countries and people representing the developed countries. Characterize the arguments of the two sides in terms of the issues surrounding population growth, energy use, resource use, and sustainable development. Then describe what should be the common interests between the two.

2. Some people say that the concept of sustainable development either is an oxymoron or represents going back to some kind of primitive living. Develop an argument contradicting this opinion, and give your own opinion on sustainable development.

3. Study Table 1–1, pick one of the vital ecosystem services listed there, and investigate the reasons why it is placed where it is by the Millennium Ecosystem Assessment.

4. Set up a debate between proponents and opponents of building a new coal-fired power plant. Use the strategies of sound science and junk science to conduct the debate, and show how a perverted use of these terms can help to obscure the real issues. Other interesting debate options include the "ozone hoax" and global warming; the Internet can provide ample material for both sides of the debate.

MAKING A DIFFERENCE

1. Find out whether there is an environmental organization or club on your campus or in your community. Join it if there is, or help to create one if there isn't. Once in the organization, join efforts to maintain awareness of important environmental issues and problems.

2. Use letters to the local newspapers or visits to town meetings to raise awareness in your community about sustainable systems and sustainable societies. Examples of issues to raise include smart growth; cluster zoning, which provides a healthy mix of open space and buildings; recycling; and ways to encourage public transit systems.

3. Encourage sound science on a national level by researching and voting in favor of political candidates who support your environmental viewpoint. Share your views with your friends and relatives.

4. Investigate the placement of waste treatment facilities and industrial plants in your city or community, especially if they are in the planning stage. If they are aimed at poorer or largely nonwhite communities, join the local population in protesting these situations.

Mastering ENVIRONMENTAL SCIENCE™

Go to **www.masteringenvironmentalscience.com** for practice quizzes, Pearson eText, videos, current events, and more.

An observer looks out over the Benxi Iron and Steel Group in Liaoning, China.

CHAPTER 2

Economics, Politics, and Public Policy

KEY TOPICS

1. **Economics and the Environment:** Economic activity depends on environmental goods and services, and it should be controlled by effective public policies and institutions.

2. **Resources in a Sustainable Economy:** The wealth of a nation is the human and natural resources a country draws on to establish and maintain an economy and may best be measured by the genuine progress indicator, not the gross domestic product.

3. **Environmental Public Policy:** Policies are essential for regulating the use of natural resources and dealing with pollution; the policies may be market based or regulatory, and they often take on a typical "life cycle."

4. **Benefit-Cost Analysis of Environmental Public Policy:** Benefit-cost analysis is an economic tool that, by law, is applied to environmental policy regulations.

5. **Politics and the Environment:** Partisan politics has played an important role in recent environmental public policy, sometimes interfering with the work of scientists.

B Y ALL MEASURES, China has become the economic giant of the early 21st century. Take your pick: umbrellas, toys, shoes, appliances, electronics, souvenirs, clothes—chances are that they were made in China. China has become the factory to the world. The Chinese economy has doubled in the last decade, growing at close to 10% per year. A new middle class numbers between 100 and 150 million. These people commonly own a car and an apartment and have opportunities to travel and purchase pricey consumer goods. With a population of 1.3 billion, China has a huge pool of labor. Now that the control of private life is a thing of the past, people are able to choose where they live, work, and travel. Millions of rural Chinese move to urban centers each year, attracted by the jobs created by factories that spring up almost overnight. Chinese schools have created a workforce with a literacy rate that is over 90%. Because of the accelerating pace of cultural life, children are now guiding their parents into the realities of modernity. Traditions are being overthrown by the Internet, by the fierce drive to learn English, and by the social mobility that accompanies the economic boom.

TOXIC AIR. There are clear signs that China's race to achieve economic superpower status carries with it some costs that go well beyond a stretching of the social fabric. Sixteen of the 20 cities with the world's most polluted air are in China.

The air is loaded with sulfur dioxide from coal-burning power plants and coal stoves, and emissions from vehicles are now adding to the toxic mix of air pollutants. Up to 750,000 premature deaths a year are believed to be traceable to air pollution. In the city of Linfen in Shangxi province, the hospital treats countless numbers of patients with respiratory disease each day. The city is dotted with coke ovens, coal-processing plants, factories, and a coal-fired power plant. It has been at the top of China's pollution charts for years. One of its hamlets, Beilu, is known as "the cancer village" because of the extremely high death rate for those 55 and older. Beijing, site of the 2008 summer Olympics, had to temporarily shut down hundreds of factories and ban millions of cars in order to bring air quality to "tolerable" levels for the games.

TOXIC WATER. Xiao Sizhu, a farmer in the village of Xiaojiadian, died at 55 of esophageal cancer. In an interview shortly before his death, Xiao spoke of swimming and fishing in the tributary of the Yellow River that runs through the village. However, he says, "Now I never go close to the water because it smells awful and has foam on top." One oncologist calls the area "the cancer capital of the world" because of esophageal and stomach cancer rates that are 25 times higher than the national average. The Yellow River itself is being sucked dry by irrigation projects stemming from some 20 dams, and the remaining water is grossly polluted with sewage and industrial wastes. Three chemical spills in 2006 turned the Yellow River red (Fig. 2–1), and another spill turned it white. More than 500 million rural Chinese are forced to use water that is contaminated by human and industrial waste; it is no wonder that cancer rates across China have been rising rapidly.

ENVIRONMENTAL LAWS? The Chinese economy has flourished in large degree because of the decentralization of power in the 1980s, where local governments and private industries were encouraged to develop their resources relatively free of the restraints of central planning. The booming economy has also put great power in the hands of the local governments and industries and has enabled them to flaunt existing environmental laws and policies. China does have its State Environmental Protection Agency (SEPA), similar to the U.S. Environmental Protection Agency (EPA), but the country is so large that SEPA has

Figure 2–1 The Yellow River turns red. This vital lifeline to central China turned red from pollutants in 2006. It is now one of the most polluted rivers on Earth.

been unable to keep up with the need for enforcement. In response, the Chinese government has been turning to international nongovernmental organizations (NGOs) and has encouraged concerned citizens to form homegrown NGOs that focus attention on pollution and other environmental issues. At least 2,000 such groups are now registered, and they are beginning to make a difference. Typically, a group—like Green Camel Bell in the western Chinese city of Lanzhou—identifies a case of blatant pollution by some factory, documents the pollution, and reports it to the central government. Green Camel Bell was successful in shutting down over 30 factories dumping wastes into tributaries of the Yellow River.

BOTTOM LINE. China's example demonstrates that economic growth can lift millions out of poverty and establish a country's strong place in the world market. But when this growth is at the cost of natural resources and the health of many of its people, it is clearly unsustainable. What is happening in China has already taken place in the industrialized countries of Japan and the West, but in these countries, many of the worst problems have been addressed by environmental laws and regulations. The Chinese example shows us that economic growth must be accompanied by the development and enforcement of just and effective environmental public policies.

We look first at economics, the subject historian Thomas Carlyle called—perhaps unfairly—"the dismal science."

2.1 Economics and the Environment

Economics is *the social science that deals with the production, distribution, and consumption of goods and services and with the theory and management of economies or economic systems.* Because economic goods and services are usually connected to—and indeed depend on—environmental goods and services, economics needs to be grounded in an understanding of the biological and physical world. Simply put, no environmental goods and services, no economy.

Relationships Between Economic Development and the Environment

In a human society, an **economy** is *the system of exchanges of goods and services worked out by members of the society.* Goods and services are produced, distributed, and consumed as people make economic decisions about what they need and want and what they will do to become players in the system—what they might provide that others would need and want. As societies develop, economic activity assumes increasingly broader dimensions, with increasingly pervasive impacts on the whole society. In future chapters, we will see numerous instances in which economic activity can damage the environment and human health. In addition, an unregulated economy can make intolerable inroads into natural resources. For example, during the latter part of the 19th century in the United States, private enterprise had unrestrained access to forests, grazing lands, and mineral deposits. This unsustainable exploitation of resources began to be addressed at the turn of the century, as government rules and regulations imposed necessary limits.

Fortunately, economic activity in a nation can also provide the resources needed to solve the problems that very activity creates. A strong relationship exists between the level of development of a nation and the effectiveness of its environmental public policies. Figure 2–2 shows a number of

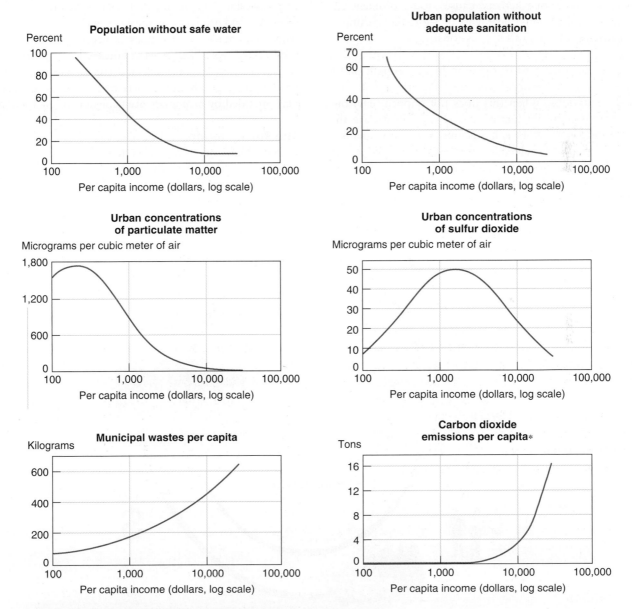

Note: Estimates are based on cross-country regression analysis of data from the 1980s.
* Emissions are from fossil fuels.

Figure 2–2 Environmental indicators and per capita income. Some of the most serious environmental problems can be improved with income growth; other problems get worse and then improve, while some problems just worsen. (*Source:* World Bank, *World Development Report 1992* (New York: Oxford University Press, 1992), p. 11.)

environmental indicators in relation to *per capita* (per individual in the population) income levels. (The data were combined from studies of different countries.) Several patterns emerge:

- Many problems decline (for example, sickness due to inadequate sanitation or water treatment) as income levels rise because the society decides to apply the resources available—usually from taxes—to address the problems with effective technologies.

- Some problems increase and then decline (for example, urban air pollutants like sulfur dioxide and carbon monoxide) when the serious consequences of the problem are recognized and public policies are developed to address them.

- Increased economic activity causes some problems to increase without any clear end in sight (for example, suburban sprawl and CO_2 emissions).

The key to solving all of these problems brought on by economic activity is the *development of effective public policies and institutions*. When this does not happen, environmental degradation and human disease and death are the inevitable outcomes. What is happening today in China is a sad case in point. To understand this relationship better, let us look at how economic systems work.

Economic Systems

Economic systems are *social and legal arrangements people construct in order to satisfy their needs and wants and improve their well-being.* Economic systems range from the communal barter systems of primitive societies to the current global economy, which is so complex that it defies description. In the organized societies of today's world, two kinds of economic systems have emerged: the **centrally planned economy**, characteristic of dictatorships, and the **free-market** (or **capitalist**) **economy**, characteristic of democracies.

In any kind of economy, there are basic components that determine the economic flow of goods and services. Classical economic theory considers **land** (natural resources), **labor**, and **capital** as the three elements that constitute the "factors of production." *Economic activity* involves the circular flow of money and products, as shown in Figure 2–3. Money flows from households to businesses as people pay for goods and services and from businesses to households as people are paid for their labor. Labor, land, and capital are invested in the production of goods and services by businesses, and the products provided by businesses are "consumed" by households.

The Rulers Decide. The cycles shown in Figure 2–3 apply to both kinds of economic systems in existence today; the two systems differ mostly in how economic decisions are

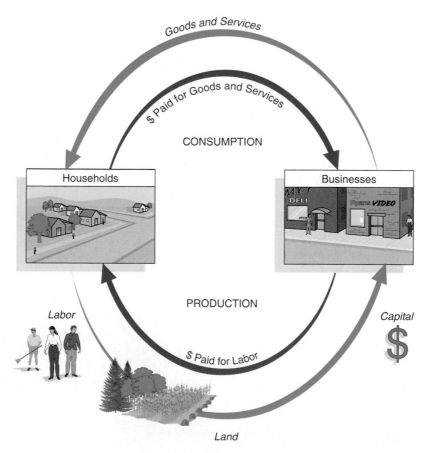

Figure 2–3 Classical view of economic activity. Land (natural resources), labor, and capital are the three elements constituting the "factors of production" (bottom green arrow). Economic activity involves the circular flow of money (blue arrows) and products (top green arrow).

made. In a centrally planned economy, the rulers make all the basic decisions regarding what and how much will be produced where and by whom. The government owns or manages most industries, and the workers often lack the freedom to choose or change jobs. This system was most prominent during the 20th century in the communist regimes of the Soviet Union and its satellites and in mainland China and some smaller Asian countries. It proved to be a disaster, both economically and environmentally. The failure of the centrally planned economy in the former Soviet Union led to economic chaos in the countries that it once comprised or dominated. North Korea, Cuba, and a few African countries remain as the last holdouts of this economic model. Indeed, it was the decision by the Chinese government to abandon this economic model that laid the foundation for China's remarkable economic growth.

The Market Decides. In a pure free-market economy, the market itself determines what will be exchanged. Goods and services are offered in a market that is free from government interference. The system is completely open to competition and the interplay of supply and demand. If supply is limited relative to demand, prices rise; if there is an oversupply, prices fall because people and corporations will economize and pay the lowest price possible. The whole system is in private hands and is driven by the desire of people and businesses to acquire goods, services, and wealth as they act in their self-interest. All "players" have free access to the market. Competition spurs efficiency as inferior products and services are forced out of the market. People can make informed decisions about their purchases because there is sufficient information about the benefits and harms associated with economic goods. The free-market economic system is thought to be at its best when left completely alone.

Governments Step In. The preceding discussion describes only the very basics of the two economic systems. In reality, no country has a pure form of either economy. The developed countries and most developing countries function with market economies, but government involvement occurs at many levels. Governments can, for example, own and operate different facets of the country's infrastructure, such as postal services, power plants, and public transport systems. Governments may also control interest rates, determine the amount of money in circulation, and adopt policies that stimulate economic growth in times of slowdown. Governments also maintain surveillance over financial processes such as stock markets and bank operations (although not too well, based on the 2008 market meltdown). On the downside, powerful business interests can manipulate a market economy, and unscrupulous people can exploit the freedom of the market to defraud people. Only governments can provide the policing that the society needs.

Is There a Conscience? Many people believe that the conventional capitalist economic system lacks a "conscience." They argue that many workings of the market economy create hardships for people (and countries) at the bottom of the economic ladder. For example, if population growth is rapid and there are not enough jobs to go around, the result often is exploitation of workers. Furthermore, a market economy only offers people *access* to goods and services; if people lack the means to pay, access alone will not meet their needs. In this situation, a society could employ a "safety net" to provide food security for impoverished people (see Chapter 12), something the market will not do. Also, it is too easy for self-interest to lead businesses and individuals to exploit natural resources or avoid costs by polluting the environment instead of producing a "clean" product.

Presumably acting in the self-interest of their people, governments also find ways to manipulate the market in their favor. This is especially true for international trade, where protective barriers and subsidies prevent the free flow of goods and services. The World Trade Organization (WTO) is supposed to enforce trade rules between nations, but the wealthy nations have been able to dominate the WTO to the detriment of the other countries.

International Trade and the World Trade Organization

Trade is a fundamental economic activity. The exchange of goods drives the engine of the free-market economy. In today's world, trade has become increasingly globalized, and in the process, world trade has become one of the most dominant forces in society. However, many question whether this is good, and their concerns become headline news every time the WTO meets. For example, in Seattle, Washington, in December 1999, tens of thousands of protesters (called "globophobes" by some) took to the streets to block the delegates from assembling—and were the recipients of tear gas, pepper spray, and rubber bullets as things began to turn ugly. The protesters were an unusual alliance of environmental activists and organized labor, groups that are often at odds over timber harvests and fishing. What were their issues?

The WTO was created in 1993 by trade ministers from many countries meeting to build on the foundation for global trade laid by the General Agreement on Tariffs and Trade (GATT). The WTO was given the responsibility for implementing numerous trade rules and, in doing so, was empowered to enforce the rules, with the option of imposing stiff trade penalties on violators. In effect, the WTO became a force for the globalization of trade, facilitating it by encouraging the reduction or elimination of trade barriers (tariffs) between countries and of subsidies to domestic industries that tended to discriminate against foreign-produced goods. Those who support the WTO claim that free trade between countries is the key to global economic growth and prosperity. This all sounds very good, but as always, the devil is in the details.

Free-Trade Rules. Since its creation, the WTO has adjudicated a number of international trade issues and, in the process, has established a reputation for elevating free trade over substantial human rights and environmental resource concerns. Furthermore, many trade agreements and the dispute rulings are carried out in closed sessions. For example, the WTO ruled against a U.S. law requiring any imports of shrimp to be certified as turtle safe (that is, the shrimp were to be certified as harvested by trawlers employing "turtle exclusion devices," or TEDs). Even though TEDs have been proven to reduce turtle mortality by 97%, the WTO ruled that the restriction was a violation of WTO rules. Other issues that have come before the WTO are a U.S. ban on the import of wooden crates from China because they harbored the Asian long-horned beetle (a serious threat to forest trees) and bans against developing countries' products that utilize child labor.

More Talk. In the Seattle meetings, the talks broke down not because of the protests, but because of a refusal of the key players—the United States, Japan, and Europe—to move from their positions. In the WTO meetings held in Cancún, Mexico, in September 2003, the talks broke down again before negotiations could proceed. This time a coalition of 20 developing nations (G-20) representing 60% of the world's people simply walked out of the meetings. Their issue was primarily the import barriers and subsidies that distort global trade in agricultural products. The developed nations are major exporters of agricultural products, and they subsidize their farmers heavily; the result is depressed world markets for these products. The developing nations want to protect their farmers from competition with these low-priced farm products, so they erect tariffs.

The future of the WTO is unclear. The **Doha Development Round** of WTO negotiations began in 2001 with the objective of lowering trade barriers around the world, but the subsequent WTO meetings in Seattle, Cancún, Hong Kong (2005), Geneva (2006), Davos (2007), and Geneva (2008) have failed to reach agreement on the issues of farming subsidies and trade barriers. In the latest Geneva talks (2008), the impasse hinged on the insistence of India, China, and other developing countries on a "special safeguard mechanism" that would allow them to raise farm tariffs if rising food imports threatened their subsistence farmers; the United States would not agree to serious cuts in its farm subsidies as long as the developing countries could impose tariffs when needed.

The WTO supposedly retains the power already delegated to it, but the continued failure to reach agreements over tariffs and subsidies signals a serious weakening in global trade governance. The likely outcome is a proliferation of bilateral arrangements between countries, which may not be all bad, given the tendency of the developed countries to dominate WTO issues in the past.

The Need for a Sustainable Economy

The global economy has proven to be quite successful in mobilizing people to engage in economic activity. The Chinese experiment is a case in point. The global economy has also been successful in improving the well-being of people: Life expectancy has never been higher; labor-saving devices remove much of the toil in wealthy countries; and aircraft, cars, computers, and cell phones open up lifestyles and work opportunities unimaginable a few decades ago. However, the global economy has also helped to create environmental problems that we documented in Chapter 1: continued poverty and hunger in developing countries, the decline of ecosystem goods and services, rising global temperatures and seas, and the loss of biodiversity. Given the vital connections between the environment and the economy, there is a clear need for a different kind of economy; "business as usual" does not work. The path to a sustainable future will require a sustainable economy.

What will such an economy look like? Instead of promoting growth, the primary goal of the economy will be improving human well-being. Instead of drawing down natural capital, the economy will value and preserve ecosystem services and goods. Instead of plunging ahead with potentially damaging technologies, societies will embrace the *precautionary principle* (see Chapter 12, p. 309) to minimize risks to human health and the environment. A rising consumer demand for green products—products like compact fluorescent bulbs, hybrid and electric vehicles, solar and wind power systems, and organic foods—will be met by businesses eager to provide them. To quote *State of the World 2008*, "a new kind of economy—a sustainable economy—is struggling to be born."

2.2 Resources in a Sustainable Economy

What are the resources on which a country draws to establish and maintain an economy? The classic economic paradigm would say that *land* (which represents the environment), *labor*, and *capital* are the essential resources needed for a country to be able to mount its economy. However, a new breed of economists, often called **ecological economists**, has emerged in recent years and has taken issue with this view. These economists point out that the classic view sees the environment simply as one set of resources within the larger sphere of the human economy (Fig. 2–3). The environment's vital role in supplying the goods and services on which human activities depend is largely ignored. They argue that the classical approach looks at things backward and that the natural environment actually *encompasses* the economy, which is constrained by the limits of resources in the environment (Fig. 2–4). The production and consumption that define the classical economic view are just part of the system. Without the vital raw materials provided by the environment and without the capacity of the environment to absorb wastes (pollution), there is no economy. Thus, the ecosystems and natural resources found in a given country provide the context for that country's economy. Similarly, the total global biosphere is the context for the global economy.

As described by Thomas Prugh, Robert Costanza, and Herman E. Daly in *The Local Politics of Global Sustainability*, **economic production** can thus be seen as it really is: *the process of converting the natural world to the manufactured world.*

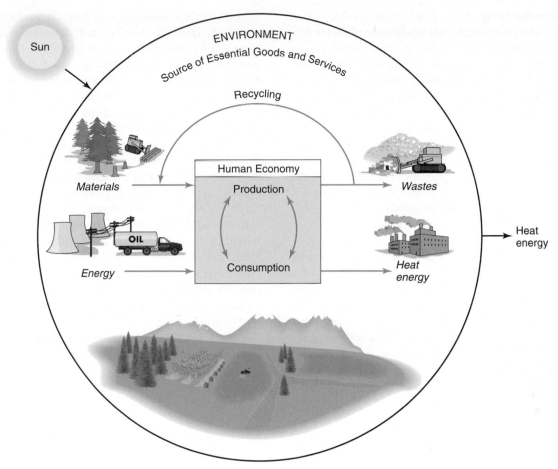

Figure 2–4 Ecological economists' view of economic activity. The natural environment encompasses the economy, which is constrained by the resources found within the environment.

Renewable and nonrenewable resources and the ecosystems containing them are turned into cars, toys, books, food, buildings, highways, computers, and so forth. Thus, to quote these authors, "Resources flow into the economy from the enfolding ecosystem and are transformed by labor and capital (using energy, also a resource), and then pass out of the economy and back into the ecosystem in the form of wastes."[1] (See Figure 2–4.)

The ecological economists' view emphasizes the role of natural ecosystems as essential life-support elements—*ecosystem capital*, as we referred to it in Chapter 1. The approach brings sustainability sharply into focus. If the economy continues to grow, the natural world continues to shrink. In time, economic growth must come to a steady state, and to be sustainable, this must happen before our natural capital is consumed beyond its ability to support our economy. If we add to ecosystem capital the nonrenewable mineral resources, such as fossil fuels and metal ores, we can use the more inclusive term **natural capital**. Thus, *natural capitalism* would be an economic system that promotes sustainability and recognizes the central role of natural ecosystems (and needs to replace *conventional capitalism*). The transition to a sustainable economic system is absolutely essential, but it will be difficult and will take years. It is the linchpin of sustainable development (Chapter 1 describes this transition in more detail).

Without a doubt, the ecosystems and mineral resources of a given country—its *natural capital*—are a major element in the wealth of that country. In recent years, the World Bank's Environment Department has been working on ways of measuring the wealth of nations and has produced some insightful analyses, which we consider next.

A New Measure of the Wealth of Nations

Environmentally sustainable development means improving human well-being over time, a process that means everything to the world's impoverished people. It is a process that requires societies to manage all of their assets that can contribute to improving human well-being. What are the components of a nation's wealth—its assets? In a classic document entitled *Where Is the Wealth of Nations? Measuring Capital for the 21st Century*,[2] the World Bank indicates three major components of capital that determine a nation's wealth: *produced capital, natural capital,* and so-called *intangible capital.*

[1]Thomas Prugh, Robert Costanza, and Herman E. Daly, *The Local Politics of Global Sustainability* (Washington, D.C.: Island Press, 2000), 19.

[2]World Bank, *Where Is the Wealth of Nations? Measuring Capital for the 21st Century* (Washington, D.C.: World Bank, 2006).

Produced capital (Fig. 2–5a) is the human-made buildings and structures, machinery and equipment, vehicles and ships, monetary savings and stocks, highways and power lines, and so forth that are essential to the production of economic goods and services. Produced capital is often the major focus of national economic planning, and it is the most easily measured of the three components of a nation's wealth. Produced capital assets have very clear income-earning potential, but they may also be subject to obsolescence and must be renewed continually. For example, a clothing factory produces a flow of goods destined for consumers, but the machinery wears out over time, and the building itself ages. Thus, we have income—a flow of goods—and depletion, which is referred to as capital consumption.

Natural capital (Fig. 2–5b) refers to the goods and services supplied by natural ecosystems and the mineral resources in the ground. Natural capital represents an indispensable set of resources, some of which are *renewable* and some *nonrenewable*. Renewable natural capital (*ecosystem capital*) is represented by forests, fisheries, agricultural soil, water resources, and the like, which can be employed in the production of a flow of goods (lumber, fish, grain). Often overlooked is the fact that this same natural capital also provides a flow of services in the form of waste breakdown, climate regulation, oxygen production, and so forth. (See Chapter 7, Figure 7–1.)

As we will see in subsequent chapters, renewable natural capital is subject to depletion, but it also has the capacity to yield a sustained income if it is managed wisely. Nonrenewable natural capital, such as oil and mineral deposits, also is subject to depletion and provides no services unless it is extracted and converted into some useful form. It is part of the wealth of a nation, however, as long as it is in the ground. Some components of natural capital are easier to measure than others. For instance, the lumber of a forest can be reckoned in board feet, but how does one measure the impact of a forest on the local climate and on its capacity to sustain a high level of biodiversity?

Intangible capital (Fig. 2–5c) is divided into three elements. The first is **human capital**, which refers to the population and its physical, psychological, and cultural attributes (innate talents, competencies, abilities, etc.). To this is added the value imparted by education, which enables members of a population to acquire skills that are useful in an economy. Education can be either formal or informal; the point is that people acquire a productive capacity through education. Investments in health and nutrition also contribute to increases in human capital.

The second element of intangible capital is what the World Bank has called **social capital**—the social and political environment that people create for themselves in a society. Social capital includes more formal structures and relationships as defined by government, the rule of law, the court system, and civil liberties. It also includes the horizontal relationships of people as they associate into religious or ethnic affiliations or join organizations in which they have a common interest. Social capital is considered a vital element of a nation's wealth because social relationships clearly affect, and are affected by, economic processes.

A third element of intangible capital is **knowledge assets**—the codified or written fund of knowledge that can be readily transferred to others across space and time. Although these assets constantly grow, they are useful only if people have access to them (hence, the vital importance of libraries, schools, universities, and the Internet).

Complementary Assets.   The three classes of assets usually complement each other as they are called on to improve human well-being. For instance, the natural assets of clean water and healthy food can improve human health and the capability to work on produced assets. Knowledge assets can provide the information for building effective social capital in the form of laws, government agencies, and interest groups. As these components of a nation's wealth are understood and managed, they enhance human well-being in many ways (Fig. 2–6). Their influence can be direct, as in the effects of a beautiful natural setting or a network of trusted friends, or indirect, as in the contribution they make to economic production and its eventual consumption.

(a) Produced capital

(b) Natural capital

(c) Intangible capital

Figure 2–5 The wealth of nations. Three major components of a nation's capital determine its wealth: (a) produced capital, (b) natural capital, and (c) intangible capital, which largely consists of human capital and the quality of formal and informal institutions.

Figure 2–6 How society's assets complement each other.
Human resources (knowledge assets, human assets), natural assets, social assets and human-made assets (produced capital) interact to enhance human well-being. (*Source: World Development Report 2003,* the World Bank, and Oxford University Press. Used with permission.)

Calculations. The World Bank's Environment Department devised various means of measuring each of these components of the wealth of nations and then set about evaluating the wealth of different nations (their assets). Some of the results of the World Bank's analysis are shown in Table 2–1. One finding that should come as no surprise is the notable differences in per capita wealth among groups of nations ranked by income.

Interestingly, the dominant source of wealth for most nations is human resources, but natural capital ranks high in a number of countries. Although natural capital often ranks third in most countries, this does not mean that it is less important than human resources or produced assets. (The World Bank authors offer a caution in interpreting their work: They did not assess the service components of natural capital because of difficulties in measuring them and calculated mainly the actual resource inputs into production—that is, forests as a source of timber, mineral assets, etc.) In particular, natural capital plays a more important role in low-income countries (Table 2–1), a signal that as they develop,

TABLE 2–1 The Per Capita Wealth of Nations by Income Group, 2000

Income Group	$ Per Capita				% Share of Total Wealth		
	Natural Capital	Produced Capital	Intangible Capital	Total Wealth	Natural Capital Share	Produced Capital Share	Intangible Capital Share
Low-income countries	1,925	1,174	4,434	7,532	26%	16%	59%
Middle-income countries	3,496	5,347	18,773	27,616	13%	19%	68%
High-income countries	9,531	76,193	353,339	439,063	2%	17%	80%
World	4,011	16,850	74,998	95,860	4%	18%	78%

Source: World Bank, *Where Is the Wealth of Nations? Measuring Capital for the 21st Century* (Washington, D.C.: World Bank, 2006).

they are more dependent on their natural resources than are the wealthier countries.

Measuring Economic Progress. The dominance of the intangible (human resources and social assets) component emphasizes the importance of investing in health, education, and nutrition in a society that needs to move further along in development. It also emphasizes the importance of having an efficient judicial system, clear property rights, and an effective and uncorrupt government—elements of social capital. In this key work on defining the wealth of nations, the World Bank has provided the basics of a tool for measuring true economic progress. It is the *growth in wealth, and all the components of that wealth*—not the growth in that once commonly employed indicator of economic status known as the gross national product (GNP) or in its now more commonly used alternative, gross domestic product (GDP).

GDP or GPI to Measure Progress?

The gross national product, or GNP, refers to *the sum of all goods and services produced (really, consumed) by a country in a given time frame.* The GNP was once the most commonly used indicator of the economic health and wealth of a country. Most countries now use the comparable *gross domestic product*, or GDP. The GDP is the GNP minus net income from abroad. As a *per capita* index, the GDP is often used to compare rich and poor countries and to assess economic progress in the developing countries. An important element in calculating a *net* GDP is accounting for the assets that are used in the production processes. Buildings and equipment, for example, are essential to production, but they gradually wear out or depreciate. Their depreciation is charged against the value of production (called capital depreciation); that is, an accounting of capital depreciation is routinely subtracted from the production of goods and services in order to generate a net GDP. This measurement is flawed, however, because it does not include goods created and services performed by a family for itself. Thus, subsistence farming in developing countries is not measured, and this omission surely produces a falsely lower GDP for these countries. The GDP also omits the natural services provided by ecosystems. For example, even though clean air is as important as asthma medicine for asthmatics, only the medicine is included in the GDP.

Natural Capital Depreciation. The economists who invented the GNP as a measuring device years ago never took into account the depreciation of natural capital—an omission that recently has received a great deal of criticism from ecological economists. Such concerns are regarded as external to a country's balance sheet. Following the classical view of economics, the natural resources and their associated natural services are considered a "gift from nature." Given this state of affairs, it is possible for a nation to cut down a million acres of forest and count the sale of the timber on the income side of the GDP ledger, whereas on the expense side, only the depreciation of chain saws and trucks would be included. Completely hidden from accounting is the loss of all the natural services once performed by the forests, and incredibly, the disappearance of the million acres of forests is shown as an economic asset! As long as this discrepancy remains, it is possible for nations to underestimate the value of their natural resources. They are able to deplete fisheries, lose soil by intensive farming, remove forests, and degrade rangelands by overgrazing and still *account for these activities as economic productivity*. The GDP can increase due to production by industries that pollute the environment and can increase again with cleanups of that pollution—a double benefit!

Correcting this system of accounting would require calculating the current market value of natural assets and considering it to be part of the stock of a nation's wealth, as recommended by the World Bank. To this value would be added the value due to the natural services performed by the ecosystems in which the resources are found (omitted by the World Bank analysis). The natural assets and the services they perform are the nation's natural capital. When the natural resources are drawn down, the depreciation represented by the loss of natural capital would be entered into the ledger as the GDP is calculated. Similar arguments can be made for human resources as a measure of a nation's wealth. A more accurate GDP would have to take all of these depreciations into account, and that doesn't happen.

Environmental Accounting. *Agenda 21*—one of the major documents that came out of the 1992 Earth Summit in Rio de Janeiro—recognized the need for "a program to develop national systems of *integrated* environmental and economic accounting in all countries." The UN's Statistical Division, which coordinates the accounting procedures of member countries, has prepared a manual designed to accomplish *Agenda 21*'s objective (*Handbook of National Accounting: Integrated Environmental and Economic Accounting*). In the manual, countries are encouraged to perform environmental accounting by putting their environmental assets and services in monetary units and keeping a parallel account of their *net domestic product* (similar to the GDP) as it is affected by environmental accounting. The document includes case studies and step-by-step instructions and strongly advocates the adoption of integrated environmental and economic accounting by UN member nations. In recent years, some 25 countries have experimented with different forms of environmental accounting, indicating that some form may become broadly adopted before long.

Genuine Progress Indicator. If the GDP is simply unable to measure true economic progress, some other index might be necessary. The nonprofit organization *Redefining Progress* has proposed the **Genuine Progress Indicator (GPI)** as a measure of economic progress. This index is calculated by assuming that some kinds of economic activity are positive and sustainable and others are not, even if the latter are included in the GDP. Some of the positive factors in the GPI—such as labor that goes into housework, parenting, and volunteering—are not calculated in the GDP. Negatives in the GPI include factors like the costs of crime, loss of leisure time, costs of pollution, depletion of nonrenewable resources, and loss of farmland. Thus, for the United States in 2004, positive contributions to the GPI totaled $11.06 trillion,

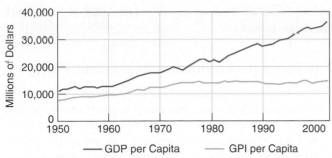

Figure 2–7 GDP and GPI per capita, 1950–2004 (in year 2000 dollars). The GDP (gross domestic product) is the conventional measure of economic progress used by economists. The GPI (Genuine Progress Indicator) is a more realistic measure of progress. Note how it is falling behind the GDP over the years. (*Source:* John Talberth, Clifford Cobb, and Noah Slattery, *The Genuine Progress Indicator 2006: A Tool for Sustainable Development*, Redefining Progress, 2006, with permission.)

while negative contributions were $6.64 trillion, yielding a net GPI of $4.42 trillion. In contrast, the 2004 GDP was $11.71 trillion. Over the years, the per capita GPI has remained fairly constant, while the GDP has risen sharply (Fig. 2–7), suggesting that the benefits of GDP growth are increasingly offset by the rising environmental and social costs of economic activity. The thinking behind the GPI has been carefully reviewed in the scientific literature, and several countries and Canadian provinces have made a first attempt to measure and apply the GPI as an index of sustainability.

Resource Distribution

None of the resources that are crucial to an economy are evenly distributed among nations. Table 2–1 shows that great differences in wealth exist among the nations of the world, something we will see many times in other chapters. It is significant that the greatest differences occur in the two categories of assets that are most closely tied to the process of development: the human-resource-dominated *intangible capital* and *produced capital*. With the exception of the oil-rich Middle East, the per capita differences in natural capital are unremarkable. Thus, *economic development*, which fosters growth in human resources and produced capital, is the pathway to achieving more equity in the distribution of resources.

Essential Conditions. There are many reasons why some countries have undergone rapid development and others have not; population growth is one that is cited in this text (Chapter 9). Perhaps the most important reasons relate to the major institutions that coordinate social and political life and that make up the social capital of a society. For example, the definition of rights, the enforcement of those rights, and the facilitation of economic exchange are all considered essential to the successful development of human capital and produced capital. A well-developed body of law, an honest legal system, inclusiveness, broad civic participation, and a free press are essential for maintaining rights within a society. A well-developed market economy and free entry into and exit from markets for people and businesses are also conducive to development. Functioning communication and transportation networks and viable financial markets are needed to sustain a

market economy. History tells us that if these resources are in place, a society will make progress in the development of human resources and produced assets. In societies where such resources are not in place, corruption, inefficiency, banditry, injustice, and intolerance often prevail; continuing poverty and environmental degradation are the predictable outcome. For example, one of the major components of intangible capital as measured by the World Bank is the "rule of law index." Switzerland scores 99.5 out of 100 on this index, while Nigeria, with its corruption and human rights abuses, scores 5.8. In fact, some countries are so poorly run that they have negative intangible capital; their people are becoming more impoverished.

Intergenerational Equity. What we have just discussed is a matter of *intragenerational equity*—the golden rule of *making possible for others what is possible for you*. All forms of current assets need to be managed well to improve the well-being of people who already exist. Sustainable development, by contrast, is more about *inter*generational equity—*meeting the needs of the present without compromising the ability of future generations to meet their needs*. Thus, we can conserve resources and avoid long-term pollution partly out of a concern for future generations.

Economists point to some problems with this perspective, however. Economics deals with future resources by applying what is called the **discount rate**. By this, we mean the *rate* used for finding the present value of some future benefit or cost. The discount rate often approximates current interest rates. For example, a dollar is worth more to us today than it will be five years from now, even without inflation, because we can put the dollar to use now and can earn interest on that dollar if we invest it. At a 5% rate of interest, in five years our dollar would be worth $1.28. Turning this idea around, a dollar five years from now is worth only $0.78 today.

Use It Up! Following this reasoning, it can be concluded that some resources or benefits are worth more now than they will be in the future. This concept can be applied to a stand of trees. The owner has several options: (1) cut all the trees and sell the timber today; (2) hold onto the trees in the hope that they will bring a better price at some time in the future; or (3) harvest the trees on a sustainable basis, spreading out profits over time. If it turns out that the trees grow more slowly than current interest rates, the economically profitable decision would be to cut them now and invest the proceeds. This is a problem for natural capital, since it would appear to argue in many cases for maximizing short-term profit over sustainable, long-term use of the resource. In other words, we should reap the economic value now and let future generations bear the ecological costs (in reality, the resource is likely to have additional value in the form of services that is not included in the economic assessment of its worth). Similarly, if we let future generations cope with global climate change, we can continue to enjoy the present benefits of the unrestrained use of fossil fuels. The same can be said about other resources, such as swordfish: We can harvest them now, and when they're gone, we can switch to some other species of fish.

This approach presents a problem for intergenerational equity. The *self-interest* of present individuals and generations is at odds with the longer term, *sustainable interests* of a community or a society. To make things a bit more challenging, some have argued that conserving resources for the future could actually be thought of as putting the interests of future generations above those of, say, the poor today. In answer, it can be said that both sets of interests are legitimate, but given a choice, many people might want to see their children and grandchildren enjoy an improved well-being and might be more inclined to seriously consider making present sacrifices in order to protect a more hopeful future.

We turn our attention now to environmental public policy, which makes possible both present and future well-being for human societies.

2.3 Environmental Public Policy

Environmental public policy includes all of a society's laws and agency-enforced regulations that deal with that society's interactions with the environment. Public policies are developed at all levels of government: local, state, federal, and international.

The Need for Environmental Public Policy

The purpose of environmental public policy is *to promote the common good.* Just what the common good consists of may be a matter of debate, but at least two goals stand out: *the improvement of human welfare* and *the protection of the natural world.* Environmental public policy addresses two sets of environmental issues: (1) the prevention or reduction of air, water, and land pollution and (2) the use of natural resources like forests, fisheries, oil, and land. All public policy is developed in a sociopolitical context that we will simply call *politics.*

What are the consequences of not having an effective environmental public policy? Human societies and their economic activities have the potential for doing great damage to the environment, and that damage has a direct impact on present and future human welfare (Table 2–2). The effects of pollution and the misuse of resources are seen most clearly in those parts of the world where environmental public policy

---

TABLE 2–2   Principal Health and Productivity Consequences of Poor Environmental Management

Environmental Problem	Effect on Health	Effect on Productivity
Water pollution and water scarcity	More than 3 million deaths and billions of illnesses a year are attributable to water pollution and poor household hygiene caused by water scarcity.	Fisheries are declining; rural household time (time spent fetching water) and municipal costs of providing safe water are increasing; water shortages constrain economic activity.
Air pollution	Many acute and chronic health impacts exist: Excessive levels of urban particulate matter and smoky indoor air are responsible for 2 million premature deaths annually.	Acid rain and ozone have harmful impacts on forests, agricultural crops, bodies of water, and human artifacts.
Solid and hazardous wastes	Diseases are spread by rotting garbage and blocked drains. Risks from hazardous wastes are typically local, but often acute.	Groundwater resources are polluted.
Soil degradation	Effects include reduced nutrition for poor farmers on depleted soils and greater susceptibility to drought.	Some 23% of land used for crops, grazing, and forestry has been degraded by erosion. Field productivity losses in the range of 0.5–1.5% of the gross national product are common on tropical soils.
Deforestation	Localized flooding leads to death and disease.	Effects include reduced potential for sustainable logging and for prevention of erosion, increased watershed instability, and diminished carbon storage capability for forests. Nontimber forest products are also lost.
Loss of biodiversity	Effects include the potential loss of new drugs.	Ecosystem adaptability is reduced, and genetic resources are lost.
Atmospheric changes	Such changes result in possible shifts in vector-borne diseases, risks from climatic natural disasters, and skin cancers attributable to depletion of ozone shield.	Coastal investments are damaged by the rise in sea level; regional changes in agricultural productivity occur.

Sources: World Bank, *Wo1rld Development Report, 1992* (New York: Oxford University Press, 1992); *Millennium Ecosystem Assessment,* 2005; UNEP Global Environment Outlook GEO$_4$, 2007.

is often not well established and implemented—the developing world. As Table 2–2 shows, millions of deaths and widespread disease are directly traceable to degraded environments. The costs to human welfare are felt in the areas of health, economic productivity, and the ongoing ability of the natural environment to support human life needs. Therefore, laws to protect the environment are not luxuries to be tolerated only if they do not interfere with individual freedoms or economic development; they are part of the foundation of any well-organized human society and essential for achieving sustainability.

Policy in the United States

The U.S. Constitution is widely regarded as responsible for establishing the most effective form of government on Earth. The convenors of the Constitution decided on three branches of government, all of which play a role in establishing and enforcing environmental public policy (Fig. 2–8). The **legislative branch**, or the Congress, consists of the Senate and the House of Representatives. All of these legislators are elected by the people from the 50 states. The Congress has the power to establish laws, which must pass both houses of the legislature and then proceed to the **executive branch** of the government. The president is elected by the people every four years (actually, by the electors, which can result in a president being elected without having received a majority, as in the year 2000). A host of administrative agencies, most headed by members of the president's cabinet, report to the president. We will encounter many of these agencies as we examine different environmental problems. The **judicial branch** of government is responsible for interpreting the law, from the Constitution and its amendments to the many pieces of legislation passed by the Congress.

Rules and Regulations. Environmental *public-policy law* is the responsibility of the Congress. Once a piece of legislation is passed by the House of Representatives and the Senate and is signed into law by the president, it is implemented by the appropriate government agencies. The EPA and most other federal agencies are required to develop *rules and regulations* for issues only broadly covered in the laws, and it is these rules that put the bite in the laws passed by Congress. In making rules, agencies are required to solicit public comments, and these most often come from major stakeholders interested in or affected by the rules—business and public interest groups. The agencies then evaluate the comments and may or may not adjust the rules, depending on the validity of the comments. Finally, agencies publish their new rules in the *Federal Register*, where they are "filed" for all interested parties to see.

One final role the Congress must play is to pass appropriations for the various programs that have already been authorized by law. The executive branch (the president) draws up a budget asking for appropriations for all of the government agencies and programs, and the Congress decides how much of what the president asks for will actually get funded. Thus, the two times the Congress can influence a given issue are when it is formulated into law and when it receives appropriations. The appropriations must be authorized every year, and new laws can always amend or abolish existing laws. Given this basic process, there are many opportunities for the political system to influence environmental public policy.

State and Local Levels. Some policies are developed at the state and local levels to solve more-local problems. States replicate much of the structure of the federal government, with legislatures, governors, a judiciary, and various agencies. Cities and towns employ conservation commissions, planning boards, health boards, water and sewer commissions, and the like. Cities and towns, for example, establish zoning regulations to protect citizens from haphazard and incompatible land uses. Some municipalities and regions go further and establish broader policies like "smart growth" to address the problems of urban sprawl. Cities and towns are adopting mandatory recycling and pay-as-you-throw regulations for municipal solid-waste management. Many states are responding to global climate change with policies aimed at reducing greenhouse gas emissions and mandating renewable energy portfolios. Eleven states have bottle laws that facilitate recycling. Indeed, states and municipalities are often out

Capitol Building	White House	Supreme Court Building

Legislative	Executive	Judicial
Congress	President	Supreme Court
	Vice-President	Federal Courts
House of Senate	Cabinet	Circuit and District Courts
Representatives	Administrative Agencies	

Figure 2–8 The three branches of government in the United States. The legislative, executive, and judicial branches provide a balance of power as laws are enacted, administered, and sometimes reviewed.

in front of the federal government on a number of environmental issues, often initiating valuable experiments in environmental public policy that could result in national policy changes.

Policy Options: Market or Regulatory?

Suppose that a legislative or administrative decision has been made to develop a policy to deal with an environmental problem. Now what? The objective of environmental public policy is to *change the behavior of polluters and resource users so as to benefit public welfare and the environment.* The two main ways to accomplish these behavior changes are by using a direct regulatory approach, in which standards are set and technologies are prescribed—the so-called *command-and-control* strategy—and by using basic *market* approaches to set prices on pollution and the use of resources. Both approaches have their advantages and disadvantages. Many newer policies include elements of each.

Command and Control. Traditionally, most of the laws that give the EPA its authority have prescribed command-and-control solutions. The laws direct the EPA to regulate pollution problems, often specifying many of the details to be used. For example, if they know the health effects of pollutants, the legislators may choose to set standards that protect the health of the most vulnerable members of the population. This is the case with the basic criteria covering air pollutants (see Chapter 19). To meet these standards, regulations are established, and polluters are required by law to comply with the regulations. The regulatory approach also works well with land use issues in which certain values are upheld that will not necessarily be protected by a straightforward market approach. Indeed, many environmental problems are not readily amenable to market-based policies.

One of the shortcomings of the regulatory approach is that it practically guarantees a certain sustained level of pollution. If a polluter is told to use a particular technology or is given a cap on emissions, the policy gives the polluter no incentive to invest in technologies that would keep pollution at lower levels than allowed. A better approach might be to set a standard (for air quality or water quality, for example) and let the polluter decide how best to achieve the standard. Here there is a command, but control is in the hands of the polluter. Pollution *prevention*, as discussed in Chapter 22, is the preferred course of action, and it is encouraging to see the EPA moving deliberately in that direction.

Market Approaches. Market-based policies have the virtues of simplicity, efficiency, and (theoretically) equity. All polluters are treated equally and will choose their responses on the basis of economic principles having to do with profitability. With a market-based approach, there are strong incentives to reduce the costs of using resources or of generating pollution. The cap-and-trade system for SO_2 emissions under the Clean Air Act of 1990 is a good example of the market approach to pollution control (see Chapter 19). User

fees (pay as you throw) for the disposal of municipal solid waste are another (Chapter 21). Several countries have restored and sustained fishery stocks by assigning individual quotas that add up to a sustainable fishing take—the total allowable catch (see Chapter 7). A new market-based initiative for conserving the services provided by ecosystems is **Payments for Ecosystem Services (PES)**. Forest, farm, and wetlands owners hold properties that provide valuable services to societies as they sequester carbon, preserve biodiversity, and prevent soil erosion. They are seldom compensated for these services. A PES program would establish a monetary value for a given ecosystem service, and those who benefit from the service would pay the landowner to maintain the service over the years and not convert the land to some environmentally suboptimal use.

Whether regulation based or market based, public policy does not appear out of nowhere. We will now examine how public policy is developed.

Public-Policy Development: The Policy Life Cycle

Environmental public policy is developed in a sociopolitical context, usually in response to a problem. As we saw (Fig. 2–2), less developed countries or regions face a group of poverty-related environmental issues, such as a lack of sanitation facilities or unsafe water supplies. As these countries develop, they become better able to deal with these basic issues through effective public policies. However, more affluence can bring new problems (like industrial air pollution) that require further public-policy development, and sometimes unexpected problems, such as stratospheric ozone depletion, arise. When specific problems are addressed through public policy in democratic societies, the development of that policy often takes a predictable course, which we will call the *policy life cycle.*

The typical policy life cycle has four stages: *recognition, formulation, implementation,* and *control* (Fig. 2–9). Each of the stages carries a certain amount of political "weight," which varies over time and is represented in the figure by the course of the life-cycle line. We will use the discovery of ozone depletion in the stratosphere as a case study to illustrate the policy life cycle.

Recognition Stage. In the early 1970s, chlorofluorocarbons (CFCs) were hailed as miracle chemicals: virtually inert, nontoxic, nonflammable, noncorrosive. They appeared in thousands of products and applications, from computers, telecommunications, pharmaceuticals, and refrigerators to food processing and cosmetics. They were ideal aerosol propellants, were extremely useful as refrigerants, and were effective as cleaning agents for electronics. Physical chemist Sherwood Rowland, while working at the University of California at Irvine in 1972, read a paper by English scientist James Lovelock, who proposed using CFCs as markers to measure atmospheric winds because of their increasing presence in the atmosphere. Rowland and new postdoctoral

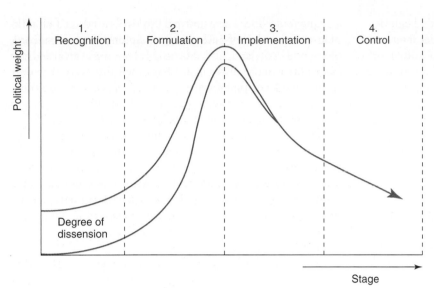

Figure 2–9 The policy life cycle. Most environmental issues pass through a policy life cycle, and each issue is accorded a different degree of political "weight" as it moves through each stage of the cycle. The final result is a policy that has been incorporated into the society and a problem that is under control.

student Mario Molina (Fig. 2–10) decided to investigate the question of what happens to CFCs in the atmosphere. The two scientists reasoned that CFCs would not break down unless they entered the stratosphere, where solar radiation is strong enough to break the tough molecules. They hypothesized that CFC breakdown would release chlorine and that the chlorine would in turn react with ozone molecules in the stratosphere in such a way that 1 chlorine molecule could destroy 100,000 ozone molecules (see Chapter 19 for details).

Molina and Rowland suddenly realized that the process they had uncovered had the potential to reduce the stratospheric ozone layer that blocked ultraviolet radiation from entering the lower atmosphere, where it could seriously damage biological tissues. They published their results in 1974.[3] The immediate response was unremarkable, so the pair began calling public attention to the dangers of CFCs. The chemical industry (especially Dow and DuPont, major CFC

[3]Mario Molina and FS Rowland. Stratospheric Sink for Chlorofluoromethanes: Chlorine Atom-Catalysed Destruction of Ozone. *Nature 249.* 810–2. June 28, 1974.

Figure 2–10 Mario Molina and Sherwood Rowland. These two physical chemists discovered the threat posed by chlorofluorocarbons to the ozone layer in the stratosphere. Their work led to the Montreal Protocol, an international agreement to phase out the use of ozone-destroying chemicals around the world.

manufacturers) responded by attacking them wherever they spoke, ridiculing their work, and even accusing them of being agents of the Soviet KGB intent on destroying capitalism.

Illustrated by these events, the *recognition stage* of public policy is low in political weight. The stage begins with the early perceptions of an environmental problem, often coming as a result of scientific research. Scientists then publish their findings, and the media pick up the information and popularize it. When the public becomes involved, the political process is under way. During the recognition stage, dissension is high. Opposing views on the problem surface as businesses or technological industries respond to the bad news that they are at the root of some new environmental problem. Eventually, the problem gets attention from some level of the government, and the possibility of addressing it with public policy is considered.

Formulation Stage. A turning point came in 1976, when the National Academy of Sciences appointed a committee to look into the claims of Rowland and Molina. The committee issued a report that confirmed the dangers of CFCs and recommended banning the chemicals. The EPA considered the evidence and banned the use of CFCs in spray cans in 1978. Environmental groups applauded the decision. CFCs continued to be used in increasing amounts, however, in other applications. Then, in 1985, a British team working in Antarctica reported a huge "hole" in the ozone layer over that continent, and further research showed that the ozone hole was due to unique conditions that led to high release of chlorine from CFCs and to subsequent ozone destruction. Ozone destruction was seen to be occurring more widely, and the world scientific and environmental organizations began calling for global action. This led to a meeting convened by the United Nations in Montreal, Canada, in 1987. There member nations agreed to scale back CFC production 50% by 2000.

Thus, the *formulation stage* is a stage of rapidly increasing political weight. The public is now aroused, and debate about policy options occurs in the corridors of power. The political battles may become fierce, as questions dealing with regulation and who will pay for the proposed changes are addressed. Media coverage is high, and politicians begin to hear from their constituencies. Lobbyists for special interests or environmental groups pressure legislators to soften or harden the policy under consideration. During the formulation stage, policy makers consider the "Three E's" of environmental public policy: **effectiveness** (whether the policy accomplishes what it intends to do in improving the environment), **efficiency** (whether the policy accomplishes its objectives at the least possible cost), and **equity** (whether the policy parcels out the financial burdens fairly among the different parties involved).

Implementation Stage. The ozone hole continued to grow, however, and the signers of the Montreal Protocol agreed to speed up the phaseout of CFCs, moving the target date for a complete phaseout to 1996. The major chemical companies capitulated and were already developing more benign substitutes for CFCs. In fact, DuPont effectively opposed an attempt in the 104th Congress to terminate U.S. participation in the Montreal Protocol. Several other chemicals in wide use that also had ozone-destroying potential were added to CFCs as substances to be regulated. It fell to regulatory agencies in the various countries to enforce the Montreal Protocol agreements.

At this point, the policy has reached the *implementation stage*, in which its real political and economic costs are exacted. The policy has been determined, and the focal point moves to regulatory agencies. During the implementation stage, public concern and political weight are declining. By now, the issue is not very interesting to the media, and the emphasis shifts to the development of specific regulations and their enforcement. Industry learns how to comply with the new regulations. Over time, greater attention may be given to efficiency and equity as all the players in the process gain experience with the policy.

Control Stage. CFCs are no longer being manufactured and released into the atmosphere. Substitutes for them are readily available. The size of the ozone hole has stopped increasing, and the amount of CFCs in the stratosphere is starting to decline. The Montreal Protocol stands as a remarkable achievement of international diplomacy and concord. Sherwood Rowland and Mario Molina were awarded the Nobel Prize in chemistry in 1995 for their discovery. The only disturbing developments are continued pressure to use other ozone-depleting chemicals like methyl bromide, which is a useful soil fumigant, and a persistent black market in CFCs.

The final stage in the policy life cycle is the *control stage*. By this point, years have passed since the early days of the recognition stage. Problems are rarely completely resolved, but the environment is improving, with things moving in the right direction. Policies (and their derived regulations) are broadly supported and often become embedded in the society, although their vulnerability to political shifts continues. Regulations may become more streamlined. The policy makers must now see that the problem is kept under control, and in due time, the public often forgets that there ever was a serious problem.

The policy life cycle is a simplified view of what is frequently a highly complex and contentious political process. At any one time, different problems will be in different stages of the life cycle, as shown in Figure 2–11 for a number of environmental problems in the industrialized countries. Also, countries in different stages of economic development will be in different stages of the policy life cycle for a given problem. For example, sickness and death due to polluted water are still serious problems in many developing countries because public policies have not yet caught up with the need for water treatment and sewage treatment. In contrast, this problem is in the control stage in the industrialized societies. The reason for the discrepancy can often be traced directly to the costs that lie behind the development and implementation of public policy.

Economic Effects of Environmental Public Policy

What are the relationships between a country's economy and its environmental public policy? Are there some policies that are too costly? How should a society parcel out its limited

Figure 2–11 Environmental problems in the policy life cycle. Different environmental problems are in different stages of the policy life cycle in industrial societies.

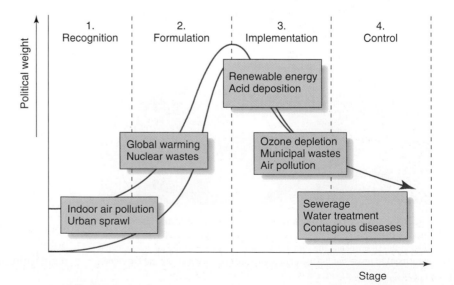

resources to address environmental problems? Are environmental regulations a burden on the economy? Let us begin to sort out these questions by looking at the relationship between environmental policies and costs.

Costs of Policies.

As we have seen, policies do not just appear out of thin air. Instead, they are the result of some version of the policy life cycle. Some policies have relatively little or no direct *monetary cost*—they do not require major investments of administration or resources. Removing subsidies to special interests and denying special access to national resources can result in a more efficient and equitable operation of the economy and can protect the environment. For example, the use of national lands for cattle grazing (Fig. 2–12a) and timber harvesting (Fig. 2–12b) is subsidized in the United States. As a result, the real costs—and environmental consequences—of these activities are borne not by the special interest groups that have access to these resources, but by all taxpayers. Removing such subsidies, however, can have very real *political* costs, as powerful interests do everything they can to hold onto their privileges.

Most environmental policies impose some very real economic costs that must be paid by some segment of society. For instance, we establish policies that control pollution because their benefits are judged to outweigh their costs—that is, the public welfare is improved, and the environment is protected. Who pays the costs? The equity principle implies that those benefiting from the policies—the customers of businesses whose activities are regulated and the people whose health is protected by laws against polluting—should pay, through either taxes or higher charges. (When businesses are taxed or industries are forced to add new technologies, they pass their costs along to consumers, so the costs of public policies are borne by the public in one way or another.)

Impact on the Economy.

For years, special interest groups and conservative think-tanks (e.g., the Competitive Enterprise Institute) have argued that environmental regulations are excessive and bad for the economy—that our concern for environmental protection costs hundreds of thousands of jobs, reduces our competitiveness in the marketplace, drives up the price of products and services, and, in general, imposes costs on the economy that are nonproductive. These views are widespread and are routinely used to try to roll back the laws and regulations that represent environmental public policy.

How sound are these arguments? This concern has been addressed by many economists in recent years, and the general findings of their studies indicate that the "environment versus economy" trade-off is a myth. Even without considering the environmental or health benefits, economists found that environmental protection has no significant adverse effect on the economy and in fact often has a positive effect.

Anecdotal Evidence. First, we must recognize that much of the evidence used to support the charges that have been made is anecdotal. For example, according to the American Petroleum Institute, "environmental restrictions on energy extraction and production have caused the loss of 400,000 jobs," and according to Florida sugar growers, "Measures to protect the Everglades will cause the loss of 15,000 jobs." Indeed, numerous cases can be cited where *some* jobs were lost because of environmental regulations.

Anecdotal evidence can also be used in *favor* of environmental policy. According to the California Planning and Conservation League Foundation, "Recycling has created 14,000 jobs in California," and according to the EPA, "The Clean Air Act of 1990 will generate 60,000 new jobs." Indeed, *many* sectors of the U.S. economy that are most subject to environmental regulations—plastics, fabrics, and so forth—have improved their efficiency and their competitiveness in the international market.

(a)

(b)

Figure 2–12 Subsidized access to pubic resources. (a) Cattle grazing in BLM-managed land near Baker City, Oregon. (b) Timber harvesting in the Olympic National Forest, Washington. Both of these activities are subsidized by taxpayers because the special interests with access to the resources pay only a fraction of the market value of the resource.

Careful Studies.   Is there something more substantial to help us with this important controversy? The best evidence comes from careful studies of what is actually happening. These studies reveal the following interesting findings:

- Estimated pollution-control costs are only 1.72% of **value added.** (Value added is the increase in value from raw materials to final production.) Industries are not going to shut down, move overseas, or lay off thousands of workers for costs of this size.

- Only 0.1% of job layoffs were attributed by employers to environment-related causes, according to a study by the U.S. Bureau of Labor Statistics.

- In general, states with the strictest environmental regulations also had the highest rates of job growth and economic performance. California instituted energy-efficiency policies in 1978, and between 1978 and 2007, those policies created almost 1.5 million jobs, while only 25,000 jobs were lost.

In short, the results of careful studies of the relationship between environmental protection and economic growth refute the suggestion that environmental regulation is bad for jobs and the economy. In fact, economic performance is highest where environmental public policy is most highly developed. Healthier workers are more productive and spend less on medical care. It may be no coincidence that in the former communist countries in Europe, where environmental public policy was deliberately suppressed in order to favor industry, both the economy and the environment have been disaster areas. The evidence indicates that concerns over energy efficiency, pollution control, and conservation of resources have encouraged businesses to modify their technologies in ways that make them *more* competitive, not less, in the national and international marketplace. The evidence also suggests that at least as many jobs have been created by environmental protection as have been lost.

The Environmental Protection Industry.   Taken as a whole, environmental protection is a huge industry. In a landmark 1990 study[4], the EPA calculated that the estimated cost of complying with environmental regulations by 1997 would be about 2.7% of the GDP—about $210 billion. Admittedly, this estimate was very rough—some believe it was too low, while others think it was too high. To put the percentage into perspective, in 1997 the United States spent an estimated 10.6% of the GDP on health care and 4.3% for national defense. Is it worth 2.7% of the GDP to provide healthy air to breathe and clean water and ecosystems that can continue to provide essential goods and services if we are willing to spend 15% on our personal health and national security?

A study of total expenditures for the environmental protection industry for one year revealed that it was responsible for the following:

- $355 billion in total industry sales;
- $14 billion in corporate profits;
- $63 billion in federal, state, and local government revenues; and
- 4 million jobs.

In sum, we can draw the following conclusions from our examination of the impact of environmental policy on the economy:

- Environmental public policy does not *diminish* the wealth of a nation; rather, it *transfers* wealth from polluters to pollution controllers and to less polluting companies.

- The "environmental protection industry" is a major job-creating, profit-making, sales-generating industry.

- The argument that environmental protection is bad for the economy is simply unsound. Not only is environmental protection good for the economy, but also environmental public policy is responsible for a less hazardous, healthier, and more enjoyable environment.

We turn next to the details of *benefit-cost analysis*, a specific tool that is employed in developing environmental public policy. In doing so, we will address the question of how to measure the cost-effectiveness of regulating pollution.

2.4 Benefit-Cost Analysis of Environmental Public Policy

The president of the United States has the power to issue executive orders, which are usually directives to government agencies or departments designed to accomplish a presidential initiative. President Ronald Reagan, concerned that the U.S. economy and business interests might be overregulated, issued Executive Order 12291 in 1981. This order required that all major regulations be supported by a *benefit-cost analysis*. A major regulation is one that imposes annual costs of at least $100 million. Presidents Bill Clinton and George W. Bush issued similar executive orders that basically continued most of the policies of Reagan's order.

A **benefit-cost analysis** begins by examining the need for the proposed regulation and then describes a range of alternative approaches. Afterward, it compares the estimated costs of the proposed action and the main alternatives to the benefits that will be achieved. All costs and benefits are given monetary values (where possible) and compared by means of what is commonly referred to as a *benefit-cost* (or *cost-benefit*) *ratio*. A favorable ratio for an action means that the benefits outweigh the costs, and the action is said to be cost effective. There is an economic justification for proceeding with it. If costs are projected to outweigh benefits, the project may be revised, dropped, or shelved for later consideration.

Benefit-cost analysis of environmental regulations is intended to build efficiency into policy so that society does not have to pay more than is necessary for a given level of environmental quality. If the analysis is done properly, it will consider *all* of the costs and benefits associated with a regulatory option. In so doing, it must address the problem of *external costs*.

---

[4]*Environmental Investments: The Cost of a Clean Environment*, Report of the Administrator of the Environmental Protection Agency to the Congress of the United States. EPA-230-1 1-90-083. November, 1990. National Center for Environmental Economics, USEPA.

External Costs

In the language of economics, an **external cost** is some effect of a business process that is not included in the usual calculations of profit and loss. For example, when a business pollutes the air or water, the pollution imposes a cost on society in the form of poor health or the need to treat water before using it. This is an *external bad*. When workers improve their job performance as a result of experience and learning, the improvement is not credited in the company's ledgers, although it is considered an *external good*.

Amenities such as clean air, uncontaminated groundwater, and a healthy ozone layer are not privately owned. In the absence of regulatory controls, there are no direct costs to a business for degrading these amenities. (In other words, they are *externalities*.) Therefore, there are no economic incentives to refrain from polluting the air or water. By including *all* of the costs and benefits of a project or a regulation, benefit-cost analysis effectively brings the externalities into the economic accounting. One suggestion for accomplishing this is the use of green fees or taxes (see Sustainability, "Green Fees and Taxes").

Environmental Regulations Impose Real Costs

The costs of pollution control include the price of purchasing, installing, operating, and maintaining pollution-control equipment and the price of implementing a control strategy. Even the banning of an offensive product costs money because jobs

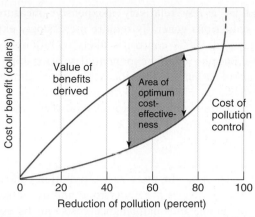

Figure 2–13 The benefit-cost ratio for reducing pollution. The cost of pollution control increases exponentially with the degree of control to be achieved. However, benefits to be derived from pollution control tend to level off and become negligible as pollutants are reduced to near or below threshold levels. The optimum cost-effectiveness is achieved at less than 100% control.

may be lost, new products must be developed, and machinery may have to be scrapped. In some instances, a pollution-control measure may result in the discovery of a less expensive way of doing something. In most cases, however, pollution control costs money. Thus, the effect of most regulations is to prevent an external bad by imposing economic costs that are ultimately shared by government, industry, and consumers.

Costs Go Up. Pollution-control costs tend to increase exponentially with the level of control to be achieved (Fig. 2–13). That is, a partial reduction in pollution may be

Sustainability
Green Fees and Taxes

Benefit-cost analysis works well only if all of the costs and benefits of a polluting activity are included in the calculations. Let us assume that this is possible. What would happen if everyone had to pay the true costs of their use of the environment as they consumed resources or degraded the air or water? Such charges are often called **green fees** (nothing to do with golf!). Oddly enough, two groups that frequently are battling—economists and environmentalists—agree that such a policy would go a long way toward solving many of our environmental problems.

Economists point out that in a free-market economy resources tend to be used in the most efficient way because people will economize. This means that most businesses and people who use, say, 100 gallons of gasoline and have to pay a levy that reflects the true costs of that gasoline and the pollution it produces will be motivated to keep their use of gasoline to an absolute minimum. They will adopt more efficient vehicles or alternative modes of transportation. Including all of the external costs of the gasoline in its price

would involve adding together the costs of maintaining our military presence in Iraq and the Persian Gulf, the costs of air pollution, and even the expected costs of coping with global climate change. The green fee could be imposed as a tax on gasoline, and to ease its impact on the economy, the fee could be implemented gradually until it brings the price of gasoline to its true level.

Internalizing all these costs would make $5 for a gallon of gasoline at the pumps seem a bargain. Many observers have suggested that a tax of, say, $100 per ton of fossil fuel carbon would go a long way toward accomplishing the objective of paying the real costs of the fuel. Great, say the environmentalists. This would mean that the people who do the most damage pay for their environmental impact, something environmental groups have been calling for all along. They insist that some of the revenues collected should be used to mitigate the impacts of pollution and resource use. If that were to be accomplished, the outcome would move our society in the direction of sustainability.

Pollution fees, gasoline taxes, and user fees all have the potential for "internalizing" the external costs of environmental degradation and the use of resources. If such green fees could become public policy, other taxes (e.g., personal income taxes) might be able to be lowered, since these taxes are often the sources of the revenue that is paying much of the external costs.

Would such a policy be politically feasible? Actually, it is already at work at the local level in thousands of communities with "pay as you throw" garbage collection. Communities simply charge a fixed rate per bag of trash collected at the curb, and the proceeds pay for the costs of municipal trash collection. If people are asked whether they *like* higher taxes, the answer is a highly predictable *no*. But if they are asked whether they would rather be taxed on their energy use, their trash production, and their purchases of high-impact goods than be taxed on their incomes and investments, they might well choose the former, especially if they can appreciate the environmental benefits that accrue from such a policy.

achieved by a few relatively inexpensive measures, but further reductions generally require increasingly expensive measures, and 100% control is likely to be impossible at any cost. Because of this exponential relationship, regulatory control often has to focus on an optimum cost-effectiveness, as Figure 2–13 indicates. In any case, the costs of pollution control constitute a powerful incentive to make substitutions, to recycle materials, or to redesign industrial processes.

Costs Go Down. In most cases, pollution-control technologies and strategies are understood and available. Thus, equipment, labor, and maintenance costs can be estimated fairly accurately. Unanticipated problems that increase costs may occur, but as technology advances and becomes more reliable, experience is gained, and lower-cost alternatives frequently emerge. The costs of pollution control are likely to be highest at the time they are initiated; then they decrease as time passes (Fig. 2–14a). The importance of this trend will become evident when we consider the time span over which costs and benefits are compared.

How accurate are the cost estimates? A look at some historical examples reveals that industry reports predict higher costs of regulations than government studies do, and surprisingly, even government analysts often overestimate the costs of regulations. For example, industry sources claimed that the costs of eliminating chlorofluorocarbons from automobile air conditioners (in order to protect the ozone layer) would increase the price of new cars by anywhere from $650 to $1,200. In reality, the price increase turned out to be much lower—from $40 to $400. A recent study of benefit-cost analyses revealed that overestimates of costs occurred in 12 of 14 regulations. The primary reason given for the overestimates is a failure to anticipate the impact of technological innovation, admittedly a factor that is difficult to predict, although another (dishonest) reason might be a deliberate ploy to project costs so high that the regulators will back down.

The costs of improving the quality of the environment represent a major economic outlay. What benefits have we received in return?

The Benefits of Environmental Regulation

The Regulatory Right to Know Act directs the Office of Management and Budget (OMB) to submit an annual report to the Congress containing, among other things, an estimate of the costs and benefits of federal rules. The 2008 report includes an up-to-date study of costs and benefits over the October 1, 1997–September 30, 2007 decade. The OMB found that the benefits (those that could be measured in monetary units) far exceeded the costs. The benefits were estimated at between $122 and $656 billion, while industries, states, and municipalities spent an estimated $46 to $54 billion to comply with the rules. The OMB report stated that the benefits of major regulations issued between 1992 and

(a)

(b)

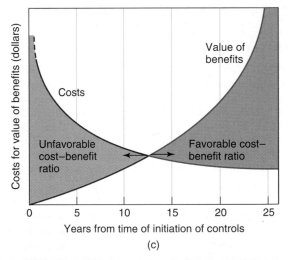

(c)

Figure 2–14 Cost-effectiveness of pollution control over time. (a) Pollution-control strategies generally demand high initial costs. The costs then generally decline as those strategies are absorbed into the overall economy. (b) Benefits may be negligible in the short term, but they increase as environmental and human health recovers from the impacts of pollution or is spared increasing degradation. (c) When the two curves are compared, we see that what may appear as cost-ineffective expenditures in the short term (5–10 years) may in fact be highly cost-effective expenditures in the long term.

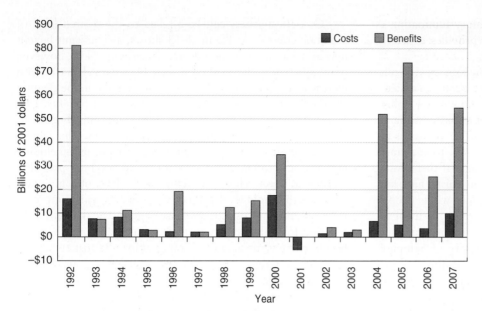

Figure 2–15 Benefits and costs of major regulations, 1992–2007. Clearly, the benefits of regulations far outweigh the costs of implementing the rules (*Source:* Office of Management and Budget, *2008 Draft Report to Congress on the Benefits and Costs of Federal Regulations and Unfunded Mandates on State, Local and Tribal Entities,* September 15, 2008.)

2007 exceeded the costs by more than fourfold (Fig. 2–15). The lion's share of both the costs and the benefits for 2007 was attributed to the EPA's regulation of particulate matter in the air. Fewer hospital visits and reductions in both premature deaths and lost workdays were benefits associated with reductions in these air pollutants mandated by the Clean Air Act.

Calculating Benefits. Such benefits of regulatory policies are seldom as easy to calculate as the costs. Estimating benefits is often a matter of estimating the costs of poor health and damages that *would have* occurred if the regulations had not been imposed (Fig. 2–14b). For example, the projected environmental damage that would have resulted from a given level of sulfur dioxide emissions from a coal-fired power plant (an external bad) becomes a benefit (an external good)

when a regulatory action prevents half of those emissions. Benefits include such things as improved human health, reduced corrosion and deterioration of materials, reduced damage to natural resources, preservation of aesthetic and spiritual qualities of the environment, increased opportunities for outdoor recreation, and continued opportunities to use the environment in the future. The dollar value of these benefits is derived by estimating, for instance, the reduction in health care costs, the reduction in maintenance and replacement costs, and the economic value generated by the enhanced recreational activity. Examples of benefits are listed in Table 2–3.

The values of some benefits can be estimated fairly accurately. For example, air-pollution episodes increase the number of people seeking medical attention. Since the medical

TABLE 2–3 Benefits That May Be Gained by the Reduction or Prevention of Pollution	
1. Improved human health	Reduction and prevention of pollution-related illnesses and premature deaths Reduction of worker stress caused by pollution Increased worker productivity
2. Improved agricultural and forest production	Reduction of pollution-related damage More vigorous growth by removal of stress due to pollution Higher farm profits, benefiting all agriculture-related industries
3. Enhanced commercial or sport fishing	Increased value of fish and shellfish harvests Increased sales of boats, motors, tackle, and bait Enhancement of businesses serving fishermen
4. Enhanced recreational opportunities	Direct uses, such as swimming and boating Indirect uses, such as observing wildlife Enhancement of businesses serving vacationers
5. Extended lifetime of materials and less necessity for cleaning	Reduction of corrosive effects of pollution, extending the lifetime of metals, textiles, rubber, paint, and other coatings Reduction of cleaning costs
6. Enhanced real estate values	Increased property values as the environment improves

attention provided has a known dollar value, eliminating a certain number of air-pollution episodes provides a health benefit of that value.

Shadow Pricing. Many benefits, however, are difficult to estimate. Accurate benefit-cost analysis depends on assigning monetary values to every benefit, but how can a dollar value be put on maintaining the integrity of a coastal wetland or the enjoyment of breathing cleaner air? The answer is to find out how much people are willing to pay to maintain these benefits. How can this be done, though, if there is no free market for the benefits? Again, economists have an answer: *Shadow pricing* involves assessing what people *might* pay for a particular benefit if it were up to them to decide. For example, to evaluate the costs of noise nuisance from aircraft (an external bad) in the vicinity of Schiphol Airport in Amsterdam, researchers evaluated the housing prices in the vicinity of many airports around the world and found that the closer the airport, and therefore the greater the noise, the lower the house value when compared with similar houses farther away. In this case, the *benefit* is seen as being free from the nuisance of low-flying aircraft, and people's willingness to pay is seen in the price of houses.

The Value of Human Life. Shadow pricing becomes difficult when the analysis has to place a value on human life. Many of the pollutants to which people are exposed are hazardous, exacting a toll on health and life expectancy. To estimate the benefits of regulating such pollutants, it is necessary to calculate how many lives will be saved or how many people will enjoy better health. The less the value that is put on a human life, the less the likelihood that a regulation will pass the benefit-cost test. Finding a value for these benefits is fraught with ethical difficulties. For example, the EPA calculated that new clean-air standards for ozone and particulates would prevent 18,700 premature deaths annually, and the monetary benefit of this prevention was valued at $110 billion (close to $6 million per life saved). Different approaches have resulted in a range of values for human life from tens of thousands of dollars to $10 million per life saved. Any benefit-cost analysis that must factor in the risk to human life requires that human lives be valued somewhere within this very broad range. The OMB valued each cancer case at $5 million, each traffic fatality at $3 million, and each lost-workday injury at $50,000.

How does shadow pricing work for the nonhuman components of natural environments—for a population of rare wildflowers, for example, or a wilderness site? The values of these things depend entirely on how willing people are to pay for their preservation. Again, monetary values must be assigned to their existence, and again, the outcome will be strongly influenced by a highly subjective element in the analysis.

Cost-Effectiveness Analysis

Cost-effectiveness analysis is an alternative option for evaluating the costs of regulations. Here the merits of the goal are accepted, and the question is: How can that goal be achieved at the least cost? To find out, alternative strategies for reaching the goal are analyzed for costs, and the least costly method is adopted.

Cost-effectiveness can be applied to the desired level of pollution control. As Figure 2–13 shows, a significant benefit may be achieved by modest degrees of cleanup. Note, however, how differently the cost and benefit curves behave with increasing reduction of pollution. At some point in the cleanup effort, the lines cross, and the costs exceed the benefits. Few additional benefits are realized when cleanup begins to approach 100%, yet costs increase exponentially. This behavior follows from the fact that living organisms—including humans—can often tolerate a threshold level of pollution without ill effect. Therefore, reducing the level of a pollutant below its threshold level will not yield an observable improvement. Optimum cost-effectiveness that meets the efficiency criterion for public policy is achieved at the point where the benefit curve is the greatest distance above the cost curve.

However, the perspective of time should be considered in calculating cost-effectiveness (Fig. 2–14c). A situation that appears to be cost ineffective in the short term may prove extremely cost effective in the long term. This is particularly true for problems such as acid deposition and groundwater contamination from toxic wastes.

Progress

What has been the result of environmental regulation to date? Pollution of air and surface water reached critical levels in many areas of the United States in the late 1960s, and since that time, huge sums of money have been spent on pollution abatement. Benefit-cost analysis shows that, overall, these expenditures have paid for themselves many times over in decreased health care costs and enhanced environmental quality. Consider the phaseout of leaded gasoline as just one example. The project cost about $3.6 billion, according to an EPA benefit-cost report. Benefits were valued at over $50 billion, $42 billion of which were for medical costs that were avoided!

The EPA Checks In. The following are some of the accomplishments of environmental public policy, as reported by the EPA:

- From 1980 to 2007, total emissions of five principal air pollutants have decreased by a total of 57%.

- Since 1976, average levels of lead in the blood of children have declined by 90%.

- Between 1988 and 2007, releases of toxic chemicals, as reported to the Toxics Release Inventory, have decreased 61%.

- In 2007, states reported that 92% of the population on community water systems had drinking water that met all of the standards, as opposed to 79% in 1993.

- Acid deposition has decreased by 33% over the past 15 years.

- More than 377,000 leaking underground storage tanks have been cleaned up since 1988.

- Between 1980 and 2008, 1,063 Superfund sites on the National Priorities List have been completely cleaned up.

- Recovery of municipal solid wastes by recycling has increased from 7% in 1970 to 33.4% in 2007, diverting 8.5 million tons away from disposal in 2007.

Between 1980 and 2007, these improvements occurred even as the GDP grew by 124%, the population rose by 33%, and the number of motor vehicle miles driven increased by 103%.

A benefit-cost analysis accompanies every new regulatory rule from the EPA and other federal agencies. Even when a rule is not classified as a major regulation, the accompanying documentation must demonstrate, by benefit-cost analysis, that the rule does not qualify as significant. Like it or not, benefit-cost analysis is now part of public policy in the United States.

2.5 Politics and the Environment

Given that environmental public policies exact a cost and can seriously impact businesses and private life, establishing and defending these policies can be a contentious affair. In Chapter 1, we described the modern environmental movement and its corollary, environmentalism. Our brief look at this movement pointed out that political battles have accompanied almost every environmental issue, and that is still the situation. Politics—the process by which groups of people make decisions—always accompanies policy.

Political Parties and the Battle for Control

In the first decades of the modern environmental movement, pollution-control policies were enacted by broad, bipartisan coalitions and signed and implemented by presidents of both parties. Grassroots citizen involvement played a key role in mobilizing this political accomplishment. By the 1990s, however, a very well articulated environmental backlash emerged. The conservative mood of the country that elected Ronald Reagan embraced a large conservative contingent that called for policy reform to reduce the regulation of industry so that market forces and private enterprise could be freed up "to move the economy forward." The same contingent also defended private property rights threatened by programs like the Endangered Species Act and wetlands regulations. This movement viewed environmentalists as the enemy of free enterprise, responsible for promoting a false view of Earth as fragile and in crisis because of human mismanagement. The antienvironmental movement received the support of many conservative Republican (and some Democratic) legislators, and the result was gridlock in Congress.

The 104th Congress (1995–1996) began with an effort to dismantle many key laws, especially the Clean Water Act, the Clean Air Act, and Superfund. When it became apparent what was happening, heavy grassroots work by many environmental organizations and extensive media coverage made it clear that the public did not support this antienvironmental thrust. As a result, most of the antienvironmental legislation died in the Senate or was vetoed by President Clinton. The more recent Congresses saw some of the same antienvironmental forces attempting once more to get their legislation through.

However, a bipartisan coalition was able to hold off new attempts to weaken environmental laws. One effect of these battles was a failure to reauthorize a number of important laws, such as the Endangered Species Act and Superfund. The laws continued in force nevertheless.

Special Interest Politics. The role and power of special interest groups became evident during these struggles. In the 1970s and 1980s, special interests (such as the coal industry and the fisheries industry) would hire lobbyists who would use their influence to try to convince congressional committee members to act in their favor on issues that were important to them. Environmental NGOs would then use the media to expose the issues to the public and call on their constituencies to telephone and send mail to their representatives in Congress. The antienvironmental interests caught on to the NGO strategy and began to mimic what they were doing, but more effectively because they had deeper pockets. Soon a constant battle between special interest groups emerged in which each would use the media and telephone calls (enhanced by phone banks, which can generate thousands of calls to legislators in a short period) to pressure members of Congress. Added to this picture was the massive explosion in campaign contributions, largely from businesses and industry.

What all this means is that environmental concerns are often a battle of special interest groups—that environmental protection is seen as one among many interests in American politics. The issues are up for grabs, and protecting the environment often depends largely on the strength of environmental interest groups. James Skillen, executive director of the Center for Public Justice, says that this is a seriously flawed development. In this setting, ecological well-being not only must have scientific proof of its validity, but also has to intrude into the political arena in competition with opposing special interests and *win*. Skillen believes that environmental justice—meaning the protection of land, air, water, and the associated biota—should not have to depend on an interest group for its advocacy. Rather, it is a matter of the *public good* and, as such, should be seen as a basic responsibility of government. To quote Skillen, "Recognition in basic law of the necessity of ecological health as the precondition of all public and market relations should become as fundamental as the recognition that certain human rights exist as the precondition of all public rules and market regulations."[5] The NGOs stepped into the gap because they found that doing so was necessary.

The Bush Years. George W. Bush identified himself as an environmentalist during the 2000 presidential campaign and, at the time, promised to reduce carbon dioxide emissions in an effort to combat global warming. Shortly after his election, President Bush began a process of environmental policy redirection (called "a perfect storm" by the *New York Times*) designed to wreck environmental programs. With the support of a host of antienvironmental interests—such as the energy, timber, and mining industries; the construction

[5]James Skillen, "How Can We Do Justice to Both Public and Private Trusts?" (paper presented at the Global Stewardship Initiative National Conference, Gloucester, Mass., October 1996). Available at cesc.montreat.edu/GSI/GSI-Conf/discussion/Skillen.html.

industry; road builders; and "free-market" think tanks—many existing environmental policies were weakened or canceled. Most prominently, the Bush administration withdrew the United States from the Kyoto Protocol, canceled support for international family planning, softened air and water quality standards, and crafted an energy policy that favored heavy exploitation of fossil fuels.

Regulatory reform? Most of this was accomplished outside the usual legislative process. Instead, the Bush administration used the regulatory process to make changes in environmental policies. The gridlock in environmental legislation enabled this development. Presidents appoint agency heads who are loyal to their administration, and the rules that emerge from federal agencies understandably reflect the ideological makeup of the party in the White House. If business lobbyists have access to important members of the administration (and they did, especially in George W. Bush's administration), they will have undue influence over agency rule making. For example, the Bush-Cheney energy plan (see Chapter 14) clearly favored the petroleum and automotive industries, and it is no secret that members of an industry task force served as Cheney's primary advisors. One of the most disturbing chapters in this story is the way that scientists within government agencies were pressured to suppress findings unfavorable to business interests or inconsistent with White House ideology and were muzzled or censored by administrators when they attempted to communicate their findings.

A New Broom. In November 2008, a majority of the American electorate voted Senator Barack Obama, a Democrat, to the office of president. With the support of a decisive Democratic majority in the House and Senate, Obama proceeded to reverse the Bush-era tide of antienvironmental policies. Following are some of the most prominent actions:

- Under Bush, federal scientists often had their work censored if it failed to conform to White House policy, especially in the realm of climate change; Obama issued a directive aimed at preventing any political interference in science—to ensure that "we make scientific decisions based on facts, not ideology," in his words.

- The Bush EPA refused to act after the Supreme Court ruled that CO_2 was a pollutant under the Clean Air Act and should be regulated if it is a health threat; the Obama EPA issued a finding that CO_2 and other greenhouse gases pose a serious threat to public health and the environment because of their clear link to climate change.

- A last-minute Bush rule allowed mountaintop coal debris to be dumped into stream beds if it was the most convenient option; new Interior Secretary Ken Salazar has requested that the U.S. District Court in Washington DC vacate this rule on the basis of its violation of the Clean Water Act.

- Another last-minute Bush rule abolished the requirement that federal agencies consult with wildlife experts before taking actions that could harm endangered species; this requirement was reinstated by the Obama administration.

- After eight years of denial and stalling on climate change by the Bush administration, the Obama administration is moving toward a new climate policy that includes a heavy stimulus for renewable energy and highly visible U.S. leadership on forging a new global agreement on climate change.

President Obama has appointed an exemplary team of scientists to the Departments of Energy and the Interior, the EPA, and the National Oceanic and Atmospheric Administration and as the President's Science Advisor and has made it clear that scientific integrity is his foremost consideration when it comes to developing and implementing environmental policies. It is fair to call this a sea change in environmental policy making, and the public has every reason to expect America to be moving in a much more environmentally sound, sustainable direction.

Citizen Involvement

How is the general public involved in environmental public policy? The public is involved directly in public policy because not only are we the recipients of the outcome of the policies, but also we pay for the benefits we receive through higher taxes and higher costs of the products we consume. We should care deeply about public policies. Another way to ask this question is: What can you do to have an impact on environmental public policy? Here are some suggestions:

- Become involved in local environmental problems. Local jurisdictions often have the power to make crucial decisions affecting the environment, and groups of citizens working together can often make a difference in how the decisions go.

- Become a member of an environmental NGO that informs its constituencies of environmental concerns around the country and lobbies for those concerns. There is a list of these organizations in Appendix A in the back of this book.

- Contact your legislators to inform them of your support for a particular environmental public policy under consideration by the Congress. You will very likely receive an acknowledgment of your concern, and if enough people do the same, your legislator is likely to reflect your views on the issue.

- Take note of the viewpoint of any political candidate on matters of environmental policy. Public concern about environmental policy can play an important role in the election or defeat of political candidates (check out what happened to Representative Richard Pombo (R–CA) in 2006).

- Stay informed on environmental affairs; only when the public is informed is it likely that grassroots support will be maintained and environmental policy will really reflect public opinion. Taking a course in environmental science is an excellent way to start getting informed.

Revisiting the Themes

Sound Science

One component of the wealth of nations is *intangible capital*. Two elements of that capital are essential to maintaining science as a key to coping with environmental issues: education and the accumulation of scientific information as part of the fund of human knowledge that can be transmitted across boundaries of space and time. A truly wealthy nation will maintain a high level of scientific competence in its populace. The work of Sherwood Rowland and Mario Molina demonstrates the vital role science can play in the development of public policy—in this case, leading to protection of Earth's ozone shield. However, scientific integrity can be corrupted by politics, as seen in the workings of the George W. Bush administration, where scientists were silenced or had their work censored by politicians. The "new broom" of the Barack Obama administration promises to restore scientific integrity in our federal offices.

Sustainability

The ecological economists have made sustainability central to their economic theories. They rightly point out that the natural environment encompasses the economy. Therefore, there are limits to global economic growth that are set by the environment, and the economy must come to a steady state at some point. A nation must manage its assets, but it can't do that unless it knows what they are and how to measure them. Depleting assets of any kind are a clear signal of unsustainable practice. A future sustainable economy must focus on human well-being, not growth. It will value and preserve ecosystem services and goods. It will develop a host of "green" products that will revolutionize our technologies. It will discard the GDP as a measure of economic progress. Instead, it may even embrace the GPI as a better measure of economic well-being. Without question, societies must develop effective environmental public policies if they are to prevent the millions of deaths and widespread disease that exact terrible tolls. Laws to protect the environment are essential for achieving sustainability. This is well illustrated by the accomplishments of environmental public policy as documented by the EPA.

Stewardship

Table 2–2 demonstrates the consequences of poor environmental stewardship; laws to protect the environment are not a luxury, but an essential foundation for a society. Then EPA Administrator Christine Todd Whitman introduced the agency's *Draft Report on the Environment 2003* with these words: "We are all stewards of this shared planet, responsible for protecting and preserving a precious heritage for our children and grandchildren. As long as we work together and stay firmly focused on our goals, I am confident we will make our air cleaner, our water purer, and our land better protected for future generations." Environmental public policy over the past 40 years has accomplished much in achieving a safer and healthier environment. This represents our society's clear commitment to planetary stewardship.

REVIEW QUESTIONS

1. What is happening these days in the Chinese economy and environment?

2. Three patterns of environmental indicators are associated with differences in the level of development of a nation. What problems decline, what problems increase and then decline, and what problems increase with the level of development?

3. Name the two basic kinds of economic systems and explain how they differ.

4. What is the role of the World Trade Organization, and how has this role broken down recently?

5. What are some characteristics of a sustainable economy?

6. Describe the three components of a nation's wealth. What insights emerge from the World Bank's analysis of the wealth of different groups of nations?

7. Why is the gross national product an inaccurate indicator of a nation's economic status? How is it corrected by the Genuine Progress Indicator?

8. What conditions are necessary for a country to make progress in the development of human resources and produced assets?

9. Explain the importance of considering both intragenerational equity and intergenerational equity in addressing resource allocation issues.

10. What is the overall objective of environmental public policy, and what are the objective's two most central concerns?

11. Describe the various bodies responsible for environmental public policy at the federal, state, and local levels.

12. What are the advantages and disadvantages of the regulatory approach versus a market approach to policy development?

13. List the four stages of the policy life cycle, and show how the discovery of ozone layer destruction and the subsequent responses to it illustrate the cycle.

14. What is the conclusion of careful studies regarding the relationship between environmental policies, on the one hand, and jobs and the economy, on the other?

15. Define the term *benefit-cost analysis* as it relates to environmental regulation. How does this analysis method address external costs? Distinguish it from cost-effectiveness analysis.

16. Discuss the cost-effectiveness of pollution control. How much progress have we made in pollution control in recent years?

17. How has the political arena affected environmental public policy in the last few years? What has been the role of special interest groups? Political parties? Presidents?

THINKING ENVIRONMENTALLY

1. Consider the ultimate impact that environmental law has on the economy. Do short-term drawbacks justify long-term goals? If there is a conflict, which should come first—the economy or the environment?

2. Research the Doha Round of meetings of the World Trade Organization. Analyze the positions of the developing countries and the industrial countries, and propose a solution that might meet the concerns of both groups.

3. Examine the three types of capital that make up the wealth of nations. Which do you think is the most valuable? Justify your answer.

4. What different roles do you think the federal and state governments should have in environmental policy?

5. Suppose it was discovered that the bleach that is commonly used for laundry is carcinogenic. Referring to the policy life cycle, describe a predictable course of events until the problem is brought under control.

MAKING A DIFFERENCE

1. Our government makes the decisions that have the biggest long-term effects on the environment. Find a cause that you feel strongly about, and find a way to support it. Demonstrating is easier than it sounds—you don't necessarily have to drive a hundred miles to join in some kind of march. It's easy to join an online movement (for example, visit www.StopGlobalWarming.org) or write a letter to a senator. It doesn't matter how you participate, just that you do.

2. Help call attention to instances of pollution in your community. You can talk to regulators in the county health department, the EPA, or whoever is in charge of monitoring pollution in your region. It's their job to listen.

3. Instead of taking a long-distance vacation and relaxing during your break, join a volunteer group, and get something done. There are many opportunities to help out.

4. Find out the costs and benefits of your own lifestyle on a purely economic level. Decide what other changes you could make to save you money and save the environment at the same time.

Mastering**ENVIRONMENTALSCIENCE**™

Go to **www.masteringenvironmentalscience.com** for practice quizzes, Pearson eText, videos, current events, and more.

Tropical rain forest.

PART TWO

Ecology:
The Science of Organisms and Their Environment

TROPICAL RAIN FORESTS seem humid, warm, dense, and full of unusual plants and animals when you first step into them. You may be unaware, however, that they are home to a greater diversity of living things than anywhere else on Earth, that they are the result of millions of years of adaptive evolution, and that they store more carbon than is in the entire atmosphere. Nor can you sense how energy flows through these forests or how nitrogen and phosphorus are cycled and recycled. This information comes from the work of many scientists who have been asking basic questions about how such natural systems function, how they came to be, and how they relate to the rest of the world.

HOW IMPORTANT IS the information that comes from these scientists? In this second part of the text, we hope to show you that information about how the natural world works is absolutely crucial to understanding how to manage and protect the natural world. In the next three chapters, we will look at the science underlying how living things survive in the world and how matter and energy move throughout systems. We will use a hierarchical approach. That is, we will start small and move large—from the molecules necessary for life and the solar energy that sustains it to how plants return to a habitat after a fire and why large regions of the world have different vegetation (such as forests, desert, and grassland). Understanding environmental science begins with an understanding of how living organisms are structured, survive, relate to their environment (and each other), and change over time.

CHAPTER 3

Basic Needs of Living Things

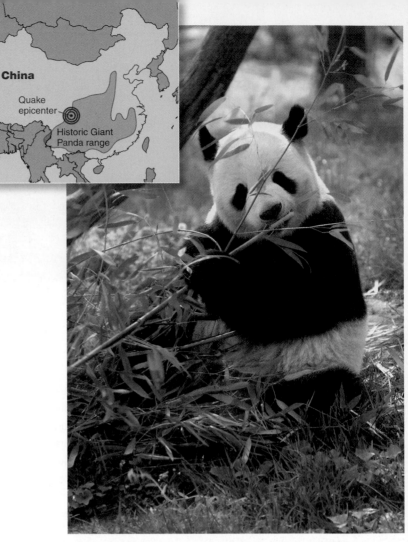

A giant panda in its native habitat, with map inset: China with Sichuan province highlighted. The current habitat for the giant panda is too small to display on this map.

KEY TOPICS

1. **Organisms in Their Environment:** The science of ecology can be described as a hierarchy of questions; different types of ecologists ask different questions, and perform different experiments, in the same natural systems.

2. **Environmental Factors:** Organisms need specific conditions and resources to survive; these define an organism's niche.

3. **Matter in Living and Nonliving Systems:** Living things need materials to build tissues and energy to carry out life processes such as photosynthesis and cellular respiration.

4. **Matter and Energy:** Very simple laws control the transformation of energy from one type to another and its flow through Earth's systems and back into space.

5. **The Cycling of Matter in Ecosystems:** Matter—particularly the important elements of carbon, phosphorus, and nitrogen—cycles through living and nonliving spheres.

O N MAY 12, 2008, a massive earthquake shook the Sichuan Province, P.R. China. It closed roads, caused buildings to collapse, and triggered mudslides that formed a lake by making a massive and unstable dam. The earthquake and the estimated 11,600 aftershocks over the next several weeks left more than 88,000 people dead or missing and up to 5 million homeless.

The earthquake also damaged two reserves for China's giant pandas: the Wolong Giant Panda Reserve Center in southwest Sichuan Province and the nearby Chengdu Panda Breeding and Research Center. Enclosures were destroyed, pandas were frightened and escaped, and food was difficult to obtain. Most of the pandas were safe, but one, an adult female named Maomao, was found dead, and another was missing. Two weeks after the earthquake, to the relief of keepers and the great rejoicing of panda conservationists, a pregnant female gave birth. Eventually, the pandas in the earthquake area had to be moved to other facilities.

China's sorrow and concern for the pandas made world news. The finding of Maomao and her funeral were on the Internet and television. Why so much concern about pandas in a disaster that harmed so many humans? The giant panda is a cultural icon of China, beloved and found in books and stories. It is endangered, with only about 1,600 still alive in the wild, many in small reserves. There are another approximately 270 in captivity. Despite their endangered species status, they remain under pressure from habitat loss and poaching. To many Chinese, the loss of a giant panda and the fear that they might be closer to extinction were another sorrow added to the loss of human life, and the survival of most of the protected pandas brought joy, much as the rescue of many humans did.

The giant panda lives on bamboo in broad-leaved forests in a decreasing habitat in China. Care for the declining pandas involves not only care for their individual health and the interbreeding population, but also protection of their habitat, including their food source. Often populations decline because damage is done to their habitat or food resources rather than to the individuals directly. The panda and its food, bamboo, give us an opportunity to look at the fundamental science behind the survival of organisms in their natural spaces.

In this chapter, we will follow the panda and the bamboo as examples of organisms living in their environment. All organisms get nutrients and energy from the environment and use them to build their bodies and carry out activities like growing, reproducing, and getting rid of wastes. The panda and bamboo both need ways to deal with the conditions under which they live—the amount of light, food, and water available, for example. We will see how the basic functions of organisms fit into the scientific discipline of ecology. We will explore basic physical rules about matter and energy and how they control the lives of organisms in their environments. To do this, we will need to understand some basic chemistry and physics. At the chapter's end, we will look at the global movements of important nutrients and how those huge cycles both affect individual organisms and are affected by human activities.

3.1 Organisms in Their Environment

Bamboo is a group of about 1,000 related types in the grass family, but unlike most grasses you may be familiar with, bamboos are woody and evergreen. As a group, bamboo is the fastest-growing plant on Earth. The areas in which bamboo is found are a function of several factors—its habitat requirements, its evolutionary history (where it first developed), and its interactions with other organisms, including humans. Just as bamboo needs certain resources (like sunlight and water) to survive, a panda needs bamboo (as a food source) to survive. The distribution and abundance of the giant panda are likewise a function of its requirements for survival, its growth, and the history of interaction between pandas and other species, especially humans. To understand why giant pandas are in decline and what conditions bamboo requires for growth, we turn to the science of ecology.

The Hierarchy of Ecology

Ecology (from *Oikos*, meaning "house or home") is *the study of all processes influencing the distribution and abundance of organisms and the interactions between living things and their environment.* A very basic understanding of ecological terms and concepts will help us here and in later chapters as we look at how changes in the environment affect living things.

Ecologists looking at the conservation of giant pandas could ask an array of different questions. They could investigate what type of soil bamboo requires to grow most rapidly or look for the factors that influence the successful breeding of giant pandas. They could also study the land use around Sichuan Province to see how it might affect giant panda habitat. Any of these inquiries could be an ecological question and different branches of ecology would address them. Therefore, ecology can be described as a hierarchy of studies. Scientists in different branches of the discipline might operate at different scales and would ask different questions about the same system. This concept is depicted in Figure 3–1.

As we investigate the science of ecology, we will go through these different levels. We will start by looking at the level of individual organisms that can be identified as belonging to species. We will see how species are grouped into larger categories and how these groups interact. Species exist in certain habitats because those places provide the nutrients and other factors that these organisms need to survive, so we will look at how those needs are met. Lastly, we will take a broader view of matter, energy, and how energy and nutrients cycle through the various systems on Earth. All of these different processes are interlinked, and ecologists study the connections as well as the whole.

Species. **Species** are the different kinds of living things in the community. A given species includes all those individuals that are like one another and that are distinct from other such groups (robins versus redwing blackbirds, for example). Species are classified together into groups of closely related species called *genera* (singular *genus*), and they in turn are grouped into *families*. These are grouped into larger groups: *orders*, *classes*, *phyla* (singular *phylum*), *kingdoms*, and *domains*. The official scientific name of any species is always Latin and has two parts—the genus name and the species epithet, or descriptive term. For example, the giant panda's name, *Ailuropoda melanoleuca* ("black-and-white cat-foot"), tells us first the genus (*Ailuropoda*) and second a descriptive word (the species epithet, *melanoleuca*). Although the giant panda is in the bear family (*Ursidae*), it is considered so different from other bears that it is the only species in this genus. In contrast, when we say, "bamboo," we actually mean a group of about 1,000 related species in the same subcategory of the grass family. The bamboo eaten by the giant panda is composed of members of two of these bamboo species.

Sometimes it is difficult to determine exactly what constitutes a species. The biological definition of a species is *all members that can interbreed and produce fertile offspring*, whereas members of different species generally do not

Figure 3–1 The hierarchy of life. Organisms can be seen as a hierarchy of complex systems from cells to tissues to organs to systems to whole individuals. Likewise, individuals function in a set of increasingly complex systems that make up the disciplines of ecology. These include populations, communities, ecosystems, and landscapes. The very largest regions are called biomes. Together all living things make up the biosphere.

interbreed. This definition is good for many organisms, but does not work for those that do not need to mate to produce offspring. In those cases, scientists use a variety of other classification methods. New species can arise over periods of time due to evolution, and species classifications are occasionally altered to reflect this.

Populations. Each species in a biotic community is represented by a certain **population**—that is, by a certain number of individuals that make up the interbreeding, reproducing group. The distinction between population and species is that *population* refers only to those individuals of a certain species that live within a given area, such as all of the gray wolves in Yellowstone National Park, whereas *species* is all inclusive, such as all of the gray wolves in the world.

Biotic Communities. The grouping of populations we observe when we study a forest, a grassland, a pond, or some other natural area is referred to as the area's **biota** (*bio* means "living") or **biotic community**. The biota includes all vegetation, from large trees through microscopic algae, and all animals, from large mammals, birds, reptiles, and amphibians through earthworms, tiny insects, and mites. It also includes microscopic creatures such as bacteria, fungi, and protozoans.

The particular kind of biotic community found in a given area is, in large part, determined by **abiotic** (nonliving, chemical, and physical) factors, such as the amount of water or moisture present, the climate, the salinity (saltiness), and the type of soil in the area. Sometimes we quickly refer to a community by its most obvious members, such as the dominant plants. For example, we might describe a "hemlock/beech forest" or an "oak/hickory forest" (Fig. 3–2). This is because vegetation is easily observed (it doesn't usually run away) and is a strong indicator of the environmental conditions of a site.

The species within a community depend on one another. For example, certain animals will not be present unless certain plants that provide their necessary food and shelter are present. Thus, the plant community supports (or limits by its absence) the animal community. In addition, every plant and animal species is adapted to cope with the abiotic factors of the region. For example, every species that lives in a temperate region is adapted in one way or another to survive the winter season, which includes a period of freezing temperatures (Fig. 3–3). These interactions among organisms and their environments will be discussed later. For now, keep in mind that the populations of different species within a biotic community are constantly interacting with each other and with the abiotic environment.

Ecosystems. An ecosystem is *an interactive complex of communities and the abiotic environment affecting them within a particular area.* A forest, a grassland, a wetland, a marsh, a pond, a sand dune, and a coral reef, each with its respective species in a particular spatial area, can be studied as distinct ecosystems. Humans are parts of ecosystems, too, and some branches of ecology, such as urban ecology, focus on human-dominated systems. Other ecologists study the

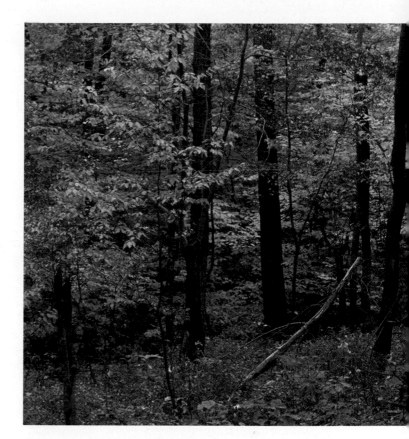

Figure 3–2 Predictable vegetation. An oak/hickory forest is dominated by oak and hickory trees, which define the type of forest habitat other species experience.

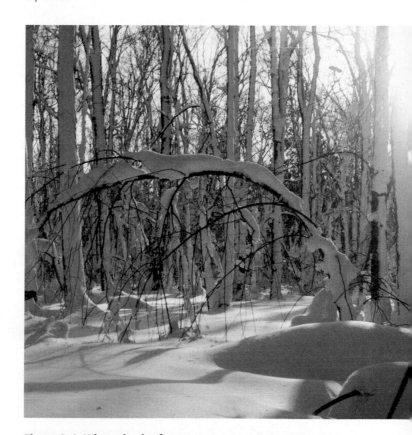

Figure 3–3 Winter in the forest. Many trees and other plants of temperate forests are so adapted to the winter season that they actually require a period of freezing temperature in order to grow again in the spring.

interaction between humans and systems such as salt marshes or the open ocean, which are deeply affected by humans but inhabited primarily by other organisms.

While it is convenient to divide the living world into different ecosystems, you will find that there are seldom distinct boundaries between ecosystems and that they are never totally isolated from one another. Many species will occupy (and thus be a part of) two or more ecosystems, or they may move from one ecosystem to another at different times, as in the case of migrating birds.

In passing from one ecosystem to another, the one may grade into the other through a transitional region, known as an **ecotone**, that shares many of the species and characteristics of both ecosystems Like the ecotone between grassland and forest depicted in Figure 3–4a, ecotones may lack unique species and contain fewer species than either bordering ecosystem. Alternatively, ecotones may, like the marshy ecotone that often occurs between the open water of a lake and dry land (Figure 3–4b), house more species and support species that do not occur in either the land ecosystem or the aquatic ecosystem.

Landscapes and Biomes. A cluster of interacting ecosystems such as a forest, an open meadow, and a nearby river together constitute a **landscape**. A landscape ecologist might do

Figure 3–4 Ecotones. (a) Ecosystems are not isolated from one another. One ecosystem blends into the next through a transitional region—an ecotone—that contains many species common to both systems, though in lower numbers than in those systems. (b) An ecotone may create a unique habitat that harbors specialized species not found in either of the ecosystems bordering it. For example, cattails, reeds, and lily pads typically grow in the ecotone between land and freshwater ponds.

(a)

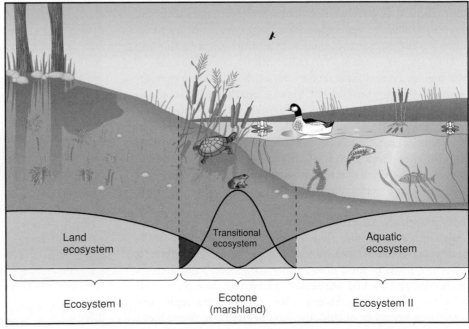

(b)

a study to find out how much suitable habitat will be left for fish to lay eggs in the riverbed if the surrounding forest is clear-cut and soil erodes or how far squirrels will be able to travel if the forest is broken into small segments. Landscapes are usually on a scale a human can envision—like looking out at the horizon and seeing a forest merge into a field or meadow. But an observer could also pull back from Earth and look at large areas of its surface that have similar characteristics.

Tropical rain forests, grasslands, and deserts are **biomes** (discussed in Chapter 5). In contrast to the much smaller landscape, a biome is a large area of Earth's surface with the same climate and similar vegetation. For example, grasslands, such as prairies, are geographical regions that can be predicted by their rainfall and temperatures. On different continents, they may have different particular species, but they are still grasslands, described by the dominant type of vegetation. As with ecosystems, there are generally no sharp boundaries between biomes. Instead, one grades into the next through transitional regions. While the term *biome* is limited to terrestrial systems, there are major categories of aquatic and wetland ecosystems that are determined primarily by the depth, salinity, and permanence of water in them. These will be discussed in Chapter 5.

Biosphere.   Regardless of how we choose to categorize and name different ecosystems, they all remain interconnected and interdependent. Terrestrial ecosystems are connected by the flow of rivers and by migrating animals. Sediments and nutrients washing from the land may nourish or pollute the ocean. Seabirds and marine mammals connect the oceans with the land, and all biomes share a common atmosphere and water cycle. Therefore, all the species on Earth live in ecosystems, and the ecosystems are interconnected, so that all living things form one huge system called the **biosphere.** The concept is analogous to the idea that the cells of our bodies form tissues, tissues form organs, organs are connected into organ systems, and all are interconnected to form the whole body.

3.2 Environmental Factors

Let's look back at our giant panda eating bamboo at the reserve. Why are the pandas living there—and not in another part of the world? The answer is the environment and history. Both the bamboo and the panda have evolved to live in the environment where they occur, with the physical, chemical, and biological factors around them. We describe these factors as either biotic (living) or abiotic (nonliving). Such factors can be characterized as either conditions or resources. **Conditions** are any factors that vary in space and time, but that are not used up or made unavailable to other species. These include temperature (extremes of heat and cold as well as average temperature), wind, pH (acidity), salinity (saltiness), and fire. **Resources** are any factors—biotic or abiotic—that are consumed by organisms. These include water, chemical nutrients (like nitrogen and phosphorus), light (for plants), oxygen, food (like the bamboo for the panda), and space for resting or nesting.

Some factors can be conditions in one situation and resources in another. For example, plants in the desert compete for water as a *resource*; however, in a pond, water is a *condition* for which plants like water lilies and cattails need special adaptations and for which they are not competing.

The degree to which each factor is present or absent profoundly affects the ability of organisms to survive. However, each species may be affected differently by each factor, and its response determines whether or not it will occupy a given region.

Optimums, Zones of Stress, and Limits of Tolerance

Different species thrive with different levels of environmental factors. This principle applies to all living things, both plants and animals. Bamboo thrives in full sun and warm temperatures, with plenty of water and well-drained, nitrogen-rich soils. In contrast, desert cacti are generally adapted for low amounts of water, poor soil fertility, and extremes of heat and cold. Aquatic systems also vary in their environmental factors and residents—for example, the amounts of freshwater and saltwater determine the identities of fish and other organisms present.

Organismal ecologists study the adaptations of organisms to specific environmental factors. Organisms can be grown under controlled conditions in which one factor is varied, while other factors are held constant. Such experiments demonstrate that, for every factor, there is an **optimum** (plural optima)—a certain level at which the organisms grow or survive most. At higher or lower levels, the organisms do less well, and at further extremes, they may not be able to survive at all (Fig. 3–5). Temperature is shown as the variable in the figure, but the idea pertains to almost any abiotic factor that might be tested.

The point at which the best response occurs is called the optimum, but this may be a range of several degrees (or other units), so it is common to speak of an *optimal range*. The entire span that allows any growth at all is called the **range of tolerance.** The points at the high and low ends of the range of tolerance are called the **limits of tolerance.** Between the optimal range and the high or low limit of tolerance are **zones of stress.** That is, as the factor is raised or lowered from the optimal range, the organisms experience increasing stress, until, at either limit of tolerance, they cannot survive.

A Fundamental Biological Principle.   Based on the consistency of observations and experiments, the following is considered to be a fundamental biological principle: *Every species has an optimum range, zones of stress, and limits of tolerance with respect to every abiotic factor, and these characteristics vary between species.* For instance, some plants cannot tolerate any freezing temperatures; others can tolerate slight, but not intense, freezing; and some actually require several weeks of freezing temperatures in order to complete their life cycles. Also, some species have a very broad range of tolerance, whereas others have a much narrower range.

Figure 3–5 Survival curve. For every factor influencing growth, reproduction, and survival, there is an optimum level. Above and below the optimum, stress increases until survival becomes impossible at the limits of tolerance. The total range between the high and low limits is the range of tolerance.

The range of tolerance for any factor affects more than just the growth of individuals. As the health and vigor of individuals affect reproduction and the survival of the next generation, the population is also influenced. Consequently, the population density (individuals per unit area) of a species is greatest where all conditions are optimal, and it decreases as any one or more conditions depart from the optimum. That is, you expect to see the most bamboo in the places where all conditions are best for bamboo and less where conditions are not so favorable.

Law of Limiting Factors. In 1840, Justus von Liebig studied the effects of chemical nutrients on plant growth. He observed that restricting any one of the many different nutrients at any given time had the same effect: It limited growth. A factor that limits growth is called (not surprisingly) a **limiting factor.** Any one factor being outside the optimal range will cause stress and limit the growth, reproduction, or even survival of a population. This observation is referred to as the **law of limiting factors** or Liebig's Law of the Minimum.

The limiting factor may be a problem of *too much* as well as a problem of *too little*. For example, plants may be stressed or killed not only by underwatering or underfertilizing, but also by overwatering or overfertilizing, which are common pitfalls for beginning gardeners. Note also that the limiting factor may change from one time to another. For instance, temperature may be limiting in the early spring, nutrients may be limiting in the winter, and water may be limiting if a drought occurs. Also, if one limiting factor is corrected, growth will increase only until another factor comes into play. The organism's genetic potential is the ultimate limiting factor. A mouse will never grow to the bulk of an elephant, no matter how much you feed it.

While one factor may be determined to be limiting at a given time, several factors outside the optimum may combine to cause additional stress or even death. In particular, pollutants may act in a way that causes organisms to become more vulnerable to disease or drought. Such cases are examples of **synergistic effects,** or **synergisms,** which are defined as two or more factors interacting in a way that causes an effect much greater than one would anticipate from each of the two acting separately.

Habitat and Niche. **Habitat** refers to the kind of place—defined by the plant community and the physical environment—where a species is biologically adapted to live. For example, a deciduous forest, a swamp, and a grassy field are types of habitats. Different types of forests (for instance, coniferous versus tropical) provide markedly different habitats and support different species of wildlife. Because some organisms operate on very small scales, we use the term *microhabitat* for things like puddles, sheltered spaces by rocks, and holes in tree trunks that might house their own small community.

Even when different species occupy the same habitat, competition may be slight or nonexistent because each species has its own niche. An animal's ecological **niche** refers to what the animal feeds on, where it feeds, when it feeds, where it finds shelter, how it responds to abiotic factors, and where it nests. Basically, the niche is the sum of all of the conditions and resources under which a species can live.

Similar species can coexist in the same habitat, but have separate niches. Competition is minimized because potential competitors are using different resources. For example, woodpeckers, which feed on insects in deadwood, do not compete with birds that feed on seeds. Bats and swallows both feed on flying insects, but they do not compete because bats feed on night-flying insects and swallows feed during the day.

In the next section, we will see how organisms take these materials from the environment and use them to build their bodies and do activities necessary for survival.

3.3 Matter in Living and Nonliving Systems

Living organisms need to be able to take in matter and energy from their environment in order to grow and function. For example, the bamboo that grows in the panda reserve can survive only because it is able to take the Sun's energy and combine it with materials from the soil, air, and water around it to make the basic building blocks of its own tissues and to power its cellular activities. In turn, the giant panda will eat the bamboo and get energy and matter necessary to build its own tissues and to do the work it needs to do in each of its cells. In order to understand this simple ecological relationship, we need to know some basic chemistry and physics. Later, we will explore what happens to energy and matter as they move through the biosphere.

Basic Units of Matter

Matter is defined as anything that occupies space and has mass. This definition covers all solids, liquids, and gases and all living as well as nonliving things. Matter is composed of **atoms**—very small pieces—that are in turn combined to form molecules, and these in turn can be combined into more complex structures.

Atoms. The basic building blocks of all matter are atoms. Only 94 different kinds of atoms occur in nature, and these are known as the naturally occurring **elements**. Atoms are made up of protons, neutrons, and electrons, which in turn are made up of still smaller particles. For example, lead (Pb) is an element. An atom of lead is the smallest piece you could have that has the characteristics of lead. A lead atom has a characteristic number of protons (positive particles), neutrons (neutral particles), and electrons (negative particles).

How can these relatively few building blocks make up the countless materials of our world, including the tissues of living things? Picture each kind of atom as a different-sized Lego® block. Like Legos, atoms can build a great variety of things. Also like Legos, natural materials can be taken apart into their separate constituent atoms, and the atoms can then be reassembled into different materials. All chemical reactions, whether they occur in a test tube, in the environment, or inside living things (and whether they occur very slowly or very fast), involve rearrangements of atoms to form different kinds of matter.

Atoms do not change during the disassembly and reassembly of different materials. A carbon atom, for instance, will always remain a carbon atom. In chemical reactions, atoms are neither created nor destroyed. The same number and kind of different atoms exist before and after any reaction. This constancy of atoms is regarded as a fundamental natural law, the **Law of Conservation of Matter.** Nuclear reactions differ from chemical reactions and could result in the splitting of an atom of one element, for example, into multiple atoms of another element. However, this is a very specific, rare instance and is not a chemical reaction. Nuclear reactions will be discussed in Chapter 15 when we talk about nuclear energy.

Now, we turn our attention to the ways atoms are put together, which atoms make up the bulk of the bodies of living organisms, and how atoms are incorporated into organisms.

Molecules and Compounds. A **molecule** consists of two or more atoms (either the same kind or different kinds) bonded in a specific way. The properties of a material depend on the specific way in which atoms are bonded to form molecules as well as on the atoms themselves. A **compound** consists of two or more *different* kinds of atoms that are bonded. For example, the fundamental units of oxygen gas, which consists of two bonded oxygen atoms, are molecules (O_2), but not compounds. By contrast, when an oxygen atom binds with hydrogen atoms to create water, it is both a molecule and a compound (H_2O).

On the chemical level, the cycle of growth, reproduction, death, and decay of organisms is a continuous process of using various molecules and compounds from the environment (food), assembling them into living organisms (growth), disassembling them (decay), and repeating the process. Driving the cycle is the genetically programmed urge living things have to grow and reproduce.

Four Spheres

During growth and decay, atoms move from the environment into living things and then return to the environment. To picture this process, think of the environment as three open, nonliving systems, or "spheres," interacting with the biosphere (Fig. 3–6). The **atmosphere** is the thin layer of gases (including water vapor) separating Earth from outer space. The **hydrosphere** is water in all of its liquid and solid compartments: oceans, rivers, ice, and groundwater. The **lithosphere** is Earth's crust, made up of rocks and minerals. Matter is constantly being exchanged within and between these four spheres.

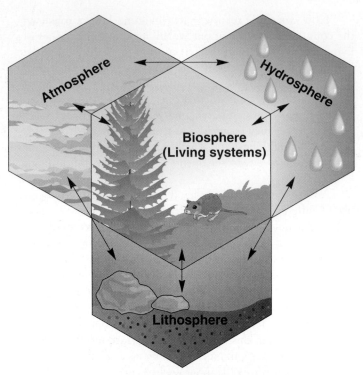

Figure 3–6 The four spheres of Earth's environment. The biosphere is all of life on Earth. It depends on, and interacts with, the atmosphere (air), the hydrosphere (water), and the lithosphere (soil and rocks). (*Source*: Adapted from *Geosystems*, 5th ed., by Robert W. Christopherson, © 2005 by Pearson Education, Inc.)

Atmosphere. The lower atmosphere is a mixture of molecules of three important gases—oxygen (O_2), nitrogen (N_2), and carbon dioxide (CO_2)—along with water vapor and trace amounts of several other gases (Fig. 3–7). The gases in the atmosphere are normally stable, but under some circumstances, they react chemically to form new compounds (for example, ozone is produced from oxygen in the upper atmosphere, as described in Chapter 18). Plants take carbon dioxide in from the atmosphere, usually through their leaves. Animals usually take oxygen in through some type of specialized organ such as a lung, but some, like earthworms, can simply absorb oxygen through their skin.

Hydrosphere. While the atmosphere is a major source of carbon and oxygen for all organisms (and a source of nitrogen for a few of them), the hydrosphere is the source of hydrogen. Each molecule of water consists of two hydrogen atoms bonded to an oxygen atom, so the chemical formula for water is H_2O. A weak attraction known as *hydrogen bonding* exists between water molecules.

Water is an important molecule for living things and usually needs to be available in liquid form. Water occurs in three different states. At temperatures below freezing, hydrogen bonding holds the molecules in position with respect to one another, and the result is a solid crystal structure (ice or snow). At temperatures above freezing, but below vaporization, hydrogen bonding still holds the molecules close, but

Clean, dry air is a mixture of molecules of three important gases.

Clean, dry air	Percent[a]		Chemical formula	Chemical diagram
	78.08	Nitrogen gas	N_2	N≡N
	20.95	Oxygen gas	O_2	O=O
	0.035	Carbon dioxide	CO_2	O=C=O

[a]The remaining 0.94 percent is composed of inert gases, which have no biological importance.

Figure 3–7 The major gases of clean, dry air. From a biological point of view, the three most important gases of the lower atmosphere are nitrogen, oxygen, and carbon dioxide. (Note that the proportion of carbon dioxide is deliberately overrepresented in the diagram.)

Water molecule
(a)

(b)

Gas
(water
vapor)

Sublimation
Deposition

Vaporization/
evaporation
Condensation

Freezing

Solid
(ice)

Liquid
(water)

Melting

(d)

(c)

Figure 3–8 Water and its three states. (a) Water consists of molecules, each of which is formed when two hydrogen atoms bond to an oxygen atom (H$_2$O). (b) In water vapor, the molecules are separate and independent. (c) In liquid water, the weak attraction between water molecules known as hydrogen bonding gives the water its liquid property. (d) At freezing temperatures, hydrogen bonding holds the molecules firmly, giving the solid state—ice.

allows them to move past one another, producing the liquid state. Vaporization (evaporation) occurs as hydrogen bonds break and water molecules move into the air independently. As temperatures are lowered again, all of these changes of state go in the reverse direction. Generally, water undergoes melting and evaporation, but sometimes water molecules leave snow or ice and go directly into the air. This is *sublimation* (Fig. 3–8). Despite the changes of state, the water molecules themselves retain their basic chemical structure of two hydrogen atoms bonded to an oxygen atom. Only the *relationship* between the molecules changes.

One reason that the states of water matter so much to organisms is that moving from one state to another either releases or requires a great deal of energy. You take advantage of this when you sweat; you can easily get rid of a lot of heat by causing water on the surface of your skin to evaporate. The heat from your body acts as energy that moves the water from a liquid to a gas state, and because the heat is gone from your body, you feel cooler.

Lithosphere. All the other elements that are required by living organisms, as well as the 72 or so elements that are not required by them, are found in the lithosphere, in the form of rock and soil minerals. A **mineral** is a naturally occurring solid, made by geologic processes; it is a hard, crystalline structure of a given chemical composition. Most rocks are made up of relatively small crystals of two or more minerals, and soil generally consists of particles of many different minerals. Each mineral is made up of dense clusters of two or more kinds of atoms bonded by an attraction between positive and negative charges on the atoms, as explained in Appendix C and shown in Figure 3–9.

Interactions. Air, water, and minerals interact with each other in a simple, but significant, manner. Gases from the air and ions (charged atoms) from minerals may dissolve in water. Therefore, natural water is inevitably a *solution* containing variable amounts of dissolved gases and minerals. This solution is constantly subject to change because various processes may remove any dissolved substances in it or additional materials may dissolve in it. Molecules of water enter the air by evaporation and leave it again via condensation and precipitation (see the hydrologic cycle, Chapter 10). Thus, the amount of moisture in the air fluctuates constantly. Wind may carry dust or mineral particles, but the amount changes constantly because the particles gradually settle out from the air. The various interactions are summarized in Fig. 3–10. The materials in these three spheres interact with the biosphere as living organisms take materials from these spheres and use them to build complex molecules in their bodies. We will discuss the process of building these molecules next.

Cl$^-$

Na$^+$

Figure 3–9 Minerals. Minerals (hard, crystalline compounds) are composed of dense clusters of atoms of two or more elements. The atoms of most elements gain or lose one or more electrons, becoming negative (−) or positive (+) ions. Salt (sodium chloride, NaCl) is held together by the attraction between the positive sodium (Na+) and negative chloride (Cl−) charges.

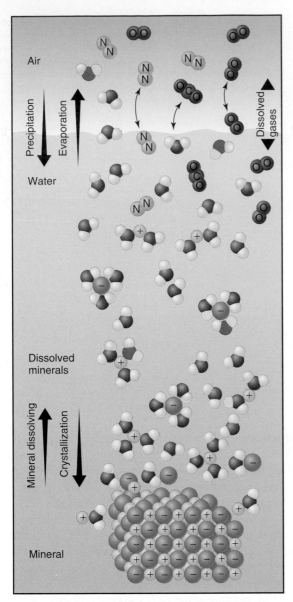

Figure 3–10 Interrelationship among air, water, and minerals. Minerals and gases dissolve in water, forming solutions. Water evaporates into air, causing humidity. These processes are all reversible: Minerals in solution recrystallize, and water vapor in the air condenses to form liquid water.

Organic Compounds

Your body, like that of the panda, or even the bamboo it eats, is composed of relatively large chemical compounds in a number of broad categories, such as proteins, carbohydrates (sugars and starches), lipids (fatty substances), and nucleic acids (DNA and RNA). These compounds usually contain six key elements: **carbon (C)**, **hydrogen (H)**, **oxygen (O)**, **nitrogen (N)**, **phosphorus (P)**, and **sulfur (S)** (Table 3–1). Living things need other elements as well, sometimes in small, but important, doses.

Unlike the relatively simple molecules that occur in the environment (such as CO_2, H_2O, and N_2), the key chemical elements in living organisms bond to form very large, complex molecules. The chemical compounds making up the tissues of living organisms are referred to as **organic**. Some of these molecules may contain millions of atoms. These molecules are constructed mainly from carbon atoms bonded into chains, with hydrogen and oxygen atoms attached. Nitrogen, phosphorus, or sulfur may be present also, but the key common denominator is carbon–carbon and carbon–hydrogen or carbon–oxygen bonds. **Inorganic**, then, refers to molecules or compounds with neither carbon–carbon nor carbon–hydrogen bonds. While this is the general rule, some exceptions occur; by convention, a few that contain carbon bonds, such as carbon dioxide, are still considered inorganic.

All plastics and countless other human-made compounds are also based on carbon bonding and are, chemically speaking, organic compounds. To resolve any confusion this may cause, the compounds making up living organisms are referred to as **natural organic compounds** and the human-made ones as **synthetic organic compounds**. In Chapter 12, we will discuss organic farming. In that context, the term *organic* will have a different meaning.

In conclusion, the elements essential to life (C, H, O, and so on) are present in the atmosphere, hydrosphere, or lithosphere in relatively simple molecules. In the living organisms of the biosphere, on the other hand, these elements are organized into highly complex organic compounds. In the case of the panda and the bamboo, the simple molecules and atoms in the soil, air, and water are combined to form the tissues of the bamboo and eaten, digested, and recombined to make the tissues of the giant panda. When the bamboo or panda dies, the reverse process occurs through decomposition and decay. Each of these processes is discussed in more detail later in the chapter. Next, we will discuss energy, the force that helps change chemical matter into substances that support life.

3.4 Matter and Energy

The universe is made up of matter and energy. Recall that **matter** is *anything that occupies space and has mass*. By contrast, light, heat, movement, and electricity do not have mass, nor do they occupy space. These are the common forms of energy with which you are probably familiar. What do the various forms of energy have in common? They affect matter, causing changes in its *position* or its *state*. For example, the release of energy in an explosion causes things to go flying—a change in position. Heating water causes it to boil and change to steam, a change of state. On a molecular level, changes of state are actually also movements of atoms or molecules. For instance, the degree of heat energy contained in a substance is a measure of the relative vibrational motion of the atoms and molecules of the substance. Therefore, we can define **energy** as *the ability to move matter*.

Energy Basics

Energy can be categorized as either kinetic or potential (Fig. 3–11). **Kinetic energy** is *energy in action or motion*. Light, heat, physical motion, and electrical current are all forms of kinetic energy. **Potential energy** is *energy in storage*. A substance or system with potential energy has the capacity,

TABLE 3–1   Elements Found in Living Organisms and the Locations of Those Elements in the Environment

| Element (Kind of Atom) | Biologically Important Molecule or Ion in Which the Element Occurs[a] | | | Location in the Environment[b] | | |
	Symbol	Name	Formula	Atmosphere	Hydrosphere	Lithosphere
Carbon	C	Carbon dioxide	CO_2	X	X	X(CO_3)
Hydrogen	H	Water	H_2O	X	(Water itself)	
Atomic oxygen (required in respiration)	O	Oxygen gas	O_2	X	X	
Molecular oxygen (released in photosynthesis)	O_2	Water	H_2O		(Water itself)	
Nitrogen	N	Nitrogen gas	N_2	X	X	Via fixation
		Ammonium ion	NH_4		X	X
		Nitrate ion	NH_3		X	X
Sulfur	S	Sulfate ion	SO_4^2		X	X
Phosphorus	P	Phosphate ion	PO_4^3		X	X
Potassium	K	Potassium ion	K^+		X	X
Calcium	Ca	Calcium ion	Ca^2		X	X
Magnesium	Mg	Magnesium ion	Mg^2		X	X
Trace Elements[c]						
Iron	Fe	Iron ion	Fe^2, Fe^3		X	X
Manganese	Mn	Manganese ion	Mn^2		X	X
Boron	B	Boron ion	B^3		X	X
Zinc	Zn	Zinc ion	Zn^2		X	X
Copper	Cu	Copper ion	Cu^2		X	X
Molybdenum	Mo	Molybdenum ion	Mo^2		X	X
Chlorine	Cl	Chloride ion	Cl^-		X	X

Note: These elements are found in *all* living organisms—plants, animals, and microbes. Some organisms require certain elements in addition to the ones listed. For example, humans require sodium and iodine.

[a]A molecule is a chemical unit of two or more atoms that are bonded. An ion is a single atom or group of bonded atoms that has acquired a positive or negative charge as indicated.

[b]X means that element exists in the indicated sphere.

[c]Only small or trace amounts of these elements are required.

or potential, to release one or more forms of kinetic energy. A stretched rubber band has potential energy; it can send a paper clip flying. Numerous chemicals, such as gasoline and other fuels, release kinetic energy—heat, light, and movement—when ignited. The potential energy contained in such chemicals and fuels is called **chemical energy.**

Energy may be changed from one form to another in innumerable ways (Fig. 3–12). Besides understanding that potential energy can be converted to kinetic energy, it is especially important to recognize that kinetic energy can be converted to potential energy. (Consider, for example, charg-

ing a battery or pumping water into a high-elevation reservoir.) We shall see later in this section that photosynthesis does just that.

Because energy does not have mass or occupy space, it cannot be measured in units of weight or volume, but it can be measured in other kinds of units. One of the most common units is the **calorie,** which is defined as the amount of heat required to raise the temperature of 1 gram (1 milliliter) of water 1 degree Celsius. This is a very small unit, so it is frequently more convenient to use the kilocalorie (1 kilocalorie = 1,000 calories), the amount of heat required to raise 1 liter

Figure 3–11 Forms of energy. Energy is distinct from matter in that it neither has mass nor occupies space. It has the ability to act on matter though, changing the position or the state of the matter. Kinetic energy is energy in one of its active forms. Potential energy is the potential that systems or materials have to release kinetic energy.

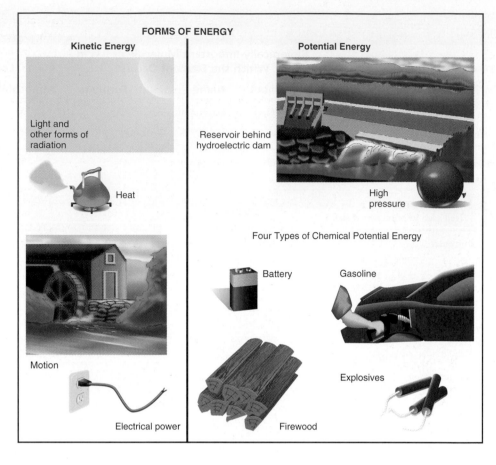

Figure 3–12 Energy conversions. Any form of energy except heat can spontaneously transform into any other form. Heat is a form of energy that flows from one system or object to another because the two are at different temperatures; therefore, heat can spontaneously transform only to something cooler.

(1,000 milliliters) of water 1 degree Celsius. (Kilocalories are sometimes denoted as "Calories" with a capital "C." Food Calories, which measure the energy in given foods, are actually kilocalories.) *Temperature* measures the molecular motion in a substance caused by the kinetic energy present in it.

If energy is defined as the ability to move matter, then no matter can be moved *without* the absorption or release of energy. Indeed, no change in matter—from a few atoms coming together or apart in a chemical reaction to a major volcanic eruption—can be separated from its respective change in energy.

Laws of Thermodynamics. Because energy can be converted from one form to another, numerous would-be inventors over the years have tried to build machines that produce more energy than they consume. A common idea is to use the output from a generator to drive a motor that in turn drives the generator to keep the cycle going and yields additional power in the bargain (a perpetual-motion machine). Unfortunately, all such devices have one feature in common: They don't work. When all the inputs and outputs of energy are carefully measured, they are found to be equal. There is no net gain or loss in total energy. This observation is now accepted as a fundamental natural law, the **Law of Conservation of Energy**. It is also called the **First Law of Thermodynamics**, and it can be expressed as follows: *Energy is neither created nor destroyed, but may be converted from one form to another*. This law really means that you can't get something for nothing.

Imaginative "energy generators" fail for two reasons: First, in every energy conversion, a portion of the energy is converted to heat energy. Second, there is no way of trapping and recycling heat energy without expending even more energy in doing so. Consequently, in the absence of energy inputs, any and every system will sooner or later come to a stop as its energy is converted to heat and lost. This is now accepted as another natural law, the **Second Law of Thermodynamics**, and it can be expressed as follows: *In any energy conversion, some of the usable energy is always lost*. Thus, you can't get something for nothing (the first law), and in fact, you can't even break even (the second law)!

Entropy. Underlying the loss of usable energy to heat is the principle of increasing entropy. **Entropy** is *a measure of the degree of disorder in a system*, so increasing entropy means increasing disorder. Without energy inputs, everything goes in one direction only—toward increasing entropy.

The conversion of energy and the loss of usable energy to heat are both aspects of increasing entropy. Heat energy is the result of the random vibrational motion of atoms and molecules. It is the lowest (most disordered) form of energy, and its spontaneous flow to cooler surroundings is a way for that disorder to spread. Therefore, the Second Law of Thermodynamics may be more generally stated as follows: *Systems will go spontaneously in one direction only— toward increasing entropy*. The second law also says that systems will go spontaneously only toward lower potential energy, a direction that releases heat from the systems (Fig. 3–13).

It is possible to pump water uphill, charge a battery, stretch a rubber band, and compress air. All of these require

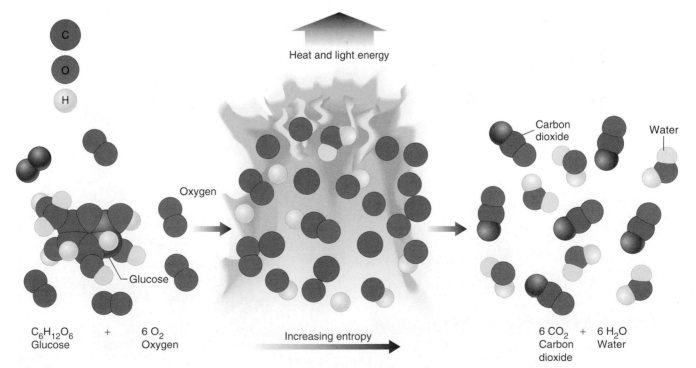

Figure 3–13 Entropy. Systems go spontaneously only in the direction of increasing entropy. When glucose, a major constituent of wood, is burned, heat is released, and the atoms become more disordered. Both of these phenomena are aspects of increasing entropy.

the input of energy in order to increase the potential energy of a system. The verbs *pump*, *charge*, *stretch*, and *compress* reflect the fact that energy is being put into the system. In contrast, flow in the opposite direction (which releases energy) occurs spontaneously (Fig. 3–14).

Whenever you see something gaining potential energy, therefore, keep in mind that the energy is being obtained from somewhere else (the first law). Moreover, the amount of energy lost from that "somewhere else" is greater than the amount gained (the second law). Let us now relate these concepts of matter and energy to organic molecules, organisms, ecosystems, and the biosphere.

Energy Changes in Organisms

Picture someone building a fire in a fireplace. The wood is largely composed of complex organic molecules. These molecules required energy to form, and therefore, their chemical bonds contain high potential energy. When these molecules are broken, the energy contained in those bonds is released and can be used to do work. This is done by **oxidation**, a loss of electrons, usually accomplished by the addition of oxygen (which causes *burning*). The heat and light of the flame are the potential energy being released as kinetic energy. In contrast, most inorganic compounds, such as carbon dioxide, water, and rock-based minerals, are nonflammable because

they have very low potential energy. Thus, the production of organic material from inorganic material represents a gain in potential energy. Conversely, the breakdown of organic matter *releases* energy. Inside the cells of the giant panda eating bamboo, the reaction that takes place is similar to that of burning wood in a fireplace, except that it occurs in a set of small steps and is much more controlled.

Producers and Photosynthesis.   This relationship between the formation and breakdown of organic matter, where energy is gained and released, forms the basis of the energy dynamics of ecosystems. **Producers** (primarily green plants) make *high-potential-energy organic molecules* for their needs from *low-potential-energy raw materials* in the environment— namely, carbon dioxide, water, and a few dissolved compounds of nitrogen, phosphorus, and other elements (Fig. 3–15). This "uphill" conversion is possible because producers use chlorophyll to absorb light energy, which "powers" the production of the complex, energy-rich organic molecules. On the other hand, **consumers** (all organisms that live on the production of others) obtain energy for movement and growth from *feeding on and breaking down organic matter made by producers* (Fig. 3–16).

Green plants use the process of **photosynthesis** to make sugar (glucose, which is stored chemical energy) from carbon dioxide, water, and light energy (Fig. 3–15). This process,

Figure 3–14 Storage and release of potential energy. (a) A simple physical example of the storage and release of potential energy. (b) The same principle applied to ecosystems.

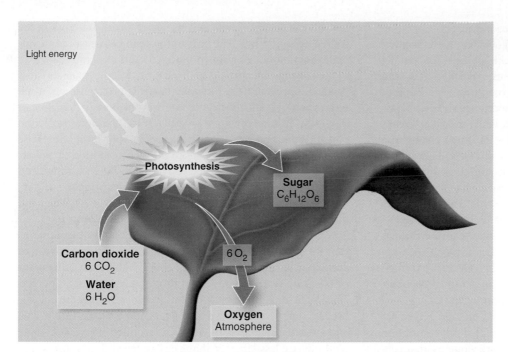

Figure 3–15 Producers as chemical factories. Using light energy, producers make glucose from carbon dioxide and water, releasing oxygen as a by-product. Breaking down some of the glucose to provide additional chemical energy, they combine the remaining glucose with certain nutrients from the soil to form other complex organic molecules that the plant then uses for growth.

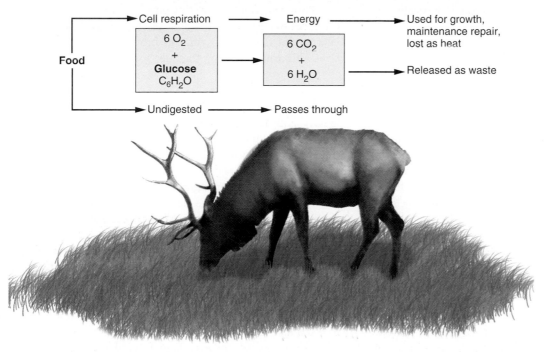

Figure 3–16 Consumers. Only a small portion of the food ingested by a consumer is assimilated into body growth and repair. Some food provides energy in chemical bonds which are broken by the process of cell respiration: this is the reverse process of photosynthesis, and waste products are carbon dioxide, water, and various mineral nutrients. A third portion is not digested and becomes fecal waste.

which also releases oxygen gas as a by-product, is described by the following chemical equation:

Photosynthesis
(An energy-demanding process)

$6\,CO_2$ + $12\,H_2O$ $\longrightarrow$ $C_6H_{12}O_6$ + $6\,O_2$ + $6H_2O$

carbon dioxide   water   light energy   glucose   oxygen   water

(gas)   input   (gas)

(low potential energy)   (high potential energy)

Chlorophyll in the cells of the plant absorbs the kinetic energy of light and uses it to remove the hydrogen atoms from water (H_2O) molecules. The hydrogen atoms combine with carbon atoms from carbon dioxide to form a growing chain of carbons that eventually becomes a glucose molecule. After the hydrogen atoms are removed from water, the oxygen atoms that remain combine with each other to form oxygen gas, which is released into the air. Water appears on both sides of the equation because 12 molecules are consumed and 6 molecules are newly formed during photosynthesis.

The key energy steps in photosynthesis take the hydrogen from water molecules and join carbon atoms together to form the carbon–carbon and carbon–hydrogen bonds of glucose. These steps convert the low-potential-energy bonds in water and carbon dioxide molecules to the high-potential-energy bonds of glucose.

Within the Plant.   The glucose produced in photosynthesis serves three purposes in the plant: (1) Glucose, with the addition of other molecules, can be the backbone used for making all the other organic molecules (proteins, carbohydrates, and so on) that make up the stem, roots, leaves, flowers, and fruit of the plant. (2) It takes energy to run the cell activities of plants such as growth. This energy is obtained when the plant breaks down a portion of the glucose to release its stored energy in a process called *cell respiration*, which is discussed later. (3) A portion of the glucose produced may be stored for future use. For storage, the glucose is generally converted to starch, as in potatoes or grains, or to oils, as in many seeds (Fig. 3–16). This stored energy is what is available to be eaten by consumers.

None of these reactions, from the initial capture of light by chlorophyll to the synthesis of plant structures, takes place automatically. Each step is catalyzed by specific **enzymes**, *proteins that promote the synthesis or breaking of chemical bonds*. The same is true of cell respiration.

Cell Respiration.   Inside each cell, organic molecules may be broken down through a process called **cell respiration** to release the energy required for the work done by that cell. Most commonly, cell respiration involves the breakdown of glucose, and the overall chemical equation is basically the reverse of that for photosynthesis:

Cell Respiration
(An energy-releasing process)

$$C_6H_{12}O_6 \ + \ 6O_2 \ \rightarrow \ 6CO_2 \ + \ 6H_2O \ + \ energy$$

glucose   oxygen   carbon dioxide   water
(high potential energy)   (low potential energy)

The purpose of cell respiration is to release the potential energy contained in organic molecules to perform the activities of the organism. Note that oxygen is *released* in photosynthesis, but *consumed* in cell respiration to break down glucose to carbon dioxide and water. All organisms need to do cell respiration, even producers like bamboo. In order to have the oxygen they need, animals like a giant panda eating the bamboo and a deer grazing on grass (Fig. 3–16) must obtain oxygen through some type of respiratory organ. This means that their ability to use food depends on the availability not only of food, but also of oxygen. In the deer, oxygen is absorbed through the lungs with every inhalation and is transported to all the body's cells via the circulatory system. Carbon dioxide, which is formed as a waste product, moves from the cells into the circulatory system and is eliminated through the lungs with every exhalation. Oxygen is more limited in water, and fish use gills to obtain it, but many aquatic places are severely oxygen limited.

In keeping with the Second Law of Thermodynamics, converting the potential energy of glucose to the energy required to do the body's work is not 100% efficient. The energy released from glucose is stored in molecules that work like compressed springs—that is, they store the energy so that the body can use it in cells later. But the amount of "energy units" that come from respiration is not nearly as much as the amount of solar energy required for the photosynthesis of glucose. Cell respiration is only 40–60% efficient. This is a great example of the Second Law of Thermodynamics at work. The rest of the energy is released as waste heat, and this is the source of body heat. This heat output can be measured in animals (cold-blooded or warm-blooded) and in plants.

Gaining Weight.   The basis of weight gain or loss becomes apparent here. Organic matter is broken down in cell respiration only as it is needed to meet the energy demands of the body. This is why your breathing rate, the outer reflection of cell respiration, varies with changes in your level of exercise and activity. If you consume more calories from food than your body needs, the excess may be converted to fat and stored, and the result is a gain in weight. In contrast, the principle of dieting is to eat less and exercise more, to create an energy demand that exceeds the amount of energy contained in your food.

The overall reaction for cell respiration is the same as that for simply burning glucose. Thus, it is not uncommon to speak of "burning" our food for energy. There are other ways of releasing stored energy from food that do not require oxygen. Such anaerobic respiration includes fermentation, the process that yeast cells use to convert grape sugars into usable energy, releasing alcohol as a by-product. However, these methods are less efficient than oxidation, and organisms that use them do not thrive unless oxygen is severely limited.

Photosynthesis and cell respiration are necessary for life on Earth to survive. In the next section, we will see how the energy that fuels photosynthesis moves through Earth's systems after traveling from the Sun.

One-Way Flow of Energy

What happens to all the solar energy entering ecosystems? Most of it is absorbed by the atmosphere, oceans, and land, thus heating them in the process. The small fraction (2–5%) captured by living plants is either passed along to the consumers that eat them or degraded into the lowest and most disordered form of energy—heat—as the plant decomposes. Eventually, all of the energy entering ecosystems escapes as heat. According to the laws of thermodynamics, no energy will actually be lost. So many energy conversions are taking place in ecosystem activities, however, that entropy is increased, and all the energy is degraded to a form unavailable to do further work. That heat energy may stay in the atmosphere for some period of time, but will eventually be re-radiated out into space. The ultimate result is that *energy flows in a one-way direction through ecosystems* and eventually leaves Earth; it is not lost to the universe, but is no longer available to ecosystems here, so it must be continually resupplied by sunlight.

As a basic energy source, sunlight is highly sustainable because it is both nonpolluting and nondepletable on a timescale meaningful to humans. Light from the Sun is a form of pure energy; it contains no substance that can pollute the environment. The Sun's energy output is remarkably constant—the radiant energy from the Sun (called the solar constant) strikes Earth's atmosphere at about 2 calories per square centimeter (0.3 square inches) per minute. How much or how little of this energy is used on Earth will neither influence nor deplete the Sun's output. Of course, the Sun's energy is depletable on a long enough scale, but relative to human time frames, solar energy is nondepletable.

Energy flow is one of the two fundamental processes that make ecosystems work. In contrast with the *flow* of energy, we talk about the *cycling* of nutrients and other elements, which are continually reused from those already available on Earth.

3.5 The Cycling of Matter in Ecosystems

Earlier we saw that the molecules that make up tissues contain certain key elements. In the bamboo and panda, for example, all the organic molecules contain carbon, the photosynthesis the bamboo is doing requires water (hydrogen and oxygen) and carbon dioxide (carbon and oxygen), potassium is part of the energy-holding mechanism in each cell, nitrogen is contained in every protein, and sulfur bonds help proteins stay in the right shape. There are many other necessary nutrients. Some are necessary in only very small amounts. For example, to make the protein hemoglobin, which carries oxygen in your blood, you need to have small amounts of iron.

According to the Law of Conservation of Matter, atoms cannot be created or destroyed, so recycling is the only possible way to maintain a dynamic system. To see how recycling takes place in the biosphere, we now focus on the pathways

of three key elements heavily affected by human activities: carbon, phosphorus, and nitrogen. Because these pathways are circular and involve biological, geological, and chemical processes, they are known as **biogeochemical cycles**. The water, or **hydrologic**, cycle is just as important, but it will be covered in Chapter 10.

The Carbon Cycle

The global carbon cycle is illustrated in Figure 3–17. Boxes represent major "pools" of carbon, and arrows represent the movement or flux of carbon from one compartment to another. The movement of carbon is measured in gigatons (1 gigaton (Gt) = 1 billion metric tons (1.1 billion US tons), sometimes referred to as 1 petagram). For descriptive purposes, it is convenient to start the carbon cycle with the "reservoir" of carbon dioxide (CO_2 molecules present in the air). Through photosynthesis and further metabolism, carbon atoms from CO_2 become the carbon atoms of the organic molecules making up a plant's body. The carbon atoms can then be eaten and become part of the tissues of all the other organisms in the ecosystem. About half is respired by plants and animals, and half is deposited to the soil (a large reservoir) in the form of **detritus** (dead plant and animal matter). Respiration by organisms in the soil that eat dead matter returns more carbon to the

Figure 3–17 The global carbon cycle. Boxes in the figure refer to pools of carbon, and arrows refer to the movement, or flux, of carbon from one pool to another.

atmosphere (as CO_2). The cycle is different in the oceans: Photosynthesis by phytoplankton and macroalgae removes CO_2 from the huge pool of inorganic carbonates in seawater, and feeding moves the organic carbon through marine food webs. Respiration by the biota returns the CO_2 to the inorganic carbon compounds in solution.

Some processes other than the transfer of carbon among producers, consumers, and organisms eating dead material are significant. The figure indicates two in particular: (1) diffusion exchange between the atmosphere and the oceans and (2) the combustion of fossil fuels, which releases CO_2 to the atmosphere. Some geological processes of the carbon cycle are not shown in Figure 3–17. For example, the fossilization of dead plants and animals produced coal deposits on many areas of Earth. This process removed vast amounts of carbon dioxide from the atmosphere and trapped it underground. Burning coal and oil releases the CO_2 to the atmosphere. For another example, limestone (such as that formed by ancient corals) also keeps carbon out of circulation; however, the weathering of exposed limestone releases carbon into the aquatic system.

Because the total amount of carbon dioxide in the atmosphere is about 765 Gt and photosynthesis in terrestrial ecosystems removes about 120 Gt/year, a carbon atom cycles from the atmosphere through one or more living things and back to the atmosphere about every six years.

Human Impacts. Human intrusion into the carbon cycle is significant. As we will see shortly, we are diverting or removing 40% of the photosynthetic effort of land plants in order to support human enterprises. By burning fossil fuels, we have increased atmospheric carbon dioxide by 35% over preindustrial levels, a topic discussed in Chapter 18. In addition, the Millennium Ecosystem Assessment reports that until the mid–20th century, deforestation and soil degradation released significant amounts of CO_2 to the atmosphere. However, more recent reforestation and changed agricultural practices have improved this somewhat. (See Sustainability, "Should We Seed the Oceans?")

The Phosphorus Cycle

The phosphorus cycle is representative of the cycles of all the biologically important mineral nutrients—those elements that have their origin in the rock and soil minerals of the lithosphere, such as iron, calcium, and potassium. We focus on phosphorus because its shortage tends to be a limiting factor in a number of ecosystems and its excess can seriously stimulate unwanted algal growth in freshwater systems.

The phosphorus cycle is illustrated in Figure 3–18. Like the carbon cycle, it is depicted as a set of pools of phosphorus and fluxes to indicate key processes. Phosphorus exists in various rock and soil minerals as the inorganic ion phosphate (PO_4^{3-}). As rock gradually breaks down, phosphate and other ions are released. This slow process is the normal means of replenishing phosphorus that is lost because of runoff and leaching. Plants absorb PO_4^{3-} from the soil or from a water solution, and once the phosphate is incorporated into organic compounds by the plant, it is referred to as

Figure 3–18 The global phosphorus cycle. Phosphorus is a limiting factor in many ecosystems. Note that the cycle is not connected to the atmosphere, which limits biosphere recycling.

organic phosphate. Moving through food chains, organic phosphate is transferred from producers to the rest of the ecosystem. As with carbon, at each step it is highly likely that the organic compounds containing phosphate will be broken down in cell respiration or by decomposers, releasing PO_4^{3-} in urine or other waste material. The phosphate may then be reabsorbed by plants to start the cycle again.

Phosphorus enters into complex chemical reactions with other substances that are not shown in this simplified version of the cycle. For example, PO_4^{3-} forms solid, insoluble compounds with a number of cations (positively charged ions), such as iron (Fe^{3+}), aluminum (Al^{3+}), and calcium (Ca^{3+}). If these cations are in sufficiently high concentration in soil or aquatic systems, the phosphorus can be bound up in chemical precipitates (solid, insoluble compounds) and rendered largely unavailable to plants. The precipitated phosphorus can slowly release PO_4^{3-} as plants withdraw naturally occurring PO_4^{3-} from soil, water, or sediments.

Human Impacts.   The most serious human intrusion into the phosphorus cycle comes from the use of phosphorus-containing fertilizers. Phosphorus is mined in several locations around the world (e.g., Florida in the United States) and is then made into fertilizers, animal feeds, detergents, and other products. Phosphorus is a common limiting factor in soils and, when added to croplands, can greatly stimulate production.

Unfortunately, human applications have tripled the amount of phosphorus reaching the oceans, from the natural level of around 8 teragrams (8 million metric tons) of phosphorus per year (Tg/yr) to the present flux of 22 Tg/yr. This increase of 14 Tg/yr is roughly equal to the global use of phosphorus fertilizer in agriculture. We are accelerating the natural phosphorus cycle as we mine it from the earth and as it subsequently moves from the soil into aquatic ecosystems, creating problems as it makes its way to the oceans. There is essentially no way to return this waterborne phosphorus to the soil, so the bodies of water end up overfertilized. This leads in turn to a severe water pollution problem known as eutrophication (Chapter 17). Eutrophication can cause the overgrowth of algae, too many bacteria, and the death of fish.

The Nitrogen Cycle

The nitrogen cycle (Fig. 3–19) has similarities to both the carbon cycle and the phosphorus cycle. Like carbon, nitrogen possesses a gas phase; like phosphorus, it acts as a limiting factor. The nitrogen cycle is otherwise unique. Most notably, unlike in the other cycles, bacteria in soils, water, and

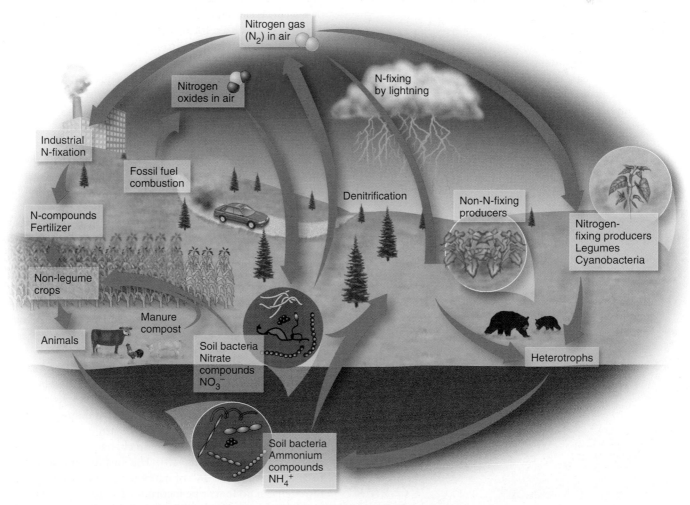

Figure 3–19 The global nitrogen cycle. Like phosphorus, nitrogen is often a limiting factor. Its cycle heavily involves different groups of bacteria.

sediments perform many of the steps of the nitrogen cycle. Like phosphorus, nitrogen is in high demand by both aquatic and terrestrial plants.

The main reservoir of nitrogen is the air, which is about 78% nitrogen gas (N_2). This form of nitrogen is called **nonreactive nitrogen**; most organisms are not able to use it in chemical reactions. The remaining forms of nitrogen are called **reactive nitrogen** (**Nr**) because they can be used by most organisms and make chemical reactions more easily.

Plants in terrestrial ecosystems ("non-N-fixing producers" in Fig. 3–19) take up Nr as ammonium ions (NO_4^+) or nitrate ions (NO_3^-). The plants incorporate the nitrogen into essential organic compounds such as proteins and nucleic acids. The nitrogen then moves from producers through consumers and, finally, to decomposers (organisms that live on dead plant and animal matter, referred to as "heterotrophs" in Fig. 3–19). At various points, nitrogen wastes are released, primarily as ammonium compounds. A group of soil bacteria, the *nitrifying bacteria*, oxidizes the ammonium to nitrate in a chemosynthetic process that yields energy for the bacteria. At this point, the nitrate is once again available for uptake by green plants—a local ecosystem cycle within the global cycle. In most ecosystems, the supply of Nr is quite limited, yet there is an abundance of nonreactive N—if it can be accessed. The nitrogen cycle in aquatic ecosystems (not shown in Fig. 3–19) is similar.

Nitrogen Fixation.   A number of bacteria and cyanobacteria (chlorophyll-containing bacteria, formerly referred to as blue-green algae) can use nonreactive N through a process called biological **nitrogen fixation**. In terrestrial ecosystems, the most important among these nitrogen-fixing organisms are bacteria in the genus *Rhizobium*, which live in nodules on the roots of legumes, the plant family that includes peas and beans (Fig. 3–20). The legume provides the bacteria with a

place to live and with food (sugar) and gains a source of nitrogen in return. From the legumes, nitrogen enters the food web. The legume family includes a huge diversity of plants, ranging from clovers (common in grasslands) to desert shrubs and many trees. Without them, plant production would be sharply impaired due to a lack of available nitrogen. In aquatic ecosystems, cyanobacteria are the most significant nitrogen fixers.

Three other important processes also "fix" nitrogen. One is the conversion of nitrogen gas to the ammonium form by discharges of lightning in a process known as *atmospheric nitrogen fixation*; the ammonium then comes down with rainfall. The second is the *industrial fixation* of nitrogen in the manufacture of fertilizer; the Haber-Bosch process converts nitrogen gas and hydrogen to ammonia. The third is a consequence of the *combustion of fossil fuels*, during which nitrogen from coal and oil is oxidized; some nitrogen gas is also oxidized during high-temperature combustion. These last two processes lead to nitrogen oxides (NO_x) in the atmosphere, which are soon converted to nitric acid and then brought down to Earth as acid precipitation. Such acid rain is discussed in Chapter 19.

Denitrification.   **Denitrification** is a microbial process that occurs in soils and sediments depleted of oxygen. A number of microbes can take nitrate (which is highly oxidized) and use it as a substitute for oxygen. In so doing, the nitrogen is reduced (it gains electrons) to nitrogen gas and released back into the atmosphere. In sewage treatment systems, denitrification is a desirable process and is promoted to remove nitrogen from the wastewater before it is released in soluble form to the environment (Chapter 20).

Human Impacts.   Human involvement in the nitrogen cycle is substantial. Many agricultural crops are legumes (peas, beans, soybeans, alfalfa), so they draw nitrogen from the air, thus increasing the rate of nitrogen fixation on land. Crops that are nonleguminous (corn, wheat, potatoes, cotton, and so on) are heavily fertilized with nitrogen derived from industrial fixation. Both processes benefit human welfare profoundly. Also, fossil fuel combustion fixes nitrogen from the air. All told, these processes add some 165 teragrams of Nr to terrestrial ecosystems annually (Fig. 3–19). This is approximately 1.5 times the natural rate of nitrogen fixation. In effect, we are more than doubling the rate at which nitrogen is moved from the atmosphere to the land.

The consequences of this global nitrogen fertilization are serious. Nitric acid (and sulfuric acid, produced when sulfur is also released by burning fossil fuels) has destroyed thousands of lakes and ponds and caused extensive damage to forests (Chapter 19). Nitrogen oxides in the atmosphere contribute to ozone pollution, global climate change, and stratospheric ozone depletion (Chapter 18). The surplus nitrogen has led to "nitrogen saturation" of many natural areas, whereby the nitrogen can no longer be incorporated into living matter and is released into the soil. There it leaches cations (positively charged mineral ions) such as calcium and magnesium from

Figure 3–20 Nitrogen fixation. Bacteria in root nodules of legumes convert nitrogen gas in the atmosphere to forms that can be used by plants.

Sustainability

Should We Seed the Oceans?

Ecosystems run on solar energy, and they recycle nutrients. The major limiting factors in marine ecosystems are the availabilities of light and nutrients. Light diminishes as water depth increases because even clear water absorbs light. The layer of water where there is adequate light for photosynthesis is known as the **euphotic zone**. Below the euphotic zone, photosynthesis does not occur. In clear water, the euphotic zone may be as deep as 180 meters (197 yards), but in turbid (cloudy) water, it may be a meter or less. In coastal waters, where the euphotic zone extends to the bottom, the bottom may support abundant plant life in the form of aquatic vegetation attached to or rooted in sediments. If the euphotic zone does not extend to the bottom, the bottom will be barren of plant life, though it may have organisms that live on falling dead matter or chemicals from deep ocean vents.

If ocean water contains too few dissolved nutrients such as phosphorus or nitrogen compounds, the growth of phytoplankton (the tiny photosynthesizers of aquatic systems) will be limited, as will the rest of the ecosystem. When organisms die in the ocean, they sink to the bottom. Nutrients carried to the bottom with this settling debris dissolve into the water and are carried along by deep-running ocean currents. Where the currents hit underwater mountains or continental rims, the nutrient-rich water is forced to the surface. These areas of **upwelling** (rising) support a rich diversity of fish and marine mammals. Consequently, the areas of the ocean supporting the most abundant marine life of all sorts are generally found within 320 kilometers (200 mi) of shore. In contrast, in the open ocean, there is less marine life the farther you move from shore.

Oceanographer John Martin of the Moss Landing Marine Laboratory in California hypothesized that we could seed the open ocean with iron, found to be the limiting nutrient in some areas. This would increase productivity by phytoplankton and thereby pull carbon dioxide out of the atmosphere. The plankton would need to fall to the bottom without being eaten and there build up deep sediments, keeping the carbon from being used by other living things (just as if we were building up new fossil deposits). Some research has shown iron supplementation to increase photosynthesis and trigger phytoplankton "blooms," growing the biomass of a redwood forest in just a few weeks. In some studies, however, such supplementation has been accompanied by a bloom in zooplankton, which return much of the carbon dioxide to the atmosphere. This occurrence might not be entirely bad, however, because even though the carbon would not be sequestered as hoped, the zooplankton could serve as food for other larger consumers.

Right now, there is tremendous interest in the science of ocean nutrient supplementation and its effects. However, near-shore dumping of nutrients such as the nitrogen and phosphorus pollution entering the Gulf of Mexico can cause "dead zones," areas largely devoid of oxygen and fish. Because of this and other concerns, some marine scientists are cautious about the use of nutrient supplements to reduce carbon in the atmosphere.

---

the soil, which leads to mineral deficiencies in trees and other vegetation. Washed into surface waters, the nitrogen makes its way to estuaries and coastal areas of oceans, where, just like phosphorus, it triggers a series of events leading to eutrophication (Chapter 20), resulting in dead seafood, harmful effects on human health, and areas of the ocean that are unfit for fish. This complex of ecological effects has been called the **nitrogen cascade**, in recognition of the sequential impacts of Nr as it moves through environmental systems, creating problems as it goes.

Comparing the Cycles

The three cycles we have looked at in depth differ in some important ways (Table 3–2). Carbon is found in large amounts in the atmosphere and in a form that can be directly taken in by plants, so carbon is rarely the limiting factor in the growth of vegetation. Both nitrogen and phosphorus are often limiting factors in ecosystems. Phosphorus has no gaseous atmospheric component (though it can be in airborne dust particles) and thus, unless added to an ecosystem artificially, enters very slowly. Nitrogen is unique because of the importance of bacteria in driving the cycle forward.

All three cycles have been sped up considerably by human actions. Nitrogen compounds in the atmosphere contribute to acid rain, and carbon dioxide is being moved from underground storage in carbon molecules to the

TABLE 3–2   Major Characteristics of the Carbon, Phosphorus and Nitrogen Cycles

Nutrient	Major Source	Interesting Feature	Human Impact
Carbon	Air	Taken in directly by plant leaves	Burning fuel moves it to air from underground
Phosphorus	Rock	No atmospheric component	Fertilizer use adds it to waterways
Nitrogen	Soil	Bacteria drive the cycle	Fertilizer moves it to soil, burning moves it to air

atmosphere, where it is acting as a greenhouse gas. Both nitrogen and phosphorus are put on soil as fertilizer or get into water from sewage and runoff. Both act in water to promote overgrowth of algae, an effect we will see in Chapter 20.

Although we have focused on carbon, phosphorus, and nitrogen, cycles exist for oxygen, hydrogen, iron, sulfur, and all the other elements that play a role in living things. Also, while the routes taken by distinct elements may differ, all of the cycles are going on simultaneously, and all come together in the tissues of living things. As the elements cycle through ecosystems, energy flows in from the Sun and through the living members of the ecosystems. The links between these two fundamental processes of ecosystem function are shown in Figure 3–21.

In this chapter, we introduced the various fields of ecology and showed that the science of ecology encompasses living things and their relationships with each other and the environment. We looked at two ecological levels in greater detail in this chapter—what is happening with individuals (organismal ecology) and what is happening to the large-scale movement of matter and energy through the biosphere (ecosystem ecology). To understand the flow of matter and energy, we also included some background on the basics of atoms, molecules, and the laws of energy. In the next two chapters, we will investigate the other subdisciplines of ecology introduced here, especially population ecology and community ecology.

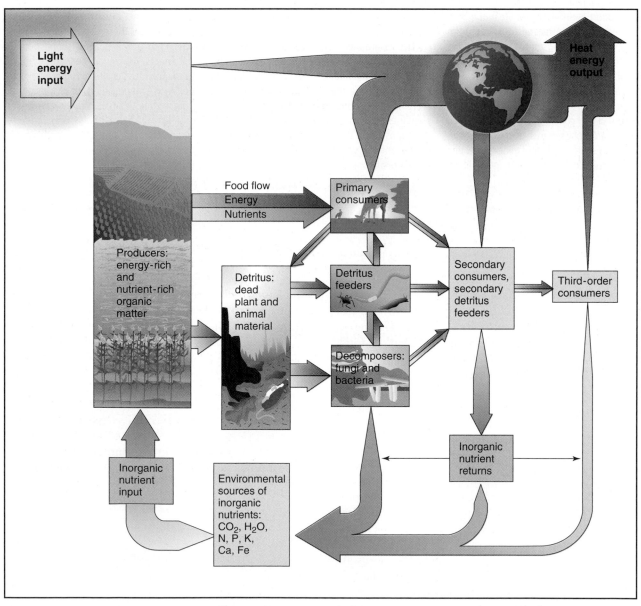

Figure 3–21 Nutrient cycles and energy flow. The movement of nutrients (blue arrows), energy (red arrows), and both (brown arrows) through the ecosystem. Nutrients follow a cycle, being used over and over. Light energy absorbed by producers is released and lost as heat energy as it is "spent."

Revisiting the Themes

Sound Science

Ecology is the science of the study of living things and the abiotic and biotic components interacting with them. Ecologists can study on a variety of levels. In this chapter, we looked at organismal ecology and ecosystem ecology—that is, how materials work in individual animals and how matter and energy flow throughout the world.

Much of this chapter is about how the basic processes of the natural world work, focusing on matter and energy. Both matter and energy are limited. Matter cycles in the form of molecules (sets of atoms) through living things and the abiotic environment. Energy flows in one direction—from the Sun, through activities on Earth, and back out into space. Using the giant panda and its food (the bamboo plant) as our examples, we demonstrated basic principles of organization within their bodies and the flow of energy and materials through their systems. The bamboo plant has a specific niche, or set of parameters, within which it needs to live. These resources and conditions include the amount of rainfall and nutrients it needs, for example. Because it has chlorophyll, bamboo is able to undergo photosynthesis. In this chemical process, inorganic building blocks are turned into more complex organic molecules whose bonds contain more chemical energy than their original components do. This energy can be released when the panda eats the bamboo and its body burns those molecules through a controlled process of cellular respiration. Energy is lost at each step of this energy path. In order to build the bodies of the bamboo and its consumer, the panda, a number of nutrients need to be available. The cycles for three of these (carbon, phosphorus, and nitrogen) were described. Humans affect these biogeochemical cycles by actions such as the burning of fossil fuels and the fertilization of agricultural land.

Sustainability

The science in this chapter that leads to specific management actions is primarily that surrounding the needs of individual species in terms of habitat and niche, the consequences of limits on nutrients and energy, and the impacts of human effects on biogeochemical cycles. These ideas will be more thoroughly explored in other chapters. For instance, the science surrounding habitat and niche underlies our understanding of why some types of organisms are particularly vulnerable to extinction (Chapter 4). The science underlying the nitrogen and phosphorus cycles is important to management decisions about water pollution (Chapter 20), and the science about carbon and nitrogen cycles aids in our understanding of air pollution and global climate change (Chapters 18 and 19). The sustainability essay ("Should We Seed the Oceans?") provides a good example of how science is combined with management to come up with solutions—in this case, possibly seeding the ocean with iron. As we saw, however, there are reasons to be cautious about large-scale manipulations.

The giant panda helps us look at basic processes within organisms and the flow of matter and energy. In the next chapter, we will look at ways individuals relate to members of their own species (population ecology) and ways species relate to other species (community ecology).

Stewardship

The giant panda focused on throughout this chapter is in decline. It has a particular habitat, which is shrinking because of competition with human activities. As stewards, the panda conservationists in China and around the world are protecting not only the panda but its habitat. Human activity also speeds up biogeochemical cycles, and this can have global effects. All of us have to be stewards not only of the nutrients in our world, but also of the natural processes that move chemicals from one part of a cycle to another. That is, stewardship involves protecting not only resources but ecosystem functions. We will look at some of the ethical and justice implications of these scientific realities in other chapters. In Chapter 6, for example, we will ask, "Why protect other species?" In Chapter 8, we will look at the implications of energy loss at every transfer and apply those principles to human population growth and the need for food.

REVIEW QUESTIONS

1. Distinguish between the biotic community and the abiotic environmental factors of an ecosystem.

2. Define and compare the terms *species*, *population*, and *ecosystem*.

3. Compared with an ecosystem, what are an ecotone, a landscape, a biome, and a biosphere?

4. How do the terms *organic* and *inorganic* relate to the biotic and abiotic components of an ecosystem?

5. What are the six key elements in living organisms?

6. What features distinguish between organic and inorganic molecules?

7. In one sentence, define *matter* and *energy*, and demonstrate how they are related.

8. Give four examples of potential energy. In each case, how can the potential energy be converted into kinetic energy?

9. State the two energy laws. How do they relate to entropy?

10. What is the chemical equation for photosynthesis? Examine the origin and destination of each molecule referred to in the equation. Do the same for cellular respiration.

11. Describe the biogeochemical cycle of carbon as it moves into and through organisms and back to the environment. Do the same for phosphorus and nitrogen.

12. What are the major human impacts on the carbon, phosphorus, and nitrogen cycles?

THINKING ENVIRONMENTALLY

1. From local, national, and international news, compile a list of ways humans are altering abiotic and biotic factors on a local, regional, and global scale. Analyze ways that local changes may affect ecosystems on larger scales and ways that global changes may affect ecosystems locally.

2. Use the laws of conservation of matter and energy to describe the consumption of fuel by a car. That is, what are the inputs and outputs of matter and energy? (*Note*: Gasoline is a mixture of organic compounds containing carbon–hydrogen bonds.)

3. Using your knowledge of photosynthesis and cell respiration, draw a picture of the hydrogen cycle and the oxygen cycle. (*Hint*: Consult the three cycles in the book for guidance.)

4. Look up everything you can find about iron seeding of the oceans, and see if you can decide what you think we should do.

MAKING A DIFFERENCE

1. Find the closest farm that does not use artificial fertilizer. Many are listed on web sites under the terms "community supported agriculture" or "CSA." See if you can arrange a visit. See what they do to lower their impact on the nitrogen and phosphorus cycles and still maintain fertile soil.

2. Visit the EPA's web site to find out how you can help protect your local watershed. Their project database can let you

get involved in monitoring, cleanups, and restorations in your area.

3. Take one of the biogeochemical cycles and try to track your impact on it. This is most easily done with carbon. Search on the web under "carbon footprint."

Mastering**ENVIRONMENTALSCIENCE**™

Go to **www.masteringenvironmentalscience.com** for practice quizzes, Pearson eText, videos, current events, and more.

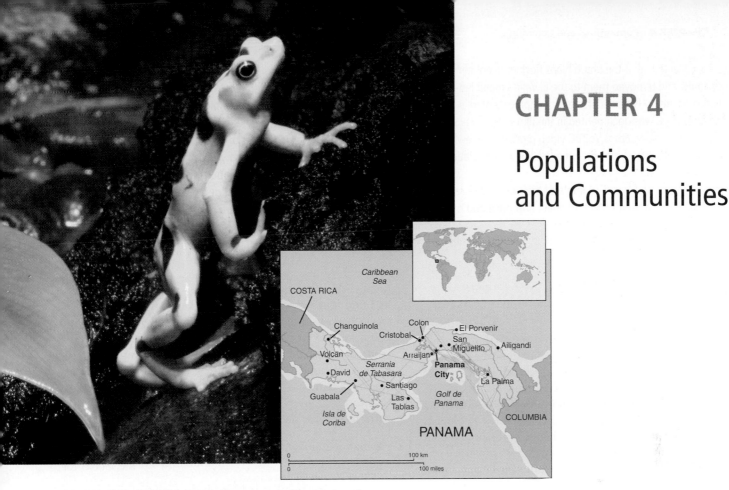

CHAPTER 4

Populations and Communities

The golden frog (*Atelopus zeteki)* of Panama, a highly endangered species, now lives only in captivity. The last known populations were removed from an area of Panama (inset) where a deadly fungus was known to be invading.

I MAGINE SITTING OUTSIDE on a summer evening, listening to a chorus of frogs. In North America, you might hear bullfrogs or spring peepers through the coming dusk. Their calls mark the breeding grounds of these species and are common sounds of spring and summer. In the rain forest of Central America, you might hear a chorus of an entirely different set of amphibians—many that are small and beautifully colored. However, one you won't hear is the golden frog (also called the golden toad) of Panama, *Atelopus zeteki.*

The golden frog, revered as a national symbol in Panama, is depicted in local mythology, sculpture, toys, lottery tickets, posters, and calendars. It is famous for its poisonous skin, which in adulthood has a bright black and yellow warning coloration. Unlike most other frogs and toads (of the group Anura, a subgroup of amphibians), it doesn't sing. Instead, the golden frog, a rare anuran that lacks eardrums, waves an arm to communicate with others of its species. Unfortunately, habitat loss over the years (through devastation of Panama's tropical forests) and overcollection for the pet trade have left golden frog populations at low levels. A new threat has also appeared—a chytrid fungus (*Batrachochytrium dendrobatidis),* which is decimating frog populations around the globe and threatening many with extinction. In 2007, concerned that the fungus was moving through areas of Panama and leaving dead and dying amphibians in its wake, conservationists made a bold move. They went into a remote

KEY TOPICS

1. **Dynamics of Natural Populations: Populations grow and decline, sometimes in predictable ways.**

2. **Limits on Populations: Populations are limited by a number of factors, including predators, resources, and sometimes factors that are unrelated to the population.**

3. **Community Interactions: Population growth can be limited by other populations in the community.**

4. **Evolution as a Force for Change: Forces in the environment drive survival of some population members over others and eventually cause evolutionary change.**

5. **Implications for Human Management: Humans alter a number of features in the environment that change population growth patterns.**

jungle area, collected all of the amphibians (including golden frogs) that they could, and removed them to the El Valle Amphibian Conservation Center in Panama. This effort, unprecedented in conservation history, was documented in the film *Leap of Faith: Saving Panama's Golden Frogs.* A BBC video production crew also filmed the removal of what was then the last known population of golden frogs in the wild.

The golden frog is a good example of the problems a rare species can face when multiple limits affect it at once and of the extreme measures that might be necessary for its survival and preservation. Under changing environmental conditions, species can often adapt or find some new niche or area to exploit. However, as seen with the golden frog, environmental changes can also put species in danger of extinction. In this chapter, we will look at how populations grow, what types of limitations populations can face, how populations of different species interact with each other, how species change over time, and what changes humans can impose on populations and groups of species.

4.1   Dynamics of Natural Populations

Like the golden frogs in their rain-forest habitats, each species in an ecosystem exists as a population; that is, each exists as a reproducing group. As we discussed in Chapter 3, a **population** is a group of members of the same species living in an area, and the populations of different species living together in one area are known as a **community.** Ecologists often want to know the pattern of change in the numbers of individuals in a population. In any population, this is easy to figure out if you have enough information. Births and immigration cause the population to grow, and deaths and emigration cause the population to shrink. This can be described by a simple equation:

(Births + Immigration) − (Deaths + Emigration)
= Change in population number

This equation shows the *change in population*, or **population growth.** If births plus immigration are more or less equal to deaths plus emigration over time, the population is said to be in **equilibrium.** But often, populations are either increasing or decreasing—that is, their population growth is *not zero*. If you take the amount the population has changed and divide it by the time it had to change, you get the **population growth rate.** Imagine there is a very small population just coming into a new ecosystem. A scientist might be interested in figuring out how fast a population could grow, how many individuals there are right now, or even what the future population size would be, using the current population and the growth rate. In order to calculate these parameters, scientists use *population growth curves*—graphs of how populations grow. Here we will look at several examples that represent broad types of growth patterns.

Population Growth Curves

Suppose for a moment you are searching the Internet and come upon a great new get-rich-quick scheme—worm farming. You order a kit, get a small number of starter worms, and plan to raise them in buckets and sell the ones you raise for composting.

In the first month, there are two worms (worms are hermaphroditic—that is, they contain both male and female organs—but they do mate). After a week, you sift the soil and find four worms, after two weeks six worms, and every week, regardless of the number of worms you started with, you have two more worms. This would be an example of a *constant* population growth rate. This is the simplest type of growth to model. However, this pattern would be unlikely. Constant growth is not something you would expect to find in nature, but it serves as a good comparison to other growth patterns because the equation is so simple:

Population number at the start + (A constant × Time)
= Population number at the end

That is, the number at the end of a period of time is equal to the number at the beginning plus a constant multiplied by the number of time units. This is illustrated by the straight orange line in Figure 4–1. Using this equation, if we started with 2 worms and displayed a constant growth rate of 2 worms per week for four weeks, the population number at the end would be 10 worms (2 + (2 × 4) = 10). At the end of 24 weeks, we would have 50 worms (2 + (2 × 24) = 50).

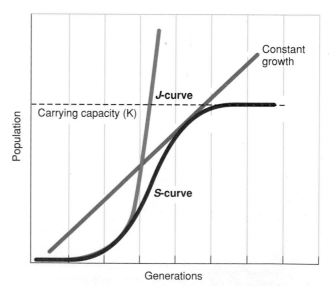

Figure 4–1 Three simple models of population growth. Constant growth is represented by the orange line, exponential growth by the blue line , and logistic growth by the red line. The carrying capacity is represented by the black line.

Exponential Increase.   Every species has the capacity to increase its population when conditions are favorable. For example, your worms could double every week (the first week 2, then 4, then 8, then 16). Such a series is called an **exponential increase.** This sort of growth results in a **population explosion.** In contrast to our first example of constant growth, exponential growth does not add a constant number of individuals to the population for each time period. Instead, the *doubling time* of the population remains constant. That is, if it takes two days to go from 8 worms to 16 worms, it will take two days to go from 1,000 worms to 2,000 worms. This is the reason such growth is described as an "explosion."

If we plot numbers over time during an exponential increase, the pattern produced is commonly called a *J*-curve because of its shape (Fig. 4–1). To calculate the growth rate of a population undergoing exponential growth, you need to know the number in the population right now, the time period, and the number of offspring individuals can produce in a given amount of time if resources are unlimited. This last value is often designated r_{max}, or just r.

The equation for exponential growth can be described as follows:

Starting population × A constant (e) multiplied
   by itself a certain number of times = Ending population

The number of times you multiply e by itself is equal to the value of r (the potential to produce offspring) multiplied by time.

Using this equation, our two starter worms produce very different results. Suppose our two worms have an $r = 0.60$ and a reproductive time period of one week. (Such an r value is probably higher than worms have in real life, but it works for our example. Many organisms have r values below 0.1

because most females do not reproduce in each time period.) Over four weeks, the population size is not that much different than it is with constant growth: $2 \times e^{0.60 \times 4} = 22$ worms. However, 24 weeks of exponential growth would yield an enormous ending population of over 3.5 million ($2 \times e^{0.60 \times 24} = 3,588,150$ worms).

This sounds complicated, but here is the point: Under unlimited conditions, organisms with a high value of r will have more rapid population growth than will those with a low r value. Exponential growth could only continue indefinitely if there were no limits to the size the population could reach.

There are times when exponential growth occurs in nature. For example, beetles getting into a big sack of rice might experience a population explosion. However, exponential growth cannot continue indefinitely. There is an upper limit to the population of any particular species that an ecosystem can support. This limit is known as the **carrying capacity** (*K*) (Fig. 4–1). More precisely, the **carrying capacity** is the maximum population of a species that a given habitat can support without being degraded over the long term—in other words, a sustainable system.

Logistic Growth.   Eventually, as a population increases, two outcomes occur: (1) The population keeps growing until it exhausts essential resources, often food, and then dies off precipitously due to starvation or other problems (this results in a crash following the *J*-curve, Fig. 4–2). Alternatively, (2) some process can slow population growth so that it levels off at or near carrying capacity. The simplest version of this latter pattern is known as **logistic growth.** Plotting this type of growth on a graph results in an *S*-shaped curve and thus is called an **S-curve** (Fig. 4–1). The point at which it levels off is

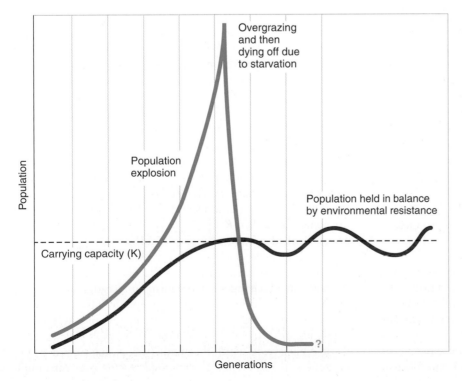

Figure 4–2 Real-life growth. In real populations, exponential growth is often followed by a crash (blue line), and logistic growth is often modified so that the population fluctuates around *K* (red line). Real-life populations often do not follow idealized equations, but such simple models help us to understand possible outcomes.

the carrying capacity (K). The equation for logistic growth can be explained this way:

Starting population + (Reproductive capacity (r)
× Population × A number that represents how far
the population is from the carrying capacity)
= Ending population

Logistic growth for our worms would work something like this. Suppose you started the population with two worms where $r = 0.60$ and $K = 2,000$ worms. Four weeks after starting the worm colony, you would have 33 worms. (Each week is calculated separately—by the fourth week, the calculation is $20.8 + (0.60 × 20.8 × ((2,000 − 20.8)/2000)) = 33$ worms). Twenty-four weeks after the worm colony started you would have 2,000 worms. They would have rapidly increased and then slowed, reaching carrying capacity in week 22 and remaining there.

There are two things to notice about the logistic growth curve. First, as the population nears the carrying capacity, the growth of the population slows until the population size remains steady and the population growth equals zero. Second, the maximum rate of population growth occurs halfway to carrying capacity. If you wanted, for example, to harvest the maximum number of worms, you would do better to keep your worm population at half of carrying capacity, not at a larger or smaller number (we will discuss this concept further in Chapter 7).

The population growth figures calculated using these simple models rarely match the exact growth figures found in real life, but many populations show *J*-shaped explosions followed by crashes (Fig. 4–2). *J*-curves come about when there are unusual disturbances, such as the introduction of a foreign species, the elimination of a predator, the sudden alteration of a habitat, or the arrival in a new section of habitat, all of which can allow rapid growth of a population. This growth is, however, followed by some type of crash. In some cases, huge overshoots of K by one population result in a collapse of the habitat for many species.

Other populations, often those controlled by complex relationships between species, show some type of *S*-curve, followed by cycles of lower and higher numbers roughly around K (Fig. 4–2). Many shoot somewhat above carrying capacity and then a little below, and eventually they cycle around the carrying capacity. This is illustrated in Figure 4–2.

Biotic Potential Versus Environmental Resistance

The rate at which members of the species reproduce, if unlimited (which we saw in the equations above as r), is a measure of its **biotic potential.** This is the number of offspring (live births, eggs laid, or seeds or spores set in plants) that members of species may produce under ideal conditions. The biotic potentials of different species vary tremendously, averaging from less than one birth per year in certain mammals and birds to millions per year for many plants and some invertebrates. To have any effect on the size of subsequent generations, however, the young must survive and reproduce in

turn. Survival through the early growth stages to become part of the breeding population is called **recruitment.**

Environmental Resistance. Unlimited population growth is seldom seen in natural ecosystems because biotic and abiotic factors cause *mortality* (death) in populations. Among the biotic factors that cause death are predators, parasites, competitors, and lack of food. Among the abiotic factors are unusual temperatures, moisture, light, salinity, and pH; lack of nutrients; and fire. The combination of all the biotic and abiotic factors that may limit a population's increase is referred to as **environmental resistance** (Fig. 4–3).

Sometimes environmental resistance lowers reproduction as well as causing mortality directly. For example, the loss of suitable habitat often prevents animals from breeding. Also, certain pollutants affect reproduction adversely. These situations are still defined as environmental resistance because they either block a population's growth or cause its decline. Additionally, environmental resistance can cause migration patterns to change, as when animals leave a drought-stricken area. Migration can lower or increase local populations.

Reproductive Strategies. The interplay of environmental resistance and biotic potential drives the success of two common **reproductive strategies** in the natural world. The first is to produce massive numbers of young, but then leave survival to the whims of nature. This strategy often results in very low recruitment. Animals with this strategy have life histories with rapid reproduction, rapid movement, and often a short life span. This strategy is highly successful if a species is adapted to an environment that can suddenly change and become very favorable, like a rain-fed temporary pond. Organisms with this strategy are usually small and tend to have huge boom-and-bust populations. Because these organisms usually have a high r, they are sometimes called ***r*-strategists** or *r-selected species.* They are also often called "weedy" or "opportunistic" species. A familiar example of this type of species is the housefly, which multiplies quickly, but also has a high mortality rate.

In contrast, the second strategy is to have a much lower reproductive rate (that is, a lower biotic potential), but then care for and protect the young until they can compete for resources with adult members of the population. An elephant or a bird like the California condor is a good example. This strategy works best where the environment is stable and already well populated by the species. Organisms with such a strategy are larger, longer lived, and well adapted to normal environmental fluctuations. Because their populations are more likely to fluctuate around the carrying capacity, they are sometimes called ***K*-strategists** or *K-selected species.* They are also called *equilibrial* species (Table 4–1).

Characteristics such as the age at first reproduction and the length of life are a part of an organism's **life history** (the progression of changes an organism undergoes in its life). Species such as the housefly and the elephant have very different life histories that reflect two extreme sets of adaptations. These life histories can be visualized in a graph called a survivorship curve, in which the number remaining from a group all born at the same time is shown decreasing over time until

Population density

Critical number

0 Increasing potential

Increasing resistance

Figure 4–3 Biotic potential and environmental resistance. A stable population in nature is the result of the interaction between factors tending to increase population and factors tending to decrease population.

Biotic Potential

- Reproductive rate
- Ability to migrate (animals) or disperse (seeds)
- Ability to invade new habitats
- Defense mechanisms
- Ability to cope with adverse conditions

Environmental Resistance

- Lack of food or nutrients
- Lack of water
- Lack of suitable habitat
- Adverse weather conditions
- Predators
- Disease
- Parasites
- Competitors

TABLE 4-1 General Characteristics of *r*-strategists and *K*-strategists

	r-strategists	*K*-strategists
Environment	Advantage if less stable	Advantage if more stable
Size	Smaller	Larger
Life span	Shorter	Longer
Age at first reproduction	Younger	Older
Offspring	More	Fewer
Parental care	Little or none	Long and involved
Population stability	Wild fluctuations	Mostly stable

the maximum life span for the species is met. Some species, like humans, have relatively low mortality in early life, and most live the bulk of a life span for that species. This survivorship pattern is depicted as Type I in Figure 4–4. Some species have many offspring that die young, and only a few live to the end of a life span. Oysters and dandelions are examples of this Type III survivorship. Generally, *r*-strategists show the Type III pattern and *K*-strategists experience the Type I pattern. Some species are not easily defined as either *r*-strategists or

K-strategists. They may also have an intermediate survivorship pattern, depicted as Type II. Squirrels and corals are likely to fall into this intermediate category. Species actually form a continuum of these survivorship curves, much as they do in terms of the characteristics of *r*- and *K*-strategists.

In several chapters, we examine species in decline, such as the giant panda of Chapter 3, and species in rapid increase, such as pest species overtaking habitats (see Chapter 13). The reason we looked at population growth equations was to show that there is a predictable connection between species characteristics, such as biotic potential (*r*), and their patterns of growth, lived out in different life histories. As we will see in Section 4.4, there are predictable ways that environmental pressures work on species with different biotic potentials.

There is also a predictable pattern to the way species are affected by large-scale human activities. While there are many exceptions, it is often the case that *r*-strategists become pests when humans alter the environment and that *K*-strategists become rarer, endangered, or extinct. Houseflies, dandelions, and cockroaches increase with human activity, while eagles, bears, and oaks decline. Some exceptions to this rule include rare opportunistic species (*r*-strategists) that are so separated from new habitat that in spite of a high biotic potential, they cannot succeed.

In sum, whether a population grows, remains stable, or decreases is the result of interplay between its biotic potential and environmental resistance. In general, a population's biotic potential remains constant; it is changes in environmental resistance that allow populations to increase or cause

Figure 4–4 Survivorship curves. Some organisms, like humans (Type I), experience low mortality throughout their lives, and most live to old age. Others (Type III), represented by an oyster, have many offspring, most of which die early. Still others are intermediate (Type II), here represented by a squirrel. Type I and Type III roughly correspond to the *K*-strategists and *r*-strategists.

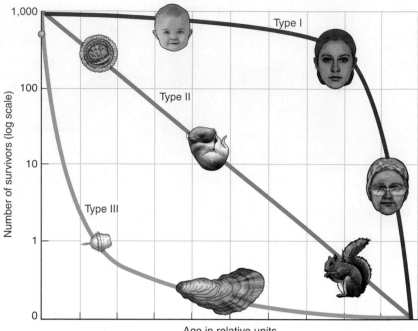

them to decrease. Population balance is a *dynamic balance,* which means that additions (births and immigration) and subtractions (deaths and emigration) are occurring continually, and the population may fluctuate around a median. Some populations fluctuate very little, whereas others fluctuate widely. As long as decreased populations restore their numbers and the ecosystem's capacity is not exceeded, the population is considered to be at equilibrium.

4.2 Limits on Populations

The carrying capacity of a habitat reflects how large a population can be sustained there, but it may not explain what is actually limiting the population. In the case of the golden frog, habitat loss forced the frogs to compete for living space and resources, destruction of streams killed their eggs, and the fungus spread rapidly through an already weakened population. Sometimes these limiting factors, such as crowding or disease, have a greater effect when there are more individuals in the population. The higher the **population density** (the number of individuals per unit area) is, the more that factor plays a role. In other cases, a factor like a tornado may keep a population from increasing, but the strength of the tornado has nothing to do with the density of the organisms in the population. We will see both of these types of limits in the next section.

Density Dependence and Density Independence

Rabbits thrive in a field with lush grass. As rabbits become more plentiful, foxes and coyotes may find it easier to catch and eat more of them and less of something else. This could slow the population growth of the rabbits. In this example,

predation acts as a **density-dependent** limit on a population. A density-dependent limit is one that increases as population density increases—such as a disease or a food shortage. The environmental resistance caused by this limit increases mortality. Conversely, as population density decreases, the environmental resistance lessens, allowing the population to recover.

Other factors in the environment that cause mortality are **density-independent** limits, those whose effects are *independent* of the crowding of the population. A sudden deep freeze in spring, for example, can kill many early germinating plants, regardless of their density. Similarly, a fire that sweeps through a forest may kill all small mammals in its wake. Although density-independent factors can be important sources of mortality, they are not involved in maintaining population equilibrium in the way we saw in the logistic equation. Because logistic growth occurs when populations become more crowded (and as they approach their carrying capacity), logistic growth is sometimes referred to as *density-dependent* population growth.

In the natural world, a population is subject to the total array of all the biotic and abiotic environmental factors around it. Many of these factors may cause mortality in a population, but only those that are density dependent are capable of *regulating* the population, or keeping it in equilibrium.

Environmental scientists distinguish between top-down and bottom-up regulation. *Top-down regulation* is the control of a population (or species) by predation, such as the control of rabbits by their predators, coyotes. In *bottom-up regulation*, the most important control of a population occurs as a result of the scarcity of some resource, as would occur if the rabbits ran out of food (Fig. 4–5). Because species interact in communities, the factor that controls a population will determine what effects adding or removing a species may have on an ecosystem. Such relationships can

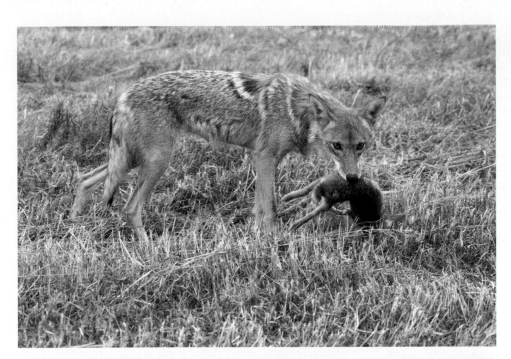

Figure 4–5 Top-down and bottom-up control. Without population controls, rabbits will rapidly multiply. Coyotes can limit them (top-down control), or they can be limited by a lack of food (bottom-up control).

cascade down a chain, so that the removal of one species has effects even on species that don't directly interact with it. This phenomenon is attributed to keystone species, as we will see in Section 4.3.

Critical Number

We've talked about mechanisms for keeping populations around carrying capacity rather than growing indefinitely, but some populations fall very low instead. There are no guarantees that a population will recover from low numbers. Extinctions can and do occur in nature. The survival and recovery of a population depend on a certain minimum population base, which is referred to as the population's **critical number.** You can see the idea of critical number at work in a pack of wolves, a flock of birds, or a school of fish. Often, the group is necessary to provide protection and support for its members. In some cases, the critical number is larger than a single pack or flock because interactions between groups may be necessary as well. In any case, if a population is depleted below the critical number needed to provide such supporting interactions, the surviving members become more vulnerable, breeding fails, and extinction is almost inevitable.

In Chapter 1, the loss of biodiversity was cited as one of the most disturbing global environmental trends. Human activities are clearly responsible for the decline, and even the extinction, of many plants and animals. This is happening because many human impacts, such as altering habitats, introducing alien species, polluting, hunting, and engaging in other forms of exploitation, are not density dependent; they can even intensify as populations decline. Concern for these declines eventually led to the Endangered Species Act, which calls for the recovery of two categories of species. Species whose populations are declining rapidly are classified as **threatened.** If the population is near what scientists believe to be its critical number, the

species may be classified as **endangered.** These definitions, when officially assigned by the U.S. Fish and Wildlife Service, set in motion a number of actions aimed at the recovery of the species in question (see Chapter 6).

4.3 Community Interactions

Now that we have described population growth as a dynamic interplay between biotic potential and environmental resistance, we can look at the ways populations interact in a whole community. Picture all of the possible relationships between two species (A and B) as being positive (helpful), negative (harmful), or neutral for each species. These relationships are depicted in Table 4–2.

The name of each type of interaction is given in the table. For example, a relationship in which one member benefits and the other is harmed $(+-)$ is **predation.** Parasitism is a subset of predation, as is herbivory (animals eating plants). A $(--)$ relationship is one in which both species are harmed. This would include *competitive* relationships, where both species use a scarce resource. **Competition** between different species is called **interspecific** competition, and competition between members of the same species is called **intraspecific** competition. In contrast, a relationship between species in which both benefit $(++)$ is called a **mutualism.** **Commensalism** occurs when one species is benefited and the other is not affected $(+0)$.

The most important interactions are predation, competition, mutualism, and commensalism, but there are others. One species might be unaffected and the other harmed (amensalism, $0-$) through an accidental interaction such as an elephant stepping on a flower or through a plant making a poison as a part of its normal metabolism that harms another organism. Similarly, it is theoretically possible to have a (00)

TABLE 4-2   The Major Types of Interactions Between Two Species. Interactions can be defined by whether the impact on each species is positive, negative, or neutral.

Interaction Name	Effect on Species A	Effect on Species B	Example
Predation	+	−	Wolves eat moose
Competition	−	−	Two rabbit species eat the same grass
Mutualism	+	+	Lichens (fungus/alga)
Commensalism	+	0	Water buffalo/egret
Amensalism	0	−	A plant makes a poison that accidentally harms another plant
No effect	0	0	Very little interaction

interaction, but there is no name for this, and presumably the interaction would be a weak one. The following looks in depth at the most important species interactions.

Predation

In any relationship in which one organism feeds on another, the organism that does the feeding is called the **predator,** and the organism that is fed on is called the **prey.** Predation is a conspicuous process in all ecosystems. Predation ranges from the classic predator-prey interactions between **carnivores** (meat eaters) and **herbivores** (plant eaters) to herbivores feeding on plants and **parasites** feeding on their hosts.

Parasites. *Parasites* are organisms—either plants or animals—that become intimately associated with their "prey" and feed on it over an extended period of time, typically without killing it. However, sometimes they weaken it, so that it becomes more prone to being killed by predators or adverse conditions. The plant or animal that is fed on is called the **host.** Parasites include a range of species, such as tapeworms, microscopic disease-causing bacteria, viruses, protozoans, and fungi (Fig. 4–6). Parasites can feed on their hosts in different ways. Tapeworms take energy from the intestine of a host animal, whereas parasitic plants tap into the sap of other plants.

Parasitic organisms affect the populations of their host organisms in much the same way that predators affect their prey—in a density-dependent manner. As the population density of the host increases, parasites and their *vectors* (agents that carry the parasites from one host to another, such as disease-carrying insects) have little trouble finding new hosts. Therefore, infection rates increase, causing higher mortality. Conversely, when the population density of the host is low, the transfer of infection is less efficient. This reduces the levels of infection and allows the host population to recover.

A tremendous variety of organisms may be parasitic. Various worms are well-known examples, but certain protozoans, insects, and even mammals (vampire bats) and plants (dodder) are also parasites. Many plant diseases and some animal diseases (such as athlete's foot, Fig. 4–6) are caused by parasitic fungi. Indeed, virtually every major group of organisms has at least some members that are parasitic. Parasites may live inside or outside their hosts, as the examples shown in Figure 4–6 illustrate.

In medicine, a distinction is generally made between bacteria and viruses that cause disease (known as **pathogens**) and parasites, which are usually larger organisms. Ecologically, however, there is no real distinction. Bacteria are foreign organisms, and viruses are organismlike entities

Figure 4–6 Several types of parasites. Nearly every biological group has at least some members that are parasitic. Shown here are (a) mistletoe, a plant parasite; (b) nematode worms (*Ascaris lumbricoides*), a human intestinal parasite; (c) lampreys, an aquatic organism that parasitizes many types of fish; and d) athlete's foot fungus on a human foot.

(a)

(b)

(c)

(d)

feeding on, and multiplying in, their hosts over a period of time. Some fungi are parasites, and some are also referred to as disease causing. Therefore, disease-causing bacteria and viruses can be considered highly specialized parasites.

Regulation of Prey. As some populations are limited by parasites, many studies have shown that herbivores are often regulated by their predators, a type of top-down control. A well-documented example is the interaction between wolves and moose on Isle Royale, a 45-mile-long island in Lake Superior that is now a national park. During a hard winter around 1900, a small group of moose crossed lake ice to the island and stayed. Their population grew considerably in the absence of predators. Then, in 1949, a small pack of wolves also managed to reach the island. The isolation of the island provided an ideal opportunity to study a simple predator-prey system, and in 1958, wildlife biologists began carefully tracking the populations of the two species (Fig. 4–7). As seen in the figure, a rise in the moose population is usually followed by a rise in the wolf population, followed by a decline in the moose population and then a decline in the wolf population. The data can be interpreted as follows: Fewer wolves represent low environmental resistance for the moose, so the moose population increases. Then the abundance of moose represents optimal conditions (low environmental resistance) for the wolves, and the wolf population increases. More wolves means higher predation on the moose (high environmental resistance); again the moose population falls. The decline in the moose population is followed by a decline in the wolf population because now there are fewer prey (high environmental resistance) for the wolves.

However, the dramatic fall in the moose population in 1996 cannot be attributed entirely to predation by the small number of wolves. Deep snow, which makes it more difficult to get food, and an infestation of ticks in 1996 caused substantial mortality. The sharp decline in moose is thought to be responsible for keeping the wolf population low, as there were few calves for them to catch. Wolves are incapable of bringing down an adult moose in good physical condition. The animals they kill are the young and those weakened by another factor, such as sickness or old age.

The observation that wolves are often incapable of killing moose that are mature and in good physical condition is extremely significant. This is what often prevents predators from eliminating their prey. As the prey population is culled down to those healthy individuals who can escape attack, the predator population will necessarily decline unless it can switch to other prey. The predators are limited by the availability of their crucial food resource. Meanwhile, the survivors of the prey population are healthy and can readily reproduce the next generation. Thus, a predator-prey relationship involves both top-down (on the prey) and bottom-up (on the predator) population regulation. Furthermore, some factors such as weather can be both density independent (when a deep freeze occurs, regardless of the number in the population) or density dependent.

A parasite can also work in conjunction with a larger predator to control a given herbivore population. When parasitic infection breaks out in a dense population of herbivores, individuals weakened by infection are more easily removed by predators, leaving a smaller, but healthier, population. Relationships between a prey population and several natural enemies are generally much more stable and less prone to wide fluctuations than are those that involve only a single predator or parasite because different predators or parasites come into play at different population densities.

Plant-Herbivore Dynamics. Herbivores are predators on plants. Just as too many wolves can bring the moose population dangerously low, too many herbivores can do the same to their plant food. In turn, whenever the predator populations are low, the prey populations encounter low environmental resistance and can overshoot their carrying capacity. These imbalances can lead to problems in the surrounding ecosystem.

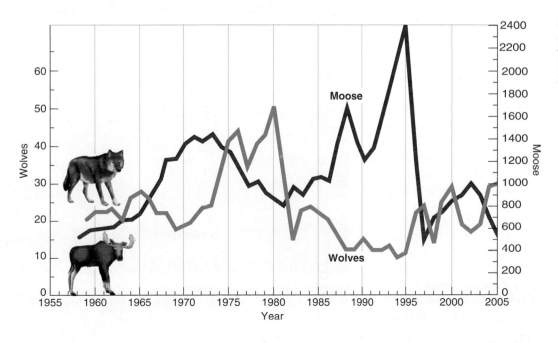

Figure 4–7 Predator-prey relationships. The fluctuations in the wolf and moose populations on Isle Royale are shown from 1955 until 2005.

Figure 4–8 Plant-herbivore interaction. In 1944, a population of 29 reindeer was introduced to St. Matthew Island, where they increased exponentially to about 6,000 and then died off due to overgrazing.

Overgrazing. If herbivores eat plants faster than the plants can grow, the plants will eventually be depleted, and the animals will suffer. A classic example is the case of reindeer on St. Matthew Island, a 128-square-mile island in the Bering Sea, midway between Alaska and Russia. In 1944, a herd of 29 reindeer was brought onto the island, where they had no predators. From these 29 animals, the herd multiplied some two-hundredfold over the next 19 years. Early in the cycle, the animals were observed to be healthy and well nourished, as supporting vegetation was abundant. By 1963, however, the size of the herd had reached an estimated 6,000, and the animals were malnourished. Lichens, an important winter food source, had been virtually eliminated and replaced by unpalatable sedges and grasses. During the winter of 1963–1964, the absence of sufficient lichens, combined with harsh weather, resulted in death by starvation of nearly the entire herd; there were only 42 surviving animals in 1966 (Fig. 4–8).

The reindeer on St. Matthew Island demonstrate that no population can escape ultimate limitation by environmental resistance, although the form of environmental resistance and the consequences may differ. If a population is not held in check, it may explode, overgraze, and then crash (exhibiting a *J*-curve and subsequent drop) as a result of starvation.

Predator Removal. Eliminating predators or other natural enemies of herbivores upsets basic plant-herbivore relationships in the same way as introducing an animal without natural enemies. In much of the United States, for example, white-tailed deer populations were originally controlled by wolves, mountain lions, and bears, most of which were killed because they were believed to be a threat to livestock and even humans. At present, deer populations in many areas would increase to the point of overgrazing if humans didn't hunt them in place of these natural predators. Indeed, population explosions do occur where hunting is prevented. If the St. Matthew reindeer population had been kept in check by predators, for example, their population might not have increased to the point at which they overgrazed the vegetation on the island.

Keystone Species. In the case of white-tailed deer, the removal of a number of predator species allowed the deer population to explode. However, sometimes the removal of a single species can cause a cascade of effects that impact far more than the just the other species with which they primarily interact. For example, in the West Coast rocky intertidal zone, a sea star species, *Pisaster ochraceus,* feeds on mussels, thus keeping the mussels from blanketing the rocks (Fig. 4–9). As a result, barnacles, limpets, anemones, whelks, and other invertebrates are able to colonize the rich intertidal habitat. When ecologist Robert Paine experimentally removed *P. ochraceus* from a rocky shoreline, the mussels crowded everything else out, and species diversity in the ecosystem was greatly reduced. Paine referred to the sea star as a **keystone**

Figure 4–9 Keystone species. The sea star *Pisaster ochraceus* is crucial for maintaining a diverse functioning community in the rocky intertidal zone of the Pacific Northwest.

species, in recognition of its crucial role in maintaining ecosystem biotic structure (in architecture, the keystone is a fundamental part of the support and structure of a building). The presence of the keystone species moderates other species that would otherwise take over the area (the mussels) and allows less-competitive species to flourish as well, thereby increasing overall diversity. Keystone species do not have to be predators, as shown in this example, but they often are.

Competition

Recall from Table 4–2 that there are species interactions in which both species are harmed. This result usually means that species are competing for some scarce resource. In Chapter 3, we discussed the concept of each species having a *niche* (the conditions and resources under which a species can live). Species are said to have overlapping ecological niches when they compete with each other. Competition can also occur between members of the same species. In the case of both types of competition, between and within species, you might expect that, over time, there would be a pressure to lower the overlap and have less negative interaction. Indeed, this is the case.

Intraspecific Competition. When members of the same species compete with each other, it is called *intraspecific competition*. This type of competition can occur over different types of resources. In many species, some members defend a **territory**. The males of many species of songbirds claim a territory, or limited space, that they will defend vigorously at the time of nesting. Their song warns other males to keep away (Fig. 4–10). The males of many carnivorous mammals, including dogs, stake out a territory by marking it with urine. The smell warns others to stay away. If other dogs do encroach, there may be a fight, but in most species, a large part of the battle is intimidation—an actual fight rarely results in serious

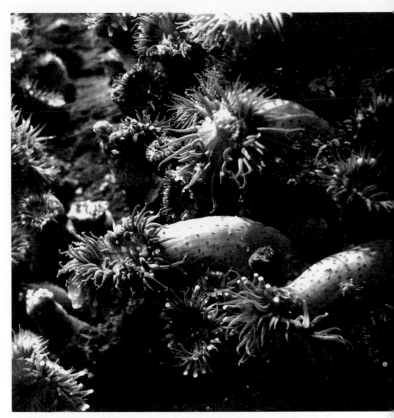

Figure 4–11 Competition. Some sea anemones, even though they are slow moving, have been found to fight in colonies, attacking animals from another colony and hitting each other with special tentacles.

physical harm. Sometimes, as in sea anemones, individuals can band together to defend a territory for a colony (Fig. 4–11).

In territoriality, what is really being protected by the defender or sought after by the invader is the claim to an area suitable for nesting, space for establishing a harem, or adequate food resources. Hence, the territory is defended only against others that would cause the most direct competition for those resources—members of the same species. Because territoriality is competition for space with resources, a lack of suitable territories is a density-dependent limitation on a population—the larger the population grows, the more intense the competition for space is.

Territoriality as an Advantage. Territoriality may seem like a negative factor, but it actually protects a population from the possibility that all of its members will get some resources, but none will get enough to survive. What if the population had no territories? If so, when resources were scarce, two things might happen: First, every encounter with another member of the species could end in a potentially lethal fight. Second, instead of some members getting enough resources to raise the next generation, it might be possible for all members to get only a portion of what they need, and a large percentage—or all—of the population would die.

Whether between species or between members of the same species, competition harms both participants. It lowers their fitness and the production of offspring. Territoriality is an adaptation that lowers the direct effects of competition on members of the same species. Individuals unable to claim a territory are usually the young of the previous generations.

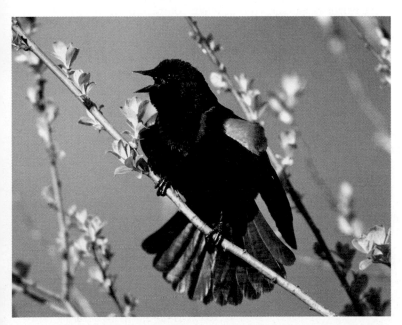

Figure 4–10 Territoriality. A red-winged blackbird announcing its claim on a territory.

Some may hang out on the fringes and will get an opportunity to hold a territory later. Others may be driven to disperse, which may open new habitats to the species. Some may die without ever getting territories, but as a whole, territories protect the population from dying out. As you might expect, territoriality is most likely in species with long life spans, multiple years to reproduce, parental care of young, and other characteristics of *K*-strategists. Territoriality is an adaptation that helps organisms disperse and stabilizes populations.

Impact on the Species.   The observation that individuals compete for scarce resources led Charles Darwin to identify the *survival of the fittest* as one of the forces in nature leading to evolutionary changes in species. Those individuals in a competing group of young plants, animals, or microbes that are able to survive and reproduce, while others are not, demonstrate superior **fitness** to the environment. Indeed, every factor of environmental resistance is a selective pressure resulting in the survival and reproduction of those individuals with a genetic endowment that enables them to better cope with their surroundings. This is the essence of **natural selection,** which is discussed in Section 4.4.

Intraspecific competition, therefore, has two distinct kinds of impact on the population of a species. In the short term, it can lead to the density-dependent regulation of a species population through such factors as territoriality and *self-thinning,* which is a phenomenon whereby crowded organisms such as trees become less numerous as they grow bigger. Intraspecific competition can also lead, however, to long-term changes as the species adapts to its environment. Those better able to compete are the ones who survive and reproduce, and their superior traits are passed on to successive generations.

Interspecific Competition.   Now we turn to the effects of competition between different species, or *interspecific competition.* Consider a natural ecosystem that contains hundreds of species of green plants, all competing for nutrients, water, and light. What prevents one plant species from driving out others? How do species survive competition? The most obvious answer is that sometimes they do not. If two species compete directly in many respects, as sometimes occurs when a species is introduced from another continent, one of the two species generally perishes in the competition. This is the *competitive exclusion principle.* For example, the introduction of the European rabbit to Australia (described later in Section 4.5) has led to the decline and disappearance of several small marsupial animal species due to direct competition for food and burrows.

In fact, *competitive exclusion* would be expected to be the norm in circumstances where a habitat is very simple and species require the same resources. However, in nature, we often see many species occupying the same habitats without becoming extinct. This has intrigued scientists. G. E. Hutchinson, a famous limnologist (a scientist who studies freshwater systems), once wrote a paper asking about this phenomenon: "A homage to Santa Rosalia or Why are there so many kinds of animals?" (Hutchinson, 1958)[1]. The answer lies, ecologists believe, in variability over space and time.

Differences in topography, type of soil, and so on mean that the landscape is far from uniform, even within a single ecosystem. Instead, it is composed of numerous microclimates or microhabitats. The specific abiotic conditions of moisture, temperature, light, and so on differ from location to location and through the course of time. This is true even in lakes, where tiny pockets of water may have different temperatures, chemistry, or light conditions. When one competitor begins to get the upper hand, often the season will change, and a different species will benefit. Thus, the adaptation of a species to specific conditions enables it to thrive and overcome its competitors in one location or time, but not in another.

These examples simply mean that ecosystems are *heterogeneous* (variable) enough to support species with many different niches. Similarly, changes over time can allow multiple species to live in the same place. For example, the spring wildflowers of temperate deciduous forests (lady slippers, hepatica, trillium) sprout from perennial roots or bulbs in the early part of that season. While they would not be able to outcompete trees for light during the summer, these plants take advantage of the light that can reach the forest floor before the trees grow leaves.

Seeming competitors can coexist in the same habitat, but have separate niches. Competition is minimized because potential competitors are using different resources. For example, woodpeckers, which feed on insects in deadwood, do not compete with birds that feed on seeds. Bats and swallows both feed on flying insects, but they do not compete because bats feed on night-flying insects and swallows feed during the day. Sometimes the "resource" can be the space used by different species as they forage for food. This is illustrated in Figure 4–12, which shows five species of warblers that coexist in the spruce forests of Maine. The birds feed at different levels in the trees and on different parts of the trees. This is called *resource partitioning,* the division of a resource and specialization in different parts of it. Subsequent study of these warblers has shown that when competition is more intense, they separate themselves even more.

All of these examples show that species can live in the same habitat because their adaptations, including behavior, *limit* competition. By limiting those competitive interactions, each species can put more energy into reproduction. However, the reason that many species can share a habitat is not simply because their niches prevent them from competing. Instead, when species live in the same place, competition can change some of their characteristics, so that they use only a portion of their potential niche (which they could have used all of when living alone). Consequently, we see species with a different niche when they are alone than when they are living together with a closely competitive species. For example, when the finch *Geospiza magnirostris* arrived on the same island as the similar species *Geospiza fortis,* the two species competed heavily for food seeds. Over 22 years of study, scientists showed that the beak sizes of both species changed, and competition was reduced as the finches began to eat slightly different-sized seeds. In populations where each species was alone, the beaks remained in a middle size. This is called *character displacement,* a physical change that lessens competition when two species co-occur (Fig. 4–13).

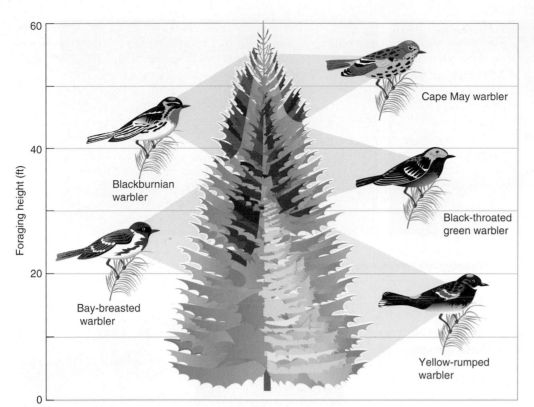

Figure 4–12 Resource partitioning. Five species of North American warblers experience less competition than it might first appear because they feed at different heights and in different parts of trees.

Cape May warbler

Blackburnian warbler

Black-throated green warbler

Bay-breasted warbler

Yellow-rumped warbler

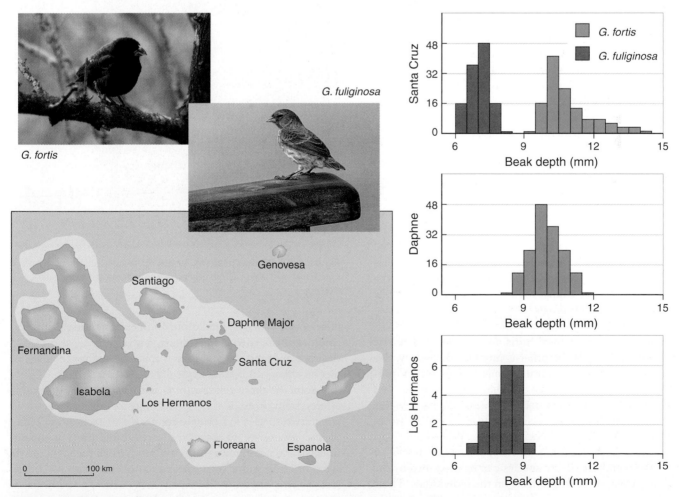

Figure 4–13 Character displacement. Finches of two species have different beak sizes when they occur together, but do not when the species occur separately.

(a)   (b)

Figure 4–14 Mutualism. (a) Pollinators and their flowering plants are mutualists. (b) A lichen is a mutualism between a fungus and an alga living symbiotically.

Mutualism

In contrast to competition, **mutualism** is an arrangement between two species whereby both benefit. There are many mutualistic relationships. For example, the relationship between pollinators (such as bees, which receive nutrition from flowers) and plants they pollinate (which receive pollination) is an example of mutualism (Fig. 4–14a). Some estimates infer that 70% of plant species have mutualistic relationships with fungi connected to their roots. The fungi benefit from the plant nutrition and in exchange make it easier for the plants to take in nitrogen and other soil nutrients. Lichens are actually a mutualistic relationship between a fungus and an alga (Fig. 4–14b). In animals, one example of a mutualism is the relationship of the anemone fish and the sea anemone. The fish protects the anemone from predation by the butterfly fish, and in turn, the anemone's stinging tentacles (to which the anemone fish is immune) protect the fish from other predators.

Commensalism

Commensalism is more rare. It occurs when two species interact, and one is benefited, while the other is unaffected. An example would be cattle egrets and water buffalo—the grazer stirs up insect prey, and the bird follows it and eats the insects (Fig. 4–15). Commensalism also includes animals that hitch rides on other species without harming them or grow on them for support. Many orchids exist in this type of relationship—they grow as vines on large trees, but do not harm or feed off of them like parasitic plants.

In contrast to commensalism is **amensalism,** where one species is *harmed*, while the other is unaffected. Usually, this is accomplished by the natural chemical compounds produced by an organism, which are harmless to itself, but harmful to other organisms around it. An example is the black walnut tree, which secrets a chemical compound from its roots that can harm or kill other plants in the area.

In summary, the relationships between two species or between two members one species can be described by the positive, negative, or neutral effects on the individuals. These potential outcomes are studied by ecologists. Not only do

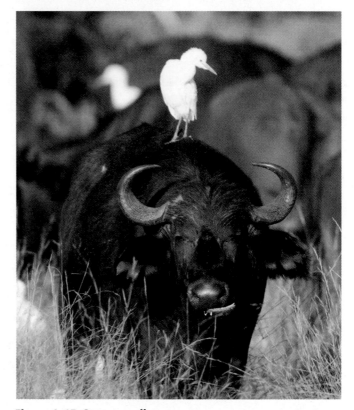

Figure 4–15 Commensalism. A cattle egret eats insects stirred up by a grazing water buffalo, which is neither harmed nor benefited by the egret's presence.

relationships such as predation, competition, mutualism, and commensalism affect individuals, but also they are part of the constant pressure that causes species to adapt to environmental change.

Some of these relationships with positive and negative effects stem from a type of **symbiosis**—a relationship in which two species live closely connected to each other. Many of the examples of parasites, predators, prey, commensalisms, and most mutualisms we have already seen would be symbiotic. As we've seen, these relationships can be beneficial (as when mutualisms benefit both parties) or detrimental (as in the case of parasites and host).

4.4 Evolution as a Force for Change

Predation and competition are two important mechanisms that keep natural populations under control. They also support the survival of individuals with qualities that can lower the impacts of negative interactions. Predators and prey become well adapted to each other's presence. Predators are rarely able to eliminate their prey species, largely because the prey has various defenses against its predators. Intraspecific competition also represents a powerful force that can lead to improved adaptations of a species to its environment. Interspecific competition, by contrast, promotes adaptations in the competitors that allow them to specialize in exploiting a resource. This specialization can lead to resource partitioning, which allows many potential competitors to share a basic resource—such as water in the soil, as illustrated by prairie grasses with long tap roots and those with short fibrous roots, shown in Figure 4–16. In this section, we discuss how these various adaptations have come about.

Selective Pressure

Most young plants and animals in nature do not survive, instead falling victim to various environmental resistance factors. These factors—such as predators, parasites, drought, lack of food, and temperature extremes—are known as **selective pressures**. These pressures can affect which individuals survive and reproduce and which are eliminated. If a predator is present, for example, prey animals with traits that protect them or that allow them to escape from their enemies (such as coloration that blends in with the background) tend to survive and repro-

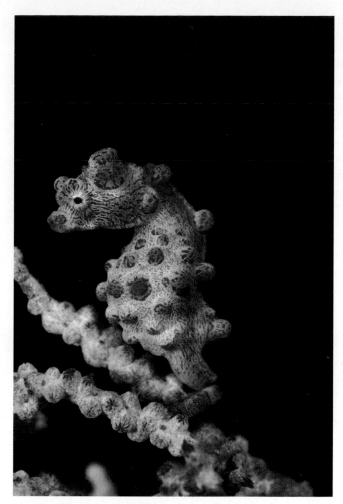

Figure 4–17 Cryptic coloration. The ability to hide (as this seahorse does) is a defense against predation, but it also can be an adaptation that allows predators to sneak up on or wait for prey.

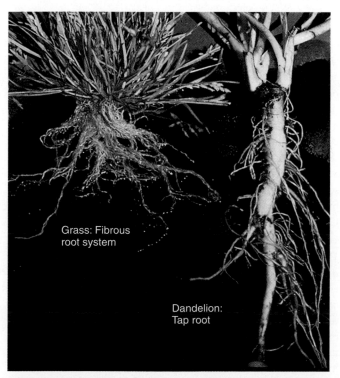

Grass: Fibrous root system

Dandelion: Tap root

Figure 4–16 Resource partitioning in plants. Prairies have plants with short fibrous roots and others with long tap roots that use different areas of water in the soil.

duce (Fig. 4–17). Those without such traits tend to become the predator's dinner. Any individual with a genetic trait that slows it down or makes it conspicuous will tend to be eaten. Thus, predators function as a selective pressure, favoring the survival of traits that enhance the prey's ability to escape or protect itself and causing the elimination of any traits handicapping the organism's escape. The need for food can also be seen as a selective pressure acting on the predator. Enhancing characteristics, such as keen eyesight and swift speed, benefit survival and reproduction. Sometimes predators use the ability to blend in as an advantage, too. The process of specific traits favoring the survival of certain individuals is known as **natural selection.**

Discovered independently by Charles Darwin and Alfred Russell Wallace, these concepts were first presented in detail by Darwin in his book *On the Origin of Species by Means of Natural Selection* (1859). The modification of the gene pool of a species by natural selection over the course of many generations is the sum and substance of **biological evolution.** Darwin and Wallace deserve tremendous credit for constructing their theory purely from their own observations, without any knowledge of genetics. The genetic information wasn't discovered until several decades later. Our modern understanding of DNA, mutations, and genetics supports the theory of evolution by natural selection.

Adaptations to the Environment

The gene pool of each population is continually tested by the selective pressures exerted by environmental resistance. Indeed, virtually all traits of any organism can be seen as features that adapt the organism for survival and reproduction, or, in Darwinian terms, **fitness.** Such characteristics of organisms include the adaptations needed for coping with climatic and other abiotic factors; obtaining nutrients, energy, and water; defending against predation and disease-causing or parasitic organisms; finding or attracting mates (or pollinating and setting seed); and migrating and dispersing. These are summarized in Figure 4–18.

The fundamental question about any trait is this: Does it increase survival and reproduction of the organism? If the answer is yes, the trait will be maintained through natural selection. Consequently, various organisms have evolved different traits to accomplish the same function. For example, the ability to run fast, to fly, or to burrow and protective features such as quills, thorns, and an obnoxious smell or taste all help reduce predation and can be found in various organisms.

The Limits of Change. As you saw in the opening essay, the Panamanian golden frog is facing a new, powerful selective pressure—a disease-causing fungus. In such a position, a species has three alternatives:

1. **Adaptation.** The population of survivors may gradually adapt to the new condition through natural selection.

2. **Migration.** The surviving population may migrate and find an area where conditions are suitable.

3. **Extinction.** Failing the first two possibilities, extinction is inevitable.

Migration and extinction need no further explanation. The critical question is, What factors determine whether a species will be able to adapt to new conditions, instead of becoming extinct?

Recall that adaptation occurs as selective pressures eliminate those individuals that cannot tolerate the new condition. For adaptation to occur, there must be some individuals with traits (*alleles*—variations of genes or new combinations of genes) that enable them to survive and reproduce under the new conditions. Even with such genetic variation, there must also be enough survivors to maintain a viable breeding population. If a breeding population is maintained, the process of natural selection should lead to increased adaptation over successive generations. If a viable population is *not* maintained at any stage, extinction results. In the case of the golden frog, the species has not yet been able to adapt to the fungus, in part because the frog was already facing pressure from habitat loss. The frog had to be removed to captivity by humans to prevent extinction. However, there are some golden frogs that have survived the fungus. If their genes can be preserved, it is possible that some day the frog will come back from the brink of extinction and be reintroduced to the wild.

In a similar example, the California condor (a very large scavenger) had declined to about 20 birds by the early 1980s because of deaths from poaching, lead poisoning, and habitat destruction. Only nine condors were left in the wild. To save the species from extinction, wildlife biologists captured the remaining wild condors and initiated a breeding program. As of July 2008, there were 320 condors, 180 of which are in captivity. This is still not a viable breeding population, but the wild condors are now forming pairs and nesting. Wildfires in the summer of 2008 made it a hard year for condors and other wildlife, so it will be quite a while before they are no longer in danger of extinction.

Keys to Survival. There are four key variables among species that will affect whether or not a viable population of individuals is likely to survive new conditions: (1) geographical distribution, (2) specialization to a given habitat or food supply, (3) genetic variation within the gene pool of the species, and (4) the reproductive rate relative to the rate of environmental change.

Consider how these factors work. It is unlikely that any change will affect all locations uniformly or equally. Therefore, a species such as the housefly—which is present over most of Earth, has no requirements for a specialized habitat or food supply, and possesses a high degree of genetic variation—is likely to survive almost any conceivable alteration in Earth's environment. The California condor, on the other hand, requires an extensive habitat of mountains, ravines, and the carcasses of large animals to eat, but has little genetic variation in its population. The less vulnerable species are likely to be *r*-strategists, described in Section 4.1. The more vulnerable species are likely to be *K*- strategists, although many species fall somewhere in between these extremes, and in some cases, even very rapidly reproducing species can be vulnerable if there is enough environmental change. The factors affecting survival of these two species types are summarized in Figure 4–19 (p. 92).

Genetic Change. Assuming the survival of a viable population, how fast can that population evolve further adaptations to enable it to better cope with new or changing conditions? Over the lifetime of the *individual*, there is no genetic change—and hence no genetic adaptation. In the population, genetic variation is brought about over generations by a number of mechanisms, including hybridization, mutation, and crossover (a process by which sections of chromosomes are swapped). Bacteria, for example, readily exchange small bits of genetic material with other bacteria, and recent studies have shown that microbes sometimes insert their genetic material into other organisms, including mammals, bringing new DNA variations into the genome. Genetic variation is the norm in living things.

The Evolution of Species. There may be as many as 14 million species of plants, animals, and microbes currently in existence, all living and functioning in ecosystems and all contributing to an amazing biodiversity. That biodiversity is diminishing, however, because present species are becoming extinct faster than new species are appearing. In order to understand why, we need to understand how new species appear.

The infusion of new variations from mutations and the pressures of natural selection serve to adapt a species to the biotic community and the environment *in which it exists.* Over

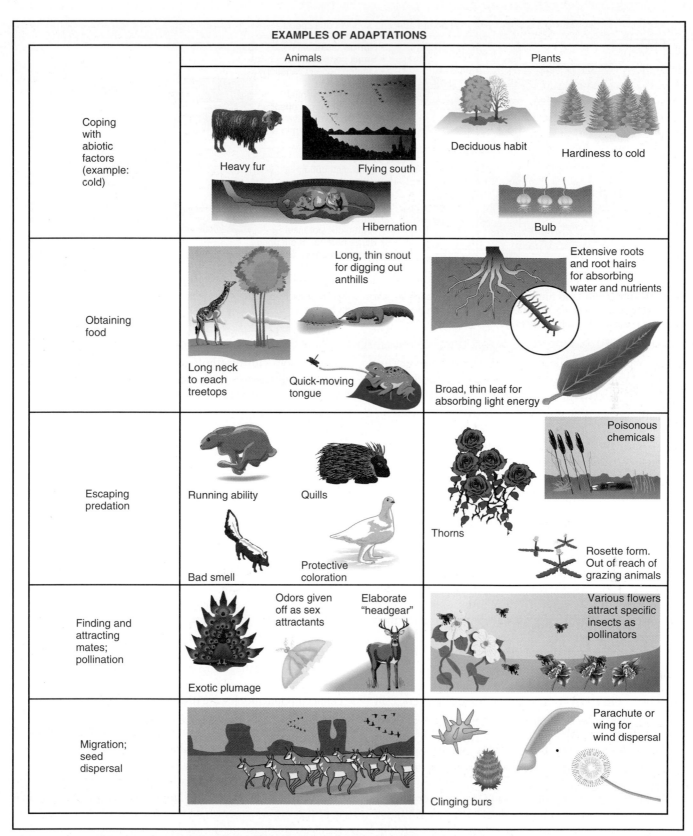

Figure 4–18 Adaptation for survival and reproduction. The five general features listed at the left are essential for the continuation of every species. Each feature is found in each species as particular adaptations that enable the species to survive or reproduce. Across the various species, a multitude of adaptations will accomplish the same function. Thus, a tremendous diversity of species exists, with each species adapted in its own special way.

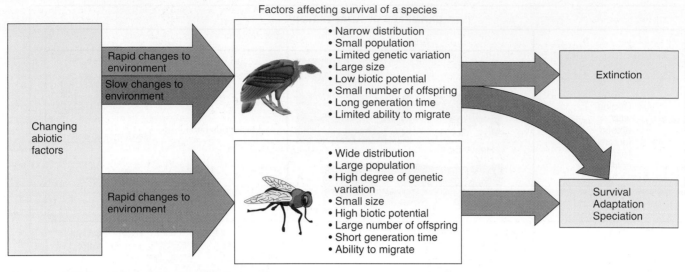

Factors affecting survival of a species

Changing abiotic factors

Rapid changes to environment
Slow changes to environment

- Narrow distribution
- Small population
- Limited genetic variation
- Large size
- Low biotic potential
- Small number of offspring
- Long generation time
- Limited ability to migrate

Extinction

Rapid changes to environment

- Wide distribution
- Large population
- High degree of genetic variation
- Small size
- High biotic potential
- Large number of offspring
- Short generation time
- Ability to migrate

Survival
Adaptation
Speciation

Figure 4–19 Vulnerable and highly adaptive species. Species differ broadly in their vulnerability to environmental changes. When confronted with rapid change in the environment, a housefly population is likely to explode, and a condor population is likely to decline.

time, in this process of adaptation, the final "product"—giraffe, anteater, redwood tree—may be so different from the population that started the process that it is considered a different species. This is one aspect of the process of speciation.

Two Species from One. The same process also may result in two or more species developing from one original species. There are only two prerequisites. The first is that the original population must separate into smaller populations that do not interbreed with one another. This **reproductive isolation** is crucial because if the subpopulations continue to interbreed, all the genes will continue to mix through the entire population, keeping it as one species. The second prerequisite is that separated subpopulations must be exposed to different selective pressures. As the separated populations adapt to these pressures, they may gradually become so different that they are not considered the same species. Thus, they are unable to interbreed with one another, even if they come together again later.

Consider, for example, the arctic and gray foxes of today. It is believed that thousands of years ago a northern subpopulation became separated from the main population, perhaps due to the effects of glaciation in North America. In the Arctic, selective pressures favor individuals that have heavier fur (to protect them from the cold); a shorter tail, legs, ears, and nose (the shortness helps conserve body heat); and white fur (which helps the animals hide in the snow). In the southern regions, selective pressures favor individuals with the opposite traits. In warmer climates, animals need a thinner coat to dissipate excessive body heat, and white fur would make the animals more conspicuous. The geographic isolation of the two subpopulations over many generations, however, made it possible for one ancestral fox population to develop into two separate species (Fig. 4–20).

Darwin's Finches. Darwin himself first observed evidence of past speciation. In 1835, Darwin collected a number of finches from different islands of the Galápagos archipelago during his famous voyage on HMS *Beagle*. Later, in consultation with ornithologist John Gould, Darwin speculated that differences in size and beak structure of the finches could have happened after subpopulations of an original parental species were isolated from one another on separate islands (Fig. 4–21).

Although Darwin himself did not work out the details of how the different Galápagos finches evolved, modern scientists have found what looks to be the most likely explanation. It is likely that at some time in the past, a few finches from the South American mainland were blown westward by a strong storm and became the first terrestrial birds to inhabit the relatively new (10,000-year-old) volcanic islands of the Galápagos. As the initial population grew and members faced increased intraspecific competition, some birds dispersed to nearby islands, where they were separated from the main population. These subpopulations encountered different selective pressures and became specialized for feeding on different things (cactus fruit, insects, and seeds, to name a few). In time, when the changed populations dispersed back to their original islands, they were different enough from the parent species that interbreeding among them did not occur, and the changed populations were distinguishable as new species.

In the finch example, Darwin did not directly observe speciation. Rather, he saw a pattern of finch species distribution, and he and subsequent scientists suggested what they thought was the best explanation for that pattern. One way to test whether an interpretation of a historical pattern is likely is to see what happens in other examples right now. As we might expect, we also see new species arise in real time today, and we have examples of species forming rapidly. The saltmarsh cordgrass, *Spartina angelica*, is one such species. It developed as a hybrid of two other related species: *Spartina alterniflora*, a native of North America, and *Spartina maritima*, a native of England. The two do not usually interbreed. *Spartina alternifora*, accidentally introduced to England's mud flats, did not thrive there, but when it bred with *Spartina maritima*, the hybrid daughter species (*S. angelica*) spread rapidly. It has a

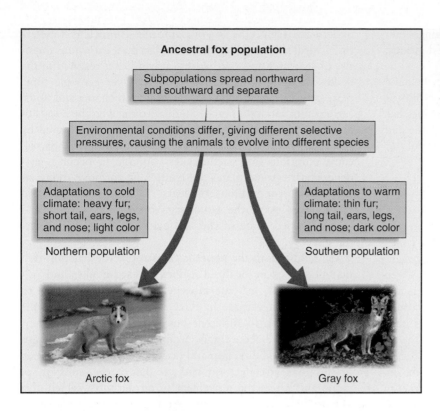

Ancestral fox population

Subpopulations spread northward and southward and separate

Environmental conditions differ, giving different selective pressures, causing the animals to evolve into different species

Adaptations to cold climate: heavy fur; short tail, ears, legs, and nose; light color

Northern population

Adaptations to warm climate: thin fur; long tail, ears, legs, and nose; dark color

Southern population

Arctic fox

Gray fox

Figure 4–20 Evolution of new species. A population spread over a large area may face various selective pressures. If the population splits so that interbreeding does not occur, the different pressures may result in the subpopulations evolving into new species. Two species of foxes show the pattern we would expect if this had occurred.

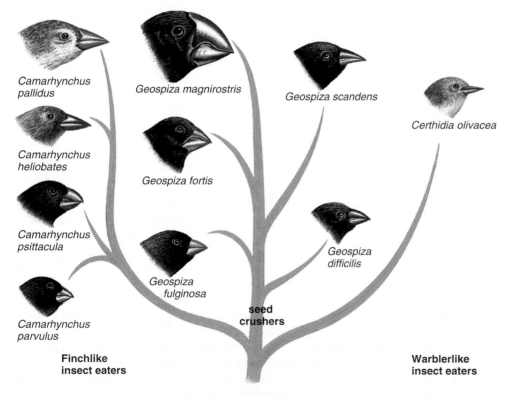

Camarhynchus pallidus

Geospiza magnirostris

Geospiza scandens

Certhidia olivacea

Camarhynchus heliobates

Geospiza fortis

Camarhynchus psittacula

Geospiza difficilis

Camarhynchus parvulus

Geospiza fulginosa

seed crushers

Finchlike insect eaters

Warblerlike insect eaters

Ancestral South American finch

Figure 4–21 Some of Darwin's finches. The similarities among these birds attest to their common ancestor. Selective pressures to feed on different foods have caused modification and speciation in adapting subpopulations. (*Source:* From *Biology: Life on Earth,* 4th ed., by Teresa Audesirk and Gerald Audesirk, copyright © 1994 by Prentice Hall, Inc., reprinted by permission of Pearson Education, Inc., Upper Saddle River, NJ 07458.)

different number of chromosomes from either parent and does not interbreed successfully with either parent species. Now, where there were two similar species, there are three, and the daughter species has a different set of characteristics and can live in a slightly different niche than either parent species.

In sum, new species are not formed from scratch; they are formed only by the modification of existing species.

Different selective pressures act on the constant variation of genetic material that naturally occurs. Combined with isolation and other processes (such as the unusual hybridization that occurred in the genus *Spartina*), new adaptations can arise. This concept is consistent with the observation that groups of closely related species are generally found in nature, instead of distinct species without close relatives.

Patterns among larger groups suggest the same thing—that organisms can be grouped into large, closely related groups, differentiated by older or newer characteristics.

Although it can be difficult to imagine, overwhelming evidence from today and patterns from the past inform us that the present array of plants, animals, and microbes has developed through evolution over long periods of time and in every geographic area on Earth. This is the source of our current biodiversity.

Drifting Continents

Geographic isolation of populations is foundational to the process of speciation. The slow movement of Earth's continents has allowed the isolation of broad groups of organisms. Using evidence from geology, German scientist Alfred Wegener proposed in 1915 that about 225 million years ago the continents were connected into a giant land mass called Pangaea. This is part of a theory that the continents have been in motion (and continue to move). Wegener's theory was hotly debated for decades, until, by the mid-20th century, the evidence for drifting continents had become irrefutable. His theory has blossomed into a grand theory known as **plate tectonics.** This theory helps us to understand earthquakes and volcanic activity and is key to understanding the geographic distribution of present-day biota.

The interior of Earth is molten rock, kept hot by the radioactive decay of unstable isotopes remaining from the time when the solar system was formed (about 5 billion years ago). Earth's crust, the lithosphere—which includes the bottoms of oceans as well as the continents—is a relatively thin layer (ranging from 10 to 250 kilometers, or 16 to 160 miles) that can be visualized as huge slabs of rock floating on an elastic layer beneath. This is much like crackers floating next to each other in a bowl of soup. These slabs of rock are called **tectonic plates.** Some 14 major plates and a few minor ones make up the lithosphere (Fig. 4–22a).

Within Earth's semimolten interior, hot material rises toward the surface and spreads out at some locations, while cooler material sinks toward the interior at other locations. Riding atop these convection currents, the plates move slowly, but inexorably, with respect to one another. The spreading process of the past 225 million years has broken up the original mass of Pangaea and brought the continents to their present positions. It also accounts for the other interactions between tectonic plates, such as volcanoes and earthquakes. The average rate of a plate's movement is about 6 centimeters (2.4 inches) per year, but over 100 million years, this adds up to almost 8,000 kilometers (5,000 miles) in the fastest-moving segments. The plate boundaries are locked by friction and hence are regions of major disturbances.

Adjacent tectonic plates move with respect to each other by separating (as in mid-ocean ridges), sliding past each other (at fault lines like the San Andreas Fault in California), or colliding. These plate movements have large environmental effects. The catastrophic **tsunami** (tidal wave) in Indonesia on December 26, 2004, resulted from the Indo-Australian plate sliding under the Eurasian plate, uplifting an underwater region the size of California as much as 8 meters (8.75 yards). The sudden uplift displaced an enormous quantity of seawater and sent waves outward that resulted in the deaths of more than 225,000 people. Over geologic time, plate collisions produce volcanic mountain chains and uplift regions into mountain ranges similar to the way the hoods of colliding cars crumple. An example of mountains caused by this type of movement is the Himalayas. The fact that volcanic eruptions and earthquakes continue to occur is evidence that tectonic plates are continuing to move today, as they have over millions of years. In fact, volcanoes and earthquakes mark the boundaries between the plates and have provided some of the best evidence for the plate tectonic theory.

In addition to the periodic catastrophic destruction that may be caused in localized regions by earthquakes and volcanic eruptions, tectonic movement may gradually lead to major shifts in climate in three ways. First, as continents gradually move to different positions on the globe, their climates change accordingly. Second, the movement of continents alters the direction and flow of ocean currents, which in turn has an effect on climate. Third, the uplifting of mountains alters the movement of air currents, which also affects climate. (For example, see the rain-shadow effect in Fig. 7–7.)

Large-scale changes in Earth's crust have been part of evolutionary pressures that have increased diversity. As continents moved to their present position, they isolated organisms into broad regions. Sometimes scientists divide the world into six *zoogeographic regions*—areas of the earth's surface with distinctive animal groups. Australia, for example, is part of the Australasian region and is home to almost all of the animals with pouches—the marsupials. Australia is located on the Indo-Australian tectonic plate. Its separation from other continents appears to be very ancient, which explains why so many species in the Australasian region are unique: They have had millions of years to evolve separately from species on other continental plates.

Sometimes the place where several tectonic plates meet will have an unusual mix of species. For example, the country of Indonesia, which consists of about 17,000 islands, is located on the junction of three tectonic plates—the Pacific, Eurasian, and Indo-Australian plates. Papua New Guinea is part of the Indo-Australian plate and has plants and animals similar to those on Australia. Other islands have biota more like parts of Asia. The island of Sulawesi lies right on a line between plates and has been formed by islands being pushed together. It is home to an odd combination of animals and plants with relatives in far distant parts of the world.

Because they have been isolated on islands, many Indonesian species have changed over time, developing adaptations not seen in their Asian or Australian relatives. Now, both Australia, long separated from other continents, and Indonesia have a high number of species that occur nowhere else in the world. Australia and Indonesia serve as examples of how long-term, large-scale changes, such as the movement of continents and the rise of islands, can change the environment for organisms and increase their diversity.

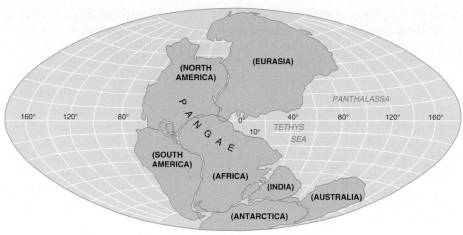

(a) 225 million years ago

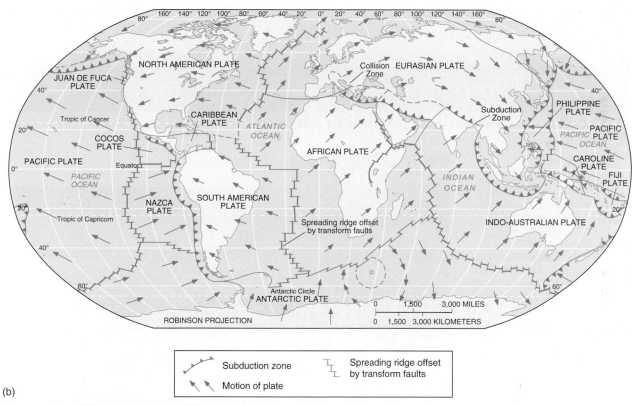

(b)

Figure 4–22 Drifting continents and plate tectonics. (a) Similarities in the types of rock, the distribution of fossil species, and other lines of evidence indicate that 225 million years ago all the present continents were formed into one huge landmass that we now call Pangaea. (b) The 14 major tectonic plates making up Earth's crust and their directions of movement. (*Source:* From *Geosystems: An Introduction to Physical Geography,* 5th ed., by Robert W. Christopherson, copyright © 2005 by Prentice Hall, Inc., reprinted by permission of Pearson Education Inc., Upper Saddle River, NJ 07458.)

4.5 Implications for Human Management

Ecologists study populations and communities for a number of reasons. One is simply to further our understanding of the world. Another is to better manage natural resources by protecting declining species such as the condor and the golden frog and by controlling pest species such as the chytrid fungus and the spotted knapweed. It is also important for us to study our own impacts on populations, as many of the causes of explosive growth or decline in populations stem from human actions (such as the removal of an important species or the introduction of non-native species). Pests will be covered in Chapter 13 and biodiversity in Chapter 6.

The keystone species mentioned in Section 4.3 are such an integral part of their ecosystems that their removal, as we discussed, can cause the collapse of the whole ecosystem. For instance, beavers were once one of the dominant ecological forces in eastern North American, creating openings in forests and wetlands that became habitats for other species and allowed them to survive. The loss of beavers (due to the fur industry and habitat destruction) removed them from much of the North American landscape, to the detriment of species that rely on small wetland meadows.

Introduced Species

Humans have impacted population equilibriums by introducing species from foreign ecosystems. These introductions alter community and population ecology relationships such as competition and predation. Over the past 500 years, and especially now that there is a vast global commerce, thousands of species of plants, animals, and microbes have been accidentally or deliberately introduced onto new continents and islands (see Sound Science, "The Village Weaverbird: Marvel or Menace?"). A recent attempt to quantify the economic losses due to introduced species in the United States estimated the total cost to be in excess of $138 billion per year.

Rabbits in Australia.   A classic example of the impact an introduced species can have on a native ecosystem comes from Australia. In 1859, rabbits were introduced into Australia from England to be used for sport shooting. The Australian environment proved favorable to the rabbits and contained no predators or other natural enemies capable of controlling them. As a result, the rabbit population exploded and devastated vast areas of rangeland by overgrazing. The devastation was extremely damaging to both native marsupials and ranchers' sheep. The problem was temporarily brought under control by introducing a disease-causing virus in rabbits. Over time, however, the rabbits adapted to the virus and began to repeat their explosive growth. In spite of the recent release of a different virus, rabbits are still Australia's most destructive pest animal, costing farmers more than $100 million each year. The impact of rabbits on vegetation can be seen on Philip Island, a 260-hectare (642-ac) island halfway between Australia and Fiji (Fig. 4–23a). The rabbits on the island were

Sound Science

The Village Weaverbird: Marvel or Menace?

DAVID LAHTI

After attending Gordon College, David Lahti earned a Ph.D. in philosophy at the Whitefield Institute at Oxford in 1998 for his work on the relationship between natural science and morality. In 2003 he received a Ph.D. from the University of Michigan in ecology and evolutionary biology, and then became a Darwin Fellow at the University of Massachusetts. He is now an Assistant Professor of Biology at Queens College CUNY in New York. He has studied weaverbirds for the last ten years in Africa, the Indian Ocean, and the Caribbean with his wife April.

A lone African tree flickers yellow and white as if on fire. Someone approaching it from a distance eventually makes out the forms of dozens of birds, hanging upside-down by their feet from globular nests, swaying frantically as they babble a complex song and flutter their outstretched wings. They are male village weaverbirds, advertising their nest-building accomplishments to a nearby flock of females. The males do have something to boast about, as weavers are considered to be the most proficient nest builders in the bird world. Once a female has chosen a nest, she lays eggs of nearly identical appearance throughout her lifetime, but the eggs of the population as a whole have among the widest variety of colors and spotting patterns of any bird. Laying a distinctive egg helps a female ensure that the eggs in her nest are her own, and not those of some African cuckoos, which mimic the eggs of other birds and lay eggs in their nests.

For most people who live with the village weaver, however, appreciating its nest-building

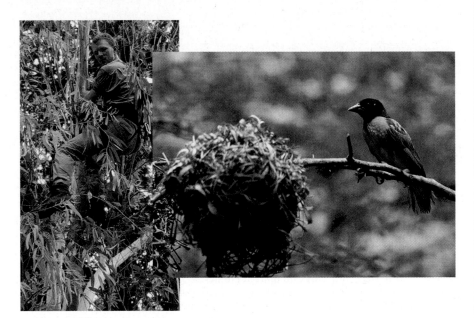

David Lahti and village weaverbird.

and egg artistry can be difficult. This species is a serious agricultural pest throughout most of its range. A margin of trees near a field of rice or sorghum can be filled with village weaver colonies, with as many as three hundred birds nesting in a single tree. This bird and one other weaver species (the quelea) pose the most significant threat to agriculture in West Africa. Rafael Jawo, a farmer in The Gambia, loses up to a third of his annual yield from dense flocks of village weavers descending on his rice fields.

The village weaver has been introduced to islands in the Caribbean and the Indian Ocean, where it has become invasive and a threat to native bird species. It is the most damaging agricultural pest in Mauritius and Haiti. The village weaver has many traits that make it an excellent invader of new areas, such as the ability to raise several broods of young per season, a generalized diet, and a preference for human-altered habitats. It is regularly transported to new areas in the cagebird trade and often escapes from captivity. In the last two decades, the village weaver has been sighted in the wild for the first time in North and South America and Europe. Many of these birds attempted to breed, and some were successful.

Throughout the world, humans have unwittingly made a menace out of the village weaver. In its natural range and with only wild food sources, it would have little adverse impact on human economies. Even today in some areas, such as central Uganda and eastern South Africa, the village weaver is a functioning element in the ecosystem without undergoing population explosions or creating serious problems for agriculture. In both of these areas, small grains suitable for the weaver's diet are not in widespread cultivation.

What can be done to remedy the situation elsewhere in Africa and on the islands? Methods of physical control such as fire, scarecrows, rattles, shooting, nest-robbing, and felling trees have been attempted, but have met with little success. Poison kills large numbers of individuals, but this does not affect long-term population sizes. Long-term alternatives may exist, however. One possibility is economic and agricultural diversification. Large stretches of land in many African nations are dedicated to the production of a single kind of crop. If a network of rice fields in central Gambia was converted into a patchwork of rice, maize, groundnuts, mangos, and palm oil, only a portion of the area would supply food for village weavers, and the more diverse ecosystem might support their competitors and predators.

For places not yet invaded, prevention is the best strategy. Transport and housing of caged birds are still largely unregulated in some European countries, such as France, which is the largest importer of weavers. If fewer birds were allowed to escape, the likelihood of a population becoming established would decrease.

The village weaver is a species with some remarkable traits. More importantly, it has been an integral part of the biological communities of sub-Saharan Africa for millions of years. To view this bird solely as an agricultural pest and an invasive species is an incomplete view, although these problems are very real and troubling. The village weaver is named for its preference for living near humans. Our challenge is to find a way to make this a peaceful coexistence.

(a)

(b)

Figure 4–23 Rabbit eradication on Philip Island. (a) In this 1978 photo, the island is largely devoid of vegetation and heavily eroded. (b) Following eradication of the rabbits in 1988, the island vegetation recovered spectacularly.

eradicated in 1988 with a combination of viruses, poisons, trapping, gassing, and shooting. The result was a spectacular recovery of the island vegetation (Fig. 4–23b).

Plants.   Introduced species can have a devastating effect on plants as well as animals. Prior to 1900, the dominant tree in the eastern deciduous forests of the United States was the American chestnut, which was highly valued for both its high-quality wood and its prolific production of chestnuts, eaten by wildlife and people alike. In 1904, however, a fungal disease called the chestnut blight was accidentally introduced when some Chinese chestnut trees carrying the disease were planted in New York. The fungus spread through the forests, killing nearly every American chestnut tree by 1950. Although oaks filled in where the chestnuts died, the ecological and commercial loss was incalculable. There is hope, however, because researchers have recently crossbred the American and Chinese chestnuts, creating a hybrid that is both 94% native and resistant to the blight. However, it will be several decades before enough of the hybrids are produced to make them available for sale to nurseries and the general public.

There are also examples of introduced plant species in the United States that have caused ecological damage. Kudzu, a vigorous vine introduced from Japan in 1876, is a famous example. It was widely planted on farms throughout the southeastern United States so it could be used for cattle fodder and erosion control. From wherever it was planted, however, kudzu invaded and climbed over adjacent forests. The species now occupies more than 3 million hectares (7 million ac) of the Deep South (Fig. 4–24). Because it is a threat to forest growth, considerable efforts are being exerted by the U.S. Forest Service to contain it. The list of nuisance introduced plants is almost endless: The spotted knapweed mentioned earlier has ruined millions of acres of rangelands; purple loosestrife, an invader of wetlands, has replaced plants more edible to wildlife; medusahead rye, a rangeland grass invader, is inedible

Figure 4–24 Introduced plant species. Kudzu overgrowing forests.

to grazing animals; and beach vitex, a coastal shrub, entangles nest-building turtles and outcompetes native dune plants.

Pests.   Most of the important insect pests in croplands and forests—Japanese beetles, fire ants, and gypsy moths, for example—are species introduced from other continents. Because of their voracious appetites, goats and pigs introduced onto islands can become pest animals, devastating both native plants and the animals that depend on that vegetation. While they may not be seen as pests, domestic cats have often proved to be effective predators and have decimated small animal populations where they have been allowed to roam free. They are responsible for greatly diminished songbird populations in urban and suburban areas, including parks (Chapter 6).

Despite all that is known about introduced species, the problem is increasing, not decreasing. This increase is mainly due to expanding world trade and travel. In the 1980s, for example, the zebra mussel and the closely related quagga mussel were introduced into the Great Lakes with the discharge of ballast water from European ships (Fig. 4–25a). Within 10 years, the mussels had colonized the Great Lakes river basins. They have continued to spread and are found in California, Colorado, and Virginia as well as other places. Zebra mussels have been found to reach densities of 700,000 individuals per square meter. They have caused untold ecological and commercial damage by displacing native mussel species and clogging water-intake pipes.

The problems of introduced species may also be exported as well as imported. In 1982, several species of jellyfishlike animals known as ctenophores were similarly transported from the East Coast of the United States to the Black Sea and the Sea of Azov in eastern Europe. The ctenophores have cost Black Sea fisheries an estimated $250 million and have completely shut down fisheries in the Sea of Azov because they kill larval fish directly and deprive larger fish of food (Fig. 4–25b).

Lessons.   There are two ecological lessons to be learned from the introduction of undesirable species. First, the regulation of populations is a matter of complex interactions among the members of the biotic community. Second, the relationships are specific to the organisms in each particular ecosystem. Therefore, when a species is transported over a physical barrier from one ecosystem to another, it is unlikely to fit into the framework of relationships in the new biotic community. In most cases, it finds the environmental resistance of the new system too severe and dies out. No harm is then done. In some instances, the introduced species simply joins the native flora or fauna and does no measurable harm. The new species has then become *naturalized*. Most of our food crop and ornamental plants have turned out to be noninvasive and will grow only under carefully controlled conditions.

(a)

(b)

Figure 4–25 Introduced species. (a) Zebra mussels introduced from Europe are now spread throughout the Great Lakes and the Mississippi Valley. The only benefit from these aliens is that they can dramatically increase water clarity where they are numerous, but that is far outweighed by the harm they do. (b) Ctenophores, originating along the Atlantic Coast in the southern United States and introduced into Europe, have destroyed fishing in the Black Sea and the Sea of Azov.

In the worst cases, the transported species becomes *invasive*. It finds physical conditions and a food supply that are hospitable, together with an insufficient number of natural enemies to stop its population growth. Then its population explodes, and it drives out native species by out-competing them for space, food, or other resources (if not by predation). Fortunately, only a small percentage of success-fully transplanted species become invasive.

Remedies. Invasive species are such a concern for the health of ecosystems that there are state and federal agencies working to slow their entry and spread. The National Invasive Species Information Center, for example, acts as a clearinghouse of information on invasive species in the United States. There are also international efforts to slow species invasions, including checkpoints at the entry into countries (such as customs checks in airports). Concern about pests is the reason people are asked in customs whether they have been to a farm or are carrying any fruits or vegetables when they enter a country. The United States has a number of regulations about invasive species, including the Nonindigenous Aquatic Nuisance Prevention and Control Act of 1990 and the National Invasive Species Act of 1996. To prevent the import of small aquatic species such as zebra mussels that can arrive in ballast water, scientists are testing the efficacy of ultraviolet light, filters, and chemicals at killing larvae. Ships entering the United States with ballast water in their tanks have to comply with a number of rules, such as exchanging their ballast water out at sea in the hopes of killing freshwater larvae before coming into local rivers.

Another solution to the takeover by an invasive species is to introduce a natural enemy. Indeed, this approach has been used in a number of cases, including in the case of the spotted knapweed (see Sound Science, "The Biological Detective: The Case of Spotted Knapweed"). Other approaches

Sound Science

The Biological Detective: The Case of Spotted Knapweed

Millions of acres in the western United States are infested with a plant that produces more seeds than others and is barely edible to grazers such as cattle, whose growth is harmed by the plant's toxins. Spotted knapweed (*Centaurea bieber-* *steinii*), a member of the Aster family, was acci-dentally introduced into the United States as seeds from Europe in the late 1800s. It thrives here not only because its seeds can remain in the ground for five years, sprouting whenever there is an opening (high biotic potential), but also because it has few natural predators (low envi-ronmental resistance).

Many herbivores eat specific plants and have adaptations that allow them to get around the

(Continued)

(Continued)

plant's defenses, but when plants are moved from one place to another, their accompanying herbivores are sometimes not moved. Suddenly without the limit of environmental resistance, the plant outcompetes other plants for light and water.

For years, scientists searched for a control for knapweed, and in the early 1970s, they thought they had the ideal solution. They introduced a gallfly that specializes in eating spotted knapweed, hoping to avoid the problems that occur when generalist herbivores are moved from one part of the globe to another and eat numerous species. It should have worked. The gallfly lays its eggs in the knapweed flowers. Larvae hatch and trigger the plant tissues to grow a gall, which protects the insect. The larvae overwinter in the galls and emerge in the spring. Because spotted knapweed plants have to put a great deal of energy into making the woody tissue in the galls, they make fewer seeds. This should have made the population decline, but the decline was not as high as expected. Why? Recently, scientists have discovered part of the answer to the mystery.

In a study reported in 2008 in the journal *Ecological Applications*, researcher Dean Pearson and his colleagues at the U.S. Forest Service's Rocky Mountain Research Station found that part of the answer lies in the relationship between the gallfly and local deer mice.* Gallfly larvae are eaten by native deer mice, which also eat seeds and other insects. Because there are so many spotted knapweed plants, there are plenty of gallflies. Because there are plenty of gallflies, there are plenty of deer mice. But the deer mice also eat the seeds of other native plants. Since the native plants make a lot fewer seeds than the introduced spotted knapweed, the knapweed seeds remaining can still outgrow the native plants. This unexpectedly complex set of relationships means that even a biological control agent chosen because of its specificity to one host may have unintended consequences.

───────────

*D. E. Pearson and R. M. Callaway, "Weed Biocontrol Insects Reduce Native Plant Recruitment Through Second-Order Apparent Competition," *Ecological Applications* 18 (2008): 1489–1500.

Spotted knapweed.

are discussed in Chapter 13 in connection with the biological control of pests. Unfortunately, control with a natural enemy is more easily said than done. Recall that control is a consequence of many factors of environmental resistance, which often include several natural enemies as well as *all* the abiotic factors. The natural enemy that is introduced could even end up as an invasive pest itself. In short, to prevent doing more harm than good, a great deal of research needs to be done before a natural enemy is introduced.

Most invasive species are not problems in their native lands. Their impact is different in a new setting, however, because ecosystems on different continents or remote islands have been isolated by physical barriers for millions of years. Consequently, the species within each ecosystem have developed adaptations to other species *within their own ecosystem*, and these do not prepare them to interact with species that have developed in other ecosystems.

Revisiting the Themes

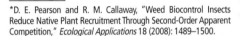

Sound Science

The golden frog of Panama is in decline. The frog's population is controlled by a disease-causing fungus as well as by human impacts like habitat loss and degradation. The chytrid fungus itself is not well controlled because it is an introduced exotic species, occurring in places where organisms have not evolved adaptations to it. This example illustrates some of the ideas of the chapter, such as the limits to population growth, the relationships between parasites (or pathogens) and hosts, the problem of introduced species, and the role humans can play in managing ecosystems.

The science of population and community ecology allows people to better understand why species live

where they do and what drives their abundances. Populations can be controlled by abiotic factors such as precipitation or by a complexity of biotic factors such as competition, predation, mutualism, and commensalism—all important relationships. Species typically have one of two reproductive strategies in their life history. Opportunistic or *r*-selected species make many small offspring, reproduce early, and do not care for their young, most of which die. In contrast, equilibrial or *K*-selected species have fewer offspring at older ages and care for the young. These differences correlate to differences in the ways in which populations grow: *r*-selected species often overshoot their carrying capacity and crash, while *K*-selected species are more likely to rise slowly and then fluctuate around the carrying capacity. Either way, all species are limited. When a few individuals start a new population, the growth may initially be exponential, but the population cannot maintain continual growth. Finally, sound science shows us that biological systems change because they are constantly varying genetically and because the environment changes over space and time to exert different selection pressures.

Sustainability

To follow the best management practices for sustainability can sometimes be difficult. This was shown with numerous examples of human actions resulting in negative unexpected consequences, particularly with the introduction of rabbits into Australia and with the attempts to slow the spread of spotted knapweed. However, in this chapter, we see that best management means paying attention to the relationships in a community that already exist and to constraints such as carrying capacity. The movement of species from a place where they had coevolved in a community to other places lacking those evolutionary relationships was highlighted as a particular problem that often results in either the explosion of an unwanted pest or the collapse of some other species. We have had to develop agencies and policies to minimize problems from the transport of species. We also saw, in the case of the golden frog, that drastic action in the short term may be required to protect species until longer-term solutions can be found.

Stewardship

When questions arise about what's best for populations and communities, the answers indicate that what seems to benefit one species (importing a plant or animal, for example) may not benefit another. A species decline can result from activities that disrupt its environment on scales of time or space that it cannot adjust to. This is definitely the case with the golden frog, which has had to adapt to too many environmental changes in a short time span. The destruction of its forest habitat, for example, has benefited human enterprises, but has also harmed other populations. Human impacts are not always to blame, however. Disruptive changes can occur simply through natural shifts in an ecosystem (such as a forest fire). One might ask, "If change is natural, why are we concerned about species loss?" This will be discussed further in Chapter 6. Throughout much of this book, we will ask how we can continue to support the needs of people for agriculture and other activities, while still maintaining healthy community structures in our ecosystems.

This chapter also shows that interactions such as predation and competition can be positive parts of ecosystems even when we might have a gut feeling that they are negative. They act as checks and balances on populations. For example, we saw that natural predators and competition keep species in check in the places where they have evolved in a community. Likewise, it might seem like territoriality would be a bad thing, but these behaviors actually lower competition and ensure that some members of the species survive.

REVIEW QUESTIONS

1. What are three basic population growth curves? When might you see them in nature?

2. Define *biotic potential* and *environmental resistance*, and give factors of each. Which generally remains constant, and which controls a population's size?

3. Differentiate between the terms *critical number* and *carrying capacity*. What is density dependence?

4. Explain the difference between *r*- and *K*-strategists. Where do these terms come from, and what are the characteristics of each broad category?

5. Describe the predator-prey relationship between the moose and wolves of Isle Royale. What other factors influence these two populations?

6. Distinguish between intraspecific and interspecific competition. How do they affect species as a form of environmental resistance?

7. What is meant by territoriality, and how does it limit the effects of competition in nature?

8. What problems arise when a species is introduced from a foreign ecosystem? Why do these problems occur?

9. What are selective pressures, and how do they relate to natural selection?

10. Describe several types of adaptations a species might have that would allow it to survive in a dry environment, escape predation, or lower competition with another species.

11. What factors determine whether a species will adapt to a change or whether the change will render it extinct?

12. How may evolution lead to the development of new species (speciation)?

13. What is plate tectonics, and how does this theory explain past movements of the continents? How have past tectonic movements affected the present-day distribution of plant and animal species on Earth's surface?

THINKING ENVIRONMENTALLY

1. Describe, in terms of biotic potential and environmental resistance, how the human population is affecting natural ecosystems.

2. Explain the key differences between *r*- and *K*-strategists, their life histories, and their reactions to environmental changes. Where do the *r* and *K* come from in population equations? What do they stand for?

3. Consider the various kinds of relationships humans have with other species, both natural and domestic. Give examples of relationships that (a) benefit humans, but harm other species; (b) benefit both humans and other species; and (c) benefit other species, but harm humans. Give examples in which the relationship may be changing—for instance, from exploitation to protection. Discuss the ethical issues involved in changing relationships.

4. Consider the village weaverbird (see Sound Science, "The Village Weaverbird: Marvel or Menace?") and describe how its various traits support its survival and reproduction. How could these traits have been affected by natural selection?

5. Describe how a human action such as removing a top predator or adding a species can have an impact on many species in an ecosystem.

MAKING A DIFFERENCE

1. If you can, plant native plants instead of alien ones in your garden. Otherwise, seeds from introduced plants could escape into the surrounding ecosystems. Native grasses, flowers, shrubs, and trees are more likely to attract native birds, butterflies, and other insects.

2. Visit a nearby national park or nature reserve. Talk to the rangers to find out about any threatened species and how they are being protected. Get involved by volunteering at your local nature center or wildlife refuge. Go wildlife or bird watching in nearby parks. Wildlife-based recreation creates millions of jobs and supports local businesses.

3. If you are a hunter or fisherman, use non-lead shot and fishing gear so birds eating spent shot or lost sinkers will not suffer lead poisoning. Follow the regulations regarding seasons and catch limits and report poachers to your local wildlife agency.

4. When choosing outdoor recreational activities, consider cross-country skiing and canoeing instead of snowmobiling or motor boating. These activities are quieter, and they increase your chances of seeing wildlife. Loud noises, especially in winter, disturb animals when they need to rest and conserve energy.

Go to **www.masteringenvironmentalscience.com** for practice quizzes, Pearson eText, videos, current events, and more.

Yellowstone National Park. Windswept wildfires consumed 165,000 acres of forest on August 20, 1988. Firefighters could only watch in awe.

KEY TOPICS

1. **Characteristics of Ecosystems:** Matter and energy flow through ecosystems by moving from one broad feeding (trophic) level to another.

2. **The Flow of Energy Through the Food Web:** Trophic (feeding) pyramids reflect the relative numbers and biomasses of different levels in a food chain.

3. **From Ecosystems to Global Biomes:** Ecosystems are connected into broad regions called biomes, and certain factors determine their placement on the globe.

4. **Ecosystem Reponses to Disturbance:** Ecological disturbances, such as a fire, are normal in ecosystems and can even be beneficial.

5. **Human Values and Ecosystem Sustainability:** Humans impact ecosystem services both positively and negatively; therefore, we need to manage ecosystems to protect their components from overuse.

I N MAY AND JUNE OF 1988, following an extended drought, lightning started a large number of fires in Yellowstone National Park, the crown jewel of America's national park system. The fires burned slowly for a while, and then, in the words of Mike Stark of the *Billings Gazette*, "Hell rode in on the wind the morning of August 20, 1988." Gale-force winds fanned the many smaller fires into conflagrations that consumed 165,000 acres in one day. Prior to the 1988 fires, the National Park Service policy on fires was to allow them to burn their course unless they were near human habitations. This was a reversal of the Smokey the Bear policy of earlier years, during which all forest fires were to be extinguished. The Yellowstone fires touched off a great political controversy over Park Service policy and led to fruitless efforts to put the fires out. In spite of the largest firefighting effort in U.S. history, the fires were finally put out by a snowfall in September. Born of weather, the fires were put out by the same force.

RECOVERY. The fires burned 36% of the park area, leaving behind a patchy landscape, with heavily or lightly burned areas interspersed among untouched areas. Within two weeks, however, grasses and herbaceous vegetation began sprouting from the nutrient-rich ashes, and a year later, abundant vegetation covered the burned areas (Fig. 5–1a). Herbivores such as bison and elk fed on the

(a)

(b)

Figure 5–1 Recovery from fire. Many observers thought that Yellowstone National Park was ruined for generations by the devastating fires of 1988. (a) One year after the fires, ground vegetation was well established. (b) Thirteen years after the fires, lodgepole pines were replacing the herbaceous vegetation.

lush new growth. Thirteen years later, lodgepole pines carpeted much of the burned-over area (Fig. 5–1b). It is expected that, some 25 years after the fires, the diversity of plants and animals in the burned areas will have completely recovered; in the meantime, the total diversity of the park has been unaffected.

Ecologists have studied the 1988 Yellowstone fires as a natural experiment and found some surprises: Even though wood buildup in the forest does matter to fire intensity, it is actually the climate that primarily drives large fires like these. The fires' impact was also much more heterogeneous (variable) than expected. Unburned portions of forest were close to very severely burned portions and provided the seeds for recolonization in burned patches. This patchiness is the norm rather than the exception in fires. Researchers also found that such large fires may even have an impact on the evolution of species, possibly by altering the survival of certain genetic strains.

While it might be hard to imagine anything more devastating to a forest ecosystem than a roaring crown fire, as events at Yellowstone have shown, the total landscape recovers. Populations of herbivores and predators flourish in the aftermath, and biodiversity can even be enhanced by natural disturbances. The park, which one senator predicted would be "a blighted wasteland for generations to come,"[1] has demonstrated that fire is vital to the ecology of natural landscapes, particularly to these northern conifer forests, which naturally experience infrequent, but severe, fires.

In Chapters 3 and 4, the focus was on the components of ecosystems. We considered the environmental conditions and resources required by living things, the basic movement of energy

and nutrients that sustain living things, and how populations work and relate to populations of other species. Now, we will look at how basic processes such as photosynthesis and respiration in individuals play out in complex systems and how energy moves from one part of the community to another. We will see how concepts like ranges of tolerance and limiting factors, which we applied to individuals in Chapter 3, apply to large groups and control their distributions over large portions of Earth. We will also look at how ecosystems change over time as a result of disturbances. We'll see how vital ecosystems are to human well-being and look at efforts to assign value to the goods and services we derive from them. We will also see that ecosystem sustainability is being severely challenged by human use.

It is a popular belief that nature will be in balance if left alone. You will learn in this chapter, however, that an ecosystem is a dynamic system in which changes are constantly occurring. The notion of a balanced ecosystem, then, must be carefully defined. We will ask questions of balance, or equilibrium, in ecosystems: What does it mean for an ecosystem to be "in balance"? Are populations normally in a state of equilibrium? What happens when a major disturbance (like a fire) interrupts the "balance" in an ecosystem? Can ecosystems and the natural populations in them change over time, yet maintain the important processes that make them sustainable?

The objective in this chapter is to examine more of the basic mechanisms that underlie the sustainability of all ecosystems, including those heavily influenced by human use. We will look at the values, goods, and services of ecosystems and their importance to humans. The chapter concludes with some perspectives on the challenge of managing ecosystems and a summary of ecosystem sustainability issues.

[1]Todd Wilkinson. "When it comes to taking care of Yellowstone, Mother Nature knows best." *National Wildlife*, August 1998, http://findarticles.com/p/articles/mi_m1169/is_ 1998_August/ai_54012275/?tag=content;col1.

5.1 Characteristics of Ecosystems

Yellowstone National Park is historically and ecologically important. It was established in 1872 as the world's first national park, and it is one of the most geologically diverse areas in the world. Located over an extinct volcano, the Yellowstone region contains 10,000 geysers and hot springs, an extensive petrified forest, and other unique natural features. The 18-million-acre (7.3-million-hectare) Greater Yellowstone Ecosystem, much larger than the park itself, is a large forested area with the headwaters of three river systems, the Yellowstone, the Snake, and the Green, which run with trout and other fish. The area is used for mineral extraction, logging, and animal grazing as well as preservation of wildlife habitat. The park and the surrounding region provide habitat for moose, elk, grizzly, bison, and migratory birds such as the rare trumpeter swan. Because of these features, in the 1970s the United Nations recognized the park as a World Heritage Site and an International Biosphere Reserve.

We can look at the Yellowstone area to better understand the characteristics of ecosystems. The Greater Yellowstone Ecosystem, like all ecosystems, contains communities of interacting species and their abiotic factors. However, the species living in this ecosystem, or any other ecosystem, may function on very different scales, which makes it difficult to delineate a fixed boundary for ecosystems. For example, the Yellowstone ecosystem contains migratory birds, which live there only seasonally. Some species inhabiting the ecosystem may also have little to no interaction with each other, such as the hot springs bacteria in Yellowstone, which have little interaction with grizzly bears. Ecosystems that have a similar type of vegetation and similar climactic conditions are grouped into broader areas called **biomes**. For example, the Greater Yellowstone Ecosystem is part of the northern temperate forest biome. We will look at biomes in more depth in Section 5.4.

Just as the study of communities raises questions about relationships that cannot be answered by looking at species separately, the study of ecosystems allows us to understand concepts that could not be predicted by understanding community interactions alone. For example, community relationships such as predation and parasitism (discussed in Chapter 4) provide the mechanisms for the flow of energy, as well as the flow of carbon, nitrogen, and phosphorus, through the natural world (discussed in Chapter 3), but these community relationships may not tell us about *ecosystem properties*, such as how much photosynthesis occurs in all plants or how much water can be absorbed by all of the plants at once. In this chapter, we will pull together the concepts presented in the previous two chapters, abiotic forces (Chapter 3) and biotic interactions (Chapter 4), to look at characteristics of ecosystems—specifically, **trophic levels**, **productivity**, and **consumption**.

Trophic Levels, Food Chains, and Food Webs

Imagine a grassy meadow containing a variety of plants. As discussed in previous chapters, plants use energy from the Sun through the process of *photosynthesis* to produce complex chemical chains from more simple molecules such as carbon dioxide and water. These plants can then be eaten by their predators, such as a grasshopper, a mouse, or even an elk (remember that herbivory is a subset of predation). In turn, the grasshopper could be eaten by its predator, a bird; the mouse eaten by an owl; and the elk fed on by a parasite or eaten by a wolf. Energy in the form of chemical bonds and nutrients in the form of molecules in the bodies of the prey are then used by the next **trophic level** (feeding level)—the predators—for energy and other physiological activities. This breakdown of chemical bonds to release energy is called *cellular respiration*. Recall that plants need to respire, too.

You can imagine two things from this mental image. First, organisms can be linked into feeding chains—the grass, to the mouse, to the snake, to the eagle, for example. Such **food chains** describe where the energy and nutrients go as they move from one organism to another. We usually refer to this as energy moving "up" the food chain. Second, you can imagine that the energy and nutrients are not all passed up the chain from one level to the next. Because of the energy laws we discussed in Chapter 3, there is inefficiency every time one organism eats another. While it is interesting to trace these feeding pathways, it is important to recognize that food chains seldom exist as isolated entities. Mice feed on several kinds of plants, are preyed on by several kinds of animals, and so on. Some organisms, including humans, eat at more than one level in the food chain. Consequently, virtually all food chains are interconnected and form a complex *web* of feeding relationships—the **food web** (Fig. 5–2).

Rather than simply asking what predators are eating the snake and how they affect the snake population or what mice are eating and how food availability affects mice, we ask different types of questions in ecosystem ecology—those that focus on trophic levels in general in order to study bigger trends in the movement of energy and materials. Examples of these questions include the following: How much of the Sun's energy do the plants in the meadow trap? How much do they use, and how much is available to plant eaters? How many predators can the ecosystem support? These questions emphasize the broader aspects of the flow of energy and materials through ecosystems, not through individual species. In order to answer these questions, ecologists divide members in the food web into categories.

Trophic Categories

All organisms in the biosphere can be categorized as either *autotrophs* or *heterotrophs*, depending on whether they produce the organic compounds they need to survive and grow. Green plants, photosynthetic single-celled organisms, and chemosynthetic bacteria are **autotrophs** (*auto* = self; *troph* = feeding) because they produce their own organic material from inorganic constituents in their environment through the use of an external energy source. They are also referred to as producers. Organisms that must consume organic material to obtain energy are **heterotrophs** (*hetero* = other; *troph* = feeding). Heterotrophs may be divided into numerous categories, the two major ones being consumers (which eat living prey) and decomposers (which break down dead organic material and can include scavengers, detritus feeders, and *chemical decomposers*).

(a)

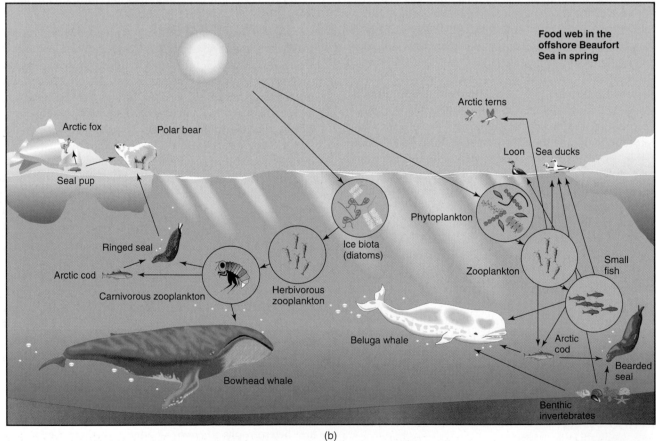

(b)

Figure 5–2 Food webs. (a) The arrows show herbivores feeding on plants and carnivores feeding on herbivores. Specific pathways, such as that from nuts to squirrels to foxes, are referred to as *food chains.* A food web is the collection of all food chains, which are invariably interconnected. (b) The marine food web in the offshore Beaufort Sea (north of Alaska and Canada) in spring.

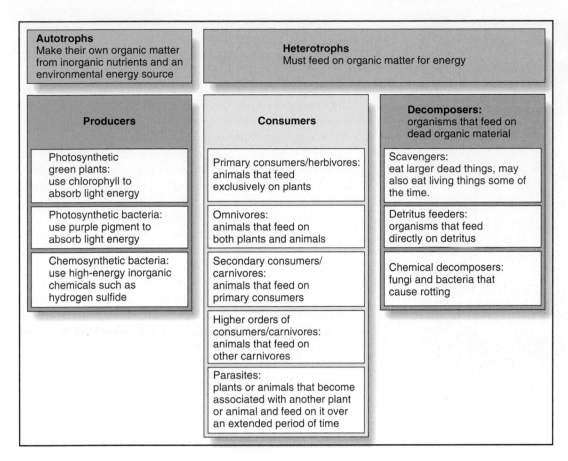

Together producers, consumers, and decomposers produce food, pass it along food chains, and return the starting materials to the abiotic parts of the environment (Fig. 5–3).

Producers. As discussed in Chapter 3, producers are organisms that capture energy from the Sun or from chemical reactions to convert carbon dioxide (CO_2) to organic matter. Most producers are green plants. Through the process of photosynthesis, these plants use light energy to convert CO_2 and water to organic compounds such as glucose (a sugar) and then release oxygen as a by-product. Plants use a variety of molecules to capture light energy in photosynthesis, but the most predominant of these is **chlorophyll**, a green pigment. Hence, plants that photosynthesize are easily identified by their green color. In some producers, additional red or brown photosynthetic pigments (in red and brown algae, for example) may mask the green. Producers range in diversity from microscopic photosynthetic bacteria and single-celled algae through medium-sized plants such as grass, daisies, and cacti to gigantic trees. Every major ecosystem, both aquatic and terrestrial, has its particular producers, which are actively engaged in photosynthesis.

Interestingly, there are bacteria, including some in the hot springs at Yellowstone National Park, that are able to use the energy in some inorganic chemicals to form organic matter from CO_2 and water. This process is called **chemosynthesis**, and these organisms are producers, too. Producers are absolutely essential to every ecosystem. The photosynthesis and growth of producers such as green plants constitute the **primary production** of organic matter, which sustains all other organisms in the ecosystem.

Consumers. All organisms in the ecosystem other than the producers feed on organic matter as their source of energy. These heterotrophs include not only all animals, but also fungi (mushrooms, molds, and similar organisms) and most bacteria. For ease of discussion, heterotrophs are often divided into *consumers*, which eat living prey, and *decomposers*, which break down dead organic material. Consumers encompass a wide variety of organisms, ranging in size from plankton to blue whales. Among consumers are such diverse groups as protozoans, worms, fish, shellfish, insects, reptiles, amphibians, birds, and mammals (including humans).

For the purpose of understanding ecosystem structure, consumers are divided into various subgroups according to their food source. Animals—as large as elephants or as small as mites—that feed directly on producers are called **primary consumers** or **herbivores** (*herb* = grass). Animals that feed on primary consumers are called **secondary consumers**. Thus, mice, which feed on vegetation, are primary consumers, whereas snakes are secondary consumers because they feed on mice (Fig. 5–4). There may also be third (tertiary), fourth (quaternary), or even higher levels of consumers. Many animals occupy more than one position on the consumer scale. For instance, humans are primary consumers when they eat vegetables, secondary consumers when they eat beef, and tertiary consumers when they eat fish that feed on smaller fish that feed on algae. Secondary and higher-order consumers are also called **carnivores** (*carni* = meat). Consumers that feed on both plants and animals are called **omnivores** (*omni* = all).

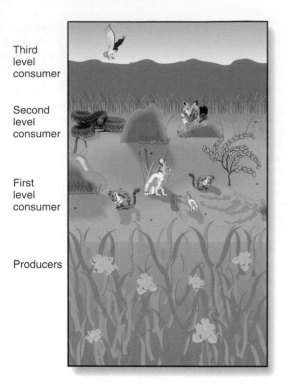

Figure 5–4 A grassland food chain. A grassland food chain, showing the producers (plants) and the first, second, and third-order consumers (mouse, snake, and hawk, respectively).

Third level consumer

Second level consumer

First level consumer

Producers

Decomposers. A very large proportion of the primary producer trophic level is not consumed in the grazing food web. As this material dies (leaves drop, grasses wither, and so on), it is joined by the fecal wastes and dead bodies from higher trophic levels and is referred to as *detritus*. This represents the starting point for a separate food web. In many cases, most of the energy in an ecosystem flows through the detritus food web (Fig. 5–5).

Detritus is composed largely of cellulose because it consists mostly of dead leaves, the woody parts of plants, and animal fecal wastes. Nevertheless, it is still organic and high in potential energy for those organisms that can digest it—namely, the decomposers. Scavengers, such as vultures, help break down large pieces of organic matter; detritus feeders, such as earthworms, eat partially decomposing organic matter; and chemical decomposers (various species of fungi and bacteria as well as a few other microbes) break down dead material on the molecular scale. Decomposers act as any other consumer does, using the organic matter as a source of both energy and nutrients. Some, such as termites, can digest woody material because they maintain decomposer microorganisms in their guts in a mutualistic, symbiotic relationship. The termite (a detritus feeder) provides a cozy home for the microbes (chemical decomposers) and takes in the cellulose, which the microbes digest for both their own and the termites' benefit (Fig. 5–6). The decomposers, breaking down detritus, serve as the primary consumers in this food chain. The detritus is similar to the "producer" trophic level, and the decomposers have their own predators that form higher trophic levels.

Most decomposers use oxygen for cell respiration, which breaks the detritus down into carbon dioxide, water, and mineral nutrients. Likewise, there is a release of waste heat, which you may observe as the "steaming" of a manure or compost pile on a cold day. The release of nutrients by decomposers is vitally important to the primary producers because it is the major source of nutrients in most ecosystems. However, some decomposers (certain bacteria and yeasts) meet their energy needs through the partial breakdown of glucose, which can occur in the absence of oxygen. This modified form of cell respiration, called **fermentation**, results in end products such as ethyl alcohol, methane gas, and acetic acid. In nature, **anaerobic**, or *oxygen-free*, environments commonly exist in the sediments of

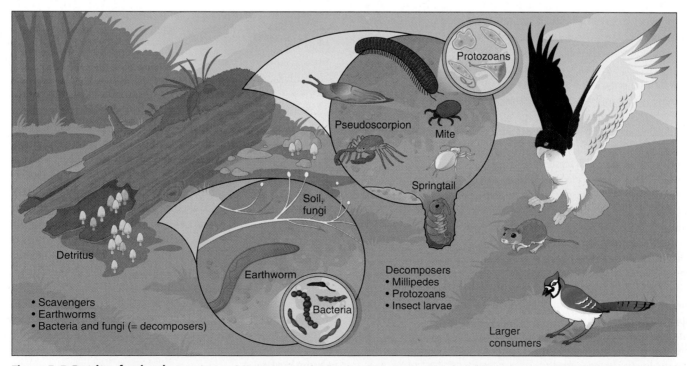

Figure 5–5 Detritus food web. Several types of decomposers work to break up large, small, and tiny bits of dead organic matter. These organisms in turn are eaten by other, higher-level consumers. In the detritus food web, the role of the producer is played by dead plant and animal matter.

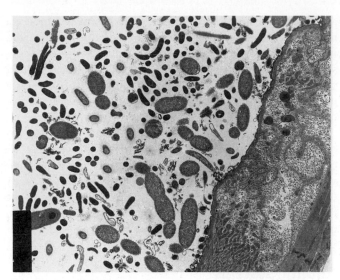

Figure 5–6 Termite gut. Termites can live on cellulose-based woody matter because their guts contain a consortium of symbiotic microbes able to digest cellulose. In this electron micrograph, the symbionts are on the left, and the termite intestinal wall is on the right.

lakes, marshes, and swamps and in the guts of animals, where oxygen does not penetrate readily. Methane gas is commonly produced in these locations. A number of large grazing animals, including cattle, maintain fermenting bacteria in their digestive systems in a mutualistic, symbiotic relationship similar to that just described for termites. As a result, both cattle and termites release methane.

Limits on Trophic Levels. How many trophic levels are there? Usually, there are no more than three or four in terrestri-

al ecosystems and sometimes five in marine systems (Fig. 5–2). This answer comes from straightforward observations. If you were to capture all of the organisms on each trophic level over a particular area or volume, dry them, and weigh them, you would find a pattern: The **biomass**, or total combined (net dry) weight, would be roughly 90% less at each higher trophic level for a terrestrial ecosystem (Fig. 5–7). For example, if you dried and weighed all of the grass and other producers in an acre of grassland, you might have 907 kilograms (1 ton or 2,000 lbs) per acre. If you could capture, dry, and weigh all of the herbivores (everything from grasshoppers to bison) in the same acre, you would find the biomass of herbivores would be about 90.7 kilograms (200 lbs) per acre. Following that pattern, the same acre would yield about 9.7 kilograms (20 lbs) of primary carnivores (animals living on grasshoppers and bison), if you captured, dried, and weighed them. At this rate, you can't go through very many trophic levels before the biomass approaches zero. If you graphed the different levels of producer and consumer mass, you would have what is commonly called a **biomass pyramid** (Fig. 5–7). We'll see the reason for this pattern as we look at where energy and nutrients go.

5.2 The Flow of Energy in Ecosystems

In most ecosystems, sunlight (or solar energy) is the initial source of energy absorbed by producers through the process of photosynthesis. (The only exceptions are rare ecosystems, such as those near the ocean floor, where the producers are chemosynthetic bacteria.) Primary production—the production of organic

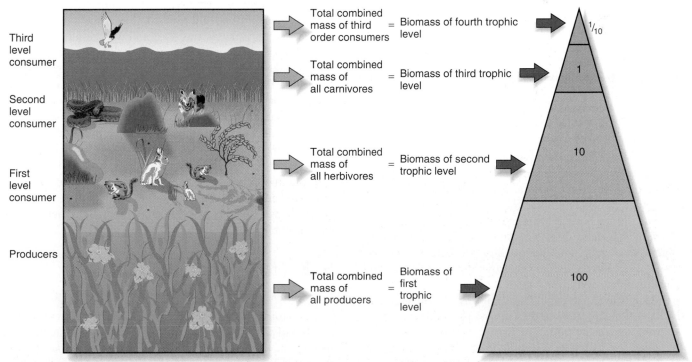

Figure 5–7 A biomass pyramid. A graphical representation of the biomass (the total combined mass of organisms) at successive trophic levels has the form of a pyramid, with about 10% of the energy contained in one trophic level being incorporated into the bodies of the next trophic level, giving each level a smaller biomass.

Segments of pyramid show relative biomass at each trophic level molecules from the Sun's energy through photosynthesis—captures only about 2%, at most, of incoming solar energy. Even though this seems like a small fraction, the resulting terrestrial net production—estimated at some 120 Gigatons (billion tons, 1.1 billion US tons) of organic matter per year—is enough to fuel all of the life in the biomes. In a given ecosystem, the actual biomass of primary producers at any given time is referred to as the *standing-crop biomass*. Both biomass and primary production vary greatly in different ecosystems. For example, a forested ecosystem maintains a very large biomass compared with a tropical grassland because there is so much carbon captured in the wood of trees. However, the rate of primary production could be higher in the grassland; grazing animals might simply eat the produced organic matter. Consequently, standing biomass is not always a good measure of productivity.

The Fate of Food.   Whereas 60–90% of the food that consumers eat, digest, and absorb is oxidized (burned) for energy, the remaining 10–40%, which is converted to the body tissues of the consumer, is no less important. This is the fraction that enables the body to grow and to maintain and repair itself. A portion of what is ingested by consumers is not digested, but simply passes through the digestive system and out as fecal wastes. For consumers that eat plants, this waste is largely **cellulose**, the material of plant cell walls. It is often referred to as *fiber*, *bulk*, or *roughage*, and some of it is a necessary part of the diet. The intestines need to push some fiber through them so that they can stay clean and open. Waste products can also include compounds of nitrogen, phosphorus, and any other elements present, in addition to the usual carbon dioxide and water. These by-products are excreted in the urine (or as similar waste in other kinds of animals) and returned to the environment. This process was illustrated in Chapter 3 (Figure 3–12).

Energy Flow and Efficiency

As the biomass pyramid suggests, when energy flows from one trophic level to the next, only a small fraction is actually passed on. This is due to three things: (1) Much of the preceding trophic level is biomass that is not consumed by herbivores, (2) much of what is consumed is used as energy to fuel the heterotroph's cells and tissues, and (3) some of what is consumed is undigested and passes through the organism as waste (e.g., feces). Thus, there is a tremendous inefficiency at each trophic level, and energy is lost to the food web. In an ecosystem, therefore, only the portion of food that becomes the body tissue of the consumer can become food for the next organism in the food chain. Incorporating matter and energy from a lower trophic level into the body of a consumer is often referred to as *secondary* production, and like primary production, it can be expressed as a rate (the amount of growth of the consumer, or consumer trophic level) over time.

The inefficiency of trophic levels means two things. First, individuals at higher levels in the biomass pyramid represent a greater amount of the Sun's energy for the same amount of body tissue. For example, a pound of hamburger requires a great deal more photosynthesis to create than a pound of soybeans. This may be counterintuitive because we just saw that there is less energy at each trophic level, but it requires a great deal more energy to produce a top-order consumer than

a mid-level consumer and likewise, more energy to produce a mid-level consumer than a producer of the same body weight. It usually also takes a longer time to produce a top-order consumer than a mid-level consumer or a producer. The top-order consumer also requires more water and other resources for a unit of body weight. In later chapters, as we look at food production in a crowded world, we will see why this matters.

Second, some materials are difficult to get rid of (such as chemicals that dissolve in fat) and remain throughout an organism's life. These chemicals remain in the bodies of predators at higher rates than in their prey and thus *biomagnify* or *bioaccumulate* (build up in the tissues) as you go up the food chain. This is the reason, as we will see in a later chapter, that some toxins build up in the tissues of higher-order consumers such as eagles and tuna.

Aquatic Systems

For simplicity, the focus of this chapter is on terrestrial ecosystems. Keep in mind, though, that exactly the same processes occur in aquatic ecosystems. As aquatic plants and algae absorb dissolved carbon dioxide and mineral nutrients from the water, they use photosynthesis to produce food and dissolved oxygen that sustain consumers and other heterotrophs. Likewise, aquatic heterotrophs return carbon dioxide and mineral nutrients to the aquatic environment.

There are two differences between aquatic and terrestrial systems in this regard, however. First, the transfer of energy is often more efficient in an aquatic system. There are also more cold-blooded animals, and less energy is required to run the bodies of cold-blooded creatures than warm-blooded creatures. It also takes less energy to support body weight in water than on land or in the air. Consequently, when less energy is used at each level, more energy is available to the next level, and food chains can be longer.

Second, aquatic systems do not always result in the same kind of biomass pyramid as the terrestrial systems. If you went out to a lake and captured all of the algae, all of the zooplankton eating the algae, all of the fish that live on zooplankton, and all of the fish that live on fish and if you dried and weighed them in their group, you might find a reverse biomass pyramid (Fig. 5–8). The reason for this upside-down pyramid is that the fish that live on fish are the oldest and largest and the algae have very short life spans and turn over rapidly. Thus, there might be more biomass in the fish, but they had to eat many times that amount of algae to form that tissue. The algae actually in the water at any one time can be small, but they have an extremely rapid rate of new production.

5.3 From Ecosystems to Global Biomes

Just as photosynthesis concepts and basic energy laws translate into broad patterns in food chain length in an ecosystem, broad patterns in ecosystems translate into a pattern we see all over the globe: There is a predictable set of organisms that live under particular conditions, such as deserts occurring in dry areas and rain forests occurring in very wet areas, for example. We can now use the concepts of optimums and limiting factors and the

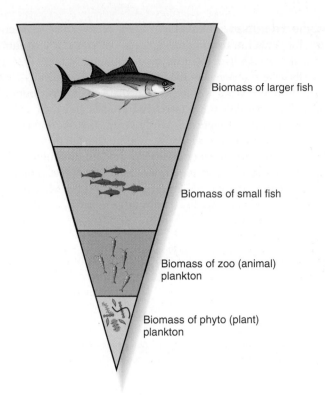

Figure 5–8 Reverse pyramid in aquatic systems. In aquatic systems such as this lake, there is a small amount of algae and a much greater biomass of large fish at any one moment. This is because the turnover is very high and life spans are so short at the lower trophic levels. The longer life spans of the upper trophic levels mean that the biomass of a fish, for example, includes all of its growth over a period of time.

concepts of differences in light and productivity to gain a better understanding of why different regions (or even localized areas) may have distinct biotic communities, creating an amazing variety of ecosystems, landscapes, and biomes.

A **biome** can be defined as a large geographical biotic community, controlled by climate. They are usually named after the dominant type of vegetation, such as deciduous forest or grassland. While the Greater Yellowstone region is an ecosystem (and possibly could be viewed as more than one), it is contained within a vast geographical region, or global biome, called the northern temperate forest. Like ecosystems, biomes have fuzzy edges. Aquatic areas such as the tidal zone and the open ocean can also be characterized by temperature and light and divided into categories. These are not typically called biomes, but function similarly.

The Role of Climate

The climate of a given region is a description of the average temperature and precipitation—the weather—that may be expected on each day throughout the entire year. Climates in different parts of the world vary widely. Equatorial regions are continuously warm, with high rainfall and no discernible seasons. Above and below the equator, temperatures become increasingly seasonal (characterized by warm or hot summers and cool or cold winters); the farther we go toward the poles, the longer and colder the winters become, until at the poles it is perpetually cold. Likewise, colder temperatures are found at higher elevations, so that there are even snow-capped mountains on or near the equator (Fig. 5–9).

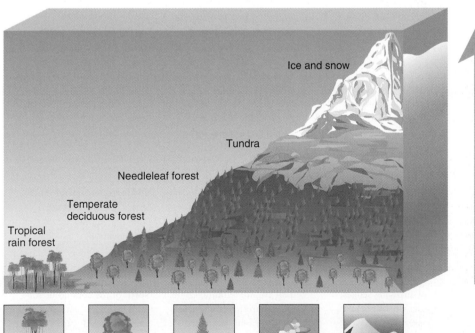

Figure 5–9 Effects of latitude and altitude. Decreasing temperatures, which result in the biome shifts depicted, occur with both increasing latitude (distance from the equator) and increasing altitude. (*Source*: Redrawn from *Geosystems*, 5th ed., by Robert W. Christopherson. Copyright © 2005 by Pearson Prentice Hall, Inc. Upper Saddle River, NJ 07458.)

Increasing altitude from sea level

Ice and snow

Tundra

Needleleaf forest

Temperate deciduous forest

Tropical rain forest

Tropical rain forest

Temperate deciduous forest

Needleleaf forest

Tundra

Ice and snow

Increasing latitude from the equator

Annual precipitation in an area also may vary greatly, from virtually zero to well over 250 centimeters (100 in.) per year. Precipitation may be evenly distributed throughout the year or concentrated in certain months, dividing the year into wet and dry seasons. A given climate will support only those species that find the temperature and precipitation levels within their ranges of tolerance. As indicated in Chapter 3 (Figure 3–5), population densities will be greatest where conditions are optimal and will decrease as any condition departs from the optimum. A species will be excluded from a region (or local areas) where any condition is beyond its limit of tolerance. How will this variation affect the biotic community?

Biome Examples.   Individual ranges of tolerance, particularly for temperature and precipitation, determine where dominant species can live. In turn, the distribution of these species describes the placement of the biomes. To illustrate, let us consider six major types of biomes and their global distribution. Table 5–1 describes these terrestrial biomes and their major characteristics, and Figure 5–10 shows the distribution of the biomes as they occur globally. Within the temperate zone (between 30° and 50° of latitude), the amount of rainfall is the key limiting factor. The **temperate deciduous forest biome** is found where annual precipitation is 75–200 centimeters (30–80 in.). Where rainfall tapers off or is highly seasonal

---

TABLE 5–1 Major Terrestrial Biomes

Biome	Climate and Soils	Dominant Vegetation	Dominant Animal Life	Geographic Distribution
Deserts	Very dry; hot days, cold nights; rainfall less than 10 in. (25 cm)/yr; soils thin and porous	Widely scattered thorny bushes and shrubs, cacti	Rodents, lizards, snakes, numerous insects, owls, hawks, small birds	North and southwest Africa, parts of the Middle East and Asia, southwestern U.S., northern Mexico
Grasslands and Prairies	Seasonal rainfall, 10 to 60 in. (25–152 cm)/yr; fires frequent; soils rich and often deep	Grass species, both tall and short; bushes and woodlands in some areas	Large grazing mammals (bison, wild horses; kangaroos); prairie dogs, coyotes, lions; termites important	Central North America, central Asia, subequatorial Africa and South America, southern India, northern Australia
Tropical Rain Forests	Nonseasonal; annual average temperature 28°C; rainfall frequent, heavy; average over 95 in. (240 cm)/yr; soils thin, poor in nutrients	High diversity of broad-leafed evergreen trees, dense canopy, vines	Enormous biodiversity; exotic, colorful insects, amphibians, birds, snakes; monkeys, tigers, jaguars	Northern South America, Central America, western central Africa, islands in Indian and Pacific oceans, southeast Asia
Temperate Forests	Seasonal; temperature below freezing in winter; summers warm, humid; rainfall from 30–80 in. (76–203 cm)/yr; soils well developed	Broad leafed deciduous trees, some conifers; shrubby undergrowth, ferns, mosses	Squirrels, raccoons, opossums, skunks, deer, foxes, black bears, snakes, amphibians, many soil organisms, birds	Western and central Europe, eastern Asia, eastern North America
Coniferous Forests	Seasonal; winters long, cold; precipitation light in winter, heavier in summer; soils acidic, with leaf litter	Coniferous trees (spruce, fir, pine, hemlock), some deciduous trees (birch, maple)	Large herbivores (mule deer, moose, elk); mice, hares, squirrels; lynx, bears, foxes; nesting area for migrating birds	Northern portions of North America, Europe, Asia, extending southward at high elevations
Tundra	Bitter cold, except for short growing season with long days and moderate temperatures; precipitation low, soils thin—and below, permafrost	Low-growing sedges, dwarf shrubs, lichens, mosses, and grasses	Year round: lemmings, arctic hares, arctic foxes, lynx, caribou; summers: abundant insects, many migrant shorebirds.	North of the coniferous forest in northern hemisphere, extending southward at elevations above the coniferous forest

Note: See http://earthobservatory.nasa.gov/Laboratory/Biome/ for biome photos and climatic data.

Figure 5–10 World distribution of the major terrestrial biomes. Photos and climate data for these biomes are maintained by the National Aeronautics and Space Administration's Earth Observatory.

Legend:
- Equatorial and tropical rain forest
- Tropical seasonal forest and scrub
- Tropical savanna (grassland)
- Temperate deciduous forest
- Conifer forest
- Temperate rain forest
- Mediterranean shrubland (chaparral)
- Tall grass prairie
- Short grass prairie
- Warm desert and semidesert
- Cold desert and semidesert
- Arctic tundra
- Alpine tundra
- Ice cap

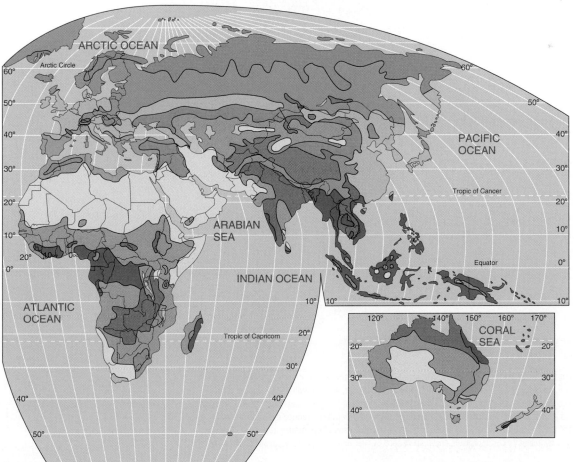

(25–150 cm or 10–60 in. per year), the **grassland** and **prairie biomes** are found, and regions receiving an average of less than 25 centimeters (10 in.) per year are occupied by a **desert biome**.

The effect of temperature, the other dominant parameter of climate, is largely superimposed on that of rainfall. That is, 75 centimeters (30 in.) or more of rainfall per year will usually support a forest, but temperature will determine the *kind* of forest. For example, broad-leaved evergreen species, which are extremely vigorous and fast growing, but cannot tolerate freezing temperatures, predominate in the **tropical rain forest**. By dropping their leaves and becoming dormant each autumn, deciduous trees are well adapted to freezing temperatures. Therefore, wherever rainfall is sufficient, deciduous forests predominate in temperate latitudes. Most deciduous trees, however, cannot tolerate the extremely harsh winters and short summers that occur at higher latitudes and higher elevations. These northern regions and high elevations are occupied

by the **coniferous forest biome**, because conifers are better adapted to those conditions.

Temperature by itself limits forests only when it becomes low enough to cause **permafrost** (permanently frozen subsoil). Permafrost prevents the growth of trees because roots cannot penetrate deeply enough to provide adequate support. However, a number of grasses, clovers, and other small flowering plants can grow in the topsoil above permafrost. Consequently, where permafrost sets in, the coniferous forest biome gives way to the **tundra biome** (Table 5–1). At still colder temperatures, the tundra gives way to permanent snow and ice cover.

The same relationship of rainfall effects being primary and temperature effects secondary applies in deserts. Any region receiving less than about 25 centimeters (10 in.) of rain per year will be a desert, but the unique plant and animal species found in hot deserts are different from those found in cold deserts. A summary of the relationship between biomes and temperature and rainfall conditions is given in Figure 5–11. The average

Figure 5–11 Climate and major biomes. Moisture is generally the overriding factor determining the type of biome that may be supported in a region. Given adequate moisture, an area will likely support a forest. Temperature, however, determines the kind of forest. The situation is similar for grasslands and deserts. At cooler temperatures, there is a shift toward less precipitation because lower temperatures reduce evaporative water loss. Temperature becomes the overriding factor only when it is low enough to sustain permafrost. (*Source*: Redrawn from *Geosystems*, 5th ed., by Robert W. Christopherson. Copyright © 2005 by Pearson Prentice Hall, Inc. Upper Saddle River, NJ 07458.)

temperature for a region varies with both *latitude* and *altitude*, as shown in Figure 5–9.

Examples of Aquatic Systems. The biome category is reserved exclusively for terrestrial systems. However, there are major categories of aquatic and wetland ecosystems that are determined primarily by the depth, salinity, and permanence of water in them, which would be similar to biomes. Among these ecosystems are lakes, marshes, streams, rivers, estuaries, bays, and ocean systems. As units of study, these aquatic systems may be viewed as ecosystems, as parts of landscapes, or as major biome-like features such as seas or oceans. Table 5–2 lists the six major aquatic systems and their primary characteristics.

TABLE 5–2 Major Aquatic Systems

Aquatic Systems	Major Environmental Parameters	Dominant Vegetation	Dominant Animal Life	Distribution
Lakes and Ponds (freshwater)	Bodies of standing water; low concentration of dissolved solids; seasonal vertical stratification of water	Rooted and floating plants, phytoplankton	Zooplankton, fish, insect larvae, turtles, ducks, geese, swans, wading birds	Physical depressions in the landscape where precipitation and groundwater accumulate
Streams and Rivers (freshwater)	Flowing water; low level of dissolved solids; high level of dissolved oxygen, often turbid from runoff	Attached algae, rooted plants	Insect larvae, fish, amphibians, otters, raccoons, wading birds, ducks, geese	Landscapes where precipitation and ground-water flow by gravity toward oceans or lakes
Inland Wetlands (freshwater)	Standing water, at times seasonally dry; thick organic sediments; high nutrients	Marshes (grasses, reeds); cattail swamps (water-tolerant trees); bogs (sphagnum moss, shrubs)	Amphibians, snakes, numerous invertebrates, wading birds, alligators, turtles	Shallow depressions, poorly drained, often occupy sites of lakes and ponds that have filled in
Estuaries (mixed)	Variable salinity; tides create two-way currents, often rich in nutrients, turbid	Phytoplankton, rooted grasses (salt-marsh grass), mangrove swamps in tropics with salt-tolerant trees, shrubs	Zooplankton, rich shellfish, worms, crustaceans, fish, wading birds, sandpipers, ducks, geese	Coastal regions where rivers meet the ocean; may form bays behind sandy barrier islands
Coastal Ocean (saltwater)	Tidal currents promote mixing; nutrients high	Phytoplankton, large benthic algae, turtle grass, symbiotic algae in corals	Zooplankton, rich bottom fauna of worms, shellfish, crustaceans; coral colonies, jellyfish, fish, turtles, gulls, terns, sea lions, seals, dolphins, penguins, whales	From coastline outward over continental shelf; coral reefs abundant in tropics
Open Ocean (saltwater)	Great depths (to 11,000 m (36,100 ft)); all but upper 200 m (656 ft) dark and cold; poor in nutrients except in upwelling regions	Phytoplankton, sargassum weed	Zooplankton and fish adapted to different depths; seabirds, whales, tuna, sharks, squid, flying fish	Covering 70% of Earth, from edge of continental shelf outward

Figure 5–12 Microclimates. Abiotic factors such as terrain, wind, and type of soil create different microclimates by influencing temperature and moisture in localized areas.

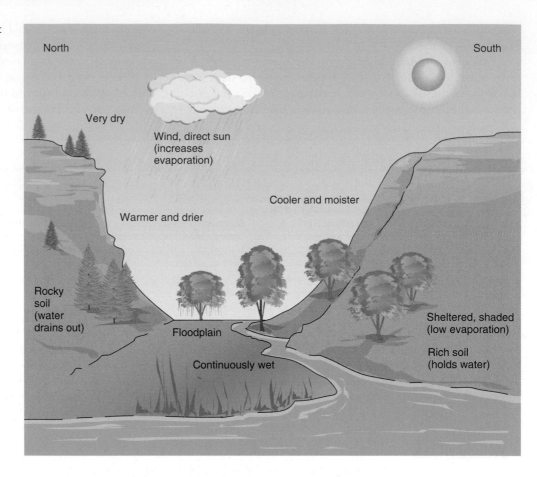

Microclimate and Other Abiotic Factors

A specific site may have temperature and moisture conditions that are significantly different from the overall climate of the region in which it is located. For example, a south-facing slope, which receives more direct sunlight in the northern hemisphere, will be relatively warmer and hence also drier than a north-facing slope (Fig. 5–12). Similarly, the temperature range in a sheltered ravine will be narrower than that in a more exposed location, and so on. The conditions found in a specific localized area are referred to as the **microclimate** of that location. In the same way that different climates determine the major biome of a region, different microclimates result in variations of ecosystems within a biome. For example, while the Greater Yellowstone region may be in the northern temperate forest biome, it includes areas of grassland and, above the tree line on mountains, areas of permafrost. This pattern of changing habitat with elevation means that the tops of mountains function ecologically as if they were more polar than their actual latitude (Fig. 5–9).

Soil type and topography also contribute to the diversity found in a biome because these two factors affect the availability of moisture. In the eastern United States, for example, oaks and hickories generally predominate on rocky, sandy soils and on hilltops, which retain little moisture, whereas beeches and maples are found on richer soils, which hold more moisture,

and red maples and cedars inhabit low, swampy areas. In the transitional region between desert and grassland, with 25–50 centimeters (10–20 in.) of rainfall per year, a soil capable of holding water will support grass, but a sandy soil with little ability to hold water will support only desert species.

Biome Productivity. All biomes have varying levels of primary productivity. Those that have high productivity may help support organisms from other biomes. For example, many sea birds migrate through salt marshes and live off the vast productivity there before traveling to other places. Highly productive biomes, like rain forests, also remove carbon dioxide from the atmosphere and trap it, something we will discuss more in Chapter 18. Biomes with high productivity aren't better than others though. Low-productivity biomes can have other roles, such as habitat for rare species.

Why are some biomes or types of ecosystems more productive than others? Figure 5–13 presents (a) the average annual net primary productivity, (b) the percentage of different ecosystems over Earth's surface, and (c) the percentage of global net primary productivity attributed to 19 of the most important ecosystems. Some key relationships between abiotic factors and specific ecosystems can be seen in the data. Tropical rain forests both are highly productive and contribute considerably to global productivity; they cover a large area of the land and are characterized by ideal climatic

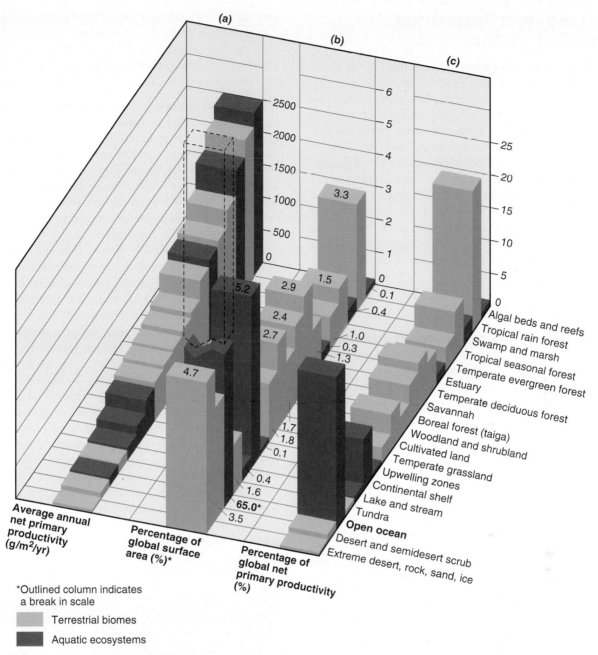

Figure 5–13 Productivity of different ecosystems. (a) The annual net primary productivity of different ecosystems, (b) the percentage of different ecosystems over Earth's surface area, and (c) the percentage of global net primary productivity. The outlined column for open ocean in (b) shows that the open ocean is such a high percentage of Earth's surface that it is off the scale relative to all of the other ecosystems.

conditions for photosynthesis—warm temperatures and abundant rainfall. The open oceans cover 65% of Earth's surface, so they account for a large portion of global productivity, yet their actual rate of production is low enough that they are veritable biological deserts. Although light, temperature, and water are abundant, primary production in the oceans is limited by the scarcity of nutrients—a good lesson in the significance of limiting factors. (See Sustainability, "Should We Seed the Oceans?" in Chapter 3.) The seasonal effects of differences in latitude can also be seen by comparing productivity in tropical, temperate, and boreal (coniferous) forests.

The biosphere consists of a great variety of environments, both aquatic and terrestrial. In each environment, plant, animal, and microbial species are adapted to all the abiotic factors. In addition, they are adapted to each other, in various feeding and nonfeeding relationships. Each environment supports a group of organisms interacting with each other and with the environment in a way that perpetuates or sustains the entire group. Every ecosystem is tied to others through species that migrate from one system to another and through exchanges of air, water, and minerals common to the whole planet.

5.4 Ecosystem Responses to Disturbance

People often picture ecosystems at equilibrium, as stable environments in which species interact constantly in well-balanced predator-prey and competitive relationships. This idea has led to the popular idea of "the balance of nature": Natural systems maintain a delicate balance over time that lends them great stability. A more recent understanding of ecosystems, however, shows that they operate in a dynamic, changing way. Species' abundances and ecosystem characteristics are not unchanging. Rather, the landscape comprises a shifting mosaic of patches, much like the burned and unburned areas of Yellowstone after a fire. The fires in Yellowstone give us a great example of how ecosystems respond to a **disturbance**—a significant change that kills or displaces many community members.

Ecological Succession

After the Yellowstone fires, new plants grew from the soil, new trees sprouted, and eventually, the original forest stands were replaced by new and different stands. First, however, these communities went through a series of intermediate stages. This phenomenon of transition from one biotic community to another is called **ecological succession**. Ecological succession occurs because the physical environment may be gradually modified by the growth of the biotic community itself, such that the area becomes more favorable to another group of species and less favorable to the current occupants. *Pioneer* species, the first to colonize a newly opened area in the aftermath of a fire, flood, volcanic lava flow, or other large disturbance, start the process. As these grow, they create conditions that are favorable to longer-lived colonizers, and succession is driven forward by the changing conditions that pave the way for other species. Driving succession forward by improving conditions for subsequent species is known as **facilitation**.

The succession of species does not go on indefinitely. A stage of development is eventually reached in which there appears to be a dynamic balance between all of the species and the physical environment. In the past, this final state was called a **climax ecosystem** because the assemblage of species continues on in space and time. The term *climax ecosystem* is still used, but many ecologists are moving away from its use because they believe it implies more stability than actually occurs in nature. In the real world, even climax communities experience change. All balances are relative to the current biotic community and the *existing* climatic conditions. Therefore, if climatic conditions change, if new species are introduced, or if old ones are removed, even climax ecosystems can change. Also, within the area of a climax ecosystem, there may be patches of disturbance that open space for new growth. Some parts of the ecosystem may be in an earlier stage of succession; sometimes there are several "final" stages, with adjoining ecosystems in the same environment at different stages.

Some biomes, such as prairies, are dependent on fires and would undergo succession to some other stage if there weren't periodic fires. We call these *fire-adapted* ecosystems and will discuss them later in the chapter. In this case, succession occurs until the successional "clock" is reset by a fire. Three classic examples of succession are presented next: primary and secondary succession on land and succession in aquatic ecosystems.

Primary Succession. If an area lacks plants and soil, such as might occur after volcanic eruption or the recession of a glacier, the process of initial invasion and progression from one biotic community to the next is called *primary succession*. An example is the gradual invasion of a bare rock or gravel surface by what eventually becomes a climax forest ecosystem. Bare rock is an inhospitable environment. It has few places for seeds to lodge and germinate, and if they do, the seedlings are killed by lack of water or by exposure to wind and sun on the rock surface. However, certain species of moss are able to exploit this environment. Their tiny spores (specialized cells that function reproductively) can lodge and germinate in minute cracks, and moss can withstand severe drying simply by becoming dormant. With each bit of moisture, moss grows and gradually forms a mat that acts as a sieve, catching and holding soil particles as they are broken from the rock or as they blow or wash by. Thus, a layer of soil gradually accumulates, held in place by the moss (Fig. 5–14).

The mat of moss and soil provides a suitable place for seeds of larger plants to lodge, and the greater amount of water held by the mat supports their germination and growth. The larger plants in turn collect and build additional soil, and eventually, there is enough soil to support shrubs and trees. In the process, the fallen leaves and other litter from the larger plants smother and eliminate the moss and most of the smaller plants that initiated the process. Thus, there

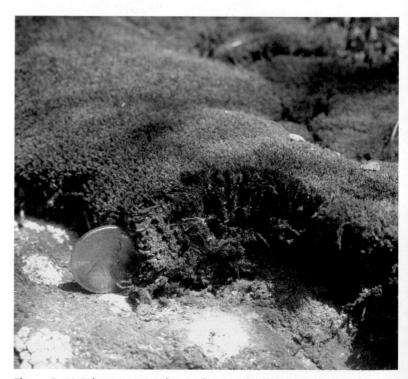

Figure 5–14 Primary succession on bare rock. Moss invades bare rock and acts as a collector, accumulating a layer of soil sufficient for additional plants to become established.

is a gradual succession from moss through small plants and, finally, to trees. The nature of the climax ecosystem differs according to the prevailing abiotic factors of the region, yielding the biomes typical of different climatic regions.

Secondary Succession. When an area has been cleared by fire, humans, or floods and then left alone, plants and animals from the surrounding ecosystem may gradually reinvade the area—not at once, but through a series of distinct stages called *secondary succession.* The major difference between primary and secondary succession is that secondary succession starts with preexisting soil. Thus, the early, prolonged stages of soil building are bypassed. Hence, the process of reinvasion begins with different species. The steps leading from abandoned agricultural fields in the eastern United States back to deciduous forests provide a classic example of secondary succession (Fig. 5–15a).

For example, in an abandoned agricultural field, crabgrass predominates among the initial invaders. Crabgrass is particularly well adapted to invading bare soil. Its seeds germinate in the spring, and it grows and spreads rapidly by means of runners; moreover, it is exceptionally resistant to drought. Despite its vigor on bare soil, crabgrass is easily shaded out by taller plants. Consequently, taller weeds and grasses, which take a year or more to develop, eventually

take over from the crabgrass. Next, young pine trees, which are well adapted to thrive in the direct sunlight and heat of open fields, gradually move in and shade out the smaller, sun-loving weeds and grasses, eventually forming a pine forest. But pine trees also shade out their own seedlings, which need bright sun to grow. Thus, the seedlings of deciduous trees, not pines, develop in the cool shade beneath the pine trees (Fig. 5–15b). Moreover, the pine trees have a limited life span and are eventually replaced by oaks, hickories, beeches, maples, and other species of hardwoods. The seedlings of these deciduous trees continue to flourish beneath the cover of their parents, providing a stable balance—the climax forest ecosystem.

Aquatic Succession. Natural succession also takes place in lakes or ponds. Succession occurs because soil particles inevitably erode from the land and settle out in ponds or lakes, gradually filling them. Aquatic vegetation produces detritus that also contributes to the filling process. As the buildup occurs, terrestrial species from the surrounding ecosystem can advance in the ecotone, and aquatic species must move farther out into the lake. In time, the shoreline advances toward the center of the lake until, finally, the lake disappears altogether (Fig. 5–16). Here the climax ecosystem may be a bog or a forest. Just as disturbances affect terrestrial ecosystems,

	Year	
Crab-grass	0–1	
Tall grass-herba-ceous plants	1–3	
Pines come in	3–10	
Pine forest	10–30	
Hard-woods come in	30–70	
Hard-wood forest climax	70+	

(a)

(b)

Figure 5–15 Secondary succession. (a) Reinvasion of an agricultural field by a forest ecosystem occurs in the stages shown. (b) Hardwoods (in this case, species of oak) are growing up underneath and displacing pines in earthern Maryland.

Figure 5–16 Aquatic succession. In this photograph, taken in Banff National Park in the Canadian Rockies, you can visualize the lake that used to exist in the low-level area. It is now filled with sediment and covered by scrub willow. Spruce and fir forest is gradually encroaching.

disturbances such as droughts—or even large floods and erosion—can send aquatic succession back to an earlier stage.

Disturbance and Resilience

In order for natural succession to occur, the spores and seeds of the various invading plants and the breeding populations of the various invading animals must already be present in the vicinity. Ecological succession is not a matter of new species developing or even old species adapting to new conditions; it is a matter of populations of existing species taking advantage of a new area as conditions become favorable. Where do these early-stage species come from if their usual fate is to be replaced by late-stage or climax species? They come from surrounding ecosystems in the early stages of succession. Similarly, the late-stage species are recruited from ecosystems in the later stages of succession. In any given landscape, therefore, all stages of succession are likely to be represented in the ecosystems. This is the case because disturbances constantly create gaps or patches in the landscape. When a variety of successional stages is present in a landscape, as opposed to one single climax stage, a greater diversity of species can be expected; in other words, biodiversity is enhanced by disturbance!

If certain species have been eliminated, natural succession will be blocked or modified. For example, when Iceland was colonized by Norsemen in the 11th and 12th centuries, the forests were cut for fuel. This process only accelerated with further European colonization in the 18th and 19th centuries. By 1850, not a tree was left standing, and Iceland had become a barren, tundralike habitat. Natural regeneration was prevented due to sheep grazing and the lack of a remaining source of seeds (Fig. 5–17). Tree seedlings are now being imported and planted in Iceland (4.5 million are planted annually) in the hope that natural succession may be reestablished.

Fire and Succession. Fire is an abiotic factor that has particular relevance to succession. It is a major form of disturbance common to terrestrial ecosystems, as we saw in the Yellowstone National Park example. About 80 years ago, forest managers interpreted the potential destructiveness of fire to mean that all fire was bad, whereupon they embarked on fire prevention programs that eliminated fires from many areas. Unexpectedly, fire prevention did not preserve all ecosystems in their existing state. In pine forests of the southeastern United States, for instance,

economically worthless scrub oaks and other broad-leaved species began to displace the more valuable pines. Grasslands were gradually taken over by scrubby, woody species that hindered grazing. Pine forests of the western United States that were once clear and open became cluttered with the trunks and branches of trees that had died in the normal aging process. This deadwood became the breeding ground for wood-boring insects that proceeded to attack live trees. In California, the regeneration of redwood seedlings began to be blocked by the proliferation of broad-leaved species.

Scientists now recognize that fire, which is often started by lightning, is a natural and important abiotic factor. As with all abiotic factors, different species have different degrees of tolerance to fire. In particular, the growing buds of grasses and pines are located deep among the leaves or needles, where they are protected from most fires. By contrast, the buds of broad-leaved species, such as oaks, are exposed, so they are sensitive to damage from fire. Consequently, in regions where these species coexist and compete, periodic fires tip the balance in favor of pines, grasses, or redwood trees. In relatively dry ecosystems, where natural decomposition is slow, fire may also help release nutrients from dead organic matter.

Figure 5–17 Iceland. Forests that originally covered much of this island nation were totally stripped for fuel in the 18th and 19th centuries. With natural succession impossible, Iceland has remained barren and tundralike, as seen here.

Figure 5–18 Ground fire. Periodic ground fires are necessary to preserve the balance of pine forests. Such fires remove excessive fuel and kill competing species.

Some plant species even depend on fire. The cones of lodgepole pine, for example, will not release their seeds until they have been scorched by fire (Fig. 5–1b).

Fire Climax Ecosystems. Ecosystems that depend on the recurrence of fire to maintain their existence are referred to as **fire climax ecosystems**. The category includes various grasslands and pine forests. Fire is being increasingly used as a tool in the management of such ecosystems. In pine forests, if ground fires occur every few years, relatively little deadwood accumulates. With only small amounts of fuel, fires usually just burn along the ground, harming neither pines nor wildlife significantly (Fig. 5–1b).

Crown fires, such as the fierce Yellowstone fires, do occur naturally because not every area is burned on a regular basis. In fact, exceedingly dry conditions can make a forest vulnerable to crown fires even when no large amounts of deadwood are present. Even crown fires, however, serve to clear deadwood and sickly trees, release nutrients, and provide for a fresh ecological start. Burned areas, such as those in Yellowstone, soon become productive meadows as secondary succession starts anew. Thus, periodic crown fires create a patchwork of meadows and forests at different stages of succession that lead to a more varied, healthier habitat that supports a greater diversity of wildlife than does a uniform, aging conifer forest. Sometimes people think that logging will reduce the number of fires or their severity in a forest. Unfortunately, because logging removes the largest trees, which are most resistant to burning, and leaves behind dead branches and smaller trees, logging can increase the potential for large fires.

Resilience. The ability of an ecosystem to return to normal functioning after a disturbance, such as a fire or flood, is called its *resilience*. Fire can appear to be a highly destructive disturbance to a forested landscape. Nevertheless, fire releases nutrients that nourish a new crop of plants, and in a short time, the burned area is repopulated with trees and is indistinguishable from the surrounding area. The processes of replenishment of nutrients, dispersion by surrounding plants and animals, rapid regrowth of plant cover, and succession to a forest can all be thought of as **resilience mechanisms** (Fig. 5–19). Resilience, therefore, helps maintain the sustainability of ecosystems.

Figure 5–19 Resilience in ecosystems. Disturbances in ecosystems are usually ameliorated by a number of resilience mechanisms that restore normal ecosystem function in a short time.

Ecosystem resilience has its limits, however. If forests are removed from a landscape by human intervention and the area is prevented from undergoing reforestation by overgrazing, the soil may erode, leaving a degraded state that carries out few of the original ecosystem's functions. As in Iceland, some disturbances may be so profound that they can overcome normal resilience mechanisms and create an entirely new and far less productive ecosystem. This new ecosystem can also have its own resilience mechanisms, which may resist any restoration to the original state. We can view this as a "tipping point"—for an ecosystem, a situation that can move the ecosystem in one direction or another (see Sustainability, "Environmental Tipping Points").

You might wonder why it matters if an ecosystem such as a forested Iceland moves into an alternative state, such as a treeless Iceland, and stays there. After all, don't treeless, cold places exist? As we will see below, ecosystems provide a number of services valuable to humans and other species. These include large-scale effects such as lowering flood surges, maintaining soil, maintaining even ground temperatures, and taking up carbon dioxide.

If just a few parts of the planet moved from one type of ecosystem to another, it really wouldn't matter. The problem is one of frequency. Large-scale human actions tend to lower ecosystem services such as flood control, wind erosion control, and carbon dioxide storage. They tend to produce ecosystems either with very low productivity or with high primary productivity, but low food chain length (such as a lake with too many nutrients). Such ecosystems tend to have less physical complexity and lower biological diversity. We will see this more when we discuss the problem of desertification (Chapter 11).

5.5 Human Values and Ecosystem Sustainability

Ecosystems have existed for millions of years, maintaining natural populations of biota and the processes that they carry out (which in turn sustain the ecosystems). One of the reasons for studying natural ecosystems is that they are models of sustainability. In this chapter, we started with trophic levels and then examined the flow of energy and nutrients in an ecosystem. These discussions have focused on how natural ecosystems work in *theory*. In *reality*, however, it is Earth's specific ecosystems that we depend on for goods and services (also known as *ecosystem capital*). Is our use of natural and managed ecosystems a serious threat to their long-term sustainability? Many ecosystem scientists believe that it is.

Appropriation of Energy Flow

As we have seen, it is the Sun that energizes the processes of energy flow and nutrient cycling. Humans make heavy use of the energy that starts with sunlight and flows through natural and agricultural ecosystems. Agriculture, for example, provides most of our food. To produce food, we have

Sustainability
Environmental Tipping Points

Apo Island, a small island ringed with coral reefs near the coast of Negros in the Philippines, has accomplished a major turnaround in the well-being of its people. This fishing community was experiencing a long decline in the fish stocks its people depended on, in spite of (and because of) newer and more destructive methods like dynamiting and the use of cyanide poisons. Rapidly rising populations were pressuring coastal fisheries all over the Philippines. The villages were locked into what could be called a *vicious cycle*, where the scarcity of local fish pushed the fishers farther offshore and forced them to use more destructive practices, which in turn led to further declines in the fish stocks, and so on.

Into this scene stepped Angel Alcala, a fisheries biologist from nearby Silliman University. Alcala proposed banning all fishing from 10% of the island's shores outward, with the purpose of creating a refuge for fish that would enable them to grow to spawning size and repopulate nearby fishing grounds. The villagers accepted his idea, began to guard the coastline, and soon prohibited all destructive fishing methods around their island. The results quickly demonstrated the wisdom of Alcala's proposal: Aquatic life on the protected reefs exploded, and within 10 years, fishers could pull in a day's catch within a few hundred yards of the shore. More than 400 other Philippine villages have established marine sanctuaries, in recognition of the success of the Apo Islanders. The villagers had turned a destructive cycle into a restorative cycle, one where the system keeps improving and returns to a healthy state. They had triggered a **tipping point**, *a situation in a human-impacted ecosystem where a small action catalyzes a major change in the state of the system.*

Tipping points act as levers that can change a system to be better able—or less able—to function as an ecosystem. They stimulate the action of **feedback loops**, where system dynamics impact a system in ways that amplify (positive loop) or dampen (negative loop) the behavior of the system. For example, predator-prey cycles function as negative feedback loops that keep both populations from exploding in terms of numbers and then crashing. Removal of predators (to favor the prey species) would be a tipping point that could allow the positive feedback of prey reproduction to bring devastation to their plant food supply.

The resilience mechanisms of ecosystems represent feedback loops that promote their health, and interference with these mechanisms can "tip" an ecosystem into a cycle that makes it difficult or impossible to restore the ecosystem to its former state. We will see why this matters on a large scale in Chapter 18, when we look at global climate change.

converted almost 11% of Earth's land area from forest and grassland biomes to agricultural ecosystems. Grasslands sustain our domestic animals for labor, meat, wool, leather, and milk. Forest biomes provide us with 3.3 billion cubic meters (117 billion cubic feet) of wood annually for fuel, building material, and paper. Finally, some 15% of the world's energy consumption is derived directly from plant material, like firewood and peat.

Calculations of the total annual global net primary production of land ecosystems range around 120 petagrams (1 petagram = 1 billion metric tons) of dry matter, including agricultural as well as the more natural ecosystems (Fig. 5–20). In 1986, researchers calculated that humans appropriate around 30% of this total production for agriculture, grazing, forestry, and human-occupied lands.[2] More recent calculations have supported this figure.[3] These estimates indicate that humans are using a large fraction of the whole. Further, because humans convert many natural and agricultural lands to urban and suburban housing, highways, dumps, factories, and the like, we cancel out an additional 8% of potential primary production. Thus, we appropriate about 40% of the land's primary production to support human needs. In using this much, we have become the dominant biological force on Earth, in spite of representing about 5% of the biomass. As ecologist Stuart Pimm put it, "Man eats Planet! Two-fifths already gone!"[4]

---

[2]Peter Vitousek, Paul R. Ehrlich, Anne H. Ehrlich, and Pamela Matson, "Human Appropriation of the Products of Photosynthesis," *BioScience* 36, no. 6. (June 1986): 368–373.
[3]N. Ramankutty, A. T. Evan, C. Monfreda, and J. A. Foley, "Farming the Planet: 1. Geographic Distribution of Global Agricultural Lands in the Year 2000," *Global Biogeochemical Cycles* 22 (2008): GB1003.
[4]Pimm, Stuart. 2001. The World According to Pimm: a scientist audits the Earth. McGraw Hill.

Involvement in Nutrient Cycling

Nutrients are replenished in ecosystems through the breakdown of organic compounds and the release of the chemicals that make them up—processes we have described as biogeochemical cycles. This maintains the sustainability of ecosystems indefinitely. However, human intrusion into these natural cycles is substantial, and it is bound to increase. As discussed in Chapter 3, our alterations of the nutrient cycles have long-term consequences. Climate change is one consequence of these alterations.

Currently, global climate change is one of the most serious environmental issues facing society. The burning of fossil fuels has increased the level of carbon dioxide in the atmosphere to the point where, by the mid–21st century, it will have doubled since the beginning of the Industrial Revolution. Methane is also released from the drying and decomposition of areas that used to be permafrost and from cattle. Because both carbon dioxide and methane are powerful greenhouse gases (see Chapter 20), a general warming of the atmosphere is occurring, leading to many other effects on Earth's climate. Numerous ecosystems are being affected by the changes, and it is questionable whether they can function sustainably in the face of such rapid changes. The threat of disturbance on such an enormous scale is a major reason why many nations of the world have begun to take steps to reduce emissions of greenhouse gases.

Value of Ecosystem Capital

How much are natural ecosystems worth to us? To the poor in many developing countries, the products of natural ecosystems keep life going and may enable them to make economic gains that can lift them from deep poverty. Theirs is a direct

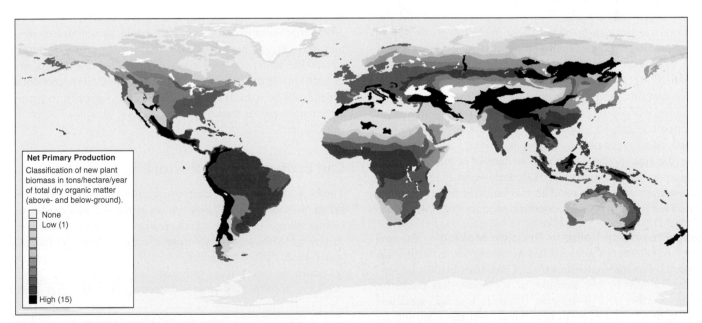

Net Primary Production

Classification of new plant biomass in tons/hectare/year of total dry organic matter (above- and below-ground).

☐ None
☐ Low (1)

■ High (15)

Figure 5–20 Global primary productivity in terrestrial ecosystems. The primary production values range from 1 ton/hectare/year (yellow) to 15 tons/hectare/year (magenta) of dry organic matter. The global annual total is 120 billion metric tons (1 billion metric tons = 1petagram). (*Source:* Bazilevich global primary productivity map, NOAA Satellite and Information Service.)

and absolute dependence. For example, 1.6 billion people, many of them the world's poorest, depend on forests for much of their sustenance (in the form of fuelwood, construction wood, wild fruits, herbs, animal fodder, and bush meat). Their access to these ecosystem goods is often uncertain. Those of us in the developed world, however, are susceptible to a very different perspective: We often assign little thought or value to natural ecosystems and are unconcerned about (and insulated from) human impacts on natural ecosystems. This perspective can have dangerous consequences.

Recall from Chapter 1 that we defined the goods and services we derive from natural systems as *ecosystem capital.* In a first-ever attempt of its kind, a team of 13 natural scientists and economists collaborated in 1997 to produce a report entitled "The Value of the World's Ecosystem Services and Natural Capital."[5] Their reason for making such an effort was that the goods and services provided by natural ecosystems are not easily seen in the market (meaning the market economy that normally allows us to place value on things) or may not be in the market at all. Thus, clean air to breathe, the formation of soil, the breakdown of pollutants, and similar items never pass through and thus are not given a value by the market economy. People are often not even aware of their importance. Because of this, these things are undervalued or not valued at all.

The team identified 17 major ecosystem goods and services that provide vital functions we depend on. The team also identified the ecosystem functions that actually carry out the vital human support and gave examples of each (Table 5–3), making the point that it is useless to consider human welfare without these ecosystem services, so, in one sense, their value as a whole is infinite. However, the **incremental value** of each type of service can be calculated. That is, changes in the quantity or quality of various types of services may influence human welfare, and an economic value can be placed on that relationship. For example, removing a forest will affect the ability of the forest to provide lumber in the future, as well as to perform other services, such as soil formation and promotion of the hydrologic cycle. The economic value of this effect can be calculated. By calculating incremental values, making many approximations, and collecting data from other researchers who have worked on individual processes, the research team tabulated the annual global value of ecosystem services performed by all types of ecosystems. According to their calculations, the total value to human welfare of a year's services amounts to $41 trillion (in 2004 dollars)—and that is considered a conservative estimate! This is close to the $55 trillion calculated for the gross world economic product in 2004.

Using Ecosystem Value in Decision Making. The real power of the team's analysis lies in its use for making local decisions. For instance, the value of wetlands cannot be represented solely by the amount of soybeans that could be grown on the land if it were drained. Instead, wetlands provide other vital ecosystem services, and these should be

balanced against the value of the soybeans in calculating the costs and benefits of a proposed change in land use. The bottom line of their analysis is, in their words, "that ecosystem services provide an important portion of the total contribution to human welfare on this planet." For this reason, the *ecosystem capital stock* (the ecosystems and the populations in them, including the lakes and wetlands) must be given adequate weight in public-policy decisions involving changes to that stock. However, because these ecosystem services are outside the market and uncertain, they are too often ignored or undervalued, and the net result is human changes to natural systems whose social costs far outweigh their benefits. For example, coastal mangrove forests in Thailand are often converted to shrimp farms, yet an analysis showed that the economic value of the forests (for timber, charcoal, storm protection, and fisheries support) exceeded the value of the shrimp farms by 70%.

A New Look.   In 2002, a new team looked at the 1997 team's calculation of value and examined the benefit-cost consequences of converting ecosystems to more direct human uses (e.g., converting a wetland to a soybean field).[6] In every case, the net balance of value was a loss—that is, services lost outweighed services gained. The team examined five different biomes and found consistent losses in ecosystem capital, running at −1.2% per year. Based on the calculated total ecosystem value of $41 trillion, this percentage represents an annual loss of $250 billion through habitat conversion alone. If this is so, the authors asked, why is this conversion still happening? They suggested that the benefits of conversion were often exaggerated through various government subsidies, seriously distorting the market analysis. Thus, the market does not adequately measure many of the benefits from natural ecosystems, nor does it deal well with the fact that many major benefits are realized on regional or global scales, while the benefits of conversion are narrowly local. Other analyses, including a National Research Council report published in 2005[7] that investigated the value of wild nature and ecosystem services, support the idea that conservation is more achievable when the real costs of harming ecosystems are included in our thinking. Beyond conservation, our tremendous dependence on natural systems should lead, logically, to proper *management* of those systems, a topic discussed later in this chapter.

Can Ecosystems Be Restored?

The human capacity for destroying ecosystems is well established. To some degree, however, we also have the capacity to restore them. In many cases, restoration simply involves stopping the abuse. For example, after pollution is curtailed, water quality often improves, and fish and shellfish gradually return to previously polluted lakes, rivers, and bays. Similarly, forests may gradually return to areas that have

[5]Robert Costanza et al., "The Value of the World's Ecosystem Services and Natural Capital," *Nature* 387 (1997): 253–260.

[6]Andrew Balmford et al., "Economic Reasons for Conserving Wild Nature," *Science* 297 (August 9, 2002): 950–953.
[7]National Research Council, *Valuing Ecosystem Services: Toward Better Environmental Decision-Making* (Washington, D.C.: National Academies Press, 2005).

TABLE 5–3 Ecosystem Services and Function

Ecosystem Service	Ecosystem Functions	Examples
Gas regulation	Regulation of atmospheric chemical composition	CO_2–O_2 balance, O_3 for UVB protection, and SO_x levels
Climate regulation	Regulation of global temperature, precipitation, and other biologically mediated climatic processes at global or local levels	Greenhouse gas regulation, dimethylsulfoxide production affecting cloud formation
Disturbance regulation	Capacitance, damping, and integrity of ecosystem response to environmental fluctuations	Storm protection, flood control, drought recovery, other aspects of habitat response to environmental variability controlled mainly by vegetation structure
Water regulation	Regulation of hydrological flows	Provisioning of water for agricultural (such as irrigation) or industrial (such as milling) processes or transporta-
Water supply	Storage and retention of water	Provisioning of water by watersheds, reservoirs, and aquifers
Erosion control and sediment retention	Retention of soil within an ecosystem	Prevention of loss of soil by wind, runoff, or other removal processes; storage of silt in lakes and wetlands
Soil formation	Soil-formation processes	Weathering of rock and the accumulation of organic material
Nutrient cycling	Storage, internal cycling, processing, and acquisition of nutrients	Nitrogen fixation, N, P, and other elemental or nutrient cycles
Waste treatment	Recovery of mobile nutrients and removal or breakdown of excess nutrients and compounds	Waste treatment, pollution control, detoxification
Pollination	Movement of floral gametes	Provisioning of pollinators for the reproduction of plant populations
Biological control	Trophic-dynamic regulations of populations	Keystone predator to control prey species, reduction of herbivory by top predators
Refuges	Habitat for resident and transient populations	Nurseries, habitat for migratory species, regional habitats for locally harvested species, or overwintering grounds
Food production	The portion of primary production extractable as food	Production of fish, game, crops, nuts, and fruits by hunting, gathering, subsistence farming, or fishing
Raw materials	That portion of primary production extractable as raw materials	The production of lumber, fuel, or fodder
Genetic resources	Sources of unique biological materials and products	Medicine, products for science, genes for resistance to plant pathogens and crop pests, ornamental species (pets and horticultural varieties of plants)
Recreation	Provision of opportunities for recreational activities	Ecotourism, sport fishing, and other outdoor recreational activities
Cultural	Provision of opportunities for noncommercial uses	Aesthetic, artistic, educational, spiritual, or scientific values of ecosystems

Source: Reprinted, by permission, from Robert Costanza, Ralph d'Arge, Rudolf de Groot, Stephen Farber, Monica Grasso, Bruce Hannon, Karin Limburg, Shahid Naeem, Robert V. O'Neill, Jose Paruelo, Robert G. Raskin, Paul Sutton, and Marjan van den Belt, "The Value of the World's Ecosystem Services and Natural Capital," *Nature* 387 (1997): 253–260.

been cleared. Humans can speed up the process by seeding and planting seedling trees and by reintroducing populations of fish and animals that have been eliminated.

In some cases, however, specific ecosystems have been destroyed or disturbed to such an extent that they require the efforts of a new breed of scientist: the *restoration ecologist*. The potential for restoration of any ecosystem rests on the following three assumptions: (1) Abiotic factors must have remained unaltered or, if not, can at least be returned to their original state; (2) viable populations of the species formerly inhabiting the ecosystem must still exist; and (3) the ecosystem must not have been upset by the introduction of one or more foreign species that cannot be eliminated and that may preclude the survival of reintroduced native species. If these conditions are met, revival efforts have the potential to restore the ecosystem to some semblance of its former state.

Unfortunately, sometimes the ecosystem is too greatly disturbed. In a prairie, for example, lack of grazing by native herbivores, overgrazing by cattle or sheep, and suppression of regular fires can lead to much woody vegetation. Exotic species may abound in the region and can continuously disperse seeds on the experimental prairie, and there may be no remnants of the original prairie grasses and herbs on the site.

There is a whole branch of ecology focused on these issues. Restoration ecologists use a set of actions described in a primer on ecological restoration published by the Society for Ecological Restoration in 2004.[8] The basic processes include taking an inventory, inventory, developing a *model* of the desired ecosystem in structural and functional terms, setting goals for the restoration efforts, designing an *implementation plan*, and carrying it out. For example, it may be necessary to remove all herbaceous vegetation and then plow and plant the land with native grasses. To maintain the prairie, it may be necessary to burn it regularly or, in the case of larger landholdings, to introduce bison. Clear *performance standards* should be stated, with the means of evaluating progress. Finally, the results of the implementation plan should be monitored and strategies developed for the long-term protection and maintenance of the restored ecosystem.

Why should we restore ecosystems? Restoration ecologists Steven Apfelbaum and Kim Chapman cite several reasons.[9] First, we should do so for aesthetic reasons. Natural ecosystems are often beautiful, and the restoration of something beautiful and pleasing to the eye is a worthy project that can be uplifting to many people. Second, we should do so for the benefit of human use. The ecosystem services of a restored wetland, for example, can be enjoyed by present and future generations. The degradation of ecosystems robs them of their ecosystem capital, and restoring them returns their value. Finally, we should do so for the benefit of the species and ecosystems themselves. It can be argued that nature has value separate from human use and that people should act to preserve and restore ecosystems and species in order to preserve that value.

The Future

On a global scale, the future growth of the human population and its rising consumption levels will severely challenge ecosystem sustainability. Over the next 50 years, the world's population is expected to increase by at least 2.5 billion people (see Chapter 8). The needs and demands of this expanded human population will create unprecedented pressures on the ability of Earth's systems to provide goods and services. Estimates based on current trends indicate that impacts on the nitrogen and phosphorus cycles will raise the amounts of these nutrients being added to the land and water to two to three times current levels. Demands on agriculture will likely require at least 15% more agricultural land; irrigated land area is expected to double, greatly stressing an already strained hydrologic cycle. Such growing demands cannot be met without facing serious trade-offs between different goods and services. More agricultural land means more food, but less forest—and, therefore, less of the important services forests perform. More water for irrigation means more rivers will be diverted, so less water will be available for domestic use and for sustaining the riverine ecosystems. These dire projections are not necessarily predictions, and they don't have to be inevitable outcomes. By looking ahead, we may choose other alternatives if we clearly understand the consequences of simply continuing present practices. There is a powerful need to understand how essential natural and managed ecosystems are for human well-being and to bring to bear the wisdom and political will necessary to promote their sustainability.

Ecosystems are resilient, living units of the landscape, often subject to natural disturbances and capable of changing over time. They also provide essential goods and services that improve the quality of human life and, in many areas, allow those in poverty to improve their lot. However, we have seen that ecosystem goods and services are routinely undervalued, with the result that, as the Millennium Ecosystem Assessment found, ecosystem capital is being degraded and overexploited in many areas of the world. To turn human efforts in a sustainable direction, we need to protect or manage the natural environment in a way that maintains the goods and services vital to the human way of life, and we need to manage ourselves. The two aspects are interrelated.

Managing Ecosystems

Virtually no ecosystem can escape human impact. We depend on ecosystems for vital goods and services—and by using ecosystems, we are in fact managing them. Forests are a good example. Management can be a matter of fire suppression (or, more recently, controlled burns), clear-cutting, or selective harvesting. The objective of management can be to maximize profit from logging or to maintain the forest as a sustainable and diverse ecosystem that yields multiple products and services. Good ecosystem management is based on understanding how ecosystems function, how they respond

---

[8]Society for Ecological Restoration, International Science and Policy Working Group, *The SER International Primer on Ecological Restoration (2004)*. Available at www.ser.org.

[9]Steven I. Apfelbaum and Kim Alan Chapman, "Ecological Restoration: A Practical Approach," in *Ecosystem Management: Applications for Sustainable Forest and Wildlife Resources*, ed. Mark S. Boyce and Alan Haney (New Haven, Conn.: Yale University Press, 1997).

Stewardship

Ecosystem Stakeholders

The pressures on ecosystems illustrate why it is crucially important to understand how ecosystems work and how human societies interact with them. Although many pressures are large-scale—even global—the decisions (and decision makers) that will most directly determine ecosystem sustainability are local or regional. The Millennium Ecosystem Assessment identifies a number of "drivers of change," factors that bring about changes in ecosystems, such as changes in land use, species introductions, and resource harvesting. Decision makers are those who are responsible for these "drivers." To understand how and why they make the decisions they do, it is useful to employ the concept of *stakeholders.* People who have an interest in or may be impacted by a given approach to environmental management, including government decision makers, are all considered to be **stakeholders**. They have a *stake* in the ecosystem resource and what happens to it. *Primary stakeholders* are those most dependent on the resource; *secondary stakeholders* may be those living near the resource, but not greatly dependent on it; government officials; scientists studying the resource; and conservation organizations.

People living within and adjacent to an ecosystem are usually the primary stakeholders. In ecosystems of the developing world, these are among the world's poorest people, and their dependence on what has been called *environmental* income is almost absolute. Primary stakeholders may remove vegetation, harvest wood, withdraw water, harvest game animals, and so on. They may use these products in their own households, or where markets exist, they may sell them for cash or exchange them for services. Their local decisions will have both intended and unintended consequences. If they harvest firewood from a forest (intended), soil erosion may increase (unintended). If many local decision makers make similar decisions, the impacts may have cumulative undesirable regional and, eventually, global consequences. Preventing such outcomes requires effective management of the ecosystem so as to maintain its sustainability and at the same time meet the needs of the various stakeholders.

One way to promote good stewardship of resources is to involve the stakeholders in management. One example of a case where poor rural communities were able to pursue community-based natural resource management and restore and manage crucial environmental resources is documented in the World Resources Institute report entitled *The Wealth of the Poor: Managing Ecosystems to Fight Poverty.** The Shinyanga region of Tanzania was once an abundant woodland environment supporting the Sukuma people with subsistence agriculture and forage for livestock. An indigenous natural resource management system of maintaining protected vegetation enclosures was removed by government fiat in the 1970s, and the result was devastating to the region. Land and soil were overused, wood was difficult to find, and traditional wild fruit and plants became scarce. In 1986, the Tanzanian government reversed its approach and promoted the revival of the original indigenous scheme of enclosures. As a result, people exerted local ownership over the natural resources, and gradually, the landscape changed from an eroded, dry land to one where vegetation and wildlife have been restored on more than 350,000 hectares (860,000 acres). The outcome was greatly improved livelihoods for the Sukuma, with dividends in fuelwood, construction timber, fodder for animals, and traditional foods. In this situation, the involvement of the primary stakeholders in decisions affecting local ecosystems was the key to restoration. One conclusion of the World Resources Institute report was that democratic reform may be the most important step toward sustainable use of ecosystem goods and services, particularly in the developing countries.

---

*World Resources Institute, *World Resources 2005: The Wealth of the Poor: Managing Ecosystems to Fight Poverty* (Washington, D.C.: World Resources Institute, 2005).

---

to disturbances, and what goods and services they can best provide to the human societies living in or around them. Since 1992, the U.S. Forest Service, the U.S. Bureau of Land Management, the U.S. Fish and Wildlife Service, the U.S. National Park Service, and the U.S. Environmental Protection Agency have all officially adopted ecosystem management as their management paradigm. What, then, is ecosystem management?

Ecosystem management comprises several main principles. First, it looks at ecosystems on both small and large scales. For example, the impact of logging on a single river and the impact of that same logging on the entire watershed are both considered in managing the ecosystem. Good management preserves the range of possible landscapes in an ecosystem instead of allowing one type to take over the area. In addition, management is not static—managers are given the opportunity to learn from experiments and to employ new knowledge from forestry and landscape science in their practice. Such "managers" can be volunteer land stewards or neighborhood groups, not just government agencies. As part of adapting to changing circumstances, management must also look at the human element of these ecosystems. The valuable goods and services of ecosystems are recognized, and through active involvement in monitoring and management, all stakeholders (all people with a stake in the ecosystem's health) are included as important elements in stewardship of the resources (see Stewardship, "Ecosystem Stakeholders"). Above all, the paradigm incorporates the objective of ecological sustainability.

The U.S. Forest Service is responsible for managing some 192 million acres (77.7 million hectares) of national forests and is a key player in overseeing public natural resources. A decade ago the Forest Service convened an interdisciplinary committee of scientists to advise the service on what sound science would recommend for managing national lands. According to the committee's report, "the first priority for management is to retain and restore the ecological sustainability of watersheds, forests, and rangelands for present and future generations. The Committee believes that *the policy of sustainability should be the guiding star for stewardship of the national forests and grasslands* to assure the continuation of this array of benefits" (referring to the goods and services provided by the national lands).[10] These were fine words, but have they been followed?

The Governing Board of the Millennium Ecosystem Assessment process examined the findings of the more than 1,350 experts who produced the many reports of the assessment and issued its own statement of the key messages to be derived from the process. The title of this work is significant: *Living Beyond Our Means: Natural Assets and Human Well-Being.*[11] As stated in the first few lines of this document: "At the heart of this assessment is a stark warning. Human activity is putting such strain on the natural functions of Earth that the ability of the planet's ecosystems to sustain future generations can no longer be taken for granted." The following are the key messages from the Governing Board:

- Everyone in the world depends on nature and ecosystem services to provide the conditions for a decent, healthy, and secure life.

- Humans have made unprecedented changes to ecosystems in recent decades to meet growing demands for food, fresh water, fiber, and energy.

- These changes have helped to improve the lives of billions, but at the same time, they have weakened nature's ability to deliver other key services such as the purification of air and water, protection from disasters, and the provision of medicines.

---

[10]Committee of Scientists, Department of Agriculture, *Sustaining the People's Lands: Recommendations for Stewardship of the National Forests and Grasslands into the Next Century* (1999). Available at www.fs.fed.us/news/science.
[11]Millennium Ecosystem Assessment, *Living Beyond Our Means: Natural Assets and Human Well-Being* (Statement of the Millennium Assessment Board) (2005). Available at www.millenniumassessment.org/en/Products.BoardStatement.aspx.

- Among the outstanding problems identified by this assessment are the dire state of many of the world's fish stocks; the intense vulnerability of the 2 billion people living in dry regions to the loss of ecosystem services, including water supply; and the growing threat to ecosystems from climate change and nutrient pollution.

- Human activities have taken the planet to the edge of a massive wave of species extinctions, further threatening our own well-being.

- The loss of services derived from ecosystems is a significant barrier to the achievement of the Millennium Development Goals to reduce poverty, hunger, and disease.

- The pressures on ecosystems will increase globally in coming decades unless human attitudes and actions change.

- Measures to conserve natural resources are more likely to succeed if local communities are given ownership of them, share the benefits, and are involved in decisions.

- Even today's technology and knowledge can reduce considerably the human impact on ecosystems. They are unlikely to be deployed fully, however, until ecosystem services cease to be perceived as free and limitless and their full value is taken into account.

- Better protection of natural assets will require coordinated efforts across all sections of governments, businesses, and international institutions. The productivity of ecosystems depends on policy choices on investment, trade, subsidy, taxation, and regulation, among others.

We are living off the future, in other words. Our current well-being is apparently dependent on drawing down the ecosystem capital that provides the goods and services: As a result, agricultural soils erode, fish stocks decline, forests shrink, and pollution of land, water, and air increases. This situation is unsustainable.

Revisiting the Themes

Sound Science

Ecosystems have natural processes that drive their functions. They are not static entities—change is a fundamental property of ecosystems. They can stand alone or be grouped into large global areas called biomes, which are predicted largely by precipitation and temperature. Ecosystems can also include sub-ecosystems, caused by variations or microclimates. Ecosystems can be characterized by their primary productivity and how that energy and associated nutrients move up through a food chain or food web. These structures contain several trophic levels of organisms that gain their energy from their prey (or in

the case of producers, primarily from the Sun). Ecosystems pull together several concepts we saw in earlier chapters. The concepts of energy and matter (and the inefficiency of their transfer) play out in forming the biomass pyramid and determining the length of food chains. Individual limits on distribution of species are the cause of large predictable areas of Earth with similar ecosystems—the biomes. The relationships between species determine the trophic levels in ecosystems. A functioning ecosystem and its food webs are resilient to disturbance—indeed, some even depend on disturbances to reset succession and allow diversity. However, extremely large or unusual disturbances might tip an ecosystem into an altered state from which it might not recover. In some instances, this leads to much lower productivity or diversity, and in others, sound environmental management can tip a degraded ecosystem back into a restored state.

Sustainability

The message of the Millennium Ecosystem Assessment is that human impacts on this resilient natural world are so unprecedented and extensive that we crossed the line into unsustainable consumption some time ago and are now depleting ecosystem capital stock instead of living off its sustainable goods and services. Converting ecosystems into farms and other human uses has frequently depleted their ability to provide us with pivotal assets. We are just beginning to calculate the lost economic value of these altered landscapes, not to mention their aesthetic and natural value. Any attempt at sustainability will need to include monitoring the world's ecosystems, setting international policies to protect them, and involving all stakeholders in their management. The basic science in Chapters 3 through 5 will inform much of the policy

making and other solutions for sustainability discussed in later chapters.

Stewardship

Ecosystem management has shown us that we need to monitor the impacts we have on our ecosystems because, while they are resilient, large disturbances can do irreparable damage. Indeed, stewardship means, first of all, understanding how ecosystems, such as the Greater Yellowstone Ecosystem, do just what they do and then stepping back and restraining our activities to allow ecosystems to continue to provide us with essential goods and services. Human agency has become the dominant force on Earth, affecting living systems all over the planet. In the process, we have the ability to drastically reduce biodiversity, or we have the ability to act as responsible stewards of our environment. Ecosystem management is, above all, a stewardship approach to the natural world. As the U.S. Forest Service's Committee of Scientists put it, "the policy of sustainability should be the guiding star" for our actions, fittingly linking sustainability and stewardship.

REVIEW QUESTIONS

1. What were the biggest surprises from the Yellowstone fires of 1988?

2. Name and describe the attributes of the two categories into which all organisms can be divided based on how they obtain nutrition.

3. Name and describe the roles of the three main trophic categories that make up the biotic structure of every ecosystem. Give examples of organisms from each category.

4. Give four categories of consumers in an ecosystem and the role that each plays.

5. Describe different members of the decomposition food web.

6. Differentiate among the concepts of *food chain*, *food web*, and *trophic levels*.

7. Relate the concept of the biomass pyramid to the fact that all heterotrophs depend on autotrophic production.

8. Describe how differences in climate cause Earth to be partitioned into major biomes.

9. What are three situations that might cause microclimates to develop within an ecosystem?

10. Identify and describe the biotic and the abiotic components of the biome of the region in which you live.

11. Define the terms *ecological succession* and *climax ecosystem*. How do disturbances allow for ecological succession?

12. What role may fire play in ecological succession, and how may fire be used in the management of certain ecosystems?

13. What is meant by ecosystem *resilience*? What can cause it to fail? How does this relate to environmental tipping points?

14. What is meant by stakeholders? How does ecosystem management involve stakeholders?

15. Succinctly describe ecosystem management.

16. Can ecosystems be restored? What has to happen for that to work?

17. How much of Earth's primary productivity is used or pre-empted by humans?

18. Examine the key messages from the Millennium Ecosystem Assessment's Governing Board, and evaluate how these points affect you now and will impact you and your children in the future.

THINKING ENVIRONMENTALLY

1. Write a scenario describing what would happen to an ecosystem or to the human system in the event of one of the following: (a) All producers are killed through a loss of fertility of the soil or through toxic contamination or (b) decomposers and detritus feeders are eliminated. Support all of your statements with reasons drawn from your understanding of the way ecosystems function.

2. Look at the following description of a broad global region, and describe what the biome name and main biota would be and how you know: 40 centimeters (16 inches) of precipitation a year, seasons, frozen much of the year, high winds.

3. Consider the plants, animals, and other organisms present in a natural area near you, and then do the following: (a) Imagine how the area may have undergone ecological succession and

(b) analyze the population-balancing mechanisms that are operating among the various organisms. Choose one species, and predict what will happen to it if two or three other native species are removed from the area. Then predict what will happen to your chosen species if two or three foreign species are introduced into the area.

4. Consult the Web site www.ecotippingpoints.org, and read several of the success stories from the site. Compare the tipping points as presented, and explain the common properties of the tipping points.

5. Explore how the human system can be modified into a sustainable ecosystem in balance with (i.e., preserving) other natural ecosystems without losing the benefits of modern civilization.

MAKING A DIFFERENCE

1. Find out where the nearest wilderness area or national park is to where you live. If possible, make a plan to visit it. Find out what management issues the park has. Do they have advisories about fire risk, invasive species, or other ecosystem management issues?

2. Using an online map, find out what river drains your watershed, and where that water goes. Is there a watershed stewardship group for the river? Who might be looking out for the protection of the ecosystem? If you are interested, organize a group volunteer activity.

3. In your area, is a there a citizens group for restoring any degraded ecosystems? There are many such small groups of citizens in the US and elsewhere. See if you can link to a citizens group. Volunteer if you can. One such activity is the annual CoastSweep, usually held in September, in which people all over the world clean up beaches. If you are coastal, see if you can help.

Go to **www.masteringenvironmentalscience.com** for practice quizzes, Pearson eText, videos, current events, and more.

CHAPTER 6

Wild Species and Biodiversity

The Atlantic puffin. Dedicated researchers are helping to return puffins to the northeastern United States.

Puffins are small seabirds that live in the cold coastal waters on both sides of the North American continent. The coast of Maine once enjoyed flourishing puffin colonies on a number of offshore islands, but the birds were hunted for their eggs, meat, and feathers from colonial times until 1900, when only one very small colony remained. Protective laws were passed, but the islands once inhabited by the puffins began to host large colonies of predatory gulls; the puffins never returned. National Audubon Society biologist Steven Kress was teaching ornithology to summer visitors on the Maine coast in the early 1970s and began thinking about bringing the puffins back to Maine. Against the advice of many biologists, Kress launched a program to accomplish his dream—Project Puffin. His plan was to obtain puffins from Newfoundland, where they were still numerous, and install them on Eastern Egg Rock, a five-acre island seven miles off the Maine coast. This proposition was no easy task.

First, the predatory gulls had to be removed, with the approval of the U.S. Fish and Wildlife Service. Then, starting with four chicks in 1973, Kress and a team of volunteers and interns installed the chicks in burrows and fed them daily until they were ready to go to sea. (Puffins spend most of their time at sea.) Each year, more chicks were brought to the island. Kress's team had to convince the puffins that the island was a thriving colony, so they installed numerous painted wooden puffin

KEY TOPICS

1. **The Value of Wild Species and Biodiversity: Ethicists believe wild species have both instrumental and intrinsic value.**

2. **Biodiversity and Its Decline: There is a tremendous loss of species occurring today, primarily from human activities; such loss harms poor people more than others.**

3. **Saving Wild Species: Cutting-edge science, policies for protection, and changes in the way people think are all required to protect wild species.**

4. **Protecting Biodiversity Internationally: Wild species do not recognize political boundaries, and therefore, efforts to protect them have to be international.**

131

decoys, played tape-recorded puffin noises, and installed bogus eggs until the first few birds began to return. Finally, in 1981, four pairs of puffins set up housekeeping on the island and raised the first Maine-bred chicks. The number of nesting pairs has grown steadily, and the island now has a thriving colony of 101 pairs.

The method used by Kress—called *social attraction*—was innovative, and his success has blossomed into an enterprise now known as the Seabird Restoration Project of the National Audubon Society. Using similar methods, Kress has presided over a number of seabird reintroductions, including several species of terns on Egg Rock and other islands in the Gulf of Maine, petrels in the Galápagos Islands, and Laysan albatrosses in Hawaii. The Fish and Wildlife Service has adopted Kress's methods to introduce seabirds to other islands where they once bred. Kress says, "Seabirds are ideal symbols for ocean conservation. Because they live in a variety of habitats—including oceans, islands, estuaries and beaches—they demonstrate how all coastal life is connected. Long-term conservation success hinges on public understanding and community support."[1]

The focus in this chapter is *wild species.* The world is experiencing a dramatic decline in wild organisms. While scientists can study individual species such as the puffin, they can also work on species decline in the context of *biodiversity*—the variety of life on Earth. In this chapter, we consider the importance of biodiversity as a whole, as well as that of individual species, and look at the public policies that seek to protect wild species and biodiversity.

[1]See http://www.projectpuffin.org/PPVC.html.

6.1 The Value of Wild Species and Biodiversity

Recall from Chapter 1 that *ecosystem capital* is the sum of all the goods and services provided to human enterprises by natural systems. This capital is enormously valuable to humankind. The 1997 mean estimate of ecosystem capital by Robert Constanza and his coauthors (see Chapter 5) was $46.7 trillion (adjusted for 2009 dollars). Such a value is so difficult to pull together that there is no subsequent estimate of the same set of parameters. However, in 2008, economist Pavan Sukhdev reported that capital loss from the worldwide financial crisis stood at $1–$1.5 trillion and capital loss from ecosystem degradation was estimated to be $2–$4.5 trillion per year.[2] The basis of much of our natural capital is ecosystems, and the basis of ecosystems is the plants, animals, and microbes—the wild species—that make them work. To maintain the sustainability of these ecosystems, their *integrity* must be preserved; this means maintaining their resilience, natural processes, and biodiversity, among other things. Placing a high priority on protecting wild species and the ecosystems in which they live is therefore crucial.

Even those who agree on the need to protect wild species, however, may disagree on the kind of protection that should be afforded to them. Some want wildlife protected so as to provide recreational hunting. Others feel strongly that hunting for sport should be banned. Many have a deeper concern for the ecological survival of wild species, and they view the ongoing loss of biodiversity as an impending tragedy. In the developing world, however, many collect or kill wild species in order to eat or make money, so their personal survival is at stake. All of these attitudes stem from the values we place on wild species. Can the differing values be reconciled in a way that still leads to sustainable management of these important natural resources?

Biological Wealth

The Millennium Ecosystem Assessment states that about 2 million species of plants, animals, and microbes have been examined, named, and classified, but that many more species have not yet been found or catalogued. Most estimates of the total number of species on Earth range from 5 to 30 million. These natural species of living things, collectively referred to as **biota,** are responsible for the structure and maintenance of all ecosystems. They, and the ecosystems that they form, represent wealth—the **biological wealth**—that makes up most of the ecosystem capital that sustains human life and economic activity with goods and services. From this perspective, the biota found in each country can represent a major component of the country's total wealth. More broadly, this richness of living species constitutes Earth's **biodiversity.**

Humans have always depended on Earth's wild species for food and materials. Approximately 12,000 years ago, however, our ancestors began to select certain plant and animal species from the natural biota and propagate them, and the natural world has never been the same. Forests, savannas, and plains were converted to fields and pastures as the human population grew and human culture developed. In time, many living species were exploited to extinction, and others disappeared as their habitats were destroyed. The Center for Biological Diversity has reported that 631 species and subspecies went extinct in North America between 1642 and 2001. A 2008 report said that global biodiversity loss costs as much as $4.5 trillion a year.[3] We have been drawing down our biological wealth, with unknown consequences.

For many Americans who live in cities and suburbs and get all their food from supermarkets, the connection between everyday life and nature may seem remote. That is an illusion, however. Our interactions with the natural world may have changed, but we still depend on biological wealth. Moreover, millions of our neighbors in the developing world are not so insulated from the natural world; they draw sustenance

[2]Pavan Sukhdev, *The Economics of Ecosystems and Biodiversity* (Interim Report funded by the German Ministry for Environment and the European Commission on the Economics of Ecosystems and Biodiversity). Presented at the UN's Convention on Biological Diversity, November 2008.

[3]Ibid.

and income directly from forests, grasslands, and fisheries. For them, "environmental income" is often the safety net that sustains them during lean times and can enable them in good times to generate wealth to improve their well-being. However, these people, too, are often engaged in practices that draw down their biological wealth. A root source of this problem is the way we regard and value wild nature.

Two Kinds of Value

In the 19th century, hunters on horseback could ride out to the vast herds of bison roaming the North American prairies and shoot them by the thousands, often taking only the tongues for markets back in the east. The passenger pigeons that once darkened the skies in huge flocks were killed at their roosts to fill a lively demand for their meat, until the species was gone (Fig. 6–1). Plume hunters decimated the populations of egrets and other shorebirds to satisfy the demands of fashion in the late 1800s. Appalled at this wanton destruction, 19th-century naturalists called for an end to the slaughter, and the U.S. public became sensitized to the losses. People began to see natural species as worthy of preservation, and naturalists began to look for ways to justify their calls to conserve nature. Thus, there was an emerging sense that species should not be hunted to extinction. But why not? Their problem then, and ours now, is to establish the thesis that wild species have some *value* that makes it essential to preserve them. If we can identify that value, then we will be able to assess what our moral duties are in relation to wild species.

Figure 6–1 The extinct passenger pigeon. Clouds of passenger pigeons darkened American skies during the 18th and 19th centuries, but relentless hunting extinguished the species by the early 20th century.

Instrumental Value.   Philosophers who have addressed this problem identify two kinds of value that should be considered. The first is **instrumental value**. A species or individual organism has instrumental value if its existence or use benefits some other entity (by providing food, shelter, or a source of income). This kind of value is usually *anthropocentric*; that is, the beneficiaries are human beings. Many species of plants and animals have instrumental value to humans and will tend to be preserved (conserved, that is) so that we can continue to enjoy the benefits derived from them.

Intrinsic Value.   The second kind of value that must be considered is **intrinsic value**. Something has intrinsic value when it has value *for its own sake*; that is, it does not have to be useful to us to possess value. How do we know that something has intrinsic value? This is a philosophical question and comes down to a matter of moral reasoning. People often disagree about intrinsic value, as illustrated by the controversies about the rights of animals. Some people argue that animals have certain rights and should be protected, for example, from being used for food, fur and hides, or experimentation. On the other side, some people hold that animals are simply property and have no rights at all. Others, probably the majority, hold an intermediate position: They agree that cruelty to animals is wrong, but they do not believe that animals should be given rights approaching those afforded to humans.

Many people believe that no species on Earth except *Homo sapiens* has any intrinsic value. If this is true, then they find it difficult to justify preserving species that are apparently insignificant. Still, in spite of the problems of establishing intrinsic value for species, support is growing in favor of preserving species that not only may appear useless to humans, but also may never be seen by anyone except a few naturalists or taxonomists (biologists who are experts on classifying organisms).

The value of natural species can be categorized as follows:

- Value as sources for food and raw materials,
- Value as sources for medicines and pharmaceuticals,
- Recreational, aesthetic, and scientific value, and
- Value for their own sake (intrinsic value).

Sources for Food and Raw Materials

Most of our food comes from agriculture, rather than being collected directly from wild populations, so it is tempting to falsely believe that food is independent of natural biota. In nature, both plants and animals are continuously subjected to the rigors of natural selection. Only the fittest survive. Consequently, wild populations have numerous traits for competitiveness, resistance to parasites, tolerance to adverse conditions, and other aspects of vigor.

In contrast, populations grown for many generations under the "pampered" conditions of agriculture tend to lose these traits because they are selected for production, not resilience. For example, a high-producing plant that lacks resistance to drought is irrigated, and the resistance to drought is ignored. Also, in the process of breeding plants for maximum production, virtually all genetic variation is eliminated. Indeed, the cultivated population is commonly called a *cultivar* (for *culti*vated *var*iety), indicating that it is a highly selected strain of the original species, with a *minimum* of genetic variation. When provided with optimal water and fertilizer, cultivars produce outstanding yields under the specific climatic conditions to which they are adapted. With their minimum genetic variation, however, they have virtually no capacity to adapt to any other conditions. For example, winter wheat varieties with high yields may be less winter hardy or more prone to pests and disease.

Wild Genes.   To maintain vigor in cultivars and to adapt them to different climatic conditions, plant breeders comb wild populations of related species for the desired traits. When found, these traits are introduced into the cultivar through crossbreeding or, more recently, biotechnology (Chapter 12). For example, in the 1970s, the U.S. corn crop was saved from blight by genes from a wild strain of maize. The point is that this trait came from a related *wild* population—that is, from natural biota. Another area in which wild species have instrumental value to humans is pest control. In Chapter 13, we will discuss the tremendous and invaluable opportunities there are to control pests by increasing genetic resistance. Genes for increasing this resistance can come only from natural biota. If wild populations are lost, the options for continued improvements in food plants will be greatly reduced.

New Food Plants.   The potential for developing *new* agricultural cultivars would also be lost if wild populations were destroyed. From the hundreds of thousands of plant species existing in nature, humans have used perhaps 7,000 in all, and modern agriculture has tended to focus on only about 30. Of these, three species—wheat, maize (corn), and rice—fulfill about 50% of global food demands. This limited diversity in agriculture makes modern plants ill suited to production under many different environmental conditions. For example, arid regions are generally considered to be unproductive without irrigation, but there are wild plants, including many wild species of the bean family, that produce abundantly under dry conditions.

Scientists estimate that 30,000 plant species with edible parts could be brought into cultivation. Many of these could increase production in environments that are less than ideal. For example, every part of the winged bean, a native of New Guinea (Fig. 6–2), is edible: pods, flowers, stems, roots, and leaves. Recently introduced into many developing countries, this legume has already contributed significantly to improving nutrition. Loss of biological diversity undercuts similar future opportunities.

Woods and Other Raw Materials.   Because species are selected from nature for animal husbandry, forestry, and aquaculture, essentially all the same arguments apply to

Figure 6–2 The winged bean. A climbing legume with edible pods, seeds, leaves, and roots, this tropical species demonstrates the great potential of wild species for human use.

those important enterprises, too. For example, nearly 3 billion people rely on wood for fuel for heating and cooking. Wood products rank behind only oil and natural gas as commodities on the world market. Demand for wood is increasing so rapidly that scientists are predicting a "timber famine" or "fuelwood crisis" in the next decade. Products such as rubber, oils, nuts, fruits, spices, and gums also come from forests. All of these products are part of the value of wild species for humans, and forests are only one example of ecosystems that provide wild and natural living resources.

Banking Genes.   Living organisms can be thought of as a bank in which the gene pools of all the species involved are deposited. Thus, biota is frequently referred to as a **genetic bank**. Because of concerns about agriculture, many scientists suggest that wild relatives of cultivated crops be conserved. In Kew, England, the Royal Botanic Gardens house the Millennium Seed Bank (MSB). In 2007, the MSB banked the billionth seed in its vast collection of the genetic diversity of world plants—the largest wild seed collection in the world. There is also the massive Svalbard Global Seed Vault, built into the permafrost deep in a mountain on a remote arctic island in Norway. This ultimate seed bank officially opened in 2008 and holds seeds that will act as a backup for seed banks around the world. Similarly, zoos act as a type of genetic bank for animals, and many zoos are active in conservation and captive breeding. In the United Kingdom, the Frozen Ark Project is collecting cells and DNA from species likely to go extinct in the near future. This collaborative effort with organizations around the world is designed to preserve genetic diversity while we try to stabilize species and slow currently high extinction rates.

Sources for Medicine

Earth's genetic bank is also important for the field of medicine. For thousands of years, the indigenous people of the island of Madagascar used an obscure plant, the rosy periwinkle, in their folk medicine (Fig. 6–3). If this plant, which grows only on Madagascar, had become extinct before 1960, hardly anyone outside Madagascar would have cared. In the 1960s, however, scientists extracted two chemicals with medicinal properties—vincristine and vinblastine—from the plant. These chemicals have revolutionized the treatment of childhood leukemia and Hodgkin's disease. Before their discovery, leukemia was almost always fatal in children; today, with vincristine treatment, there is a 99% chance of remission.

The story of the rosy periwinkle is just one of hundreds. Chinese star anise, a tree in the magnolia family, yields star-shaped fruits that are harvested and processed to produce shikimic acid, the raw material for making Tamiflu®, an effective treatment against influenza viruses. Paclitaxel (trade name Taxol®), an extract from the bark of the Pacific yew, has proven valuable for treating ovarian, breast, and small-cell cancers. For a time, this use threatened to decimate the Pacific yew because six trees were required to treat one patient for a year. Now, the substance is extracted from the leaves of the English yew tree, a horticultural plant that is easy to maintain. Today Taxol® is the most prescribed antitumor drug.

Stories like these have created a new appreciation for the field of *ethnobotany*, the study of the relationships between plants and people. To date, more than 3,000 plants have been identified as having anticancer properties. Drug companies have been financing field studies of the medicinal use of plants by indigenous peoples (called *bioprospecting*) for years. The search for beneficial drugs in the tropics has led to the creation of parks and reserves to promote the preservation of natural ecosystems that are home to both the indigenous people and the plants they use in traditional healing. In Surinam, for example, a consortium made up of drug companies, the U.S. government, and several conservationist and educational groups has created the Central Surinam Reserve, a 4-million-acre

Figure 6–3 The rosy periwinkle. This native to Madagascar is a source of two anticancer agents that are highly successful in treating childhood leukemia and Hodgkin's disease.

(1.64-million-hectare) protected area. Table 6–1 lists a number of well-established drugs that were discovered as a result of analyzing the chemical properties of plants used by traditional healers. Who knows what other medicinal cures are out there?

Recreational, Aesthetic, and Scientific Value

The species in natural ecosystems also provide the foundation for numerous recreational and aesthetic interests, ranging from sportfishing and hunting to hiking, camping, bird-watching,

---

TABLE 6–1   Modern Drugs from Traditional Medicines

Drug	Medical Use	Source	Common Name
Aspirin	Reduces pain and inflammation	*Filipendula ulmaria*	Queen of the meadow
Codeine	Eases pain; suppresses coughing	*Papaver somniferum*	Opium poppy
Ipecac	Induces vomiting	*Psychotria ipecacuanha*	Ipecac
Pilocarpine	Reduces pressure in the eye	*Pilocarpus jaborandi*	Jaborandi plant
Pseudoephedrine	Reduces nasal congestion	*Ephedra sinica*	*Ma-huang* shrub
Quinine	Combats malaria	*Cinchona pubescens*	Cinchona tree
Reserpine	Lowers blood pressure	*Rauwolfia serpentina*	Rauwolfia
Scopolamine	Eases motion sickness	*Datura stramonium*	Jimsonweed
Theophylline	Opens bronchial passages	*Camellia sinensis*	Tea
Tubocurarine	Relaxes muscles during surgery	*Chondrodendron tomentosum*	Curare vine
Vinblastine	Combats Hodgkin's disease	*Catharanthus roseus*	Rosy periwinkle

photography, and more (Fig. 6–4). Periodically, the U.S. government conducts a survey on outdoor recreation; the latest was carried out from 2000 to 2001. Of all people age 16 and older who were surveyed, an estimated 202 million were involved in some form of outdoor recreation. The largest activity was viewing natural scenery, an activity reported by 125 million people. The survey included opinions on environmental issues, and people's attitudes clearly reflected both interest in and concern about the environment, an indication that their contact with the outdoors had sensitized them. For example, 82% of respondents agreed or strongly agreed with the statement that "humans are severely abusing the environment," and 80% agreed or strongly agreed that, "when humans interfere with nature, it often produces disastrous consequences." Very likely, much of the broad public support for preserving wild species and habitats stems from the aesthetic and recreational enjoyment people derive from them. While hunting has declined in recent decades, other outdoor activities remain popular. A 2006 U.S. Fish and Wildlife Service survey on fishing, hunting, and wildlife found that 87.5 million U.S. resident adults (age 16 and older) participated in

wildlife-centered recreation. This number included 30 million people who fished, 12.5 million who hunted, and 71.1 million who did some type of wildlife observation such as bird-watching. Hunting, fishing, and wildlife-watching result in employment for 2.6 million people, nearly as many as employed by the computer industry, and produce $108 billion in revenues in the United States.

Recreational and aesthetic values constitute a very important source of support for maintaining wild species because activities inspired by these values also support commercial interests. **Ecotourism**—whereby tourists visit a place in order to observe wild species or unique ecological sites—represents the largest foreign-exchange-generating enterprise in many developing countries. Because some percentage of recreational dollars will be spent on activities related to the natural environment, any degradation of that environment affects commercial interests. As the national surveys indicated, recreational, aesthetic, and scientific activities involve a great number of people and represent a huge economic enterprise.

Scientific value comes from our need to have living things available so we can learn basic laws of nature, the way

Figure 6–4 Recreational, aesthetic, and scientific uses. Natural biota provides numerous values, a few of which are depicted here.

ecosystems function, and the way the world works generally. Societies have long held that such "pure science" is valuable. Many scientists are engaged in this enterprise by sheer curiosity about the world, including the functioning of individual types of organisms, and biota provides the nature that we get to study. Realistically, though, most of the science done with these species aims for more pragmatic goals—new medicines, protection from gene loss in agriculture and forestry, and other outcomes already discussed.

A Cautionary Note. The use of wild species and biodiversity for instrumental value can raise some problems of justice as we navigate the needs of people with competing claims. For example, while the rosy periwinkle was a success story because of its effectiveness in combating cancer, little of the money returned to Madagascar, a very poor country. In other cases, such as the case of turmeric (a spice commonly used in India), ancient herbal remedies have been patented by large companies. While the patent on turmeric was eventually overturned, the patent on the rosy periwinkle was not, and money from the rosy periwinkle came to Madagascar only when the patent ran out. The indigenous people in the areas where biodiversity is highest may not be the ones to benefit from their resources. Likewise, ecotourism may bring money to poor areas of the world, but such tourism can also increase pollution, directly harm wildlife, or alter local societies in negative ways. For example, whale-watching boats on the St. Lawrence River disrupt the feeding of whales. Tourist boats on Mexican lakes frighten flocks of flamingoes and reduce their feeding time. In the Caribbean, coral reefs are damaged by divers, boat anchors, and shopkeepers who remove coral to sell. For indigenous people, ecotourism can marginalize them as part of the scenery, or they can remain poor, while tourism operators gain wealth. Nonetheless, there is sustainable tourism, and people intending to travel in the hopes of promoting conservation can find tourism that is done well.

The Loss of Instrumental Value. Overall, a loss of biodiversity has a tremendous negative effect on the world in real, practical ways. The Economics of Ecosystems and Biodiversity (TEEB) report, released in May 2008, was the first international report to detail the economic and life-quality effects of biodiversity loss. Researchers have estimated that the loss of ecosystem services and biodiversity costs $78 billion annually and, even more importantly, that these costs are higher for the world's poorest people. Continued loss of biological diversity, the report said, could halve the income of the poorest billion and a half people. Because the people who benefit most from the rapid use of resources are not the people who are most harmed by the loss of those resources, many ethicists believe such an outcome to be morally wrong. Thus, even the instrumental values of Earth's resources have an ethical component.

Value for Their Own Sake

The usefulness (instrumental value) of many wild species is apparent. But instrumental value alone may not be a compelling reason to protect many wild species. What about the species that have no obvious value to anyone—including

many plant and animal species? Another strategy for preserving all wild species is to emphasize the *intrinsic* value of species rather than their unknown or uncertain ecological and economic values. From this viewpoint, the extinction of any species is an irretrievable loss of something valuable.

Environmental ethicists argue that the long-established existence of species means they have a right to continued existence. Living things have ends and interests of their own, and the freedom to pursue them is often frustrated by human actions. Do humans have the right to terminate a species that has existed for thousands or millions of years? Environmental philosopher Holmes Rolston III puts it this way: "Destroying species is like tearing pages out of an unread book, written in a language humans hardly know how to read, about the place where they live."[4] Some support this view by arguing that there is value in every living thing and that one kind of living thing (e.g., humans) has no greater value than any other. This argument, however, can lead to some difficulties, such as having to defend the rights of pathogens and parasites. A more common viewpoint held by ethicists is that, because humans have the ability to make moral judgments, they also have a special responsibility toward the natural world, and that this responsibility includes concern for other species.

Religious Support. Many ethicists find their basis for intrinsic value in religious thought. For example, in the Jewish and Christian traditions, Old Testament writings express God's concern for wild species when He created them. Judeo-Christian scholars maintain that by declaring His creation good and giving it His blessing, God was saying that all wild things have intrinsic value and, therefore, deserve moral consideration and stewardly care. In the story of Noah, described in the Hebrew Bible, God makes a covenant not only with humans, but also with all creatures, "every living thing." Similarly, in the Islamic tradition, the Quran (Koran) proclaims that the environment is the creation of Allah and should be protected because its protection praises the Creator. Many Native American religions have a strong environmental ethic (a system of moral values or beliefs); for example, the Lakota hold that humans and other forms of life should interact like members of a large, healthy family. Drawing on Hindu philosophy, the Chipko movement of northern India represents a strong grassroots environmentalism. Ethical concern for wild species underlies several religious traditions and represents a potentially powerful force for preserving biodiversity. From many backgrounds, even if species have no demonstrable use to humans, it can still be argued that they have a a value and a right to exist and that humans thus have a responsibility to protect them from harm that results from human actions.

The Land Ethic. Further, there is an ethic about the preservation not just of species, but also of their ecosystems. This view of nature was most famously described by conservationist Aldo Leopold in an essay published in 1949 entitled "The Land Ethic." Leopold is known for being one of the first scientists to understand the importance of both fire and

---

[4]Holmes Rolston III, *Environmental Ethics: Duties to and Values in the Natural World* (Philadelphia: Temple University Press, 1988), p. 129.

predators in maintaining ecosystem health. He advocated for protection of wild places—*wilderness*—as well as for better care of agricultural and other human-dominated land. Leopold wrote a number of essays and is well known in the literary community as well as among conservationists. *A Sand County Almanac* sold more than 2 million copies and is printed in nine languages. "The Land Ethic" was instrumental in generating interest in the protection of wilderness. In it, Leopold describes his viewpoint: "A thing is right when it tends to preserve the integrity, stability, and beauty of the biotic community. It is wrong when it tends otherwise."[5]

6.2 Biodiversity and its Decline

In the introduction to this chapter, biodiversity was defined as simply *the variety of life on Earth*. We can be more specific than that. The dimensions of biodiversity include the genetic diversity within a species and the variety of species as well as the range of communities and ecosystems. When scientists use the term *biodiversity*, however, sometimes they mean a

characteristic that they can calculate and compare between habitats. In this definition, scientists use two measures to calculate biodiversity—the number of species and a measure of how "even" the species are. A habitat has low biodiversity if it is dominated by only one species, with few members of other species, because those few might easily leave or die out. Given the same number of species, diversity is considered to be higher in habitats where the dominance by any single species is low. In this chapter, our main focus is the diversity of species. Recall from Chapter 4 that, over long periods of time, natural selection leads to speciation (the creation of new species) as well as extinction (the disappearance of species). Now, we are interested in how humans interact with wild species and impact biodiversity over a relatively short time.

How Many Species?

Almost 2 million species have been discovered and described, and certainly, many more than that exist. Most people are completely unaware of the great diversity of species that occur within any given taxonomic category. Groups especially rich in species are the flowering plants (270,000 species) and the insects (950,000 species), but even less diverse groups, such as birds or ferns, are rich with species that are unknown to most people (Table 6–2). Groups that are conspicuous or

---

[5]Aldo Leopold, "The Land Ethic," in *A Sand County Almanac and Sketches Here and There* (Oxford, England: Oxford University Press, 1949). Available at http://www.luminary.us/leopold/land_ethic.html

---

TABLE 6–2 Known and Estimated Species on Earth

Domain	Eukaryote kingdoms	No. of described species	Estimated total
Archaea		175	?
Bacteria		10,000	?
Eukarya			
	Animalia		1,320,000
	Craniata (vertebrates), total	52,500	55,000
	Mammals	4,630	
	Birds	9,750	
	Reptiles	8,002	
	Amphibians	4,950	
	Fishes	25,000	
	Mandibulata (insects and myriapods)	963,000	8,000,000
	Chelicerata (arachnids, etc.)	75,000	750,000
	Mollusca	70,000	200,000
	Crustacea	40,000	150,000
	Nematoda	25,000	400,000
	Fungi	72,000	1,500,000
	Plantae	270,000	320,000
	Protoctista	80,000	600,000
Total		1,750,000	14,000,000

Notes: This table presents recent estimates of the number of species of living organisms in the high-level groups recognized, and in some selected groups within them. Vertebrate classes are distinguished because of the general interest in these groups. The described species column refers to species named by taxonomists. Most groups lack a formal list of species. All estimates are approximations. They are inevitably inaccurate because new species will have been described since publication of any checklist and more are continually being described, and other names turn out to be redundant synonyms. In general, the diversity of microorganisms, small-sized species, and those from habitats difficult to access are likely to be seriously underestimated. Among Archaea and Bacteria the figures of 175 and 10,000 are very rough estimates of 'species' defined on features shown in culture; there appears to be no sound estimate of the total amount of prokaryote diversity. The estimated total column includes provisional working estimates of the number of described species plus the number of unknown species; the total figure is highly imprecise. Only a small selection of animal phyla is shown, but the figure for Animalia applies to all. Figures in the total row are for all species in all domains.

Source: B. Groombridge and M. D. Jenkins, *World Atlas of Biodiversity* (Berkeley: University of California Press, 2002).

commercially important, such as birds, mammals, fish, and trees, are much more fully explored and described than, say, insects and very small invertebrates (such as soil nematodes, fungi, and bacteria). Taxonomists are aware that their work in finding and describing new species is incomplete. Trained researchers who can identify major groups of organisms are few, and the task of fully exploring the diversity of life would require a major sustained effort in systematic biology. For example, Costa Rica, less than half the size of New York State, is estimated to contain at least 5% of all living species. Estimates of unknown species keep rising as taxonomists explore the rain forests more and more. Whatever the number of species, the planet's biodiversity represents an amazing and diverse storehouse of biological wealth.

The Decline of Biodiversity

Biodiversity is declining in the United States as well as around the world. Because U.S. ecosystems are well studied, the trends that show themselves here (including the decline of many plant and animal species) may represent larger global trends. **Endemic** species (those found in only one habitat, and nowhere else) are especially at risk. Some areas of the country and of the world are particularly vulnerable to species loss. These areas are often the focus of special conservation efforts.

North America. The biota of the United States are as well known as any. Even so, not much is known about most of the 200,000-plus species of plants, animals, and microbes estimated to live in the nation. At least 500 species native to the United States, including at least 100 vertebrates, are known to have gone extinct (or have gone missing and are believed to have gone extinct) since the early days of colonization. An inventory of 20,897 wild plant and animal species in the 50 states, carried out in the late 1990s by a team of biologists from the Nature Conservancy and the National Heritage Network (and more recently NatureServe), concluded that one-third are vulnerable (threatened), imperiled (endangered), or already extinct (Fig. 6–5). Mussels, crayfish, fishes, and amphibians—all species that depend on freshwater habitats—are

at greatest risk. In part, this potential for loss stems from the high number of *endemic* aquatic species in the American Southeast. For example, there are close to 200 species of freshwater mussels and clams in North America, giving these bivalves the greatest species diversity of any freshwater bivalve group in the world. Flowering plants are also of great concern, with one-third of their numbers in decline in North America.

Species *populations* are a more important element of biodiversity than even the species' *existence*. Populations of species that occupy different habitats and ranges contribute to biological wealth, as they provide goods and services important in ecosystems. Across North America, general population declines of well-studied wild species are also occurring. Commercial landings of many species of fish are down, puzzling reductions in amphibian populations are occurring all over the world, and many North American songbird species, such as the Cerulean warbler, wood thrush, and scarlet tanager, have been declining and have disappeared entirely from some local regions. These birds are neotropical migrants; that is, they winter in the neotropics of Central and South America and breed in temperate North America. According to a report from the National Audubon Society, more than one-fourth of North American birds are already declining or are in danger of doing so.

Global Outlook. Worldwide, the loss of biodiversity is even more disturbing. One of the most recent global analyses of biodiversity is the Millennium Ecosystem Assessment. Figure 6–6, from this assessment, shows a comparison of extinction rates from the fossil record, from the recent past, and as projected into the future. For mammals, the background (past) extinction rate is less than one extinction every thousand years, with the exception of the higher rates during five great extinction events. (Scientists estimate the background extinction rate from the fossil record of marine animals and from rates of change in DNA.) This rate can be compared with known extinctions of the last several hundred years (for mammals and birds, 20 to 25 species per hundred years and, for all groups, 850 species over about 500 years), indicating that the current rate is 100 to 1,000 times higher than past rates.

In 2008, the International Union for Conservation of Nature (IUCN) reported that some 23% (1,141) of mammal species and 12% (1,226) of bird species were globally threatened. These two groups have been thoroughly studied.[6] The status of all other groups of organisms is much less well known. The majority of threatened species are concentrated in the tropics, where biodiversity is so rich as to be almost unimaginable. Biologist Edward O. Wilson identified 43 species of ants on a single tree in a Peruvian rain forest, a level of diversity equal to the entire ant fauna of the British Isles.[7] Other scientists found 300 species of trees in a 1-hectare (2.5-acre) plot and almost 1,000 species of beetles on

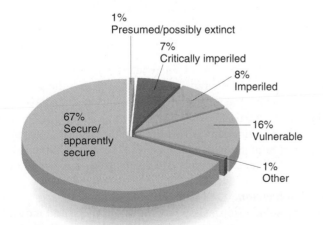

1%
Presumed/possibly extinct

7%
Critically imperiled

8%
Imperiled

16%
Vulnerable

1%
Other

67%
Secure/
apparently
secure

Figure 6–5 The state of U.S. species. Fully one-third of over 20,000 species of plants and animals surveyed were found by biologists to be at risk of extinction. (*Source:* Data from the Nature Conservancy and NatureServe.)

[6]Birdlife International Report. *The State of the World's Birds: Indicators for our changing world* (2008). Available at http://www.biodiversityinfo.org/sowb/default.php?r=sowbhome.
[7]Edward O. Wilson, "The Arboreal Ant Fauna of Peruvian Amazon Forests: A First Assessment," *Biotropica* 19 (1987): 245–251.

Figure 6–6 Species extinction rates. The distant past rate refers to average extinction rates from the fossil record; the recent past rate refers to known extinctions over the last few hundred years; the future rate is based on models that assess the trends of change in factors affecting biodiversity from 1970 to 2050. The height of each box represents the possible range of values. (*Source:* Millennium Ecosystem Assessment, *Ecosystems and Human Well-Being: Biodiversity Synthesis* (Washington, D.C.: World Resources Institute, 2005).)

nineteen specimens of a species of tree in Panama.[8] Unfortunately, the same tropical forests that hold such high biodiversity are also experiencing the highest rate of deforestation. Because the inventory of species in these ecosystems is so incomplete, it is virtually impossible to assess extinction rates in such forests. Regardless of the exact rates, numerous species are in decline, and some are becoming extinct. No one knows how many, but the loss is real, and species loss represents a continuing depletion of the biodiversity of our planet.

Reasons for the Decline

Extinctions of the distant past were caused largely by processes of climate change, plate tectonics, and even asteroid impacts. Current threats to biodiversity are sometimes described with the acronym **HIPPO**, which refers to five factors: **habitat destruction, invasive species, pollution, population,** and **overexploitation**. We will discuss each factor in more detail below. The losses in the future will be greatest in the developing world, where biodiversity is greatest and human population growth is highest. Africa and Asia have lost almost two-thirds of their original natural habitat. People's desire for a better way of life, the desperate poverty

of rural populations, and the global market for timber and other natural resources are powerful forces that will continue to draw down biological wealth on those continents.

One key to slowing the loss of biodiversity lies in bringing human population growth down, as we will see in Chapter 8. If the human population increases to 10 billion, as some demographers believe that it will, the consequences for the natural world may be frightening. Another key is to pull individual consumption down to sustainable levels in parts of the world with high consumption of energy and materials. These two concepts will be vital to solving several environmental issues and will reoccur throughout the rest of the book.

Habitat Change. By far the greatest source of biodiversity loss is the physical alteration of habitats through the processes of conversion, fragmentation, simplification, and intrusion. Habitat destruction has already been responsible for 36% of the known extinctions and is the key factor in the currently observed population declines. Natural species are adapted to specific habitats, so if the habitat changes or is eliminated, the species go with it.

Conversion. Natural areas are converted to farms, housing subdivisions, shopping malls, marinas, and industrial centers. When a forest is cleared, for example, destruction of the trees is not the only result; every other plant or animal that occupied the destroyed ecosystem, either permanently or temporarily (e.g., migrating birds), also suffers. Global forest

---

[8]T. L. Erwin, "The Tropical Forest Canopy: The Heart of Biotic Diversity." In *Biodiversity*, edited by E. O.Wilson, National Academy Press, Washington, D.C., pp.123–129.

Figure 6–7 The border of Haiti and the Dominican Republic. Severe deforestation in Haiti, on the left, leads to habitat loss for many species. It also causes a number of problems for humans. The forested area on the right is the Dominican Republic.

cover has been reduced by 40% already, and the decline continues. For example, the decline in North American songbird populations has been traced to the loss of winter forest habitat in Central and South America and the increasing fragmentation of summer forest habitat in North America. Agriculture already occupies 38% of land on Earth; habitats like croplands that replace natural habitats such as grasslands are quite inhospitable to all but a few species that tend to be well adapted to the new managed landscapes. Haiti provides one extreme example of habitat loss. Figure 6–7 shows the border between Haiti and the Dominican Republic, with Haiti on the left. Land use in Haiti has resulted in a tremendous loss of forests, something important not only in the context of loss of species, but also as a driver of soil erosion and water quality problems (as we will see in later chapters). In the Dominican Republic, social trends such as greater stability and democratization have helped the country avoid unrest and environmental destruction.

Fragmentation. Natural landscapes generally have large patches of habitat that are well connected to other similar patches. Human-dominated landscapes, however, consist of a mosaic of different land uses, resulting in small, often geometrically configured patches that frequently contrast highly with neighboring patches (Fig. 6–8). Small fragments of habitat can support only small numbers and populations of species, making them vulnerable to extermination. Species that require large areas are the first to go; those that grow slowly or have naturally unstable populations are also vulnerable.

Reducing the size of a habitat creates a greater proportion of **edges**—breaks between habitats that expose species to predators. Increased edge favors some species, but may be detrimental to others. For example, the Kirtland's warbler is an endangered species that is highly dependent on large patches of second-growth jack pines in Michigan. The species is endangered because its habitat has been greatly fragmented, creating edges that favor the brown-headed cowbird, a nest parasite that can invade the forest and lay its eggs in the nest of the rare warbler. Edges also favor species that prey on nests, such as crows, magpies, and jays. These predators are thought to be partly responsible for declines in many neotropical bird populations.

Simplification. Human use of habitats often simplifies them. Removing fallen logs and dead trees from woodlands for firewood, for example, diminishes an important microhabitat on which species ranging from woodpeckers to germinating forest trees depend. When a forest is managed for the production of one (or a few) species of tree, tree diversity declines, and as a result, so does the diversity of the plant and animal species that depend on the less favored trees. Streams are sometimes "channelized"—their beds are cleared of fallen trees and stony *riffle* areas (places with more rapidly moving water), and sometimes the stream is straightened out by dredging the bottom. Such alterations inevitably reduce the diversity of fish and invertebrates that live in the stream.

Intrusion. Birds use the air as highways, especially as they migrate north and south in the spring and fall. Recently, telecommunications towers have presented a new hazard to birds. Although television towers have been around for decades, new towers are sprouting up on hilltops and countrysides in profusion. Cell phone signal towers are also common. The lights often placed on these towers can attract birds, which usually migrate at night, and the birds simply collide with the towers and supporting wires. In 2000, a

Figure 6–8
Fragmentation.
Development usually leads to
the breaking up of natural areas,
resulting in a mosaic of habitats
that may not support local
populations of species.

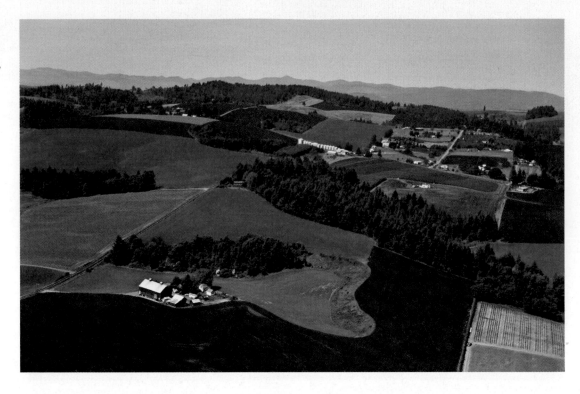

study estimated that the towers kill somewhere between
5 and 50 million birds a year.[9] Three organizations petitioned
the Federal Communications Commission (FCC) to incorporate bird mortality in its decisions about new towers in 2002,
but nothing was done. The scientific research from the study
gave environmentalists the power to go to court to force the
FCC to consider risk to birds in decision making. Finally, a
federal appeals court ruled in 2008 that the FCC did in fact
have to come up with a plan to protect birds. This ruling
resulted in a meeting of five agencies with the FCC in June of
2008. The problem remains unsolved, but new tower designs
or lighting may help. Ornithologist Daniel Klem estimated
that between 100 million and 1 billion birds die every year
from crashing into glass windows. Birds are just one example of organisms harmed by the intrusion of structures into
habitats. Any solution to wild species' decline must include
creative ways to lower the impact of human structures on
other organisms.

Invasive Species.   An exotic, or alien, species is one that is
introduced into an area from somewhere else, often a different continent (see Chapter 4). Because the species is not
native to the new area, it is frequently unsuccessful in establishing a viable population and quietly disappears. Such is the
fate of many pet birds, reptiles, and fish that escape or are
deliberately released. Some exotic species establish populations that remain at low levels without becoming pests.
Occasionally, however, an alien species finds the new environment to be very favorable, and it becomes an *invasive
species*, thriving, spreading out, and perhaps eliminating

native species by predation or competition for space or food.
In Chapter 13, you will learn that most of the insect pests
and plant parasites that plague agricultural production were
accidentally introduced. Invasive species are major agents in
driving native species to extinction and are responsible for an
estimated 39% of all animal extinctions since 1600.

Accidental Introductions.   The brown tree snake was
introduced onto the island of Guam as a stowaway on cargo
ships during World War II (Fig. 6–9). This snake is mildly poisonous, grows to a length of 8 feet, and eats almost anything
smaller than itself. In a little over 50 years, the snakes have
eliminated 9 of Guam's 12 native species of birds. They have no

Figure 6–9 The brown tree snake. This snake has invaded Guam and
other countries. It is causing the decline of many native birds.

[9]G. Shire, *New Report Documents Hundreds of Thousands of Birds Killed at
Communications Towers* (2000). Available from the American Bird Conservancy,
at www.abcbirds.org/newsandreports/releases/000629.html.

natural predators on the island, and sometimes they even invade local houses. Currently, wildlife officials are concentrating their efforts on preventing the snakes from spreading to other islands in the Marianas chain. Many other species were also introduced accidentally. Sahara mustard, an invasive species in New Mexico and other states, most likely came to the area in shipments of date palms from the Middle East.

May I Introduce... Deliberate introductions sanctioned by the U.S. Natural Resources Conservation Service (in the interest of "reclaiming" eroded or degraded lands) have brought us kudzu (Chapter 4) and other runaway plants, such as the autumn olive and multiflora rose. Saltcedar was introduced in the American Southwest for erosion control and has taken over riverbanks there. Many other exotic plants are introduced as horticultural desirables. Brazilian pepper, an ornamental shrub resembling holly, has become the most widespread exotic in Florida. It is an aggressive colonizer, taking over many native plant communities and fundamentally changing ecosystems like the Everglades (Fig. 6–10). The plant now occupies more than 700,000 acres (283,279 hectares) in central and southern Florida. A task force recently concluded, "The uncontrolled expansion of Brazilian pepper constitutes one of the most serious ecological threats to the biological integrity of Florida's natural systems."[10]

There are specific instances of both deliberate and accidental introductions in **aquaculture**—the farming of shellfish, seaweed, and fish—which produces one-third of all seafood consumed worldwide. Over 100 species of aquatic plants and animals are raised in aquaculture in the United States, most of which are not native to the farming locations. Parasites, seaweeds, invertebrates, and pathogens have been introduced together with the desired aquaculture species, often escaping from the pens or ponds and entering the sea or nearby river

system. For example, Asian black carp are used in raising catfish because the carp eat snails that harbor parasites capable of ruining the catfish meat. So far, none is known to have escaped in the United States, but everywhere else that they have been used, they have escaped and spread. Sometimes the aquaculture species itself is a problem. Atlantic salmon, for example, are now raised along the Pacific Coast, and many have escaped from the pens and have been breeding in some West Coast rivers, where they compete with native species and spread diseases.

Over Time. The transplantation of species by humans has occurred throughout history, to the point where most people are unable to distinguish between the native and exotic species living in their lands. European colonists brought literally hundreds of weeds and plants to the Americas, and now most of the common field, lawn, and roadside plant species in eastern North America are exotics. For example, almost one-third of the plants in Massachusetts are alien or introduced—about 725 species, of which 66 have been specifically identified as invasive or likely to become so. Among the animals introduced to North America, the most notorious have been the house mouse and the Norway rat; others include the wild boar, the starling, the horse, and the nutria. One of the most destructive exotics is the house cat: Studies in the 1990s showed that the millions of domestic and feral cats in the United States are highly efficient at catching small mammals and birds. They kill an estimated 1 billion small mammals (chipmunks, deer mice, and ground squirrels, as well as rats) and hundreds of millions of birds annually. One report calculates the annual cost of invasive species in the United States to be $137 billion.[11]

Species originating in the Americas can cause havoc in other places, too. For example, the North American gray

---

[10] Amy Ferriter, *Brazilian Pepper management Plan for Florida* (1997). Available at http://www.fleppc.org/Manage_Plans/schinus.pdf.

[11] D. Pimental, L. Lach, R. Zuniga, and D. Morrison, "Environmental and Economic Costs of Nonindigenous Species in the United States," *BioScience* 50 (2000): 53–65.

(a)

(b)

Figure 6–10 The Brazilian pepper bush. (a) Shown is a typical Everglades landscape, with wooded hammocks surrounded by marshes. (b) A virtual monoculture of Brazilian pepper covers this Everglades marsh area, obliterating the native biota.

Figure 6–11 The gray squirrel. While native to North America, the gray squirrel is an introduced invasive species in Europe, where it causes the decline of other species.

squirrel is invasive in Europe, where it outcompetes the native red squirrels and may be infecting them with a fatal virus. The gray squirrel is one of the 100 most invasive species worldwide (Fig. 6–11).

Invasive Species and Trophic Levels. Plants provide food for herbivores that in turn provide food for carnivores. Non-native plants, because they may have different resistance mechanisms, may be difficult for herbivores to eat and thus may keep energy and materials from passing up the food chain, even if a species has been in a new ecosystem for a long time. For example, the Norway maple was introduced to North America in 1756—but in spite of being in the United States for hundreds of years, these maples provide much less food up the food chain for herbivores (like caterpillars) and their predators (like song birds) than native trees do.

Pollution.   Another factor that decreases biodiversity is pollution, which can directly kill many kinds of plants and animals, seriously reducing their populations. For example, nutrients (such as phosphorus and nitrogen) that travel down the Mississippi River from the agricultural heartland of the United States have created a "dead zone" in the Gulf of Mexico, an area of more than 10,000 square miles (as of 2008) where oxygen completely disappears from depths below 20 meters every summer[12] (see Chapter 20). Shrimp, fish, crabs, and other commercially valuable sea life are either killed or forced to migrate away from this huge area along the Mississippi and Louisiana coastline.

Pollution destroys or alters habitats, with consequences just as severe as those caused by deliberate conversions. Every oil spill kills seabirds and often sea mammals, sometimes by the thousands. Some pollutants such as DDT can travel up food chains and become more concentrated in higher consumers (Chapter 5). Acid deposition and air pollution kill

[12]B. Walsh. *The Gulf's Growing "dead zone,"* (June 17, 2008). Available at http://www.time.com/time/nation/article/0,8599,1815305,00.html.

forest trees; sediments and large amounts of nutrients kill species in lakes, rivers, and bays; and global climate change, brought on by greenhouse gas emissions, is already having an impact on many species. Climate change is likely to accelerate in the next 50 years and is discussed in depth in Chapter 18.

Pollution can spread disease. Human wastes can spread pathogenic microorganisms to wild species, a threat called *pathogen pollution.* For example, manatees (endangered aquatic mammals) have been infected by human papillomavirus, cryptosporidium, and microsporidium, with fatal outcomes. The cause of a mysterious rash of deformities in many amphibian species in the 1990s has finally been identified as the larval stage of a flatworm that invades the tadpoles and induces the deformities (Fig. 6–12). The recent and rapid rise in the incidence of these deformities has been traced to habitats that have been altered by human use; high nutrient pollution has led to large snail populations, which are intermediate hosts of the flatworm.

Population.   Human population growth puts pressure on wild species through several mechanisms—through direct use of wild species, through conversion of habitat into agriculture or other use, through pollution, and through having large numbers of people using resources that wild species also need. Human population trends will be described in depth in Chapters 8 and 9. Overconsumption is tied to overpopulation and to the concept of *overexploitation*—the last letter in the HIPPO acronym—which will be discussed next. Even if each person used a modest amount of each resource, large numbers of people might use too much, leading to a total overconsumption. Thus, overpopulation is a problem even if people use few resources individually. However, a small group of people can also overuse resources. Indirectly, overconsumption can drive pollution, conversion of habitat, and other effects that harm wildlife. In a world in which some individuals own literally billions of times the money

Figure 6–12 Deformed frogs. Mysterious outbreaks of deformed amphibians have now been traced to a flatworm parasite that derails normal development.

and resources that others own, people with the most consumptive lifestyles have a disproportionate effect on the environment. Differences in level of consumption and in sheer numbers of people drive some of the tensions among the high-, middle-, and low-income countries (see Chapter 8).

Overexploitation. The overharvest of a particular species is known as *overexploitation*. Just as the overharvest of puffins led to their decline, removing whales, fish, or trees faster than they can reproduce will lead to their ultimate extinction. In addition, overuse of individual species harms ecosystems. Overexploitation is estimated to have caused 23% of recent extinctions and is often driven by a combination of greed, ignorance, and desperation. Poor resource management often leads to a loss of biodiversity. Forests and woodlands are overcut for firewood, grasslands are overgrazed, game species are overhunted, fisheries are overexploited, and croplands are overcultivated. These practices not only deplete the resource in question, but also often set into motion a cycle of erosion and desertification, with effects far beyond the exploited area. This problem is explored in depth in Chapter 7.

Trade in Exotics. One prominent form of overuse is the trafficking in wildlife and in products derived from wild species. Much of this "trade" is illegal. Worldwide, the International Criminal Police Organization (Interpol) estimates that it generates at least $12 billion a year, making it the third largest illicit income source, after drugs and guns. This illegal activity flourishes because some consumers are willing to pay exorbitant prices for furniture made from tropical hardwoods (like teak), exotic pets, furs from wild animals, traditional medicines from animal parts, and innumerable other "luxuries," including polar-bear rugs, ivory-handled knives, and reptile-skin handbags. For example, some Indonesian and South American parrots sell for up to $10,000 in the United States, and a panda-skin rug can bring in $25,000.

One example of illegal wildlife trade that still continues is that of *shahtoosh* shawls (Figure 6–13a), which can bring

thousands of dollars and can be purchased in many large cities in India and elsewhere, in spite of a trade ban since 1975. The shawls, extremely soft and warm, are made of the wool of the Tibetan antelope, or chiru (Figure 6-13b). They became the "must-have" luxury item for the wealthy elite of London, New York, Hong Kong, and other large cities in the 1980s when fur became unpopular. Their popularity remained in spite of the negative effect on the animals—in part, because people who bought them, and even some of the Kashmiri weavers who created them, did not know that the animals are killed to get the wool. In fact, three antelopes are needed to make one shawl. Some 20,000 chirus were killed each year, and the population of the species has declined to less than 50,000. Retraining Kashmiri weavers and providing them with other sustainable fine wools, as well as several public court cases that made the shawls less popular in the West, have countered this trend, but not entirely.

Greed. The long-term prospect of extinction does not curtail the activities of exploiters because, to them, the prospect of a huge immediate profit outweighs it. Even when the species is protected, the economic incentive is such that poaching and black-market trade continue. Particularly severe is the growing fad for exotic pets—fish, reptiles, birds, and houseplants. In many cases, these plants and animals are taken from the wild; often, only a small fraction even survives the transition. When animals and plants are removed from their natural breeding populations, these individuals are no better than dead when it comes to maintaining the species. At least 40 of the world's 330 known species of parrots are threatened because of this fad. Many pet stores carry imported species. Not purchasing wild-caught species is one way individuals can help preserve biodiversity.

In 1992, the United States, the largest importer of exotic birds, adopted the *Wild Bird Conservation Act*, a law designed to stop wild capture of declining birds, uphold international treaties in wildlife trade, and support sustainable breeding programs. In 2004, more than 200 nongovernmental organizations

Figure 6–13 Shahtoosh shawl and Tibetan antelope (chiru). (a) Shahtoosh shawls are prized items for the wealthy elite. (b) This species of antelope is hunted for its delicate fur, which is used to make shahtoosh shawls.

(NGOs) signed the European Union Wild Bird Declaration, a call for wild bird protection in the European Union (EU). In a huge step forward in bird conservation, in 2007, the EU decided unanimously to prohibit the import of wild birds.

HIPPO represents the major reasons for species decline. Habitat is disrupted or degraded. Species are introduced intentionally or accidentally and disrupt relationships in communities. Pollution harms species directly and can also harm them indirectly, as in the case of the frogs and parasites. Both human population growth and overconsumption by individuals exacerbate the decline of wild species. Finally, overexploitation of a wild species used as a resource can lead to its decline. All of these problems have solutions, however. Such solutions demand both individual efforts and corporate decision making.

Consequences of Losing Biodiversity

Why is biodiversity loss a concern? The Millennium Ecosystem Assessment team had this answer in their report: because biodiversity is essential for the ecosystem services and goods that human societies derive from the natural world. For example, richly diverse mangrove forests and coral reefs are buffers against coastal storms and floods. The well-being of millions of rural poor in the developing world is tied to the provision

of products of natural ecosystems, especially during economic hard times. In fact, biodiversity loss threatens the achievement of virtually all of the Millennium Development Goals (see Chapter 8). Nature-based tourism would not happen without the rich diversity of wildlife and flora that attract and delight people; for many countries of sub-Saharan Africa, nature-based tourism represents the largest earner of foreign currency.

Can ecosystems lose some species and still retain their functional integrity? Research evidence indicates that the dominant plants and animals determine major ecosystem processes such as energy flow and nutrient cycling. (See Sound Science, "Biodiversity: Essential or Not?") Thus, simplifying ecosystems by driving rarer species to extinction will not necessarily lead to ecosystem decline. This has its risks, however: It is possible to lose *keystone species*—species whose role is absolutely vital to the survival of many other species in an ecosystem (see Chapter 4). For example, through felling trees and building dams, beavers make local habitats that undergo succession as beavers move throughout the landscape, increasing ecosystem diversity in a region. Alternatively, the keystone species in an ecosystem may be a predator at the top of a food chain that keeps herbivore populations under control. Sometimes keystone species are the largest animals in the ecosystem, and because of their size,

---

Sound Science

Biodiversity: Essential or Not?

A fierce debate is raging among ecologists, focused on a very important question that has serious political and economic ramifications: Will the loss of species in an ecosystem bring it closer to collapse? What is the role of biodiversity in promoting normal functioning, such as productivity, in an ecosystem? On one side of this controversy are some of the leading ecologists in the United States (such as David Tilman) and the United Kingdom (John Lawton), who have published experimental work indicating that biodiversity is essential to the healthy functioning of an ecosystem.

In one line of experimental work—the Ecotron Experiment—researchers added various plants, insects, and worms to identical plots in enclosed chambers. They found that the more species they added to a plot, the more biomass the plot produced and thus the greater the productivity of the plot. Other work carried out in Minnesota showed that plots with a greater diversity of species were more resistant to drought. Tilman, Lawton, and others authored an important *Issues in Ecology* paper published by the Ecological Society of America in 1999. Citing their experimental work, the authors concluded, "Given its importance to human welfare, the maintenance of ecosystem functioning should be

included as an integral part of national and international policies designed to conserve local and global biodiversity." *

But wait a minute, say Michael Huston of Oak Ridge National Laboratory in Tennessee, Phil Grime of the University of Sheffield in the United Kingdom, and others.[†] There is a good chance that these experimental studies and the conclusions drawn from them were "politically manipulated." Not only that, but in some important cases, the experiments were flawed and therefore inconclusive. Years ago theoretical ecologist Robert May concluded, on the basis of mathematical modeling, that biodiversity has no consistent effect on the functioning of an ecosystem. What is more important, the critics say, is how individual organisms respond to environmental change. Citing experimental evidence in support of this view, these researchers note that, if particular species are present, they will often dominate ecosystem functioning because of their superior adaptations to a

given environment; the presence of many other species can be irrelevant to ecosystem functions.

This evidence does not indicate that biodiversity is unimportant—in fact, greater biodiversity will make the presence of key species more likely in an ecosystem. In a sense, biodiversity is an insurance policy against environmental variability such as droughts or harsh winter weather. But to say that biodiversity is *the* key to ecosystem functioning and to transfer this conclusion to policy recommendations goes beyond what the science supports, say critics (including Ross Thompson and Brian Starzomski, authors of a 2007 review of the subject[‡]). These critics believe that the science surrounding ecosystem function and biodiversity is still being sorted out and suggest that the positive value society places on biodiversity needs to be justified with other reasons. Because the scientific connection between the number of species in an ecosystem and its health is not always clear, it is even more important for people to think through the instrumental and intrinsic values of species.

*S. Naeem, et al. "Biodiversity and Ecosystem Functioning: Maintaining Natural Life Support Processes." *Issues in Ecology*. Ecological Society of America. p. 2.
[†]J. Kaiser, "Ecology: Rift Over Biodiversity Divides Ecologists," *Science* 289, 5483(2000): pp. 1282–1283.

[‡]R. Thompson and B. Starzomski, "What Does Biodiversity Actually Do? A Review for Managers and Policy Makers," *Biodiversity and Conservation* 16, no. 5 (2007): 1359–1379.

they have the greatest demands for unspoiled habitat. Thus, wolves, elephants, tigers, moose, and other large animals are considered "umbrella" species for whole ecosystems.

As we have seen, one thing many declining species have in common is that they are often long-lived, large, and older at first reproduction and have high parental care of offspring (as *K*-strategists are). Thus, they are vulnerable to rapid changes in the environment. Sometimes species go into decline although they are quite common. Species such as the passenger pigeon, once the most common bird in North America, can leave a huge ecological gap when they decline or disappear. Those species least likely to be harmed by human actions include widely distributed species, with small size, rapid reproduction, short generation time, low parental care of offspring, and an ability to migrate (characteristics of *r*-strategists). Such species (roaches, houseflies, and others) are most likely to become pest species. Thus, the overarching effect of human disturbance of the environment is not only to decrease species that we may like, but also to increase a number that will become major pests.

Moving Forward. What will happen as rare or unknown species go extinct because of human activities? Some people will mourn their passing. We do not really know, however, what we are losing when species become extinct. One thing is certain: The natural world is less beautiful and less interesting as a result. We might be inclined to feel overwhelmed by these numbers and the significant decline in species and in abundances of organisms. But there are glimmers of hope. Sometimes species we thought were extinct are found not to be. Two of these are the Beck's petrel, not seen for 80 years before being found by an ornithologist in 2008, and the pygmy tarsier, not seen since 1921 and thought extinct, but found on the Indonesian island of Sulawesi in 2008. Sometimes new populations of very rare species are discovered, as happened in the case of the highly threatened greater bamboo lemur. A new population of this rare species was found in 2007, around 400 kilometers (240 mi) from the only previously known population. As we saw in the opening story, sometimes the hard work of scientists and conservation volunteers pays off in the recovery of a rare and declining species such as the puffin. We will see a number of these success stories as we look at policies protecting species.

Sometimes new protections emerge from a change in policy. Ironically, the conservation of birds was enhanced in the EU when concern over avian flu made it possible to pass regulations limiting the import of wild birds. Sometimes scientific accomplishments make conservation doable. For example, 2008 saw the first live rhinoceros birth from frozen semen, and black-footed ferrets, an extremely rare weasel-like animal, have been born from sperm stored from fathers who had died eight years before! These breakthroughs in captive breeding may save species. Once in awhile, conservationists catch a break, as they did in 2008 when a rare Vietnamese turtle, one of the last four of its species in the world, was swept from a lake by a flood and caught by a fisherman who was going to sell the reptile to a restaurant for soup. Conservationists were able to purchase the 68 kilogram (150 lb) turtle and return it to the lake. In the next section, we will see more of the efforts scientists and others are putting into saving wild species.

6.3 Saving Wild Species

If wild species and biodiversity as a whole have so much value and if they are in such decline, there must be efforts in the works to solve the problem. Indeed, there are. Those efforts involve basic science, public policy, and decisions at the local, national, and international levels. Scientists are on the front lines of biodiversity protection, as they often have a better knowledge of what is out there in the world and whether or not it is declining. However, scientists cannot stop biodiversity loss. That requires laws, and it demands enforcement of those laws. Protection of biodiversity also requires stepping back to look at the big picture, seeing how people relate to the rest of the natural world, and examining how we can do so in more sustainable ways.

The Science of Conservation

The example of the scientific work that was used to save the puffins is just one of many in *conservation biology*, the branch of science that is most focused on the protection of populations and species. Conservation biologists use techniques like captive breeding, telemetry (transmitting data through remote communication), and tracking devices in order to learn more about the natural history of particular species and protect those that are vanishing.

There is another critical element in the science of protecting species—**taxonomy**. Taxonomy is the cataloging of species and the naming of new ones. It allows us to understand what species are out there and to identify those in the most trouble. Right now, the field of taxonomy is suffering from a lack of experts, and this lack is harming biodiversity conservation. By 2018, there may be no specialists in fungi in the United Kingdom, in spite of the fact that fungi provide medicines, can be important crop pests, and are important parts of soil health.[13] Because most people cannot even identify the species we are dealing with, it is difficult to promote solutions. The loss of expertise as taxonomists retire and the pressure from international treaties to protect species may require increased funding to train new taxonomists.

Policy and Political Structure

Public policies, and the agencies that make and support them, are necessary for the protection of species. For example, if we want to save wild species, we first need to ask who *owns* wildlife. In the United States, property owners do not own the wildlife living on their lands; wildlife resources are public resources, protected under the Public Trust Doctrine. The government holds these resources *in trust* for all the people and is obliged to provide protection for these resources. In the United States, this protective role is usually exercised by state fisheries and wildlife agencies. Sometimes the law mandates federal jurisdiction, as in the case of endangered

---

[13]S. Connor. "Fungi scientists are endangered species: Budget cuts threaten research into basics that propagate life on Earth," *The Independent* (November 2008). Available at http://www.independent.co.uk/news/science/fungi-scientists-are-endangered-species-1038862.html.

species. Game animals were some of the first animals people recognized as having the potential to be overharvested; consequently, they are a group that receives special oversight.

Game Animals in the United States.

Game animals are those animals traditionally hunted for sport, meat, or pelts. In the early days of the United States, there were no restrictions on hunting, and a number of species were hunted to extinction (the great auk, heath hen, and passenger pigeon) or near extinction (the bison and wild turkey). As game animals became scarce in the face of unrestrained hunting, regulations were enacted. State governments, backed up by the federal government, enacted laws establishing hunting seasons and bag limits and hired wardens to enforce the laws. Some species were given complete protection in order to allow their populations to build up to numbers that would once again allow hunting.

Turkey Boom. Consider the wild turkey. A favorite game species, this bird was hunted to the brink of extinction, but was making a slow comeback by the 1930s as a result of hunting restrictions. At that time, there was a total population of about 30,000 birds in a few scattered states. After World War II, state and federal programs addressed the need for protecting turkey habitats. The birds were reintroduced into areas they had once inhabited, and hunting quotas were strictly limited. The turkey is now found in 49 states, and its population has risen to almost 7 million as a result of these measures (Fig. 6–14). Unfortunately, the turkey was also introduced to western states, where it was not native, and it is now a pest in those areas. Ideally in the future, it would be protected in states where it is native and removed from states where it is invasive.

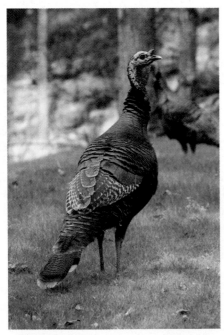

Figure 6–14 The wild turkey. Restoration efforts have so successfully reestablished the wild turkey throughout the United States that hunting within carefully controlled limits now takes place.

Hunting and Conservation.

Using hunting and trapping fees as a source of revenue, state wildlife managers enhance the habitats that support important game species. In spite of what seems on the surface to be a destruction of wildlife, hunting has many positive aspects. Besides the fees hunters pay, many hunters belong to organizations dedicated to the game they are interested in hunting. Organizations such as Ducks Unlimited and the National Wild Turkey Federation raise funds that are used for the restoration and maintenance of natural ecosystems vital to the game they are interested in hunting. For example, Ducks Unlimited announced in 2008 that more than 12 million acres (4.86 million hectares) of vital wetland habitat for wildfowl have been conserved in North America. The organization's activities have helped stem the tide of loss that has reduced North America's wetlands by more than 50% over the years. These activities are in great measure responsible for the remarkable yearly hunting harvest of ducks and geese in North America—some 14.5 million ducks and more than 3.7 million geese in the 2007–2008 season.[14]

Defenders of hunting and trapping argue that their prey are often animals that lack natural predators and would increase to the point of destroying their own habitat, something that frequently happens in the United States with larger animals such as deer and elk. For example, there are so many deer in Pennsylvania that, in 2004, they were estimated to be at three times the natural carrying capacity. The deer eat crops, damage cars in collisions, and seriously harm forest understories through extensive browsing. It might seem like hunting is the answer, but sometimes it is not that simple. In spite of high deer numbers, some hunters have actually lobbied the state to take action to *increase* the deer herd. Why? These hunters are hunting in the northern and western part of the state, where deer populations were once heavy. Unfortunately, the deer have left these forests and have moved to the southern and eastern parts of the state, where they eat agricultural crops and browse in suburban developments. They wreak economic damage, but are very difficult to hunt. Thus, the perception of hunters is that deer populations are declining, while the perception of many other people (including wildlife managers) is that there are too many deer for ecosystems to support.

In Australia, kangaroos are so common that professional sharpshooters now cull more than 3.5 million kangaroos a year (out of an estimated 35–50 million), mostly for the pet food and leather industries. Here the kangaroo acts as a wild "range" animal. It is better adapted to the grasslands than cattle might be and provides a source of regional income, while keeping people from switching to animals like domestic cattle or sheep, which are harder on the soil. However, many members of the nonhunting public object to the killing of wildlife, and some groups, such as the Humane Society of the United States and People for the Ethical Treatment of Animals (PETA), actively campaign to limit or end hunting and trapping. Some practices that are regarded as especially cruel,

---

[14]U.S. Fish and Wildlife Service, *New Report Shows 2007–2008 Waterfowl Harvest* (2008). Available at www.fws.gov/news/NewsReleases/showNews.cfm?newsId=705E9F08-C75D-6CC7-15088A1A4B67E811.

such as the use of leghold steel traps, have been banned in several states as a result of ballot initiatives.

Even people who are not opposed to hunting may feel that encouraging hunting because of a lack of other predators may not be getting at the fundamental problems with species loss. They argue for the return of predators such as grizzly bears, mountain lions, and wolves in order to restore more natural checks and balances to ecosystems. Because humans do not always get along well with top predators, such reintroductions are unlikely to occur on a large scale.

Backyard Menagerie. As in the case of the deer in Pennsylvania, many species of animals are frequently found in urban and suburban areas. Common game animals, such as rabbits, doves, and squirrels, are well adapted to suburban environments. Due to protection from overhunting, viable populations of these animals are being maintained. Serious problems, however, have emerged as a result.

The number of animals killed on roadways now far exceeds the number killed by hunters. As rural areas are developed, the increasing numbers of animals found on roadways are a serious hazard to motorists. With 5.7 million miles (9.2 million km) of paved roadways in the United States, more than a million animals a day become roadkill. Overpasses and tunnels are increasingly being built to provide wildlife with safe corridors (Fig. 6–15). A 2008 study at Purdue University identified amphibians as especially affected by road mortality.[15] More than 95% of the roadkill that researchers identified over 11 months along a stretch of road in the Midwest were amphibians.

Many nuisance animals are thriving in highly urbanized areas, creating various health hazards. Opossums, skunks, and deer are attracted to urban areas by opportunities for food;

---

[15]D. J. Glista, T. L. DeVault, and J. A. DeWoody, "Vertebrate Road Mortality and Its Impact on Amphibians," *Herpetological Conservation and Biology* 3 (2008): 77–87.

Figure 6–15 Highway overpass. This wildlife overpass in Banff, British Columbia, facilitates wildlife traffic between the forests intersected by the highway.

unsecured garbage cans and pet food left outside will ensure visits from raccoons. In 2007, the Centers for Disease Control and Prevention reported 7,258 cases of rabid animals in the United States, 57% of which were raccoons and skunks.[16]

Because wild animals are in the public trust, their numbers are managed by government agencies. Now that many populations are increasing, the government sometimes has to limit them. In recent years, suburbanites have been increasingly attacked by cougars, bears, and alligators as urbanization encroaches on the wild (Fig. 6–16), with resulting

---

[16]J. D. Blanton, et al. *Rabies Surveillance in the United States During 2007.* Available at http://74.125.45.104/search?q=cache:_-mdy6Rbj9gJ:www.cdc.gov/rabies/docs/rabies_surveillance_us_2007.pdf+cdc+number+rabid+animals+in+US2007&hl=en&ct=clnk&cd=1&gl=us.

Figure 6–16 Cougar on the roof. As suburbs encroach on cougar habitats in the western United States, attacks on humans have become more numerous. This particular cougar on the roof of a home in a California suburb has been shot with a tranquilizer dart.

increases in contact between people and these large predators. Coyotes, which once roamed only in the Midwest and western states, are now found in every state and are increasing in numbers. A highly adaptable predator, the coyote will eat almost anything. In suburban neighborhoods, they often dine on cats, small dogs, and human garbage. (One enterprising pair had a den on the median strip of Boston's beltway, where they lived off roadkill.) Finding a way to protect predators and human safety at the same time is a part of our modern stewardship task.

One highly controversial effort to control unwanted animals is carried out by an agency of the U.S. Department of Agriculture. In 1998, the agency changed its name from **Animal Damage Control** to **Wildlife Services** in an effort to soften its image. This agency responds to requests from livestock owners, farmers, homeowners, and others to remove nuisance animals and birds. "Removal" virtually always means killing, and Wildlife Services routinely uses poisons, traps, and other devices to kill around 2.5 million animals yearly. Of these, most are native species such as raccoons or skunks, but others, such as wild hogs and starlings, are exotic invasive species. Wildlife Services plays the important role of keeping both invasive species and native species from overtaking human areas and limiting negative human-wildlife interactions.

Protecting Endangered Species

In colonial days, huge flocks of snowy egrets inhabited the coastal wetlands and marshes of the southeastern United States. In the 1800s, when fashion dictated fancy hats adorned with feathers, egrets and other birds were hunted for their plumage. By the late 1800s, egrets were almost extinct. In 1886, the newly formed National Audubon Society began a press campaign to shame "feather wearers" and end the practice. The campaign caught on, and gradually, attitudes changed; new laws followed. Government policies that protect animals from overharvesting are essential to keep species from the brink of extinction. Even when cultural standards change due to the efforts of individual groups (such as the National Audubon Society), laws and policy measures must follow to ensure that endangered populations remain protected. Since the 1800s, several important laws have been passed to protect a wide variety of species.

Lacey Act. Florida and Texas were the first states to pass laws protecting plumed birds. Then, in 1900, Congress passed the **Lacey Act**, forbidding interstate commerce in illegally killed wildlife—and therefore making it more difficult for hunters to sell their kill. Since then, numerous wildlife refuges have been established to protect the egrets' breeding and migratory habitats. With millions of people visiting these refuges and seeing the birds in their natural locales, attitudes have changed significantly. Today the thought of hunting these birds would be abhorrent to most of us, even if official protection were removed. Thus protected, egret populations were able to recover substantially. In the meantime, the Lacey Act and its amendments have become the most important

piece of legislation protecting any wildlife from illegal killing or smuggling. Under the act, the U.S. Fish and Wildlife Service can bring federal charges against anyone violating a number of wildlife laws.

Special agents of the U.S. Fish and Wildlife Service worked on over 12,177 violations of the Lacey Act and other wildlife-related laws in 2007. Their efforts resulted in fines of more than $14 million and a total of 31 years of jail time for those convicted. Following are some examples of their work: (1) Owners of an animal park in South Dakota were caught trafficking bear gall bladders and buying and selling threatened grizzly bears; (2) a Utah man was fined for having 106 illegally caught bobcat skins that he had intended to sell in Europe (where their value was $38,000); and (3) the poaching of thousands of undersized leopard sharks in California waters was stopped, the criminals were prosecuted, and nearly $1 million in fines were used to create a fund to restore leopard shark habitat in the San Francisco Bay area. In 2007, the Combat Illegal Logging Act of 2007 (S. 1930) was proposed to expand the Lacey Act to cover rare timber and timber products. It is still waiting to move forward.[17]

Endangered Species Act. Congress took another major step when it passed a series of acts to protect endangered species. The most comprehensive and recent of these acts is the **Endangered Species Act** of 1973 (reauthorized in 1988). Recall from Chapter 4 that an *endangered species* is a species that has been reduced to the point where it is in imminent danger of becoming extinct if protection is not provided (Fig. 6–17). Genetically distinct populations (subspecies) such as the Florida panther and the peninsular bighorn sheep may also be protected. Further, the act provides for the protection of *threatened species*, which are judged to be in jeopardy, but not on the brink of extinction. When a species is officially recognized as being either endangered or threatened, the law specifies substantial fines for killing, trapping, uprooting (in the case of plants), modifying significant habitat of, or engaging in commerce in the species or its parts. The legislation forbidding commerce includes wildlife threatened with extinction anywhere in the world.

The Endangered Species Act is administered by the U.S. Fish and Wildlife Service for terrestrial and freshwater species (and a few marine mammals) and by the National Oceanic and Atmospheric Administration (NOAA) Fisheries Service for marine and anadromous species (those that migrate into freshwater). There are three crucial elements in the process of designating a species as endangered or threatened:

1. **Listing.** Species may be listed by the appropriate agency or by petition from individuals, groups, or state agencies. Listing must be based on the best available information and must not take into consideration any economic impact the listing might have. Current listed species are summarized in Table 6–3.

---

[17]USFWS Actions. *US FWS Office of Law Enforcement.* Annual Report FY 2007 (2008). Available at http://www.fws.gov/le/aboutle/annual.htm.

(a) Male Pine Barrens Tree Frog   (b) Silversword   (c) Red-cockaded Woodpecker

(d) Karner Blue Butterfly   (e) Swamp Pink   (f) Whooping Cranes

(g) Devil's Hole Pupfish   (h) White Oryx   (i) Manatee

Figure 6–17 Endangered species. Shown are some examples of endangered species—species whose populations in nature have dropped so low that they are in imminent danger of becoming extinct unless protection is provided. Go to **GRAPHit!** at www.masteringenvironmentalscience.com.

2. **Critical Habitat.** When a species is listed, the agency must also designate as *critical habitat* the areas where the species is currently found or where it could likely spread as it recovers. A 1995 Supreme Court decision made it clear that federal authority to conserve critical habitats extended to privately held lands.

3. **Recovery Plans.** The agency is required to develop recovery plans that are designed to allow listed species to survive and thrive.

As Table 6–3 indicates, as of November 2008, 1,327 U.S. species were currently listed for protection under the act;

TABLE 6–3 Federal Listings of Threatened and Endangered U.S. Plant and Animal Species, 2008

Summary of Listed Species, Listed Populations,[a] and Recovery Plans[b] as of 11/21/2008

	United States			Foreign			Total Listings (U.S. and Foreign)	U.S. Listings with Active Recovery Plans[c]
	Endangered	Threatened	Total Listings	Endangered	Threatened	Total Listings		
Mammals	70	14	84	256	20	276	360	57
Birds	75	15	90	179	6	185	275	85
Reptiles	13	24	37	66	16	82	119	38
Amphibians	13	11	24	8	1	9	33	17
Fishes	74	65	139	11	1	12	151	102
Clams	62	8	70	2	0	2	72	70
Snails	64	11	75	1	0	1	76	70
Insects	47	10	57	4	0	4	61	39
Arachnids	12	0	12	0	0	0	12	12
Crustaceans	19	3	22	0	0	0	22	18
Corals	0	2	2	0	0	0	2	0
Animal Subtotal	449	163	612	527	44	571	1,183	508
Flowering Plants	572	143	715	1	0	1	716	631
Conifers and Cycads	2	1	3	0	2	2	5	3
Ferns and Allies	24	2	26	0	0	0	26	26
Lichens	2	0	2	0	0	0	2	2
Plant Subtotal	600	146	746	1	2	3	749	662
Grand Total	1,049	309	1,358	528	46	574	1932	11,70

Source: Department of the Interior, U.S. Fish and Wildlife Service, Division of Endangered Species, November 21, 2008.
[a]A listing has an E or a T in the "status" column of the tables in 50 C.F.R. §17.11(h) or 50 C.F.R. §17.12(h) (the "List of Endangered and Threatened Wildlife and Plants").
[b]Twelve animal species (7 in the United States and 5 in foreign countries) are counted more than once in the above table, primarily because these animals have distinct population segments (each with its own individual listing status).
[c]Recovery plans pertain only to U.S. species.

recovery plans are in place for 1,170 of them, but only 526 species have designated critical habitat because of lack of data and lack of resources. In addition, 253 candidate species are waiting to be listed.

Alternatives and Roadblocks. The Endangered Species Act was scheduled for reauthorization in 1992, but because of political battles, it has been operating on year-to-year budget extensions. It has been a political lightning rod since its enactment. Opposition to the act comes from development, timber, recreational, and mineral interests as well as other lobby groups. Organizations such as the American Land Rights Association and the American Forest Resource

Council continually lobby for weakening or abolishing the Endangered Species Act because they believe it limits their property rights to log, build buildings, and carry out other activities. They have allies in the Congress—including former Congressman Richard Pombo, who once chaired the House Resources Committee. In 2005, Pombo sponsored H.R. 3824, the Threatened and Endangered Species Recovery Act (TESRA).

TESRA made a number of changes that were considered by conservation scientists to lower the likelihood of species recovery. These included eliminating protection of critical habitat for species, lessening consultation between the U.S. Fish and Wildlife Service and the Environmental Protection

Agency (EPA) on the impacts of pesticides on wildlife, and bypassing review of federal actions by wildlife agencies. It would hamper the ability of scientists to use modeling and government reports in their assessments of extinction risk. The objections of scientists to TESRA were so strong that six scientific societies signed a joint statement outlining their concerns in 2006.[18] TESRA passed the House in the 109th Congress, but failed to make it through the Senate and was not reintroduced subsequently.

Conflicting Values. Some critics claim that the Endangered Species Act has been a failure because only 10 of the 1,300 listed species have recovered and been taken off the list. In response, proponents state that the two main causes of extinction—habitat loss and invasive species—are on the increase. Furthermore, species are listed only when they have already reached dangerously low populations, which means that they almost certainly would have gone extinct without the protection. If success were measured by the number of species that have stabilized or increased their populations, then the act might be judged a success, since some 41% of listed species have responded this way after being listed.

Other critics of the Endangered Species Act believe that it does not go far enough. A major shortcoming is that protection is not provided until a species is officially listed as endangered or threatened and a recovery plan is established. For half of the protected vertebrate species, fewer than 1,000 individuals remained in each population at the time of listing; for half of the listed plant species, fewer than 120 individuals remained. Currently, the list's backlog includes some 253 "candidate species," all of which are acknowledged to be in need of listing. In the first 20 years following the passage of the Endangered Species Act, 114 U.S. species went extinct (or "missing," which means that, despite at least 10 years of survey, they haven't been seen). The majority of these never made it to the list and became extinct during a delay in listing procedures.

Another contentious issue is the establishment of critical habitat. Opponents believe that the critical habitat designation puts unwanted burdens on property owners and assert that the critical habitat findings have little merit in conserving species. TESRA would have completely eliminated critical habitat

protections and replaced them with a statement that recovery plans must identify areas of "special value" to the protection of the species. However, nothing in this designation is intended to be mandatory. Critics of TESRA answer that critical habitat works; species with critical habitat have been twice as likely to be on the road to recovery as those without it.

Future Legislation. TESRA was one of a number of attempts to weaken the Endangered Species Act. It had the strong support of political and commercial interest groups, but was opposed by a host of environmental organizations and scientists. In March 2006, a letter signed by almost 6,000 scientists from around the country was delivered to the U.S. Senate, urging them to pass legislation that maintains and strengthens the provisions and funding of the Endangered Species Act and to reject the proposals from the House that would weaken the act.

In the final analysis, the Endangered Species Act is a formal recognition of the importance of preserving wild species, regardless of whether they are economically important. Species listed as endangered or threatened have legal rights to protection under the law. The act is something of a last resort for wild species, but it embodies an encouraging attitude toward nature that has now become public policy. Some changes may make the act function better. These might include tax breaks as incentives to landowners caring for endangered species or other measures to lift the burden of critical habitat designations. In 2007, a bipartisan group of 17 senators introduced another bill—the Endangered Species Recovery Act (ESRA) of 2007 (S.700)—to suggest just that. The reauthorization of a strong Endangered Species Act, possibly with support from other laws like this, will be a major signal of our continuing commitment to consider the value of wild species on grounds other than economic or political.

Seeing Success

Efforts to protect species, such as the Lacey Act, the Marine Mammals Protection Act, the Endangered Species Act, and many other laws designed to keep our biotic heritage from going extinct, have resulted in some successes. An excellent example of a species restored to many former areas of its range because of these protection and recovery efforts is the gray wolf (see Sustainability, "Return of the Gray Wolf"). Other species have also benefited significantly from these laws.

---

[18]*Scientific Societies' Statement on the Endangered Species Act* (February 27, 2006). Available at www.esa.org/pao/policyStatements/pdfDocuments/2 2006_finalStatement_Scientific%20Societies%20ESA.pdf.

Sustainability
Return of the Gray Wolf

There has been a profound shift in public attitude toward wolves over the past few decades. Prior to the Endangered Species Act, wolves had been exterminated from all but 2 of the lower 48

states. Currently, gray wolf populations are on the increase. In fact, more than 3,900 of them are now found in 9 states. Whereas states once poisoned and trapped wolves and paid bounties for

them, a nonprofit organization called *Defenders of Wildlife* now pays ranchers for any losses they suffer from wolf predation. (In the last 20 years, Defenders has compensated ranchers nearly

(Continued)

$700,000 for such losses.) The data from Idaho, home of more than 500 wolves, show that domestic dogs kill 20 times more sheep than wolves do.

In a highly publicized program, the U.S. Fish and Wildlife Service began releasing captured Canadian wolves in 1995 into Wyoming's Yellowstone National Park and central Idaho. The 14 wolves released in Yellowstone in 1995 produced 9 pups, successfully preying on the abundant elk and bison in the park and rewarding visitors with views of their activities. Encouraged by their success, the wildlife biologists added 17 more wolves to Yellowstone in 1996 and then halted the reintroductions. In 2006, the Yellowstone wolf population numbered 118. Recent observations suggest that the wolves are bringing greater ecological health to areas of Yellowstone, largely through their impact on the overabundant elk herds in the park. Streamside willows have been restored, much to the delight of beavers and moose; the elk have abandoned feeding on the willows because wolves can easily catch them in those habitats.

Wolf populations in the Great Lakes and Rocky Mountain areas were delisted in 2008 after achieving population goals set by the U.S. Fish and Wildlife Service for recovery. However,

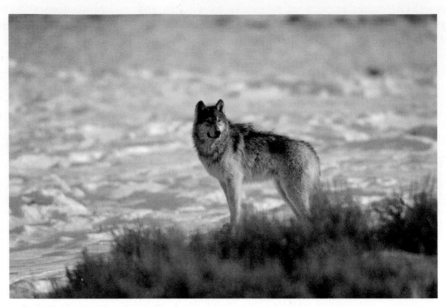

Gray wolf.

the Rocky Mountain population was relisted in an emergency court ruling after 37 wolves were shot in one month. The relisting was a contentious move, promoted by environmentalists and decried by ranchers and farmers. The conflicting interests of protectionists on the one hand and farmers and ranchers on the other will have to be worked out through a careful program of regulation of wolf populations—just as many other large animals, such as bears, cougars, and coyotes, are kept under control by regulations and hunting quotas.

Birds of Prey.   One of the species recently removed from the list because its numbers have recovered significantly is the American bald eagle, the U.S. national symbol (Fig. 6–18). The peregrine falcon, another previously endangered species, has also been removed for the same reason. Both are large raptors—predatory birds at the top of the food chain. Both the bald eagle and the peregrine falcon were driven to extremely low numbers because of the use of DDT as a pesticide from the 1940s through the 1960s. Carried up to these predators through the food chain, DDT caused a serious thinning of the birds' eggshells that led to nesting failures in the two species and in numerous other predatory birds (see Chapter 5). DDT use was banned in both the United States and Canada in the early 1970s, and the stage was set for the birds to recover. On August 20, 1999, the peregrine falcon was delisted, and on June 28, 2007, the Interior Department took the American bald eagle off the Endangered Species List. Both raptors will still be protected by the Migratory Bird Treaty Act, and the eagle will be protected by the Bald and Golden Eagle Protection Act.

Fly Away Home.   A few species have gained exceptional public attention, and heroic efforts have been mounted to save them. Efforts to save the whooping crane, for example, include virtually full-time monitoring and protection of the single remaining flock, which for years numbered only in the teens. The historic migratory wild flock is increasing steadily and reached a population of 266 individuals in 2008 from a low of

Figure 6–18 The American bald eagle. The Endangered Species Act worked well in bringing this magnificent national symbol to the point where it was dropped from the list.

14 cranes in 1939. This flock migrates from their wintering grounds on the Texas Gulf Coast to Wood Buffalo National Park in Canada. A nonmigratory flock has been reintroduced in central Florida. Some 31 birds now make up the flock, but breeding has been tough because of bobcat predation and an extended drought in Florida; releases into that flock have been

Figure 6–19 Whooping cranes and pilot. A pilot in an ultralight aircraft leads a flock of whooping cranes in their migration from Wisconsin to Florida.

stopped. A spectacular effort has established a new migratory flock of 72 cranes that summer in Wisconsin and migrate to Florida. This flock was first "taught" its migratory path by following researchers in ultralight aircraft (Fig. 6–19). Unfortunately, each of these flocks is extremely vulnerable to environmental changes. For example, a difficult winter in 2008 left the wild flock declining for the first time since 2001. The Texas flock lost 20% of its numbers in twelve months. However, whooping cranes have suffered setbacks before, and conservationists remain hopeful.

The Spotted Owl.   Critics of the Endangered Species Act claim that it sometimes goes too far to protect a species. A high-profile case was the northern spotted owl. This species became a focal point in the battle to save some of the remaining old-growth forests of the Pacific Northwest. The owl is found only in these forests and had dwindled to a population estimated at 2,400 pairs. In June 1990, the U.S. Fish and Wildlife Service listed the owl as a threatened species. Years of bitter controversy followed until a compromise—the **Northwest Forest Plan**—was worked out in 1994. The plan, employing the ecosystem management approach now adopted by the Forest Service (see Chapter 7), set aside 7.4 million acres (3 million hectares) of federal land in California, Oregon, and Washington where logging is prohibited in stands of trees older than 80 years. In 2005, chapters of the National Audubon Society filed notice of intent to sue the U.S. Fish and Wildlife Service, claiming that the service has never completed a final recovery plan for the northern spotted owl. In 2006, the service convened a team to make a recovery plan for the owl. They wrote a draft recovery plan, which was reviewed by the public in 2007 and released in 2008.

6.4 Protecting Biodiversity Internationally

Serious efforts are being made to preserve biodiversity around the world, especially in the tropics, where so much of the world's biodiversity exists. These protective efforts require an immense amount of coordination among local, state, and federal authorities. The National Biological Information

Infrastructure is a collaborative program that makes protecting species easier for agencies in the United States, but the United States also has to coordinate with the rest of the world. These partnerships not only create treaties, but also monitor species, share scientific advances, and find solutions to the needs of people who are clashing with wild species.

International Developments

There are numerous international developments for the protection of species. Some, like the **International Union for Conservation of Nature** (IUCN, formerly called the World Conservation Union), are organizations that monitor the successes and failures of our conservation efforts. Others are organizations that help coordinate scientists or policy makers around the globe. For example, the Invasive Species Specialist Group (ISSG) is a part of the Species Survival Commission of IUCN. The ISSG is made up of invited world experts who act as advisors and who maintain a global invasive species database. The policy and treaty makers have to hammer out documents like the Convention on Biological Diversity. Finally, there has to be funding for species protection, some of which is provided by the Critical Ecosystem Partnership Fund and other foundations.

The Red List.   The IUCN maintains a "Red List" of threatened species.[19] Similar to the U.S. endangered species list, the Red List uses a set of criteria to evaluate the risk of extinction for thousands of species throughout the world. The list is updated frequently, and there were 16,928 species of plants and animals on it in 2008. Once published in book form, the Red List is now available in searchable electronic format on the Internet. Each species on the list is given its proper classification, and its distribution, documentation, habitat, ecology, conservation measures, and data sources are described.

The IUCN is not actively engaged in preserving endangered species in the field, but its findings are often the basis of conservation activities throughout the world, and it provides crucial leadership to the world community on issues involving biodiversity. One of its efforts was to help establish the *Trade Records Analysis for Flora and Fauna in International Commerce* (TRAFFIC) network, which monitors wildlife trade and helps to implement the decisions and provisions of the *Convention on International Trade in Endangered Species of Wild Fauna and Flora* (CITES).

CITES.   Endangered species outside the United States are only tangentially addressed by the Endangered Species Act. However, under the leadership of the United States, the *Convention on International Trade in Endangered Species of Wild Fauna and Flora*, or CITES, was established in the early 1970s. CITES is not specifically a device to protect rare species—instead, it is an international agreement (signed by 169 nation-states) that focuses on trade in wildlife and wildlife parts. The treaty recognizes three levels of vulnerability of species—the highest being species threatened with extinction— and covers 5,000 animals and 28,000 plants, fungi, and other

[19]http://www.iucnredlist.org/.

species. Restrictive trade permits and agreements between exporting and importing countries are applied to those species, sometimes resulting in a complete ban on trade if the nations agree. Every two or three years the signatory countries meet at a Conference of the Parties. The latest such conference was held in The Hague, The Netherlands in June 2007.

Perhaps the best-known act of CITES was to ban the international trade in ivory in 1989 in order to stop the rapid decline of the African elephant (from 2.5 million animals in 1950 to about 609,000 in 1989 and only 470,000 in 2008). Recently, several countries have applied to CITES to resume ivory sales and conduct a limited harvest of elephants; permission was given for sales of stocks of ivory at several conferences, and each time poaching of elephants resumed. Unfortunately, legal sales of ivory from the few stable populations may make poaching in other populations worse. In 2006, 3.9 tons of ivory shipped from Cameroon and seized in Hong Kong were shown (via DNA testing) to have come from elephants in Gabon. Any plan to protect African elephants needs to enable people in Africa to manage wildlife without overexploitation, persuade others to stop using ivory, and use science to stop ivory crime. These changes require a world outcry against ivory collection and an expanded effort to stop ivory crime to save the African elephant from extinction.

Convention on Biological Diversity. Although CITES provides some protection for species that might be involved in international trade, it is inadequate to address broader issues pertaining to the loss of biodiversity. With the support of the United Nations Environment Programme (UNEP), an ad hoc (formed for this purpose) working group proposed an international treaty that would move to conserve biological diversity worldwide. After several years of negotiation, the **Convention on Biological Diversity** was drafted and became one of the pillars of the 1992 Earth Summit in Rio de Janeiro. The *Biodiversity Treaty,* as the convention is called, was ratified in December 1993 and is now in force. As originally organized, the treaty addresses three complementary objectives: (1) the conservation of biodiversity, (2) the sustainable use of biodiversity services, and (3) the equitable sharing of the use of genetic resources found in a country. The treaty establishes a secretariat and designates the Conference of the Parties (the CITES treaty members) as the governing body. Nine conferences have been held, the latest in Bonn, Germany, in May 2008. At the 2002 meeting, delegates adopted a *Strategic Plan*, with the mission to "achieve, by 2010, a significant reduction of the current rate of biodiversity loss at the global, regional and national level, as a contribution to poverty alleviation and to the benefit of all life on Earth."

The secretariat of the Convention on Biological Diversity and the UNEP jointly published the report *Global Biodiversity Outlook 1* in 2001 and then *Global Biodiversity Outlook 2* in 2006. *Global Biodiversity Outlook 3* is expected to be launched in 2010, which has been declared the International Year of Biodiversity. *Outlook 2* lays out in detail the prospects for achieving the 2010 goal of reducing the current rate of loss of biodiversity. The bottom line: Although the targets laid out in the *Strategic Plan* are achievable and the tools for achieving them are largely in place, it will take an unprecedented and costly effort to reach success. As it surveys current trends, the publication shows that the loss of biodiversity is continuing, at all levels. Deforestation continues; population trends of key land, freshwater, and marine species all show declines; and more species are threatened with extinction. One positive note is that the area of protected spaces is increasing.

Even though President Clinton signed the Convention on Biological Diversity in 1993, lobbying by a number of key organizations—especially those concerned about property rights—prevented ratification of the treaty by the U.S. Senate, leaving U.S. involvement in a state of limbo. However, many of the provisions of the treaty are being carried out by different U.S. agencies. The National Biological Service, for example, is working on several of the treaty's provisions. The United States continues to send delegations to the Conference of the Parties and participates in other related meetings, such as those dealing with genetically modified organisms.

Critical Ecosystem Partnership Fund. A new player on the international scene, the Critical Ecosystem Partnership Fund, emerged in August 2000. Jointly sponsored by the World Bank, Conservation International, the government of Japan, the MacArthur Foundation, and the Global Environment Facility, the fund provides grants to NGOs and community-based groups for conservation activities in biodiversity "hot spots"—34 regions, making up just 2.3% of Earth's land surface (Fig. 6–20), in which 75% of the most threatened mammals, birds, and amphibians are located. By 2008, the fund had already provided $102 million in grants to some 1,300 partners to work on preserving biodiversity in the "hot spots."

Stewardship Concerns

What we do about wild species and biodiversity reflects on our wisdom and our values. Wisdom dictates that we take steps to protect the biological wealth that sustains so many of our needs and economic activities. Our values come from our view of species and their instrumental and intrinsic worth. Our values also reflect what we think about the importance of justice—whether we believe, for example, that the people who most benefit economically from the use of an ecosystem should be the same people who bear the risks of its decline or whether we value maintaining choices and options for future generations.

Wisdom. The scientific team that put together the UN Global Biodiversity Assessment focused on four themes in its recommendations:

1. **Reforming policies that often lead to declines in biodiversity.** Many governments (including that of the United States) subsidize the exploitation of natural resources and agricultural activities that consume natural habitats. These policies are frequently counterproductive, leading to the

Figure 6–20 Biodiversity "Hot Spots." These regions are considered the "emergency rooms" of planetary biodiversity, and represent most of the "hot spots" being funded by the Critical Ecosystem Partnership Fund. (*Source:* Redrawn from various sources.)

decline of important natural goods and services (such as soil fertility and water cycling), which tend to be undervalued economically because they do not enter the market.

2. **Addressing the needs of people who live adjacent to or in high-biodiversity areas or whose livelihood is derived from exploiting wild species.** Although protecting biodiversity benefits entire societies, the people who are closest to the protected areas are key to the success of efforts to maintain sustainability of natural resources. These people must be given ownership rights or at least communal-use rights to natural resources, and they must be involved in the protection and management of wildlife resources. Experience has shown that when local people benefit from the protection of wildlife, they will more readily take up stewardship responsibility for the wildlife. Recently, community-based natural resource management emerged as a strategy for wildlife-rich African countries to preserve their biodiversity.

3. **Practicing conservation at the landscape level.** Too often, we tend to believe that a given refuge or park will protect the biodiversity of a region when, in reality, many wild species require larger contiguous habitats or a variety of habitats. For example, to address the needs of migratory birds, summer and winter ranges and flyways between them must be considered. Buffer zones should be created between major population centers and wildlife preserves. This approach requires cooperation among governments, private-property owners, and corporations.

4. **Promoting more research on biodiversity.** Far too little is known about how to manage diverse ecosystems and landscapes, and much remains to be learned about the

species within the boundaries of different countries. For example, Costa Rica is engaged in a scientific inventory of its biodiversity, with an eye especially on the development of new pharmaceutical or horticultural products. Toward that end, a new field called *biodiversity informatics* has been invented. Its goal is to put taxonomic data and other information on species online—that is, to "turn the internet into a giant global biodiversity information system," as one observer put it.[20] Species 2000, a program with the "objective of enumerating all known species of organisms on Earth," is one such effort.[21] Another is the Global Biodiversity Information Facility, whose express purpose "is to make the world's biodiversity data freely and universally available."[22]

Values. Our values also are demonstrated by our approach to wild species. Is the natural world simply grist for our increasing consumption, or is nature something to be managed sustainably and in a stewardly fashion and then passed on to our children and their children?

Here is an appropriate word from the conclusion of *Global Biodiversity Outlook 2* (2006):

> In our everyday choices, we all have direct impacts on biodiversity and the state of our planet's ecosystems. . . . We all derive benefits from biodiversity, and we will all

---

[20]F. A. Bisby. "The quiet revolution: biodiversity informatics and the internet," *Science*, 89, 5488 (2000): pp. 2309–12.
[21]Goddard Space Flight Center, An Online Catalogue of the World's Known Species. http://gcmd.nasa.gov/records/Species 2000-99.html.
[22] Global Biodiversity Information Facility. http://www.gbif.org/GBIF_org/joingbif .htm#4.

suffer from its loss . . . [but] this loss will not be borne equally. Failure to deal with the biodiversity crisis will most severely affect the poor of the developing world. Proof of the compassion and care of the global community for those less fortunate lies in ensuring that the basis for their livelihoods is conserved and used sustainably and that the benefits of this use are shared equitably.[23]

At the start of the chapter, we saw puffins return to a small island off Maine because of the efforts of conservation biologists. The success of Project Puffin shows us that we can indeed stem the tide of species declines with hard work.

---

[23]Global Biodiversity Outlook 2. "Conclusions: Actions to Achieve the 2010 Target," 2006: pp.76–77.

Further strategies to address the loss of biodiversity must focus on preserving the natural ecosystems that sustain wild species. Ecosystems are discussed in detail in Chapter 7, including how we depend on them, what we are doing to them, and what we must do to resist the forces that seem to be leading to a future in which all that is left of wild nature is what we have managed to protect in parks, preserves, and zoos.

Thomas Jefferson once said, "For if one link in nature's chain might be lost, another might be lost, until the whole of things will vanish by piecemeal." As this quote suggests, loss of wild species and their associated genetic and ecosystem diversity could lead to a breakdown in the natural functioning of ecosystems. Indeed, as we look at the world around us, we see that Jefferson was correct. Currently, many animal and plant populations are in decline. However, efforts made in science and policy and the political will of all of us can result in successes such as the comeback of the puffin.

Revisiting the Themes

Sound Science

Sometimes we do not know enough science to manage species well. Such science is critical to upholding the Endangered Species Act, the most powerful policy in the United States for the protection of endangered and threatened species. Fortunately, many universities provide excellent programs in wildlife management. The Millennium Ecosystem Assessment and the Heinz State of the Nation's Ecosystems, as well as annual reports by the IUCN, pull together the solid work of thousands of scientists. Conservation biology, like restoration ecology, is a science that is focused on the protection of living things. The work of conservation biologists helps us to understand why some species are going extinct, where they travel, when they are recuperating, and sometimes when wildlife-related crimes have occurred. The science of taxonomy is foundational to the protection of species, and making sure we have scientists ready to study has to be part of the solution to biodiversity decline.

Sustainability

Because ecosystem capital is of such high value to human enterprises, the pathway to a sustainable future depends on protecting the wild species that make ecosystems work. Such protection is afforded by local and federal policies and international treaties. Sustainability is enhanced when we take steps to encourage healthy populations of organisms; if they thrive, so do their natural ecosystems. Sometimes, however, it is necessary to hunt or otherwise remove animals (such as deer) from populations when their numbers increase too much. Long-term plans for sustainability have to address the root issues that cause biodiversity declines, including those caused by overpopulation and overconsumption.

Stewardship

The reestablishment of puffins off the coast of Maine is good stewardship at work. Stewardship places a high value on wild species. Although instrumental value is sufficient reason for preserving many species, perceiving intrinsic value might be the nearest to a pure stewardship ethic because it works out primarily for the sake of the species. Of all the policies dealing with wild species, the Endangered Species Act comes closest to supporting the intrinsic value of living things.

The protection of biodiversity is tied to the ways people value wild species. These values can include instrumental or utilitarian values as well as intrinsic values. Wild species and their genes represent biological wealth, providing the basis for the goods and services coming from natural ecosystems and sustaining human life and economies. Species can also be thought of as a genetic bank, capable of being drawn on when new traits are needed. Losing biodiversity eventually leads to a biologically impoverished future.

REVIEW QUESTIONS

1. Define *biological wealth* and apply the concept to the human use of that wealth.

2. Compare instrumental value and intrinsic value as they relate to determining the worth of natural species. Where does Leopold's idea of the land ethic fit into these two categories?

3. What are the four categories into which the human value of natural species can be divided? Give examples of each one.

4. What means are used to protect game species, and what are some problems emerging from the adaptations of many game species to the humanized environment?

5. What is the Lacey Act, why was it needed, and how is it used to protect wild species?

6. How does the Endangered Species Act preserve threatened and endangered species in the United States? Give some examples of how the act has been implemented.

7. What are the most critical controversies surrounding the Endangered Species Act? What alternatives have been proposed?

8. What is biodiversity? What do scientists need to know to calculate it for a habitat?

9. What do the letters in the acronym HIPPO stand for? What is an example of each?

10. Give several examples of ways habitat change can either harm species or aid them.

11. What are IUCN, CITES, and the Convention on Biological Diversity? How do their roles differ?

12. What are the major recommendations of the UN Global Diversity Assessment for the protection of biodiversity?

THINKING ENVIRONMENTALLY

1. Some people argue that each individual animal has an intrinsic right to survival. Should this right extend to plants and microorganisms? What about the *Anopheles* mosquito, which transmits malaria, or tigers that sometimes kill people in India? What about the bacteria that cause typhoid fever? Defend your position.

2. Choose an endangered or threatened animal or plant species, and use the Internet to research what is currently being done to preserve it. What dangers is your chosen species subject to?

3. The Endangered Species Act is the most important legislation in the United States for the protection of species. However, there are critics who think it is too strict and those who think it is not strict enough. Investigate one example of the application of the law—for example, to gray wolves—and decide what you think ought to be done.

4. Log onto Conservation International's biodiversity hot spots Web page and research one of the 34 hot spots. What is being done to protect biodiversity in this location?

MAKING A DIFFERENCE

1. Instead of devoting your whole yard to grass, convert some of your yard to native plants. The National Wildlife Federation has a great deal of information on turning your backyard into wildlife habitat, and even a certification program. They can be found on the web.

2. Support groups that conserve and protect wildlife through the political process. Groups like the Center for Biological Diversity and others will alert members when there are political actions that they can support.

3. Consider volunteering or interning for a program that conserves wildlife. Project Puffin, mentioned at the chapter's start, has volunteer opportunities, as do many wildlife conservation groups. These opportunities are best found on the Internet.

4. Be careful about purchasing products that might come from a wild population. For example, it is better to buy pets bred in captivity rather than wild-caught pets.

Mastering ENVIRONMENTALSCIENCE™

Go to **www.masteringenvironmentalscience.com** for practice quizzes, Pearson eText, videos, current events, and more.

CHAPTER 7

The Use and Restoration of Ecosystems

Halibut fishing. Fishermen harvest Pacific halibut on board the *Mar del Norte*, off Kodiak Island, Alaska. The 2008 catch was 57.1 million pounds, valued at $200 million.

KEY TOPICS

1. **Ecosystem Capital: Natural ecosystems have great economic value, as they provide goods and services vital to human well-being.**

2. **Conservation, Preservation, Restoration: Ecosystems are put to both consumptive and productive uses—and occasionally need to be restored to a more healthy state.**

3. **Biomes and Ecosystems Under Pressure: Forest and ocean ecosystems are under major pressures, but there are sustainable ways to fill the demands.**

4. **Protected Lands: Public and private management of lands is key to keeping habitats both protected and productive.**

THE PACIFIC HALIBUT is a large flatfish (up to 8 ft in length and 800 lb in weight) found across the northern parts of that ocean in bottom waters of the continental shelf. It is a prized table fish and has supported an active commercial fishery since the late 1800s. The eastern Pacific fishery has yielded annual catches as high as 100 million lbs and has been heavily regulated since the 1920s by the International Pacific Halibut Commission. Typically, the commission would set an annual *total allowable catch* (TAC), based on its assessment of the halibut population, and then the fishing fleet would fish until the TAC was reached. Soon this took on all the aspects of a fishing derby, with increasingly shorter seasons, until, in 1990, the season lasted only six days. Almost all of the fish caught in that short time had to be frozen, which lowers the quality of the fish and the price received by the fishers. The intensive fishing led to lost and damaged gear and endangered the crew and vessels as they fished in spite of adverse weather conditions.

Canada was the first country to move away from this fishing derby approach in 1991, followed by the United States in 1995. Both countries adopted a management strategy called the individual quota (IQ) system, whereby owners of fishing vessels were allocated a percentage of the annual TAC based on the size of their vessel and their recent fishing performance. Then they could decide when to fish within a season that lasted as long as eight months. Entry into the fishery was strictly limited—the number of fishing vessels declined from 3,500 to 1,200 by 2007. The shift proved highly successful. The fishers saw an increase in their income, as they could sell their fish in the more lucrative fresh-fish market. Also, they could fish in better weather, and they lost less fishing gear. Indeed, to everyone's surprise, fish stocks improved so well that, from 1995 to 1999, the TAC increased by 52% (from 48.6 million lbs to 74.1 million lbs) and has averaged 70.8 million lbs since then. The change in management strategy took place with the full involvement of the fishing community, and even though the fishery is now more thoroughly regulated than ever (fishers must report when they intend to go out, fish catches are carefully monitored, and observers are present on many vessels), the fishery is healthy, and the fishers are better off.

Recall from Chapter 6 that one key to saving wild species is to preserve their habitats—the ecosystems in which they are found. Populations of wild species continue to decline, however, because ecosystems are also in decline throughout the world. In addition to the loss of species, the natural goods and services that ecosystems provide are degraded or lost. Not just the natural world, but also human well-being is threatened in the face of this unsustainable decline of ecosystems.

The focus in this chapter is particularly on the natural ecosystems that directly sustain human life and economy. We will pay special attention to those ecosystems that are in the deepest trouble. These ecosystems, the species in them, and the goods and services they generate are identified as *ecosystem capital*, a major component of the *natural capital* wealth of nations (see Chapter 2).

7.1 Ecosystem Capital

Earth consists of ecosystems that vary greatly in their species makeup, but that exhibit common functions, such as energy flow and the cycling of matter. The major types of terrestrial and aquatic ecosystems—the *biomes*—reflect the response of plants, animals, and microbes to different climatic conditions. Earth's land area can be divided into five broad categories for the purposes of this chapter (Table 7–1): **forests and woodlands, grasslands and savannas, croplands, wetlands,** and **desert lands and tundra.** The oceanic ecosystems can be categorized as **coastal ocean and bays, coral reefs,** and **open ocean.** These eight terrestrial and aquatic systems provide all of our food; much of our fuel; wood for lumber and paper; fibers—leather, furs, and raw materials for fabrics; oils and alcohols; and much more. The world economy and human well-being directly depend on the exploitation of the **natural goods** (referred to in Table 1–1 by the Millennium Ecosystem Assessment team as *provisioning services*) that can be extracted from ecosystems.

Ecosystems also perform a number of other valuable **natural services** (*regulating* and *cultural services* in Table 1–1) as they process energy and circulate matter in the normal course of their functioning. Some of the services are general

TABLE 7–1　Ecosystems on Earth's Surface

Ecosystem	Area (million km^2)	Area (million mi^2)	Percentage of Total Land Area
Forests and woodlands	47.3	18.2	31.7
Grasslands and savannas	30.2	11.6	20.2
Croplands	16.1	6.2	10.7
Wetlands	5.2	2.0	3.5
Desert lands and tundra	51.0	19.6	34.0
Total land area	**149.8**	**57.6**	**100.0**
Coastal ocean and bays	21.8	8.4	
Coral reefs	0.52	0.2	
Open ocean	426.4	164.0	
Total ocean area	**448.7**	**172.6**	

and pertain to essentially all ecosystems, whereas others are more specific. Figure 7–1 shows combinations of services provided to societies by a spectrum of ecosystems. A thorough listing of them can be found in Table 5–3. In their normal functioning, all natural and altered ecosystems perform some or all of these natural services, free of charge, year after year. Recall from Chapter 5 that a team of natural scientists and economists calculated the total value of a year's global ecosystem goods and services to be $41 trillion!

To illustrate the value of an ecosystem, consider a mangrove wetland in South Asia that is converted to a shrimp farm (Fig. 7–2). Studies of shrimp aquaculture in Thailand show that the economic value of the shrimp is $3,375 per acre over five years, following which the site is abandoned as yields decline sharply. Left intact, the mangroves provide services not fully captured in the marketplace: storm protection for coastline communities, forest products from mangrove harvest, and spawning ground for local fisheries. These services have a sustained estimated value of $14,450 per acre. In addition, there are net social costs of the shrimp farm

in the form of pollution and land degradation that would require expensive restoration, calculated at −$2,200 per acre. Thus, the net income from the shrimp farming, when social costs are subtracted, is lowered to $1,175, which is only 8% of the value of services provided by the intact mangrove wetland.

The benefit is local, short term, and specific (to the shrimp farmer, who is likely to be from outside the community), while the loss of services is more regional, long term, and diffuse (to all those living near the coast). Many areas of Southeast Asia hit by the 2004 tsunami had been deprived of their protective mangrove forests and were more heavily damaged than those with intact mangroves.

Ecosystems as Natural Resources

If the natural services performed by ecosystems are so valuable, why are we still draining wetlands and removing forests? The answer is clear: *A natural area will receive protection only if the value a society assigns to services provided in its*

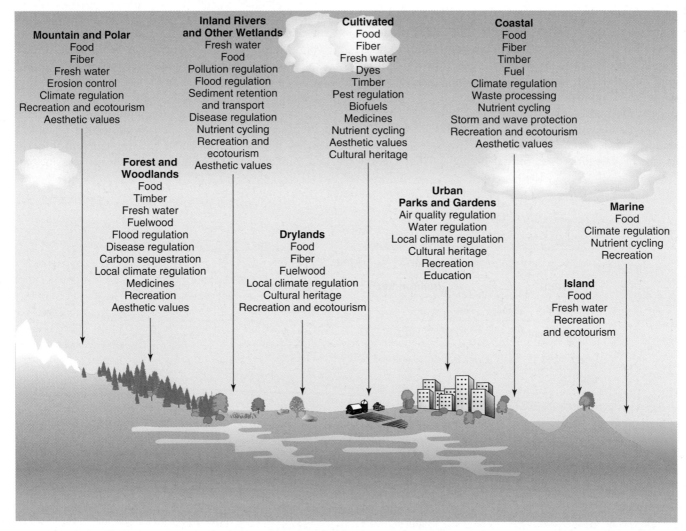

Figure 7–1 Services from various types of ecosystems. Different ecosystems deliver different combinations of services that support human well-being. (*Source:* Millennium Ecosystem Assessment, *Living Beyond Our Means: Natural Assets and Human Well-Being* (Statement of the Millennium Assessment Board) (2005).)

Figure 7–2 Shrimp aquaculture farms in Thailand. Carved out of mangrove forest, these shrimp farms provide only a fraction of the value of the services performed by the intact mangroves.

natural state is higher than the value the society assigns to converting it to a more direct human use.

Natural ecosystems and the biota in them are commonly referred to as **natural resources.** As resources, they are expected to produce something of value, and the most commonly understood value is *economic* value. By referring to natural systems as *resources*, we are deliberately placing them in an economic setting, and it becomes easy to lose sight of their ecological value. Instead of *natural resources*, we use the concept of **ecosystem capital** in order to avoid making this mistake.

Valuing. Markets provide the mechanism for assessing economic value, but do poorly at placing monetary values on ecosystem services. People and institutions may express their preferences for different ecosystem services via the market. (Recall from Chapter 6 that this expression of value is *instrumental* in nature.) Instrumental valuation works fairly well for the *provisioning* services provided by ecosystems (goods such as timber, fish, and crops). However, the *regulating* and *cultural services* provided by natural ecosystems—the more ecological values—are far more difficult to put prices on. These services are *public goods*, not usually provided by markets, but contributing greatly to people's well-being. In fact, they are absolutely essential.

Regulating and cultural services are also less local. Large forested areas can modify the climate in a region and absorb and hold water in the hydrologic cycle. A small forest can be cut and the timber sold, with no noticeable impact on the weather, but with a definite economic value in a market setting, providing income to the owners. If the larger area is also cut, it will have both an immediate economic benefit and a long-term loss in regulating and supporting services. Thus, it will have an impact on people's well-being in the formerly forested area and well beyond, as short-term profits are made at the long-term environmental expense of many. As discussed

in Chapter 4, when the benefits and losses in such conversions are reasonably evaluated economically, losses inevitably outweigh benefits.

Private Versus Public Lands. Some natural ecosystems are maintained in a natural or seminatural state because that is how they provide the greatest economic (direct-use) value for their owners. For example, most of the state of Maine is owned by private corporations that exploit the forests for lumber and for paper manufacturing. However, if Maine experienced a population explosion and corporations could sell their land to developers for more money than could be gained from harvesting timber, forested lands would quickly become house lots. Similar conversions can be pictured: tropical forests to grasslands for grazing, agricultural fields to suburbs, and wetlands to vacation developments.

Some natural ecosystems either are publicly owned (state and federal lands) or cannot be owned (ocean ecosystems). These ecosystems are still considered natural resources and may be subject to economically motivated exploitation. *Sustainable* exploitation of such systems will maintain the natural services they perform. Fortunately, ecosystems not only can sustain continued exploitation, but also can be restored from destructive uses that have degraded them and removed their natural biota.

Domesticated Nature. Whether private or public, human domination of the landscape is pervasive, and ecosystems have been converted or domesticated in vast areas of the planet. What we call natural may be only an illusion because humankind has so converted and domesticated natural systems that it is difficult to find areas of the world without human impact. Even where systems appear natural, it is likely that humans have killed larger animals, suppressed wildfires, controlled rivers, introduced exotic species, and harvested wood. Elsewhere, we have converted grasslands and forests to grazing lands and crop cultivation, so that almost 50% of the land surface is now used in the production of crops and animal products and over half of the forests have been lost in the process. To be sure, the domesticated ecosystems still provide provisioning and regulating and cultural services, and may do so sustainably, but they do so at our consent and largely for our benefit.

Future Pressures. As population continues its relentless rise, pressures on ecosystems to provide resources will increase: More food, more wood, more water, and more fisheries products will be needed. Maintaining sustainable exploitation of ecosystem capital will become more difficult, in view of the already intense pressure being put on those ecosystems. The main objective of the thorough work of the Millennium Ecosystem Assessment was to evaluate the health of the services that people derive from the natural world. As described in Chapter 1, the assessment team found that 15 of the 24 services they evaluated were in decline worldwide, five were in a stable state, and four were increasing (and doing so at the expense of other services). The governing board of the assessment was so concerned about these findings that it published a separate document for a "nonspecialist" readership— *Living Beyond Our Means: Natural Assets and Human Well-Being.* This brief report begins: "At the heart of this assessment is a stark warning. Human activity is putting such strain on the natural functions of Earth that the ability of the planet's ecosystems to sustain future generations can no longer be taken for granted. . . . This assessment shows that healthy ecosystems are central to the aspirations of humankind."

7.2 Conservation, Preservation, Restoration

An ecosystem's remarkable natural capacity to regenerate makes it (and its biota) a **renewable resource**. In other words, an ecosystem has the capacity to replenish itself despite certain quantities of organisms being taken from it, and this renewal can go on indefinitely—it is sustainable. Recall from Chapter 5 that every species has the biotic potential to increase its numbers and that, in a balanced population, the excess numbers fall prey to parasites, predators, and other factors of environmental resistance. It is difficult to find fault with activities that effectively put some of this excess population to human use. The trouble occurs when users (hunters, fishers, loggers, whalers) take more than the excess and deplete the breeding population.

Conservation Versus Preservation

Conservation of natural biotas and ecosystems does not imply *no* use by humans whatsoever, although this may sometimes be temporarily expedient in a management program to allow a certain species to recover its numbers. Rather, the aim of conservation is to *manage or regulate use* so that it does not exceed the capacity of the species or system to renew itself. Conservation is capable of being carried

Figure 7–3 Muriqui monkey. Protection of this species of monkey in Brazil involves allowing some cutting of the forest in order to provide second-growth forests.

out sustainably, and when sustainability is adopted as a principle, conservation has a well-defined goal.

Preservation is often confused with conservation. The objective of the preservation of species and ecosystems is to *ensure their continuity, regardless of their potential utility.* Effective preservation often precludes making use of the species or ecosystems in question. For example, it is impossible to maintain old-growth (virgin) forests and at the same time harvest the trees. Thus, a second-growth forest can be *conserved* (trees can be cut, but at a rate that allows the forest to recover), but an old-growth forest must be *preserved* (it must not be cut down at all).

There are times when conservation and preservation come into conflict. The Muriqui monkey of Brazil was once thought to require virgin forests, leading to a concern for protecting such forests for the sake of the species (Fig. 7–3). Recent research has shown, however, that the monkeys actually do better in second-growth forests, which support a greater range of vegetation on which the monkeys feed. Indefinite preservation of the virgin forests would lead to a decline in the population of the Muriqui monkey, a seriously endangered species. Thus, *conservation* of the forests is essential for *preservation* of this New World monkey.

Patterns of Human Use of Natural Ecosystems

Before examining the specific biomes and ecosystems that are under the highest pressure from human use, we need to study a few general patterns of such use to understand what is happening in these natural systems and why it is happening.

Consumptive Use. People in more rural areas make use of natural lands or aquatic systems that are in close proximity to them. When people harvest natural resources in order to provide for their needs for food, shelter, tools, fuel, and clothing, they are engaged in **consumptive use.** This kind of exploitation

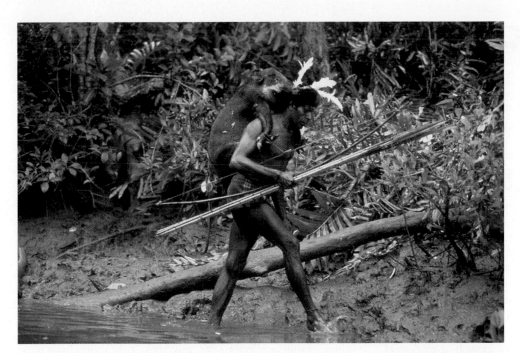

Figure 7–4 Consumptive use. A hunter from the Asmat tribe, in Irian Jaya, Indonesia, brings home game. Wild game is an important food source for millions in many parts of the world.

usually does not appear in the calculations of the market economy of a country. People may barter or sell such goods in exchange for other goods or services in local markets, but most consumptive use involves family members engaged in hunting and gathering to meet their own needs. Thus, people are hunting for game (Fig. 7–4), fishing, or gathering fruits and nuts in order to meet their food needs, or else they are gathering natural products like firewood, forage for animals, or wood and palm leaves to construct shelters or to use as traditional medicines. This "wild income," as it has been called, is highly important to the poor; some 90% of the 1.2 billion poorest people depend on forests for some of their income.

Bush Meat. Wild game, or "bush meat," is harvested in many parts of Africa and provides a large proportion of the protein needs of people. In Congo, for example, it is as high as 75% of a person's protein intake. The estimated take in the Congo basin alone is somewhere between 1 and 3 million metric tons annually. Without access to these resources, people's living standard would decline, and their very existence could be threatened. Unfortunately, this is a largely unregulated practice and often involves poaching in wildlife parks. According to the Convention on International Trade in Endangered Species (CITES), this practice has contributed to the decline of 30 endangered species. Making matters worse, commercial hunters have recently turned what was once a matter of local people using local wildlife for food into an unsustainable industry that supplies cities with tons of "fashionable" meat. For example, primate meat (e.g., chimpanzee) has been found in markets in New York, London, Paris, Brussels, and others.

Dependence on consumptive use is most commonly associated with the developing world, but also exists in the more rural areas of the developed world, where people depend on wood for fuel or wild game for food. Their well-being would be diminished if they were denied access to these "free" resources.

Productive Use. In contrast to consumptive use, **productive use** is the exploitation of ecosystem resources for economic gain, as when products like timber and fish are harvested and sold for national or international markets (Fig. 7–5). Productive use is an enormously important source of revenue and employment for people in every country. The commercial trade in wood products for pulp, lumber, and fuel and in nonwood forest products amounted to $468 billion in 2006, employing 13.7 million people. The global fisheries catch in 2006, which reached 92 million metric tons (84 million tons), was valued at $91.2 billion. Another productive use is the collection of wild species of plants and animals for cultivation or domestication. The wild species may provide the initial breeding stock for commercial plantations or

Figure 7–5 Productive use. Loggers load poplar logs to be transported to a lumber mill, where they will be made into wafer board for the building industry.

ranches, or they may be used as sources of genes to be introduced into existing crop plants or animals to improve their resistance or to add other desirable genetic traits. Finally, as discussed in Chapter 6, wild species continue to provide sources of new medicines.

Tenure. Consumptive and productive uses of natural ecosystem resources are the consequence of the rights of tenure, or property rights, over the land and water holding those resources. Four kinds of tenure are recognized: (1) *private ownership*, which restricts access to natural resources; (2) *communal ownership*, which permits use of natural resources by members of the community; (3) *state ownership* of natural resources, which implies regulated use; and (4) *open access*, where natural resources can be used by anyone. In all of these arrangements—private, communal, state, and open access—there is the potential either for sustainable use of the resources or for unsustainable use leading to the loss of the resources and the services provided by the natural ecosystem being exploited. As a result, the perspective of stewardship is vital. Do the resource managers and harvesters see themselves as caretakers, as well as exploiters, of the natural resource? Are public policies that encourage stewardship and sustainability in place? Is the local or national government presence weak or corrupt? Are sound scientific principles employed in management strategies? Keep these questions in mind as we next examine two patterns that are commonly seen when people exploit natural resources.

Maximum Sustainable Yield. The central question in managing a renewable natural resource is this: How much continual use can be sustained without undercutting the capacity of the species or system to renew itself? A scientific model employed to describe this amount of use is the

maximum sustainable yield (MSY): *the highest possible rate of use that the system can match with its own rate of replacement or maintenance.* Besides pertaining to the harvesting of natural biota, the concept of MSY can apply to the maintenance of parks, air quality, water quality and quantity, and soils—indeed, the entire biosphere. *Use* can refer to the cutting of timber, hunting, fishing, park visitations, the discharge of pollutants into air or water, and so on. Natural systems can withstand a certain amount of use (or abuse, in terms of pollution) and still remain viable. However, a point exists at which increasing use begins to destroy the system's regenerative capacity. Just short of that point is the MSY.

Optimal Population. An important consideration in understanding how the MSY works is the **carrying capacity** of the ecosystem—the maximum population the ecosystem can support on a sustainable basis. If a population is well below the carrying capacity of the ecosystem (Figure 7–6a), then allowing that population to grow will increase the number of reproductive individuals and thus the yield that can be harvested. However, as the population approaches the carrying capacity of the ecosystem, new individuals must compete with older individuals for food and living space. As a result, recruitment may fall drastically (Figure 7–6b). When a population is at or near the ecosystem's carrying capacity, production—and hence sustainable yield—can be increased by thinning the population so that competition is reduced and optimal growth and reproductive rates are achieved. Thus, the MSY cannot be obtained with a population that is at the carrying capacity. Theoretically, the **optimal population** for harvesting the MSY is just halfway to the carrying capacity (Fig. 7–6).

The practical use of this model is complicated by the fact that the carrying capacity—and, therefore, the optimal population—is variable. Carrying capacity may vary from

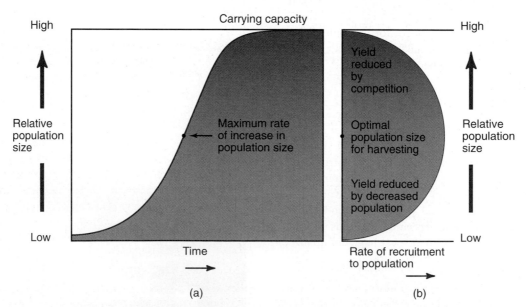

Figure 7–6 Maximum sustainable yield. The maximum sustainable yield occurs when the population is at the optimal level (meaning the rate of increase in population is at a maximum). (a) The logistic curve of population size in relation to carrying capacity. (b) Recruitment plotted against population size, showing the effects of competition and decreased population levels.

year to year as the weather fluctuates. Replacement of harvested individuals may also vary from year to year because some years are particularly favorable to reproduction and recruitment, while others are not. Human impacts, such as pollution and other forms of altering habitats, adversely affect reproductive rates, recruitment, carrying capacity, and, consequently, sustainable yields. For these reasons alone, managing natural populations to achieve the MSY is fraught with difficulties. Furthermore, accurate estimates of the size of the population and the recruitment rate must be made continually, and these data are hard to come by for many species.

Precautionary Principle. The conventional approach has been to use the estimated MSY to set a fixed quota—in fishery management, the **total allowable catch (TAC)**. If the data on population and recruitment are inaccurate, it is easy to overestimate the TAC, especially when there are pressures from fishers (and the politicians who represent them) to maintain a high harvest quota. In the face of such uncertainties, and also in response to repeated cases of overuse and depletion of resources, resource managers have been turning to the **precautionary principle**. That is, where there is uncertainty, resource managers must favor the protection of the living resource. Thus, the exploitation limits must be set well enough below the MSY to allow for uncertainties. Those using the resource ordinarily want to push the limits higher, however, so the result is frequent conflicts between users and managers. These conflicts are addressed in greater detail when the problems of fisheries are discussed later in this section.

Using the Commons. Where a resource is owned by many people in common or by no one (open access), it is known as a **common-pool resource** Strictly speaking, a **commons** refers to a system with open access where use by one does not subtract from use by others; for example, *knowledge* is a commons. Examples of natural common-pool resources include federal grasslands, on which private ranchers graze their livestock; coastal and open-ocean fisheries used by commercial fishers; groundwater drawn for private estates and farms; nationally owned woodlands and forests harvested for fuel in the developing countries; and the atmosphere, which is polluted by private industry and everyone who drives motor vehicles.

The exploitation of such common-pool resources presents some serious problems and can lead to the eventual ruin of the resource—a phenomenon called the *tragedy of the commons*, taken from the title of a 1968 essay by the late biologist Garrett Hardin.[1] Sustainability implies that common-pool resources will be maintained so as to continue to yield benefits not just for the present, but also for future users.

Tragedy of the Commons. As described by Hardin, the original "commons" were pastures around villages in England. These pastures were provided free by the king to anyone who wished to graze cattle. Herders were quick to realize that whoever grazed the most cattle stood to benefit the most. Even if they realized that a particular commons was being overgrazed, those who withdrew their cattle simply sacrificed personal profits, while others went on using that commons. One herder's loss became another's gain, and the commons was overgrazed in any case. Consequently, herders would add to their herds until the commons was totally destroyed. The lesson of this story is that where there is no management of the commons, everyone pursuing their own interests will lead to a tragedy for all.

Hardin's essay applies to a limited, but significant, set of problems that arise where there is *open access* to a common-pool resource, but where there is *no regulating authority* (or there is one, but it is ineffective) and *no functioning community*. Exploitation of the resource then becomes a free-for-all in which profit is the only motive. Coastal and offshore fisheries have consistently demonstrated the reality of the tragedy of the commons, with stocks of desirable fish declining all over the world. The tragedy can be avoided only by limiting freedom of access.

Limiting Freedom. One arrangement that can mitigate the tragedy is *private ownership*. When a renewable natural resource is privately owned, access to it is restricted, and in theory, it will be exploited in a manner that guarantees a continuing harvest for its owner(s). This theory does not hold, however, when an owner maximizes immediate profit and then moves on. Modern corporations lend themselves to this strategy: For example, corporate raiders can take over a natural resource—say, a timber company and its holdings—and clear-cut the forest in order to pay off debts incurred in the process of the takeover. Many owners, though, will be in for the long run and manage the resources more responsibly because it is in their best interests to do so.

Where private ownership is unworkable or lacking, the alternative is to *regulate access* to the commons. Regulation should allow for (1) protection, so that the benefits derived from the commons can be sustained; (2) fairness in access rights; and (3) mutual consent of the regulated. Such regulation can be the responsibility of the state, but it does not have to be. In fact, the most sustainable approach to maintaining the commons may be local community control, wherein the power to manage the commons resides with those who directly benefit most from its use and there are strong social ties and customs that can function well over time to protect the commons (see Stewardship, "The Turtle Island Commons"). By contrast, in many situations, state control of a commons has accelerated its ruin and has led to an associated social breakdown and the impoverishment of people.

Maine Lobsters. The lobster is a tasty crustacean found in the colder marine waters of the Northern Hemisphere, especially where there are shallow-water rocky subtidal seascapes interspersed with sandy patches. The state of Maine is famous for its lobsters, and management of the lobster fishery—a common-pool resource—has evolved into a workable combination of state and local regulation. The state has established conservation laws that protect juvenile and breeding female lobsters and limit the number of lobster

---

[1]Garrett Hardin, "The Tragedy of the Commons," *Science* 162 (1968): 1243–1247.

Stewardship

The Turtle Island Commons

Located at the southwestern tip of the Philippine Islands, the Turtle Island municipality is a system of six small islands that harbor the most important green turtle rookery in Southeast Asia. Each year thousands of female green turtles come ashore to lay eggs on the beaches that surround each island, and each year the local people collect eggs, which they sell to dealers that market the eggs, considered a delicacy, in Taiwan and other Asian cities. For decades, conservation of the eggs was not intentionally practiced, and the fishery was in danger of a "tragedy of the commons" outcome. However, a small group of conservationists from the Philippine Department of Environment and Natural Resources came to the islands in 1984 to establish a program that would not only allow the local residents to continue to collect eggs, but also provide for the sufficient survival of hatchlings to ensure continuity of the turtle population. The program, called the Pawikan Conservation Project (PCP), included scientific data collection as well as recruitment of local egg collectors to serve as wardens.

Working with the local communities, PCP evolved a system of permits, where each family on the islands would be entitled to harvest eggs on a limited number of days. Rules included limiting harvesting to 60% of the eggs on four of the islands, while on the remaining two islands, no harvesting was allowed. The PCP staff worked within the social structures of the islands, building trust with the authorities and developing a plan that met with broad acceptance. The result was a steady improvement in the percentage of

eggs conserved, from an initial 50% up to 80% within 15 years.

A New Law

Then, in 2001, the Philippine Congress passed the Wildlife Resources and Conservation Act, understood as a major victory for conservation of natural resources. The act prohibited the taking of marine turtle eggs, and the Turtle Islands were declared a wildlife sanctuary. Shortly after the law was passed, PCP was disbanded, and turtle egg conservation on the islands effectively ceased. Within two years, survival of eggs dropped to 40%. The reason, of course, was poaching. The local island government refused to acknowledge the new law, viewing it as an intrusion into its territorial rights. Department of Environment and Natural Resources personnel were physically prevented from visiting the islands and monitoring activities there. By imposing an external (and unenforceable) rule on the Turtle Island residents, the government effectively ruined years of work building an effective local arrangement that conserved the turtle population.

The take-home lesson is clear: Sustainable stewardship of a common-pool resource must involve a program that works with the local community, one that enables the people to benefit from the resource, while at the same time protecting the productive base of that resource. Several of the principles listed in Table 7–2 were violated, and the result was a threatened collapse of the turtle egg fishery. The good news in this case is that the PCP and the islanders are

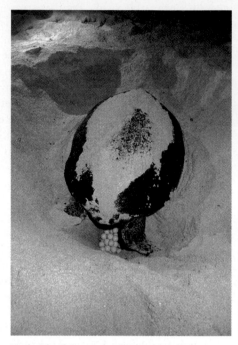

Turtle Island Commons. Green turtle laying eggs on a beach.

talking again, and the islanders are beginning to return to the system of harvesting that existed prior to the new national conservation law.

Source: Raul P. Lejano and Helen Ingram, "Place-Based Conservation: Lessons from the Turtle Islands," *Environment* 49 (November 2007): 18–26.

---

traps (Fig. 7–7). The lobster fishers, however, have developed a second set of local rules that determine who can fish a given harbor and coastline area. It is an informal system, but it works because the fishers are part of a functioning community where they know each other and are monitoring each other's harvesting. The result is a successful outcome where the state's lobster fishery is thriving even though the number of fishers has more than doubled since 1973.

Public Policies. In sum, to achieve the objectives of conservation when harvesting living resources, it is important to consider both the concept and limitations of the MSY and the social and economic factors causing overuse and degradation that diminish the sustainable yield. After reviewing these factors, public policies can be established and enforced that protect natural resources effectively. Some principles

that should be embodied in these policies are presented in Table 7–2. When policies based on principles like these are in place, it is possible for people to continue to put natural resources to sustainable consumptive and productive uses. The vast harvest taken from the natural world each year is proof that sustainable use not only is possible, but also actually occurs with many resources and in many locations around the world.

In some situations, however, exploitation and degradation have gone too far, and the resilience of an ecosystem has been overcome. In those cases, natural services and uses can be restored only if the habitats are deliberately restored as well. In recent years, the need for the restoration of damaged ecosystems has become clear, and a new subdiscipline with that objective—called **restoration ecology**—has been developed.

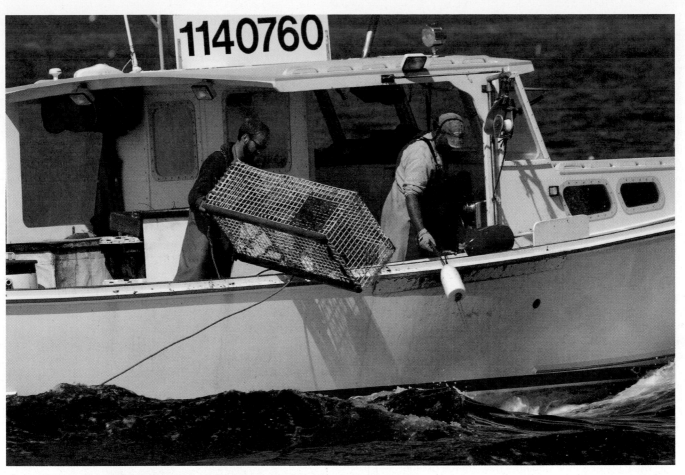

Figure 7–7 Exploiting a commons. Lobster boat unloading a lobster pot in Maine waters.

TABLE 7–2   Principles That Should Be Incorporated into Public Policies to Protect Natural Resources
1. Natural resources cannot be treated as open-access commons. Habitats and species should be held by private, communal, or state entities that are responsible for their exploitation and sustainability.
2. Scientific research should be employed to assess the health of the resource and to set sustainable limits on its use.
3. To accommodate uncertainty, the precautionary principle should be used in setting limits for exploitation.
4. Regulations should be enforced; the resource must be protected from poaching.
5. Economic incentives that encourage the violation of regulations should be eliminated.
6. Subsidies that encourage exploitation of the resource should be removed.
7. All stakeholders should be involved in resource-related decisions; those living close to the resource should share in profits from its use.
8. Security of tenure is crucial for poor people who depend on the resource for their well-being.

Ecosystem Restoration

A great increase in restoration activity has occurred during the last 30 years, spurred on by federal and state programs and the growing science of *restoration ecology*. The intent of ecosystem restoration is to repair the damage to specific lands and waters so that normal ecosystem integrity, resilience, and productivity return. It has become a worldwide, $70 billion industry. Because of the complexity of natural ecosystems, however, the practical task of restoration is often difficult. Frequently, soils have been disturbed, pollutants have accumulated, important species have disappeared, and other—often exotic—species have achieved dominance.

For these reasons, a thorough knowledge of ecosystem and species ecology is essential to the success of restoration efforts. The ecological problems that can be mended by restoration include those resulting from soil erosion, surface strip mining, the draining of wetlands, hurricane damage, agricultural use, deforestation, overgrazing, desertification, and the eutrophication of lakes. Every case is different. Some cases involve working with several different ecological problems at the same time, such as the huge restoration project under way in the (formerly) vast Everglades of southern Florida.

Everglades Restoration. A much-publicized plan to restore the Everglades received congressional approval in 2000. The plan, the **Comprehensive Everglades Restoration Plan (CERP)**, is expected to take 36 years and almost $11 billion to complete (Figure 7–8). The federal government

◆ **Surface Water Storage Reservoirs**
181,300 acres of above and in-ground reservoirs are planned to store millions of gallons of water.

● **Aquifer Storage and Recovery**
More than 300 underground water storage wells are proposed to store up to 1.6 billion gallons of treated water a day in confined aquifers.

🌢 **Stormwater Treatment Areas**
35,600 acres of manmade wetlands will be constructed to remove pollutants and other harmful contaminants from water before it is discharged to the Everglades.

■ **Wastewater Reuse**
Two advanced treatment plants are proposed to recycle more than 220 million gallons of wastewater a day, adding a new source of high quality water for the southern Everglades.

★ **Seepage Management**
Barriers are proposed to be built to stop the rapid underground seepage of water out of the Everglades, which today results in the loss of millions of gallons of water each year.

▲ **Removing Barriers to Sheetflow**
More than 240 miles of canals and levees may be removed to restore the historic overland sheetflow through the Everglades wetlands. Sections of Tamiami Trail will be elevated to handle increased water flows contributed by CERP project features.

✦ **Operational Changes**
Changes will be made in the regional water management system to benefit Lake Okeechobee, the Everglades, and the coastal estuaries.

Figure 7–8 The Everglades restoration plan. The goal of this plan is to restore historical water flow patterns so that ecological restoration can occur. The seven major features of the plan are described at the left of the map. The lower three maps illustrate the historical, current, and future water flow through southern Florida. (*Source:* Comprehensive Everglades Restoration Plan, www.evergladesplan.org.)

promised to provide half of the funds and state, tribal, and local agencies the other half. CERP is managed by the U.S. Army Corps of Engineers, the federal agency that was actually responsible for creating much of the system of water control that made the restoration necessary.

The Everglades is a network of wetland landscapes that once occupied the southern half of Florida. Now reduced to half its original size through development and wetlands draining, it is still an amazing and unique ecosystem, harboring a great diversity of species and many endangered and threatened species. It is home to 16 wildlife refuges and 4 national parks, the crown jewel being Everglades National Park.

In Bondage. Because Florida is so flat, the general topography promoted a slow movement of water south from Lake Okeechobee by way of a 40-mile-wide, 100-mile-long shallow "river of grass" ending in Florida Bay. Floridians once viewed it as an unproductive and forbidding swamp, so they brought the water flow under human control with a system of 1,000 miles of canals, 720 miles of levees, and hundreds of water control structures (locks, dams, and spillways). The system is bounded on the east by a string of cities and their suburbs. In its center, 550,000 acres of croplands were created, which form a lucrative sugar cane industry. Now, much of the water that originally made its way through the "river of grass" is diverted to irrigate the croplands and to feed the thirsty suburbs and cities. Water shortages in the winter leave too little for the natural systems, and in the summer rainy season, too much water is diverted to the Everglades. The water quality, once pristine, is now degraded because of nutrients (especially phosphorus) from agricultural runoff. The changed nutrient regime promotes the growth of invasive species and has caused a decline in the native vegetation.

Water Release. A broader restoration program was originally begun in 1996, coordinated by the South Florida Ecosystem Restoration Task Force. (CERP is now a major element of the task force's plans.) The task force has the following goals: (1) "getting the water right"; (2) restoring, protecting, and preserving natural habitats and species; and (3) promoting the compatibility of the natural and built (human) systems. The total task force project is expected to cost close to $15 billion. CERP's role is the first goal: "getting the water right." The plan calls for removing 240 miles of levees and canals and creating a system of reservoirs and underground wells to capture water for release during the dry season. The new flowage is designed to restore the river of grass, thereby restoring the 2.4 million acres of Everglades—not to the original state, but at least to a healthy system that promotes the second goal of the task force. If CERP does its job, close to 2 billion gallons of water that flow out of the Everglades each day will be redirected to flow southward. The plan calls for 80% of this water to feed the Everglades and 20% to feed the thirst of South Florida's cities and farms.

Land Buyout. A major step in the restoration project was announced in 2008, when U.S. Sugar agreed to sell its land and facilities to the state of Florida for $1.7 billion. This sale involved 187,000 acres of land to the south and east of

Lake Okeechobee and could recreate wetlands to store up to 1.2 cubic kilometers (0.28 cubic miles) of water that otherwise would need to be pumped into reservoirs and underground wells during the summer. Currently, the excess water is pumped eastward and westward out to sea, disturbing the salinity and nutrient regimes in the receiving estuaries. The water stored in the new wetlands would then be free to flow south into the Everglades during the dry season, once the path to the Everglades is made clear by other planned projects. Unfortunately, this planned sale has been scaled way back because of Florida's financial uncertainties, a result of the 2008–2009 national recession. The new proposal would invest $530 million for 72,500 acres of land south of Lake Okeechobee and leave open future options for the state to buy more land.

There is no blueprint for a project like the Everglades restoration. The long time frame, the need for a continued infusion of federal and state money, and the great difficulties that will arise in carrying out the restoration guarantee that the project will attract skepticism as well as effusive praise. It will be in the news for a long time.

Pending Restorations. As the values of natural ecosystems become more recognized, efforts to restore damaged or lost ones will become increasingly important. Many other large systems, such as the California Bay Delta, Chesapeake Bay, Platte River Basin, and Upper Mississippi River System, have suffered from human abuse and ignorance and are now the focus of restoration projects. The Mississippi Delta, battered by hurricanes Katrina and Rita in 2005, is in need of a huge effort to restore the damaged wetlands and barrier islands. Restoration is under way at many sites that are ecological gems, including the Galápagos Islands, under siege from many exotic species; the Illinois River, confined by levees; the Brazilian Atlantic forest, severely deforested; and Tampa Bay, affected by its urban surroundings.

The next section discusses a number of still-functioning biomes and ecosystems that are in trouble because of their great economic importance.

7.3 Biomes and Ecosystems Under Pressure

The biomes and ecosystems of the world provide us with priceless services and economically valuable goods. Although human activities affect virtually all biomes and ecosystems, some are under more pressure than others. This section begins with a look at the forest biomes, the most important of all biomes in terms of their economic significance and potential for active human management.

Forest Ecosystems

Forests are the normal ecosystems in regions with year-round rainfall that is adequate to sustain tree growth. Forests are also the most productive systems the land can support, and they are self-sustaining. Forests and woodlands (ecosystems with mixed trees and grasses) perform a number of vital natural services. Among other things, they conserve biodiversity, moderate regional climates, prevent erosion, store carbon and nutrients, and provide recreational opportunities. They provide a number of vital goods, too, such as lumber, the raw material for making paper, fodder for domestic animals, fibers, gums, latex, fruit, berries, nuts, and fuel for cooking and heating. In fact, forests provide fuel, food, medicine, and income for over 1.6 billion people, mostly in the developing countries. In spite of this value, the major threat to the world's forests is not simply their *exploitation*, but rather their *total destruction*.

Forest Resources Assessments. The UN's Food and Agricultural Organization (FAO) conducts periodic assessments of forest resources and compares the data with previous information in order to evaluate trends in forested areas. In 2001, the agency published its most comprehensive analysis ever, the *Global Forest Resources Assessment 2000* (also referred to as the FRA 2000). One of its products was a series of global forest maps, based on satellite images with 1-kilometer resolution. Figure 7–9 shows the locations of the four major world forest biomes. The latest FRA (in 2005) did not repeat the satellite forest mapping, but did provide information on changes since 2000 and was distinguished by its focus on sustainable forest management (discussed later in this section). This assessment was the basis for FAO's report titled *State of the World Forests 2009*. Major findings of the FRA assessments are as follows:

1. In 2005, the world's forest cover was 3.95 billion hectares (9.8 billion acres), or 30% of total land area. The FAO defines "forests" as lands where trees are the dominant life form and produce a canopy greater than 10% cover. An additional 1.4 billion hectares (3.5 billion acres) are classified as "other wooded land," sometimes referred to as "woodlands," where the tree cover is less than 10% or the trees and/or shrubs are species that do not grow to a height above 5 meters.

2. Deforestation continues to occur, primarily in the developing countries. Forest cover in the developed countries has remained stable. The FAO defines **deforestation** as *the removal of forest and replacement by another land use.* Thus, logging per se does not count as deforestation if the forests are allowed to regenerate. The current global rate of net deforestation is estimated at 7.3 million hectares (18 million acres) per year, some 18% better than in the 1990s. The annual net changes in forest area in major world regions are shown in Figure 7–10.

3. Throughout the world, the most important forest product is wood for industrial use; half of the forest lands are designated for "production," where the wood is harvested for pulp (paper source), lumber, and fuelwood. The FRA 2005 reported that there have not been any drastic changes in forest production over the past 15 years, except for a steady increase in plantation forests.

4. Worldwide, about 9% of the forests are protected as national parks or reserves. An additional 65% of the world's forests have protection as one of their designated functions. These proportions are on the increase, a recognition by many countries of the important nontimber services of forests.

5. The role of forests in climate change was formally acknowledged in November 2001 at a meeting of the

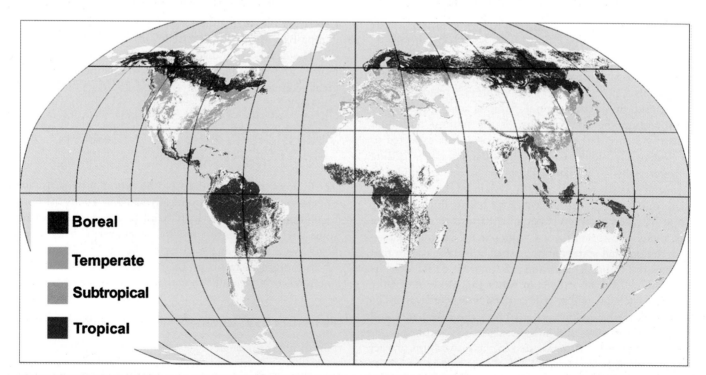

Boreal

Temperate

Subtropical

Tropical

Figure 7–9 World forest biomes. The FAO *Global Forest Resources Assessment 2000* produced this map showing the locations of the major world forests.

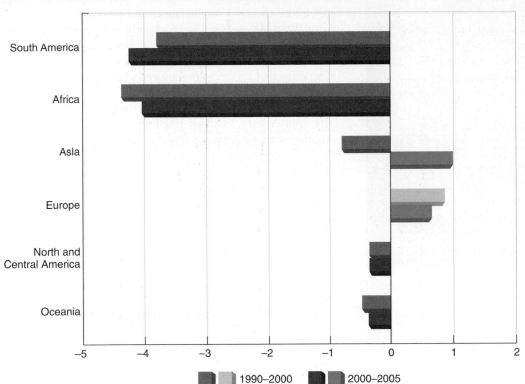

Annual net change in forest area by region 1990–2005 (million ha per year)

Figure 7–10 Net changes in forest area. The annual net change of forests from 1990 to 2000 and from 2000 to 2005 in the different regions of the world. Virtually all of the losses occurred in tropical forests of South America and Africa. Go to **GRAPHit!** at www.masteringenvironmentalscience.com. (*Source:* Food and Agricultural Organization of the United Nations, *State of the World's Forests 2009.*)

signers of the Kyoto Protocol in Marrakech, Morocco. The outcome of this accord is a much more thorough inventory of the role of forests as carbon stores, sources of carbon emissions, and carbon sinks. The FRA 2005 estimated that forest ecosystems in 2005 contained 638 gigatons (Gt) of carbon, more than the amount of carbon as CO_2 in the atmosphere (see Chapter 18). Half of this carbon is in living wood and deadwood, and half is in the forest soils.

The next *Forest Resources Assessment* is scheduled to be completed by 2010. As part of this assessment, countries have agreed to conduct a global remote-sensing survey of forests, using high-resolution satellite imagery, which is far superior to the relatively inaccurate field sampling and low-resolution imagery on which current forest resource management is based. The assessment will, for the first time, cover the entire land surface of Earth and will provide countries with the tools to monitor and manage forests and report on land use changes far more accurately than ever before.

Forests as Obstacles. Even though the rate of deforestation is lower than in the 1990s, it is still a major cause for concern. Why are forests being cleared when they could be managed for wood production on a sustainable basis? Even though forests are highly productive systems, it has always been difficult for humans to exploit them for food. According to the FRA 2005 data, most of the deforestation was due to *conversion into pastures and agricultural land.*

Unlike forests, grasslands have short food chains in which herbaceous growth supports large herbivores that can yield meat and other animal products. Alternatively, forests can be replaced by cultivated plants that are used directly for food. Thus, the forests have always been an obstacle to conventional animal husbandry and agriculture. Indeed, the first task the early European colonists faced when they came to the Western Hemisphere was to clear the forests so they could raise crops. Once the forests were cleared, continual grazing or plowing effectively prevented the trees from regrowing.

Consequences. Clearing a forest has the following immediate consequences for the land and its people:

- The overall productivity of the area is reduced.
- The standing stock of nutrients and biomass, once stored in the trees and leaf litter, is enormously reduced.
- Biodiversity is greatly diminished.
- The soil is more prone to erosion and drying.
- The hydrologic cycle is changed, as water drains off the land instead of being released by transpiration through the leaves of trees or percolating into groundwater.
- A major carbon dioxide sink (removal of CO_2 from the air) is lost.
- The land no longer yields forest products.
- People who depend on harvesting forest products lose their livelihood.

Figure 7–11 Logging operations. A forest in New Guinea is logged for its tropical wood.

- The trade-off, of course, is that now the land will yield crops or support livestock, more people may be fed, and (usually) more profit will accrue to the owners.

Unless forests are cleared and converted to other uses, they can yield a harvest of wood for fuel, paper, and building materials. Approximately 3 billion cubic meters (3.7 billion yd^3) of wood are harvested annually from the world's forests for fuel, wood, and paper (Fig. 7–11). It is unreasonable to expect the countries of the world—especially the developing countries—to forgo making use of their forests. However, forest-management practices vary greatly in their impact on the forest itself and on surrounding ecosystems.

Types of Forest Management. The practice of forest management, usually with the objective of producing a specific crop (hardwood, pulp, softwood, wood chips, etc.), is called **silviculture.** Several practices are employed, with quite different impacts on the total forest ecosystem involved. Because trees take from 25 to 100 years to mature to harvestable size, the normal process of management is a *rotation*, which is a cycle of decisions about a particular stand of trees from its early growth to the point of harvest. Two options are open to the forester: *even-aged management* and *uneven-aged management*. With even-aged management, trees of a fairly uniform age are managed until the point of harvest, cut down, and then replanted, with the objective of continuing the cycle in a dependable sequence.

Clear-Cutting. Typically, fast-growing, but economically valuable, trees are favored for even-aged management. (Others are culled from the stand.) In the Pacific Northwest, Douglas fir is frequently managed this way. Harvesting is often accomplished by *clear-cutting*—removing an entire stand at one time. This management strategy has been severely criticized because it creates a fragmented landscape with serious impacts on biodiversity and adjacent ecosystems; also, a clear-cut leaves an ugly, unnatural-looking site for many years. Large timber companies often employ the practice because it is efficient at the time of harvest and it does not involve much active management and planning.

Other Methods. Uneven-aged management of a stand of trees can result in a more diverse forest and lends itself to different harvesting strategies. For example, in *selective cutting*, some mature trees are removed in small groups or singly, leaving behind a forest that continues to maintain a diversity of biota and normal ecosystem functions. Replanting is usually unnecessary, as the remaining trees provide seeds. Another strategy is *shelter-wood cutting*, which involves cutting the mature trees in groups over a period of, say, 10 or 20 years, such that at any time there are enough trees both to provide seeds and to give shelter to growing seedlings. Like selective cutting, this method takes more active management and skill, but unlike clear-cutting, it leaves a functional ecosystem standing. In effect, it should lead to a sustainable forest, an objective that the entire wood and paper industry is currently focusing on—at least in principle.

Sustainable Forestry. In forestry, sustainability must be carefully defined. A common objective of forestry management is **sustained yield**, which means that the production of wood is the primary goal and the forest is managed to harvest wood continuously without being destroyed. It is similar in concept to MSY, in that its objective is to maximize harvest rates, but it ignores the ecosystem properties of forests. **Sustainable forest management,** by contrast, was defined at the United Nations Conference on Environment and Development (UNCED) of 1992: *Forests are to be managed as ecosystems, with the objectives of maintaining the biodiversity and integrity of the ecosystem, but also to meet the social, economic, cultural and spiritual needs of present and future generations.* FAO presented an updated sustainable forest management concept in the FRA 2005, which included the following key elements:

1. **Forest resources**—to maintain adequate forest cover to support the social, economic, and environmental dimensions of forestry; in particular, to reduce deforestation, restore degraded forest landscapes, maintain forests for carbon storage, and support conservation efforts.

2. **Biological diversity**—to conserve and manage biodiversity at all its levels—species, genetic, and landscape—and especially to protect fragile ecosystems, with the aim of maintaining the diversity of life and allowing the possibility of developing new products in the future, such as medicines.

3. **Forest health and vitality**—to protect forests from wildfire, pests, diseases, pollution, and other damaging agents in order to maintain and improve long-term forest health and productivity in support of the range of forest functions.

4. **Productive functions of forest resources**—to manage forests so as to produce a wide range of wood and non-wood forest products, ensuring that harvesting and production are sustainable.

5. **Protective functions of forest resources**—to promote the role of forests in providing a host of valuable services

such as maintaining soil and aquatic systems; reducing the risks of floods, avalanches, and drought; and maintaining clean water.

6. **Socioeconomic functions**—to address the important economic contributions of forest resources for employment and the many values that come from processing and marketing forest products; to protect the forests, as they provide cultural, spiritual, or recreational value; and to enhance opportunities for indigenous and community management systems.

7. **Legal, policy, and institutional framework**—to provide the framework necessary to support the six themes already listed, including participatory decision making, fair and just use of forest resources, and proper law enforcement.

The FRA 2005 is organized according to these thematic elements; the assessment reported significant progress since 1990 in sustainable forest management on regional and global scales. The tropical forests are especially critical.

Tropical Forests. Because of their continued deforestation, tropical forests are of great concern. They are the habitat for millions of plant and animal species and a storehouse of biodiversity. These forests are also crucial in maintaining Earth's climate, serving as a major sink for carbon and restraining the buildup of global carbon dioxide. Nevertheless, tropical forests continue to be removed at a rapid rate. Between 1960 and 1990, 20% of the tropical forests (over 445 million hectares, or 1.1 billion acres)—equivalent to two-fifths of the land area of the United States—were converted to other uses. According to the FRA 2005, the annual global deforestation rate from 1990 through 2005 was about 13 million hectares (32 million acres), and there has been little sign of decrease. Most of this loss occurred in the tropical forests (Fig. 7–10).

Reasons for Removal. Typically, an area to be cleared is cut and then allowed to dry for a few weeks before being burned. The unmistakable signal of tropical deforestation is the plumes of smoke from burning vegetation, which can often be detected from satellites or the space shuttles. Deforestation is caused by a number of factors, all of which come down to the fact that the countries involved are in need of greater economic development and have rapid population growth. The FRA 2000 study concluded that the current major cause of deforestation is *conversion to pastures and agriculture*. Many governments in the developing world are promoting deforestation by encouraging the colonization of forested lands. Indonesia, for example, has embarked on a program of resettlement and intends to convert 20% of its remaining forests to agricultural production. The clearing of forests for subsistence agriculture is especially intense in Africa, where population growth is unrelenting. In Brazil's Amazon Basin, commercial ranches, home to some 74 million cattle, are replacing rain forests, enabling Brazil to dominate the beef export market. The Brazilian government has introduced antideforestation measures, which had some success in reducing the Amazon deforestation for several years (2004–2007), but recent satellite-based reports indicate that the rate of deforestation is rising again.

New Trends. Left to grow, the forests of the developing world are an important resource that can generate much-needed revenue. Developing countries account for 26% of the international trade in forest products. Often, concessions are sold to multinational logging companies, which reap high profits by exploiting the economic desperation of the developing countries and harvest the timber with little regard for regenerating the forest. Chinese and other Asian companies are logging millions of acres of tropical forest in smaller countries like Belize, Suriname, and Papua New Guinea, where regulation is weak and corruption is strong. More important is the consumptive use of forests by millions of people living in the forests or on their edges. Owning few private possessions and often little land, they absolutely depend on foraging in and extracting goods from the forests. Among the encouraging trends in forest management in the developing world—trends that indicate that those in power are paying more attention to the needs of indigenous people and the importance of forest goods and services (other than just wood)—are the following:

1. **Practicing sustainable forest management.** A growing number of developing countries are joining the developed countries in promoting sustainable forest management. An estimated 6% of the tropical forests are now under a formal forest management plan.

2. **Designating forest area for conservation.** A number of developing countries are increasing the area of their forests designated for conservation. Globally, more than 11% of forest area is so designated.

3. **Establishing plantations of trees for wood or other products (cacao, rubber, etc.).** More than 40% of industrial harvests originate in plantations (Fig. 7–12). Although

Figure 7–12 Plantation forest. A rubber tree plantation in Vietnam.

biodiversity is lower than in natural forests, plantations can continue to recycle nutrients, hold soil, and recycle water. Plantations also take the pressure off existing natural forests, allowing them to be maintained for biodiversity conservation.

4. **Setting aside extractive reserves that yield nontimber goods.** Among these goods are latex, nuts, fibers, and fruits. Recent calculations show that some forests are worth much more as extractive reserves than as sources of timber. Some 3.6 million hectares (9 million acres) of the Brazilian Amazon have been set aside as extractive reserves.

5. **Preserving forests as part of a national heritage and putting them to use as tourist attractions.** This practice can often generate much more income than logging can.

6. **Putting forests under the control of indigenous villagers.** The villagers can then collectively use the forest products in traditional ways. Given tenure over the land, the villagers tend to exercise stewardship over their forests in a way that is sustainable. Where this practice has been implemented, the forests have fared better than where they have been placed under state control.

Certification. The *Forest Stewardship Council* was the result of a recent initiative directed toward the certification of wood products. This alliance of nongovernmental organizations such as the World Wide Fund for Nature and National Wildlife Federation, industry representatives, and forest scientists has developed into a major international organization with the mission of promoting sustainable—or as they call it, "responsible"—forestry by certifying forest products for the consumer market. By 2009, more than 110 million hectares (242 million acres) in 81 countries had been certified. Certification makes it possible for consumers to choose wood products that have been harvested sustainably.

Ocean Ecosystems

Covering 75% of Earth's surface, the oceans and coastal ecosystems provide priceless goods and services that enable commerce and enhance human well-being. As vast as the oceans are, human activities have impacted them, often severely. A team of British, Canadian, and American scientists recently constructed a global map that pictures the severity of the effects of such human impacts as commercial shipping, resource extraction, nutrient runoff from land, fishing, species invasion, and greenhouse gas emissions (Fig. 7–13). The team's work provides a first attempt at documenting the human impacts throughout the oceans and is expected to stimulate strategies for minimizing the ecological impact of human activities in locations where that impact is especially severe. One of those activities is commercial fishing.

Marine Fisheries. Fisheries provide employment for at least 200 million people and account for more than 15% of the total human consumption of protein. (The term **fishery** refers either to a limited marine area or to a group of fish or shellfish species being exploited.) For years, the oceans beyond a 12-mile limit were considered international commons. By the end of the 1960s, however, numerous regions of the sea were being seriously depleted of many species by overfishing on the part of international fleets equipped with factory ships and modern fish-finding technology. In the mid-1970s, as a result of agreements forged at a series of UN Conferences on the Law of the Sea, nations extended their limits of jurisdiction to 200 miles offshore. The United States accomplished this with the Magnuson Act of 1976. Since

Figure 7–13 Human impact on ocean ecosystems. This map indicates highest-impact areas in the Eastern Caribbean (B), the North Sea (C), and Japanese waters (D) and a least-impact area in the vicinity of northern Australia (E).

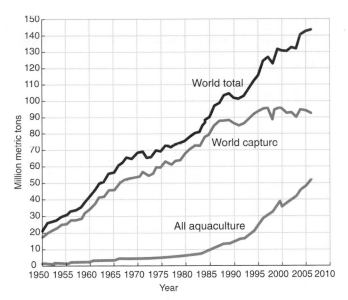

Figure 7–14 **The global fish harvest.** The global fish catch plus aquaculture equals the world total. Note how aquaculture is providing all of the increase since 1995. Shown are data for 1950–2006. (*Source:* Data from FAO FAOSTAT.)

many prime fishing grounds are located between 12 and 200 miles from shore, this action effectively removed most fisheries from the international commons and placed them under the authority of particular nations. As a result, some fishing areas recovered, while nationally based fishing fleets expanded to exploit the fisheries.

The Catch. The total recorded harvest from marine fisheries and fish farming is reported annually by the FAO. The harvest has increased remarkably since 1950, when it was just 20 million metric tons (Fig. 7–14). By 2006, it had reached almost 144 million metric tons. Aquaculture accounted for 51.6 million metric tons, or 36% of world fish supplies that year. Based on the trends shown in the figure, the "capture" fisheries leveled off in the 1990s, and the continued rise in fish production is due to aquaculture (see Sustainability, "Aquaculture Filling the Gap"). Of the total of 144 million metric tons, 92 million metric tons were the freshwater and marine fisheries catch, to which should be added some 7 million tons of bycatch (fish caught and discarded). Over 37% of the capture fisheries are species like whiting, herring, and anchovy that are reduced to fish products

Sustainability

Aquaculture Filling the Gap

Modern aquaculture, which is the farming of aquatic organisms, includes species ranging from seaweed to bivalve mollusks to freshwater and marine fish. Globally, aquaculture production—particularly the farming of marine and estuarine (coastal) species—has increased dramatically in the past 20 years. (See Figure 7–14.) Without a doubt, this production is offsetting the plateau in the capture fisheries, but coastal aquaculture is not without its problems and critics. For example, shrimp farming in many tropical areas has been criticized for destroying mangrove habitats. In this case, shrimp aquaculture is being blamed, at least in part, for accelerating the decline of fisheries for some species that depend on the mangroves for a part of their life cycle. Another concern is that farming carnivorous species such as shrimp and salmon puts an even greater demand on the capture-fish species that are used to make the feed for the farmed species (although less valuable species like anchovies and herring are used for that purpose). Atlantic salmon are farmed worldwide, and there is concern that, as they escape from the net pens (which they frequently do), they will spread diseases and hybridize with the wild populations, which are often highly threatened already.

Coastal aquaculture sites are typically in estuaries and lagoons or in ocean waters very close to the shoreline. These are the waters of the coastal zone that are the most used by humans for many other purposes, such as sportfishing, boating, and waterskiing. The nearshore coastal waters are also the most likely to be affected by sewage effluents and land runoff carrying many kinds of pollutants. The result in many cases is conflict among the various users, and the long-term outlook for coastal aquaculture is that nearshore waters are becoming less and less available for farming. There does not seem to be much argument about the need for expanded farming of coastal species to increase world fish and shellfish supplies, but the industry appears to be facing more problems than ever before. So what can be done to make aquaculture more sustainable in the long term?

One approach is open-ocean aquaculture, which is the farming of coastal species in ocean

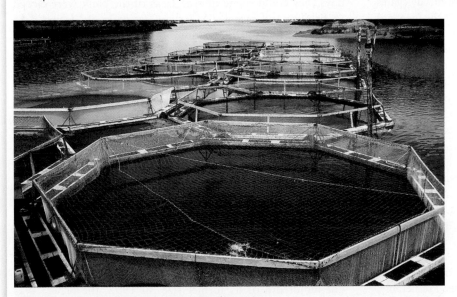

Aquaculture. Coastal saltwater salmon pens in Lubec, Maine.

(Continued)

(Continued)

waters several kilometers from the shoreline. This is a new practice in North America, and even in Europe and other areas where it has been long practiced, total production levels are far lower than they could be. The most polluted areas in many coastal zones are the estuaries and embayments very close to shore. Hence, open-ocean waters are relatively clean. High water quality is important for aquaculture, but it is especially critical for species such as bivalve mollusks, which concentrate pathogens and other pollutants that can be passed on to human consumers. Also, new satellite-based remote-sensing efforts are revealing previously unknown nutrient-rich upwelling zones. In some areas in the Gulf of Maine, for example, phytoplankton concentrations can exceed the levels thought to occur only in enriched estuarine waters. This means that open-ocean waters may be able to support a high productivity of farmed species such as the blue mussel.

Mussels and other bivalve mollusks are herbivorous animals, feeding lower on the food web. A long-term sustainable approach to aquaculture would see more use of herbivorous species, reducing the impact on the wild fish stocks. Indeed, herbivorous fish like carp, tilapia, and milkfish, together with the marine mollusks, already make up 80% of the global output of aquaculture.

The problems facing aquaculture today are not unusual because aquaculture is yet another complex practice that consists of competing demands on our environment. The problems can be solved only by approaches that are truly sustainable in the long term. In this case, it means bringing policy makers and the aquaculture industry together to work toward reducing the environmental costs of aquaculture.

Source: Contributed in part by Ray Grizzle, Jackson Estuarine Laboratory, University of New Hampshire.

---

to feed the farm-raised fish and animals. The salmon and shrimp aquaculture farms consume several times as much fish protein as they actually produce.

The Limits. The world fish catch may appear stable, but many species and areas are overfished. The FAO has concluded that 52% of fish stocks are fully exploited (at their maximum sustainable yield), 16% are overexploited and will likely decline further, and 8% are depleted and are much less productive than they once were. Much of the catch (one-third) is of such low commercial value that it is used only for fish meal and oil production. Aquaculture is growing at a rate of 9% per year in the developing countries, helping to alleviate poverty and improve the well-being of rural people.

The FAO has also concluded that the capture fisheries are at their upper limit, as indicated by the long-term catch data. Fisheries scientists reason that, if overexploited and depleted fisheries were restored, the catch could increase by at least 10 million metric tons. That could happen, however, only if fishing for many species was temporarily eliminated or greatly reduced and the fisheries were managed on a more sustainable basis. At the 2002 Johannesburg World Summit, delegates agreed (one of the few agreements to come out of the meeting) to restore depleted global fisheries to achieve an MSY "on an urgent basis" by 2015 and to establish a network of protected marine areas by 2012. Recent studies have shown that marine reserves provide sheltered nurseries for surrounding areas and can supply world-record-size fish to adjacent fisheries.

Georges Bank. Events on Georges Bank, New England's richest fishing ground, illustrate the fisheries' problems well. Cod, haddock, and flounder (called groundfish because they feed on or near the bottom and are caught by bottom trawling) were the mainstay of the fishing industry for centuries. In the early 1960s, these species accounted for two-thirds of the fish population on the bank. After 1976, fishing on Georges Bank doubled in intensity, resulting in a decline in the desirable species and a rise in the so-called rough species—dogfish (sharks) and skates. As the rough species increased their numbers, fishers focused their attention on them, even though they were less valuable. Predictably, dogfish and skate populations have also been overfished and are now subject to regulation.

Management Councils. The prized species declined on Georges Bank because the fishery was poorly managed. The Magnuson Act established eight regional management councils made up of government officials and industry representatives. The councils are responsible for creating management plans for their regions, while the National Marine Fisheries Service (NMFS) provides advice and scientifically based information on assessing stocks. The NMFS also has the authority to reject elements of the plans submitted to it by the councils. The New England Fishery Management Council (NEFMC) began its task by setting TAC quotas, but fishers claimed that the TACs were set too low and successfully argued for an indirect approach that employed mesh with openings of a size that allowed smaller fish to escape. In less than 10 years, the number of boats fishing Georges Bank doubled, with disastrous results, as indicated by the cod-landing data (Fig. 7–15).

The NEFMC discarded the mesh-size approach and initiated a number of regulations designed to take some pressure off the Georges Bank cod. Vessels were restricted in the days they could spend at sea, areas of the bank were closed to fishing, and target TACs were set at levels designed to reduce the harvest. In addition, the NMFS has been buying back fishing permits and paying fishers to scrap their boats. In spite of these measures, cod landing consistently exceeded the TAC for years, by 150–200%, until 2004–2005, when the stock had plummeted to its lowest level ever and restrictions finally led to a much reduced harvest. Recently, the New England groundfishery (with the blessing of the NEFMC) began to operate a community-based quota system; this has been so successful that 17 additional groups of eastern fishers have petitioned to start similar systems.

Other Cod Fisheries. The Grand Banks off the coast of Newfoundland were once the richest fishing grounds in the world. The cod, however, were decimated by Canadian fishers, who were encouraged by their government to exploit the fishery and were given little warning of its imminent collapse. By 1991, the cod population had dropped to one-hundredth of its former size. In 1992, the fishery was closed, costing the jobs of 35,000 fishers. So far the cod population has not rebounded, and cod are at their lowest levels

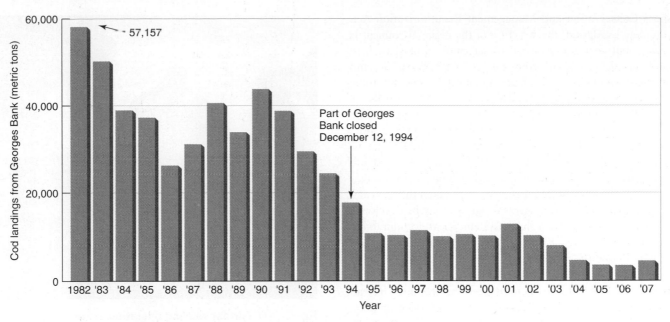

Figure 7–15 Fishery in distress. Data showing cod landings from Georges Bank, 1982–2007.
(*Source:* Northeast Fisheries Science Center, Woods Hole Laboratory, 2009.)

in a century. Across the water, Europe's cod fishery is experiencing the same decline: The North Sea's cod stocks are now only 15% of what they were in the 1970s. Proposals to reduce harvests have met with fierce resistance from the European Union fishing nations, who argue that the reductions would destroy them.

The basic problem of the fisheries is that there are too many boats, rigged with high technology that gives the fish little chance to escape, chasing too few fish. Equipped with sophisticated electronic equipment that can find fish in the depths and with trawlers with 200-foot-wide openings that scoop up everything on and close to the bottom, the fishing industry has become too efficient (Fig. 7–16). A better system of management is needed, and the size of the fishing fleet must be reduced so that the remaining fishers are better able

to make a living without overfishing. A limited system of access is needed, too, such as the IQ system used to manage the Pacific halibut fishery that we discussed at the beginning of the chapter. This system, also called **catch shares**, effectively gives fishers the equivalent of property rights in the fishery. The rights are transferable and allow the fishers to determine how and when they will harvest their catch. Coupled with accurate population assessments by fishery scientists, the catch shares system has produced an economically stable and environmentally sustainable Pacific halibut fishery, and there is no reason that it should not be applied to fisheries everywhere.

Fisheries Law Reauthorized. In December 2006, Congress passed the second reauthorization of the Magnuson Act, with strong bipartisan support: the **Magnuson-Stevens**

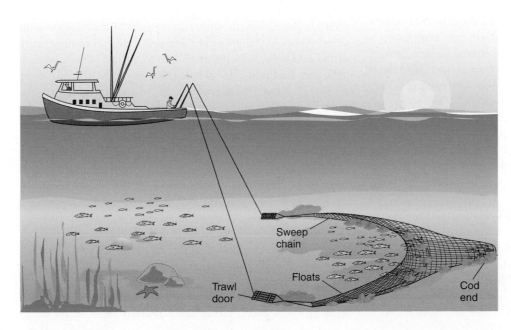

Figure 7–16 Bottom trawling. Because of the way the ocean bottom is degraded, this method, used to harvest groundfish, has been compared to clear-cutting forests. The steel trawl doors weigh up to 5 tons, and the distance between the doors is as much as 200 feet.

Fishery Conservation and Management Reauthorization Act. In the new legislation, the structure of the regional councils is retained, but they are required to set catch limits based on sound scientific advice from the council's Scientific and Statistical Committee (SSC). The annual catch limits are to be set below the maximum sustainable yield, and the councils are to employ precautionary principles in setting those limits. The law also requires councils to end overfishing within two years after a species is judged to be overfished (it used to be 10 years!). As with the previous version of the act, depleted fish stocks must be rebuilt and maintained at biologically sustainable levels. The law also requires steps to be taken to assess and minimize bycatch. Limited access programs may be developed along the lines of the catch shares system. Clear accountability measures are required to ensure that the councils and their scientists follow the mandates of the new law. Some fisheries scientists were disappointed that the new law does not require an ecosystem-based approach to fisheries management, instead maintaining the existing species-by-species approach. It does, however, strengthen the role of the SSCs, which improves the science-based approach to fisheries management.

Other Issues. Many other issues pertaining to fisheries need to be addressed. These include the banning of shark finning, a practice in which just the fins of sharks are taken to satisfy the demand for shark-fin soup; the accurate certification of "dolphin-safe" tuna (to discourage the practice of fishing that sets nets on dolphin schools because tuna are usually feeding below them); shrimping in the Gulf of Mexico that kills endangered sea turtle species and generates a huge bycatch; declining stocks of swordfish and tuna even as fishing intensity increases; and aquaculture demands placed on ocean fish because the ground meal made from ocean fish harvests is fed to pen-raised fish.

Marine Reserves. Marine Protected Areas (MPAs) are areas of the coasts and sometimes open oceans that have been closed to all commercial fishing and mineral mining. They often allow less-harmful use like small-scale fishing, tourism, and recreational boating, although some MPAs prohibit all such activities. The number of MPAs is growing, now numbering more than 4,000 of varying sizes. Over 18% of coral reefs are covered by MPAs, but not all of these are rigorously managed. The benefits of MPAs are well known and include basic protection of these vital ecosystems, increased tourism, and the spillover of fish and larvae from the MPA to surrounding fisheries. The most direct path to the restoration of a fishery or a damaged habitat is likely to be the creation of a marine reserve. In 2009 President George W. Bush established three huge MPAs in waters surrounding several South Pacific islands, totaling 505,000 square kilometers (195,000 mi^2) in size.

International Whaling. Another fishery of great concern to the international community is the great whales. Whales, found in the open ocean and in coastal waters, were heavily depleted by overexploitation until the late 1980s. Whales once were harvested for their oil, but are now prized for their meat, considered a delicacy in Japan and a few other countries. In 1974, the International Whaling Commission (IWC), an organization of nations with whaling interests, decided to

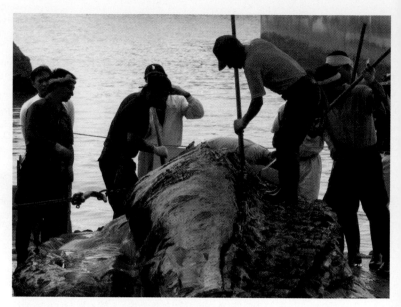

Figure 7–17 Current whaling operation. Butchers slice into a Baird's beaked whale at a pierside slaughterhouse in Wada, Japan.

regulate whaling according to the principle of maximum sustainable yield. Whenever a species of whale dropped below the optimal population for such a yield, the IWC instituted a ban on hunting that species in order to allow the population to recover. At that time, three species (the right whale, bowhead whale, and blue whale) were at very low levels and were immediately protected. Because of difficulties in obtaining reliable data on and enforcing catch limits, the IWC took more drastic action and placed a moratorium on the harvesting of all whales beginning in 1986. The moratorium has never been lifted; however, some limited whaling by Japan, Iceland, and Norway continues, as does harvesting by indigenous people in Alaska, the Russian Federation, and Greenland (Fig. 7–17).

Table 7–3 lists the status of 13 whale species or populations that are of commercial interest. Estimates of the numbers remaining range from 100 Northwest Pacific gray whales to up to 2 million sperm whales. Their main threat today is entanglement with fishing gear and collisions with ships. Although reliable data are still hard to acquire, many of the whale species appear to be recovering. The bowhead, gray and Southern right whales, for example, recently were upgraded from "Vulnerable" to "Lower Risk" by the International Union for Conservation of Nature (IUCN). In the time between the IWC moratorium and the present, however, the basis of the ethical controversy over whaling has shifted from conservation to animal rights. Many people worldwide simply believe that it is wrong to kill and eat such large and unique mammals. People from whaling nations counter with the argument that their culture includes eating whales, just as other cultures include eating cows or turkeys.

Whale Stakes. There has been heavy pressure from three members of the IWC—Japan, Iceland, and Norway—to reopen whaling. The interests of these countries focus on the small minke whale, although the Japanese have begun taking the much larger fin whales. In 1993, Norway resumed whaling for minkes, citing the country's right to refuse to

TABLE 7–3 Remaining Numbers of Large Whales

Endangered	Vulnerable	Lower Risk or Insufficient Data
Blue whale (3,000–5,000)	Sperm whale (1 million–2 million)	Minke whale (935,000)
North Atlantic right whale (300–350)		Bryde's whale (40,000–80,000)
North Pacific right whale (under 1,000)		Bowhead whale (11,750)
Fin whale (50,000–90,000)		Southern right whale (7,500)
Sei whale (50,000)		Humpback whale (63,600)
Northwest Pacific gray whale (100)		Gray whale (26,300)

Sources: WWF, IUCN *Red Data List* (2008), and International Whaling Commission.

accept specific IWC rulings. Norway based its decision on information submitted by the Scientific Committee of the IWC, which judged the minke population to be numerous enough to absorb a sustainable exploitation. Currently, Norway takes up to 650 and Japan over 1,000 minke whales per year, the Norwegians for human consumption and the Japanese for "scientific purposes." It is no secret that the Japanese "scientific" harvest ends up in commercial markets; the Japanese people, however, are losing their taste for whale meat, and much of it ends up as pet food or in school dinners. The newest Japanese rationale for whaling, totally without scientific merit, is that the whales eat too many fish that humans should be catching!

The IWC is right in the middle of this controversy. At its annual meetings, prowhaling members push for a resumption of whaling with quotas, and antiwhaling countries oppose any such resumption. The 2007 meeting of the IWC reaffirmed the moratorium on commercial whaling, and the 2008 meeting closed without any vote taken. At this meeting, however, the IWC set up a new working group to study compromises on the divisive issues of whale sanctuaries and research whaling. The sanctuary issue has made headlines lately because of Australia's claim that Japanese whalers are killing whales in the Australian Whale Sanctuary. Australia declared the sanctuary based on its sovereignty in Antarctica. Boats from Greenpeace and Sea Shepherd Conservation Society have made a practice of harassing Japanese whaling vessels as they encounter them in the Australian Whale Sanctuary. There seems to be little chance for an end to the whaling controversy as long as Japan, Norway, and Iceland persist in killing whales. Perhaps the best hope is the rising sentiment in these countries against whaling, fueled by opportunities to see whales on whale-watching trips.

Whale Watching. Whale watching has become an important tourist enterprise in coastal areas (Fig. 7–18). Stellwagen Bank, within easy reach of boats from Boston,

Figure 7–18 Whale watching. A group of humpback whales entertains boatloads of whale watchers on Stellwagen Bank just north of Provincetown, Cape Cod. Whale watching has replaced whaling in New England waters that were once famous as whaling centers.

Cape Ann, and Cape Cod, Massachusetts, has become the center of a whale-watching industry estimated to be worth more than $30 million annually in New England alone. From spring through fall, scores of boats venture offshore daily to watch the whales that congregate over the bank. Many of the humpback whales seem to enjoy entertaining the visitors and often frolic alongside the boats for hours. Recently, the International Fund for Animal Welfare reported that whale watching now flourishes in 87 countries and generates more than $1 billion in business annually.

Besides having aesthetic and entertainment value, whale watching is of scientific value. Whale-watching tour boats usually carry a biologist along who identifies the whale species and interprets the experience for the visitors. The biologists are often associated with groups such as the Cetacean Research Unit and the Whale Conservation Institute, which have studied the whales of Stellwagen Bank since 1979 and have published many papers on the humpback whale. Largely in response to its importance for whales, Stellwagen Bank was designated a National Marine Sanctuary in 1993.

Coral Reefs. In a band from 30° north to 30° south of the equator, coral reefs occupy shallow coastal areas and form *atolls*, or islands, in the ocean. Coral reefs are among the most diverse and biologically productive ecosystems in the world. The reef-building coral animals live in a symbiotic relationship with photosynthetic algae called **zooxanthellae** and, therefore, are found only in water shallower than 75 meters (245 ft). The corals build and protect the land shoreward of the reefs and attract tourists, who enjoy the warm, shallow waters. In addition, because the reefs attract a great variety of fish and shellfish, they are important sources of food and trade for local people, generating an estimated $30 billion per year in tourism and fishing revenues. In recent years, coral reefs have been the subject of concern because they have shown signs of deterioration in areas close to human population centers. The year 1998 was a watershed: Before then, an estimated 11% of coral reefs had been destroyed by human activities. Since then, the percentage of destroyed reefs has increased to 20%. The threats are both global and local.

Global Threats. The Intergovernmental Panel on Climate Change (IPCC), in its 2007 assessment (see Chapter 18), forecast that the coming global temperature increases would bring on "more frequent coral bleaching events and widespread mortality." Extensive *coral bleaching* (wherein the coral animals expel their symbiotic algae and become white in appearance) severely damaged an estimated 16% of the world's reefs (Fig. 7–19) during the record-high sea surface temperatures accompanying the 1997–1998 El Niño. Some coral bleaching is an annual phenomenon related to the normal high temperatures and light intensities of summer, but the most recent incidents are unprecedented in scope. Studies have shown that increased sea surface temperatures can bring on 100% bleaching; indeed, 2005 brought the highest water temperatures on record in the Caribbean, bleaching 90% of corals in many locations (and contributing

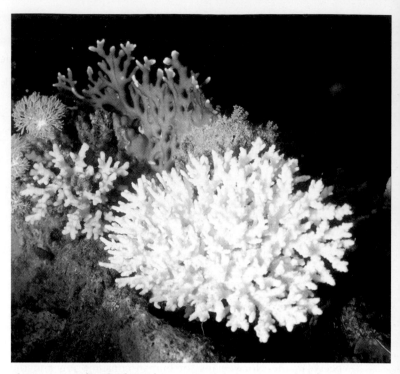

Figure 7–19 Bleached coral. This coral in the Red Sea shows the effects of excessively warm temperature.

to a record number of intense hurricanes). Corals are resilient, recovering from bleaching if sea surface temperatures decline in a short time, and some species are more tolerant of bleaching than others. Repeated bleaching, however, weakens the corals and eventually kills them.

The other global threat—one that is more long-term—is *ocean acidification.* As the surface oceans take up increasing amounts of the CO_2 pumped into the atmosphere by humans, sea water becomes more acid, and the carbonate ion concentration is reduced. This in turn makes it more difficult for coral animals to build their calcium carbonate skeletons. If we allow atmospheric CO_2 to rise above 800 ppm (it is now at 380 ppm and rising), by the end of the century the carbonate ion concentration will be reduced by half. Scientists point out that persistent high summer temperatures and the gradual acidification of the oceans could permanently wipe out coral reefs over vast areas of the shallow tropical oceans—an enormous loss of biodiversity and economic value.

Exploitation. A more local source of coral damage stems from poverty and greed. Islanders and coastal people in the tropics often scour the reefs for fish, shellfish, and other edible sea life. Some exploitation of the reefs, however, is driven by the lucrative trade in tropical fish for food and the pet trade. Local residents sometimes use dynamite and cyanide to flush the fish out of hiding in the coral, a practice that is not only destructive to the coral, but also dangerous to the perpetrators. "Fishing" with cyanide is a growing (illegal) practice that has been encouraged by exporters in the Indo-Pacific region. Halting this practice will take a concerted effort on the part of the Asian countries involved to police the reefs and work toward retraining the cyanide fishers to use nondestructive fishing methods. It will also require countries that import tropical fish to monitor and certify cyanide-free imports.

One solution to this destructive trade is to foster effective, sustainable management of the coral reef resources. Ideally, such management would be community based so that local people benefit from the exploitation. Often, though, communities in the developing world lack the capacity to carry out this kind of management, in which case outside assistance is required (as described in Stewardship, "The Turtle Island Commons," p. 168). Another solution is to incorporate coral reefs in Marine Protected Areas (MPAs). A recent study found that 980 MPAs are already providing 19% of the world's coral reefs with some level of protection. Again, effective management and enforcement are essential, and even this is not always effective.

Mangroves. Often, just inland of the coral reefs is a fringe of mangrove trees. These trees have the unique ability to take root and grow in shallow marine sediments. There they protect the coasts from damage due to storms and erosion and form a rich refuge and nursery for many marine fish. Despite these benefits, mangroves are under assault from coastal development, logging, and shrimp aquaculture. Between 1983 and the present, half of the world's 45 million acres (18 million ha) of mangroves were cut down, with percentages ranging from 40% (in Cameroon and Indonesia, for example) to nearly 80% (in Bangladesh and the Philippines, for instance).

Although logging (for paper pulp and chipboard) and coastal land development are believed to be responsible for most of the mangrove deforestation, the development of ponds for raising shrimp is proving destructive, too. In this practice, the mangrove forests are cleared, and the land is bulldozed into shallow ponds, where black tiger prawns are fed with processed food. The aquaculture of prawns is now a multibillion-dollar industry, satisfying the great demand for shrimp in the developed countries. Unfortunately, in many cases the ponds become polluted and no longer support shrimp, so new ponds are then created by clearing more mangroves. The regular flushing of the ponds, required to remove excess nutrients and wastes, often degrades adjacent coastal environments. The massive removal of mangroves has brought on the destabilization of entire coastal areas, causing erosion, the siltation of sea grasses and coral reefs, and the ruin of local fisheries. Developing nations that once considered the mangroves to be useless swampland are now beginning to recognize the natural services they perform. Local and international pressure to stop the destruction of mangroves is growing.

We have investigated how the management (or lack thereof) of natural resources has affected ecosystems around the world. In the next section, we will focus on the connection between public policy and natural resource management as it is reflected in public and private lands in the United States.

7.4 Protected Lands

The time has passed when the loss of ecosystems could be excused on the grounds that there are suitable substitute habitats just over the hill. With the rising human population, industrial expansion, and pressure to convert natural resources to economic gain, there will always be reasons to exploit natural ecosystems. The last resort for many species and ecosystems is protection by law in the form of national parks, wildlife refuges, and reserves. The World Database on Protected Areas (WDPA) recognizes 106,900 "protected areas," sites whose management is aimed at "achieving specific conservation objectives," according to the Convention on Biological Diversity. These sites, representing 13% of Earth's land area, cover a spectrum of protection from strict nature preserves to managed resource use. However, establishment of a protected area is not always followed by effective management and control over misuse.

Public and Private Lands in the United States

In March 2009, the Omnibus Public Lands Management Act was signed into law by President Obama. This sweeping legislation protected an additional 2.1 million acres of wilderness, increased the wild and scenic river system by 50%, established new national parks and monuments, and provided for a new federal land system, the National Landscape Conservation System. The United States is unique among the countries of the world in having set aside a major proportion of its landmass for public ownership. Nearly 40% of the country's land is publicly owned and is managed by state and federal agencies for a variety of purposes, excluding development. The distribution of federal lands, shown in Figure 7–20, is greatly skewed toward Alaska and other western states, a consequence of historical settlement and land distribution policies. Nonetheless, although most of the East and Midwest are in private hands, forests and other natural ecosystems still function all across these regions.

Wilderness. Land given the greatest protection (preservation) is designated as **wilderness**. Authorized by the Wilderness Act of 1964, this classification includes 109 million acres in more than 700 locations—over 4% of the land area of the United States. The act provides for the permanent protection of these undeveloped and unexploited areas so that natural ecological processes can operate freely. Permanent structures, roads, and motor vehicles and other mechanized transport are prohibited. Timber harvesting is excluded. Some livestock grazing and mineral development are allowed where such use existed previously; hiking and other similar activities are also allowed. Areas in any of the federally owned lands can be designated as wilderness; the Bureau of Land Management, National Park Service, Forest Service, and Fish and Wildlife Service all manage wilderness lands.

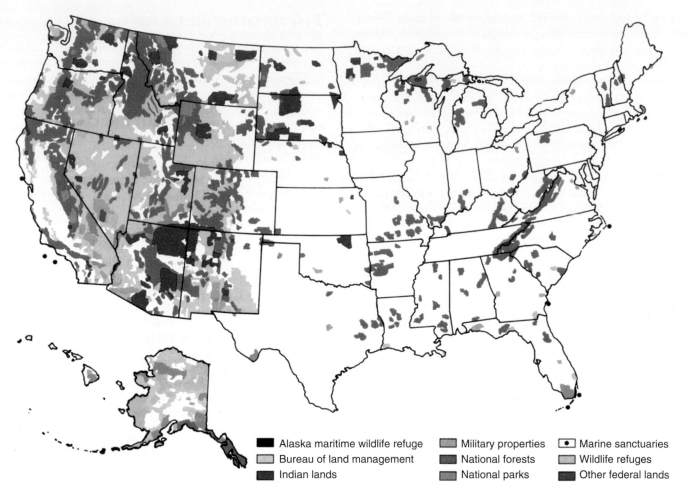

Figure 7–20 Distribution of federal lands in the United States. Because the East and Midwest were settled first, federally owned lands are concentrated in the West and Alaska. (*Source:* Council on Environmental Quality, *Environmental Trends*, 1989.)

Legend:
- Alaska maritime wildlife refuge
- Bureau of land management
- Indian lands
- Military properties
- National forests
- National parks
- Marine sanctuaries
- Wildlife refuges
- Other federal lands

National Parks and National Wildlife Refuges. The **national parks**, administered by the National Park Service, and the **national wildlife refuges**, administered by the Fish and Wildlife Service, provide the next level of protection to 84 million and 96 million acres, respectively. Here the intent is to protect areas of great scenic or unique ecological significance, protect important wildlife species, and provide public access for recreation and other uses. The dual goals of protecting and providing public access often conflict with each other because the parks and refuges are extremely popular, drawing so many visitors (more than 310 million visits a year) that protection can be threatened by those who want to see and experience the natural sites. At a number of the most popular parks, auto traffic has been replaced by shuttle vehicles because of the problems of parking and habitat destruction accompanying the use of motor vehicles within the parks. All of the federal agencies involved are trying to cope with the rising tide of recreational vehicles like snowmobiles and off-road vehicles used on most federal lands. The Blue Ribbon Coalition, a lobbying group representing owners and manufacturers of off-road vehicles, fiercely defends keeping federal lands accessible to these vehicles.

Increasingly, agencies, environmental groups, and private individuals are working together to manage natural sites as part of larger ecosystems. For example, the Greater Yellowstone Coalition has been formed to conserve the larger ecosystem that surrounds Yellowstone National Park (Fig. 7–21). Yellowstone and Grand Teton National Parks form the core of the ecosystem, to which seven national forests, three wildlife refuges, and private lands are added to form the greater ecosystem of 18 million acres. The coalition acts to restrain forces threatening the ecosystem. Thus, it has challenged the Forest Service's logging activities, addressed residential sprawl and road building on the private lands, acted to protect crucial grizzly bear habitat, and curtailed the senseless shooting of bison that wander off the national parklands in search of winter grazing.

This cooperative approach is important for the continued maintenance of biodiversity because so much of the nation's natural lands remain outside of protected areas. Cooperation may also help restrict development up to the borders of the parks and refuges, keeping them from becoming fairly small natural islands in a sea of developed landscape.

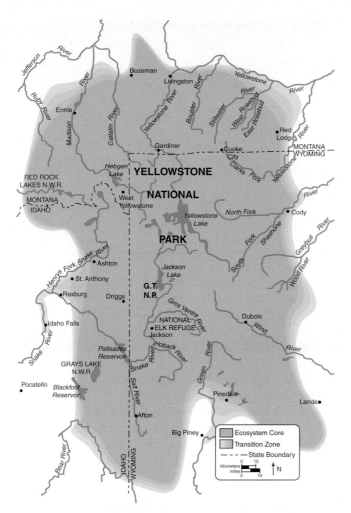

Figure 7–21 Greater Yellowstone ecosystem. Yellowstone National Park is the center of a much larger ecosystem, which has received attention from the Greater Yellowstone Coalition.

National Forests. Forests in the United States represent an enormously important natural resource, providing habitat for countless wild species, as well as supplying natural services and products. These forests range over 740 million acres, of which about two-thirds are managed for commercial timber harvest. Almost three-fourths of the managed commercial forestland is in the East and privately owned; the remainder is mainly in the West and is administered by a number of government agencies, primarily the Forest Service and the Bureau of Land Management.

Deforestation is no longer a problem in the United States. Although we have cut all but 5% of the forests that were here when the colonists first arrived, second-growth forests have regenerated wherever forestlands have been protected from conversion to croplands or house lots. There are more trees in the United States today than there were in 1920. In the East, some second-growth forests have aged to the point where they look almost as they did before their first cutting, except for the absence of the American chestnut.

The Forest Service is responsible for managing 193 million acres of national forests. The Bureau of Land

Management oversees 258 million acres that are a great mixture of prairies, deserts, forests, mountains, wetlands, and tundra. Most federal timber production comes from the national forest lands.

Multiple Use. The Forest Service's management principle in the 1950s and 1960s was *multiple use*, which allowed for a combination of extracting resources (grazing, logging, and mining), using the forest for recreation, and protecting watersheds and wildlife. Although the intent was to achieve a balance among these uses, multiple use actually emphasized the extractive uses; that is, it was output oriented and served to justify the ongoing exploitation of public lands by private interest groups (ranchers, miners, and especially the timber industry).

New Forestry. A forest-management strategy called **New Forestry** was introduced in the late 1980s. This practice is directed more toward protecting the ecological health and diversity of forests than toward producing a maximum harvest of logs. New Forestry involves cutting trees less frequently (at 350-year intervals instead of every 60 to 80 years), leaving wider buffer zones along streams to reduce erosion and protect fish habitats, leaving dead logs and debris in the forests to replenish the soil, and protecting broad landscapes—up to 1 million acres—across private and public boundaries, while involving private landowners in management decisions. The Forest Service began adopting some of these management principles in the early 1990s, and they formed the core of what is now the official management paradigm of the Forest Service: **ecosystem management**.

To harvest timber, government foresters select tracts of forest they judge ready for harvesting and then lease the tracts to private companies, which log and sell the timber. The timber harvest from national forests has been a controversial issue. Historic data on timber sales (Fig. 7–22) reflect major shifts in logging. Note the rapid rise following World War II to satisfy the housing boom that was under way at the time, the drop reflecting rising environmental concern in the 1970s, increased logging during the Reagan-Bush years, and a steep drop with the Clinton administration as a result of renewed environmental concern. According to a recent Forest Service analysis, growth of forests has exceeded harvests for several decades in both national and private forests, indicating that the forests are being managed sustainably.

The Roadless Controversy. The Clinton administration's Forest Service chief, Mike Dombeck, embraced the ecosystem management paradigm, and one of his first acts was to initiate a moratorium on building new logging roads, which became the Roadless Area Conservation Rule. The Forest Service had been criticized for its road building because the existing roads had contributed to the destruction of forests and streams. The timber industry and many members of Congress from states with national forests objected to the moratorium. They believed that it would halt timber harvests and prevent access by off-road vehicles in roadless areas (they were right!). Some 58 million acres of forestland were affected by the moratorium. The Bush administration,

Figure 7–22 National forest timber sales. Timber sales over the years reflect changes in resource needs and political philosophy.

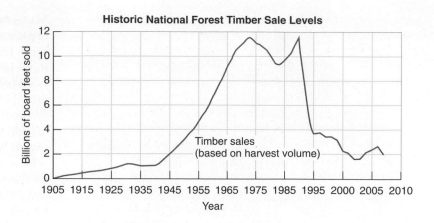

unhappy with the roadless rule, replaced the road ban with a ruling in 2005 that allows governors to ask the Secretary of Agriculture to open lands to roads and development. In 2006, this ruling was canceled by a federal district court order to reinstate the roadless rule. Another district court tossed out the roadless rule in 2008, creating conflicting court decisions and initiating a lengthy appeal process. In his campaign for the presidency, Barack Obama promised to establish the roadless rule as the law of the land, making it likely that this rule will continue to protect 58 million acres of forests.

Fires. The years 2006 and 2007 set new records for forest fires, with 9.9 million acres burning the former year (a record) and 9.3 million acres the latter. These continue a trend begun in 2004, with burned acreages double the 10-year average. With a decade of drought in the western states and serious drought in the South, it is no surprise that forest fires have become more extensive. Many scientists believe that this drought pattern is a consequence of the buildup of greenhouse gases and its impacts on climate. A hot debate has focused on why so much forest land bears what is called *fuel loads*—meaning dead trees and dense understories of trees and other vegetation—and what to do about it. Most observers believe that a combination of fire suppression and common logging practices are responsible for the high fuel loads. It is also true that drought dries out vegetation, and when combined with lightning to start fires and high winds to spread them, most forests will burn, regardless of what is done to avoid wildfires. As we saw in Chapter 5, wildfires can also be seen as natural disturbances, rather than ecological disasters, a confirmation of the dynamic nonequilibrium nature of ecosystems.

There is little solid research on what to do about reducing the threat of wildfires. However, the Bush administration put in place two strategies to address the issue: *fuel reduction* and *salvage logging*. The Healthy Forests Restoration Act of 2003 (H.R. 1904) addressed fuel reduction by providing incentives to logging companies to thin fireprone forests, especially near populated areas. In return for this work, the loggers get to harvest larger, commercially valuable trees at a lower cost. In addition, arguing that citizens' appeals were stalling forest-thinning projects, the law ruled that environmental reviews and judicial oversight would be restricted.

Salvage logging, on the other hand, involves harvesting timber from recent burns (fires often consume only the outer branches of trees). The Forest Service authorizes sales of timber on recently burned national forest lands, but is required to follow conservation safeguards mandated by the National Environmental Policy Act and the Endangered Species Act. This takes time, but in recent years, this kind of logging has been profitable to private operators in national forests. A 2006 study of postwildfire logging in an Oregon forest showed, however, that salvage logging not only added to the continued fire risk (loggers left branches and brush behind), but also actually killed up to three-fourths of the seedlings that had regenerated on the burned land.[2] Many scientists believe that the best approach to regenerating a forest is to do nothing; removing the dead trees kills seedlings, disturbs the soil, removes nesting sites for wildlife, and reduces nutrients and shade needed for new trees to grow.

Protecting Nonfederal Lands. Between 1996 and 2005, voters approved more than $25 billion through state, county, and town ballots to protect land for open space and parks. The voters decided to tax themselves to pay for open-space preservation. The largest measure was passed in California, where voters approved $2 billion for land acquisition. In Maine, the Nature Conservancy bought 185,000 acres along the St. John River for $35 million, and the New England Forestry Foundation paid Pingree Associates $28 million for development rights on 760,000 acres of forests. Pingree will continue to harvest trees, but only in a manner that is certified by the Forest Sustainability Council. These acquisitions are indicative of a groundswell of public and private support for maintaining open space. Outright purchase by municipalities is one option, but not the only one.

Land Trusts. With so much land in private hands, the future of natural ecosystems in the United States is always uncertain, particularly when the land has special appeal for recreation and aesthetic enjoyment, because there the potential for development is great. Often, landowners and towns-people want to protect natural areas from development, but

---

[2]D. C. Donato, J. B. Fontaine, J. L. Campbell, W. D. Robinson, J. B. Kauffman, and B. E. Law. "Post-Wildfire Logging Hinders Regeneration and Increases Fire Risk," *Science* 311 (January 20, 2006): 352.

are wary of turning the land over to a government authority. One creative option is the **private land trust**, a nonprofit organization that will accept either outright gifts of land or **easements**—arrangements in which the landowner gives up development rights into the future, but retains ownership of the parcel. The land trust may also purchase land to protect it from development. The land trust movement is growing considerably: There were 429 land trusts in the United States in 1980 and more than 1,667 in 2005, as reported by the Land Trust Alliance, an umbrella organization in Washington, D.C., serving local and regional trusts. These trusts protect almost 12 million acres of land. The Nature Conservancy and other national organizations have helped to protect an additional 25 million acres in the United States. The Nature Conservancy also protects 102 million acres in other countries.

Land trusts are proving to be a vital link in the preservation of ecosystems. In 1999, alarmed about signs of development in their region, 30 residents of Grand Lake Stream in northeastern Maine (pop. 150) met to share their concerns. Out of that meeting grew the Downeast Lakes Land Trust, which, over time, has carried out a $30 million conservation effort that has protected 342,000 acres of woodlands, 60 lakes, and 1,500 miles of riverfront. These lands connected several other conservation holdings in eastern Maine and New Brunswick, and the result (to the surprise of the original 30 residents) is that more than 1 million contiguous acres are protected against development. The Downeast Land Trust is busy now constructing hiking trails and canoe-access campsites so that the public will be able to enjoy hiking, canoeing, and fishing in the conserved landscape.

Final Thoughts

Ecosystems everywhere are being exploited for human needs and profit. In addition to all the examples discussed in this chapter, other areas that are in trouble include wetlands drained for agriculture and recreation, overgrazed rangelands, and rivers that are overdrawn for irrigation water. The purpose of this chapter has been to highlight some of the most critical problems and to indicate steps being taken to correct them. It is certain that greater pressures will be put on natural ecosystems as the human population continues to rise. These pressures must be met with increasingly effective protective measures if we want to continue to enjoy the goods and services provided by ecosystems. The problems are most difficult in the developing world, where poverty forces people to take from nature in order to survive. If natural areas are to be preserved, the needs of people must be met in ways that do not involve destroying ecosystems. People must be provided with alternatives to overexploitation, a situation requiring both wise leadership and effective international aid.

As nations formulate policies for dealing with the environment in sustainable ways, they will need better information on what is happening and what the consequences of those policies will be. Sound scientific information is essential, and in many areas, there are large gaps in our knowledge. The Millennium Ecosystem Assessment, a four-year, $20 million effort described in Chapter 1, has produced a series of publications directed toward finding out what is happening to the world's ecosystems and helping countries to both deal with ecosystems that are in trouble and meet the goals of the numerous international environmental conventions now in place. It is without doubt the most accessible and rigorous data set on world ecosystems ever assembled.

In April 2000, when he was UN Secretary-General, Kofi Annan included in his millennium report to the General Assembly a stern warning that far too little concern was being given to the sustainability of our planet: "If I could sum it up in one sentence, I should say that we are plundering our children's heritage to pay for our present unsustainable practices. . . . We must preserve our forests, fisheries, and the diversity of living species, all of which are close to collapsing under the pressure of human consumption and destruction." "In short," Annan stated, "we need a new ethic of stewardship." Indeed, under UN sponsorship, the Billion Tree Campaign, begun in 2006, overran its goal in 18 months and has now become the Seven Billion Tree Campaign. From individuals to heads of states, people in 155 countries are planting trees, acting in stewardship concern for the climate and for peoples' needs.

Revisiting the Themes

Sound Science

Calculating a total allowable catch (TAC) requires solid scientific data on fish populations and an understanding of the MSY model. These limits are difficult, if not impossible, to accomplish for mobile creatures such as fish and whales. Fisheries scientists are constantly challenged to assess fish stocks accurately, and their recommendations can often be trumped by political and economic decisions, as is well illustrated in the plight of the New England fisheries.

Ensuring sustainable harvests in forests requires thorough scientific assessments of tree growth and a balancing of the needs of ecosystems with those of the harvesters. The restoration of ecosystems requires a great deal of sound science in the form of species and ecosystem ecology. It also requires the ability to make midcourse corrections, which projects like the Everglades restoration will require in the future. Scientific research indicates that salvage logging is a bad idea, but loggers and some politicians may not want to hear that. Finally, the Millennium Ecosystem Assessment is an excellent example of enlisting the world's best scientists in gathering and distributing vital information on the state of the world's ecosystems.

Sustainability

The management of the Pacific halibut industry demonstrates how to achieve a sustainable harvest of a desirable natural resource with the use of catch shares. Sustainable exploitation of natural resources makes it possible for them to continue to produce the essential goods and services that support human well-being. Unfortunately, there are many cases of unsustainable use: the taking of "bush meat" by commercial hunters, the cyanide poisoning of coral reefs, the exploitation of a commons where there is open access to all, and the deforestation in tropical rain forests, to name a few. The principle of sustainability is built into the model of the maximum sustainable yield, but for fisheries, it turns out to be harder to accomplish in practice than in theory. For this reason, it is wise to use the precautionary principle in attempting to set such a threshold for a resource. Sustainable forestry, however, is not so difficult, as long as it is focused on managing the forests as functioning ecosystems and as long as biodiversity is respected. Where resources are being exploited in much of the developing world, sustainability is unworkable without

taking into account the needs of local people and involving them in management decisions.

Stewardship

Stewardly care of ecosystems means consciously managing them so as to benefit both present and future generations. Sustainable consumptive and productive use of ecosystem resources involves tenurial rights over them; exercising these rights should be a matter of stewardship embodied in rules and policies. Table 7–2 presents a set of principles that promote such stewardship. When forests and fisheries are under the control of the people who most directly depend on their use, stewardship can flourish. The Turtle Island fishery (actually, eggery!) demonstrates how this can work, even while sometimes needing the help of outside organizations. Management of the fisheries exhibits both good and poor stewardship in different regions: Compare the Pacific Northwest and Georges Bank fisheries.

It is encouraging to see how much land (and sometimes water) is set aside for protection in perpetuity. Finally, it is heartening to see the former UN Secretary-General endorsing a stewardship ethic as an important step in preserving the natural resources that are under tremendous pressure from present generations. The Seven Billion Tree Campaign is a tangible demonstration of that ethic.

REVIEW QUESTIONS

1. What were the problems in the Pacific halibut fishery, and how were they resolved?

2. What are some goods and services provided by natural ecosystems?

3. Compare the concept of *ecosystem capital* with that of *natural resources*. What do the two reveal about values?

4. Compare and contrast the terms *conservation* and *preservation*.

5. Differentiate between *consumptive use* and *productive use*. Give examples of each.

6. What does *maximum sustainable yield* mean? What factors complicate its application?

7. What is the *tragedy of the commons*? Give an example of a common-pool resource, and describe ways of protecting such resources.

8. When are restoration efforts needed? Describe efforts under way to restore the Everglades.

9. Describe some of the findings of the most recent FAO Global Forest Resources Assessments. What are the key elements of sustainable forest management?

10. What is deforestation, and what factors are primarily responsible for deforestation of the tropics?

11. What is the global pattern of exploitation of fisheries? Compare the yield of the capture fisheries with that of aquaculture.

12. Compare the objectives of the original Magnuson Act with those of the 2006 Magnuson-Stevens Fishery Conservation and Management Reauthorization Act.

13. What is the current status of the large whales? Discuss the controversy over continued whaling by some countries.

14. How are coral reefs and mangroves being threatened, and how is this destruction linked to other environmental problems?

15. Compare the different levels of protection versus use for the different categories of federal lands in the United States.

16. Describe the progression of the management of our national forests during the last half century. What are current issues, and how are they being resolved?

17. How do land trusts work, and what roles do they play in preserving natural lands?

THINKING ENVIRONMENTALLY

1. It is an accepted fact that both consumptive use and productive use of natural ecosystems are necessary for high-level human development. To what degree should consumptive use hold priority over productive use? Think about more than one resource (lumber, bush meat, etc.).

2. Consider the problem presented by Hardin of open access to the commons without regulation. To what degree should the freedom of the use of these areas be limited by the authorities? Make use of Table 7–2 when you defend your position.

3. Consider the benefits and problems associated with coastal and open-ocean aquaculture. Is it a useful practice overall? Justify your answer.

4. Kofi Annan stated that we are in need of a "new ethic of stewardship." What principles should this new ethic be built on?

MAKING A DIFFERENCE

1. To determine the impact you have on the ecosystems around you, calculate your ecological footprint (several sites on the Internet can help with this). What can you do to lower your ecological footprint?

2. Consider Michael Pollan's work *The Omnivore's Dilemma*. The book deals with America's eating habits, government regulations regarding food production, farming techniques, and similar topics. Decide which eating habits are ecologically sustainable and which are not. Go to the Internet and search for sites that offer ecosystem-friendly recipes.

3. By transporting firewood from its native area, you may run the risk of spreading diseases and alien insect species that can damage trees and overrun whole areas of forest. You can avoid this by using firewood harvested locally or, if you need to carry firewood a distance, by finding wood marked with a U.S. Department of Agriculture (USDA) tag confirming that the logs are safe to move.

4. Organize a volunteer tree-planting group. You can acquire young trees at a nursery, college agricultural department, or city government for a price. To raise money for a large project, try to find individual or corporate sponsors.

5. When you visit parks and other natural areas, stay on the designated trails. Avoid making new trails that might lead to erosion, and never disturb nesting birds and wild animals raising their young.

Mastering**ENVIRONMENTALSCIENCE**™

Go to **www.masteringenvironmentalscience.com** for practice quizzes, Pearson eText, videos, current events, and more.

PART THREE

The Human Population and Essential Resources

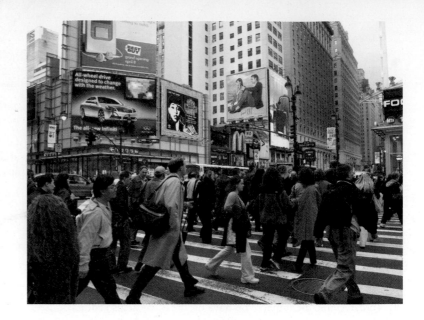

From A POPULATION PERSPECTIVE, the 20th century was remarkable: We started the century with fewer than 2 billion people on Earth and ended with 6 billion and counting. The United Nations Population Division estimated in 2008 that our numbers may reach 9.1 billion by the middle of the 21st century—almost a 50% increase in 50 years. Virtually all of the increase will be in the developing countries, which are already densely populated and straining to meet the needs of their people for food, water, health care, shelter, and employment.

YET IN RECENT DECADES, a population decline has been occurring in some industrialized countries. Populations are aging in these countries, bringing serious social, economic, and political consequences such as a shrinking workforce, a rising retiree population, rising health care costs, and a rising "replacement migration" of workers from developing countries.

IN CHAPTER 8, you will learn how human populations are similar to and different from populations of other species. We will see how historic changes in technology have affected our population growth, and how that growth can affect the rest of the world. We will also look at the demographic transition, which has brought an end to population growth in much of the industrialized world.

IN CHAPTER 9, we will discuss what needs to be done to bring the developing countries through the demographic transition, and look at efforts being made to change consumption and population patterns.

IN CHAPTER 10–13, we will look at the critical resources humans need, how we manage and use these resources, and their relationship to a growing population. We need to look at population and consumption concerns in tandem with resource use and pollution, because population growth and its twin, consumption, affect every environmental issue.

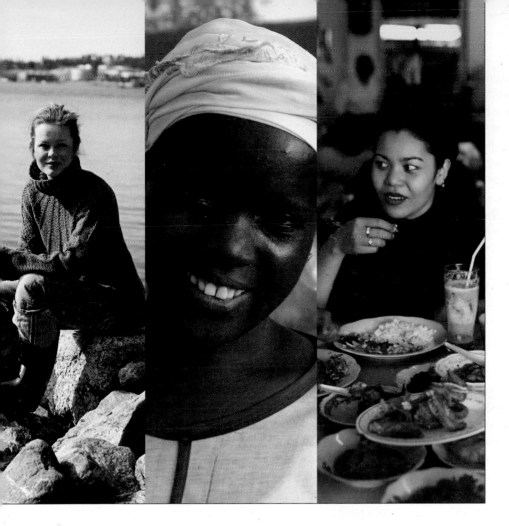

KEY TOPICS

1. **Humans and Population Ecology:** Humans, like other organisms, are subject to natural laws and ecological processes such as the law of limiting factors, but show some significant differences from other creatures in their ability to change their own world.

2. **Population and Consumption: Different Worlds:** The world can be divided into countries with very high incomes and low fertility, middle incomes and usually lower fertility, and low incomes but very high fertility, although there can be significant ranges within countries and within each group.

3. **Consequences of Population Growth and Affluence:** Unlimited population growth or unlimited use of natural resources is unsustainable; both population growth and consumption patterns must be addressed for stewardship of resources to occur.

4. **Projecting Future Populations:** Social scientists use a number of tools to understand populations and project likely future outcomes, such as the demographic transition—a shift from high fertility and mortality to low fertility and mortality.

THE COLD WINTER wind cuts through the streets of Stockholm, Sweden, as Anje, a 19-year-old Swedish university student, hurries home to the warmth of her apartment. There she will settle in for an evening at the computer and her books, studying for a degree in the banking industry. She will stop for a fast-food meal on the way. Like many Swedes, Anje plans to travel extensively when her studies are over and work for an international firm. While she is dating, Anje has no plans to marry in the near future. She is likely to marry late, if she does, and have children later than many people in the world. Like many in the European Union (EU),

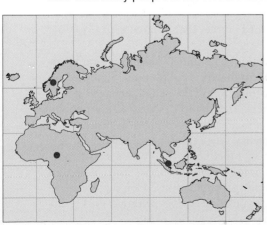

Anje gets a free university education. She works hard because jobs will be scarce and she wants, someday, to be established. When she does want to have children, Anje is likely to have good benefits, with medical care and maternity leave. She is likely to care for her elderly parents in a society in which the number of elderly people is increasing, but good social and medical services will keep them active long into their seventies, maybe their eighties.

A world away, in the heat of bustling Jakarta, Indonesia, 19-year-old Atin rides a motorcycle home with her brother. They will stop at a market for chicken and noodles and come home to the small house they share with their parents and three siblings. Atin and her family have a small television and a few other luxuries. A medical emergency could set them back, though, and Atin is worried about her father's coughing. Like millions of Indonesian men, he smokes local clove and tobacco cigarettes. Their family, like many others, was hurt by the Asian financial crisis in the 1990s. Atin's father lost a more stable job and took a series of temporary jobs, and their mother began to work part-time. Atin dropped out of high school to help her family. Now Atin and her brother are working to help support the family. Atin hopes to marry a young man her family knows if she and he can get the money together. Like many Indonesians, she will most likely have a smaller family than her parents did.

In the dry season, Burkina Faso (in West Africa) is a place of dust and sun. Awa, 19-year-old married mother of two, walks with her children along a dusty road toward her home. She is carrying a large container of water on her head. Awa's husband is working in agriculture in neighboring Cote d'Ivoire (Ivory Coast), where he makes a small amount of money and sends it home. Neither Awa nor her husband can read, and there is no nearby medical facility. She is worried because she is pregnant and this time has been feeling very ill. Awa and her husband had another child who died shortly after birth, and Awa is afraid that this coming baby will not survive either, as she is sick and still nursing another child. Her feet ache and she is very tired, but there is little to be done. She cares some of the time for her mother and for the children of her sister, who has died of AIDS. Her four surviving siblings have similar lives and live nearby. They all farm during the rainy season, growing cassava, yams, peanuts, and maize, and a few members of the extended family work on a cotton farm or have other relatives working in the Ivory Coast.

Anje, Atin, and Awa are three of millions of young people living around the world. Their life circumstances give them different health concerns, access to education, and medical care. They are also likely to have different family sizes and to use natural resources differently. In the 20th century, the global human population experienced an unprecedented explosion, more than tripling its numbers. Now, the rate of growth is slowing down, but the increase in absolute numbers continues to be substantial—and in some parts of the world, there isn't much slowing.

Remarkable changes in technology and substantial improvements in human well-being have accompanied this growth. In just the last 50 years, average income per capita has nearly tripled, global economic output has risen sevenfold (from $7 to $46 trillion), life expectancy has risen from 46 to 68 years, and infant mortality has been cut by two-thirds (from 157 to 52 deaths per thousand live births). Even so, Anje, Atin, and Awa do not live in a world where people experience average conditions. The remarkable improvements in living conditions are not experienced equally. Extreme poverty is still widespread. An estimated 1.2 billion people live on less than $1 per day, and the income gap between the richest and the poorest countries is enormous.

Humans are animals, and population ecologists can study human populations and ask many of the same questions they would if they were studying wolves or elk. But humans are also exceptional. They can think about and make decisions about their own fertility, they have ethical systems that inform their decisions, and they can use technology to prolong life and increase food production. Even more unusual, humans have cultural constructs like money. Money and its economic institutions mean that survival is driven not only by evolutionary fitness, but also by the economic circumstances into which people are born.

In this chapter, we will put the stories of people like Anje, Atin, and Awa into the context of the dynamics of population growth and the related concept of consumption and its social and environmental consequences. A continually growing population is unsustainable, as is a continually increasing consumption of natural resources, so the focus in this and the next chapter is population and consumption stability—and what is required to get there.

8.1 Humans and Population Ecology

Humans are part of the natural world, and human populations are subject to processes such as birth and death. Figure 8–1 shows human population growth over the last 2,000 years. The long slow incline is followed by a rapid rise, giving what looks like the *J*-shaped curve we saw in Chapter 4. However, on closer inspection, the last few decades have shown decreasing population growth rates, and projections (possible future scenarios determined by assumptions about changes we will see) suggest that the global human population may level off, so that the pattern we will see by 2100 will be more like the *S*-shaped curve that characterizes logistic growth (Chapter 4).

The field of collecting, compiling, and presenting information about human populations like that presented in Figure 8–1 is called **demography;** the people engaged in this work are **demographers.** Demographers may study population processes such as migration or changes in fertility and mortality. In environmental science, biological science quickly becomes intertwined with other fields, and some social demographers include economic, cultural, social, and biological factors in their analyses of populations. Consequently, we cannot discuss human population ecology without quickly discussing differences in wealth or health care.

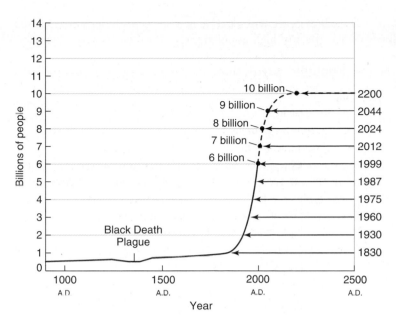

Figure 8–1 World population over the centuries. Population grew slowly for most of human history, but in modern times, it has expanded greatly. (*Sources:* Basic plot from Joseph A. McFalls, Jr., "Population: A Lively Introduction," *Population Bulletin* 46, no. 2 (1991): 4; updated from UN Population Division, medium projection, 2008.)

Demographers use a set of specialized terms, described in Table 8–1, which we will use throughout the chapter.

r- or *K-* Strategists

In Chapter 4, we contrasted organisms like cockroaches, which breed rapidly and have boom-and-bust populations, with others like elephants, which maintain smaller, but more stable population sizes. The cockroaches were described as *r-selected species*, or *r-strategists*, because they have a high *r*, or reproductive potential. They were contrasted with *K-selected species*, or *K-strategists*, which remain close to *K* (carrying capacity) and are characterized by long life spans, older age at first reproduction, high parental care of offspring, and fewer offspring. We also saw that these characteristics form a continuum, with some organisms not easily placed in either group.

TABLE 8–1   Demographic Terms

Term	Definition
Growth Rate (annual rate of increase or decrease)	The rate of growth of a population, as a percentage. Multiplied by the existing population, this rate gives the net yearly increase (or decrease, if negative) for the population.
Total Fertility Rate	The average number of children each woman has over her lifetime, expressed as a rate based on fertility occurring during a particular year.
Replacement-Level Fertility	A fertility rate that will only replace a woman and her partner, theoretically 2.0, but adjusted slightly higher because of mortality and failure to reproduce.
Infant Mortality	The number of infant deaths per thousand live births.
Population Profile (age structure)	A bar graph plotting numbers of males and females for successive ages in the population, starting with the youngest at the bottom.
Population Momentum	The effect of current age structure on future population growth. Young populations will continue growing even after replacement-level fertility has been reached, due to reproduction by already existing age groups.
Crude Birthrate	The number of live births per thousand in a population in a given year.
Crude Death Rate	The number of deaths per thousand in a population in a given year.
Epidemiologic Transition	The shift from high death rates to low death rates in a population as a result of modern medical and sanitary developments.
Fertility Transition	The decline of birthrates from high levels to low levels in a population.
Demographic Transition	The tendency of a population to shift from high birth and death rates to low birth and death rates as a result of the epidemiologic and fertility transitions. The result is a population that grows very slowly, if at all.

When we ask "Are humans *r*- or *K*-strategists?" it becomes clear that humans are like other organisms, but also significantly different. Just looking at the population growth curve of the last thousand years (Fig. 8–1), we might be tempted to conclude that we are looking at a *J*-shaped curve (Chapter 4) representing exponential growth typical of weedy species that move from place to place quickly and cannot outcompete other organisms over the long term (*r*-strategists). However, humans have characteristics such as high parental care and late reproduction that we would expect of the more stable *K*-strategists.

How can humans have the rapid population growth curve of a weedy, explosive species, but the life history of a slower growing, *equilibrial* species (those with population numbers remaining near the carrying capacity)? The answer lies in the fact that in some ways humans are critically different from other organisms. Even though the Millennium Ecosystem Assessment recognizes that "humans are an integral part of ecosystems," there are some differences between humans and other organisms that make our role in ecosystems different from that of others.

Remember that a *population* is an interbreeding group of a particular species in the same area. We do not typically talk about a widespread species as having a *global population* because local populations rise and decline according to local conditions. Drought in one area might be offset by good conditions in another. In contrast, we do talk about "human population" and mean the whole global population. Prior to the modern era, human populations acted more separately—disease, plenty, and scarcity were local effects, and populations rose and fell with surrounding conditions. Several changes, due to society and culture, have altered the way human populations interact and have changed the way human population ecology works. In the modern age, what happens in one part of the world affects humans in other parts of the world. Humans act both as local populations, with population characteristics such as family size and life expectancy, and as part of a larger global whole. Humans are unique in the biosphere in our ability to regulate our reproduction, use fire, store food for longer times, and adapt our environment with technology so we can live in more places. Consequently, population ecology can be applied to humans, but not in exactly the same way it is applied to other organisms.

Revolutions

The changes that have allowed humans to become such a dominant part of the landscape are so extreme that they can be considered revolutionary. Here we will look at a number of the revolutions that have brought humans from a number of isolated populations—ruled (like other species) primarily by local conditions—to the global human population today. In the past, there have been several large-scale upheavals in the way humans do things. These include the development of agriculture (which allowed settling into cities), the Industrial Revolution, modern medicine, the green agricultural revolution of the 1970s, and the current environmental revolution. In each case, we will see how new ways of doing things altered the effect of high populations.

Neolithic Revolution. Natural ecosystems have existed and perpetuated themselves on Earth for hundreds of millions of years, while humans are relative newcomers to the scene. Anatomically modern humans most likely arose in Africa around 200,000 years ago. The earliest fully modern humans in Europe appeared some 40,000 years ago, during the end of a period known as the Upper **Paleolithic** (around 50,000 to 10,000 years ago). Archeological evidence suggests that Paleolithic humans survived in small tribes as hunter-gatherers, catching wildlife and collecting seeds, nuts, roots, berries, and other plant foods (Fig. 8–2). Settlements were never large and were of relatively short duration because, as one area was picked over, the tribe was forced to move on. As hunter-gatherers, these people were much like other omnivorous consumers in natural ecosystems. Populations could not expand beyond the sizes that natural food sources supported, and deaths from predators, disease, and famine were common.

About 12,000 years ago, a highly significant change in human culture occurred when humans in the Middle East began to develop animal husbandry and agriculture. The development of agriculture provided a more abundant and reliable food supply, but it was a turning point in human history for other reasons as well. Because of its profound effect, it is referred to as the **Neolithic Revolution**. Agriculture allows permanent (or at least long-term) settlements and the specialization of labor. Some members of these settlements specialize in tending crops and producing food, freeing others to specialize in other endeavors. With this specialization of labor, there is more incentive and potential for technological development, such as better tools and better means of transporting water and other materials. Trade with other settlements begins, and commerce is born. In addition, living in cities allows a greater storage of food and the development of preservation techniques, and food storage means that food stored from one year might be able to help people in years with less food production, evening out the risk from year to year. A consequent reduced mortality rate, coupled with more reliable food production, supports population growth, which in turn supports (and is supported by) expanding agriculture. In short, modern civilization and population growth had their origins in the invention of agriculture about 12,000 years ago.

Industrial Revolution. For about another 11,000 years, the human population increased and spread out over the planet, but those increases were not as dramatic as what we see today. Agriculture and natural ecosystems supported the growth of civilizations and cultures that increased their knowledge and mastery over the natural world. With the birth of modern science and technology in the 17th and 18th centuries, the human population—by 1800, almost a billion strong—was on the threshold of another revolution: the **Industrial Revolution** (Fig. 8–3). The Industrial Revolution and its technological marvels were energized by fossil fuels—

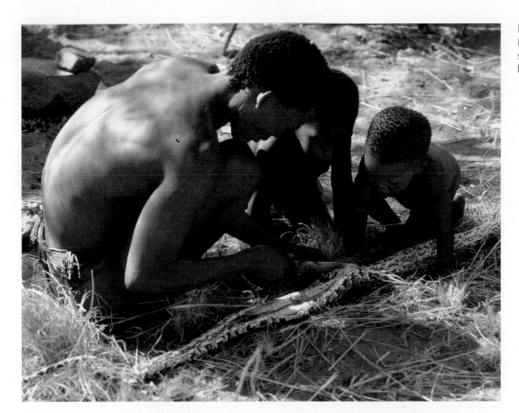

Figure 8–2 Hunter-gatherer culture. Before the advent of agriculture, all human societies had to forage for their food, as these bushmen from Namibia are doing.

Figure 8–3 Industrial Revolution. The Industrial Revolution began in England in the 1700s. Coal was the energy source, and economic growth and pollution were the consequences.

first coal, then oil and gas. In the past, people had used other fuels, particularly peat (compacted plant material in wetlands) and wood. These fuels do the same thing that fossil fuels do—they allow people to use energy that was trapped by plants from the sun a long time ago. The use of coal in particular opened up vast resources of trapped energy from other eras. This energy was used by humans to do work that we would never have been able to do if we had relied on human power—or even on the power of domesticated animals. The burning of fossilized plants and animals, or of wood, is similar to the process of cellular respiration (Chapter 3): It breaks chemical bonds to release stored energy and releases water and carbon dioxide as waste products. Fossil fuels contain other chemicals as well, such as sulfur and nitrogen compounds, mercury, lead, and arsenic, all of which can contribute to the pollution of land, air, and water and will be discussed in other chapters.

The harnessing of this extra energy allowed people to produce more food. However, the Industrial Revolution came with some unexpected costs. Pollution and exploitation took on new dimensions as the industrial powers turned to the extraction of raw materials from all over the world. In time, every part of Earth was affected by this revolution and continues to be affected today. As a result, we now live in a time of population growth and economic expansion, but unfortunately with all of the environmental problems outlined in Chapter 1.

Medical Revolution. One of the main reasons for the slow and fluctuating population growth prior to the early 1800s was the prevalence of diseases that were often fatal, such as smallpox, diphtheria, measles, and scarlet fever. These diseases hit infants and children particularly hard, and it was not uncommon for a woman who had seven or eight live births to have only two or three children reach adulthood. In addition, epidemics of diseases such as the black plague of the 14th century, typhus, and cholera eliminated large numbers of adults. Prior to the 1800s, therefore, the human population was essentially in a dynamic balance with natural enemies—mainly diseases—and other aspects of environmental resistance. High reproductive rates were largely balanced by high mortality, especially among infants and children. With high birth and death rates, the population growth rate was low in these preindustrial societies.

In the late 1800s, Louis Pasteur and others demonstrated that many diseases were caused by infectious agents (now identified as various bacteria, viruses, and parasites) and that often these organisms were transmitted via water, food, insects, and rodents. Soon vaccinations against different diseases were developed, and whole populations were immunized against such scourges as smallpox, diphtheria, and typhoid fever. At the same time, cities and towns began treating their sewage and drinking water. Later, in the 1930s, the discovery of penicillin (the first in a long line of antibiotics) resulted in cures for often-fatal diseases such as pneumonia and blood poisoning. Improvements in nutrition began to be

significant as well. In short, better sanitation, medicine, and nutrition brought about spectacular reductions in mortality, especially among infants and children, while birthrates remained high. From a biological point of view, the human population began growing almost exponentially, as does any natural population once it is freed from natural enemies and other environmental restraints.

The Green Revolution. After the industrial and medical revolutions, human populations increased dramatically. Concerns about whether we would be able to feed the rapidly burgeoning global population led to intense efforts to find new ways to increase agricultural efficiency. The development of chemical pesticides in World War II, along with an increase in irrigation and fertilizer use, dramatically increased crop yields. These trends were enhanced by the development of high-yield crops such as specialized rice, which yielded more per acre than other crops. These advances in agriculture did increase crop yields dramatically and allowed many countries to provide food for their rapidly growing populations (Fig. 8–4). However, as we will see in subsequent chapters, the industrialized agriculture that allowed for such high yields came at significant costs, much as the Industrial Revolution did. Increased erosion, soil and water pollution, and the loss of native (often hardier) plant varieties were some of those costs of industrial agriculture.

Just as the use of fossil fuels allowed people to use energy trapped over millions of years to do the activities of a single year, industrialized agriculture allowed people to use resources, particularly soil and underground water, more rapidly than they were replaced. This problem will be discussed in some detail in chapters on soil, water, and food production. In addition, the use of pesticides initially saved many human lives and protected crops. In the long term, however, the overuse of pesticides resulted in **resistance**, the ineffectiveness of a pesticide when the target organisms are

Figure 8–4 Green Revolution. Indian farmers harvest high-yield wheat. Different varieties of crops, fertilizer, irrigation, and pesticides are all part of industrialized agriculture, which helped feed growing populations but led to the pollution of soil and water.

no longer affected by it. A current worldwide interest in sustainable agriculture is an attempt to rectify some of these problems. However, as we will see in Section 8.3, population growth is pressuring us to increase, rather than decrease, high-yield agriculture.

The Newest Revolution. In the past, humans experienced revolutions with new technological breakthroughs. Now, we have the Internet, computers, nanotechnology, robotics, and solar and other technologies. Many of these can be put to use in what some hope will be a real "green" revolution—an **environmental revolution**. More-efficient technologies, better urban and regional planning, policy and industrial changes, and the personal decisions of billions of people will be required to drive this revolution.

Humans are part of natural ecosystems, but they are different enough from other organisms to make their population ecology global and to allow them to change their world more extensively. They have characteristics of both *r*- and *K*-strategists. We see the *J*-shaped curve for human population growth because at different times in history we have increased the number of people the world can support by changing the way we do things. Our past growth might seem to suggest that natural laws do not apply to humans. In the next section, we will look at carrying capacity and limits and ask whether they apply to human populations.

Do Humans Have a Carrying Capacity?

One of the most notable features of today's human population growth is that it is difficult to determine a carrying capacity. If people were close to starving many years ago, and yet today there are many more people, people live longer, and proportionately fewer are extremely poor, how can we talk about humans reaching a carrying capacity? In fact, some people claim that the whole idea of a carrying capacity does not apply to humans. Economist Julian Simon, the author of *The Ultimate Resource*,[1] claimed that human ingenuity was the ultimate resource and that there was no such thing as a limit for humans. He believed we would colonize space and that human populations really have no upper limit. There are plenty of people who would agree with him, in part because every time in the past that you might think humans had hit a carrying capacity, they have solved some problem and passed it.

Most ecologists and demographers, however, would disagree. In their view, an escape into space is unlikely to solve the world's problems for the vast majority of humans, and humans are subject to limits and natural laws of population growth. How then do we reconcile natural laws with the pattern of human population growth in the last 2,000 years? Throughout history, humans have improved their survival rate, increased their populations, and increased their life span by doing some things other organisms cannot do (or cannot do on the scale we can). These improvements have essentially increased the carrying capacity for our species. For example, we were once limited by what we could capture or hunt in one season, but the development of agriculture and trade removed that limit. We were limited by our energy requirements, but the use of fossil fuels gave us a new resource; we were limited by child mortality, but medical breakthroughs lowered child mortality and populations exploded.

This pattern of population growth is actually exactly what you would expect to see under the law of limiting factors—each time a limit is removed, population increases until there is another limiting factor. Julian Simon represented a group that believes that we will always be able to eliminate limiting factors. However, in the past, increases have come at the expense of using up nonrenewable resources (like fossil fuels). Increases in population have also come at the expense of creating another set of limits—this time, pollutants. The limiting factors that humans face in the 21st century include pollution from our own activities, a limit on agricultural land, a limit on ocean fish, and trade-offs between ecosystems we need for their services (such as photosynthesis, flood control, and habitat) versus land and water for human activities. The argument that Simon makes also includes no value for nonhuman parts of nature except as resources for humans. That is, the rest of nature has no *intrinsic* value (Chapter 6).

Since humans seem able to increase their carrying capacity in the short term—while potentially making trade-offs that will be paid for in the long term—determining a carrying capacity for the global human population is difficult. Demographer Joel Cohen explained the reason in his book *How Many People Can the Earth Support?*[2] Cohen explained that there isn't a clear-cut answer to the question of carrying capacity because, with humans, some of the answer depends on what living standard we are willing to accept. For example, if we never ate meat, we did not rely on wild organisms, and we were willing to never travel, heat our homes, or use paper, Earth could support more people. On the other hand, if everyone on Earth wanted a 21st-century American way of life, we would need multiple Earths to support the global population we already have. Because so much of the answer depends on standard of living expectations, Cohen argued, it is very difficult to determine an actual carrying capacity. Some people have used Cohen's argument to say that there is no carrying capacity at all. That was not Cohen's meaning, as he has made clear in other writings. Cohen simply meant that we can't figure out a carrying capacity for humans in the same way we do for fish, ducks, or flies because we have (and want) more choices.

While it may be difficult to find a single number for human carrying capacity because we may not all agree on what living standard we are willing to accept, the world's scientific societies disagree with Julian Simon and believe that there is some type of limit to human population growth. In 1993, representatives of the National Academies of Sciences from 58 countries met in New Delhi, India, to agree on and

---

[1]Julian Simon, *The Ultimate Resource* (Princeton, NJ: Princeton University Press, 1983).

[2]Joel Cohen, *How Many People Can the Earth Support?* (New York: Norton, 1995).

sign a statement about world population. They said, among other things, that "the earth is finite and that natural systems are being pushed ever closer to their limits."[3]

Scientists believe that, even if a concrete number may be hard to agree on, it is important to try to estimate what carrying capacity limits might be from an ecosystem perspective. In a 2004 paper in the journal *BioScience*, two scientists evaluated 69 papers on world population and carrying capacity, looking to see if there were any central trends. They performed a meta-analysis (an analysis of multiple other analyses) and identified a number of factors that could limit human populations, such as "water availability, energy, carbon, forest products, nonrenewable resources, heat removal, photosynthetic capacity, and the availability of land for food production."[4] They concluded that the best estimates for global human carrying capacity centered on 7.7 billion people. One important thing the analysis showed is that the best predictions of how large our human populations will get by 2050 are greater than the best estimates of our carrying capacity.

From the dawn of human history until the beginning of the 1800s, the population increased slowly and variably, with periodic setbacks. It was roughly 1830 before world population reached the 1 billion mark. By 1930, however, just 100 years later, the population had doubled to 2 billion. Barely 30 years later, in 1960, it reached 3 billion, and in only 15 more years, by 1975, it had climbed to 4 billion. Thus, the population doubled in just 45 years, from 1930 to 1975. Then, 12 years later, in 1987, it crossed the 5 billion mark! In 1999, world population passed 6 billion, and it is expected to hit 7 billion in 2011 or 2012.

On the basis of current trends (which assume a continued decline in fertility rates), the UN Population Division (UNPD) medium projection predicts that world population will pass the 7 billion mark around 2012, the 8 billion mark in 2024, and the 9 billion mark in 2047 and will reach 9.1 billion around 2050. At that point, world population will still be increasing by 34 million per year. After that, it is all guesswork (Fig. 8–1). Future growth and projections will be described in Section 8.4.

8.2 Population and Consumption: Different Worlds

To understand what is likely to happen in the next decades, we need to understand the lives of people like Anje, Atin, and Awa. In spite of the fact that globalization is part of the modern world, people in wealthy, middle-income, and poor countries live in radically different economic and demographic conditions. In fact, even within each country there can be tremendous disparities in wealth and in quality of life.

Rich Nations, Middle-Income Nations, Poor Nations

The World Bank, an arm of the United Nations, divides the countries of the world into three main economic categories, according to average per capita gross national income (Fig. 8–5):

1. **High-income, highly developed, industrialized countries.** This group (1.056 billion in 2007) includes the United States, Canada, Japan, Korea, Australia, New Zealand, the countries of western Europe and Scandinavia, Singapore, Taiwan, Israel, and several Arab states (2007 gross national income per capita, $11,456 and above; average of $37,566). Anje in Sweden is in this group.

2. **Middle-income, moderately developed countries.** This group (4.26 billion) includes mainly the countries of Latin America (Mexico, Central America, and South America), northern and southern Africa, China, Indonesia and other southeastern Asian countries, many Arab states, eastern Europe, and countries of the former U.S.S.R. It is further divided by the World Bank into lower middle-income and upper middle-income countries. (2007 gross national income per capita, $936–$3,705 for the former category, $3,706–$11,455 for the latter). Indonesia, home to Atin and her family, is in the lower middle income group.

3. **Low-income, developing countries.** This group (1.296 billion) comprises the countries of eastern, western, and central Africa, India and other countries of southern Asia, and a few former Soviet republics. (2007 gross national income per capita, less than $936; average of $578). Awa and her family in Burkina Faso are members of this group.

The high-income nations are commonly referred to as **developed countries,** whereas the middle- and low-income countries are often grouped together and referred to as **developing countries.** Of course, all countries are in the process of developing in some sense, but here the term refers to economic changes. The terms *more developed countries* (MDCs), *less developed countries* (LDCs), and *third-world countries* are being phased out, although you may still hear them used. (The first world was the high-income countries, whereas the second world was the former Communist bloc, which no longer exists. Therefore, referring to the developing countries as the third world is obsolete.)

The disparity in distribution of wealth among the countries of the world is mind-boggling. The high-income countries make up just 15% of the world's population, yet they control about 80% of the world's wealth, calculated on the basis of gross national income. The low-income, developing countries, with 37% of the world's population, control only 3.0% of the world's gross national income. This amounts to a difference in per capita income of 63 to 1! The distribution of wealth within each country is also disproportionate, so the poorest people living in these poor nations are extremely badly off.

---

[3]Population Summit of the World's Scientific Academies, 1993. National Academy of Sciences, National Academy of Engineering, Institute of Medicine (SEM) page 7. Available at http://books.nap.edu/openbook.php?record_id=9148&page=7.
[4]J. Van Den Bergh and P. Rietveld, "Reconsidering the Limits to World Population: Meta-analysis and Meta-prediction," *BioScience* 54, no. 3 (2004): 195.

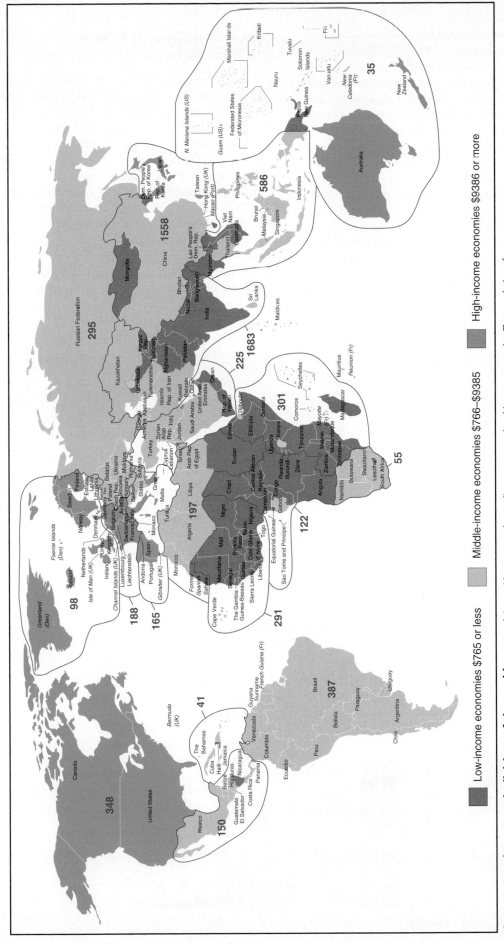

Figure 8–5 Major economic divisions of the world. Nations of the world are grouped according to gross national income per capita. The population of various regions (in millions) is also shown by magenta lines and numbers. (*Sources: World Development Report, 2005* (New York: Oxford University Press, Inc.), Copyright © 2005 by the International Bank for Reconstruction and Development/The World Bank. Populations from *World Population Data Sheet* (Washington, D.C., Population Reference Bureau, 2008).)

Low-income economies $765 or less

Middle-income economies $766–$9385

High-income economies $9386 or more

The disparity of wealth is difficult to understand just by looking at general income figures. The UN Development Program (UNDP) uses the *Human Development Index* as a measure of general well-being, based on life expectancy, education, and per capita income. The UNDP also uses the *Human Poverty Index*, based on additional information about literacy and living standards, to make a more direct measurement of poverty in both low- and high-income countries (Fig. 8–6). Between 10% and 15% of the people in developed countries are unable to afford adequate food, shelter, or clothing, compared with about 45% of those in developing countries. Anje, in Sweden, lives in the country with the lowest human poverty index, while Awa, in Burkino Faso, lives in the country with the highest. These data, available yearly in the UNDP Human Development Report, have been used to focus attention on the most deprived people in a country, to help countries develop appropriate policies, and to mark progress toward sustainable development. Clearly, there is work to do in all countries in order to address the general problem of poverty.

Population Growth in Rich and Poor Nations

The population growth shown in Figure 8–1 is for the world as a whole. If you look at population growth in developed versus developing countries, you find a discrepancy that parallels the great difference in wealth between these groups of countries. According to the UNPD, the high-income countries such as Anje's Sweden, with a combined population of 1.2 billion in mid-2008, are growing at a rate of 0.1% per year. These countries will add less than 1 million to the world's population in a year. The middle-income countries, such as Atin's Indonesia, and the low-income countries, such as Awa's Burkino Faso, together had a mid-2008 total population of 5.6 billion. The populations of these countries are increasing at a rate of 1.5% per year, adding 75 million in a year. Consequently, more than 98% of world population growth is occurring in the developing (middle- and low-income) countries. In fact, population growth is highest in

Land area

Technical notes
- Data are from the United Nations Development Programme's 2004 Human Development Report.
- The United Nations Development Programme uses one Human Poverty Index for poorer territories and another for richer territories. The scores of the latter can be divided by 10 so the indices are comparable.

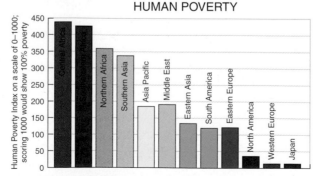

HUMAN POVERTY

Human Poverty Index on a scale of 0–1000; scoring 1000 would show 100% poverty

(Central Africa, Southern Africa, Northern Africa, Southern Asia, Asia Pacific, Middle East, Eastern Asia, South America, Eastern Europe, North America, Western Europe, Japan)

Figure 8–6 Another way to picture poverty. The territory size shows the number of poor people in the country. On the human poverty index, Burkina Faso ranks highest and Sweden ranks lowest. (*Source:* SASI Group (University of Sheffield) and Mark Newman (University of Michigan).)

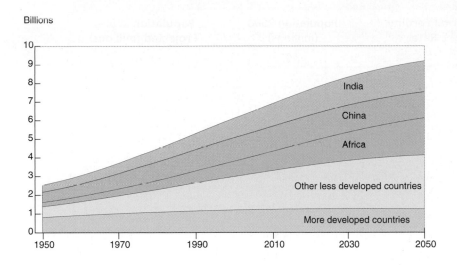

Billions

Figure 8–7 Population growth in different world regions. Because of larger populations and higher birthrates, developing countries represent an ever-growing share of the world's population. This trend is expected to continue. (*Source:* UN Population Division, *World Population Prospects: The 2008 Revision* (New York: United Nations, 2008).)

the least developed countries (Fig. 8–7). What lies behind this discrepancy?

Population growth occurs when births outnumber deaths. In the absence of high mortality, the major determining factor for population growth is births, conventionally measured using the **total fertility rate**—the average number of children each woman in a population has over her lifetime. Theoretically, a total fertility rate of 2.0 will give a stable population because two children per woman will replace a couple when they eventually die. A total fertility rate greater than 2.0 will give a growing population because each generation is replaced by a larger one, and, barring immigration, a total fertility rate less than 2.0 will lead to a declining population because each generation will eventually be replaced by a smaller one. Given that infant and childhood mortality is not in fact zero and that some women do not reproduce, **replacement-level fertility**—the fertility rate that will just replace the population of parents—is 2.1 for developed countries and slightly higher for developing countries, which have higher infant and childhood mortality.

Total fertility rates have dropped all over the world, but most dramatically in the high-income countries. Total fertility rates in developed countries have declined over the past two centuries to the point where they now average 1.6, with some as low as 1.1 (Table 8–2, Fig. 8–8). The one major exception is the United States, with a total fertility rate of 2.1 in 2008. When countries industrialize and have modern medicine, they typically undergo these changes, described as the *demographic transition*. These changes will be described in Section 8.4 when we discuss how demographers project future population trends.

In developing countries, fertility rates have come down considerably in recent years, but they still average 2.8. Some rates are as high as 6.8 or more, however (Table 8–2). Thus, the populations of developing countries, half of which are the poor (low-income) countries, will continue growing, while the populations of developed countries will stabilize or

decline. As a consequence, the percentage of the world's population living in developing countries—already 82%—is expected to climb steadily to more than 90% by 2075 (Fig. 8–7). Nevertheless, it is not just the developing countries that have problems.

Different Populations, Different Problems

Just as we saw in the discussion of carrying capacity, both the number of people and their standards of living play a role in whether we achieve sustainability or experience environmental degradation. Some time ago ecologist Paul Ehrlich and physicist John Holdren proposed a formula to account for the human factors that contribute to environmental deterioration and the depletion of resources. They reasoned that human pressure on the environment was the outcome of three factors: *population*, *affluence*, and *technology*. They offered the following formula:

$$I = P \cdot A \cdot T$$

According to this equation, called the *IPAT formula*, environmental impact (*I*) equals population (*P*), multiplied by affluence and consumption patterns (*A*), and multiplied by the level of technology of the society (*T*). The *IPAT* formula fits well with the concept of carrying capacity. The equation is a simplification, and cannot be used for multiple environmental problems at once. However, there is broad agreement that these three factors play the major role in putting pressure on the environment. Given the high level of technology in the industrialized countries and the affluent lifestyle that accompanies it, a fairly small population can have a very large impact on the environment.

This reality causes tension between affluent countries with lower population growth and poorer countries with higher population growth, as they try to negotiate solutions to environmental problems. Sometimes poorer countries see that nations like the United States achieved wealth by doing activities that were not environmentally sound, only later

TABLE 8–2 Population Data for Selected Countries

Country	Total Fertility Rate	Population 2008 (millions)	Population 2050— Projected (millions)
World	2.6	6,705	9,352
Developing Countries:			
Average	2.8		
(excluding China)	3.2		
Egypt	3.1	74.9	117.9
China	1.6	1,324	1,437
Afghanistan	6.8	32.7	81.9
India	2.9	1,150	1,755
Iraq	4.8	29.5	61.9
Burkino Faso	6.2	15.2	37.5
Haiti	4.0	9.1	15.1
Indonesia	2.6	239.9	343.1
Developed Countries:			
Average	1.6		
United States	2.1	305	438
Canada	1.6	33.3	41.9
Japan	1.3	127.7	95.1
Sweden	1.9	9.2	10.4
Italy	1.3	59.9	61.7
Spain	1.4	46.5	43.9

Source: Data from *2008 World Population Data Sheet* (Washington, D.C.: Population Reference Bureau, 2008).

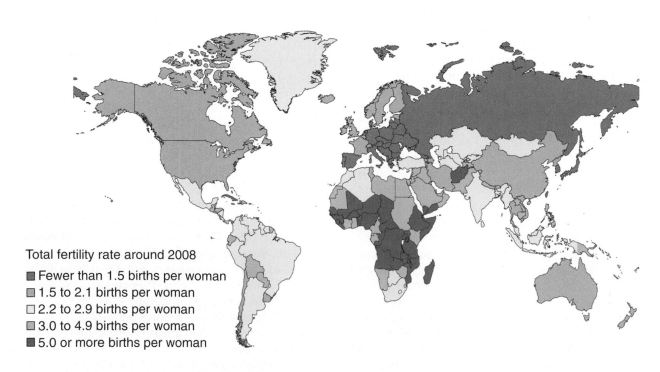

Total fertility rate around 2008

- Fewer than 1.5 births per woman
- 1.5 to 2.1 births per woman
- 2.2 to 2.9 births per woman
- 3.0 to 4.9 births per woman
- 5.0 or more births per woman

Figure 8–8 Total fertility rates around the world. Fertility rates vary around the world. (*Source:* UN Population Division, *World Population Prospects: The 2008 Revision* (New York: United Nations, 2008).)

developing policies for environmental protection. Such nations may resent being told to protect their own resources. Wealthier nations are often called on to give development aid to poorer nations in the hopes of pulling them out of poverty, slowing population growth, and promoting sustainable solutions to global problems. Affluent nations are also often asked to develop more efficient technologies and lower resource use to protect the rest of the globe from environmental degradation.

Impact of Wealth. Sometimes people think that having wealth will solve environmental problems—that is, the rich can afford the luxury of caring for the environment. Wealthy countries, the thinking goes, can more easily afford the technologies to lower air pollution, clean water, and take care of sewage, for example. Give the poor quick ways to improve their economies, and eventually, they will also take care of the environment. It is true that the environment has improved in the United States and western Europe with the passage of many environmental laws and that some wealth is required to build technologies to deal with the needs of large groups of people with minimal environmental degradation. But the relationship between economic wealth and environmental health is not that clear-cut for two reasons. First, while issues like infectious diseases in water might get better, other problems like municipal waste increase with wealth. The second reason is that wealth allows people to care for the environment in their area, but doing so often pushes environmental problems to other, poorer places. For example, the high-income countries may be able to clean up their own sewage because they have the money for treatment plants. However, the extraction of resources and the manufacture of goods in other countries for use in these high-income countries have impacts in many other parts of the world.

It is for this reason—environmental impacts are not always experienced at the same place where resources are consumed—that arguments that overpopulation is not a problem fall flat. Some people argue that because the highest-density places—like the Netherlands, Singapore, Hong Kong, and Japan—are places with healthy, long-lived, wealthy human populations, overpopulation is not real. If they can be so crowded, why can't we have a whole world like that? But high-density, wealthy areas such as Singapore use resources and affect ecosystems over a much wider area. People living in high-density areas have impacts in the far reaches of the seas and land. This is why we talk about an environmental (or ecological) **footprint,** or effect of something we might be doing. An environmental footprint is an estimate of the amount of land and ocean required to provide the resources you need or absorb the wastes you produce. The environmental footprint of Hong Kong, for example, is many times greater than the size of the metropolitan region.

Some estimates conclude that the average American places at least 20 times the demand on Earth's resources—including its ability to absorb pollutants—as does the average person in Bangladesh, a poor Asian country. Major world pollution problems—such as the depletion of the ozone layer, the impacts of global climate change, and the accumulation of toxic wastes in the environment—are largely the consequence of the high consumption associated with affluent lifestyles in the developed countries. For instance, the United States, with only 4.5% of the world's population, is currently responsible for more than 22% of the total global emissions of carbon dioxide, the major greenhouse gas. Likewise, much of the global deforestation and loss of biodiversity is due to consumer demands in developed countries. Anje in Sweden probably uses fewer resources and has a smaller footprint than her cousin in the United States because the EU has more regulations on products and because Europe is so tightly packed and distances between cities are much smaller. Public transportation, biking, smaller vehicles, and other lower-impact services are the norm. However, Europeans still have very high impacts on the world in comparison to citizens of low-income countries.

The developing countries, especially the low-income group of countries, do have a population problem, and it is making their progress toward sustainable development that much more difficult. Their needs are great: economic growth, more employment, wise leaders, effective public policies, fair treatment by other nations, and, especially, technological and financial help from the wealthy nations. Meeting these needs is a matter of justice, one of the key components of stewardship. Even this description is too simplistic because within each group of nations, there is a tremendous difference in wealth and poverty levels and individual experiences. There is a rising middle class in some middle-income countries with consumption patterns similar to those in the United States, Japan, and Europe. This group is called "the consumer class." For example, India has millions of extremely wealthy people and an emerging middle class (as wealthy as Europeans, Japanese, or Americans), while millions more live in extreme poverty. In China, there are 240 million people in this rising middle- and upper-class group. Even in low-income countries, the top leaders may be extremely wealthy, while the rest of the citizens are poor.

On the other hand, the United States, an extremely wealthy country, also has large disparities between groups in terms of health care and access to services. Because of these disparities, it ranks lowest among the developed countries in life expectancy in spite of spending more per person on health care than any other country. It has more children in poverty and more people in prison than any other high-income country. Consequently, simple statements about the life experiences of high-, middle-, and low-income countries may not apply to every person. The disparities between rich and poor within a country make it difficult for the poor. Differences between countries are not the only differences that justice requires us to address.

Enter Stewardship. Despite having a fairly stable population, the developed countries have their own problems with consumption, affluence, damaging technologies, and burgeoning wastes. These issues must be addressed to achieve sustainability, and this, too, requires wise leaders and effective public policies. It also requires willpower, both individually and as a society, and an ethical basis for caring for others and for the rest of the natural world.

Fortunately, the environmental impacts of affluent lifestyles may be moderated to a large extent by practicing environmental stewardship. For example, suitable attention to wildlife conservation, pollution control, energy conservation and efficiency, and recycling may offset (to some extent) the negative impact of a consumer lifestyle. In fact, a life devoted to conservation or other aspects of environmental stewardship might entirely offset the negatives and have a highly positive effect overall. As a result, the *IPAT* formula might be modified to

$$I = \frac{P \cdot A \cdot T}{S}$$

where *S* stands for stewardly concern and practice.

Some people argue that population growth is the main problem, others claim that our highly consumption-oriented lifestyle is chiefly to blame, and still others maintain that it is our inattention to stewardship that is the prime shortcoming. In order to make the transition to a sustainable future, however, all three areas (and more) must be addressed. That is, the population must stabilize (a demographic transition), consumption must decrease (a resource and technology transition), and stewardly action must increase (a political/sociological transition).

8.3 Consequences of Population Growth and Affluence

Expanding populations and increasing affluence—it sounds like trouble for the environment, and it is. It also means trouble for people in high-, middle-, and low-income countries.

In countries experiencing rapid population growth, people overuse land, and many flee to cities in the hopes that there will be new opportunities there. These hopes are not necessarily fulfilled. In high-income countries, increases in industrialization have resulted in huge amounts of wealth, but in problems as well, including the transfer of environmental degradation to places where those most responsible for generating it do not have to experience its price.

Countries with Rapid Growth

Prior to the Industrial Revolution, many humans survived through subsistence agriculture. Families lived on the land, raised livestock, and produced enough crops for their own consumption and perhaps some extra to barter for other essentials. Forests provided firewood and structural materials for housing. With small, isolated, and relatively stable populations, this system was basically sustainable. As the older generation passed away, the land and natural systems could still support the next generation. Indeed, many cultures sustained themselves in this way over thousands of years.

After World War II, modern medicines—chiefly vaccines and antibiotics—were introduced into developing nations, whereupon death rates plummeted and populations grew rapidly. Today 75% of the world's poor in developing countries live in rural areas, and most are engaged in small-scale agriculture (Fig. 8–9), but only 4% of official development assistance goes to agriculture. Rapid growth has negative consequences on a population that is largely engaged in farming a few acres or less of land. When farms become too small to support the next generation, six basic options are

Figure 8–9 Small-scale agriculture. Seventy percent of the world's poor live in rural areas and are heavily dependent on small-scale agriculture. These Philippine women are planting rice seedlings.

possible, all of which are being played out to various degrees by people in these societies:

1. Reform the system of land ownership.

2. Intensify the cultivation of existing land to increase the production per unit area.

3. Open up new land to farm.

4. Engage in illicit activities for income.

5. Emigrate to other countries, either legally or illegally.

6. Move to the cities and seek employment.

As we will see, each of these options has consequences.

Land Ownership Reform. Rising population growth in rural developing countries has increased the need to reform the system of land ownership. Collectivization (the gathering of farmers into group farms) and ownership by the wealthy few are two patterns of agricultural land ownership that have historically kept rural peoples in poverty. Collectivization originated within 20th-century communism and was one of the great failures of the former Soviet Union. Today state-owned land is being privatized, with the result that agricultural production is on the rise. When China abandoned collective agriculture in 1978 and assigned most agricultural land to small-scale farmers, farm output grew more than 6% a year for the next 15 years.

Ownership by the wealthy few was a common result of colonialism in the 19th and 20th centuries. For example, under the apartheid system in South Africa, 87 percent of South Africa's land was available only to whites or was owned by the government, leaving only a small amount of poor-quality land for use by blacks. Since apartheid ended, the South African government has been slowly trying to solve the problems brought about by such land inequality. However, resolving these kinds of inequities can disrupt the

social order, as events in Zimbabwe have shown. In that country, a chaotic land reform program, which was supposed to return land to landless people, has virtually destroyed the country's agriculture and helped to plunge that country into economic decline.

Intensifying Cultivation. The introduction of more highly productive varieties of basic food grains in the Green Revolution has had a dramatic beneficial effect in supporting the growing population, but is not without some concerns (see Chapter 9). For example, intensifying cultivation means working the land harder. Traditional subsistence farming in Africa involved rotating cultivation among three plots and allowing some of the land to lie fallow. With pressures to increase productivity, plots have been put into continuous production with no time off. The results have been a deterioration of the soil, decreased productivity (ironically), and erosion. Given the countertrends of rapidly increasing population and the deterioration of land from overcultivation, food production per capita in Africa is currently decreasing and may decrease in China in the near future.

Opening Up New Lands. Opening up new lands for agriculture may sound like a good idea, but there is really no such thing as "new land." Most good agricultural land is already in production. Opening up new land means converting natural ecosystems to agricultural production, which means losing the goods and services those ecosystems were contributing. Even then, converted land is often not well suited for agriculture. Most of the tropical deforestation in South and Central America has occurred in order to increase agricultural production (Fig. 8–10). Much of this deforestation is done by poor, young people who are seeking an opportunity to get ahead. Unfortunately, in a short time, the cleared land becomes unproductive for agriculture, leaving the people again in poverty.

Figure 8–10 Deforestation in the tropics. Millions of acres of rain forest in Central and South America are being cut down each year to make room for agriculture, as shown in this photograph from Peru.

Figure 8–11 Migration. Ecuadorans seeking visas to work in Spain line up at the Spanish Embassy in Quito, Ecuador.

Illicit Activities. Anyone who doesn't have a way to grow sufficient food must gain enough income to buy it—and sometimes desperate people break the law to do this. Although it is difficult to draw the line between the need and the greed that also draws people into illicit activities, it is undeniable that the shortage of adequate employment exacerbates the problem. In many developing countries, income is obtained from illegal activities such as raising drug-related crops and poaching wildlife.

Migration Between Countries. The gap between high- and low-income countries is reflected in the perception of many in the poorer countries who believe they can improve their well-being by migrating to a wealthier country. The fact that populations in the wealthy countries are aging suggests a strategy that, on the surface, looks appealing. That is, the wealthy countries need more and younger workers, so they should welcome the migrants from the low-income countries who are looking desperately for work. Such migration does happen. Each year many millions migrate to the United States and Europe to find a better life or to escape civil wars and ethnic persecution (Fig. 8–11).

Immigration, however, has its problems, too. Prejudice against foreigners is common, especially in countries with strong ethnic and cultural homogeneity. Even though the United States is a nation largely built on immigration, opposition to foreigners has intensified since the September 11, 2001, terrorist attacks. (See Stewardship, "Immigration: An American Tradition.")

Stewardship

Immigration: An American Tradition

For people trapped by poverty or lack of opportunity in their homeland, emigration to another country has always seemed a way to achieve a better life. As the New World opened up, many millions of people realized this dream by emigrating to the United States and other countries. The United States and Canada are countries composed largely of immigrants and their descendants. Until 1875, all immigration into the United States was legal, so all who could manage to arrive could stay and become citizens. This openness was inscribed on the Statue of Liberty. The inscription reads, in part, "Give me your tired, your poor, your huddled masses yearning to breathe free, the wretched refuse of your

teeming shore. . . . Send these, the homeless, tempest-tossed to me."

In the past, emigration from the Old World created a flood of migrants immigrating to the New World. This emigration relieved population pressures in European countries and aided in the development of the New World. A totally open policy toward immigration today, however, would be untenable. The United States, with its current population of 308 million (2009), is no longer a vast, open land awaiting development. Current U.S. citizens are stewards of the resources of the country and have an obligation to future inhabitants. They have to make good choices not only about their own resource use,

but also about immigration policies. How much immigration should be permitted, and should some groups be favored over others? Current immigration policy makes it easier for skilled people to gain citizenship and relatively difficult for unskilled people to do so. This policy has created what is commonly referred to as a "brain drain" for other countries. Brainpower, many point out, is the "export" that developing nations can least afford.

Current immigration laws officially admit around 900,000 new immigrants per year, a number larger than we have received at any time since the 1920s and larger than the number accepted by all other countries combined.

At present, immigration accounts for about 35% of U.S. population growth.

A recent report by the National Research Council examined the economic impacts of our immigration policies and concluded that legal immigration basically benefits the U.S. economy and has little negative impact on native-born Americans with the exception of a few areas of the country where immigrants are especially numerous. In 1990, Congress established the U.S. Commission on Immigration Reform and directed it to submit a report (The Jordan Commission Report) at the end of 1997. The Commission recommended decreasing total immigration, increasing visas annually for spouses and minor children of legal permanent residents, and discontinuing the immigration of unskilled laborers, while continuing to welcome more highly skilled immigrants. These recommendations have not all been implemented. If they are, they will result in only a small reduction in the rate of legal immigration.

Unofficially, hundreds of thousands of would-be migrants that are unable to gain access through legal channels seek ways to enter the country illegally. The United States maintains an active border patrol, especially along the border with Mexico, which several thousand people try to cross each night. Most are caught and returned, but an undetermined number—estimated at 200,000 to 300,000 per year—slip through. The Illegal Immigration Reform Act of 1996 addressed this problem by strengthening the border patrol and stepping up efforts to locate and deport illegal aliens. However, these plans have not been wholly successful.

The Jordan Commission Report also noted the negative impacts of illegal immigration on the United States. In a further attempt to halt illegal immigration, construction was begun on a long fence on the border of Mexico in the latter years of the Bush administration. The proposed fence was hotly contested by environmentalists, local landowners, the country of Mexico, Native Americans in the region, and other groups, but erection of the fence commenced anyway. The effectiveness of the fence is still debated. It appears that immigration, either legal or illegal, will continue to be a large and increasing part of growth in the United States well into the future.

Much migration is also not from a poor country to a wealthy one, but from one low- or middle-income country to another, especially in areas with civil strife. Refugee immigration leads to temporary refugee camps, where diseases and hunger often take a terrible toll on human life. Some "migrants" are easy targets for exploitation. Recruiters from plantations in the Ivory Coast pay parents in neighboring Burkina Faso (families like Awa's) to send their children to work on the plantations, where they are used as virtual slaves.

Migration to Cities. Faced with the poverty and hardship of the countryside, many hundreds of millions of people in developing nations continue to migrate to cities in search of employment and a better life. This urbanization is one of the biggest trends in population today. In 2008, the world passed a landmark—more than half of the population lived in cities; this trend is expected to continue, so that by 2050 urban dwellers will account for 70% of the global population. In 2007, there were 19 metroregions that qualified as "megacities" of 10 million or more, the top ten of which are shown in Fig. 8–12a, but most urbanites live in cities of less than half a million. Most of the net growth of the next 50 years in the developing countries will be absorbed in urban areas. About one-third of that urban growth is expected to occur in China and India.

Challenges to Governments. Population growth and migration to urban areas are outpacing economic growth and the provision of basic services in many developing countries. The most rapidly expanding cities, especially in sub-Saharan Africa, have fallen so far behind in providing basic services that they are getting worse, not better. Streets are potholed, sanitation and drinking water are poor, electricity and telephone service are erratic, and crime is rampant. Many are forced to live in sprawling, wretched shantytowns and slums that do not even provide adequate water and sewers, much less other services (Fig. 8-12b). Diseases like malaria and malnutrition are rampant, and the incidence of HIV/AIDS is much higher in the cities than in the countryside, a consequence of much higher numbers of single men.

People need basic services like education, housing, health care, and transportation, and many developing countries are simply unable to provide these fast enough to keep up with population growth and migration to urban areas. For example, Pakistan's population increased more than fourfold between 1950 and 2005 as a result of decades of 3% annual growth. According to Pervez Musharraf, former president of Pakistan, that country's population growth was "the main factor retarding economic growth, poverty alleviation, and action on joblessness."[5] Although there are connections between rapid population growth in the developing world and environmental decline, the most pressing issue is the massive poverty that afflicts the low-income countries where population growth is still very high. A completely different set of issues connects population and environmental decline in the affluent countries.

Countries with Affluence

The United States has the dubious distinction of leading the world in the consumption of many resources. We consume the largest share of 11 of 20 major commodities: aluminum, coffee, copper, corn, lead, oil, oilseeds, natural gas, rubber, tin, and zinc. We lead in per capita consumption of many other items, such as meat. The average American eats about twice the global average of meat. We lead the world in paper consumption, at 725 pounds (330 kg) per person per year. All of these factors (and many more like them) contribute to the unusually high environmental impact each of us makes on the world.

Despite the adverse effects of affluence, increasing the average wealth of a population can affect the environment positively. An affluent country such as ours provides amenities like safe drinking water, sanitary sewage systems and sewage treatment, and the collection and disposal of refuse. Thus, many forms of pollution are held in check, and the

---

[5]"The President of Pakistan on the Need to Slow Population Growth in the Muslim World," *Population and Development Review* 31, no. 2(2005): 399.

Figure 8–12 Growing cities. (a) The top 10 world metropolitan areas in 2005. Since 1970, cities in the developing world have grown phenomenally, and a number of them are now among the world's largest. (b) Slums on the outskirts of São Paulo, Brazil. Thirty-two percent of the city's population lives in these blighted areas. (*Source:* Data for part (a) from UN Population Division, *World Urbanization Prospects: The 2007 Revision* (New York: United Nations, 2007).)

(a)

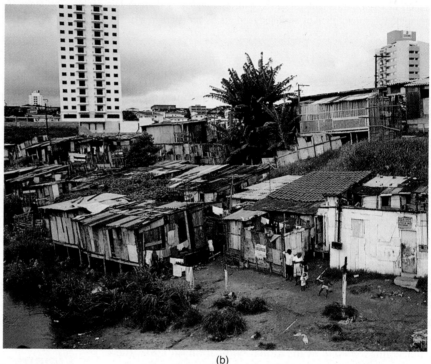

(b)

environment improves with increasing affluence. In addition, if we can afford gas and electricity, we are not destroying our parks and woodlands to obtain firewood. In short, we can afford conservation and management, better agricultural practices, and pollution control, thereby improving our environment.

Still, because the United States consumes so many resources, we also lead the world in the production of many pollutants. For example, by using such large quantities of fossil fuel (coal, oil, and natural gas) to drive our cars, heat and cool our homes, and generate electricity, the United States is responsible for a large share of the carbon dioxide produced. As mentioned, with about 4.5% of the world's population, the United States generates 22% of the carbon dioxide emissions that may be changing global climate. Similarly, emissions of chlorofluorocarbons (CFCs) that have degraded the ozone layer, emissions of chemicals that cause acid rain, emissions of hazardous chemicals, and the production of nuclear wastes are all largely the by-products of affluent societies.

High individual consumption places enormous demands on the environment. The world's wealthiest 20% of people (some located in low- and middle-income countries) are responsible for 86% of all private consumption and 80% of world trade. As a consequence, by 1995, 60% of major world fisheries were either fully exploited or overexploited, and old-growth forests in southern South America are being clear-cut and turned into chips to make fax paper. Oil spills are a "by-product" of our appetite for energy. Tropical forests are harvested to satisfy the desires of the affluent for exotic wood furnishings. Metals are mined, timber harvested, commodities grown, and oil extracted, all far from the industrialized countries where these goods are used. Every one of these activities has a significant environmental impact. As increasing numbers of people strive for and achieve greater affluence, it seems more than likely that such pressures, and other ones like them, will mount.

One way of generalizing the effect of affluence is to say that it enables the wealthy to clean up their immediate environment by transferring their wastes to more distant locations. It also allows them to obtain resources from more distant locations, so they neither see nor feel the impacts of getting those resources. However, affluence also provides people with opportunities to exercise lifestyle choices that are consistent with the concerns for stewardship and sustainability.

With this picture of population growth and its impacts in view, the next section discusses future population trends, how demographers identify them, and the biggest population trend—the demographic transition.

8.4 Projecting Future Populations

During the 1960s, the world population growth rate peaked at 2.1% per year, after having risen steadily for decades. Following this peak, it began a steady decline (Fig. 8–13). Twenty years later the number of persons added per year peaked at 87 million. The reason for the difference is that even though the population growth rate was in decline, there were increasing numbers of people who could reproduce. The total number of persons added every year did not go down until the growth rate fell low enough to offset the increase in reproducing people.

The declines in both the population growth rate and the number of persons added per year are primarily a consequence of the decline in the *total fertility rate*—that is, the average number of babies born to a woman over her lifetime. In the 1960s, the total fertility rate was an average of 5.0 children per woman; it has since steadily declined to its present value of 2.6 children per woman. In addition to the total fertility rate, the age at which a woman first reproduces also controls how rapidly a population grows. A population in which women have children at young ages will grow more quickly than one in which women are slightly older when they reproduce, even if women have the same number of children. That is, there are more generations alive at the same time when women reproduce young. Thus, total fertility rate is the most important, but not the only, factor determining population growth.

Extrapolating the trend of lower fertility rates leads to the UN Population Division's projection that the global human population will reach 9.2 billion by 2050; it will not have leveled off at that time, but is likely to level off sometime in the 22nd century. This projection is the UN's *medium* scenario; other scenarios are based on different fertility assumptions. Figure 8–14 depicts the UN Population Division's projections until 2050 under four different sets of

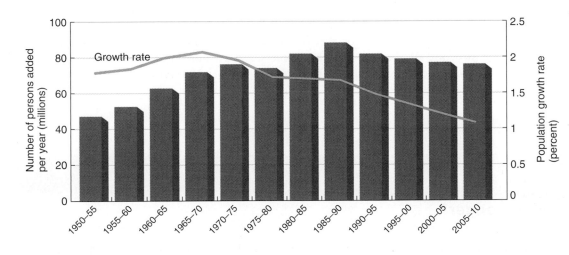

Figure 8–13 World population growth rate and absolute growth. Declining fertility rates in the last three decades have resulted in a decreasing rate of population growth. Absolute numbers, however, are still adding 78 million persons per year. (*Source:* Data from Shiro Horiuchi, "World Population Growth Rate," *Population Today* (June 1993): 7 and (June/July 1996): 1; updated from UN Population Division, *World Population Prospects: The 2008 Revision* (New York: United Nations, 2008).)

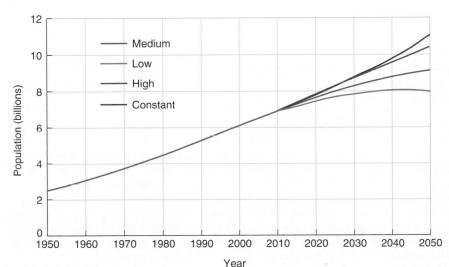

Figure 8–14 United Nations population projections. The most recent population projections demonstrate the vital role played by different fertility assumptions. The *Constant* projection assumes that fertility rates remain at their present level (2.6 children per woman). The *Medium* projection assumes a gradual decline in fertility in the developing countries, where 85% drop below replacement-level fertility by 2050 and the world total fertility rate is 2.02 children per woman. This projection is given the highest probability by the UN Population Division. The *High* projection assumes fertility rates a ½ child greater than the *Medium* projection, and the *Low* projection assumes fertility rates a ½ child lower than the *Medium* projection. All of these projections include the impact of the AIDS epidemic on increasing mortality. (*Source:* UN Population Division, *World Population Prospects: The 2008 Revision* (New York: United Nations, 2008).)

assumptions. The assumption of declining fertility rates is crucial; if current fertility rates remain unchanged (*constant* in the figure), the 2050 population will be 11.9 billion. To achieve the medium scenario, people will need to continue to purposefully lower their fertility in the high fertility areas of the world, at a rate we are not seeing yet. The UN doesn't usually project past 50 years out. The projected leveling off over 9 billion raises serious concerns about whether Earth can sustain such numbers.

Population Profiles

In studying population growth, we must consider more than just the increase in numbers, which is simply births minus deaths. We must also consider how the number of births ultimately affects the entire population over the **longevity**, or lifetimes, of the individuals (See Sound Science, "Are We Living Longer?"). A **population profile** is a bar graph showing the number or proportion of people (males and females separately) at each age for a given population. Figure 8–15 shows examples of such graphs. The data are collected through a census of the entire population using a combination of household questionnaires and estimates for groups such as the homeless. In the United States and most other countries, a detailed census is taken every 10 years. Between censuses, the population profile may be adjusted using data regarding births, deaths, immigration, and the aging of the population.

A population profile shows the **age structure** of the population—that is, the number of people in each age group at a given date. It is a snapshot of the population at a given time. For example, population profiles of the United States for 1985 and 2005 and a projected future profile for 2050 are shown in Figure 8–15. Leaving out the complication of migration for the moment, each bar in the profile represents one *cohort* (group of the same age) of the population. As you look from one graph to the next, you can see a particular group as it ages. In high-income countries such as the United States, the proportion of people who die before age 60 is relatively small. Therefore, the population profile below age 60 is an "echo" of past events that affected birthrates. Figure 8-15a shows, for example, that smaller numbers of people were born between 1931 and 1935 (ages 54–50 in 1985). This drop was a reflection of lower birthrates during the Great Depression. The dramatic increase in people born in 1946 and for 14 years thereafter (ages 39–25 in 1985) is a reflection of returning veterans and others starting families and choosing to have relatively large numbers of children following World War II—the "baby boom." A number of industries expanded and then contracted as the baby-boom generation moved through a particular age range, and this phenomenon is not yet past. In sequence, schools, then colleges and universities, and then the job market were affected by the large influx of baby boomers. As the large baby-boom generation, now in middle age (and projected to be extremely elderly in 2050), moves up the population profile (Fig. 8–15a–c), any business or profession that provides goods or services to seniors is looking forward to a period of growth. Planners also use population profiles to understand the probable needs for social security programs in the future.

Sound Science

Are We Living Longer?

Most of us would like to live a long and prosperous life, hopefully just slipping away in our sleep at some advanced age. With the advent of modern medicine and disease control, this pleasant goal has become possible for more and more people. Average **life expectancy** (the number of years a newborn can expect to live under current mortality rates) has been increasing dramatically. Over the past two centuries, world life expectancy increased from about 25 years to 65 for men and 69 for women. Life expectancy and mortality rates are linked, and as infant mortality in particular has declined in the developing countries, their average life expectancy has climbed sharply. Life expectancies in those countries are now within 10 years of life expectancies in the developed countries (73 for men and 80 for women). What is not well known is that life expectancy has also continued to climb about

2½ years every decade for the low-mortality developed countries. Mortality rates have declined for every age.

Longevity

Can we hope that this trend will continue until, perhaps a century from now, human life expectancies will be approaching or going beyond 100? There is intense debate over this question. First, however, we must differentiate between life expectancy and longevity. **Longevity** refers to the *maximum life span for a species*. The verifiable record for humans is held by Madame Jeanne Louise Calment, who died in France on August 4, 1997, at 122 years of age. The record for the United States is that of a man who died at 115 years of age in 1882! These figures suggest that human longevity has not really increased over the years. Indeed, many workers in

the field of gerontology (the study of aging) are convinced that, in humans (as in all species), there is an aging process that cannot be prevented—that built into our cells are the biochemical seeds of destruction. Cells, tissues, and organs wear out at variable rates in different people. Thus, we will not likely find life expectancies rising beyond about 85 years anywhere or anytime in the future.

Other workers argue that even though absolute longevity seems to be a limiting process, we can expect life expectancies to continue the increase they have shown in recent years. They ask, Why should life expectancies stop at 85? The trend of a 2½-year increase per decade is based on reducing many of the factors that cause mortality at different ages. This trend should continue, given the medical prowess of our 21st-century science.

a. 1985

b. 2005

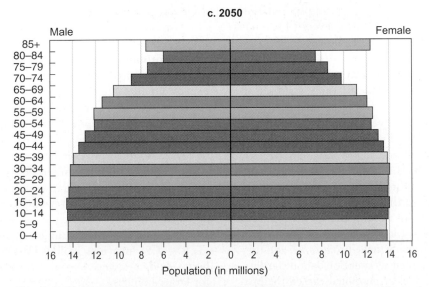

c. 2050

Figure 8–15 Population profiles of the United States. The age structure of the U.S. population (a) in 1985, (b) in 2005, and (c) projected to 2050. Go to GRAPHit! at www.masteringenvironmentalscience.com. (*Source:* U.S. Census Bureau, International Data Base.)

Figure 8–16 Total fertility rate, United States, 1920–2008. The changes in fertility led to the baby boom and the baby bust in the United States. The rate now hovers about the replacement level. (*Sources*: Population Reference Bureau, "What Drives U.S. Population Growth?" *Population Bulletin* 57, no. 4 (December 2002); *2008 World Population Data Sheet* (Washington, D.C.: Population Reference Bureau, 2008).)

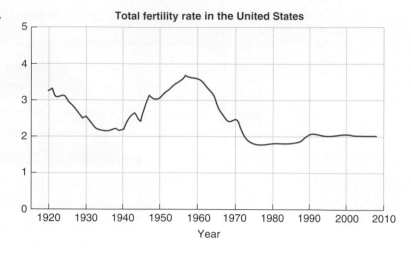

Total fertility rate in the United States

The baby boom is not the only phenomenon we see in U.S. population profiles. The general drop in numbers of people born from 1961 to 1976 (ages 15–24 in 1985) resulted from declining fertility rates—the "baby bust." The rise in numbers of people born in more recent years (ages 15–24 in 2005) is termed the "baby boom echo" and is due to the large baby-boom generation producing children, even though the actual total fertility rate remained near 2.0. These changes in fertility are shown in Figure 8–16.

Predicting Populations

Current population growth in a country is calculated from three vital statistics: births, deaths, and migration. The population of the United States grew by more than 2.7 million people during 2008. This growth was the result of the combination of these three factors. There were 4.32 million births (15,000 more than the highest previous record in 1957). Immigration was projected to add one person every 36 seconds to the population, and in January 2009, the projection was one birth every 8 seconds and one death every 12 seconds in the United States. Thus, the total population was increasing by one person approximately every 14 seconds (U.S. Census Bureau News), although these numbers are imprecise because of illegal immigration.

Demographers used to make population *forecasts*. Most often, they took current birth and death rates, factored in expected immigration, and then extrapolated the data into the future. They were virtually always wrong. For example, in 1964, the total fertility rate (TFR) in the United States was around 3.0. The U.S. Census Bureau forecast that the TFR would range between 2.5 and 3.5 up to 2000 and projected a 2000 population of 362 million under the higher fertility assumption—81 million more than the actual number. The agency totally missed the baby bust of the late sixties and seventies, although the trend in 1964 was definitely downward.

Nowadays demographers only make *projections*, and they cover their bases by making their assumptions about fertility, mortality, and migration very clear. As noted earlier, the projection that the world population will reach 9.2 billion in 2050 is based on the assumption that fertility rates will continue their decline. The United Nations gives three different projections of future world population (Fig. 8–14). The medium scenario assumes that world fertility will drop from 2.55 children per woman to slightly above 2 children per woman by 2050, the high-fertility scenario assumes a total fertility rate of 2.5 by 2050, and the low-fertility scenario assumes that the total fertility rate will reach 1.5 by 2050. Recall that the constant fertility line (red) shows what will happen if the fertility rates around the world do not change. Note how each fertility assumption generates profoundly different world populations. Notice also that none of these projections, even the one based on a drop in total fertility to 1.5 children per woman, will lead to a population smaller than today's population by 2050.

Population Projections for Developed Countries.
The 2005 population profile of Italy, a developed country in southern Europe, shown in Figure 8–17a, reflects the fact that Italian women have had a low fertility rate for some time. If the 2005 total fertility rate of 1.3 remains constant for the next 20 years, the profile presented in Figure 8–17b is obtained. This profile shows a dramatic increase in the number of older people, a great reduction in the number of children and young people, and a net decline of 830,000.

Graying of the Population. For the next 20 years, Italy's population will be **graying**, a term used to indicate that the proportion of elderly people is increasing. In 2005, one-fourth of Italians were aged 60 and above; by 2050, 41% are expected to be 60 or older. Overall, a net population decrease is expected to occur. What opportunities and risks does the changing population profile imply for Italy? If you were an advisor to the Italian government, what would your advice be for the short term? For the longer term? Unless a smaller population is the goal, it might be wise to encourage, and to provide incentives for, Italian couples to bear more children. In fact, many European governments have policies to encourage women to have more children, with mixed results so far. France, with one of the highest fertility rates in Europe (TFR = 1.9), provides child allowances, tax credits for in-home child care, and discounts on many goods and services for larger families.

The very low fertility rate and the expected declining population seen in Italy are typical of an increasing number of highly developed nations. Europe as a whole is on a trajectory

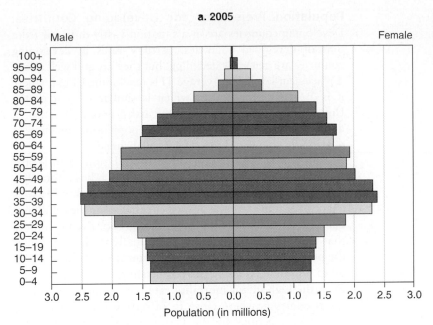

a. 2005

Male / Female

Population (in millions)

Figure 8–17 Projecting future populations: A developed country. A population profile of Italy, representative of a highly developed country, (a) in 2005 and (b) projected to 2025. Note how larger numbers of persons are moving into older age groups and the number of children is diminishing. (*Source:* U.S. Census Bureau, International Data Base.)

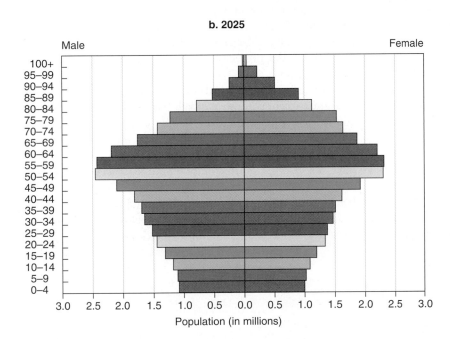

b. 2025

Male / Female

Population (in millions)

of population decline if only natural increase (rather than immigration) is considered. Who will produce the goods and services needed by the European countries' aging populations? Will their economies remain competitive on the world markets? The surge of older people retiring may overwhelm government pension systems; life expectancies near 80 and retirement around 60 mean decades of retirement income.

One obvious solution is to allow more immigration. However, allowing the declining numbers of Italian or French people to be replaced by immigrant populations has implications for art, culture, and religion. Such changes are sometimes difficult for the host society. In fact, Europe is in the midst of a migration wave. More than 20 million immigrants now live and work in the countries of western Europe, many of whom do not have resident status. In some of these countries, fear and mistrust of these "guest workers" have led to violent attacks on them. Many Europeans do not want more immigration, even to prevent the loss of the labor force and population decline. Yet just to maintain their current populations, most of these countries will have to triple their current immigration levels in the near future.

Less Graying Here. In contrast to other developed countries, the fertility rate in the United States reversed directions in the late 1980s and started back up. On the basis of the lower fertility rate, the U.S. population had been projected to stabilize at between 290 and 300 million toward the middle

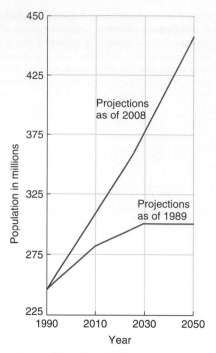

Figure 8–18 Population projections for the United States.
Projections shift drastically with changes in fertility. Contrast the 1989 projection, based on a fertility rate of 1.8, with the 2008 projection, based on an increased fertility rate of 2.1 and current immigration. (*Source:* U.S. Census Bureau, U.S. Population Projections.)

of the next century. With the higher fertility rate of 2.1 and with an influx of immigrants, the U.S. population is now projected to be 439 million by 2050 (Fig. 8–15c) and to continue growing indefinitely (Fig. 8–18). For this projection, immigration is assumed to remain constant at current levels.

These projections for the United States show how differences in the total fertility rate and high immigration profoundly affect population estimates when they are extrapolated forward. In light of the concerns for sustainable development, the population policy of the United States (and scenarios that could lead to lower future growth) are under heavy debate.

Population Projections for Developing Countries.
Developing countries are in a situation vastly different from that of developed countries. Fertility rates in developing countries are generally declining, but they are still well above replacement level. The average TFR (excluding China, where it is 1.6) is currently 3.2, which is similar to that of the United States at the peak of the baby boom. Because of decades of high fertility rates, the population profiles of developing countries have a pyramidal shape.

Burkina Faso. The 2008 population profile for Awa's home country of Burkina Faso, which has a total fertility rate of 6.3, is shown in Figure 8–19a. Burkina Faso is a good example of the lowest-income countries and has demographics that are typical of that group. Even assuming that this high fertility rate will gradually decline by 2025, the population is likely to increase from 15.3 million (2008) to 25.4 million, yielding the profile shown in Figure 8–19b. Even the high infant mortality rate (currently 91 per thousand) comes nowhere close to offsetting the high fertility rate. In projections by the UN, the impact of AIDS also is included. AIDS has increased child and young adult mortality and lowered average life spans in a number of countries. The pyramidal form of the profile remains the same in both 2008 and 2025 because, for many years, the rising generation of young adults produces an even larger generation of children. Thus, the pyramid gets wider and wider until the projected declining fertility rate begins to take effect.

While highly developed countries are facing the problems of a graying population, the high fertility rates in developing countries maintain an exceedingly young population. An "ideal" population structure, with equal numbers of persons in each age group and a life expectancy of 75 years, would have one-fifth (20%) of the population in each 15-year age group. By comparison, 40% to 50% of the population is below 15 years of age in many developing countries, whereas less than 20% of the population is below the age of 15 in most developed countries (Table 8–3).

TABLE 8–3 Populations by Age Group

Region or Country	Percent of Population in Specific Age Groups			
	<15	15 to 65	>65	Dependency Ratio*
Sub-Saharan Africa	43	54	3	85
Latin America	30	64	6	56
Asia	27	66	7	54
Indonesia	29	65	6	54
Europe	16	68	16	47
Sweden	17	65	18	54
China	19	73	8	37
United States	20	67	13	49

Source: Data from *2008 World Population Data Sheet* (Washington, D.C.: Population Reference Bureau, 2008).
*Number of individuals below 15 and above 65, divided by the number between 15 and 65 and expressed as a percentage.

a. 2005

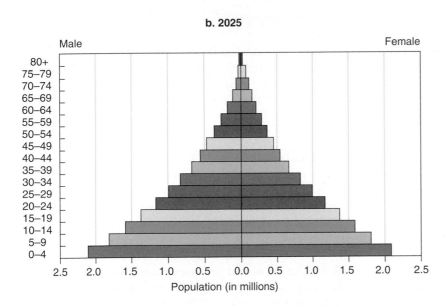

b. 2025

Figure 8–19 Projecting future populations: A developing country. (a) The 2005 population profile of Burkina Faso, a developing country. (b) Projection for 2025, based on the assumption that the total fertility rate of 4.8 will decline to 2.7 by 2025. (*Source*: U.S. Census Bureau, International Data Base.)

Growth Impacts. What do these differing population structures mean in terms of the need for new schools, housing units, hospitals, roads, sewage facilities, telephones, and so forth? One of the things they mean is that if a country such as Burkina Faso is simply to maintain its current standard of living, the amount of housing and all other facilities (not to mention food production) must be almost doubled in as little as 25 years. In particular, jobs must be available for a large number of young people. Unfortunately, the population growth of a developing country can easily cancel out its efforts to get ahead economically.

Present and projected population profiles for developed and developing countries are compared in Figure 8–20. The figure shows that, while little growth will occur in developed countries over the next 50 years, enormous growth is in store for the developing world, even assuming that fertility rates in

the developing world continue their current downward trend. Even bringing the fertility rates of developing countries down to 2.0 right now will not immediately stop population growth in those countries because of a phenomenon known as population momentum.

Population Momentum

Population momentum refers to the effect of current age structures on future populations. In a young population, such as Burkino Faso's, momentum is *positive* because such a small portion of the population is in the upper age groups (where most deaths occur) and many children are entering their reproductive years. Even if these rising generations have only two children per woman (replacement-level fertility), the number of births will far exceed the number of deaths. This imbalance will

Figure 8–20 Comparing projected populations. 2000 population profiles for developed and developing countries, projected to 2050. (*Source:* U.S. Census Bureau, International Data Base.)

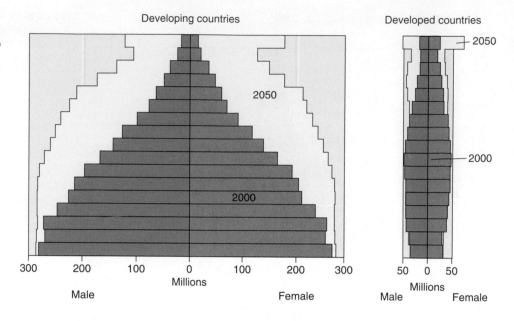

Developing countries

Developed countries

continue until the current children reach the Burkina Faso limits of life expectancy—50 to 60 years. In other words, only a population that has been at or below replacement-level fertility for decades will stop growing (achieve a stable population).

Despite population momentum, efforts to stabilize population are not fruitless. It just means that population growth cannot be halted quickly. Like a speeding train, there is a long time between applying the brakes and stopping completely. The earlier the fertility rates are reduced, the greater the likelihood is of ending population growth and achieving a sustainable society in the near future.

Europe's population, on the other hand, will soon begin to experience *negative* population momentum as a consequence of the low fertility of the past three decades. It has been calculated that if the current TFR of 1.5 births continues for 15 more years, negative momentum will shrink the EU population by some 88 million by 2100. It is this prospect (and all the social consequences accompanying an aging population) that has prompted several European countries (e.g., France) to adopt pro-natal policies—policies that make it easier to have and raise children.

The Demographic Transition

The concept of a stable, nongrowing global human population based on people freely choosing to have smaller families can become a reality because it is already happening in developed countries. Early demographers observed that the modernization of a nation brings about more than just a lower death rate resulting from better health care. A decline in the fertility rate also occurs as people choose to limit the size of their families. Thus, as economic development occurs, human societies move from a primitive population stability, in which high birthrates are offset by high infant and childhood mortality, to a modern population stability, in which low infant and childhood mortality are balanced by low

birthrates. A gradual shift in birth and death rates from the primitive to the modern condition in the industrialized societies is called the **demographic transition** (Fig. 8–21). The basic premise of the demographic transition is that there is a causal link between modernization and a decline in birth and death rates.

Birth and Death Rates. To understand the demographic transition, you need to understand the **crude birthrate (CBR)** and the **crude death rate (CDR).** The CBR and the CDR are *the number of births and deaths, respectively, per thousand of the population per year.* By giving the data per thousand of the population, populations of different countries can be compared, regardless of their total size. The term *crude* is used because no consideration is given to what proportion of the population is old or young, male or female. Subtracting the CDR from the CBR gives the increase (or decrease) per thousand per year. Dividing this result by 10 then yields the *percent* increase (or decrease) of the population.

| Number of births per 1,000 per year (CBR) | − | Number of deaths per 1,000 per year (CDR) | = | Natural increase (or decrease) in population per 1,000 per year | ÷ 10 = | Percent increase (or decrease) in population per year |

A zero-growth population is achieved if, and only if, the CBR and CDR are equal and there is no net migration.

Epidemiologic Transition. Throughout most of human history, crude death rates were high—40 or more per thousand for most societies. By the middle of the 19th century, however, the epidemics and other social conditions responsible for high death rates began to recede, and death rates in Europe and North America declined. The decline was gradual in the now-developed countries, lasting for many decades and finally stabilizing at a CDR of about 10 per thousand (Fig. 8–21). At present, cancer, cardiovascular disease, and other degenerative diseases account for most mortality, and many people survive to old age. This pattern of change in mortality factors has been called the *epidemiologic* transition and represents one element of the demographic transition. (Epidemiology is the study of diseases in human societies.)

Fertility Transition. Another pattern of change over time can be seen in crude birthrates. In the now-developed countries, birthrates have declined from a high of 40 to 50 per thousand to 9 to 12 per thousand—a *fertility* transition. As Figure 8–21 shows, this fertility transition did not happen at the same time as the epidemiologic transition; instead, it was delayed by decades or more. Because net growth is the difference between the CBR and the CDR, the time during which these two patterns are out of phase is a time of rapid population growth. The developed countries underwent such growth during the 19th and early 20th centuries and during the baby-boom years of the mid–20th century.

Phases of the Demographic Transition. The demographic transition is typically presented as occurring in the four phases shown in Figure 8–22. **Phase I** is the primitive stability resulting from a high CBR being offset by an equally high CDR. **Phase II** is marked by a declining CDR—the epidemiologic transition. Because fertility and, hence, the CBR remain high, population growth accelerates during Phase II. The CBR declines during **Phase III** due to a declining fertility rate, but population growth is still significant. Finally, **Phase IV** is reached, in which modern stability is achieved by a continuing low CDR and an equally low CBR.

Developed countries have generally completed the demographic transition, so they are in Phase IV. Developing countries, by contrast, are still in Phases II and III. In most of these countries, death rates declined markedly as modern medicine and sanitation were introduced in the mid–20th century. Fertility and birthrates are declining, but remain considerably above replacement levels (Fig. 8–21). Therefore, populations in developing countries are still growing rapidly.

In light of these circumstances, the key question is: What must happen in the developing countries so that they complete their demographic transition and reach Phase IV? Is economic development essential to undergoing the transition, or may other factors, such as family planning programs, come into play?

In this chapter, we looked at three young people—Anje in Sweden, Atin in Indonesia, and Awa in Burkino Faso. We saw that their life experiences were dramatically different from each other, in part because of broad differences in wealth, health care, and other goods and services between countries. Even within countries, great variability exists. Those differences in life experiences drive differences in consumption patterns and fertility—two parameters that need to be brought into line to have a sustainable world.

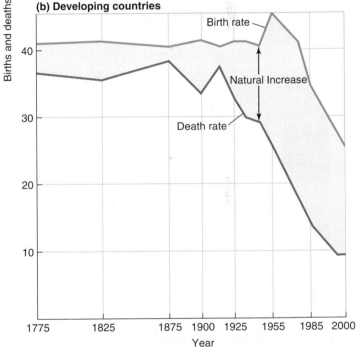

Figure 8–21 Demographic transition in developed and developing countries. (a) In developed countries, the decrease in birthrates proceeded soon after, and along with, the decrease in death rates, so very rapid population growth never occurred. (b) In developing countries, both birth and death rates remained high until the mid-1900s. Then the sudden introduction of modern medicine caused a precipitous decline in death rates. Birthrates remained high, however, resulting in very rapid population growth. (Redrawn with permission of Population Reference Bureau.)

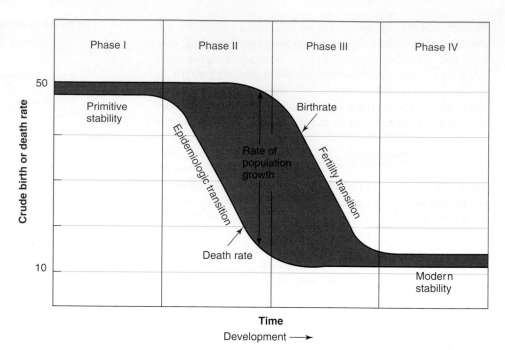

Figure 8–22 The demographic transition. The epidemiologic transition and the fertility transition combined to produce the demographic transition in the developed countries over many decades. (*Source:* Adapted from Joseph A. McFalls, Jr., "Population: A Lively Introduction," *Population Bulletin* 53, no. 3 (1998): 39, courtesy of Population Reference Bureau, 2000.)

In Chapter 9, we will investigate the factors that influence birth and death rates, and you will explore the relative contributions of economic development and family planning in bringing about the fertility transition. In the process, we can begin to picture what sustainable development might look like for the developing world, in particular, and what it will take to get there.

Revisiting the Themes

Sound Science

Demography, the science of human population phenomena, reflects the diligent work of many scientists whose concern is to document data and interpret it. Their work forms much of the data and information presented in this chapter. Demography can be viewed in the context of population ecology as a whole. Human populations can be studied like other populations and are subject to natural laws such as the law of limiting factors and carrying capacity. In spite of some opinions, there is consensus in the scientific community that population growth and increased demand for resources cannot continue indefinitely. However, humans are also quite different from other organisms. It is difficult to classify them as either r- or K-strategists because their life history of slow growth and high parental care does not easily match their current rapid global population growth and opportunistic use of new habitats. In the past, the carrying capacity of humans has been increased by using technology to trap energy from other eras and open up new habitats. However, many of the things we did to achieve modern life have prices, as we will see in later chapters.

Demographers have clarified the nature of the demographic transition, and their population projections enable us to see what the future will look like if current trends continue.

Sustainability

A sustainable world is a world whose population is generally stable—it cannot increase indefinitely, as the world's population currently is doing. World population may level off at 10 billion or even 9 billion, but we may be exceeding Earth's carrying capacity. Two or three billion more people will need to be fed, clothed, housed, and employed, all in an environment that continues to provide essential goods and services. Many agree that sustainability will be possible only if the additional people maintain a modest, low-consumption lifestyle, much less than that of the developed countries. People already in developed countries will need to make changes to lower their own consumption so that others may have access to resources. These changes involve not only myriad personal decisions, but also coordinated action and the development of more efficient technologies.

In a sense, the United States is playing a small role in the process by accepting so many immigrants from developing countries—about 1% of the annual population growth in these countries (hardly a significant percentage). More significant is the impact of immigration on the United States itself, which is currently on a trajectory of indefinite population growth because of immigration. Sustainability requires that we somehow stabilize population both nationally and globally.

Stewardship

In order to achieve the goal of sustainability, humans need to steward resources by stabilizing both population growth and resource consumption. Such stewardship requires us to promote justice and may require difficult decisions about what is right. Calling on the developing countries to keep down their future demands on the environment could be considered unjust, as so many people live in abject poverty, while others consume resources at high rates. Still, there is no way for these countries to meet their future needs by mimicking what the developed world is currently doing with resources and the environment. Justice demands (and sustainability requires) that both the rich and the poor countries reduce their environmental impact.

Stewardship requires both individual efforts and corporate action through policy. If enough people in the high- and middle-income countries practice environmental stewardship, the pathway to such a future will be easier. Stewardly concern and action can do much to alleviate the negative impacts of affluence and technology, prominent items in the *IPAT* formula. Population growth has to slow as well. Achieving that end without trampling on the rights of others is part of the job ahead of us. Immigration also offers ethical dilemmas as immigrants move to places that may or may not be prepared to receive them.

REVIEW QUESTIONS

1. In what ways is human population ecology similar to and different from that of other organisms? Why is it difficult to determine a carrying capacity for humans?

2. How has the global human population changed from prehistoric times to 1800? From 1800 to the present? What is projected over the next 50 years?

3. How does the World Bank classify countries in terms of economic categories?

4. What three factors are multiplied to give total environmental impact? Are developed nations exempt from environmental impact? Why or why not?

5. What are the environmental and social consequences of rapid population growth in rural developing countries? In urban areas?

6. Describe the negative and positive impacts of affluence (high individual consumption) on the environment.

7. What information is given by a population profile?

8. How do the population profiles and fertility rates of developed countries differ from those of developing countries?

9. Compare future population projections, and their possible consequences, for developed and developing countries.

10. Discuss the immigration issues pertaining to developed and developing countries. Which countries are sending the most immigrants and which are receiving the most?

11. What is meant by population momentum, and what is its cause?

12. Define the crude birthrate (CBR) and the crude death rate (CDR). Describe how these rates are used to calculate the percent rate of growth and the doubling time of a population.

13. What is meant by the demographic transition? Relate the epidemiologic transition and the fertility transition (two elements of the demographic transition) to its four phases.

14. How do the current positions of the developed and developing nations differ in the demographic transition?

THINKING ENVIRONMENTALLY

1. Consider how humans are different from and similar to other species. In what ways does it seem obvious that population ecology principles apply to humans? Why do some people see humans as not subject to the law of limiting factors?

2. List the parameters you think you would need to keep track of to make accurate predictions about future human population growth. For example, what conditions might drive mortality and fertility trends?

3. Consider the population situation in Italy or Germany: very low birthrates, an aging population, and eventual decline in overall numbers. What policies would you recommend to these countries, assuming their desire to achieve a sustainable society?

4. Consider the populations of Sweden, Indonesia, and Burkina Faso. How could people in each of these countries help those in the other countries to steward their resources better or to promote a sustainable future?

MAKING A DIFFERENCE

1. Calculate your ecological (environmental) footprint using a site such as http://www.myfootprint.org.

2. Come up with two things you can do to lower your resource consumption. They might include buying less of something you don't need or buying a product with less packaging.

Agree to do it for one month, and decide whether you could continue longer.

3. Educate yourself on immigration policy. Decide whether you support immigration reform and what policies you would like to see in place. Then write to an elected official about it.

Go to **www.masteringenvironmentalscience.com** for practice quizzes, Pearsone Text, videos, current events, and more.

New and old India. (above) Heritage City, an apartment complex in Gurgaon, on the southwest fringe of Delhi, India. Gurgaon has become a corporate and industrial center. (below) Another side of Gurgaon: This is the old India. Inset: Fisherman in Kerala.

CHAPTER 9

Population and Development

I F YOU WERE TO CALL JC PENNEY FOR A CREDIT APPLICATION, the chances are that you would be talking to a young Indian person, one of 13,000 working for Genpact, a call center in Gurgaon, a small city on the outskirts of Delhi, India. Genpact recruits some 1,100 new employees a month from a pool of well-educated young people, who aspire to live in apartments like those of Heritage City. Outside this high-tech enclave, however, is the old India, with families living in shantytowns, sacred cows roaming the streets, garbage and trash everywhere, and shoeless children begging for rupees. This is the stark contrast found in many countries with a history of rapid population growth.

In the 60 years since India gained its independence, its population has tripled. With 1.15 billion people already, India is now growing at a rate of 18 million per year and is expected to surpass China's population by 2030. In spite of many decades of family-planning programs, India's population has continued its relentless growth, canceling out much of the country's gains in food production, health care, and literacy. Almost half of India's children under the age of three are malnourished, 75% of its population lives on less than $2 a day, and almost half of its women remain illiterate. With a per capita gross national income of $950 per year,

221

India has a very long way to go to achieve the kind of development that characterizes the industrialized countries. It might be tempting to consider this a hopeless situation were it not for Kerala, one region of India where things are quite different.

Kerala, a state in the south of India, has a population of 33 million and an area of 39,900 square miles (about the size of Kentucky). This makes Kerala the second most densely populated state in India and close to the most densely populated region in the world (Fig. 9–1). Situated only 10° north of the equator, Kerala is tropical, with lush plantations of coffee, tea, rubber, and spices in the highlands and rice, coconut, sugarcane, tapioca, and bananas in the lowlands. With the Arabian Sea lapping at Kerala's shore, fishing is an important industry. Kerala is very much like the rest of India in that it is crowded, per capita income is low, and food intake is around 2,200 calories per day (considered adequate, but on the low end). Here, however, the similarities end.

The people of Kerala have a life expectancy of 73 years, compared with 65 for all of India. Infant mortality in the state is 12 per thousand versus 57 per thousand for India. The fertility rate is 1.8 (below replacement level) compared, to 2.8 for India. Literacy is near 90%, almost all villages in the state have access to schools and modern health services, and women have achieved high offices in the land and are as well educated as men. Even though Kerala must be considered a poor region by every economic measurement, its people are well on the way to achieving a stable population.

What is different about Kerala? For one thing, there is a strong public-policy commitment to health care and education. Also, land distribution is relatively equitable, food distribution is efficient, and India's old caste system has all but disappeared. These investments

Figure 9–1 Kerala. This state in southwestern India has achieved remarkable progress toward a stable population.

in social policy have paid off. Kerala's total fertility rate dropped from 3.7 to 1.8 in just two decades; in the last few years, Kerala has been making significant progress in per capita income growth. In short, Kerala shows that it is possible to bring a developing region from the middle phase of the demographic transition to the threshold of its completion without the long-term economic development that characterized the industrial countries as they underwent their demographic transition.

This chapter follows up on the demographic information presented in Chapter 8. We begin with a reassessment of the demographic transition.

9.1 Reassessing the Demographic Transition

Recall from Chapter 8 that the *demographic transition* describes the shift from high birth and death rates to low birth and death rates over time in the industrialized countries. Using the concept of the demographic transition, you can plot the current birth and death rates of the major regions of the world to visualize where they are in the transition (Fig. 9–2). The vertical dividing line on this plot separates the regions (to the right of the line) that are on a fast track to completing the demographic transition or are already there from the regions (to the left of the line) that are in the middle of the demographic transition. These regions (half of the world) are mostly in their fifth decade of rapid population increase. It is as if they are trapped there, with serious consequences.

Different Ways Forward

The countries in these regions would all like to experience the economic development that has brought South Korea, Indonesia, Malaysia, Brazil, and others into the middle- and even high-income nation groups. If they did so, it is likely that they would then move through the demographic transition. There are great disparities in economic progress among developing countries, however. A recent World Bank study of developing economies found that growth in per capita income from 1990 to 2005 was strongly correlated with technological progress; both of these measures were at their lowest in three regions with continued high population growth and poverty: sub-Saharan Africa, the Middle East, and Latin America (Fig. 9–3).[1]

[1]Kavita Watsa, *Technology & Development: Findings from a World Bank Report: Global Economic Prospects 2008: Technology Diffusion in the Developing World* (Washington, D.C.: World Bank, 2008).

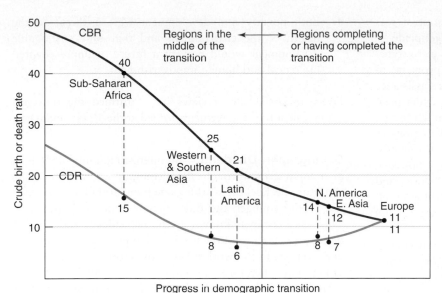

Figure 9–2 World regions in the process of demographic transition. Crude birthrates (CBRs, red) and crude death rates (CDRs, green) are shown for major regions of the world. A dividing line separates countries at or well along in the demographic transition from those still in the middle of the transition. (*Source:* Data from *2008 World Population Data Sheet* (Washington, D.C.: Population Reference Bureau, 2008).)

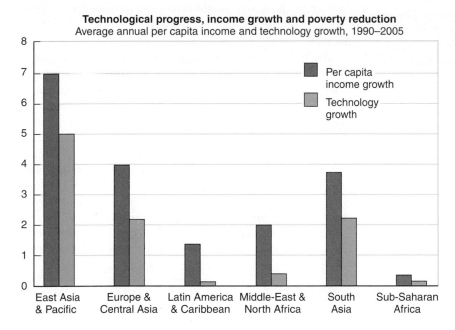

Figure 9–3 Technological progress, income growth, and poverty reduction. Average annual per capita income and technology growth, 1990–2005, in selected regions. (*Source: Technology & Development (Findings from a World Bank Report). Global Economic Prospects 2008: Technology Diffusion in the Developing World* (Washington, D.C.: World Bank, 2008).)

How did the now-developed nations come through the demographic transition without getting caught in the poverty-population trap? The improvements in disease control that lowered death rates (the epidemiological transition) occurred gradually through the 1800s and early 1900s. Industrialization, which introduced factors that lowered fertility, occurred over the same period. Therefore, there was never a huge discrepancy between birth and death rates (see Figure 8–22). In contrast, modern medicine was introduced into the developing world relatively suddenly, bringing about a precipitous decline in death rates, while the fertility-lowering effects of development have been slow in coming.

What do the developing countries need to do to fast-forward their way through the demographic transition? From early on, two basic conflicting theories offered answers to this question:

1. We need to speed up economic development in the high-growth countries, and population growth will slow down "automatically," as it did in the developed countries.

2. We need to concentrate on population policies and family-planning technologies to bring down birthrates.

Population Conferences. These two schools of thought were reflected at two UN population conferences—one in Bucharest, Romania, in 1974 and a second in Mexico City in 1984. At the Bucharest conference, the United States was a strong advocate of population control through family planning, while the developing nations argued that "development was the best contraceptive." Their resistance to family planning was also bolstered by feelings that the developed world's promotion of population control was another form of economic imperialism or even genocide.

At the conference in Mexico City, the sides were somewhat reversed. Developing countries facing real problems of excessive population growth were asking for more assistance with family planning, whereas the United States, under pressure from right-to-life advocates, took the position that development was the answer and terminated all contributions to international family-planning efforts, a policy that remained in effect until 1993. The other developed countries, however, remained convinced that family planning was essential and continued to support international efforts to help the developing countries implement policies designed to bring fertility rates down.

End of Debate. The most recent population conference, the *International Conference on Population and Development (ICPD)*, was held in Cairo in 1994. Poverty, population growth, and development were clearly linked at this conference, and resources and environmental degradation were an important focus. The ideological split between developed and developing countries was a thing of the past—all agreed that population growth must be dealt with in order to make progress in reducing poverty and promoting economic development. The responsibility for bringing fertility rates down was placed firmly on the developing countries themselves, although the developed countries pledged to help with technology and other forms of aid. There was also broad agreement that

1. Women's rights to health care, education, and employment were foundational to achieving slower population growth;

2. Development must be linked to a reduction in poverty;

3. The existing poverty in the developing countries was an affront to human dignity that should not be tolerated; and

4. Both poverty and development were a threat to the health of the environment and that only sustainable development would prevent a future of unprecedented biological and human impoverishment.

Five-year anniversary reviews have affirmed the universal commitment to the agreements forged at the 1994 conference.

Demographic Dividend. An interesting thing happens as a country goes through the demographic transition. As birthrates decline, the working-age population increases relative to the younger and older members of the population. This relationship is known as the **dependency ratio** and is defined as the ratio of the nonworking population (under 15 and over 65) to the working-age population (see Table 8–3). For a time, therefore, the society can spend less on new schools and old-age medical expenses and more on factors that will alleviate poverty and generate economic growth—a *demographic dividend*. This demographic window is open only once, however, as the numbers of younger children decrease, and stays open only for a few decades, until the numbers of older people increase.

Figure 9–4 shows how dependency ratios have changed over time—and will likely change in the future—for several world regions. A number of East Asian and Latin American countries, such as South Korea, Brazil, and Mexico, have taken advantage of this window. They have put in place population and economic policies that are consistent with poverty reduction and economic expansion, and as a result, they have experienced rapid economic development. The countries of South Asia (for example, Pakistan, India, and Bangladesh) are currently approaching this window of opportunity. Will they make the necessary investments in health, education, and economic opportunities to take full advantage of the demographic window? To do so, they must directly help the poorest people in their societies.

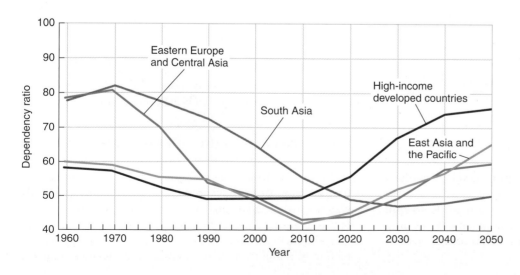

Figure 9–4 The demographic window. As countries experience gradually decreasing fertility, the dependency ratio declines and opportunities for development increase. The plot compares several developing regions with some developed countries. (*Source: World Development Report 2003.*)

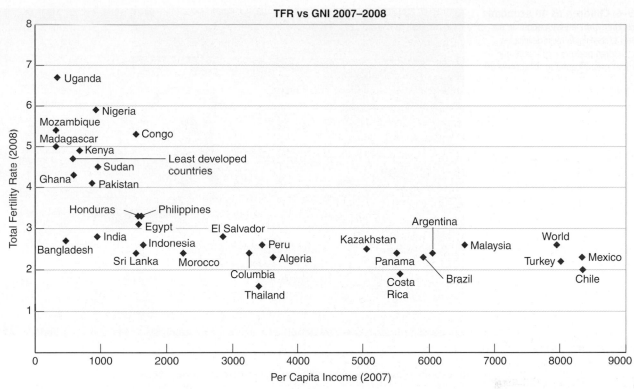

Figure 9–5 Fertility rate and income in selected developing countries. There is a correlation between income (gross national income in purchasing power parity per capita) and lower total fertility. Factors that affect fertility more directly are health care, education for women, and the availability of information and services related to contraception. (*Source:* Data from 2008 *World Population Data Sheets* (Washington, D.C.: Population Reference Bureau, 2008); *World Development Report 2009* (Washington, D.C.: World Bank, 2009).)

Large and Small Families

The fertility transition is the most vital element in the demographic transition. Many studies have shown that high fertility and poverty are linked, including the correlation between fertility rate and gross national income per capita (Fig. 9–5). Countries on the far left side of this plot are all in the early or middle stages of the demographic transition; most are low-income economies. Those in the middle and to the right are all middle-income economies and are moving through the transition.

To the observer in a developed country, it may seem strange that so many poor women in developing countries have large families. It is obvious from our perspective that more children could spread a family's income more thinly and handicap efforts to get ahead economically. What we fail to recognize, however, is that the poor in developing countries live under very different sociocultural conditions. Numerous studies and surveys reveal the following as primary reasons that the poor in developing countries have large families:

1. **Security in one's old age.** A traditional custom and need—because there is no "safety net" in many developing countries—is that old people will be cared for by their children. Therefore, a primary reason given by poor women in developing nations for desiring many children is "to ensure my care in old age."

2. **Infant and childhood mortality.** Closely coupled with the desire for security in one's old age is the experience of high infant and childhood mortality. The common and often personal experience of children dying leads people to try to make sure that some of their children will survive to adulthood. In Nigeria, for example, 100 infants die out of every 1,000 born.

3. **Helping hands.** In subsistence-agriculture societies, women do most of the work relating to the direct care and support of the family. A child as young as 5 can begin to help with many of the chores, and 12-year-olds can do an adult's work (Fig. 9–6). In short, children are seen as a productive asset.

4. **Status of women.** The traditional social structure in many developing countries still discourages women from obtaining higher education, owning businesses or land, and pursuing many careers. They are forced to stay at home and work for no pay. Often, respect for a woman grows higher the more children she bears.

5. **Availability of contraceptives.** Poor women often lack access to reproductive health information, services, and facilities. Providing contraceptives to women is a major facet of family planning. Studies show a strong correlation between lower fertility rates and the percentage of couples using contraception (Fig. 9–7). In fact, each 16% increase in contraceptive use translates into one

Figure 9–6 Children as an economic asset. Children working with adults in the fields in Bali, Indonesia. In most developing countries, children perform adult work and thus contribute significantly to the income of the family.

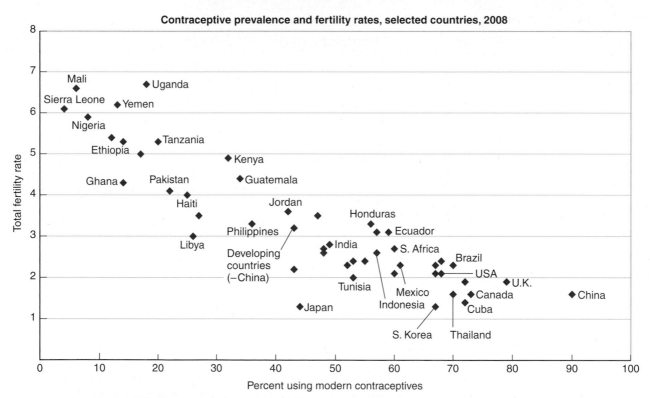

Figure 9–7 Prevalence of contraception and fertility rates. More than any other single factor, lower fertility rates are correlated with the percentage of the population using contraceptives. (*Source:* Data from *2008 World Population Data Sheets* (Washington, D.C.: Population Reference Bureau, 2008).)

less child. The decline in fertility rates in the 1980s and 1990s is directly correlated with a rise in the percentage of couples using contraception (rising from 38% in 1980 to 60% in the late 1990s).

Perhaps the most profound finding in surveys of women in the developing world is that large numbers state that they

want to delay having their next child or that they do not want any more children. Many of these women are not using contraceptives, however, because contraceptives are frequently unavailable or too expensive.

Conclusions. The five factors supporting large families are common to preindustrialized, agrarian societies. With

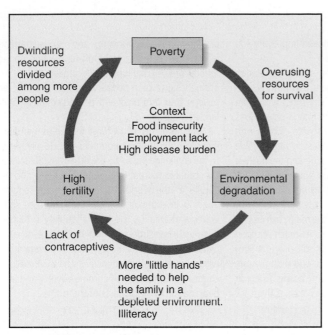

Figure 9–8 The poverty cycle. Poverty, environmental degradation, and high fertility rates become locked in a self-perpetuating vicious cycle.

industrialization and development, however, come conditions conducive to having small families. These conditions include the relatively high cost of raising children, the existence of pensions and a Social Security system, opportunities for women to join the workforce, free access to inexpensive contraceptives, adequate health care, wide educational opportunities and high educational achievement, and an older age at marriage.

Vicious Cycle. Fertility rates in developing countries remain high not because people in those countries are behaving irrationally or irresponsibly, but because the sociocultural climate in which they live has favored high fertility for years and because contraceptives are often unavailable. Furthermore, these are usually impoverished people in a context of low food productivity, a heavy disease burden, and lack of employment. Their poverty, environmental degradation, and high fertility drive one another in a vicious cycle (Fig. 9–8): Increasing population density leads to a greater depletion of rural community resources like firewood, water, and land, which encourages couples to have more children to help gather resources, and so on. They are in a poverty trap (See Stewardship, "Poverty Traps and Natural Resources Management").

Stewardship

Poverty Traps and Natural Resources Management

CHRIS BARRETT

Cornell University

Dr. Chris Barrett is Professor of Applied Economics and Management and Co-Director of the African Food Security and Natural Resources Management Program at Cornell University. He holds degrees from Princeton, Oxford, and Wisconsin–Madison and previously served on the faculty at Utah State University and as a staff economist with the Institute for International Finance. He is co-editor of the American Journal of Agricultural Economics *and serves as an associate editor or editorial board member of* Environment and Development Economics, *the* Journal of African Economies, *and* World Development.

The poorest people in the world rely disproportionately on the natural resource base, earning their living from agriculture, fishing, forestry, and hunting. Thus, questions of human development and the alleviation of poverty cannot be separated from issues of environmental management. The two are inextricable.

Some poverty, like some environmental problems, is short-lived. People lose a job unexpectedly, suffer from a drought, or endure some other shock that sets them back temporarily. In most of these cases, though, people can and usually do recover. There is remarkable inherent resilience in people, just as there is in many ecosystems.

Many of the rural poor are not so lucky. Their poverty is chronic, not transitory, reflecting low productivity due to their lack of productive assets such as land, livestock, and capital and the meager returns they earn on the few assets they do own. Lacking access to credit, their only path out of long-term poverty would require unrealistic levels of personal sacrifice and savings in order to make productivity-enhancing investments in livestock, new technologies, or education. Indeed, in many cases, poor farmers cannot even afford to maintain current productivity by investing in replenishment of the soils on which they heavily depend. Instead, they must rely ever more heavily on the natural resource base, mining it for current subsistence at the cost of future resource productivity and continued poverty. Partly as a consequence, nearly two-fifths of the world's agricultural land is seriously degraded, with this figure higher and growing in many of the poorest areas of the world. The chronic poor thus appear trapped in a downward spiral of poverty and resource degradation.

Poverty traps ultimately depend on the existence of threshold effects. Threshold effects exist when people with very similar initial positions follow different trajectories. While farmers above a threshold can accumulate capital and grow their way out of poverty over time, those caught below the threshold are unable to accumulate productive capital or to adopt improved tech-

Witchweed (Striga hermonthica), a parasitic weed that can devastate maize crops in Africa. (*Source:* L. J. Musselman from the Striga Photo Gallery of Species, www.science.siu.edu/parasitic-plants/Scrophulariaceae/Striga.Gallery.html.)

nologies or natural resources management practices. They are then more likely to exit farming or to be forced to farm ever more marginal or fragile land. A vicious cycle of resource degradation and productivity decline ensues.

We see such patterns among farmers in Madzuu, a poor community in the Vihiga District of western Kenya, where high population density—more than 1,100 people/km²—and a lack of

(Continued)

(Continued)

employment options outside of farming have decreased average farm size to only 0.3 hectares (about three-quarters of an acre). Seventy-one percent of Madzuu households survived on the equivalent of $.50/day per person or less in 2002. Madzuu maize farmers harvest only about one-seventh as much corn per acre, on average, as U.S. farmers. This is largely due to soils that have become heavily degraded through excessive, continuous cultivation and few nutrient amendments through inorganic fertilizers and the application of manure. Another reason for the low yields is infestation by a parasitic plant, Witchweed. This weed infests nutrient-depleted unirrigated soils in western Kenya.

The poor cannot afford to make such investments. Income distribution data indicate that the incomes of households suffering soil degradation between 1989 and 2002 were much lower than the incomes of those who enjoyed improving soil quality over the same period.

Part of this is due to the strong correlation between educational attainment and both household income and farm productivity. Those with more education are more likely to find good-paying jobs off-farm that provide a steady source of cash with which to buy fertilizers and invest in higher-value products such as exotic dairy cattle and tea bushes. As compared to other Madzuu households, those who had completed secondary school were 5 times more likely to own improved dairy stock, were 18 times more likely to grow tea (the main cash crop in the region), and applied 4 times as much nitrogen fertilizer per hectare (although still only 19 kilograms/hectare (17 lb/acre), about one-sixth the rate of American farmers). With scant access to credit with which to invest in key assets such as livestock or productivity-enhancing and soil-sustaining inputs such as fertilizer, the off-farm labor market for educated, skilled workers becomes a key source of financing to keep people from falling below the poverty trap threshold and degrading their farmland in Madzuu.

The problems of natural resources management in poor rural communities are closely linked to the challenge posed by poverty traps. Sustaining the valuable natural capital on which the poorest people in the world especially depend requires ongoing investment in maintaining forests, soils, water, and wildlife. The poor's inability to undertake such investments—and the failure of compensatory action by governments, private businesses, and nonprofit agencies to fill the breach—creates a vicious cycle of chronic poverty and land degradation in many parts of the tropics.

Thus, it is not just economic development that leads to declining fertility rates. Rather, fertility rates decline insofar as development provides (1) security in one's old age apart from the help of children, (2) lower infant and childhood mortality, (3) universal education for children, (4) opportunities for higher education and careers for women, and (5) unrestricted access to contraceptives and reproductive health services. These are all outcomes of public policies, and countries and regions (like Kerala) that have made them a high priority have moved into and through the demographic transition.

9.2 Promoting Development

The good news is that many developing countries have made remarkable *economic* progress. The gross national products of some countries have increased as much as fivefold, bringing them from low- to medium-income status, and some medium-income nations have achieved high-income status. Although the world economy is still strongly dominated by the developed countries, the developing countries have become more and more involved in what is now an integrated global economy. At the heart of this economy is technology, and as Figure 9–3 indicates, those developing regions showing the highest growth in income also show the greatest technological progress. There is still a great gap between rich and poor countries in income, as well as technology, but the percentage of impoverished people in the world—those living on less than **$1.25 per day** (the new *extreme poverty* criterion)—has dropped from 42% to 21.3% of the developing world's population in the 19 years since 1990. These percentages correspond to 1.8 billion people in 1990 and 1.2 billion in 2009.

Tremendous *social* progress has been made in many developing countries, too. Efforts from many branches of the United Nations have aided the work of various government programs. Private charitable organizations and nongovernmental organizations (NGOs) have also played a large role. Literacy rates, the percentage of the population with access to clean drinking water and sanitary sewers, and other social indicators of development have generally improved. Furthermore, the fertility rates of most developing countries have declined, although they are still far from the replacement level (Table 9–1).

The successes just described, however, are dulled by some sobering facts. A sixth of the world's population (1.2 billion people) still lives on less than $1.25 a day, nearly 1 billion lack access to clean water, over 1 billion are living in urban slums under deplorable conditions, and over 1 billion are

TABLE 9–1 Decline in Total Fertility Rates

	1985	1990	1995	2000	2005	2008
Africa	6.3	6.2	5.8	5.3	5.1	4.9
Latin America and the Caribbean	4.2	3.5	3.1	2.8	2.6	2.5
Asia (excluding China)	4.6	4.1	3.5	3.3	3.0	2.8
China	2.1	2.3	1.9	1.8	1.6	1.6
Developed countries	2.0	2.0	1.6	1.5	1.5	1.6

Source: Data from various *World Population Data Sheets* (Washington, D.C.: Population Reference Bureau).

malnourished. These are the world's poorest people. As a consequence of the deficiencies they face, the poor suffer from higher rates of disease, infant mortality, and illiteracy; lower life expectancy; poor living conditions; social exclusion; and powerlessness. As Nelson Mandela, former leader of South Africa, put it, "Massive poverty and obscene inequality are such terrible scourges of our times—times in which the world boasts breathtaking advances in science, technology, industry and wealth accumulation—that they have to rank alongside slavery and apartheid as social evils."[2]

Is the solution to these problems simply to facilitate economic growth and the alleviation of poverty will follow? The United Nations Fund for Population Activities (UNFPA) puts it the following way in *State of the World Population 2002*: "Overall economic growth is not enough: it requires directing development efforts to the poor.... Economic growth will not by itself end poverty.... Ending extreme poverty calls for commitment to the task, and specific action directed to it." This is the explicit objective of the Millennium Development Goals.

Millennium Development Goals

In 1997, representatives from the United Nations, the World Bank, and the Organisation for Economic Co-Operation and Development (OECD) met to formulate a set of goals for international development that would reduce the extreme poverty in many countries and its various impacts on human well-being. The goals were sharpened and then adopted as a global compact by all UN members at the UN Millennium Summit in 2000. Table 9–2 lists the 8 *Millennium*

[2]Speech in London's Trafalgar Square for the Campaign to End Poverty in the Developing World. BBC News, February 3, 2005.

TABLE 9–2   Millennium Development Goals and Targets (Conditions in 1990 Are Compared with Those in 2015, the Year for Reaching the Targets)

Goal 1. Eradicate extreme poverty and hunger.
- **Target 1:** Reduce by half the proportion of people whose income is less than $1 (now $1.25) a day.
- **Target 2:** Reduce by half the proportion of people who suffer from hunger.

Goal 2. Achieve universal primary education.
- **Target 3:** Ensure that all children, boys and girls alike, are enrolled in primary school.

Goal 3. Promote gender equality and empower women.
- **Target 4:** Eliminate gender disparities in primary and secondary education by 2005 and in all levels of education by 2015.

Goal 4. Reduce child mortality.
- **Target 5:** Reduce mortality rates by two-thirds for all infants and children under five.

Goal 5. Improve maternal health.
- **Target 6:** Reduce maternal mortality rates by three-quarters.
- **Target 7:** Achieve, by 2015, universal access to reproductive health.

Goal 6. Combat HIV/AIDS, malaria, and other diseases.
- **Target 8:** Have halted, by 2015, and begun to reverse the spread of HIV/AIDS.
- **Target 9:** Achieve, by 2010, universal access to treatment for HIV/AIDS for all those who need it.
- **Target 10:** Have halted, by 2015, and begun to reverse the incidence of malaria and other major diseases.

Goal 7. Ensure environmental sustainability.
- **Target 11:** Integrate the principles of sustainable development into country policies and programs, and reverse the loss of environmental resources.
- **Target 12:** Reduce by half the proportion of people without sustainable access to safe drinking water and basic sanitation.
- **Target 13:** Achieve a significant improvement in the lives of at least 100 million slum dwellers by 2020.

Goal 8. Forge a global partnership for development.
- **Target 14:** Develop a trading system that is open, rule based, and nondiscriminatory—one that is committed to sustainable development, good governance, and a reduction in poverty.
- **Target 15:** With Official Development Assistance (ODA), address the special needs of the least developed countries (including the elimination of tariffs and quotas for their exports and an increase in programs for debt relief).
- **Target 16:** Deal with the debt problems of all developing countries by taking measures that will make debt sustainable in the long run.
- **Target 17:** In cooperation with the private sector, make available the benefits of new technologies, especially information and communications technologies.

Sources: United Nations, World Bank, Organisation for Economic Co-Operation and Development, Millennium Ecosystem Assessment.

Development Goals (MDGs) and 17 targets. As well as targets, each goal has definite indicators for monitoring progress. The indicators are measurable and provide feedback to all who are concerned about whether progress is being made toward the different goals.

The goals reinforce each other and should all be worked on together. For example, the indicators of progress for goal 4 (*reduce child mortality*) are the infant mortality rate, the under-five mortality rate, and the proportion of one-year-old children immunized against measles. Achieving goals 5 and 6 (*improve maternal health* and *combat HIV/AIDS, malaria, and other diseases*) should go far toward accomplishing goal 4.

Most importantly, the MDGs are a clear set of targets for the developing countries themselves. Each country is expected to work on its own needs, but it is expected that all of the countries will do so in partnership with developed countries and development agencies, consistent with goal 8 (*forge a global partnership for development*). The low-income countries simply lack the resources to go it alone. It is important to note that the MDGs are not focused on the economic development normally tracked by measures like gross national income. They specifically address extreme poverty wherever it is found. As the goals are achieved, however, their impacts will certainly improve economic development in the different countries.

The Millennium Project and MDG Support Team.

An unprecedented amount of global attention has been focused on the MDGs. Virtually every UN agency, scores of NGOs, and government agencies of many countries have produced strategies and directives that demonstrate how they are mobilizing to achieve the MDGs. In the midst of this activity, the **Millennium Project** was given the mandate to develop a coordinated action plan that other agencies and organizations can consult as they address the MDGs. The Millennium Project, commissioned by the UN Secretary-General in 2002, was an independent advisory body headed by economist Jeffrey Sachs, a world expert in poverty, and involved some 250 development experts from around the world. After working for three years, the final report—*Investing in Development: A Practical Plan to Achieve the Millennium Development Goals*—was released in January 2005. The plan spells out many objectives for practical projects that have been shown to work, such as promoting microlending (see Section 9.3), supplying fertilizer for farmers, eliminating school fees, and establishing school meal programs.

The successor to the Millennium Project is the **MDG Support Team**, operating under the mandate of the United Nations Development Program (UNDP). The MDG Support Team works with countries by invitation, helping them to prepare and implement national strategies for achieving the MDGs, developing tools and methods specific for that country, coordinating technical expertise across the UN family of agencies, and spotlighting successful MDG initiatives. The UNDP also works on obtaining the financing needed for support of MDG work in the recipient countries.

Progress Toward the MDGs.

A recent summary of global and regional progress toward the MDGs is found in *The Millennium Development Goals Report 2008*, produced by statisticians from a host of UN agencies. Now that we are more than halfway to the goals' target date of 2015, how is it going?

Extreme Poverty. The first goal (*eradicate extreme poverty and hunger*) aims to reduce by half the proportion of people whose income is less than $1 (now $1.25) a day. This goal has almost been met at the global level, but the 2008–2009 worldwide recession is expected to cause an additional 55 to 90 million people to slip below the extreme poverty level and push the number of hungry people to over a billion. Furthermore, there are major regions—in particular, sub-Saharan Africa and Southern Asia—that are far behind on this goal. Ironically, even if the MDG is achieved, there will still be almost a billion people living in extreme poverty and hunger.

Goals Affecting Children. Another goal that will apparently be met (because of progress already made) is goal 2 (*achieve universal primary education*). The goal that seems least likely to be achieved is goal 4 (*reduce child mortality*), with the target of reducing under-five mortality rates by two-thirds (Fig. 9–9). More than one-third of these deaths are associated with malnutrition, a clear link to goal 1, target 2 (*reduce by half the people who suffer from hunger*). In 2007, half of the 9.2 million child deaths occurred in sub-Saharan Africa. Many other regions are falling short of the target; of the major regions, only Latin America and the Caribbean, Eastern Asia, and Northern Africa seem to be on track. Of course, reaching the target means that 4 million children will still die of preventable causes every year, instead of the present 9.2 million.

Other Goals. *The Millennium Development Goals Report 2008* lists a number of additional successes:

- In all but two regions, primary school enrollment is 90% or higher.

- The number of deaths from measles has fallen dramatically; about 80% of children in developing countries now are immunized.

- At least 1.6 billion people have gained access to safe drinking water since 1990.

- The number of deaths from AIDS has declined, as has the number of those newly infected with HIV virus.

The report also lists areas where greater effort is required if goals are to be met:

- One-fourth of all children in developing countries are underweight, caused by undernourishment.

- More than 500,000 women die annually in childbirth or from complications from pregnancy.

- Some 2.5 billion people (almost half of the developing world) live without adequate sanitation.

- Foreign aid expenditures by the developed countries have been in decline for two years and risk falling short of commitments made in 2005.

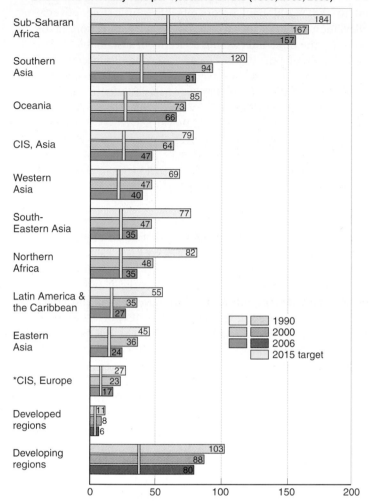

Under-five mortality rate per 1,000 live births (1990, 2000, 2006)

Sub-Saharan Africa: 184, 167, 157

Southern Asia: 120, 94, 81

Oceania: 85, 73, 66

CIS, Asia: 79, 64, 47

Western Asia: 69, 47, 40

South-Eastern Asia: 77, 47, 35

Northern Africa: 82, 48, 35

Latin America & the Caribbean: 55, 35, 27

Eastern Asia: 45, 36, 24

*CIS, Europe: 27, 23, 17

Developed regions: 11, 8, 6

Developing regions: 103, 88, 80

Legend:
- 1990
- 2000
- 2006
- 2015 target

Figure 9–9 Millennium Development Goal 4, Target 5. As a result of efforts toward this goal—*reduce child mortality*—deaths among children under five finally dropped below 10 million. About half of these deaths occur in sub-Saharan Africa. Death from measles has dropped dramatically; 80% of children in developing countries now receive a measles vaccine.
(*Source: Millennium Development Goals Report 2008* (New York: United Nations, 2008).)
*Commonwealth of Independent States (CIS) countries are Russia, Ukraine, Georgia, and other countries formerly in the Soviet Union.

Country-Level Reports. Global and regional reports are important in providing a general picture of progress being made toward the MDGs, but the real action is in the individual countries. Toward that end, some 140 countries now produce reports, coordinated and often funded by the UNDP. For example, Albania is carrying the MDG process to the local level and reporting results on the Internet. The country report highlights significant disparities among the country's different regions, and each region has addressed the MDGs in detail, providing strategies for achieving the goals and estimates of cost.

Paying for It. How much will it cost to achieve the MDGs by 2015, and how will these costs be met? It is the responsibility of the developing countries themselves to manage their own economic and social development, and therefore, they must take full responsibility for creating a climate that will facilitate the partnership envisioned in goal 8 of the MDGs. They must also devote significant financial resources toward achieving the goals. The best estimates of the Millennium Project suggest that for the lowest-income countries, the costs will have to be split evenly between domestic sources and Official Development Aid (ODA) from the wealthy donor countries. Both developing countries and donor countries will have to commit major new outlays of funds if the MDGs are to be met. At a September 2008 "high-level event" at UN Headquarters, a coalition of governments, foundations, and businesses pledged an estimated $16 billion in new commitments to meet the MDGs. Equally important is foreign investment and participation in global markets, which directly stimulate and help grow the economies of these countries. More important yet to the poorest people, however, is the work of various UN and private organizations that directly provide assistance and funnel aid to the developing countries. These agencies are discussed next, followed by the issues of development aid and the debt crisis.

World Agencies at Work

In 1944, during World War II, delegates from around the world met in New Hampshire and conceived a vision of development for the countries ravaged by World War II. They established the International Bank for Reconstruction and Development (now called the **World Bank**). The World Bank functions as a special agency under the UN umbrella, owned by the countries that provide its funds. As the developed countries of Europe and Asia got back on their feet, the World Bank directed its attention more toward the developing countries. With deposits from governments and commercial banks in the developed world, the World Bank now lends money to governments (and only governments) of developing nations for a variety of projects at interest rates somewhat below the going market rates. In effect, the World Bank helps governments borrow large sums of money for projects they otherwise could not afford.

Annual loans from the World Bank have climbed steadily, from $1.3 billion in 1949 to $24.7 billion in 2008. Through its power to approve or disapprove loans and the amount of money it lends, the World Bank has become the major single agency providing aid to developing countries for the past 50 years. How does the agency carry out its mission, and how does its record look?

What Is the World Bank? The World Bank is actually five closely associated agencies. The most prominent ones are the International Bank for Reconstruction and Development (IBRD) and the International Development Association (IDA). The IBRD aims at reducing poverty in middle-income and some of the stronger low-income countries through loans and advisory services. The IDA enables the World Bank to provide interest-free loans to the poorest countries. These countries are so poor that they would be unable to borrow on market terms. The International Finance Corporation (IFC) works through the private sector to promote economic development in countries that are investment risks. For example, the IFC loaned $11 million in 2006 to a Mauritanian company to finance the building of a modern ready-mix cement production plant in that country. The IFC also provides substantial advice and guidance to governments and businesses.

Bad Bank, Good Bank. It is inaccurate either to credit the World Bank for all the progress made in development or to blame it for all areas in which progress has been lacking. However, critics cite many examples wherein the bank's projects have actually exacerbated the cycle of poverty and environmental decline.

Bad Bank. For example, the World Bank loaned India nearly a billion dollars to create a huge electric-generating facility consisting of five coal-burning power plants at Singrauli and to develop six open-pit coal mines to support the plants. The new power goes to distant cities, however, and has done little for the poor, who cannot afford electrical hookups. Worse, the project displaced over 200,000 rural poor people who had farmed the fertile soil of the region for generations, and it moved them to a much less fertile area, but allowed them no say in the matter and gave them little, if any, compensation. In addition, the project has caused extensive air and water pollution. Hydroelectric dams in a number of countries have similarly displaced people and exacerbated their poverty.

This and many other past projects funded by the World Bank have been destructive to the environment and often to the poorer segments of society. For example, the World Bank funneled $1.5 billion into Latin America from 1963 to 1985 to clear millions of acres of tropical forests. Most of the cleared land was given over to large cattle operations that produced beef for export. In other countries, projects have emphasized growing cash crops for export and fostering huge mechanized plantations, while leaving the poor marginalized. At that time, the World Bank procedures considered environmental concerns as an add-on rather than as an integral component of its policies.

Good Bank. There has been a sea change in recent years in how the World Bank does business with the developing world. In the words of former World Bank President James D. Wolfensohn, "Poverty amidst plenty is the world's greatest challenge, and we at the Bank have made it our mission to fight poverty with passion and professionalism. This objective is at the center of all of the work we do." Toward that end, the bank helped initiate the Millennium Development Goals. It now requires *Poverty Reduction Strategy Papers* for all countries receiving IDA loans. The World Bank's *Country Assistance Strategy* identifies how its assistance reduces poverty and establishes and maintains sustainable development.

Environmental Strategy. As evidence of a change in policies, the World Bank adopted a new environmental strategy, *Making Sustainable Commitments: An Environmental Strategy for the World Bank*, in 2002. The stated goal of the new strategy "is to promote environmental improvements as a fundamental element of development and poverty reduction strategies and actions." The strategy "recognizes that sustainable development, which balances economic development,

social cohesion, and environmental protection, is fundamental to the World Bank's core objective of lasting poverty alleviation." Thus, the World Bank's new direction takes seriously the three intersecting concerns in sustainable development: development that balances economic, social, and environmental factors (see Figure 1–11). A new analysis of more than 11,000 World Bank projects funded to address poverty reduction or biodiversity found that these goals were often mutually reinforcing. The World Bank uses a rigorous process to rate its projects for overall performance and sustainability; the analysis showed that, depending on the different objectives, some 60 to 85% of all projects were rated as satisfactory or highly satisfactory.

Other UN Agencies. To quote its mission statement, the "UNDP is the UN's global development network, advocating for change and connecting countries to knowledge, experience and resources to help people build a better life." The agency has offices in 131 countries and works in 166. It helps developing countries attract and use development aid effectively. As we have seen, its network also coordinates global and national efforts to reach the MDGs. The UN Food and Agricultural Organization (FAO) and the World Health Organization (WHO) are other important agencies whose work will be discussed in future chapters.

Should We Keep the UN? The United Nations and its various agencies frequently come under attack in the United States, where critics charge that the organization is useless to us. These critics would like to reduce or eliminate foreign aid. The fact is that, without the work of the World Bank and UN agencies like the UNDP and the FAO, the world would be in much worse shape than it is. We are fortunate that we do not need development aid, but billions of others around the world who are much worse off depend on such aid for continued economic development, especially as it is reflected in the MDGs. This is not a luxury, but a necessity in this globally connected world where so many suffer from the effects of poverty.

We turn now to the debt crisis. To some extent, this crisis is an outcome of World Bank policies, but it is also a consequence of development aid in general.

The Debt Crisis

The World Bank, many private lenders, and wealthy nations lend money to developing countries. Theoretically, development projects are intended to generate additional revenues that would be sufficient for the recipients to pay back their development loans with interest. Unfortunately, however, a number of things have gone wrong with this theory, such as corruption, mismanagement, and honest miscalculations.

Over time, developing countries as a group have become increasingly indebted (Fig. 9–10). Their total debt reached $4.0 trillion in 2007. Interest obligations climb accordingly, and any failure to pay interest gets added to the debt, increasing the interest owed—the typical credit-debt trap. In 2007, the developing countries received $969 billion in new loans and paid back $672 billion in principal and interest on existing loans.

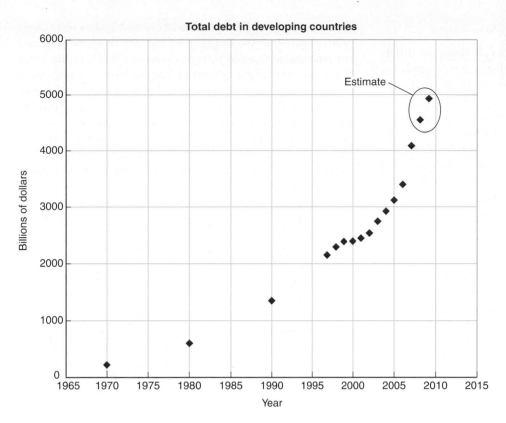

Total debt in developing countries

Figure 9–10 Total developing country debt. The developing countries have built up a debt of $4 trillion in the past 38 years. Figures for 2008 and 2009 are estimates. (*Source: World Economic Outlook 2008* (Washington, D.C.: International Monetary Fund, April, 2008).)

A Disaster. The debt situation continues to be an economic, social, and ecological disaster for many developing countries. In order to keep up even partial interest payments, poor countries often do one or more of the following:

1. **Focus agriculture on growing cash crops for export.** This clearly impacts the country's ability to feed its people.

2. **Adopt austerity measures.** Government expenditures are reduced so that income can go to pay interest. But what is cut? Usually, it is funds for schools, health clinics, police protection in poor areas, building and maintenance of roads in rural areas, and other goods and services that benefit not only the poor, but also the country as a whole.

3. **Invite the rapid exploitation of natural resources (for example, logging of forests and extraction of minerals) for quick cash.** With the emphasis on quick cash, few, if any, environmental restrictions are imposed. Thus, the debt crisis has meant disaster for the environment.

In essence, these measures are all examples of liquidating ecosystem capital to raise cash for short-term needs. They do not represent sustainability. As is typical with the credit trap, many countries have paid back in interest many times what they originally borrowed, yet the debt remains. Since the beginning of the global debt crisis in 1979, the developing countries have paid a cumulative $8.345 trillion in debt service and still have a current debt of $4.043 trillion and rising (Fig. 9–10). However, an important mitigating fact about this debt is that only 10% of this debt is owed by the low-income countries (those with gross national income per capita of

$936 or less), a consequence of recent debt cancellation. It had become obvious that many countries would never be able to repay their debts.

Debt Cancellation. The World Bank is addressing the problems of debt and poverty directly through two initiatives: the *Consultative Group to Assist the Poorest* (CGAP) and the *Heavily Indebted Poor Country* (HIPC) initiative. The CGAP is designed to increase access to financial services for very poor households through what is called *microfinancing* (to be discussed shortly). The HIPC initiative addresses the debt problem of the low-income developing countries (mostly from sub-Saharan Africa). To qualify, countries have to demonstrate a track record of carrying out economic and social reforms that lead to greater stability and alleviate poverty. By 2008, 33 countries had received debt relief totaling $105 billion, leaving a remaining debt of $9 billion in those countries. The intention of HIPC is to release resources that can then be used for development. In most countries that qualified for debt relief, spending on education and health has risen substantially.

Zambia. To demonstrate how HIPC has worked, consider the case of Zambia. Before HIPC, Zambia's debt was $7.1 billion. Through HIPC, that country's total debt was reduced to $500 million. Although this sounds like a tremendous break for Zambia, the country routinely experiences food shortages, it has a gross national income (GNI) of $800 per capita, and servicing this debt was costing the country twice what it spends on education, more than 7% of its GNI. Thus, the news of debt cancellation was received with joy in Zambia, and the government responded by introducing free health care for people in rural areas of the country.

Development Aid

College students and their parents know all about the differences among scholarships, grants, and loans as they apply for financial aid. Paying back college loans is no picnic for the thousands of students who accept them in order to afford their education. It is all called "aid," but the best kind of aid is the grant or scholarship, which does not have to be repaid. The developing countries are well aware of the differences, too, and have been receiving grant-type aid from private and public donors that is essential to their continued development. How much aid have they been given?

Official Development Assistance.   The best records are for official development assistance (ODA), coming from donor countries (Fig. 9–11). The total amount in 2007 was $93 billion, part of an encouraging trend following a decade of declining aid in the 1990s, but down from a record high in 2005. The 1990s decline likely occurred as a result of the end of the Cold War, since the West and the Communist states were no longer competing for influence in the developing countries. Studies have indicated that the allocation of aid often follows the interests of the donor countries.

As Figure 9–11 shows, aid as a percentage of the GNI of donor countries declined to 0.22% and then began to rise to its present rate of .28%. At the Rio Earth Summit of 1992, the aid target for donor countries was 0.7% of GNI. If aid had achieved this target in 2007, the amount given would have been $259 billion. Only Denmark, Sweden, Norway, the Netherlands, and Luxembourg have met or exceeded the 0.7% target. At the 2005 Paris Declaration on Aid Effectiveness, representatives from donor countries pledged to raise their ODA to .35% by 2010, well short of the 0.7% target, but a significant increase. They will have to get cracking, however, and increase ODA aid by $18 billion per year from 2008 to 2010 in order to reach that goal (Fig. 9–11).

Some of the shortfall in aid is explained as "donor fatigue." That is, rich countries have tired of handing out multimillion-dollar rescue packages, only to have a large percentage go to fill the pockets of corrupt leaders or to cover administrative costs. During the Cold War days, aid was used to erect palaces, build torture chambers, and line the Swiss bank accounts of many corrupt leaders. However, steps have recently been taken to minimize this misdirection of funds. Allocation of aid is increasingly being tied to policy reform in the recipient countries. The *Poverty Reduction Strategy Paper (PRSP)* approach of the World Bank is one example. In order to continue receiving aid, governments must prepare reports on policies that show real reform in addressing poverty and corruption.

Migration and Remittances.   Some 175 million people live outside the country of their birth, and many more second-generation immigrants maintain ties to their native countries. Most of these migrants left their home country looking for employment; wage levels in developed countries are five times those of the countries of origin of most migrants. The migrants typically send much of their money home—**remittances**, as they are called. This is a surprisingly large flow of funds—estimated at $240 billion in 2007 and more than double the size of international aid flows. There is no doubt that these funds alleviate the severity of poverty in the developing countries and can be thought of as a "do it yourself" kind of development aid. Of course, this flow of funds represents a money drain on the economy of the developed countries, but the migrant and immigrant workers often take jobs no one else wants.

Aid from the United States.   How much aid does the United States give? Our ODA donations were $21.8 billion in 2007, the largest of any donor, but only 0.16% of our GNI. Norway, by contrast, gives the equivalent of .95% of its GNI in development aid. To put it in its proper perspective, though, contributions from religious organizations, foundations, corporations, nongovernmental organizations, and individuals in the United States provided some $36.9 billion in aid in 2007, outstripping our ODA. When remittances and private capital investments are added to these,

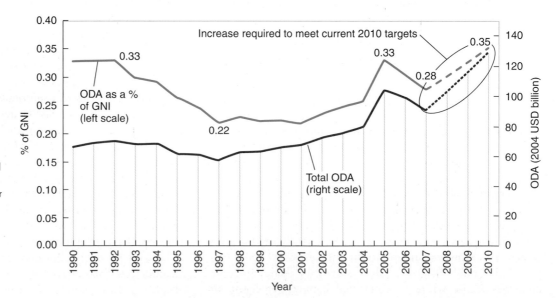

Figure 9–11 Official development assistance. Development assistance from donor countries in relation to their gross national income, 1990–2010 (projected). On the left axis, Official Development Assistance (ODA) is plotted as a percentage of gross national income of the donor countries. The right axis shows ODA in billions of dollars. The figures beyond 2007 are projections of what it would take to reach the 2010 targets of the donor countries (*Source:* Organisation for Economic Co-Operation and Development, *Aid Targets Slipping Out of Reach?* (Paris: OECD, December 2008).)

TABLE 9–3 U.S. Economic Contributions to Developing Countries, 2007

Type of Funding	Billions of $
Official Development Assistance (ODA)	21.8
Private Philanthropy	36.9
Foundations	3.3
Corporations	6.8
Private and Voluntary Organizations	10.8
Volunteerism	3.5
Universities and Colleges	3.9
Religious Organizations	8.6
Remittances	79.0
Private capital investments	97.5
U.S. Total Economic Engagement	235.2

Source: The Index of Global Philanthropy 2009, Executive Summary. The Center for Global Prosperity. Hudson Institute, 2009 (www .global-prosperity.org)

the total economic contributions from the United States to developing countries amount (for 2007) to $235 billion (Table 9–3), which is actually 1.35% of our 2007 GNI. However, the global financial crisis that began in late 2007 will make it unlikely that the United States and other developed countries will be increasing their aid anytime soon. The financial crisis will likely impact the developing countries in other ways.

Global Financial Crisis. Developing world economies experienced record economic growth during 2002–2007, a result of increased export earnings and substantial foreign direct investment. The collective GNI of developing countries grew more than 5% per year, peaking at 8% in 2006 and well outpacing the developed countries. However, in late 2007, the U.S. housing bubble burst, leading to a collapse of financial markets that spread to other developed countries. The United States fell into a severe recession, and the recession spread globally. This global crisis will likely impact the developing countries in a number of ways, even though they are not as vulnerable as the developed countries to the meltdown of financial markets. There will be a reduction in their exports, there will be much less foreign investment in their markets, and the GNI growth they had recently experienced will likely decline. The developed countries may be tempted to reduce development aid and raise barriers to trade from developing countries. Also, as unemployment grows, foreign-born workers are likely to be impacted and will be less able to send remittances to their countries of origin.

Even though foreign assistance is vital, the developing countries themselves must also address the root causes of poverty, and this usually means public policy changes at more local levels. To illustrate, the discussion in the next section assesses the relative roles played by development and family planning.

9.3 A New Direction: Social Modernization

As the case of Kerala demonstrates, demographic experts are recognizing that the shift from high to low fertility rates in the poorer developing countries does not require the economic trappings of a developed country. Instead, what is needed are efforts within the country made on behalf of the poor, with particular emphasis on the following points:

1. Improving education—especially literacy and the education of girls and women;

2. Improving health—especially lowering infant mortality and improving life expectancy;

3. Making family planning accessible (that is, both available and affordable);

4. Enhancing income through employment opportunities; and

5. Improving resource management (reversing environmental degradation).

In all of these areas, the focus should be on women because they not only bear the children, but also are the primary providers of nutrition, child care, hygiene, and early education. It is women who are most relevant in determining both the numbers and the welfare of subsequent generations.

Fortunately, the world is by no means starting from scratch in any of these areas. Numerous programs, both private and government funded, have been in existence for many years, and a wealth of experience and knowledge has been gained. The result of success in these efforts is called **social modernization**. We will look at each area in more detail.

Improving Education

The education in question is not college or graduate school—or even advanced high school. It is basic literacy—learning to read, write, and do simple calculations (Fig. 9–12). Illiteracy rates among poor women in developing countries are commonly 50–70%, in part because the education of women is not considered important and in part because expanding populations have overwhelmed school systems and transportation systems. Providing basic literacy will empower people to glean information from pamphlets on everything from treating diarrhea, to conditioning soils with compost, to baking bread. An educated populace is an important component of the wealth of a nation.

Investing in the education of children is a key element of the public-policy options of a developing country, one that returns great dividends. In 1960, for example, Pakistan and

Figure 9–12 Education in developing countries. Providing education in developing countries can be quite cost effective because any open space can suffice as a classroom and few materials are required. This is a class in Mumbai, India.

South Korea both had similar incomes and population growth rates (2.6%), but very different school enrollments— 94% in Korea versus 30% in Pakistan. Within 25 years, Korea's economic growth was three times that of Pakistan's, and now the rate of population growth in Korea has declined to 0.5% per year, while Pakistan's is still 2.2% per year. Kerala, where literacy is greater than 90%, is another case in point. The lowest net enrollment of children in primary schools (about 71%) is in sub-Saharan Africa, where 38 million children are still out of school. Remember, Millennium Development Goal 2 is to ensure that children everywhere receive a primary-school education. This is a worthwhile goal, but in some ways, it is insufficient. Evidence shows that achieving at least 50% secondary schooling will help large segments of a poor country's population out of poverty and will have a strong impact in decreasing fertility and improving the health of children.

Improving Health

The health care needed most by poor communities in the developing world is not high-tech bypass surgery or colonoscopy. Instead, it is the basics of good nutrition and hygiene—steps such as boiling water to avoid the spread of disease and properly treating infections and common ailments such as diarrhea. (In developing countries, diarrhea is a major killer of young children, but it is easily treated by giving suitable liquids, a technique called *oral rehydration therapy.*) The discrepancy in infant mortality rates (infant deaths per 1,000 live births) between the developed countries (6/1,000) and developing countries (54/1,000) speaks for itself.

Life Expectancy. One universal indicator of health is human life expectancy. In 1955, average life expectancy globally was 48 years. Today, it is 68 years and rising; it is

expected to reach 73 years by 2025. This progress is the result of social, medical, and economic advances in the latter half of the 20th century that have substantially increased the well-being of large segments of the human population. Recall from Chapter 8 the *epidemiologic transition*: the trend of decreasing death rates that is seen in countries as they modernize. Recall also that, as this transition occurred in the now-developed countries, it involved a shift from high mortality rates (due to infectious diseases) toward the present low mortality rates (due primarily to diseases of aging, such as cancer and cardiovascular diseases).

Two Worlds. The countries of the world have undergone the epidemiologic transition to different degrees, with very different consequences. Despite the progress indicated by the general rise in life expectancy, there is a 10-year difference in life expectancy in the developed countries (77 years) and that in the developing countries (67 years). The so-called least developed countries, totaling 1.14 billion people, have a life expectancy of only 55 years. The discrepancy between the most developed and least developed countries can be seen quite clearly in the distribution of deaths by main causes, shown in Figure 9–13. Infectious diseases were responsible for 47% of mortality in the least developed countries, compared with only 6% in the developed countries.

The solution is very evident: Primary health services need to be centered on people's needs and expectations, they must provide health security in the context of people's communities, and there must be universal access to primary health care. Disease prevention and the promotion of health are critical components of this primary health care. Clear focus on these issues has come from the MDGs, and the attention given to achieving the MDGs is improving the delivery of health care to people in the low-income countries.

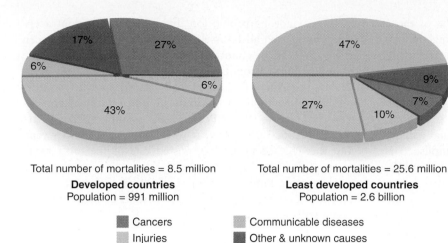

Total number of mortalities = 8.5 million
Developed countries
Population = 991 million

Total number of mortalities = 25.6 million
Least developed countries
Population = 2.6 billion

■ Cancers   ■ Communicable diseases
■ Injuries   ■ Other & unknown causes
■ Cardiovascular

Figure 9–13 Leading causes of mortality in the most developed and least developed countries. Infectious diseases are predominant in the least developed countries, while cancer and diseases of the cardiovascular system predominate in the highly developed countries. (*Source:* Data from World Health Organization, *Projections of Mortality and Burden of Disease, 2002–2030* (2008).)

Health care in the developing world must especially emphasize pre- and postnatal care of the mother as well as that of the children. Maternal mortality is more than 900 per 100,000 live births in a number of developing countries where maternal health care is almost nonexistent. Many government, charitable, and religious organizations are involved in providing basic health care, and when this is extended to rural countrysides in the form of clinics, it is one of the most effective ways of delivering primary health care to families. When women are better off, whole populations are improved. This is the focus of reproductive health.

Reproductive Health. A fairly new concept in population matters, *reproductive health* was strongly emphasized at the International Conference on Population Development held in Cairo, Egypt, in 1994. Reproductive health focuses on women and infants and includes the following major elements:

• Prenatal care;

• Safe childbirth and postnatal care;

• Information and services pertaining to contraception;

• Prevention and treatment of sexually transmitted diseases (STDs);

• Abortion services (where legal) and care afterward;

• Prevention and treatment of infertility; and

• Elimination of violence against women (coercive sex, rape), sexual trafficking, and female circumcision and infibulation (traditional practices in some African societies).

Reproductive health care underpins virtually all of the MDGs, and especially goals 4, 5, and 6. Goal 5 (*improve maternal health*) was expanded in 2005 to include reproductive health services, and the new target is to achieve universal access to reproductive health by 2015. It must be addressed particularly to the needs of young people. This is often a controversial issue, but sex education for adolescents is a very real need, and not just in the developing

world. Premarital sex is on the rise in all regions of the world, resulting in unplanned pregnancies, out-of-wedlock births, unsafe abortions, and STDs. The chilling impact of AIDS has given a new urgency to health care and sex education.

AIDS. One of the greatest challenges to health care in the developing countries is the sexually transmitted disease known as **acquired immune deficiency syndrome (AIDS)**. Unfortunately, the global epidemic of AIDS is most severe in many of the poorest developing countries, which are least able to cope with the consequences. In 2007, more than 90% of the 33 million people infected with the human immunodeficiency virus (HIV) lived in the developing countries. In sub-Saharan Africa, the virus is spread primarily by heterosexual contact, resulting in high incidences of female infection, which often gets transmitted to their children. The incidence of infected adults is more than 18% in countries such as Botswana, Namibia, and South Africa. The worst case is Swaziland, a small country in southern Africa, where the incidence is 26%. There, life expectancy is only 33 years as a consequence of the AIDS epidemic. The news is not all bad, however. In 2007, the number of new infections remained steady at 2.7 million, the number of deaths dropped to 2 million, and the number of HIV-positive people in the developing countries who are receiving anti-HIV drug treatment rose to more than 3 million.

These positive results are the outcome of a remarkable increase in funding from the developed countries. Recall that target 8 of the MDGs is to have halted and begun to reverse the spread of HIV/AIDS by 2015. Major sources of funding—some $10 billion in 2007 for the developing countries—are the World Bank; the Global Fund to Fight AIDS, Tuberculosis and Malaria; the U.S. President's Emergency Plan for AIDS Relief (PEPFAR); and the local country budgets. PEPFAR has provided $15 billion over five years, and in 2008, Congress increased funding for the program (PEPFAR II) to $39 billion for the next five years. The funds are used for a combination of testing, treatment, and prevention. The 3 million receiving treatment in the low- and middle-income countries are only 30% of those who need it, however, which suggests that closing the gap would require many more billions. A better approach, according to those in the field, is to place greater emphasis on prevention.

Most prevention approaches involve getting people to change their sexual behavior—to practice "safe sex," limit their partners, avoid prostitution, and delay sexual activity until marriage. That these things can be accomplished is proven by the case of Uganda, where a 70% decline in HIV prevalence has been registered in recent years, linked to a 60% reduction in casual sex. Education and literacy are important parts of combating this social problem. The Ugandan government has mounted a national AIDS control program that includes a major education campaign. Recent evidence has revealed another effective preventive measure— namely, male circumcision. Studies have shown a 60% reduction in HIV risk in circumcised males involved in heterosexual relationships. Indirectly, male circumcision also protects more women than most other preventive strategies. The WHO is strongly recommending that countries offer male circumcision as one of their anti-AIDS services.

Family Planning

For those who can pay, information on contraceptives, related materials, and treatments are readily available from private doctors and health care institutions. The poor, however, must depend on family-planning agencies, which are supported by a combination of private donations, government funding, and small amounts the clients may be able to afford. The stated policy of family-planning agencies (or similar private services) is, as the name implies, to *enable people to plan their own family size—that is, to have children only if and when they want them*. In addition to helping people avoid unwanted pregnancies, family planning often involves determining and overcoming fertility problems for those couples who are having reproductive difficulties. Family planning is a critical component of reproductive health care.

More specifically, family-planning services include the following:

- Counseling and education for singles, couples, and groups regarding human reproduction, the hazards of sexually transmitted diseases (AIDS, in particular), and the benefits and risks of various contraceptive techniques;

- Counseling and education on achieving the best possible pre- and postnatal health for mother and child, with the emphasis on good nutrition, sanitation, and hygiene;

- Counseling and education to avoid high-risk pregnancies— those that occur when a woman is too young or too old or that follow too closely on a previous pregnancy— which seriously jeopardize the health, and even the life, of the mother; and

- Contraceptive materials or treatments after people have been properly instructed about all alternatives.

The vigorous promotion and provision of contraceptives has proven all by itself to lower fertility rates, as seen in Fig. 9–7. Those countries that have implemented effective family-planning programs have experienced the most rapid decline in fertility. For example, Thailand initiated a vigorous family-planning program as part of a national population policy in 1971. Population growth subsequently declined from 3.1% per year to the present 0.5% per year, and the Thai economy has posted one of the most rapid rates of increase over the same years. Encouraging and implementing family planning is the first and most important step a country can take to improve its chances of developing economically.

Unmet Need. A society's fertility is determined not only by the choices couples make to have children, but also by births that are unplanned. Women who are not currently using contraception, but who want to postpone or prevent childbearing, are said to have an *unmet need*. Thus, both the spacing of children and the limiting of family size are involved. One major goal of family-planning policies and agencies is to ensure that all those who want or need reproductive health services actually have access to them. This goal is far from achieved in many parts of the developing world. The unmet need for family planning ranges from 10% to 50% of women in the different developing countries and affects an estimated 120 million women. Where women's unmet need is minimal, fertility rates are invariably lower. The governments of countries that want to lower their fertility know that meeting the unmet need for family-planning services is perhaps their most crucial objective. Indeed, much international aid is targeted at helping these countries provide reproductive health services to their families.

Abortion. Unfortunately, to many people, family planning conjures up images of the abortion clinic. Abortions, by definition, are terminations of unwanted pregnancies. Nearly everyone agrees that an abortion is the least desirable way to avoid having an unwanted child—especially in view of the other alternatives that are available. The document resulting from the Cairo population conference explicitly states that abortions should *never* be used as a means of family planning. Therefore, it is particularly important to understand that the primary functions of family planning are education and the provision of services directed at avoiding unwanted or high-risk pregnancies. If family-planning services were universally available and people availed themselves of these services, there would be far fewer unwanted pregnancies and, hence, fewer abortions.

How Many? According to the WHO, 219 million women become pregnant each year (Fig. 9–14). Of these, 46 million (21%) resort to abortion. More than one-third (20 million) of these abortions are unsafe, performed by people lacking the necessary skills or in medically deficient settings. As a result, more than 95% of the 77,000 women who die annually from such unsafe abortions are in the developing countries. This disturbing recourse to abortions

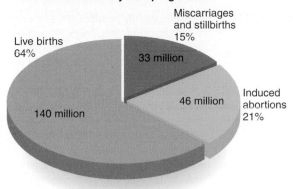

The outcomes of a year's pregnancies

Miscarriages and stillbirths 15%
33 million

Live births 64%
140 million

Induced abortions 21%
46 million

Figure 9–14 Outcome of the world's pregnancies. Abortions represent 21% of the pregnancies around the world. (*Source: World Health Report 2005, Updated*).

can be seen as a consequence of the lack of family-planning education and services. All studies show that cutbacks in family-planning services result in *more* unwanted pregnancies and *more* demand for abortions, not *less*. Indeed, after the Reagan administration cut off family-planning monies in 1984, abortions in the United States increased an estimated 70,000 a year.

In the United States, the legal abortion rate declined in the late 1990s and is currently at 19.4 per 1,000 women of childbearing age (1.2 million abortions in 2005), the lowest level since 1976. Much of this decline is traced to the increased availability of *emergency contraception*—methods of preventing pregnancy within a few days after unprotected

sexual intercourse. The most common technique is a pill based on the same hormones used in ordinary birth control. These pills (Plan B and Preven®) are far less controversial than the drug mifepristone (formerly known as RU-486), a nonsurgical alternative to early abortions. Plan B and Preven® were approved by the U.S. Food and Drug Administration (FDA) in the late 1990s for prescription use. An appeal to make Plan B available over the counter (OTC) led in 2005 to a controversial FDA decision to deny permission. The FDA has since reversed that decision, and now women 17 years of age or older have OTC access to the drug. Younger women must consult a health care professional to get access.

Family-Planning Agencies. Planned Parenthood, which operates clinics throughout the world, is probably the best known family-planning agency. Another significant player is UNFPA, which provides financial and technical assistance to developing countries at their request. The emphasis of UNFPA is on combining family-planning services with maternal and child health care and expanding the delivery of such services in rural and marginal urban areas. Support for this UN agency and other family-planning agencies has become a political football in the United States. The Clinton administration—through the Agency for International Development—tried to continue their funding, and many Republican members of the House opposed it. The argument, used since the Reagan years, is that these agencies promote abortions or, in the case of China, condone government efforts that have forced couples to have no more than one child. (See Sustainability, "China's Population Policies.")

Sustainability

China's Population Policies

Here are a few options associated with the limits of Earth's carrying capacity for the human species. What are the ethical and moral implications of each option?

1. You can argue, as some optimists do, that the problems do not really exist—that technology will always have a solution to make life better and better for more and more people.

2. Perhaps a disease, a famine, or a natural disaster—or even war—will come along and take care of the situation for us in "nature's way."

3. Some consider forced sterilization or abortion as a viable solution to the population problem.

Would you agree that these options lack credibility or compassion and are unacceptable to most people?

4. The most acceptable option—the one that has been the theme of this chapter—is to create an economic and social climate in which people, of their own volition, will desire to have fewer children. Along with such a climate, society will provide the means (family planning) to enable people to exercise that choice. This option has been realized "unconsciously" in developed countries, but it raises certain ethical dilemmas when you attempt to apply it to developing countries. For example, how far can you go in manipulating the social or economic environment before choice becomes coercion?

With its current population of 1.3 billion (a fifth of the world's people), China provides the most comprehensive example of a country that offers extensive economic incentives and disincentives aimed at reducing population growth. Some years ago China's leaders recognized that, unless population growth was stemmed, the country would be unable to live within the limits of its resources. Because of inevitable population momentum, the leaders felt that the country could not even afford a total fertility rate of 2.0. Thus, they set a goal of one child per family, and to achieve that goal, they instituted an elaborate array of incentives and deterrents. The prime *incentives* were as follows:

- Paid leave to women who have fertility-related operations—namely, sterilization or abortion procedures;

(Continued)

(Continued)

- A monthly subsidy to one-child families;
- Job priority for only children;
- Additional food rations for only children;
- Housing preferences for single-child families; and
- Preferential medical care to parents whose only child is a girl. (There is a strong preference for sons in China, and parents generally wish to have children until at least one son is born.)

Penalties for an excessive number of children in China included the following:

- Bonuses received for the first child must be repaid to the government if a second is born.
- A tax (now called a "social compensation fee") must be paid for having a second child.

- Higher prices must be paid for food for a second child.
- Maternity leave and paid medical expenses apply only to the first child.

Besides improving economic opportunities, these incentives and deterrents have helped China achieve a precipitous drop in its total fertility rate, from a rate of about 4.5 in the mid-1970s to a rate of 1.6 in 2008. The policy is unpopular, but without it, China's population would be 400 million larger. In practice, the one-child policy has been rigorously pursued only in urban areas, home to some 36% of the population. People in rural areas and minorities (who are exempted from the policy) have a higher fertility rate. In the last decade, rapid economic growth and more personal freedoms have made the Chinese people harder to control. Many are

willing to pay the penalties. It was the continued enforcement of the policy (including some enforced abortions and sterilizations) that prompted the Bush administration to withdraw funding from the UNFPA.

Despite these efforts, the population of China is still growing due to momentum. (A large percentage of the population is still at or below reproductive age.) Recently, the nation's top population official affirmed that the current policy would need to continue for at least 10 more years, when almost 200 million people will enter childbearing age. One disturbing consequence of China's policy is the skewed ratio of males to females. Males are preferred in Chinese families, so some parents try to make sure that their one child is a male, through such extreme measures as abortion and even infanticide of female children.

Starting in 2002, the Bush administration withheld funding to UNFPA (even though Congress voted $235 million in allocations) because of unsupported claims that funds were being used in China to support forced abortions and sterilizations. According to Americans for UNFPA, the funds withheld from UNFPA could have prevented 294,000 maternal deaths and allowed 82 million women to delay their first or next pregnancy. The last word in this battle comes from President Obama, who has taken steps to restore funding for UNFPA and overturn the Global Gag Rule.

The Global Gag Rule. The first act of President George W. Bush in 2001 was to reinstate the Mexico City Policy, known less formally as the "Global Gag Rule." This policy was first imposed by President Reagan in 1984. The gag rule prohibits any U.S. government aid from being given to foreign family-planning agencies if they provide abortions, counsel women about abortion if they are dealing with an unwanted pregnancy, or advocate for abortion law reforms in their own country. The rule is directed at NGOs like the International Planned Parenthood Federation. This rule is viewed by many as a serious blow to attempts to increase the availability of contraceptive services in the developing countries because the United States had been the largest international donor to family-planning programs. More recently, the Bush administration withdrew its support for the entire 1994 ICPD Program of Action because it uses terms such as *reproductive services* and *reproductive health care*. These are deemed to imply a right to abortion. Critics of this move pointed out (unsuccessfully) that the conference documents made clear that "in no case should abortion be promoted as a method of family planning" and that the Cairo agreement was critical to the fight against HIV/AIDS and prevented the very unwanted pregnancies that could lead to abortions.

Employment and Income

The bottom line of any economic system is the exchange of goods and services. At its simplest level, this entails a barter economy in which people agree on direct exchanges of certain goods or services. Barter economies are still widespread in the developing world.

The introduction of a cash economy facilitates the exchange of a wider variety of goods and services, and everyone may prosper, as they have a wider market for what they can provide and a wider choice of what they can get in return. In a poor community, everyone may have the potential to provide certain goods or services and may want other things in return, but there may be no money to get the system off and running. In a growing economy, people who wish to start a new business venture generally begin by obtaining a bank loan to set up shop. The poor, however, are considered high credit risks. Furthermore, they may want a smaller loan than a commercial bank wants to deal with, and many of the poor may be women, who are denied credit because of gender discrimination alone. For these three reasons, poor communities have trouble getting start-up capital. Fortunately, however, the situation is changing.

Grameen Bank. In 1976, Muhammad Yunus (Fig. 9–15), an economics professor in Bangladesh, conceived and created a new kind of bank (now known as the Grameen Bank) that would engage in **microlending** to the poor. As the name implies, microloans are small—they averaged just $330 in 2007—and are short-term loans, usually just four to six months. Nevertheless, they provide such basic things as seed and fertilizer for a peasant farmer to start growing tomatoes, some pans for a baker to start baking bread, a supply of yarn for a weaver, some tools for an auto mechanic, an inventory of goods to start a grocery store, and so on.

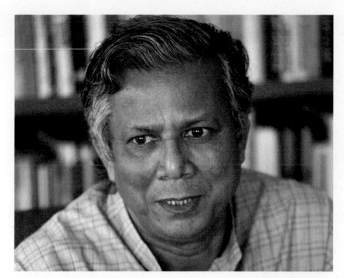

Figure 9–15 Muhammad Yunus. An economics professor in Bangladesh, Yunus created the Grameen Bank, which initiated the microlending movement. The Grameen Bank has lent over $6.6 billion over the past 35 years, and the microlending model has been duplicated in more than 40 countries. Yunus received the 2006 Nobel Peace Prize in recognition of the global impact of his work.

Yunus secured his loans by having the recipients form **credit associations**—groups of several people who agreed to be responsible for each other's loans. With this arrangement, the Grameen Bank has experienced an exceptional rate of payback—greater than 97%. Small-scale agriculture loans from the Grameen Bank have had outstanding results. In a rural area of Bangladesh, small loans, along with horticultural advice, are now enabling peasant farmers to raise tomatoes and other vegetables for sale to the cities. These people doubled their incomes in three years.

Microlending has been found to have the greatest social benefits when it is focused on women because, as Yunus observed, "When women borrow, the beneficiaries are the children and the household. In the case of a man, too often the beneficiaries are himself and his friends." The credit associations also create another level of cooperation and mutual support within the community, particularly when the loans are directed toward women. There are now 2,481 branches of the Grameen Bank in Bangladesh, covering over 80,000 villages.

More Microlending. The unqualified success of microlending in stimulating economic activity and enhancing the incomes of people within poor communities has been so remarkable that the Grameen concept has been adopted in more than 40 countries around the world (including the United States) with various modifications. Recently, the World Bank dramatically increased its support of microfinancing through the CGAP, which coordinates the work of some 54 different funding agencies around the world. A major development in this arena has been the work of the Microcredit Summit Campaign, an outgrowth of the 1995 UN Fourth Conference on Women in Beijing. An update from this campaign indicates that, by the end of 2007, some

3,300 institutions had granted loans to 154 million "poorest clients," 85% of whom were women. These efforts may be doing more than anything else to meet the MDGs, as they empower the poor to improve their own well-being.

Resource Management

The world's poor depend on local ecosystem capital resources—particularly water, soil for growing food, and forests for firewood. Many lack access to enough land to provide an income and often depend on foraging in woodlands, forests, grasslands, and coastal ecosystems. This activity generates income and is vital to those in extreme poverty; 90% of the world's poorest people depend on forests for some of their sustenance (by extracting fuelwood, construction wood, wild fruits and herbs, fodder, and "bush meat" for both subsistence and for cash). Forests, fisheries, reefs, grasslands, and waterways all can be resources for the poor—so-called common-pool resources (see Chapter 7). These can be a safety net, as well as an employment source, but if not managed in some way, common-pool resources are liable to overuse, especially when populations are increasing.

Effective strategies for managing natural resources often can be found in traditional knowledge, which can be shared broadly with the help of appropriate organizations (Fig. 9–16). "Farmer-to-farmer" advisory and training services are spreading with the help of NGOs and UN agencies. Empowering the poor to manage community- or state-owned lands is one approach that has worked in Nepal. There forestry user groups are given the right to own trees, but not the land, and there are now some 6,000 user groups managing 450,000 hectares (1.1 million acres). The groups develop forest-management plans, set timber sale prices, and manage the surplus income from the operations. Protecting forest resources and tackling land degradation in drylands are two sustainable steps that can be addressed by country policies.

Putting It All Together

Each of the five components of social modernization—education, health, family planning, employment and income, and resource management—depends on and supports the other components. For example, better health and nutrition support better economic productivity, better economic productivity supports obtaining a better education, and a better education leads to a delay in marriage and the desire to have fewer children. The availability of family-planning services is essential to realizing the desire of parents to have fewer children. All the components work together to alleviate the conditions of poverty, reverse environmental degradation, and reduce population growth. Conversely, the lack of any component—especially family-planning services—will undercut the ability to achieve all the other components.

The elements of social modernization are entirely compatible with the MDGs (Table 9–2). Although the MDGs do not directly address slowing population growth or making

Figure 9–16 Improving resource management. A major step in enhancing incomes and protecting the environment is to encourage better resource management. Here Eritrean women are planting mangrove seedlings in a restoration project to reclaim coastal land, generate wood for harvest, and provide leaves for animal fodder. (*Source:* Gordon Sato, by permission.)

the demographic transition, every one of the goals will be more reachable with continued progress in social modernization. The MDGs particularly address the needs of the poorer developing countries—the very ones that still have high fertility and stagnant economies.

One reality that the developing countries must confront in their progression toward economic and social development is globalization. What role can globalization play in bringing about social modernization, the demographic transition, and economic growth in the developing countries?

Globalization. Defined as *the accelerating interconnectedness of human activities, ideas, and cultures*, globalization is profoundly changing many human enterprises, especially in its economic impacts. A small number of developing countries (China, Malaysia, and Indonesia, for example) have successfully entered the global market, and more strive to do so. Their exports have paved the way for economic progress, and these countries are becoming more connected to the economic- and information-rich high-tech world.

The poorer countries, however, are mostly populated by self-employed peasants who have not made the transition to becoming wage earners in a market economy. This is not easy, and they can't do it by themselves. The jobs have to be there, and that often requires investment from outside agencies—businesses and NGOs with the knowledge and capital needed. Also, there has to be some hope of marketing the products, which means entering the global market on a level playing field. At present, however, the playing field is far from level because developed countries protect their exports with subsidies, quotas, and tariffs, and all of these trade barriers work against the developing countries (Chapter 2). Toward that end, goal 8 of the MDGs (*forge a global partnership for development*) calls for eliminating these barriers and encouraging the poor countries themselves to develop a market-based trading system committed to sustainable development and attractive to foreign investment.

Alternatively, many people become small-scale entrepreneurs with the aid of microlending agencies and never worry about the global market. This is an attractive alternative and one that is growing exponentially. Indeed, globalization can be positively harmful to the poor if it opens markets for agricultural products from the developed countries; it can ruin subsistence farming. Globalization can also widen the gap between the rich and poor within a society, as the wealthier members capitalize on the information flow and quick transfer of goods and services, leaving the poor farther behind. And as we have seen, when there is a global meltdown in financial markets, the developing countries are impacted in a number of ways.

Globalization is often presented as an unquestionable good, but critics claim that it doesn't seem capable of promoting some other goods, such as financial stability and social justice. It also has the potential to dilute and even destroy cultural and religious ideals and norms when it is uncritically welcomed into developing countries. Whether helpful or harmful, globalization is taking place, making it essential that the developing countries confront it head-on and adapt its connections to their needs and cultures.

Revisiting the Themes

Sound Science

Technological progress is one of the keys to economic development, but it depends on educating the population beyond the secondary level, something that many developing countries do not do well. The science of demography has brought us the insights of the demographic dividend, based on changes in dependency ratios. Medical advances and health care have helped bring many countries through the epidemiologic transition—however, there are still great needs for basic health care in the poorer countries. Sound science has brought us improved AIDS treatment through anti-HIV medicines, but so far has not come up with an effective vaccine.

Sustainability

The demographic transition is essential for a sustainable future. This chapter is about what has to be done economically and socially to bring about development that is sustainable. Economic progress is being made in many developing countries, and it is accompanied by most of the elements of social modernization. However, economic progress is still lacking in many others. For these, social modernization is the most feasible approach, as it will bring fertility rates down and enable countries to open the demographic window that may make economic progress possible.

An element of development activities that is unsustainable is the debt built up over the years by many poorer developing countries. Although this has been helped by much debt cancellation, the overall debt of developing countries continues to build up, and the repayment of these debts forces many developing countries to liquidate ecosystem capital in order just to make their payments.

One of the key MDGs is goal 7: *Ensure environmental sustainability*. This goal is facilitated by integrating principles of sustainability into country policies and programs and by reversing the loss of environmental resources. Improving the management of these resources is an important component of social modernization.

Stewardship

Justice for the developing world was one of the key ethical issues identified in the discussion of stewardship in Chapter 1. *Distributive justice* is an ethical ideal that considers the gross inequality of the human condition as unjust—the kind of inequality represented by the billion-plus people who live on less than $1.25 a day versus typical people in the wealthy countries. Although equal distribution of the world's goods and services is neither attainable nor necessary, the existing poverty in the poor countries is an affront to human dignity. In order to address this injustice, the alleviation of poverty must be made a priority of public policy. Aid from the rich countries to the poor countries is not an opportunity for advancing political goals, but rather a part of the obligation we have as human beings to look after and care for each other. As you read each of the MDGs, think about those whose well-being will be improved as the goals are being met.

REVIEW QUESTIONS

1. What are the two basic schools of thought regarding the demographic transition? How were these reflected in the three most recent global population conferences?

2. Discuss the five specific factors that influence the number of children a poor couple may desire.

3. What are the MDGs? Where did they come from? Cite an example of one goal and the target used to measure progress in attaining it.

4. What are two major agencies that promote development in poor nations, and how do they carry out their work?

5. What is meant by the debt crisis of the developing world? What is being done to help resolve this crisis?

6. What is development aid, and how does it measure up against the need for such aid?

7. What are the five interdependent components that must be addressed to bring about social modernization?

8. Define *family planning*, and explain why it is critically important to all other aspects of development.

9. What is meant by an unmet need? What is the Global Gag Rule?

10. What is microlending? How does it work?

11. What is the significance of globalization for economic and social development in the developing countries?

THINKING ENVIRONMENTALLY

1. Is the world population below, at, or above the optimum? Defend your answer by pointing out things that may improve and things that may worsen as the population increases.

2. Suppose you are the head of an island nation with a poor, growing population and the natural resources of the island are being degraded. What kinds of policies would you initiate, and what help would you ask for to try to provide a better, sustainable future for your nation's people?

3. List and discuss the benefits and harms of writing off debts owed by developing nations.

4. Imagine you are an official in a third-world country working to deal with an epidemic such as AIDS. Do you emphasize prevention or treatment? What priority does this take over other government tasks?

5. Do you believe contraceptives should be made available free of charge by governments, since this would have the potential to curb population growth and/or abortion rates? Defend your answer.

MAKING A DIFFERENCE

1. Think carefully about your own reproduction. What concerns will you and your mate weigh as you plan your family?

2. Become involved in the abortion debate, pro or con. Whether the United States supports international family planning, whether abortions remain legal in your state, and other issues will be determined by votes cast by you or your representatives.

3. Become involved in, and support, programs promoting effective sex education and responsible sexual behavior. Consider the advantages of abstinence and monogamy as ways to avoid needing an abortion or contracting a sexually transmitted disease.

4. If you are not convinced of the profound impacts of poverty in the developing world, log on to the Millennium Project Web site and read about the Millennium Village movement (www.unmillenniumproject.org/mv/index.htm).

5. Encourage sustainability by buying products that originate from the developing world.

Go to **www.masteringenvironmentalscience.com** for practice quizzes, Pearson eText, videos, current events, and more.

Satellite views of the Aral Sea: left, 1989; right, 2008. The southern sea is getting smaller every year and has split into two basins; the smaller northern sea is on the way to recovery. (*Source:* NASA Earth Observatory.)

KEY TOPICS

1. **Water: A Vital Resource: In the developed countries, a huge infrastructure of dams, irrigation systems, treatment plants, and pipelines has brought fresh water under control, while in developing countries, many people still lack access to safe drinking water.**

2. **Hydrologic Cycle: Natural Cycle, Human Impacts: Water circulates through the atmosphere, oceans, and land through evaporation, condensation, precipitation, and gravitational flow; human activities can create serious disruptions of this cycle.**

3. **Water: A Resource to Manage, a Threat to Control: Surface waters are dammed to meet the needs of irrigation, industry, and domestic uses and also to prevent devastating floods.**

4. **Water Stewardship: Supply and Public Policy: Human societies are placing increasing demands on water resources; five options for meeting these demands are explored.**

THE FORMER SOVIET UNION was responsible for one of the world's worst environmental disasters: the death of the Aral Sea, located in Central Asia and once the fourth largest lake on Earth. Soviet planners in the 1930s determined that the area surrounding the sea could grow cotton if it were irrigated. In turn, the cotton would earn valuable hard currency as an export crop. In time, the Amu Darya and the Syr Darya, the two large rivers flowing into the Aral, were tapped for irrigation water. By 1960, millions of hectares of land came under irrigation, the Soviet Union became the world's second largest cotton exporter, and the project was judged a huge success. However, from the 50 cubic kilometers of water the sea was receiving in 1965 (an already reduced amount from what it had received in the 1930s), inflowing water was reduced to zero by the early 1980s. Predictably, the sea, once 68,000 square kilometers (26,000 mi^2) in area, began to shrink.

IMPACTS. The most immediate impact was felt in the large fishery, an industry that had employed 60,000 people. Because essentially no new water was entering the lake, the salinity began to increase as water evaporated, devastating the lake's ecology. The commercial fishery collapsed by the early 1980s. In time, the lake lost 90% of its original volume, the water level dropped more than 53 feet, and as the photos here show, the surface area of the lake shrank greatly. The Aral Sea has separated into the Big Aral to the south (itself separated into a shallow east and deeper west basin) and the Small Aral to the north. The Big Aral is now three times as salty as the ocean and no longer has fish. There were additional impacts as well. The exposed lake bed (more than 54,000 km², or 21,000 mi²) now contains accumulated salts, and the dry winds of the area pick up the salt and dust, laden with fertilizers, herbicides, and pesticides, dropping them on the irrigated land and choking people for miles around. Many health impacts have been reported from the area, from chronic bronchitis to increased cancer rates. The local climate changed; summers are hotter, and winters are colder. Many animal species unique to the Aral Sea have become extinct.

REGIME CHANGE. With the collapse of the Soviet Union in 1991, the future of the Aral Sea is now in the hands of the republics of Uzbekistan and Kazakhstan. Turkmenistan, Tajikistan, and Kyrgyzstan take water from the Amu Darya and the Syr Darya for irrigation, so they are also involved. These five republics are all struggling economically, and the diverted waters support billions of dollars worth of agriculture annually, employing millions of people. There is little hope for restoring the Aral Sea to its former status because it would take stopping all irrigation from the two rivers for half a century just to increase the Big Aral's area to twice its present size. In the process, the surrounding countries' economies would be destroyed.

The Small Aral, however, is on the way to recovery as a result of an $85 million restoration project funded by the World Bank. The project erected a dam between the Small Aral and the Big Aral in 2005, and in less than a year, the northern portion filled with water from the Syr Darya, years ahead of expectations. The water's rise is covering some 1,000 square kilometers (390 mi²) of dry seabed. With the salinity falling to one-third that of seawater, freshwater fish have returned, and fishers caught 2,000 tons of fish in 2007. And with the early filling, water is now beginning to flow over the dam into the parched Big Aral. There is little hope, however, for recovery of the Big Aral anytime soon—too much water is still being diverted for irrigation. The eastern basin is now so salty that only brine shrimp can live in it.

OTHER DRY RUNS. The Aral Sea's fate is not unique. The Rio Grande, the second longest river in the United States, disappears about 300 feet from the Gulf of Mexico because its waters are drawn down all along its course for domestic use and irrigation. The mighty Colorado River is almost completely drained by the United States before it can reach Mexico, leaving a delta in the Gulf of California that has become a moonscape of baked mud. The Dead Sea is "dying" because its inflow of fresh water is diverted by both Israel and Jordan. The Dead Sea may disappear by 2050 unless seawater is pumped into it from the Red Sea, a prohibitively expensive project. In these and many other parts of the world, domestic use, industrial use, and especially agricultural use for irrigation are competing for an increasingly scarce supply of life-giving water. Chapter 10, the first of four chapters on essential resources, focuses on water.

10.1 Water: A Vital Resource

Water is absolutely fundamental to life as we know it. Happily, Earth is virtually flooded with water. A total volume of some 1.4 billion cubic kilometers (325 million mi³) of water covers 75% of Earth's surface. About 97.5% of this volume is the salt water of the oceans and seas. The remaining 2.5% is **fresh water**—water with a salt content of less than 0.1% (1,000 ppm). This is the water on which most terrestrial biota, ecosystems, and humans depend. Of the 2.5%, though, two-thirds are bound up in the polar ice caps and glaciers. Thus, only 0.77% of all water is found in lakes, wetlands, rivers, groundwater, biota, soil, and the atmosphere (Fig. 10–1). Nevertheless, evaporation from the oceans combines with precipitation to resupply that small percentage continually through the solar-powered hydrologic cycle, described in detail in Section 10.2. Thus, *fresh water is a continually renewable resource.*

Streams, rivers, ponds, lakes, swamps, estuaries, groundwater, bays, oceans, and the atmosphere all contain water,

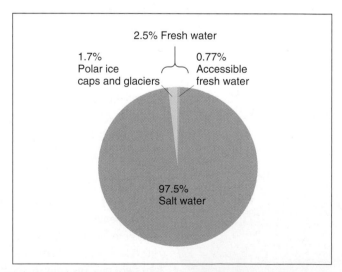

Figure 10–1 Earth's water. The Earth has an abundance of water, but terrestrial ecosystems, humans, and agriculture depend on accessible fresh water, which constitutes only 0.77% of the total.

and they all represent ecosystem capital—goods and services vital to human interests. (See Table 5–2 for the array of aquatic ecosystems and their characteristics.) They provide drinking water, water for industries, and water to irrigate crops. Bodies of water furnish energy through hydroelectric power and control flooding by absorbing excess water. They provide transportation, recreation, waste processing, and habitats for aquatic plants and animals. Fresh water is a vital resource for all land ecosystems, modulating the climate through evaporation and essential global warming (when water is in the atmosphere as water vapor). During the last two centuries, many of these uses (and some threats to them) have led us to construct a huge infrastructure designed to bring water under control. We have built dams, canals, reservoirs, aqueducts, sewer systems, treatment plants, water towers, elaborate pipelines, irrigation systems, and desalination plants. As a result, waterborne diseases have been brought under control in the developed countries, vast cities thrive in deserts, irrigation makes it possible to grow 40% of the world's food, and one-fifth of all electricity is generated through hydropower. These great benefits are especially available to people in the developed countries (Fig. 10–2a).

In the developing world (Fig. 10–2b), by contrast, 1.1 billion people still lack access to safe drinking water, 2.6 billion do not have access to adequate sanitation services, and more than 1.6 million deaths each year are traced to waterborne diseases (mostly in children under five). All too often in developing countries, water is costly or inaccessible to the poorest in society, while the wealthy have it piped into their homes. In addition, because of the infrastructure that is used to control water, whole seas are being lost, rivers are running dry, millions of people have been displaced to make room for reservoirs, groundwater aquifers are being pumped down, and disputes over water have raised tensions from local to international levels. Fresh water is a limiting resource in many parts of the world and is certain to become even more so as the 21st century unfolds.

Feeding a growing world population will challenge our water management infrastructure, especially as cities and industries compete with agriculture for scarce water resources. The growing impact of climate change on the hydrologic cycle is in turn worsening the problems brought on by droughts and floods. Rainfall variability can reduce the economic growth of nations as floods devastate some croplands and droughts dry them up in other parts of the world. The UN Environment Program Director, Achim Steiner, on World Water Day 2007, said, "If we want to avoid water scarcity as the permanent theme for the 21st century, a big part of the solution is cuts in greenhouse gas emissions of 60 to 80 percent."[1]

There are two ways to consider water issues. The focus in this chapter is on *quantity*—that is, on the global water cycle and how it works, on the technologies we use to control water and manage its use, and on public policies we have put in place to govern our different uses of water. Chapter 20 focuses on water *quality*—that is, on water pollution and its consequences, on sewage-treatment technologies, and on public policies for dealing with water pollution issues. Water—so vital to life—moves through the land, oceans, and atmosphere in a cycle.

---

[1]As reported in *Europaworld*, March 23, 2007.

(a)

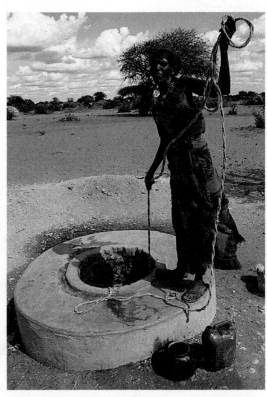

(b)

Figure 10–2 Differences in availability. Human conditions range from (a) being able to luxuriate in water to (b) having to walk long distances to obtain enough simply to survive. In many cases, however, the abundance of water is illusory because water supplies are being overdrawn. Go to **GRAPHit!** at www. masteringenvironmentalscience.com.

10.2 Hydrologic Cycle: Natural Cycle, Human Impacts

Earth's **water cycle**, also called the **hydrologic cycle**, is represented in Figure 10–3. The basic cycle consists of water rising to the atmosphere through *evaporation* and *transpiration* (the loss of water vapor as it moves from the soil through green plants and exits through leaf pores) and returning to the land and oceans through *condensation* and *precipitation*. Modern references to the hydrologic cycle distinguish between *green water* and *blue water*. Green water is water in vapor form, while blue water is liquid water wherever it occurs. Green water and blue water are vitally linked in the hydrologic cycle. Terms commonly used to describe water are listed and defined in Table 10–1.

Evaporation, Condensation, and Purification

Recall from Chapter 3 that a weak attraction known as hydrogen bonding tends to hold water molecules (H_2O) together. Below 32°F (0°C), the kinetic energy of the molecules is so low that the hydrogen bonding is strong enough to hold the molecules in place with respect to one another, and the result is *ice*. At temperatures above freezing, but below boiling (212°F or 100°C), the kinetic energy of the molecules is such that hydrogen bonds keep breaking and reforming with different molecules. The result is *liquid water*. As the water molecules absorb energy from sunlight or an artificial source, the kinetic energy they gain may be enough to allow them to break away from other water molecules entirely and enter the atmosphere. This process is known as **evaporation**, and the result is **water vapor**—water molecules in the gaseous state.

You will learn in Chapter 18 that water vapor is a powerful greenhouse gas. It contributes about two-thirds of the total warming from all greenhouse gases. The amount of water vapor in the air is the **humidity**. Humidity is generally measured as **relative humidity**, the amount of water vapor as a percentage of what the air can hold *at a particular temperature*. For example, a relative humidity of 60% means that the air contains 60% of the maximum amount of water vapor it could hold at that particular temperature. The amount of water vapor air can hold increases and decreases with the temperature. When warm, moist air is cooled, its

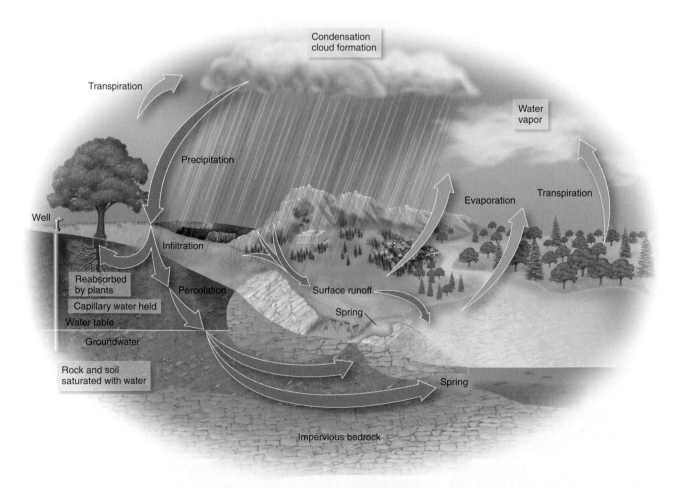

Figure 10–3 The hydrologic cycle. The Earth's fresh water is replenished as water vapor enters the atmosphere by evaporation and transpiration from vegetation, leaving salts and other impurities behind. As precipitation hits the ground, three additional pathways are possible: surface runoff, infiltration, and reabsorption by plants. Green arrows depict green water, present as vapor in the atmosphere; blue arrows show liquid water.

TABLE 10–1   Terms Commonly Used to Describe Water

Term	Definition
Water quantity	The amount of water available to meet demands.
Water quality	The degree to which water is pure enough to fulfill the requirements of various uses.
Fresh water	Water having a salt concentration below 0.1%. As a result of purification by evaporation, all forms of precipitation are fresh water, as are lakes, rivers, groundwater, and other bodies of water that have a throughflow of water from precipitation.
Salt water	Water that contains at least 3% salt (30 parts salt per 1,000 parts water), typical of oceans and seas.
Brackish water	A mixture of fresh water and salt water, typically found where rivers enter the ocean.
Hard water	Water that contains minerals, especially calcium or magnesium, that cause soap to precipitate, producing a scum, curd, or scale in boilers.
Soft water	Water that is relatively free of minerals.
Polluted water	Water that contains one or more impurities, making the water unsuitable for a desired use.
Purified water	Water that has had pollutants removed or is rendered harmless.
Storm water	Water from precipitation that runs off of land surfaces in surges.
Green water	Water in vapor form originating from water bodies, the soil, and organisms—the source of water for precipitation.
Blue water	Precipitation, renewable surface water runoff, and groundwater recharge—the focus of management and the main source of water for human withdrawals and natural ecosystems.

relative humidity rises until it reaches 100%, and further cooling causes the excess water vapor to condense back to liquid water because the air can no longer hold as much water vapor (Fig. 10–4).

Condensation is the opposite of evaporation. It occurs when water molecules rejoin by hydrogen bonding to form liquid water. If the droplets form in the atmosphere, the result is fog and clouds. (Fog is just a very low cloud.) If the droplets form on the cool surfaces of vegetation, the result is dew. Condensation is greatly facilitated by the presence of *aerosols* in the atmosphere. **Aerosols** are microscopic liquid or solid particles (like hair spray!) originating from land and water surfaces. They provide sites that attract water vapor and promote the formation of droplets of moisture. Aerosols may originate naturally, from sources such as volcanoes, wind-stirred dust and soil, and sea salts. Anthropogenic sources—sulfates, carbon, and dust—contribute almost as much as the natural sources (more in some locations) and can have a significant impact on regional and global climates.

Purification occurs when water is separated from the solutes and particles it contains. The processes of evaporation and condensation purify water naturally. When water in an ocean or a lake evaporates, only the water molecules leave the surface; the dissolved salts and other solids remain behind in solution. When the water vapor condenses again, it is thus purified water—except for the pollutants and other aerosols it may pick up from the air. The water in the atmosphere turns over every 10 days, so water is constantly being purified. (The most chemically pure water for use in laboratories is obtained by distillation, a process of boiling water and condensing the vapor.)

Thus, evaporation and condensation are the source of all natural fresh water on Earth. Fresh water that falls on the land as precipitation gradually makes its way through aquifers, streams, rivers, and lakes to the oceans (blue water flow). In the process, it carries along salts from the land, which eventually accumulate in the oceans. Salts also accumulate in

Figure 10–4 Condensation. The amount of water vapor that air can hold increases and decreases with the temperature. The red line follows an air mass that starts with 40% relative humidity (RH) at 30°C and is cooled. When the mass of air cools to 17°C, it has reached 100% RH, and condensation begins, forming clouds. If we started with a higher RH, clouds would form sooner when the air is cooled. Further cooling and condensation results in precipitation.

inland seas or lakes, such as the Great Salt Lake in Utah. The salinization of irrigated croplands (see Chapter 11) is a noteworthy human-made example of this process. Most of the fresh water from precipitation, however, is returned directly to the atmosphere by evaporation and transpiration.

Precipitation

Warm air rises from Earth's surface because it is less dense than the cooler air above. As the warm air encounters the lower atmospheric pressure at increasing altitudes, it gradually cools as it expands—a process known as *adiabatic cooling*. When the relative humidity reaches 100% and cooling continues, condensation occurs and clouds form. As condensation intensifies, water droplets become large enough to fall as precipitation. *Adiabatic warming*, by contrast, occurs as cold air descends and is compressed by the higher air pressure in the lower atmosphere.

Precipitation on Earth ranges from near zero in some areas to more than 2.5 meters (100 in.) per year in others. (Figure 10–5 shows the pattern over land.) The distribution depends on patterns of rising and falling air currents. As air rises, it cools, condensation occurs, and precipitation results. As air descends, it tends to become warmer, causing evaporation to increase and dryness to result. A rain-causing event that you see in almost every television weather report is the movement of a cold front. As the cold front moves into an area, the warm, moist air already there is forced upward because the cold air of the advancing front is denser. The rising warm air cools, causing condensation and precipitation along the leading edge of the cold front.

Two factors—global convection currents and rain shadow—may cause more or less continuously rising or falling air currents over particular regions, with major effects on precipitation.

Convection Currents.   Global convection currents occur because the Sun heats Earth most intensely over and near the equator, where rays of sunlight are almost perpendicular to Earth's surface. As the air at the equator is heated, it expands, rises, and cools; condensation and precipitation occur. The constant intense heat in these equatorial areas ensures that this process is repeated often, thus causing high rainfall, which, along with continuous warmth, in turn supports tropical rain forests.

Rising air over the equator is just half of the convection current, however. The air, now dry, must come down again. Pushed from beneath by more rising air, it literally "spills over" to the north and south of the equator and descends over subtropical regions (25° to 35° north and south of the equator), resulting in subtropical deserts. The Sahara of northern Africa and the Kalahari of southern Africa are prime examples. The two halves of the system composed of the rising and falling air make up a **Hadley cell** (Fig. 10–6a and b). Because of Earth's rotation, winds are deflected from the strictly vertical and horizontal paths indicated by a Hadley cell and tend to flow in easterly and westerly directions—the *trade winds* (Fig. 10–6c), which blow almost continuously from the same direction.

Rain Shadow.   The second situation that causes continually rising and falling air occurs when moisture-laden trade winds encounter mountain ranges. The air is deflected

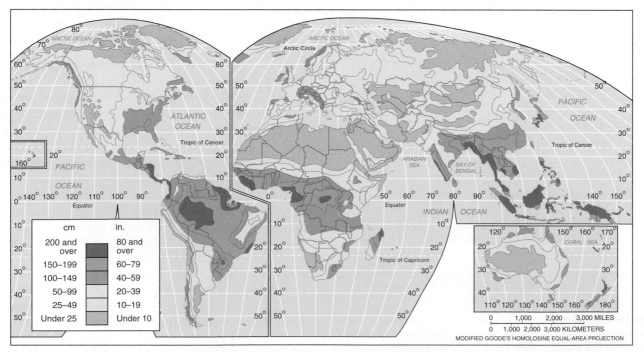

Figure 10–5 Global precipitation. Note the high rainfall in equatorial regions and the regions of low rainfall to the north and south. (*Source:* Robert W. Christopherson, *Geosystems: An Introduction to Physical Geography,* 5th ed., Pearson/Prentice Hall, 2005, Upper Saddle River, NJ.)

Figure 10–6 Global air circulation. (a) The two Hadley cells at the equator. (b) The six Hadley cells on either side of the equator, indicating general vertical airflow patterns. (c) Global trade-wind patterns, formed as a result of Earth's rotation.

upward, causing cooling and high precipitation on the wind-ward side of the range. As the air crosses the range and descends on the other side, it becomes warmer and increases its capacity to pick up moisture. Hence, deserts occur on the leeward sides of mountain ranges. The dry region down-wind of a mountain range is referred to as a **rain shadow** (Fig. 10–7). The severest deserts in the world are caused by the rain-shadow effect. For example, the westerly trade winds, full of moisture from the Pacific Ocean, strike the Sierra Nevada mountains in California. As the winds rise over the mountains, large amounts of water precipitate out, supporting the lush forests on the western slopes. Immediately east of the southern Sierra Nevada, however, lies Death Valley, one of the driest regions of North America.

Figure 10–5 shows general precipitation patterns for the land; the general atmospheric circulation (Fig. 10–6) and the effects of mountain ranges (Fig. 10–7) combine to give great variation in the precipitation reaching the land. The distribution of precipitation in turn largely determines the biomes and ecosystems found in a given region.

Groundwater

As precipitation hits the ground, it may either soak into the ground (**infiltration**) or run off the surface. The amount that soaks in compared with the amount that runs off is called the **infiltration-runoff ratio**.

Runoff flows over the surface of the ground into streams and rivers, which make their way to the ocean or to inland seas. All the land area that contributes water to a particular stream or river is referred to as the **watershed** for that stream or river. All ponds, lakes, streams, rivers, and other waters on the surface of Earth are called **surface waters**.

Water that infiltrates has two alternatives (Fig. 10–3). The water may be held in the soil in an amount that depends on the water-holding capacity of the soil. This water, called **capillary water**, returns to the atmosphere either by evaporation from the soil or by transpiration through plants (a green water flow). The combination of evaporation and transpiration is referred to as **evapotranspiration**.

The second alternative is **percolation** (a blue water flow). Infiltrating water that is not held in the soil is called **gravitational water** because it trickles, or percolates, down through pores or cracks under the pull of gravity. Sooner or later, however, gravitational water encounters an impervious layer of rock or dense clay. It accumulates there, completely filling all the spaces above the impervious layer. This accumulated water is called **groundwater**, and its upper surface is the **water table** (Fig. 10–3). Gravitational water becomes groundwater when it reaches the water table in the same way that rainwater is called lake water once it hits the surface of a lake. Wells must be dug to depths below the water table. Groundwater, which is free to move, then seeps into the well and fills it to the level of the water table.

Recharge. Groundwater will seep laterally as it seeks its lowest level. Where a highway has been cut through rock layers, you can frequently observe groundwater seeping out. Layers of porous material through which groundwater moves are called **aquifers**. It is often difficult to determine the location of an aquifer. Many times these layers of porous rock are found between layers of impervious material, and the entire formation may be folded or fractured in various ways. Thus, groundwater in aquifers may be found at various depths between layers of impervious rock. Also, the **recharge area**—the area where water enters an aquifer—may be many miles away from where the water leaves the aquifer. Underground aquifers hold some 99% of all *liquid* fresh water; the rest is in lakes, wetlands, and rivers.

Underground Purification. As water percolates through the soil, debris and bacteria from the surface are generally filtered out. However, water may dissolve and leach out certain

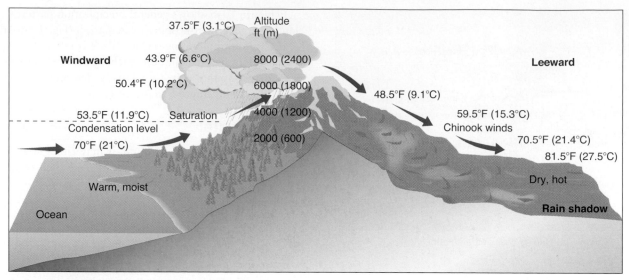

Figure 10–7 Rain shadow. Moisture-laden air in a trade wind cools as it rises over a mountain range, resulting in high precipitation on the windward slopes. Desert conditions result on the leeward side as the descending air warms and tends to evaporate water from the soil.

minerals. Underground caverns, for example, are the result of the gradual leaching away of limestone (calcium carbonate). In most natural situations, the minerals that leach into groundwater are harmless. Indeed, calcium from limestone is considered beneficial to health. Thus, groundwater is generally high-quality fresh water that is safe for drinking. A few exceptions occur when the groundwater leaches minerals containing sulfide, arsenic, or other poisonous elements that make the water unsafe to drink.

Drawn by gravity, groundwater may move through aquifers until it finds some opening to the surface. These natural exits may be seeps or springs. In a **seep**, water flows out over a relatively wide area; in a **spring**, water exits the ground as a significant flow from a relatively small opening. As seeps and springs feed streams, lakes, and rivers, groundwater joins and becomes part of surface water. A spring will flow, however, only if it is *lower* than the water table. Whenever the water table drops below the level of the spring, the spring will dry up. Groundwater movement actually constitutes one of the loops in the hydrologic cycle.

Loops, Pools, and Fluxes in the Cycle

The hydrologic cycle consists of four physical processes: evaporation, condensation, precipitation, and gravitational flow. There are three principal loops in the cycle: (1) In the *evapotranspiration loop* (consisting of green water), the water evaporates and is returned by precipitation. On land, this water—the main source for natural ecosystems and rain-fed agriculture—is held as capillary water and then returns to the atmosphere by way of evapotranspiration. (2) In the *surface runoff loop* (containing blue water), the water runs across the ground surface and becomes part of the surface water system. (3) In the *groundwater loop* (also containing blue water), the water infiltrates, percolates down to join the groundwater, and then moves through aquifers, finally

exiting through seeps, springs, or wells, where it rejoins the surface water. The surface runoff and groundwater loops are the usual focus for human water resource management.

In the hydrologic cycle, there are substantial exchanges of water among the land, the atmosphere, and the oceans; such exchanges are the cycle's "fluxes." Figure 10–8 shows estimates of these fluxes and of the "pools" that hold water globally. Every year the hydrologic cycle circulates approximately 570,000 cubic kilometers of water, the equivalent of 25 times the water in the Great Lakes. Absent from Figure 10–8, however, are the human impacts on the cycle, which are large and growing.

Human Impacts on the Hydrologic Cycle

A large share of the environmental problems we face today stems from direct or indirect impacts on the water cycle. These impacts can be classified into four categories: (1) changes to Earth's surface, (2) changes to Earth's climate, (3) atmospheric pollution, and (4) withdrawals for human use.

Changes to the Surface of Earth. Recall from previous chapters that the direct loss of forests and other ecosystems to various human enterprises diminishes the goods and services these systems have been providing. However, the effects of human-caused changes on the water cycle may be even more profound. In most natural ecosystems, precipitation is intercepted by vegetation and infiltrates into porous topsoil. From there, water provides the lifeblood of natural and human-created ecosystems. The evapotranspiration that takes place in these ecosystems—the green water flow—not only sustains the ecosystems, but also recycles the water to local rainfall. Some of the water that infiltrates percolates down to recharge the groundwater reservoir. Then its gradual release through springs and seeps maintains the flow of streams and rivers at relatively uniform rates. The reservoir

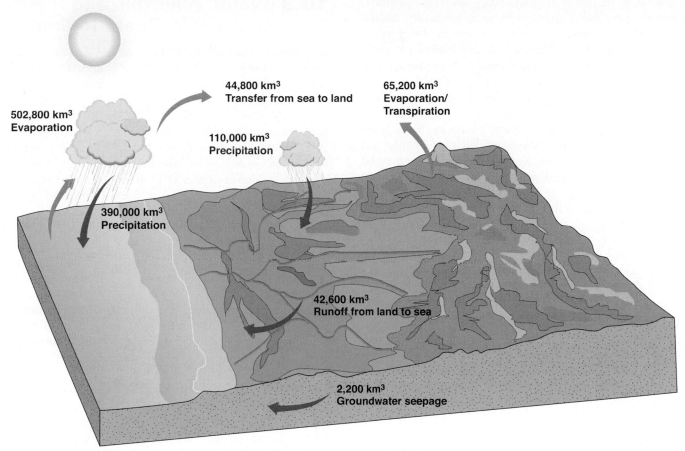

Figure 10–8 Yearly water balance in the hydrologic cycle. The data show (1) the contribution of water from the oceans to the land via evaporation and then precipitation, (2) the movement of water from the land to the oceans via runoff and groundwater seepage, and (3) the net balance of water movement between terrestrial and oceanic regions of Earth. Values are in cubic kilometers per year. (*Source:* Figure 4.2 from *Earth Science*, 8th ed., by Edward Tarbuck and Frederick K. Lutgens, copyright © 1997 by the authors, reprinted by permission of Pearson Education, Inc., Upper Saddle River, NJ; revised with data from *GEO Yearbook 2003*, United Nations Environment Programme.)

of groundwater may be sufficient to maintain a flow even during a prolonged drought. In addition, dirt, detritus, and microorganisms are filtered out as the water percolates through soil and porous rock, resulting in groundwater that is drinkable in most cases.

As forests are cleared or land is overgrazed and plants are not there to intercept rainfall, the pathway of the water cycle is shifted from infiltration and groundwater recharge to runoff, and the water runs into streams or rivers almost immediately. This sudden influx of water into waterways may not only cause a flood, but also bring along sediments and other pollutants via surface erosion. Increased runoff necessarily means less infiltration and, therefore, less evapotranspiration and groundwater recharge. Lowered evapotranspiration means less moisture for local rainfall. Groundwater may be insufficient to keep springs flowing during dry periods. Dry, barren, and lifeless streambeds are typical of deforested regions—a tragedy for both the ecosystems and the humans who are dependent on the flow.

Wetlands function to store and release water in a manner similar to the way the groundwater reservoir does. Therefore, the destruction of wetlands has the same impact as deforestation: Flooding is exacerbated, and waterways are polluted during wet periods and dry up during droughts. For example, the spring 2008 flooding in Iowa occurred in a landscape that has been greatly manipulated. Ninety percent of the state's wetlands have been filled, and the tallgrass prairies have been converted to plowed fields (corn alone covers two-thirds of the state). The fields have been provided with underground drainage pipes to lower the water table, and streams and creeks have been straightened. The result was predictable: With nothing but plowed land to receive it, rainwater moved quickly from the land to the rivers. Much of the flooding was perhaps inevitable, but the human changes to the landscape certainly contributed to the record heights reached by Iowa's rivers. Just as in the Aral Sea, humans here have altered the landscape for intense agricultural production and financial gain. The same tug of war occurs between agriculture and natural consequences (even in a developed country).

Climate Change. There is now unmistakable evidence that Earth's climate is warming because of the rise in greenhouse gases (see Chapter 18), and as this occurs, the water cycle is being altered. A warmer climate means more evaporation from land surfaces, plants, and water bodies because evaporation increases exponentially with temperature. A wetter atmosphere means more and, frequently, heavier precipitation and more flood events. In effect, *global warming is believed to be speeding up the hydrologic cycle.* Already, the United States and Canada have experienced a 10–15% increase in precipitation over the past 50 years (the same time frame for significant global warming to be felt). Regional and local changes are difficult to project; different computer models predict different outcomes for a given region. However, global climate models (see Chapter 18) project a radically changing hydrological world as the 21st century unfolds. The climatologists calculate that a warmer climate will likely generate more hurricanes and more droughts; the extended drought in the western United States and the record hurricane years of 2005 and 2008 may not be a coincidence. They also calculate that many of the currently water-stressed areas like East Africa and the Sahel region (see Chapter 12) will get less water, which will bring large agricultural losses.

According to the Millennium Ecosystem Assessment, "A changing climate can modify all elements of the water cycle, including precipitation, evapotranspiration, soil moisture, groundwater recharge, and runoff. It can also change both the timing and intensity of precipitation, snowmelt, and runoff."

Atmospheric Pollution. Aerosol particles form nuclei that enable water to condense into droplets. The more such particles there are, the greater is the tendency for clouds to form. Anthropogenic aerosols are on the increase, primarily in the form of sulfates (from sulfur dioxide in coal), carbon (as soot), and dust. They form a brownish haze that is associated with industrial areas, tropical burning, and dust storms. Their impact on cloud formation is substantial, and where these aerosols occur, solar radiation to Earth's surface is reduced (so they have a cooling effect). Their most significant impact, however, is on the hydrologic cycle. The unique size spectrum of the anthropogenic aerosols (they promote smaller than normal droplets) causes them to actually *suppress* rainfall where they occur in abundance, even though they encourage cloud formation. As they do so, the atmospheric cleansing that would normally clear the aerosols is suppressed, and they remain in the atmosphere longer than usual. With suppressed rainfall come drier conditions, so more dust and smoke (and more aerosols) are the result. Note that this works just opposite to the impact of the greenhouse gases on climate and the hydrologic cycle. Again, there are significant regional differences; the aerosol impact is more local, whereas the impact of greenhouse gases on climate and the hydrologic cycle is more global. In addition, the greenhouse gases are gradually accumulating in the atmosphere, while the aerosols have a lifetime measured in days.

Finally, there are many problems related to the fourth impact, withdrawals of water for human use (as we make use of surface waters and groundwater). Humans are major players in the hydrologic cycle, as is discussed in the next section.

10.3 Water: A Resource to Manage, a Threat to Control

According to the Millennium Ecosystem Assessment, human use now appropriates about 3,600 cubic kilometers of water per year, which represents 27% of the accessible freshwater runoff on Earth. Annual global withdrawal is expected to rise by about 10% every decade. The U.S. Geological Service prepares estimates of major uses of fresh water in the United States every five years (Table 10–2). It is significant that Americans use less water now than they did in 1980, even with population increases (Fig. 10–9). Most of the water used in homes and industries is for washing and flushing away unwanted materials, and the water used in electric power production is for taking away waste heat. These are **nonconsumptive** uses because the water, though now contaminated with wastes, remains available to humans for the same or other uses if its quality is adequate or if it can be treated to remove undesirable materials. In contrast, irrigation is a **consumptive** use because the applied water does not return to the water resource. It can only percolate into the ground or return to the atmosphere through evapotranspiration. In either case, the water does reenter the overall water cycle but is gone from human control.

Worldwide, the largest use of water is for irrigation (70%), second is use for industry (20%), and third is for direct human use (10%). These percentages vary greatly from one region to another, depending on natural precipitation and the degree to which the region is developed (Fig. 10–10). Most increases in withdrawal are due to increases in irrigated lands. In the United States, irrigation accounts for 65% of freshwater consumption.

Sources. About 37% of domestic water comes from groundwater sources and 63% from surface waters (rivers, lakes, reservoirs) in the United States. Before municipal water supplies were developed, people drew their water from whatever source they could, such as wells, rivers, lakes, or

TABLE 10–2 U.S. Demands on Fresh Water	
Use	**Gallons (Liters) Used per Person per Day**
Consumptive	
Irrigation and other agricultural use	485 (1835)
Nonconsumptive	
Electric power production	477 (1805)
Industrial use	72 (273)
Public supply and self-supply	164 (621)

Source: S. S. Hutson, N. L. Barber, J. F. Kenny, K. S. Linsey, D. S. Lumia, and M. A. Maupin, *Estimated Use of Water in the United States in 2000* (Circular 1268) (Reston, Va.: U.S. Geological Survey, 2004).

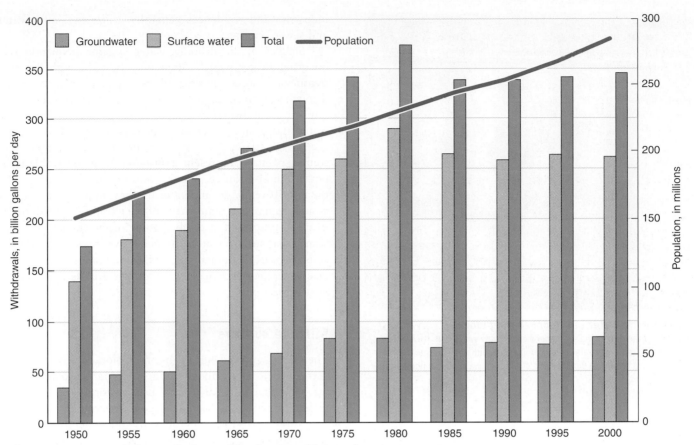

Figure 10–9 U.S. trends in population and freshwater withdrawals. Since 1980, Americans have been using less water, even though the population has increased substantially. (*Source:* S. S. Hutson, N. L. Barber, J. F. Kenny, K. S. Linsey, D. S. Lumia, and M. A. Maupin, *Estimated Use of Water in the United States in 2000* (Circular 1268) (Reston, Va.: U.S. Geological Survey, 2004).)

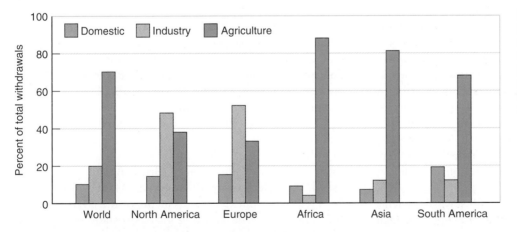

Figure 10–10 Regional usage of water. The percentage used in each category varies with the climate and relative development of the region. A less developed region with a dry climate (e.g., Africa) uses most of its water for irrigation, whereas an industrialized region (e.g., Europe) requires the largest percentage for industry.

rainwater. This approach is still used in rural areas in the developing countries. Women in many of these countries walk long distances each day to fetch water. Because surface waters and shallow wells often receive runoff, they frequently are polluted with various wastes, including animal excrement and human sewage likely to contain pathogens (disease-causing organisms). In fact, an estimated 90% of wastewater in developing countries is released to surface waters without any treatment. Yet, unsafe as it is, this polluted water is the only water available for an estimated 1.1 billion people in

less developed countries (Fig. 10–11). It is commonly consumed without purification, but not without consequences: According to the UN World Health Organization, contaminated water is responsible for the deaths of more than 1.6 million people in developing countries each year, 90% of whom are children. Recall from Chapter 9 that Millennium Development Goal 7 is to reduce by half the proportion of people without access to safe drinking water.

Technologies. In the industrialized countries, the collection, treatment, and distribution of water is highly developed.

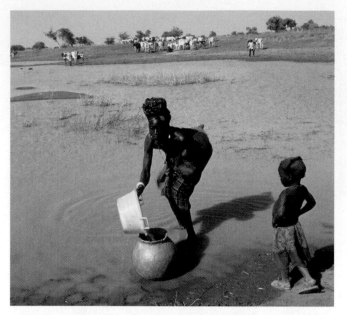

Figure 10–11 Water in the developing world. In many villages and cities of the developing world, people withdraw water from rivers, streams, and ponds. These sources are often contaminated with pathogens and other pollutants. This woman in Burkina Faso is collecting water from a pond used by people and animals.

Larger municipalities rely primarily on surface-water sources, while smaller water systems tend to use groundwater. In the former, dams are built across rivers to create reservoirs, which hold water in times of excess flow and can be drawn down at times of lower flow. In addition, dams and reservoirs may offer power generation, recreation, and flood control. Reservoirs created by dams on rivers are also major sources of water for irrigation. In this case, no treatment is required. However, water for municipal use is piped from the reservoir to a treatment plant, where it is treated to kill pathogens and remove undesirable materials, as shown in Figure 10–12. After treatment, the water is distributed through the water system to homes, schools, and industries. Wastewater, collected by the sewage system, is carried to a sewage-treatment plant, where it is treated before being discharged into a natural waterway. (See Chapter 20.) Often, wastewater is discharged into the same river from which it was withdrawn, but farther downstream.

Whenever possible, both water and sewage systems are laid out so that gravity maintains the flow through the system. This arrangement minimizes pumping costs and increases reliability. On major rivers, such as the Mississippi, water is reused many times. Each city along the river takes water, treats it, uses it, treats it again, and then returns it to the river. In developing nations, the wastewater frequently is discharged with minimal or no treatment. Thus, as the water moves downstream, each city has a higher load of pollutants to contend with than the previous city had, and ecosystems at the end of the line may be severely affected by the pollution. Pollutants include industrial wastes, as well as pollutants from households, because

industries and residences generally utilize the same water and sewer systems.

Smaller public drinking-water systems frequently rely on groundwater, which often may be used with minimal treatment because it has already been purified by percolation through soil. More than 1.5 billion people around the world rely on groundwater for their domestic use. Both surface water and groundwater are replenished through the water cycle (Fig. 10–3). Therefore, in theory at least, these two sources of water represent a sustainable or renewable (self-replenishing) resource, but they are not inexhaustible. Over the last 50 years, groundwater use has outpaced the rate of aquifer recharge in many areas of the world. The death of the Aral Sea (discussed at the beginning of the chapter) demonstrates that surface waters are being overdrawn as well. The consequences of human use and manipulation of both surface water and groundwater resources are discussed in the next subsection.

Surface Waters

The Glen Canyon Dam spans the Colorado River at Lee's Ferry, Arizona, just above Grand Canyon National Park. It was completed in 1963 and began storing excess water in Lake Powell, the second largest reservoir in the United States (Fig. 10–13). Operated by the federal Bureau of Reclamation, the dam generates hydropower, and the reservoir stores water for distribution to Arizona, California, Nevada, and Mexico. During the dam's first two decades of operation, the water released was strictly designed to maximize hydropower; it could fluctuate many-fold in flow in the course of a day.

Dams and the Environment. An environmental impact study in the late 1980s and early 1990s concluded that the operation of the dam had seriously damaged the downstream ecology of the Colorado River and its recreational resources (hiking, river rafting). In response, Congress passed the Grand Canyon Protection Act in 1992, which required a process of "adaptive management" that would adjust water-release rates in a manner designed to enhance the ecological and recreational values of the river. Secretary of the Interior Bruce Babbitt then established the Glen Canyon Adaptive Management Work Group to carry out the requirements of the Grand Canyon Protection Act. The working group is made up of representatives from six Native American tribes, seven states in the river basin, several power contractors, and numerous federal agencies and environmental groups. Many natural and social science projects have been conducted subsequently, and the outcome has been a public-policy success, with the dam now meeting the ecological and social needs of the river ecosystem and the people who live around it or visit it.

To trap and control flowing rivers, more than 45,000 large dams (more than 50 feet high) have been built around the world. Some 3,000 of these dams contain reservoirs with

Figure 10–12 Municipal water use and treatment. (a) Water is often taken from a river or reservoir, piped to the treatment plant, treated, used, and then returned to the source. (b) At the treatment plant, (1) chlorine is added to kill bacteria, (2) alum (aluminum sulfate) is introduced to coagulate organic particles, and (3) the water is put into a settling basin for several hours to allow the coagulated particles to settle. The water is then (4) filtered through sand, (5) treated with lime to adjust the pH, and (6) put into a storage water tower or reservoir for distribution to your home.

Figure 10–13 Glen Canyon Dam. Glen Canyon Dam blocks the Colorado River just north of Grand Canyon National Park. Built to provide hydroelectric power and water to states in the Colorado River basin, the dam holds back Lake Powell, the second largest reservoir in the United States.

storage volumes greater than 95 billion liters (25 billion gal), inundating 49 million hectares (120 million acres) of land and containing more than 6,250 cubic kilometers (1,500 mi³) of water. Of the 42,600 cubic kilometers of annual runoff in the hydrologic cycle (Fig. 10–8), only 31% is accessible for withdrawal because many rivers are in remote locations and because much of the remaining water is required for navigation, flood control, and the generation of hydroelectric power. The large dams have an enormous direct social impact, leading to the displacement of at least 40 million people worldwide and preventing access by local people to the goods and services of the now-buried ecosystems.

Dam Impacts. The United States is home to some 75,000 dams at least six feet in height and an estimated 2 million smaller structures. These have been built to run mills (now an obsolete usage), control floods, generate electric power, and provide water for municipal and agricultural use. These dams have enormous ecological impacts. When a river is dammed, valuable freshwater habitats, such as waterfalls, rapids, and prime fish runs, are lost. When the river's flow is diverted to cities or croplands, the waterway below the diversion is deprived of that much water. The impact on fish and other aquatic organisms is obvious, but the ecological ramifications go far beyond the river. Wildlife that depends on the water or on food chains involving aquatic organisms is also adversely affected. Wetlands occupying floodplains along many rivers, no longer nourished by occasional overflows, dry up, resulting in frequent die-offs of waterfowl and other wildlife that depended on those habitats (Fig. 10–14). Fish such as salmon, which swim from the ocean far upriver to spawn, are seriously affected by the reduced water level and have problems getting around the dam, even

one equipped with fish ladders (a stepwise series of pools on the side of the dam, where the water flows in small falls that fish can negotiate). If the fish do get upriver, the hatchlings have similar problems getting back to the ocean. On the Columbia and Snake rivers, juvenile salmon suffer 95% mortality in their journey to the sea as a result of negotiating the dams and reservoirs that block their way.

Dams Gone. Increasingly, people are recognizing the costs of dams and finding them to be unacceptable. More and more, existing dams are being challenged: More than 714 have already been dismantled in the United States, and others await the same fate. Removing a dam is not easy, however. Legal complexities abound where long-existing uses of a dam (to control floods, or provide irrigation water or lake-like still water for recreation) conflict with the expected advantages of removal (reestablishing historic fisheries such as salmon and steelhead runs and restoring the river for recreational and aesthetic use). Practical problems also exist. Often, the reservoir behind a dam has received massive amounts of sediment from upriver, and removing the dam exposes the sediments and washes them downstream. For example, a proposal to remove low dams on the Blackstone River in Massachusetts was abandoned in the early 1990s because it was found that sediments built up behind the dams were contaminated with heavy metals released from upstream factories. Yet, most often the removal of a dam brings major benefits, as seen with the Edwards Dam on Maine's Kennebec River. After the dam was removed in 1999, the river saw the return of striped bass, salmon, alewives, and shad, as the river water remained cool and well aerated. Canoers, kayakers, and anglers once again made recreational use of the additional 17 miles of restored river

(a)   (b)

Figure 10–14 Ecological effects of diversion. (a) Snow geese and sandhill cranes in Bosque del Apache Wildlife Refuge, New Mexico. (b) What happens to these birds when wetlands dry up because of water diversion or a falling water table?

habitat. This dam removal has served as a model for dam removal across the United States.

Impacts on Estuaries. When river waters are diverted on their way to the sea, serious impacts are seen in **estuaries**—bays and rivers in which fresh water from a river mixes with seawater. Estuaries are among the most productive ecosystems on Earth; they are rich breeding grounds for many species of fish, shellfish, and waterfowl. As a river's flow is diverted (say, to irrigated fields), less fresh water enters and flushes the estuary. Consequently, the salt concentration increases, profoundly affecting the estuary's ecology. This is exactly what is happening in San Francisco Bay. Over 60% of the fresh water that once flowed from rivers into the bay has been diverted for irrigation in the Central Valley (2.5 million acres irrigated) and for municipal use in Southern California (25 million people served). In dry years (which are becoming more common), the farms and municipalities remove a third of the water that would normally flow through the delta into San Francisco Bay. Without the freshwater flows, salt water from the Pacific has intruded into the bay, with devastating consequences. Chinook salmon runs, for example, have almost disappeared. Sturgeon, Dungeness crab, and striped bass populations are greatly reduced. Exotic species of plants and invertebrates have replaced many native species, and the tidal wetlands have been reduced to only 8% of their former extent. The delta smelt, an endangered species, is in steep decline.

Upriver from the bay, over a thousand miles of levees protect farmlands that have sunk 20 feet or more below sea level due to compacting. If the levees collapsed during an earthquake, massive amounts of salt water would be sucked upriver and would shut down the huge pumps that divert fresh water to the cities and farms. Given these major problems, the CALFED Bay-Delta Program was established, according to its mission statement, to "develop and implement a long-term comprehensive plan that will restore ecological health and improve water management for beneficial uses of the Bay-Delta System." However, this is not working, as major stakeholders (farmers, government agencies, environmental groups) have not been able to agree on changes to the delta recommended by scientists.

The problem is not limited to the United States. The southeastern end of the Mediterranean Sea was formerly flushed by water from the Nile River. Because this water is now held back and diverted for irrigation by the Aswan High Dam in Egypt, that part of the Mediterranean is suffering severe ecological consequences (such as severe coastal erosion) as sediments normally flowing out to the coast are kept behind the dam. Recall, too, the profound consequences of diverting the waters that once fed the Aral Sea.

Floods. One of the critical reasons for dams is to prevent devastating floods. Floods have always been common. In many parts of the world, however, the frequency and severity of flooding are increasing—not because precipitation is

Figure 10–15 Flooding of the Koshi River. When the river burst its banks in August 2008, residents of Sunsari were forced to wade to safety. The river changed course, threatening more than a million people in northern India.

greater, but because both deforestation and cultivation have increased, and they cause erosion and reduce infiltration. A recent study of floods in 56 developing countries has confirmed the "sponge" effect of forests, concluding that continued loss of forests will lead to greater loss of lives and economies for the developing countries where floods are frequent. For example, extreme flooding in Bangladesh (a very flat country only a few feet above sea level) is now common because the Himalayan foothills in India and Nepal have been deforested. Due to sediment deposited from upriver erosion, the Ganges River basin has risen 5–7 meters (15–22 ft) in recent years!

The worst monsoon in 25 years ravaged Bangladesh, India, and Pakistan in the summer of 2007, killing more than 2,200 people and leaving some 31 million others homeless. The monsoon season comes every summer, bringing vital rain to the farmers of South Asia, but also often causing massive destruction. Sometimes the devastation is more a matter of human mismanagement, as in the Indian state of Bihar in August 2008. There the Koshi River, carrying a flow only a tenth of its stated capacity, broke through dirt levees constructed to contain the river and the 80 million tons of silt carried yearly down the river from the Himalayas in Nepal. The breach changed the course of the river, and it flooded half of the impoverished state of Bihar, stranding more than 1 million people (Fig. 10–15). Some villages were under 10 feet of water; the Indian government mounted a rescue and relief effort, but was unable to address the break until late in the fall. Experts argue that the breach came because the levees allowed massive silt deposition, a consequence of flawed design of the levees.

Groundwater

Groundwater Levels. Because groundwater supplies are vital for domestic use and especially for agriculture, they are a resource that requires management. Within the seven-state High Plains region of the United States, the Ogallala aquifer supplies irrigation water to 4.2 million hectares

(a)

(b)

(c)

Figure 10–16 Exploitation of an aquifer. (a) Pumping up water from the Ogallala aquifer has made this arid region of the United States into some of the most productive farmland in the country. (b) Water is applied by means of center-pivot irrigation, in which the water is pumped from a central well to a self-powered boom that rotates around the well, spraying water as it goes. (c) An aerial photograph shows the extent of center-pivot irrigation throughout the region. Groundwater depletion will bring an end to this kind of farming.

(10.4 million acres), one-fifth of the irrigated land in the nation (Fig. 10–16). This aquifer is probably the largest in the world, but it is mostly "fossil water," recharged during the last ice age. The recent withdrawal rate has been about 28 billion cubic meters (23 million acre-ft) per year, two orders of magnitude higher than the recharge rate. Water tables have dropped 30–60 meters (100–200 ft) and continue to drop at the rate of 6 feet per year. Irrigated farming has already come to a halt in some sections, and it is predicted that over the next 20 years another 1.2 million hectares (3 million acres) in this region will be abandoned or converted to dryland farming (ranching and the production of forage crops) because of water depletion.

As mentioned earlier, some 99% of all liquid fresh water is in underground aquifers. Of this groundwater, more than three-fourths is like the Ogallala aquifer—with a recharge rate of centuries or more. Such groundwater is considered

nonrenewable. Renewable groundwater is replenished by the percolation of precipitation water, so it is vulnerable to variations in precipitation. In tapping groundwater, we are tapping large, but not unlimited, natural reservoirs, contained within specific aquifer systems. Its sustainability ultimately depends on balancing withdrawals with rates of recharge. For most of the groundwater in arid regions, there is essentially no recharge. The resource must be considered nonrenewable. Like oil, it can be removed, but any water removed today will be unavailable for the future. Groundwater depletion has several undesirable consequences.

Falling Water Tables. The simplest indication that groundwater withdrawals are exceeding recharge is a falling water table, a situation that is common throughout the world. Because irrigation consumes far and away the largest

Figure 10–17 Sinkhole. The removal of groundwater may drain an underground cavern until the roof, no longer supported by water pressure, collapses. The result is the sudden development of a sinkhole, such as this one, which consumed a home in Frostproof, Florida, on July 12, 1991.

amount of fresh water, depleting water resources will ultimately have its most significant impact on crop production. Although running out of water is the obvious eventual conclusion of overdrawing groundwater, falling water tables can bring on other problems.

Diminishing Surface Water. Surface waters are affected by falling water tables. In various wetlands, for instance, the water table is essentially at or slightly above the ground surface. When water tables drop, these wetlands dry up, with the ecological results described earlier. Further, as water tables drop, springs and seeps dry up as well, diminishing even streams and rivers to the point of dryness. Thus, excessive groundwater removal creates the same results as the diversion of surface water.

Land Subsidence. Over the ages, groundwater has leached cavities in the ground. Where these spaces are filled with water, the water helps support the overlying rock and soil. As the water table drops, however, this support is lost. Then there may be a gradual settling of the land, a phenomenon known as **land subsidence**. The rate of sinking may be 10–15 centimeters (6–12 in.) per year. In some areas of the San Joaquin Valley in California, land has settled as much as 9 meters (29 ft) because of groundwater removal. Land subsidence causes building foundations, roadways, and water and sewer lines to crack. In coastal areas, subsidence causes flooding unless levees are built for protection. For example, where a 10,000 square kilometer (4,000 mi²) area in the Houston–Galveston Bay region of Texas is gradually sinking because of groundwater removal, coastal properties are being abandoned as they are gradually being inundated by the sea. Land subsidence is also a serious problem in New Orleans, sections of Arizona, Mexico City, and many other places throughout the world. When it happens in cities, underground pipes break, causing leaks and wastage of domestic water. Sewage pipes can also fracture, contaminating the groundwater aquifer.

Subsidence is one of the serious problems in many cities in developing countries, where the public water supply is not dependable. It is estimated that more than 1 billion people in Asian cities alone depend on groundwater—in most cases, drilling their own wells. For example, in Shanghai, China, overuse of groundwater has brought on land subsidence averaging 1.8 meters (6 ft) over an area of 400 square kilometers (150 mi²). It is a classic case of tragedy of the commons (see Chapter 7), where an open-access resource is subject to indiscriminate use. The result is subsidence, contamination by intruding salt water (see below) and sewage, and ultimately, depletion of the resource.

Another kind of land subsidence, a **sinkhole**, may develop suddenly and dramatically (Fig. 10–17). A sinkhole results when an underground cavern, drained of its supporting groundwater, suddenly collapses. Sinkholes may be 300 feet (91 m) across and as much as 150 feet deep. The problem of sinkholes is particularly severe in the southeastern United States, where groundwater has leached numerous passageways and caverns through ancient beds of underlying limestone. An estimated 4,000 sinkholes have formed in Alabama alone, some of which have "consumed" buildings, livestock, and sections of highways.

Saltwater Intrusion. Another problem resulting from dropping water tables is **saltwater intrusion**. In coastal regions, springs of outflowing groundwater may lie under the ocean. As long as a high water table maintains a sufficient head of pressure in the aquifer, fresh water will flow into the ocean. Thus, wells near the ocean yield fresh water (Fig. 10–18a). However, lowering the water table or removing groundwater at a rapid rate may reduce the pressure in the aquifer, permitting salt water to flow back into the aquifer and hence into wells (Fig. 10–18b). Saltwater intrusion is a serious problem in many European countries along the Mediterranean coast.

Figure 10–18 Saltwater intrusion. (a) Where aquifers open into the ocean, fresh water is maintained in the aquifer by the pressure of fresh water inland. (b) Excessive removal of water may reduce the pressure, so that salt water moves into the aquifer.

(a)

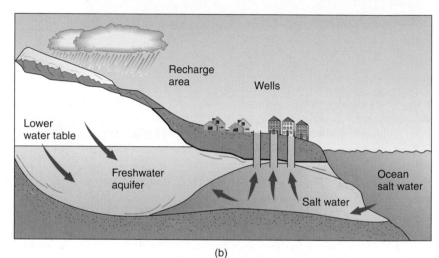

(b)

10.4 Water Stewardship: Supply and Public Policy

The hydrologic cycle is entirely adequate to meet human needs for fresh water because it processes several times as much water as we require today. However, the water is often not distributed where it is most needed, and the result is persistent scarcity of water in many parts of the world (see Stewardship, "Water—Key to Life and Progress in Darewadi"). In the developing world, there is still a deficit of infrastructure, such as wells, water-treatment systems, and (sometimes) large dams, for capturing and distributing safe

Stewardship

Water—Key to Life and Progress in Darewadi

Drylands, defined by low, often seasonal, rainfall between 25 and 75 centimeters (10 and 30 in.) per year, are home to more than 2 billion people. Darewadi village in western India's state of Maharashtra is a typical example. Home to some 800 people, the village occupies a watershed of 1,540 hectares (3,800 acres) fed by an average rainfall of 45 centimeters (18 in.) per year. In 1995, the rural village's economy was largely based on subsistence agriculture for three to four months of the year, with the main dryland crop of millet. Many of the villagers kept sheep and goats, which they grazed on watershed land. After raising crops, many of the village men left in search of seasonal work. Water was trucked in for the four driest months (April–July).

Ten years later the village no longer needs truck delivery of water; farms are worked for nine to ten months of the year; many more crops are raised, yielding five times the former income; trees and grasses are growing on all nonagricultural land; and a large number of village families have new television sets, bicycles, and motorcycles. The key to this successful transformation? Water management, aided by a very effective development program funded by the German government and implemented by a statewide nongovernmental organization (NGO), the Watershed Organization Trust (WOTR). WOTR personnel worked directly with the villagers, who formed a village watershed committee representing all stakeholders (including women and landless poor) to oversee the work done (by villagers) to restore the watershed. They built simple water harvesting systems such as rainwater-catchment dams (check dams) and hillside contour trenches; they effectively banned grazing on hillsides and planted trees and grasses. The results began to be evident as groundwater aquifer levels (tapped by wells for drinking and irrigation water) rose 3 meters (10 ft), in spite of several drought years. Irrigated acreage has doubled, and the water trucks have stopped coming. This village-based program succeeded because the NGO did some excellent capacity building and because villagers worked hard to regenerate their watershed.

To quote the WOTR, "A watershed is not only a geographic area, but also a living space. It is basically the area of survival of the community living within it and drawing its sustenance from it." The NGO is taking this perspective of a watershed to other areas of the state of Maharashtra, implementing a number of projects that are designed to improve the well-being of its inhabitants:

- Implementing public-private community partnership projects in association with the government of Maharashtra in 50 villages;
- Promoting and implementing efforts to provide drinking water and sanitation to individual households and schools;
- Providing training programs for participatory watershed and rural development at a training center in Darewadi village and other sites;
- Promoting the empowerment of women through self-help groups that encourage their active participation in village development and entrepreneurship; and
- Monitoring the growth and promoting the health of children and building the capacity of women health workers.

In all of these projects, WOTR puts the local community at the center of the development process. Gradually, a village at a time, local people are being encouraged and empowered to take control of their future, and in so doing, they are not only working their way out of a cycle of poverty, but also teaching other communities how to do it.

Source: World Resources Institute, *World Resources 2005: The Wealth of the Poor. Managing Ecosystems to Fight Poverty* (Washington, D.C.: World Resources Institute 2005); available on the Watershed Organization Trust Web site, www.wotr.org).

---

drinking water. Despite the growing negative impacts of overdrawing water resources, expanding populations create an ever-increasing demand for additional water for agriculture, industry, and municipal use. Water for food dominates the current water demand, estimated at 6,800 cubic kilometers per year.[2] Of this, some 5,000 cubic kilometers are used as green water flow in the world's rain-fed agriculture and 1,800 cubic kilometers as blue water flow in the form of irrigation water withdrawn from rivers, lakes, and groundwater. It is estimated that 4,500 cubic kilometers (of the total 6,800 km^3) are attributed to the developing countries, to fulfill current food needs.

If we assume that virtually all of the future population increase will be in the developing countries, then so also will be the increased need for food. To eradicate hunger and reach an adequate dietary demand of 3,000 kilocalories/day by 2030, water resource planners[3] estimate that an additional 4,200 cubic kilometers of water for irrigation will be required. This almost doubles the current water use for food production in the developing countries and, if provided by surface water and groundwater, would more than double the current blue water flow—an unacceptable and probably impossible scenario. And this says nothing about future increased domestic and industrial water use.

What are the possibilities for meeting existing needs and growing demands in a sustainable way? Actually, there are only about five options: (1) capture more of the runoff water, (2) gain better access to existing groundwater aquifers, (3) desalt seawater, (4) conserve present supplies by using less water, and (5) make food production more efficient. Making matters more difficult, there is no single U.S. agency responsible for forming public policy on such water supply issues.

More Dams to Capture Runoff

Some 260 new dams become operational each year around the world, a figure that is down from 1,000 per year in the mid–20th century. Certainly, there are opportunities to capture more of the seasonal floodwaters that rush to the sea in a short period of time, and a combination of flood control, water storage, and hydropower can be a powerful argument for dam construction.

Three Gorges Dam. A spectacular example is the Three Gorges Dam, constructed across a scenic stretch of the Yangtze River in China (Fig. 10–19). The project is the centerpiece of the Chinese government's effort to industrialize and join the modern age. Completed in 2006, the dam is the largest hydroelectric project in the world, generating 22,000 megawatts of electricity when the reservoir is filled. It is also expected to provide control over the disastrous flooding of a river that took more than 300,000 lives in the 20th century. More than 1.2 million people—including entire cities, farms, homes, and factories—have been displaced and relocated to make way for the 600-kilometer-long (370-mi-long) reservoir. Because of unforeseen landslides along the reservoir's banks, an additional 3 to 4 million people will be relocated in coming years. As the reservoir has been filling, lowering the Yangtze River flow by 50%, downstream impacts have been severe. Salt water is creeping upriver at the mouth of the river, ships have been stranded, pollutants have accumulated, and many fish populations have been reduced due to interference with breeding cycles. One fear is that the dam could be breached by a major earthquake, as the dam reservoir sits on two major geologic faults. In spite of the many drawbacks, Chinese officials stifled internal criticism and went ahead with the project. On the positive side, the power generated will contribute greatly to China's renewable energy resources, raising the proportion of electricity generated from renewable sources from 7% to 15%. And it will, for a time,

---

[2]M. Falkenmark and J. Rockström, "The New Blue and Green Water Paradigm: Breaking Ground for Water Resources Planning and Management," *Journal of Water Resources Planning and Management* 132 (May/June 2006): 129–132.
[3]Ibid, p. 131.

Figure 10–19 Three Gorges Dam. This dam on the upper Yangtze River in China is the world's largest hydroelectric project. Although the dam is completed, the reservoir is still filling. (Aerial photo taken November 2008).

bring much of the major flood potential of the Yangtze under control.

Wild Rivers. Protection has been accorded to some rivers with the passage of the Wild and Scenic Rivers Act of 1968, which keeps rivers designated as "wild and scenic" from being dammed or affected by other harmful operations. Some 11,300 miles of rivers have been protected, but more than 60,000 miles qualify for protection under the act. The numerous large and small dams across the country have impacted some 600,000 miles of our rivers, by comparison. Those rivers designated as wild and scenic are in many ways equivalent to national parks; thus, like national parks, they need public supporters and defenders, such as the organization American Rivers. There are consistent attempts on the part of special interests (the hydropower industry, agriculture) to undo the recent work establishing wild and scenic rivers.

Tapping More Groundwater

Already, more than 2 billion people depend on groundwater supplies. In many areas, groundwater use exceeds aquifer recharge, leading to shortages as the water table drops below pump levels. Groundwater depletion is considered the single greatest threat to irrigated agriculture, and it is happening in many parts of the world. India, China, West Asia, countries of the former Soviet Union, the American West, and the Arabian Peninsula are all experiencing declining water tables. Calculations by the International Water Management Institute suggest that an expansion of irrigation on the order of 20% might be possible, but this will not meet the food needs in the developing countries by 2030.[4]

Exploiting renewable groundwater will continue to be an option, but it is unlikely to provide great increases in water supply because it is these same sustainable supplies, recharged by annual precipitation, that are being increasingly polluted. Agricultural chemicals such as fertilizers and pesticides, animal wastes, and industrial chemicals readily enter groundwater, making groundwater pollution as much of a threat to domestic water use as depletion. For example, some 60,000 square kilometers of aquifers in western and central Europe will likely be contaminated with pesticides and fertilizers within the next 50 years. Also, India and Bangladesh are plagued with arsenic-contaminated groundwater, which has extracted a heavy toll in sickness and death for millions in those countries.

Desalting Seawater

The world's oceans are an inexhaustible source of water, not only because they are vast, but also because any water removed from them will ultimately flow back in. Not many plants can be grown in full seawater, however, and although researchers are developing salt-tolerant plants via genetic modification, we are not there yet. Thus, with increasing water shortages and most of the world's population living near coasts, there is a growing trend toward **desalination** (desalting) of seawater for domestic use. More than 13,000 desalination plants have been installed in some 120 countries, especially in the Middle East, providing more than 12 billion gallons of drinking water per day.

Two technologies—microfiltration (reverse osmosis, or RO) and distillation—are commonly used for desalination. Smaller desalination plants generally employ microfiltration, in which great pressure forces seawater through a reverse osmosis membrane filter fine enough to remove the salt. Larger facilities, particularly those that can draw from a source of waste heat (for example, from electrical power plants), generally use distillation (evaporation and condensation of vapor). Efficiency is gained by using the heat given off by condensing water to heat the incoming water. Even where waste heat is used, however, the costs of building and maintaining the plant, which is subject to corrosion from seawater, are considerable. Under the best of circumstances, the production of desalinized water costs about $2 per 3,800 liters (1,000 gal). This is two to four times what most city dwellers in the United States currently pay, but it is still not a high price to pay for drinking water (think of bottled water prices).

---

[4]Ibid, p. 131.

Figure 10–20 Desalination plant. The Tampa Bay seawater desalination plant came on line in 2008 and now delivers 25 million gallons of drinking water a day. It is the country's first such plant built to be a city's primary water source.

Tampa Bay Leads the Way (in the United States).

In January 2008, the Tampa Bay seawater desalination plant was completed, signaling a new day for desalination (Fig. 10–20). The plant is the country's first ever built to serve as a primary water source. The plant uses salty cooling water from a power plant and reverse osmosis to produce 25 million gallons of drinking water per day (MGD), providing Pasco County, Pinellas County, St. Petersburg, and other cities with some 10% of their drinking water. The Tampa Bay region has experienced chronic water shortages and is exploiting an opportunity that has attracted worldwide attention. The water is expected to cost Tampa Bay Water $2.49 per 1,000 gallons, wholesale over 30 years—more than the $1 per 1,000 gallons for existing groundwater sources, but still a reasonable cost.

Although the higher cost might cause some people to cut back on watering lawns and to implement other conservation measures, most people in the United States could afford desalinized water without unduly altering their lifestyles. For irrigating croplands, however, the higher cost of desalinized seawater would probably be prohibitive, except possibly for high-value cash crops like greenhouse flowers or vegetables.

The world's largest RO desalination plant employs a unique energy recovery device (a rotary displacement pump) that is helping greatly to make desalination affordable. It was built at Hadara, between Tel Aviv and Haifa in Israel, and will produce 72 MGD, with a potential expansion to 94 MGD. The largest distillation plant is the Jebel Ali desalination plant in the United Arab Emirates, producing 215 MGD, or 2,500 gallons per second.

Using Less Water

A developing-nation family living where water must be carried several miles from a well finds that one gallon per person per day is sufficient to provide for all of its essential needs, including cooking and washing. Yet, a typical household in the United States consumes an average of 380 liters (100 gal) per person per day. If all indirect uses are added (especially irrigation), this figure increases to 4,900 liters (1,300 gal) per person per day. Similarly, a peasant farmer may irrigate by carefully ladling water onto each plant with a dipper, while typical modern irrigation floods the whole field. The way water resource planners think is beginning to change, however: Instead of asking *how much water we need and where can we get it*, people are now asking *how much water is available and how can we best use it*. The good news is that the rate of water use per capita has actually begun to drop, which can be attributed to some well-needed water conservation strategies already put in place. Some specific measures that are being implemented to reduce water withdrawals are described in the next several subsections.

Agriculture. Agriculture is far and away the largest consumer of fresh water—some 40% of the world's food is grown in irrigated soils. Most present-day irrigation wastes huge amounts of water, and in fact, half of it never yields any food. Where irrigation water is applied by traditional flood or center-pivot systems, 30–50% is lost to evaporation, percolation, or runoff. Several strategies have been employed recently to cut down on this waste. One is the *surge flow* method, in which computers control the periodic release of water, in contrast to the continuous-flood method. Surge flow can cut water use in half.

Drip, Drip. Another water-saving method is the *drip irrigation* system, a network of plastic pipes with pinholes that literally drip water at the base of each plant (Fig. 10–21). Although such systems are costly, they waste much less water. Also, they have the added benefit of retarding salinization (see Chapter 11). Studies have shown that drip irrigation can reduce water use 30–70%, while actually increasing crop yields, compared with traditional flooding methods. Although drip irrigation is spreading worldwide, especially in arid lands such as Israel and Australia, 97% of the irrigation in the United States, and 99% throughout the world, is still done by traditional flood or center-pivot methods.

Figure 10–21 Drip irrigation. Irrigation consumes the most water. Drip irrigation offers a conservative method of applying water.

Figure 10–22 Treadle pump. This treadle pump in Bangladesh is operated like a step exercise machine. Millions of these pumps now allow rural farmers to raise crops during dry seasons.

The reason that so few farms have changed over to drip irrigation systems is that it costs about $1,000 per acre to install them. In comparison, water for irrigation is heavily subsidized by the government, so farmers pay next to nothing for it. Therefore, it makes financial sense to use the cheapest system for distributing water, even if it is wasteful. Calculations of construction costs and energy subsidies to provide irrigation water to farmers indicate an annual subsidy of $4.4 billion for the 11 million acres of irrigated land in the western United States, an average of $400 an acre. Governments collect an average of 10% or less of the actual cost in water fees charged to farmers. Reducing this subsidy would greatly encourage water conservation through the use of more efficient irrigation technologies.

Treadle Pumps. In the developing countries, irrigation often bypasses the rural poor, who are at the greatest risk for hunger and malnutrition. Affordable irrigation technologies can be remarkably successful, as in Bangladesh, where low-cost treadle pumps (Fig. 10–22) enable farmers to irrigate their rice paddies and vegetable fields at a cost of less than $35 a system. The pump works like a step exercise machine, is locally manufactured, and has been adopted by the millions. The irrigation it affords makes it possible for the farmers to irrigate small plots during dry seasons from groundwater lying just a few feet below the surface. Over 1.5 million treadle pumps have been sold in Bangladesh alone, increasing the productivity of more than 600,000 acres of farmland there. The pumps are now also in use in India, Nepal, Myanmar, Cambodia, Zambia, Kenya, and South Africa.

Municipal Systems. The 100 gallons of water each person consumes per day in modern homes are used mostly for washing and removing wastes—that is, for flushing toilets (3–5 gallons per flush), taking showers (2–3 gallons per minute), doing laundry (20–30 gallons per wash), and so on.

Watering lawns, filling swimming pools, and other indirect consumption only add to this use.

Water conservation has long been promoted as a "save the environment" measure, though without much effect. Now, numerous cities are facing the stark reality that it will be extremely expensive—and in many cases, impossible—to increase supplies by the traditional means of building more reservoirs or drilling more wells. The only practical alternative, they are discovering, is to take real steps toward reducing water consumption and waste. A considerable number of cities have programs whereby leaky faucets will be repaired and low-flow showerheads and water-displacement devices in toilets will be installed free of charge. Cities in Arizona are paying homeowners to replace their lawns with **xeriscaping**—landscaping with desert species that require no additional watering—and the business is thriving. Many cities and towns ban certain uses of water when droughts reduce the available supply. Brown lawns and dusty automobiles are the result, signs that a city's water supply is threatened.

A Flush World. In 1997, the last phase of a regulation authorized by the 1992 National Energy Act took effect, and it became illegal to sell 6-gallon commodes. In their place is the new wonder of the flush world: the 1.6-gallon toilet (Fig. 10–23). When the new toilets first came into use, homeowners found that they could no longer depend on an easy flush; in fact, plumbers would usually offer a free plunger with installation of the early 1.6-gallon models. Newer versions, however, use even less water, work perfectly well, and save 10 gallons or more a day per person. In fact, the newest models are dual flush, with options for a 1.28 gallon (solids) or a 0.8 gallon (liquids) flush. The 50 million low-flow toilets now in place in the United States save an estimated 600 million gallons of water a day. New York City is providing

Figure 10–23 A WaterSense commode. New, improved, high-efficiency commodes receive the EPA's WaterSense label and could save a family of four $90 a year on water bills—or $2,000 over the lifetime of the toilet. This one has a 0.8 gallon option for liquid wastes and a 1.6 gallon option for solids.

significant rebates to people who replace their old toilets with the new models, and Los Angeles is offering the low-flow toilets free as part of its efforts to restore Mono Lake.

Gray Water. It seems wasteful to raise all domestic water to drinking-water standards and then use most of it to water lawns and flush toilets. Increasingly, gray water recycling systems are being adopted in some water-short areas. **Gray water**, the slightly dirtied water from sinks, showers, bathtubs, and laundry tubs, is collected in a holding tank and used for such things as flushing toilets, watering lawns, and washing cars. Going further, a number of cities are using treated wastewater (sewage water) for irrigating golf courses and landscapes, both to conserve water and to abate the pollution of receiving waters (see Chapter 20). Residents of California use more than 160 billion gallons of treated wastewater for such purposes, and in Israel, 70% of treated wastewater is reused for irrigation of nonfood crops. If the idea of reusing sewage water turns you off, recall that *all* water is recycled by nature. There is hardly a molecule of water you drink that has not moved through organisms—including humans—numerous times. A number of communities are already treating their wastewater so thoroughly that its quality exceeds what many cities take in from lakes and rivers.

Making Food Production and Use More Efficient

We have already addressed the need and options for improvement in irrigation technology. However, most of the agriculture in developing countries is rain-fed, depending on green water flow from soil and resulting in evapotranspiration. Further, most of the soils are dryland soils (see Chapter 11), receiving only 10 to 30 inches of rainfall per year. Crops raised on these soils are characteristically low in yield. Yields can be improved by employing water harvesting techniques, such as storing the seasonal rainfall with "check dams" (small rock or raised soil dams across seasonal water courses). These dams recharge aquifers, restrain soil erosion, and provide for some irrigation water. Another strategy is to enhance rainfall retention, maintain soil fertility, and restrain soil erosion through conservation tillage, where crop residues are left in place and new crops are planted with little or no tilling.

A great deal of water goes into the manufacture of food. One pound of wheat, for example, takes about 1,000 pounds (500 quarts) of water to grow. Whenever food is wasted, water is wasted, too. Some food is lost in storage, but much is lost in food processing and wasteful food habits at the consumer level. It is estimated that at least 50% of edible food is lost or wasted between the field and the table. Reducing food wastage, therefore, conserves water. Another option for reducing water loss in water-stressed regions is to import food, or ship *virtual water*, as it is called. Many arid countries avoid using their own water supplies by importing farm products from regions that have excess fresh water. Indeed, an estimated 800 cubic kilometers of virtual water are traded yearly in the form of foodstuffs. However, many of the arid countries are also economically poor and have a limited ability to purchase great quantities of food.

Even given some improvements in yields from better water harvesting and a limited increase in groundwater withdrawal, there will still be a gap between crop production and the food necessary to meet the needs of developing countries by 2030. This gap will have to be met by converting more forests and grasslands to agricultural lands, according to Falkenmark and Rockström.[5]

Public-Policy Challenges

The hydrologic cycle provides a finite flow of water through each region of Earth. When humans come on the scene, this flow of water is inevitably divided between the needs of the existing natural ecosystems and the agricultural, industrial, and domestic needs of humans, with the latter usually being met first. In fact, recent calculations indicate that humans now use 26% of total terrestrial evapotranspiration and 54% of accessible precipitation runoff. We are major players in the water cycle, and as we have seen, many facets of our water use are unsustainable. It takes public policy to strike a sustainable balance between competing water needs, and this is often sadly lacking.

Water Wars. Maintaining a supply of safe drinking water for people is a high-priority issue. In addition, water for irrigation is vital to food production across much of the world. Often, these two demands come into conflict, generating what have been referred to as "water wars." California has been the scene of many water conflicts. For example, Southern California depends heavily on Colorado River water for irrigation of the extensive Imperial Valley farms, as well as for drinking water for its large cities. Recently, an agreement was struck that diverts 65 billion gallons of Colorado River irrigation water a year to the city of San Diego for domestic use. To purchase the water and also pay farmers to take their land out of production, the city is investing $2 billion. The deal will fallow some 30,000 acres out of a total of 450,000 acres of irrigated farmland. Normally, the irrigation water drains into the Salton Sea from the farmlands, providing the major freshwater source for the sea, which is California's largest lake and the site of a highly productive fishery. Now, it is clear that the diversion is causing the sea to shrink because the lost water represents some 25% of its intake. As the sea shrinks, the exposed seabed will dry up and will be scoured by winds, spreading salt and chemicals widely over nearby crops. The sea is also a haven for wildfowl and migrating waterbirds. If the salinity increases (it is already 30% higher than seawater), it could wipe out the fishery and also threaten the birds.

The Southern California scene demonstrates that water for people can often mean less water for natural ecosystems as rivers, wetlands, and groundwater are exploited and freshwater ecosystems are deprived of their life-sustaining water. This is not simply a matter of aesthetics, however, because these ecosystems provide a wide variety of vital goods and services. They also are host to many threatened and endangered species. A way must be found to manage water for both people and the natural world. How can this be accomplished?

---

[5]Ibid, p. 130.

National Water Policy. Many countries have addressed their water resources and water needs with a national policy; not so the United States. The Clean Water Act and subsequent amendments authorize the U.S. Environmental Protection Agency (EPA) to develop programs and rules to carry out its mandate for oversight of the nation's water quality (considered in Chapter 20). The EPA, however, does not deal with water quantity. There is no federal agency to provide similar oversight for the water quantity issues discussed in this chapter.

Peter Gleick, a leading expert on global freshwater resources, addresses this public-policy need,[6] pointing out that most water policy decisions in the United States are made at local and regional levels. National policies, however, are needed to guide these lower-level decisions and also to deal with interstate issues and federally managed resources. The last time a water policy report was issued was in 1950, when President Harry S. Truman established a water resources policy commission and the commission gave its report. This commission needs to be revived or a new one needs to be established, and in either case, it must be given a mandate to collect data on water resources and problems and issue recommendations on how the federal government could facilitate water stewardship in the 21st century.

Key Issues. The following issues need to be addressed by a water resources policy commission:

1. Water efficiency must be promoted as the primary strategy for meeting future water needs. In the United States, scarcity and rising costs have led to significant gains in efficiency. As a result, water withdrawals peaked in the early 1980s and have leveled off since then (Fig. 10–9). This change has been attributed to steps taken in all three sectors (domestic, industry, and irrigation) to reduce water losses.

2. Water subsidies need to be reduced or eliminated. Sustainable water management will never happen until water is priced according to its real costs. Policy makers have persistently subsidized water resources, especially for agricultural use, and the result has been enormous waste and inefficiencies. Instead of subsidies, caps must be placed on withdrawals from groundwater and surface water in order to protect ecosystem services.

3. Polluters must be charged according to their effluents. When municipalities, farmers, and industries are allowed to pollute waterways without any accountability, the result is that everyone downstream is forced to pay more to clean up the water and aquatic ecosystems are degraded accordingly. Various approaches are possible, such as permit fees, "green taxes," direct charges, and market-based trading (see Chapter 20).

4. Watershed management must be integrated into the pricing of water. Watersheds hold the key to water purity, aquifer recharge, and water storage, as well as maintaining natural wetland and riparian habitats that are vital to many plants and animals. The city of New York found out that every dollar it invested in watershed management paid off handsomely, saving as much as $200 for new treatment facilities.

5. Water authorities must regulate dam operations so that river flow is maintained in a way that simulates natural flow regimes. When this happens, rivers below dams can sustain recreational fisheries and other uses, and the natural biota flourishes.

6. The United States must respond to the global water crisis with adequate levels of international development aid. Achieving the Millennium Development Goal of reducing by half the 1.1 billion people without access to safe drinking water will save millions of lives and billions of dollars annually.

7. Much more research and monitoring are needed to provide the basic data necessary to make informed policy decisions. It is fair to say that if you can't measure it, you can't manage it. Essential data on the recharge and withdrawal of water from groundwater aquifers, on stream flow, on river runoff, and on water uses are poorly collected in most developing countries. The Millennium Ecosystem Assessment has provided several outstanding scientific assessments of water needs and the goods and services associated with natural waters.[7]

International Action. On the international front, the World Commission on Water for the 21st Century sponsors the World Water Forum, which convened for the fifth time in 2009. The findings of the forum relate directly to solving the great water-related needs of Earth, especially in the developing countries. With a large proportion of the developing world already experiencing shortages of clean water or water for agriculture, the population increases and pressure to develop these countries' resources that are certain to come in the next few decades will undoubtedly subject hundreds of millions to increased water stress. The problem is not that Earth contains too little fresh water; rather, it is that we have not yet learned to manage the water that our planet provides. To quote the World Water Council's *World Water Vision* report, "There is a water crisis, but it is a crisis of management. We have threatened our water resources with bad institutions, bad governance, bad incentives, and bad allocations of resources. In all this, we have a choice. We can continue with business as usual, and widen and deepen the crisis tomorrow. Or we can launch a movement to move from vision to action—by making water everybody's business."[8]

[6]Peter H. Gleick, "Global Water: Threats and Challenges Facing the United States," *Environment* 43 (March 2001): 18–26.

[7]Millennium Ecosystem Assessment, *Ecosystems and Human Well-Being: Wetlands and Water Synthesis* (Washington, D.C.: World Resources Institute, 2005); Millennium Ecosystem Assessment, *Ecosystems and Human Well-Being*, vol. 1, *Current State and Trends* (Washington, D.C.: World Resources Institute, 2005), Chapter 7, "Fresh Water," and Chapter 20, "Inland Water Systems"; Millennium Ecosystem Assessment, *Ecosystems and Human Well-Being*, vol. 3, *Policy Responses* (Washington, D.C.: World Resources Institute, 2005), Chapter 7, "Freshwater Ecosystem Services."

[8]*World Water Vision: Making Water Everybody's Business*. World Water Council, 2000.

Revisiting the Themes

Sound Science

What is known about the hydrologic cycle is the result of sound scientific research—research that is needed to make progress in managing water resources. Work performed under the Millennium Ecosystem Assessment has given us a better understanding of such matters as groundwater recharge, essential river flows, and the interactions between different land uses and water resources. The techniques of water treatment have been fine-tuned by scientific work, as well as the development of reverse osmosis for desalination. It was the lack of sound science that led to the breach of the Koshi River levees in Bihar, India, in 2008.

Sustainability

The opening story showed how unsustainable practices in water diversion led to the destruction of the Aral Sea and the degradation of rivers and other aquatic ecosystems. Still, the hydrologic cycle is a remarkable global system that renews water and does so sustainably. Groundwater—especially the fraction that is nonrenewable because it was formed centuries ago—is being used in an unsustainable manner. The consequences are immediate: Water tables disappear into the depths of Earth, farms dry up because there is no water, and land subsidence ruins properties and groundwater storage itself. Conserving water through xeriscaping, low-flush toilets, gray water use, and more efficient food production and consumption is a sustainable path.

Stewardship

Exchanges among land, water, and atmosphere represent a cyclical system that has worked well for eons and that can continue to supply all our needs into the future—if we manage water wisely. However, we are changing the cycle in ways that reflect a lack of wisdom and a lack of resolve to meet people's needs in a just way. In particular, we are refusing to restrain our release of greenhouse gases, thus ensuring that global climate change will continue and intensify. Stewardship of water resources means that we consider the needs of the natural ecosystems—and especially endangered species—when we make policy decisions. It means that we address the wastefulness that characterizes so much of our water use. It means in particular that we show concern and help the millions who lack adequate clean water and are suffering because of it. Some examples of good stewardship are the Glen Canyon Dam operation, the removal of outdated dams to restore river flow, the use of treadle pumps in many developing countries, and water management in Darewadi village. In the United States, developing a national water policy would be a huge step in stewardly management.

REVIEW QUESTIONS

1. What are the lessons to be learned from the Aral Sea story?

2. Give examples of the infrastructure that has been fashioned to manage water resources. What are the challenges related to developing countries?

3. What are the two processes that result in natural water purification? State the difference between them. Distinguish between green water and blue water.

4. Describe how a Hadley cell works, and explain how Earth's rotation creates the trade winds.

5. Why do different regions receive different amounts of precipitation?

6. Define *precipitation*, *infiltration*, *runoff*, *capillary water*, *transpiration*, *evapotranspiration*, *percolation*, *gravitational water*, *groundwater*, *water table*, *aquifer*, *recharge area*, *seep*, and *spring*.

7. Use the terms defined in Question 6 to give a full description of the hydrologic cycle, including each of its three loops—namely, the evapotranspiration, surface runoff, and groundwater loops. What is the water quality (purity) at different points in the cycle? Explain the reasons for the differences.

8. How does changing Earth's surface (for example, by deforestation) change the pathway of water? How does it affect streams and rivers? Humans? Natural ecology?

9. Explain how climate change and atmospheric pollution can affect the hydrologic cycle.

10. What are the three major uses of water? What are the major sources of water to match these uses?

11. How do dams facilitate the control of surface waters? What kinds of impacts do they have?

12. Distinguish between renewable and nonrenewable groundwater resources. What are the consequences of overdrawing these two kinds of groundwater?

13. What are the five options for meeting existing water scarcity needs and growing demands?

14. Describe how water demands might be reduced in agriculture, industry, and households.

15. What is the status of water policy in the United States? Cite some key issues that should be addressed by any new initiatives to establish a national water policy.

THINKING ENVIRONMENTALLY

1. Imagine that you are a water molecule, and describe your travels through the many places you have been and might go in the future as you make your way around the hydrologic cycle time after time. Include travels through organisms.

2. Suppose some commercial interests want to create a large new development on what is presently wetlands. Describe a debate between those representing the commercial interests and those representing environmentalists who are concerned about the environmental and economic costs of development. Work toward negotiating a consensus.

3. Describe how many of your everyday activities, including your demands for food and other materials, add pollution to

the water cycle or alter it in other ways. How can you be more stewardly with your water consumption?

4. An increasing number of people are moving to the arid southwestern United States, even though water supplies are already being overdrawn. If you were the governor of one of the states in this region, what policies would you advocate to address this situation?

5. Suppose some commercial interests want to develop a golf course on what is presently forested land next to a reservoir used for city water. Describe the impacts this development might have on water quality in the reservoir.

MAKING A DIFFERENCE

1. Since 1992, showerheads and faucets have been required to use no more than 2.5 gallons of water per minute. If you have older plumbing, it's possible that you may not meet these standards. Equipment such as aerating showerheads can help you reduce your water use (and your water bill) without decreasing effectiveness too much.

2. Consider taking a "navy" shower (so-called because ships at sea must conserve water): Wet down, then turn the shower off, then soap down, and then rinse down. You will save both water and energy this way.

3. Americans drink billions of single-use water bottles per year. In addition to the carbon miles it takes to ship these

products, most Americans simply throw the recyclable containers away. A much more sustainable method of drinking is simply to fill reusable bottles with filtered tap water.

4. Old water heaters are most likely running at less than 50% efficiency. If this is the case for you, you can replace it with a newer model or simply use a tankless heater, which heats water only when you need it.

5. A dishwasher can save more than 5,000 gallons of water per year as well as hundreds of hours of your time. Scraping your dishes first, running the washer only for full loads, and using the energy-saver settings can reduce water and energy usage even more.

Mastering**ENVIRONMENTALSCIENCE**™

Go to **www.masteringenvironmentalscience.com** for practice quizzes, Pearson eText, videos, current events, and more.

Farming in Guatemala. Maize is a mainstay of nutrition for the people of Guatemala. It is possible to dramatically improve maize production with proper treatment of the soil.

CHAPTER 11

Soil: The Foundation for Land Ecosystems

THIRTY-FIVE YEARS AGO GABINO LÓPEZ attended classes on *sustainable agriculture* in San Martín Jilotepeque, Guatemala, and then returned to his village to apply what he had learned to his own small farm. He planted vegetation barriers along the contours of his land to stop erosion and applied cow manure to enrich the soil. His maize harvest increased from 0.8 to 1.3 tons/hectare (1,050 lb/acre). López's harvest jumped to 3.4 tons/ hectare when he plowed under his crop residues, as opposed to when he burned them or rotated crops. Aware of his success, his friends asked him for advice, and before long, Gabino López became an international consultant in *ecological agriculture*, traveling from Mexico to Ecuador and even India. He and his fellow farmers employ a number of practices that follow the "five golden rules of the humid tropics": (1) keeping the soil covered, (2) using minimal or even no tillage, (3) using mulch to provide nutrients for crops, (4) maximizing biomass production, and (5) maximizing biodiversity. As a result, the maize harvest has averaged 4.3 tons/ hectare in a number of local villages.

Recently, López reported on the concept of *farmer experimenters* in the context of the work of COSECHA, a Honduran nongovernmental organization (NGO). To overcome the problem of irregular rainfall, the farmers tested a number of different contraptions for trapping and holding the water for small-scale irrigation. The farmers decided that 1- to 2-cubic-meter

KEY TOPICS

1. **Soil and Plants: Soils vary greatly in texture and vertical profile, providing attributes that determine much of the success of plant growth.**

2. **Soil Degradation: The three practices that lead to erosion and soil degradation are overcultivation, overgrazing, and deforestation.**

3. **Soil Conservation: Soil conservation requires action at two levels: individual landholders and public policy.**

catchments were ideal and found many additional uses for the water. Water harvesting is joining green manures and natural pest control in the growing reach of agricultural development that involves villagers experimenting and teaching others what they learn. It is an ongoing process, and it is leading to sustainable agriculture for people in serious need of all the food they can raise. On the other side of the world, a very different story must be told.

RUSSIAN DESERT. In 1997, in Komsomolsky, located in southeastern Russia, Alexandra Bakhtayeva told a *Boston Globe* reporter that she had to dig herself out every day from the sand blowing in from surrounding dry land that was once prime grazing land. The windblown sand means that the area has undergone severe desertification—conversion from stable, rolling grasslands to drifting sand. The desert area is growing, expanding by thousands of acres annually in this Russian southland. What used to be soil is gone and has been replaced by sand. This is a human-made desert, the result of failed communist agricultural policies and, earlier than that, the removal of the local Kalmyk people by Stalin as he forced collectivization of the rural areas of the Soviet Union.

Before their removal by Stalin, the Kalmyks very carefully avoided plowing the land and, instead, rotated the land being grazed by their cattle and sheep. Shepherds even wore flat-bottomed shoes to avoid breaking up the topsoil, and their fat-tailed sheep were a breed with flat-bottomed hooves. The combination of plowing the thin soil in an attempt to raise crops and grazing millions of Caucasian sheep with sharp hooves gradually broke down the soil and gave it to the winds. Now, the sand has already buried 25 towns, with more on the way; nearly half of the territory of Kalmykia is severely desertified. Farther to the east, encroaching deserts already occupy 2.6 million square kilometers (1 million square miles) of China, over one-fourth of China's landmass (Fig. 11–1). There is hope, however. As Figure 11-1 illustrates, workers are building barriers to stabilize sand dunes in China's Dong Shawo Desert. In Kalmykia, with the help of the United Nations Environment Program (UNEP), the local forest service has begun planting shrubs and grasses on 32,000 acres of sand in an effort to halt sand movement. Also, a regional representative has traveled to Mongolia to investigate importing the kind of sheep that once grazed the area. These and other efforts are intended to limit the area of moving sand and reclaim some of the degraded land.

PAST NEGLECT. Ninety percent of the world's food comes from land-based agricultural systems, and the percentage is growing as the ocean's fish and natural ecosystems are increasingly being depleted. Protecting and nurturing agricultural soils, which are the cornerstone of food production, must be a central feature of sustainability. Yet it is a feature that has been overlooked repeatedly

Figure 11–1 Desertification in China. Workers outside Wuwei, Gansu Province, are building barriers by using hay to create grid patterns that will keep the sand from further encroachment.

in the past. The story of Easter Island in Chapter 1 is just one of many. In his book *A Green History of the World*,[1] Clive Ponting documents how the fall of the ancient Greek, Roman, and Mayan empires was more the result of a decline in agricultural productivity due to soil erosion than of outside forces. The Dust Bowl of the 1930s in the United States, brought on by widespread conversion of dry prairies to crops followed by an extended drought, saw the out-migration of 2.5 million people from the Great Plains states.

As the world's population edges up to 7 billion, croplands and grazing lands are being increasingly pressed to yield more crops and other products. The landmark 1991 Global Assessment of Soil Degradation (GLASOD) study reported that 15% of Earth's land was degraded, mostly due to soil erosion. Yet a new study by the Land Degradation Assessment in Drylands (LADA) found that **land degradation** is worse than that.[2] Over 20% of cultivated areas, 30% of forests, and 10% of grasslands were degraded between 1981 and 2003. Some 1.5 billion people depend directly on the land undergoing degradation. Throughout the world, soils have been (and continue to be) degraded by erosion, the buildup of salts, and other problems that can only undermine future productivity.

Our objective in this chapter is to develop an understanding of the attributes of soil required to support good plant growth, of how these attributes may deteriorate under various practices, and of what is necessary to maintain a productive soil. We will then look at what is being done nationally and internationally to rescue this essential resource and to establish practices that are sustainable. In Section 11.1, we present some characteristics of soil and then examine the relationships between soil and plants that sustain productivity.

[1]Clive Ponting, *A Green History of the World* (New York: Penguin Books, 1991).
[2]Bai, Z.G. et al, 2008. Global Assessment of Land Degradation and Improvement 1: Identification by Remote Sensing. Report 2008/01, FAO/ISRIC — Rome/Wageningen.

11.1 Soil and Plants

Soil can be defined as *solid material of geological and biological origin that is changed by chemical, biological, and physical processes, giving it the ability to support plant growth*. A rich soil is much more than the dirt you might get out of any hole in the ground. Indeed, agriculturists cringe when anyone refers to soil as *dirt*. The quality of soil can make the difference between harvesting an abundant crop and abandoning a field to weeds. You have already learned

DETRITUS

Leaves, stems, flowers, seeds
derived from plants

Continual
deposition

**DYNAMIC
INTERACTIONS =
TOPSOIL**

MINERAL PARTICLES

• Parent materials

• Leached minerals

• Weathered clay,
  sand, silt from bedrock,
  glacial deposits,
  wind and alluvial
  deposits

DETRITUS FOOD WEB

• Biological breakdown
  of detritus to humus

• Bioturbation and mixing
  with mineral particles

• Nitrogen
  fixation

Figure 11–2 Topsoil formation. Soil production involves a dynamic interaction among mineral particles, detritus, and members of the detritus food web.

(Chapter 4) that various detritus feeders and decomposers feed on organic detritus in an ecosystem. Nutrients from the detritus are released into the soil and absorbed by producers, thus recycling the nutrients. In a productive soil, the detritus feeders and decomposers constitute a biotic community of organisms that not only facilitates the transfer of nutrients, but also creates a soil environment that is most favorable to the growth of roots. In short, a productive topsoil involves dynamic interactions among the organisms, detritus, and mineral particles of the soil (Fig. 11–2).

Soil Characteristics

Most soils are hundreds of years old and change very slowly. *Soil science* is an integrative science that is at the heart of agricultural and forestry practice. Many years of study of soils have resulted in a system of classification of soil profiles and soil structure and a taxonomy of soil types from all over the world—as well as a large body of scientific literature investigating relevant topics. Section 11.1 presents just enough of this information to enable you to open the door to soil science; you may even be motivated to follow up on your own in this fascinating field.

Soil Texture. The mineral material of soil, or **parent material,** has its origin in the geological history of an area. Parent material could be rock or sediments deposited by wind, water, or ice.

Sooner or later parent material is broken down by natural **weathering** (gradual physical and chemical breakdown) to the point where often it is impossible to tell just what kind of rock the soil came from. As rock weathers, it breaks down into smaller and smaller fragments. Below the size of small stones, these fragments (called *soil separates*) are classified as *sand, silt,* and *clay.* **Sand** is made up of particles from 2.0 to 0.063 millimeters in size, **silt** particles range from 0.063 to 0.004 millimeters, and **clay** is anything finer than 0.004 millimeters. Particles larger than sand are called gravel, cobbles, or boulders, depending on their size. You can see the individual rock particles in sand, and you may be familiar with the finer particles called silt, but you would need a good microscope to see clay particles. If you wash clay in water, the water immediately takes on a cloudy or muddy appearance because the clay particles are suspended in it. The moldable, "gooey" quality of clay so necessary for pottery appears when just enough water is added for the particles to slide about one another on a film of water, but still cling together. On drying further, the clay particles adhere in hard clods (or elegant pots!).

Proportions. The sand, silt, and clay particles constitute the mineral portion of soil. **Soil texture** refers to the relative proportions of each type of particle in a given soil. If one predominates, the soil is said to be sandy, silty, or clayey. A proportion that is commonly found in soil consists of roughly 40% sand, 40% silt, and 20% clay. A soil with these

Figure 11–3 The soil texture triangle. Relative proportions of sand, clay, and silt are represented on each axis. Major classes of soil are indicated on the triangle. For clay, read across horizontally; for silt, read diagonally downward; for sand, read diagonally upward to the left. The texture content of any soil should total 100% if the triangle is read properly.

proportions is called a **loam.** You can determine the texture of a given soil by shaking a small amount of it with water in a large test tube to separate the particles and then allowing them to settle. Because particles settle according to their weight, sand particles settle first, silt second, and clay last. The rough proportion of each can then be seen. (More precise measurements require a laboratory analysis.) Soil scientists classify soil texture with the aid of a triangle that shows the relative proportions of sand, silt, and clay in a given soil (Fig. 11–3).

Properties. Three basic considerations determine how several important properties of the soil are influenced by its texture:

1. Larger particles have larger spaces separating them than smaller particles have. (Visualize the difference between packing softballs and packing golf balls in containers of the same size.)

2. Smaller particles have more surface area relative to their volume than larger particles have. (Visualize cutting a block in half again and again. Each time you cut it, you create two new surfaces, but the total volume of the block remains the same.)

3. Nutrient ions and water molecules tend to cling to surfaces. (When you drain a nongreasy surface, it remains wet.)

These properties of matter profoundly affect such soil properties as infiltration, nutrient- and water-holding capacities, and aeration (Table 11–1). Note how the soil properties correspond to particle size (sand, silt, and clay) in the table.

Soil texture also affects **workability**—the ease with which a soil can be cultivated. Workability has a tremendous impact on agriculture. Clayey soils are very difficult to work with because even modest changes in moisture content can turn them from too sticky and muddy to too hard and even bricklike. Sandy soils are very easy to work because they become neither muddy when wet nor hard and bricklike when dry.

Soil Profiles. The processes of soil formation create a vertical gradient of layers that are often (but not always) quite distinct. These horizontal layers are known as **horizons,** and a vertical slice through the different horizons is called the **soil profile** (Fig. 11–4). The profile reveals a great deal about the factors that interact in the formation of a soil. The topmost layer, the **O horizon,** consists of dead organic matter (detritus) deposited by plants: leaves, stems, fruits, seeds, and so forth. Thus, the O horizon is high in organic content and is the primary source of energy for the soil community. Toward the bottom of the O horizon, the processes of decomposition are well advanced, and the original materials may be unrecognizable. At this point, the material is dark and is called **humus.**

Subsurface Layers. The next layer in the profile is the **A horizon,** a mixture of mineral soil from below and humus from above. The A horizon is also called **topsoil.** Fine roots from the overlying vegetation permeate this layer. The A horizon is usually dark because of the humus that is present and may be shallow or thick, depending on the overlying ecosystem. Vital to plant growth, topsoil grows at the rate of an inch or two every hundred years. In many soils, the next layer is the **E horizon,** where *E* stands for **eluviation**—the process of *leaching* (dissolving away) many minerals due to the downward movement of water. This layer is usually paler in color than the two layers above it.

Below the E horizon is the **B horizon.** This layer is characterized by the deposition of minerals that have leached

Soil Texture	Water Infiltration	Water-Holding Capacity	Nutrient-Holding Capacity	Aeration	Workability
Sand	Good	Poor	Poor	Good	Good
Silt	Medium	Medium	Medium	Medium	Medium
Clay	Poor	Good	Good	Poor	Poor
Loam	Medium	Medium	Medium	Medium	Medium

TABLE 11–1 Relationship Between Soil Texture and Soil Properties

O Horizon: Humus. (surface litter, decomposing plant matter)

A Horizon: Topsoil. (mixed humus and leached mineral soil)

E Horizon: Zone of leaching. (less humus, minerals resistant to leaching)

B Horizon: Subsoil. (accumulation of leached minerals like iron and aluminum oxides)

C Horizon: Weathered parent material. (partly broken-down minerals)

Figure 11–4 Soil profile. Major horizons from the surface to the parent material in an idealized soil profile.

from the A and E horizons, so it is often high in iron, aluminum, calcium, and other minerals. Frequently referred to as the **subsoil**, the B horizon is typically high in clay and is reddish or yellow in color from the oxidized metals present. Below the B horizon is the **C horizon**, which is the parent mineral material that originally occupied the site. This layer represents weathered rock, glacial deposits, or volcanic ash and usually reveals the geological process that created the landscape. The C horizon is affected little by the biological and chemical processes that go on in the upper layers.

Because there are innumerable soils across the diverse landscapes of the continents, soil profiles will differ in the thickness and the content of the layers. However, all soils exist in layers and can be characterized by their texture.

Soil Classification. Soils come in an almost infinite variety of vertical structures and textures. To give some order to this diversity, soil scientists have created a taxonomy, or classification system, of soils. It works much like the biological taxonomy with which you may be familiar (see Chapter 6). The most inclusive group in the taxonomy is the soil *order*. If you are classifying a soil, you find the order first and then work your way downward through the taxonomic categories (*suborders, groups, subgroups, families*) until you come to the soil *class* that best corresponds to the soil in question. There are literally hundreds of soil classes. Figure 11–5 displays the worldwide distribution of the 12 major orders.

☐ Alfisols (High-Nutrient Soils)
■ Andisols (Volcanic Soils)
■ Aridisols (Desert Soils)
■ Entisols (New Soils)

☐ Gelisols (Permafrost Soils)
■ Histosols (Organic Soils)
■ Inceptisols (Young Soils)
■ Mollisols (Prairie Soils)

■ Oxisols (Tropical Forest Soils)
■ Spodosols (Conifer Forest Soils)
☐ Ultisols (Low-Nutrient Soils)
■ Vertisols (Swelling Clay Soils)

■ Rock Land
☐ Shifting Sands
☐ Ice/Glacier

Figure 11–5 Global map of soil orders. The map shows the worldwide distribution of the 12 soil orders. (*Source:* After the U.S. Department of Agriculture's Natural Resources Conservation Services, World Soil Resources staff.)

Following are the characteristics of the four major soil orders that are most important for agriculture, animal husbandry, and forestry.

Mollisols. Mollisols are fertile, dark soils found in temperate grassland biomes. They are the world's best agricultural soils and are encountered in the midwestern United States; across temperate Ukraine, Russia, and Mongolia; and in the pampas (a broad region of grassy plains) in Argentina. They have a deep A horizon and are rich in humus and minerals; precipitation is insufficient to leach the minerals downward into the E horizon.

Oxisols. These are soils of the tropical and subtropical rain forests. They have a layer of iron and aluminum oxides in the B horizon and have little O horizon, due to the rapid decomposition of plant matter. Most of the minerals are in the living plant matter, so oxisols provide limited fertility for agriculture. If the forests are cut (as in "slash-and-burn" agriculture, a common practice in the tropics), a few years of crop growth can be obtained, but in time, the intense rainfall leaches the minerals downward, forming an impervious hardpan that resists further cultivation.

Alfisols. Alfisols are widespread, moderately weathered forest soils. Although not deep, they have well-developed O, A, E, and B horizons. Alfisols are typical of the moist, temperate forest biome and are suitable for agriculture if they are supplemented with organic matter or mineral fertilizers to maintain soil fertility.

Aridisols. These are the very widespread soils of **drylands** and deserts (drylands are *arid, semiarid, and seasonally dry areas*). The paucity of vegetation and precipitation leaves aridisols relatively unstructured vertically. They are thin and light colored and, in some regions, may support enough vegetation for rangeland animal husbandry. Irrigation used on these soils usually leads to salinization, as high evaporation rates draw salts to surface horizons, where they accumulate to toxic levels.

The descriptions of these four classes of soil demonstrate that different soils vary significantly in their properties. Soil scientists speak of "soil constraints," meaning the various characteristics of different soils that may or may not lead to successful agriculture. What are the major characteristics that enable soils to support plant growth, especially agricultural crops?

Soil and Plant Growth

For their best growth, plants need a root environment that supplies optimal amounts of mineral nutrients, water, and air (oxygen). The pH (relative acidity) and salinity (salt concentration) of the soil are also critically important. **Soil fertility**—the soil's ability to support plant growth—often refers specifically to the presence of proper amounts of nutrients, but it also includes the soil's ability to meet all the other needs of plants. Farmers speak of a given soil's ability to support plant growth as the **tilth** of the soil.

Mineral Nutrients and Nutrient-Holding Capacity.
Mineral nutrients—phosphate (PO_4^{3-}) potassium (K^+), calcium (Ca^{2+}), and other ions—are present in various rocks,

along with non-nutrient elements. Minerals initially become available to roots through the weathering of rock. Weathering, however, is much too slow to support normal plant growth. The nutrients that support plant growth in natural ecosystems are supplied mostly through the breakdown and release (recycling) of nutrients from detritus.

Leaching. Nutrients may literally be washed from the soil as water moves through it, a process called **leaching**. Leaching not only lessens soil fertility, but also contributes to pollution when materials removed from the soil enter waterways. Consequently, the soil's capacity to bind and hold nutrient ions until they are absorbed by roots is just as important as the initial supply of those ions. This property is referred to as the soil's **nutrient-holding capacity**.

Fertilizer. In agricultural systems, there is an unavoidable removal of nutrients from the soil with each crop because nutrients absorbed by plants are contained in the harvested material. Therefore, agricultural systems require inputs of nutrients to replace those removed with the harvest. Nutrients are replenished with applications of **fertilizer**—material that contains one or more of the necessary nutrients. Fertilizer may be organic or inorganic. **Organic fertilizer** includes plant or animal wastes or both; manure and compost (rotted organic material) are two examples. Organic fertilizer also includes leguminous fallow crops (alfalfa, clover) or food crops (lentils, peas), which fix atmospheric nitrogen. **Inorganic fertilizers** are chemical formulations of required nutrients, without any organic matter included. Inorganic fertilizers are much more prone to leaching than organic fertilizers.

Water and Water-Holding Capacity.
As with nutrients, the supply of water in the soil is crucial to plant growth. Water is constantly being absorbed by the roots of plants, passing up through the plant and exiting as water vapor through microscopic pores in the leaves—a process called **transpiration** (Fig. 11–6). The pores, called *stomata* (singular, *stoma*), are essential to permit the entry of carbon dioxide and the exit of oxygen in photosynthesis; however, the plant's loss of water via transpiration through the stomata is dramatic. A field of corn, for example, transpires an equivalent of a field-sized layer of water 43 centimeters (17 in.) deep in a single growing season. Inadequate water results in *wilting*, a condition that conserves water, but that also shuts off photosynthesis by closing the stomata and preventing gas from being exchanged. If the wilted condition is too severe or too prolonged, the plants die.

Three Soil Attributes. Water is resupplied to the soil naturally by rainfall or artificially by irrigation. Three attributes of the soil are significant in this respect. First is the soil's ability to allow water to **infiltrate**, or soak in. If water runs off the ground surface, it won't be useful to the plants. Worse, it may cause erosion, which is discussed shortly.

Second is the soil's ability to hold water after it infiltrates, a property called **water-holding capacity**. Poor water-holding capacity implies that most of the infiltrating water percolates on down below the reach of roots—not very far in the case of seedlings and small plants—again becoming useless. What is desired is a good water-holding capacity—the

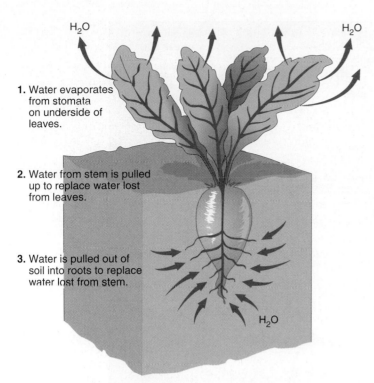

Figure 11–6 Transpiration. When water evaporates from the leaves of a plant, a vacuum is created that pulls the water up through the plant tissues. The roots draw water from the soil to replenish the evaporated water. The Sun's energy drives transpiration via the initial evaporation process.

H₂O

1. Water evaporates from stomata on underside of leaves.

2. Water from stem is pulled up to replace water lost from leaves.

3. Water is pulled out of soil into roots to replace water lost from stem.

ability to hold a large amount of water like a sponge—providing a reservoir from which plants can draw between rains. If the soil does not have such water-holding capacity, the plants will have to depend on frequent rains or irrigation or else suffer the consequences of drought. Sandy soils are notorious for their poor water-holding capacity, whereas clayey and silty soils are good at holding water.

The third critical attribute of the soil is its ability to minimize **evaporative water loss**. This kind of evaporation depletes the soil's water reservoir without serving the needs of plants. The O horizon functions well to reduce evaporative water loss by covering the soil. These aspects of the soil-water relationship are summarized in Figure 11–7.

Aeration. Novice gardeners commonly kill plants by overwatering, or "drowning," them. Roots need to "breathe." Basically, they are living organs and need a constant supply of oxygen for energy via metabolism. Land plants depend on the soil being loose and porous enough to allow the diffusion of oxygen into, and carbon dioxide out of, the soil, a property called **soil aeration**. Overwatering fills the air spaces in the soil, preventing adequate aeration. So does **compaction**, or packing of the soil, which occurs with excessive foot or vehicular traffic. Compaction also reduces infiltration and increases runoff. Again, soil texture strongly influences this property, as indicated in Table 11–1.

Relative Acidity (pH). The term **pH** refers to the acidity or alkalinity of any solution. A solution that is neither acidic nor alkaline is said to be neutral and has a pH of 7. The pH

scale, which runs from 1 to 14, is discussed more fully in Chapter 19. For now, it is important to know that different plants are adapted to different pH ranges. Most plants (as well as animals) do best with a pH near neutral, but many plant species not only tolerate, but also do better in distinctly acidic or alkaline soils (blueberries, for example, are notorious acid lovers).

Salt and Water Uptake. A buildup of salts in the soil makes it impossible for the roots of a plant to take in water. Indeed, if salt levels in the soil get high enough, water can be drawn *out* of the plant (by osmosis), resulting in dehydration and death. Only plants with special adaptations can survive saline soils, and to date, none of those are crop plants (but researchers are making progress in bioengineering salt-tolerant genes into crop plants). The importance of the problem is explained in Section 11.2, when we discuss how irrigation may lead to the accumulation of salts in soil (salinization).

The Soil Community

As we've seen, to support desired plant growth, the soil must (1) have a good supply of nutrients and a good nutrient-holding capacity; (2) allow infiltration, have a good water-holding capacity, and resist evaporative water loss; (3) have a porous structure that permits good aeration; (4) have a pH near neutral; and (5) have a low salt content. Moreover, these attributes must be sustained. How does a soil provide and sustain such attributes? Which is the best soil?

Recall the principle of limiting factors from Chapter 3: The poorest attribute is the limiting factor. The poor water-holding capacity of sandy soil, for example, may preclude agriculture altogether because the soil dries out so quickly. The inability of clay soils to allow water infiltration or

Figure 11–7 Plant-soil-water relationships. Water lost from the plant by transpiration must be replaced from a reservoir of water held in the soil. In addition to the amount and frequency of precipitation, the size of this reservoir depends on the soil's ability to allow water to infiltrate, to hold water, and to minimize direct evaporation.

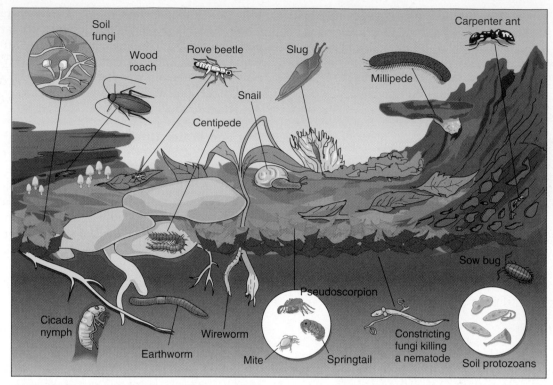

Figure 11–8 Soil as a detritus-based ecosystem. A host of organisms, major examples of which are shown here, feed on detritus and burrow through the soil, forming a humus-rich topsoil with a loose, clumpy structure.

aeration prevents these from being good soils. As Table 11–1 indicates, the best soil textures are silts and loams because limiting factors are moderated in these two types of soil. Even though silts and loams are rated only "medium," their soil texture limitations are greatly improved by the organic parts of the soil ecosystem—the detritus and soil organisms.

Organisms and Organic Matter in the Soil. The dead leaves, roots, and other detritus accumulated on and in the soil support a complex food web, including numerous species of bacteria, fungi, protozoans, mites, insects, millipedes, spiders, centipedes, earthworms, snails, slugs, moles, and other burrowing animals (Fig. 11–8). The most numerous and important organisms are the smallest—the bacteria. With the use of fluorescent staining (a process that makes bacteria glow when viewed in a fluorescence microscope), literally millions of bacteria can be seen and counted in a gram of soil (Fig. 11–9).

Humus. As all these organisms feed, the bulk of the detritus is consumed through their cell respiration, and carbon dioxide, water, and mineral nutrients are released as by-products (see Chapter 4). However, each organism leaves a certain portion undigested; that is, a portion resists breakdown by the organism's digestive enzymes. This residue of partly decomposed organic matter is **humus**, found in high concentrations at the bottom of the O horizon. A familiar example is the black or dark brown spongy material remaining in a dead log after the center has rotted out. **Composting**

is the process of fostering the decay of organic wastes under more or less controlled conditions, and the resulting compost is essentially humus. (Composting as a means of processing and recycling various wastes is discussed further in Chapter 21.) Humus is good for plant growth because it has an extraordinary capacity for holding both water and nutrients—as much as 100 times greater than the capacity of clay, on the basis of weight.

Figure 11–9 Soil bacteria. This photomicrograph shows soil bacteria (the bright spots) stained with acridine orange, a fluorescent dye. The bar scale indicates a length of 10 microns.

Figure 11–10 Humus and the development of soil structure. On the left is a humus-poor sample of loam. Note that it is a relatively uniform, dense "clod." On the right is a sample of the same loam, but rich in humus. Note that it has a very loose structure, composed of numerous aggregates of various sizes.

Soil Structure and Topsoil. As animals feed on detritus on or in the soil, they often ingest mineral soil particles as well. For example, it is estimated that as much as 37 tons per hectare (15 tons per acre) of soil pass through earthworms each year in the course of their feeding. As the mineral particles go through the worm's gut, they actually become "glued" together by the indigestible humus compounds. Thus, earthworm excrements—or **castings**, as they are called—are relatively stable clumps of inorganic particles plus humus. The burrowing activity of organisms keeps the clumps loose. This loose, clumpy characteristic is a desirable **soil structure** (Fig. 11–10). Whereas soil *texture* describes the *size* of soil particles, soil *structure* refers to their *arrangement*. A loose soil structure is ideal for infiltration, aeration, and workability. The clumpy, loose, humus-rich soil that is best for supporting plant growth is the topsoil, represented in a soil profile as the A horizon (see Figure 11–4).

If plants are grown on adjacent plots, one of which has had all its topsoil removed, the results are striking: The yield from plants grown on subsoil is only 10% to 15% of that from plants grown on topsoil. In other words, a loss of all the topsoil would result in an 85% to 90% decline in productivity. The development of topsoil from subsoil or parent material is a process that takes hundreds of years or more. Figure 11–11 shows a site where the topsoil was removed 50 years ago to provide fill for a highway. Although a lichen cover and some stunted trees have developed on the exposed gravel, there is still no topsoil there.

Interactions. There are some important interactions between plants and soil biota. A highly significant one is the symbiotic relationship between the roots of some plants and certain fungi called **mycorrhizae.** Drawing some nourishment from the roots, mycorrhizae penetrate the detritus, absorb nutrients, and transfer them directly to the plant. Thus, there is no loss of nutrients to leaching. Another important relationship is that of certain bacteria that add nitrogen to the

soil, as discussed in Chapter 3. Not all soil organisms are beneficial to plants, however. For example, *nematodes*, small worms that feed on living roots, are highly destructive to some agricultural crops. In a flourishing soil ecosystem, however, nematode populations may be controlled by other soil organisms, such as a fungus that forms little snares to catch and feed on the worms (Fig. 11–12).

Soil Enrichment or Mineralization. You can now see how the aboveground portion and the soil portion of an ecosystem support each other. The bulk of the detritus, which supports the soil organisms, is from green-plant producers, so green plants support the soil organisms. In turn, by feeding on detritus, the soil organisms create the chemical and

Figure 11–11 The results of removing topsoil. This "soil" is nothing but gravel with topsoil removed. After 50 years, only lichens and a few stunted trees have developed. The white object is a ballpoint pen.

Figure 11–12 Predatory fungus. A soil nematode (roundworm), a root parasite, captured by the constricting rings of the predatory fungus *Arthrobotrys anchonia.*

physical soil environment that is most beneficial to plant growth.

Green plants protect the soil and, consequently, themselves in two other important ways as well: The cover of living plants and detritus (1) protects the soil from erosion and (2) reduces evaporative water loss. Thus, it is desirable

to maintain an organic mulch around any garden vegetables that do not maintain a complete cover themselves.

Unfortunately, the mutually supportive relationship between plants and soil can be broken all too easily. The maintenance of topsoil depends on additions of detritus in sufficient quantity to balance losses. Without continual additions of detritus, soil organisms will starve, and their benefit in keeping the soil loose and nutrient rich will be lost. Additional consequences can occur as well. Although resistant to digestion by soil organisms, humus does decompose at the rate of about 2% to 5% of its volume per year, depending on the surrounding conditions. (The rate is most rapid in the tropics.) As the soil's humus content declines, the clumpy aggregate structure created by soil particles glued together with the humus breaks down. Water- and nutrient-holding capacities, infiltration, and aeration decline correspondingly. This loss of humus and the consequent collapse of topsoil are referred to as the **mineralization** of the soil because what is left is just the gritty mineral content—sand, silt, and clay—and no humus.

Thus, topsoil is the result of a dynamic balance between detritus additions and humus-forming processes, on the one hand, and the breakdown and loss of detritus and humus (Fig. 11–13), on the other. If additions of detritus are insufficient, there will be a gradual deterioration of the soil. Conversely, mineralized soils can be revitalized through generous additions of compost or other organic matter.

Figure 11–13 The importance of humus to topsoil. Topsoil is the result of a balance between detritus additions and humus-forming processes as well as their breakdown and loss. If additions of detritus are insufficient, the soil will gradually deteriorate.

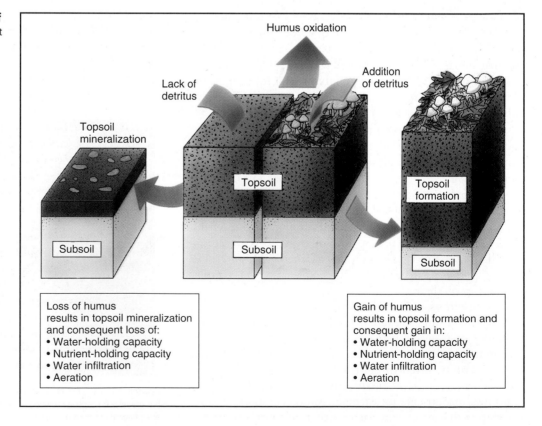

11.2 Soil Degradation

In natural ecosystems, there is always a turnover of plant material, so new detritus is continuously supplied. However, when humans come on the scene and cut forests, graze livestock, or grow crops, the soil is at the mercy of our management or mismanagement. When key soil attributes required for plant growth or for other ecosystem services deteriorate over time, the soil is considered *degraded*. How serious is this problem?

The GLASOD report on degraded land quoted earlier (15% degraded) was for a long time the only truly global assessment. The data for the report came from questionnaires sent to soil experts around the world. A global map was produced; it was not a measure of land degradation, but an estimate. Very little of the information was validated by collecting data on actual soil conditions or information on crop productivity. The authors actually pointed this out in the report. The GLASOD map began to be challenged in places like Burkina Faso, one of the poorest and most densely populated countries in the dryland region of West Africa. The 2001 World Economic Forum calculated that Burkina Faso ranked the highest of any African dryland country in soil degradation (based on GLASOD). Other reports indicated that up to 75% of the land was degraded. In "Soil Degradation in the West African Sahel: How Serious Is It?"[3] however, David Niemeijer and Valentina Mazzucato reported that agricultural yields for virtually all crops have increased in the last 40 years, during which time the population more than doubled. Using subsistence agriculture, the local farmers showed a remarkable ability to manage their land by employing soil and water conservation practices appropriate to drylands. As a result of this and other recent reports, the subsequent World Economic Forum declared that the GLASOD data are suspect and certainly out of date and pointed to the great need for a new assessment of land degradation.

To provide a much more accurate picture of this problem, a project called the Land Degradation Assessment in Drylands (LADA) was launched by the Food and Agricultural Organization of the United Nations (FAO). LADA defines land degradation as "a reduction in the capacity of land to perform ecosystem functions and services that support society and development." This project has produced a new assessment of global land degradation based on satellite remote-sensing techniques: Global Assessment of Land Degradation and Improvement (GLADA).[4] The GLADA assessment employed data from a 23-year time series by National Oceanic and Atmospheric Administration (NOAA) satellites and precipitation data from over 9,300 land stations. Their most incisive analysis combines plant productivity with rain-use efficiency to produce a map of regions that are worsening, calling them "hot spots," and regions that are showing improving trends, calling them "bright spots." Their data indicate that over the time frame of 1981–2003, some 24% of global land area showed a worsening trend. Of this, almost 20% is cropland, representing more than 20% of all cultivated areas. The good news is that some 16% of the land area revealed an improving trend, of which 18% is cropland.

For example, one region with worsening conditions is South Africa. The GLADA analysis revealed that 29% of the country suffered land degradation (Fig. 11–14a) from 1981 to 2003, and this covered 41% of all cropland. Some 38% of the South African population depends on (and has exacerbated) the degrading areas (Fig. 11–14b). LADA is partnering with countries like South Africa not only to produce a satellite-based assessment, but also to work at local levels to promote the adoption of sustainable land and ecosystem management. In South Africa, the degrading areas are suffering from poor management, and poor management can lead to another serious problem: soil erosion.

Erosion

How is topsoil lost? The most pervasive and damaging force is **erosion**, the process of soil and humus particles being picked up and carried away by water or wind. Erosion follows anytime soil is bared and exposed to the elements. The removal may be slow and subtle, as when soil is gradually blown away by wind, or it may be dramatic, as when gullies are washed out in a single storm.

In most natural terrestrial ecosystems, a vegetative cover protects against erosion. The energy of falling raindrops is intercepted by the vegetation, and the water infiltrates gently into the loose topsoil without disturbing its structure. With good infiltration, runoff is minimal. Any runoff that does occur is slowed as the water moves through the vegetative or litter mat, so the water has too little energy to pick up soil particles. Grass is particularly good for erosion control because, when runoff volume and velocity increase, well-anchored grass simply lies down, forming a smooth mat over which the water can flow without disturbing the soil underneath. Similarly, vegetation slows the velocity of wind and holds soil particles.

Splash, Sheet, and Gully Erosion. When soil is left bare and unprotected, however, it is easily eroded. Water erosion starts with what is called **splash erosion**, as the impact of falling raindrops breaks up the clumpy structure of the topsoil. The dislodged particles wash into spaces between other aggregates, clogging the pores and thereby decreasing infiltration and aeration. The decreased infiltration results in more water running off and carrying away the fine particles from the surface, a phenomenon called **sheet erosion**. As further runoff occurs, the water converges into rivulets and

---

[3]David Niemeijer and Valentina Mazzucato, "Soil Degradation in the West African Sahel: How Serious Is It?" *Environment* 44, no. 2 (2002): 20–31.
[4]Bai, Z.G., D. L. Dent, L. Olsson, and M. E. Schaepman,. *Global Assessment of Land Degradation and Improvement 1: Identification By Remote Sensing.* (Report 2008/01), FAO/ISRIC,. Rome/Wageningen. 2008).

Figure 11–14 Land degradation and population in South Africa. Across South Africa, 29% of the country suffered degradation, including 41% of all croplands. Some 17 million people depend on these degrading areas. Go to **GRAPHit!** at www.masteringenvironmentalscience.com. (*Source:* After *Global Assessment of Land Degradation and Improvement 1: Identification by Remote Sensing*, by Z. G. Bai, D. L. Dent, and M. E. Schaepman, FAO, 2008.)

streams, which have greater volume, velocity, and energy and hence greater capacity to pick up and remove soil. The result is the erosion into gullies, or **gully erosion** (Fig. 11–15). Once started, erosion can readily turn into a vicious cycle if it is not controlled. Eroded soil is less able to support the regrowth of vegetation and is exposed to further erosion, rendering it even less able to support vegetation, and so on.

Desert Pavement and Cryptogamic Crusts.   Another very important and devastating feature of wind and water erosion is that both always involve the *differential* removal of soil particles. The lighter particles of humus and clay are the first to be carried away, while rocks, stones, and coarse sand remain behind. Consequently, as erosion removes the finer materials, the remaining soil becomes progressively coarser—sandy,

Figure 11–15 Erosion. Severe erosion due to poor farming practices, near Bosencheve, Mexico.

Figure 11–16 Formation of desert pavement. (a) As wind erosion removes the finer particles, the larger grains and stones are concentrated on the surface. (b) The result is desert pavement, which protects the underlying soil from further erosion. Traffic such as off-road vehicles breaks up the desert pavement and initiates further erosion.

stony, and finally, rocky. Such coarse soils frequently reflect past or ongoing erosion. Did you ever wonder why deserts are full of sand? The sand is what remains after the finer, lighter clay and silt particles have blown away. In some deserts, the removal of fine material by wind has left a thin surface layer of stones and gravel called a **desert pavement**, which protects the underlying soil against further erosion (Fig. 11–16). Damaging this surface layer with vehicular traffic allows another episode of erosion to commence.

Water erosion can cause other changes in the soil composition. Often, when rainfall clogs soils, the soils harden when they dry. Following this, the soil may be colonized by different kinds of primitive plants (algae, lichens, and mosses) called cryptogams. Their growth and colonization create a crust on the soil, called a **cryptogamic crust**. Such crusts can stabilize the soil and add nutrients through nitrogen fixation, but they can also inhibit water infiltration and seed generation. They are readily broken up by livestock trampling or human intrusion, and if the soil below is loosened, it can again be subject to wind and water erosion.

Drylands and Desertification

Recall that clay and humus are the most important components of soil for both nutrient- and water-holding capacity. As clay and humus are removed, nutrients are removed as well because they are bound to those particles. The loss of water-holding capacity is even more serious, however. Regions that have sparse rainfall or long dry seasons support grasses, scrub trees, or crops only insofar as soils have good water-holding and nutrient-holding capacity. As these soil properties are diminished by the erosion of topsoil, such areas become deserts, both ecologically and from the standpoint of agricultural production. Indeed, the term **desertification** is used to denote this process. Note that desertification doesn't mean advancing deserts; rather, the term refers to *a permanent reduction in the productivity of arid, semiarid, and seasonally dry areas* (traditionally called *drylands*). It is a process of land degradation with a number of possible causes, including extended droughts, overgrazing, erosion, deforestation, and overcultivation. It is extremely serious because it is so permanent.

Dryland ecosystems cover 41% of Earth's land area. They are defined by precipitation, not temperature. Beyond the relatively uninhabited deserts, much of the land receives only 25 to 75 centimeters (10 to 30 in.) of rainfall a year, a minimal amount to support rangeland or nonirrigated cropland (Fig. 11–17). Droughts are common features of the climate in drylands, sometimes lasting for years. When the

rainfall returns, however, vegetation often returns, indicating that such lands are not desertified. However, according to the Millennium Ecosystem Assessment, 10% to 20% of these lands suffer from some form of land degradation. These drylands, occupying some 6 billion hectares, are home to more than 2 billion people and are found on every continent except Antarctica. Since these dryland areas are commonly inhabited by some of the poorest people on Earth, promoting sustainable land management is essential for improving standards of living and meeting the Millennium Development Goals (MDGs) successfully.

Recognizing the severity of this problem, the United Nations established the Convention to Combat Desertification (UNCCD), which was signed and officially ratified by over 100 nations in late 1996. The first Conference of the Parties was held in October 1997 in Rome. Conferences were held annually until 2001 and, since then, biennially. These conferences and other regular UNCCD meetings have been concerned with such issues as the funding of projects to reverse land degradation, "bottom-up" programs that enable local communities to help themselves, the gathering and dissemination of traditional knowledge on effective drylands agricultural practices, and the development of a 10-year strategic plan. With the support of the UNCCD, affected countries have been developing National Action Programs. In 2005, a UNCCD meeting launched a new alliance to fight land degradation in sub-Saharan Africa, **TerrAfrica**. The alliance consists of several UN agencies and many African countries and is designed to coordinate efforts at all levels aimed toward arresting land degradation and promoting sustainable land management. For example, the government of Madagascar has been working on sustainable land management with several TerrAfrica partners, especially the World Bank, combating land degradation in its upland watersheds and southern arid lands.

Even though variations in climate often play a role in the processes of desertification, it is human impacts that pose the greatest threat to the health of dryland ecosystems. Accordingly, let us now take a closer look at the practices that lead to erosion, one of the main elements of desertification.

Causing and Correcting Erosion

The three major practices that expose soil to erosion and lead to soil degradation are *overcultivation*, *overgrazing*, and *deforestation*, all of which are a consequence of unsustainable management practices. All of these are practices with impacts that can be reversed, as we will see.

Overcultivation. Traditionally, the first step in growing crops has been (and, to a large extent, still is) plowing to control weeds. The drawback is that the soil is then exposed to wind and water erosion. Further, it may remain bare for a considerable time before the newly planted crop forms a complete cover. After harvest, much of the soil again may be left exposed to erosion. Runoff and erosion are particularly

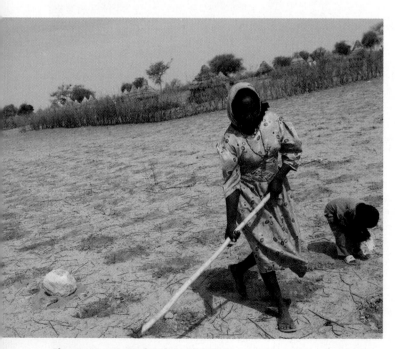

Figure 11–17 Drylands cultivation. Displaced people in West Darfur, Sudan, planting a crop in the semi-desert soil, in anticipation of some rainfall.

severe on slopes, but wind erosion may extract a heavy toll on any terrain with minimal rainfall.

Plowing is frequently considered necessary to loosen the soil to improve aeration and infiltration through it, yet all too often the effect is just the reverse. Splash erosion destroys the soil's aggregate structure and seals the surface, so that aeration and infiltration are decreased. The weight of tractors used in plowing adds to the compaction of the soil. In addition, plowing accelerates the oxidation of humus and evaporative water loss.

Despite the harmful impacts inherent in cultivation, systems of crop rotation—a cash crop such as corn every third year, with hay and clover (which fixes nitrogen as well as adding organic matter) in between—have proved sustainable. However, as food or economic demands cause farmers to abandon rotation, degradation and erosion exceed regenerative processes, and the result is a gradual decline in the quality of the soil. This is the essence of overcultivation.

No-Till. A technique that permits continuous cropping, yet minimizes soil erosion, is **no-till agriculture**, which is now routinely practiced over much of the United States. According to this technique, the field is first sprayed with herbicide to kill weeds, and then a planting apparatus is pulled behind a tractor to accomplish several operations at once. A steel disk cuts a furrow through the mulch of dead weeds, drops seed and fertilizer into the furrow, and then closes it (Fig. 11–18). At harvest, the process is repeated, and the waste from the previous crop becomes the detritus and mulch cover for the next. Thus, the soil is never left exposed, erosion and evaporative water loss are reduced, and there is enough detritus, including roots from the previous crop, to maintain the topsoil. A variation on this theme is **low-till farming**, which is catching on in Asia. For example, to plant wheat after a rice crop has been harvested, only one pass over a field is made—to plant seed and spread fertilizer. Previously, farmers would plow 6 to 12 times to break up the muddy soil and dry it out for the wheat, and then they would irrigate the land.

Fertilizer. Another aspect of overcultivation involves the use of inorganic versus organic fertilizer. There is no question that optimal amounts of required nutrients can be efficiently provided by the suitable application of inorganic chemical fertilizer. The failing of chemical fertilizer is its lack of organic matter to support soil organisms and build soil structure. Under intensive cultivation—one or more cash crops every year—nutrient content may be kept high with inorganic fertilizer, but mineralization, and thus soil degradation, proceeds in any case. Then, with the soil's loss of nutrient-holding capacity, applied inorganic fertilizer is prone to simply leach into waterways, causing pollution.

This is not to say that chemical fertilizers do not have a valuable place in agriculture—after all, the exclusive use of organic material may not provide enough of certain nutrients to support optimal plant growth. What is required is for growers to understand the different roles played by organic material and inorganic nutrients and then to use each as necessary.

NRCS. Regardless of cropping procedures and the use of fertilizer, a number of other techniques are widely employed to reduce erosion. Among the most conspicuous are **contour strip cropping** (plowing and cultivating at right angles to contour slopes) and **shelterbelts** (protective belts of

Figure 11–18 Apparatus for no-till planting. This 12-row planter is planting no-till cotton with minimal disturbance to the protective mulch layer on the surface of the soil. The apparatus creates furrows with cutting discs, applies herbicide and fertilizer, drops in seeds, and closes the furrows.

Figure 11–19 Contour farming and shelterbelts. (a) Cultivation up and down a slope encourages water to run down furrows and may lead to severe erosion. The problem is reduced by plowing and cultivating along the contours at a right angle to the slope. In this photograph, strip cropping is also being utilized: The light green bands are one type of crop, the dark bands another. (b) Shelterbelts—belts of trees around farm fields—break the wind and protect the soil from erosion.

(a)

(b)

trees and shrubs alongside plowed fields) (Fig. 11–19). The **U.S. Natural Resources Conservation Service** (NRCS), formerly the Soil Conservation Service, was established in response to the Dust Bowl tragedy of the early 1930s. Through a nationwide network of regional offices, the NRCS provides information to farmers or other interested persons regarding soil and water conservation practices. (The regional offices can be found through county government telephone listings. Federal extension service offices will test soil samples, provide an analysis thereof, and make recommendations.) The NRCS performs a regular inventory of erosion losses in the United States. (See Sound Science, "Erosion by Equation.")

Sound Science

Erosion by Equation

By any name or measure, soil erosion is a highly undesirable process. It is a highly visible one, too. Ugly gashes, for example, can be seen in fields and slopes that are not well vegetated (see Fig. 11–15), streams and rivers run brown with sediment after a heavy rain, deltas form where rivers empty into larger bodies of water, and reservoirs fill up with sediment over time. Eroded soil represents lost agricultural productivity, and the process can certainly lead to desertification if it is left unchecked. Erosion is also a natural geological process, wherein land surfaces undergo weathering and sediments form over geological time, eventually becoming the thick layers that form sedimentary rock. Thus, some erosion is bound to occur, no matter what farming methods are practiced.

The National Resources Inventory reports erosion rates on a 5-year basis. For 2003, they amounted to about 1.7 billion tons of eroded material. Much erosion is judged to be "excessive," originating from 100 million acres of cropland deemed vulnerable. How is erosion measured, especially if measurements are taken for the entire United States? In most cases, it is done via map reading, aerial photography, and the judicious use of a *universal soil loss equation* and a *wind erosion equation*. As a result, the figures that are obtained are *estimates* rather than *measurements*. The actual rates are not measured and, indeed, are probably not measurable for the whole country.

What happens when someone actually measures erosion and compares it with estimated rates? UCLA geographer Stanley W. Trimble asked this question in a study of the Coon Creek basin in Wisconsin, published in 1999.* He had the good fortune of having available a previous historical study of the basin by the Soil Conservation Service, which left many markers in the sediments of the basin, and a huge amount of archived data. Trimble took soil profiles all around the basin, dated the layers of sediment, and was able to work back to the original surface of organic soil of the prairie, which dated to about 1850, when the basin was first farmed. He found rates of erosion that were high in the late 19th century, went even higher in the 1920s and 1930s, and then began to decline as farmers adopted conservation practices. Most remarkably, the rates from the 1970s to the 1990s were only 6% of their peak rate.

Although the nationally measured erosion rates showed a slight decline during the period of the 1970s to the 1990s compared with earlier rates, the low rates found by what can be called "ground truth" in Trimble's work do not come close to the much higher estimates. Trimble's work has come under some criticism, but his response is that "the burden of proof is on those who have been making these pronouncements about big erosion numbers.... They owe us physical evidence. For one big basin, I've measured the sediment and I'm saying, I don't see it."[†] At the very least, this kind of work suggests that more such actual measurements should be made and that the published erosion estimates should be taken with, well, a grain of silt.

*From "Decreased Rates of Alluvial Sediment Storage in the Coon Creek Basin, Wisconsin," by Stanley Trimble, *Science* 285 (August 20, 1999): 1975–99, copyright © by American Assn. for the Advancement of Science.
[†]As quoted in Glanz, James. "Sharp Drop Seen in Soil Erosion Rates." *Science* 285 (August 20, 1999): 1187–1189.

Estimates indicate that soil erosion on croplands has lessened in recent years, from 2.1 billion tons in 1992 to 1.7 billion tons in 2003. This is likely a consequence of improved conservation practices, such as windbreaks, grassed waterways, and field border strips of perennial vegetation to filter runoff.

Overgrazing. Grasslands that receive too little rainfall to support cultivated crops have traditionally been used for grazing livestock. In fact, some 65% of dryland areas are rangelands. Likewise, forested slopes that are too steep for cropping are commonly cleared and put into grass for grazing. Unfortunately, such lands are too often overgrazed. As grass production fails to keep up with consumption, the land becomes barren. Wind and water erosion follows, and soils become degraded.

Overgrazing is not a new problem. In the United States in the 1800s, the American buffalo (bison) was slaughtered to starve out Native Americans and allow for stocking the rangelands with cattle. Overgrazing was rampant, leading to erosion and encroachment by hardy desert plants such as sagebrush, mesquite, and juniper, which are not palatable to cattle. Western rangelands now produce less than 50% of the livestock forage they produced before the advent of commercial grazing. Yet according to a U.S. General Accounting Office (GAO) study, 20% of the rangelands remain overstocked.[5]

The broader ecological impact of overgrazing should not go unnoticed either. The National Public Lands Grazing Campaign documents the many harmful effects of livestock grazing on public lands. Livestock compete with native animals for forage, reducing the latter's numbers substantially (about one-third of the endangered species in the United States are in jeopardy because of overgrazing or other practices associated with raising cattle, such as predator-control programs and the suppression of fire). Particularly hard hit are wooded zones along streams and rivers, zones that are trampled by cattle seeking water. The resulting water pollution by sediments and cattle waste is high on the list of factors making fish species the fastest-disappearing wildlife group in the United States. A recent study of rangeland management concluded that (1) desertification impacts some 85% of North America's drylands and (2) the most widespread cause of rangeland desertification is livestock grazing.[6]

Public Lands. In many cases, overgrazing occurs because the rangelands are public lands not owned by the people who own the animals. Where this is the case, herders who choose to withdraw their livestock from the range sacrifice income, while others continue to overgraze the range. Therefore, the incentive is for all to keep grazing, even though the range is being overgrazed. This problem, known as the "tragedy of the commons," was discussed in Chapter 7.

In a variation of the tragedy of the commons, the U.S. Bureau of Land Management (BLM) and the Forest Service lease grazing rights for 2 million square kilometers of federal (i.e., taxpayer-owned) lands for a nominal fee of $1.35 (in 2009) per animal unit (one cow-and-calf pair or five sheep) per month, less than one-tenth of what a private owner would be paid for the same rights. In 2006, over 14 million animal unit months were used on BLM and Forest Service lands. The fee structure was established by the Taylor Grazing Act of 1934, which effectively prohibited any reduction in grazing levels and kept the fees well below market value. The income to the government is far less than the costs of administering the program: The grazing program lost $115 million in 2004, after subtracting fee receipts. A worse problem is the ecological cost exacted by livestock to watersheds, streams, wildlife, and endangered species, considered to be as high as $500 million. Attempts in Congress and the Department of the Interior (which houses the BLM) to raise grazing fees are consistently met with opposition from western congressmen, who threaten to cut the BLM budget.

[5]Rangeland Management: More Emphasis Needed on Declining and Overstocked Grazing Allotments (GAO/RCED-88-80, June 10, 1988).

[6]Bowker, M. A., J. Belnap, M. E. Miller. 2006. "Spatial Modeling of Biological Soil Crusts to Support Rangeland Assessment and Monitoring." *Rangeland Ecology and Management* 59: 527.

One solution lies in better management. Because range-lands are now producing at only a 50% level or less, their restoration could benefit both wildlife and cattle production. The NRCS has a program (the **Conservation Stewardship Program**) that provides information and support to enable ranchers who own their lands to burn unwanted woody plants, reseed the land with perennial grass varieties that hold water, and manage cattle so that herds are moved to a new location before overgrazing occurs. One solution gaining favor among ranchers and environmental groups is for the government to buy up some of the 26,000 existing grazing permits and "retire" rangelands. The ranchers would be offered a generous payment for the permits they hold, and the land would be used for wildlife, recreation, and watershed protection.

Deforestation. Forest ecosystems are extremely efficient systems both for holding and recycling nutrients and for absorbing and holding water because they maintain and protect a very porous, humus-rich topsoil. Investigators at Hubbard Brook Forest in New Hampshire found that converting a hillside from forest to grass doubled the amount of runoff and increased the leaching of nutrients manyfold.

Much worse is what occurs if the forest is simply cut and the soil is left exposed. The topsoil becomes saturated with water and slides off the slope in a muddy mass into waterways, leaving barren subsoil that continues to erode. Forests continue to be cleared at an alarming rate—about 13 million hectares (32 million acres) per year in 2005, most of which are in the developing countries, according to the *Global Forest Resources Assessment 2005* from the FAO. The problem is particularly acute when tropical rain forests are cut. The soils of equatorial regions have been subjected to heavy rains and leaching for millennia. Parent materials are already maximally weathered, and any free nutrients have long since been leached away. Consequently, tropical soils (oxisols) are notoriously lacking in nutrients. When the forests are cleared, the thin layer of humus with nutrients readily washes away. Only the nutrient-poor subsoil, which is very poor for agriculture, is left.

The Other End of the Erosion Problem. Water that is unable to infiltrate flows over the surface immediately into streams and rivers, overfilling them and causing flooding. Eroding soil, called **sediments**, is carried into streams and rivers, clogging channels and intensifying flooding, filling reservoirs, killing fish, and generally damaging the ecosystems of streams, rivers, bays, and estuaries. Around the world, coral reefs are dying because of sediments and other pollutants carried with them. Indeed, excess sediments and nutrients resulting from erosion are recognized as the greatest pollution problem of surface waters in many regions of the world. Meanwhile, groundwater resources are also depleted because the rainfall runs off rather than refilling the reservoir of water held in the soil or recharging groundwater.

Irrigation and Salinization

Irrigation—supplying water to croplands by artificial means—has dramatically increased crop production in regions that typically receive low rainfall. However, irrigation

Figure 11–20 Flood irrigation. The traditional method of irrigation is to flood furrows between rows with water from an irrigation canal. This method is extremely wasteful because most water evaporates or percolates beyond the root zone or because the water table rises to waterlog the soil and prevent the crop from growing.

is also a major contributor to land degradation. Traditionally, water has been diverted from rivers through canals and flooded through furrows in fields, a technique known as **flood irrigation** (Fig. 11–20). In recent years, **center-pivot irrigation**, in which water is pumped from a central well through a gigantic sprinkler that slowly pivots itself around the well, has become much more popular (see Figures 10–16b and c). In the United States, the Bureau of Reclamation is the major government agency involved with ongoing supplies of irrigation water to the western regions. More than 800 dams and reservoirs have been constructed by the bureau since its inception in 1902, and it now provides irrigation water for 4 million hectares (10 million acres) of farmland. Total irrigated land acreage in the United States is 22 million hectares (55 million acres), one-sixth of all cultivated cropland, according to the U.S. Department of Agriculture.

Worldwide, irrigated acreage has risen dramatically in the last few decades. The FAO estimates that total irrigated land now amounts to 277 million hectares (684 million acres), a 54% increase over the past 30 years. New irrigation projects continue to be built, but the expansion in acreage is now being offset in large part by another ominous trend: *salinization*.

Salinization is the accumulation of salts in and on the soil to the point where plant growth is suppressed (Fig. 11–21). Salinization occurs because even the freshest irrigation water contains at least 200 to 500 ppm (0.02% to 0.05%) of dissolved salts. Adding water to dryland soils also dissolves the high concentrations of soluble minerals that are often present in those soils. As the applied water leaves by evaporation or transpiration, the salts remain behind and gradually accumulate as a precipitate. Because it happens in drylands and renders the land less productive or even useless, salinization is considered a form of desertification.

Figure 11–21 Salinization. Millions of acres of irrigated land are now worthless because of the accumulation of salts left behind as water evaporates.

Lost land is an enormous problem. Worldwide, an estimated 1.5 million hectares (3.7 million acres) of agricultural land—or the equivalent of Connecticut and Rhode Island—is lost yearly to salinization and waterlogging. Asia alone has lost an estimated 130 million hectares (320 million acres) of farmland to salinization. In the United States, the problem is especially acute in the lower Colorado River basin and in the San Joaquin Valley of California, areas in which a total of 160,000 hectares (400,000 acres) has been rendered unproductive, representing an economic loss of more than $30 million per year. Adding to the problems, water supplies are fast being depleted by withdrawals for irrigation. This and related problems were discussed in Chapter 10.

Salinization can be avoided, and even reversed, if sufficient water is applied to leach the salts down through the soil. Unless there is suitable drainage, however, the soil will become a waterlogged quagmire in addition to being salinized. Drainage pipe may be installed under the surface, but only at great expense, and then attention must be paid to where the salt-laden water is drained. For example, the Kesterson National Wildlife Refuge in California was destroyed by pollution from irrigation drainage from selenium-enriched soils of the San Joaquin Valley. After receiving the salty drainage water for only three years, birds, fish, insects, and plants began to die off. The site was declared a toxic waste dump and has been drained and capped with soil. At least 14 other locations in the western United States have been identified as having the potential to accumulate toxic irrigation water, now known as the Kesterson Effect.

11.3 Soil Conservation

In the previous sections, we examined the nature of soils and how they support the growth of plants. We saw that healthy soils are essential for maintaining agricultural production. We examined the human activities that lead to erosion and desertification—overcultivation, overgrazing, and deforestation. Along the way, we encountered some agencies and organizations working to preserve and protect the soils.

There is broad consensus that soils are vital to human societies and that they should be preserved. This is encouraging, and so long as this consensus translates into action, we will be moving in the direction of sustainability. Because topsoils grow so slowly (an inch every hundred years at best), sustainability means doing all we can to reduce soil erosion.

Soil conservation must be practiced at two levels. The most important is the level of the *individual landholder*. As the chapter-opening story of Gabino López demonstrated, those working on the land are best situated to put into practice the conservation strategies that not only enhance their soils, but also bring better harvests and improve their way of life. A great deal of traditional knowledge exists with landholders, who routinely practice soil conservation techniques. The increased production in Burkina Faso is a prime example. This is stewardship in action.

The second level is the level of *public policy*. In particular, farm policies play a crucial role in determining how soils are conserved. Policies governing the grazing of public lands and forestry practices also lead to either good or poor stewardship of the soils. The story of Komsomolsky, at the beginning of the chapter, demonstrated how agricultural and human-resource policies can lead to a soil-related disaster on an epic scale. The two levels must work in harmony to bring about effective, sustainable stewardship of soil resources. Let us examine briefly what is being done in these two areas, beginning with individual landholders.

Helping Individual Landholders

In the developing world as well as the developed world, it is the individual landholders, farmers, and herders who hold the key to sustainable soil stewardship. They must be convinced that what they do will work, that it is affordable, and that, in the long run, their own well-being will improve. Frequently, this requires taking a number of small, realistic steps. It may also require help in the way of microlending (see Chapter 9), sound advice (often best coming from their peers), or simply encouragement to experiment (see Sustainability, "Three-Strata Forage System for Mountainous Drylands").

Sustainability

Three-Strata Forage System for Mountainous Drylands

The Hindu Kush Himalayas (HKH) are a mountain chain straddling eight Asian countries from China and Myanmar in the east to Afghanistan and Pakistan in the west. The world's highest mountain region, it is also the most populous. The HKH region occupies 4.29 million square kilometers (1.65 million mi²) and is home to more than 155 million people. Farming, forestry, and livestock sustain most of the people in the region. The land is fragile and is now threatened with an imbalance between population and available productive land. Rapid population growth and intensifying agriculture have led to, and are still producing, great increases in erosion and landslides as people clear forested slopes in a desperate attempt to meet their basic needs for fuelwood, timber, animal fodder, and cultivation. Aware of this intensifying problem, the **International Centre for Integrated Mountain Development** has promoted studies on how such a system can be brought into sustainable use. What is needed for the

HKH region is an approach that will rehabilitate degraded lands, provide a sustainable source of fodder for animals, and also furnish fuelwood for domestic use. The *three-strata forage system* (TSFS) can accomplish all of these aims and has been shown to work in other dryland environments.

The three strata are (1) grass and legumes, (2) shrubs, and (3) fodder trees. Given the right combination of plants, this vegetation will prevent soil erosion, yield fodder and human food and medicines, and create a pleasant environment where some greenery will persist all year. Such a system has worked well in the dry regions of Bali and is in use by almost all of the farmers there. In the HKH region, researcher Shaheena Hafeez has recommended a number of plants (grasses, legumes, shrubs, and trees) to use in a TSFS application for the mountainous terrain. One of the most interesting varieties is a tall shrub commonly found throughout the region: sea buckthorn (*Hippophae* spp.). This plant can

root and thrive in poor soils because it is a nitrogen fixer. Its root system is highly developed and holds the soil well. Sea buckthorn supplies palatable forage for animals all year and produces up to 1,300 pounds of berries per acre. The berries are a rich source of vitamins and can be made into jams, juices, and beverages. Oil from the berries is valued for its medicinal properties. A six-year-old sea buckthorn forest can provide more than 7 tons of fuelwood per acre on a sustainable basis.

The TSFS is only one of a number of strategies needed for meeting what are called the "silent emergencies" of the remote HKH region—poverty, energy and water shortages, scarcity of fuelwood and fodder, and poor access to sanitation, health, and education—as opposed to the "loud emergencies" of global climate change, loss of biodiversity, and the like. The TSFS is an important strategy because it promises to provide most of the basic goods and services essential to life in these arid mountain lands.

Toward those ends, the UN Convention to Combat Desertification is a large step in the right direction. Other efforts include the new *TerrAfrica Alliance* and the *Sustainable Agriculture and Rural Development Initiative* (SARD) facilitated by the FAO. The latter is a kind of umbrella organization that coordinates efforts to reach small farmers with sound agricultural practices; SARD puts the efforts in the context of encouraging sustainable development and achieving the MDGs. An example is found in the Bobo Diaulasso region of Burkina Faso, a rural community in Africa's drylands. Here intensive farming, driven by population growth, has led to a loss of soil fertility and declining crop yields. An FAO project has stepped in to enable a group of farmers to organize and adopt a range of sustainable practices, such as mixing legume and cereal grain plots to preserve soil fertility, cultivating soybeans to diversify income sources, and producing forage to feed livestock during the extended dry season. FAO brought the landholders and other stakeholders together to consolidate these findings into a set of established practices that could be implemented in other similar situations.

The Keita Project.   Preventing and reversing the degradation of dryland soils remains a key need for helping the more than 2 billion people who live in drylands, often among the poorest on Earth and subject to climatic variations that can destroy their livelihood and well-being. Niger is one of the hottest countries of the world, with much of its land in the

Sahara Desert itself. Below the Sahara, Niger is part of the Sahel, an ecoclimatic belt that runs across Africa from the Atlantic Ocean to the Red Sea. It is a semiarid grassland, receiving 6–20 inches of rainfall a year, but often less in drought years. In 1982, the Italian government initiated a project in the Keita district of Niger, over 4,000 square miles of plateaus, rocky slopes, and grasslands inhabited by 230,000 people. The project targeted the serious degradation brought on by overpopulation and the effects of soil and wind erosion. Some 41 dams were established to catch water from the monsoonal summer rains, more than 18 million trees were planted, and countless check dams (see Chapter 10, p. 243) were constructed to stop rainwater sheet erosion and provide soil and moisture for crops. Most of the work was carried out by local people, especially the women, who were paid with food. The antierosion land management scheme took many years, but it worked. The Keita district is now a flourishing environment for crops (Fig. 11–22), livestock, and people, with better roads, schools, and other social services. Desertification has been halted. The people now have control over their plots of land, an important step toward long-term stewardship of local resources.

Two Paradigms.   Desertification has been happening in many places and, when made worse by climate change or extended drought, can lead to permanent degradation and to greater poverty and misery. The human factors include population growth and immigration, unjust land tenure policies,

Figure 11–22 Land in the Keita district of Niger. This district was the recipient of a remarkable land restoration effort headed by the government of Italy.

and economic pressure to produce cash crops, all of which intensify the pressure on the land to increase its ecosystem services. The productivity of the land declines as overgrazing and overcultivation lead to increased erosion. These results are only made worse in the face of extended drought and some impacts of climate change. The final result seems practically irreversible and plunges local populations into deep poverty, disease, and malnutrition.

This paradigm is not inevitable, however, and a counter-paradigm involves a chain of events that may better describe what is happening on the ground in many dryland regions, such as Burkina Faso and Niger. Dryland populations are capable of responding to the evidence of degradation and taking corrective measures, including better managing livestock and grazing, adopting soil and water conservation measures, and diversifying their income in ways that relieve pressure on the land. The result is a sustainable relationship with dryland ecosystems that results in improving human well-being. Some of the processes of desertification, although distressing, may be signs of the cyclical nature of dryland ecosystems; they simply mean that in time things are likely to change and that what seemed irreversible was in fact not.

If there is a bottom line here, it is that dryland soils are resilient and that desertification is not the inevitable outcome of human occupation, which reminds us that humans are also resilient and capable of learning to adapt to the challenges of drought and population increase.

Public Policy and Soils

Despite the crucial importance of saving agricultural land, forests, and rangeland, soil degradation in the form of erosion, desertification, and salinization continues to occur because of human activities. The losses are on a collision course with sustainability. Who is responsible for making changes? At the first level, it is the individual landowner, but landowners are subject to public policies, the second level of decision making for soil conservation.

Subsidies. Farm policy in the United States was originally focused solely on increasing production. This goal has been achieved, as is demonstrated by continuing high yields and surpluses. The federal government supports U.S. agriculture through a number of programs; the farm lobby is a powerful one that has successfully maintained its support through the years. In more recent years, farm policy has emphasized maintaining farm income and support of farm commodities—in other words, subsidies. Farmers are guaranteed price levels for many products: grains, cotton, sugar, peanuts, dairy products, soybeans, and so on. In their recent book *Perverse Subsidies*, Norman Myers and Jennifer Kent[7] point out how subsidies are bad news for the economy and the environment. Because of government subsidies, U.S. taxpayers supported agricultural prices to the tune of $11 billion in 2009; crop insurance and loan programs added another $10 billion in support. These subsidies occurred in spite of record farm income for the year. Myers and Kent also argue that subsidies are bad news for the environment because they encourage excessive use of pesticides and fertilizers, they reduce crop rotation by locking farmers into annual crop support subsidies, and they promote the continued drawdown of groundwater aquifers through irrigation. Recall from Section 11.2 how similar subsidies to ranchers for grazing on federal lands have undercut the objectives of soil conservation for those lands.

Sustainable Agriculture. In comparison, the goals of sustainable agriculture are to (1) maintain a productive topsoil, (2) keep food safe and wholesome, (3) reduce the use of chemical fertilizers and pesticides, and last, but far from least, (4) keep farms economically viable. Professional agriculturists are becoming increasingly aware of the shortfalls of modern farming and are at work developing alternatives. Many options actually mimic practices of the past, such as contouring, crop rotation, terracing, the use of smaller equipment, and the application of a reduced amount of chemicals or no chemicals at all. In 1988, the U.S. Department of Agriculture started the **Sustainable Agriculture Research and Education (SARE)** program, which provides funding for investigating ways to accomplish all of the goals of sustainable agriculture just listed. The program has continued to receive funding of $5 to $12 million annually, a small, but encouraging amount earmarked for building and disseminating knowledge about sustainable agricultural systems.

Farm Legislation. The **Federal Agricultural Improvement and Reform Act (FAIR)** of 1996 attempted to bring needed reforms. Subsidies and controls over many farm commodities were reduced or eliminated, giving farmers greater flexibility in deciding what to plant, but also forcing them to rely more heavily on the market to guide their decisions (and earn their income!). Unfortunately, declining commodity prices prompted yearly emergency farm aid packages,

---

[7]Norman Myers and Jennifer Kent, *Perverse Subsidies: How Tax Dollars Can Undercut the Environment and the Economy* (Washington, D.C.: Island Press, 2001).

essentially maintaining the subsidies and controls targeted by FAIR.

The 2002 **Farm Security and Rural Investment Act** succeeded FAIR. It continued to subsidize a host of farm products, maintaining price supports and farm income for American farmers. The latest farm bill is the **Food, Conservation and Energy Act of 2008**, passed by Congress over President George W. Bush's veto. The president objected to the continued high level of support to big farms, arguing that payment limitations and subsidies for crops were too high. Nevertheless, Congress passed a bill that maintains the high levels of farm support and, because of subsidies for corn ethanol production, encourages farmers to plow under remaining native grasslands. It does, however, continue a number of programs aimed at soil and wetland conservation.

Conservation Programs. In 1985, Congress passed the **Conservation Reserve Program (CRP)**, under whose authority highly erodible cropland could be established as a "conservation reserve" of forest and grass. Farmers are paid about $125 per hectare ($50 per acre) per year for land placed in the reserve. As of 2008, 14 million hectares (34.7 million acres) had been placed in the reserve, saving an estimated 454 million tons of topsoil per year from erosion—at a cost of $1.8 billion in "rental payments" to the 430,800 farms enrolled in the program. Unfortunately, the 2008 farm bill lowered the 2009 acreage for the CRP to 32 million acres.

The new farm bill continued conservation funding for a number of other programs. In 1985, conservation activities supported by the U.S. Department of Agriculture received $1 billion, whereas these programs now receive more than $5 billion annually. Funding was provided for two newer programs, the **Wildlife Habitat Incentives Program (WHIP)** and the **Environmental Quality Incentives Program (EQIP)**, both of which encourage conservation-minded landowners to set aside portions of their land or address pollution problems. The 2002 farm bill initiated the **Conservation Security Program (CSP)**, a voluntary program that encourages stewardship of private farmlands, forests, and watersheds by providing landowners incentives to conserve soil and water resources on their lands. This was renamed the **Conservation Stewardship Program** in the 2008 farm bill; over $1 billion was provided for the new CSP. Over 12 million acres of farmlands may be enrolled each year.

Public policies aside, farmers should have incentives to nurture the soil on lands under cultivation. The more farmers incorporate conservation practices in their operations, the more productive their farms will become in the long run. The same applies to those who care for livestock that graze open rangelands. Most observers agree that the most significant obstacle to establishing sustainable soil conservation is a lack of knowledge about what such conservation can do. In particular, this applies to the developing countries, where agricultural policies often are poorly developed and the majority of landholders have little access to agricultural extension services like those provided in the United States by the NRCS. In cases where they are provided with such services, as with Gabino López, they often accomplish surprising gains in agricultural production.

Revisiting the Themes

Sound Science

Soil science uses the principles of sound science to understand how soils are formed, how they are structured, and how they can be sustained. As with any science, however, it is possible to draw erroneous conclusions from the data, as seen with the GLASOD data on world soils and soil degradation and with the data on soil erosion in the United States. The truth is that more field measurements are desperately needed, and the UNCCD's Land Degradation Assessment in Drylands (LADA) is a big step in that direction. In the United States, soil erosion is measured with map reading and equations, and when some scientists actually measured erosion (see p. 286), they found that the equation approach grossly overestimated erosion rates.

Sustainability

Topsoils take hundreds of years to form and represent the most vital component of land ecosystems. Under most conditions, they are highly sustainable. Because of heavy human use for agriculture, grazing, and forestry, however, lands are being degraded, causing the productivity of those ecosystems to decline. In particular, the drylands have been recognized as extremely vulnerable, and the UNCCD has been helping farmers to adopt new approaches to farming. Programs like COSECHA and TerrAfrica are spreading the word on these new approaches as well as on many sound traditional approaches that are not widely known. The three-strata forage system of the mountainous drylands of Asia (see p. 290) represents an amazing strategy for restoring soils and providing animal fodder and fuelwood that, if practiced, will promote long-term sustainability in those lands. On the other hand, the fees charged for grazing livestock on public lands in the United States are woefully inadequate and promote poor and unsustainable use of public lands.

Stewardship

Those who work the land or graze their livestock are stewards of the soil. They are temporary tenants whose lives will be far shorter than the time it took to form the soil. Their stewardship can lead to sustainable soil health, or it can lead to further erosion and degradation of the soil. A few short years of poor management can lead to the loss of hundreds of years worth of topsoil. Soil is cared for at the local level. Global figures may frighten or impress, but soil stewardship takes place one field at a time, by the people who are most familiar with the land. In many countries, soil erosion is being reversed with practices like no-till agriculture and contour strip cropping. In the United States, the Conservation Stewardship Program encourages farmers to care for the soils on their lands. This and other conservation programs under the U.S. Department of Agriculture have been adequately supported by the Food, Conservation and Energy Act of 2008.

REVIEW QUESTIONS

1. Compare the land use decisions and their consequences in Guatemala and Kalmykia, as found in the opening story.

2. Name and describe the properties of the three main components of soil texture.

3. What are the different layers, or horizons, that compose the soil profile? What makes each one distinct from the others?

4. Name and describe the four major soil orders listed in this chapter. Where are they mainly found?

5. What are the main things that plant roots must obtain from the soil? Name and describe a process (natural or unnatural) that can keep plants from obtaining the amounts of each of these things they need for survival.

6. Describe, in terms of both content and physical structure, the soil environment that will best fulfill the needs of plants.

7. State and define the relationships among humus, detritus, and soil organisms. Describe the role each of these factors plays in creating a soil favorable to plants.

8. What is meant by mineralization? What causes it? What are its consequences for soil?

9. What are some problems with the sweeping claims made about soil erosion and soil degradation? How have these claims been challenged?

10. What is the impact of water and wind on bare soil? Define and describe the process of erosion in detail.

11. What is meant by desertification? Describe how the process of erosion leads to a loss of water-holding capacity and, hence, to desertification.

12. What are drylands? Where are they found, and how important are they for human habitation and agriculture?

13. What are the three major cultural practices that expose soil to the weather?

14. What is meant by salinization, and what are its consequences? How does salinization result from irrigation?

15. What are the two levels at which soil conservation must be practiced? Give an example of what is being done at each level.

THINKING ENVIRONMENTALLY

1. Why is soil considered to be a detritus-based ecosystem? Describe how the aboveground portion and the belowground portion of an ecosystem act as two interrelated and interdependent communities.

2. Suppose you have uncovered an excellent new technique for keeping dryland soils moist and fertile. How would you get your message to the subsistence farmers?

3. Critique the following argument: Erosion is always with us; mountains are formed and then erode, and rivers erode canyons. You can't ever eliminate soil erosion, so it is foolish to ask farmers to do so.

4. The current system of management of grazing lands by the BLM is clearly unsustainable. What would you suggest in the way of public policy to make it sustainable?

MAKING A DIFFERENCE

1. To prevent soil erosion in your gardens, mulching twice a year can hold moisture in the soil during the summertime and keeps plants warmer during the winter. It also keeps rain from washing away your precious topsoil.

2. Don't use a leaf blower. It can erode topsoil more than even some of the most violent winds. It also burns fossil fuels.

3. Try mowing your fallen leaves to make shreds. Leave the shredded leaves in place and cover some with bark chips in garden beds to generate soil that is aerated, is rich with nutrients, and holds moisture better. Both of these practices combat erosion.

4. Composting is extremely sustainable and a very good method to practice. Two-thirds of solid waste is made up of purely organic material such as table scraps, trimmings from the yard, and paper products. Information on how to compost is available on the Internet. If you don't have enough space for traditional composting, you could try vermicomposting, which utilizes the surprising sustainable powers of worms.

5. When you use fertilizers, use organic products and ones labeled "slow-release," since "quick-release" fertilizer can be easily washed out and pollutes groundwater.

Go to **www.masteringenvironmentalscience.com** for practice quizzes, Pearson eText, videos, current events, and more.

A terraced rice plantation in Bali, Indonesia.

CHAPTER 12

The Production and Distribution of Food

HALF THE PEOPLE in the world—most of them in the developing countries—eat rice daily. Although an excellent food for energy, rice is a poor source of vitamins and other nutrients; hence, it must be supplemented with other foodstuffs for a balanced diet. Unfortunately, millions cannot afford other foods and depend heavily on rice for their nutrition. The result is two of the most common nutritional deficiencies: vitamin A deficiency, which leads to blindness and immune-system failures, and iron deficiency, which ultimately causes anemia and also immune-system problems. According to the World Health Organization, more than 195 million children suffer from vitamin A deficiency, and among these, over 5 million are afflicted with night blindness. Iron deficiency causes more than 100,000 maternal deaths a year during pregnancy and childbirth. If rice could somehow be supplemented with iron and vitamin A, two serious nutrient disorders could be conquered.

Biochemist Ingo Potrykus, working at the Swiss Federal Institute of Technology in Zurich, believed that it could be done and, in the early 1990s, began a program of research that would lead to a remarkable accomplishment of plant biotechnology: golden rice (Fig. 12–1). Over a period of seven years,

Figure 12–1 Golden rice. The yellow rice kernels contain beta carotene, inserted into them with biotechnology to enable the rice to help meet nutritional needs for vitamin A. The early golden rice 1 (GR1) is on the upper right, GR2 is on the left, and normal rice on the lower right. (*Source:* Martin Enserink, "Tough Lessons from Golden Rice," *Science* 320 (April 25, 2008): 470.

using the techniques of gene splicing, Potrykus, colleague Peter Beyer of the University of Freiburg in Germany, and their coworkers incorporated several genes into a rice genome that would enable the plant to synthesize beta carotene (which the body can use in turn to synthesize vitamin A) and iron in the body of the grain. (Both beta carotene and iron are normally present in the leaves of the plant, but when the rice is milled, only the grain is saved.) The new rice, called GR1, was named golden rice because of the color of the beta carotene in the grains. However, the amount of beta carotene in GR1 was so low that a person would have to eat many pounds of the rice daily to get to the recommended intake.

Since its development, golden rice has been improved to the point where a new variety (GR2) contains enough beta carotene to satisfy vitamin A needs at typical developing country diet levels. GR2 was created by Syngenta, a major agrochemical company. Some genetic lines of golden rice are well along in the regulatory approval process. Continued work on golden rice is now taking place in six laboratories in the Philippines, India, and Vietnam. The researchers are crossing GR1 and GR2 lines with popular rice varieties that are more suitable for warm climates in Asia.

FRANKENFOODS. The golden rice concept has met with a less-than-universal welcome because it is a genetically modified (GM) crop, containing genes from corn and beans. The nongovernmental organization (NGO) Greenpeace has waged an unrelenting campaign against golden rice, claiming that it is a ploy on the part of Syngenta to open the door to other GM crops (however, Syngenta has turned over the rights to GR2 to a humanitarian board, which Potrykus chairs). Genetically modified organisms are called "Frankenfoods" by some because they contain foreign genes and were created in a laboratory. The international debate over these foods has injected a new set of fears into the food production and distribution industry, and the debate is far from settled. However, if society is to meet the challenge of feeding another 3 or 4 billion people in this century and if we are to do so with better nutrition, many believe that genetic-engineering technology will have to play a key role.

Throughout the world, three decades of rapid population growth have left hundreds of millions dependent on imported food or food aid, while one in five in the developing world remains undernourished. As the world population continues its relentless rise (Fig. 1–3), no resource is more vital than food. Can the world's farmers and herders produce food swiftly enough to keep up with population growth? After all, Earth holds only a finite amount of the resources needed for food production—namely, suitable land, water, energy, and fertilizer.

SUCCESSES. By many measures, human societies have done very well at putting food on the table. More people are being fed than ever before, with more nutritional food. In the last 30 years, world food production has more than doubled, rising even more rapidly than population. In addition, the daily amount of food available in the developing countries has increased by 25%. A lively world trade in foodstuffs forms the bulk of economic production for many nations. Incredibly, more people are now overweight (1.3 billion) than are hungry. Optimists would say that, although regions of chronic hunger and malnutrition—and occasional famines—exist, these are the exceptions to an otherwise remarkable accomplishment. They claim that, when the world population levels off (at 10 billion?), enough food will still be produced to sustain it. Others question whether that is at all possible. There is real concern whether it will be possible to meet the Millennium Development Goal (MDG) of reducing by half the proportion of people suffering from hunger.

This chapter examines food-production systems, problems surrounding the lack of food, and the global effort being made to reduce hunger and malnutrition.

12.1 Crops and Animals: Major Patterns of Food Production

Some 12,000 years ago the Neolithic Revolution saw the introduction of agriculture and animal husbandry, which probably did more than anything else to foster the development of human civilization. Virtually all of the major crop plants and domestic animals were established in the first thousand years of agriculture. Between 1450 and 1700, world exploration and discovery led to an exchange of foods that greatly influenced agriculture and nutrition in the Western world. From the New World came potatoes, maize (corn), beans, squash, tomatoes, pineapples, and cocoa. Rice came from the Orient. In exchange, Europeans brought to the New World wheat, onions, sugarcane, and a host of domestic animals—horses, pigs, cattle, sheep, and goats. With the advancing capabilities of science and technology in the 1800s, the stage was set for a remarkable change in agriculture.

The Development of Modern Industrialized Agriculture

Until 150 years ago, the majority of people in the United States lived and worked on small farms. Human and animal labor turned former forests and grasslands into systems that produced enough food to supply a robust and growing nation (Fig. 12–2a). Farmers used traditional approaches to combat pests and soil erosion: Crops were rotated regularly, many different crops were grown, and animal wastes were returned to the soil. The land was good, and farming was efficient enough to allow an increasing number of people to leave the farm and join the growing ranks of merchants and workers living in cities and towns. Then, in the mid-1800s,

the Industrial Revolution came to the United States, and it had a major impact on farming.

The Transformation of Traditional Agriculture. The Industrial Revolution transformed agriculture so profoundly that today, in the United States, some 3 million farmers and farm workers produce enough food for all the nation's needs, plus a substantial amount for trade on world markets (Fig. 12–2b). This revolution increased the efficiency of farming remarkably. Since the mid-1930s, the number of farms has decreased by two-thirds (from 6.8 million to 2.1 million), while the size of farms has grown fourfold. (They now average 449 acres, or 182 hectares) Farming is a large and important business in the United States. All farm and farm-related employment (all jobs involved in producing the goods and services needed to satisfy the demand for farm products) accounts for 15% of the total U.S. workforce. Indeed, the revolution has achieved such gains in production that the United States frequently has had to cope with surpluses of many crops.

Virtually every industrialized nation has experienced this agricultural revolution. The pattern of developments in U.S. agriculture could just as well describe that in France, Australia, or Japan. Crop production has been raised to new heights, doubling or tripling yields per acre (Fig. 12–3). To understand how this tremendous increase in production was made possible, you need to understand the components of the agricultural revolution. These in turn reveal some of the problems brought on by the revolution.

Infrastructure. It is not enough to simply allow farmers to produce crops on their land. The transformation of agriculture was greatly facilitated by developing a supporting infrastructure. This involved rural electrification and road building, agricultural programs in state colleges and universities, the establishment of markets and efficient transportation

(a)

(b)

Figure 12–2 Traditional versus modern farming. (a) Traditional farming practiced in Arkansas. For hundreds of years, traditional practices on American farms supported the growing U.S. population. (b) Modern agricultural practice, illustrated by a combine harvesting wheat in America's Midwest.

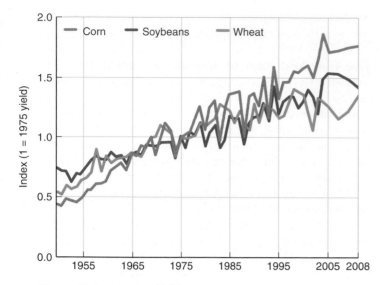

Figure 12–3 U.S. crop yields. Corn, wheat, and soybean yields (measured as tons or bushels per acre), indexed to 1975 as the base year. A value above 1.0 indicates a higher yield than in 1975; a value below 1.0 indicates a lower yield than in 1975. (*Source:* USDA National Agricultural Statistics Service.)

of farm goods, banks and credit unions to handle farm loans, agricultural and soil extension services for farmers, irrigation facilities, the formation of farm cooperatives, price and income support programs, and many other forms of government subsidies of farming. Most of these developments were essential in achieving the agricultural revolution, but many believe that the long-term maintenance of farm subsidies simply favors large corporate farms and works against the interests of farmers in developing countries.

Machinery. An incredible array of farm machinery handles virtually every imaginable need for working the soil: seeding, irrigating, weeding, and harvesting. The list is impressive: tractors, combines, handlers, harrows, mowers, toppers, tedders, balers, spreaders, foragers, plows, tillers, disc harrows, scufflers, strippers, seed drills, swathers, and more. This machinery has enabled farmers to cultivate far more land than ever. However, the shift from animal labor to machinery has created a dependency on fossil fuel energy. When the price of oil rises, so does the cost of food production.

Land Under Cultivation. In the United States, cropland and pastureland comprise some 987 million acres (400 million ha). This is about one-third of the total land area of the lower 48 states. Agriculture, therefore, has an enormous impact on the landscape. Before 1960, much of the increased production in the United States came from bringing new land into production. Since then, increases in crop yields and consistent crop surpluses have taken the pressure off to convert additional land to cropland, enabling farmers to be selective in the land they cultivate. The surpluses have also given farmers the opportunity to take erosion-prone land out of production. Under current farm policy, the **Conservation Reserve Program** reimburses farmers for "retiring" erosion-prone land and planting it with trees or grasses (see Chapter 11). It is likely that some of this land will soon be brought back into production to meet a rising demand for corn used in ethanol production.

Essentially all of the good cropland in the United States is now under cultivation or held in short-term reserve. Globally, agriculture occupies 38% of the land, and the net rate of growth of this land has been constant at around 0.3% per year over the past 30 years. The expansion of cropland comes at the expense of forests and wetlands, however, which are both economically important and ecologically fragile.

Fertilizers and Pesticides. Farmers long ago learned that animal manure and other organic supplements to soil could increase their crop yields. When chemical fertilizers first became available for use, however, farmers discovered that they could achieve even greater yields. Moreover, the fertilizers were more convenient to use and more available than manure. When fertilizers were first employed, 15 to 20 additional tons of grain were gained from each ton of fertilizer used. Between 1950 and 1990, worldwide fertilizer use rose tenfold. Then, following a decline traced to hard times in the countries of the former Soviet Union after it broke up in the 1990s, the worldwide use of fertilizer resumed its rise and is higher than ever (206 million tons in 2008). The high demand is tied to high prices of food on the global market and has led to a doubling of world fertilizer prices in 2007. Most of the current increase is in "underfertilized" countries such as China, India, and Brazil, but some is due to the rise in biofuel crops.

Chemical pesticides have provided significant control over insect and plant pests, but due to natural selection, the pests have become resistant to most of the pesticides. As a result, pesticide use has tripled since 1970, but the percentage of crops lost to pests has remained constant. Also, efforts to reduce the use of pesticides have begun because of side effects to human and environmental health. Many of the chemicals that are in use have not been adequately tested for human concerns and their potential for causing genetic defects. As is discussed in Chapter 13, progress is being made toward developing natural means of control that are environmentally safe, but these new methods will be unlikely to increase yields. And as with the use of farm machinery, fertilizer and pesticide manufacture depends on fossil fuel energy.

Irrigation. Worldwide, irrigated acreage doubled between 1961 and 2003. By 2003 (the latest year with data), irrigated acreage represented 18% of all cropland—some 683 million acres (277 million ha)—and produced 40% of the world's food. Irrigation is still expanding, but at a much slower pace because of limits on water resources. (Irrigation represents 70% of all water use.) Recall from Chapter 10 that much current irrigation is unsustainable because groundwater resources are being depleted. In addition, production is being adversely affected on as much as one-third of the world's irrigated land because of waterlogging and salinization—consequences of irrigating where there is poor drainage.

High-Yielding Varieties of Plants. Several decades ago plant geneticists developed new varieties of wheat, corn, and rice that gave yields double to triple those of traditional varieties. This feat was accomplished by selecting strains that diverted more of the plant's photosynthate (photosynthetic product) to the seeds and away from the stems, leaves, and roots. The seeds of modern wheat, rice, and corn now receive more than 50% of the photosynthate, close to the calculated physiological limit of 60%. As these new varieties were grown in the developed countries and then introduced throughout the world, production soared, and the Green Revolution was born.

The Green Revolution

The same technologies that gave rise to the agricultural revolution in the industrialized countries were eventually introduced into the developing world. There, they gave birth to the remarkable increases in crop production called the **Green Revolution**.

In 1943, the Rockefeller Foundation sent agricultural expert Norman Borlaug and three other U.S. agricultural scientists to Mexico, with the objective of exporting U.S. agricultural technology to a less developed nation that had serious food problems. Their aim was to improve the traditional crops grown in Mexico, especially wheat. Mexican wheat was well adapted to the subtropical climate, but it gave low yields and responded to fertilization by growing very tall stalks that were easily blown over. Using wheat from other areas of the world, Borlaug and his coworkers bred a dwarf hybrid with a large head and a thick stalk. The hybrid did well in warm weather when provided with fertilizer and sufficient water (Fig. 12–4). The program was highly successful: By the 1960s, Mexico had closed the gap between food production and food needs, wheat production had tripled, and Mexican wheat appeared on the export market.

Nobel Effort. Research workers with the Consultative Group on International Agricultural Research (CGIAR) extended the work done in Mexico, introducing modern varieties of high-yielding wheat and rice to other developing countries. In one success story, Borlaug induced India to import hybrid Mexican wheat seed in the mid-1960s, and in six years, India's wheat production tripled. Within a few years, many of the world's most populous countries turned the corner from being grain importers to achieving stability and, in some cases, even becoming grain exporters. Thus, while world population was increasing at its highest rate (2% per year), rice and wheat production increased 4% or more per year. In 35 years, global food production doubled. The Green Revolution has probably done more than any other single scientific or other achievement to prevent hunger and malnutrition. Borlaug was awarded the Nobel Peace Prize in 1970 in recognition of his contribution.

Panacea? The high-yielding modern varieties are now cultivated throughout the world and have become the basis of food production in China, Latin America, the Middle East, southern Asia, and the industrialized nations. Because the technology raises yields without requiring new agricultural lands, the Green Revolution has also held back a significant amount of deforestation in the developing world. But as remarkable as it was, the Green Revolution is not a panacea for all of the world's food-population difficulties. The high-yielding grains do best on irrigated land, and water shortages have begun to occur as a result of this dependence (Chapter 10). Also, the modern varieties require constant inputs of fertilizer, pesticides, and energy-using mechanized labor, all of which can be in short supply in developing countries. Further, Green Revolution agriculture did not eradicate hunger or poverty. Although India has been self-sufficient in terms of food since 1990, one-fifth of the population still suffers from malnutrition because they can't afford to purchase the food they need and there is no effective safety net.

Impacts. The CGIAR recently sponsored a study of the impact of the Green Revolution in the developing world

(a)

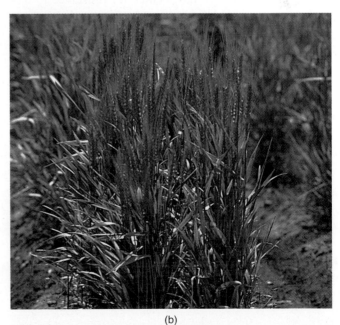
(b)

Figure 12–4 Traditional versus high-yielding wheat. A comparison of (a) an old variety of wheat, shown growing in Rwanda, with (b) a new, high-yielding variety of dwarf wheat growing in Mexico.

Figure 12–5 The yield gap for cereals. The difference in yield in terms of tons per acre between sub-Saharan Africa and other regions continues to widen. (*Source:* FAOSTAT.)

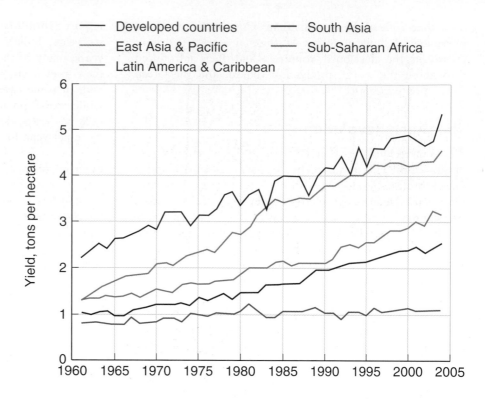

between 1960 and 2000.[1] The study's findings have shed light on that important agricultural development:

- The early years of the Green Revolution (1960–1980) were only the beginning; research on high-yielding crops has continued, and more varieties continue to be developed, released, and adopted by farmers.

- More-recent research has focused on resistance to diseases, pests, and climatic stresses. This work has been the key to the increased adoption of modern varieties.

- The early Green Revolution contributed greatly to expanded food production in Asia and Latin America; the later years of the Green Revolution have benefited Africa and the Middle East as a greater variety of crops was developed.

- If there had been no Green Revolution, crop yields in developing countries would certainly have been lower, leading to higher food prices and an increase in lands brought under cultivation everywhere. Less food would have been available, leading to higher levels of hunger and malnutrition and of infant mortality.

Even with the modern varieties, agricultural production in sub-Saharan Africa has lagged behind that of the rest of the developing world (Fig. 12–5). The reason for this gap is the dominance of subsistence agriculture in that region.

---

[1]R. E. Evenson and D. Gollin, "Assessing the Impact of the Green Revolution, 1960 to 2000," *Science* 300 (May 2, 2003): 758–762.

Subsistence Agriculture in the Developing World

In most of the developing world, plants and animals continue to be raised for food by *subsistence farmers*, using traditional agricultural methods. These farmers represent the great majority of rural populations. **Subsistence farmers** live on small parcels of land that provide them with the food for their households and, it is hoped, a small cash crop. From the point of view of the modern world, such farmers are very poor, although many do not consider themselves to be so. Like past agricultural practice in the United States, subsistence farming is labor intensive and lacks practically all of the inputs of industrialized agriculture. Also, it is often practiced on marginally productive land (Fig. 12–6).

Typically, a family owns a small parcel of land for growing food and maintains a few goats, chickens, or cows. The system may be sustainable if crop residues are fed to livestock, the livestock manure is used as a fertilizer, and the family's nutrition is adequate. Such a family is making the best use of very limited resources. Keep in mind, however, that subsistence agriculture is practiced in regions experiencing the most rapid population growth, even though that kind of agriculture is best suited for low population densities. An estimated 1.4 billion people in Latin America, Asia, and Africa—more than one-third of the people there—depend on subsistence agriculture. It has been called the "silent giant" that feeds most of the world's rural poor.

Problems in Africa. Low yields, rapid population growth, poverty, hunger, and high child mortality are the rule in sub-Saharan Africa. Two-thirds of the region's people live

Figure 12–6 Subsistence farming.
Subsistence farming feeds more than 1.4 billion
people in the developing world.

in rural lands and are dependent on agriculture for their livelihood. Because of the failures of agriculture, many in the region are dependent on imported food, and world food prices have been soaring. The World Bank's *World Development Report 2008: Agriculture for Development* states that agriculture alone carries the potential for lifting rural Africa out of its poverty. To raise crop yields, an *African Green Revolution* is needed, one that duplicates the success of the Green Revolution in Asia and Latin America. Most of the rural farmers lack the basic resources—fertilizer and high-yield seeds, in particular—to make improvements in crop yields. African leaders are waking up to this need after decades of neglecting funding for rural agricultural development. In Malawi, for example, government subsidies for fertilizer and high-yield seeds led to a doubling in food production in three years.

To facilitate more rapid progress, the Rockefeller and Gates foundations have initiated the *Alliance for a Green Revolution* in Africa and have appointed Kofi Annan, former secretary-general of the United Nations, as chairman of the alliance. Starting with foundation funding at $150 million, the alliance will coordinate financial aid from the wealthy countries toward achieving this sorely needed African Green Revolution. Already, African crop scientists are working with conventional breeding techniques to improve traditional African food crops. The alliance will also work throughout the subcontinent to help improve irrigation, soil health, and local and regional agricultural markets. The basic goal of this work is to enable the millions of subsistence farmers in sub-Saharan Africa to grow enough food to sustain their families and to produce enough to encourage the economic development of their communities and their countries.

Successes. Because subsistence agricultural practice varies with the local climate and with local knowledge, there are regions of the world where it works well. In some areas, subsistence agriculture involves shifting cultivation within tropical forests—often called slash-and-burn agriculture (Fig. 12–7). Research has shown that this practice can be sustainable. The cultivators create highly diverse ecosystems in which the cleared land supports a few years of crops and gradually shifts into agroforestry—a system of tree plantations with different ground crops employed as the trees grow.

Animal Farming and Its Consequences

Foot-and-mouth disease and mad-cow disease have recently plagued the livestock industry from Europe to South America. Australia, however, has been free of these diseases, and the cattle business there is booming. Australia is home to 28 million cattle, making it the world's top beef exporter. These cattle are ranch raised, grazing over vast areas of the Australian rangelands (Fig. 12–8). One cattle company alone manages a herd of 400,000 animals and would like to expand it to a million. The country's reputation as a provider of "disease-free" beef has brought good times for Australian cattlemen.

Raising livestock—sheep, goats, cattle, buffalo, and poultry—has many parallels to raising crops, and there are

Figure 12–7 Slash-and-burn agriculture.
Countryside in Mindanao, Philippine Islands, showing the results of sustainable slash-and-burn agriculture. Cultivated fields are interspersed with trees and natural areas in a diverse ecosystem.

Figure 12–8 Cattle ranching in Australia.
Jackaroos (Australian cowboys) drive a herd of cattle in the Injune area, Queensland. Go to **GRAPHit!** at www.masteringenvironmentalscience .com.

many connections between the two. Fully one-fourth of the world's croplands are used to feed domestic animals. In the United States alone, 70% of the grain crop—one half the cultivated acreage—goes to animals. The care, feeding, and "harvesting" of some 4 billion four-footed livestock and almost 18 billion birds constitute one of the most important economic activities on the planet. The primary force driving this livestock economy is the large number of the world's people who enjoy eating meat and dairy products—primarily, most of the developed world and growing numbers of people in less developed nations. As with crop farming, however, there are two patterns: (1) In the developed world and increasingly in the developing world, animals are raised in large numbers and often in confinement (the so-called confined animal feeding operations, or CAFOs); (2) in rural societies in the developing world, livestock and poultry are raised on family farms or by pastoralists who are subsistence farmers.

CAFOs. Industrial-style animal farming can damage the environment and human health in a host of ways. Because so much of the plant crop is fed to animals, all of the problems of industrialized agriculture apply to animal farming. Another serious problem is the management of animal manure. In developing countries, manure is a precious resource that is used to renew the fertility of the soil, build shelters, and provide fuel. In developed countries, it is a wasted resource. Close to 460 million tons of animal waste is produced each year in the United States, some of which leaks into surface waters and contributes to die-offs of fish, contamination with pathogens, and a proliferation of algae. CAFOs now produce more than 50% of the animals raised for meat in the United States. With factory farms on the sharp increase, wastes from manure and meat processing are either bypassing or overwhelming the often inadequate treatment systems available and polluting the nation's waterways. The Environmental Protection Agency (EPA) asserts that animal-based agriculture is the most widespread source of pollution in the nation's rivers.

The crowded factory farms are perfect conditions for diseases to incubate and spread among the animals and from animals to humans. Some very serious public health concerns have emerged, none more potentially devastating than **avian flu.** At least 12 years ago this disease (which has been around for centuries) mutated to a virulent form that has devastated poultry flocks in Asia—more than 140 million poultry have died or been purposely destroyed. Avian flu is spreading around the world, a result of wild birds that can carry the disease, but are symptom free. The virus has begun to spread to humans from birds and has infected more than 400 people in the last several years, often with deadly results. Because influenza viruses are capable of mutating and recombining with each other, there is great concern in world health organizations that avian flu will precipitate a devastating global human flu pandemic. Further, many bacteria associated with animal farming are human pathogens, causing food-borne human diseases. The bacterium *Salmonella* alone is estimated to cause losses of $2.5 billion per year in the United States. Because antibiotics are heavily used in animal feed, many of the bacterial contaminants have developed resistance to antibiotics, and these are far more difficult to treat when humans are infected.

Rain Forest Crunched. In Latin America, more than 58 million acres (23 million ha) of tropical rain forests have been converted to cattle pasture. Even though most of this land is best suited for growing rain-forest trees, some of it does support a rural population of subsistence farmers producing a diversity of crops. Much of the land, however, is held by relatively few ranchers who own huge spreads. At first, the conversion of land from forest to pasture was the outcome of government policies meant to encourage the colonization of unoccupied land, and most of the meat was for domestic use. However, today cattle production in the Amazon basin has expanded and is export driven; Brazilian beef exports bring in $1.5 billion annually.

Climate Change. According to the UN's intergovernmental Panel on Climate Change, deforestation and other changes in land use in the tropics release an estimated 1.6 billion tons of carbon to the atmosphere annually, contributing a significant amount of carbon dioxide to the greenhouse effect. Also, because their digestive process is anaerobic, cows and other ruminant animals annually emit some 100 million tons of methane, another greenhouse gas, through belching and flatulence. The anaerobic decomposition of manure leads to an additional 30 million tons of methane per year. All this methane released by livestock makes up about 3% of the gases causing global warming (Chapter 18).

Good Cow. Even though their animals also contribute to the methane problem, it is callous to fault the subsistence farmers whose domestic animals enhance their diet and improve their quality of life. In fact, one of the most important kinds of sustainable development aid brought to rural families is the gift of a cow or a few goats or rabbits, as carried out by Heifer Project International (Fig. 12–9). Working in 115 countries, the project has distributed large animals, beehives, fowl, and fish fingerlings to families. The organization works with grassroots groups of local people who oversee a given project. Each project is chosen to meet a genuine need and to improve the environment. Those receiving the animals must agree to be trained in how to raise them properly and must be committed to passing on the gift in the form of donating offspring of their livestock to other needy people.

The well-being of millions in the developing world is tied directly to the animals they raise, and the impact of these animals on the environment is often sustainable. Animals that are well managed can enhance the soil and enable rural farmers to maintain a balanced farm ecosystem. The rise in animal husbandry in the developing world—farmers in the developing countries now own two-thirds of the world's domestic animals—has made an important contribution to nutrition, especially for women and small children. In sum, animal farming is far more likely to be sustainable in the rural farms and pastoral herds of the developing world than in the CAFOs of the developed world, where there is a pressing need to address problems of pollution and disease.

Figure 12–9 Heifer Project International. An East African woman gives a goat to a neighbor as part of the project's "passing on the gift" ceremony. Participants agree to pass on offspring of their livestock to others.

Biofuels and Food Production

Because of the accumulation of greenhouse gases in the atmosphere, global climate change is happening, with consequences that are impacting virtually every country (Chapter 18). The most basic cause of climate change is the burning of fossil fuels, which has driven atmospheric CO_2 (a greenhouse gas) more than 35% above preindustrial levels. One of the important ways to mitigate climate change is to burn **biofuels** (ethanol and oils derived from agricultural crops) instead of fossil fuels. Because there is no new CO_2 being released to the atmosphere, biofuels are a form of renewable energy. But there is another, more basic reason for using biofuels. In the past, fossil fuel oil was less expensive than biofuels, but in recent years, oil prices have fluctuated wildly, and biofuels have become very attractive.

The primary biofuel in production today is ethanol, derived from the fermentation of cornstarch (in the United States) or cane sugar (in Brazil). Ethanol is a suitable substitute for gasoline, and with inflated gasoline prices, ethanol manufacture is quite competitive. The corn farmers in Iowa are happy to sell their corn to the highest bidder, so a rising fraction (one-third of the 2008 crop) of U.S. corn is devoted to meeting the growing demand from ethanol distilleries. At the same time—and in the minds of many, directly connected—food prices have risen worldwide to the point of crisis.

Consequences. Some observers are accusing the United States and the European Union (EU) of embarking on "a criminal path" because they divert corn from hungry people to ethanol production. The United States produces 40% of the world's corn, a staple food for millions, and 55–60% of the corn offered on the world food market. A nutritionist with the UN World Food Program stated that "global price rises mean that food is literally being taken out of the mouths of hungry children whose parents can no longer feed them."[2] How accurate is this accusation?

[2]World Food Program News. "Price rises eating away at children's futures: an interview with WFP nutritionist Andrew Thorne-Lyman." April 28, 2008.

While it is certainly true that biofuel production has created a new demand for corn, it must be noted that the prices of wheat, rice, and soybeans have also risen dramatically—in fact, more than the price of corn. So far the new demand for corn is being met with larger corn plantings; the increase was accomplished by taking land out of soybean production. Ethanol production uses field corn, which is usually fed to livestock. Only the cornstarch is used for ethanol production; the proteins, vitamins, and fiber are converted to high-energy animal feed, sweeteners, and corn oil.

Common factors that are contributing to the high food prices are (1) the price of oil (used by farm machinery and in fertilizer manufacture), which reflects a growing competition from developing countries like China and India for a resource that is limited; (2) bad weather and poor harvests in many parts of the world, and especially in Australia, which is normally a major wheat-exporter; and (3) the rising demand for meat and animal products, again from emerging economies like China. Research by the International Food Policy Research Institute (IFPRI) suggests that perhaps 30% of the grain price increases between 2000 and 2007 are linked to biofuel production. Many observers suggest that some limit be put on corn ethanol production, while making a greater investment in a new generation of biofuels based on grasses and timber waste (see Chapter 16).

In the meantime, there is the need to meet the food needs of a growing world population. Looking ahead to 2020, will we be able to feed the additional 1½ billion people and also make significant progress on meeting the Millennium Development Goal of reducing by half existing hunger and malnutrition in the world? What are the prospects?

Future Prospects. A team from the IFPRI looked at the prospects a few years ago, asking those very questions. As a baseline, the IFPRI team expected the demand for grains to be much as it has been, slightly outpacing population increase (Fig. 12–10). But it also expected a big jump in the world's consumption of meat and projected a 50% rise in

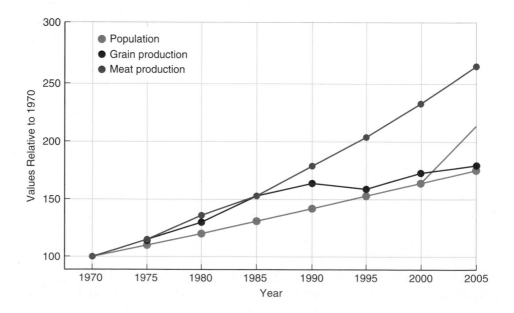

Figure 12–10 Global population and grain and meat production. Grain production has kept up with the population increase, and meat production has greatly outpaced it over the last 35 years. (*Source:* FAOSTAT.)

consumption by 2020. This increase is traced to rising economies in the developing world. The major consequence of this rise in demand for meat is a shift from grains for human consumption to the production of feed grains and soybeans.

The IFPRI was pessimistic about the ability of developing countries to meet the rise in demand for grains. The team pointed out that only about 40 million more hectares are likely to be converted to crops, due to major constraints on suitable land; most of the prime arable land in the world is now in cultivation. Team members pointed out that increases in yields for grains are slowing down and believe that rising demand for grains will have to be met by trade—with grain imports by the developing countries to double by 2020. They questioned whether the major grain exporters would be able to meet that demand and suggested that only countries with significant economic development would be able to afford the imports. In particular, the IFPRI focused on sub-Saharan Africa, which it believes will be a "hot spot" for hunger and malnutrition for years to come.

Is the IFPRI scenario realistic? The team members clearly did not anticipate the food crisis of 2007–2008 or the surge in biofuel production. And although they spotlighted sub-Saharan Africa, they did not seem to think there was much hope for an African Green Revolution. As we have seen, there are new prospects that indicate this might actually happen.

Global Picture. Given the limits on suitable land for agriculture, there seem to be only two prospects for increasing food production: (1) continue to increase crop yields and (2) grow food crops on land that is now used for feedstock crops, biofuels, or cash crops. As mentioned earlier, a dramatic rise in crop yields in the developing countries was the major accomplishment of the Green Revolution: Wheat yields tripled, rice and corn yields more than doubled, potato yields rose 78%, and cassava yields rose 36%, according to the UN Food and Agricultural Organization (FAO). Interestingly, grain yields have continued to rise in some of the *developed* countries, too. In France, for example, wheat yields have quadrupled since 1950, reaching more than 6 tons per hectare. Rice yields in Japan have risen 67% in the same period. Actually, the genetic potential exists for wheat yields of 14 tons per hectare or higher. Can we expect yields to continue to increase up to their genetic potential?

It turns out that the great differences in grain yields among regions (Table 12–1) have less to do with the genetic strains used and more to do with the weather. Egypt and Mexico, for example, achieve much higher yields than does the United States because most of their wheat is irrigated, while U.S. wheat is rain-fed and grown on former grasslands. Australian wheat yields are even lower, the result of the country's sparse rainfall. Once the agricultural land is planted with high-yielding strains and fertilized to the maximum, other factors—soil, rainfall, and available sunlight—limit productivity. These environmental limits are a reminder that agricultural sustainability is highly dependent on soil and water conservation (Chapters 10 and 11) and on the weather. Variations in climate affect harvests locally and globally,

Country	Yield (tons/hectare)
United Kingdom	7.3
Egypt	6.5
France	6.3
Mexico	5.0
China	4.8
Zimbabwe	4.6
Poland	3.9
Pakistan	2.8
India	2.7
United States	2.6
Argentina	2.5
Canada	2.4
Ukraine	2.3
Russian Federation	2.0
Kenya	1.8
Kazakhstan	1.3
Australia	1.1

TABLE 12–1   Annual Wheat Yield in Various Countries, 2007

Source: FAOSTAT Agricultural Data, FAO, 2008.

injecting significant instability in food production, which makes future planning extremely chancy. Another imponderable in future food production is global climate change: As the predicted warming of the 21st century occurs, it is impossible to project how rainfall patterns will change. However, scientists at NASA's Goddard Institute of Space Studies estimate that developing countries could lose 334 million acres (135 million ha) of essential farm land in the next 50 years due to temperature increases in the tropics.

Less Meat and Biofuels. What of the possibility of switching from the production of feed grain and cash crops to food for people? Almost 70% of domestic grain in the United States is used to feed livestock. The percentages drop over other regions of the world in proportion to the economic level of the region. Sub-Saharan Africa and India, for example, use only 2% of their grain to feed livestock. Feed grain can be considered a buffer against world hunger; if the food supply becomes critical, it might force more of the world's people to eat lower on the food chain (less meat, more grain). The trend, however, is in exactly the opposite direction. Of course, converting land use from cash crops to food crops also is possible, but it is a complex undertaking because it involves such issues as instituting land reform and maintaining a balance of trade.

The surge in biofuel production in the United States is not likely to subside anytime soon. In 2008, more than 4 billion bushels of corn, almost one-third of the crop, were diverted to ethanol production. This demand is expected to rise to 5 billion bushels by 2010 as more ethanol plants come on line. More corn production will come, but at the risk of using marginal croplands and continuing the shift of soybean production to corn. Since more corn will be diverted to ethanol production (currently offering the best prices to corn growers), less will be available for animal feed and export.

In the face of these constraints, Gordon Conway, president of the Rockefeller Foundation, has called for a new Green Revolution. He has dubbed it a **Doubly Green Revolution**—"a revolution that is even more productive than the first Green Revolution and even more 'green' in terms of conserving natural resources and the environment. During the next three decades, it must aim to repeat the successes of the Green Revolution on a global scale in many diverse localities and be equitable, sustainable, and environmentally friendly."[3] Conway believes that this new revolution is possible in large part because of the emergence of biotechnology and the kinds of genetic manipulations demonstrated by the development of golden rice.

12.2 From Green Revolution to Gene Revolution

Genetic engineering makes it possible to combine characteristics from genetically different plants and to incorporate desired traits into crop lines and animals, producing so-called transgenic, or genetically modified (GM), varieties. This is a radically different technology from the genetic research that

[3]Gordon Conway. *"The Doubly Green Revolution: Food for All in the 21st Century."* 1999. Cornell University Press. Ithaca, NY.

produced the Green Revolution: Researchers are no longer limited to genes that already exist within a species or that could arise via mutation; it is now possible to exchange genes among bacteria, animals, and plants.

This kind of technology can help the developing world to produce more food. Genetically modified organisms are under intense research and development, with many new GM organisms coming out of the lab. Still, in spite of the obvious potential of these organisms, there are concerns about their development and use. We need to look at both sides of this controversy.

The Promise

The earliest and still the most widely adopted genetically altered products to be marketed are (1) cotton plants with built-in resistance to insects that comes from genes taken from a bacterium (so-called Bt, standing for *Bacillus thuringiensis,* the name of the bacterium) and (2) corn and soybeans resistant to the herbicide Roundup® (chemically, *glyphosate*), allowing farmers to employ no-till techniques. More recently, biotechnology has developed sorghum (an important African crop) resistant to a parasitic plant known as witchweed, which infests many crops in Africa; corn, potatoes, and cotton resistant to insects; rice resistant to bacterial blight disease; and trees and salmon that grow very rapidly, to name a few. In just 13 years since bioengineered seeds became commercially available, farmers have overwhelmingly turned to the transgenic breeds of soybeans, cotton, and corn. Worldwide, 282 million acres (114 million ha) in 23 countries were planted with bioengineered crops in 2007 (Fig. 12–11); 99% of this area is accounted for by just four crops (corn, soybeans, canola, and cotton). A rapidly increasing trend is toward the so-called stacked products, containing two or more biotech genes in a crop (e.g., traits directed toward different insects, or insect resistance plus herbicide tolerance).

Figure 12–11 Global area of genetically modified crops. The plot shows the rapid increase in global areas planted with genetically modified crops, primarily cotton, soybeans, canola and corn. (*Source:* Clive James. "Global Status of Commercialized Biotech/GM Crops." Brief 37. ISAAA, 2007, p. 3.)

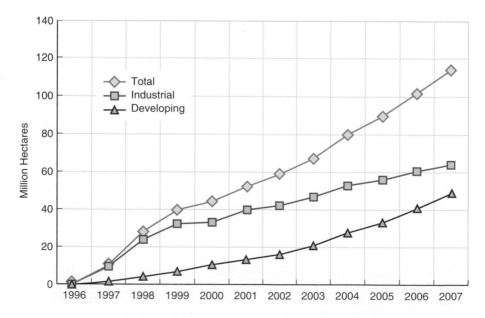

Biotech crop research that can benefit the developing countries is proceeding at a rapid pace, often in those very countries. China, in particular, has emerged as a leader in plant biotechnology, having recently sequenced the rice genome. Among other aims, the objectives of agricultural biotechnology are (1) to incorporate resistance to diseases and pests that attack important tropical plants; (2) to increase tolerance to environmental conditions, such as drought and a high salt level, which stress most plants; (3) to improve the nutritional value of commonly eaten crops (like the work with golden rice); and (4) to produce pharmaceutical products in ordinary crop plants ("pharma" crops). Thus, products under development include virus-resistant cassavas, sweet potatoes, melons, squash, tomatoes, and papayas (Fig. 12–12); protein-enhanced corn and soybeans; fungus-resistant bananas; bananas and tomatoes that contain an antidiarrheal vaccine; and drought-tolerant sorghum and corn. The potential for transgenic crops and animals seems almost unlimited.

Among the important environmental benefits of bioengineered crops are reductions in the use of pesticides because the crops are already resistant to pests, less erosion because no-till cropping is facilitated by the use of herbicide-resistant crops, and less environmental damage associated with bringing more land into production because existing agricultural lands will produce more food. When Chinese cotton farmers began turning to Bt cotton, for example, they found that they used much less pesticide, saving money, time, and their own health in the process.

Marker-Assisted Breeding. The new technologies also involve crop improvements without resorting to transgenic traits. For example, a technique called *marker-assisted breeding* identifies desirable traits in crop plants or their wild ancestors and locates the genes with DNA sequencing. Next, plants with the desired gene are crossbred with a modern crop breeding line. Then DNA screening of seedlings for the desired gene bypasses the need to grow plants to maturity to see if they were successfully transformed. Using this technique, a gene from a wild rice ancestor was incorporated into a modern high-yield rice variety, raising the yield by 18%. Other crops being developed with this approach are carrots (a high-calcium "super-carrot"), potatoes (more starch), and pigeon peas (higher yields). This approach has the distinct advantage of not requiring the special testing and permits needed with transgenic crops, greatly reducing costs and time delays. It also bypasses the objections raised to transgenic crops in many countries.

The Problems

Concerns about genetic engineering technology involve three considerations: environmental problems, food safety, and access to the new techniques.

Environmental Concerns. A major environmental concern focuses on the pest-resistant properties of the transgenic crops. If pests have a broad exposure to the toxin or some other resistance incorporated into the plant, it is possible that

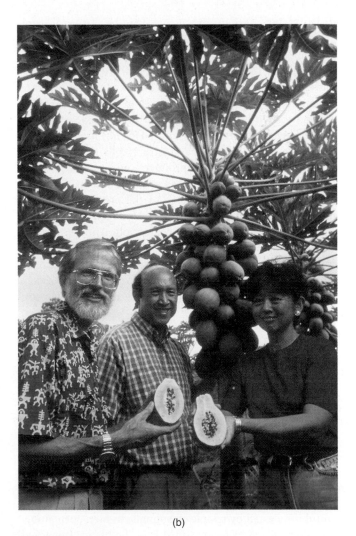

(a)   (b)

Figure 12–12 Papaya plants in Hawaii. (a) The healthy plants have been genetically altered to prevent ring spot, a viral disease of papayas; the plants on the right are traditional ones. (b) The developers of a papaya that is resistant to the deadly virus: from left, Dr. Richard Manshardt, Dr. Dennis Gonsalves, and Maureen Fitch, a Hawaiian graduate student.

they will develop resistance to the toxin and thus render it ineffective as an independent pesticide. The transgenic crop then loses its advantage. Indeed, glyphosate resistance has been found in a number of weeds that infest soybean, corn, and cotton fields. The scale of this resistance is currently small. However, because so much acreage worldwide is now planted with glyphosate-resistant crops, the resistant weeds could spread widely and cause economic disaster for farmers. Another concern is the ecological impact of the crops. For example, pollen from Bt corn (resistant to the corn borer) can be spread by the wind to adjacent natural areas where beneficial insects may pick it up and be killed by the toxin. This effect was shown to occur in laboratory tests with monarch butterflies. Recent studies indicate, however, that the Bt corn poses no risk to the butterflies, and the EPA has cleared the corn for continued planting. A third environmental concern arises because genes for herbicide resistance or for tolerance to drought and other environmental conditions can also spread by pollen to ordinary crop plants or their wild relatives, possibly creating new or "super" weeds. Field studies indicate that this concern is valid.

Safety Issues. Food safety issues arise because transgenic crops contain proteins from different organisms and might trigger an unexpected allergic response in people who consume the food. Tests have shown that a Brazil nut gene incorporated into soybeans was able to express (or manufacture) a protein in the soybeans that induced an allergic response in individuals already allergic to Brazil nuts. Also, antibiotic-resistance genes are often incorporated into transgenic organisms in order to provide a way to trace cells that have been transformed. The transgenic product could then convey the resistance to antibiotics to pathogens in human systems by a separate process of gene transfer, or the product could prevent the antibiotic from being used effectively in an individual who has consumed the transgenic food. Another concern relates to the possibility that plants could produce new toxic or allergenic substances in their tissues in response to the presence of the foreign genes. To date, however, none of these concerns has become evident in countries where GM crops have been grown over the past 13 years.

The biotech industry has recognized the advantages of producing pharmaceuticals by engineering genes for desirable products into common crop plants. More than 200 permits have been granted by the U.S. Department of Agriculture (USDA) for field tests of such applications, involving hormones, enzymes, diagnostic drugs, and other substances. The Union of Concerned Scientists (UCS) recently brought together a panel of experts to examine the risks of this technology, and their report led UCS to ask the USDA to halt the outdoor production of these crops. The primary risk is that these "pharma" crops could contaminate ordinary food crops—some of the compounds being developed could be harmful if ingested by people or animals. In fact, corn altered to produce a pig vaccine spread in fields in Iowa and Nebraska in 2002, and only a last-minute intervention by the USDA kept the altered corn out of food. Since it seems unlikely that a foolproof system could be established, UCS is urging an alternate strategy: Why not produce the pharmaceuticals with noncrop plants?

Access in the Developing World. The problems concerning access to the new technologies relate to the developing world. For the first few years, almost all genetically modified organisms were developed by large agricultural-industrial firms, with profit as the primary motive. Accordingly, farmers have been forbidden by contract from simply propagating the seeds themselves and must purchase seeds annually. Farmers in the developing countries are less able to afford the higher costs of the new seeds, which must be paid up front each year. Fortunately, some noncommercial and donor-funded laboratories are taking aim at this problem. Further, biotech research is booming in China, India, and the Philippines, where most of the fruits of the research are being made available to developing-world farmers. In addition, the genetically modified seeds are spreading rapidly through seed "piracy."

In 2004, the UN FAO gave its qualified approval to agricultural biotechnology in *The State of Food and Agriculture*,[4] but concluded that the promise of the technology to alleviate global hunger and improve the well-being of farmers in developing countries was still more theory than reality. However, in 2007 some 7.5 million subsistence farmers benefited from insect-resistant cotton in China that was developed in Chinese biotech laboratories, and some 3.8 million small farmers in India are now experiencing cotton yield increases of 50%. Two indications of the potential of developing countries to produce their own applications can be seen in Iran and South Africa. Iran began in 2006 to deploy a Bt rice developed in that country and hopes within a few years to eliminate its need to import rice (currently 1 million tons a year). South African researchers have developed corn resistant to the maize streak virus, which devastates corn crops throughout sub-Saharan Africa (Fig. 12–13).

All of these concerns, combined with the general fear of what unknown technologies can bring, have stimulated a rather heated controversy over the spreading use of genetically modified organisms by farmers and the foods developed from such organisms. Activists around the world have rallied to protest the development of such organisms and the "Frankenfoods" derived from them. The protests have been strongest in Europe, where concerns over food have been exacerbated by other recent food scares, such as mad-cow disease. Some suspect, however, that foreign governments are using the controversy to prevent imports of less expensive foods. Far less concern has arisen so far in the United States. Because corn and soybeans are used in so many processed foods, an estimated 60% of all foods in the United States are thought to contain some genetically modified substances. Some consumer groups would like to see mandatory labeling of all such foods, but U.S. policy does not require such labeling.

---

[4]United Nations, Food and Agricultural Organization, *The State of Food and Agriculture 2003–2004: Agricultural Biotechnology: Meeting the Needs of the Poor?* (Rome: UN FAO, 2004).

Figure 12–13 Maize streak virus. Leaves are affected by the virus, but the most important impact is to cause the plant to be stunted and the corn ear to fail to develop. A heavy infection can cause 100% crop failure. This virus is widely spread throughout sub-Saharan Africa.

The Policies

Biotechnology and its application to food crops do not exist in a regulatory vacuum. In the United States, the EPA, the USDA, and the Food and Drug Administration (FDA) all have regulatory oversight of different elements of the application of biotechnology to food crops. Concern over their oversight led to a report from the National Academy of Sciences' National Research Council, a prestigious and influential body. The report, *Genetically Modified Pest-Protected Plants: Science and Regulation*, concluded that transgenic crops have been adequately tested for environmental and health effects, but that the three agencies involved need to do a better job of coordinating their work and sharing their information with the public. Overall, the report endorsed the science, technology, and regulation, stating, "The committee is not aware of any evidence that foods on the market are unsafe to eat as a result of genetic modification."[5] At the same time, the report called for more research on the environmental and safety issues.

Cartagena Protocol. On the international level, the UN Convention on Biodiversity sponsored a key conference in Montreal, Canada, in January 2000 to deal with trade in genetically modified organisms. After heated negotiations, the conference reached an agreement, called the **Cartagena Protocol on Biosafety.** Surprisingly, the agreement was welcomed by governments, the private sector, and environmental groups alike. At issue was a fundamental philosophy about how technologies, particularly new ones, should be regulated. Those skeptical of the claims of biotech companies about the safety of their products wanted *proof that the genetically modified organisms were safe* before they were allowed into their countries. Those advocating the use of such organisms wanted to see the *suspected dangers proven* before genetically modified organisms could be denied access to overseas markets.

Precautionary Principle. The Cartagena Protocol states that the "lack of scientific certainty due to insufficient relevant scientific information and knowledge. . . . shall not prevent [a country] from taking a decision" on the import of genetically modified organisms. The protocol puts the right to deny entry of any of these organisms in the hands of the importing country, but its decision must be based on sound science (involving an assessment of the risks involved) and the broad sharing of information about the products. Thus, the agreement makes operational, for the first time in international environmental affairs, the so-called **precautionary principle**, which states that *where there are threats of serious or irreversible damage, lack of scientific certainty should not be used as a reason for failing to take measures to prevent potential damage.* According to another stipulation in the protocol, shipments that contain food commodities made with genetically modified organisms must be labeled to say that they *may* contain them. This wording was a compromise that enabled opposing parties to reach a consensus on the entire protocol. The protocol has been ratified by 147 countries (but not the United States) and became legally binding in 2003.

A regulatory framework is being erected around the use of this new technology. The public response is still negative in Europe and positive in the United States, although activists have been quite vocal on both sides of the Atlantic and are being heard in the developing countries. The EU has lifted a ban on genetically modified foods, but is requiring clear labeling on any such food or animal feed containing more than 0.9% of genetically modified ingredients. Countries everywhere are cautiously examining the technology, but the trend is in favor of its adoption; in 2007, 23 countries, including several in the EU, were growing biotech crops. Clearly, the future almost certainly will have to include major advances in food production from biotechnology, and many of these advances are likely to benefit the developing countries. A new plant revolution is under way, and it is unlikely to be stopped. Former U.S. president Jimmy Carter put it well: "Responsible biotechnology is not the enemy; starvation is."[6] If food production is to keep pace with population growth, biotechnological advances will be essential, in the view of most observers.

[5]National Research Council, Committee on Genetically Modified Pest-Protected Plants, *Genetically Modified Pest-Protected Plants: Science and Regulation* (Washington, D.C.: National Academy Press, 2000), 8.

[6]Jimmy Carter. "Forestalling Famine with Biotechnology." *The Washington Times*, July 11, 1997.

12.3 Food Distribution and Trade

For centuries, the general rule for basic foodstuffs—grains, vegetables, meat, and dairy products—was *self-sufficiency*. Whenever climate, blight (as in the 19th-century Irish potato famine), or war interrupted the agricultural production of a nation or region, the inevitable result was famine and death, sometimes on the scale of millions. Once colonies were established in the New World, timber, furs, tobacco, fish, and, later, sugar, coffee, cotton, and other raw materials began to flow back to the Old World. In turn, the Old World exported manufactured goods, which helped transform the colonies into societies much like the European ones that had given birth to them. With the Industrial Revolution, trade between nations intensified, and soon it became economically feasible to ship basic foodstuffs around the world. A lively and important world trade in foodstuffs arose, and as it did, the need for self-sufficiency in food diminished. Like other sectors of the economy, food has become globalized. World trade in agricultural products was worth $1.035 trillion in 2008.

Patterns in Food Trade

Given this global pattern, agricultural production systems do much more than supply a country's internal food needs. For some nations (such as the United States and Canada), the capacity to produce more basic foodstuffs than the home population needs represents an extremely important economic enterprise. For many other countries (especially those of the developing world), special commodities such as coffee, fruit, sugar, spices, palm oil, cocoa, and nuts provide their only significant export products. This trade clearly helps the exporter, and it allows importing nations to enjoy foods that they are unable to raise themselves or that are out of season. Given the realities of a market economy, the exchange works well only as long as the importing nation can pay cash for the food. Cash is earned by exporting raw materials, fuel, manufactured goods, or special commodities. In this way, for example, Japan imports $48 billion worth of food and livestock feed (representing 60% of total consumption), but more than

makes up for it by exporting $596 billion in manufactured goods (cars, electronic equipment, and so on) annually.

Grain on the Move. The most important foodstuffs on the world market are grain (wheat, rice, corn, barley, rye, and sorghum) and oilseeds, such as soybeans. Some of this is imported by high- and middle-income countries to satisfy the rising demand for animal protein. It is instructive to examine the pattern of global trade in grain over the past seven decades (Table 12–2). In 1935, only western Europe was importing grain; Asia, Africa, and Latin America were self-sufficient. By 1950, new patterns were emerging, and today the trade in grains—as well as other basic foodstuffs—represents a development with great economic and political implications. As the table shows, North America has become the major source of exportable grains—the world's "breadbasket" or, in another sense, the world's "meat market." The trade in oilseeds is similarly lopsided. In this case, the United States, Argentina, and Brazil have a corner on the export trade, and Europe and China are major importers. Most of this is used as a protein supplement in animal feeds.

Over the past 50 years, Asia and Africa have shown an increasing dependence on imported grain (and for China, imported soybeans). These regions have also experienced 50 years of continued population growth. Although much of the food needs of these regions are met by internal production (Asia, for example, produced 940 million tons of grain and 123 million tons of oilseeds in 2007), the persistent dependency is an ominous signal. If, as IFPRI projects, the developing countries import twice as much grain by 2020, the grain-exporting countries will have to either greatly increase domestic production or use less grain (not likely). And the rising use of corn for biofuel production goes, well, against the grain.

At no time in recent history has the world grain supply run out. The global amount of grain produced in 2008 was 2.127 billion tons, and carryover stocks at year's end amounted to 409 million tons. Production was up by 46 million tons, but total use exceeded production by 80 million tons. However, with some use of carryover stocks, enough grain is available to satisfy the world's food needs—and enough to feed millions of animals and continue to keep some in storage (Fig. 12–14).

TABLE 12–2   World Grain Trade Since 1935

Region	Amount Exported or Imported (Million Metric Tons)[1]									
	1935	**1950**	**1960**	**1970**	**1980**	**1990**	**1995**	**2000**	**2004**	**2007**
North America	5	23	39	56	131	123	108	98	101	120
Latin America	9	1	0	4	−10	−12	−4	−16	−8	−7
Western Europe	−19	−22	−25	−30	−62	−10	21	13	10	0.2
Africa	1	0	−2	−5	−15	−31	−31	−44	−42	−51
Asia	2	−6	−17	−37	−63	−82	−78	−74	−79	−72
Australia and New Zealand	3	3	6	12	19	15	13	21	25	9

Source: FAO FAOSTAT, FAO Food Outlook, June 2008.

[1] A minus sign in front of a figure indicates a net import; no sign indicates a net export.

Cereal production and utilization

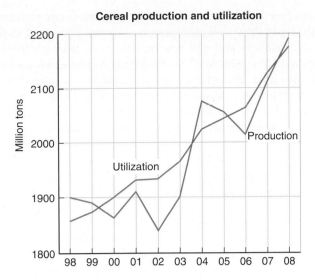

Figure 12–14 Production and utilization of world grains, 1998–2008. When utilization is greater than production, grains are supplied from carryover stocks. (*Source*: GIEWS, "FAO Food Outlook," June 2008.)

TABLE 12–3	Rise in World Market Prices of Selected Commodities		
		Average	
Commodity	Cost Basis	2002–2006 Price	2007–2008 Price
Wheat	$/ton	$168	$319
Coarse grains	$/ton	113	181
Rice	$/ton	262	361
Oilseeds	$/ton	293	486
Vegetable oils	$/ton	587	1,015
Butter	$/100 kg	162	294
Whole milk powder	$/100 kg	192	417

Source: OECD/FAO Agricultural Outlook 2008–2017.

In spite of this, the world food situation entered a period in 2007 and 2008 that became known as "the global food crisis."

The Global Food Crisis

Between 2006 and 2008, the prices of food commodities on the world market rose precipitously—on average, 100%

(Table 12–3)—triggering food riots and emergency measures in many countries. In response, the United Nations addressed this issue at a scheduled World Food Summit in Rome in June 2008 (see Sustainability, "2008 World Food Summit"). If nothing else, this crisis reflects the interconnected global trade in food. What was behind this unusual rise in prices? What have been the consequences?

Sustainability

2008 World Food Summit

Global conferences have a reputation for being long on promises and short on delivery. However, one basic function of such conferences is to draw the attention of national leaders to international issues. The World Food Summit, held in Rome in June 2008, was originally scheduled as the "High-Level Conference on World Food Security: The Challenges of Climate Change and Bioenergy" and was sponsored by the UN FAO as the latest of series of meetings held every six years. Although climate change and bioenergy were addressed, the conference shifted much of its focus to the existing world food crisis. Attended by heads of state and high-level ministers from 181 countries, the conference did indeed get the attention of national leaders as details of food shortages, high food prices, and shortfalls in aid were presented. This was the third World Food Summit; earlier summits in 1996 and 2002 led to verbal commitments to eradicate hunger and malnutrition, at least reaching the Millennium

Development Goal of a 50% reduction in hungry people by 2015.

The meeting eventually developed a set of short- and long-term plans in its final declarations. A major focus was the need for investing in agriculture in ways that would help small-scale farmers in developing countries to accomplish dramatic increases in food production. In the short term, this means developed countries providing technical assistance, and in the long term, it means developed countries urging the governments of developing countries to increase their investment in agriculture. Another focus was the critical need in the short term for "seeds, fertilizer and irrigation," the main components of the first Green Revolution, so that a second Green Revolution would be accomplished, especially in Africa. Addressing the immediate food crisis, the summit approved plans to enlist more support and resources for UN agencies to provide food assistance. Pledges of $6 billion in new funding for the

food crisis were made by the developed countries, adding to an existing $7 billion in pledges.

Two contentious issues at the summit were trade and biofuels. Some 40 food-exporting countries impose different restrictions on trade—taxes, quotas, or actual bans. And the perpetual problem of agricultural subsidies in the developed countries was addressed by participants, but did not appear in the final declarations. Also, a number of participants challenged the biofuel-producing countries to declare a moratorium on the use of grains and oilseeds for biofuel production, claiming that this practice was responsible for much of the rise in food prices. In the end, the declarations promised only to "study" this issue in depth.

Perhaps the clearest message of the summit was that feeding the world population over the next 20 years is as important and challenging an issue as addressing global climate change, and we are already falling behind.

A number of factors conspired to create the situation where supply was unable to keep up with demand:

1. higher production costs due to the rising cost of petroleum and fertilizer;

2. as we have seen already, the diversion of corn and other commodities to the production of biofuels;

3. a strong demand for higher-quality diets in the emerging economies of Asia (e.g., more meat, dairy products, vegetable oils);

4. weather-related shortfalls of harvests in key exporting countries (Australia, Canada); and

5. a background of declining carryover stocks of many commodities.

Several of these factors are transitory, suggesting that some relief to the food crisis is on the way. Better weather can cure the exporting countries' harvest shortfalls, farmers will respond to higher prices by investing in available technologies to increase yields and croplands, and the current severe downturn in the global economy has already brought on a sharp decline in petroleum prices. However, the new biofuel economy and the demand for better diets in Asian emerging economies are here to stay. The outlook, according to the UN FAO, is for a decline in food prices in the short run, but a decline to new, higher levels than those of the past 10 years.

One definite consequence of the high commodity prices is that farmers are getting much higher prices for their crops; this is especially true for the farmers in the developed countries that dominate the food export business. However, farmers in the developing countries are not connected to the world markets and will benefit little from the high market commodity prices. And the developing countries whose food deficits force them to import commodities are faced with prices they can't afford. For most people in these countries, food expenditures are over 50% of their income, and the result is a significant increase in undernourishment and malnourishment. This brings us to the topic of food security.

Food Security

The Bread for the World's Hunger Institute defines **food security** as "assured access for every person to enough nutritious food to sustain an active and healthy life." Who is responsible for food security? Figure 12–15 displays *three major levels of responsibility for food security:* the **family**, the **nation**, and the **global community**. At each level, the players are part of a market economy as well as a sociopolitical system. In a market economy, *food flows in the direction of economic demand*. Need is not taken into consideration. If there are hungry pets and hungry children, the food will go to the pets if their owners have money and the children's parents don't. In other words, the cash economy, following the rules of the market, provides the *opportunity* to purchase food, but not the food itself. Where the economic status of the player is very low (a jobless parent, a poor country), the next level may be willing and able to provide the needed purchasing power or the food (but maybe not).

Family	Nation	Global community

Goal: Personal and family food security
Policies:
—Employment security
—Adequate land or livestock
—Good health and nutrition
—Adequate housing
—Effective family planning
—Access to food

Goal: Self-sufficiency in food and nutrition
Policies:
—Just land distribution
—Support of sustainable agriculture
—Effective family planning
—Promotion of market economy
—Avoidance of militarization
—Effective safety net to provide food when needed

Goal: Sustainable food and nutrition for all countries
Policies:
—Food aid for famine relief
—Appropriate technology in development aid
—Aid for sustainable agricultural development
—Debt relief
—Fair trade
—Disarmament
—Family planning assistance

Figure 12–15 Responsibility for food security. Goals and strategies for meeting food needs at three levels of responsibility: the family, the nation, and the global community.

Family Level. The most important level of responsibility is the family. The goal at this level is to meet the nutritional needs of everyone in the family to an extent that provides freedom from hunger and malnutrition. For an individual, there are four legitimate options for attaining food security: You can purchase the food, raise the food, gather the food from natural ecosystems, or have the food provided by someone, usually in the context of the family. In the event of economic or agricultural failure at the family level, the fourth option implies that there is an effective *safety net*—that at some national, state, or local level there are policies or programs whose objective is to meet the food security needs of all individuals in the society.

National Level. Safety-net policies and programs can be of two types: (1) official policies, represented by a variety of welfare measures, such as the Food Stamp Program and the Supplemental Security Income program, and (2) voluntary aid through hunger-relief organizations. For example, Feeding America, a network of food banks throughout the country, forms part of the voluntary safety net. More than 25 million Americans receive emergency food help from the organization. Many more of the hungry are working poor, often living in the suburbs, who must choose between paying for food and paying for heat in their homes; or they may be the elderly, who must choose between paying for their prescription drugs and paying for food. These people often depend on one of the more than 50,000 local food charities for at least one good meal a day. Feeding America retrieves food from food processors and suppliers and distributes it to the organization's network of food banks. A recent law, the *Charity, Aid, Recovery, and Empowerment Act of 2003 (CARE Act)*, allows family farmers, ranchers, and restaurant owners to deduct the costs of food donated to agencies such as Feeding America as a charitable contribution.

For developing countries, an appropriate goal at the national level is *self-sufficiency in food*—enough food to satisfy the nutritional needs of all of a country's people. To meet this goal, a nation can either produce all the food its people need or buy the food on the world market. This goal implies the existence of policies to eliminate chronic hunger and malnutrition among its people, such as a just land-distribution policy and a functioning market economy. Many nations, though, are not self-sufficient in food, and those that cannot afford to buy all they need must turn to the global community for *food aid* and other forms of technical development assistance. Even a poor country like Sierra Leone can establish a food security program; there it is called Operation Feed the Nation, and the program, supported with funds from many external sources, has become a mainstay of agricultural development and household support in that African country.

Global Level. Although a substantial amount of food aid flows from rich countries to food-poor countries, much more food is imported and paid for by the developing countries. The World Trade Organization (WTO) is the body that governs international trade. Meeting every few years, the WTO brings together developed and developing member countries for negotiations and policy decisions (see Chapter 2). The wealthy developed countries continue to maintain policies that are highly protective of their agricultural sectors, while calling on the developing countries to liberalize their trade policies. This hypocrisy works itself out in the *tariffs*—taxes the developed countries set on imported agricultural products—and in the *subsidies* given to the agricultural sector. For example, Haiti cut its tariff on rice from 50% to 3% in 1995, opening the country's markets to subsidized rice from the United States (called "dumping"). By 2000, rice production in Haiti was cut in half, leaving thousands of small rice farmers without their means of livelihood. Now, Haiti is the third largest importer of U.S. rice, at 240,000 tons per year. At the same time, rice production is subsidized to the tune of $1 billion per year in the United States.

Recall from Chapter 11 that the latest U.S. farm bill continued the subsidies of most agricultural products, even adding some new ones. The EU is notorious for its high tariff barriers and large subsidies for its agricultural sector and doles out $41 billion a year to its farmers. Even though the EU has enacted agricultural reforms that change the rules for subsidies, farmers still receive a one-time annual payment based on previous subsidies and the size of their farm.

Other Needs. There are other initiatives that can help the poorer countries to become self-sufficient in food. One of the most important is relieving the debt crisis in developing countries, an issue addressed in Chapter 9. Another factor in meeting global nutrition needs is the *trade imbalance* between the industrial and the developing countries. The developing countries typically export raw goods such as cash crops, mineral ores, and petroleum and import more-sophisticated manufactured products such as aircraft, computers, and machinery. Prices for the latter have risen, while those for the commodities from developing countries have only recently begun to rise. To some extent, the increasing use of labor in the developing countries in electronic assembly and in the manufacture of clothing has offset this imbalance. Indeed, the globalization of markets is working to the advantage of the developing countries, which have a surplus of labor and low labor costs. As a result, many industrial corporations have outsourced their manufacturing to developing countries. What the poor need is jobs, and even if the pay is far below what workers in industrial countries receive, it is often enough to help workers in developing countries improve their well-being.

Appropriate global food security goals for the wealthy nations are listed in Fig. 12–15. These are policies that promote self-sufficiency in food production and sustainable relationships between the poorer nations and their environments as well as addressing the trade imbalance and human exploitation problems.

12.4 Hunger, Malnutrition, and Famine

At a UN World Food Conference in 1974, delegates from all nations subscribed to the objective "that *within a decade* no child will go to bed hungry, that no family will fear for its next day's bread, and that no human being's future and capacities will be stunted by malnutrition." In 2008, 34 years later, the United Nations held a World Food Summit and addressed the food crisis that was responsible for adding 75 million to the 848 million people still suffering from hunger and malnutrition in the world (see Sustainability, "2008 World Food Summit," p. 311). With the MDGs, a new, more modest target has been proposed: *to cut world hunger in half by 2015* in the developing countries. This means a reduction in the proportion of hungry people from 20% to 10%. In light of the evident availability of enough food to feed everyone and in view of the global market in foodstuffs, why is there still so much hunger and malnutrition in the world?

Malnutrition Versus Hunger

Hunger is the general term referring to a lack of basic food required to provide energy and to meet nutritional needs such that the individual is unable to lead a normal, healthy life.

Malnutrition is the lack of essential nutrients, such as specific amino acids, vitamins, and minerals, and **undernourishment** is the lack of adequate food energy (usually measured in Calories). Another nutritional disorder that is becoming epidemic, even in many developing countries, is **overnourishment**. Worldwide, more than 1.3 billion people overeat and are overweight. Fully one-third of the U.S. population is clinically obese (more than 30 pounds overweight).

To address this and other nutritional problems in the United States, the USDA established six food groups and arranged them in a pyramid. The original pyramid, with its one-size-fits-all approach that identified six food groups and a range of servings, has been replaced by an Internet-based pyramid that tailors the food groups and amounts to age and physical activity (Fig. 12–16). With its stress on physical activity and 12 different age groups, the new pyramid provides a better understanding of nutritional needs, but it only does this if the user employs the Internet. Following the Web-based pyramid should facilitate weight control and cut down

MyPyramid
STEPS TO A HEALTHIER YOU

Based on the information you provided, this is your daily recommended amount from each food group.

GRAINS 7 ounces	VEGETABLES 3 cups	FRUITS 2 cups	MILK 3 cups	MEAT & BEANS 6 ounces
Make half your grains whole	**Vary your veggies**	**Focus on fruits**	**Get your calcium-rich foods**	**Go lean with protein**
Aim for at least **3 1/2 ounces** of whole grains a day	Aim for these amounts **each week:** **Dark green veggies** = 3 cups **Orange veggies** = 2 cups **Dry beans & peas** = 3 cups **Starchy veggies** = 6 cups **Other veggies** = 7 cups	Eat a variety of fruit Go easy on fruit juices	Go low-fat or fat-free when you choose milk, yogurt, or cheese	Choose low-fat or lean meats and poultry Vary your protein routine– choose more fish, beans, peas, nuts, and seeds

Find your balance between food and physical activity

Be physically active for at least **30 minutes** most days of the week.

Know your limits on fats, sugars, and sodium

Your allowance for oils is **6 teaspoons a day.**

Limit extras–solid fats and sugars–to **290 calories a day.**

Your results are based on a 2200 calorie pattern.

Name: _____

This calorie level is only an estimate of your needs. Monitor your body weight to see if you need to adjust your calorie intake.

Figure 12–16 The new food guide pyramid. The new pyramid is a guide produced by the U.S. Department of Agriculture to help people evaluate their food intake and exercise regimen so as to provide adequate nutrition and keep their weight under control. It requires visiting a Web site, www.mypyramid.gov, or gaining access to printed materials that present and explain the way the pyramid works. Physical activity and age are used to give 12 variations of the pyramid and the recommended balance of calories and food types. The figure represents the recommended food intake for a 20-year-old female who engages in 30–60 minutes of physical activity each day. The Web site also provides a meal-tracking worksheet to help in achieving the diet goals.

on the chances of developing diabetes, heart disease, high blood pressure, and some cancers, typical problems of the overnourished. Bottom line? The best way to combat obesity is to increase the amount of exercise you get and to lower your total calorie intake.

Although overnourishment is occurring in many developing countries, undernourishment is even more prevalent, especially in sub-Saharan Africa and South Asia. There the major nutritional problems are a lack of proteins and some vitamins (malnutrition), especially in children, and a lack of food for energy (undernourishment) for all ages.

Extent and Consequences of Hunger

Accurate, reliable figures on the worldwide extent of hunger are unavailable, mainly because few governments make any effort to document such figures. On the basis of surveys of available food in the society, with accessibility to food factored in, the FAO estimated in mid-2009 that 1 billion people in the developing world and an additional 15 million in the rest of the world are underfed and undernourished.

Where? Almost two-thirds of the undernourished (642 million) live in Asia and the Pacific. The countries most seriously affected are India, Bangladesh, Pakistan, and China. Sub-Saharan Africa has the highest percentage of undernourished, with one-third of the population, some 265 million people, afflicted. Hunger is a massive problem, but it affects individual people with unique faces and personalities. Consider the following quotes from the World Bank's *Voices of the Poor*. From Senegal: "Your hunger is never satisfied, your thirst is never quenched; you can never sleep until you are no longer tired." And from a 10-year-old child from Gabon: "When I leave for school in the mornings, I don't have any breakfast. At noon there is no lunch, in the evening I get a little supper, and that is not enough. So when I see another child eating, I watch him, and if he doesn't give me something, I think I'm going to die of hunger."[7]

Consequences. The effects of malnutrition and undernutrition are greatest in preschool children and next greatest in women. Hunger can prevent normal growth in children, leaving them thin, stunted, and often mentally and physically impaired (Fig. 12–17), a condition known as protein-energy malnutrition. Research has shown that undernutrition in early childhood can seriously limit growth and intellectual development throughout life. Women in the developing world are especially vulnerable because they are subject to widespread patterns of low status and because they work hard. Entire families are held down by hunger, unable to produce or work for the food they need.

Sickness and death are companions of hunger. Because poor nutrition lowers a person's resistance to disease, measles, malaria, and diarrheal diseases become common and are

Figure 12–17 Malnourished children. These orphaned children are often left to their own devices in this makeshift refugee camp in Sudan.

major causes of death in the malnourished and undernourished. Eleven million children under the age of five die each year in the developing countries; at least half of these deaths are caused either directly or indirectly by hunger and malnutrition. Rarely do the children actually starve to death; instead, death comes from the inadequate food that leaves them weak, underweight, and vulnerable to disease. Hunger is often a seasonal phenomenon in rural areas that are supported by subsistence agriculture, as people are forced to ration their stored food in order to survive until the beginning of the next harvest. Anyone who travels in the developing world cannot help but notice that few people in the rural areas look well fed. Most are thin and spare from a lifetime of hard work and limited access to food.

Root Cause of Hunger

In the view of most observers, *the root cause of hunger is poverty*. Our planet produces enough food for everyone alive today. However, hungry and malnourished people lack either the money to buy food or enough land to raise their own. A recent FAO report on food insecurity states: "Most of the widespread hunger in a world of plenty results from grinding, deeply rooted poverty."[8]

Inadequate food is only one of many consequences of poverty. As mentioned in Chapter 9, 1.4 billion people live in absolute poverty today, with incomes of less than $1.25 a day.

[7]Narayan, Deepa with Raj Patel, Kai Schafft, Anne Rademacher and Sarah Koch-Schulte. 2000. *Voices of the Poor: Can Anyone Hear Us?* New York, N.Y.: Published for the World Bank, Oxford University Press.

[8]FAO. "State of Food Insecurity in the World 2002." Rome, Italy.

Most of these people reside in rural villages, are illiterate, spend half or more of their income on food, and represent races, tribes, or religions that suffer discrimination. It is to address this kind of poverty and its consequences that the MDGs (see Table 9–2) were articulated and promoted around the world. Addressing hunger and the poverty that causes it will be essential to making progress toward all the other goals. One measure of progress toward reducing hunger is the proportion of children under five who are underweight because of chronic or acute hunger. Figure 12–18 compares data on underweight proportions from 1990 and 2006, showing that only a few regions—namely, eastern Asia, Latin America and the Caribbean, and northern Africa—are on target to reduce this measure of hunger 50% by 2015. The number of underweight children in the developing countries is still a staggering 140 million. It is feared that the current global food crisis will push this number even higher.

Hunger and poverty do not always go from bad to worse. A number of Asian countries, including China, Indonesia, and Thailand, significantly reduced the extent of poverty and hunger in them during the 1980s and 1990s. Indonesia has benefited from oil exports and Green Revolution technology, and it continues to put a major emphasis on rural development and the country's social infrastructure. Malawi has been able to double its agricultural productivity through a voucher program for fertilizer and seed, helping poor farmers to feed their families and produce crops for the market. It is possible, therefore, for societies to address the needs of the hungry poor with appropriate public policies and to make progress in reducing the extent of absolute poverty and hunger. In contrast, in any given year there are countries and sometimes whole regions that face exceptional food emergencies. It is here that international responsibility comes most sharply into focus.

Famine

A **famine** is a severe shortage of food accompanied by a significant increase in the death rate. Famine is a clear signal that a society is either unable or unwilling to distribute food to all segments of its population. Two factors—*drought* and *conflict*—have been immediate causes of famines in recent years.

Drought. Drought is blamed for famines that occurred in 1968–1974 and again in 1984–1985 in the Sahel region of Africa. The Sahel, a broad belt of drylands south of the Sahara Desert, is occupied by 70 million people who practice subsistence agriculture or tend cattle, sheep, and goats. Although the region normally has enough rainfall to support dry grassland or savanna ecosystems, the rainfall is seasonal, undependable, and prone to failure. Beginning in 1965, the region experienced 20 years of subnormal rainfall, with tragic results. Crops withered, forage for livestock declined, watering places dried up, and livestock died. Both farmers and pastoralists began abandoning their land and migrating toward urban centers, where they were often herded into refugee camps. Unsanitary conditions in the camps and the

Figure 12–18 Underweight children. The figure shows the percentages of children under age five who were underweight in 1990 and 2006. Millennium Development Goal 1 aims at reducing the 1990 proportion by 50% by 2015. (*Source: Millennium Development Goals Report 2008* (New York: United Nations, 2008), 10.)

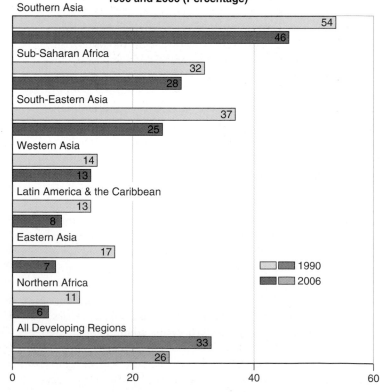

already weakened condition of the refugees led to the spread of infectious diseases such as dysentery and cholera, and many thousands died before effective aid could be organized. The 1984–1985 Sahelian famine is thought to have been responsible for almost a million deaths in Ethiopia alone. The number would have been higher if not for aid extended by Africans and numerous international agencies.

Since 1999, periods of drought and normal rainfall have alternated in the western Sahel. Sometimes, as in 2003–2004, favorable weather led to desert locust infestations, which devastated crops in many regions. Drought and deforestation have brought on catastrophic desertification, which in turn has triggered massive dust storms that spread clouds of Sahara dust across the Atlantic Ocean and into the Northern Hemisphere.

Although rainfall is normal in the western Sahel at present, the eastern Sahel—the "Horn of Africa"—remains in the grip of an extended drought that has put more than 13 million people at risk. Cattle are dying, and malnutrition is on the increase.

Warning Systems. In order to prevent droughts from leading to famines, two major efforts provide crucial warnings about food insecurity in different regions of the world. One, the FAO's *Global Information and Early Warning System (GIEWS)*, monitors food supply and demand all over the world, watching especially for those places where food supply difficulties are expected or are happening. The GIEWS issues frequent reports and alerts that are usually the result of its field missions or on-the-ground reports from other UN agencies and NGOs. The agency also relies on satellite imagery for crop-monitoring activities.

The second major effort is the *Famine Early Warning System Network (FEWS NET)*, funded by the U.S. Agency for International Development. FEWS NET is a partnership of agencies involved in satellite operations focusing on sub-Saharan Africa. It continually measures rainfall and agricultural conditions in the region and, like the GIEWS, issues regular bulletins and special reports on the Internet, giving governments and relief agencies accurate and timely assessments of the status of food security in African countries.

Conflict. Famines in which the causative factor was not drought, but war, threatened Ethiopia, Eritrea, Somalia, Rwanda, Sudan, Mozambique, Angola, and Congo in the 1990s. Devastating and prolonged civil warfare put millions of Africans at risk of famine. The civil wars disrupted the farmers' normal planting and harvesting and displaced millions from their homes and food sources. Governments in power maintained control over food and relief supplies; relief agencies operated under dangerous conditions and frequently incurred casualties. In some areas, the problem was made worse by persistent droughts. In Mozambique alone, 900,000 people died from direct military action or from indirect effects of the war there. At least 15 countries in sub-Saharan Africa have experienced food emergencies in recent years because of internal conflicts and their effects on

neighboring countries. Continuing conflicts in Afghanistan and Sri Lanka interfere with access to food, but actual famine is not occurring there or anywhere else.

Famines from drought and war are preventable. India, Brazil, Kenya, and southern Africa have coped with droughts in recent years by mobilizing effective relief in the form of food, clothing, and medical assistance. Indeed, the drought of the early 1990s accelerated the peace process in Mozambique and helped change the political landscape in southern Africa. Cooperation between South Africa and the 10 nations of the Southern Africa Development Community to prevent famine lowered barriers between South Africa and its neighbors prior to the coming of democracy to that troubled nation.

Hunger Hot Spots

Much of Africa has experienced long-term and severe droughts as well as widespread civil conflict. Erratic weather patterns and a general warming trend thought by some scientists to be the outcome of global climate change have caused crop harvests to oscillate between average and poor. And with the recent precipitous rise in food prices, many millions in the region are experiencing food insecurity. The FAO GIEWS identifies 20 African countries currently considered to be "in crisis and requiring external assistance."

Drought has once again brought Ethiopia to the brink of a food crisis. Some 11 million need aid in order to prevent famine. Government land policies are a problem: The government owns all of the land, yet 85% of the people in Ethiopia are farmers. History has shown that if farmers own their own land, they will care for it much more effectively. Chronic food insecurity has led to increasing malnutrition, and the failure of seasonal rains has caused wide-scale death of livestock. Somalia is considered to be the worst humanitarian disaster in the world, according to the FAO. Over 3 million, some 45% of the population, require emergency food, and because of the lawlessness of that country, armed groups attack aid workers and hijack food. Zimbabwe, once a major breadbasket for southern Africa, is now a humanitarian disaster area, brought on largely by President Robert Mugabe's "land reform" policies. Once-thriving farms owned by white settlers were seized and resettled by Mugabe supporters who had little knowledge of agriculture. The combination of recurrent drought, internal unrest, and a refusal by Mugabe to allow aid workers into the country has pushed over 5 million (40% of its people) into a state of desperate hunger where, for many, the only food is ground-up tree roots.

In Sudan, a 21-year civil war and drought combined to put almost 1 million people in peril, triggering a massive airdrop by the World Food Program to provide almost 10,000 tons of food a month to the southern parts of Sudan (Fig. 12–19). The Sudanese government systematically blocked aid shipments in order to pacify (and starve) the rebels in the south and west. A peace agreement in 2005 has led to massive population movements as refugees return to their

Figure 12–19 Food aid. Women receive aid from the World Food Program in Sudan.

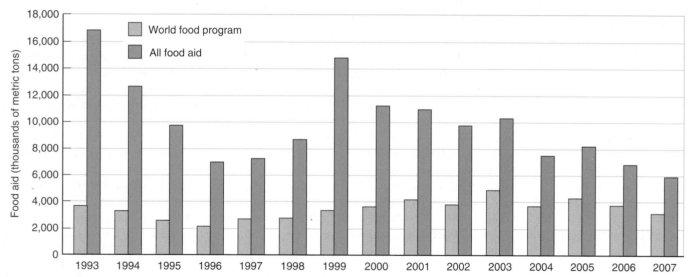

Figure 12–20 Global food aid. Global food aid from 1993 to 2007. Yellow indicates food aid delivered by the World Food Program; green indicates food aid from all sources. (*Source:* World Food Program statistics, FAIS, UNFAO.)

devastated homelands. Continued civil conflict still occurs in the Darfur region, however. Together with the failures of seasonal rains, the conditions in Sudan will require continued deliveries of thousands of tons of emergency food aid.

Other countries with recurrent hunger and food insecurity include North Korea, Haiti, and Bangladesh. The FAO in early 2009 identified 32 countries, including the 20 African countries in need, as being "in crisis and requiring external assistance."[9] The "external assistance" here is *food aid*; the reasons for the need vary from conflict to drought to floods, hurricanes and cyclones to localized crop failures. Widespread, severe hunger and famine are being averted today only

because of food aid. What, then, is the function of food aid and just how effective is it?

Food Aid

The World Food Program (WFP) of the United Nations coordinates global food aid, receiving donations from the United States, Japan, and the countries of the European Union. In addition, many nations conduct their own programs of bilateral food aid to needy countries. Although international food aid was cut in half (to 6.2 million metric tons) between 1993 and 1996, it reached 15 million metric tons in 1999; it has since dropped to its lowest level in 30 years (Fig. 12–20). Much of the increase in 1999 was the result of a major shortfall in grain

[9]FAO. Crop Prospects and Food Situation. No. 1. February 2009.

production in the Russian Federation and a consequent extension of food aid primarily from the United States (admittedly, to boost U.S. farm exports, which had been falling).

Food aid is distributed to countries all over the world, not just where famines are threatened. This raises the question: What is the proper role of food aid? Aid is vital in saving lives where food emergencies occur, but what about the people who suffer from chronic hunger and malnutrition? When should food be given to those in need, instead of being distributed according to market economics?

When Aid Doesn't Help.   Numerous humanitarian campaigns to end world hunger have been mounted in the last 50 years. The United States and Canada have been world leaders in giving away food (which is first purchased from farmers and, therefore, represents a subsidy). A number of serious famines have been moderated or averted by these efforts, and certainly the need for ongoing emergency food aid is undeniable. However, routinely supplying food aid in an attempt to alleviate chronic hunger in developing countries may be the worst thing to do.

The problem is that people will not pay more than they have to for food. Therefore, free or very cheap foreign food undercuts the local market. In effect, local farmers must compete economically with free or low-cost imported food. When they cannot earn a profit, they stop producing and eventually enter the ranks of the poor. The cycle continues, with people who sell goods to the farmers also suffering when the farmers lose buying power. In the long run, the entire local economy deteriorates. Hence, the donation of food, while well intended, often aggravates the very conditions that it is meant to alleviate. Meanwhile, population pressures continue to build, and the magnitude of the problem increases.

When It Does.   The WFP is aware of this problem and targets most of its food aid to emergency situations. Some aid does go to "development projects," benefiting poor people in countries where food security is at risk and regions where malnutrition is severe. The intention of this aid is to free poor people of the need to provide food for their families and allow them to devote their efforts to other development activities. As an important component of this strategy, the WFP now makes 80% of its food purchases in the developing countries themselves. This is a "win-win" solution, where small-scale farmers receive income for their crops and needy people are fed. Unfortunately, the U.S. Congress recently refused to change a law requiring food aid from the United States to be American-grown, an obvious concession to the farm lobby.

Food aid will undoubtedly continue to be an international responsibility. It is an important buffer against famine, and it can become part of local development aid and an assault against childhood malnutrition. As part of the solution to the chronic hunger and malnutrition among the poor, however, simple food aid is not enough. Much more will be accomplished by restructuring the economic arrangements between rich and poor nations and by extending development aid aimed at fostering self-sufficiency in food and sustainable interactions with the environment.

Closing Thoughts on Hunger

In 1980, a Presidential Commission on World Hunger delivered its report to President Jimmy Carter, with the major recommendation "that the United States make the elimination of hunger the primary focus of its relations with the developing world." The thrust of the report was that if we respect human dignity and have a sense of social justice, we must agree that hunger is an affront to both. The right to food must be considered a basic human right. It follows, then, that we as a nation have a moral obligation to respond to world hunger, the report concluded. Has that moral obligation made its way into public policy in the ensuing years?

Although food is our most vital resource, we do not treat it as a commons (free to all that need it). Indeed, the production and distribution of food constitutes one of the most important economic enterprises on Earth. Alleviating hunger is primarily a matter of addressing the absolute poverty that afflicts one of every five people on the planet. To treat food as a commons would be to treat wealth as a commons. While this is one of the tenets of socialist systems, this notion has proved to be unworkable. We are part of the world economy now, and it is a market economy—for the perfectly good reason that nothing else seems to work, given the realities of human nature.

What has not been done, however, is to bring this market economy under the discipline of sustainability. Short-term profit crowds out long-term sustainable management of natural resources. National self-interest in the rich countries prevents them from leveling the playing field for agricultural commodities. Why are the rich nations content to maintain the current high subsidies to agriculture that allow them to flood markets with underpriced products? Why is hunger not at the top of the policy agenda of the United States?

In the developing countries, there is a critical need for enlightened leadership. Sub-Saharan Africa has been plagued with corrupt and inept elites who are more interested in staying in power than in reducing poverty in their countries. Tribal conflicts have taken a terrible toll in lives and hunger. The AIDS epidemic is ravaging many of the countries, and in some, the leaders have been reluctant to admit that they have a problem.

We understand the situation quite well. No new science or technology is needed to alleviate hunger and at the same time promote sustainability as we grow our food. The solutions lie in the realm of political and social action at all levels of responsibility. Given the current groundswell of concern about the environment and the global attention to the Millennium Development Goals, there may never be a better time to turn things around and take more seriously our responsibilities as stewards of the planet and as our brothers' (and sisters') keepers.

Revisiting the Themes

Sound Science

The new high-yield grains that led to the Green Revolution are sound science in practice. So is agricultural research that continues to seek new ways to improve crop yields. Although many might disagree, the new biotech work all over the world producing genetically modified crops—especially the work that is designed to help farmers in developing countries improve their ability to grow food—is sound science in action. The example of golden rice highlights the potential for this technology. Sound science is also needed to provide an accurate evaluation of all genetically modified crops and foods before they are implemented. The satellite imagery used by the GIEWS and FEWS NET to assess hunger hot spots is also a commendable application of science to human needs.

Sustainability

Global land that is used for agricultural crops and animal husbandry represents one of the most crucial and irreplaceable components of ecosystem capital. However, this land must be nourished; it is a renewable resource, but it can be drawn down by unsustainable practices that lead to erosion and other forms of land degradation. As populations increase, pressure will mount to turn more forests and wetlands into croplands, a transformation with trade-offs that are not usually beneficial due to the loss of the goods and services of the natural lands.

Subsistence agriculture has the potential to operate sustainably as pastoralists and small farmers tend their crops and flocks, recycling manure and feeding crop residues to their animals. Even slash-and-burn cultivation can be practiced sustainably. In contrast, factory animal farming (as currently practiced) wastes manure and pollutes the air and water in the name of economic efficiency. Many consider the diversion of food crops to biofuel production unsustainable and are urging countries to put a cap on this process. Some food aid is harmful to sustainability in developing countries, where self-sufficiency in food is essential to establishing a sustainable society.

Stewardship

The remarkable progress made in keeping the world's people fed is certainly good stewardship in action. This is especially true where efforts are being made, such as those by the WFP, FEWS NET, the GIEWS, and the many NGOs involved in hunger and development aid, to identify those suffering from food insecurity and to ensure that they have sufficient food for their needs. Thousands of organizations and people show that they care as this work goes on. The work of Borlaug and his team, of the CGIAR, and of the IRRI is particularly noteworthy because they have sought to improve the yields of the more important grains and make the new high-yielding varieties available to the developing countries. The efforts to improve agricultural production in sub-Saharan Africa are laudable, as are applications of genetic engineering to developing-country crops. The Feeding America organization plays an important role in the United States in keeping people from going hungry in a land of plenty. Meeting the Millennium Development Goal of reducing the proportion of hungry people in the world by half is a worthy objective, but it still leaves hundreds of millions to suffer from the effects of hunger.

REVIEW QUESTIONS

1. Describe the major components of the agricultural revolution. What are the environmental costs of each?

2. What is the Green Revolution? What have been its limitations and its gains?

3. Describe subsistence agriculture, and discuss its relationship to sustainability.

4. How do sustainable animal farming and industrial-style animal farming affect the environment on different scales?

5. What are biofuels, and what are their impacts on food production?

6. What are some viable options for increasing food production in the future?

7. What are the major advantages and problems associated with using biotechnology in food production?

8. Describe the patterns in grain trade between different world regions over the last 70 years. How does the food crisis of 2007–2008 relate to these patterns?

9. Describe the three levels of responsibility for meeting food needs. At each level, list several ways food security can be improved.

10. Define *hunger, malnutrition,* and *undernourishment.* What is the extent of these problems in the world? What is their root cause?

11. Discuss the causes of famine and severe hunger, and name the geographical areas most threatened by them.

12. How is relief for famine and severe hunger accomplished? Why does food aid sometimes aggravate poverty and hunger?

THINKING ENVIRONMENTALLY

1. Imagine that you have been sent as a Peace Corps volunteer to a poor African nation experiencing widespread hunger. Design a strategy for assessing the needs of the people and for contacting appropriate sources for help.

2. Log on to www.mypyramid.gov and compare your diet with that recommended by the USDA program. Use the meal tracking worksheet to record your food intake over a two- to three-day period. Which nutrients are lacking? Which are in excess? What changes in your diet would reconcile these differences?

3. Use the Web to evaluate the work of FEWS NET (www.fews .net) and the GIEWS (www.fao.org/GIEWS). How do these organizations function in bringing food aid to people with the greatest need?

MAKING A DIFFERENCE

1. Try to buy cloth bags for food shopping instead of paper or plastic. These are reusable, as opposed to the disposable varieties, and are very sustainable. Even though paper bags come from trees (and are both renewable and recyclable), they generate a huge amount of air pollutants and water pollutants in their manufacture, more so than plastic.

2. If you are cautious about genetically altered foods and you want to know whether a product is organic or not, check the bar code number on the package: If the first digit is 9, the food is organic. If that number is 4, the item was a result of conventional farming. If the number is 8, then it has been genetically engineered.

3. Plant your own garden; you control all of the chemical inputs (if any), and no carbon dioxide is emitted in getting the goods onto your plate.

4. Bring your lunch to work; you'll save your own money as well as the environment.

5. Demonstrate your concern for the hungry and homeless in your city by giving your time (or donating foodstuffs) to a local soup kitchen or food pantry.

Mastering**ENVIRONMENTALSCIENCE**™

Go to **www.masteringenvironmentalscience.com** for practice quizzes, Pearson eText, videos, current events, and more.

CHAPTER 13

Pests and Pest Control

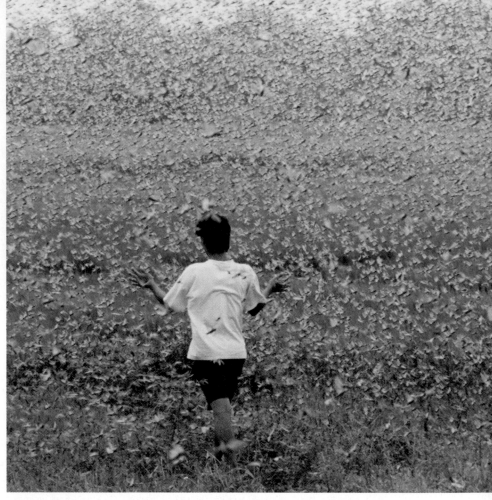

Locusts surround a Filipino farmer in Tarlae province, where they caused crop losses of $690,000 in 1994.

KEY TOPICS

1. **The Need for Pest Control: What are pests, and what are the different methods we use to control them?**

2. **Chemical Treatment: Promises and Problems: As new and more effective chemical pesticides, such as DDT, are employed, some serious problems begin to appear.**

3. **Alternative Pest Control Methods: There are numerous alternatives to using pesticides to control pests.**

4. **Socioeconomic Issues in Pesticide Use: Integrated pest management (IPM) is an important way to reduce the use of pesticides.**

5. **Pesticides and Policy: Public policy for controlling pesticides is based on several laws.**

By morning the wind had brought the locusts; they invaded all Egypt and settled down in every area of the country in great numbers. Never before had there been such a plague of locusts, nor will there ever be again. They covered all the ground until it was black. They devoured all that was left . . . everything growing in the fields and the fruit on the trees. Nothing green remained on tree or plant in all the land of Egypt.

Exodus 10:13–15.

THIS ACCOUNT OF THE DESERT LOCUST is found in the Old Testament book of Exodus. It was one of the plagues that Moses announced as he tried to get the pharaoh to release the Israelite people from bondage. It could very well be a description of a modern locust plague. Ordinarily shy, solitary creatures, desert locusts tend to stay put, feeding on available vegetation in the drylands of Africa and West Asia (Figure 13–1). However, conditions of ample rain and vegetation can trigger a remarkable switch from the *solitary* to the *gregarious* phase as numbers of locusts in a site increase. They undergo explosive population growth, forming a swarm of up to 30 million locusts per square mile (80 million per square kilometer) and eating almost all vegetation in their path before they take wing again to find new food sources.

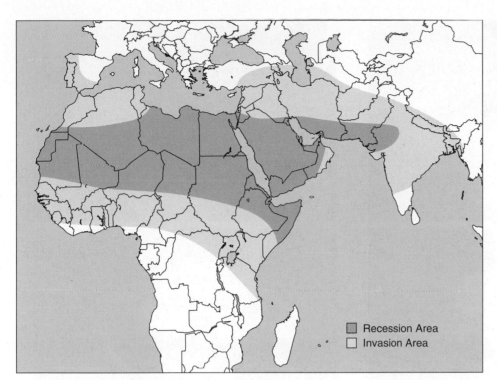

Figure 13–1 Desert locust range. The locusts usually live solitary lives in the green areas shown on the map (recession area). But with the right conditions, they come together in swarms that can migrate for thousands of kilometers, shown in yellow on the map (invasion area). The migrating swarms are the basis of locust plagues. (*Source:* "Locust!" by Rebecca Lindsey, Earth Observatory, NASA.)

In October 2004, some 7 to 10 million acres (3 to 4 million hectares) of land in a band across Africa were covered with migratory locusts, the worst plague in 15 years. The locusts are a serious pest for the subsistence farmers who work the difficult dryland soils for cereal crops. It is especially frustrating when rainfall is better than average and locust swarms fly in and devastate an expected abundant harvest. Sometimes yellow aircraft swoop down on the swarm and spray pesticides on the locusts, stopping the swarm in its tracks. This is the outcome of satellite surveillance and the Desert Locust Information Service,

provided by the UN Food and Agricultural Organization (FAO). Sometimes, however, funds and personnel are not up to the task, and the locust swarms devastate local economies in struggling countries like Senegal and Mauritania. Lacking human intervention, the farmers hope for the kind of strong winds that can blow the aerial swarms thousands of miles away (as in the case of the ancient Egyptians, where we read that when the pharaoh repented, "the Lord changed the wind to a very strong west wind, which caught up the locusts and carried them into the Red Sea" (Exodus 10:19)).

13.1 The Need for Pest Control

Since earliest times, humans have suffered frustration and losses brought on by destructive pests. To this day, farmers, herdsmen, and homeowners wage a constant battle against the insects, plant pathogens, and unwanted plants that compete with their human rivals for the biological use of crops, animals, and homes.

One dictionary defines a **pest** as "any organism that is noxious, destructive, or troublesome."[1] This definition includes a broad variety of organisms that interfere with humans or with our social or economic endeavors: pathogens, nuisance wild animals, annoying insects, molds, and so forth. Other chapters consider nuisance wild animals (Chapter 6) and pathogens (Chapter 17); the emphasis in this chapter is on those pests that interfere with crops, grasses, domestic

animals, and structures—primarily the agricultural pests and weeds.

Agricultural pests are organisms that feed on agricultural crops or ornamental plants or animals. The most notorious of these organisms are various insects, but certain fungi, viruses, worms, snails, slugs, rats, mice, and birds also fit into the category (Fig. 13–2 a–f). **Weeds** are plants that compete with agricultural crops, forests, and forage grasses for light and nutrients (Fig. 13–2 g–h). Some unwanted plants poison cattle or have other serious effects; others simply detract from the appearance of lawns and gardens. Bringing these pests under control has three main purposes: to *protect our food*, to *protect our health*, and *for convenience*.

Part of the credit for modern human prosperity can be attributed to pest control. For example, pesticides are vital elements in the prevention of the diseases that kill and incapacitate humans. In addition to having an agricultural use, pesticides have become important public-health tools used to combat diseases such as malaria and sleeping sickness.

[1] Oxford English Dictionary, 2nd ed. Oxford University Press, 1989.

Figure 13–2 Important pests. (a) Fire ants. (b) Cotton boll weevil. (c) *Aedes aegypti* mosquito. (d) Male medfly. (e) Colorado potato beetle. (f) Formosan subterranean termites. (g) Vampire weed. (h) Crabgrass.

The Importance of Pest Control

Insects, plant pathogens, and weeds destroy an estimated 37% (before and after harvest) of potential agricultural production in the United States, at an estimated yearly loss of $122 billion to consumers and producers. Elsewhere, pests also destroy important crops: (1) In the late 1990s, Russia suffered a 70% yield loss from a potato fungus, *Phythophtora infestans*, considered to be global agriculture's worst crop disease. (2) The "vampire weed," *Striga hermonthica*, is a root parasite that causes several billion dollars in losses each year in East Africa. (3) Soybean rust, *Phakopsora pachyrhizi*, has spread throughout South America, causing $2 billion in losses in Brazil alone in 2004. It was blown into the southern United States by Hurricane Ivan in 2004 and is spreading northward into the Midwest, which is very bad news for a crop worth almost $27 billion in 2007.

The Environmental Protection Agency (EPA) estimated that efforts to control these losses in the United States in 2001 (the latest available survey) involved the use of 888 million pounds (404,000 metric tons) of **herbicides** (chemicals that kill plants) and other **pesticides** (chemicals that kill animals and insects considered to be pests) annually (Fig. 13–3), at a direct cost of $11.1 billion. According to the EPA, 5.05 billion pounds (2.30 million metric tons) of pesticides were used worldwide in 2001, at a cost of $32 billion. Many of the changes in agricultural technology, such as monoculture and the widespread use of genetically identical crops, have boosted yields and also brought an increase in the proportion of crops lost to pests—from 31% in the 1950s to 37% today. During the last half-century, the use of herbicides and other pesticides multiplied manyfold, leading to a disturbing and unsustainable dependency on them.

Different Philosophies of Pest Control

Medical practice employs two basic means of treating infectious diseases. One approach is to give the patient a massive dose of antibiotics, hoping to either eliminate the pathogen or stop it before it can get established. The other approach is to stimulate the patient's immune system with a vaccine to produce long-lasting protection against any future invasion. In practice, both means are often used to keep a particular pathogen under control.

The same basic philosophies govern the control of agricultural pests. The first is **chemical treatment**. Like the use of antibiotics, chemical treatment seeks a "magic bullet" that will eradicate or greatly lessen the numbers of the pest organism. Although it has had much success, this approach gives only short-term protection. Furthermore, the chemical often has side effects that are highly damaging to other organisms.

The second philosophy is **ecological control**. Like stimulating the body's immune system, this approach seeks to give long-lasting protection by developing control agents on the basis of knowledge of the pest's life cycle and of ecological relationships. Such agents, which may be other organisms or chemicals, work in one of two ways: Either they are highly specific for the pest species being fought, or they manipulate one or more aspects of the ecosystem. Ecological control emphasizes the protection of people and domestic plants and animals from damage from pests rather than eradication of the pest organism. Thus, the benefits of pest control can be obtained, while maintaining the integrity of the ecosystem.

These two philosophies are actually combined in the approach called **integrated pest management (IPM)**. IPM is *an approach to controlling pest populations by using all suitable methods—chemical and ecological—in a way that brings*

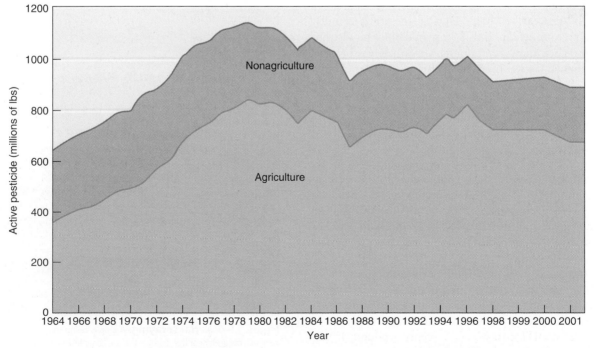

Figure 13–3 Pesticide use in the United States. Nonagricultural and agricultural uses of pesticides (active ingredients) are shown for the period 1964–2001. (*Source:* Data from the Environmental Protection Agency, Office of Pesticide Programs.)

about long-term management of pest populations and also has minimal environmental impact. This approach is increasing in usage, especially where pesticides are seen as undesirable because of health risks and in developing countries where the cost of pesticides is often prohibitive. We will discuss IPM further in Section 13.4.

13.2 Chemical Treatment: Promises and Problems

Pesticides are categorized according to the group of organisms they kill. There are insecticides (for insects), rodenticides (for mice and rats), fungicides (for fungi), herbicides (for plants), and so on. None of these chemicals, however, is entirely specific to the organisms it was designed to control; they can all pose hazards to other organisms, including humans.

Development of Chemical Pesticides and Their Successes

Finding effective materials to combat pests is an ongoing endeavor. The early substances (frequently referred to as **first-generation pesticides**) used to control agricultural pests included toxic heavy metals such as lead, arsenic, and mercury. Scientists now recognize that these substances may accumulate in soils, inhibit plant growth, and poison animals and humans. In addition, toxic heavy metals lose their effectiveness as pests become increasingly resistant to them. For example, in the early 1900s, citrus growers were able to kill 90% of injurious **scale insects** (minute insects that suck the juices from plant cells) by placing a tent over an infested tree and piping in deadly cyanide gas for a short time. By 1930, this same technique killed as little as 3% of these pests.

The next step had its origins in the science of organic chemistry in the early 1800s. During the 19th century, chemists synthesized thousands of organic compounds, but for the most part, these compounds sat on shelves because uses for them had not yet been found. By the 1930s, however, with agriculture expanding to meet the needs of a rapidly increasing population and with first-generation pesticides failing, farmers needed something new. In time, **second-generation pesticides**, as they came to be called, were developed as a result of synthetic organic chemistry.

The DDT Story. In the 1930s, a Swiss chemist named Paul Müller began systematically testing some organic chemicals for their effect on insects. In 1938, he hit on the chemical dichlorodiphenyltrichloroethane (DDT), a chlorinated hydrocarbon that had first been synthesized some 50 years earlier. Just traces of DDT killed flies in Müller's laboratory.

DDT appeared to be nothing less than the long-sought "magic bullet," a chemical that was extremely toxic to insects and yet seemed nontoxic to humans and other mammals. It was very inexpensive to produce. At the height of its use in the early 1960s, DDT cost about 20 cents a pound. It was **broad spectrum**, meaning that it was effective against a multitude of insect pests. It was also **persistent**, meaning that it

did not break down readily in the environment and, hence, provided lasting protection. This last attribute provided additional economy by eliminating both the material and the labor costs of repeated treatments.

In War. . . . DDT quickly proved successful in controlling important insect disease carriers. During World War II, for example, the military used DDT to control body lice, which spread typhus fever among the men living in dirty battlefield conditions. As a result, World War II was one of the first wars in which fewer people died of typhus than of battle wounds. DDT sprayed over the island of Saipan enabled the U.S. Marines to defeat both dengue fever and the Japanese army. The World Health Organization (WHO) of the United Nations used DDT throughout the tropical world to control mosquitoes and greatly reduced the number of deaths caused by malaria. (See Stewardship, "DDT for Malaria Control," p. 332) There is little question that DDT saved millions of lives. In fact, the virtues of DDT were so outstanding that Müller was awarded the Nobel Prize for medicine in 1948 for his discovery.

And in Peace. . . . Postwar uses of DDT expanded dramatically. The chemical was sprayed on forests to control defoliating insects such as the spruce budworm. It was routinely sprayed on salt marshes to kill nuisance mosquitoes. It was sprayed on suburbs to control the beetles that spread Dutch elm disease. DDT even proved highly effective in controlling agricultural insect pests (Fig. 13–4). Indeed, DDT was so effective that many crop yields increased dramatically. Growers could ignore other, more painstaking methods of pest control such as crop rotation and the destruction of old crop residues. They could grow varieties that were less resistant, but more productive. They could grow certain crops in warmer or moister regions, where formerly the damage from pests had been devastating. In short, DDT gave growers more options for growing the most economically productive crop.

Figure 13–4 Aerial spraying. A crop is being sprayed with pesticides to keep insects under control. Spraying is the basic technique still used in most control programs, although nowadays the pesticides in use do not persist in the environment for more than a week or two.

TABLE 13–1   Characteristics of Some Pesticides

Insecticide Class	Examples		Toxicity to Mammals	Persistence
Organophosphates	Parathion, malathion, phorate, chloropyrifos		High	Moderate (weeks)
Carbamates	Carbaryl, methomyl, aldicarb, aminocarb, carbofuran		High to moderate	Low (days)
Chlorinated hydrocarbons	DDT, toxaphene, dieldrin, chlordane, lindane		Relatively low	High (years)
Pyrethroids	Permethrin, bifenthrin, esfenvalerate, decamethrin		Low	Low (days)
Herbicide Class	**Examples**		**Effects on Plants**	
Triazines	Atrazine, simazine, cyanizine		Interfere with photosynthesis, especially in broadleaf plants	
Phenoxy	2-, 4-D; 2-, 4-, 5-T; methylchlorophenoxybutyrate (MCPB)		Cause hormonelike effects in actively growing tissue	
Acidamine	Alachlor, propachlor		Inhibit germination and early seedling growth	
Dinitroaniline	Trifluralin, oryzalin		Inhibit cells in roots and shoots; prevent germination	
Thiocarbamate	Ethylpropylthiolcarbamate (EPTC), cycloate, butylate		Inhibit germination, especially in grasses	
Phosponate	Glyphosate		Inhibit amino acid synthesis; kill a wide variety of plants	

Source: Private Pesticide Applicator Training Manual, 1993, U.S. Environmental Protection Agency, Region VIII.

The success of DDT led to the development of a great variety of synthetic organic pesticides. Currently, the U.S. Environmental Protection Agency (EPA) Pesticide Program regulates more than 18,000 pesticide products, most of which are synthetic organic pesticides. Examples and some characteristics of these insecticides and herbicides are listed in Table 13–1.

Problems Stemming from Chemical Pesticide Use

The problems associated with the use of synthetic organic pesticides can be categorized as follows:

- development of resistance by pests,
- resurgences and secondary-pest outbreaks, and
- adverse environmental and human health effects.

These problems can be extremely widespread and persistent and have led to an increased need for alternative methods of pest control.

Development of Resistance by Pests. The most fundamental problem for growers is that chemical pesticides gradually lose their effectiveness. Over the years, it becomes necessary to use larger and larger quantities, to try new and more potent pesticides, or to do both to obtain the same degree of control. Synthetic organic pesticides fared no better than first-generation pesticides in this respect. For example, in 1946, 2.2 pounds (1 kg) of pesticides provided enough protection to produce about 60,000 bushels of corn. By 1971, it took 141 pounds (64 kg) to produce the same amount, and losses due to pests actually *increased* during the intervening years.

Evolution at Work. Resistance builds up because pesticides destroy the sensitive individuals of a pest population, leaving behind only those few that already have some resistance to the pesticide. Resistant insect populations develop rapidly because insects have a phenomenal reproductive capacity. A single pair of houseflies, for example, can produce several hundred offspring that may mature and reproduce themselves only two weeks later. Consequently, repeated pesticide applications result in the unwitting selection and breeding of genetic lines that are highly, if not totally, resistant to the chemicals that were designed to eliminate them. Cases have been recorded in which the resistance of a pest population has increased as much as 25,000-fold. Recent work on mosquito resistance to a group of pesticides has shown that resistance genes can be spread globally in a brief time. In short, pesticide resistance simply confirms that "natural" selection is a powerful evolutionary force.

The Last Roundup®. When one pesticide is used extensively, resistance is virtually inevitable. A case in point is the herbicide glyphosate (trademark Roundup®), which is used widely on corn, soybeans, and cotton that have been genetically modified to be herbicide tolerant (HT). These crops contain a bacterial gene that, when inserted, makes them resistant to glyphosate. HT crops can be planted without tilling and then after emergence can be safely sprayed with glyphosate, wiping out all encroaching weeds. Use of the HT crops has led to enormous increases in the use of the herbicide, to the exclusion of most other herbicides. As expected, several weeds—e.g., mare's tail, water hemp, Palmer pigweed, and ryegrass—have developed resistance to glyphosate (Fig. 13–5); agricultural scientists have found an 8- to 12-fold resistance of Palmer pigweed to glyphosate in the southern United States. One agricultural expert predicted that in several years there will be an epidemic of glyphosate-resistant weeds. If

Figure 13–5 Mare's tail (*Conyza canadensis*) resistant to glyphosate. This plant is growing well alongside herbicide-tolerant soybeans, after treatment with herbicide. Hundreds of thousands of acres of soybeans and other crops are now infested with resistant mare's tail. (*Source:* Purdue University Extension Weed Science.)

that happens, farmers will be spending much more per acre on herbicides, and they will have to return to conventional tillage, which will greatly increase soil erosion.

Over the years of pesticide use, the number of resistant species has climbed steadily. Many major pest species are resistant to all of the principal pesticides. Indeed, over a thousand species of insects, plant diseases, and weeds have shown resistance to pesticides. Interestingly, as a pest population becomes resistant to one pesticide, it also may gain resistance to other, unrelated pesticides, even though it has not been exposed to the other chemicals.

Resurgences and Secondary-Pest Outbreaks. The second problem with the use of synthetic organic pesticides is that, after a pest has been virtually eliminated with a pesticide, the pest population not only recovers, but also explodes to higher and more severe levels. This phenomenon is known as a **resurgence**. To make matters worse, small populations of insects that were previously of no concern because of their low numbers suddenly start to explode, creating new problems. This phenomenon is called a **secondary-pest outbreak**.

A recent study in California listed a series of 25 major pest outbreaks, each of which caused more than $1 million worth of damage. All but one involved resurgences or secondary-pest outbreaks. Moreover, the species appearing in secondary-pest outbreaks quickly became resistant to pesticides, thus compounding the problem. At first, pesticide proponents denied that resurgences and secondary-pest outbreaks had anything to do with the use of pesticides. However, careful investigations have shown that resurgences and secondary-pest outbreaks occur because the insect world *is part of a complex food web.*

Thus, the chemical approach breaks down because it ignores basic ecological principles. It assumes that the ecosystem is a static entity in which one species, the pest, can simply be eliminated. In reality, the ecosystem is a dynamic system of interactions, and a chemical assault on one species will inevitably perturb the system and produce other, undesirable effects. Populations of plant-eating insects are frequently held in check by other insects that parasitize or prey on them (Fig. 13–6). Pesticide treatments often have a greater impact on these natural enemies than on the plant-eating insects they are meant to control. Consequently, with their natural enemies suppressed, the populations of both the original target pest and other plant-eating insects explode. To achieve sustainability, therefore, we must understand how ecosystems work and adapt our interventions accordingly.

Treadmill. The late entomologist Robert van den Bosch coined the term **pesticide treadmill** to describe attempts to eradicate pests with synthetic organic chemicals. It is an apt term because the chemicals do not eradicate the pests. Instead, they increase resistance and secondary-pest outbreaks, which lead to the use of new and larger quantities of chemicals, which in turn lead to more resistance and more secondary-pest outbreaks, and so on (Fig. 13–7). The use of glyphosate on HT crops is a good example. The process is an unending cycle, constantly increasing the risks to human and environmental health, and is clearly not sustainable. In the words of one observer, "Modern agriculture has a serious chemical dependency—an addiction to pesticides."[2]

Adverse Human Health Effects. As with all toxic substances, pesticides can be responsible for both acute and chronic health effects.

Acute Effects. According to the American Association of Poison Control Centers, more than 96,300 persons suffered acute poisonings from pesticides in the United States during 2007. Most were farm workers or employees of pesticide companies who came in direct contact with the chemicals. Different pesticides affect different human systems: glyphosate poisoning, for example, can lead to nausea, abdominal pain, shock, and respiratory failure; pyrethrin poisoning leads to allergic reactions, seizures, pneumonia, and coma.

There is no global documentation of pesticide poisoning; however, the WHO has recently proposed standards for diagnosis and identification of acute pesticide poisoning cases. Anecdotal cases and some regional documentation of pesticide poisonings suggest that as many as 39 million people may suffer from acute poisoning each year, according to the Pan American Health Organization. Most of these cases occur in the developing countries, where regulations and training are often lax and information on pesticides may be poorly understood. The use of pesticides by untrained persons is considered to be the major cause of these poisonings, but in many cases, children and families come in contact with the pesticides through aerial spraying, the dumping of pesticide wastes, or the use of pesticide containers to store drinking water.

[2]Anne McGinn. *Pop Stoppers.* www.tcetoday.com/MagPDFs/716sustainag.pdf. July 13, 2009.

Insect food chains

(a)

Figure 13–6 Insect food chains.
(a) Food chains exist among insects, just as they do among higher animals. (b) Aphid lion (on right) impaling and eating a larval aphid (a small insect that sucks sap from plants).

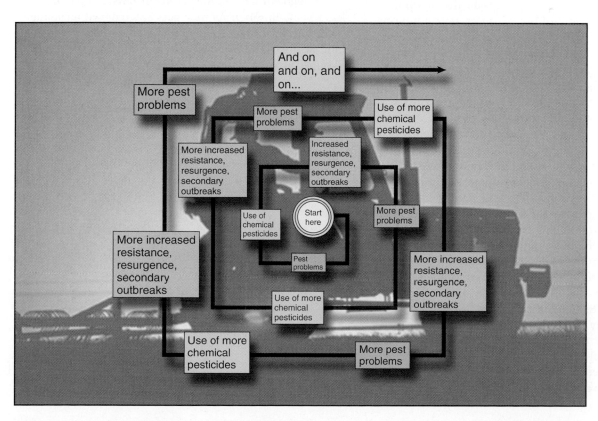

(b)

Figure 13–7 The pesticide treadmill. Treatment with chemical pesticides aggravates many pest problems. The continued use of these products demands ever-increasing dosages of pesticides, which further aggravate pest problems and produce more contamination of food-stuffs and ecosystems.

Chronic Effects. Pesticides are applied to fields and orchards to reduce pest damage to crops. They are also used to protect harvested food so that it is brought undamaged to market. Because of the wide-ranging use of pesticides by most farmers, consumers are inevitably exposed to pesticide residues on their food. Also, many occupational exposures to pesticides are subacute. The public-health concern is that the pesticides might have chronic effects, even at low levels of exposure. Among the chronic effects of pesticides is the potential for causing cancer, as indicated by animal testing. In fact, evidence of carcinogenicity has been observed for 20 pesticides. Epidemiological evidence has implicated organochlorine pesticides in various cancers, including lymphoma and breast cancer.

Other chronic effects include dermatitis, neurological disorders, birth defects, and infertility. For example, male sterility has been clearly linked to dibromochloropropane (DBCP), once used to control nematodes and now banned. Two disorders recently associated with pesticide use are suppression of the immune system and disruption of the endocrine system. New epidemiological evidence suggests that long-term exposure to pesticides can trigger Parkinson's disease in some people. Persistent exposure to pesticides among factory workers in India led to abnormally low white blood cell counts. (White blood cells are crucial elements of the immune system.) Similar results were found for agricultural workers in the former Soviet Union. Laboratory studies show many forms of immune suppression by different types of pesticides.

Endocrine Disruptors. Laboratory tests have shown that a number of pesticides, including atrazine and alachlor (herbicides) and DDT, endosulfan, diazinon, and methoxychlor (insecticides), interfere with reproductive hormones. Both a rise in the incidence of breast cancer among humans and reports of abnormal sexual development in alligators and other animals in the wild have given credence to the possibility that very low levels of a number of chemicals are able to mimic or disrupt the effects of estrogenic hormones (sexual hormones that are highly potent at low concentrations). It is well documented that farmworkers and herbicide sprayers have defective sperm counts. A recent study confirmed that men exposed to low levels of several pesticides, probably through drinking water, were many times more likely to have defective sperm or low sperm counts than were those with little or no exposure.

Concern about this reached Congress, and the Food Quality Protection Act (FQPA) of 1996 directed the EPA to develop procedures for testing chemicals for endocrine disruption activity. An EPA-established advisory committee strongly recommended proceeding with a screening program, and in 1999, the EPA established the **Endocrine Disruptor Screening Program** (EDSP). Since that time, the program has selected several assay methods and will use them in a two-tiered screening and testing process. The EPA has selected 73 chemicals, virtually all pesticides, and began issuing testing orders for the chemicals in 2009 (13 years after the FQPA authorized action). Selection of the chemicals was based on consideration of their pathways to human exposure (in food, drinking water, home use of pesticides, and so forth). Eventually, all pesticides are to be tested in the EDSP, but at the current pace of progress, it will be decades before this happens.

Adverse Environmental Effects. Aerial spraying for insect control has pointed most clearly to the adverse environmental effects of pesticides. The story of DDT, used so widely during the 1940s, 1950s, and early 1960s (80,000 metric tons in 1962 alone), illustrates the hazards.

DDT, Continued. In the 1950s and 1960s, ornithologists (people who study birds) observed drastic declines in populations of many species of birds that fed at the tops of food chains. Fish-eating birds such as the bald eagle and osprey (Fig. 13–8) were so affected that their extinction seemed imminent. Investigators at the U.S. Fish and Wildlife National Research Center near Baltimore, Maryland, showed that the problem was reproductive failure: Eggs were breaking in the nest before hatching. The investigators also found that the fragile eggs contained high concentrations of dichlorodiphenyldichloroethylene (DDE), a product of the partial breakdown of DDT by the animal's body. DDE interferes with calcium metabolism, causing birds to lay thin-shelled eggs. Further study revealed that birds were acquiring high levels of DDT and DDE by bioaccumulation and biomagnification, the process of accumulating higher and higher doses through the food chain.

Because of accumulation, small, seemingly harmless amounts received over a long period of time may reach toxic levels. This phenomenon—referred to as **bioaccumulation**—can be understood as follows: Many synthetic organic chemicals are highly soluble in lipids (fats or fatty compounds), but less soluble in water. As they pass through cell membranes, which are lipid, they come out of water solution and enter into the lipids of the body. Thus, traces of synthetic organics like pesticides and their breakdown products that are

Figure 13–8 Osprey. Populations of fish-eating birds such as the osprey, brown pelican, and bald eagle were decimated in the 1950s and 1960s by the effects of widespread spraying with DDT. With the banning of the pesticide, these populations have greatly recovered. The bald eagle was taken off the endangered species list in 1994.

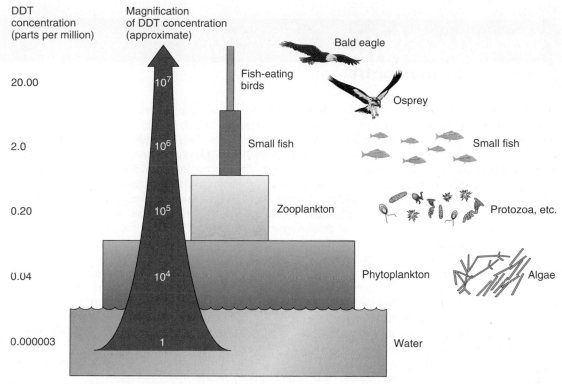

Figure 13–9 Biomagnification. Organisms on the first trophic level absorb the pesticide from the environment and accumulate it in their bodies (bioaccumulation). Then each successive consumer in the food chain accumulates the contaminant to yet a higher level. Thus, the concentration of the pesticide is magnified manyfold throughout the food chain. Organisms at the top of the chain are likely to accumulate toxic levels.

absorbed with food or water are trapped and held by the body's lipids, while the water and water-soluble wastes are passed in the urine. Because synthetics are unnatural compounds, the body cannot fully metabolize them and has no mechanism to excrete them. Accordingly, trace levels consumed over time gradually accumulate in the body and may produce toxic effects sooner or later.

Up the Chain. Bioaccumulation, which occurs in the individual organism, may be compounded through a food chain. Each organism accumulates the contamination from its food, so the concentration of contaminant in its body is many times higher than that in its food. In effect, the next organism in the food chain now has a more contaminated food and accumulates the contaminant to yet a higher level. Eventually, all the contaminant accumulated by the large biomass at the bottom of the food pyramid is concentrated, through food chains, into the smaller and smaller biomass of organisms at the top of the food pyramid. This multiplying effect of bioaccumulation that occurs through a food chain is called **biomagnification**. One of the most distressing aspects of bioaccumulation and biomagnification is that there are no warning symptoms until concentrations of contaminant in the body are high enough to cause problems. Figure 13–9 shows how DDT and its metabolic products worked their way up the food chain to the top predators.

Silent Spring. In the 1950s, Rachel Carson, a U.S. Fish and Wildlife Service biologist and an accomplished science writer, began reading the disturbing scientific accounts of the effects of DDT and other pesticides on wildlife. Carson was finally galvanized into action by a letter from a friend distressed over the large number of birds killed when the friend's private bird sanctuary was sprayed for mosquito control. By 1962, Carson finished a book-length documentation of the effects of the almost uncontrolled use of insecticides across the United States. The book, *Silent Spring*, became an instant best-seller. Its basic message was that, if insecticide use continues as usual, there might someday come a spring with no birds—and with ominous consequences for humans as well.

Silent Spring triggered a debate that has not ended. Representatives of the agricultural and chemical industries claimed that the book was an unreasonable and unscientific account that, if taken seriously, would halt human progress. At the same time, however, the book was hailed as an unparalleled breakthrough in environmental understanding. Eventually, concerns about environmental and long-term health effects led to the banning of DDT in the United States and most other industrialized countries in the early 1970s. Numerous other chlorinated hydrocarbon pesticides (for example, chlordane, dieldrin, endrin, and heptachlor) were also banned because of their propensity for bioaccumulation in the environment and suspected potential for causing cancer. In the years since the banning of DDT, the bird species that were adversely affected have recovered. Today the debate centers on the continued use of DDT against malaria in developing countries. (See Stewardship, "DDT for Malaria Control.")

Stewardship

DDT for Malaria Control

The use of DDT for controlling malaria is one of those situations where stewardship decisions involve conflicting views of the common good. Malaria exacts a terrible toll in death and disease: 350–500 million bouts of sickness and a million deaths a year. Fifty years ago malaria seemed to be on the way out as new synthetic drugs killed the protozoan parasites that caused the disease and DDT wiped out mosquitoes everywhere. But resistance emerged with a vengeance; now, the parasites are highly resistant to most common malaria drugs, and the mosquitoes are resistant to many pesticides. Although DDT has been banned in developed countries because of its human health and environmental dangers, it is still used in many developing countries, largely because it is so inexpensive. DDT is one of a group of 12 persistent organic pollutants targeted for phasing out by the United Nations Environment Program.

In malarial regions in the developing world, villagers are exposed to mosquitoes because they cannot afford screens on their homes, mosquito nets to sleep under, or repellent to spray on their children, four of whom die from malaria every minute. The anopheles mosquito, which carries malaria, is a night hunter, biting people as they rest or sleep. In 22 countries, DDT is still the main line of defense against the mosquito. The typical application involves spraying inside houses and eaves at a concentration of 2 g/m^2, once or twice a year. Even where the mosquitoes are resistant to its killing effects, the DDT acts as a repellent or irritant that keeps the mosquitoes from staying around. In 2006, the WHO reversed a 30-year policy and officially put DDT on a par with bed nets and antimalarial drugs as major tools for controlling malaria.

Those in favor of pushing forward with the ban cite the problems of keeping DDT out of agricultural fields, where it can enter food chains and poison top predators. (Once the pesticide is made available, controlling its use is difficult.) Alternative pesticides, such as synthetic pyrethroids, are already available (but resistance to these is appearing, too). An integrated public-health approach is needed wherein mosquito breeding places are eliminated, bed nets are employed, and malaria cases are treated promptly to intercept the life cycle of the parasite. However, these steps are seldom all present in many developing-country locales. Until they are, organizations such as the Malaria Foundation International and the Malaria Project are actively campaigning against the proposed action for a worldwide ban on DDT. To date, these organizations have obtained a special exemption for countries to continue to use DDT for controlling disease-carrying insects.

Almost 50 years later, Rachel Carson is credited with stimulating the start of the modern environmental movement and the creation of the EPA. *Silent Spring* has become a classic, and the regulation of insecticides and other toxic chemicals is in a sense a monument to Rachel Carson, who died of cancer only two years after her book was published.

Nonpersistent Pesticides: Are They the Answer? A key characteristic of chlorinated hydrocarbons is their persistence; that is, they take a long time to break down. Lingering in the environment for years, they can contaminate many organisms and become biomagnified. This persistence results because these chemicals have such complex structures that soil microbes are unable to fully break them down. DDT, for example, has a half-life on the order of 20 years.

Because the persistent pesticides have been banned, the agrochemical industry has substituted *nonpersistent* pesticides for the banned compounds. The synthetic organic phosphates (for example, malathion, parathion, and chlorpyrifos), as well as carbamates such as aldicarb and carbaryl, have been used extensively in place of chlorinated hydrocarbons (see Table 13–1). These compounds are potent inhibitors of the enzyme cholinesterase, which is essential for proper functioning of the nervous system in all animals. They break down into simple nontoxic products within a few weeks after their application. Thus, there is no danger of their migrating long distances through the environment and affecting wildlife or humans long after being applied.

Toxicity. For several reasons, however, nonpersistent pesticides are not as environmentally sound as they might appear. First of all, they are persistent enough to "ride the food supply from farmer to consumer," as Anne Platt McGinn put it in her Worldwatch paper, *Why Poison Ourselves?*[3] The total environmental impact of a pesticide is a function not only of its persistence, but also of its toxicity, its dosage, and the location where it is applied. Many of the nonpersistent pesticides are far more toxic than DDT. This higher toxicity, combined with the frequent applications needed to maintain control, presents a significant hazard to agricultural workers and others exposed to these pesticides. For instance, the organophosphates are responsible for an estimated 70% of all pesticide poisonings. The FQPA requires the EPA to develop new health-based standards that address the risk of children's exposure to such pesticides. The fallout from the new standards has led to the banning of any use of parathion on produce and of chlorpyrifos for all over-the-counter uses, such as on flea collars and in lawn products. Chlorpyrifos may no longer be used on tomatoes and is prohibited for use on apples after blooming has ceased. Diazinon, a widely used agricultural insecticide and a common household product once marketed as Spectracide®, has been banned for many agricultural uses as well as all indoor and outdoor residential uses.

---

[3]Anne Platt McGinn, *Worldwatch Paper #153: Why Poison Ourselves? A Precautionary Approach to Synthetic Chemicals* (Washington, D.C.: Worldwatch Institute, November 2000).

Bird Kill. Nonpersistent pesticides may still have far-reaching environmental impacts. For example, carbofuran is a pesticide widely used on alfalfa, rice, and soybeans. However, one application to a sunflower field in Colorado in 2006 led to the deaths of 2,200 birds of several species. In another incident, 800–1,200 ducks and geese died after treatment of an alfalfa field with carbofuran. The EPA estimates that as many as 2 million birds have been killed annually by carbofuran. This pesticide is also toxic to humans; as little as ¼ teaspoon can be fatal. Because of these risks, the EPA announced in 2008 that it would no longer allow residues of carbofuran on any food, essentially revoking all uses. The EPA decision was based on the grounds that the pesticide posed an unacceptable risk to toddlers.

Another hazard of nonpersistent pesticides is that desirable insects may be just as sensitive to them as pest insects are. Bees, for example, which play an essential role in pollination, are highly sensitive to them. Thus, the use of these compounds creates an economic problem for beekeepers and jeopardizes pollination. Also, regular spraying of neighborhoods with malathion to control mosquitoes leads inevitably to great declines in butterflies and fireflies.

Finally, nonpersistent chemicals are just as likely to cause resurgences and secondary-pest outbreaks as persistent pesticides are, and pests become resistant to nonpersistent pesticides just as readily.

13.3 Alternative Pest Control Methods

Numerous ecological and biological factors affect the relationship between a pest and its host. *Ecological control* seeks to manipulate one or more of these natural factors so that crops and people are protected without jeopardizing environmental and human health. This natural approach, unlike the chemical treatment approach, depends on an understanding of the pest and its relationship with its host and with its ecosystem. The more we know about the organisms involved, the greater are our opportunities for ecological control.

The life cycle typical of moths and butterflies is shown in Figure 13–10. Many groups of insects have a similarly complex life cycle. The development of each stage may be influenced by numerous abiotic factors, and at each stage, the insect may be vulnerable to attack by a parasite or predator. The proper completion of each stage depends on internal chemical signals provided by insect hormones. Locating mates, finding food, and other behaviors depend on external chemical signals. All these findings suggest ways in which pest populations may be controlled without resorting to synthetic chemical pesticides.

The four general categories of ecological pest control are (1) **cultural control**, (2) **control by natural enemies**, (3) **genetic control**, and (4) **natural chemical control**. We will consider each of these in turn.

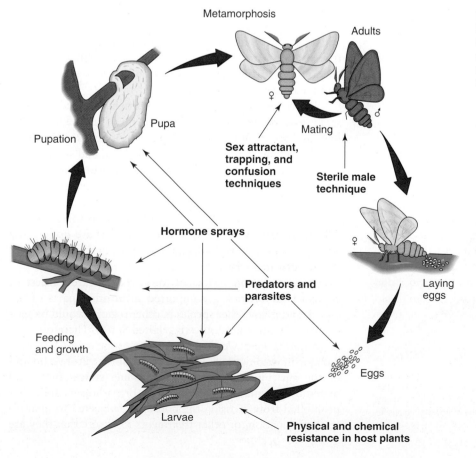

Figure 13–10 Complex life cycle of insects. Like the moth shown here, most insects have a complex life cycle that includes a larval stage and an adult stage. Biological control methods recognize the different stages and attack the insect, using knowledge of its needs and life cycle.

Metamorphosis

Adults

Pupa

Pupation

Mating

Sex attractant, trapping, and confusion techniques

Sterile male technique

Hormone sprays

Predators and parasites

Laying eggs

Feeding and growth

Eggs

Larvae

Physical and chemical resistance in host plants

Cultural Control

A cultural control is a nonchemical alteration of one or more environmental factors in such a way that the pest finds the environment unsuitable or is unable to gain access to its target. This type of control can be practiced on all scales—by a homeowner, a farmer, or even a government.

Control of Pests Affecting Crops, Gardens, and Lawns. Homeowners often use excessive amounts of herbicides to maintain a weed-free lawn. Weed problems in lawns are frequently a result of cutting the grass too short. If grass is left at least 3 inches (8 cm) high, it will usually maintain a dense enough cover to keep out most crabgrass and other noxious weeds. Thus, monitoring the height of grass is a form of cultural weed control. A gardener may also control pests by avoiding plants that act as attractants or by planting ones that act as repellants for certain pests. Roses are attractants, while marigolds and chrysanthemums are justly famous for repelling insects.

Some parasites require an alternative host, so they can be controlled by eliminating that host (Fig. 13–11). Also, hedgerows, fencerows, and shelterbelts can provide refuges where natural enemies of pests (birds, amphibians, praying mantises, and so on) can be maintained.

Crops. Managing any crop residues that are not harvested is important. Spores of plant disease organisms and insects may overwinter or complete part of their life cycle in the dead leaves, stems, or other plant residues that remain in the fields after harvesting. Plowing under or burning the material may be quite effective in keeping pest populations to

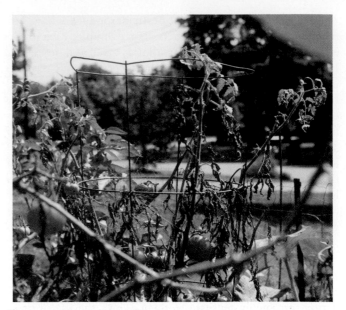

Figure 13–12 Tomato wilt. This tomato plant has been devastated by tomato wilt. Growing tomatoes in the same plot maintains wilt spores in the soil, guaranteeing that wilt will strike the plants every year. (*Source:* Author's garden.)

a minimum. In gardens, a clean mulch of material such as grass clippings or hay will keep down the growth of weeds and protect the soil from drying and erosion.

Growing the same crop on a plot of land year after year keeps the pest's food supply continuously available (Fig. 13–12). Hence, crop rotation—the practice of changing crops from one year to the next—may provide control because pests of the first crop cannot feed on the second crop and vice versa. Crop rotation is especially effective in controlling root nematodes (roundworms that live in the soil and feed on roots) and other pests that do not have the ability to migrate appreciable distances.

For economic efficiency, agriculture has moved progressively toward monoculture; witness the Corn Belt, Cotton Belt, and so on. When a pest outbreak occurs, monoculture is most conducive to its rapid multiplication and spread. The spread of a pest outbreak is impeded, and other natural controls may be more effective, if there is a mixture of crop species, some of which are not vulnerable to attack. One approach that works well in the British Isles is to intersperse cultivated strips with uncultivated ones that are not treated with pesticides. Natural enemies of pests are maintained in the uncultivated strips.

Border Patrol. Most of the pests that are hardest to control were unwittingly imported from other parts of the world, and many other species in other regions would be serious pests if introduced into the United States. Therefore, it is important to keep would-be pests out of the country. This is a major function of U.S. Customs and Border Protection and of the agriculture departments of some states. Biological materials that may carry pest insects or pathogens are either prohibited from crossing the border or subjected to quarantines, fumigation, or other treatments to ensure that they are

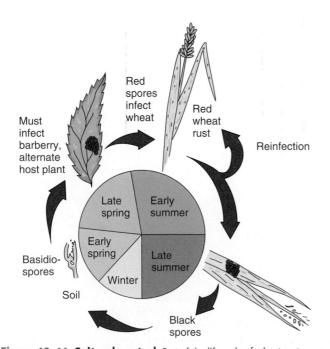

Figure 13–11 Cultural control. Part of the life cycle of wheat rust, a parasitic fungus that is a serious pest on wheat, requires that the rust infest barberry, an alternate host plant. The elimination of barberry in wheat-growing regions has been an important cultural control.

free of pests. A recent example is the Asian longhorned beetle, which arrived in packing crates from China. The beetle infests healthy maple, birch, poplar, and ash trees and eventually kills them. As a result of this pest, new regulations require that wooden crates and pallets from China be kiln-dried or chemically treated.

Cultural Control of Pests Affecting Humans. We routinely practice many forms of cultural control against diseases and parasitic organisms. Some of these practices are so familiar and well entrenched in our culture that we no longer recognize them for what they are. For instance, disposing properly of sewage and avoiding drinking water from unsafe sources are cultural practices that protect against waterborne disease-causing organisms. Combing and brushing the hair, bathing, and wearing clean clothing are cultural practices that eliminate head and body lice, fleas, and other parasites. Changing bed linens regularly protects against bedbugs. Properly and systematically disposing of garbage and keeping a clean house with all cracks sealed and with good window screens are effective methods for keeping down populations of roaches, mice, flies, mosquitoes, and other pests. Sanitation requirements in handling

and preparing food are cultural controls designed to prevent the spread of disease. The refrigeration, freezing, canning, and drying of foods are cultural controls that inhibit the growth of organisms that cause rotting, spoilage, and food poisoning.

If these practices are compromised, as they usually are in any major disaster, there is the very real danger of additional widespread mortality resulting from outbreaks of parasites and diseases. As discussed in Chapter 17, sickness and death are the consequences where these practices are not broadly pursued in a society, as in many of the less developed countries.

Control by Natural Enemies

There are four well-defined types of natural enemies: **predators, parasitoids, pathogens,** and **plant-eaters.** This type of control is used all over the world. In the eastern United States, tiny predator beetles have been imported in an attempt to control the (also imported) hemlock wooly adelgid, which is killing off hemlock trees. In Africa, mealybugs are controlled by parasitic wasps (see Sound Science, "Wasps 1, Mealybugs 0"); these types of wasps are also effective in

Sound Science

Wasps 1, Mealybugs 0

The cassava (manioc) plant originated in South America, but has been cultivated throughout the tropical world. Currently, in sub-Saharan Africa, it is the primary food for more than 200 million people, one-third of the human population in that region. Cassava is a high-yielding crop that requires no modern technology and, accordingly, is raised by subsistence farmers everywhere.

The mealybug (see photo (a)), an insect not previously seen in Africa, appeared in Congo in the early 1970s and spread across sub-Saharan Africa, leaving a trail of ruined harvests and hunger in its wake. Zaire and Congo, unable to afford pesticides, turned for help to the International Institute of Tropical Agriculture in Nigeria. The group formed the Biological Control Program and began to look for natural enemies of the mealybug.

Returning to the land of origin of the cassava, a researcher found the mealybug in Paraguay and observed that the insect was kept under control by natural predators and parasites. After extensive testing, researchers identified a parasitic wasp, *Anagyrus lopezi*, as the prime candidate for controlling the bugs. No larger than a comma on this page (see photo (b)), the female

(a)   (b)

Control by natural enemies. (a) Swarm of mealybugs. (b) Parasitoid wasp, highly magnified.

wasp seeks out mealybugs, paralyzes them with a sting, and deposits eggs that hatch inside the mealybug and eat their way through the insect. Once this wasp-bug relationship was identified, the wasps were reared by the thousands on captive mealybugs and then spread across the continent over a period of eight years. The results were a spectacular success; the tiny wasps have brought the mealybug under control in almost all

of Africa. It is estimated that every dollar invested in the control program has yielded $149 in crops saved from destruction. The best news is that the control is permanent and does not require the repeated application of expensive and environmentally damaging pesticides. In the wake of the project, national biological control programs have been established and are now focusing on pests of other crops of sub-Saharan Africa.

(a)

(b)

(c)

Figure 13–13 Parasitic wasps. (a) The life cycle of the parasitic wasp that uses the gypsy moth as its host. (b) A wasp depositing eggs in a gypsy moth larva. (c) Another insect parasite, a braconid wasp, lays its eggs on the pest known as the tomato hornworm, which is the larva of the sphinx moth. The wasp larvae feed on the caterpillar, and shortly before the caterpillar dies, they emerge and form the cocoons seen here.

controlling caterpillar populations (Fig. 13–13). In sub-Saharan Africa, the swarming desert locust was completely controlled by spraying "Green Muscle," a mix of dry spores of the fungus *Metarhizium anisopliae* with oil, on infested fields. Water hyacinth, which has blanketed many African lakes, is coming under control following the introduction of Brazilian weevils. More than 30 weed species worldwide are now limited by insects introduced into their habitats.

Protect the Natives. The problem with using natural enemies lies in finding organisms that provide control of the target species without attacking other, desirable species.

Entomologists estimate that of the 50,000 known species of plant-eating insects that have the potential for being serious pests, only about 1% actually are. The populations of the other 99% are held in check by one or more natural enemies. Therefore, the first step in using natural enemies for control should be *conservation*—protecting the natural enemies that already exist. This means avoiding the use of broad-spectrum chemical pesticides, which may affect natural enemies of the target species even more than they do the target pests themselves. Eliminating or placing considerable restrictions on the use of broad-spectrum chemical pesticides will, in many cases, allow natural enemies to reestablish themselves and

(a)

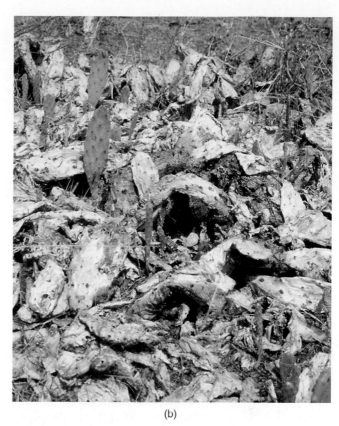

(b)

Figure 13–14 Prickly pear cactus. (a) Prickly pear cactus was a devastating weed in Australia. It was brought completely under control by the cactus moth, here shown as larvae. (b) The same moth was used to help control prickly pear in Krueger National Park, South Africa. However, the cactus-eating moth is now invading the southern United States and threatening a number of useful prickly pear species that are not at all considered pests.

control *secondary pests*—those that become serious problems only after the use of pesticides.

Import Aliens as a Last Resort. However, effective natural enemies are not always readily available. In many cases, the absence of a pest's natural enemies is the result of accidentally importing the pest without also importing its natural enemies. Quite often, effective natural enemies have been found by systematically combing the home region of an introduced pest and finding its various predators or parasites. To date, the USDA has released almost 1,000 types of insects to control pests, with generally favorable results. Identifying the enemy species is vital, and the species must be carefully tested before intentionally releasing it, something that has not always been done well. Some 16% of 313 parasitoid wasp species introduced into North America have attacked native species. The cactus moth (Fig. 13–14), so successful in Australia, was introduced to a Caribbean island to control prickly pear in 1956. Unfortunately, it has made its way to the United States and is now "eating its way westward in the U.S.,"[4] as one report put it. The moth eats many prickly pear species and is a serious threat to desert ecosystems as well as a thriving prickly pear agricultural crop in Mexico. It has most recently been found just off the Yucatan Peninsula, and the Mexican government is taking steps to eradicate the moth there.

Genetic Control

Most plant-eating insects and plant pathogens attack only one species or a few closely related species. This specificity implies a genetic incompatibility between the pest and any species that are not attacked. Most genetic-control strategies are designed to develop genetic traits in the host species that provide the same incompatibility—that is, resistance to attack by the pest. The technique has been used extensively in connection with plant diseases caused by fungal, viral, and bacterial parasites. For example, in the years 1845–1847, the potato crop in Ireland was devastated by late blight, a fungal parasite. Nearly a million people starved, and another million people emigrated to escape the same fate. Nowadays, protection against such disasters is provided in large part by growing varieties of potato that are resistant to the blight. It is no overstatement to say that the world owes much of its production of potatoes, corn, wheat, and other cereal grains to the painstaking work of plant geneticists who selected and bred disease-resistant varieties. The battle between agricultural scientists and plant diseases never ends; before the 1950s, wheat stem rust used to devastate wheat fields until breeders developed wheat varieties resistant to the rust. However, a new strain of the fungus appeared in Africa in the

---

[4]M. Alma Solis, Stephen D. Hight, and Doria R. Gordon. "Tracking the Cactus Moth Cactoblastis cactorum Berg., as it flies and eats its way westward in the U.S." *News of the Lepidopterists' Society* 46: 3–5. Spring 2004.

Figure 13–15 Hessian fly. This pest of wheat is controlled by maintaining wheat plants that produce a chemical that is toxic to the fly.

early 2000s and is spreading eastward into Asia, infecting many varieties of wheat in the developing world.

Control with Chemical Barriers. A chemical barrier is a chemical produced by the plant we want to protect; the substance is lethal or at least repulsive to the would-be pest. Once they have identified such a barrier, plant breeders can use selection and crossbreeding to enhance the trait in the desirable plant. The relationship between wheat and the Hessian fly provides an example (Fig. 13–15). The fly lays its eggs on wheat leaves, and the larvae move down the leaves and into the main stem as they feed. The weakened stem either dies or is broken in the wind. The Hessian fly was introduced into the United States in the straw bedding of Hessian soldiers during the Revolutionary War. The fly eventually spread throughout much of the Midwest, causing widespread devastation until scientists at the University of Kansas developed a variety of wheat that produces a chemical that is toxic to the insect; the chemical kills the larvae when they feed on the leaves. Increasing resistance through breeding has reduced losses to this pest to less than 1% where resistant varieties have been planted for several years.

Control with Physical Barriers. Physical barriers are structural traits that impede the attack of a pest. For example, leafhoppers are significant worldwide pests of cotton, soybeans, alfalfa, clover, beans, and potatoes, but they can damage only plants that have relatively smooth leaves. Hooked hairs on the leaf surfaces of some plants tend to trap and hold immature leafhoppers until they die. Similarly, glandular hairs that exude a sticky substance fatally entrap alfalfa weevil larvae (Fig. 13–16). Such traits can be enhanced in vulnerable plants through selective breeding.

Control with Sterile Males. Another genetic-control strategy involves flooding a natural population with sterile males that have been reared in laboratories. Combating the screwworm fly provides a prime illustration. This fly, which is closely related to the housefly and looks much like it, lays its eggs in open wounds of cattle and other animals. The larvae feed on blood and lymph, keeping the wound open and festering. Secondary infections frequently occur and often lead to the death of the animal.

Early in the 20th century, the problem became so severe that cattle ranching from Texas to Florida and northward was becoming economically impossible. In studying the situation during the 1940s, Edward Knipling, an entomologist with the USDA, observed two essential features of screwworm flies: (1) Their populations are never very large, and (2) the female fly mates just once, lays her eggs, and then dies. Knipling reasoned that if the female mated with a sterile male, no offspring would be produced. His hypothesis was correct, and today sterile males are routinely used to control this pest. After huge numbers of screwworm larvae are grown on meat in laboratories, the resulting pupae are subjected to just enough high-energy radiation to render them sterile. These sterilized pupae are then air-dropped into the infested area. Ideally, 100 sterile males are dropped for every normal female in the natural population, giving a 99% probability that wild females will mate with one of the sterile males. The technique proved so successful that it eliminated the screwworm fly from Florida in 1958–1959 and was subsequently employed to eradicate the screwworm fly in Mexico and most of Central America. The sterile-insect technique has saved billions of dollars for the cattle industry and has been used successfully on a number of agriculturally important pests.

Recently, the technique was used to eradicate the tsetse fly from the island of Zanzibar, thus eliminating the disease trypanosomiasis (sleeping sickness), which the tsetse fly carries. Indeed, now herdsmen can raise their cattle without the devastating losses the disease causes elsewhere in sub-Saharan Africa, where more than 10 million square miles (26 square kilometers) are inhabited by tsetse flies. Efforts are under way to extend this success to other parts of the continent, through the Pan African Tsetse and Trypanosomiasis Eradication Campaign.

Strategies Using Biotechnology. Biotechnology has multiplied the potential for genetic control. More complex than basic plant breeding, genetic engineering makes it possible to introduce genes into crop plants from other plant species, bacteria, and viruses. The new transgenic crops (see Chapter 12) are undergoing rapid development and

Figure 13–16 Alfalfa glandular hairs. This scanning electron micrograph shows an immature potato leafhopper trapped by sticky glandular hairs on the stem of a resistant alfalfa strain.

testing. More than 280 million acres (113 million ha) of transgenic crops are planted globally, the most important being soybeans, cotton, and corn. This approach is not without its problems, however: Recall from Chapter 12 that public acceptance of genetically modified organisms is lagging behind the exploding technology, especially in Europe.

One promising strategy is to incorporate the protein coat of a plant virus into the plant itself. When the plant "expresses" (that is, manufactures) the virus's protein coat, the plant becomes resistant to infection by the real virus. In this way, crop plants have been made resistant to more than a dozen plant viruses. Another strategy is to craft a chemical—in this case, a specific ribonucleic acid (RNA) structure—that interferes with an insect's normal RNA, silencing a gene that controls molting and sidetracking that process—thus, a gene-silencing pesticide.

Bt. Endowing plants with resistance to insects has been engineered by incorporating a potent protein produced by a bacterium, *Bacillus thuringiensis* (Bt). This protein kills the larvae of a number of plant-eating insects and is harmless to mammals, birds, and most other insects. Scientists have engineered the gene into a number of plants, including cotton, potatoes, and corn (Fig. 13–17). This development alone is expected to cut in half the use of pesticides on cotton—a crop that consumes more than 10% of the pesticides used throughout the world. The protection comes from a naturally occurring protein used for three decades by home gardeners, organic growers, and other farmers. Wide-scale use of the Bt cotton (Monsanto's Bollgard® cotton) revealed that it is effective against two of three pests, but farmers had to employ pesticides to control the bollworm (a kind of moth larva) in much of the acreage planted with the new crop. In India, farmers planting Bt cotton have been surprised by mealybug infestations and have had to resort to expensive pesticides for control.

In 2007, 49% of corn acreage and 59% of cotton acreage in the United States were planted using Bt varieties. This represents an enormous exposure of one basic "pesticide" to the insects, and the question is not *if* insect resistance will develop, but *when*. Anticipating this resistance problem, the EPA has required seed companies to verify that farmers plant 20% of their fields with non-Bt versions of the crop. The non-Bt fields provide a refuge for pests that are susceptible to the Bt toxin, allowing for these insects to interbreed with potentially resistant colleagues from the Bt acreage and, hopefully, dilute resistance genes in the insect populations. So far the strategy seems to have worked.

Roundup® Again. One biotech strategy has been to give a crop species resistance to a broad-spectrum herbicide. Although this might appear to encourage the use of herbicides, it is actually expected to make it possible to use smaller total amounts of herbicide. The herbicide glyphosate (Roundup®), first discussed in Section 13.2, inhibits an amino acid pathway that occurs only in plants and not in animals, biodegrades rapidly, and can control both grasses and broadleaf plants. Genes for glyphosate resistance have been successfully introduced into cotton, canola, soybeans, and corn. Farmers save money on herbicides because they need to spray Roundup® only once to control weeds in their soybean fields. (Almost 90% of the U.S. soybean crop is now planted with Roundup Ready® soybeans.)

Not So Fast. . . . There is a downside, however, to these current biotech innovations. First, they are not well suited to subsistence agriculture in developing countries, as they foster a dependence on a costly annual supply of seeds under current corporation practices. Also, there is the definite possibility of developing "superweeds" that are resistant to the herbicides or plant viruses, as resistance genes are shared with genetically close wild species. Furthermore, insects are likely to develop resistance to Bt plants much more readily than they would to occasional spraying of the Bt insecticide directly. Nevertheless, it is hoped that the applications of biotechnology will gradually reduce the burden of pesticides in the environment and their potential impact on human health. One sign that this strategy is working is the rapid decline in the crop-duster industry (Fig. 13–4), which is in a tailspin because of the growing use of genetically modified crops.

Natural Chemical Control

Hormones are chemicals produced in humans and other animals that provide "signals" that control developmental processes and metabolic functions. In addition, insects produce many **pheromones**—chemicals secreted by one individual that influence the behavior of another individual of the same species.

The aim of natural chemical control is to isolate, identify, synthesize, and then use an insect's own hormones or pheromones to disrupt its life cycle. Two advantages of natural chemicals are that they are nontoxic and that they are highly specific to the pest in question. (They do not affect the natural enemies of the pest to any appreciable extent.) If the affected pest is eaten by another organism, it is simply digested.

Scientists have discovered that caterpillar pupation is triggered by a decrease in the level of the chemical called **juvenile hormone**. If this chemical is sprayed on caterpillars, pupation does not occur. The larvae simply continue to feed and grow, become grossly oversized, and eventually die. One newly developed insecticide, called Mimic®, is a synthetic variation of **ecdysone**, the molting hormone of insects.

Figure 13–17 Bioengineered potatoes. The center row shows potato plants that have the Bt gene incorporated into their genome. These plants are protected from the Colorado potato beetle, while surrounding rows show typical devastation of nonengineered potato plants by beetles.

Figure 13–18 Trapping technique. Adult Japanese beetles are lured into a trap by scented bait they find attractive. Once inside the trap, baffles prevent the beetles from escaping. These traps are available in hardware stores and are quite effective, although some users find that they are too effective, luring beetles to their yards and gardens without actually catching them. (Japanese beetles are notorious lawn grubs as larvae; as adults, they are voracious consumers of garden flowers and vegetables.)

Mimic begins the molting process in insect larvae, but doesn't complete it. As a result, the larva is trapped in its old skin and eventually starves to death. Mimic is specific to moths and butterflies and is viewed as a potent agent in the control of the spruce budworm, the gypsy moth, the beet armyworm, and the codling moth—all highly devastating pests. Mimic is now registered for use throughout North America and has proven effective against the hemlock looper in Newfoundland and the gypsy moth in the United States.

Adult insects secrete pheromones to attract mates. Once identified and synthesized, these pheromones may be used in either the trapping technique or the confusion technique. In the trapping technique, the pheromone is used to lure males or females into traps or into eating poisoned bait (Fig. 13–18). In the confusion technique, the pheromone is dispersed over the field in such quantities that males become confused, cannot find females, and thus fail to mate.

The enormous potential of natural chemicals for controlling insect pests without causing ecological damage or disruption has been recognized for at least 30 years. Research has identified a large number of pheromones, and these natural chemicals are now key tools employed in IPM on many insects, such as the boll weevil and the pink bollworm for cotton, the codling moth for pears and apples, the tomato pinworm for tomatoes, and several bark beetles that infest forest trees, among others.

13.4 Socioeconomic Issues in Pesticide Use

The increasing availability of alternative pest control methods and the accumulating evidence of the failures of chemical treatment have led to a growing movement toward avoiding the use of pesticides, particularly on foods. This movement must contend with a strong tendency on the part of farmers to employ pesticides.

Pressures to Use Pesticides

A species becomes a pest only when its population multiplies to the point of causing significant damage. Natural controls are generally aimed at keeping pest populations below damaging levels, not at total eradication of the populations. By keeping pest populations down, natural controls avert significant damage, while preserving the integrity of the ecosystem.

Therefore, the question to be asked when facing any pest species is: Is the species causing significant damage? Damage should be deemed significant only when the economic losses due to the damage considerably outweigh the cost of applying a pesticide. This point is called the **economic threshold** (Fig. 13–19). If significant damage is not occurring, natural controls are already operating, and the situation is probably best left as is. Spraying with synthetic chemicals at this stage is more than likely to upset the natural balance and make the situation worse through resurgences. It is also not cost effective. If, however, significant damage *is* occurring, a pesticide treatment may be in order.

Even if there is no evidence of immediate damage from a pest, a grower who believes that his or her plantings are at risk is likely to resort to **insurance spraying**—the use of

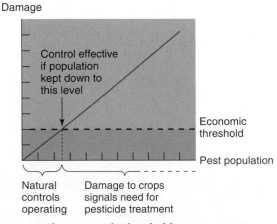

Figure 13–19 The economic threshold. The objective of pest control should not be to eradicate a pest totally. All that is needed is to keep population levels below the economic threshold.

pesticides to prevent losses to pests. For example, pesticides are heavily used in European apple crops, where the most important threats are mildew and scab—airborne diseases that, once established, cause significant economic damage. Farmers typically employ insurance spraying to control these diseases.

Consumers also put indirect pressure on growers to use pesticides. From customers to supermarket chains to canneries, there is a tendency to select the best-looking fruits and vegetables, leaving the remainder to be sold at lower prices or trashed. Like Snow White, we all tend to reach for the apparently perfect, unblemished apple. Growers know that blemished produce means less profit, so they indulge in **cosmetic spraying**—the use of pesticides to control pests that harm only the item's outward appearance. Cosmetic spraying accounts for a significant fraction of pesticide use, does nothing to enhance crop yield or nutritional value, and results in an increase in pesticide residues remaining on the produce.

Integrated Pest Management

IPM aims to minimize the use of synthetic organic pesticides without jeopardizing crops. This is made possible by addressing all the interacting sociological, economic, and ecological factors involved in protecting crops. With IPM, the crop and its pests are seen as part of a dynamic ecosystem. The goal is not eradicating pests, but maintaining crop damage below the economic threshold. The EPA uses a four-tiered approach in describing IPM:

1. **Set action thresholds.** With the economic threshold in mind, IPM identifies a point at which pest populations or environmental conditions indicate that some control action is needed.

2. **Monitor and identify pests.** Pest populations are monitored, often by persons employed by local agricultural extension services or farm cooperatives or by persons acting as independent consultants. These **field scouts** are trained in identifying and monitoring pest populations (traps baited with pheromones are used for this purpose)

and in determining whether the population exceeds the economic threshold.

3. **Prevent pests.** Cultural and biological control practices form the core of IPM techniques. Practices such as crop rotation, polyculture (instead of monoculture), the destruction of crop residues, the maintenance of predator populations, and carefully timed planting and fertilizing are basic to IPM. "Trap crops" are often used. That is, early strips of a crop such as cotton are planted to lure existing pests and then the pests are destroyed by hand or by the limited use of pesticides before they can reproduce.

4. **Control pests.** If the preceding steps indicate that pest control is needed in spite of the preventive methods, measures are taken to decrease the pest population. Here selected brands of pesticides may be used in quantities that do the least damage to the natural enemies of the pest.

Making the economic benefits of natural controls known to growers is an important aspect of IPM. Sponsored by the University of Minnesota, Radcliffe's IPM World Textbook is an outstanding online resource for IPM information. Because pesticides increased yields and profits when they were first used, many farmers still cling to them, believing that they offer the only way to bring in a profitable crop. In some instances, however, the rising costs of pesticides, along with their tendency to aggravate pest problems, have eliminated their economic advantage. **Pest-loss insurance**, which pays the farmer in the event of loss due to pests, has enabled insured growers in developed countries to refrain from unnecessary and costly "insurance spraying."

Indonesia. Particularly in the developing world, government agricultural policy often determines the extent to which IPM is adopted. Governments and aid agencies usually subsidize the purchase of pesticides, thereby strongly encouraging growers to step onto the pesticide treadmill. By contrast, Indonesia has provided a viable IPM model for other rice-growing countries. Faced with declining success in controlling the brown plant hopper (Fig. 13–20), Indonesian

Figure 13–20 Brown plant hoppers. Immature brown plant hoppers, shown on the stem of a rice plant.

Figure 13–21 Integrated pest management. IPM has helped Indonesian rice farmers bring the brown plant hopper under control after years of frustration on the pesticide treadmill.

rice growers have switched from heavy pesticide spraying to a light spraying regime that preserves the natural enemies of the insect. The success of the program can be traced to close cooperation between the Indonesian government and the FAO (Fig. 13–21). FAO workers conducted training sessions for farmers that lasted throughout the growing season. Eventually, more than 200,000 farmers were taught about the rice agroecosystem and IPM techniques, and these farmers formed a corps that could in turn teach others.

The economic and environmental benefits of the Indonesian IPM program have been remarkable. The government has saved millions of dollars annually by not purchasing pesticides, farmers have not had to invest in pesticides and spraying equipment, the environment has been spared the application of thousands of tons of pesticide, fish are once again thriving in the rice paddies, and the health benefits of reduced pesticide use have been spread from applicators to consumers and wildlife. The FAO is sponsoring similar IPM training programs in many other developing countries.

Organically Grown Food

For consumers concerned about pesticides in their food, **organic** food is an attractive option. Walk into any supermarket and you will find counters of fresh vegetables, meats, prepared foods, and Fair Trade foods, all bearing the label of *organic*. Many farmers are turning away from the use of pesticides, chemical fertilizers, antibiotics, and hormones as they raise grains, vegetables, and livestock for the organic food trade. Typically, organic farms are small, employ traditional farming methods with diverse crops, and are tied to local economies. For the farmers, the organically raised crops often represent lower crop yields, but also lower expenses. The soil in their fields and orchards is richer, and they have a product that is increasingly in demand. Sales of organics have increased 20% a year over the past decade, and organically grown food is now a $14 billion enterprise in the United States and more than $40 billion worldwide. The major reason for this expanding market is concern over health and safety, although consumers also report that the organic foods are usually tastier. They are also more costly, but recent tests have shown that organic foods are far less likely to have any pesticide residues, and those that do have residues contain much lower amounts than conventionally grown foods. Most residues in organic foods came from older, persistent pesticides, such as DDT, lingering in the soil.

USDA Organic. The certification of organic foods has been a contentious process, and eventually, the industry requested help from the government. In response, in 1990 Congress passed the **Organic Foods Protection Act**, which established the National Organic Standards Board (NOSB), under USDA auspices. After many years of work, proposed guidelines for certification from the NOSB were aired in

Figure 13–22 Organic foods. This symbol on food signals that the food or crop has been produced on a certified organic farm.

early 1998. Possibly in response to lobbying from the food industry, the board allowed genetically engineered foods, irradiated foods (foods subjected to radiation to kill bacteria), and crops fertilized with sewage sludge to be included as organic. A firestorm of protest from the organic-food community ensued. The NOSB went back to work, and in late 2002, the USDA published its final standards for certifying organic foods. As a result, no product involving genetically engineered or irradiated foods or foods fertilized with sewage sludge can be certified as 100% organic. The standards also prohibit the use of conventional pesticides, antibiotics, growth hormones, and chemical fertilizers. To market certified organic foods, farmers now must have their operations scrutinized by USDA-approved inspectors. Only then may they use the USDA seal (Fig. 13–22) on their products. If the food is at least 95% organic, it may also display the seal, but may not claim to be 100% organic.

13.5 Pesticides and Policy

In regulating the use of pesticides, three concerns are paramount: (1) Pesticides must be evaluated both for their intended uses and for their impacts on human health and the environment; (2) those who use the pesticides—especially agricultural workers—must be appropriately trained and protected from the risks of close contact; and (3) because most of the agricultural applications involve food, the public must be protected from the risks of pesticide residues on food products.

FIFRA

The **Federal Insecticide, Fungicide, and Rodenticide Act,** commonly known as FIFRA, addresses the first two of the preceding concerns. This law, established by Congress in

1947 and amended in 1972, is administered by the EPA, which has total jurisdiction over the manufacture, sale, use, and testing of pesticides. FIFRA requires manufacturers to register pesticides with the EPA before marketing them. The registration procedure includes testing for toxicity to animals (and, by extrapolation, to humans). From the test results, usage standards are set. For example, chlordane, which was used for 40 years as a pesticide on crops, lawns, and gardens, was tested and shown to be acutely toxic to animals in high doses. Chronic exposure led to damaging effects on most vital systems and caused liver cancer in animals. Acute and chronic exposure of humans to chlordane showed similar results, except for the development of cancer. On the basis of these results, the EPA canceled all uses of chlordane for food crops in 1978, but allowed continued use against termites until 1988, when all uses of the chemical in the United States were banned. The law also stipulates that the EPA must determine whether a product "generally causes unreasonable adverse effects on the environment," and if it so determines, the product may be restricted or banned. This was the basis for the banning of DDT.

The EPA's Pesticide Program places high priority on registering "reduced-risk" pesticides, especially those that promise to replace more toxic chemicals. Most are **biopesticides**—microbial pesticides like Bt, plant-incorporated protectants such as corn bioengineered to manufacture Bt in its tissues, or biochemical substances that are naturally occurring, like the pheromones (e.g., Mimic). These pesticides often clear the registration process within a year, compared with the typical three years a conventional pesticide might take. In 2006, for example, the EPA registered 26 new products, 15 of which were reduced-risk pesticides. Once registered, a pesticide must have an EPA-approved label that lists its active ingredients, instructs the public on its proper use, and informs users of the risks involved with its use.

In protecting workers from risks posed by occupational exposure to pesticides, the EPA works with state environmental agencies, which have primary responsibility for enforcement. Under its Worker Protection Standard, the EPA has issued requirements for safety training, protective equipment, emergency assistance, and permitting standards. The states must then demonstrate to the EPA that, on the basis of the federal regulation, they have adequate regulation and enforcement mechanisms.

FFDCA

Protecting consumers from exposure to pesticides on food was a contentious and confusing process for many years. Three agencies are involved: the EPA, the Food and Drug Administration (FDA), and the Food Safety and Inspection Service of the USDA. Basically, the EPA sets the allowable tolerances for pesticide residues, based on its assessment of toxicity, and the FDA monitors and enforces the tolerances on all foods except meat, poultry, and egg products, which are managed by the USDA.

The original enabling act was the **Federal Food, Drug, and Cosmetic Act (FFDCA)** of 1938. For many years, one clause of the FFDCA, the so-called **Delaney clause**, was highly controversial. The clause stated, "No [food] additive shall be deemed to be safe if it is found to induce cancer when ingested by man or animal." Since pesticides represent one of the largest categories of toxic chemicals to which people are exposed, this clause prohibited many pesticides from being used on foodstuffs when those pesticides had been found to cause cancer in laboratory tests with animals. In essence, the law stated that, if a given pesticide presents *any* risk of cancer, *no detectable residue* may remain on the food. As analytical chemical techniques became more and more sensitive, extremely low traces of a pesticide could be measured, and according to the Delaney clause, if that pesticide caused cancer in experimental animals, farmers could not use it.

FQPA

The National Research Council examined this dilemma and recommended in 1987 that the anticancer clause be replaced with a "negligible risk" standard, based on risk-analysis procedures that had been developed subsequent to the Delaney clause. After years of debate, Congress passed the **Food Quality Protection Act (FQPA)** in 1996 and did away with the Delaney clause and amended both FIFRA and FFDCA. The following are the major requirements of the act:

*   The new safety standard is "a reasonable certainty of no harm" for substances applied to foods.

*   Special consideration must be given to the exposure of young children to pesticide residues.

*   Pesticides or other chemicals are prohibited if, when consumed at average levels over the course of a lifetime, they can be shown to carry a risk of more than one case of cancer per million people.

*   All possible sources of exposure to a given pesticide must be evaluated, not just exposure from food.

*   All older products on the market before 1996 must be reassessed according to the new requirements, within 10 years.

*   The same standards are to be applied to raw and processed foods (previously regulated separately).

Care for Kids. The focus on children is the outcome of a 1993 report by the National Research Council, "Pesticides in the Diets of Infants and Children."[5] The focus is on children both because they typically consume more fruits and vegetables per unit of body weight than do adults and because studies have shown that children are frequently more susceptible to carcinogens and neurotoxins. Accordingly, the FQPA requires the EPA to add a 10-fold safety factor in assessing children's risks from pesticides.

The USDA Pesticide Data Program (PDP) provides annual analyses of pesticides in fresh and processed foods. The EPA

[5]Committee on Pesticides in the Diets of Infants and Children. "Pesticides in the Diets of Infants and Children." National Research Council. National Academy Press, Washington, D.C. 1993.

evaluates toxicity data on individual pesticides and then, using the USDA's PDP analyses and other information on risk assessment (see Chapter 17), sets *limits on the amounts of a pesticide that can remain in or on foods*—called **tolerances**. To date, the EPA is on target with the directives of the FQPA, having published tolerance reassessments for more than 9,637 pesticide products (out of a total of 9,721). More than 1,900 tolerances were rejected. As mentioned earlier, two pesticides—methyl parathion and chlorpyrifos—recently were banned on the basis of these reevaluations because they were found to pose unacceptable risks to children.

For an example of a "tolerance," another less toxic organophosphate, malathion, is listed at 8 ppm tolerance for many crops, meaning that no more than 8 ppm of malathion may be found on or in the product. For milk, however, the tolerance is set at 0.5 ppm, an indication of milk's special place in children's diets. USDA and FDA inspectors constantly monitor foods in interstate commerce to enforce these tolerance limits.

Pesticides in Developing Countries

The United States currently exports thousands of tons of pesticides (active ingredients) each year. Some of these are products banned in the United States itself. A number of factors—low literacy, weak laws, unfamiliarity with equipment—increase the potential for serious exposure to toxic pesticides in the developing countries. FIFRA requires "informed permission" prior to the shipment of any pesticides banned in the United States. This permission comes from the purchaser, which can often be a foreign subsidiary of the exporting company. The EPA must then notify the government as to the identity of the importing company.

PIC. Fortunately, the international community has erected a more effective system There is now a process of **prior informed consent (PIC)**, whereby exporting countries inform all potential importing countries of actions they have taken to ban or restrict the use of pesticides or other toxic chemicals. Governments in the importing countries respond to the notifications via UN agencies, which then disseminate all the information they receive to the exporting countries and follow up by monitoring the export practices of those countries. The PIC process was officially approved at a UN conference in 1998 and went into force as the **Rotterdam Convention** in February 2004; more than 70 countries have ratified the agreement.

Code of Conduct. The problem of unsafe pesticide use in the developing countries has also been addressed by the FAO. In spite of early opposition from the pesticide industry and exporting countries, an international *Code of Conduct on the Distribution and Use of Pesticides*, which addressed in detail conditions of safe pesticide use, was approved by the FAO in November 2002. The code makes clear the responsibilities of both private companies and countries receiving pesticides in promoting their safe use. Unlike the PIC Convention, this code calls for voluntary action and so is not legally binding, but it has already proven useful in holding private industry and importing countries to standards of safe use. In the view of some observers, both PIC and the FAO Code of Conduct are not as strong as they should be. Nevertheless, their very existence represents great progress in addressing a difficult problem in the developing countries.

Final Thoughts

Some years ago a group of leading entomologists called for a 50% reduction in pesticide use in the United States, supporting their recommendation by detailing the economic and environmental benefits of such a reduction. Shortly afterward, the heads of the EPA, FDA, and USDA issued a joint announcement of their commitment to reducing pesticide use in the United States. Yet, as Figure 13–3 indicates, we have a long way to go to achieve a 50% reduction from the 1979 high point of pesticide use. As with many environmental issues, significant progress in pesticide control has benefited from grassroots action and pressure from public-interest groups and NGOs, such as the Consumers Union and the Pesticide Action Network. These and other groups are pressing for continued movement in the direction of ecological pest management and continued progress in keeping food free of pesticide residues.

Revisiting the Themes

Sound Science

It was sound science that uncovered the connections between DDT and the decline of many predatory birds—biomagnification was a great surprise to everyone. Some amazing applications of sound science are seen in the imaginative work on alternative pest controls, such as Knipling's sterile-male technique, the use of pheromones to control pests, and the many studies that have uncovered natural enemies of pests, leading to their control. Agricultural scientists are constantly challenged to keep up with new strains of plant pests by developing resistant varieties of the crops they attack.

Sustainability

Ecological control is a sustainable approach to controlling pests, while the chemical treatment approach guarantees that pesticides will need to be used indefinitely. Pest resistance, resurgences, and secondary outbreaks have been the norm. The ecological control methods provide the means to achieve control of most pests in a sustainable way, especially when they are controlled by natural enemies. IPM presents a reasonable alternative for managing pests by limiting the use of pesticides. The organic food movement is proof that food can be grown without chemicals and, for consumers and the environment, is a great way to keep pesticides out of the natural world (and the refrigerator). Reducing or eliminating the use of pesticides and keeping food safe and healthy are two of the pillars of the sustainable-agriculture movement. (See Chapter 11.)

Stewardship

The widespread use of DDT during and just after World War II saved millions of lives. However, the continued, often indiscriminate, use of this persistent pesticide and others like it led to trouble in the environment, and the careful work of Rachel Carson in documenting this trouble in the 1960s was a landmark in environmental stewardship. Yet the continued use of DDT to control malaria-bearing mosquitoes shows that discerning the common good is not always easy. Because stewardship also includes human welfare and health, the focus of the FQPA on children was a major event in stewardly care. It can be argued that the work of the EPA in complying with this important legislation is stewardship in action. The agency is an aggressive promoter of reduced use of the more toxic pesticides.

REVIEW QUESTIONS

1. Define *pests*. Why do we control them?

2. Discuss two basic philosophies of pest control. How effective are they?

3. What were the apparent virtues of the synthetic organic pesticide DDT?

4. What adverse environmental and human health effects can occur as a result of pesticide use?

5. Define *bioaccumulation* and *biomagnification*.

6. Why are nonpersistent pesticides not as environmentally sound as first thought?

7. Describe the four categories of natural, or biological, pest control. Cite examples of each and discuss their effectiveness.

8. Define the term *economic threshold* as it relates to pest control. What are cosmetic spraying and insurance spraying?

9. What are the four steps in IPM? Explain how IPM worked in Indonesia.

10. What is organic food, and how is it now certified?

11. How does FIFRA attempt to control pesticides? What new perspective does the FQPA bring to the policy scene?

12. Discuss recent policy regarding the export of pesticides to developing countries.

THINKING ENVIRONMENTALLY

1. Consider the ethics of the United States' exportation of pesticides. A large portion of this exported material is banned in the United States itself. If the choice was up to you to continue this or put it to an end, what would you do? Remember that such pesticides have at least a positive immediate benefit to these countries. Are they a necessary evil?

2. Almost one-third of the chemical pesticides bought in the United States are for use in houses and on gardens and lawns. What should manufacturers and users do to ensure the limited and prudent use of pesticides?

3. Should the government give farmers economic incentives to switch from pesticide use to IPM? How might that work?

4. Investigate how bugs and weeds are controlled on your campus. Are IPM techniques being used? Organic fertilizers? If not, consider lobbying for their use.

5. Imagine you're the mayor of a mid-sized suburban town where mosquitoes invade from neighboring marshes. What kind of restrictions would you place on local pesticide use? Defend your answer.

MAKING A DIFFERENCE

1. When you plant a garden, select hardy varieties that are known to flourish in your particular region. Pests are less likely to destroy these types, since they grow much faster than they can be eaten, even without pesticides.

2. Try planting some herbs in your garden, since this often has many pest-oriented benefits. They are very likely to attract the good, predator bugs that will naturally take out the pests. Sage, basil, chives, yarrow, cilantro, and garlic are all great choices.

3. Another way to bring in the predatory insects is to buy them. A great number of companies now breed and sell species such as praying mantises, lacewings, and ladybugs, all of which help get rid of your pest problems.

4. To help keep a pesticide-free lawn looking green, let the grass grow a little longer than usual. This way it will need less water and food. Select a mower that mulches grass and leaves; lawn clippings provide nitrogen, and fallen leaves are great for conditioning soil.

Mastering**ENVIRONMENTALSCIENCE**™

Go to **www.masteringenvironmentalscience.com** for practice quizzes, Pearson eText, videos, current events, and more.

PART FOUR

Harnessing Energy for Human Societies

I N JUNE 2008, gasoline prices in the United States averaged $4 a gallon, and many drivers in the rural South spent 15% or more of their income on gasoline to get to their jobs. Detroit declared the sport-utility vehicle to be dead as sales of these gas guzzlers plummeted and buyers turned to more efficient, smaller cars. After a summer of high energy prices, however, an economic crisis plunged the country and the world into a recession, gasoline prices dropped below $2 a gallon, and for a time at least, drivers could fill their tanks for less than $50. Hopefully, the high gasoline prices of the summer of 2008 will be remembered as shoppers make their new vehicle purchases. One thing is certain: As the world pulls out of the 2008–2009 recession, energy costs will rise again. These developments provide clear evidence of the crucial link between energy and the economy.

THE INDUSTRIAL COUNTRIES are so deeply committed to fossil fuel and nuclear energy that we cannot imagine an economy apart from these energy sources. Developing countries are following the same example, hoping to enhance the well-being of their people. A recent analysis of the world's oil reserves suggests that oil production may peak within the next decade and then begin a decline that will drive prices up to new heights. Will we invade wildlife refuges and fragile coastal waters to withdraw every last drop of oil from the ground? Looming in the background is the specter of global climate change and the struggle of dealing with this threat, which is tied so directly to the use of fossil fuels.

THE UNITED STATES has only 2% of the world's oil reserves, but consumes 25% of its oil, leaving us at the mercy of a global oil market. The Organization of Petroleum Exporting Countries (OPEC) cartel and the unpredictable Persian Gulf countries possess more than 80% of the planet's oil reserves, so we will have to do business with them for the foreseeable future. The intensity of our interest is reflected by two wars since 1991 and a constant military presence in the Persian Gulf countries and their surrounding waters. Can we chart a path to a secure energy future? One approach is "business as usual"— pump oil, drive gas-guzzling vehicles, gouge mountains for coal, and lace the countryside with natural-gas wells and pipelines—but there are serious consequences to this approach. An alternative—the sustainable and more secure future—is to focus our energy policy on greater energy efficiency and a move toward renewable energy sources. A post-election poll (November 2008[1]) found that 78% of voters believed that investing in clean energy is vital to the recovery of the American economy. The challenge is there, and the choice is ours to make. Chapters 14, 15, and 16 describe what's going on in this important area.

---

[1]Zogby Newswire, November 7, 2008. www.zogby.com/news/readnews.cfm?ID=1637.

CHAPTER 14

Energy from Fossil Fuels

Hobet 21, a strip mine in West Virginia. The mine, owned by Arch Coal, covers 12,000 acres that were once a mountain in this Appalachian region. The coal industry has stripped hundreds of thousands of acres to mine the low-sulfur bituminous coal in this region.

KEY TOPICS

1. **Energy Sources and Uses:** Our civilization depends heavily on coal, natural gas, and petroleum for energy; these three primary energy sources are coupled in different ways to major end uses.

2. **Exploiting Crude Oil:** The energy source for transportation, oil reserves, exploration, and trade have a strong impact on the U.S. economy now that we import two-thirds of our oil.

3. **Other Fossil Fuels:** Natural gas and coal are also vital fuels in the United States with significant reserves, heavily used in power plants to generate electricity.

4. **Energy Policy and Security:** Supply-side and demand-side energy policies for fossil fuels have been legislated in recent years, yielding different pathways to the future.

THE WHITE MOUNTAINS of New Hampshire are a magnet for tourists and climbers, with nearly 50 mountain peaks above 4,000 feet. A major part of this northern extension of the Appalachian mountain chain is included in the White Mountain National Forest, some 777 acres of hardwood and coniferous forests and an extensive alpine area (Fig. 14–1). Those who live in and visit the White Mountains would likely be experiencing a very different terrain if large coal seams were discovered buried down in the geological strata of the mountains. Since mining is one of the extractive activities permitted in the country's national forests, the White Mountain National Forest would begin to look like the Appalachians of West Virginia. To get at the coal, huge machines and dynamite would be used, stripping trees and topsoil and removing the tops of the mountains, dumping the fill in the valleys and obliterating streams and valleys and small towns. It is hard to imagine New Englanders allowing this to happen, but our country's thirst for coal would likely make it inevitable.

Mountaintop removal mining, as this process is called, began in the southern Appalachians in the late 1970s as a more economical way to get at the valuable low-sulfur coal found in seams up to 1,000 feet down. The Hobet 21 mine, shown at the beginning of the chapter, is only one of many that have devastated

Figure 14–1 The White Mountains. Mount Washington from the Appalachian Trail in the White Mountains.

parts of West Virginia, Kentucky, Tennessee, and Virginia. Roughly 7%—more than 470 mountains—of the 12 million acres of mountainous land in these states has been or will soon be disturbed by mountaintop removal. Already, over 700 miles of streams have been completely buried, and another 800 miles have been degraded by the waste products that leach from the mine waste. Whole towns have been bought out by the powerful mining companies, and residents who refuse to sell have been threatened by those who support the mining operations because they provide jobs.

Lawsuits in the late 1990s over the impact of mine waste held the practice down, but a rule by the Bush administration changed all that. In 2002, mining debris was reclassified from "waste" to "fill," thus allowing companies free rein to dump the debris into stream valleys. A number of grassroots organizations, including the Ohio Valley Environmental Coalition, Mountain Justice, Appalachian Voices, and Christians for the Mountains, have been formed to combat mountaintop removal mining. They face a formidable array of politicians and coal industrialists who argue that the procedure is legal, it generates thousands of jobs, and it provides the country with coal it desperately needs. Indeed, West Virginia supplies about 15% of U.S. coal production. One miner from the small West Virginia town of Twilight stated, "I know the country needs coal, but they don't worry about the people who live in these areas. They just don't care, I reckon."[2]

OIL SPILLS AND DRILLS. Another major environmental issue is the oil spills that seem to happen on a regular basis. Oil for the global economy is moved from oil fields and ports and transported by oil tankers to refineries around the world. Millions of gallons are in transit at any given time, and it is inevitable that oil spills will occur. In 1989, the *Exxon Valdez* spilled 11 million gallons of crude oil into pristine Prince Edward Sound, and the oil is still leaching into the food chain of Alaskan waters. Spills occur every year. In 2002, the tanker *Prestige* sank off the Spanish coast

[2]Steve Webb, as reported by Clem Gutatta in *West Virginia Blue*, April 1, 2008, www.wvablue.com/showDiary.do?diaryId=1798.

with 21 million gallons of fuel oil, which continue to leak out of the sunken ship. An oil tanker collided on the Mississippi River with a barge filled with fuel oil in July 2008, spilling 420,000 gallons of fuel oil from the barge.

Drilling for oil and gas in the Arctic National Wildlife Refuge (ANWR), shown in Figure 14–2, has become a political firefight in recent years. The Bush administration made opening up the ANWR to oil development a major piece of its national energy policy. Opposing this were most of the Democratic members of Congress and a host of environmental organizations. The oil industry is predictably very interested in developing the ANWR oil because U.S. oil reserves are continuing to decline and developing the refuge's oil would maintain thousands of jobs and generate a great deal of profit for the industry. The ANWR survived numerous congressional battles through the end of 2006, and with the Democratic capture of the Congress in 2006 and the presidency in 2008, the refuge will likely remain closed for the moment.

The environmental consequences of oil spills, drilling operations, and mountaintop mining are some of the many costs of doing business in a fossil-fuel-driven economy. In future chapters, you will learn some of the other costs, such as air pollution, which is impacting human and environmental health, and rising greenhouse gases, which are leading to global climate change.

The objective in this chapter is to gain an overall understanding of the sources of fossil fuel energy that support transportation, homes, and industries. The consequences of our dependence on fossil fuels are discussed, as are public policy and options for the future.

Figure 14–2 Caribou in the Arctic National Wildlife Refuge. The Porcupine herd, some 123,000 strong, migrates on tundra that sits on an unexploited oil field. The refuge is pristine wilderness, and the intrusion of oil rigs, roads, pipelines, and villages would not only threaten the wildlife, but also clash with the attractions and aesthetic value of the landscape.

14.1 Energy Sources and Uses

Throughout human history, the advance of technological civilization has been tied to the development of energy sources. In early times (and even today in less developed regions), the major energy source was muscle power. Some people lived in relative luxury by exploiting the labor of others—slaves, indentured servants, and minimally paid workers. Human labor was supplemented by the use of domestic animals for agriculture and transportation, by water or wind power for milling grain, and by the Sun.

By the early 1700s, inventors had already designed many kinds of machinery. The limiting factor was a continual power source to run them. The breakthrough that launched the Industrial Revolution was the development of the steam engine in the late 1700s. In a steam engine, water is boiled in a closed vessel to produce high-pressure steam, which pushes a piston back and forth in a cylinder. Through its connection to a crankshaft, the piston turns the drive wheel of the machinery. Steam engines rapidly became the power source for steamships, steam shovels, steam tractors, steam locomotives, and stationary engines to run sawmills, textile mills, and virtually all other industrial plants.

The Three Kings: Coal, Oil, and Gas

Coal. At first, the major fuel for steam engines was firewood. Then, as demands for energy increased and firewood around industrial centers became scarce, coal was substituted. By the end of the 1800s, coal had become the dominant fuel, and it remained so into the 1940s. In addition to being used as fuel for steam engines, coal was widely used for heating, cooking, and industrial processes. In the 1920s, coal provided 80% of all energy used in the United States.

Although coal and steam engines powered the Industrial Revolution, which greatly improved life for most people, there were many drawbacks. The smoke and fumes from the numerous coal fires made air pollution in cities far worse than anything seen today. Writers of the time recorded that they often could not see as far as a city block away because of the smoke. Coal is also notoriously hazardous to mine and dirty to handle, and burning coal results in large quantities of ash that must be removed. As for steam engines, because of the size and bulk of the boiler, the engines were heavy and awkward to operate (Fig. 14–3). Frequently, the fire had to be started several hours before the engine was put into operation in order to heat the boiler sufficiently.

Oil Rules. In the late 1800s, the simultaneous development of three technologies—the internal combustion engine, oil-well drilling, and the refinement of crude oil into gasoline and other liquid fuels—combined to provide an alternative to steam power. The replacement of coal-fired steam engines and furnaces with petroleum-fueled engines and oil furnaces was an immense step forward in convenience. Also, the air quality improved greatly as cities were gradually rid of the smoke and soot from burning coal. (It was only in the 1960s, with the tremendous proliferation of cars, that pollution from gasoline engines became a problem.) Further, the gasoline-driven internal combustion engine provided a valuable power-to-weight advantage that allowed rapid advances in technology. A 100-horsepower gasoline engine weighs only a tiny fraction of what a 100-horsepower steam engine and its boiler weigh, and jet engines have an even greater power-to-weight ratio. Automobiles and other forms of ground transportation would be cumbersome, to say the least, without this power-to-weight advantage, and airplanes would be impossible.

Economic development throughout the world since the 1940s has depended largely on technologies that consume gasoline and other fuels refined from crude oil. By 1951, crude oil became the dominant energy source for the nation. Since then, it has become the mainstay not only of the United States (making up 40% of total U.S. energy consumption),

Figure 14–3 Steam power. Steam-driven tractors marked the beginning of the industrialization of agriculture. Today tractors half the size do 10 times the work and are much easier and safer to operate.

Figure 14–4 Global primary energy supply. The world consumes some 11,741 million tons of oil equivalent. The sources of the total primary energy supply are shown here. Note that 81% comes from fossil fuels. (*Source:* International Energy Agency, 2008.)

but also of most other countries, both developed and developing. Crude oil currently provides about 34% of the total annual global energy production (Fig. 14–4). In the countries of eastern Europe and in China, coal is still the dominant fuel. Coal ranks second globally, at 26%, and provides 23% of U.S. demand.

Gas, Naturally. Natural gas, the third primary fossil fuel, is found in association with oil or during the search for oil. Natural gas is largely methane, which produces only carbon dioxide and water as it combusts; thus, it burns more cleanly than coal or oil. In terms of pollution, it is the most desirable fuel of the three. Despite the obvious fuel potential of natural gas, at first there was no practical way to transport it from wells to consumers. Any gas released from oil fields was (and in many parts of the world still is) flared—that is, vented and burned in the atmosphere, a tremendous waste of valuable fuel. Gradually, the United States constructed a network of pipelines connecting wells with consumers. With the completion of these pipelines, the use of natural gas for heating, cooking, and industrial processes escalated rapidly because gas is clean, convenient (no storage bins or tanks are required on the premises), and relatively inexpensive. Currently, natural gas satisfies about 24% of U.S. energy demand and 20.5% of the global demand.

Thus, three fossil fuels—crude oil, coal, and natural gas—provide 86% of U.S. energy consumption and 81% of the world's consumption. The remaining percentages are furnished by nuclear power, hydropower, and renewable sources. The changing pattern of energy sources in the United States is shown in Figure 14–5. The picture is similar for other developed countries of the world, although the percentages differ somewhat, depending on the energy resources of the country. With fossil fuel energy, people and goods have become so mobile that the human community has in fact become a global community. Adding to this linkage in a powerful way is the explosion in information storage and flow, made possible by digital devices and now the Internet—all powered by electricity.

Electrical Power Production

A considerable portion of the energy we gain from fossil fuels is used to generate **electrical power**, defined as the amount of work done by an electric current over a given time. The electricity itself is an **energy carrier**—it transfers energy from a **primary energy source** (coal or waterpower, for instance) to its point of use. Electricity now defines our modern technological society, powering appliances and computers in our homes, lighting up the night, and enabling the global Internet and countless other vital processes in business and industry to operate. More than 33% of fossil fuel production is now used to generate electricity in the United States; in 1950, the figure was only 10%.

Generators. Electric generators were invented in the 19th century. In 1831, English scientist Michael Faraday discovered that passing a coil of wire through a magnetic field

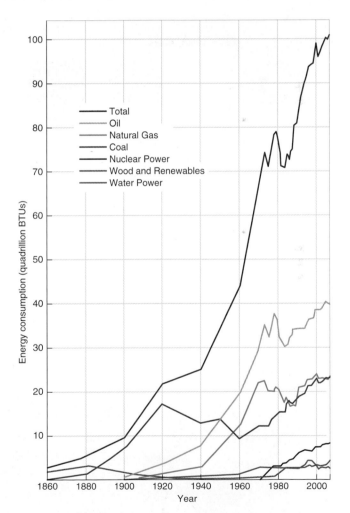

Figure 14–5 Energy consumption in the United States. Total consumption and major primary sources from 1860 to 2007 are shown here in quadrillion Btus (one Btu is the amount of heat required to raise one pound of water one degree Fahrenheit). Note how the mix of primary sources has changed over the years and how the total amount of energy consumed has continued to grow. Note also the skyrocketing increase in the use of oil after World War II (1945) as private cars and car-dependent commuting became common. (*Source:* Data from Energy Information Administration, U.S. Department of Energy, *Annual Energy Review 2007*, June 2008.)

causes a flow of electrons—an electrical current—in the wire. An **electric generator** is basically a coil of wire that rotates in a magnetic field or that remains stationary while a magnetic field is rotated around it. The process converts mechanical energy into electrical energy. As discussed in Chapter 4, any time energy is converted from one form to another, some is lost through resistance and heat (in accordance with the Second Law of Thermodynamics). Also, some energy is lost in transmitting the electricity over wires from the generating plants to the end users. In the end, it takes three units of primary energy to create one unit of electrical power that actually is put to use. This apparently inefficient proposition is justified because electrical power is so useful and indispensable.

Turbogenerators. In the most widely used technique for generating electrical power, a primary energy source is employed to boil water, creating high-pressure steam that drives a **turbine**—a sophisticated paddle wheel—coupled to the generator (Fig. 14–6a). The combined turbine and generator are called a **turbogenerator.** Any primary energy source can be harnessed for boiling the water. Coal, oil, and nuclear energy are most commonly used at present, but burning refuse, solar energy, and geothermal energy (heat from Earth's interior) may be more widely used in the future.

In addition to steam-driven turbines, gas- and water-driven turbines are used. In a gas turbogenerator, the high-pressure gases produced by the combustion of a fuel (usually natural gas) drive the turbine directly (Fig. 14–6b). For hydroelectric power, water under high pressure—at the base of a dam or at the bottom of a pipe beginning at the top of a waterfall—is used to drive a hydroturbogenerator (Fig. 14–6c).

Figure 14–6 Electrical power generation. Electricity is produced commercially by driving generators with (a) steam turbines, (b) gas turbines, and (c) water turbines. The percentage of the U.S. electricity supply derived from each source (as of 2007) is indicated. Smaller amounts of power (2.5%) are now also coming from wind energy, solar thermal energy, and photovoltaics (solar cells). (d) Each generator in this array in a hydroelectric plant is attached by a shaft to a turbine propelled by water.

TABLE 14–1	Energy Sources for Generating Electrical Power (% of total)		
Source	**New England**	**United States**	**World**
Nuclear power	28.5	19.4	14.8
Natural gas	40.3	21.9	20.1
Coal	14.5	48.7	41.0
Oil	3.0	1.6	5.8
Hydro	7.3	6.0	16.0
Wood, refuse, renewables	6.4	2.5	2.3

Sources: U.S. Energy Information Administration, 2008; International Energy Agency, 2008.

Wind turbines are also coming into use (see Figure 16–17.) A utility company supplying electricity to its customers will employ a mix of power sources, depending on the cost and availability of each and the demand cycle.

Fluctuations in Demand. Where does your electricity come from? It depends. Most utility companies are linked together in what are called *pools.* In New England, for example, ISO New England ties together some 350 electric-power-generating plants through 8,000 miles of transmission lines. The utility is responsible for balancing electricity supply and demand, regardless of daily or seasonal fluctuations. The generating capacity of the ISO New England utilities is currently 33,336 megawatts (MW), produced by a variety of energy sources, as shown in Table 14–1. One MW is enough electricity to power 800 homes, on average. The greatest demands on electrical power in New England come in the summer months, when air conditioning use is at its highest. The all-time peak load (highest demand) record of 28,130 MW was set on August 2, 2006, a very hot day in New England. A power pool must accommodate daily and weekly electricity demand as well as seasonal fluctuations.

Demand Cycle. The typical pattern of daily and weekly electrical demand in the United States is shown in Figure 14–7. The baseload represents the constant supply of power provided by the large coal-burning and nuclear plants with capacities as high as 1,000 MW. As demand rises during the day, the utility draws on additional plants that can be turned on and off—the intermediate and peak-load power sources that represent the utility's reserve capacity. These are the utility's gas turbines and diesel plants, or the utility may use pumped-storage hydroelectric power (water is pumped to a reservoir using off-peak power and is allowed to flow through turbines when needed). Occasionally, a deficiency in available power will prompt a brownout (a reduction in voltage) or a blackout (a total loss of power), disrupting whole regions temporarily. Such events occur most likely during times of peak power demand and may be precipitated by a sudden loss of power accompanying the shutdown of a baseload plant.

In our current economy, demand for electricity has been rising faster than supply, resulting in a decline in reserve capacity from 25% in the 1980s to the current level of 15%. Summer heat waves represent the greatest source of sudden pulses in demand, and as the climate continues to grow

Figure 14–7 Weekly electrical demand cycle. Electrical demand fluctuates daily, weekly, and seasonally. A variety of generators must be employed to meet base, intermediate, and peak electricity needs.

warmer in many parts of the United States, more and more regional utilities are being pushed to the edge of their ability to satisfy summer electricity demand. An even more serious problem is the antiquated system that controls the power transmission grid connecting power sources to users.

Blackout.   Power brownouts and blackouts are a serious threat to an economy that is now highly dependent on computers and other electronic devices. On August 14, 2003, the largest blackout in U.S. history hit eight eastern states and parts of Canada, imposing a $30 billion cost on the economies of the two countries. The blackout began when several power lines in Ohio brushed against trees and shut down. Alarm software failed, leaving controllers unaware of the problem. Transmission lines around the failure cut out, a local power plant shut down, more and more lines and plants dropped out, and within 8 minutes, 50 million people across eight states and two Canadian provinces were without power. What is needed to prevent such major blackouts is a "self-healing **smart grid**," one that is able to monitor problems in real time, react to trouble quickly, and isolate trouble areas so that a cascading failure like the one in 2003 would be impossible. The technology for this fix is available, but funding is a huge problem, as the electric utility industry tends to run its business on very slim profit margins. Because of the possibility of terrorist attacks on the power grid, the U.S. Department of Homeland Security has declared a "self-healing infrastructure" as one of its highest priorities in a new national plan. Fixing this problem may require some form of nationalization of the transmission grid.

Clean Energy?

There was a time when electrical power was promoted as being the ultimate clean, nonpolluting energy source. This is true *at the point of its use*: Using electricity creates essentially no pollution. For a time, it was also relatively inexpensive, and electric heating systems were installed in many new homes. This is no longer true. The public knows that electricity is (1) an expensive way to heat homes and (2) generated from fossil fuels and nuclear energy. Coal-burning power plants are the major source of electricity in the United States, but are heavily implicated in acid deposition and global climate change (Chapters 18 and 19). Nuclear power, discussed in Chapter 15, is a much-mistrusted technology because of the potential for accidents and the problem of disposal of the nuclear waste products. Additional adverse effects of using electrical energy apply to other parts of the fuel cycle, such as mining coal and processing uranium ore. The pollution is simply transferred from one part of the environment to another. The only nonpolluting energy sources are the renewables—hydroelectricity, solar, wind, and biomass—and these have their own limitations (Chapter 16).

Efficiency.   The problem of transferring pollution from one place to another becomes even more pronounced when you consider efficiency because the production of electricity from burning fossil fuels has an *efficiency of only 30–35%*. The energy loss is accounted for partly by some heat energy from the firebox going up the chimney and partly by a large amount of heat energy remaining in the spent steam as it comes out the end of the turbine. These unavoidable energy losses, called

Figure 14–8 Cooling towers. These structures, the most conspicuous feature of both coal-fired and nuclear power plants, are necessary to cool and recondense the steam from the turbines before returning it to the boiler. Air is drawn in at the bottom of the tower, over the condensing apparatus, and out the top by natural convection. Cooling is aided by the evaporation of water that condenses in a plume of vapor as the moist air exits the top of the tower, as seen in this photo.

conversion losses, are a consequence of the need to maintain a high temperature difference between the incoming steam and the receiving turbine so as to maximize efficiency. Heat energy will go only toward a cooler place, so it cannot be recycled into the turbine. The most common practice is to dissipate it into the environment by means of a **condenser** (a device that turns turbine exhaust steam into water). In fact, the most conspicuous features of coal-burning and nuclear power plants are the huge cooling towers used for that purpose (Fig. 14–8). Another significant energy loss occurs as the electricity is transmitted and distributed to end users by connecting wires.

An alternative to cooling towers is to pass water from a river, a lake, or an ocean over the condensing system (Fig. 14–6a). With this approach, however, the waste heat energy is transferred to the body of water, with the result that all the small planktonic organisms drawn through the condensing system with the water are cooked and the warm water added back to the waterway may have deleterious effects on the aquatic ecosystem. Thus, waste heat energy discharged into natural waterways is referred to as **thermal pollution**.

Electricity is a form of energy that is indispensable to modern societies as they are currently structured. Worldwide, the electrical power industry produces over 17,000 billion kilowatt-hours and revenues valued at more than $800 billion a year and, at the same time, generates a challenging array of environmental problems.

Matching Sources to Uses

In addressing whether energy resources are sufficient and where additional energy can be obtained, it is necessary to consider more than just the source of energy because some forms of energy lend themselves well to one use, but not to others. For example, the cars and trucks that travel the U.S. system of highways depend virtually 100% on liquid fuels. Almost all other machines used for moving, such as tractors, airplanes, locomotives, and bulldozers, depend likewise on liquid fuels. Crude oil, together with much smaller amounts of biofuels, is needed to support this transportation system. The nuclear industry has promoted nuclear power as a solution to the energy shortage. Nuclear power can do little to mitigate the demand for crude oil, however, because it is suitable only for boiling water to drive turbogenerators, which will do essentially nothing for our transportation system. The same can be said for coal. Although this situation might change with more widespread use of electric vehicles, their practicality for long distances still requires a technological breakthrough in the form of a low-cost, lightweight battery that can store large amounts of power.

Energy Flow.   Primary energy use is commonly divided into four categories: (1) transportation, (2) industrial processes, (3) commercial and residential use (heating, cooling, lighting, and appliances), and (4) the generation of electrical power, which goes into categories (2) and (3) as a secondary energy use. The major pathways from the primary energy sources to the various end uses are shown in Figure 14–9. Transportation, which represents 29% of energy use, depends entirely on petroleum, whereas nuclear power, coal, and water power are limited to the production of electricity. Natural gas and oil are more versatile energy sources, and significant shifts have

occurred in the use of these sources in recent years. For example, today only 2% of U.S. electricity is generated by oil, whereas 30 years ago it was about 17%.

Figure 14–9 also indicates the high proportion of consumed energy that goes directly to waste heat rather than being used for its intended purpose. Some waste is inevitable, as dictated by the Second Law of Thermodynamics, but the current losses are much greater than necessary. The efficiency of energy use can be at least doubled for some uses, including cars that get twice as many miles per gallon and appliances that consume half as much power. For uses like electric power generation, however, somewhat smaller increases in efficiency can be expected. In general, increasing the efficiency of energy use would have the same effect as increasing the energy supply, at far less expense. Such increases in efficiency represent a largely untapped "reserve" of energy that is saved, the **conservation reserve** discussed in more detail in Section 14.4.

In contemplating the energy future, we need to consider more than gross supplies and demands. That is, we must also examine how particular primary sources can be matched to particular end uses. Furthermore, what are the most cost-effective ways of balancing supplies and demands—by increasing supplies or by decreasing consumption through conservation, efficiency, and demand management?

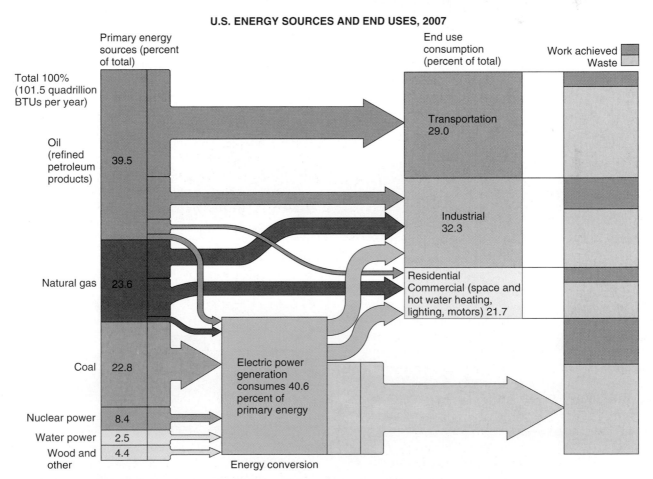

Figure 14–9 Pathways from primary energy sources to end uses in the United States.
Only major pathways are shown. Note that end uses are connected to primary sources in specific ways. Note also the large percentage of energy that is wasted as it is converted to heat and lost. (*Source:* Data from Energy Information Administration, U.S. Department of Energy, *Annual Energy Review 2007*, June 2008.)

14.2 Exploiting Crude Oil

The United States has abundant reserves of coal, adequate reserves of natural gas (at least for the time being), and the potential for the expansion of nuclear power. Therefore, no shortages loom in these areas now. (Note that the environmental concerns regarding coal and nuclear power are being set aside for the moment.) Petroleum to fuel our transportation system, however, is another matter. The U.S. crude-oil supplies fall far short of meeting demand; we now depend on foreign sources for 66% of our crude oil, and this dependency is steadily growing as U.S. reserves diminish. Foreign sources come at increasing costs, such as trade imbalances, military actions, economic disruptions, and coastal oil spills.

Let us examine the situation, starting with where crude oil comes from and how it is exploited.

How Fossil Fuels Are Formed

The reason crude oil, coal, and natural gas are called *fossil fuels* is that all three are derived from the remains of living organisms (Fig. 14–10). In the Paleozoic and Mesozoic geological eras, 100 to 500 million years ago, freshwater swamps and shallow seas, which supported abundant vegetation and phytoplankton, covered vast areas of the globe. Anaerobic conditions in the lowest layers of these bodies of water impeded the breakdown of detritus by decomposers. As a result, massive quantities of dead organic matter accumulated.

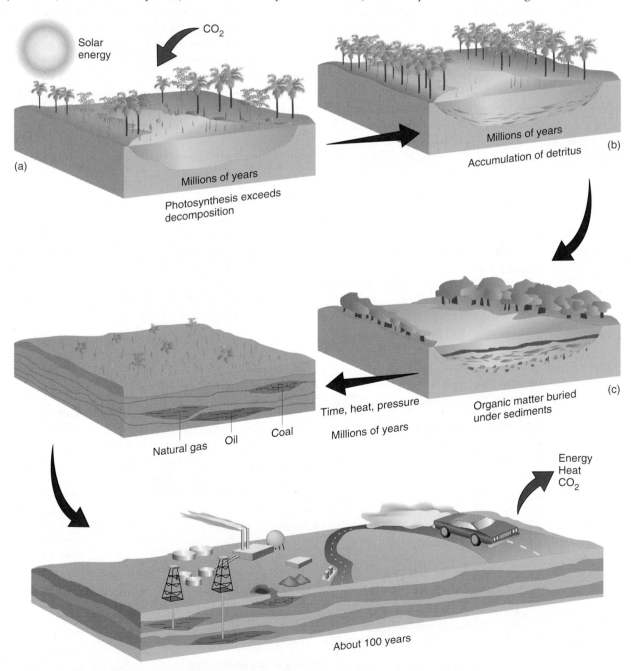

Figure 14–10 Energy flow through fossil fuels. Coal, oil, and natural gas are derived from biomass that was produced many millions of years ago. Deposits are finite, and because formative processes require millions of years, they are nonrenewable.

Over millions of years, this organic matter was gradually buried under layers of sediment and converted by pressure and heat to coal, crude oil, and natural gas. Coal is highly compressed organic matter—mostly leafy material from swamp vegetation—that decomposed relatively little. Sediments buried 7,500 to 15,000 feet are in a temperature range that breaks down large organic molecules to smaller carbon molecules, producing crude oil. Below 15,000 feet, molecular breakdown leads to the one-carbon compound methane, or natural gas.

Renewal. Additional fossil fuels may still be forming through natural processes to this day. However, they cannot be considered renewable resources, as we are using fossil fuels far faster than they ever formed. In fact, it takes 1,000 *years* to accumulate the amount of organic matter that the world now consumes each *day*. Because fossil fuels are a finite resource, the pertinent question is not *whether* they will run out, but *when* they will.

Crude-Oil Reserves Versus Production

The total amount of crude oil remaining represents Earth's oil resources. Finding and exploiting these resources is the problem. The science of geology provides information about the probable locations and extents of ancient shallow seas. On the basis of their knowledge and their field experience, geologists make educated guesses as to where oil or natural gas may be located and how much may be found. These educated guesses are the world's **estimated reserves**. The estimates may be far off the mark because there is no way to determine whether estimated reserves actually exist except by exploratory drilling.

If exploratory drilling strikes oil, further drilling is conducted to determine the extent and depth of the **oil field**. From that information, a fairly accurate estimate can be made of how much oil can be economically obtained from the field. That amount then becomes **proved reserves**. Amounts are given in barrels of oil (1 barrel = 42 gallons), and the content of each reserve field is expressed in probabilities. Thus, a 50% probability, or P50, of 700 million barrels for a field means that a total of 700 million barrels is just as likely as not to come from the field. Proved reserve estimates can range from P05 (a 5% probability) to P90. A P90 estimate is more reliable, but oil producers are often vague about probabilities and might prefer to use a P05 or a P10 estimate in order to give the impression of a large reserve for political or economic reasons. Also, proved reserves hinge on the economics of extraction. Thus, such reserves may actually increase or decrease with the price of oil because higher prices justify exploiting resources that would not be worth extracting for lower prices. The final step, the withdrawal of oil or gas from the field, is called **production** in the oil business. *Production*, as used here, is a euphemism. "It is production," said ecologist Barry Commoner, "only in the sense that a boy robbing his mother's pantry can be said to be producing the jam supply."[3] In reality, this step is *extraction* from Earth.

---

[3]Commoner, Barry. "Toward a Sustainable Energy Future." *E Magazine*. May/June 1991, p. 55.

Recovery. Production from a given field cannot proceed at a constant rate because crude oil is a viscous fluid held in pore spaces in sedimentary rock, just as water is held in a sponge. Initially, the field may be under so much pressure that the first penetration produces a gusher. Gushers, however, are generally short lived, and the oil then seeps slowly from the sedimentary rock into a well from which it is pumped out. Ordinarily, conventional pumping (**primary recovery**) can remove only 25% of the oil in an oil field. Further removal, of up to 50% of the oil in the field, is often possible, but it is more costly because it involves manipulating pressure in the oil reservoir by injecting brine or steam (**secondary recovery**), which forces the oil into the wells. In some fields, even greater recovery (**enhanced recovery**) is obtained by injecting carbon dioxide, which breaks up oil droplets and enables them to flow again.

Economics—the price of a barrel of oil—determines the extent to which reserves are exploited. No oil company will spend more money to extract oil than it expects to make selling the oil. At $10 per barrel of crude oil, the price in the late 1990s, it was economical to extract no more than 25–35% of the oil in a given field. Thus, an increase in price makes more reserves available. In the 1970s and early 1980s, higher oil prices justified reopening old oil fields and created an economic boom in Texas and Louisiana. Then an economic bust occurred in the late 1980s, and a drop in oil prices caused the fields to be shut down again. Following that, oil prices began to rise in 1998, reached a peak of $145 a barrel in mid-2008, and then began a sharp decline in September of 2008 as global oil demand dropped—a consequence of the severe global economic crisis. Although lower oil prices are good news for consumers, the volatility of the market is bad news for oil producers, who need a more stable market in order to make future plans for oil exploitation.

Declining U.S. Reserves and Increasing Importation

Understanding the relationships among production, price, and the exploitation of reserves can help clarify the U.S. energy dilemma. Up to 1970, the United States was largely oil independent. That is, exploration was leading to increasing proved reserves, such that production could basically keep pace with growing consumption (Fig. 14–11). In 1970, however, new discoveries fell short, so production decreased, while consumption continued to grow rapidly. This downturn vindicated the predictions of the late geologist M. King Hubbert, who proposed that oil exploitation in a region would follow a bell-shaped curve. Hubbert observed the pattern of U.S. exploitation and predicted that U.S. production would peak (now called the *Hubbert peak*) between 1965 and 1970. At that point, about half of the available oil would have been withdrawn, and production would decline gradually as reserves were exploited. Figure 14–11 demonstrates that a Hubbert curve can be applied to U.S. oil production.

To fill the energy gap between rising consumption and falling production, the United States depended increasingly on imported oil, primarily from the Arab countries of the

Figure 14–11 Oil production and consumption in the United States. Four stages can be seen: (1) Up to 1970, the discovery of new reserves allowed production to parallel increasing consumption. (2) From 1970 to 1973, a lack of new oil discoveries caused production to decline, while consumption continued to climb, resulting in a vast increase in oil imports. (3) In the late 1970s to early 1980s, high oil prices promoted both lowered consumption and increased production—which included bringing the Alaskan oil field into production. (4) From 1986, when oil prices declined sharply, until the present, consumption has again increased, while production has resumed its decline, making us increasingly dependent on foreign oil. Go to **GRAPHit!** at www.masteringenvironmentalscience.com.
(*Source:* Energy Information Administration, U.S. Department of Energy, *Annual Energy Review 2007*, June 2008.)

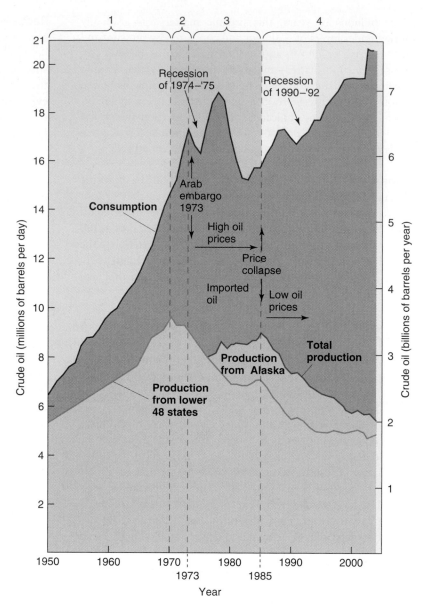

Middle East. European countries and Japan did likewise. Because imported oil cost only about $2.30 per barrel ($12.16 in 2007 dollars) in the early 1970s and Middle Eastern reserves were more than adequate to meet the demand, this course seemed to present few problems.

The Oil Crisis of the 1970s. Because of the low price of oil from Middle Eastern countries, the United States and most other highly industrialized nations were becoming more and more dependent on oil imports. A group of predominantly Arab countries known as the Organization of Petroleum Exporting Countries (OPEC) formed a cartel and agreed to restrain production in order to get higher prices. Imported oil began to cost more in the early 1970s, and then, in conjunction with an Arab-Israeli war in 1973, OPEC initiated an embargo of oil sales to countries, such as the United States, that gave military and economic support to Israel. The effect was almost instantaneous because we depend on a fairly continuous flow from wells to points of consumption. (Compared with total use, little oil was held in storage.) Spot shortages occurred, which quickly escalated into widespread panic and long lines at gas stations because of our widespread dependence on cars and trucks. We were willing to pay almost anything to have oil shipments resumed—and pay we did: OPEC resumed shipments at a price of $12 a barrel ($55 in 2007 dollars), four times the previous price.

Then, by continuing to limit oil production all through the 1970s, OPEC was able to keep supplies tight enough to force prices higher and higher. In the early 1980s, a barrel of oil delivered to the United States cost $36 ($75 in 2007 dollars), and the United States was paying accordingly. The high energy costs were devastating to the economy as inflation and unemployment rose and the country experienced several recessions. The purchase of foreign oil became, and remains, the single largest factor in our **balance-of-payment deficit**— where we buy more from other countries than we sell as exports (Fig. 14–12a).

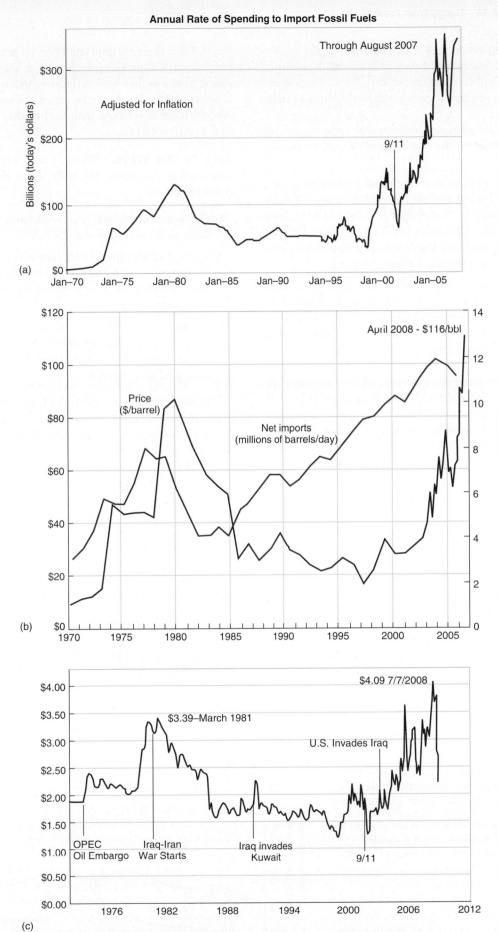

Annual Rate of Spending to Import Fossil Fuels

(a)

Through August 2007

Adjusted for Inflation

9/11

Billions (today's dollars)

$300

$200

$100

$0

Jan–70   Jan–75   Jan–80   Jan–85   Jan–90   Jan–95   Jan–00   Jan–05

(b)

$120

$100

$80

$60

$40

$20

$0

April 2008 - $116/bbl

Price
($/barrel)

Net imports
(millions of barrels/day)

1970   1975   1980   1985   1990   1995   2000   2005

14

12

10

8

6

4

2

0

(c)

$4.09 7/7/2008

$4.00

$3.50

$3.00

$2.50

$2.00

$1.50

$1.00

$0.50

$0.00

$3.39–March 1981

U.S. Invades Iraq

OPEC
Oil Embargo

Iraq-Iran
War Starts

Iraq invades
Kuwait

9/11

1976   1982   1988   1994   2000   2006   2012

Figure 14–12 The cost of fossil fuel imports. (a) The inflation-corrected cost of total oil imported
to the U.S. economy; (b) the cost in dollars per barrel (today's dollars) and net imports; (c) gasoline prices in
today's dollars. (*Source:* zFacts.com, using data from U.S. Department of Energy sources.)

359

Adjusting to Higher Prices. In response to the higher prices, the United States and other industrialized countries made a number of adjustments during the 1970s. The following are the most significant changes made by the United States:

- We increased domestic production by stepping up oil exploration, building the Alaska pipeline, and reopening old oil fields.

- Congress took steps to decrease consumption by setting new standards for automobile fuel efficiency at 27.5 miles per gallon or mpg (cars averaged 13 mpg in 1973), lowered speed limits to 55 miles per hour, promoted higher efficiencies for building insulation and appliances, and began to develop alternative energy sources.

- To protect against another OPEC boycott, we created a strategic oil reserve in underground caverns in Louisiana. The current stockpile is 702 million barrels of oil, equivalent to about 33 days of consumption at 21 million barrels per day.

With the exception of the stockpile, all of these steps are economically appealing only when the price of oil is high. If the price is low, the economic returns or savings may not justify the necessary investments (even though the steps are environmentally desirable).

The results of the new actions were not immediate, nor could quick returns have been expected. For example, it takes at least three to five years to design a new car model and start production. Then another six to eight years pass before enough new models are sold and old ones retired to affect the overall highway "fleet."

Recovery. As we entered the 1980s, however, these efforts were definitely having an impact. The United States continued to depend substantially on foreign oil, but consumption was declining, and production, up after the Alaska pipeline was completed, was holding its own. (See Figure 14–11.) Elsewhere in the world, significant discoveries in Mexico, Africa, and the North Sea made the world less dependent on OPEC oil. As a result of all these factors, plus OPEC's inability to restrain its own members' production, world oil production exceeded consumption in the mid-1980s, and there was an oil "glut."

Supply exceeded demand, with predictable results. In 1986, world oil prices crashed from the high $20s to $14 per barrel. From that time until late 1999, oil prices fluctuated in the range from $10 to $20 per barrel and, due to inflation, were almost as low as they were in the early 1970s (Fig. 14–12b). The lower oil prices apparently benefited consumers, industry, and oil-dependent developing nations, but they hardly solved the underlying problem.

Back to Old Ways. Whereas high oil prices stimulated constructive responses, the collapse in oil prices undercut those responses:

- Oil exploration was curtailed, and production from older fields was terminated.

- Conservation efforts and incentives were abandoned. Government standards for automobile fuel efficiency were frozen at 27.5 mpg. Speed limits were raised to 65 and even 75 mph.

- Congress terminated tax incentives and other subsidies for the development or installation of alternative energy sources, destroying many new businesses engaged in solar and wind energy.

Production of crude oil in the United States continued to drop as proved reserves were drawn down. The area of declining production now includes Alaska, which reached its peak production in 1988. At the same time, consumption of crude oil rose again because both the number of cars on the highways and the average number of miles per year that each car is driven increased. Adding to that consumption, consumers began buying vehicles that were less fuel efficient. Minivans, four-wheel-drive sport-utility vehicles (SUVs), light trucks, and other less-fuel-efficient models constituted more than 50% of the new-car market. Our dependence on foreign oil has grown steadily since the mid-1980s, and it is now at 66% and rising (Fig. 14–13).

Back to the Future. The collapse in oil prices of the 1990s set the stage for another developing crisis: a rise in oil prices that began in late 1998. OPEC was concerned about the continuing oil glut and its low prices, so it decided to cut production just as East Asia was coming out of a serious recession. The rapid rise in demand brought on a shortfall in the oil supply, and the price of crude oil rose precipitously (Fig. 14–12b). With demand outpacing supply, oil prices kept

Figure 14–13 Consumption, domestic production, and imports of petroleum products. After the oil crises of the 1970s, the United States became less reliant on foreign oil. Since the mid-1980s, however, consumption of foreign oil has for the most part been rising steadily. (*Source:* Energy Information Administration, U.S. Department of Energy, *Annual Energy Review 2007,* June 2008.)

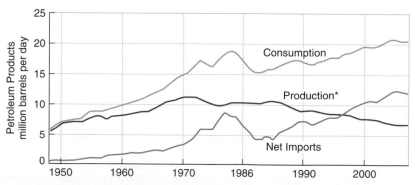

*Crude oil and natural gas plant liquids production.

rising, reaching a peak of over $140 a barrel in mid-2008. Home owners were strapped with high fuel-oil prices, commuters faced gasoline prices in the $3–$4 range (Fig. 14–12c), people cut back on driving, economic growth began to slacken, and only the oil industry prospered as record profits poured in. Internationally, food prices rose to unprecedented levels, forcing many in the developing world into dependency and hunger. In the United States, shoppers purchased smaller vehicles, sales of hybrid vehicles blossomed, and the auto industry began closing factories that produced SUVs, vans, and trucks. Once again, Congress responded, raising the fuel efficiency standard, mandating the increasing use of renewable fuels, and establishing new standards for energy efficiency.

Recession. Then the bottom dropped out of the stock market in September 2008, a consequence of overpriced housing, mortgage excesses, and uncontrolled stock market speculation. A global economic meltdown followed, during which major financial institutions failed or were bought out by governments, credit suddenly became almost unavailable, and job losses accelerated. In short, the countries of the world slid into a recession that was called the worst in 75 years. The impact on oil prices was profound: Oil imports dropped, supply exceeded demand, and the price dropped back below $70 a barrel by October of the same year. Gasoline prices dropped down into the $2 a gallon range, and home owners facing the winter could anticipate much lower fuel-oil prices. Oil-exporting countries suddenly faced a sharp decline in their revenues, and OPEC members began to consider cutting their output in order to stabilize prices. Nevertheless, the rest of the world, including developed and developing countries alike, remained dependent on oil from those relatively few nations with surplus production.

The Consequences of U.S. Dependency

What problems does this growing dependency on foreign oil present? As the U.S. dependency on foreign oil rises, we are faced with problems at three levels: (1) the costs of purchasing oil, (2) the risk of supply disruptions due to political instability in the Middle East, and (3) ultimate resource limitations in any case.

Costs of Purchase. The yearly cost to the United States of purchasing foreign oil and gas is shown in Figure 14–12a. The cost dropped considerably from its high of more than $130 billion (corrected for inflation) in the early 1980s, due to falling prices and conservation efforts. It began a rapid rise in 2000, however, and exceeded $300 billion per year. In seven years, our petroleum imports increased by 24%, and the price per barrel rose almost fivefold!

The cost for crude oil represents about 36% of our current balance-of-payment deficit ($815 billion in 2007), which has significant consequences for our economy. The price we pay at the pump is basically the same whether the oil is produced here or abroad. If the oil is produced at home, however, the money we pay for it circulates within our own economy, providing jobs and producing other goods and services. When we buy oil from a foreign country, we are draining our own economy.

Finally, there are the ecological costs of oil spills such as the *Exxon Valdez* incident (estimated at $15 billion) and of other forms of pollution from the drilling, refining, and consumption of fuels—although estimating total damage in dollars is sometimes difficult. There are also military costs to consider when protecting our sources of foreign oil.

Persian Gulf Oil. The Middle East is a politically unstable region of the world. As noted earlier, it was the unexpected Arab boycott that plunged us into the first oil crisis in 1973. Maintaining access to Middle Eastern oil is also a major reason for our efforts toward negotiating peace agreements between various parties in that region. Recognizing this political instability, the United States maintains a military capability to ensure our access to Persian Gulf oil. This capability was tested in the fall of 1990 when Saddam Hussein of Iraq invaded Kuwait, then producing 6 million barrels of oil per day. The U.S.-led Persian Gulf War of early 1991 threw Hussein's armies out of Kuwait and gave the United States an ongoing presence in the Persian Gulf region. It was this presence, however, that eventually angered the radical Islamic terrorist Al Qaeda organization and led to the terrible September 11, 2001, attack on the World Trade Center and the Pentagon.

Two More Wars. Within a month, U.S. and British forces invaded Afghanistan, which had been harboring Osama bin Laden and elements of Al Qaeda. The purpose of the war was to capture bin Laden, destroy the Al Qaeda training camps, and overthrow the ruling Taliban regime. Later, North Atlantic Treaty Organization forces joined in to help stabilize the country. Although Afghanistan is not an oil-producing country, it can be argued that the Afghan war was a consequence of our presence in the Persian Gulf region. That war is still in progress, as the Taliban have sustained an insurgency that continues to destabilize the country and bin Laden is still at large.

Then, in 2003, a U.S.-British coalition invaded Iraq in order to overthrow Saddam Hussein's rule and eliminate suspected weapons of mass destruction (Fig. 14–14). As the momentum was building for the invasion, protesters around the world rallied to the cry "No blood for oil," implying that the real motivation behind the pending war was to gain access to Iraq's rich oil reserves. Some hindsight suggests that oil was part of the picture. For example, coalition forces were prepared to protect Iraq's oil fields and did so immediately after the invasion. However, there were no plans to protect the Iraqi National Museum, which was looted of priceless historical artifacts. Although it is clear that the United States and Britain did not simply commandeer the Iraqi oil, they have spent billions to restore Iraq's oil industry—and the United States and other oil-hungry Western countries are eager customers for the oil now that it is flowing again (at a rate of 2½ million barrels per day, close to prewar levels). Those countries rely on Persian Gulf oil for half of their oil imports. It is fair to say that military costs associated with maintaining access to Persian Gulf oil represent a substantial U.S. government subsidy of our oil consumption, one that has been costly also in human lives.

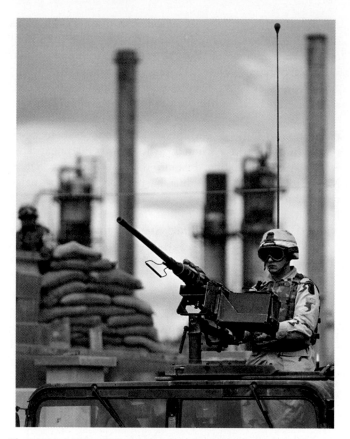

Figure 14–14 The U.S.-Iraqi war and oil. A soldier from the U.S. 173rd Airborne stands guard at a refinery near the Baba Gurgur oil field in Kirkuk, Iraq.

Other Sources. The United States has turned more and more to non–Middle Eastern oil suppliers. Recently, Canada replaced Saudi Arabia as the number-one exporter of oil to the United States, with about 17% of the market. Mexico, Colombia, and Russia are also major sources of oil. Western Hemisphere sources now provide about 44% of our imports. The relative political stability and proximity of these sources are good reasons for the United States to turn to them and away from the Middle East. Russia, a number of former Soviet republics, and several African countries (especially Nigeria) are important new sources of oil. However, most of these other sources, like the United States, are on the downside of the Hubbert curve and are seeing production declining. The conclusion? We will continue to depend on OPEC and the Persian Gulf for a substantial share of our oil imports for the present and even more so in the future.

Oil Resource Limitations and Peak Oil

Crude-oil production in the United States is decreasing because of diminishing domestic reserves. Moreover, on the basis of the "Easter-egg hypothesis," there is dim hope for major new U.S. finds. According to this hypothesis, as your searching turns up fewer and fewer eggs, you draw the logical conclusion that nearly all the eggs have been found. In terms of oil, North America is already the most intensively explored of any continental landmass. The last major find was the Alaskan oil field in 1968. Discoveries since then have

been small in comparison, with increasing numbers of dry holes in between. Indeed, most of the "new" oil in the United States is coming from innovative computer mapping of geological structures and horizontal drilling technology that enable oil to be identified and withdrawn from small isolated pockets that were previously missed within old fields. Thus, an increasing percentage of our needs must continue to be met by imported oil.

How much oil is still available for future generations? About 1,050 billion barrels (BBs) of oil have already been used, almost all in the 20th century. Currently, the nations of the world are using about 85 million barrels per day, or some 31 BBs a year. Rising demand by the developing countries and the continued rise in oil demand by the developed countries point to an increase of about 2% per year, so that demand could reach 40 BBs a year by 2020. How long can the existing oil reserves provide that much oil?

Hubbert's Peak. As with many issues, it depends on whom you believe. Recently, oil geologists Colin Campbell and Jean Laherrère provided an analysis that puts the world's proved reserves at about 850 BBs.[4] This is much lower than the 1,238 BBs claimed by oil industry reports (Table 14–2). Campbell and Laherrère derived their estimate from the P50 values of *proved* reserves and discounted some reports of reserves that were inflated for political reasons (for example, those of OPEC countries in the mid-1980s). On the basis of their estimates and the known amount already used, the two geologists calculated a Hubbert curve indicating that peak oil production will occur sometime in the current decade (Fig. 14–15). When that happens, production will begin to taper off in spite of rising demand, and the world will enter a new era of rising oil prices. Princeton professor and former oil geologist Kenneth Deffeyes used the Hubbert technique in his book *Beyond Oil: The View from Hubbert's Peak*[5] to draw the same conclusions. Even if these scientists are wrong and the oil industry estimates are right, it would delay the peak by only a decade at most. Deffeyes predicts: "Global oil production will probably reach a peak sometime during this decade. After the peak, the world's production of crude oil will fall, never to rise again. . . . Ignoring the problem is equivalent to wagering that world oil production will continue to increase forever. My recommendation is for us to bet that the prediction is roughly correct."[6]

U.S. Geological Survey. In 2000, on the basis of extensive geological studies, the U.S. Geological Survey (USGS) released an assessment of world oil and gas that included "*undiscovered* reserves"—that is, reserves from fields that have not yet been found. According to this assessment, there are undiscovered reserves of 732 BBs of oil remaining from unknown fields and an additional 688 BBs in known fields that could be unaccounted for by conventional analyses.

[4]Colin J. Campbell and Jean H. Laherrère, "The End of Cheap Oil." *Scientific American*, March 1998.
[5]Kenneth Deffeyes, *Beyond Oil: The View from Hubbert's Peak* (New York: Hill and Wang, 2005).
[6]Kenneth, Deffeyes, *Hubbert's Peak: The Impending World Oil Shortage* (Princeton, N.J.: Princeton University Press, 2001), 1.

TABLE 14–2 Proved World Reserves and Annual Use of Fossil Fuel Reserves, 2007

Region	Reserves		
	Petroleum (billion barrels)*	Natural Gas (trillion ft³)*	Coal (billion metric tons)*
North America	69.3**	281.6	250.5
South and Central America	111.2	272.8	16.3
Europe and Eurasia	143.7	2,097.9	272.2
Middle East	755.3	2,583.3	1.4
Africa	117.5	514.9	49.6
Asia Pacific	40.8	510.7	257.5
Total	1,237.8	6,261	847.5

	Use		
	Petroleum	Natural Gas	Coal
World use, 2007	31 billion barrels	103.8 trillion ft³	6.69 billion metric tons
World use, 2007 (quads)	177 quads	106 quads	118 quads

Source: BP Statistical Review of World Energy, 2008.
*The fuels may be compared by calculating their energy content in British thermal units (Btus) and then converting these figures to quadrillion Btus or quads, where 1 quad = 10^{15} Btus. One billion barrels of petroleum yield about 5.7×10^{15} Btus, or 5.7 quads. One trillion cubic feet of natural gas yield about 1.02×10^{15} Btus, or 1.02 quads. One billion metric tons of coal yield about 17.65×10^{15} Btus, or 17.65 quads.
**Does not include an estimated oil equivalency of 152 billion barrels in Canadian oil sands.

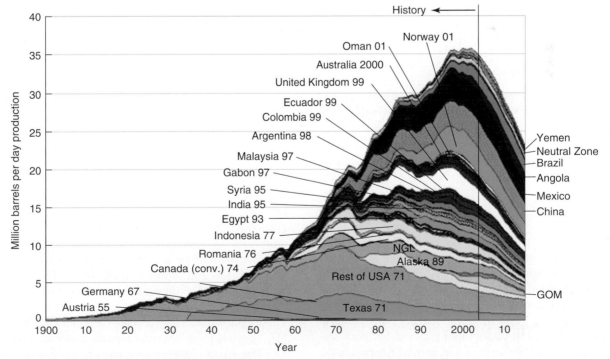

Figure 14–15 Hubbert curves of oil production. Oil production in a region follows a bell-shaped curve. Dates next to countries refer to when oil production peaked in those countries. The combined curves are for petroleum sources outside OPEC and the former Soviet Union. The evidence indicates that world oil production from these sources may have peaked already. (*Source: BP Statistical Review of World Energy 2004*, as presented in W. Zittel and J. Schindler, October 15, 2004, EnergyBulletin.net.)

Campbell believes that the USGS estimates are overly optimistic, based as they are on oil yet to be found (not *proved* reserves!).[7] Even if much of this oil is found, it will at best move peak oil production back a few years or a decade, given the rising consumption rates.

Downward Trajectory. At the current rate of use (which is predicted to rise), the proved reserves can supply only 40 years of oil demand. New discoveries continue to add to proved reserves, but they are only a fraction of our annual use. Thus, the 21st century will become known as an era of declining oil use, just as the 20th century saw continuously rising use. The National Petroleum Council, an advisory group comprised of oil industry executives and other energy experts, reported in 2007 to then Energy Secretary Samuel Bodman that global crude oil production will not keep up with the growing demand in the next 25 years. The German-based Energy Watch Group stated that global oil production already peaked in 2006 and would begin a downward trajectory.

As oil becomes more scarce, the OPEC nations will once again dominate the world oil market. The Middle East possesses 61% of the world's proved reserves, and the developed world—in particular, the United States—will have to rely more and more on that region for its oil imports. In light of this fact, and in light of the reality of terrorist activities, it would seem wise to begin to prepare for that day by reducing our dependency on foreign oil now. We can become more independent in three ways: (1) We can increase the fuel efficiency of our transportation system, which uses most of the oil; (2) we can use the other fossil fuel resources we have to make fuel for vehicles; and (3) we can develop alternatives to fossil fuels, many of which are ready now.

14.3 Other Fossil Fuels

Unlike crude oil, considerable reserves of natural gas and comparatively vast reserves of coal and oil absorbed in sand and rock are still available in the United States and elsewhere in the world. Natural gas is gradually coming into use as a fuel for vehicles, and there is some potential in this respect for coal and oil.

Natural Gas

In the United States, natural gas has been rising as a fossil fuel of choice—so much so that 16.5% of our annual use now comes from imports, almost all by pipeline from Canada. Almost equal quantities of this fuel go to industrial use, residential use, and electrical power generation. As with oil, the cost of natural gas is subject to market fluctuations in supply and demand, and prices often rise and fall with seasonal changes. At the 2007 rate of use, proved reserves in the United States hold only a nine-year supply of gas. This is

somewhat deceiving, however, because natural gas is continually emitted from oil- and gas-bearing geological deposits and new drilling is opening up more deposits. The expected long-term recovery of natural gas in the United States is significantly greater than the proved reserves, putting the supply closer to 50 years.

Worldwide, the estimated natural-gas resource base is even more plentiful (Table 14–2). Almost four times as much equivalent gas (that is, equivalent in energy) is likely to be available as oil. Much of it, however, is relatively inaccessible. Gas must either be pumped through a pipeline (a process limited by the length of the line) or be subjected to high pressures so that it remains a liquid at room temperature (liquefied natural gas, or LNG). It can then be shipped in large tankers; currently, eight LNG terminals in the United States handle these ships. Thirteen more have been approved, but are not yet built. Constructing new LNG terminals has proven to be difficult, as the public views the terminals and tankers as major security and safety hazards.

Natural-Gas-Run Cars. Natural gas can also meet some of the fuel needs for transportation. With the installation of a tank for compressed gas in the trunk and some modifications of the engine fuel-intake system—costing several thousand dollars—a car will run perfectly well on natural gas. Natural gas is a clean-burning fuel, producing carbon dioxide and water, but virtually no hydrocarbons or sulfur oxides. Natural gas is in use in the United States by many buses and a number of private and government car fleets, but use is restricted due to the limited number of natural-gas service stations. Detroit automakers have stopped selling natural-gas-powered cars, but Europeans can buy them from eight different automakers. There the fuel is taxed at a much lower rate than gasoline, and the result is a growing fleet—800,000 strong—of natural-gas-powered vehicles in Europe. Germany alone has almost 800 service stations that provide natural gas. It will take stronger public-policy support for this option to grow in the United States; until that happens, the natural-gas option will likely stay on the back burner.

Synthetic Oil. With the aid of the Fischer-Tropsch process, known since the 1920s, natural gas can be converted to a hydrocarbon that is liquid at room temperature and pressure—basically, a synthetic oil. Researchers are hotly pursuing ways to convert natural gas to synthetic oil more efficiently. Presently, the fuel produced is only about 10% more expensive than oil. Several major oil companies are spending billions on gearing up refineries to convert gas to liquid diesel and home-heating fuels (Fig. 14–16), and researchers are seeking an appropriate catalyst that can convert natural gas (methane) to methanol (methyl alcohol). These options promise to make the abundant, but inaccessible, "stranded" natural-gas deposits suddenly very valuable.

The Costs. Developing natural gas has definite trade-offs. For example, to develop the vast northern Canadian and Alaskan gas deposits, a 3,600-mile gas pipeline will be needed to conduct the gas to the U.S. Midwest. Current gas prices already justify building the pipeline, and several

[7]Campbell, Colin. Peak Oil: An Outlook on Crude Oil Depletion. 2002. www.greatchange.org/ov-campbell,outlook.html.

Figure 14–16 Gas-to-liquid fuel plant. This plant in Bintulu, Malaysia, captures "waste" natural gas and converts it into 12,500 barrels/day of liquid fuels. Royal Dutch/Shell is building a plant 11 times as large in Qatar.

competing plans are being evaluated by the state of Alaska, the Canadian government, and the U.S. government. The cost is expected to run upward of $25 billion. Needless to say, such a pipeline would carry an enormous environmental cost. Substantial gas reserves exist in the lower 48 states, especially in the Gulf Coast and the Rocky Mountain states. Prompted by Bush administration directives, gas-drilling rigs are going full tilt on many federal lands, causing environmental damage from roads, wells, and pipelines. Some are in sensitive wildlife areas, such as the Padre Island National Seashore and the Powder River Basin in Montana and Wyoming.

Applying natural gas to the transportation sector—both directly as a fuel and indirectly as a synthetic oil—may extend the oil phase of our economy, but it will not be a sustainable solution because the natural-gas reserves are also limited. What about coal, then, the most abundant fossil fuel?

Coal

The world's leading producer of coal is China (2.8 billion tons in 2007), where a booming economy has stimulated energy demand. Most of this is burned to produce electricity as the country builds more than two coal power plants a week. Coal provides 80% of China's electricity. In the United States, 49% of electricity comes from coal-fired power plants. In cost per British thermal unit (Btu), coal is much less expensive than natural gas or oil. Current U.S. reserves are calculated to be 264 billion tons, and at the 2007 rate of production (1.14 billion tons/year), the supply could last 230 years. Unlike oil and natural gas, we produce more coal each year than we use, and we export about 5% of coal production annually.

The mining work can be hazardous, and injuries and fatalities are not uncommon—in 2007, 33 fatal injuries occurred. More serious, however, is the occupational disease called black lung, or coal workers' pneumoconiosis (CWP), acquired by inhaling coal dust. Yearly fatalities from this disease, mostly males of advanced age, have been declining, but are still over 700.

Coal can be obtained by surface mining (69%) or underground mining (31%). Both methods have substantial environmental impacts. Land subsidence and underground fires often occur in conjunction with underground mines. The town of Centralia, Pennsylvania, for example, has been abandoned and bought by the government because of a coal fire that started 40 years ago and could burn for another 100 years (Fig. 14–17). Coal fires around the world produce almost as much carbon dioxide as do all the cars and trucks in the United States, contributing significantly to the world's greenhouse gas emissions. In underground mining, at least 50% of the coal must be left in place to support the roof of the mine, which is one reason why mining companies have turned to strip mining.

Strip Mining. In the more common surface, or strip, mining, dynamite is used to break overlying layers, and then gigantic power shovels turn aside the rock above the coal seam and remove the coal (see chapter opening photo). This method totally destroys the ecology of the region, as forests are removed and streams are buried by the coal wastes. Although federal regulations require strip-mined areas to be reclaimed—that is, graded and replanted—it takes many decades before an ecosystem remotely resembling the original one can develop, and the topography is in shambles. In arid areas of the West, water limits may prevent the ecosystem from ever being reestablished. Consequently, surface-mined areas may be turned into permanent deserts. Furthermore, erosion, acid leaching, and mine waste have numerous adverse effects on waterways and groundwater in the surrounding area. For example, Massey Energy Co., the largest coal producer in the Appalachian region, recently reached a $30 million settlement agreement with the Environmental Protection Agency (EPA). The coal company was alleged to

Figure 14–17 Coal fire. Smoke seeps from an underground coal mine fire in Centralia, Pennsylvania. The town has been abandoned because of the fire.

have violated the Clean Water Act on at least 4,500 occasions while discharging mining waste and sediment into streams in connection with its mountain-top-removal mining operations.

Coal Power. Ninety-one percent of the coal the United States produces is used to generate electricity. Once it is mined, coal is transported to large power plants. A typical 1,000-MW plant burns 8,000 tons of coal a day—a mile-long train's worth of coal every day. Combustion of this much coal releases 20,000 tons of carbon dioxide and 800 tons of sulfur dioxide (most of which is prevented from entering the atmosphere if the power plant is well provided with scrubbers—see Chapter 19). Then there is the waste: 800 tons of fly ash and 800 tons of boiler residue ash daily, requiring a special landfill.

The government- and industry-sponsored Clean Coal Power Initiative (CCPI) program is the successor to the Clean Coal Technology (CCT) program. The CCT supported a spectrum of experimental technologies applicable to existing power plants and to a new generation of coal-fired power plants. The objectives of both programs have been to remove pollutants from the coal before or after burning and to achieve higher efficiencies so that greenhouse gas emissions are reduced. To date, some of the projects have helped industry meet the sulfur dioxide (SO_2) emission limits of the Clean Air Act Amendments of 1990. Another technology developed under the program is the **integrated gasification combined cycle (IGCC) plant**, where coal is mixed with water and oxygen and heated under pressure to produce a synthetic gas (**syngas**) that is burned in a gas turbine to produce electricity. Sulfur and other contaminants are removed before burning. The CO_2 stream from the process is concentrated; one syngas plant in North Dakota pipes the CO_2 to an oil field, where it is injected into oil wells to enhance recovery.

The CCPI is providing a new round of government-industry projects designed to meet the challenges of coal use in the 21st century. Clean coal technologies would be an improvement over the current ones. But unless they include carbon capture technology, they will still involve substantial greenhouse gas emissions and all of the drawbacks of mining the coal. Until a change in policy requires carbon capture from coal-burning power plants, it will not happen on a commercial scale because the technology will impose a significant cost (estimated at $40 per ton).

Oil Shale and Oil Sand

The United States has extensive deposits of oil shale in Colorado, Utah, and Wyoming. **Oil shale** is a fine sedimentary rock containing a mixture of solid, waxlike hydrocarbons called *kerogen*. When shale is heated to about 1,100°F (600°C), the kerogen releases hydrocarbon vapors that can be recondensed to form a black, viscous substance similar to crude oil, which can then be refined into gasoline and other petroleum products. However, about a ton of oil shale yields little more than half a barrel of oil. The mining, transportation, and disposal of wastes necessitated by an operation

producing, for instance, a million barrels a day (just 5% of U.S. demand) would be a Herculean task, to say nothing of its environmental impact. Nevertheless, because the deposits contain an estimated 800 BBs of oil, the rising costs of importing oil make it inevitable that every domestic source will be considered. In November 2008, the Interior Department's Bureau of Land Management (BLM) finalized rules for how commercial oil-shale leases will be regulated, opening the way for oil companies to begin serious consideration of developing oil from the shale deposits. The process faces stiff opposition because of its potential to damage air quality and groundwater; any oil-shale development is subject to National Environmental Protection Act analysis (environmental impact statements) and must get local and state approval.

Oil sand is a sedimentary material containing bitumen, an extremely viscous, tarlike hydrocarbon. When oil sand is heated, the bitumen can be "melted out" and refined in the same way as crude oil can. Northern Alberta, in Canada, has the world's largest oil-sand deposits (152 BBs as proved reserves), and Canada is commercially exploiting them because the cost has been competitive with oil prices. Oil sands yielded 41 million barrels of oil in 2007, worth $1.4 billion, and production is expected to double by 2010. The environmental damage is substantial; 82,000 acres of boreal forest and wetlands have already been heavily disturbed. The extraction process generates enormous quantities of greenhouse gases. The United States is eager to buy the oil, which represents 10% of our imported oil. It is transported via a pipeline from Alberta to Port Arthur, Texas, where several refineries are located.

Worldwide, the combination of oil shale and oil sand has been estimated to hold several trillion barrels of oil. Large-scale commercial development of these resources will not occur until oil prices rise dramatically—something that may already be under way.

14.4 Energy Policy and Security

According to the Department of Energy, the period from 1973 to 1995 saw an 18% reduction in energy growth, saving the economy $150 billion a year in total energy expenditures. This was due to the steps taken in response to the oil crisis of the 1970s. Thus, public policy can bring about major improvements in energy use. Unfortunately, we also slipped back into old ways when the crisis faded into history. The state of affairs clearly demanded a new energy policy. In 2001, a task force chaired by then Vice President Cheney produced *Reliable, Affordable, and Environmentally Sound Energy for America's Future: Report of the National Energy Policy Development Group* (**The National Energy Policy Report**), signaling the Bush administration's broad intentions for America's energy policy. Then, after four years of debate, Congress passed the **Energy Policy Act of 2005**, which intended to establish energy policies for the United States for years to come. However, with the control of the 110th Congress passing to the Democratic Party, a sea change began in environmental policy, producing the **Energy Independence and Security Act of 2007** at the end of 2007.

In this section, we examine the concerns involved in moving the country toward a secure energy future as raised by the Union of Concerned Scientists in their report *Energy Security: Solutions to Protect America's Power Supply and Reduce Oil Dependence.* We then examine the government's latest energy policies in light of present problems and future needs.

Security Threats

There are several security threats that the United States incurs through its dependence on fossil fuel energy. We will focus here on the three most pressing threats.

Oil Dependence.   The United States imported $300 billion worth of oil in 2007, representing 36% of our balance-of-payment deficit and 66% of our oil consumption. The global oil market is basically beyond our control, although we continue to invest vast sums to maintain stability in the Persian Gulf region, where 61% of the world's proved reserves are found. The U.S. (and global) economy demonstrated its sensitivity to three sudden oil price shocks over the past 30 years. Each time, although the supply dropped only 5%, the price rose rapidly and brought on inflation, a drop in the gross national product, and a recession. The years 2006–2008 brought oil prices to a new high, threatening a repeat of this economic roller coaster. Relying increasingly on oil from the OPEC cartel and the volatile Persian Gulf states is simply asking for economic and political trouble in the future.

Terrorism.   Nuclear power plants, hydropower dams, oil and gas pipelines, refineries, tankers, and the electrical grid—all of these present attractive targets for terrorists intent on doing damage to the people and the economy of the United States. Our continued involvement in the Middle East has done much to inflame the anger and frustration that can provide sufficient motivation for terrorist acts, as was

Figure 14–18 Hunter bags pipeline. The Trans-Alaska pipeline was shut down for three days when a hunter punctured it with a bullet.

TABLE 14–3	CO$_2$ Emissions per Unit of Energy for Fossil Fuels (Natural Gas = 100%)
Natural Gas	100%
Gasoline	134%
Crude Oil	138%
Coal	178%

Source: Data from Energy Information Administration, U.S. Department of Energy.

demonstrated on September 11, 2001. Nuclear power plants are particularly vulnerable to a terrorist attack, and the consequences of crashing an airliner into one could be devastating. It would be relatively easy to disrupt energy flow to major areas with an attack on the infrastructure for the distribution of oil, gas, and electricity. The Trans-Alaska pipeline, for example, was shut down for three days after a hunter shot a hole in it (Fig. 14–18).

Global Climate Change.   The long-term prospect of using fossil fuels to meet our energy needs poses a huge risk to our economic, environmental, and national security because of global climate change. Recall that all fossil fuels are carbon-based compounds and none can be burned without releasing carbon dioxide (CO$_2$) as a by-product. Of the top three fossil fuels—coal, oil, and natural gas—coal produces the most CO$_2$ and natural gas produces the least (Table 14–3). There is overwhelming evidence that the greenhouse gas emissions of the 20th century, especially CO$_2$, have brought on a rise in global temperatures, among other observable impacts. (See Chapter 18.) Most of these emissions are the consequence of our heavy use of fossil fuels. If greenhouse gases continue their steady rise in the atmosphere, climate changes will melt more of the polar ice caps, causing sea level to rise. In addition, there will be major shifts in precipitation patterns; more extreme storms, droughts, and heat waves; ecosystem instability; and disruptions in agriculture. Because the United States is one of the world's leading producers of greenhouse gases and also far outpaces other countries in per capita consumption of fossil fuels (Fig. 14–19), what we do will profoundly affect not only our own security, but also that of the entire planet.

Energy Policies

Both *supply-side* and *demand-side* policy options are available. Because energy supply and demand is such a nationwide matter, only public policy has the sweeping power to effect changes that will make a difference. The Cheney report (CR) estimated that, over the next 20 years, oil consumption will increase 30%, natural-gas consumption will increase 50%, and electricity (and, therefore, coal) demand will increase 45%. The Cheney report recommended ways to

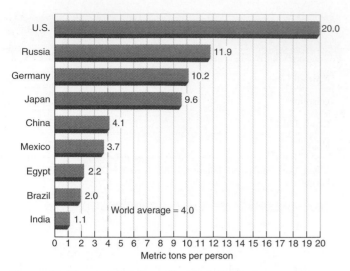

Figure 14–19 Annual carbon dioxide emissions per capita from burning fossil fuels, selected countries, 2005. (*Source:* Energy Information Agency, U.S. Department of Energy, *Annual Energy Review 2008*, June 2009.)

meet these rising demands, largely with fossil fuels, and the Energy Policy Act (PA) acted on most of these recommendations, but not always favorably. The newer Energy Independence and Security Act (EI), however, mainly addresses the demand side.

Supply-Side Policies

- **Exploring and developing domestic sources of oil and gas**
  CR: Recommended opening the ANWR and offshore locations to oil and gas exploration and production.
  PA: Promotes taking an inventory of offshore oil and gas resources, but no drilling in ANWR.

- **Increasing the use of the vast coal reserves for energy**
  CR: Recommended adding between 1,300 and 1,900 new coal-fired electric power plants in the next 20 years.
  PA: Promotes energy from coal via clean coal technologies, loan guarantees, and research incentives.

- **Continuing subsidies to oil and nuclear industries**
  CR: Recommended tax incentives to encourage energy production from fossil and nuclear fuels.
  PA: Guarantees $6.9 billion in tax incentives and billions more in loans for these industries.
  EI: Included a broad repeal of tax incentives for oil and gas in the original legislation, designed to offset the cost of many of the new elements of the law, but Republican opposition to the repeal forced the removal of this option. A limited repeal was enacted, enough to offset some estimated costs.

- **Removing environmental and legal obstacles to energy development**
  CR: Recommended streamlining the permitting process for drilling and hydropower licensing and employing eminent domain to establish new electrical transmission lines.

PA: Provides tax credits for hydropower plants and streamlines permitting for drilling.

- **Providing access to remote sources of natural gas**
  CR: Recommended expediting the construction of a natural-gas pipeline to bring Alaskan natural gas to the lower 48 states.
  PA: Requires periodic reporting of progress in pipeline construction, which now depends on state and private monies, and establishes federal control over the siting and permitting of new LNG terminals.

Business as Usual? Supply-side policies, especially those dealing with fossil fuel energy, are "business-as-usual" solutions to our energy security issues. They do little or nothing to reduce our vulnerability to oil price disruptions, terrorist actions, and global climate change. *Demand-side* policies, by contrast, have the advantage of reducing our energy needs and making it more possible to move to a future in which renewable energy can take over. At the same time, these policies will decrease our vulnerability to terrorists and to disruptions of the oil market. They will also save money and reduce pollution from burning fossil fuels.

Demand-Side Policies. In rethinking our energy strategies, we need to realize that energy has no real meaning or value by itself. Its only worth is in the work it can do. Thus, what we really want is not energy, but comfortable homes, transportation to and from where we need to go, manufactured goods, computers that work, and so forth. As a result, many energy analysts argue that we should stop thinking in terms of where we can get additional supplies of the old fuels to keep things running as they are. Instead, we should be thinking in terms of how we can satisfy these needs with a minimum expenditure of energy and the least environmental impact.

Imagine discovering a new oil field having a production potential of at least 6 million barrels a day—three times the capacity of the Alaskan field even at its height. Furthermore, assume that this newfound field is inexhaustible and that exploiting it will not adversely affect the environment. The oil field does not exist, but as a **conservation reserve**, it demonstrates what can be achieved through greater energy efficiency. The Cheney report (CR), the Energy Policy Act (PA), and the Energy Independence and Security Act (EI) addressed ways to lower energy demand:

- **Increasing the mileage standards for motor vehicles**
  CR: Recommended that the Secretary of Transportation study and provide recommendations for increasing the **Corporate Average Fuel Economy** (CAFE) standards for motor vehicles, but in ways that would avoid any negative impact on the U.S. automotive industry.
  PA: Authorizes continued "study" of the CAFE standards after the House and Senate overwhelmingly opposed their increase. The act did, however, grant *consumer tax credits* for fuel-efficient hybrid and other advanced technology vehicles.

EI: Raises the CAFE standards for the combined fleet of cars and light trucks gradually, so as to meet a target of 35 mpg by model year 2020. This will save about 1.1 million barrels of oil per day, half of what we import from the Persian Gulf. This goal has been superseded by rules from the Obama administration that mandate a target of 39 mpg for cars and 30 mpg for light trucks by 2016.

As noted in Section 14.2, this *conservation reserve* has already been tapped to some extent, especially with the doubling of the average fuel efficiency of cars from 13 to 27.5 mpg. To put it bluntly, *raising the CAFE requirements for all vehicles is the key to greater oil independence and energy security.*

- **Increasing the energy efficiency of lighting, appliances, and buildings**

  CR: Recommended promoting higher efficiency standards for appliances when possible and applying these standards to additional appliances.

  PA: Provides tax breaks for manufacturers of energy-efficient appliances; encourages the EPA to continue its Energy Star program (see Chapter 16), which also addresses energy-efficient buildings; and provides tax breaks for those making energy efficiency improvements to their homes.

  EI: Requires a 30% efficiency increase for lightbulbs. Compact fluorescent lightbulbs (CFLs) are between 20% and 25% efficient (Fig. 14–20), whereas incandescent lightbulbs are only 5% efficient. The act does not ban the incandescent bulb; however, the long-lasting CFLs already comply with the required efficiency increase and will likely become the light standard of the future. This act also sets new efficiency standards for appliances: refrigerators, freezers, washers, dishwashers, dehumidifiers, and residential boilers.

  The act also addresses energy efficiency in buildings and industry, encouraging the development of energy-efficient "green" commercial buildings and supporting a weatherization program sponsored by the Department of Energy.

Figure 14–20 Energy-efficient lightbulbs. Replacing standard incandescent lightbulbs with screw-in fluorescent bulbs (shown here) can cut the energy demand for lighting by 75–80%. The higher cost of the fluorescent bulbs is more than offset by the greater efficiency and much longer lifetime of the bulb.

Building codes currently require improved insulation and double-pane windows for new homes and buildings. Interestingly, some utility companies are subsidizing the installation of fluorescent lightbulbs and energy-efficient appliances for consumers as a cost-effective alternative to building additional power plants.

- **Encouraging industries to use combined heat and power (CHP) technologies**

  CR: Recommended providing tax credits and more flexibility in permitting for CHP projects.

  PA: Takes no significant positive action on CHP, but amends a previous act (PURPA) that required utilities to buy power from CHP plants.

  EI: Defines this technology as a suitable form of energy savings, making it eligible for energy efficiency grants and research proposals.

According to the CHP methodology (see Sustainability, "CHP: Industrial Common Sense"), a factory or large building installs a small power plant that produces electricity and heats the building with the "waste" heat. Such systems can

---

Sustainability

CHP: Industrial Common Sense

In many situations, a single energy source can be put to use to produce both electrical and heat energy. This process, called *combined heat and power* (CHP), or *cogeneration*, can be employed to conserve energy in two fundamental ways. First, a power plant whose primary function is to generate electricity also can be used to heat surrounding homes and buildings—for example, the Consolidated Edison electric utility has been using steam from waste heat in New York City for years. Second, a facility whose primary function is to produce heat or mechanical energy can also generate electricity, which is used by the facility and can also be sold to the local utility. The system can work for all sorts of fuels and heat engines. This is by far the most common application of CHP.

The arrangement makes perfect sense. Every building and factory requires both electricity and heat. Traditionally, these two forms of energy come from separate sources—the electricity

(continued)

from power lines brought in from some central power station and the heat from an on-site heating system, usually fueled by natural gas or oil. Ordinary power plants are, at best, 30% efficient in their use of energy. Combining the two sources in CHP can provide an overall efficiency of fuel use of more than 80%. Besides providing savings on fuel, cogenerating systems have at least 50% lower capital costs per kilowatt generated than conventional utilities have. This cost savings makes CHP quite competitive with utility-produced power.

For example, the Santa Barbara County Hospital in California recently installed a cogeneration system based on natural gas as the fuel source. Natural gas drives a turbine with the capacity to generate 7 MW of electricity, much of which is used by the hospital. Boilers use the waste heat from the turbine to produce high-pressure steam, which provides all of the hospital's heating and cooling needs. Any steam not required by the hospital is directed back to the turbine to increase its power. Electrical power not used by the hospital is sold to the local utility. The system has the added advantage of protecting the hospital against wide-scale blackouts or brownouts that may occur in the region's centralized utility system.

The enactment of the **Public Utility Regulatory Policies Act (PURPA)** in 1978 was essential for stimulating the use of CHP. Under this act, a facility with cogenerative capacity (1) is entitled to use fuels that are not usually available to electricity-generating plants (for example, natural gas), (2) must produce above-standard efficiency of fuel use, and (3) is entitled to have a utility buy the cogenerated electricity

(a) TRADITIONAL WAY

Fossil fuel to generate electricity → Central power station — Waste heat — Transmission losses — 500 kW

Fossil fuel for heating and hot water — 2 fuel units to generate 500 kW and heat building → Heat

(b) COGENERATION

Fossil fuel to generate electricity on premises — Electrical generator — 500 kW

$1\frac{1}{3}$ fuel units to generate 500 kW and heat building — Turbine — Waste heat supplies heating needs

Combined heat and power facility. (a) The traditional method uses two forms of energy from separate sources for buildings. (b) In cogeneration, the heat lost from an on-site power-generating system provides heating needs.

at a fair price. The Energy Policy Act of 2005 amended PURPA to remove this last provision as a favor to the utilities industry. CHP facilities must now compete with other energy suppliers and providers as they sell or purchase electricity.

Cogeneration is becoming increasingly popular in U.S. industries as a major strategy for cutting energy costs. Given the rising prices of

petroleum and natural gas, pressures favoring energy conservation will increase, and cogeneration will continue to grow. Its impact can be seen in the electric power sector of the economy: CHP electricity contributes almost 8% of the total electricity generated in the United States and saves building and industry owners over $5 billion a year in energy costs.

---

achieve an efficiency of 80%! In another energy-saving step, a new technology—the *combined-cycle natural-gas unit*—has been increasingly employed to generate electricity. Two turbines are employed in this unit: The first turbine burns natural gas in a conventional gas turbine (Fig. 14–6b); the second turbine is a steam turbine (Fig. 14–6a) that runs on the excess heat from the gas turbine. The use of this technology boosts the efficiency of the conversion of fuel energy to electrical energy from around 30% to as high as 50%. The cost of combined-cycle systems is half that of conventional coal-burning plants, and the pollution from such systems is much lower. A large proportion of new units being built or planned for the future consists of combined-cycle systems.

- **Promoting greater use of non-fossil-fuel sources of energy (nuclear and renewable energy)**

 CR: Recommended building more nuclear power plants and increasing and subsidizing alternative energy sources.

 PA: The act included a wide range of incentives to stimulate nuclear power: tax credits for generated electricity, insurance against regulatory delays, and federal loan

guarantees. These incentives amount to $13 billion. The act also authorized $4.5 billion in support of renewable energy and efficiency, with tax credits for wind, solar, and biomass energy and major funding for research and development of renewable energy initiatives. The Energy Policy Act will be reviewed in Chapters 15 and 16.

EI: Establishes a renewable fuel standard, encouraging the production of biofuels. However, a more extensive renewable energy portfolio standard, requiring the electric utilities to obtain 15% of their energy supply from renewable sources by 2020, was withdrawn in the end because of Republican opposition in the Senate. These and other renewable energy provisions will be covered in Chapter 16.

Final Thoughts

Conserving energy is extremely important, but keep in mind that *reducing* our use of fossil fuels is not *eliminating* that use. There are two major pathways for developing non-fossil-fuel energy alternatives: pursuing nuclear power and promoting renewable-energy applications. Nuclear power has been in use for 50 years, and various renewable-energy systems have also been developed. Thus, both nuclear power and renewable alternatives are at a stage where they could be expanded to provide further energy needs if suitable technological solutions

and public acceptance (in the case of nuclear power) or pressure and government support (in the case of renewable energy) are provided. Chapter 15 focuses on the nuclear option, and Chapter 16 focuses on renewable options. Broad public policy needs are summarized at the end of Chapter 16.

Revisiting the Themes

Sound Science

Science and technology have flourished in the age of fossil fuels, not least because they have devoted much research toward developing new and often ingenious ways of extracting the energy from those fuels. The need today, however, is for scientists and engineers to point the way toward using fossil fuels with greater efficiency. Two examples are the CHP technology that is being increasingly employed by factories and institutions and the IGCC plants developed under the Clean Coal Technology Program. Another key research effort is improving ways to convert natural gas to a liquid so that remote gas reserves can be accessed and substituted for the more polluting coal and oil resources. The greatest need is for renewable energy technologies, a subject addressed in Chapter 16.

Sustainability

The 19th, 20th, and 21st centuries have benefited immensely from the fossil fuels buried in the Earth over past eons. These are, however, nonrenewable resources. They will not be available for very long, as human history goes. We are currently traveling on an unsustainable track, powered by fuels that are

running out. Sustainable options are available, but we are not putting our best efforts there. The most serious consequence of our current track is that we continue to pump greenhouse gases, especially CO_2, into the atmosphere, where they are already affecting the global climate and promise to do so at an accelerating rate as we burn ever more fossil fuels. Eventually, human civilization will have to make a transition to power by renewable energy. The sooner this becomes a reality, the better off the entire planet will be.

Stewardship

What concern, if any, should we have for future generations? During the 20th century, we used half of the available oil, and our use is accelerating. What will we leave for our grandchildren and great-grandchildren? Making the transition to a renewable-energy future will be difficult, and it cannot happen overnight. A thoughtful stewardship of available resources should motivate us to refrain from pumping all the oil and natural gas to satisfy present demands and to plan for the future generations that will be facing far greater energy difficulties than we are as they struggle to complete the transition to renewable energy.

REVIEW QUESTIONS

1. How have the fuels to power homes, industry, and transportation changed from the beginning of the Industrial Revolution to the present?

2. What are the three primary fossil fuels, and what percentage does each contribute to the U.S. energy supply?

3. Electricity is a secondary energy source. How is it generated, and how efficient is its generation?

4. Name the four major categories of primary energy use in an industrial society, and match the energy sources that correspond to each.

5. What is the distinction between estimated reserves and proved reserves of crude oil, and what factors cause the amounts of each to change?

6. How does the price of oil influence the amount produced?

7. What were the trends in oil consumption, in the discovery of new reserves, and in U.S. production before 1970? After 1970? What is a Hubbert curve?

8. What did the United States do in the early 1970s to resolve the disparity between oil production and consumption? What events caused the sudden oil shortages of the mid-1970s and then the return to abundant, but more expensive, supplies?

9. Why did an oil glut occur in the mid-1980s, and what were its consequences in terms of oil prices and the continuing pursuit of the responses to the crisis of the 1970s?

10. What have been the directions of U.S. production, consumption, and importation of crude oil since the mid-1980s?

11. Describe the situation with respect to natural gas and coal for the United States in terms of domestic supplies, the consequences of developing and using the fuels, and the ability of those sources to supplement oil.

12. What are the security risks and consequences of our growing dependence on foreign oil? How do they relate to the Persian Gulf War of 1991? To terrorism? To the 2003 U.S.-Iraqi war?

13. What phenomenon may restrict the consumption of fossil fuels before resources are depleted?

14. Compare the Bush administration's energy policy with the Energy Policy Act of 2005 and the Energy Independence and Security Act of 2007 relative to supply- and demand-side strategies.

15. To what degree can energy conservation serve to mitigate energy dependency? What are some prime examples of energy conservation? To what degree may conservation alleviate future energy shortages and the cost of developing alternative sources of energy?

THINKING ENVIRONMENTALLY

1. Statistics show that less developed countries use far less energy per capita than do developed countries. Explain why this is so.

2. Between 1980 and 1999, gasoline prices in the United States declined considerably relative to inflation and then rose dramatically in the 2000s. Predict what they will do in the next 10 years, and rationalize your prediction.

3. Suppose your region is facing power shortages (brownouts). How would you propose solving the problem? Defend your proposed solution on both economic and environmental grounds.

4. List all the environmental impacts that occur during the "life span" of gasoline—that is, from exploratory drilling to the consumption of gasoline in your car.

5. Explore the positions of the following regarding an increase in CAFE requirements: the fossil fuel industry, the automobile-manufacturing companies, the steel industry, Congress, the Union of Concerned Scientists, and the National Academy of Sciences. How do they differ, and why?

MAKING A DIFFERENCE

1. An easy first step for reducing your energy is replacing your conventional lightbulbs with compact fluorescents. You've probably heard that before, but have you done it? Switch them today, and you'll save big on your electric bills. Fluorescent bulbs do contain a small amount of mercury, so if you accidentally break one, it would be wise to open your windows; and when you clean up the glass, don't use your bare hands. The mercury levels are low enough that the last traces will vaporize in a matter of hours.

2. The 275 million television sets in the United States use over 50 billion kilowatt hours of electricity a year. That's the same as about 10 coal power plants. As the screens get larger, power consumption will increase. Right now, there are three major HDTV technologies—plasma, LCD, and DLP (digital light processing). Of this group, DLP is by far the most efficient (0.13 watts per inch versus 0.33 watts for

plasma). Also, look for Energy Star models; across the board, they use 30% less electricity than standard models.

3. When you are in the market for transportation, choose a vehicle that gets high mileage per gallon, in anticipation of the higher fuel costs that will certainly occur before the vehicle wears out. If you can afford it, purchase a hybrid vehicle.

4. To further cut down on power, turn your thermostat down a few degrees in the winter and up a few degrees in the summer. Wear warm or light clothes to help make up for the temperature difference. One-sixth of power use in the United States is due to air conditioning.

5. Buy a power strip. This makes it easy to turn things off without having to go to the trouble of unplugging them. Some strips can even detect idle current and shut them off without you having to even flip any switches.

Mastering**ENVIRONMENTALSCIENCE**™

Go to **www.masteringenvironmentalscience.com** for practice quizzes, Pearson eText, videos, current events, and more.

CHAPTER 15
Nuclear Power

Watts Bar Nuclear Generating Station, Spring City, Tennessee. Unit 1 was the last reactor to come online in the United States, in 1996. Unit 2 construction was halted in 1988, but has since resumed, with completion expected in 2013.

O N JULY 16, 2007, a magnitude-6.8 earthquake struck an area just offshore of the city of Kashiwazaki, Japan, some 455 kilometers northwest of Tokyo. The earthquake killed 10 people, injured 1,800, and left over 10,000 homeless. Earthquakes are common events in Japan, which sits on the Pacific Rim, an area well known for its seismic activity. This one, however, was centered only 10 miles (16 km) from the world's largest nuclear facility, the Kashiwazaki-Kariwa Nuclear Power Plant. The power plant is home to seven nuclear reactors, generating almost 8,000 megawatts (MW) of electricity. When the earthquake hit, the four reactors that were online were shut down automatically by built-in safety mechanisms. Although the reactors and their containment buildings were undamaged, the quake started a fire in a transformer building (Fig. 15–1), knocked over 400 barrels of low-level nuclear waste, broke pipes, and caused some leakage of radioactive water into the Sea of Japan. The earthquake significantly exceeded the level of seismic activity for which the nuclear power plant was designed, and the plant was shut down for over a year in order to inspect all possible damage and install greater earthquake-proofing.

KEY TOPICS

1. **Nuclear Energy in Perspective:** Nuclear power was born in the United States, and nuclear energy now generates 19% of U.S. electricity.

2. **How Nuclear Power Works:** The objective of nuclear power technology is to use a controlled nuclear reaction to drive a turbogenerator.

3. **Nuclear power plants generate highly radioactive wastes and have the potential to cause catastrophic accidents.**

4. **More Advanced Reactors:** Fast-neutron reactors are used to reprocess spent nuclear fuel.

5. **The Future of Nuclear Power:** Global climate change has made the nuclear energy option more attractive.

Figure 15–1 Fire at Kashiwazaki-Kariwa Nuclear Power Plant.
On July 16, 2007, an earthquake shook Japan's largest nuclear power facility, starting a fire in a transformer building and knocking over 400 barrels of nuclear waste.

Even with its violent introduction into the nuclear age with the bombing of Hiroshima and Nagasaki in 1945, Japan is determined to increase its nuclear power capacity. Japan's 55 reactors now supply 30% of the country's electrical power, and the goal is to reach 43% by 2010. The Kashiwazaki earthquake is only the latest in a series of problems experienced by Japan's nuclear industry. The worst accident happened on September 30, 1999, at a nuclear power plant in Tokaimura, a village 120 kilometers northeast of Tokyo. There some workers were mixing some uranium fuels in buckets (a violation of regulations), and the mixture reached critical mass and began a nuclear-fission chain reaction, emitting high levels of gamma rays and neutrons. By the time the reaction was shut down, enough gamma radiation had escaped to expose townspeople and workers to varying levels of radiation. Two of the workers died of radiation poisoning. Then, in August 2004, five workers were killed at the Mihama Nuclear Power Plant west of Tokyo as a result of a ruptured steam pipe. Fortunately, there were no radiation leaks.

These incidents have led to a groundswell of public protest in Japan against the planned expansion of nuclear power and have embarrassed the government and nuclear industry officials, who were not always forthright in their release of information following the incidents. The controversy over nuclear energy in Japan is only one of many in the troubled history of nuclear power. In this chapter, we will delve into that history through the lenses of science and public policy as the many concerns about nuclear power are raised and discussed.

15.1 Nuclear Energy in Perspective

Rising natural-gas prices, continued emissions of sulfur dioxide and nitrogen oxides from coal, and the long-term threat of climate change all suggest that an energy future based on fossil fuels is not an option. In light of these serious energy-related problems, it seems prudent to do everything possible to develop non-fossil-fuel sources of energy. Nuclear power is an alternative that contributes few emissions to global warming, and there is sufficient uranium to fuel nuclear reactors well into the 21st century, with the possibility of extending the nuclear fuel supply indefinitely through reprocessing technologies. Thirty-one countries now have nuclear power plants either in place or under construction, and in some of these countries, nuclear power generates more electricity than any other source. Is nuclear energy the key to the future of global energy—the best route to a sustainable relationship with our environment?

The History of Nuclear Power

For years, geologists have known that fossil fuels would not last forever. Sooner or later, other energy sources would be needed. It was anticipated that nuclear power could produce electricity in such large amounts and so cheaply that we would phase into an economy in which electricity would take over virtually all functions, including the generation of other fuels, at nominal costs. Following World War II, determined to show the world that the power of the atom could benefit humankind, the U.S. government embarked on a course to lead the world into the Nuclear Age.

The Nuclear Age. Thus did the government move into the research, development, and promotion of commercial nuclear power plants (along with the continuing development of nuclear weaponry). Using this research, companies such as General Electric and Westinghouse constructed nuclear power plants that were ordered and paid for by utility companies, with assurances from the federal government, via the Price-Anderson Act of 1957, that these corporations and utilities would be covered by insurance for any legal liabilities incurred. The Atomic Energy Commission, now the Nuclear Regulatory Commission (NRC), an agency in the U.S. Department of Energy (DOE), set and enforced safety standards for the operation and maintenance of the new plants, as it does today.

In the 1960s and early 1970s, utility companies moved ahead with plans for numerous nuclear power plants (Fig. 15–2). By 1975, 53 plants were operating in the United States, producing about 9% of the nation's electricity, and another 170 plants were in various stages of planning or construction. Officials estimated that by 1990 several hundred plants would be online and by the turn of the century as many as a thousand would be operating. A number of other industrialized countries got in step with their own programs, and some less developed nations were going nuclear by purchasing plants from industrialized nations.

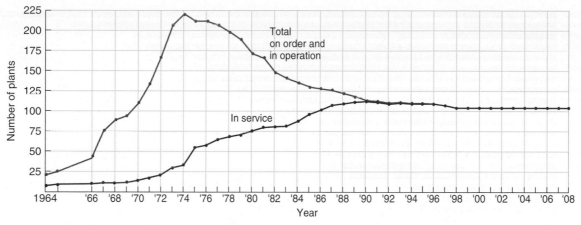

Figure 15–2 Nuclear power in the United States. Since the early 1970s, when orders for plants reached a peak, few utilities called for new plants, and many have canceled earlier orders. Nevertheless, the number of plants in service increased steadily as plants under construction were completed. The number of operating plants peaked at 112 and is holding steady at 104. (*Source:* Data from U.S. Department of Energy.)

Curtailed. After 1975, however, the picture changed dramatically. Utilities stopped ordering nuclear plants, and numerous existing orders were canceled. The last order for a plant in the United States that was not subsequently canceled was placed in 1978. In some cases, construction was terminated even after billions of dollars had already been invested. Most strikingly, the Shoreham Nuclear Power Plant on Long Island, New York, after being completed and licensed at a cost of $5.5 billion, was turned over to the state in the summer of 1989, to be dismantled after generating electricity for only 32 hours (Fig. 15–3). The reason was that citizens and the state deemed that there was no way to evacuate people from the area should an accident occur. Similarly, just a few weeks earlier, California citizens voted to shut down the Rancho Seco Nuclear Power Plant, located near Sacramento, after a 15-year history of troubled operation. In the United States, 28 units (separate reactors) have been shut down permanently for a variety of reasons.

Currently.... At the end of 2008, 104 nuclear power plants were operating in the United States. The Watts Bar Unit 1 reactor in Tennessee was the last unit to come online, beginning operations in February 1996; Unit 2—partially completed, but halted in 1988—is again under construction and will be the first reactor to be completed in the 21st century in the United States. The 104 existing plants (the most in any country) generate 19% of U.S. electrical power. However, in looking ahead, the U.S. Energy Information Administration projects an *increase* in nuclear generating capacity, from the present level of 100 gigawatts (GW) to 115 GW in 2030, from new power plants expected to open as a result of the Energy Policy Act of 2005 (this act provides many incentives to stimulate the nuclear power industry).

Global Picture. After the catastrophic accident at Chernobyl in the Soviet Union in April 1986, nuclear power was being rethought in many countries. Yet the demand for electricity is more robust than ever, and the other means of generating electricity have their own problems. Coal-powered electricity produces more greenhouse gases and other pollutants than does any other form of energy. Oil and natural-gas supplies are more limited, and oil is vital to transportation and home heating. Hydroelectric power is already heavily developed. Wind and solar power, although increasing

Figure 15–3 The Shoreham Nuclear Plant on Long Island, New York. With little electricity produced, this plant was closed because of concerns about whether surrounding areas could be evacuated in case of an accident. The plant now hosts two wind turbines that generate about 1/8,000 of the power the nuclear plant would have produced.

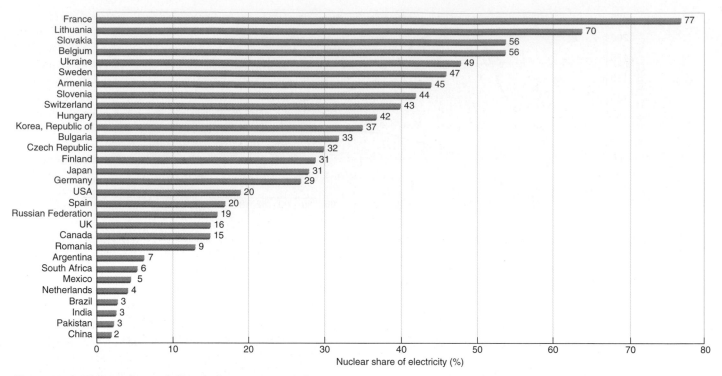

Figure 15–4 Nuclear share of electrical power generation. In 2008, those countries lacking fossil fuel reserves tended to be the most eager to use nuclear power. (*Source:* Data from International Atomic Energy Agency.)

rapidly, is still only a small fraction of the total electricity generated.

Therefore, nuclear power plants continue to be built. Including those in the United States, the world has a total of 439 operating nuclear plants, with an additional 42 under construction. Globally, electricity generated from nuclear power has increased at a rate of 3.2% per year for the last decade. Nuclear power now generates about 14% of the world's electricity, but dependence on nuclear energy varies greatly among the 31 countries operating nuclear power plants (Fig. 15–4). France now produces 77% of its electricity with nuclear power and has plans to go to more than 80%. China and India, the world's population giants, also are investing heavily into nuclear energy, each with six new units under construction. If the safety and economic issues can be resolved, can nuclear power supply future energy needs? In order to react intelligently to this issue, we must start with a clear understanding of what nuclear power is and what its pros and cons are.

15.2 How Nuclear Power Works

The objective of nuclear power technology is to control nuclear reactions so that energy is released gradually as heat. As with plants powered by fossil fuels, the heat energy produced by a nuclear plant is used to boil water and produce steam, which then drives conventional turbogenerators. Nuclear power plants are baseload plants (see Chapter 14); they are always operating unless they are being refueled, and they are large plants, generating up to 1,400 MW.

From Mass to Energy

The release of nuclear energy is completely different from the burning of fuels or any other chemical reactions discussed so far. To begin with, materials involved in chemical reactions remain unchanged at the atomic level, although the visible forms of these materials undergo great transformation as atoms are rearranged to form different compounds. For example, the organic carbon in gasoline is still carbon after it is burned and comes out of the tailpipe as carbon dioxide. Nuclear energy, however, involves changes at the atomic level through one of two basic processes: *fission* and *fusion*. In **fission**, a large atom of one element is split to produce two smaller atoms of different elements (Fig. 15–5a). In **fusion**, two small atoms combine to form a larger atom of a different element (Fig.15–5b). In both fission and fusion, the mass of the product(s) is less than the mass of the starting material, and the lost mass is converted to energy in accordance with the law of mass-energy equivalence ($E=mc^2$) first described by Albert Einstein. Because the small loss of mass is multiplied by the speed of light *squared*, the amount of energy released by this mass-to-energy conversion is tremendous: The sudden fission or fusion of just one kilogram of material, for example, releases the devastatingly explosive energy of a nuclear bomb.

The Fuel for Nuclear Power Plants. All current nuclear power plants employ the fission (splitting) of uranium-235. The element uranium, which occurs naturally in various minerals in Earth's crust, exists in two primary forms, or isotopes: uranium-238 (^{238}U) and uranium-235 (^{235}U). **Isotopes**

(a) FISSION

Free
neutrons

+ ENERGY

Uranium-235

Radioactive
by-products
(mostly wastes)

(b) FUSION

Hydrogen-2

Free
neutron

Helium

+ ENERGY

Hydrogen-3

○ Neutrons
● Protons

Figure 15–5 Nuclear fission and fusion. Nuclear energy is released from either (a) fission, the splitting of certain large atoms into smaller atoms, or (b) fusion, the fusing together of small atoms to form a larger atom. In both cases, some of the mass of the starting atom(s) is converted to energy.

of a given element contain different numbers of neutrons, but the same number of protons and electrons.

The number 238 or 235 that accompanies the chemical name or symbol for uranium is called the **mass number** of the element. It is the sum of the number of neutrons and the number of protons in the nucleus of the atom. All atoms of any given element must contain the same number of protons, so variations in mass number represent variations in numbers of neutrons. ^{238}U contains 92 protons and 146 neutrons; ^{235}U contains 92 protons and 143 neutrons. Although all isotopes of a given element behave the same chemically, their other characteristics may differ profoundly. In the case of uranium, ^{235}U atoms will readily undergo fission, but ^{238}U atoms will not. Like uranium, many other elements exist in more than one isotopic form.

It takes a neutron hitting the nucleus at just the right speed to cause ^{235}U to undergo fission. Because ^{235}U is an unstable isotope, a small, but predictable number of the ^{235}U atoms in any sample of the element undergo radioactive decay and release neutrons, among other things. If one released neutron moving at just the right speed hits another ^{235}U atom, the latter becomes ^{236}U, which is highly unstable and undergoes fission immediately into lighter atoms (**fission products**). The fission reaction gives off several more neutrons and releases a great deal of energy (Fig. 15–5a). Ordinarily,

these neutrons are traveling too fast to cause fission, but if they are slowed down and then strike another ^{235}U atom, they can cause fission to occur again. In this way, more neutrons and more energy are released, with the potential to repeat the process. A domino effect, known as a *chain reaction*, may thus occur (Fig. 15–6a).

Nuclear Fuel.   To make nuclear "fuel," uranium ore is mined, purified into uranium dioxide (UO_2), and enriched. Significant uranium deposits in the United States are found in Wyoming and New Mexico and are mined underground or in open pits. The ore is then subjected to **milling**, where it is crushed, treated chemically, and rendered into **yellowcake**, so called because of its yellowish color. The yellowcake, about 80% UO_2, is then delivered to a conversion facility for purification and enrichment. Because 99.3% of all uranium found in nature is ^{238}U (thus, only 0.7% is ^{235}U), **enrichment** involves separating ^{235}U from ^{238}U to produce a material containing from 3% to 5 % of ^{235}U (and the rest ^{238}U). Since these two isotopes are chemically identical, enrichment is based on their slight difference in mass. The technical difficulty of enrichment is the major hurdle that prevents less developed countries from advancing their own nuclear capabilities.

Nuclear Bomb.   When ^{235}U is *highly* enriched, the spontaneous fission of an atom can trigger a chain reaction. In nuclear weapons, small masses of virtually pure ^{235}U or other fissionable material are forced together so that the two or three neutrons from a spontaneous fission cause two or three more atoms to undergo fission; each of these in turn triggers two or three more fissions, and so on. The whole mass undergoes fission in a fraction of a second, releasing all the energy in one huge explosion (Fig. 15–6b).

The Nuclear Reactor.   A nuclear reactor for a power plant is designed to sustain a continuous chain reaction (Fig. 15–6c), but not allow it to amplify into a nuclear explosion. Control is achieved by enriching the uranium to 3–5% ^{235}U. This modest enrichment will not support the amplification of a chain reaction into a nuclear explosion. However, in the process of fission, some of the faster neutrons are absorbed by ^{238}U atoms, converting them into ^{239}Pu (plutonium), which then also undergoes fission when hit by another neutron. So much plutonium is produced in this way that at least one-third of the energy of a nuclear reactor comes from plutonium fission.

Moderator.   A chain reaction can be sustained in a nuclear reactor only if a sufficient mass of enriched uranium is arranged in a suitable geometric pattern and is surrounded by a material called a *moderator*. The **moderator** slows down the neutrons that produce fission so that they are traveling at the right speed to trigger another fission. In slowing down the neutrons, the moderator gains heat. In nuclear plants in the United States, the moderator is near-pure water, and the reactors are called light-water reactors (LWRs). (The term *light water* denotes ordinary water (H_2O), as opposed to the substance known as *heavy water*, or deuterium oxide (D_2O). Deuterium is an isotope of hydrogen.)

Fuel Rods.   To achieve the geometric pattern necessary for fission, the enriched uranium dioxide is made into pellets that are loaded into long metal tubes. The loaded tubes are

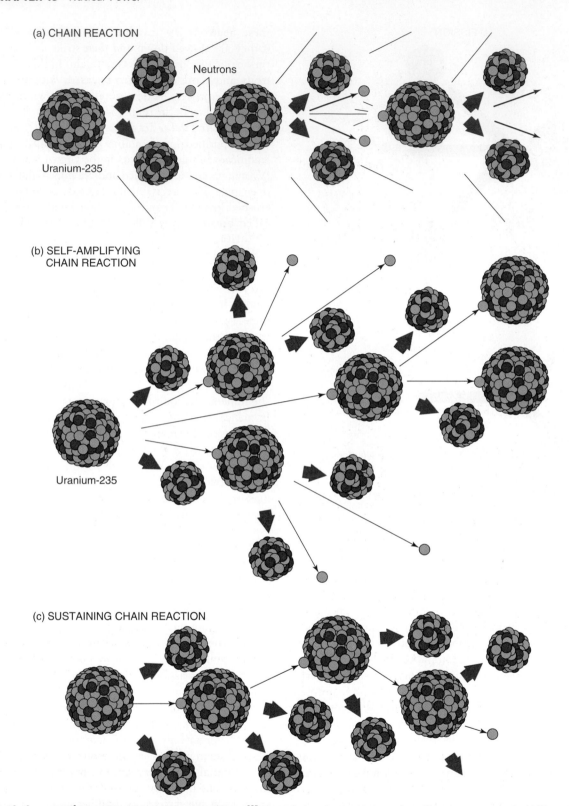

(a) CHAIN REACTION

Neutrons

Uranium-235

(b) SELF-AMPLIFYING
CHAIN REACTION

Uranium-235

(c) SUSTAINING CHAIN REACTION

Figure 15–6 Fission reactions. (a) A simple chain reaction. When a ²³⁵U atom fissions, it releases two or three high-energy neutrons, in addition to energy and split "halves." If another ²³⁵U atom is struck by a high-energy neutron, it fissions, and the process is repeated, causing a chain reaction. (b) A self-amplifying chain reaction leading to a nuclear explosion. Since two or three high-energy neutrons are produced by each fission, each may cause the fission of two or three additional atoms. (c) In a sustaining chain reaction, the extra neutrons are absorbed in control rods, so that amplification does not occur.

called **fuel elements** or **fuel rods** (Fig. 15–7a). Many fuel rods are placed close together to form a reactor core inside a strong reactor vessel that holds the water, which serves as both moderator and coolant (Fig. 15–7b). Over time, fission products that also absorb neutrons accumulate in the fuel

rods and slow down the rate of fission and heat production. The highly radioactive spent-fuel elements are then removed and replaced with new ones.

Control Rods. The chain reaction in the reactor core is controlled by rods of neutron-absorbing material, referred

(a)

Control
rods

Fuel
elements

(b)

H_2O moderator

Uranium-235

Neutron-absorbing
material

(c)

Figure 15–7 A nuclear reactor. (a) In the core of a nuclear reactor, a large mass of uranium is created by placing uranium in adjacent tubes, called fuel elements. The rate of the chain reaction is moderated by inserting or removing rods of neutron-absorbing material (control rods) between the fuel elements. (b) The fuel and rods are surrounded by the moderator fluid, near-pure water. (c) Technicians ready the core housing to receive uranium fuel elements in this nuclear reactor.

to as **control rods,** inserted between the fuel elements. The chain reaction is started and controlled by withdrawing and inserting the control rods as necessary. Hence, a nuclear reactor is simply an assembly of fuel elements, moderator-coolant, and movable control rods (Fig. 15–7). As the control rods are removed and a chain reaction is initiated, the fuel rods and the moderator become intensely hot.

The Nuclear Power Plant.   In a nuclear power plant, heat from the reactor is used to boil water to provide steam for driving conventional turbogenerators. One way to boil the water is to circulate it through the reactor, as in the **boiling-water reactor.** Two-thirds of the power plants in the United

States, however, are **pressurized-water reactors** (Fig. 15–8), employing a double loop. In these, the moderator-coolant water is heated to more than 600°F by circulating it through the reactor, but it does not boil because the system is under very high pressure (155 atmospheres). The superheated water is circulated through a heat exchanger, boiling other, unpressurized water that flows past the heat exchanger tubes. This action produces the steam used to drive the turbogenerator.

LOCA.   The double-loop design of the pressurized-water reactor isolates hazardous materials in the reactor from the rest of the power plant. However, both reactor types have one serious drawback: If the reactor vessel should crack, the sudden loss of water from around the reactor,

Figure 15–8 Pressurized nuclear power plant. The double-loop design isolates the pressurized water from the steam-generating loop that drives the turbogenerator.

called a **loss-of-coolant accident (LOCA),** could result in the core's overheating. The sudden loss of the moderator-coolant water would cause fission to cease, since the moderator would no longer be present. Nevertheless, the fuel core could still overheat because 7% of the reactor's heat comes from radioactive decay in the newly formed fission products. In time, the uncontrolled decay would release enough heat energy to melt the materials in the core, a situation called a **meltdown.** Then the molten material falling into the remaining water could cause a steam explosion. To guard against all this, backup cooling systems keep the reactor immersed in water should leaks occur, and the entire assembly is housed in a thick concrete containment building (Fig. 15–8).

Comparing Nuclear Power with Coal Power

Because nuclear plants are baseload plants that provide the foundation for meeting the daily and weekly electrical demand cycle, they must be replaced with other baseload plants. If nuclear power plants are phased out, one option is replacing such large plants with coal-burning plants. The United States has abundant coal reserves, but is burning coal the course of action we wish to pursue? Nuclear power has some decided environmental advantages over coal-fired power (Fig. 15–9). Comparing a 1,000-MW nuclear plant with a coal-fired plant of the same capacity, each operating for one year, we find the following characteristics:

- **Fuel needed.** The coal plant consumes 2–3 million tons of coal. If this amount is obtained by strip mining, significant environmental destruction and acid leaching

Figure 15–9 Nuclear power versus coal-fired power. Both methods of generating electricity have their environmental advantages and disadvantages. The nuclear power option assumes perfect containment of radioactivity and the availability of some method for waste storage and disposal.

will result. If the coal comes from deep mines, there will be human costs in the form of accidental deaths and impaired health. The nuclear plant requires about 30 tons of enriched uranium, obtained from mining 75,000 tons of ore, with much less harm to humans and the environment. The fission of about 1 pound (0.5 kg) of uranium fuel releases energy equivalent to burning 50 tons of coal.

- **Carbon dioxide emissions.** The coal plant emits more than 7 million tons of carbon dioxide into the atmosphere, contributing to global climate change. The nuclear plant emits none while it is producing energy, but fossil fuel energy is used in the mining and enriching of uranium, the construction of the nuclear plant, the decommissioning of the plant after it is shut down, and the transportation and storage of nuclear waste. Since a coal plant also needs to be constructed and the coal and waste ash must be transported, much the same can be said for the coal plant.

- **Sulfur dioxide and other emissions.** The coal plant emits more than 300,000 tons of sulfur dioxide, particulates, and other pollutants, leading to acid rain and health-threatening air pollution (unless it is captured by scrubbers). The nuclear power plant produces no acid-forming pollutants or particulates.

- **Radioactivity.** A coal plant releases 100 times more radioactivity than does a nuclear power plant because of the natural presence of radioactive compounds (uranium, thorium) in the coal. The nuclear plant releases low levels of radioactive waste gases.

- **Solid wastes.** The coal plant produces about 600,000 tons of ash that require disposal. (Much of it is reused in cement production.) The nuclear plant produces about 250 tons of highly radioactive wastes that require storage and ultimate safe disposal. (Safe disposal is an unresolved problem.)

- **Accidents.** A worst-case accident in the coal plant could result in fatalities to workers and a destructive fire, a situation common to many industries. Accidents in a nuclear plant can range from minor emissions of radioactivity to catastrophic releases that can lead to widespread radiation sickness, scores of human deaths, untold numbers of cancers, and widespread, long-lasting environmental contamination.

It is *the disposal of radioactive wastes* and *the potential for accidents* that have led to public concern over nuclear power. Can these problems be overcome, or are other energy options—natural-gas-powered boilers or renewable energy—less problematic and less costly than nuclear power?

15.3 The Hazards and Costs of Nuclear Power Facilities

Assessing the hazards of nuclear power requires an understanding of radioactive substances and their danger.

Radioactive Emissions

When uranium or any other element undergoes fission, the split "halves" are atoms of lighter elements, such as iodine, cesium, strontium, cobalt, or any of some 30 other elements. These newly formed atoms—called the *direct products* of the fission—are generally unstable isotopes of their respective elements. Unstable isotopes (called **radioisotopes**) become stable by spontaneously ejecting subatomic particles (alpha particles, beta particles, and neutrons), high-energy radiation (gamma rays and X-rays), or both. Radioactivity is measured in **curies** (after Marie Curie, the scientist who first coined the term *radioactivity*); one gram of pure radium-226 gives off one curie per second, which is approximately 37 billion spontaneous disintegrations into particles and radiation. The particles and radiation are collectively referred to as **radioactive emissions.** Any materials in and around the reactor may also be converted to unstable isotopes and become radioactive by absorbing neutrons from the fission process (Fig. 15–10). These indirect products of fission, along with the direct products, are the **radioactive wastes** of nuclear power. The direct products of fission generate **high-level wastes,** so called because they are highly radioactive. The indirect products of fission, such as the reactor materials, are much less radioactive and are classified as **low-level wastes,** a category that also includes materials from hospitals and industry.

Biological Effects. A major concern of the public regarding nuclear power is that large numbers of people may be exposed to low levels of radiation, thus elevating their risk of cancer and other disorders. Is this a valid concern?

Radioactive emissions can penetrate biological tissue, resulting in radiation exposure. This exposure is measured as *absorbed dose*, using joules per kilogram (J/kg)—energy imparted (J) per mass of body tissue (kg). These units are called *Grays* (Gy) for low-level radiation and *Sieverts* (Sv) for high-level radiation; the difference relates to the fact that high-level radiation does more damage per dose. For simplicity, all dose units are often reported in Sieverts (the old unit *rem*, still in use by some agencies, is the equivalent of 0.01 Sv).

As radiation penetrates tissue, it displaces electrons from molecules, leaving behind charged particles, or *ions*, so the emissions are called *ionizing radiation*. The process of ionization may involve breaking chemical bonds or changing the structure of molecules in ways that impair their normal functions. Ordinarily, the radiation is not felt and leaves no visible mark unless the dose is high.

High Dose. In high doses, radiation may cause enough damage to prevent cell division. This is why, in medical applications, radiation can be focused on a cancerous tumor to destroy it. However, if the whole body is exposed to such levels of radiation (over 1 Sievert is considered a high dose), a generalized blockage of cell division occurs that prevents the normal replacement or repair of blood, skin, and other tissues. This result is called *radiation sickness* and may lead to death a few days or months after exposure. The two stricken Japanese workers in the Tokaimura accident received from

Figure 15–10 Radioactive wastes and radioactive emissions. Nuclear fission results in the production of numerous unstable isotopes, designated radioactive waste. These isotopes give off potentially damaging radiation until they regain a stable structure.

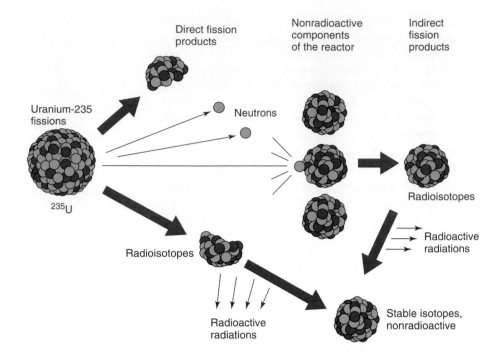

10 to 17 Sieverts of radiation—ultimately, a lethal dose. Very high levels of radiation may totally destroy cells, causing immediate death.

Low Dose. In lower doses, radiation may damage DNA, the genetic material inside cells. Cells with damaged (mutated) DNA may then begin dividing and growing out of control, forming malignant tumors or leukemia. If the damaged DNA is in an egg or a sperm cell, the result may be birth defects (mutations) in offspring. The effects of exposure to radiation may go unseen until many years after the event (10 to 40 years is typical). Other effects include the weakening of the immune system, mental retardation, and the development of cataracts.

Exposure. Health effects are directly related to the level of exposure. There is broad agreement that doses between 100 and 500 milliSieverts (1 milliSievert, or mSv = 1/1,000 of a Sievert) result in an increased risk of developing cancer. Evidence for this hypothesis comes from studies of patients with various illnesses who were exposed to high levels of X-rays in the 1930s, before the potential harm of such radiation was understood. People in these groups subsequently developed higher-than-normal rates of cancer and leukemia.

Is radiation dangerous even at doses below 100 milliSieverts? Some scientists point to the ability of living cells to repair small amounts of damage to DNA and believe that there is a threshold of radiation below which no biological effects occur. A recent study by the National Research Council (NRC)[1] found that there is no safe level of radiation, dismissing the idea of a threshold. Federal agencies already employ the concept of no threshold and assume a direct relationship between the amount of radiation and the incidence of cancer. The NRC report predicted that 1 in

1,000 individuals would develop cancer from an exposure to 10 mSv; a 10-fold increase in exposure would increase the risk to 1 in 100. Federal standards set 1.7 mSv/yr as the maximum exposure permitted for the general populace per year, except for medical X-rays.

Sources of Radiation. Nuclear power is by no means the only possible source of radiation. There is also normal background radiation from radioactive materials, such as uranium and radon gas that occur naturally in the Earth's crust and cosmic rays from outer space. For most people, background radiation is the major source of radiation exposure. In addition, we deliberately expose ourselves to radiation from medical and dental X-rays and, more and more frequently, from CT scans, by far the largest source of human-induced exposure. For the average person, these deliberate exposures equal one-fifth the exposure from background sources. The average person in the United States receives a dose of about 3.6 mSv per year (Table 15–1). Thus, the argument becomes one of relative hazards. That is, does or will the radiation from nuclear power significantly raise radiation levels and elevate the risk of developing cancer?

Getting Nuked? During normal operation of a nuclear power plant, the direct fission products remain within the fuel elements, and the indirect products are maintained within the containment building that houses the reactor. No routine discharges of radioactive materials (other than permissible minor stack emissions) into the environment occur. Even very close to an operating nuclear power plant, radiation levels from the plant are lower than normal background levels. A radiation detector will pick up more radiation from the ground or the concrete of a basement floor than it will when held within 150 yards of a nuclear power plant. Careful measurements have shown that public exposure to radiation from normal operations of a power plant is less than 1% of natural background radiation.

---

[1]National Research Council, *Health Risks from Exposure to Low Levels of Ionizing Radiation: Beir VII-Phase 2* (Washington, D.C.: National Academy of Sciences, 2005). Available at fermat.nap.edu/execsumm_pdf/11340.

TABLE 15–1   Relative Doses from Radiation Sources

Source	Dose
All sources (avg.)	3.6 mSv/year
Cosmic radiation at sea level*	0.26 mSv/year
Terrestrial radiation (elements in soil)	0.26 mSv/year
Radon in average house	2 mSv/year
X-rays and nuclear medicine (avg.)	0.50 mSv/year
Consumer products	0.11 mSv/year
Natural radioactivity in the body	0.39 mSv/year
Mammogram	0.30 mSv
Chest X-ray	0.10 mSv
CT scan of abdomen	10 mSv
Continued fallout from nuclear testing	0.01 mSv/year
Living near a nuclear power station (assuming perfect containment)	<0.01 mSv/year
Federal standards for the general populace	1.7 mSv/year

Source: Data from U.S. Nuclear Regulatory Commission.
*Increases 0.005 mSv for every 100 feet of elevation.

The main concern about nuclear power, therefore, should not focus on a plant's normal operation. Instead, the real problems arise from the storage and disposal of radioactive wastes and the potential for accidents.

Radioactive Wastes

To understand the problems surrounding nuclear waste disposal, you must understand the concept of **radioactive decay**, a process in which, as unstable isotopes eject particles and radiation, they become stable and cease to be radioactive. As long as radioactive materials are kept isolated from humans and other organisms, the decay proceeds harmlessly.

Half-Life. The rate of radioactive decay is such that half of the starting amount of a given isotope will decay within a certain period. In the next equal period, half of the remainder (half of a half, which equals one-fourth of the original) decays, and so on. The time for half of the amount of a radioactive isotope to decay is known as the isotope's **half-life**. The half-life of an isotope is always the same, regardless of the starting amount.

Each particular radioactive isotope has a characteristic half-life. The half-lives of various isotopes range from a fraction of a second to many thousands of years. Uranium fission results in a heterogeneous mixture of radioisotopes, the most common of which are listed in Table 15–2. Much of this material—in particular, the remaining ^{235}U and the

plutonium (^{239}Pu) that has been created by the neutron bombardment of ^{238}U—can be recovered and recycled for use as nuclear fuel in an operation called **reprocessing**. Although Britain and France have a large reprocessing operation, U.S. policy until recently prohibited the practice because of concerns over plutonium and atomic weapons (plutonium can be more easily fabricated into nuclear weapons than can uranium).

Disposal of Radioactive Wastes

The development of nuclear power went ahead without ever fully addressing the issue of what would be done with the radioactive wastes. Proponents of nuclear power generally assumed that the long-lived high-level wastes could be solidified, placed in sealed containers, and buried in deep, stable rock formations (geologic burial) as the need for such containment became necessary. However, this has not yet

TABLE 15–2   Common Radioactive Isotopes Resulting from Uranium Fission and Their Half-Lives

Short-Lived Fission Products	Half-Life (days)
Strontium-89	50.5
Yttrium-91	58.5
Zirconium-95	64.0
Niobium-95	35.0
Molybdenum-99	2.8
Ruthenium-103	39.4
Iodine-131	8.1
Xenon-133	5.3
Barium-140	12.8
Cerium-141	32.5
Praseodymium-143	13.6
Neodymium-147	11.1
Long-Lived Fission Products	**Half-Life (years)**
Krypton-85	10.7
Strontium-90	28.0
Ruthenium-106	1.0
Cesium-137	30.0
Cerium-144	0.8
Promethium-147	2.6
Additional Product of Neutron Bombardment	**Half-Life (years)**
Plutonium-239	24,000

happened. Thus, the current problem of nuclear waste disposal is twofold:

- **Short-term containment.** Storage during this period allows the radioactive decay of short-lived isotopes (Table 15–2). In 10 years, fission wastes lose more than 97% of their radioactivity. Wastes can be handled much more easily and safely after this loss occurs.

- **Long-term containment.** The Environmental Protection Agency (EPA) had recommended a 10,000-year minimum to provide protection from the long-lived isotopes. Government standards require isolation for 20 half-lives. (Plutonium has a half-life of 24,000 years.)

Tanks and Casks. For short-term containment, spent fuel is first stored in deep swimming pool–like tanks on the sites of nuclear power plants. The water in these tanks dissipates waste heat (which is still generated to some degree) and acts as a shield against the escape of radiation. The storage pools can typically accommodate 10 to 20 years of spent fuel. However, the storage pools at U.S. nuclear plants had been filled to 50% of their capacity by 2004 and will be at 100% by 2015. After a few years of decay, the spent fuel may be placed in air-cooled dry casks for interim storage until long-term storage becomes available. The casks are engineered to resist floods, tornadoes, and extremes of temperature. Currently, 49 facilities (mostly nuclear plants) are licensed to employ dry storage, with the prospect of many more being licensed as pool storage capacity is exhausted.

In the meantime, the wastes from the world's commercial reactors have been accumulating at a rate of 9,000 tons a year, reaching 290,000 tons at the end of 2008, all stored on-site at the power plants. Of these wastes, 57,000 tons are in the United States. Furthermore, because of neutron bombardment of the reactor walls, all nuclear power plants will eventually add to the stockpile of radioactive wastes.

Radioactivity and the Military. Some of the worst failures in handling radioactivity have occurred at military facilities in the United States and in the former Soviet Union. A recent calculation taken from government figures indicated that the U.S. nuclear program has, over the years, sickened 36,500 Americans and killed more than 4,000, most of these in connection with the manufacture and testing of nuclear weapons (most of the sickness and death is due to cancer). Further, liquid high-level wastes stored at many U.S. facilities have leaked into the environment and contaminated wildlife, sediments, groundwater, and soil. Activities at these sites have been shrouded in secrecy; only recently have documents revealing past accidents and radioactive releases been declassified and made available to the public. Deliberate releases of uranium dust, xenon-133, iodine-131, and tritium gas have been documented at Hanford, Washington; Fernald, Ohio; Oak Ridge, Tennessee; and Savannah River, South Carolina. Cleaning up these sites is now the responsibility of the DOE, which has already spent $50 billion on this problem and estimates that the final cost could run as high as $250 billion, probably the largest environmental project the country will ever undertake.

From Russia, with Curies. Russian military weapons facilities have been even more irresponsible. The worst is a giant complex called Chelyabinsk-65, located in the Ural Mountains. For at least 20 years, nuclear wastes were discharged into the Techa River and then into Lake Karachay. At least 1,000 cases of leukemia have been traced to radioactive contamination from the Chelyabinsk facility. Even today, standing for one hour (thus receiving 6 Sv of radiation) on the shore of Lake Karachay will cause radiation poisoning and death within a week. The lake dried up in the summer of 1967, and winds blew radioactive dust with 5 million curies across the countryside, contaminating hundreds of thousands of people. This lake is considered to be the most polluted lake on Earth, a legacy of the Cold War and an enormous continuing source of radioactive contamination. Recognizing the dangers of spreading contamination, Russian authorities have filled the lake with hollow concrete blocks, rocks, and soil, and plants now spread their greenery over this permanently contaminated site.

Megatons to Megawatts. The end of the Cold War has brought the welcome dismantling of nuclear weapons by the United States and the nations of the former Soviet Union. Thousands of nuclear weapons are being dismantled, many as a result of earlier agreements between U.S. President Bill Clinton and Russian President Boris Yeltsin. The agreements include a joint closure of all remaining plutonium weapons production reactors, funds to aid Russia in destroying missile silos and dismantling nuclear submarines and bombs, and vast cuts in the nuclear arsenals of the two former enemies. The radioactive components of the weapons must be handled with great care and disposed of where they are safe from illegal access and where they will pose no danger to human health for centuries to come—a daunting task.

One element of the U.S.-Russian agreements, the **Megatons to Megawatts** program, involves a government-industry partnership wherein a private U.S. company oversees the dilution of weapons-grade uranium to the much lower enrichment of power-plant uranium, sells it to U.S. nuclear power plants, and pays the Russian government a market price (to date, $5.7 billion). As a result, by September 2008, 13,795 nuclear warheads had been eliminated, and 345 tons of highly enriched uranium had been diluted to some 10,100 tons of power-plant fuel (Fig. 15–11). Fully half of the uranium fuel currently used by U.S. power plants comes from this program. The closing down of much of the nuclear warfare enterprise is certainly a welcome development, but the problems of security and of disposal of the wastes will remain as a long-term legacy of the Cold War.

High-Level Nuclear Waste Disposal. The United States and most other countries that use nuclear power have decided that geologic burial is the safest option for disposing of the highly radioactive spent fuel, but no nation has developed plans to the point of actually carrying out the burial. Many nuclear nations have not even been able to find a site that may be suitable for receiving wastes. Where sites have been selected, questions about their safety have surfaced. The basic problem is that no rock formation can be guaranteed to remain stable and dry for tens of thousands of years.

Figure 15–11 Megatons to Megawatts. Cylinders containing nuclear-warhead-derived fuel from Russia are unloaded in the United States. This program has eliminated 13,795 nuclear warheads and currently supplies half of the fuel for U.S. nuclear power plants.

Everywhere scientists look, there is evidence of volcanic or earthquake activity or groundwater leaching within the last 10,000 years or so, which is to say that it may occur again in a similar period. If such events did occur, the still-radioactive wastes could escape into the environment and contaminate water, air, or soil, with consequent effects on humans and wildlife.

Yucca Mountain. In the United States, efforts to locate a long-term containment facility have been hampered by a severe *not in my backyard* (*NIMBY*) syndrome. A number of states, under pressure from citizens, have passed legislation categorically prohibiting the disposal of nuclear wastes

within their boundaries. In the meantime, the need to select and develop a long-term repository has become increasingly critical, given the buildup of spent fuel in every nuclear power plant. The **Nuclear Waste Policy Act of 1982** committed the federal government to begin receiving nuclear waste from commercial power plants in 1998. At the end of 1987, Congress called a halt to the debate and selected a remote site, Yucca Mountain, in southwestern Nevada (Fig. 15–12), to be the nation's civilian nuclear waste disposal site. No other sites are being considered. Not surprisingly, Nevadans fought this selection, passing a law in 1989 that prohibits anyone from storing high-level radioactive waste in the state. The federal government has the power to override state prohibitions, though, and the Nevada site has undergone intensive study and exploratory construction over the past 20 years, at a cost of $4 billion.

The EPA has the responsibility to set safety standards for radiation doses, and the agency has been involved with the Yucca Mountain site from early on. The EPA has had to revise its standards several times because of court rulings. The applicable standard—15 millirems (0.15 mSv) for the first 10,000 years—establishes a limit on risk of radiation to individuals near the repository. After a 2004 court ruling that a 10,000-year period was inadequate, the EPA extended the protection standard to *1 million years*, but raised the allowable dose maximum past 10,000 years to 100 millirems (1.0 mSv), which is still below the current level of average radiation exposure. Critics of the site and DOE's plans for disposal question if any degree of safety can be guaranteed for such long periods of time, given the possibilities of earthquakes, volcanic activity, or other changes in geologic time. However, studies have indicated that the storerooms, 1,000 feet (300 m) above current groundwater levels, will be safe from invasion by groundwater for at least 10,000 years.

Figure 15–12 Yucca Mountain, Nevada. This remote ridgeline in Nevada was chosen as the nation's sole high-level nuclear waste depository. The site is on federally protected land, approximately 100 miles northwest of Las Vegas, Nevada.

Yucca Rejected. In July 2002, President George W. Bush signed a resolution (passed by Congress) voiding a move by Nevada's governor to block further development of the site. The Yucca Mountain facility was officially designated to be the nation's nuclear waste repository. The DOE submitted a license application for operation of the Yucca Mountain repository to the NRC in June 2008, and NRC staff recommended that the agency adopt DOE's environmental impact statement for the project. However, the Obama victory in November 2008 spelled doom for the repository. The Obama administration decided to terminate the Yucca Mountain program, citing health and safety concerns as the reason. Instead, the administration has appointed a Blue Ribbon Commission to explore "other nuclear waste disposal options." Needless to say, the state of Nevada is delighted that Yucca Mountain may never be host to high-level nuclear wastes. (See Stewardship, "Showdown in the New West.")

Although the waste issues are a serious concern, it is the potential for accidents that generates the greatest concern over nuclear power.

Nuclear Power Accidents

On March 28, 1979, the Three Mile Island Nuclear Power Plant near Harrisburg, Pennsylvania, suffered a partial meltdown as a result of a series of human and equipment failures and a flawed design. The steam generator (Fig. 15–8) shut down automatically because of a lack of power in its feedwater pumps, and eventually, a valve on top of the generator opened in response to the gradual buildup of pressure. Unfortunately, the valve remained stuck in the open position and drained coolant water from the reactor vessel. There were no sensors to indicate that this pressure-operated relief valve was open. Operators responded poorly to the emergency, shutting down the emergency cooling system at one

Stewardship

Showdown in the New West

In a scenario reminiscent of the Civil War, three western states have become the focus of a controversy over states' rights. On one side is the federal government, especially the DOE, which is trying to solve the problem of what to do with nuclear wastes and which thinks it has a reasonable solution. On the other side are politicians in the states designated as repository sites, who are responding to the voices of citizens of those states. At stake may be the future of nuclear energy in the United States.

WIPP

The DOE has constructed the $1.5 billion Waste Isolation Pilot Plant (WIPP) in salt caves 2,150 feet beneath the desert in southeastern New Mexico, 26 miles from the city of Carlsbad. The plant is on federal land and passed all necessary safety reviews. It was ready to open for business in October 1991. "Business," in this case, was for the plant to become the repository for up to a million barrels of plutonium wastes from nuclear weapons plants and laboratories around the country as well as for the plutonium from dismantled nuclear warheads. (No commercial nuclear wastes will go to the plant.) It is the world's first (and only) deep geological repository for high-level plutonium wastes.

New Mexico sued the DOE on the grounds that it was in violation of federal law prohibiting the opening of such a repository without congressional approval. The state of Texas and four environmental groups joined New Mexico in bringing suit. In 1998, after a lengthy review, the EPA gave the WIPP certification to begin receiving radioactive materials from nuclear weapons facilities around the country, and in 1999, the site began to receive wastes from a number of facilities. By 2009, the WIPP had received over 7,000 shipments without incident.

Yucca Mountain

Opposition to the long-term storage of nuclear wastes is even stronger in Nevada, which has the dubious distinction of having been selected for the nation's only repository for commercial nuclear wastes, at Yucca Mountain. The site is a barren ridge in the desert about 100 miles northwest of Las Vegas. All sorts of surveys have been carried out to test the mood of Nevadans regarding the long-term storage of wastes in their state. The responses have been uniformly negative: Three-fourths of Nevadans agreed that the state should continue to do all it can to prevent the establishment of the Yucca Mountain site. The state's attempt to block development of the site failed because the U.S. Supreme Court ruled that Nevada must process applications for permits to continue work there. Nevertheless, the state of Nevada is continuing its legal challenges to the facility. The latest of nine lawsuits has the state challenging the EPA's radiation exposure standards on the basis that they fail to protect the health and safety of Nevada citizens. Under George W. Bush, Yucca Mountain became the officially designated repository for commercial nuclear wastes. But as we have seen, the Obama administration has halted all funding for the continued development of the Yucca Mountain facility.

Goshute Reservation

The NRC has granted a license to the consortium Private Fuel Storage (PFS) to receive up to 44,000 tons of radioactive waste, in the form of dry casks, on the Skull Valley Goshute Reservation in Utah. This is an interim storage site intended to relieve the commercial nuclear power plants of their nuclear waste storage problem until Yucca Mountain is open for business. Terms of the deal have been kept secret, but no doubt some large sums of money are involved. State politicians have tried everything possible to stop the deal, but the site, 40 miles from Salt Lake City, is under control of the Native American tribe, which has benefited from grants from the DOE and financial backing from PFS as they have fought a nine-year battle to get the necessary approvals. However, the Department of the Interior recently issued two decisions against the proposed site, and the tribe has filed a federal lawsuit to overturn the decisions.

People in New Mexico, Nevada, and Utah are indignant that their states could become the dumping ground for nuclear wastes from around the nation, but in a sense, their concern is a proxy for what will be an enduring concern for generations yet unborn. We need to deal responsibly with the legacy of nuclear energy. In a real sense, the nation's security is also at stake because the wastes stored on the grounds of the existing and shut power plants are prime targets for terrorists.

point and shutting down the pumps in the reactor vessel. One instrument error compounded the problem: Gauges told operators that the reactor was full of water when, actually, it needed water badly. The core was uncovered for a time and suffered a partial meltdown. Ten million curies of radioactive gas were released into the atmosphere.

The drama held the whole nation—particularly the 300,000 residents of metropolitan Harrisburg, who were poised for evacuation—in suspense for several days. The situation was eventually brought under control, and no injuries or deaths occurred, but it would have been much worse if the meltdown had been complete. The reactor was so badly damaged and so much radioactive contamination occurred inside the containment building that the continuing cleanup is proving to be as costly as building a new power plant. There are no plans to restart the reactor. GPU Nuclear, operator of the plant, has since paid out $30 million to settle claims from the accident, although the company has never admitted the existence of any radiation-caused illnesses.

Chernobyl. Prior to 1986, the scenario for a worst-case nuclear power plant disaster was a matter of speculation. Then, at 1:24 A.M. local time on April 26, 1986, events at a nuclear power plant in Ukraine (then a part of the Soviet Union) made such speculation irrelevant (Fig. 15–13). Since that day, Chernobyl has served as a horrible example of nuclear energy gone awry.

Here's how it happened: While conducting a test of standby diesel generators, engineers disabled the power plant's safety systems, withdrew the control rods, shut off the flow of steam to the generators, and decreased the flow of coolant water in the reactor. However, they did not allow for the radioactive heat energy generated by the fuel core, and lacking coolant, the reactor began to heat up. The extra steam that was produced could not escape and had the effect of rapidly boosting the energy production of the reaction. In an attempt to quell the reactor, the engineers quickly inserted the carbon-tipped control rods. The carbon tips acted as moderators, slowing down the neutrons that were produced in the reaction. The neutrons, however, were still speedy enough to trigger more fission reactions, and the result was a split-second power surge to 100 times the maximum allowed level. Steam explosions then blew the 2,000-ton top off the reactor, the reactor melted down, and a fire was ignited in the graphite that was being used as a moderator, burning for days. At least 50 tons of dust and debris bearing 100–200 million curies of radioactivity in the form of fission products were released in a plume that rained radioactive particles over thousands of square miles, 400 times the radiation fallout from the atomic bombs dropped on Hiroshima and Nagasaki in 1945.

Consequences. As the radioactive fallout settled, 135,000 people were evacuated and relocated. The reactor was eventually sealed in a sarcophagus of concrete and steel. A barbed-wire fence now surrounds a 1,000-square-mile exclusion zone around the reactor site. The soil remains contaminated with radioactive compounds, yet for years 2,000 Ukrainian workers were bused in daily to work at the

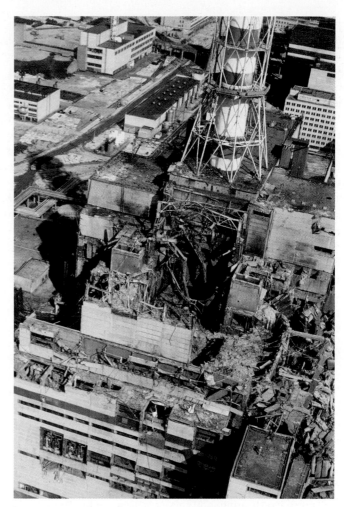

Figure 15–13 Chernobyl. An aerial view of the Chernobyl reactor on April 29, 1986, three days after the explosion. The disaster was the worst nuclear power accident, directly killing at least 31 people and putting countless others in the surrounding countryside at risk for future cancer deaths.

remaining reactor in the Chernobyl complex. In December 2000, the last reactor was permanently shut down.

Only 2 engineers were directly killed by the explosion, but 28 of the personnel brought in to contain the reactor in the aftermath of the explosion died of radiation within a few months. Over a broad area downwind of the disaster, buildings and roadways were washed down to flush away radioactive dust. Even with these precautions, many people in or near the evacuation zone were exposed to high levels of radiation, especially the short-lived radioisotope iodine-131. Predictions of hundreds of thousands of cancer deaths among those exposed followed shortly after the accident.

A recent report by a team of experts convened by United Nations agencies[2] has provided the most reliable assessment of the impacts of the accident on the surrounding populations. The main health impact has been an outbreak of thyroid cancer among children who drank milk contaminated

---

[2]The Chernobyl Forum: 2003–2005, *Chernobyl's Legacy: Health, Environmental and Socio-economic Impacts, and Recommendations to the Governments of Belarus, the Russian Federation and Ukraine,* 2nd rev. version (Vienna, Austria: International Atomic Energy Agency, 2005). Available at www.iaea.org/Publications/Booklets/Chernobyl/chernobyl.pdf.

with radioactive iodine. Thousands of cases have been diagnosed in this group, with numerous deaths. Several thousand additional deaths due to cancer are expected, but the increase will be hard to detect because of the hundreds of thousands of cancer deaths from other causes to be expected in the affected populations. However, 22 years after the accident, Russian television's *Health* program reported increases in thyroid cancer some 20 to 70 times above normal in many regions of the country and is recommending that people between the ages of 20 and 40 from regions around the Ukraine seek ultrasound thyroid examinations.

The hastily constructed shelter over the damaged power plant is in a state of decay, prompting a plan for construction of a safe structure for the long-term confinement of the reactor complex. A $1.4 billion "New Safe Confinement," funded by some 28 governments and coordinated by the European Bank for Reconstruction and Development, will consist of a steel structure large enough to house the Statue of Liberty; construction was begun in 2008 and will take five years to complete.

Could It Happen Here?   Are we in the United States in danger of such an explosion occurring at a nuclear power plant? Nuclear scientists argue that the answer is no because U.S. power plants have a number of design features that should make a repeat of Chernobyl impossible. For one thing, the Chernobyl reactor used graphite as a moderator rather than the water used in LWRs. Further, LWRs are incapable of developing a power surge more than twice their normal power, well within the designed containment capacity of the reactor vessel. Finally, LWRs have more backup systems to prevent the core from overheating, and these reactors are housed in thick, concrete-walled containment buildings designed to withstand explosions such as the one that occurred in the Chernobyl reactor, which had no containment building. LWRs are not immune to accidents, however, the most serious being a complete core meltdown as a result of a total loss of coolant. This has never happened, but Three Mile Island was a close call.

The real cost of the accidents at Three Mile Island and Chernobyl must be reckoned in terms of public trust. Public confidence in nuclear energy plummeted after the two accidents, which pointed to human error as a highly significant factor in nuclear safety—and human error is something the public understands well. Nuclear proponents suffered a serious loss of credibility with Three Mile Island, but Chernobyl was their worst nightmare come true—a full catastrophe, just what the antinuclear movement had predicted might someday occur.

Safety and Nuclear Power

As a result of Three Mile Island and other, lesser incidents in the United States, the NRC has upgraded safety standards not only in the technical design of nuclear power plants, but also in maintenance procedures and in the training of operators. Thus, proponents contend, nuclear plants were designed to be safe in the beginning, and now they are safer than ever. Some proponents of nuclear power claim that we now have the technology to build *inherently safe* nuclear reactors,

designed in such ways that any accident would result in an automatic quenching of the chain reaction and suppression of heat from nuclear decay. In reality, however, there is no such thing as an inherently safe reactor, since that concept implies no release of radioactivity under any circumstances—an impossible expectation.

Passive Safety.   Instead, nuclear scientists are proposing a new generation of nuclear reactors with built-in *passive safety* features rather than the *active safety* features found in current reactors. **Active safety** relies on operator-controlled actions, external power, electrical signals, and so forth. As accidents have shown, operators may override such safety factors, and electricity, valves, and meters can fail or give false information. **Passive safety**, by contrast, involves engineering devices and structures that make it virtually impossible for the reactor to go beyond acceptable levels of power, temperature, and radioactive emissions; their operation depends only on standard physical phenomena, such as gravity and resistance to high temperatures.

New Generations of Reactors.   Nuclear scientists distinguish four generations of nuclear reactors. *Generation I* reactors are the earliest, developed in the 1950s and 1960s; few are still operating. The majority of today's reactors are *Generation II* vintage, the large baseline power plants, of several designs. *Generation III* refers to newer designs with passive safety features and much simpler power plants—among them the so-called advanced boiling-water reactors (ABWRs) (Fig. 15–14). General Electric (GE) has submitted the Economic Simplified Boiling Water Reactor (ESBWR) design for NRC approval. Compared to GE's current ABWR, this new design has many more passive safety features. In the event of a LOCA, two separate passive systems would allow water to drain by gravity to the reactor vessel. In the United States, four utility companies have applied for a federal construction and operating license based on this new reactor; of the 26 new U.S. license applications, the majority are for the Westinghouse AP1000 Advanced Passive Reactor, a pressurized-water reactor with new passive safety features.

In East Asia, the nuclear reactor design of choice is now the ABWR, which produces 1,400 MW of electricity. Japan, with 55 nuclear plants, has 4 of these in operation; several more are under construction in Japan and Taiwan, and there are plans for many more. The new plants are designed to last for 60 years, can be constructed within 5 years, and are simple enough that a single operator can control them under normal conditions. They generate electricity at a cost competitive with the costs of fossil fuel options.

Generation IV plants are now being designed and will likely be built within the next 20 years. One example of a Generation IV reactor is the pebble-bed modular reactor (PBMR), which will feed spherical carbon-coated uranium fuel pebbles gradually through the reactor vessel, like a gumball machine. The PBMR will be cooled with fluidized helium, an inert gas, which will also spin the turbines. These reactors will be small, will produce about 160 MW of power, and are expected to be cheap to build, inherently safe, and

Figure 15–14 Advanced boiling-water reactor. The core is surrounded by three concentric structures: a reactor pressure vessel, in which heat from the reactor boils water directly into steam; a concrete chamber (outlined with heavy black line) and a water pool, which together contain and quench steam vented from the reactor in an emergency; and a concrete building, which acts as a secondary containment vessel and shield. Any excessive pressure in the reactor will automatically open valves that release steam into a quenching pool, reducing the pressure. Water from the quenching pool can, if necessary, flow downward to cool the core. (*Source:* "Advanced Light-Water Reactors," by M. W. Golay and N. E. Todreas. Copyright 1990 by *Scientific American.* All rights reserved.)

inexpensive to operate. The modules can be built in a factory and shipped to the location of the power plant.

One motive behind the new designs is to restore the public's confidence in nuclear energy. Previously, nuclear proponents had emphasized the very low probabilities of accidents. As we have seen, though, improbable events can happen, and when they happen to nuclear power plants, the consequences can be dreadful. It may be that the public was on the way to a renewed appreciation of nuclear power, but this was set back on September 11, 2001.

Terrorism and Nuclear Power. What would have happened if one of the terrorist suicide groups that took over four airliners on September 11, 2001, had targeted a nuclear power plant instead of the World Trade Center? This question has raised great concern in many venues, from the NRC, to Congress, to local hearings in municipalities that host nuclear plants. After some uncertainty following September 11, the general consensus is that a jetliner could not penetrate the very thick walls of the containment vessel protecting the reactor. It could, however, destroy the control building and bring on a LOCA.

Another terrorist method might be to attack the nuclear plant with a dedicated strike force, overcome the guards, and bring about a loss of cooling water and a core meltdown by manipulating the controls. If this happened, the reactor vessel would likely be breached and a large release of radiation would occur. (This did not happen at Three Mile Island because the operators cooled the core sufficiently before it could melt completely.) Although few, if any, immediate civilian fatalities would result, there would be the long-term effects of radiation on the incidence of cancer. Over the years, the NRC has conducted mock terrorist attacks on nuclear plants, and a relatively high number of these attacks have penetrated the plants' security.

Response. Following September 11, the NRC immediately beefed up the requirements for security around every power plant—adding more guards, more physical barriers, and vehicle checks at greater distances from the plant—as well as keeping the shoreline near the plants off limits. The Energy Policy Act of 2005 has stepped up security measures, with requirements to escalate the kinds of threat that nuclear plant security forces must be able to meet and to conduct "force-on-force" security exercises at least once every three years, among others. A recent National Academy of Sciences (NAS) report[3] concluded that the most vulnerable locations

---

[3]National Research Council, *Safety and Security of Commercial Spent Nuclear Fuel Storage: Public Report* (Washington, D.C.: National Academies Press, 2006). Available at www.nap.edu/catalog/11263.html.

for terrorist attack are the spent-fuel pools. Every power plant has a spent-fuel storage pool containing at least as much highly radioactive material as a nuclear core. These pools are not at all as heavily protected as the reactor vessels. (The pool walls may be only 18 inches thick.) If a pool lost its water, the fuel could heat up and cause a fire that could release greater amounts of cesium-137 than the Chernobyl accident. The NAS recommended further policy decisions that would protect the pools from a loss-of-coolant-pool-water event brought on by terrorist activity.

Making a Bomb. Mohamed ElBaradei, head of the International Atomic Energy Agency, reported in 2008 to the UN General Assembly that nearly 250 thefts of nuclear or radioactive material had occurred during the past year; not much of it was ever recovered. Although the sum of all the stolen material was not enough to build a nuclear device, the potential to steal enough for such a device was demonstrated. In fact, on two occasions an ABC News team successfully smuggled a soda-can-size cylinder of depleted uranium through U.S. customs radiation detectors to demonstrate the inadequacy of the detectors currently in use. Since the smuggled uranium had a radiation signature similar to that of highly enriched uranium (HEU), the news team basically showed that terrorists could smuggle enough HEU into the United States to enable the manufacture of a crude nuclear weapon. Further, since plutonium is not highly radioactive, it would be relatively easy to smuggle bomb-grade amounts of plutonium into a country. About 15 pounds of plutonium would be sufficient to make a crude bomb. Guarding against these possibilities is the responsibility of the Department of Homeland Security; the department has announced that it will beef up its radiation detection with new devices.

The potential for accidents and terrorist acts certainly raises questions about a more widespread promotion of nuclear power. However, as Figure 15–2 shows, utilities in the United States were already turning away from nuclear power before Three Mile Island and Chernobyl. The reasons were mainly economic.

Economics of Nuclear Power

What was behind the many canceled nuclear power plants? First, projected future energy demands were overly ambitious, so a slower growth rate in the demand for electricity postponed orders for all types of power plants. Second, increasing safety standards for the construction and operation of nuclear power plants caused the costs of plants to increase at least fivefold, even after inflation is considered. Adding to the rise in costs was the withdrawal of government subsidies to the nuclear industry. Third, public protests frequently delayed the construction or start-up of a new power plant. Such delays increased costs still more because the utility was paying interest on its investment of several billion dollars even when the plant was not producing power. As these costs were passed on, consumers became yet more disillusioned with nuclear power. Finally, safety systems may protect the public, but they do not prevent an accident from financially ruining the utility. Since radioactivity prevents

straightforward cleanup and repair, an accident can convert a multibillion-dollar asset into a multibillion-dollar liability in a matter of minutes, as Three Mile Island demonstrated. Thus, nuclear power involves a financial risk that utility executives have been hesitant to take.

Longevity? Originally, it was thought that nuclear plants would have a lifetime of about 40 years. It now appears that lifetimes will be considerably less for many of the older plants. For example, the government of Ontario decided in 2005 to shut down two 540-MW CANDU nuclear reactors more than a decade before their scheduled retirement. The reactors had been idle since 1997 because of corrosion in the pipes carrying heavy coolant water from the reactor core. The repair bill was estimated to be $1.6 billion, approximately the cost of a new reactor.

Worldwide, more than 119 nuclear plants have been shut down after an average operating lifetime of 20 years. This shorter lifetime substantially increases the cost of the power produced because the cost of the plant must be repaid over a shorter period. The lifetimes of nuclear power plants are shorter than originally expected due to *embrittlement* and *corrosion*. **Embrittlement** occurs as neutrons from fission bombard the reactor vessel and other hardware. Gradually, this neutron bombardment causes the metals to become brittle enough that they may crack under thermal stress—for example, when emergency coolant waters are introduced in the event of a LOCA. When the reactor vessel becomes too brittle to be considered safe, the plant must either be shut down or be repaired at great cost.

Corrosion is a normal consequence of steam generation. Very hot, pressurized water flows from the core into the steam generator through thousands of 3/4-inch-diameter pipes immersed in water (Fig. 15–8). The water inside and outside these pipes contains corrosive chemicals that, over time, cause cracks to develop in some of the pipes. If the main line conveying steam from the generator to the turbine were to rupture, the sudden increase in pressure in the generator could cause several cracked pipes to break at once. In that case, radioactive moderator-coolant water would be released and would overload safety systems, forcing the plant to vent radioactive gas to the outside. Cracked pipes are "repaired" basically by plugging them, cutting the overall plant power. In March 1995, using a new high-tech probe, officials discovered that up to half of the steam generator pipes in the Maine Yankee plant in Wiscasset, Maine, had developed cracks, some penetrating 80% of the pipes' thickness. As a result, the plant was shut down after only 24 years of operation. Recent problems of corrosion in reactor vessels have both surprised and alarmed NRC officials; apparently, the cooling water can be far more corrosive than originally expected.

Decommissioning. Closing down, or decommissioning, a power plant can be extremely costly. Decommissioning Maine Yankee (Fig. 15–15a), now complete, cost $635 million. (By comparison, the plant cost just $231 million to build in the late 1960s.) The site now houses some 64 dry casks

(a)   (b)

Figure 15–15 Maine Yankee. (a) The Maine Yankee Nuclear Power Plant in the process of being decommissioned, in 2003. The decommissioning is now complete, involving the removal of 200,000 tons of solid waste by rail, truck, and barge. (b) The site now consists of 200 acres of conservation land, 400 acres of land for economic development, and 12 acres of secured storage of dry casks containing the spent nuclear fuel. The photo shows railroad cars transporting processed soil, with dry casks in the background.

containing the spent-fuel assemblies (Fig. 15–15b), which await disposal at their final resting place, wherever that happens to be. The owners of the plant are allowed to bill former customers to recover some costs—a "stranded cost" nobody will like. Faced with these costs, some utilities are opting to repair older plants in spite of the high costs of such repairs. Other closed reactors are on sites currently occupied by active units. The old, closed-down reactors are simply defueled and allowed to sit until the day when all the reactors are closed and the entire site must be decommissioned. In two closed reactors, only the nuclear components were removed, and the plants were converted to conventional natural-gas plants.

New Life. Currently, nuclear power is being viewed more favorably by policy makers, as well as by many in the environmental community, in spite of the economic issues. Most of the operating nuclear power plants in the United States were built before global climate change was recognized as a serious problem, but concerns about the greenhouse gas emissions of conventional coal-fired power plants have brought many to reconsider the advantages of nuclear power. This has come into sharp focus as the time has been drawing near for the retirement of a number of older nuclear plants. New Jersey and Vermont are exploring extending the licenses of two reactors beyond their 40-year terms, in light of the agreement among nine northeastern states to restrict greenhouse gas emissions from power plants. Indeed, the nuclear power industry is hoping to extend the life of its fleet of aging plants, most of which will reach the end of their operating licenses within the next 10 to 15 years. The industry looks

with hope at the favorable political climate for nuclear power, anticipating the construction of a number of new nuclear plants within the decade. With 26 new license applications received and 3 site permits issued, the NRC has a renewed set of responsibilities after years of status quo. Most of the new reactors will be built on the sites of existing nuclear power plants.

Overseas, nuclear power is also looking up. The most recent nuclear reactor to be built in the United Kingdom, Sizewell B, cost almost $3,000 per kilowatt of capacity, almost 10 times as much as a gas-fired plant. Offsetting this cost, however, is the lower operating cost of nuclear energy when compared with fossil fuel–powered electricity (about two-thirds of the latter). France is building a new nuclear plant on the Normandy coast, after a 10-year respite. Italy and Sweden, decades after public referenda turned down nuclear power, are now planning to resume building nuclear plants. Thirty-one countries are still committed to maintaining current nuclear power plants, and 15 more countries are interested in developing nuclear power; 5 new nuclear plants have gone online in the last two years, and 42 are under construction.

15.4 More Advanced Reactors

Uranium—especially ^{235}U—is not a highly abundant mineral on Earth. At the height of optimism about nuclear energy in the 1960s, when as many as 1,000 plants were envisioned by the turn of the century, it was forecast that shortages of ^{235}U would develop. Breeder reactors, which utilize chain reactions, were seen as the solution to this problem.

Breeder (Fast-Neutron) Reactors

Recall from Section 15.2 that when a ^{235}U atom fissions, two or three neutrons are ejected. Only one of these neutrons needs to hit another ^{235}U atom to sustain a chain reaction; the others are absorbed by something else. The breeder reactor is designed so that (nonfissionable) ^{238}U absorbs the extra neutrons, which are allowed to maintain their high speed. (Such reactors are now called *fast-neutron reactors*.) When this occurs, the ^{238}U is converted to plutonium (^{239}Pu), which can then be purified and used as a nuclear fuel, just like ^{235}U. Thus, the fast-neutron reactor converts nonfissionable ^{238}U into fissionable ^{239}Pu, and because the fission of ^{235}U generally produces two neutrons in addition to the one needed to sustain the chain reaction, the fast-neutron reactor may produce more fuel than it consumes. Furthermore, 99.3% of all uranium is ^{238}U, so converting that to ^{239}Pu through fast-neutron reactors effectively increases nuclear fuel reserves more than 100-fold.

In the United States and elsewhere, small fast-neutron reactors are operated for military purposes. France, Russia, and Japan are currently the only nations with commercial fast-neutron reactors, and these have not been in continuous use. Fast-neutron reactors present most of the problems and hazards of standard fission reactors, plus a few more. The consequences of a meltdown in a fast-neutron reactor would be much more serious than those of a meltdown in an ordinary fission reactor, due to the large amounts of ^{239}Pu, with its half-life of 24,000 years. In addition, because plutonium can be purified and fabricated into nuclear weapons more easily than can ^{235}U, the potential for the diversion of fast-neutron reactor fuel to illicit weapons production is greater. Hence, the safety and security precautions needed for fast-neutron reactors are greater. Also, fast-neutron reactors are more expensive to build and operate. However, they have the ability to extract much more energy from recycled nuclear fuel, and they produce much less high-level waste than do conventional nuclear plants.

Reprocessing

The United States has enough uranium stockpiled to satisfy current industry needs. Although the global supply of uranium is not a problem right now, a number of countries are reprocessing spent fuel through chemical processes, recovering plutonium and mixing it with ^{238}U to produce mixed-oxide (MOX) fuel that is about 5% plutonium and suitable for further use in many nuclear power plants. In this way, some of the high-level waste is reused and does not add to the stockpile of waste. In fact, the DOE has proposed a resumption of reprocessing in the United States, in a program called the Advance Fuel Cycle Initiative (AFCI), and has received funding to "develop and demonstrate nuclear fuel cycles that could reduce the long-term hazard of spent nuclear fuel and recover additional energy," according to the budget request. One outcome of AFCI is a new plan from DOE and the Bush administration, the Global Nuclear Energy Partnership (GNEP). This initiative involves several new facilities, to be built by the DOE in the United States, that would reprocess spent nuclear fuel and supply it to other countries. Also, establishing the reprocessing facilities would prepare for a new generation of nuclear reactors that use the fast-neutron technology. Congress, however, has not been receptive to GNEP and has recommended no further funding for the program.

Fusion Reactors

The vast energy emitted by the Sun and other stars comes from fusion (Fig. 15–5b). The Sun, as well as other stars, is composed mostly of hydrogen. Solar energy is the result of the fusion of hydrogen nuclei (protons) into larger atoms, such as helium. Scientists have duplicated the process in the hydrogen bomb, but hydrogen bombs hardly constitute a useful release of energy. The aim of fusion technology is to carry out fusion in a controlled manner in order to provide a practical source of heat for boiling water to power steam turbogenerators.

The d–t Reaction.   Since hydrogen is an abundant element on Earth (there are two atoms of it in every molecule of water) and helium is an inert, nonpolluting, nonradioactive gas, hydrogen fusion is promoted as the ultimate solution to all our energy problems; that is, fusion affords pollution-free energy from a virtually inexhaustible resource: water. However, most current designs do not use regular hydrogen (^{1}H), but rather employ the isotopes deuterium (^{2}H) and tritium (^{3}H) in what is called the *d–t reaction*. Deuterium is a naturally occurring nonradioactive isotope that can be extracted in almost any desired amount from the hydrogen in seawater. Tritium, by contrast, is an unstable gaseous radioactive isotope that must be produced artificially. Radioactive and difficult to contain, tritium is a hazardous substance. As a result, fusion reactors could easily become a source of radioactive tritium leaking into the environment unless effective (and costly) designs are used to prevent that.

In the present state of the art, fusion power is still an energy consumer rather than an energy producer. The problem is that it takes an extremely high temperature—some 3 million degrees Celsius—and high pressure to get hydrogen atoms to fuse. In the hydrogen bomb, the temperature and pressure are achieved by using a fission bomb as an igniter! Getting to the high temperature is only one part of the problem; extracting the heat from the reacting region is also a daunting task, given the large amount of energy released from the fusion reactions.

Tokamak.   A major technical problem is how to contain the hydrogen while it is being heated to the tremendously high temperatures required for fusion into helium. No known material can withstand these temperatures without vaporizing, although three techniques are being tested. One is the Tokamak design, in which ionized hydrogen is contained within a magnetic field while being heated to the necessary temperature. The second is laser fusion, in which a tiny pellet of frozen hydrogen is dropped into a bull's-eye, where it is hit simultaneously from all sides by powerful laser

Figure 15–16 Laser fusion. In this experimental instrument at Lawrence Livermore Laboratory, 30 trillion watts of optical power are focused onto a tiny pellet of hydrogen smaller than a grain of sand and located in the center of a vacuum chamber. For less than a billionth of a second, the fusion fuel is heated and compressed to temperatures and densities near those found in the Sun.

beams (Fig. 15–16). The beams heat and pressurize the pellet to the point of fusion. A third design is a device called the Z machine, which delivers very short, but powerful, pulses of current through an array of wires. This concentrated electrical energy results in a blast of X-rays that recently reached 1.8 million degrees Celsius, close to the 2 to 3 million degrees it would take to initiate fusion.

Some controlled fusion has been achieved in the Tokamak and laser fusion devices, and in late 1994, a Tokamak facility at the Princeton Plasma Physics Laboratory in Princeton, New Jersey, achieved 10.7 MW of fusion power in 0.27 second of reaction. As yet, however, the breakeven point has not been reached: More energy is required to run the magnets or lasers than is obtained by fusion. (It took 39.5 MW of energy to sustain the reaction at Princeton.) In the United States, fusion energy research is proceeding, with annual funding in the hundreds of millions of dollars, an indication that this technology is considered important.

Pot of Gold? An international research team has presented a design for an International Thermonuclear Experimental Reactor (ITER), a prototype fusion reactor of the Tokamak variety. The facility, to be built in Cadarache, France, is expected to cost over $14 billion, and the reactor will produce 10 times as much energy as it consumes. Recently, the United States rejoined the project some 10 years after withdrawing because of concerns about costs and the

science itself. The most optimistic view sees the ITER coming on line by 2016. As this plan indicates, developing, building, and testing a fusion power plant will require at least another 10 to 20 years and many billions of dollars. This assumes that the engineering difficulties will be overcome, which is by no means a given. Thus, fusion is, at best, a very long-term and uncertain option. Reflecting that view, Congress made drastic cuts in the current ITER funding. Many scientists believe that fusion power will always be the elusive pot of gold at the end of the rainbow. The standard joke about fusion power is this: *It is the energy source of the future and always will be!*

15.5 The Future of Nuclear Power

With fossil fuels contributing to global climate change and with alternatives such as solar and wind power (Chapter 16) in their early stages, the long-term outlook for energy is discouraging. Using nuclear power seems to be an obvious choice; however, there continues to be organized opposition to nuclear energy, and it is still a costly technology, highly dependent on government patronage. Like Italy, Sweden and Germany decided some years ago to phase out nuclear power, but opinion polls have lately shown broad public support for keeping nuclear power in place. Russia and other eastern European countries are holding onto their nuclear power, and currently, eight more reactors are under construction in the Russian Federation. A new plant is being built in Finland, and the United Kingdom is planning a new generation of reactors to replace its aging fleet. Asian countries are welcoming nuclear power, with China, Japan, South Korea, and India seemingly intent on using it to fulfill a major share of their energy needs.

Opposition

Opposition to nuclear power is based on several premises:

- People have a general distrust of technology they do not understand, especially when that technology carries with it the potential for catastrophic accidents or the hidden, but real, capacity to induce cancer.

- Many observers are critical of the way nuclear technology is being managed. They are aware that the agency responsible for licensing and safety regulations (the NRC) is also a strong supporter of the commercial nuclear industry.

- Problems involving lax safety, operator failures, and cover-ups by nuclear plants and their regulatory agencies have occurred in the United States, Canada, and Japan.

- The problems of high construction costs and unexpectedly short operational lifetimes have already been mentioned. Thus, the economic argument can also be used to oppose nuclear power.

- The nuclear industry has repeatedly presented nuclear energy as extremely safe, arguing that the probabilities of accidents occurring are very low. However, when accidents do occur, probabilities become realities, and the arguments are moot.

- Nuclear power plants are viewed as prime targets for terrorist attacks, which could bring about a potentially devastating release of radioactivity. Critics argue that, therefore, it only makes sense to reduce the number of such targets, not add more of them.

- There remains the crucial problem of disposing of nuclear waste. All parties agree that the waste must be placed somewhere safe, but there the agreement ends, and siting a long-term nuclear waste repository has been a difficult political as well as technological problem in country after country.

Mismatch? Opponents of nuclear power also cite the basic mismatch between nuclear power and the energy problem. The main energy problem for the United States is an eventual shortage of crude oil for transportation purposes, yet nuclear power produces electricity, which is not yet used for transportation. Consequently, nuclear power simply competes with coal-fired power in meeting the demands for base-load electrical power. Given the high costs and additional financial risks of nuclear power plants, coal is cheaper, and the United States does have abundant coal reserves.

Nonetheless, there are still the environmental problems of mining and burning coal, especially global climate change. As discussed in Chapter 14, burning coal emits more CO_2 per unit of energy produced than any other form of energy does. If the long-term environmental costs were factored into the price of coal, that price would be considerably higher than it is now.

In the final analysis, nuclear power may be a unique energy option, but it is also a controversial technology. Interestingly, public opinion about nuclear power has been shifting. Numerous polls indicate that Americans view nuclear power as increasingly important, in the light of surging prices for oil and gas. Some leaders in the environmental field are viewing nuclear energy more favorably, given the serious consequences of continued use of fossil fuels for global climate change. Opposition continues, however, because of the perceived risks to both human health and the environment and because of the mismatch to our most critical energy problem: an impending shortage of fuel for transportation. Opponents don't believe that the benefits are worth the risks.

Rebirth of Nuclear Power

If nuclear energy is to have a brighter future, it will be because we have found the continued use of fossil fuels to be so damaging to the atmosphere that we have placed limits on their use, but have not been able to develop adequate alternative energy sources. Nuclear supporters point out that U.S. nuclear power plants prevent the annual release of 164 million tons of carbon, 5.1 million tons of sulfur dioxide, and 2.4 million tons of nitrogen oxides. Also, it appears that the electric car will soon be available from the major car manufacturers. If this catches on, nuclear-generated electricity can indeed be substituted for oil-based fuels.

Observers agree that, if the rebirth of nuclear power is to come, a number of changes will have to be made:

- Reactor safety concerns will have to be addressed, through the use of smaller, advanced light-water reactor designs with built-in passive safety features.

- The potential for terrorist attacks and sabotage will also have to be addressed. The public must be convinced that every possible preventative measure has been taken.

- The industry's manufacturing philosophy will have to change to favor standard designs and factory production of the smaller reactors instead of the custom-built reactors presently in use in the United States. Recently, the DOE selected three designs for development, one of which—the Westinghouse advanced pressurized-water reactor—is a 600-MW unit similar to the one pictured in Figure 15–14. The unit, designed for modular construction, is based primarily on passive safety.

- The framework for licensing and monitoring reactors must be streamlined, but without sacrificing safety concerns. This has largely been accomplished with the Energy Policy Act of 2005. The rules allow the NRC to approve a plant design prior to the selection of a specific site.

- The waste dilemma must be resolved. With the Yucca Mountain waste repository now in limbo, resolving the dilemma is becoming more difficult.

- Political leadership will be required to accomplish all these developments, and this is not a given now that Barack Obama, who has questions about the nuclear power option, has been elected president.

Nuclear Energy Policies. President George W. Bush made expanding nuclear energy a major component of his energy policy by taking the following steps:

- Proposing the *Nuclear Power 2010* program, with plans to identify new sites for nuclear power plants, to streamline regulations associated with site licensing, and to encourage industry to build new plants that could become operational by 2010;

- Moving toward a resolution of the nuclear waste problem by approving the Yucca Mountain site for a repository; and

- Signing into law the **Energy Policy Act of 2005**, which included a wide range of incentives to enable the Nuclear Power 2010 program: a tax credit for the first 6,000

MW of new nuclear capacity for the first eight years of operation, insurance to protect companies building new reactors from the risk of regulatory delays, and federal loan guarantees for up to 80% of construction costs of new power plants. The act also authorizes $1.25 billion toward the construction of a special nuclear plant that would produce hydrogen from water, a key part of a coming fuel-cell technology for automobiles (see Chapter 16). These incentives amount to $13 billion.

The newer **Energy Independence and Security Act of 2007** addressed nuclear power only tangentially; the legislation had a provision to promote American-built nuclear power plants in overseas markets.

The new Energy Secretary, Steven Chu, is a Nobel laureate atomic physicist who considers nuclear energy to be "safe and reliable." His 2010 budget reflects increased funding for nuclear research and development. Although no new plants will be built by 2010, it begins to look like nuclear power in the United States may rise like a phoenix from the ashes of its recent decline.

Revisiting the Themes

Sound Science

The science behind nuclear power was developed during World War II and was used to produce the two atomic bombs that ended the war with Japan in 1945. Nuclear physicists were eager to show that atomic energy could be put to beneficial uses, however, and with that in mind, the technology for nuclear power plants was developed. Sound science was, and continues to be, essential for this enterprise, especially in regard to improvements in safety and efficiency. Nevertheless, scientific uncertainties remain: For example, what is a safe level of exposure to radioactivity? How can nuclear wastes be stored safely for thousands of years to come? Can nuclear reactors be designed to be accident proof? Can fusion energy ever become a commercial reality? Answers to these questions will challenge scientists for years to come.

Sustainability

Looking toward a sustainable energy future, many believe that nuclear power has a role to play in getting us there. Because uranium must be mined and is not an abundant mineral, uranium supplies will not last indefinitely. Reprocessing fuel can extend the life of nuclear power, but it carries risks that may be unacceptable. Still, nuclear fission may be a way of bridging the gap between fossil fuel energy and renewable energy. The most important aspect of fission is that nuclear energy does not release greenhouse gases and would be capable of providing most of our electrical power, thus enabling us to avoid relying on coal.

The current dilemma is how to deal with the buildup of the highly radioactive spent-fuel wastes. Storing them in pools and casks on the grounds of the power plants is unsustainable. Consequently, countries using nuclear power must find ways to dispose of the wastes safely, which means making sure that they are contained for thousands of years.

The prospect of fusion energy continues to be intriguing, as fuel sources are abundant. The technology is extremely daunting, however, and expensive. At present, fusion looks like a dubious option.

Stewardship

It was a major political decision to develop nuclear power for generating electricity. The technology is so expensive and fraught with risk that only government-level agencies can manage the development and ongoing regulation of nuclear power. The roles of the NRC and the EPA are crucial in protecting the U.S. public from the hazards and risks of keeping nuclear power in operation and safely handling its wastes.

Government subsidies and guarantees remain in place for this enterprise, and given its potential for massive consequences if things go wrong, that may continue to be necessary. International agreements, such as those between the United States and Russia to dismantle nuclear warheads, point to the importance of maintaining close political ties among all countries with nuclear weapons capabilities or even with nuclear power plants (since these can be a source of weapons-grade nuclear material).

REVIEW QUESTIONS

1. Compare the outlook regarding the use of nuclear power in the United States and globally in each of the following decades: 1960s, 1980s, and early 21st century.

2. Describe how energy is produced in a nuclear reaction. Distinguish between fission and fusion.

3. How do nuclear reactors and nuclear power plants generate electrical power?

4. What are radioactive emissions, and how are most humans exposed to them?

5. How are radioactive wastes produced, and what are the associated hazards?

6. Describe the two stages of nuclear waste disposal.

7. What problems are associated with the long-term containment of nuclear waste? What is the current status of the disposal situation?

8. Describe what went wrong at Three Mile Island and Chernobyl.

9. What features might make nuclear power plants safer? What about terrorism?

10. What are the four generations of nuclear reactors? Describe the advantages of Generations III and IV.

11. Discuss economic reasons that have caused many utilities to opt for coal-burning plants rather than nuclear-powered plants.

12. How do fast-neutron and fusion reactors work? Does either one offer promise for alleviating our energy shortage?

13. Explain why nuclear power does little to address our largest energy shortfall.

14. Discuss changes in nuclear power that might brighten its future.

THINKING ENVIRONMENTALLY

1. Evaluate the risks of nuclear power. Are we overly concerned or not concerned enough about nuclear accidents? Could an accident like Chernobyl happen in the United States?

2. Would you rather live next door to a coal-burning plant or a nuclear power plant? Defend your choice.

3. Global climate change has been cited by nuclear power proponents as one of the most important justifications for further development of the nuclear power option. Give reasons for their belief, and then cite some reasons that would counter their argument.

4. Go to the website www.pbmr.com and investigate the reactor described there. Where is it being designed, and where will it likely be used first? What are its features?

MAKING A DIFFERENCE

1. Pay a visit to a nuclear power plant, and explore the way the facility (a) is protected against terrorist attack, (b) continually trains the plant's operators, (c) keeps the public informed in the event of a nuclear accident, and (d) handles its high-level wastes.

2. Visit the radiology department of your local hospital, and ask a radiologist to explain the different ways radiation is used in the hospital's medical procedures and how (or whether) the hospital tries to limit the radiation exposure of patients. Ask the radiologist what the hospital does with its low-level radioactive waste.

MasteringENVIRONMENTALSCIENCE™

Go to **www.masteringenvironmentalscience.com** for practice quizzes, Pearson eText, videos, current events, and more.

Offshore wind power. The Scroby Sands wind farm uses 30 two-MW wind turbines made by Vestas, a Danish company. It is located on a sandbank two miles east of Great Yarmouth, on the English coast, and generates power for 41,000 homes. A similar offshore wind farm has been proposed for Nantucket Sound, off the southern Massachusetts coast.

KEY TOPICS

1. **Putting Solar Energy to Work:** An enormous amount of solar energy reaches Earth, and countries are turning this energy into hot water and electricity at a great pace.

2. **Indirect Solar Energy:** Hydropower, wind power, and biomass energy are ways of harvesting solar energy indirectly, as all three are a consequence of the Sun's work.

3. **Renewable Energy for Transportation:** Renewable energy is beginning to meet the need to replace oil for transportation, currently by using biofuels and potentially by developing hydrogen and fuel-cell technologies.

4. **Additional Renewable-Energy Options:** Geothermal energy is poised to expand its share of electrical power; tidal power and wave power may be options for the future.

5. **Policies for a Sustainable-Energy Future:** A sustainable-energy future must see renewable energy replacing fossil fuel energy; our national energy policies now encourage both renewable energy and energy conservation more than ever before.

NANTUCKET SOUND, a shallow body of water south of Cape Cod, rarely sees a windless day. Indeed, coastal locations are well known for their constant and often cooling winds, attracting vacationers and year-round residents who also enjoy the beaches and scenery. Summer populations in Massachusetts towns on Cape Cod swell to many times their winter numbers, putting stress on the area's energy infrastructure. So when Cape Wind Associates proposed an offshore wind farm in Nantucket Sound that could supply 75% of Cape Cod's electricity needs, the company expected a warm welcome. Instead, its proposal ignited a firestorm of controversy. The proposed wind farm is not small. Some 130 turbines, each 417 feet tall, would be spread over 24 square miles of Nantucket Sound and generate 470 megawatts (MW) of power. Concerns range from "visual pollution" to threats to the area's tourism, navigation, fishing industry, and migrating birds. Organizations like the Alliance to Protect Nantucket Sound filed numerous lawsuits during the seven years since Cape Wind first proposed the wind farm, and all were eventually dismissed. Proponents of the project, however, view these concerns as a classic case of NIMBY (Not in My Back Yard) and point out that New England desperately needs the renewable energy of offshore winds. Opinion polls on and off Cape Cod strongly support the project. To date, every state and federal

Figure 16–1 Rooftop hot water. In warm climates, solar hot-water heaters are becoming commonplace. Note the solar water-heating panels on each of the homes in this development in southern California.

agency examining the project has given its approval. The U.S. Mineral Management Service (MMS), which has authority over all offshore developments in federal waters, issued a favorable environmental impact statement in January 2009 and is expected to give the project its lease in 2009. Power could be coming to Cape Cod and Nantucket Island from the wind farm as early as 2011.

Aware of the concern for visual pollution, Garden State Offshore Energy has received the rights for construction of a wind farm 16 to 20 miles off the coast of New Jersey. Delaware, Maryland, Rhode Island, Virginia, New York, and Texas have also received proposals for new offshore wind turbines. Belgium, France, Sweden, and the United Kingdom are developing offshore wind farms in the 100- to 300-MW range.

In the hills east of San Francisco, regiments of wind turbines standing in rows up the slopes and over the crests of the hills are producing electrical power equivalent to that produced by a large coal-fired power plant. At the end of 2008, wind power in the United States generated 23,000 MW, enough to power over 6.5 million homes. Every year installations of wind turbines set new records. Wind turbines are also becoming commonplace across northern Europe and are sprouting up as well in India, Mexico, Argentina, New Zealand, and other countries around the world.

SOLAR, TOO. Panels of solar cells are also providing electricity in both developed and developing countries around the world. Throughout Israel and other countries in warm climates, it is now commonplace to have water heated by the Sun (Fig. 16–1). Even in temperate climates, many people have discovered that proper building design, insulation, and simple solar collection devices can reduce energy bills 75% or more. In the desert northeast of Los Angeles are "farms" with rows of trough-shaped mirrors tipped toward the Sun (Fig. 16–2).

Figure 16–2 Solar thermal power in southern California. A solar-trough power plant. Sunlight striking the parabolic-shaped mirrors is reflected onto the central pipe, where it heats a fluid that is used in turn to boil water and drive turbogenerators.

These reflectors are focusing the Sun's rays to boil water or synthetic oil and drive turbogenerators.

The inertia of "business as usual" has kept the world on a track of growing dependence on fossil fuels, with nuclear power retaining its share of energy production for the present. In the meantime, as the preceding examples demonstrate, the use of energy from the Sun and wind has been making remarkable progress, to the point where these renewable sources of energy are becoming cost competitive with traditional energy sources—and are far more practical in many situations. Given the unsustainability of our present course—due to the limited supplies of fossil fuels and their polluting impacts on global climate and the air we breathe—a renewable-energy future is absolutely essential.

The objective of this chapter is to give you a greater understanding of the potential for sustainable energy from sunlight, wind, biomass, and other sources. In 2007,

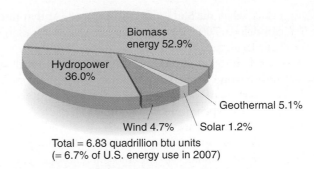

Figure 16–3 Renewable-energy use in the United States.
A mix of sources of renewable energy provided 6.7% of the nation's energy used in 2007. (*Source:* Data from Department of Energy, Energy Information Agency, *Renewable Energy Annual,* May 2009.)

renewable energy provided 6.7% of U.S. primary energy (Fig. 16–3); worldwide, renewable energy supplies 18% of energy use.

16.1 Putting Solar Energy to Work

Before turning our attention to the practical ways that solar energy is captured and used, let us consider some general concepts of solar energy.

Solar energy originates with thermonuclear fusion in the Sun. (Importantly, all the chemical and radioactive polluting by-products of the reactions remain behind on the Sun.) The solar energy reaching Earth is radiant energy, entering at the top of the atmosphere at 1,366 watts per square meter, the **solar constant**. This energy ranges from ultraviolet light to visible light and infrared light (heat energy) (Fig. 16–4). About half of this energy actually makes it to Earth's surface, 30% is reflected, and 20% is absorbed by the atmosphere. Thus, full sunlight can deliver about 700 watts per square meter to

Figure 16–4 The solar-energy spectrum.
Equal amounts of energy are found in the visible-light and the infrared regions of the spectrum. In the Sun, the nuclear fusion of hydrogen to form helium emits a spectrum of radiant energy that reaches Earth.

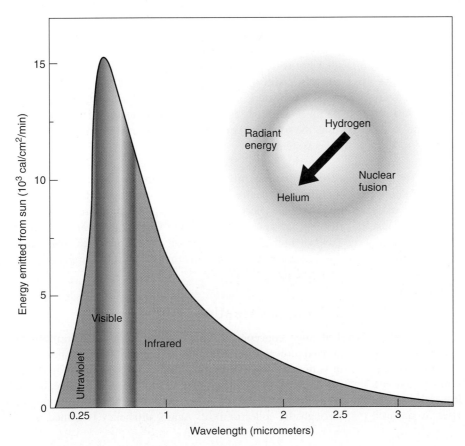

Earth's surface when the Sun is directly overhead. At that rate, the Sun can deliver 700 MW of power (the output of a large power plant) to an area of 390 square miles (1,000 km²).

The total amount of solar energy reaching Earth is vast—almost beyond belief. Just 40 minutes of sunlight striking the land surface of the United States yields the equivalent energy of a year's expenditure of fossil fuel. The Sun delivers 10,000 times the energy used by humans.

Moreover, using some of this solar energy will not change the basic energy balance of the biosphere. Solar energy absorbed by water or land surfaces is converted to heat energy and eventually lost to outer space. Even the fraction that is absorbed by vegetation and used in photosynthesis is ultimately given off again in the form of heat energy as various consumers break down food (Chapter 3). Similarly, if humans were to capture and obtain useful work from solar energy, it would still ultimately be converted to heat and lost in accordance with the Second Law of Thermodynamics (see Chapter 3). The overall energy balance would not change.

Although solar energy is an abundant source, it is also diffuse (widely scattered), varying with the season, latitude, and atmospheric conditions. The main problem associated with using solar energy is one of taking a diffuse and intermittent source and concentrating it into an amount and form, such as fuel or electricity, that can be used to provide heat and to run vehicles, appliances, and other machinery. In addition, what do you do when the Sun is not shining? These problems involve the *collection*, *conversion*, and *storage* of solar energy. Also, in the final analysis, overcoming such hurdles must be cost effective.

Consider how natural ecosystems address these three problems. Plant leaves are excellent solar *collectors*, gathering light over a wide area. With the aid of chlorophyll, photosynthesis *converts* and *stores* light energy as chemical energy—namely, starch and other forms of organic matter—in the plant. The plant biomass then "fuels" the rest of the ecosystem. In the sections that follow, you will see how we can overcome or, even better, sidestep these three problems in various ways to meet our energy needs in a cost-effective manner.

Solar Heating of Water

Solar hot-water heating is already popular in warm, sunny climates. A solar collector for heating water consists of a thin, broad box with a glass or clear plastic top and a black bottom in which water tubes are embedded (Fig. 16–5). Such collectors are called **flat-plate collectors**. Faced toward the Sun, the black bottom gets hot as it absorbs sunlight—similar to how black pavement heats up in sunlight—and the clear cover prevents the heat from escaping, as in a greenhouse. Water circulating through the tubes is thus heated and is conveyed to a tank where it is stored.

In an *active system*, the heated water is moved by means of a pump. In a *passive system*, natural convection currents are used. In a passive solar water-heating system, the system must be mounted so that the collector is lower than the tank. Thus, heated water from the collector rises by natural

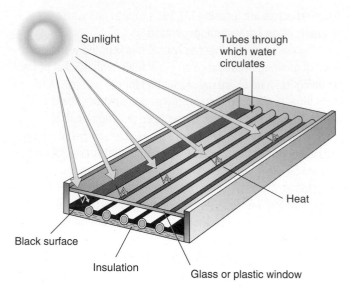

Figure 16–5 The principle of a flat-plate solar collector.
As sunlight is absorbed by a black surface, it is converted to heat. A clear glass or plastic window over the surface allows the sunlight to enter the collector, which traps the heat. Air or water is heated as it passes over and through tubes embedded in the black surface.

convection into the tank, while cooler water from the tank descends into the collector (Fig. 16–6). The tank will usually have a source of auxiliary heat (electric or gas) in order to get the temperature to a desired level or to provide heat when solar energy is insufficient. There are no pumps to buy, maintain, or remember to turn on and off.

In temperate climates, where water in the system might freeze, the system may be adapted to include a heat-exchange coil within the hot-water tank. Then antifreeze fluid is circulated between the collector and the tank. In the United States, more than 1 million residential and 200,000 commercial solar hot-water systems have been installed (see Figure 16–1), but this is still only a small fraction of the total number of hot-water heaters. The reason is the initial cost, which is 5 to 10 times as much as gas or electric heaters. However, over time, the solar system costs less to operate than an electric or a gas system.

In the United States, solar thermal systems have been used most commonly to heat swimming pools. Systems used for this purpose are inexpensive and consist of black plastic or rubberlike sheets with plastic tubing that circulates water through the system internals. In China, however, more than 10% of households are getting their hot water from solar thermal systems; China leads the world in this application of solar energy.

Solar Space Heating

The same concept for heating water with the Sun can be applied to heat spaces. Flat-plate collectors such as those used in water heating can be used for space heating. Indeed, the collectors for space heating may even be less expensive, home-made devices because it is necessary only to have air circulate through the collector box. Again, efficiency is gained if the

Figure 16–6 Solar water heaters. In nonfreezing climates, simple water-convection systems may suffice. In freezing climates, an antifreeze fluid is circulated. Solar heat is augmented by an auxiliary heat source in the hot-water tank.

(a)   (b)

Figure 16–7 Passive hot-air solar heating. (a) Many homeowners could save on fuel bills by adding homemade solar collectors such as that shown here. (b) Air heated in the collector moves into the room by passive convection.

collectors are mounted to allow natural convection to circulate the heated air into the space to be heated (Fig. 16–7).

Building = Collector. The greatest efficiency in solar space heating, however, is gained by designing a building to act as its own collector. The basic principle is to have windows facing the Sun. In the winter, because of the Sun's angle of incidence, sunlight can come in and heat the interior of the building (Fig. 16–8a). At night, insulated drapes or shades can be pulled down to trap the heat inside. The well-insulated building, with appropriately made doors and windows, would act as its own best heat-storage unit. Beyond good insulation, other systems for storing heat, such as tanks of water or masses

of rocks, have not proved to be cost effective. Excessive heat load in the summer can be avoided by using an awning or overhang to shield the window from the high summer Sun (Fig. 16–8b).

Along with design, positioning, and improved insulation, appropriate landscaping can contribute to the heating and cooling efficiency of both solar and nonsolar space-heating designs, thereby conserving energy. In particular, deciduous trees or vines on the sunny side of a building can block much of the excessive summer heat, while letting the desired winter sunlight pass through. An evergreen hedge on the shady side can provide protection from the cold.

Figure 16–8 Solar building siting.
In contrast to utilizing expensive and complex active solar systems, solar heating *can* be achieved by suitable architecture and orientation of the home at little or no additional cost. (a) The fundamental feature is large, Sun-facing windows that permit sunlight to enter during the winter months. Insulating drapes or shades are drawn to hold in the heat when the Sun is not shining. (b) Suitable overhangs, awnings, and deciduous plantings prevent excessive heating in the summer.

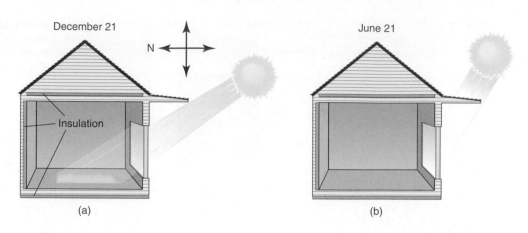

December 21   June 21

(a)   (b)

Earth-Sheltered Housing. One approach that combines insulation and solar heating is **earth-sheltered housing** (Fig. 16–9). The principle is to use earth as a form of insulation and orient the building for passive solar energy. Two strategies are employed. One is to build up earth against the building's walls (earth berms) in a more conventional house design; the other is to cover the building to varying degrees with earth, exposing the interior to the outside with windows facing southward. Earth has a high capacity for heat storage, and since the walls and floors of an earth-sheltered home are usually built with poured concrete or masonry, at night the building will radiate the heat stored during the day and warm the house. In the summer, the building is kept cool by its contact with earth, although moisture often has to be controlled with dehumidifiers.

Energy Stars. In almost any climate, a well-designed passive solar-energy building can reduce energy bills substantially with an added construction cost of only 5% to 10%. Since about 25% of our primary energy is used for space and water heating, proper solar design, broadly adopted, would create enormous savings on oil, natural gas, and electrical power. In 2001, the Environmental Protection Agency (EPA)

Figure 16–9 Earth Hall. This Earth-sheltered building at Au Sable Institute of Environmental Studies in Mancelona, Michigan, houses offices, classrooms, and laboratories.

extended its **Energy Star** program to buildings and began awarding the Energy Star label to public and corporate edifices. Buildings awarded the label use at least 40% less energy than others in their class and must meet a number of other criteria to qualify. By 2008, almost 4,100 buildings had earned the label, saving $1.5 billion annually in lower energy bills and reducing carbon dioxide emissions by 25 billion pounds—a major contribution to **energy conservation**.

A common criticism of solar heating is that a backup heating system is still required for periods of inclement weather. Good insulation is a major part of the answer to this criticism. People with well-insulated solar homes find that they have minimal need for backup heating. When backup is needed, a small wood stove or gas heater suffices. In any case, the criticism concerning the need for backup heating misses the point: The objective of solar heating is to reduce our dependency on conventional fuels. Even if solar heating and improved insulation reduced the demand for conventional fuels by a mere 20%, that would still represent sustainable savings of 20% of the traditional fuel and its economic and environmental costs.

Solar Production of Electricity

Solar energy can also be used to produce electrical power, thus providing an alternative to coal and nuclear power. Currently, a few methods are feasible, and two in particular, *photovoltaic cells* and *solar-trough collectors*, are proving to be economically viable.

Photovoltaic Cells. A solar cell—more properly called a **photovoltaic, or PV, cell**—looks like a simple wafer of material with one wire attached to the top and one to the bottom (Fig. 16–10). As sunlight shines on the wafer, it puts out an amount of electrical current roughly equivalent to that emitted by a flashlight battery. Thus, PV cells collect light and convert it to electrical power in one step. The cells, measuring about four inches square, produce about one watt of power. Some 40 PV cells can be linked to form a module that generates enough energy to light a lightbulb. Varying amounts of power can be produced by wiring modules together in panels.

How They Work. The simple appearance of PV cells belies a highly sophisticated level of science and technology.

Figure 16–10 Photovoltaic cell. Converting light to electrical energy, this cell provides enough energy to run the small electric motor needed to turn the fan blades.

Each cell consists of two very thin layers of semiconductor material separated by a junction layer. The lower layer has atoms with single electrons in their outer orbital that are easily lost. The upper layer has atoms lacking electrons in their outer orbital; these atoms readily gain electrons. The kinetic energy of light photons striking the two-layer "sandwich" dislodges electrons from the lower layer, creating a current that can flow through a motor or some other electrical device back to the upper side. Thus, with no moving parts, solar cells convert light energy directly to electrical power with an efficiency of 15–30%.

Because they have no moving parts, solar cells do not wear out. However, deterioration due to exposure to the weather limits their life span to about 20 years. The major material used in PV cells is silicon, one of the most abundant elements on Earth, so there is little danger that the production of PV cells will ever suffer because of limited resources. The cost of these cells lies mainly in their sophisticated design and construction.

Uses. PV cells are already in common use in pocket calculators, watches, and numerous toys. Panels of PV cells provide power for rural homes, irrigation pumps, traffic signals, radio transmitters, lighthouses, offshore oil-drilling platforms, Earth-orbiting satellites, and other installations that are distant from power lines. It is not hard to imagine a future in which every home and building has its own source of pollution-free, sustainable electrical power from an array of PV panels on the roof. Current installations in many countries and in some 42 states in the United States involve *net metering*, where the rooftop electrical output is subtracted from the customer's use of power from the power grid.

Cost. The cost of PV power (cents per kilowatt-hour) is the cost of the PV cells divided by the total amount of power they may be expected to produce over their lifetime (currently around 25 cents per kilowatt-hour). This cost must be compared with that of other power alternatives (6–12 cents per kilowatt-hour for residential electricity). The first PV cells had a cost factor several hundred times that of electricity from conventional power stations transmitted through the power grid, so those cells were used mainly in areas far from the grid. PV power had its first significant application in the 1950s, in the solar panels of space satellites. This application started the development cycle rolling. As more-efficient cells and less-expensive production techniques evolved, costs came down dramatically. In turn, applications, sales, and potential markets expanded, creating the incentive for further development. In response to this, it is the fastest-growing energy technology industry in the world. In 2008, the industry shipped solar panels with a total capacity of 6,900 MW; existing solar PV panels now have the capacity to turn sunlight into more than 16,000 MW (Fig. 16–11).

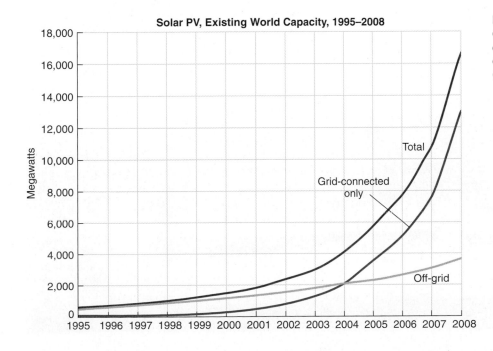

Solar PV, Existing World Capacity, 1995–2008

Figure 16–11 The market for PV panels. Global sales of PV panels have increased dramatically, exceeding 16,000 MW by 2009. Over three-fourths of the PV panels are connected to the electric grid. (*Source: Renewables 2009 Global Status Report.*)

Figure 16–12 PV system inverter. Inverter (on right) and controllers for a PV system at Channel Islands National Park, California. The inverter is a device that connects the PV panels of a rooftop system to the electric grid or devices powered by PV electricity. It converts the direct current coming from PV panels to alternating current and also acts as a control for the system.

Inverters. The most complicated and yet necessary component of a PV system is the **inverter**. The inverter (Fig. 16–12) acts as an interface between the solar PV modules and the electric grid or batteries. It must change the incoming direct current (DC) from the PV modules to alternating current (AC) compatible with the electricity coming from the grid and/or the devices that will be powered by the PV system. It must also act as a control, able to detect and respond to fluctuations in voltage or current on either DC or AC, and it must be robust enough to withstand the high temperatures of an attic (where rooftop PV systems are involved). An inverter for a rooftop system will cost several thousand dollars, but in most cases, these are also eligible for local and national subsidies. Like the cost of PV cells, the cost of inverters has also come down over time.

Utilities. The utility companies are beginning to move toward large-scale PV installations. At the end of 2008, there were 53 plants worldwide with capacities greater than 10 MW. The largest one in the United States is the 14-MW Nellis Air Force Base plant in Nevada (Fig. 16–13). Sunny Spain is building PV power plants at a furious pace; in 2008 alone, Spain erected 194 facilities with a capacity of 2 or more MW, including the world's largest solar PV power plant, with a capacity of 60 MW, at Olmadilla.

However, the most promising future for PV power may be in the installation of PV panels on rooftops, where a huge amount of unused space is readily available. California's Edison International has installed the first of 150 solar projects on warehouse and factory roofs, a 2-MW capacity array of solar panels on a distribution center in Fontana; the company anticipates spending $875 million to provide 250 MW of electricity on commercial rooftops. For residential use, utilities in California, New York, New Jersey, and Connecticut have established programs that provide incentives to customers to install 2- to 4-kilowatt (kW) PV systems on their

Figure 16–13 PV power plant. The Nellis Air Force Base plant in Nevada consists of 72,416 solar panels, each generating 200 watts. The panels are built on trackers that follow the Sun as it moves across the sky. Built in 26 weeks, the plant covers 140 acres and provides 25% of the energy consumed at the base, where 12,000 people live and work.

roofs. A 2-kW system provides about half of the annual energy needs of a residence, and in case the homeowner is generating more than needed, the utility buys the excess electricity at the retail rate. Los Angeles has a goal of 100,000 solar PV systems on rooftops by 2010. A federal investment tax credit of 30% of the system cost was enacted in the Emergency Economic Stabilization Act of 2008, providing further incentives to homeowners to go solar. A similar program in Japan has led to the installation of more than 1,500 MW of solar PV power—over 400,000 separate installations.

New PV Technologies. To compete with electricity from conventional sources, the cost of solar cells needs to keep coming down. At least three new technologies may accomplish this: (1) thin-film PV cells, in which cheap amorphous silicon or cadmium telluride is used instead of expensive silicon crystals and the film can be applied as a coating on roofing tiles or glass; (2) silicon crystals that are sliced endwise to make "SLIVERs," long, very thin slices that produce photocells with only one-tenth the silicon used in normal cells; and (3) glass coated with light-absorbing dyes that transmit energy to solar cells on the edge of the glass, while allowing much of the light to pass through to conventional solar panels. All of these techniques are either in production or close to it.

Concentrated Solar Power (CSP).

With government funding, several technologies have been developed that convert solar energy into electricity by using reflectors (or concentrators) such as mirrors to focus concentrated sunlight onto a receiver that transfers the heat to a conventional turbogenerator. These devices work well only in regions with abundant sunlight, but they are moving into commercial production for use in the U.S. Southwest and in Spain.

Solar Trough. One method that is proving to be cost effective is the **solar-trough** collector system, so named because the collectors are long, trough-shaped reflectors tilted toward the Sun (Fig. 16–2). The curvature of the trough is such that all of the sunlight hitting the collector is reflected onto a pipe running down the center of the system. Oil or some other heat-absorbing fluid circulating through the pipe is thus heated to very high temperatures. The heated fluid is passed through a heat exchanger to boil water and produce steam for driving a turbogenerator.

Nine solar-trough facilities in the Mojave Desert of California have been connected to the Southern California Edison utility grid since the early 1990s. With a combined capacity of 350 MW, these facilities have about one-third the capacity of a large nuclear power plant and are converting a remarkable 22% of incoming sunlight to electrical power at a cost of 12 cents per kilowatt-hour, not much more than the cost of power from coal-fired facilities. Solar-trough plants have experienced a renaissance in recent years, with several new plants completed by 2007 in several countries and many more in the planning or construction phase. One advantage of this technology is the capability of storing some of the thermal energy for release at night.

Power Tower. A *power tower* is an array of Sun-tracking mirrors that focuses the sunlight falling on several acres of land onto a receiver mounted on a tower in the center of the

Figure 16–14 Power tower Solar Two. Sun-tracking mirrors are used to focus a broad area of sunlight onto a molten-salt receiver mounted on the tower in the center. The hot salt is stored or pumped through a steam generator that drives a conventional turbogenerator.

area (Fig. 16–14). The receiver transfers the heat energy collected to a molten-salt liquid, which then flows either to a heat exchanger to drive a conventional turbogenerator or to a tank at the bottom of the tower to store the heat for later use. The latest pilot plant model of this technology, Solar Two, is now connected to southern California's utility grid and is generating 10 MW of electricity, enough for 10,000 homes.

Spain has CSP projects in the pipeline for some 1,000 MW, both solar-trough and power tower systems. A royal decree and government subsidies have paved the way for Spain to become a world leader in commercial CSP development (adding substantially to its rising investment in PV power plants).

Dish-Engine System. The dish-engine system (Fig. 16–15) is a smaller system, consisting of a set of parabolic concentrator dishes that focus sunlight onto a receiver. Fluid (often molten sodium) in the receiver is transferred to an engine (a Stirling engine) that generates electricity directly. Dish-engine systems have demonstrated efficiencies of 30%, higher than those of any other solar technology. They generate from 5 to 40 kW, and any number of them can be linked together to provide larger amounts of electricity. They are well suited for off-grid applications to provide power to remote, sunny areas. Stirling has worked with the Sandia National Laboratories to build an array of six dish-engine systems in Arizona, mainly as a demonstration of the potential for this technology to provide electricity as units are connected together to make a power plant.

The Future of Solar Energy.

Solar electricity is growing at a phenomenal rate—the solar PV business is a $20 billion industry, and it is growing at a rate of 40% a year (see Figure 16–11). The industry is hard-pressed to keep up with current demand. Solar energy does have certain disadvantages, however. First, the available technologies are still more expensive than conventional energy sources, although subsidies are helping to level the playing field. Second, solar energy works only during the day, so

Figure 16–15 Solar dish-engine system. An array of circular mirrors focuses solar energy on a receiver, which transmits the energy to an engine that generates electric power.

it requires a backup energy source, a storage battery, or thermal storage for nighttime power. Also, the climate is not sunny enough to use solar energy in the winter in many parts of the world. Against these drawbacks should be weighed the hidden costs that are not included in the cost of power from traditional sources: air pollution, strip mining, greenhouse gas emissions, and nuclear waste disposal, among others.

Matching Demand. The Sun provides power only during the day, but 70% of electrical demand occurs during daytime hours, when industries, offices, and stores are in operation. Thus, considerable savings still can be achieved by using solar panels just for daytime needs and continuing to rely on conventional sources at night. All the solar power that can conceivably be built in the next 20 years or more could be used to supply the daytime demand, leaving traditional fuels to satisfy nighttime demand. In particular, the demand for air-conditioning, which, after refrigeration, is the second largest power consumer, is well matched to energy from PV cells. In the long run, the nighttime load might be carried by forms of indirect solar energy, such as wind power and hydropower (discussed later in the chapter).

According to Table 14–1 (see Chapter 14), about 68% of our electrical power (56% worldwide) is currently generated by coal-burning and nuclear power plants. Therefore, the development of solar and wind electrical power can be seen as gradually reducing the need for coal and nuclear power. Unlike new nuclear or coal power plants, which take years to plan and build, solar or wind facilities can become operational within months of the decision to build them, they can be installed in small increments, and additional modules can be added as demand requires. The utility does not have to guess what power demands will be in 10 or 15 years. With solar or wind power, a utility can add capacity as it is needed, make relatively small investments at any given time, and have those investments start paying back almost immediately. Thus, this approach involves much less financial risk both for the utility and for its consumers.

One further advantage of solar and wind facilities is their relative invulnerability to terrorist attack, compared with oil and gas pipelines and large power plants, especially nuclear facilities. Further, renewables produce no hazardous wastes and are geographically dispersed.

Solar power has also proved suitable for meeting the needs of electrifying villages and towns in the developing world, where, because centralized power is unavailable, rural electrification projects based on PV cells are already beginning to spread throughout the land. (See Sustainability, "Transfer of Energy Technology to the Developing World.")

Sustainability

Transfer of Energy Technology to the Developing World

The nations of the industrialized Northern Hemisphere have achieved their level of development by using energy technologies based largely on fossil fuels (and, to a lesser extent, nuclear energy). These nations' development took place during a time when fossil fuels were inexpensive. Only as those fuels became more expensive did the nations of the North begin to get serious about technologies that would make energy use more efficient. The specter of global climate change has brought a new perspective to the need to wean the industrialized North away from fossil fuel energy in the 21st century.

The same traditional fossil fuel–based technologies have been adopted by the developing nations of the Southern Hemisphere, except that these nations are lagging behind in their ability to implement the technology on a large scale. As discussed in Chapter 14, however, the fossil fuel–based energy pattern of the 20th century will have to be replaced largely with renewable-energy systems in the 21st century. We have only one or two generations remaining for the dominance of fossil fuel energy. If the United States and other developed nations are willing to play a significant role in helping the developing nations, it is questionable whether the same developmental path the North took should be promoted. Instead, there is the opportunity to engage in some "leapfrogging" technology transfer: The North can put its development dollars into such technologies as photovoltaics, wind turbines, and efficient public transport systems for the cities.

The solar route is especially attractive for many of the climates in the developing world, where an estimated 1.6 billion people lack electricity. For example, some 200,000 PV systems installed in Kenya over a 25-year period now provide power to 5% of the rural Kenyan population. The systems, with costs ranging from $300 to $1,500, are successfully marketed to people with incomes averaging less than $100 per month. Only the solar panels are imported; local companies provide the batteries and other system components. People use the energy for lighting, television, and radio. Annual sales in Kenya exceed 30,000 systems, a sign that the technology is in great demand and is fostering appropriate development in the area. Credit and financing extended to both companies and users have been key to this success. Some of the financing originates with development agencies from the North, but the majority is Kenyan. Building the infrastructure to market and service the PV systems has also been an essential part of the success story, and that can be accomplished by a relatively small number of organizations and individuals. These solar home systems are being added in many other developing countries, especially China, Bangladesh, India, Mexico, and Morocco.

The cost of the alternative—putting roughly 1.6 billion people who are still without electricity on the **power grid** (and thus requiring central power plants, high-voltage transmission lines, poles, wires, and transformers)—is unimaginable!

Solar electric power has a bright future. Costs of solar PV systems are declining by about 5% per year, as efficiencies of PV cells improve and manufacturers achieve economies of scale with the rising demand. Homeowners are ready to spend $10,000 or more to install rooftop PV systems, especially when subsidies are now available from the federal government and many states. The technology is now mature, and many small installation companies are competing for home solar PV installations.

Of course, solar power is only one of a number of renewable forms of energy. Some of these other sources are derived from solar energy and have been around for a long time.

16.2 Indirect Solar Energy

Water, fire, and wind have provided energy for humans throughout history. Dams, firewood, windmills, and sails represent indirect solar energy because energy from the Sun is the driving force behind each. (Recall in Chapter 10 how solar energy powers the hydrologic cycle of evaporation and precipitation and also drives the convection currents that produce the wind). What is the potential for expanding these options from the past into major sources of sustainable energy for the future?

Hydropower

Early in technological history, it was discovered that the force of falling water could be used to turn paddle wheels, which in turn would drive machinery to grind grain, saw logs into lumber, and do other laborious tasks. The modern culmination of this use of water power, or hydropower, is huge hydroelectric dams, where water under high pressure flows through channels, driving turbogenerators (Fig. 16–16). The amount of power generated is proportional to both the height of the water behind the dam—that is what provides the pressure—and the volume of water that flows through. Chapter 10 describes the role played by dams in trapping and controlling rivers for flood control, irrigation, and hydropower.

About 6% of the electrical power generated in the United States currently comes from hydroelectric dams, most of it from about 300 large dams concentrated in the Northwest and Southeast. Worldwide, hydroelectric dams have a generating capacity of 840,000 MW, with more very large projects in the construction or planning stage. Hydropower generates 19% of electrical power throughout the world and is by far the most common form of renewable energy in use.

Trade-offs. Dams—in particular, the larger ones—are often controversial; they provide important benefits, but also involve some serious consequences. The most obvious *advantages* are these:

- Hydropower eliminates the cost and environmental effects of fossil fuels and nuclear power. Hydroelectric

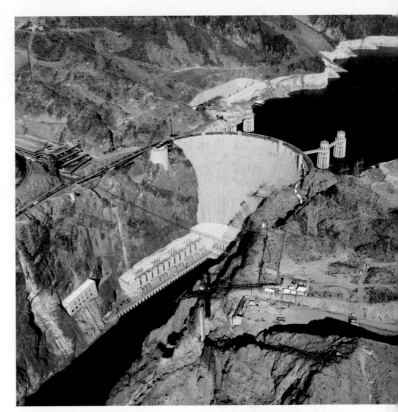

Figure 16–16 Hoover Dam. About 6.7% of the electrical power used in the United States comes from large hydroelectric dams such as this one.

plants have longer lives than fuel-fired ones; some plants have been in service for 100 years.

- The dams provide flood control on many rivers that have taken huge tolls of human lives, and they also provide irrigation water for agriculture. Some 40% of the world's food is generated from the irrigation facilitated by the reservoirs behind dams.

- Reservoirs behind the dams often provide recreational and tourist opportunities.

- Dams can be employed in **pumped-storage power plants;** during the night, when electricity demand is low, the dam's hydrogenerator pumps water up to an elevated reservoir through a tunnel. At times of high demand, the water is returned to the lower reservoir through the tunnel, where it spins turbines and generates electricity.

There are some serious *disadvantages*, however:

- The reservoir created behind the dam inevitably drowns farmland or wildlife habitats and perhaps towns or land of historical, archaeological, or cultural value. Glen Canyon Dam (on the border between Arizona and Utah), for example, drowned one of the world's most spectacular canyons.

- Dams and the large reservoirs created behind them often displace rural populations. In the last 50 years, some 40 to 80 million people have been forced to move to accommodate the rising waters of reservoirs.

- Dams impede or prevent the migration of fish, even when fish ladders are provided. Federal surveys show that fish habitats are suffering in the majority of the nation's rivers because of damming.

- Because the flow of water is regulated according to the need for power, dams wreak havoc downstream. Water may go from near flood levels to virtual dryness and back to flood levels in a single day. Other ecological factors are also affected because sediments with nutrients settle in the reservoir, meaning that smaller-than-normal amounts reach the river's mouth.

More Dams?   Even if such trade-offs were not enough, thoughts of greatly expanding water power in the United States are nullified by the fact that few sites conducive to large dams remain. Already, 75,000 dams six feet high or more dot U.S. rivers, with an estimated 2 million smaller structures. As a result, only 2% of the nation's rivers remain free flowing, and many of these are now protected by the Wild and Scenic Rivers Act of 1968, a law that effectively gives certain scenic rivers the status of national parks. The United States and most developed countries have already brought their hydropower to capacity, and the trend is in the direction of removing many of the dams that impede the natural flow of rivers.

Proposals for new dams outside of the United States are embroiled in controversy over whether the projected benefits justify the ecological and sociological trade-offs. For example, the 50-meter-high Nam Theun 2 Dam, a 1,060-MW project on a tributary of the Mekong River in Laos, is expected to help that country begin a process of economic development. On hold since the mid-1990s, the project received final approval in 2005 from the World Bank and Asian Development Bank for guarantees and loans, and construction is almost completed. The project has displaced 6,200 people and will damage or destroy 174 square miles of sensitive environments as the reservoir fills. In addition, it is basically being built to sell power to Thailand (only 75 MW are to be channeled to Laos at first). International Rivers (IR) is monitoring implementation of the project. IR is concerned that the Laos government is ill equipped to manage a project of this size and is not likely to use the revenue for poverty alleviation in that country. A 2008 report from IR indicated that, while work on the dam is on schedule, the social and environmental mitigation programs are lagging behind.

Dam Report.   Another huge dam, completed in 2006, is the Three Gorges Dam on the Yangtze River in China. (See Figure 10–19, Chapter 10.) The largest dam ever built (1.4 miles long), Three Gorges has displaced some 1.2 million people in order to generate 22,000 MW of electricity and control the flood-prone river. Offsetting the human and economic costs is the fact that it would take more than a dozen large coal-fired power plants to produce the same amount of electricity.

A special World Commission on Dams, convened to examine the impacts and controversies surrounding large dams, reported its findings in 2000. The commission began with the assertion that dams are only a means to an end, the end being *"the sustainable improvement of human welfare."*

Figure 16–17 Horse Hollow Wind Energy Center. Located in Taylor and Nolan counties, Texas, this is the world's largest wind farm. It has a 735-MW capacity, developed by 421 wind turbines spread over 47,000 acres of land. Go to **GRAPHit!** at www.masteringenvironmentalscience.com.

The report found dams to be a mixed blessing, concluding that large dams should be built only if no other options exist.[1] Many guidelines were presented for assessing costs and benefits of specific projects—in particular, for finding ways to incorporate participatory decision-making, sustainability, and accountability in the planning and implementation of projects. Many more dams will be built, largely in the developing countries (especially China, Brazil, and India), and these will certainly play a role in the energy future of those countries.

Wind Power

In 2006, the Horse Hollow Wind Energy Center in Taylor and Nolan counties, Texas (Fig. 16–17), became the world's largest wind farm. The "farm," consisting of 304 wind turbines spread over almost 47,000 acres, is capable of generating 735 MW of power, enough to provide the electricity needs of 260,000 households. In 1999, Texas established a renewable portfolio standard—a mandated target for renewables requiring that 2,000 MW of additional renewable capacity be added within a decade. Also aided by an energy production tax credit (PTC) of 1.9 cents per kilowatt-hour, the Horse Hollow Wind Energy Center and others built since 1992 brought the U.S. wind energy–generating capacity to 24,300 MW in 2008 (Fig. 16–18).

The United States is the world leader in wind energy, with Germany close behind (24,000 MW capacity in 2008). In the past seven years, global wind power capacity has increased at a rate of 28% per year, to 121,000 MW at the start of 2009. The World Wind Energy Association is predicting that global wind power will reach 170,000 MW by the end of 2010. Wind is the second largest component of renewable electricity in the world (behind hydropower) because it has become economically competitive with conventional energy sources, and countries all over the world are

---

[1]*Dams and Development: A New Framework for Decision-Making. The Report of the World Commission on Dams.* Earthscan Publications. 2000.

Cost of Energy and Cumulative Domestic Capacity

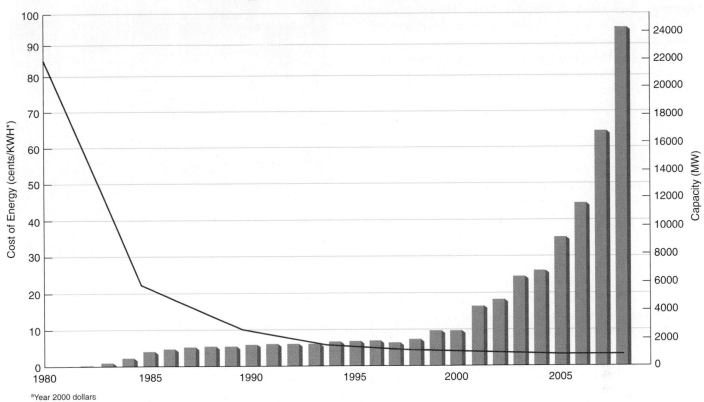

^aYear 2000 dollars

Figure 16–18 Wind energy in the United States. As the cost of wind energy has dropped to the level of being very competitive with conventional energy sources, wind turbine capacity has risen dramatically. (*Data source:* American Wind Energy Association.)

encouraging the installation of wind turbines by setting targets and providing subsidies.

Once a standard farm fixture, windmills fell into disuse as transmission lines brought abundant lower-cost power from central generating plants. Not until the energy crisis and rising energy costs of the 1970s did wind begin to be seriously considered again as a potential source of sustainable energy. Growing rapidly, wind power now supplies 1.5% of global electricity demand, but there are no technical, economic, or resource limitations that would prevent wind power from supplying at least 12% of the world's electricity by 2020, with a generating capacity of 1.2 million MW. In Denmark, which is hard-pressed by a lack of fossil fuel energy, wind now supplies more than 19% of its electricity.

Design. Many different designs of wind machines have been proposed and tested, but the one that has proved most practical is the age-old concept of wind-driven propeller blades. The propeller shaft is geared directly to a generator. (A wind-driven generator is more properly called a **wind turbine** than a windmill.) As the reliability and efficiency of wind turbines have improved, the cost of wind-generated electricity has decreased. **Wind farms**—arrays of up to several thousand such turbines—are now producing pollution-free, sustainable power for less than five cents per kilowatt-hour (with the PTC), a rate that is competitive with the rates of traditional sources. Moreover, the amount of wind that can

be tapped is immense. The American Wind Energy Association calculates that wind farms located throughout the Midwest could meet the electrical needs of the entire country, while the land beneath the turbines could still be used for farming. Hundreds of new turbines are sprouting from midwestern farms, and the farmers are paid handsome royalties ($2,000 to $3,000 per turbine per year) for leasing their land to wind developers. The U.S. Department of Energy (DOE) is promoting a scenario where wind energy will provide 20% of U.S. electricity needs by 2030—an ambitious, but by no means impossible target.

Drawbacks. Still, wind power does have some drawbacks. First, it is an intermittent source, so problems of backup or storage must be considered. Second is an aesthetic consideration: One or two wind turbines can be charming, but a landscape covered with them can be visually unappealing (to some). Third, wind turbines can be a hazard to birds. Locating wind farms on migratory routes or near critical habitats for endangered species such as condors could be a problem for soaring birds in particular. Mortality from wind turbines may be upward of 40,000 per year, but this number pales in comparison to the hundreds of millions of birds killed annually by cats, automotive traffic, and glass windows.

As discussed earlier, plans for offshore wind farms are putting a new spin on wind energy. Denmark already has many offshore turbines, as do the Netherlands, the United

Kingdom (see p. 397), and Sweden. Belgium, France, and Germany are also in the development stage. The primary lure of offshore wind is the dependability and strength of the maritime winds. Siting is perhaps easier, too, as the turbines will be far enough offshore to have much less of a visual impact than onshore sites and land does not have to be purchased. Whether offshore or onshore, the global drive to install wind power is generating electricity that is emission free; in fact, the 26,000 MW of new wind turbines installed in 2008 displaced 76 million tons of carbon dioxide, or emissions from 40 coal-fired power plants.

Wind power and hydropower are two forms of indirect solar energy; another highly important form is biomass energy.

Biomass Energy

Burning firewood for heat is the oldest form of energy that humans have used throughout history. However, there is nothing like a new name to put life into an old concept. Thus, "burning firewood" has become "utilizing biomass energy." *Biomass energy* means "energy derived from present-day photosynthesis." As Figure 16–3 indicates, biomass energy leads hydropower in renewable-energy production in the United States. (Most uses of biomass energy are for heat.)

In addition to burning wood in a stove, major means of producing biomass energy include burning municipal wastepaper and other organic waste, generating methane from the anaerobic digestion of manure and sewage sludges, and producing alcohol from fermenting grains and other starchy materials.

Burning Firewood.   Wherever forests are ample relative to the human population, firewood—or fuelwood, as it is called—can be a sustainable energy resource, and indeed, wood has been a main energy resource over much of human history. Fuelwood is the primary source of energy for heating and cooking for some 2.6 billion people, amounting to about 7% of total energy use from all sources.

In the United States, wood stoves have enjoyed a tremendous resurgence in recent years: About 5 million homes rely entirely on wood for heating, and another 20 million use wood for some heating. The most recent development in wood stoves is the *pellet stove*, a device that burns compressed wood pellets made from wood wastes (Fig. 16–19). The pellets are loaded into a hopper (controlled by computer chips) that feeds fuel when it is needed. The stove requires very little attention and burns efficiently, relying on one fan to draw air into it and another fan to distribute heated air to the room.

Fuelwood Crisis?   Recall from Chapter 7 that two patterns of use determine how forest resources will be exploited: consumptive use and productive use. Most fuelwood use in developing countries fits the *consumptive* pattern: People simply forage for their daily needs in local woodlands and forests. However, gathering wood, converting wood into charcoal, and selling the wood products in urban areas translate into important *productive* use activities for many in developing countries. All these activities have the potential to

Figure 16–19 Pellet stove. Pellet stoves use pelletized wood waste to achieve an efficient burn that requires much less care than burns from conventional woodstoves.

degrade local forests and woodlands. There was serious concern in the later decades of the 20th century about deforestation due to wood gathering, but although there are local shortages, it appears that global use of fuelwood peaked in the late 1990s and is now declining. As development progresses, people have more disposable income and shift up the "energy ladder" from wood to charcoal to fossil fuels. The Millennium Ecosystem Assessment judged that fuelwood demand has not become a significant cause of deforestation, although there are local areas of concern. This could change if the cost of fossil fuels continues to rise.

Burning Wastes.   Facilities that generate electrical power from the burning of municipal wastes (waste-to-energy conversion) are discussed in Chapter 21. Many sawmills and woodworking companies are now burning wood wastes, and a number of sugar refineries are burning cane wastes to supply all or most of their power. In Spain, wastes from olive oil production are powering three electrical facilities, providing over 32 MW of power, enough to supply 100,000 households. Power from these sources may not meet more than a small percentage of a country's total electrical needs, but it represents a productive and relatively inexpensive way to dispose of biological wastes.

Producing Methane.   Chapter 20 examines how the anaerobic digestion of sewage sludge yields *biogas*, which is two-thirds methane, plus a nutrient-rich treated sludge that is a good organic fertilizer. Animal manure can be digested likewise. When the disposal of manure, the production of energy, and the creation of fertilizer can be combined in an efficient cycle, great economic benefit can be achieved. In China, millions of small farmers maintain a simple digester in the form of an underground brick masonry fermentation chamber

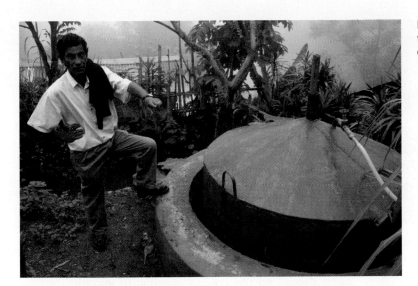

Figure 16–20 Biogas power. Animal wastes are introduced to this unit in rural India and mixed with water. The wastes then decompose under anaerobic conditions, producing biogas.

with a fixed dome on top for storing gas (Fig. 16–20). Agricultural wastes, such as pig manure and cattle dung, are put into the chamber and diluted 1:1 with water, and the anaerobic digestion produces the biogas, which is used for cooking, heating, and lighting. The residue slurry makes an excellent fertilizer. The concept has been expanded throughout the developing world to provide an alternative to fuelwood. Currently, some 25 million households employ these devices. Both India and China provide subsidies for these family-scale biogas plants.

16.3 Renewable Energy for Transportation

Declining oil reserves and the enormous impact of transportation's demands for oil on our economy and on global climate suggest that the most critical need for a sustainable energy future is a new way to fuel our vehicles. Renewable energy is already answering a small, but growing part of this need in the form of biofuels and may have the answer for the long term in the form of hydrogen and fuel cells.

Biofuels

Complex organic matter (primarily plants, but also animal wastes) can be processed to make fuels for vehicles. Currently, two fuels are seeing rapid expansion globally: ethyl alcohol (ethanol) and biodiesel. High oil prices, strong farmer support, government subsidies, and environmental concerns have led to a major expansion of global ethanol and biodiesel production since 2000. Although these two biofuels currently provide only 1% of global transportation fuels, this is just the beginning, as new technologies and government support promise to make biofuels a major competitor for fossil fuels. The current sources of biofuels are agricultural and food commodities.

Ethanol. Ethanol is produced by the fermentation of starches or sugars. The usual starting material is corn, sugarcane, sugar beets, or other grains. Fermentation is the process used in the production of alcoholic beverages. The only new part of the concept is that, instead of drinking the brew, we distill it and put it in our cars. However, production costs make ethanol more expensive than gasoline unless oil sells for more than $55 a barrel. To stimulate the industry in the United States, federal tax credits have been extended to biofuels, which make them even more competitive. *Gasohol*, 10% ethanol and 90% gasoline, has been promoted and marketed in the Midwest since the late 1970s. The enthusiastic support of gasohol in that region is not motivated purely by the desire to save fossil fuel: Gasohol also provides another market for corn crops and thereby improves the economy of the region.

Farm Products. In 2008, 10 billion gallons (240 million barrels) of ethanol for fuel, the equivalent of 160 million barrels of oil, were used in the United States (ethanol contains about two-thirds the energy of regular gasoline, gallon for gallon). One-fourth of the nation's corn crop is dedicated to this use; however, an ethanol factory also produces corn oil and livestock feed. Ethanol plants under construction will increase the production capacity to 12.6 billion gallons a year. A tax credit of 45 cents a gallon helps ethanol compete with gasoline, and a renewable fuel standard (or RFS, which requires a minimum volume of renewable fuel in gasoline) mandated a level of 9 billion gallons of renewable fuel (read, corn-based ethanol) in 2008, to rise to a limit of 15 billion gallons of corn-based ethanol by 2015. The RFS target rises to 36 billion gallons in 2022, but the increase beyond 2015 must be met by using sources other than corn.

Can we expect ethanol production from corn to make serious inroads on the oil import and greenhouse gas emissions problems? Currently, ethanol represents about 5% of U.S. gasoline consumption. This could rise to 20% if the country's entire corn crop were devoted to ethanol production—an impossible assumption, given the importance of corn for animal feed and export. With some 85 million acres devoted to corn harvest in the United States, there is little suitable farmland for significant expansion. And as we saw in Chapter 12, there are serious concerns that biofuel production has contributed significantly to recent rising

Figure 16–21 Miscanthus x giganteus. This hardy perennial grass grows to a height of four meters and is a prime candidate for a second-generation cellulosic biofuel.

prices of food grain, diverting corn from hungry people to ethanol production.

Another issue is the energy efficiency of corn-based ethanol. For corn-based ethanol, fossil fuel energy is required to produce the corn (fertilizer, pesticides, farm machinery), to transport it, and to operate the ethanol plant. A recent analysis[2] has calculated that using ethanol instead of gasoline reduces greenhouse gas emissions, mile for mile, by only 13%. However, some biofuel critics argue that if new land is cleared of its forests or grasslands, the carbon emissions from the land use change should be added to the debit from fossil fuel greenhouse gas emissions. When this is done, corn-based ethanol is a net loser in its impact on climate change. Thus, it could be reasoned that the primary strategy in using corn-based ethanol is to reduce our reliance on imported oil.

Second-Generation Biofuel. A more likely long-run technology for producing ethanol is the use of cellulosic feed-stocks such as agricultural crop residues, grasses (such as switchgrass), logging residues, fast-growing trees, and fuel-woods from forests. Ethanol produced this way is called second-generation biofuel (current starch-based processes produce first-generation biofuel). One crop with exceptional potential is Miscanthus (Fig. 16–21), a giant perennial grass that can be grown in poor soils and that outproduces corn and other grasses.

With the use of enzymes to break down the cellulose to sugars, demonstration facilities have shown that cellulosic feedstock technology can be a cheaper and more energy-efficient process than corn-based production. The technology is now being scaled up to commercial production; at least 11 plants are under construction in the United States, and 1 plant in Louisiana has begun production. Recall that the RFS requires that all of the increase in the target biofuel after 2016 be derived from second-generation biofuel sources. An available tax credit of $1.01 per gallon of cellulosic biofuel is moving this industry forward rapidly.

Brazil has also invested heavily in fuel ethanol; there the feedstock is sugarcane, which is much less costly to farm than corn. Brazil used to lead the world in ethanol production, but has been surpassed by the United States. For years, ethanol has been added at a level of 25% of fuel in Brazil, but recently, the country mandated the production of flex-fuel vehicles that can run on any mixture of alcohol and gasoline. With the lower production costs and the capacity to shift all sugarcane acreage to ethanol production, Brazil is on the road to meeting all of its domestic fuel demand with biofuels.

Air Quality. Besides substituting for gasoline, ethanol is a suitable substitute for methyl tertiary butyl ether (MTBE), a petroleum derivative that has been used as a fuel additive to make the burning of the gasoline cleaner. Both MTBE and ethanol improve air quality when they are made part of **reformulated gasoline**, and a number of metropolitan areas depend on this type of gasoline to improve their air quality in order to meet Clean Air Act requirements. MTBE, which is a carcinogen in animals, has been showing up in water supplies and precipitating legal actions against the oil industry. The Energy Policy Act of 2005 refused to grant the industry the liability protection it wanted, and therefore, it has abandoned MTBE and has turned to ethanol. Billions of gallons of the ethanol additive are involved, which means that ethanol's place in the fuel market is further established.

Biodiesel. Imagine a truck exhaust that smells like french fries. Not only is that an improvement over diesel exhaust, but also it is an indication that the truck is burning biodiesel, made largely from soybean oil. Sales have reached 600 million gallons of the fuel, which is very competitive with petroleum diesel fuel because of a hefty $1-per-gallon production tax credit. Still, this is only 1% of diesel fuel transportation use in the United States. Standard biodiesel is a 20% mix of soybean oil in normal diesel fuel, but practically any natural oil or fat can be mixed with an alcohol to make the fuel; one excellent source is recycled vegetable oil from frying. However, like corn-based ethanol, there are competing uses for the vegetable oils.

There are some offbeat technologies surfacing to produce biodiesel, such as the factory in Missouri that takes in truckloads of turkey wastes from a Butterball turkey-processing plant and turns them into biofuels. The patented process belongs to Changing World Technologies, and its founder, Brian Appel, says that it can handle almost any organic waste imaginable and turn it into useful oil. Currently, the Missouri plant turns a profit, converting 250 tons of turkey waste per day into 20,000 gallons of oil and also selling a fertilizer product from the rendered wastes.

Hydrogen: Highway to the Future?

Over the past decade, there has been increasing interest in powering cars with hydrogen. Conventional cars with internal combustion engines can be adapted to run on hydrogen gas in the same manner as they can be adapted to run on natural gas. (Hydrogen is not really a fuel, but an energy carrier, like electricity; it must be generated using some other energy

[2]Alexander E. Farrell, Richard J. Plevin, Brian T. Turner, Andrew D. Jones, Michael O'Hare, and Daniel M. Kammen,. "Ethanol Can Contribute to Energy and Environmental Goals," *Science* 311 (January 27, 2006): 506–508.

source.) Neither carbon dioxide nor hydrocarbon pollutants are produced during the burning of hydrogen. Indeed, the only major by-product of H_2 combustion is water vapor:

$$2H_2 + O_2 \rightarrow 2H_2O + energy$$

(Some nitrogen oxides will be produced because the burning still uses air, which is nearly four-fifths nitrogen.)

A Good Idea, But. . . . If hydrogen gas is so great, why aren't we using it? The answer is that there is virtually no hydrogen *gas* on Earth. Any hydrogen gas in the atmosphere has long since been ignited by lightning and burned to form water. And although many bacteria in the soil produce hydrogen in fermentation reactions, other bacteria are quick to use the hydrogen because it is an excellent source of energy. Thus, there are abundant amounts of the *element* hydrogen, but it is all combined with oxygen in the form of water (H_2O) or other compounds.

Although hydrogen can be extracted from water via **electrolysis** (the reverse of the preceding reaction), electrolysis requires an *input* of energy. Hydrogen can also be derived chemically from hydrocarbon fuels such as methane, petroleum oil, and methyl alcohol, but these processes are also energy losers; that is, more energy is put into the process than is contained in the hydrogen. Moreover, because these compounds contain carbon, it would be better to use them directly than to obtain hydrogen from them and use it. Therefore, the use of hydrogen as a truly clean energy carrier must await an energy source that is itself suitably cheap, abundant, and nonpolluting—like solar energy.

Plants Do It. For eons, the natural world has used photosynthesis to split water into hydrogen and oxygen, using light energy (see Chapter 3). To date, two ways have been proposed for mimicking this process. The first is to employ a catalyst to do what the photosynthetic pathway does. Recently, researchers from the Massachusetts Institute of Technology (MIT) described a process that employs a water-splitting catalyst made from cobalt and phosphorus, two common chemicals.[3] The process uses an electrical current that could come from solar cells and generates hydrogen and oxygen; it operates in neutral water and ambient temperature and pressure. Stay tuned!

Electrolysis. The second approach is simply to pass an electrical current through water and cause the water molecules to dissociate. Hydrogen bubbles come off at the negative electrode (the cathode), while oxygen bubbles come off at the positive electrode (the anode). (You can demonstrate this process for yourself with a battery, some wire, and a dish of water. A small amount of salt placed in the water will facilitate the conduction of electricity and the release of hydrogen.) The hydrogen gas is collected, compressed, and stored in cylinders, which can then be transported or incorporated into a vehicle's chassis. One important drawback to this technology is that it is difficult to store enough hydrogen in a vehicle to allow it to operate over long distances. Also,

compressing the hydrogen requires energy, thus reducing the energy efficiency involved. As an alternative, the hydrogen can be combined with metal hydrides, which can absorb the gas and release it when needed. Finding a hydride that is not too heavy has been a challenge, but the process is actively under development.

Solar Energy to Hydrogen. A rather elaborate plan has been suggested for taking solar energy–produced hydrogen for vehicles to its logical conclusion. We would start by building arrays of solar-trough or photovoltaic generating facilities in the deserts of the southwestern United States, where land is cheap and sunlight plentiful. The electrical power produced by these methods would then be used to produce hydrogen gas by electrolysis. The most efficient way to move the hydrogen is through underground pipelines; hundreds of miles of such pipelines already exist where hydrogen is transported for use in chemical industries. Although this would be an expensive part of a hydrogen infrastructure, the technology for it is already well known.

Model U. The Ford Motor Company unveiled a "concept car" in the United States it has dubbed the Model U. The Model U sports a 2.3-liter internal combustion engine that runs on hydrogen and is equipped with a hybrid electric drive. The car uses four 10,000-psi tanks to store hydrogen, enough to range 300 miles before a fill-up. And there's the problem: Before such cars can become feasible, there must be a hydrogen-fueling infrastructure. Once that infrastructure is in place, cars like the Model U would likely be replaced by a completely new way of moving a vehicle: by means of *fuel cells*.

Fuel Cells. An alternative to burning hydrogen in conventional internal combustion engines uses hydrogen in *fuel cells* to produce electricity and power the vehicle with an electric motor. **Fuel cells** are devices in which hydrogen or some other fuel is chemically recombined with oxygen in a manner that produces an electrical potential rather than initiating burning (Fig. 16–22a). Emissions consist solely of water and heat! Because fuel cells create much less waste heat than conventional engines do, energy is transferred more efficiently from the hydrogen to the vehicle—at a rate of 45% to 60% versus the 20% for current combustion engine vehicles. To power a vehicle, hundreds of fuel cells are combined into a fuel-cell stack. The vehicle must also include a hydrogen storage device, a cooling system (fuel cells also generate heat), and a device to force oxygen into the fuel cells. Fuel cells are now powering buses in Vancouver and Chicago, and 36 Daimler Chrysler urban buses with fuel-cell drives are now cruising the streets of 10 European cities, Beijing (China), and Perth (Australia). The fuel-cell stack develops more than 200 kW of power, from hydrogen in tanks mounted on the roof of the bus. These public transportation vehicles had logged a cumulative 1.24 million miles (2 million km) as of 2008.

Not to be outdone by Ford, GM presented a drivable car based on the fuel-cell concept—the Hy-wire—and followed this with a more advanced version, the Sequel. The vehicle has no engine compartment because the fuel cells, hydrogen tanks, and electric motor are all part of a "skateboard" chassis. Steering, braking, and other mechanical elements of a conventional

---

[3]Kanan, Matthew W. and Daniel G. Nocera. "In Situ Formation of an Oxygen-Evolving Catalyst in Neutral Water Containing Phosphate and Co^{2+}." *Science* (August 22, 2008): 1072–1074.

Figure 16–22 Hydrogen-oxygen fuel cell.
(a) Hydrogen gas flows into the cell and meets a catalytic electrode, where electrons are stripped from the hydrogen, leaving protons (hydrogen ions) behind. The protons diffuse through a proton exchange membrane, while electrons flow from the anode to the motor. As the electrons then flow from the motor to the cathode of the fuel cell, they combine with the protons and oxygen to produce water. Thus, the chemical reaction pulls electrons through the circuit, producing electricity that runs the motor. Hydrogen is continuously supplied from an external tank, oxygen is drawn from the air, and the only by-product is water. (b) Powered by a compact fuel cell stack, the Honda FCX Clarity has a range of 270 miles on a charge of hydrogen (4 kg), delivering the equivalent of 68 miles per gallon. The vehicle also has a battery for extra assist, similar to the battery in today's hybrid gas-electric vehicles. It will be available for lease at $600 per month.

(a)

(b)

vehicle are all replaced with electronically controlled units, called *by-wire systems*, that connect at a single docking port to the chassis. Thus, various bodies can be mounted on the chassis and even exchanged if desired. Many other "concept" vehicles operating on fuel cells are now on the road, developed by all the major car manufacturers. Honda, however, is the first manufacturer to commit to a dedicated fuel-cell vehicle assembly line. The Honda FCX Clarity (Fig. 16–22b) is rolling off the assembly lines, but at a pace of several dozen a year, it will be a long time before it makes an impact on the market. The Clarity will be available for lease in California regions that are supported by hydrogen refueling stations.

FreedomCAR. Major obstacles to the widespread application of fuel-cell-driven vehicles are their high cost and the lack of an infrastructure to provide hydrogen for refueling. Still, the technology is improving rapidly. It received a major boost when President Bush launched the FreedomCAR and Fuel Partnership initiative in 2003. Originally, FreedomCAR was a partnering effort between the government and the auto companies to promote research toward fuel-cell-powered vehicles, with the goal of 2010 for commercial introduction. The National Research Council (NRC) recently evaluated the program and recommended a midcourse shift toward a more rapid development of plug-in hybrid electric vehicles (PHEVs), while still supporting some research toward fuel-cell vehicles. The research will work toward developing the cars, the fuel-cell technology, and the necessary hydrogen infrastructure. Following the NRC recommendation, stimulus funding of $2.4 billion from the American Recovery and Reinvestment Act of 2009 is designed to help meet the Obama administration's goal of putting 1 million PHEVs on the road by 2015.

Once a hydrogen infrastructure is in place and vehicles powered by fuel cells are available, we will have entered a **hydrogen economy**—one in which hydrogen becomes a major energy carrier. With solar- and wind-powered electricity also in place, we would no longer be tied to nonrenewable, polluting energy sources and would have made the transition to a sustainable-energy system. Still, the obstacles in the way of the hydrogen economy are formidable, and many believe it will be decades before it even has a chance of happening. Global climate concerns and the costs of importing oil convey an urgency to resolving this issue; reducing our use of oil in the near term is crucial, and many believe that the best pathway is greater use of hybrid vehicles and other technologies already available.

16.4 Additional Renewable-Energy Options

Aside from water, wind, biomass, and hydrogen energy, there are other renewable-energy options that are worth pursuing.

Geothermal Energy

In various locations in the world, such as the northwestern corner of Wyoming (now Yellowstone National Park), there are springs that yield hot, almost boiling, water. Natural steam vents and other thermal features are also found in such areas. They occur where the hot, molten rock of Earth's interior is close enough to the surface to heat groundwater, particularly in volcanic regions. Using such naturally heated water or steam to heat buildings or drive turbogenerators is the basis of **geothermal energy**. In 2008, geothermal energy provided over 10,000 MW of electrical power (equivalent to the output of 10 large nuclear or coal-fired power stations) in countries as diverse as Nicaragua, the Philippines, Kenya, Iceland, and New Zealand. Today the largest single facility is in the United States at a location known as The Geysers, 70 miles (110 km) north of San Francisco (Fig. 16–23). With

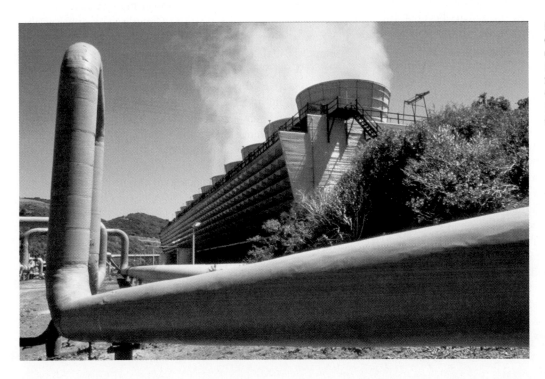

Figure 16–23 Geothermal energy. One of the 11 geothermal units operated by the Pacific Gas & Electric Company at The Geysers in Sonoma and Lake Counties, California. California geothermal energy produces 1,800 MW of power, roughly 4.3% of California's electricity needs.

Figure 16–24 Geothermal heat pump system. The pipes buried underground facilitate heat exchange between the house and the Earth. This system can either cool or heat a house and can be installed almost anywhere, although at a higher initial cost than a conventional heating, ventilation, and air-conditioning system.

Air handler, with indoor coil and fan connected to air ducts

Heat exchange pipes

over 2,900 MW from geothermal energy, the United States is the world leader in the use of this energy source; new projects in the development stage will increase this to over 6,000 MW. As impressive as this application of geothermal energy is, nearly double the amount is being used to directly heat homes and buildings, largely in Japan and China.

A recent study[4] by MIT suggested that the United States employ an emerging technology called enhanced geothermal systems (EGS). EGS involves drilling holes several miles deep into granite that holds temperatures of 400 degrees or more and injecting water under pressure into one hole where it absorbs heat from the rock and, as steam, flows up another shaft to a power plant where it generates electricity. Such plants already exist in Australia, Europe, and Japan. This underground heat is widespread in the United States, unlike the present limited geothermal sources. The MIT team proposed a $1 billion investment to exploit the EGS potential, predicting that by 2050 some 100 gigawatts (GW) of electrical power could be developed in this way.

Heat Pumps.   A less spectacular, but far more abundant, energy source than the large geothermal power plants exists anywhere pipes can be drilled into the Earth. Because the ground temperature below six feet or so remains constant, the Earth can be used as part of a heat exchange system that extracts heat in the winter and uses the ground as a heat sink in the summer. Thus, the system can be used for heating and cooling and does away with the need for separate furnace and air-conditioning systems. As Figure 16–24 shows, a geothermal heat pump (GHP) system involves loops of buried pipes filled with an antifreeze solution, circulated by a pump connected to the air handler (a box containing a blower and a filter). The blower moves house air through the heat pump and distributes it throughout the house via ductwork.

Four elementary schools in Lincoln, Nebraska, installed GHP systems for their heating and cooling. Their energy cost savings were 57% compared with the cost of conventional heating and cooling systems installed in two similar schools. Taxpayers will save an estimated $3.8 million over the next 20 years with the GHP systems. According to the EPA, these systems are by far the most cost-effective energy-saving

systems available. Although they are significantly more expensive to install than conventional heating and air-conditioning systems, they are trouble free and save money over the long run. New heat pump installations in the United States are growing at an annual rate of 20% per year, with almost 1 million installed by 2008.

Tidal and Wave Power

A phenomenal amount of energy is inherent in the twice-daily rise and fall of the ocean tides, brought about by the gravitational pull of the Moon and the Sun. Many imaginative schemes have been proposed for capturing this limitless, pollution-free source of energy. The most straightforward idea is the **tidal barrage**, in which a dam is built across the mouth of a bay and turbines are mounted in the structure. The incoming tide flowing through the turbines would generate power. As the tide shifted, the blades would be reversed so that the outflowing water would continue to generate power.

In about 30 locations in the world, the shoreline topography generates tides high enough—20 feet (6 m) or more—for this kind of use. Large tidal power plants already exist in two of those places: France and Canada. The only suitable location in North America is the Bay of Fundy, where the Annapolis Tidal Generating Station has operated since 1984, at 20-MW capacity. Thus, this application of tidal power has potential only in certain localities, and even there, it would not be without adverse environmental impacts. The barrage would trap sediments, impede the migration of fish and other organisms, prevent navigation, alter the circulation and mixing of saltwater and freshwater in estuaries, and perhaps have other, unforeseen ecological effects.

Other techniques are being explored that harness the currents that flow with the tides. For example, in 2006, Verdant Power installed six 15-foot-tall underwater turbines to harness the tidal energy of the East River in New York City; unfortunately, the currents were too strong for the blades and sheared off the tips. After another attempt also failed, Verdant has installed another turbine that has held up and is generating power for a nearby supermarket. The city plans to add many more turbines if all goes well, generating 10 MW of electricity. San Francisco is eyeing a similar scheme to tap the energy of the 400 billion gallons of water that flow under the Golden Gate Bridge with every tide.

[4]Tester, Jefferson et al. *The Future of Geothermal Energy: Impact of Enhanced Geothermal Systems (EGS) on the United States in the 21st Century.* Massachusetts Institute of Technology and the Department of Energy. 2006.

Ocean Waves. Standing on the shore of any ocean, an observer would have to be impressed with the energy arriving with each wave, generated by offshore winds. It might be possible to harness some of this energy, but the technological challenge is daunting. The ideal location is one that receives the wave's force before it hits the shoaling sea floor, is close enough to shore to facilitate hookup with transmission cables, and is deep enough so that the equipment will not crash on the sea floor during storm turbulence. In the United States, the Pacific Northwest provides these conditions, and the Pacific Gas and Electric Company has contracted with Finavera Renewables to install the nation's first, admittedly small, wave power plant off the coast of northern California. Eight buoys will generate two MW of electricity as they bob up and down with the waves, expanding and contracting a hose that pumps pressurized water through a turbine. There are other proposed mechanisms for capturing wave energy, and it is fair to say that this technology is in the early research-and-development stage. The U.S. DOE is supporting research on ocean energy on a 50% cost-share basis.

Ocean Thermal-Energy Conversion

Over much of the world's oceans, a thermal gradient of about 20°C (36°F) exists between surface water heated by the Sun and the colder deep water. **Ocean thermal-energy conversion (OTEC)** is the name of an experimental technology that uses this temperature difference to produce power. The technology involves using the warm surface water to heat and vaporize a low-boiling-point liquid such as ammonia. The increased pressure of the vaporized ammonia would drive turbogenerators. The ammonia vapor leaving the turbines would then be recondensed by cold water pumped up from as much as 300 feet (100 m) deep and returned to the start of the cycle.

Various studies indicate that OTEC power plants show little economic promise—unless, perhaps, they can be coupled with other, cost-effective operations. For example, in Hawaii, a shore-based OTEC plant uses the cold, nutrient-rich water pumped from the ocean bottom to cool buildings and supply nutrients for vegetables in an aquaculture operation, in addition to cooling the condensers in the power cycle. Even so, interest in duplicating such operations is minimal at present.

16.5 Policies for a Sustainable Energy Future

In Chapters 14–16, we have examined our planet's energy resources, requirements, and management options. The global issues are clear: Fossil fuels, especially oil and natural gas, will not last long at the current (and increasing) rate at which they are being consumed. Even more important, every use of fossil fuels produces unhealthy pollutants and increases the atmospheric burden of carbon dioxide, accelerating the already serious climate changes brought on by rising levels of greenhouse gases. Renewable energy means sustainable energy, and a sustainable-energy future does not include

significant fossil fuel energy. Global targets, therefore, are (1) *stable atmospheric levels of greenhouse gases, especially carbon dioxide*, and (2) *a sustainable-energy development-and-consumption pattern that is based on renewable energy*. Our main focus in this chapter has been on sustainable energy *development*, but there is another energy strategy that focuses on energy *consumption*—namely, **energy conservation and efficiency**. Energy saved is energy that does not have to be developed (or paid for), and in Chapter 14, we laid out two policies that have energy conservation as a goal. These were *increasing the mileage standards for motor vehicles* and *increasing the energy efficiency of lighting, appliances, and buildings*. Earlier in this chapter, we stressed the Energy Star program, as it encourages municipalities and businesses to achieve energy savings. We turn now to public policy in the United States and the recent and very significant changes that have encouraged renewable energy and energy conservation.

National Energy Policy

Nationally, the United States has a serious *oil-transportation* problem: We import two-thirds of our oil, much of which is consumed in transportation. This problem is an enormous drag on our economy. In fact, more than $500,000 leaves the United States *every minute* to pay for imported oil. Natural-gas resources are also being used increasingly to generate electrical power, prompting major price increases. Finally, we have an *energy security* problem: Not only are we subject to the whims of Organization of Petroleum Exporting Countries and other oil producers, but also we are vulnerable to terrorist attacks on various facets of our energy infrastructure.

These national issues, as well as other, global ones, prompted a policy response from the Bush administration and Congress. In the previous two chapters, we examined the policies pertaining to fossil fuels and nuclear power as they emerged in the **2005 Energy Policy Act** and the **2007 Energy Independence and Security Act**. These laws have become the major expression of federal energy policy, although many other laws promoting renewable energy and conservation have been enacted. The **American Recovery and Reinvestment Act of 2009** gave a huge boost ($43 billion) to renewable energy and energy conservation, with the intent of doubling the energy from renewable sources by 2012.

For renewable energy, supply-side policy provisions enacted into law

- Establish a renewable fuel standard (RFS) for ethanol and biodiesel that rises from 9 billion gallons in 2008 to 15 billion gallons by 2015.

- Extend the RFS to 36 billion gallons by 2022, but all the increase over 15 billion gallons must be met by feedstocks other than corn (i.e., second-generation sources).

- Continue funding for research and development on renewable energy, including major funding increases for solar energy and biomass programs.

- Extend the production tax credit for electricity generated with wind, hydropower, geothermal, and biomass energy.

- Provide investment tax credits for geothermal heat pumps, solar water heating, biofuel, and PV systems.

- Contain provisions aimed at enabling geothermal energy to compete with fossil fuels in generating electricity.

With their focus on *energy conservation and efficiency*, demand-side provisions enacted into law

- Raise the Corporate Average Fuel Economy standards for the combined fleet of cars and light trucks gradually, to meet a target of 39 miles per gallon for cars and 30 miles per gallon for light trucks by model year 2016.

- Extend the Energy Star building energy efficiency program to include schools, retail establishments, private homes, and health care facilities.

- Provide tax credits for many types of improvements to homes and commercial buildings that enhance energy efficiency.

- Require a 30% increase in efficiency for incandescent lightbulbs, which is already met by current compact fluorescent lightbulbs (CFLs). This guarantees that the CFLs will become the light standard of the future.

- Direct EPA to inventory waste-energy potential and establish a grant program to encourage waste-energy recovery (see Chapter 21).

- Develop efficiency standards for most appliances and provide tax credits for manufacturers of energy-saving appliances.

- Continue funding for development of the FreedomCAR and plug-in hybrid electric vehicles (PHEVs).

- Provide an income tax credit applicable to the purchase of new hybrid vehicles, PHEVs, or fuel-cell-powered vehicles.

These recent policies did *not* establish a federal *renewable portfolio standard (RPS)*. This would have required utilities to provide an increasing percentage of power from nonhydroelectric renewable sources (wind, solar, biomass). However, the states are setting the bar for the RPS: 26 states and the District of Columbia have implemented RPS requirements. Most states have adopted target percentages of 10% to 20%, with deadlines ranging from 2010 to 2025. Globally, at least 64 countries have a national target for renewable energy.

These policies also did not take any direct action on global warming, although they expanded DOE's research program for carbon capture and sequestration. Also, it can be argued that every provision that promotes greater energy efficiency and the use of renewable-energy sources is going to reduce greenhouse gas emissions in the United States.

Final Thoughts

Our look at current energy policy indicates that the United States is making a serious effort to address the need to develop renewable-energy sources and to improve the efficiency of current energy use. This policy also continues the effort to move toward a hydrogen economy. What about the global targets of achieving stable atmospheric levels of greenhouse gases and a long-term sustainable-energy policy? Unfortunately, current policy also promotes further use and development of fossil fuel energy, as summarized in Chapter 14. It's "business as usual" for the fossil fuel industries. How can that be changed?

One development that has caught everyone's attention in recent years is the rising price of gasoline. Until recently, the United States was the only industrialized country that has seen fit to keep gasoline prices remarkably low. Fuel in all other highly developed countries is so heavily taxed that it costs consumers upward of $6 per gallon—unlike U.S. prices, which were below $1.50 for 12 of the last 18 years. When gasoline rose to $3 a gallon in 2005, consumers began to turn away from gas-guzzling sport-utility vehicles and muscle cars and have been opting for the rising number of gasoline-electric hybrid cars appearing on the lots of many car manufacturers. This supports the suggestion of a number of economists and environmental groups that we should be paying more—much more—for gasoline (and other fossil fuels). To accomplish this, we would need a policy change: a **carbon tax**—that is, a tax levied on all fuels according to the amount of carbon dioxide that they produce when consumed. Such a tax, proponents believe, would provide both incentives to use renewable sources, which would not be taxed, and disincentives to consume fossil fuels. It is hard to imagine any step that would be more effective (or more controversial) in reducing greenhouse gas emissions in the United States. A number of European countries have already adopted such a tax.

Are the preceding developments, even the suggested carbon tax, enough to enable us to achieve a sustainable-energy system and to mitigate global climate change? Very likely, no. Yet many of them are moving us in the stewardly direction that is vital to the future of the global environment.

Revisiting the Themes

Sound Science

Some elegant technologies have been discussed in this chapter, developed by scientists in research laboratories and then applied to solving some very practical problems. PV devices, for example, are being intensely researched in order to bring down the costs and make solar energy competitive with conventional energy sources. Research at MIT has produced a real breakthrough in the discovery of a catalyst that can split water, imitating photosynthesis. The National Renewable Energy Laboratory in Golden, Colorado, was established by Congress in 1977 and has developed into a cutting-edge research institute that produced the solar-trough and solar-dish systems. It is now exploring hydrogen technologies and improved PV devices as well as a number of biomass technologies and low-wind-speed turbines and other renewable-energy options. This is science at its best, with public funding put to work to develop new ideas that may revolutionize the energy scene.

Sustainability

At the risk of repetition, global civilization is on an unsustainable energy course. Our intensive use of fossil fuels has created a remarkable chapter in human history, but we are still terribly dependent on maintaining and increasing our use of oil, gas, and coal. The results have already been seen in rising atmospheric greenhouse gases and the climate changes they are producing. Future use of fossil fuel will bring on more intense climate changes, and long before we burn all the available oil, gas, and coal, we will have been forced to turn to other energy sources. Renewable-energy sources derived from the Sun can

provide a sustainable, inexhaustible supply of energy that is virtually free from harmful environmental impacts. It is imperative that the United States, as well as other countries, embark on a course to bring about a transition to a renewable-energy economy, most likely based on hydrogen and electricity as energy carriers, to be produced from renewable sources.

Stewardship

Stewardship of energy means doing more with less and turning away from the fossil fuels that are generating the greenhouse gases that are responsible for global climate change. Government policies are rapidly moving forward in encouraging renewable energy by providing subsidies for homeowners and businesses to embrace renewable-energy options. Policies are also improving the energy efficiency and conservation picture. After all, energy saved is energy that does not have to be developed. Picture solar panels on every home, church, hospital, and school roof and wind turbines on college campuses, grazing lands, and offshore wind farms. Many nongovernmental organizations are involved with programs that distribute PV and biogas systems to people in poor, rural, developing-country communities. The impact is literally being felt around the globe.

REVIEW QUESTIONS

1. How does solar energy compare with fossil fuel energy in satisfying basic energy needs? What are the three fundamental problems in harnessing solar energy?

2. How do active and passive solar hot-water heaters work?

3. How can a building best be designed to become a passive solar collector for heat?

4. How does a PV cell work, and what are some present applications of such cells?

5. What is happening in recent years in the PV market? How are utilities encouraging PV power?

6. Describe the solar-trough system, how it works, and its potential for providing power.

7. What is the potential for developing more hydroelectric power in North America versus developing countries, and what are the impacts of such development?

8. Where is wind power being harvested, and what is the future potential for wind farms?

9. What are some ways of converting biomass to useful energy, and what is the potential environmental impact of each?

10. What biofuels are used for transportation, and what is the potential for increasing biofuels in the United States?

11. How can hydrogen gas be produced via the use of solar energy? How might hydrogen be collected and stored to meet the need for fuel for transportation in the future?

12. How do fuel cells work, and what is being done to adapt them to power vehicles?

13. What is geothermal energy, and what are two ways it is being harnessed?

14. What is the potential for developing tidal power in the United States?

15. How does renewable energy address the key global and national energy issues?

16. Describe current policy for renewable energy and energy efficiency as an outcome of the passage of the Energy Policy Act of 2005. What crucial issues were omitted from the act?

THINKING ENVIRONMENTALLY

1. What do you believe needs to happen before solar power becomes a thoroughly mainstream form of energy production? Defend your answer.

2. What do you think could be a viable solution for backup power systems for solar heaters in the event of bad weather?

3. How much promise does hydrogen power hold? Is its eventual sustainability just a myth or something to be pursued?

4. Imagine you are charged with building a facility run entirely by renewable resources (with the possible exception of backup power). How would you evaluate which sources have the most potential?

5. Audit all of the energy needs in your daily life. How much energy do you consume? Describe how each need might be satisfied by a renewable-energy option. (Check the Internet for tools for this exercise.)

MAKING A DIFFERENCE

1. Many everyday electrical products can now be powered by photovoltaic (PV) cells. You can find solar-powered chargers for MP3 players, cell phones, digital cameras, and even laptops. While this power may not be as dependable as that from a wall charger, it certainly provides a sustainable alternative while the Sun shines (especially useful on road and camping trips, where conventional power isn't available).

2. Carpool! While we wait for renewable-energy sources in transportation, it's our job to try to conserve our dwindling fossil fuels.

3. If you live close to work, consider not driving there at all and instead taking a bicycle. The benefits are endless; great exercise, low emissions, and no money toward gas are just the beginning.

4. Take advantage of the following services from your electric company: energy conservation, home-energy audits, and subsidized compact fluorescent lighting.

5. Consult the free *Greentips* newsletter of the Union of Concerned Scientists on its Web site (www.ucsusa.org), and follow the advice given in this publication for reducing your energy footprint.

6. Adopt ways to use less fuel and electricity both at home and where you work. (Turn thermostats up in the summer and down in the winter, turn off lights when you leave a room, wear sweaters, use energy-efficient lightbulbs, etc.)

Go to **www.masteringenvironmentalscience.com** for practice quizzes, Pearson eText, videos, current events, and more.

Paper mill at sunset, Georgetown, South Carolina.

PART FIVE

Pollution and Prevention

A N INDUSTRIAL PLANT sends a steady stream of exhaust gases into the air, and the wind carries the pollutants away. If it didn't, life in the vicinity of the plant would be nasty. But where is "away"? Is it the next county, state, or country? The upper atmosphere? In reality, there is no "away" for the pollution of land, water, and air resulting from the activities of the 6.7 billion people inhabiting Earth. We are fouling our own nest, our only planet.

IN HIS BOOK *Red Sky at Morning*,[1] James Gustave Speth describes **four long-term trends in pollution** that marked the 20th century:

First, *the volumes of pollutants released increased by huge amounts*—so much so that the developed countries finally began to legislate limits on many pollutants.

Second, *the pollution evolved from gross excesses to microtoxicity*, where major releases of toxic gases and waterborne pathogens were replaced by smaller, but much more toxic releases of organic chemicals and trace metals like mercury.

Third, *industrial pollutants have shifted from the first world to the third world*—from the developed to the developing countries. China, for example, is reaping the benefits of rapid industrial growth and at the same time suffering from disgusting and health-damaging pollution.

Fourth, *the effects of pollution have shifted from local to global*. Air pollution, resulting from the use of fossil fuels and industrial chemicals, has brought acid rain on a regional scale and global climate change and ozone depletion on a global scale.

In this part of the book, Chapter 17 begins with a look at the links between environmental hazards (like pollution) and human health. The chapters that follow examine the major forms of pollution, and you will discover that pollution is often a by-product of some useful activity (although sometimes intentionally released). We cover the range of pollutants, from global and regional air pollutants to water pollutants to municipal and industrial wastes. In every chapter, the public policies that have emerged to deal with each of the problems raised by pollution are discussed. Without exaggeration, the problems of pollution pose some of the most difficult and important issues facing human society. One of the essential transitions of the 21st century will be that from pollution-intensive economic production to environmentally benign processes as we work toward a sustainable future.

---

[1]Speth, James Gustave. *Red Sky at Morning*. Yale University Press, New Haven. 2004.

CHAPTER 17

Environmental Hazards and Human Health

Poultry slaughter. Romanian veterinary worker collects a duck for culling in a village quarantined because of avian flu. About 100,000 domestic birds had to be culled to keep the disease from spreading further.

KEY TOPICS

1. **Links Between the Environment and Human Health: Environmental hazards that bring on human misery, disease, and death can be cultural, biological, physical, and chemical.**

2. **Pathways of Risk: Some risks that lead to exposure to environmental hazards are poverty, tobacco use, disease transmission, toxic chemicals, and natural disasters.**

3. **Risk Assessment: Human and environmental health can be protected by evaluating risks associated with particular hazards and taking action to manage the risks.**

A 32-YEAR-OLD MAN FROM the village of Kubu Sembelang in North Sumatra, Indonesia, died on May 22, 2006, of respiratory failure. He was the seventh member of an extended family to die within the month with symptoms of a severe flu. Three of the victims spent a night in a small room with a family member who was coughing frequently, and the evidence suggests that they became infected at that time. Specimens taken from the victims confirmed that they died of an infection with the H5N1 influenza virus, better known as the *avian flu* virus. The virus from the human cases was found to be genetically similar to viruses isolated earlier from poultry in Sumatra. What marks this cluster of cases as unique is the mode of transmission of the virus. Although over 220 human cases (and more than 125 deaths) were known at the time, virtually all of these victims had contact with poultry likely to have been carrying the H5N1 virus. The North Sumatra cases, however, were apparaently the result of human-to-human transmission.

The highly virulent H5N1 avian flu virus first appeared in Hong Kong in 1997, killing entire flocks of chickens within 48 hours. Since that time, H5N1 avian flu has infected all kinds of poultry and

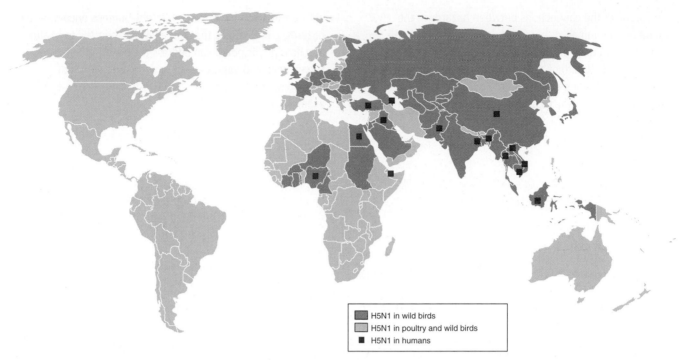

Figure 17–1 Countries with confirmed cases of avian flu (H5N1), 2008. The map shows the location of the H5N1 virus in poultry, wild birds, and humans. These data are collected by the WHO. (*Source:* Centers for Disease Control Web site.)

other birds and has spread through South Asia and into Africa and Europe—and could arrive in the Western Hemisphere at any time (Fig. 17–1). Most of this spread has been traced to the poultry business, but the virus has also been found in wild migratory fowl. The first human cases appeared in 2003, stirring concern that avian flu could at some point spread from human to human (as indicated in the North Sumatra cases) and become a global pandemic. The epidemic among poultry has devastated the poultry industry and rural farmers who depend on their birds for food and income; once a bird is found with avian flu symptoms, the whole flock must be destroyed. Hundreds of millions of poultry have been slaughtered to contain the epidemic since it began in 1997.

High concern over the potential of the H5N1 virus to trigger a human pandemic stems from the nature of influenza viruses. Their genetic material is contained on a number of strands of viral ribonucleic acid (RNA), and such viruses occasionally mix strands with other types of influenza. If a human sick with ordinary, contagious flu contracts the deadly avian flu and the viruses exchange genes, there is the possibility of creating a new, virulent form of avian flu that can spread rapidly from human to human. At present, avian flu's high mortality (262 deaths for 436 cases) in humans is the result of its novelty; there is no history of exposure to this virus and thus no built-in immunity to it.

In April 2009, an influenza outbreak was reported from North America, and because the virus contained strands of influenza from swine viruses (as well as avian and human viruses), it has become known as **swine flu**, a novel H1N1 virus. Early reports of cases in Mexico indicated a severe disease, with over 100 deaths. However, even though the disease spread rapidly throughout the world (with over 160,000 cases reported in four months), the death rate was remarkably low (0.7%). Nevertheless, because of the rapid spread of the swine flu virus, the World Health Organization (WHO) has declared a global influenza pandemic. The last three influenza pandemics occurred in 1918, 1957, and 1958; the 1918 flu killed more than 20 million people. As with avian flu, the WHO is monitoring the swine flu pandemic closely and coordinating the global response to these diseases.

The avian and swine influenzas are examples of a number of emerging new diseases that continually remind us we are not always able to control the hazards posed by our inevitable contacts with the natural environment and, particularly, with the animal world. Some 39 diseases have jumped from animals to humans in the last 25 years. Severe acute respiratory syndrome (SARS), Lassa fever, the Ebola virus, the hantavirus, and the West Nile virus all have taken human lives, and there is every reason to expect more new diseases to appear as we continue to manipulate the natural environment and change ecological relationships.

THE OLD ENEMIES.   To put these diseases in perspective, however, it is not the new, emerging diseases that represent the greatest threat to us; rather, it is the common, familiar ones that take the greatest toll of human life and health—diseases such as malaria, diarrhea, respiratory viruses, and worm infestations, which have been kept at bay in the developed countries, but continue to ravage the developing countries. In the developed countries, cancer is the killer that is most closely linked to the environment, leaving us with questions about the chemicals we are exposed to on a daily basis.

The study of the connections between hazards in the environment and human disease and death is often called **environmental health**; but it is *human* health that is the issue here, not the health of the natural environment. In this chapter, we will examine the nature of environmental hazards and the consequences of exposure to those hazards. Then we will consider some pathways whereby humans encounter the hazards, interventions that can alleviate the risks to health, and how public policy can address the need for continued surveillance and improvement in our management of environmental hazards. We begin with some basics as we examine what pollution is.

17.1 Links Between the Environment and Human Health

The U.S. Environmental Protection Agency (EPA) defines **pollution** as "the presence of a substance in the environment that, because of its chemical composition or quantity, prevents the functioning of natural processes and produces undesirable environmental and health effects." Any material that causes the pollution is called a **pollutant**.

Pollution is not usually the result of deliberate mistreatment of the environment. The pollutants are almost always the *by-products* of otherwise worthy and essential activities—producing crops, creating comfortable homes, providing energy and transportation, and manufacturing products—and of our basic biological functions. Pollution problems have become more pressing over the years because both growing population and expanding per capita use of materials and energy have increased the amounts of by-products that go into the environment. Also, many materials now widely used, such as aluminum cans, plastic packaging, and synthetic organic chemicals, are **nonbiodegradable**. That is, they resist attack and breakdown by detritus feeders and decomposers and consequently accumulate in the environment. An overview of the categories of pollutants that result from various activities is shown in Figure 17–2.

It is important to note the breadth and diversity of pollution. Any part of the environment may be affected, and almost anything may be a pollutant. The only criterion is that the addition of a pollutant results in undesirable changes. The impact of an undesirable change may be largely aesthetic—hazy air obscuring a distant view or the unsightliness of roadside litter, for instance. The impact may be on ecosystems as a whole—the die-off of fish or forests, for example. Or the impact may be on human health—such as human wastes contaminating water supplies. The impact also may range from very local—the contamination of an individual well—to global—adding ozone-depleting chemicals to the atmosphere. We tend to think of pollution as the introduction of human-made materials into the environment, but undesirable changes may be caused by the introduction of too much of what are otherwise natural compounds, such as fertilizer nutrients introduced into waterways and carbon dioxide introduced into the atmosphere.

Thus, the slogan "Don't pollute" is a gross oversimplification. The very nature of our existence necessitates the production of wastes. Our job in remediating present and future pollution problems is like the concept of sustainable development itself. We need to adapt the means of meeting our present needs so that by-products are managed in ways that will not jeopardize present and future generations. The general strategy in each case must be to

1. Identify the material or materials that are causing the pollution—the undesirable change;

2. Identify the sources of the pollutants;

3. Clean up the environment already impacted by pollution;

4. Develop and implement pollution-control strategies to prevent the pollutants from entering the environment; and

5. Develop and implement alternative means of meeting the need that do not produce the polluting by-product—in other words, avoid the pollution altogether.

Basically, the strategy for addressing pollution accomplishes one of the transitions to a sustainable society presented in Chapter 1: "A technology transition from pollution-intensive economic production to environmentally friendly processes."

Environmental Health

With our focus in this chapter on human health, a more precise definition of **environment** is *the whole context of human life—the physical, chemical, and biological setting of where and how people live.* Thus, the home, air, water, food, neighborhood, workplace, and even climate constitute elements of the human environment. Within the human environment, there are *hazards* that can make us sick, cut our lives short, or contribute in other ways to human misfortune. In the context of environmental health, a **hazard** is anything that can cause (1) injury, disease, or death to humans; (2) damage to personal or public property; or (3) deterioration or destruction of environmental components.

The existence of a hazard does not mean that undesirable consequences inevitably follow. Instead, we speak of the connection between a hazard and something happening because of that hazard as a **risk**, defined here as the *probability* of suffering injury, disease, death, or some other loss as a result of exposure to a hazard. The notion of **vulnerability** is also important; some people, especially the poor, are more vulnerable to certain risks than are others. Therefore, an appropriate notation of risk should be

$$\text{Risk} = \text{Hazard} \times \text{Vulnerability}$$

Figure 17–2 Categories of pollution. Pollution is an outcome of otherwise worthy human endeavors. Major categories of pollution and some activities that cause them are shown here.

The presence of avian flu in poultry is a *hazard* that presents the *risk* of humans contracting the disease and dying. People working with poultry are much more *vulnerable* than are others. Because risks are expressed as probabilities, the analysis of the probability of suffering some harm from a hazard becomes a problem for scientists or other experts to address, and we will deal with the assessment of risk as it becomes the basis of policy decisions.

The Picture of Health. Health has many dimensions—physical, mental, spiritual, and emotional. The WHO defines **health** as "a state of complete physical, social, and mental well-being, and not merely the absence of disease or

infirmity."[2] Unfortunately, measuring all these dimensions of health for a society is virtually impossible. Thus, to study environmental health, we will focus on *disease* and consider health to be simply *the absence of disease.* Two measures are used in studies of disease in societies: *morbidity* and *mortality.* **Morbidity** is the incidence of disease in a population and is commonly used to trace the presence of a particular type of illness, such as influenza or diarrheal disease. **Mortality** is the incidence of death in a population. Records are usually kept of the cause of death, making it possible to

---

[2]WHO, *Glossary.* Available at www.who.int/health-systemsperformance/docs/glossary.htm.

analyze the relative roles played by infectious diseases and factors such as cancer and heart disease. **Epidemiology** is the study of the presence, distribution, and prevention of disease in populations.

Public Health.   Protecting the health of its people has become one of the most important facets of government in modern societies. In the United States, the lead federal agency for protecting the health and safety of the public is the *Centers for Disease Control and Prevention (CDC)*, an agency within the Department of Health and Human Services. The CDC's mission is "to create the expertise, information, and tools that people and communities need to protect their health—through health promotion, prevention of disease, injury and disability, and preparedness for new health threats."[3] Each state also has a health department, and most municipalities have health agents. In addition, there is a huge health care industry in the United States, with federal programs such as Medicare and Medicaid, hospitals, health maintenance organizations, and local physicians and other health professionals. The delivery of health care is spread around all of these entities, but the primary responsibility for health risk management and prevention resides with the CDC and the state public-health agencies. These agencies, for example, may require public-health measures such as immunizations and quarantines, and they are responsible for monitoring certain diseases and environmental hazards and for controlling epidemics. Indeed, it was the CDC that first reported the swine flu cases. The CDC also gathers data and provides information on health issues to state and local health care providers.

Other Countries.   Virtually every country has a similar *ministry of health* that acts on behalf of its people to manage and minimize health risks. The health policies that are put into place, however, are subject to limitations of information and funding. Ideally, a health ministry should direct its limited resources toward strategies that will accomplish the greatest risk prevention. Fortunately, all countries have access to the programs and information of the WHO, a UN agency established in 1948 with the mission of enabling "all peoples to attain the highest possible level of health." The WHO is centered in Geneva, Switzerland, and maintains six regional offices around the world. The agency is staffed by health professionals and other experts and is governed by the UN member states through the World Health Assembly.

Life Expectancy.   One universal indicator of health is human life expectancy. In 1955, average life expectancy globally was 48 years. Today it is 68 years and rising gradually: It is expected to reach 73 years by 2025. This progress is the result of social, medical, and economic advances in the past 55 years that have increased the well-being of larger and larger segments of the human population. Recall from Chapter 8 the *epidemiologic transition*: the trend of decreasing death rates that is seen in countries as they modernize. Recall also that, as this transition occurred in the now-developed countries, it involved a shift from high mortality (due to infectious diseases)

toward the present low mortality rates (due primarily to diseases of aging, such as cancer and cardiovascular diseases).

Two Worlds.   The countries of the world have undergone the epidemiologic transition to different degrees, with very different consequences. Despite the progress indicated by the general rise in life expectancy, one-sixth (9.2 million) of the annual deaths in the world are children under the age of five in the developing world. As we saw in Chapter 9, a high proportion (47%) of the mortality in the less developed countries is caused by common infectious diseases (see Figure 9–13).

An entirely different perspective is brought into focus as we consider some of the environmental hazards that accompany industrial growth and intensive agriculture. In addition, some of the most lethal hazards in the developed world are the outcome of purely voluntary behavior—in particular, smoking tobacco and engaging in risky sexual activity. We will look at four classes of environmental hazards: cultural, biological, physical, and chemical.

Environmental Hazards

There are two fundamental ways to consider hazards to human health. One is to regard the lack of access to necessary resources as a hazard. For example, lack of access to clean water and nourishing food is likely to result in harm to a person. Investigating hazards from this perspective means considering the social, economic, and political factors that prevent a person from having access to such basic needs. Although this is a fundamentally important perspective, much of it is outside the scope of this chapter (and has already been discussed in Chapters 9 and 12). Instead, the focus will be primarily on *the exposure to hazards in the environment.* What is it in the environment that brings the risk of injury, disease, or death to people?

Cultural Hazards.   Many of the factors that contribute to mortality and disability are a matter of choice or at least can be influenced by choice. People engage in risky behavior and subject themselves to hazards. Thus, they may smoke cigarettes, eat too much, drive too fast, use addictive and harmful drugs, consume alcoholic beverages, sunbathe, engage in risky sexual practices, get too little exercise, or choose hazardous occupations. People generally subject themselves to these hazards because they derive some pleasure or other benefit from them. Wanting the benefit, they are willing to take the risk that the hazard will not harm them. Factors such as living in inner cities, engaging in criminal activities, and so on are cultural sources of mortality, too. As Figure 17–3 indicates, 40% of all deaths in the United States can be traced to cultural hazards, and in most cases, deaths from cultural hazards are preventable (if people refrain from their risky behaviors).

The connection between behavior and risk is especially lethal in the case of acquired immune deficiency syndrome (AIDS), caused by the human immunodeficiency virus (HIV). AIDS is one of a group of sexually transmitted diseases (STDs) that includes gonorrhea, syphilis, and chlamydia. Recall from Chapter 9 that AIDS is taking a terrible toll in the developing world, where more than 90% of HIV-infected

---

[3]CDC, from www.cdc.gov/about/organization/mission.htm.

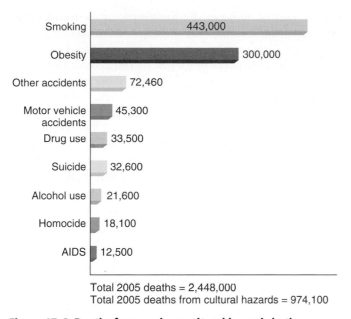

Total 2005 deaths = 2,448,000
Total 2005 deaths from cultural hazards = 974,100

Figure 17–3 Deaths from various cultural hazards in the United States, 2005. Many hazards in the environment are a matter of personal choice and lifestyle. Other hazards are due to accidents and the lethal behavior of other people. (*Sources:* Data from the National Vital Statistics Reports, April 24, 2008; smoking deaths and obesity deaths from the Centers for Disease Control.)

TABLE 17–1	Global Mortality from Major Infectious Diseases, 2004
Cause of Death	**Estimated Yearly Deaths***
Acute respiratory infections	4,259,000
Diarrheal diseases	2,163,000
HIV/AIDS	2,040,000
Tuberculosis	1,464,000
Malaria	889,000
Measles	424,000
Meningitis	340,000
Pertussis (whooping cough)	254,000
Tetanus	163,000
Hepatitis	159,000
Syphilis	99,000
Trypanosomiasis (sleeping sickness)	52,000
Leishmaniasis	47,000
Schistosomiasis	41,000
Dengue	18,000
Intestinal roundworms	6,000

*Total deaths from infectious diseases in 2004 are estimated at 13.8 million by the WHO.
Source: Data from World Health Organization, *The Global Burden of Disease: 2004* (October 2008).

people live. Wherever it occurs, AIDS is largely the consequence of high-risk sexual behavior. At present, there is no cure or vaccine for the HIV virus.

Biological Hazards. Human history can be told from the perspective of the battle with pathogenic bacteria and viruses. It is a story of epidemics such as the black plague and typhus, which ravaged Europe in the Middle Ages, killing millions in every city, and of smallpox, which swept through the New World. The story takes a positive turn in the 19th century with the first vaccinations and the "golden age" of bacteriology (a span of only 30 years), when bacteriologists discovered most of the major bacterial diseases and brought bacteria into laboratory culture. The 20th century saw the advent of virology (the study of viruses and the treatment of the diseases they cause), the great discovery of antibiotics, immunizations that led to the global eradication of smallpox and the victory over polio and many other childhood diseases, and the growing influence of molecular biology as a powerful tool in the battle against disease.

The battle is not over, however, and never will be. Pathogenic bacteria, fungi, viruses, protozoans, and worms continue to plague every society and indeed every person. They are inevitable components of our environment. Many are there regardless of our human presence, and others are uniquely human pathogens whose access to new, susceptible hosts is mediated by the environment. Table 17–1 shows the leading global diseases and the annual number of deaths they cause.

Approximately one-fourth of global deaths are due to infectious and parasitic diseases. The leading causes of death in this category are the acute respiratory infections (for example, pneumonia, diphtheria, influenza, and streptococcal infections), both bacterial and viral. Pneumonia is by far the most deadly of these (Fig. 17–4). Other bacterial and

Figure 17–4 Jim Henson. The creator of the Muppets succumbed in 1990 to a virulent pneumonia caused by the common bacterium *Streptococcus pneumoniae*.

viral respiratory infections represent the most common reasons for visits to the pediatrician in the developed countries, but it is primarily in the developing countries that the respiratory infections lead to death, mostly in children who are already weakened by malnourishment or other diseases.

Diarrheal diseases were responsible for 1.7 million deaths in children under age five in 2004. One-third of these deaths are traced to a little-known virus—**rotavirus**—that infects virtually every child between the ages of three months and five years. Rotavirus is rarely diagnosed, yet it causes most cases of severe diarrhea in young children in developed countries and many childhood deaths due to diarrhea in developing countries. The difference is treatment; the infection often leads to dehydration, which is quickly life threatening, but can be reversed with rehydration therapy (the replacement of lost fluids with balanced saline solutions). The virus is as contagious as a cold virus, and fortunately, one infection usually induces lifetime immunity. Many other cases of diarrhea are the consequence of ingesting food or water contaminated with bacterial pathogens, such as *Salmonella*, *Campylobacter*, *Shigella*, and *E. coli*, from human wastes.

Tuberculosis.   Although AIDS has overtaken tuberculosis as the disease that causes the most adult deaths, tuberculosis continues to be a major killer. Close to one-third of the world's people are infected with the microbe *Mycobacterium tuberculosis* (Fig. 17–5), and perhaps 10% of these will develop a life-threatening case during their lifetime. Complacency about treatment and prevention in recent years has led to a resurgence of tuberculosis, even though the bacterial pathogen that causes it has been known for more than 100 years and its genome has

been sequenced. Some of the resurgence can be traced to the AIDS epidemic because HIV infection eventually leads to a compromised immune system, which allows diseases like tuberculosis to flourish. Much of the disease's comeback, however, is the result of multi-drug-resistant strains of the microbe.

Malaria.   Of the infectious diseases present in the tropics, malaria is by far the most serious, accounting for an estimated 247 million cases each year and 881,000 deaths. Caused by protozoan parasites of the genus *Plasmodium*, malaria is propagated by mosquitoes of the genus *Anopheles*. The disease begins with an infected mosquito biting a person (Fig. 17–6). From there, the parasites invade the liver and then the red blood cells. After multiplying in the red blood cells, the parasites are released, destroying the cells. The parasites then invade other red blood cells. While in the liver and the red blood cells (most of the time), the parasites are hidden from immune system cells and antibodies. This entire chain of events occurs in synchronized cycles, leading to the periodic episodes of fever, chills, and malaise that are typical of malaria. With the most virulent parasite, *P. falciparum*, 60% of the red blood cells may be destroyed in one episode, producing rapid and severe anemia. The waste products of the parasites, when they break out of the cells, produce high fever, headaches, and vomiting in the victim.

Most of the diseases responsible for mortality are also leading causes of debilitation in humans of all ages. More than 4 billion episodes of diarrhea occur in the human population over the course of a year, and at any given time, more than 3.5 billion people suffer from the chronic effects of parasitic worms such as hookworms and schistosomes.

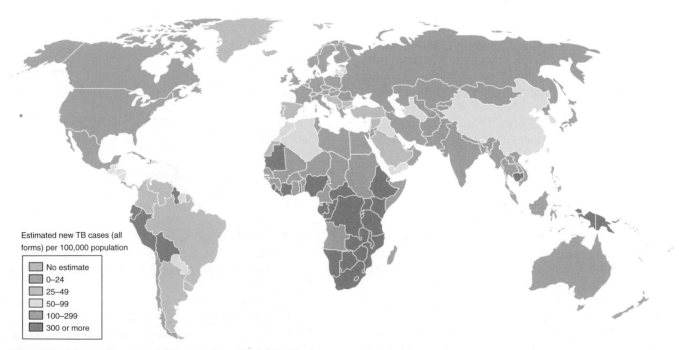

Estimated new TB cases (all forms) per 100,000 population

- No estimate
- 0–24
- 25–49
- 50–99
- 100–299
- 300 or more

Figure 17–5 Global map of tuberculosis. This disease, caused by the bacterium *Mycobacterium tuberculosis*, is far from controlled. Close to one-third of the world's people are likely infected by the pathogen. Tuberculosis (TB) incidence rates refer to numbers per 100,000 in the population. (*Source:* The World Health Organization, *Global Tuberculosis Control: Surveillance, Planning, Financing,* 2008.)

HOW MALARIA SPREADS

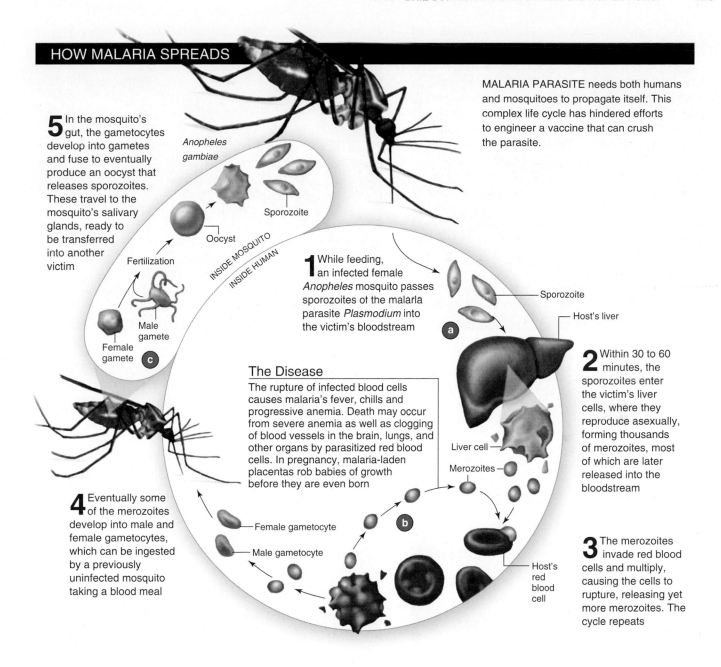

5 In the mosquito's gut, the gametocytes develop into gametes and fuse to eventually produce an oocyst that releases sporozoites. These travel to the mosquito's salivary glands, ready to be transferred into another victim

Anopheles gambiae

Sporozoite

Oocyst

Fertilization

INSIDE MOSQUITO

INSIDE HUMAN

Male gamete

Female gamete

MALARIA PARASITE needs both humans and mosquitoes to propagate itself. This complex life cycle has hindered efforts to engineer a vaccine that can crush the parasite.

1 While feeding, an infected female *Anopheles* mosquito passes sporozoites of the malarla parasite *Plasmodium* into the victim's bloodstream

Sporozoite

Host's liver

2 Within 30 to 60 minutes, the sporozoites enter the victim's liver cells, where they reproduce asexually, forming thousands of merozoites, most of which are later released into the bloodstream

Liver cell

Merozoites

The Disease

The rupture of infected blood cells causes malaria's fever, chills and progressive anemia. Death may occur from severe anemia as well as clogging of blood vessels in the brain, lungs, and other organs by parasitized red blood cells. In pregnancy, malaria-laden placentas rob babies of growth before they are even born

4 Eventually some of the merozoites develop into male and female gametocytes, which can be ingested by a previously uninfected mosquito taking a blood meal

Female gametocyte

Male gametocyte

Host's red blood cell

3 The merozoites invade red blood cells and multiply, causing the cells to rupture, releasing yet more merozoites. The cycle repeats

Figure 17–6 Life cycle of malarial parasite. The life cycle of the protozoan parasite *Plasmodium*, which causes malaria in humans, involves the *Anopheles* mosquito and a human host. (*Source:* Dunavan, Claire Panosian, "Tackling Malaria," *Scientific American* 293 (December 2005): 74–83. Used with permission.)

Physical Hazards. Natural disasters—including hurricanes, tornadoes, floods, forest fires, earthquakes, landslides, and volcanic eruptions (Fig. 17–7)—take a toll of human life and property every year. They are the outcome of hydrological, meteorological, or geological forces. However, the 12 months following November 2004 were unimaginably dreadful. During that time, three major natural disasters took nearly 300,000 lives and left millions homeless and without means of self-support: the 2004 Indian Ocean tsunami, Hurricane Katrina on the U.S. Gulf Coast on August 29, 2005, and an earthquake in Pakistan on October 8,

2005. The earthquake-spawned tsunami of December 26, 2004, claimed some 225,000 lives, more than any other tsunami on record. More recently, the year 2008 saw two tragic disasters: an earthquake in China and a tropical cyclone in Myanmar; Table 17–2 records the mortality from major natural disasters for 2006 through 2008.

Looking at the recent record, there are *two patterns* that emerge: hazards like tsunamis, tornadoes, and earthquakes that are impossible to anticipate and hazards that are largely a consequence of choices people make about where to live.

(a) (b)

(c) (d)

Figure 17–7 Dangerous physical hazards. Natural disasters bring death and destruction in their wake. (a) Devastation in Banda Aceh on the Indonesian island of Sumatra, from the tsunami that struck countries in the eastern Indian Ocean, December 26, 2004. (b) Aerial view of New Orleans under water from Hurricane Katrina, August 30, 2005. (c) A magnitude-7.9 earthquake devastated the Sichuan area of China on May 12, 2008. (d) Family and friends clean up a damaged restaurant in Newburgh, Indiana, after an F3 tornado tore through Indiana and Kentucky on November 6, 2005, killing 25 people.

Out of Nowhere. Each year in the United States, 780 tornadoes strike on average, spawned from the severe weather accompanying thunderstorms. Because of the tendency for cold, dry air masses from the north to mix with warm, humid air from the Gulf of Mexico, the central United States generates more tornadoes than anywhere else on Earth. Of short duration, they develop into some of the most destructive forces known in nature, with winds reaching as

high as 300 miles per hour. The most intense tornadoes have killed hundreds of people. On April 3, 1974, for example, a line of tornadoes from Canada to Georgia took more than 300 lives and caused untold property damage.

The Indian Ocean tsunami and the Pakistan earthquake were completely unavoidable. There is still no way to forecast earthquakes, but there are ways to mitigate their impacts. Tsunami warning systems have existed for the Pacific Ocean

TABLE 17–2 Mortality from Major Natural Disasters, 2006–2008

Country	Type of Disaster	Date	Deaths
Myanmar	Tropical cyclone	May 2, 2008	138,366
China (P. Republic)	Earthquake	May 12, 2008	87,476
Indonesia	Earthquake	May 27, 2006	5,778
Bangladesh	Tropical cyclone	November 15, 2007	4,234
Philippines	Tropical cyclone	November 30, 2006	1,399
France	Heat wave	July 15, 2006	1,388
Afghanistan	Storm	June 21, 2008	1,317
Philippines	Landslide	February 17, 2006	1,126
Bangladesh	Flood	July 21, 2007	1,110
India	Flash flood	June 11, 2008	1,063

Source: Emergency Events Database, www.emdat.be.

since 1965, but not until 2006 for the Indian Ocean, and earthquake-resistant buildings can be constructed (but were not for rural mountain villages in Pakistan). This illustrates the fact that the most devastating consequences of natural hazards usually are experienced by those who are least capable of anticipating them and dealing with their effects—the poor. Indeed, most of the deaths and loss of homes from Hurricane Katrina occurred among the lower-income residents.

In Harm's Way. Much of the loss brought on by natural disasters is a consequence of poor environmental stewardship. Hillsides and mountains are deforested, leaving the soil unprotected; people build homes and towns on floodplains; villages are nestled up to volcanic mountains; cities are constructed on known geological fault lines; and coastal marshes and mangrove forests are replaced by house lots. A general human tendency is to assume that disasters happen only to other people in other places, and even if there is a risk, say, of a hurricane striking a coastal island, many people are willing to take that calculated risk in order to enjoy life on the water's edge. Trends on the U.S. "hurricane coasts" (Gulf of Mexico and southern Atlantic) are a case in point: Coastal populations have increased 1% per year between 1990 and 2005, in spite of the many hurricane strikes. One writer suggested that town zoning boards should create a *stupid zone* for such locations. Other "stupid zone" locations would be in the middle of 25-year floodplains, in highly flammable coniferous forests, on top of earthquake faults, and in valleys below dormant volcanoes.

Certainly, part of the reason for the rising human and economic costs of natural disasters is the fact that there are more people in the world to be affected. But as one climatologist put it, if natural disasters are having an increasingly greater impact, the culprit is not Mother Nature, but human nature.

In Chapter 18, we will examine the question of whether global climate change is behind the rising tide of natural disasters, especially intense hurricanes, tornadoes, floods, and droughts. Whether it is or not, all the evidence suggests that climate change will definitely impact future natural disasters as temperatures and sea levels rise. This is a problem we are bringing on ourselves.

Chemical Hazards. Industrialization has brought with it a host of technologies that employ chemicals such as cleaning agents, pesticides, fuels, paints, medicines, and those used directly in industrial processes (Fig. 17–8). The manufacture, use, and disposal of these chemicals often bring humans into contact with them. Exposure is either through the ingestion of contaminated food and drink, the breathing of contaminated air, absorption through the skin, or direct use or by accident (discussed at length in Chapter 22). *Toxicity* (the condition of being harmful, deadly, or poisonous) depends not only on *exposure*, but also on the *dose* of the toxic substance—the amount actually absorbed by a sensitive organ. Further, for most substances, there is a threshold below which no toxicity can be detected. (See Figure 22–1.)

Making things even more difficult, different people have different thresholds of toxicity for given substances. For example, children are often at greater risk than adults because children are growing rapidly and are incorporating more of their food (and any contaminants that are in it) into new tissue. Embryos are even more sensitive. Substances transmitted across the mother's placenta can have grave impacts on the embryo's development, especially in the early stages.

Carcinogens. Many chemicals, such as heavy metals, organic solvents, and pesticides, are hazardous to human health even at very low concentrations. Episodes of acute

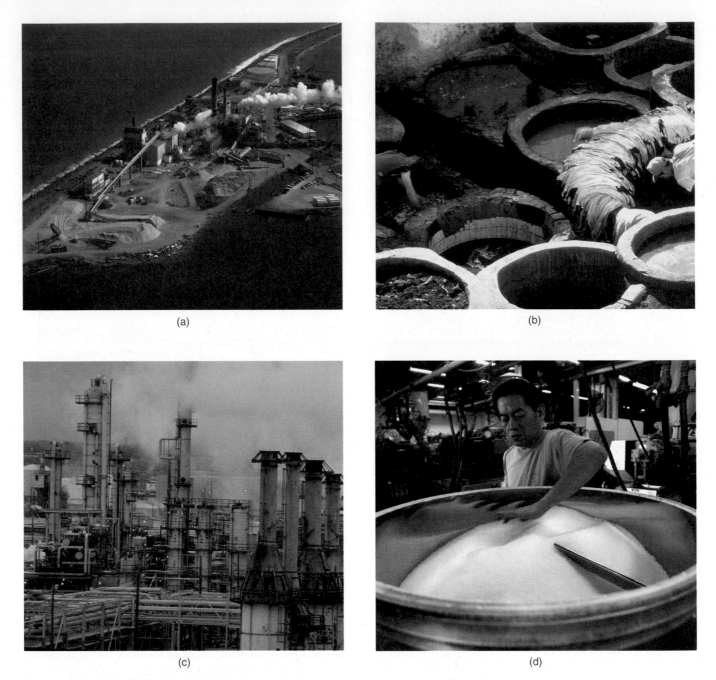

(a)

(b)

(c)

(d)

Figure 17–8 Industrial processes and hazards. Many industries produce chemicals and pollutants that contribute to the impact of toxic substances on human health. Some examples of hazardous workshops are (a) a paper mill in Port Angeles, Washington; (b) a tannery in Marrakech, Morocco; (c) an oil refinery in Philadelphia, Pennsylvania; and (d) a plastics factory in Singapore. Go to **GRAPHit!** at www.masteringenvironmentalscience.com

poisoning are easy to understand and are clearly preventable, but the long-term exposure to low levels of many of these substances presents the greatest challenge to our understanding. Some chemical hazards are known *carcinogens* (cancer-causing agents). Cancer takes a very large toll in the developed countries, accounting for one-fourth of mortality (see Chapter 9, Figure 9–13). Because cancer often develops over a period of 10 to 40 years, it is frequently hard to connect the cause with the effect. The *Eleventh Report on Carcinogens*,[4]

from the U.S. Department of Health and Human Services, lists 51 chemicals (and 3 biological agents) now known to be human carcinogens and 188 more that are "reasonably anticipated to be human carcinogens" because of animal tests. Many are substances used in both developed and developing countries. Other potential effects of toxic chemicals are impairment of the immune system, brain impairment, infertility, and birth defects.

Many developing countries are in double jeopardy because of both their high level of infectious disease and their rising exposure to toxic chemicals through industrial processes and home use. In both cases, the risks are largely preventable. In places where industrial growth is especially

---

[4]*Report on Carcinogens, Eleventh Edition*; U.S. Department of Health and Human Services, Public Health Service, National Toxicology Program.

rapid, insufficient attention is often given to the pollutants that accompany that growth. For example, 13 of the 15 cities with the worst air pollution are found in Asia, where economic growth has risen rapidly.

Cancer. In the United States, 23% of all deaths—559,300 in 2005—were traced to cancer, and even though it is largely a disease of older adults, cancer can strike people of all ages. How is it that a simple chemical like benzene can cause leukemia? How can aflatoxin cause liver cancer? To answer these questions, we must understand cellular growth and metabolism. During normal growth, cells develop into a variety of specific types that perform unique functions in the body: blood cells, heart muscle cells, liver cells, absorptive cells that line the small intestine, and so forth. Some cells, like the stem cells that form the cells of the blood, keep on dividing until death. Other cells, however, are programmed to stop dividing and simply perform their functions, often throughout life. A cancer is a cell line that has lost its normal control over growth. Recent research has revealed the presence of as many as three dozen genes that can bring about a malignant (out-of-control) cancer; these genes are like ticking time bombs that may or may not go off.

Carcinogenesis. Carcinogenesis (the development of a cancer) is now known to be a process with several steps, often with long periods of time in between. In most cases, a sequence of five or more mutations must occur in order to initiate a cancer. (A mutation is a change in the informational content of the DNA that controls processes in the cell.) *Environmental* carcinogens are either chemicals that are able to bind to DNA and prevent it from functioning properly or agents, such as radiation, that can strike the DNA and disrupt it. When such an agent initiates a mutation in a cell, the cell may go for years before the next step—another mutation—occurs. Thus, it can take upward of 40 years for some cells to accumulate the sequence of mutations that leads to a malignancy. In a malignancy, the cells grow out of control and form tumors that may spread (metastasize) to other parts of the body. Early detection of cancers has improved greatly over the years, and many people survive cancer who would have succumbed to it a decade ago. The best strategy, however, is prevention, so great attention is now given to the environmental carcinogens and habits, like tobacco use, that are clearly related to cancer development.

17.2 Pathways of Risk

The environmental hazards identified in Section 17.1 are responsible for untold human misery and death, but if many of the consequences of exposure to the different hazards are preventable, how is it that the hazards turn into mortality statistics? In other words, what are the pathways that lead from risks of infection, exposure to chemicals, and vulnerability to physical hazards to human deaths? If prevention is the goal—and it should be—this knowledge is crucial. In 2002, the WHO annual report[5] was entitled *Reducing Risks, Promoting Healthy Life.* The document is a thoroughly researched presentation of the major risk factors "for which the means to reduce them are known," as the report puts it. The report shows that a remarkably small number of risk factors are responsible for a vast proportion of premature deaths and disease. The top 10 factors (Fig. 17–9) are responsible for more than one-third of all deaths and much of the global disease burden. We will consider only a few examples of these risk factors because a thorough examination of this topic would take up volumes.

The Risks of Being Poor

One major pathway for hazards is poverty: The WHO has stated that poverty is the world's biggest killer. Environmental risks are borne more often by the poor, not only in the developing countries, but also in the developed countries of the world. The people who are dying from infectious diseases are those who lack access to adequate health care, clean water, nutritious food, healthy air, sanitation, and shelter. Because of this, they are exposed to more environmental hazards and thus encounter greater risks. The world's number one risk factor (Fig. 17–9) is called **underweight**, a problem of malnutrition that strikes primarily children less than five years of age. (See Chapter 12.) This factor is strongly related to poverty: Those households living on less than $1.25 per day (one-fifth of the world's population) show the highest prevalence of underweight in children. A 2008 report in *The*

---

[5]World Health Organization, *The World Health Report 2002: Reducing Risks, Promoting Healthy Life* (Geneva: World Health Organization, 2002). Available at www.who.int/whr/2002/en.

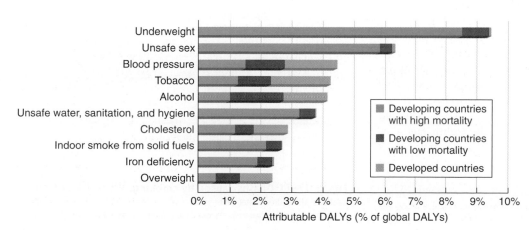

Figure 17–9 The 10 leading global risk factors. These 10 factors are responsible for more than one-third of all deaths and much of the global burden of disease. They are reported as a percentage of global DALYs (disability-adjusted life years); one DALY equals the loss of one healthy life year. (*Source:* The World Health Organization, *World Health Report 2002.*)

Figure 17–10 Public health clinic. This outreach immunization center in Bangladesh provides basic health information and services to people in the surrounding area.

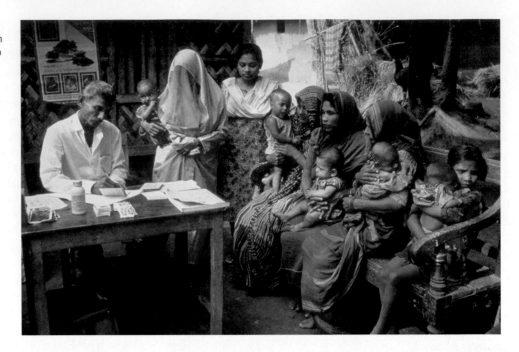

Lancet[6] found that more than one-third (2.2 million) of the childhood deaths under age five are caused by malnutrition.

Wealth. Overall, the wealthier a country becomes, the healthier is its population. However, in virtually all countries, those with wealth are able to protect themselves from many environmental hazards. People in developed countries tend to live longer, and when they die, the chief causes of death are diseases of old age—the so-called degenerative diseases, such as cancers and diseases of the circulatory system (heart disease, strokes, and so forth). By contrast, only one-third of the deaths in the developing countries are indicative of people who are living out their life span. Another indicator is age at death; people under the age of 50 account for two-fifths (more than 20 million) of the annual deaths in the world. Most of these deaths occur in the developing world, are due to infectious disease, and are preventable.

There are many reasons for this discrepancy between developed and developing nations. One is education. Advances in socioeconomic development and advances in education are strongly correlated. As people—and especially women—become more educated, they act on their knowledge to secure relief from hazards to health. They may, for example, improve their hygiene, immunize their children (Fig. 17–10), or recognize dangerous symptoms such as dehydration and seek oral rehydration therapy. Other reasons for the discrepancy between health in the developed world and that in the developing world are the lack of safe drinking water and sanitation and the high indoor air pollution associated with the use of solid fuels such as wood, dung, and coal in simple cooking stoves. This latter risk factor (the eighth factor in Fig. 17–9) leads to a high incidence of respiratory tract infections and chronic lung disease.

Priorities. Education, nutrition, and the general level of wealth in a society do not tell the whole story, however. A nation may make a deliberate policy choice to put its resources into improving the health of its population, rather than, say, militarization or the development of modern power sources. Countries such as Costa Rica, China, and Sri Lanka have a much longer life expectancy and lower infant mortality than their gross domestic product would predict. They have accomplished this enviable record by focusing public resources on such concerns as immunization, the upgrading of sewer and water systems, and land reform.

The Cultural Risk of Tobacco Use

Lifestyle choices such as refraining from exercise, overeating, driving fast, imbibing alcohol, climbing mountains, and so forth carry with them a significant risk of accident and death. However, one cultural hazard, tobacco use, is the leading cause of death in the United States (Fig. 17–11) and ranks

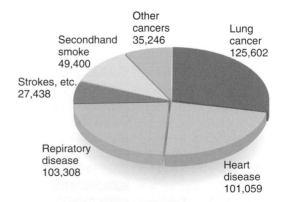

Secondhand smoke 49,400

Other cancers 35,246

Lung cancer 125,602

Strokes, etc. 27,438

Repiratory disease 103,308

Heart disease 101,059

Figure 17–11 Deaths caused by smoking. Approximately 443,000 deaths in the United States (average annual number, 2000–2004) are attributed to smoking. (*Source:* Centers for Disease Control, MMWR, November 14, 2008.)

[6]Robert E. Black et al. "Maternal" and child undernutrition: global and regional exposures and health consequences. *The Lancet* 371: 243–260. Jan. 19, 2008.

fourth in the global tally of lethal risk factors (Fig. 17–9). Although tobacco use is declining in some developed countries, it is on the increase in most developing countries and remains high in the former socialist countries of eastern Europe. Smoking often begins in childhood. Lured by the image of smoking as "cool" or by peer pressure, youngsters are very quickly addicted by the nicotine in cigarettes, an addiction that can last a (shortened) lifetime. According to the CDC, some 43.4 million adults (19.8%) in the United States smoke cigarettes, and half of these will die or become disabled because of their habit.

Marlboro Country?   Tobacco is the only known product on the market that kills half of its users. Cigarette smoking has been clearly and indisputably correlated with cancer and other lung diseases. Smoking has been shown to be responsible for 29% of cancer deaths in the United States. The WHO puts the number of smokers worldwide at 1.9 billion, with 5.4 million deaths per year due to cigarette smoking. Eighty percent of smokers are in the developing world, where the habit is increasing at a rate of 2% per year. In China, 57% of men are smokers. The WHO reported in 2008 that smoking is a growing epidemic in the developing countries, where its full impact is yet to come because tobacco-related disease and death don't begin until decades after the start of tobacco use.

Studies have shown that smokers living in polluted air experience a much higher incidence of lung disease than do smokers living in clean air—a synergistic effect. Certain diseases typically associated with occupational air pollution show the same synergistic relationship with smoking. For example, black lung disease is seen almost exclusively among coal miners who are also smokers, and lung disease predominates in smokers who are exposed to asbestos. Smoking is linked to half of the tuberculosis deaths in India, which leads the world in that category. Smoking also kills by impairing respiratory and cardiovascular functioning. The CDC estimates that smoking costs the United States $193 billion a year in health care costs and lost job productivity because of premature death and disability.

In the United States, several measures have been taken to regulate this cultural hazard. Tobacco products are highly taxed, for example, providing substantial revenues to the states ($14.5 billion) and the federal government ($7.3 billion). Raising cigarette taxes has been shown to be one of the most effective ways to reduce smoking, especially in young people. Surgeons general of the United States have issued repeated warnings against smoking since 1964, and public policy has taken them seriously by requiring warning labels on smoking materials, banning cigarette advertising on television, promoting smoke-free workplaces, requiring nonsmokers' areas in restaurants, and banning smoking on all domestic airline flights. Since the warnings began, the U.S. adult smoking population has gradually dropped from 42% to 19.8%.

Secondhand Smoke.   In January 1993, the EPA classified environmental tobacco smoke (ETS), meaning secondhand smoke, as a Class A (known human) carcinogen. In the words of then EPA Administrator Carol Browner, "Widespread exposure to secondhand smoke in the United States presents a serious and substantial public health risk." The EPA's action was unanimously supported by its science advisory board. The consequences of this action have been substantial: Smoke-free zones and outright bans in public areas such as shopping malls, bars, restaurants, and workplaces are now common. In particular, the EPA has taken specific steps to protect children from ETS in all public places, and the Occupational Safety and Health Administration is promoting policies to protect workers from involuntary exposure to ETS in the workplace. The impact of these actions on public health should be significant.

Goodbye, Joe.   Another significant development came from the Food and Drug Administration. In February 1994, citing the addictive properties of nicotine, then FDA Administrator David Kessler proposed that the agency have the authority to regulate cigarettes as a drug. At congressional hearings following the FDA's initiative, tobacco executives denied that cigarette smoking was addictive. The hearings included allegations of a cover-up of the tobacco industry's own research that revealed the addictive properties of nicotine. The FDA launched efforts to curb tobacco sales to the under-aged. One early casualty was "Joe Camel," a cartoon character that seemed to be targeted at the young. In March 2000, however, the U.S. Supreme Court ruled that the FDA did not have the authority to regulate tobacco. That authority, said the Court, belonged to Congress, and until 2009, the many bills filed in Congress to extend that authority to the FDA did not make it into law. However, in June 2009, the House and Senate voted overwhelmingly to give the FDA sweeping authority to regulate tobacco products, finally overcoming the efforts of a powerful tobacco lobby.

Legally Speaking.   On the legal front, there has been some success—and some failure. The attorneys general of a number of states brought suit to recover some of their Medicaid expenses brought about by smoking-related illnesses. In late 1998, 46 states reached a $246 billion settlement with the top four tobacco companies wherein the companies would remunerate the states over a period of years, while the states would agree to drop lawsuits asking the companies to pay for their Medicaid expenses. The money has been used not only to reimburse the states, but also to help finance programs to discourage people—especially youths—from smoking.

In 1999, the federal Justice Department filed a suit against the six largest tobacco companies, accusing them of lying about the dangers of smoking, manipulating nicotine levels, and directing huge promotional campaigns at children. The government sought $280 billion from the companies, arguing that this represents 30 years of past profits that the companies received unlawfully. That suit lasted six years—until, in August 2006, U.S. District Judge Gladys Kessler ruled that, although the tobacco companies conspired to deceive the public about the dangers of their products, an earlier court ruling prohibited her from imposing the multibillion-dollar fine requested by the Justice Department.

FCTC and MPOWER. The battle against smoking is rapidly spreading worldwide. After several years of efforts, the WHO *Framework Convention on Tobacco Control (FCTC)* was adopted in May 2003 by the World Health Assembly. This convention is a treaty that aims to reduce the spread of smoking by requiring signatory countries to raise taxes, restrict advertising, require larger health warnings on tobacco materials, and reduce ETS (steps already taken in the United States). The treaty went into effect on February 27, 2005, and has been signed by 168 countries. Many countries have already taken action called for in the treaty; Ireland, Spain, Norway, and Italy have banned smoking in indoor public places, and other countries have begun banning advertising and requiring printed warnings on cigarette packages.

The WHO, with funding of $500 million from Bill Gates and New York City's Mayor Michael Bloomberg (Fig. 17–12), initiated a campaign in 2008 to combat the global smoking epidemic in developing countries. Given the acronym MPOWER, the campaign has six components with proven strategies to reduce tobacco use:

- Monitor tobacco use and prevention policies.
- Protect people from tobacco smoke.
- Offer help to quit tobacco use.
- Warn about the dangers of tobacco.
- Enforce bans on tobacco advertising, promotion, and sponsorship.
- Raise taxes on tobacco.

MPOWER will work with the FCTC and employ WHO agencies and partners to promote the campaign policies through antismoking campaigns in the developing countries. Although the MPOWER strategies have been effective in reducing smoking in the United States, only 5% of the rest of the world uses any of these components at present. That should change, given the new funding available.

Risk and Infectious Diseases

Epidemiology has been described as "medical ecology" because the epidemiologist traces a disease as it occurs in geographic locations as well as tracing the mode of transmission and consequences of the disease in the field. Epidemiologists point to many reasons infectious diseases and parasites are more prevalent in the developing countries. One major pathway of risk is contamination of food and water due to inadequate hygiene and inferior sewage treatment. Some of the risk can be traced to the failure to educate people about diseases and how they can be prevented. Much, however, is simply a consequence of the general lack of resources needed to build effective public-health-related infrastructure in cities, towns, and villages of the developing world. According to the UN's 2008 report on the Millennium Development Goals, an estimated 2.5 billion people are without adequate sanitation, and 870,000 lack access to safe drinking water. Many diarrheal diseases are transmitted in this way, and some of these, such as cholera and typhoid fever, are deadly and capable of spreading to epidemic proportions. Any diarrheal infection in small children can be fatal if it is accompanied by malnutrition or dehydration. By making oral rehydration therapy kits readily available throughout the country, Mexico reduced mortality rates in children by 70% in less than a decade.

Even Here? These problems are not unique to the developing world. Inadequate surveillance in developed countries sometimes leads to major outbreaks of diarrheal diseases, with fatal outcomes. For instance, in 1993, the water supply

Figure 17–12 Bill Gates and Michael Bloomberg announce MPOWER. MPOWER is the new campaign to stop people around the world from smoking; Gates and Bloomberg invested $500 million in the campaign in 2008.

of Milwaukee, Wisconsin, became contaminated with *Cryptosporidium parvum*, a protozoal parasite originating in farm animal wastes. Some 370,000 people developed diarrheal illnesses from the outbreak, and more than 4,000 were hospitalized. The factorylike conditions under which food processing in the developed countries takes place often give pathogens an opportunity to contaminate food on a wide scale, resulting in diarrheal outbreaks and major recalls of food products. Peanut butter, that staple food for kids of all ages, was implicated in an outbreak of *Salmonella typhimurium* in 2008–2009. This bacterium infects the intestinal tract and brings on diarrhea, fever, and cramps. Over 500 cases were recorded, and the source was traced to a Peanut Corporation of America factory in Georgia. The occurrence of such outbreaks, the resurgence of known diseases, and the emergence of new ones like SARS, AIDS, avian flu, swine flu, and the Ebola virus are the concern of public-health agents and epidemiologists in organizations such as the WHO and the CDC. (See Sound Science, "An Unwelcome Globalization.")

Tropical Diseases. The tropics, where most of the developing countries are found, have climates ideally suited for the year-round spread of disease by insects. Mosquitoes are *vectors* for several deadly and debilitating diseases, such as yellow fever, dengue fever, elephantiasis, Japanese encephalitis, West Nile virus, and malaria. Of these, malaria is by far the most serious (Fig. 17–13). Public-health attempts to control malaria have been aimed at **vector control**—attacking the *Anopheles* mosquito with the use of insecticides—and **treatment strategies**—curing people once they have been infected. Malaria was once prevalent in the southern United States, but an aggressive campaign in the 1950s to eliminate the *Anopheles* mosquito and identify and treat all human cases of malaria led to the complete eradication of the disease there.

Unfortunately, in the tropics the mosquitoes have developed resistance to all of the pesticides employed, and eradication has remained an elusive goal. DDT is still used in some developing countries to spray the walls of huts and houses, but its use in this manner is highly controversial because of the pesticide's well-known harmful environmental and health impacts. (See Stewardship, "DDT for Malaria Control," p. 332). However, the spraying has proven to be highly effective. Resuming their use of DDT after several years with another pesticide, South Africa and Zambia were able to dramatically reduce their malaria caseloads.

Sound Science

An Unwelcome Globalization

We live in a time when globalization is rapidly encompassing travel, information, trade, and investment. The Internet ties people together in ways unimagined a few years ago. The globalization of health, however, remains an elusive goal, similar to the globalization of economic well-being. Laurie Garrett, in *The Coming Plague*,* describes an unwelcome form of globalization: the globalization of disease. Garrett examines the recent history of emerging diseases such as AIDS, Ebola, hantavirus, Rift Valley fever, Legionnaires' disease, and others. She also explains the resurgence of familiar diseases like tuberculosis, cholera, and pneumonia partly as a consequence of the widespread overuse of antibiotics. Many of the new diseases are clearly linked to changes in land use, which bring humans into close contact with rodents or other animals that harbor viruses previously unknown to medicine and often deadly to humans. Resurgent diseases, by contrast, are a creation of our medical practice. By treating people with antibiotics without restraint, we unknowingly select strains that are immune to the antibiotics and that pass on their resistant genes to unrelated bacteria by way of plasmid transfer. (Plasmids are small circular pieces of DNA that can move from one bacterium to another in a semisexual process.) The most notorious recent example of this is methicillin-resistant *Staphylococcus aureus* (MRSA). This bacterium, normally found on the skin, can thrive in many places in the body and is responsible for almost 19,000 fatalities a year in the United States alone. MRSA has become resistant to practically every antibiotic and is frequently acquired in hospitals.

The heroes of the book are the women and men on the front lines of epidemiology, people such as Joe McCormick of the CDC, who doggedly pursued lethal viruses in Africa, and Patricia Webb, who worked on the deadly viruses in maximum-containment laboratories back in the United States. Garrett makes a plea for a greater commitment from our universities, medical schools, and government agencies to train workers who will be capable of recognizing new diseases and who will be able to move about equally well in the laboratory, the hospital, and the field in pursuit of knowledge and public-health intervention around the world.

The importance of Garrett's perspective was recently emphasized with the appearance of several new and deadly viral diseases. This chapter began with the story of avian flu and swine flu, which have not yet run their course. The West Nile virus, which is carried by birds (and is fatal to them), first appeared in the United States in 1999 and has spread to almost every state. In 2008, 1,356 cases were recorded, with 44 deaths. This virus, spread by mosquitoes, may have traveled to the United States in a bird or an infected human. A particularly deadly virus causing SARS appeared in China in late 2003 and spread rapidly to several countries. SARS infected more than 8,000 people and killed over 770 before it was contained in mid-2003. The epidemic is believed to have originated in wild animals, which are often sold as food in China, and then spread human-to-human. A recent count indicated that three-fourths of emerging diseases are diseases of wild animals and that their appearance is increasing in frequency. The potential of these diseases for causing sickness and death has been proven; the one sure fact is that we have not seen the last of them.

*Laurie Garrett, *The Coming Plague: Newly Emerging Diseases in a World Out of Balance* (New York: Penguin Books, 1995).

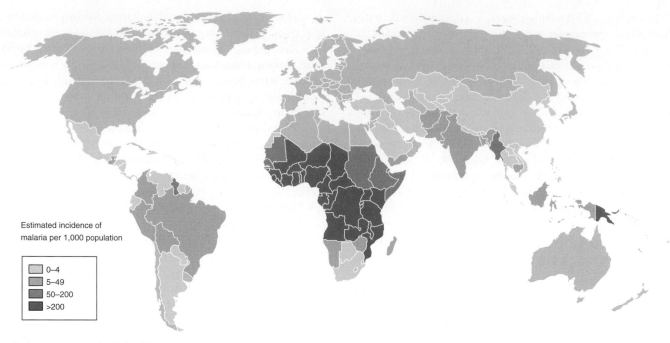

Figure 17–13 Malaria incidence per 1,000 population, 2006. There were an estimated 247 million cases of malaria in 2006, most of which were in sub-Saharan Africa. An estimated 881,000 deaths occurred in 2006 as a result of malarial infection. *(Source:* World Health Organization, *World Malaria Report 2008.*)

Net Results. One highly promising effort funded by the WHO's Tropical Disease Research Program found that children provided with insecticide-treated nets over their beds (bed nets) experienced a substantial reduction in mortality from all causes (Fig. 17–14). In Vietnam, a combination of bed nets and more-effective drugs reduced malaria deaths by 98% in seven years. This strategy is the basis of a cost-effective, large-scale intervention throughout Africa to reduce the high mortality from malaria, and it is working. Bed nets are especially effective against *Anopheles* mosquitoes, which emerge from hiding and feed primarily at night. Using a combination of long-lasting insecticide-treated bed nets, indoor DDT spraying, and quick access to drug treatment, Zambia aims to reduce malaria mortality 75% by 2009. Already, over 4 million bed nets have been distributed, and huge declines in malaria incidence have been seen.

Complicating control of the disease, the *Plasmodium* protozoans have developed resistance to one treatment drug after another. Until recently, chloroquine was quite effective against malaria in Africa. Now it is ineffective, and health workers have turned to artemisinin, a drug derived from the common Chinese weed *Artemisia annua*. This drug works quickly and effectively, but the WHO says it should be used only in combination with several other antimalarials, called artemisinin combination therapy (ACT), because of the danger of creating artemisinin-resistant strains of the malarial parasite when used alone. Indeed, ACT resistance has begun to crop up in western Cambodia and is being met with a containment plan to eliminate malaria from regions where the resistance has been found.

Good News. In 2002, molecular biologists successfully sequenced the genomes of the *Anopheles* mosquito and the most lethal malaria parasite, *P. falciparum*. Armed with this information, researchers are now able to target potential weak points in both organisms in the search for new vaccines and drugs. Research continues on the development of new, more effective antimalarial drugs and on the development of

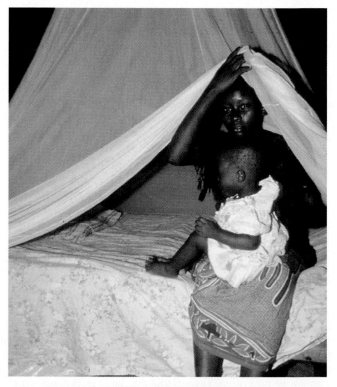

Figure 17–14 Insecticide-treated bed net. This Kenyan woman and her child were protected at night from malaria-bearing mosquitoes. Bed nets have been shown to prevent infection.

an effective vaccine. Although the parasite's complicated life cycle and the fact that malaria does not readily induce long-term immunity have made a vaccine an elusive goal, a vaccine recently developed by GlaxoSmithKline—labeled RTS,S—is promising enough to begin field tests in Africa.

A number of plans have been developed over the years aimed at conquering malaria, the latest of which is the Global Malaria Action Plan (GMAP). Unlike the others, this plan has an extremely ambitious goal: reduce malaria deaths to near zero by 2015 and then gradually eliminate malaria everywhere until it is eradicated. Initial commitments of almost $3 billion in late 2008 are promising, but full implementation of the plan will require many more billions. Even if GMAP is not completely successful, it will have very likely met the Millennium Development Goal target of having halted and begun to reverse the incidence of malaria.

Toxic Risk Pathways

How is it that people are exposed to chemical substances that can bring them harm? Airborne pollutants represent a particularly difficult set of chemical hazards to control, as they are difficult to measure and difficult to avoid. Humans breathe 30 pounds (14 kg) of air into their lungs each day. Although some of the symptoms of pollution that people suffer involve the moist surfaces of the eyes, nose, and throat, the major site of impact is the respiratory system. Three categories of impact can be distinguished:

1. **Chronic.** Pollutants cause the gradual deterioration of a variety of physiological functions over a period of years.

2. **Acute.** Pollutants bring on life-threatening reactions within a period of hours or days.

3. **Carcinogenic.** Pollutants initiate changes within cells that lead to uncontrolled growth and division (cancer), which is frequently fatal.

Indoor Air Pollution. Much attention is given to air pollutants released to the atmosphere—and rightly so. (Chapter 19 discusses these pollutants.) However, *indoor air pollution* can pose an even greater risk to health. The air inside the home and workplace often contains much higher levels of hazardous pollutants than outdoor air does! In the developed countries, the overall indoor air pollution problem is threefold. First, increasing numbers and types of products and equipment used in homes and offices give off hazardous fumes. Second, buildings have become increasingly well insulated and sealed; hence, pollutants are trapped inside, where they may accumulate to dangerous levels. Third, people are exposed more to indoor pollution than to outdoor pollution. The average person spends 90% of his or her time indoors, and the people who spend the most time inside are those most vulnerable to the harmful effects of pollution—namely, small children, pregnant women, the elderly, and the chronically ill.

The sources of indoor air pollution are quite varied, as evidenced in Figure 17–15. One of the least excusable of these is smoking, which carries a much higher health risk than any of the other sources shown in Figure 17–15. ETS

contains 40 substances that are known carcinogens and many others that are potent respiratory irritants.

Developing Countries. The most serious indoor air pollution threat is found in the developing world, where at least 3 billion people continue to rely on biofuels like wood and animal dung for cooking and heat. This threat was identified as one of the top 10 health risks by the WHO (Fig. 17–9). Fireplaces, cooking fires, and stoves are often improperly ventilated (if at all), exposing the inhabitants to very high levels of pollutants, such as carbon monoxide, nitrogen and sulfur oxides, soot, and benzene (the last two being known carcinogens). Four major problems have been associated with indoor air pollutants in the developing world: acute respiratory infections in children; chronic lung diseases, including asthma and bronchitis; lung cancer; and birth-related problems. One study in Mexico showed that women who were continually exposed to indoor smoke had 75 times the risk of developing a chronic lung disease as women who weren't exposed. Solutions to this problem include the use of ventilated stoves that burn more efficiently and the conversion from biofuels to bottled gas or liquid fuels such as kerosene.

Asthma. One common consequence of indoor air pollution is asthma, which is currently at epidemic proportions in the United States, afflicting 20 million people each year. An allergic disease, asthma attacks the respiratory system so that air passages tighten and constrict, causing wheezing, tightness in the chest, coughing, and shortness of breath. In acute attacks, the onset is sudden and sometimes life threatening. Substances that can trigger asthma attacks include dust, animal dander, mold, secondhand smoke, swimming pool chlorine, and various gases and particles. Asthma is expensive, too, with direct and indirect costs estimated at more than $11 billion a year in the United States, where more than 3,800 die from the condition annually. Through its Indoor Environments Division Web site, the EPA provides helpful guidelines for controlling many of the substances that trigger asthma in schools, homes, and other indoor environments.

Worms, Anyone? Oddly, asthma is much more common in developed countries, in spite of the greater exposure to indoor air pollution in developing countries. Recent research has suggested that frequent parasitic worm infections stimulate a well-regulated anti-inflammatory network in the immune systems of most people in the developing countries, protecting them against many allergic diseases like asthma. In the developed countries, where such parasitic infections are uncommon, this protection is lacking, so allergic disorders are much more common. Although this disparity suggests that it might be beneficial to play host to a few worms, it would be preferable to identify the parasite molecules that induce the protective immune response and to employ those molecules as vaccines.

Toxicology. Even though the substances found in indoor air are hazardous, the link between their presence in the air and the development of a health problem is harder to establish than with the infectious diseases. As with cancer, often the best that can be done is to establish a statistical correlation between exposure levels and the development of

Figure 17–15 Indoor air pollution. Pollutants originate from many sources and can accumulate to unhealthy levels, leading to asthma and "sick-building syndrome" (characterized by eye, nose, and throat irritations; nausea; irritability; and fatigue) as well as other serious problems.

an adverse effect. The science of **toxicology** studies the impacts of toxic substances on human health and investigates the relationships between the presence of such substances in the environment and health problems like cancer. Chapter 22 discusses toxic substances and toxicology in greater detail.

Disaster Risk

Each year, more than 200 million people are affected by natural disasters. These disasters inevitably trigger emergency relief efforts and subsequent humanitarian assistance and reconstruction—all good things. The global costs of weather-related disasters have increased dramatically in recent years, from $9 billion per year in the 1980s to $45 billion in the 2000s. Enormous resources are devoted to these efforts, and societies need to maintain their capacity to respond swiftly and effectively to natural disasters. They receive help with this from the UN Disaster Assessment and Coordination (UNDAC) system; in 2008, the UNDAC sent teams to 16 areas.

However, much of the human death and misery brought on by natural hazards could be prevented through rigorous *disaster risk reduction* strategies. Two resources have recently become available to assist societies to move in this direction. One is the publication *Living with Risk: A Global Review of Disaster Reduction Initiatives*,[7] produced in 2004 by the Inter-Agency Secretariat of the International Strategy for Disaster Reduction. This publication, available on the Internet, applies the concepts of hazards, risks, and risk assessment to natural disasters and goes on to recommend policies and a framework for guiding and monitoring disaster risk reduction.

A second resource was created in January 2005, when 168 country delegations met in Kobe, Japan, for the UN World Conference on Disaster Reduction. At that conference,

[7]Inter-Agency Secretariat of the International Strategy for Disaster Reduction, *Living with Risk: A Global Review of Disaster Reduction Initiatives* (Geneva: United Nations International Strategy for Disaster Reduction (UN/ISDR), 2004). https://unp.un.org/.

delegates adopted the *Hyogo Framework for Action*, a global blueprint for disaster risk reduction efforts for the coming decade. The Framework identified five priorities for action:

1. **Make Disaster Risk Reduction a Priority:** Ensure that disaster risk reduction is a national and a local priority with a strong institutional basis for implementation.

2. **Know the Risks and Take Action:** Identify, assess, and monitor disaster risks—and enhance early warning. A tsunami warning system in the Indian Ocean would have prevented thousands of deaths on December 26, 2004.

3. **Build Understanding and Awareness:** Use knowledge, innovation, and education to build a culture of safety and resilience at all levels.

4. **Reduce Risk:** Reduce the underlying risk factors. For example, in earthquakes, it is usually the houses and buildings that kill the people. Building standards in earthquake-prone areas can greatly reduce the death toll of an earthquake.

5. **Be Prepared and Ready to Act:** Strengthen disaster preparedness for effective response at all levels.

We now turn our attention to *risk assessment*, one of the most important tools of the toxicologist in the developed countries. Risk assessment is an approach to the problems of environmental health and is now a major element of public policy—but also the subject of some controversy.

17.3 Risk Assessment

Although, in the developed countries, our lives are freer of hazards and risks than ever before, our society still faces hazards and the risks that they present to our health. For our own self-interest, we should know about and evaluate the risks to which we are subjected. Such knowledge enables us to make informed choices as we consider the benefits and risks of the hazards around us. In the developing countries, the heavy burden of disease so common in many of the poorer regions calls for a medical response out of humanitarian concern. But *treating the sick*, as helpful and appropriate as it is, will go on indefinitely if *prevention* is ignored. According to the WHO 2002 report, "Focusing on risks to health is the key to preventing disease and injury." When the risks are thoroughly understood, strategies for reducing disease and death are far easier to identify and prioritize. If governments are to act as stewards of the health of their people, they desperately need the kind of information a thorough evaluation of health risks can bring them.

Risk assessment is *the process of evaluating the risks associated with a particular hazard before taking some action in a situation in which the hazard is present.* Figure 17–16

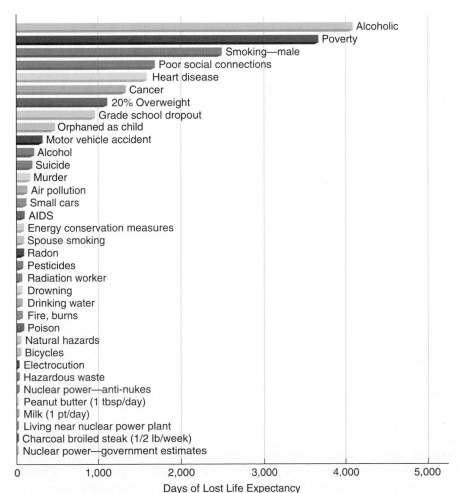

Days of Lost Life Expectancy

Figure 17–16 Loss of life expectancy from various risks. One way to illustrate the relative risks posed by different hazards is to rank them in terms of average days of lost life associated with each risk. (*Source:* R. Wilson and E. A. C. Crouch, *Risk-Benefit Analysis,* 2nd ed. Cambridge: Harvard University Press, 2001.)

TABLE 17–3	Some Commonplace Hazards, Ranked According to Their Degree of Risk

Hazardous Action	Annual Risk*		
Cigarette smokers	10.0	per	1,000
All cancers	2.0	per	1,000
Firefighters	4.0	per	10,000
Police killed in line of duty	2.9	per	10,000
Hang gliding	2.6	per	10,000
Air pollution	2.5	per	10,000
Motor vehicle accident	1.6	per	10,000
Snowmobiling	1.3	per	10,000
Home accidents	1.1	per	10,000
Firearms	1.1	per	10,000
Airline pilot	10.0	per	100,000
Mountain climbing	6.4	per	100,000
Alcohol consumption	6.4	per	100,000
Boating	5.0	per	100,000
Swimming	3.0	per	100,000
Four tablespoons peanut butter per day (cancer risk from aflatoxin)	1.3	per	100,000
Drinking water containing EPA limit of chloroform	7.0	per	1,000,000
Chest X-ray	3.0	per	1,000,000
Electrocution	1.8	per	1,000,000

*Probability of dying.
Source: R. Wilson and E. A. C. Crouch, *Risk-Benefit Analysis*, 2nd ed. (Cambridge, Mass.: Harvard University Press, 2001), copyright © 2001 by Richard Wilson and Edmund Crouch. Reprinted by permission of the authors.

shows a number of risks, expressed as days of lost life expectancy, for people in a typical developed country. Alternatively, risks may be expressed as the probability of dying from a given hazard. According to Table 17–3, for example, the annual risk of dying from a motor vehicle accident is 1.6 per 10,000. That is, in the course of a year, 1.6 out of every 10,000 who ride in or drive a motor vehicle will die as the result of an accident. Figure 17–16 suggests that this is equivalent to about 300 days of lost life expectancy. In other words, people who never enter a motor vehicle will live an average of 300 days longer than those who do.

Truthfully, not very many people actually make choices about the hazards in their lives on the basis of such risk assessments. However, risk assessment has become an important process in the development of public policy and has long been a major way of applying science to the hard problems

of environmental regulation. The WHO is now recommending risk assessment as an ideal way to promote the health of people everywhere.

Risk Assessment by the EPA

Risk assessment began at the EPA in the mid-1970s as a way of addressing the cancer risks associated with pesticides and toxic chemicals. It has developed into a highly science-based methodology directed toward understanding how pollutants may impact human health and the environment. The National Center for Environmental Assessment (NCEA) is the EPA's lead agency for risk assessment. As currently performed at the EPA, risk assessment for human health risks has four steps: hazard assessment, dose-response assessment, exposure assessment, and risk characterization. The description that follows includes elements of new guidelines for risk assessment that were updated in 2005 by the EPA.

Hazard Assessment: Is This Really a Hazard? Hazard assessment is the process of examining evidence linking a potential hazard to its harmful effects. In the case of accidents, the link is obvious. The use of cars, for example, involves a certain number of crashes and deaths. In these cases, *historical data*, such as the annual highway death toll, are useful for calculating risks.

In other cases, the link is not so clear because there is a time delay between the first exposure and the final outcome. For example, establishing a link between exposure to certain chemicals and the development of cancer some years later is often very difficult. In cases where the linkage is not obvious, data may be needed from epidemiological studies or animal tests. An **epidemiological study** tracks how a sickness spreads through a community. Thus, to find a link between cancer and exposure to some chemical, an epidemiological study would examine all the people exposed to the chemical and determine whether that population has more cancer than the general population. These findings are considered to be the best data for risk assessment and have resulted in scores of chemicals being labeled as known human carcinogens.

Ask the Mice. We do not want to wait 20 years to find out that a new food additive causes cancer, so we turn to **animal testing** (Fig. 17–17) to find out now what *might* happen in the future. A test involving several hundred animals (usually mice) takes about three years and costs more than $250,000. If a significant number of the animals develop tumors after having been fed the substance being tested, then the substance is either a possible or a probable human carcinogen, depending on the strength of the results.

Three objections to animal testing have been raised: (1) Rodents and humans may have very different responses to a given chemical; (2) the doses used on the animals are often much higher than those to which humans may be exposed; and (3) some people are opposed on ethical grounds to the use of animals for such purposes. Although there are obvious differences between rodents and humans, all chemicals shown by epidemiological studies to be human carcinogens

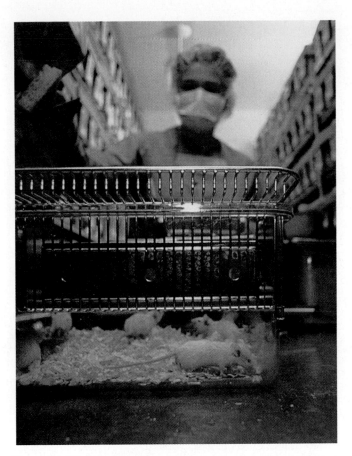

Figure 17–17 Animal testing. Laboratory mice are routinely used to test the potential of a chemical to cause cancer. The mice are an important source of information helping to assess the presence of hazards in food, cosmetics, or the workplace.

are also carcinogenic to test animals, suggesting that animal tests have some predictive value for humans.

Just Test People? In the past, chemical manufacturers that were interested in assessing the safety of pesticides (in order to have their products approved for use) paid human subjects to test the products. Such tests are far less expensive than the usual animal testing. (Volunteers may be paid only a few hundred dollars; mice apparently charge much higher fees!) A National Academy of Sciences panel addressed this topic in 1998, stating that the use of human volunteers "to facilitate the interests of industry or of agriculture" is unjustifiable. The EPA agreed and banned any use of such tests. However, the Bush administration, responding to requests from manufacturers, pressed the EPA to reevaluate its policy and allow human testing. A 2006 rule from the EPA forbids such testing on children and pregnant women (but allows it for others), satisfying the pesticide manufacturers. The Natural Resources Defense Council has challenged the rule in court, and it is hoped that the Obama administration will stop human pesticide testing altogether.

Another source of information is the chemical or process itself: What are its physical and chemical properties, and what mode of action might the chemical or process take in inducing cancer? Hazard assessment takes a "weight-of-the-evidence" approach to determining the carcinogenic potential of a chemical or process. Standard descriptors are used in its conclusions, such as "likely to be carcinogenic to humans." In the end, hazard assessment tells us that we *may* have a problem.

Dose-Response and Exposure Assessment: How Much for How Long? When animal tests or human studies show a link between exposure to a chemical and an ill effect, the next step is to analyze the relationship between the concentrations of the chemical in the test (the **dose**) and both the incidence and the severity of the response. It is at this point that vulnerability may be introduced; for example, it is known that children are more vulnerable to some cancer-causing chemicals because they are growing. From this information, projections are made about the number of cancers that may develop in humans who are exposed to different doses of the chemical. Unless there is reason to assume otherwise, a linear response is used to determine an acceptable level of exposure. This process is called **dose-response assessment**.

Following dose-response assessment is **exposure assessment**. This procedure involves identifying human groups already exposed to the chemical, learning how their exposure came about, and calculating the doses and length of time of their exposure.

Risk Characterization: What Does the Science Say? The final step, **risk characterization**, is to pull together all the information gathered in the first three steps in order to determine the risk and its accompanying uncertainties. Here is where science is able to inform the risk manager. The characterization summary should provide information about key supportive data and methods, their limitations, and, especially, the estimates of risk and the uncertainties involved. Commonly, risk is expressed as the probability of a fatal outcome due to the hazard, as in Table 17–3 for common, everyday risks. For example, the EPA expresses cancer risk as an "upper-bound, lifetime risk," meaning the highest value in the range of probabilities of the risk, calculated over a lifetime. The Clean Air Act of 1990 directs the EPA to regulate chemicals that have a cancer risk of greater than one in a million for the people who are subject to the highest doses. That is the same standard employed by the FDA in regulating chemicals in food, drugs, and cosmetics. One important limitation of risk characterization is that the analysts must work with the available information, which never seems to be sufficient. The final estimate of risk, therefore, includes a measure of the imprecision that always accompanies the use of imperfect information. Over 300 risk assessments on subjects from aerosols to watersheds are published on the EPA's NCEA Web site.

Public-Health Risk Assessment

As it was preparing the *2002 World Health Report*, the WHO was breaking new ground in the area of risk assessment. Until then, most risk assessment took place as a function of environmental regulation. The new effort began by looking at all of the risk factors commonly known to be responsible for poor health and mortality and then asking this question: "Of all the disease burden in this population, how much could be caused by this [particular] risk?" Then certain risk factors were chosen for special attention on the basis of the following criteria:

- **Potential global impact.** The risk is likely to be among the leading causes of poor health and of mortality. For example, the risks of being overweight are now well documented in many of the developed countries, where the condition is a serious problem.

- **High likelihood of causality.** It should be possible to trace the cause-effect relationships between poor health and the factor in question. Thus, overweight persons have a greater tendency to develop diabetes, high blood pressure, and high cholesterol; these in turn lead to greater risks of heart disease, stroke, and many forms of cancer.

- **Modifiability.** It should be possible to develop country-wide risk reduction policies that substantially lessen the impact of the factor. For instance, overeating is a matter of individual behavior, which can be difficult to change, but a society can draw greater attention to the epidemic of obesity, its health risks, and the steps that can be taken to prevent overeating.

- **Availability of data.** There must be reliable data on the prevalence of the risk factor and its relationship to disease and mortality. With regard to overeating, the CDC coordinates a state-based monitoring system (the Behavioral Risk Factor Surveillance System) that collects information via telephone interviews. The CDC reports information on overeating and other, similar behavior-related health risks (participating in insufficient physical activity, failing to eat fruits and vegetables, using tobacco and alcohol) to the state and local health departments.

The DALY.   The data collected must be assessed with the use of a "common currency," one that facilitates a comparison of different types of risks. The WHO report uses the measurement known as the *DALY*—the *disability-adjusted life year*. One DALY represents the loss of one healthy year of a person's life. This measure, now in common usage in the global health community, assesses the burden of a disease in terms of life lost due to premature mortality and time lived with a disability (see Figure 17–9).

The WHO report uncovered some remarkable differences in risk factors in different groups of countries (Table 17–4). Each risk factor calls for a different suite of interventions. As a country evaluates its highest health risks, it will often have to prioritize its risk prevention strategies. This is especially true of the high-mortality developing countries, where public-health resources are severely limited. It is the opinion of the WHO that the scientific bases for the main risk factors presented in the 2002 report are well understood. Also, the strategies for cost-effective risk prevention for the main risks are straightforward. There is broad agreement about the need for action, for example, to address the increase in tobacco consumption and the role of unsafe sex in the HIV/AIDS epidemic. The WHO has an important role

TABLE 17–4   Ten Leading Selected Risk Factors as Percentage Causes of Disease Burden, Measured in DALYs*

Developing Countries	
High-mortality countries	
Underweight	14.9%
Unsafe sex	10.2%
Unsafe water, sanitation, and hygiene	5.5%
Indoor smoke from solid fuels	3.7%
Zinc deficiency	3.2%
Iron deficiency	3.1%
Vitamin A deficiency	3.0%
Blood pressure	2.5%
Tobacco	2.0%
Cholesterol	1.9%
Low-mortality countries	
Alcohol	6.2%
Blood pressure	5.0%
Tobacco	4.0%
Underweight	3.1%
Overweight	2.7%
Cholesterol	2.1%
Indoor smoke from solid fuels	1.9%
Low fruit and vegetable intake	1.9%
Iron deficiency	1.8%
Unsafe water, sanitation, and hygiene	1.7%
Developed Countries	
Tobacco	12.2%
Blood pressure	10.9%
Alcohol	9.2%
Cholesterol	7.6%
Overweight	7.4%
Low fruit and vegetable intake	3.9%
Physical inactivity	3.3%
Illicit drugs	1.8%
Unsafe sex	0.8%
Iron deficiency	0.7%

*Disability-adjusted life years: one DALY = loss of one healthy year of a person's life.
Source: World Health Report, 2002.

to play in coordinating efforts to translate scientific information into action, and its 2002 report is a major step in carrying out that role.

Risk Management

In human health and environmental risk assessment, the analysis of hazards and risks is a task that falls primarily to the scientific community. The task is incomplete, however, if no further consideration is given to the information gathered. Risk management naturally follows and is the responsibility of lawmakers and administrators. EPA policy separates risk assessment and risk management within the agency.

Risk management involves (1) *a thorough review of the information available pertaining to the hazard in question and the risk characterization of that hazard* and (2) *a decision as to whether the weight of the evidence justifies a regulatory action.* Without doubt, public opinion can play a powerful role in both processes. In general, however, a regulatory decision will hinge on one or more of the following considerations:

1. **Cost-benefit analysis.** This type of analysis compares costs and benefits relative to a technology or proposed chemical product; if done fairly, the analysis can make a regulatory decision very clear cut. When the EPA issued new regulations for air-pollution particulate matter in 2006, the benefits in human health far outweighed the regulatory costs.

2. **Risk-benefit analysis.** A decision can be made with regard to the risks versus the benefits of being subjected to a particular hazard, especially when those benefits cannot be easily expressed in monetary values. The use of medical CT scans is a good example. Because they represent a radiation exposure, the scans carry a calculable risk of cancer, but the benefit derived when you scan the brain for a suspected tumor is much greater than the cancer risk involved. However, there is concern about repeated CT scans, since one abdominal scan is equivalent to 500 chest X-rays.

3. **Public preferences.** As we will see, people have a much greater tolerance for risks that they feel are under their own control or are voluntarily accepted.

Risk management has been thoroughly incorporated into the EPA's policy-making process. Its most common use has been in the design of regulations. The process has enabled the agency to target appropriate hazards for regulation and to determine where to aim the regulations—that is, at the source, at the point of use, or at the point of disposal of the hazardous agent. As was done for toxic chemicals and cancer, a given risk level may be adopted as a standard against which policy makers can measure new risks.

Risk Perception

The U.S. public's concern about environmental problems can be traced to the fear of hazards that pose a risk to human life and health. People may perceive that their lives are more hazardous than ever before, but that is not true. In fact, our society is freer from hazards than it ever has been, as is evidenced by increased longevity. Why, then, do people protest against nuclear power plants, waste sites, and pesticide residues in food when, according to experts, these hazards pose extremely small risks (Fig. 17–16)? The answer lies in people's **risk perceptions**—their intuitive judgments about risks. In short, people's perceptions are not consistent with the results coming from a scientific analysis of risk. There are some good reasons for this discrepancy—and some lessons to be learned from it.

Hazard Versus Outrage. The reason for the inconsistency between public perception and actual risk calculations, according to Peter Sandman[8] of Rutgers University, is that the public perception of risks is often more a matter of *outrage* than one of *hazard*. Sandman holds that, while the term *hazard* expresses primarily a concern for fatalities only, the term *outrage* expresses a number of additional concerns:

1. **Lack of familiarity with a technology.** Examples include how nuclear power is produced and how toxic chemicals are handled.

2. **Extent to which the risk is voluntary.** Research has shown that people who have a choice in the matter will accept risks roughly 1,000 times as great as when they have no choice. People are much more accepting of a risk if they are in control of the elements of that risk, as in driving an automobile, swimming, and eating.

3. **Public impression of hazards.** Accidents involving many deaths (as at Bhopal—see Chapter 22) or a failure of technology (for example, at Three Mile Island—see Chapter 15) are thoroughly imprinted into public awareness by media coverage and are not quickly forgotten.

4. **Overselling of safety.** The public becomes suspicious when scientists or public-relations people play up the benefits of a technology and play down its hazards.

5. **Morality.** Some risks have the appearance of being morally wrong. If it is wrong to subject humans to pesticide testing, the notion that the chemical companies clearly benefit from the testing is unacceptable. You should obey a moral imperative, regardless of the costs.

6. **Fairness.** The benefits and the risks should be connected. If the benefits go to someone else, why should you accept any risk?[9]

Media's Role. The public perception of risk is strongly influenced by the media, which are far better at communicating the outrage elements of a risk than they are at communicating the hazard elements. Public concern over mad-cow disease received extensive media coverage and had a high "outrage quotient," and beef from Canada was banned in 2003 when the disease was found in two animals. However, cigarette smoking, which causes 443,000 deaths a year in the United States alone, receives minimal media attention because it is

[8]Peter M. Sandman. *"Risky Business." Rutgers Magazine.* April/June 1989: 36–37.
[9]V. Covello and P. Sandman, "Risk Communication, Evolution and Revolution," in *Solutions to an Environment in Peril,* ed. A. Wolbarst (Baltimore, Md.: Johns Hopkins University Press, 2001), 164–178.

not "news." Hence, there is no outrage factor. Indeed, it has been suggested that, if all the year's smoking fatalities occurred on one day, the media would go ballistic, and smoking would be banned the next day!

Public Concern and Public Policy. Generally speaking, *public concern*, rather than cost-benefit analysis or risk-benefit analysis conducted by scientists, often drives public policy. The EPA's funding priorities are set largely by Congress, which reflects public concern. Is this a problem?

If public outrage is the primary impetus for public policy, some serious risks may get less attention than they deserve. In particular, environmental concerns are commonly perceived as much less important than they really are because of the public's preoccupation with other, more immediate issues (Table 17–5). This difference underscores the importance of *risk communication*, a task that should not be left to the media. For example, the value of ecosystems and their connections to human health and welfare need far greater emphasis in the public consciousness, and this responsibility falls to the scientific and educational communities, as well as to government agencies. Studies have shown that the most effective risk communication occurs by starting with what people already know, discovering what they need to know, and then tailoring the message so that it provides people with the knowledge that helps them make a more informed evaluation of risks. The process should not stop there, though: To be successful, risk communication should also involve dialogue among all the parties involved—namely, policy makers, the public, experts, and other interested groups.

The public's concern for more than the probabilities of fatalities has merit. Public concern must be heard, understood, and given a reasonable response. It may not be the best source of public policy, but it reflects certain values and concerns that could easily be omitted by an "objective," scientific assessment. The fact is that subjective judgments are going to play a role at every step in the risk-assessment process, from hazard assessment to risk perception to risk management. The uncertainties involved in risk assessment should remind us that the process is only a tool—and an imperfect one at best.

The Precautionary Principle

The use of risk assessment has come under attack in recent years, with three major concerns: (1) It does not adequately reflect the uncertainty inherent in much of the assessment process, (2) it has been largely employed in situations where a chemical or a process is already in use, and (3) the burden of proof of harm falls largely on the regulators.

In recognition of these problems, there is a strong movement toward employing the **precautionary principle** in formulating public policy to protect the environment and human health. The Wingspread Conference of 1998—a

TABLE 17–5 Top Domestic Priorities for Obama and Congress

Percent considering issue a "top priority"	Jan. 2008	Jan. 2009	2008–2009 Change
Strengthening the nation's economy	75	85	+10
Improving the job situation	61	82	+21
Defending the United States against terrorism	74	76	+2
Improving educational system	66	61	−5
Dealing with U.S. energy problems	59	60	+1
Reducing health care costs	69	59	−10
Reducing budget deficit	58	53	−5
Providing health insurance to uninsured	54	52	−2
Dealing with problems of poor	51	50	−1
Reducing crime	54	46	−8
Strengthening the military	42	44	+2
Reducing middle-class taxes	46	43	−3
Protecting the environment	56	41	−15
Dealing with illegal immigration	51	41	−10
Reducing influence of lobbyists	39	36	−3
Dealing with global trade	37	31	−6
Dealing with global warming	35	30	−5

Source: Pew Research Center, Washington, D.C., January 22, 2009.

meeting of 35 academic scientists, government researchers, labor representatives, and grassroots environmentalists from the United States, Canada, and Europe—stated the following consensus definition of the principle: "Where an activity raises threats of harm to human health or the environment, precautionary measures should be taken even if some cause-and-effect relationships are not fully established scientifically. In this context, the proponent of an activity, rather than the public, should bear the burden of proof."

The precautionary principle is well established in Europe, where it originated, and has become the basis for European environmental law. For example, the principle has been employed to restrict the import of genetically modified foods (Chapter 12). It is also an essential component of many UN environmental treaties. How it might relate to U.S. risk-based environmental policy is still under debate. The principle operates in some measure in our approach to pharmaceuticals and pesticides, according to which a substance is presumed guilty until proven innocent; that is, the substance is not permitted to be put on the market until tests show that it is safe to use.

The precautionary principle represents a potent tool for implementing stewardship in the environmental health arena. Although risk assessment is useful in many situations, it should not be used as a blanket policy. The precautionary principle enables us to act to prevent some potential environmental health problems, providing a corrective to pressures from some vested interests and enthusiasts to push technology in more risky directions. The old saying "An ounce of prevention is worth a pound of cure" captures the sense of this approach. Where uncertainty is substantial, and especially where the penalty for being wrong is great, it would seem wise to make the precautionary principle a guiding principle overseeing the entire risk assessment process.

Stewards of Health. In a very real sense, public-health ministries and agencies are the primary stewards of the health and welfare of a country's people. Public policy should clearly reflect stewardship principles that put the health of people and the environment above the economic bottom line, which often seems to drive our decisions. To quote the 2002 *World Health Report*, "Governments are the stewards of health resources. This stewardship has been defined as 'a function of a government responsible for the welfare of the population and concerned about the trust and legitimacy with which its activities are viewed by the citizenry.'" Governments can accomplish much in the arena of preventive health, but they must nurture their scarce resources carefully. It is in that sense that the WHO focuses on reducing risks in order to promote health and provides a valuable tool for accomplishing this to all countries. Beyond the government, health care providers are also stewards whose responsibility is vitally important to environmental health.

Revisiting the Themes

Sound Science

The continuing battle against pathogens and human disorders like cancer and heart disease requires the sustained attention of dedicated research laboratories around the world. This is modern science in the direct interest of humankind. Risk assessment utilizes the scientific community in the final step of the process, at which risk characterization takes place. At that point, the scientists have evaluated the risk as best they can, so it is now up to those responsible for risk management. The WHO states that the basis for the major risk factors plaguing most societies has been scientifically established. There is more work to do, of course, but the failure to prove scientific causality cannot be an excuse for inaction by those responsible for their people's health.

Sustainability

A foundational strategy of sustainable development is to reduce poverty, in part because the poor are susceptible to many hazards. Numerous risk factors are a consequence of the lack of access to clean water, healthy air, sanitation, nutritious food, and health care. The infectious diseases in particular are a tremendous impediment to the economic progress in poorer countries. It is hard to see how those countries can make much progress when HIV/AIDS, malaria, and tuberculosis take such a heavy toll on their children and young adults. The first seven Millennium Development Goals all target environmental health problems. (See Table 9–2.) If the world community takes these goals seriously, the outcome should be a revolution in public health that will move scores of countries much farther along in their quest for sustainable development.

Stewardship

Governments are the stewards of their country's health resources. Health ministries have a difficult task if they are not given the funding and personnel to carry out their mission. Fortunately, organizations such as the WHO can provide the vital information governments need to address their countries' greatest health problems. The WHO exemplifies the stewardship concept of a community of nations acting together to bring care and information to those whose health is at risk. Those who work alongside the WHO—the health ministries, health-related nongovernmental organizations, and health care providers—are the hands and hearts of the world community.

REVIEW QUESTIONS

1. Define *pollution*, *pollutant*, *nonbiodegradable*, and *environment*.

2. Differentiate between a hazard and a risk. Where does vulnerability fit in?

3. Define *morbidity*, *mortality*, and *epidemiology*.

4. Describe the public-health roles of the CDC and the WHO.

5. What are the four categories of human environmental hazards? Give examples of each.

6. List the top 10 risk factors that are responsible for global mortality and disease.

7. Document the efforts to control tobacco use both in the United States and globally.

8. What is the significance of malaria worldwide, and what are some recent developments in the battle against this disease?

9. Outline the four steps of risk assessment used by the EPA.

10. What methods are used to test chemicals for their potential to cause cancer?

11. Describe the process of risk assessment for public health as it was recently carried out by the WHO. What is a DALY?

12. Discuss the relationship between public risk perception and assessment, on the one hand, and public policy, on the other.

13. What is the precautionary principle, and how would employing the principle improve our approach to environmental public policy?

THINKING ENVIRONMENTALLY

1. Consider Table 17–4. Discuss each group's risk factors that are primarily a consequence of human choice. Pick one of these factors, and describe how you might proceed to bring it under control in an appropriate country.

2. What global accountability is there for natural disaster relief? Consider the example of devastating weather in developing countries triggered by climate change (which could be instigated by the more prosperous).

3. Suppose your town wanted to spray trees to rid them of a nuisance pest. How would you use cost-benefit analysis to determine whether spraying is a good idea? Is there a better approach?

4. Consult the CDC Web site (www.cdc.gov) and investigate one of the emerging infectious diseases. What are the risks you take and what precautions should you take if traveling to an area where the disease has been found?

MAKING A DIFFERENCE

1. Monitor the hazards in your own life, and take steps to reduce high-risk behavior.

2. Install a carbon monoxide detector if you don't already have one—and check your home's radon levels.

3. Steer clear of phthalates, found in nearly three-quarters of today's beauty products. These are endocrine disruptors, which can obstruct hormone functions; cause infertility, miscarriage, birth defects, and breast cancer; and damage organs.

4. Six percent of an average American's day is spent on the road, but that time makes for 60% of their fine particle pollution intake. Driving less is not only good for the environment, but also good for you!

5. Instead of using toxic house cleaners, consider homemade combinations of white vinegar, castile soap, baking soda, and water.

Mastering**ENVIRONMENTALSCIENCE**™

Go to **www.masteringenvironmentalscience.com** for practice quizzes, Pearson eText, videos, current events, and more.

Katrina from space. Satellite image of Hurricane Katrina in the Gulf of Mexico, August 28, 2005, as it approaches the Gulf Coast. (*Source*: NASA/Jeff Schmaltz, MODIS Land Rapid Response Team.)

CHAPTER 18

Global Climate Change

I T WAS A DISASTER WAITING TO HAPPEN. Many studies had warned of serious consequences if a major hurricane came ashore at New Orleans. There would be enormous damage as levees failed and the city flooded; the loss of life could be catastrophic. Then, on August 29, 2005, Hurricane Katrina made landfall on the Gulf Coast just east of New Orleans. The day before landfall, Katrina was a category 5 hurricane, with winds over 175 miles per hour (mph). By landfall, Katrina's intensity dropped to category 4 (131–155 mph on the Saffir-Simpson scale), but that was enough to drown New Orleans as levees failed and the waters of Lake Pontchartrain poured into the city. It was the costliest storm ever, with $134 billion in damage; it was also the deadliest U.S. storm since 1928, causing more than 1,800 deaths in four states. The storm surge swept over the Louisiana, Mississippi, and Alabama coasts, leaving behind mountains of debris, ruined homes, and broken lives. Katrina was followed in less than a month by powerful Hurricane Rita, which hit the Louisiana and Texas coast west of New Orleans. The two hurricanes turned 217 square miles of the Louisiana coast and wetlands into water.

Katrina was only one (but the worst) of a record 27 named storms in the 2005 Atlantic hurricane season. Although 2006 was

KEY TOPICS

1. **Atmosphere, Weather, and Climate:** The atmosphere is a collection of gases held in a thin envelope around Earth; the climate is the result of long-term weather patterns in a region.

2. **Climate Change Science:** The Intergovernmental Panel on Climate Change has presented clear evidence of the changes in Earth's climate and has assigned its cause to human agency.

3. **Response to Climate Change:** International steps, such as the Kyoto Protocol, have been taken to mitigate climate change, but more drastic cuts in emissions are necessary.

Figure 18–1 Number of category 4 and 5 hurricanes, by ocean basin. Every ocean basin shows an increase in the number of strong hurricanes over the most recent 15-year period. The unusually active 2005 season is not even included in the plot. (*Source:* World Resources Institute, *Hot Climate, Cool Commerce,* May 2006.)

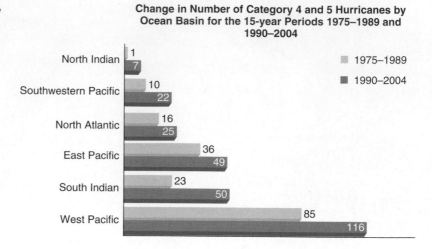

Change in Number of Category 4 and 5 Hurricanes by Ocean Basin for the 15-year Periods 1975–1989 and 1990–2004

near normal, with 10 named storms, 2007 and 2008 had 15 and 18 named storms, respectively. Before 1995, the average for named storms was about 10. Several research groups have recently published studies that suggest a link between hurricanes and global warming. They pointed out that the trends of increasing numbers of category 4 and 5 hurricanes (Fig. 18–1) and increasing numbers of named storms coincide with the trend of rising sea surface temperatures in tropical seas. The high sea surface temperatures result in more water vapor in the air, leading to conditions that create a rising column of warm, humid air that can be set spinning by Coriolis force (Earth's rotation) and lead to hurricane formation. Evidence is clear that the oceans have been absorbing an increasing amount of heat energy.

The possible connections among hurricanes, the atmosphere, oceans, and human activities raise some important questions. Could it be that the warming trend now evident in global temperatures is responsible for more intense hurricanes? What is happening with our climate? Indeed, what does control the climate?

We will seek answers to these questions as we investigate the atmosphere, how it is structured, and how it brings us our weather and climate. Then we will consider the scientific evidence for global climate change, finishing with a look at the varied responses to global climate change.

18.1 Atmosphere, Weather, and Climate

The atmosphere is a collection of gases that gravity holds in a thin envelope around Earth. The gases within the lowest layer, the **troposphere**, are responsible for moderating the flow of energy to Earth and are involved with the biogeochemical cycling of many elements and compounds—oxygen, nitrogen, carbon, sulfur, and water, to name the most crucial ones. The troposphere ranges in thickness from 10 miles (16 km) in the tropics to 5 miles (8 km) in higher latitudes, due mainly to differences in solar energy. This layer contains practically all of the water vapor and clouds in the atmosphere; it is the site and source of our weather. Except for local temperature inversions, the troposphere gets colder with altitude (Fig. 18–2). Air masses in this layer are well mixed vertically, so pollutants can reach the top within a few days. Substances entering the troposphere—including pollutants—may be changed chemically and washed back to Earth's surface by precipitation (Chapter 19). Capping the troposphere is the **tropopause**, a boundary region where air shifts from cooling with height and begins to warm.

Above the tropopause is the **stratosphere**, a layer within which temperature *increases* with altitude, up to about 40 miles above the surface of Earth. The temperature increases primarily because the stratosphere contains ozone (O_3), a form of oxygen that absorbs high-energy radiation emitted by the Sun. Because there is little vertical mixing of air masses in the stratosphere and no precipitation from it, substances that enter it can remain there for a long time. Beyond the stratosphere are two more layers, the *mesosphere* and the *thermosphere*, where the ozone concentration declines and only small amounts of oxygen and nitrogen are found. Because none of the reactions we are concerned with occur in the mesosphere or thermosphere, we shall not discuss those two layers. Table 18–1 summarizes the characteristics of the troposphere and stratosphere.

Weather

The day-to-day variations in temperature, air pressure, wind, humidity, and precipitation—all mediated by the atmosphere—constitute our **weather**. **Climate** is the result of long-term weather patterns in a region. The scientific study

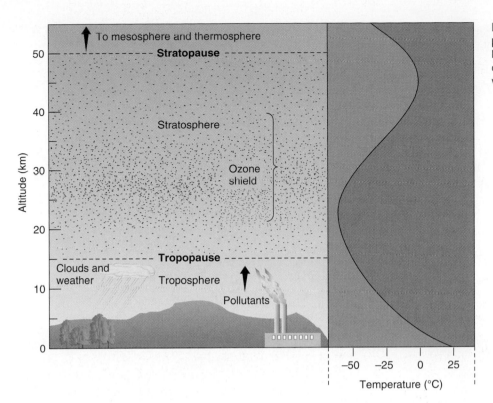

Figure 18–2 Structure and temperature profile of the atmosphere. The plot on the left shows the layers of the atmosphere and the ozone shield, while the plot on the right shows the vertical temperature profile.

of the atmosphere—both of weather and of climate—is **meteorology**. It is fair to think of the atmosphere-ocean-land system as an enormous weather engine, driven by solar power and strongly affected by the rotation of Earth and its tilted axis. Solar radiation enters the atmosphere and then takes a number of possible courses (Fig. 18–3). Some (29%) is reflected by clouds and Earth's surfaces, but most is absorbed by the atmosphere, oceans, and land, which are heated in the process. The land and oceans then release energy back upward through evaporation, convection, and reradiation of infrared energy.

Flowing Air. Some of the energy that is released from Earth's surface is transferred to the atmosphere. Thus, air masses grow warmer at the surface of Earth and will tend to expand, becoming lighter. The lighter air will then rise,

creating *vertical* air currents. On a large scale, this movement creates the major *convection currents* we encountered in Chapter 10 (Fig. 10–6). Air must flow in to replace the rising warm air, and the inflow leads to *horizontal* airflows, or wind. The ultimate source of the horizontal flow is cooler air that is sinking, and the combination produces the Hadley cell (Fig. 10–6). As discussed in Chapter 10, these major flows of air create regions of high rainfall (equatorial), deserts (25° to 35° north and south of the equator), and horizontal winds (trade winds).

Convection. On a smaller scale, **convection currents** bring us the day-to-day changes in our weather as they move in a general pattern from west to east. Weather reports inform us of regions of high and low pressure, but where do these come from? Rising air (due to solar heating) creates high pressure

TABLE 18–1 Characteristics of Troposphere and Stratosphere

Troposphere	Stratosphere
Extent: Ground level to 10 miles (16 km)	Extent: 10 miles to 40 miles (16 km to 65 km)
Temperature normally decreases with altitude, down to −70°F (−59°C)	Temperature increases with altitude, up to +32°F (0°C)
Much vertical mixing, turbulent	Little vertical mixing, slow exchange of gases with troposphere, via diffusion
Substances entering may be washed back to Earth	Substances entering remain unless attacked by sunlight or other chemicals
All weather and climate take place here	Isolated from the troposphere by the tropopause

Figure 18–3 Solar-energy balance.
Much of the incoming radiation from the Sun (29%) is reflected back to space, but the remainder (71%) is absorbed by the oceans, land, and atmosphere, where it creates our weather and fuels photosynthesis. Eventually, this absorbed energy is radiated back to space as infrared energy (heat). (*Source:* Rebecca Lindsey, Climate and Earth's Energy Budget, NASA Earth Observatory, January 14, 2009.)

up in the atmosphere, leaving behind a region of lower pressure close to Earth. Conversely, once the moist, high-pressure air has cooled by reradiating heat to space and losing heat through condensation (thereby generating precipitation), it flows horizontally toward regions of sinking, cool, dry air (where the pressure is lower). There the air is warmed at the surface and creates a region of higher pressure (Fig. 18–4). The differences in pressure lead to airflows, which are the winds we experience. As the figure shows, the winds tend to flow from high-pressure regions toward low-pressure regions.

Jet Streams.   The larger-scale air movements of Hadley cells are influenced by Earth's rotation from west to east. This creates the trade winds over the oceans and the general flow of weather from west to east. Higher in the troposphere, Earth's rotation and air-pressure gradients generate veritable rivers of air, called **jet streams,** that flow eastward at speeds of more than 300 mph and that meander considerably. Jet streams are able to steer major air masses in the lower troposphere. One example is the polar jet stream, which steers cold air masses into North America when it dips downward in latitude.

Put Together. . . .   Air masses of different temperatures and pressures meet at boundaries we call **fronts,** which are regions of rapid weather change. Other movements of air masses due to differences in pressure and temperature include hurricanes and typhoons and the local, but very destructive, tornadoes. Finally, there are also major seasonal airflows—the **monsoons**—which often represent a reversal of previous wind patterns. Monsoons are created by major

differences in cooling and heating between oceans and continents. The summer monsoons of the Indian subcontinent are famous for the beneficial rains they bring and notorious for the devastating floods that can occur when the rains are heavy. Putting these movements all together—taking the general atmospheric circulation patterns and the resulting precipitation, then adding the wind and weather systems generating them, and finally mixing all this with the rotation of Earth and the tilt of the planet on its axis, which creates the seasons—yields the general patterns of weather that characterize different regions of the world. In any given region, these patterns are referred to as the region's **climate.**

18.2 Climate Change Science

Climate was described in Chapter 5 as the average temperature and precipitation expected throughout a typical year in a given region. Recall that the different temperature and moisture regimes in different parts of the world have generated different types of ecosystems called *biomes*, reflecting the adaptations of plants, animals, and microbes to the prevailing weather patterns, or climate, of a region. The temperature and precipitation patterns themselves are actually caused by other forces—namely, the major determinants of weather previously outlined. Humans can adjust to practically any climate (short of the brutal conditions of high mountains and burning deserts), but this is not true of the other inhabitants of the particular regions we occupy. If other living organisms in a region are adapted to a particular

Figure 18–4 A convection cell. Driven by solar energy, convection cells produce the main components of our weather as evaporation and condensation occur in rising air and precipitation results, followed by the sinking of dry air. Horizontal winds are generated in the process.

climate, then a major change in the climate represents a serious threat to the structure and function of the existing ecosystems. The subject of climate change is such a burning issue today because we depend on these other organisms for a host of vital goods and services without which we could not survive (see Chapter 7). If the climate changes, can these ecosystems change with it in such a way that the vital support they provide us is not interrupted? Is Earth's climate indeed changing? If so, what is causing it to change?

Synopsis of Global Climate Change

In 2007, after sifting through thousands of studies about global warming, scientists from the Intergovernmental Panel on Climate Change (IPCC) published the Fourth Assessment Report (AR4) on climate change. They concluded that *warming of the climate system is unequivocal.* Our climate is changing, the scientists said: The atmosphere is warmer (Fig. 18–5), the oceans are warmer, there is more precipitation, there are more extreme weather events (like hurricanes), glaciers are melting, and the sea level is rising.

The IPCC report stated that it is *very likely* (i.e., there is greater than 90% probability) that the warming is caused by increased concentrations of anthropogenic (human-made) greenhouse gases (GHGs), which are trapping increased amounts of infrared radiation. The most significant anthropogenic GHG is carbon dioxide (CO_2), which has risen by 39%—from 280 parts per million (ppm) to 388 ppm—since the Industrial Revolution began (Fig. 18–6). The source of

this added CO_2 is no secret: Most of it comes from burning fossil fuels, and a smaller amount comes from burning forests (deforestation). The global use of fossil fuel is rising every year, and if nothing is done to reduce it, the 21st century is likely to see climate changes that are dangerous, possibly catastrophic. Sea levels will keep rising; there will be more-frequent heat waves, changed weather patterns, and threats to natural ecosystems; and these changes will go on for hundreds of years. Needless to say, this is not good news, and it has gotten the attention of the world. Efforts are under way to respond by way of **mitigation** (reducing GHG emissions) and **adaptation** (adjusting to coming climate changes). However, our civilization has no experience with a problem like this, and the way forward is still being debated.

IPCC. The AR4, the fourth in a series of reports from the IPCC, is an incredibly detailed and comprehensive study. In 1988, the UN Environment Program and the World Meteorological Society established the IPCC in order to provide accurate and relevant information that would lead to an understanding of human-induced climate change. The IPCC has three working groups: one to assess the scientific issues (Working Group I), another to evaluate the impact of global climate change and the prospects for adapting to it (Working Group II), and a third to investigate ways of mitigating the effects (Working Group III). The latest IPCC assessment working groups consist of more than 2,000 experts in the appropriate fields from 154 countries. These experts are unpaid and participate at the cost of their own research and

Global Mean Temperature over Land and Ocean

Figure 18–5 Annual mean global surface temperature anomalies. The baseline, or zero point, is the 20th-century average temperature. The warming trend since 1975 is conspicuous. (*Source:* National Climatic Data Center, NOAA, 2009.)

Figure 18–6 Atmospheric carbon dioxide concentrations. The concentration of CO_2 in the atmosphere fluctuates between winter and summer because of seasonal variation in photosynthesis. The average concentration is increasing due to human activities—in particular, burning fossil fuels and deforestation. (*Source:* Mauna Loa Observatory, Hawaii, NOAA Research Laboratory, Scripps Institution of Oceanography, 2009.)

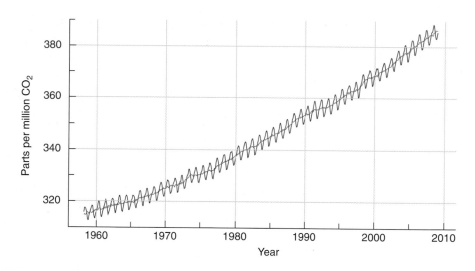

other professional activities. Each working group produced a separate report;[1] all reports are available from the IPCC Web site (www.ipcc.ch).

The work of the IPCC has been guided by two basic questions:

Risk assessment—Is the climate system changing, and what is the impact on society and ecosystems?

Risk management—How can we manage the system through adaptation and mitigation?

Third Assessment. In 2001, the IPCC released its Third Assessment Report. The major headings in Working Group I's "Summary for Policymakers" provide a picture of the consensus understanding of global climate change science at the turn of the century and are presented here:

- An increasing body of observations gives a collective picture of a warming world and other changes in the climate system.

[1]See, for example, Susan Solomon et al., eds., *Climate Change 2007: The Physical Science Basis*, Contribution of Working Group I to the Fourth Assessment Report of the Intergovernmental Panel on Climate Change (New York: Cambridge University Press, 2007).

- Emissions of GHGs and aerosols due to human activities continue to alter the atmosphere in ways that are expected to affect the climate.

- Confidence in the ability of models to project future climate has increased.

- There is new and stronger evidence that most of the warming observed over the last 50 years is attributable to human activities.

- Human influences will continue to change atmospheric composition throughout the 21st century.

- Global average temperature and sea level are projected to rise under all IPCC scenarios.

- Anthropogenic climate change will persist for many centuries.

- Further action is required to address remaining gaps in information and understanding.

A Nobel Effort. Building on past assessments and new results from the past six years of research, the three IPCC working groups released the Fourth Assessment Report in stages during 2007, the final product being *Climate Change 2007: Synthesis Report*, which was released late in the year and contained key findings of the working groups. In an unexpected move, the Nobel Committee awarded the 2007 Nobel Peace Prize to the IPCC and to former Vice President Al Gore "for their efforts to build up and disseminate greater knowledge about man-made climate change, and to lay the foundations for the measures that are needed to counteract such change,"[2] according to the Nobel Prize jury. Gore is considered to be the leading advocate of the need to take action on climate change, highlighted by his Academy Award–winning film *An Inconvenient Truth*.

In this chapter, we will explore the information on climate change in some detail, making full use of the IPCC reports. We begin with a look into the past, which harbors "records" of climate change.

Climates in the Past

Searching the past for evidence of climate change has become a major scientific enterprise, one that becomes more difficult the further into the past we try to search. Systematic records of the factors making up weather—temperature, precipitation, storms, and so forth—have been kept for little more than 100 years. Nevertheless, these records already inform us that our climate is far from constant. The record of surface temperatures, from weather stations around the world and from literally millions of observations of temperatures at the surface of the sea, tells an interesting story (Fig. 18–5). Since 1880, global average temperature has shown periods of cooling and warming, but since 1976, it has increased 0.6°C (1.1°F). During the 20th century, two warming trends

occurred, one from 1910 to 1945 and the latest dramatic increase from 1976 until the present.

Further Back. Observations on climatic changes can be extended much further back in time with the use of **proxies**—measurable records that can provide data on factors such as temperature, ice cover, and precipitation. For example, historical accounts suggest that the Northern Hemisphere enjoyed a warming period from 1100 to 1300 A.D. This was followed by the "Little Ice Age," between 1400 and 1850 A.D. Additional proxies include tree rings, pollen deposits, changes in landscapes, marine sediments, corals, and ice cores.

Some work done quite recently on ice cores has both provided a startling view of a global climate that oscillates according to several cycles and afforded some evidence that remarkable changes in the climate can occur within as little as a few decades. Ice cores in Greenland and the Antarctic have been analyzed for thickness, gas content (specifically, carbon dioxide (CO_2) and methane (CH_4), two GHGs), and **isotopes**, which are alternative chemical configurations of a given compound, due to different nuclear components. Isotopes of oxygen, as well as isotopes of hydrogen, behave differently at different temperatures when condensed in clouds and incorporated into ice.

The record based on these analyses indicates that Earth's climate has oscillated between ice ages and warm periods (Fig. 18–7a). During major ice ages, huge amounts of water were tied up in glaciers and ice sheets, and the sea level was lower by as much as 400 feet (120 m). Ice cores from Antarctica show eight glacial periods, and eight interglacial periods, over the past 800,000 years. During the glacial periods, temperatures and GHGs were low; during the warm interglacial periods, just the opposite was true. The GHG and temperature data match exactly. Carbon dioxide levels ranged between 150 and 280 ppm.

The most likely explanation for these major oscillations is the existence of known variations in Earth's orbit, where, in different modes of orbital configuration, the distribution of solar radiation over different continents and latitudes varies substantially. These oscillations take place according to several periodic intervals, called **Milankovitch cycles** (after the Serbian scientist who first described them), of 100,000, 41,000, and 23,000 years.

Rapid Changes. Superimposed on the major oscillations is a record of rapid climatic fluctuations during periods of glaciation and warmer times (Fig. 18–7b). One such rapid change, called the *Younger Dryas* event (*Dryas* is a genus of Arctic flower), occurred toward the end of the last ice age. Earth had been warming up for 6,000 years and then plunged again into 1,150 years of cold weather. At the end of this event, 11,700 years ago, Arctic temperatures rose 7°C in 50 years! The impact on living systems must have been enormous. It is unlikely that this warming was due to variations in solar output: There is simply no evidence that the Sun has changed over the last million years, let alone having undergone major changes in just a few decades. Instead, scientists have narrowed the field of possible explanations to the link between the atmosphere and the oceans.

[2]http://nobelprize.org/nobel_prizes/peace/laureates/2007/.

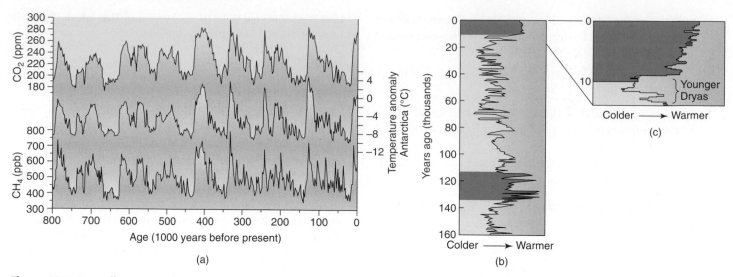

(a)

(b)

(c)

Figure 18–7 Past climates, as determined from ice cores. (a) Carbon dioxide, methane, and estimated global temperature from Antarctic ice cores, covering the past 800,000 years, compared with changes of the past century. Note how closely the temperature and GHG concentrations coincide. (b) Temperature patterns of the last 160,000 years, demonstrating climatic oscillations. (c) Higher resolution of the last 12,000 years. The Younger Dryas cold spell occurred at the start of this record. Note how rapidly the cold spell dissipated, at around 10,700 years before the present. (*Sources:* (a) The Bohr Institute; (b) & (c) Adapted from Robert W. Christopherson, *Geosystems: An Introduction to Physical Geography,* 5th ed. (Upper Saddle River, NJ: Pearson Prentice Hall, 2005).)

Ocean and Atmosphere

More than two-thirds of Earth is covered by oceans, making it mostly a water planet. Since we all live on land, it is hard for us to imagine the dominant role the oceans play in determining our climate. Nevertheless, the oceans are the major source of water for the hydrologic cycle and the main source of heat entering the atmosphere. Recall from Chapter 10 that the *evaporation of ocean water* supplies the atmosphere with water vapor, and when water vapor condenses in the atmosphere (Fig. 10–4), it supplies the atmosphere with heat (latent heat of condensation). The oceans also play a vital role in climate because of their innate *heat capacity*—the ability to absorb energy when water is heated. Indeed, the entire heat capacity of the atmosphere is equal to that of just the top 3 meters of ocean water! It is this property that makes the climate milder in coastal areas. Finally, through the movement of currents, the oceans *convey heat* throughout the globe.

Thermohaline Circulation. A thermohaline circulation pattern dominates oceanic currents, where *thermohaline* refers to the effects that temperature and salinity have on the density of seawater. This pattern, or **conveyor** system (Fig. 18–8), acts as a giant, complex conveyor belt, moving water masses from the surface to deep oceans and back again. A key area is the high-latitude North Atlantic, where salty water from the Gulf Stream moves northward on the surface and is cooled by Arctic air currents. Cooling increases the density of the water, which then sinks to depths of up to 4,000 meters—the *North Atlantic Deep Water* (NADW). This deep-water current spreads southward through the Atlantic to the southern tip of Africa, where it is joined by

cold Antarctic waters. Together, the two streams spread northward into the Indian and Pacific oceans as deep currents. Gradually, the currents slow down and warm, becoming less dense and welling up to the surface, where they are further warmed and begin to move as surface waters back again toward the North Atlantic. This movement transfers enormous quantities of heat toward Europe, providing a climate

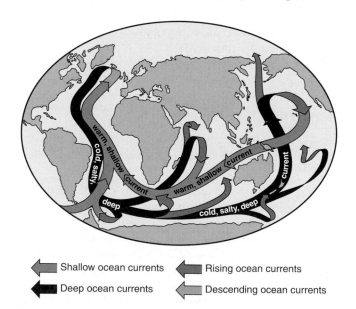

Figure 18–8 The oceanic conveyor system. Salty water flowing to the North Atlantic is cooled and sinks, forming the North Atlantic Deep Water system. This deep flow extends southward and is joined by Antarctic water, whereupon it extends into the Indian and Pacific oceans. Surface currents then proceed in the opposite direction, returning the water to the North Atlantic. (Adapted from Figure 3.15 from *Our Changing Planet,* 2nd ed., by Fred T. MacKenzie. Copyright © 1998 by Prentice Hall, Inc. Reprinted by permission of Pearson Education, Inc., Upper Saddle River, NJ 07458.)

that is much warmer than the high latitudes there would suggest. The circulation pattern completes one cycle over a period of about 1,000 years and is vital to the maintenance of current climatic conditions.

Abrupt Change. Recent evidence indicates that the conveyor system has been interrupted in the past, changing climate abruptly. One mechanism that could accomplish such a change is the appearance of unusually large quantities of fresh water in the North Atlantic. This would have the effect of lowering the density of the water (and, therefore, preventing much of the massive sinking that normally occurs there). With this stagnation occurring in surface Arctic waters, the northward movement of warmer, saltier water would slow down. North Atlantic marine sediments show evidence of the periodic invasion (six times in the last 75,000 years) of icebergs from the polar ice cap, which supplied huge amounts of fresh water as they melted—called Heinrich events, after the scientist who first described them. The evidence indicates that these invasions coincided with rapid cooling, as recorded in ice cores, and suggests that the conveyor system shifted southward, with the NADW forming nearer to Bermuda than to Greenland. When this shift occurred, a major cooling of the climate took place within a few decades. Although it is not clear how, the normal conveyor pattern returned and brought about another abrupt change, this time warming conditions in the North Atlantic. The Younger Dryas event likely involved just such a shift in the conveyor system, brought on by the sudden release of dammed-up water from glacial Lake Agassiz into the St. Lawrence drainage. Referring to these abrupt changes in a recent article, oceanographer Wallace Broecker commented, "Earth's climate system has proven itself to be an angry beast. When nudged, it is capable of a violent response."[3]

What If . . . ? Among the likely consequences of extended global warming are increased precipitation over the North Atlantic and more melting of sea ice and ice caps. If such a pattern is sustained, it could lead to a weakening in the normal operation of the conveyor and a change in climate, especially in the northern latitudes—a possibility that Broecker has called "the Achilles' heel of our climate system." Using the terminology introduced in Chapter 5, this would be a *tipping point.* Scientists with the U.S. Global Change Research Program (USGCRP) have recently concluded that the conveyer will indeed decrease as a consequence of global warming, but that the decrease will extend over the course of the 21st century, rather than occurring abruptly.[4] Even with a weakening of the conveyor system, most observers believe that expected increases in GHGs and the overall warming that this brings will prevent anything like a new ice age (as portrayed by movies like *The Day After Tomorrow*).

Ocean-Atmosphere Oscillations. Regional climates are notoriously erratic on time scales from a few days to decades. Studies of a number of ocean-atmosphere processes—processes that produce erratic climates on a global scale—have provided new insights into natural climate variability. There is, for example, the *North Atlantic Oscillation (NAO)*, where atmospheric pressure centers switch back and forth across the Atlantic, switching winds and storms in the process. Then there is the *El Niño/La Niña Southern Oscillation (ENSO)*, where major shifts in atmospheric pressure over the central equatorial Pacific Ocean lead to a reversal of the trade winds. Incredibly, these reversals dominate the climate over much of the planet, and they can last for a year or more at a time. The 1997–2000 ENSO was the most intense of recent decades, and it spawned powerful global weather patterns that caused over $36 billion in damages and over 22,000 deaths. There is also a warm-cool cycle, the *Interdecadal Pacific Oscillation (IPO)*, that swings back and forth across the Pacific over periods of several decades.

These processes are becoming better understood, and as they are, they are beginning to provide explanations for some previously puzzling developments. For example, it is clear (Fig. 18–5) that a global warming trend was halted in the mid-1940s and that it resumed 30 years later. Researchers have found evidence that this midcentury cooling was the result of shifting ocean circulation linked to the NAO. Forecasters have suggested that this and the other ocean-atmosphere oscillations have been offsetting greenhouse-based warming off and on, but that by the end of this decade, greenhouse warming will overcome these effects and intensify.

The Earth as a Greenhouse

Factors that influence the climate include interactive *internal components* (oceans, the atmosphere, snow cover, sea ice) and *external factors* (solar radiation, the Earth's rotation, slow changes in our planet's orbit, and the gaseous makeup of the atmosphere). **Radiative forcing** is the influence any particular factor has on the energy balance of the atmosphere-ocean-land system. A *forcing factor* can be *positive*, leading to *warming*, or *negative*, leading to *cooling*, as it affects the energy balance. Forcing is expressed in units of Watts per square meter (W/m^2); averaged over all of the planet, solar radiation entering the atmosphere is about 340 W/m^2. It is this radiation that is acted on by the various forcing factors.

Warming Processes. The interior of a car heats up when the car is sitting in sunlight with the windows closed. This heating occurs because sunlight comes in through the windows and is absorbed by the seats and other interior objects, thus converting light energy into heat energy, which is given off in the form of infrared radiation. Unlike sunlight, infrared radiation is blocked by glass and so cannot leave the car. The trapped heat energy causes the interior air temperature

---

[3]Wallace S. Broecker, "Does the Trigger for Abrupt Change Reside in the Ocean or in the Atmosphere?" *Science* 300 (June 6, 2003): 1519–1522.
[4]CCSP, *Abrupt Climate Change*, A report by the U.S. Climate Change Science Program and the Subcommittee on Global Change Research (Reston, Va.: U.S. Geological Survey, 2008).

to rise. This is the same phenomenon that keeps a greenhouse warmer than the surrounding environment.

Greenhouse Gases. On a global scale, water vapor, CO_2, and other gases in the atmosphere play a role analogous to that of the glass in a greenhouse. Therefore, they are called **greenhouse gases (GHGs)**. Light energy comes through the atmosphere and is absorbed by Earth and converted to heat energy at the planet's surface. The infrared heat energy radiates back upward through the atmosphere and into space. The GHGs that are naturally present in the troposphere absorb some of the infrared radiation and reradiate it back toward the surface; other gases (N_2 and O_2) in the troposphere do not (Fig. 18–9). This greenhouse effect was first recognized in 1827 by French scientist Jean-Baptiste Fourier and is now firmly established. The GHGs are like a heat blanket, insulating Earth and delaying the loss of infrared energy (heat) to space. Without this insulation, average surface temperatures on Earth would be about −19°C instead of +14°C, and life as we know it would be impossible.

Therefore, our global climate depends on Earth's concentrations of GHGs. If these concentrations increase or decrease significantly, their influence as positive forcing agents will change, and our climate will change accordingly.

Ozone in the troposphere has a positive forcing effect, whereas its impact in the stratosphere is negative. However, the positive forcing predominates. Tropospheric ozone has a strong anthropogenic component linked to pollution, as discussed in Chapter 19. It varies greatly in time and location and is highest in the vicinity of industrial societies.

Cooling Processes. Earth's atmosphere is also subject to negative forcing factors. For example, on average, clouds cover 50% of Earth's surface and reflect some 23% of solar radiation away to space before it ever reaches the ground. Sunlight reflected in this way is called the **planetary albedo**, and it contributes to overall cooling by preventing a certain amount of warming in the first place. The effect of the albedo is especially true for *low-lying* clouds and can be readily

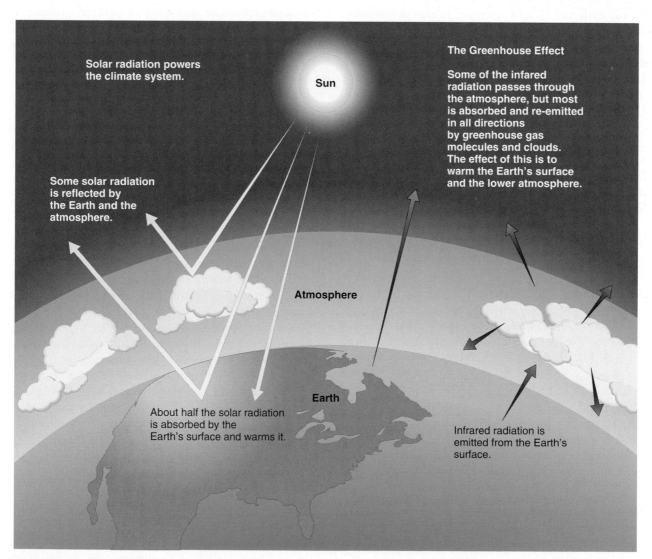

Figure 18–9 The greenhouse effect. Solar radiation that reaches Earth's surface is converted to infrared radiation and is radiated upward. Greenhouse gases in the atmosphere intercept this radiation and are warmed. The greenhouse gases in turn reradiate heat energy, and much of this is directed back toward Earth's surface. (*Source:* S. Solomon et al., eds., *Climate Change 2007: The Physical Science Basis*, Contribution of Working Group I to the Fourth Assessment Report of the Intergovernmental Panel on Climate Change (Cambridge, England: Cambridge University Press, 2007), FAQ 1.3, Figure 1.)

appreciated if you think of how you are comforted on a hot day when a large cloud passes between you and the Sun. The radiation that was heating you and your surroundings is suddenly intercepted higher up in the atmosphere, and you feel cooler. However, *high-flying*, wispy clouds have a positive forcing effect, absorbing some of the solar radiation and emitting some infrared radiation themselves. Overall, the current net impact of clouds is judged to be a slightly negative forcing.

Snow and ice also reflect sunlight, contributing to the planetary albedo. Recent work, however, suggests that this effect is reduced greatly by the widespread black carbon soot originating from anthropogenic sources. By darkening the snow and ice, the soot promotes the absorption of radiant energy rather than its reflection.

Volcanoes. Volcanic activity can also lead to planetary cooling. When Mount Pinatubo in the Philippines erupted in 1991, some 20 million tons of particles and aerosols entered the atmosphere and contributed to a significant drop in global temperature as radiation was reflected and scattered away. This global cooling effect lasted until the volcanic debris was finally cleansed from the atmosphere by chemical change and deposition, a process that took several years.

Aerosols. These are microscopic liquid or solid particles originating from land and water surfaces (Chapter 10).

Climatologists have found that industrial aerosols (from ground-level pollution) play a significant role in canceling out some of the warming from GHGs. Sulfates, nitrates, dust, and soot from industrial sources and forest fires enter the atmosphere and react with compounds there to form a high-level aerosol haze (Fig. 18–10). This haze reflects and scatters some sunlight; the aerosols also contribute to the formation of clouds, which likewise increase the planetary albedo. *Sooty* aerosol (from fires) has an overall warming effect, but industrial *sulfate* aerosol creates substantial cooling, mainly through its tendency to increase global cloud cover. The mean residence time of the sulfates and soot forming the aerosol is about a week, so the aerosol does not increase over time, as the GHGs do. However, because anthropogenic sulfate is more than double that coming from natural sources, its effect is substantial and persistent. Climatologists estimate that the cooling effect of these pollutants has counteracted much of the potential global warming in recent decades. Because of progress made in curbing industrial pollutants in the United States and Europe, this impact has been declining over those regions. However, industrial growth in China and India has created a major rise in aerosols over Asia.

Solar Variability. Since the Sun is the source of radiant energy that heats Earth, any variability in the Sun's radiation reaching Earth will likely influence the climate. Direct

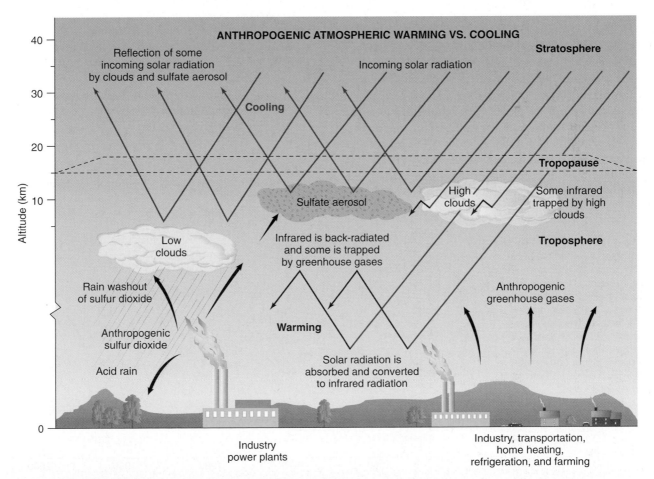

Figure 18–10 Global warming and cooling. Factors involved in atmospheric warming and cooling are illustrated. Greenhouse gases promote global warming; sulfur dioxide leads to cooling.

monitoring of solar irradiance goes back only a few decades, but it has confirmed that there is an 11-year cycle of changes involving solar radiation increases during times of high sunspot activity. Sunspots are known to block cosmic ray intensity, and since cosmic rays help seed clouds, the diminished cloud cover allows more solar radiation to reach Earth's surface. Indeed, variations in solar forcing are thought to be partly responsible for the early-20th-century warming trend. However, solar output began declining around 1985 and continued for at least 20 years, yet global temperatures were rising rapidly at that time and have continued to rise. The IPCC AR4 assigned a slight warming role to fluctuations in solar radiation, but found that GHGs were some 13 times more responsible for the rising global temperatures than were solar changes.

Thus. . . . Global atmospheric temperatures are a balance between the positive and negative forcing from natural causes (volcanoes, clouds, natural GHGs, solar irradiance) and anthropogenic causes (sulfate aerosols, soot, ozone, increases in GHGs). The net result varies, depending on one's location. The various forcing agents bring us year-to-year fluctuations in climate and make it difficult to attribute any one event or extreme season to human causation. Nevertheless, there has clearly been enough of a shift in many features of our climate in recent decades to generate international attention.

Evidence of Climate Change

There is much natural variation in weather from year to year, and local temperatures do not necessarily follow globally averaged ones. It is a fact, however, that the 10 warmest years on record occurred between 1997 and 2008 (Fig. 18–5). The year 2005 set a new record for global temperature—the warmest year since the late 1800s. Since the mid-1970s, the average global temperature has risen about 0.6°C (1.1°F), almost 0.2°C per decade. The warming is happening everywhere, but is especially rapid at high latitudes in the Northern Hemisphere. Because of the continued increases in anthropogenic GHGs in

the atmosphere, the observed warming is considered to be the consequence of an "enhanced greenhouse effect."

Satellites. An interesting and controversial aspect of the warming trend involved measurements of tropospheric temperatures (from Earth's surface to 8 km above) made by 13 different satellites over two decades. One early analysis of the data indicated virtually no temperature increase in the troposphere. Much was made of this apparent discrepancy with the surface temperature trends by skeptics eager to disprove global warming. This situation was addressed by the Climate Change Science Program (CCSP), which had created in 2002 by the Bush administration to address apparent unresolved questions about global climate change. The CCSP study, reported in May 2006, found that "the discrepancy no longer exists because errors in the satellite . . . data have been identified and corrected. New data sets have also been developed that do not show such discrepancies." Indeed, the study found a remarkable agreement between surface temperatures and tropospheric temperatures, all of which indicate a continued rise in temperature consistent with "our understanding of the effects of radiative forcing agents [showing] clear evidence of human influences on the climate system."[5] This report, the first of 21 assessments planned by the program, was approved by the White House.

Ocean Warming. The upper 3,000 meters of the ocean have warmed measurably since 1955 (Fig. 18–11), a warming that dwarfs the observed warming of the atmosphere, accounting for 90% of the heat increase of Earth systems in the last several decades. There is a fascinating story about the actual measurements of ocean heat content shown in Figure 18–11 (and described in Sound Science, "Getting It Right"). A recent analysis of Earth's energy imbalance[6]

---

[5]CCSP, *Temperature Trends in the Lower Atmosphere: Steps for Understanding and Reconciling Differences*, A Report by the Climate Change Science Program and the Subcommittee on Global Change Research (Washington, D.C.: CCSP, 2006).
[6]James Hansen et al., "Earth's Energy Imbalance: Confirmation and Implications," *Science* 308 (June 2005): 1431–1435.

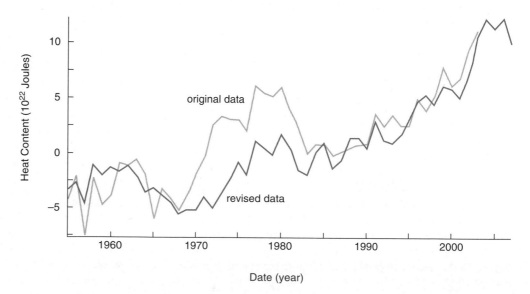

Figure 18–11 Heat capture by the oceans. The plot shows the changes in heat content of the upper 2,290 feet (700 m) of the world's oceans, as plotted by the Global Oceanographic Data Archeology and Rescue Project (GODAR). Heat content is measured in joules (J). Two plots are seen: an original data plot and a revised data plot after corrections were made. See Sound Science, "Getting It Right." (*Source:* Rebecca Lindsey, *Correcting Ocean Cooling*, Earth Observatory, November 5, 2008.)

Sound Science

Getting It Right

Josh Willis works at NASA's Jet Propulsion Laboratory and specializes in measurements of the ocean's response to global warming. He publishes his results in peer-reviewed journals like the *Journal of Geophysical Research*. In a 2004 paper, he presented work that showed the ocean heat content increasing between 1993 and 2003, in agreement with what climatologists were seeing with the atmospheric temperature measurements. This is no trivial matter; the oceans absorb some 90% of the heat from global warming. However, Willis then did a follow-up study with colleague John Lyman, updating the time series from 2003 to 2005, showing that the oceans suddenly began to cool, and in a large way. This alarmed many climatologists and, at the same time, fed the climate change skeptics, who pounced on the data as proof that global warming wasn't happening.

One climatologist who was especially dubious about Willis's new data was Takmeng Wong, who works at NASA's Langley Research Center. Wong had been measuring the exchange of energy between the Earth and space, and his data showed that the amounts of entering and outgoing energy at the top of the atmosphere were gradually climbing out of balance. More energy was entering than leaving. This was true for the 2003 to 2005 time period as well, and Wong politely asked Willis to reexamine his data, since there was a

Josh Willis, NASA Jet Propulsion Laboratory.

clear discrepancy between the two studies. When Willis measured an even deeper ocean cooling in 2006, he began to entertain serious doubts about his data.

The source of the temperature data being employed by Willis was a flotilla of Argo floats—underwater robots that would sink to 2,000 meters, drift with the ocean currents, and then pop up to the surface, taking water temperatures as they ascended. They then transmitted the data to a satellite. Willis found that, when tested,

many new Argo floats were giving bad data—in fact, indicating cooler temperatures than other devices being employed. He also consulted an independent set of data on satellite measurements of sea level. When water heats up, it expands, and the sea level data were showing a continual rise. One final factor in the data relates to a different data set on ocean temperatures from devices called XBTs (for expandable bathythermographs, which are dropped into the ocean and relay temperatures as they sink). An independent paper had shown that the XBT data from previous years were too warm. When Willis corrected his ocean warming time series with the too-warm XBT measurements and the too-cool Argo measurements, the indications of ocean cooling went away. The warm pulses of the 1970s and 1980s disappeared, and the time series showed a gradual warming (Fig. 18–11).

The conclusion to this story is that you can never have too many independent measurements if you want to measure something as huge as a warming planet. Taken together, ocean warming, sea level rise, and energy exchange with space are now telling a consistent story, one of a warming that can be attributed only to rising GHG emissions.

Source: Rebecca Lindsey, *Correcting Ocean Cooling*, NASA Earth Observatory, November 5, 2008.

makes the case for the ocean's continuing absorption of heat; over the last decade, the ocean has absorbed essentially all of the heat not yet seen in the atmosphere. This net absorption of heat energy has one long-term consequence and one more immediate.

The long-term consequence is *the impact of this stored heat* as it eventually comes into equilibrium with the atmosphere, raising temperatures over land and in the atmosphere even more. The immediate consequence is the *rise in sea level* that occurs because of thermal expansion. Sea level rose about 6 inches (15 cm) during the 20th century, a rate of 1.5 millimeters per year (Fig. 18–12). It is now rising more rapidly, at a rate of 3.3 millimeters per year. About half of this rise is due to thermal expansion, and the remainder results from the melting of glaciers and ice fields. The IPCC AR4 documents a general loss of land ice, as glaciers and ice caps have thinned and melted. Significantly, historic sea level data indicate that the sea level was unchanged until 100 years ago; this modern sea level rise is a new phenomenon and is consistent with the known pattern of anthropogenic climate warming.

Other Observed Changes.   The IPCC AR4 documents a number of climate impacts, all of which are consistent with and contribute to the picture of GHG-caused climate change (note that most of these changes have potentially serious consequences for human and natural systems):

- Broad increases in warm temperature extremes and decreases in cold temperature extremes have been documented. There are also seasonal changes: Spring comes earlier and fall is later in the Northern Hemisphere, shifting ecosystems and populations out of sync. More tree deaths and forest insect damage have also been traced to the warmer temperatures. Heat waves, like those that scorched Europe and the United States in 2006, are increasing in intensity and frequency.

- Droughts are also increasing in frequency and intensity; the fraction of land area experiencing drought has more than doubled over the last 30 years. More than 60% of the United States has been under a drought that started in the late 1990s, seriously affecting agricultural crops and water resources. The increased forest fire activity in

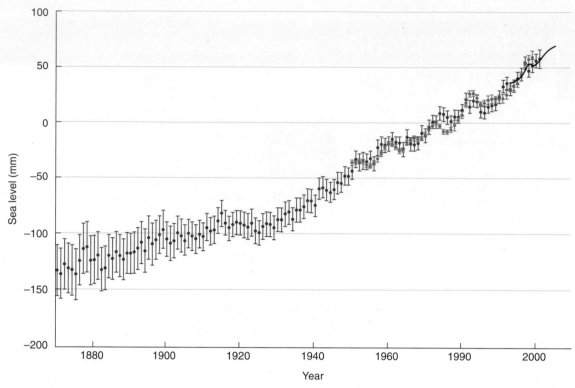

Figure 18–12 The rise in global mean sea level. Data from 1870 to 1950 reconstructed sea level from tide records in red, tide gauge measurements since 1950 in blue, and satellite altimetry since 1992 in black. (*Source:* S. Solomon et al., eds., *Climate Change 2007: The Physical Science Basis*, Contribution of Working Group I to the Fourth Assessment Report of the Intergovernmental Panel on Climate Change (Cambridge, England: Cambridge University Press, 2007), Figure 5.13.)

the western United States in recent decades is likely related to droughts, warmer temperatures, and earlier springs. Australia experienced a rash of tragic forest fires in early 2009, linked to 12 years of drought conditions.

- Rising Arctic temperatures have brought about a major shrinkage of Arctic sea ice, with decreases of 11.7% in the last decade (Fig. 18–13). While overall global temperature has increased a moderate 1°F, temperatures in Alaska, Siberia, and northwestern Canada have risen 5°F in summer and 10°F in winter. Spring comes two weeks earlier in the Arctic than it did 10 years ago. The polar ice cap has lost nearly 20% of its volume in the last two decades, and permafrost has been melting all over the polar regions—lifting buildings, uprooting telephone poles, and breaking up roads.

- The Greenland Ice Sheet, a huge reservoir of water, is melting and calving icebergs at unprecedented rates, undoubtedly a response to the rise in temperatures in the northern latitudes. A net loss of 27 cubic miles of water occurred between 2003 and 2005. Record warm summers have accelerated Greenland ice melt since then. The two-mile-high ice sheet contains enough water to raise ocean levels 23 feet (7 m).

- Temperatures in Antarctica, which holds enough ice to raise sea levels by 185 feet (57 m), have risen by more than 0.5°C over the past 50 years; West Antarctica showed the greatest warming, at 0.85°C. The West Antarctic Ice Sheet (WAIS) is the continent's most vulnerable ice mass, since it is grounded below sea level. Ocean water is constantly in contact with the edges of this ice sheet, and the ice sheet has been shrinking as this water eats away at the ice. Melting of this ice sheet would raise global sea level by 16 to 20 feet (5 to 6 m).

- Glaciers have resumed a retreat that has been documented for more than a century, but that slowed in the 1970s and 1980s. The World Glacier Monitoring Service has followed the melting and slow retreat of mountain glaciers all over the world. The response time of glaciers to climatic trends is slow—10 to 50 years—but the data collected show an acceleration of glacial "wastage" since 1990 that is likely a response to recent rapid atmospheric warming (Fig. 18–14).

- Patterns of precipitation are changing, with greater amounts from 30° N and 30° S poleward and with smaller amounts from 10° N to 30° N. Also, there have been widespread increases in heavy precipitation events, leading to more frequent flooding.

- As we saw earlier, there is a trend toward more-intense tropical hurricanes, which correlate with higher sea surface temperatures. In the Atlantic basin, the frequency of named storms has increased significantly over the past 13 years.

- Marine fish species populations are shifting northward as sea temperatures have warmed, impacting the commercial fisheries. Changing sea temperatures have also

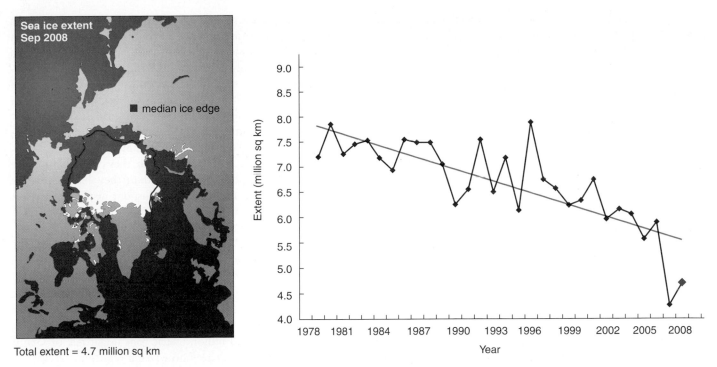

Total extent = 4.7 million sq km

Figure 18–13 Decline of Arctic sea ice. (a) The extent of sea ice in September 2008 was 4.7 million km², the second lowest measure on record. The magenta line is the median ice extent for September from 1979 to 2000. (b) Tracking with satellites shows the 30-year downward decline in the Arctic sea ice at the end of the northern summer. The decline is now 11.7% per decade. (*Source:* National Snow and Ice Data Center, University of Colorado, October 2008.)

(a)   (b)

Figure 18–14 Shrinkage of the Muir glacier. The Muir glacier, a tidewater glacier in Alaska's Glacier Bay, has retreated more than 7 miles (12 km) and thinned by more than 2,600 feet (800 m) since 1941. Once covered with ice and snow (a), the land now holds trees and shrubs (b). (*Source:* NSIDC/WDC for Glaciology, compiled 2002, updated 2006, *Online glacier photograph database,* Boulder, Colo.: National Snow and Ice Data Center/World Data Center for Glaciology, digital media.)

affected the spatial distribution of phytoplankton production over major ocean areas.

- The oceans have been soaking up some 40% of the CO_2 emissions from human sources, for which we may be thankful, since this slows the rise of atmospheric CO_2 considerably. However, in doing so, the oceans are experiencing a decrease in pH, a process known as **ocean acidification.** Although the decrease is only about 0.1 pH unit to date, there is serious concern about the future state of seawater pH in light of continued anthropogenic CO_2 production. At some level, ocean acidification will make the warm oceans inhospitable to coral reefs; the coral animals depend on an equilibrium for carbon minerals that is impacted by acidic conditions. The bottom line: Human use of fossil fuels has actually begun to change the chemistry of the surface oceans.

Given the accumulating evidence that the climate is changing and that it is already causing an array of changes in biological, physical, and chemical systems on Earth, it is time to address the causes of these changes in more detail.

Rising Greenhouse Gases

According to the IPCC AR4 Synthesis Report: "Global GHG emissions due to human activities have grown since preindustrial times, with an increase of 70% between 1970 and 2004."[7] The most important of these anthropogenic GHGs is carbon dioxide (CO_2).

Carbon Dioxide. More than 100 years ago, Swedish scientist Svante Arrhenius reasoned that differences in CO_2 levels in the atmosphere could greatly affect Earth's energy budget. Arrhenius suggested that, in time, the burning of fossil fuels might change the atmospheric CO_2 concentration, although he believed that it would take centuries before the associated warming would be noticeable. He was not concerned about the impacts of this warming, arguing that such an increase would be beneficial. (He lived in Sweden!)

Carbon Dioxide Monitoring. In 1958, Charles Keeling began measuring CO_2 levels on Mauna Loa, in Hawaii. Measurements there have been recorded continuously, and they reveal a striking increase in atmospheric levels of the gas (Fig. 18–6). The concentrations increased exponentially until the energy crisis in the mid-1970s, rose at a

rate of 1.5 ppm/year for several decades, and in the 2000s began rising at a rate of 2.0 ppm/year and higher. The data also reveal an annual oscillation of 5–7 ppm, which reflects seasonal changes of photosynthesis and respiration in terrestrial ecosystems in the Northern Hemisphere. When respiration predominates (late fall through spring), CO_2 levels rise; when photosynthesis predominates (late spring through early fall), CO_2 levels fall. Most salient, however, is the relentless rise in CO_2. As of early 2009, atmospheric CO_2 levels were at 388 ppm, 39% higher than they were before the Industrial Revolution and higher than they have been for over 800,000 years (Fig. 18–7a). Thus, our insulating blanket is thicker, and it is reasonable to expect that this will have a warming effect.

Sources. As Arrhenius suggested, the obvious place to look for the source of increasing CO_2 levels is our use of fossil fuels. Every kilogram of fossil carbon burned produces about 3 kilograms of CO_2 that enters the atmosphere. (The mass triples because each carbon atom in the fuel picks up two oxygen atoms in the course of burning and becoming CO_2.) Currently, 8.0 billion metric tons (gigatons, or Gt) of fossil fuel carbon (GtC) are burned each year (Fig. 18–15), all added to the atmosphere as CO_2 (about 3% of this total comes from cement production and is usually included in figures reported for fossil fuel emissions). At least half of this amount comes from the industrialized countries (Fig. 18–16). It is estimated that the burning of forest trees is another anthropogenic source that adds some 1.6 GtC annually to the carbon already coming from fossil fuel combustion. Over the past 50 years, the release of carbon from fossil fuel use and deforestation has almost tripled. Fortunately, more than half of the carbon emissions are removed annually by so-called *sinks*.

---

[7]IPCC, 2007: *Climate Change 2007: Synthesis Report.* Contribution of Working Groups I, II and III to the Fourth Assessment. Report of the Intergovernmental Panel on Climate Change (Core Writing Team, Pachauri, R.K and Reisinger, A. (eds.)). IPCC, Geneva, Switzerland, 104 pp.

Figure 18–15 Global carbon emissions from fossil fuel combustion. Data from 1980 to 2008, from the Carbon Dioxide Information Analysis Center (CDIAC), Energy Information Administration (EIA), and International Energy Agency (IEA). The rise in the 2000s decade is conspicuous. Go to **GRAPHit!** at www .masteringenvironmentalscience.com. (*Sources:* CDIAC, USEIA, IEA.)

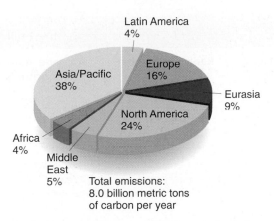

Figure 18–16 Sources of carbon dioxide emissions from fossil fuel burning. Total emissions in 2006 were approximately 8.0 GtC, or 29 billion metric tons of CO_2. Half of this (Europe, North America, Japan, South Korea, Taiwan, Saudi Arabia) comes from the industrialized countries (*Source:* Data from U.S. Energy Information Administration, December 2008.)

Sinks. Careful calculations show that if all the CO_2 emitted from burning fossil fuels accumulated in the atmosphere, the concentration would rise by 3 or more ppm per year, not the 2 ppm charted in Figure 18–6. In the 1990s, this meant an addition of approximately 8 GtC per year to the atmosphere by anthropogenic sources; yet only 3.3 GtC per year actually accumulated. Thus, there must be carbon sinks that absorb CO_2 and keep it from accumulating at a more rapid rate in the atmosphere. Recent work with stable carbon and oxygen isotopes and careful measurements of atmospheric CO_2 around the globe have brought us much closer to quantifying annual fluxes in CO_2 and identifying the sinks. There is broad agreement that the oceans serve as a sink for much of the CO_2 emitted; some of this is due to the uptake of CO_2 by phytoplankton, and some is a consequence of the undersaturation of CO_2 in seawater. There are limitations to the ocean's ability to absorb CO_2, however, because only the top 300 m of the ocean are in contact with the atmosphere. (As mentioned earlier, the deep ocean layers do mix with the upper layers, but the mixing time is over 1,000 years.)

Calculations indicate that, in the 1990s, the ocean sink accounted for an uptake of 2.2 ± 0.4 GtC annually, or about 30% of emissions.

The seasonal swings illustrated in Figure 18–6 show that the biota can influence atmospheric CO_2 levels. Measurements indicate that terrestrial ecosystems also serve as carbon sinks. Indeed, these land ecosystems apparently stored a net 1.0 ± 0.5 GtC annually during the 1990s, a figure that includes the losses from deforestation and thus implies an average gross annual uptake of some 2.5 GtC during those years. As a result, terrestrial ecosystems, and especially forests, are increasingly valued for their ability to sequester carbon. NASA prepared a satellite intended to measure CO_2 throughout the atmosphere and confirm the locations of the sinks. The February 2009 launch unfortunately failed to reach orbit, and the satellite crashed into the ocean.

A simple model of the major dynamic pools and fluxes of carbon, as presently understood, is presented in Figure 18–17.

Other Gases. Water vapor, methane, nitrous oxide, ozone, and chlorofluorocarbons (CFCs) also absorb infrared radiation and add to the insulating effect of CO_2 (Table 18–2). Most of these gases have anthropogenic sources and are increasing in concentration, raising the concern that future warming will extend well beyond the calculated effects of CO_2 alone.

Water Vapor. Water vapor absorbs infrared energy and is the most abundant GHG. Although it plays an important role in the greenhouse effect, its concentration in the troposphere is quite variable. Through evaporation and precipitation (the hydrologic cycle—see Chapter 10), water undergoes rapid turnover in the lower atmosphere. However, recent measurements have found a general increase in water vapor in the atmosphere of 400 grams per cubic meter per decade since 1988. As temperatures over the land and oceans rise, evaporation increases, and the water vapor concentration, or humidity, rises. As the humidity rises, it traps more heat in the atmosphere, which causes even more warming. This phenomenon is called **positive feedback** and is one of the more disturbing features of future warming. The recent

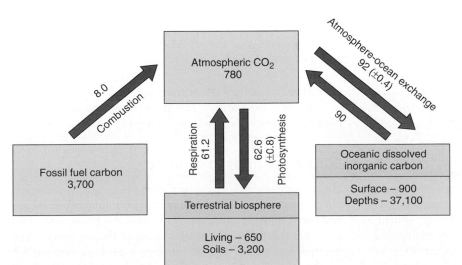

Figure 18–17 Global carbon cycle. Data are given in GtC (billions of metric tons of carbon). Pools are in the boxes, and fluxes are indicated by the arrows. (*Data sources:* IPCC AR4, 2007: CRS Report for Congress RL34059, 2007.)

TABLE 18–2   Anthropogenic Greenhouse Gases in the Atmosphere

Gas	Average Preindustrial Concentration (ppb)*	Approximate Current Concentration (ppb)	Radiative Forcing (W/m²)	Average Years in Atmosphere
Carbon dioxide (CO_2)	280,000	388,000	1.63	ca. 120
Methane (CH_4)	714	1,875	0.48	12
Nitrous oxide (N_2O)	270	322	0.16	114
Chlorofluorocarbons and halocarbons	0	1.3	0.32	50–100

Sources: IPCC AR4; recent data from CDIAC.
*Parts per billion; 1,000 ppb = 1 part per million (ppm).

rise in atmospheric humidity is entirely consistent with the warming that has accompanied the rise in anthropogenic GHGs.

Methane. Methane (CH_4), the third-most-important GHG, is 20 times more effective a GHG than is CO_2, molecule for molecule. Primarily a product of microbial fermentative reactions, methane's main natural sources are wetlands and green plants. Anthropogenic sources include livestock (methane is generated in the stomachs of ruminants), landfills, coal mines, natural-gas production and transmission, rice cultivation, and manure. Although methane is gradually destroyed in reactions with other gases in the atmosphere, it is being added to the atmosphere faster than it can be broken down. The concentration of atmospheric methane has more than doubled since the Industrial Revolution, as revealed in core samples taken from glacial ice (Table 18–2). After rising at a rate of around 1.8 parts per billion (ppb) per year, methane now appears to be increasing more slowly, at about 0.8 ppb per year. Measurements indicate that two-thirds of the present methane emissions are anthropogenic. Like CO_2, methane is more abundant than it has been for at least the last 800,000 years.

Nitrous Oxide. Nitrous oxide (N_2O) levels have increased some 18% during the last 200 years and are still rising at a rate of 0.8 ppb per year. Sources of the gas include agriculture, the oceans, and biomass burning; lesser quantities come from fossil fuel burning and industry. N_2O is produced in agriculture via anaerobic denitrification processes (see Chapter 3), which occur wherever nitrogen (a major component of fertilizers) is highly available in soils. The buildup of nitrous oxide is particularly unwelcome because its long residence time (114 years) makes the gas a problem in not only the troposphere, where it contributes to warming, but also the stratosphere, where it contributes to the destruction of ozone.

Ozone. Although ozone in the troposphere is short-lived and varies at many locations, it is a potent GHG. Some ozone from the stratosphere (where it is formed) descends into the troposphere, but the greatest source is anthropogenic, through the action of sunlight on pollutants (Chapter 19). Estimates indicate that the concentration of ozone in the atmosphere has increased 36% since 1750.

Major sources are automotive traffic and burning forests and agricultural wastes.

CFCs and Other Halocarbons. Emissions of halocarbons are entirely anthropogenic. Like nitrous oxide, halocarbons are long lived and contribute to both global warming in the troposphere and ozone destruction in the stratosphere. Used as refrigerants, solvents, and fire retardants, halocarbons have 10,000 times the capacity to absorb infrared radiation that CO_2 does. The rate of production of chlorofluorocarbons (CFCs) has declined since the Montreal Accord of 1987, and the concentration of CFCs in the troposphere leveled off in the late 1990s and is now slowly declining. However, these gases are highly stable and will continue to exert their warming effects for many decades.

Together the other anthropogenic GHGs are estimated to trap almost as much infrared radiation as CO_2 does. Although the tropospheric concentrations of some of these gases are rising, it is hoped that they will gradually decline in importance because of steps being taken to reduce their levels in the atmosphere, leaving CO_2 as the primary GHG to cope with in the future. The IPCC AR4 Synthesis Report states: "Global atmospheric concentrations of CO_2, CH_4 and N_2O have increased markedly as a result of human activities since 1750 and now far exceed pre-industrial values determined from ice cores spanning many thousands of years. . . . There is *very high confidence* that the global average net effect of human activities since 1750 has been one of warming, with a radiative forcing of +1.6 W/m²." Figure 18–18 shows the current understanding of the various radiative forcing agents, in units of W/m². It is evident that positive anthropogenic forcing far outweighs negative anthropogenic forcing.

Future Changes in Climate

Higher temperatures, rising sea levels, heat waves, droughts, intense precipitation events, shifts in seasons, melting ice sheets, and Arctic thawing—these are happening right now, and GHG levels are continuing to rise as a result of human agency. The International Energy Agency projects a growing demand for fossil fuels as populations and economies grow. If global fossil

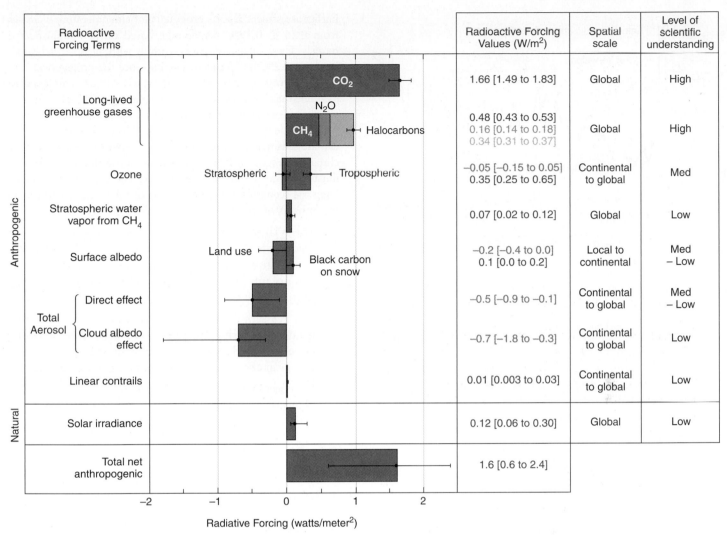

Figure 18–18 Radiative forcing. Positive and negative factors contribute to changes in energy balance within the atmosphere and at the Earth's surface. The figure presents the human factors that have led to the current sustained increase in radiative forcing (RF), in Watts per square meter. The 90% confidence limits are indicated, and on the right, the relative radiative forcing and spatial scale of effects are shown, along with the level of scientific understanding. (*Source:* S. Solomon et al. eds., *Climate Change 2007: The Physical Science Basis*, Contribution of Working Group I to the Fourth Assessment Report of the Intergovernmental Panel on Climate Change (Cambridge, England: Cambridge University Press, 2007), Figure SPM 2.)

fuel use continues on its present trajectory, emissions of GHGs will increase 35% by 2030. By 2050, this grows to over 100%. Given the huge reserves of coal, ample supplies of natural gas, and remaining oil reserves, there is no reason to doubt this projection. However, it is clear that commitment to this future use of fossil fuels also will commit us to greater changes in climate. How much change, and when, would be a matter of guesswork if it were not for the use of climate models.

Modeling Global Climate. Weather forecasting employs powerful computers that are capable of handling large amounts of atmospheric data and applying mathematical equations to model the processes taking place in the atmosphere, oceans, and land. Forecasts of conditions for 72 hours or more have become quite accurate. Even though computing power has increased greatly in recent years, the somewhat chaotic behavior of weather parameters prevents longer-term forecasting.

Modeling climate is an essential strategy for exploring the potential future impacts of rising GHGs. Climatologists employ similarly powerful computers and have combined

global atmospheric circulation patterns with ocean circulation, radiation feedback from clouds, and land surface processes to produce *atmosphere-ocean general circulation models (AOGCMs)* that are capable of simulating long-term climatic conditions. Fourteen centers around the world are now intensely engaged in exploring climate change by running models coupling the atmosphere, oceans, and land.

The best test of such models is the degree to which they can simulate present-day climate. A recent study compared some 50 climate models developed over the past two decades in order to answer this question: How well do coupled models simulate today's climate?[8] The authors concluded that the most recent models, especially those used for the IPCC AR4, have reached an unprecedented level of realism. Figure 18–19, from the IPCC AR4, shows how closely models track the observed temperature changes from 1900 to 2004, taking into consideration natural and anthropogenic forcing.

[8]Thomas Reichler and Junsu Kim, "How Well Do Coupled Models Simulate Today's Climate?" *Bulletin of the American Meteorological Society* 89, no. 3 (March 2008): 303–331.

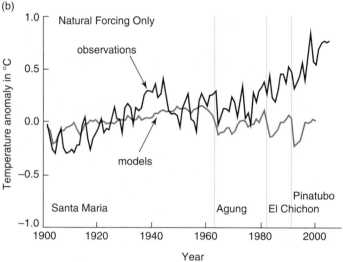

Figure 18–19 Comparison between modeled and actual data on temperature anomalies, 1900–2000. Observed values are shown by the black line, and runs from 13 models are shown in color. (a) Model results with both anthropogenic and natural forcings. (b) Model results with only natural forcings. The gray lines mark the timing of major volcanic events. (*Source:* S. Solomon et al., eds., *Climate Change 2007: The Physical Science Basis*, Contribution of Working Group I to the Fourth Assessment Report of the Intergovernmental Panel on Climate Change (Cambridge, England: Cambridge University Press, 2007), Figure 23, TS.)

Earlier assessment reports projected global temperature increases from 0.15 to 0.29°C per decade, based on less sophisticated models. These values compare favorably with the observed values of about 0.2°C per decade occurring since the projections.

Running the Models. The AOGCMs are employed to examine the changes that might take place over the 21st century. A number of assumptions about GHG emissions are combined with a range of climate responses to GHG concentrations (called **climate sensitivity**) to project some 35 scenarios of 21st-century climate changes. In Figure 18–20, we show three scenarios to illustrate a range of climate responses. The assumptions underlying the scenarios are critical (none of the scenarios considers the impacts of specific policies to reduce GHGs):

- The A1B scenario reflects a world in which population reaches its peak in midcentury and then declines, economic growth is rapid, and new and efficient technologies are introduced; there is a balance across all energy sources, with no specific reliance on any one technology.

- The A2 series reflects a heterogeneous world of independent countries going their own way, with increasing population and regionally varied economic growth; technological change is slower than in other scenarios.

- B1 reflects a more globalized world with population developments as in A1B, but with rapid changes in economies and effective global cooperation to reach sustainability in all sectors; in particular, technologies are clean and resource efficient.

Every scenario employed in the models generates a rise in CO_2 concentrations and a corresponding rise in mean global temperature. Even if there were no increase at all in radiative forcing agents, current GHG concentrations represent a *commitment* to further temperature increase, due to the delay in the release to the atmosphere of energy stored in the oceans. That commitment is on the order of about 0.1°C per decade for the next several decades, as shown in Figure 18–20 (the orange curve). The blue, green, and red curves (Fig. 18–20) represent the B1, A1B, and A2 scenarios extended to 2100. Beyond 2100, forcing is kept constant in the A1B and B1 scenarios, and what is seen is *commitment* based on accumulated GHGs.

Figure 18–20 Model projections of global mean warming. Projections are for the A2, A1B, and B1 scenarios starting in 2000 and extended to 2100. The orange curve from 2000 shows the forcing "commitment" if greenhouse gas emissions were held constant in 2000. Extensions beyond 2100 are for constant forcing, showing the commitment to further temperature increase due to the delay in energy release (*Source:* S. Solomon et al., eds., *Climate Change 2007: The Physical Science Basis*, Contribution of Working Group I to the Fourth Assessment Report of the Intergovernmental Panel on Climate Change (Cambridge, England: Cambridge University Press, 2007), Figure 32, TS.)

Projections. The AR4 examined the implications of the model runs and projected the potential future climate changes that could be expected, given the range of scenarios and also the current picture of climate change already under way. The major consequence of rising GHG levels during the 21st century is *rising temperatures*—the amount of rise depending on energy choices and other factors, such as population growth. Rising global temperatures are linked to two major impacts: *regional climatic changes* and *a rise in sea level*. Both of these effects show up in all of the models. Some of the most significant findings are presented here:

- **Equilibrium climate sensitivity**—*the sustained change in mean global temperature that occurs if atmospheric CO_2 is held at double preindustrial values, or about 550 ppm*—is in the range of 2–4.5°C, most likely 3°C. This occurs in all of the scenarios, and many observers believe it could happen by 2050.

- The actual warming of the 21st century will be a consequence of GHG emissions, whatever they are. The model scenarios project a range of surface warming in 2100 compared with 1980–2000; for the B1, A1B, and A2 scenarios, the estimates average 1.8°C, 2.8°C, and 3.4°C, respectively. The projected rate of warming for the next two decades is 0.2°C per decade.

- The warming is expected to occur in geographical patterns similar to those of recent years: warming greatest at the higher latitudes and in continental interiors and least over the North Atlantic and the southern oceans. However, it will be warmer everywhere.

- Because of the warming climate, snow cover and sea ice will decrease, possibly fully opening up the Arctic Ocean by the end of the century. Glaciers and ice caps will continue to shrink, resulting in a rise in sea level. In the Arctic, up to 90% of the upper layer of permafrost is expected to thaw.

- The sea level, currently rising at a rate of 3.3 mm per year, will certainly continue to rise as a consequence of continued thermal expansion and snow and ice melt (Fig. 18–12). The range projected by the IPCC for the end of the 21st century is 7 to 23 inches (19 to 59 cm). Because of the slow response time of thermal expansion, larger increases (30 to 80 cm) could occur in the next two centuries. Significantly, the projections did not consider any increases in the rate of melting of the Greenland and Antarctic Ice Sheets.

- As the upper layers of the North Atlantic Ocean warm and become more fresh, the thermohaline circulation is expected to slow down, but no models show it collapsing.

- Storm intensities are expected to increase, with higher wind speeds, more extreme wave heights, and more intense precipitation.

- Heat waves will become more frequent and longer lasting as the climate warms. Growing seasons will lengthen, and frost days will decrease in middle and high latitudes. Outbreaks of cold air in the Northern Hemisphere will diminish by 50% or more compared with the present.

- Precipitation will decrease in already dry regions and increase in already wet regions; there will be an increase in extremes of daily precipitation. Areas of extreme drought will expand, affecting up to 30% of the world's land surface.

- The changing climatic parameters will have profound impacts on ecosystems and species, the impacts accruing as the temperature warms over time (Fig. 18–21): damage to polar ecosystems (as we have seen), bleached and

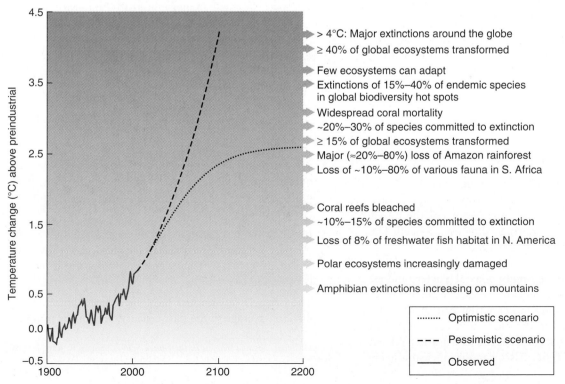

Figure 18–21 Global warming impacts on biological systems. The impacts build up with greater temperature changes over time. The optimistic scenario is B1 from IPCC AR4; the pessimistic scenario is A2 from IPCC AR4. (*Source:* Richard Kerr, "How Urgent Is Climate Change?" *Science* 318 (November 23, 2007): 1230–1231.)

dying coral reefs, major loss of Amazon rain forests, widespread extinctions of species, and loss of ecosystem resilience.

A "business-as-usual" approach to climate change is best illustrated by the A2 scenario, which projects a temperature change between 3°C and 5°C. Note that the difference in temperature between an ice age and the warm period following it is only 5°C, so in a century, unless some serious measures are taken, Earth's climate will change dramatically. Responding to these changes will likely involve unprecedented and costly adjustments. The impact on natural ecosystems could be highly destabilizing. For the agricultural community, the greatest difficulty in coping with climatic change is *not knowing what to expect*. Already, farmers lose an average of one in five crops because of unfavorable weather. As the climate shifts, the vagaries of weather will become more pronounced, and crop losses are likely to increase.

Sea Level.   There is great uncertainty about the magnitude of the rise in sea level. A half-meter rise in sea level will flood many coastal areas and make them much more prone to damage from storms, forcing people to abandon properties and migrate inland. A one-meter rise would flood 17% of Bangladesh, creating tens of millions of refugees and obliterating half of the country's rice lands. For many of the small oceanic nations, such as the Maldives, off the coast of India, and the Marshall Islands, in the Pacific, a rise in sea level means obliteration, not just alteration. Models suggest that, if the mean global temperature rises as much as 3°C above present levels, it will set up conditions that could cause widespread loss of the Greenland Ice Sheet. Recent studies have made it clear that the IPCC projections of sea level rise are *seriously on the low side*. These studies indicate that 21st-century warming would lead to a sea level rise of 0.8 to 2 meters, considering the plausible rates of melting of glaciers that run into the sea (in Greenland and Antarctica). And once atmospheric GHG levels are stabilized (assuming they are!), temperatures and sea levels will continue to rise for hundreds of years because of the slow response time of the oceans. The expected continued rise would cause disasters in most coastal areas, which are home to half the world's population and its business and commerce.

What About the Antarctic?   Some new studies indicate that the Antarctic could be a huge factor in future sea level rise. As we have seen, the Greenland-sized WAIS has been shrinking. The Ross and Weddell seas extend outward from this ice field, and the landward fringes of these seas are covered with thick floating shelves of ice. The Greenland and Antarctic Ice Sheets hold enough water to raise sea level by 230 feet (70 m). For the Antarctic, snowfall in the interior and the melting of coastal glaciers and ice sheets have long provided a delicate balance in their impact on sea level. Some recent data indicate that Antarctic snowfall has changed very little in the past 50 years, while 87% of 244 Antarctic glaciers studied have shown a net retreat over the same period of time. Also, satellite measurements

have shown that the ice sheet is losing as much as 36 cubic miles of ice a year, enough to raise sea level by 0.4 mm a year. These losses of ice are attributed to higher temperatures on the continent (almost 3°C over the last 50 years). However, most climate models project a 5% snowfall increase in Antarctica for every 1°C temperature rise. Clearly, this is a major uncertainty in any assessment of future sea level rise. Stay tuned!

Climate Change Impacts in the United States

In 2002, the Bush administration launched the Climate Change Science Program (CCSP—now called the Global Change Research Program, or GCRP) to integrate the efforts of many federal agencies involved in the scientific effort to understand global climate change. One of the tasks of the CCSP has been to communicate the findings of the scientific community to policy makers and the broader public by publishing reports on climate change issues. These reports, covering a broad range of topics, are available on the Internet (www.climatescience.gov) and are very well done. The 2009 report, *Global Climate Change Impacts in the United States*,[9] draws from 19 existing CCSP reports and other peer-reviewed scientific assessments, such as those from the IPCC. It is designated as the **United Synthesis Product** (USP) and provides a detailed look at the impacts of climate change on various *sectors* (e.g., transportation, agriculture, human health) and *regions* (e.g., Northwest, Southwest, Northeast, Alaska) of the United States. Some highlights of the report are worth examining.

National Climate Change.   Most of the global impacts of climate change—temperature increases, droughts, intensity of hurricanes, seasonal changes, increased intensity of heavy rain events, rising sea level, large wildfires, and melting glaciers, among others—have been observed in the United States. The projected future changes just reviewed will certainly affect the United States. However, national and regional climates vary more than global climate, and the different regions of the United States have experienced different impacts and will do so in the future. Following are key observations from the report:

- The average U.S. temperature has risen more than 2°F over the past 50 years and will rise more; how much more depends primarily on the amount of heat-trapping gases emitted globally.

- Precipitation has increased an average of about 5% over the past 50 years. Shifting patterns have generally made wet areas wetter, while dry areas have become drier. This is projected to continue.

- The heaviest downpours have increased approximately 20% on average in the past century, and this is projected

[9]*Global Climate Change Impacts in the United States*, Thomas R. Karl, Jerry M. Melillo, and Thomas C. Peterson, (eds.). Cambridge University Press, 2009.

to continue, with the strongest increases in the wettest places.

- Many types of extreme weather events, in addition to heavy downpours, have become more frequent and intense during the past 40 to 50 years.

- The destructive energy of Atlantic hurricanes has increased in recent decades and is projected to increase further in this century.

- Sea level has risen 2 to 5 inches during the past 50 years along many U.S. coasts and is projected to rise more in the future.

- For cold-season storms outside the tropics, storm tracks are shifting northward, and the strongest storms are projected to become stronger.

- Arctic sea ice is declining rapidly and this is projected to continue.

Some interesting graphics are used to illustrate regional climate changes. For example, for the Midwest, the summer climate in Illinois and Michigan is shown shifting southward and westward (Fig. 18–22) as temperatures increase and precipitation decreases. Significantly, the impacts of climate change in Alaska are greater than those in any other region of the United States. Average temperatures are projected to rise 4°F to 7°F by midcentury and could rise 8°F to 13°F by the end of the century if global emissions continue on their present course. The USP report's concluding thoughts emphasize the importance of involving scientists, policy makers, and the public in "designing, initiating, and evaluating *mitigation* and *adaptation* strategies. . . . The best decisions about these strategies will come when there is widespread understanding of the complex issue of climate change—the science and its many implications for our nation."

We turn now to our society's response to climate change.

18.3 Response to Climate Change

The world's industries and transportation networks are so locked into the use of fossil fuels that massive emissions of CO_2 and other GHGs seem certain to continue for the foreseeable future, bringing on significant climate change. One possible response to this situation is **adaptation**. That is, we must anticipate some harm to natural and human systems and should plan adaptive responses to lessen the vulnerability of people, their property, and the biosphere to coming changes.

Yet it seems more sensible to reduce the rate at which emissions are added to the atmosphere and, eventually, bring about a sustainable balance—although no one thinks that doing so will be easy. This is the **mitigation** response: Take action to reduce emissions.

Skeptics about global warming can be found in many sectors, from predictable sources such as the fossil fuel industry, Rush Limbaugh, and many conservative think tanks, to

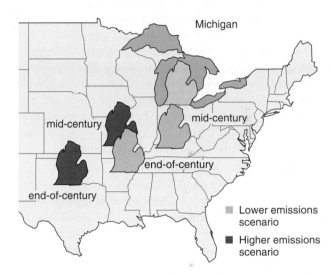

Figure 18–22 Climate on the move in the Midwest. Model projections of summer average temperature and precipitation changes in Illinois and Michigan for midcentury and end of century. The lower emissions scenario is IPPC B1, and the higher emissions scenario is IPPC A2. Note that both states are projected to get considerably warmer and have less summer precipitation. (*Source:* CSSP, *Global Climate Change Impacts in the United States: Unified Synthesis Product,* 2009.)

some members of the scientific community. John Holdren, chief scientist in the Obama administration and a leading climate change scientist, describes three stages through which climate change skeptics tend to move: Stage 1—*You're wrong about climate change, and we can prove it.* The trouble is that there are no other plausible candidates for the observed disruption of global climates. So they go to Stage 2—*OK, the climate is changing, but it isn't changing all that much. The threat is overplayed. Why take enormously costly steps to prevent something that may have positive results or may even never happen?* Faced with the overwhelming evidence, many with scientific credentials eventually move on to Stage 3—*Well, it will do some real damage, but it's too late or too costly to do anything about it. We'll just have to adjust to it.*

The skeptics can be found on talk radio and Internet blogs, in op-ed pieces and the Congress of the United States. Holdren makes the point that the unfounded skepticism is not just regrettable; it is also downright dangerous. We need to get going on shaping remedies to meet the challenges, and the widespread, but misguided skepticism has already delayed the desperately needed political consensus for action. There are ethical issues at stake, too.

Ethical Principles. Indeed, why respond at all? Here are **three principles** that speak to this question:

1. The first is the **precautionary principle**, as articulated in the 1992 Rio Declaration. The principle states, "In order to protect the environment, the precautionary approach shall be widely applied by States according to their capabilities. Where there are threats of serious or irreversible damage, lack of full scientific certainty shall not be used as a reason for postponing cost-effective measures to prevent environmental degradation." Employing this principle is like taking out insurance: We are taking measures to avoid a highly costly, but uncertain, situation (albeit one that is appearing more and more likely). The risk of serious harmful consequences is real, the scientific consensus is strong, and the political response, as we will see, is becoming robust.

2. The second is the **polluter pays principle**. Polluters should pay for the damage their pollution causes. This principle has long been part of environmental legislation. CO_2 is where it is because the developed countries (and now many developing countries) have burned so much fossil fuel since the beginning of the industrial era.

3. The third is the **equity principle**. Currently, the richest 1 billion people produce 55% of CO_2 emissions, while the poorest 1 billion produce only 3%. International equity and intergenerational equity are ethical principles that should compel the rich and privileged to care about those generations that follow—especially those in the poor countries who are even now experiencing the consequences of global climate change. Climate change is going to cause harm, as we have seen. If we are able to take action that will prevent both present and future harm, it would be wrong not to do so.

Mitigation

It will be costly to take the actions needed to mitigate the causes of climate change and to adapt to changes that are unavoidable. But the costs of inaction will be far higher. If climate change is unchecked, every country will be impacted, but it will be most disastrous for the world's poorest countries and people. Widespread droughts, rising sea levels, more intense and numerous hurricanes and typhoons, and more heat waves, wildfires, severe water shortages, crop failures, and other climatic disruptions will create hundreds of millions of climate refugees. Conflict and terrorism will surely follow in the wake of such deprivation, as people are radicalized by

their plight. The impacts on the natural world will be equally severe, as we have seen (Fig. 18–21).

Some attempts have been made to put a price tag on inaction; these are admittedly rough estimates, but ones based on sound analyses. For the United States, one study by the Natural Resources Defense Council estimates a total economic cost of global warming at 3.6% of gross domestic product (GDP) by 2100, or almost $3.8 trillion (in today's dollars). For just four impacts—hurricane damage, energy costs, real estate losses, and water costs—the price tag comes to almost $1.9 trillion. The Stern Review,[10] an economic analysis commissioned by the British government, estimated that inaction on global climate change would result in economic loss equivalent to 5% to 20% of global GDP per year, for the indefinite future. Given the economic, environmental, and social costs of the consequences of unrestrained release of GHGs, it would seem that mitigation is absolutely essential. We need to reduce emissions of GHGs in order to prevent disaster. The next question is, What should be our target?

Achieving Stabilization. The stated objective of the Framework Convention on Climate Change (FCCC) is to *stabilize the GHG content of the atmosphere at levels and on a time scale that would prevent dangerous anthropogenic interference with the climate system.* This means that GHG emissions must be brought under control to the point where they are no longer accumulating in the atmosphere, and it must be done before global temperatures rise to dangerous levels. Earth history tells us that with 3°C global warming, equilibrium sea level could be at least 80 feet (25 m) higher than today. A rise of 2°C could trigger irreversible melting of the Greenland Ice Sheet and a sea level rise of several meters. Both of these scenarios are clearly "dangerous" to human civilization and Earth systems in general and are to be avoided. However, with current warming at about 0.8°C and more warming already in the pipeline, we could get to 2°C in a couple of decades. Many feel that this could well be a *tipping point*, a point beyond which irreversible—and clearly dangerous—runaway climate change occurs no matter what we do.

So we have a possible target, a cushion of sorts, of 1–1.5°C more warming, corresponding to a stable concentration of CO_2 (with the other GHGs included) in the atmosphere somewhere between 400 and 500 ppm. To achieve stabilization at 450 ppm will require an energy system that is revolutionary because current carbon emissions of 8 GtC/year will have to fall to 2 GtC/year by 2100 and then drop even further. This is the *inconvenient truth* Al Gore referred to in his 2006 film on global warming. The IPCC AR4 Synthesis Report presented an analysis of six stabilization scenarios and their correspondence with global temperatures and sea level rise (Table 18–3). Global emissions of GHGs corresponding to the different scenarios are shown in Figure 18–23a, along with the time course of allowable emissions. The relationship between temperature increase above preindustrial levels and GHG stabilization-level

---

[10]Nicholas Stern, *The Economics of Climate Change: The Stern Review* (Cambridge, England: Cambridge University Press, 2007).

TABLE 18–3 Stabilization Scenarios and Corresponding Global Temperatures and Sea Level Rise

Category	CO₂-equivalent concentration at stabilization (including all GHGs) (ppm)	Peaking year for CO₂ emissions	Change in global CO₂ emissions in 2050 (% of 2000 emissions)	Global avg. temperature increase at equilibrium (°C)	Global avg. sea level rise above preindustrial at equilibrium from thermal expansion (m)
I	445–490	2000–2015	−85 to −50	2.0–2.4	0.4–1.4
II	490–535	2000–2020	−60 to −30	2.4–2.8	0.5–1.7
III	535–590	2010–2030	−30 to +5	2.8–3.2	0.6–1.9
IV	590–710	2020–2060	+10 to +60	3.2–4.0	0.6–2.4
V	710–855	2050–2080	+25 to +85	4.0–4.9	0.8–2.9
VI	855–1130	2060–2090	+90 to +140	4.9–6.1	1.0–3.7

Source: IPCC, *Climate Change 2007: Synthesis Report* (Geneva, Switzerland: IPCC, 2008).

(a)

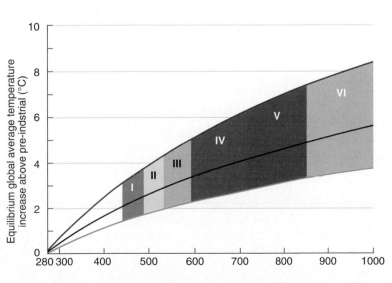

(b)

Figure 18–23 CO₂ emissions and equilibrium temperature increases for a range of stabilization levels. (a) Global CO₂ emissions for 1940 to 2000 and projected emissions ranges for six categories of stabilization scenarios from 2000 to 2100. The thickness of each band represents the uncertainty inherent in the model projections. The dashed lines (post-SRES range) show the upper and lower emissions range of more recently developed scenarios. (b) Corresponding relationship between a CO₂ stabilization target and the likely equilibrium global temperature increase above preindustrial. The black line through the middle shows the best estimate of the response of temperature to GHG concentrations (climate sensitivity); the red and blue lines show the upper and lower bounds of climate sensitivities. (*Source:* IPCC, *Climate Change 2007: Synthesis Report*, Contribution of Working Groups I, II and III to the Fourth Assessment. Report of the Intergovernmental Panel on Climate Change (Geneva, Switzerland: IPCC, 2007), Figure 5-1.)

concentration for the six scenarios is shown in Figure 18–23b. The IPCC states: "In order to stabilize the concentration of GHGs in the atmosphere, emissions would need to peak and decline thereafter. The lower the stabilization level, the more quickly this peak and decline would need to occur."[11] Clearly, the only category in Table 18–3 and Figure 18–23 that is consistent with the FCCC objective is category I. Again, from the IPCC: "Mitigation efforts over the next two to three decades will have a large impact on opportunities to achieve lower stabilization levels."

We turn now to the question of what the international community has done in response to global climate change.

What Has Been Done? One of the five documents signed by heads of state at the UN Conference on Environment and Development's Earth Summit in Rio de Janeiro in 1992 was the **Framework Convention on Climate Change (FCCC)**. This convention agreed to the goal of *stabilizing GHG levels in the atmosphere*, starting by reducing GHG emissions to 1990 levels by the year 2000 in all industrialized nations. Countries were to achieve the goal by voluntary means. Five years later it was obvious that the voluntary approach was failing. All the developed countries except those of the European Union *increased* their GHG emissions by 7% to 9% in the ensuing five years. The developing countries increased theirs by 25%!

Kyoto Protocol. Prompted by a coalition of island nations (whose very existence is threatened by global climate change), the third Conference of Parties to the FCCC met in

Kyoto, Japan, in December 1997 to craft a binding agreement on reducing GHG emissions. In Kyoto, 38 industrial and former eastern bloc nations agreed to reduce emissions of six GHGs to 5% *below* 1990 levels, to be achieved by 2012. The signatories to this agreement are called the *Annex I parties*; all others (the developing countries) are *non–Annex I parties*. The percentages of the reductions varied (Table 18–4). At the conference, the developing countries refused to agree to *any* reductions, arguing that the developed countries had created the problem and that it was only fair that the developing countries continue on their path to development as the developed countries had, energized by fossil fuels. This stance was in accord with the FCCC principle of "common but differentiated responsibilities." Under this principle, each nation must address climate change, but its priorities and efforts could differ according to national circumstances. Thus, the developing countries argued that their first priority was to alleviate poverty, whereas the developed countries had no such excuse.

The accord requires that at least 55 countries ratify it and that Annex I ratifying countries account for at least 55% of Annex I GHG emissions. More than 180 parties have ratified (many are non–Annex I parties), thus meeting the first requirement for the protocol. The second requirement for ratification was met in October 2004, when Russia signed on. The total percentage of Annex I emissions is 63.7%. The United States is the only Annex I party that has not ratified the Kyoto Protocol.

The signers of the Kyoto Protocol have a great deal of flexibility in deciding how they will achieve their GHG reductions. The most obvious choices involve renewable

[11]IPCC, *Climate Change 2007: Synthesis Report,* Contribution of Working Groups I, II and III to the Fourth Assessment. Report of the Intergovernmental Panel on Climate Change (Geneva, Switzerland: IPCC, 2007).

TABLE 18–4 Kyoto Emission Allowances and 16-Year Greenhouse Gas Emission Changes

Country	Kyoto Allowances (%) (1990–2012)	Observed Change (%) (1990–2006)	2006 Greenhouse Gas Emissions (10^6 Metric Tons of Carbon-equivalents)
United States	−7	+14.4	1,914
European Union	−8	−2.2	1,132
United Kingdom	−12.5	−15.1	179
Germany	−21	−18.2	274
Australia	+8	+28.8	146
Canada	−6	+21.7	196
Japan	−6	+5.3	365
Russian Federation	0	−34.2	597
All Annex I countries	−5.2	−1.7	4,996/4,052*
Non–Annex I countries			4,117*

Sources: UNFCCC; Carbon Dioxide Information Analysis Center.
*CO_2 emissions only.

energy, and possibly nuclear energy, and a variety of energy conservation strategies. They may also plant forests to act as carbon sinks or engage in emissions trading with other Annex I parties, either buying or selling assigned emissions amounts to meet targets. They may even implement projects that reduce emissions in other countries and earn credits toward their target emissions. Compliance with the Kyoto Protocol is monitored carefully, and there are penalties for failure to meet commitments. As Table 18–4 shows, many Annex I countries are on target for meeting their commitments, and several (the United States, Canada, Australia, and Japan) still have a long way to go.

The targeted reductions of the Kyoto Protocol will *not* stabilize atmospheric concentrations of GHGs. The IPCC calculates that it would take immediate reductions in emissions of at least 60% worldwide to stabilize GHG concentrations at today's levels. Since this has no chance of happening, the concentration of GHGs in the atmosphere will continue to rise. The major weakness of the Kyoto Protocol is that the world's largest GHG emitters—India, China, and the United States—are not participating. Nevertheless, Kyoto is an important first step toward the stabilization of the atmosphere.

Bali and Beyond. In December 2007, a UN-sponsored climate conference held in Bali addressed the need for greater cuts in GHG emissions when the Kyoto Protocol expires in 2012. At the Bali meeting, parties to the FCCC agreed to what was essentially a "road map" to future negotiations. Efforts to set new goals for cuts in emissions and targets for stable atmospheric concentrations of GHGs were thwarted by the United States and some developing countries. Instead, it was agreed that "deep cuts in global emissions will be required" to achieve the objective of avoiding dangerous climate change, and a negotiation process was established, to be implemented by the FCCC Ad Hoc Working Group as it discusses issues to be presented at the next meeting of the

parties to the FCCC in late 2009 in Copenhagen. The negotiations are intended to include developing countries in the decision making, which is to focus in particular on emissions reductions and other mitigation measures. Delegates were hopeful that a new administration in the United States would cooperate more fully with the negotiations and future meetings. This brings us to the question, What in fact has the United States been doing about global climate change and the FCCC?

U.S. Policy. Shortly after taking office in 2001, President George W. Bush stated that he opposed the Kyoto Protocol for two reasons:

1. It exempts the developing countries and thus is unfair.

2. It would cause serious harm to the U.S. economy.

He also cited the "incomplete state of scientific knowledge of the causes of, and solutions to, global climate change."[12] Later, Environmental Protection Agency (EPA) Administrator Christine Whitman officially notified the world that the United States would withdraw from the Kyoto agreement. Instead of working with the Kyoto Protocol, the Bush administration announced in 2002 the **Global Climate Change Initiative** **(GCCI).** The flagship policy of this initiative is a reduction of 18% in **emissions intensity** over the next 10 years. Note that emissions intensity is *not* the same as emissions; it is the *ratio* of GHG emissions to economic output, the latter measured as GDP.

Critics point out that the program is voluntary, and even if it works, it will simply continue a trend that has seen emissions intensity reduced by 28% from 1990 to 2007. While that happened, our total GHG emissions actually rose 17% (Fig. 18–24). It is clear that the GCCI simply will not even

---

[12]http://www.grist.org/article/gelbspan-boiling/

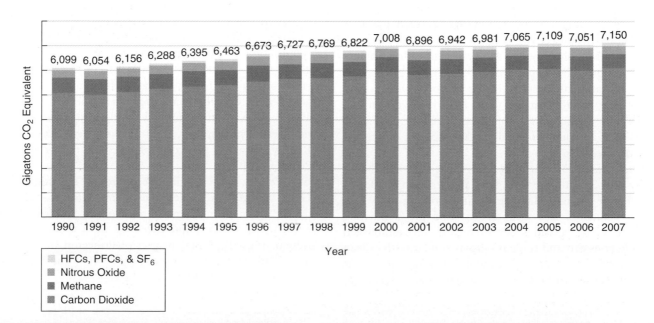

Figure 18–24 U.S. greenhouse gas emissions, 1990–2007. Four types of greenhouse gases are shown. Total emissions rose 17%, while greenhouse gas intensity fell 28%. Units are GtCO₂ equivalents. (*Source:* EPA, Inventory of U.S. Greenhouse Gas Emissions and Sinks, 1990–2007, April 2009.)

come close to a reasonable reduction on the order of those being mandated by the Kyoto Protocol nations. A more fruitful effort has been the Bush administration's **Climate Change Science Program** (CCSP), initiated in 2002, which was mentioned previously. The CCSP/GCRP now includes ongoing research on climate change, formerly in independently funded programs. This and other climate change programs have received some $37 billion in funding since 2001, expenditures that have been cited by the Bush administration as proof that the United States is not ignoring climate change.

In reality, the Bush administration in many ways placed roadblocks in the path of climate change science and policy. The U.S. role in thwarting international accord has already been mentioned. Here at home, the Bush EPA declined to act on a Supreme Court decision that ordered the agency to consider GHG pollution as a threat to human health and welfare. A House committee found that the Bush White House attempted to stifle communications from climate scientists and heavily edited reports from federal agencies, tilting them in the direction of questioning human influence on the climate by emphasizing scientific uncertainties about global climate change.[13] In what seemed to be a softening toward the issue, President Bush in 2008 proposed a new climate goal for the United States, calling for a limit to the growth of GHGs by 2025. Critics pointed out that this would allow U.S. emissions to rise at least 10% more and called the move "too little, too late."

A Sea Change. The Democratically controlled 110th Congress began to move ahead on climate change issues. The House established a Select Committee on Energy Independence and Global Warming, and numerous bills are in the pipeline covering a wide spectrum, from climate change research to GHG emissions cap-and-trade programs and emissions taxes. The 111th Congress has taken up these issues, and the new White House administration has promised to bring a sea change in U.S. climate change policy. President Obama has made climate legislation a high priority and has signaled his commitment to the issue by appointing Harvard physicist John Holdren as director of the White House Office of Science and Technology Policy. Holdren is a leading climate scientist. Other key appointments are Jane Lubchenko (another climatologist) as head of the National Oceanic and Atmospheric Administration and Nobel Prize–winning physicist Steven Chu as head of the Department of Energy. Chu has been authorized to lead the development of alternative energy sources.

States and Corporations. It is encouraging that a majority of U.S. states are taking unilateral action to address global climate change. Many states are adopting renewable portfolio standards (see Chapter 16) and are mandating reporting of GHG emissions. Several regions have established GHG emissions cap-and-trade programs similar to the Clean Air Act program that was successful in bringing down sulfur dioxide emissions. More than 300 mayors have pledged to move their cities to GHG reductions similar to those called for by Kyoto. The emerging mosaic of regional, state, and local initiatives is likely to put pressure on the federal government to enact a coherent regulatory strategy to bring down GHG emissions.

Convinced that the scientific findings call for prompt action, the heads of many of the country's largest corporations joined with environmental group leaders to urge Congress to fast-track climate legislation. They are members of the U.S. Climate Action Partnership, which has called for a reduction in GHG emissions by 80% of 2005 levels, to be achieved by 2050 by a **cap-and-trade program.** Under such a system, the government imposes strict limits on emissions from a group of industries and then grants permits for emissions. If a company comes in under its cap, it can sell its unused amount to those who emit more. Over time, the emission limit is gradually reduced. Capping and trading GHG emissions combine to create one of the most prominent of the mitigation options, one that is already in operation in the European Union and the Annex I parties.

Mitigation Tools. Many options are available for reducing GHG emissions, some of which are already in place in the United States and globally. Table 18–5 presents a summary of most of the prominent mitigation tools. Appropriately, the majority of the options address the energy sector, where most of the anthropogenic GHGs are originating. What is needed, of course, is to "decarbonize" the energy economy, so that energy needs are met without fossil fuels. As Chapter 16 pointed out, every provision that promotes greater energy efficiency and the use of renewable energy sources is going to reduce GHG emissions in the United States. Recent legislation has made progress in energy efficiency and renewable energy, as discussed in Chapters 14 and 16.

There will be economic costs and benefits to mitigation efforts. The benefits come in the form of new industries and the jobs they create, savings from efficiency measures, and the ecosystem benefits from forests that are preserved and regrown. The net costs, as calculated by the Stern Review for stabilization at 550 ppm CO_2, amount to around 2% of global GDP by 2050. These are the costs of cutting GHG emissions by reducing demand for emissions-intensive goods and services, increasing efficiencies of energy use, avoiding deforestation, and switching to lower-carbon technologies for power, heat, and transport. Recall that the Stern Review calculated the costs of inaction to be from 5% to 20% of GDP.

Wedges. An imaginative approach to mitigation and GHG stabilization—called *stabilization wedges*—was introduced by Stephen Pacala and Robert Socolow.[14] Their approach has been popularized by the Carbon Mitigation Initiative (CMI), a joint project of Princeton University, BP, and Ford Motor Company. Picture two futures—one in which emissions are allowed to double and one in which they

---

[13]U.S. House of Representatives, Committee on Oversight and Government Reform, *Political Interference with Climate Change Science Under the Bush Administration*, December 2007. Available at oversight.house.gov/documents/20071210101633.pdf.

[14]Stephen Pacala and Robert Socolow, "Stabilization Wedges: Solving the Climate Problem for the Next 50 Years with Current Technologies," *Science* 305 (August 13, 2004): 968–972.

TABLE 18–5   Options for Mitigation of Greenhouse Gas Emissions

Cap-and-trade. Place a cap on GHG emissions; for CO_2, this can be accomplished by limiting the use of fossil fuels in industry and transportation. Rights to emit GHGs would be allocated to different companies and would be tradable in a market-based system. This can work internationally, where different countries are allocated permits and can trade with other countries. It is anticipated that the cap will be reduced gradually until a stabilized atmosphere is achieved. This option generates potentially large revenues, which can be put to use in ways that can alleviate some of the problems created by the program.

Renewable energy. Invest in and deploy an increasing percentage of energy in the form of renewable-energy technologies: wind power, solar collectors, solar thermal energy, photovoltaics, biofuels, hydrogen-powered vehicles, and geothermal energy, among others.

Carbon capture and storage. Sequester CO_2 emitted from burning fossil fuels at power plants by capturing it at the site of emission and pumping it into deep, porous underground formations. This is especially vital for coal-burning power plants, as coal is the most abundant fossil fuel energy source for many major GHG emitters.

Nuclear power. Encourage the development of nuclear power, first resolving issues concerning cost-effectiveness, reliability, spent fuel, and high-level waste.

Reforestation. Stop the loss of tropical forests and encourage the planting of trees and other vegetation over vast areas now suffering from deforestation. Trees and other vegetation provide a sink for carbon that removes CO_2 from the atmosphere.

Efficiency. Make energy conservation rules much more stringent. (Tighten building codes to require more insulation, use energy-efficient lighting, and so forth.)

Mileage standards. Reduce the amount of fuels used in transportation by raising mileage standards, encouraging car pooling, and stimulating mass transit in urban areas.

Subsidies. Remove fossil fuel subsidies like depletion allowances, tax relief to consumers, support for oil and gas exploration, and the substantial costs of maintaining access to Middle Eastern oil.

Carbon tax. Levy a tax on all fuels according to the amount of carbon dioxide that they produce when consumed. Such a tax, proponents believe, would provide both incentives to use renewable sources, which would not be taxed, and disincentives to consume fossil fuels.

---

are kept at current levels (Fig. 18–25a). The doubling path would lead to certain, and possibly catastrophic, climate change. To prevent this future, GHG emissions would be kept flat through 2055 and then would be reduced. This requires cutting carbon output by some 7 billion tons per year by 2055, and the carbon savings is what is called the *stabilization triangle* (Fig. 18–25b). Rather than trying to do this with any one mitigation option, the CMI broke the triangle into seven "wedges," or mitigation strategies, each of which would reduce 1 billion tons of carbon emissions by 2055. On the CMI Web site, 15 wedge strategies are presented and described in some detail. They are presented in four categories: Efficiency and Conservation, Fossil-Fuel-Based Strategies, Nuclear Energy, and Renewables and Biostorage. Each strategy is considered to be available now and, if scaled up, can account for one full wedge of emissions reduction. The mitigation tools in Table 18–5 are all represented in the suite of wedge strategies. The significant point of the exercise is that only seven of the wedges, if brought up to scale, could in fact bring about the desired future. The CMI has developed a game out of the concept, which helps drive home "the carbon mitigation challenge and the trade-offs involved in planning climate policy," as stated in the CMI guide.[15]

As a bottom line, even with such wedges, climate change is already happening and will likely accelerate in the future. *Mitigation* is the response that is essential to reducing the magnitude of future climate change, but *adaptation* will also be needed.

Adaptation

Adaptation means making adjustments in anticipation of changes brought on by the rising sea levels and temperatures of the next decades. It also means reducing vulnerability to the inevitable impacts of climate change. What makes this process difficult is that the adaptations will be aiming at a moving target—and one that is uncertain at best. Adaptation is not cost free. Climate changes are bringing us into ever new situations, and some of them are developing rapidly. IPCC Working Group II addressed adaptation in its AR4 report, *Climate Change 2007: Impacts, Adaptation and Vulnerability*.[16] The Stern Review detailed adaptations for developed and developing countries, many of which are presented in Table 18–6.

New Funds.   In light of the risks, the IPCC stated, "The effects of climate change are expected to be greatest in developing countries in terms of loss of life and relative effects on

---

[15]Roberta Hotinski, *Stabilization Wedges: A Concept and a Game* (Princeton, NJ: Carbon Mitigation Initiative, January 2007). Available at www.princeton.edu/wedges.

[16]Martin Parry, Osvaldo Canziani, Jean Palutikof, Paul van der Linden, and Clair Hanson, eds., *Climate Change 2007: Impacts, Adaptation and Vulnerability*, Contribution of Working Group II to the Fourth Assessment Report of the Intergovernmental Panel on Climate Change (New York. Cambridge University Press, 2007).

Figure 18–25 Stabilization wedges.
(a) Two futures are pictured: an emissions-doubling path, which is the path we are on right now, and a flat path, where emissions are limited to 7 billion metric tons of carbon. The stabilization triangle is the difference between these two paths. (b) The stabilization triangle is filled with seven wedges, or mitigation strategies, each of which avoids 1 billion tons of carbon. (*Source:* Roberta Hotinski, *Stabilization Wedges: A Concept and a Game* (Princeton, NJ: Carbon Mitigation Initiative, January 2007). Available at www.princeton.edu/wedges.

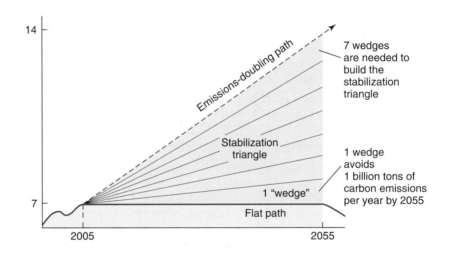

TABLE 18–6 Examples of Adaptation Strategies

All Countries

- **Agriculture:** Farmers shift to climate-resistant crops or crops more appropriate to the new climate regime. Irrigation is expanded to areas once fed only by rain.
- **Structural concerns:** Societies invest to reduce the costs and disruption from extreme weather events like storms, floods, droughts (e.g., seawalls, new reservoirs, revegetating coastal wetlands).
- **Emergency preparedness:** Societies invest in early-warning systems, high-quality climate information, intervention when needed (e.g., heat wave warnings, heavy precipitation events, shelters for those displaced or impacted by events).
- **Reducing risks:** Societies establish a financial safety net to help their poorest members, providing insurance that accurately reflects new risks to discourage poor choices.

Developing Countries

- **Promoting development:** Investments are made in health, education, economic opportunity, and food security (to lower people's vulnerability to climate impacts and help them build adaptive capacity).
- **Controlling diseases:** Malaria, AIDS, tuberculosis, and emerging diseases are brought under control so that a changing climate does not increase their threat.
- **Enhancing economic progress:** Societies promote literacy, electronic communication, road infrastructure, and diversification of economic activity.

Source: Nicholas Stern, *The Economics of Climate Change: The Stern Review* (Cambridge, England: Cambridge University Press, 2007).

investment and the economy." As a start, the FCCC established two new funds to address this concern: a **Least Developed Countries Fund** to advise countries on national adaptation strategies and a **Special Climate Change Fund** to provide additional financial assistance to developing countries affected by climate change. Subsequently, the FCCC added the **Adaptation Fund** and the **Climate Investment Funds,** the latter sponsored by the World Bank. The emphasis of all of these funds is on the developing countries, since it is assumed that the developed countries will be far more able to afford the costs of adaptation (and can also afford to finance the funds).

Poverty and Climate Change. The World Bank and a host of UN and other organizations released a report that addresses the preceding concerns. Called *Poverty and Climate Change: Reducing the Vulnerability of the Poor Through Adaptation,* the report makes some telling points:

1. Climate change impacts will be superimposed on existing vulnerabilities—water availability, food security, health and disease, and life in coastal areas. The poor already suffer disproportionately from these risks.

2. The countries with the fewest resources will likely bear the greatest burden of climate change, with greater loss of life and economic impacts.

3. Ecosystem goods and services are likely to be disrupted by climate change. The poor depend especially on these for survival, so they will be most affected.

4. Even today more than 96% of disaster-related deaths take place in developing countries. Major weather events can set back development for decades.

5. Because poverty is so severe in these countries, adaptation must be treated in the context of reducing poverty and achieving the Millennium Development Goals. (See Table 9–2.) Practically every one of the goals is threatened by the anticipated impacts of climate change, and the report addresses these impacts in detail.

Strategies. Specific adaptation strategies will vary with the different circumstances, but in general, the World Bank report suggests that "the best way to address climate change impacts on the poor is by integrating adaptation measures into sustainable development and poverty reduction strategies." Adaptations addressed in the report include improved governance, vulnerability assessments, access to accurate information on climate change, and the integration of impacts into economic processes. Those adaptations that have "no regrets" benefits—that is, those measures that would foster desirable social benefits, regardless of the intensity or existence of climate change impacts—are especially encouraged. Strategies differ for the developed and developing countries in some areas, but many adaptations are common to all countries (Table 18–6).

Geoengineering

The threat of climate change is generating a host of futuristic schemes to help the fight against global warming. Some of the schemes are clearly unworkable, while others have been floated as serious proposals. For example, there is a proposal to fertilize vast areas of the oceans with iron, thereby stimulating phytoplankton growth, which would remove CO_2 and sink to the depths. Another proposed scheme erects scrubbers, devices that would chemically absorb CO_2 in the air and then store it in deep reservoirs. Still other methods would block some of the solar radiation from reaching Earth's surface by, for example, injecting sulfate aerosols into the stratosphere or launching satellites into space that would release reflecting particles in the path of the Sun's radiation. Most climate experts dismiss these schemes on the basis of their huge costs or the likelihood of unintended consequences such as interference with weather patterns or ecosystem destruction; besides, they might not work at all.

It is fair to say that none of these schemes compares with what is already happening. Whatever the outcome, it is certain that we are conducting an enormous global experiment in geoengineering, and our children and their descendants will be living with the consequences. Global climate change is perhaps the greatest challenge facing human civilization in the 21st century.

Revisiting the Themes

Sound Science

Much of this chapter reflects some outstanding science, since our knowledge of climate change depends on the solid work of thousands of scientists. It is largely the scientists who are calling attention to the perils of climate change. They were the first to discover the risks involved, and they are now the chief advocates of effective action. The IPCC, with the thousands of scientists involved, has performed a masterful assessment of the scientific basis of climate change. In the United States, the CCSP/GCRP has also produced a portfolio of excellent

(continued)

analyses of climate change and its many dimensions. Nevertheless, more good scientific work is needed because there are still unanswered questions and uncertainties.

Sustainability

Earth is in the midst of an unsustainable rise in atmospheric GHG levels, the result of our intense use of fossil fuels. In short, we completely depend on a host of technologies that are threatening our future. All projections of future fossil fuel use and GHGs point to global consequences that are serious, but not inevitable. The United States and other developed countries will not escape these consequences, but the gravest of them affect the developing countries. A sustainable pathway is still open to us, and it involves a combination of steps we can take to mitigate the emissions and bring the atmospheric concentration of GHGs to a stable, even declining, level.

Stewardship

Stewardly care for Earth is not really an option. We must act, and several principles have been cited as arguments for effective action to prevent dangerous climate change: the precautionary principle, the polluter pays principle, and the equity principle. If we care about our neighbors and our descendants, we will take action both to mitigate the production of GHGs and to enable especially the poorer countries to adapt to coming unavoidable climate changes, even if taking such action is costly.

Ecosystems depend on climate, and if we change the climate too rapidly, many ecosystems are likely to suffer serious disruptions and will no longer provide essential goods and services societies now depend on. We allow these ecosystems to change at our own peril. In the end, ecosystems will redevelop, and many species will adapt or move to new locations, but there may well be major losses in biodiversity, and the ecosystem adaptations will not happen overnight. The short-term impacts will be especially hard on those societies that live close to the land (that is, the developing countries).

REVIEW QUESTIONS

1. Explain the possible connection between intense hurricanes and the effects of current warming.

2. What are the important characteristics of the troposphere? Of the stratosphere?

3. Provide a brief synopsis of global climate change. What are the main issues and trends?

4. What has been discovered about global climate trends in the past? What is the contemporary trend showing?

5. What is the conveyor system? How does it work? How does it connect with global warming?

6. What is radiative forcing? Describe some warming and cooling forcing agents.

7. What evidence do global land and ocean temperatures provide for a warming Earth?

8. What additional changes contribute to the picture of human-caused warming?

9. Which of the GHGs are the most significant contributors to global warming? How do they work?

10. Describe the data for atmospheric CO_2. What are significant sources and sinks for atmospheric CO_2?

11. How are models employed in climate change research? Describe several scenarios for 21st-century climate change.

12. What are the major IPCC projections for climate changes in the 21st century?

13. Summarize the key findings of the CCSP Synthesis (USP) report.

14. What mitigation steps could be taken to stabilize the GHG content of the atmosphere?

15. Trace the political history of the FCCC and the Kyoto Protocol. What is the next international step in reaching agreement on climate change mitigation?

16. Describe recent U.S. climate change policy, and compare it with those of other developed countries.

17. What are some adaptation steps that should be taken to anticipate inevitable changes in climate?

THINKING ENVIRONMENTALLY

1. What steps could you take to lower your climate impact? Check the Union of Concerned Scientists Web site for suggestions.

2. Research the skeptics of global climate change. What evidence do they offer in support of their views? How does it compare with the evidence from the IPCC? On balance, what do the data indicate?

3. Even if the human role in global climate change is highly likely, but not necessarily a certainty, should our response to it be any different?

4. What recent developments most increase the urgency of action in the international arena?

5. Log on to www.princeton.edu/~cmi/resources/stabwedge.htm, get some friends together, and play the Stabilization Wedge Game. The Web site has instructions and materials for the activity.

MAKING A DIFFERENCE

Anything you can do to cut energy use will cut your CO_2 output. In general, cutting CO_2 will also cut your bills. Listed below are just a few examples.

1. Buy items with the smallest amounts of packaging, The more packaging products have, the more space they take up in the distribution trucks, and the more CO_2 emissions they create.

2. Washing your clothes in cold water instead of hot will keep 200 pounds of carbon dioxide from entering the atmosphere a year.

3. Cutting your beef intake by just a quarter can save up to 250 pounds of CO_2 annually.

4. The worst form of transportation for the climate is driving alone in a car. Use public transportation when possible, bike, walk, carpool, buy a hybrid—there are many options! When deciding on a car, make fuel economy a high priority. Driving 25 fewer miles a week will earn you 1,500 fewer pounds of CO_2 per year.

5. When driving, keep your car speed steady and avoid sudden acceleration or braking. Driving 65 instead of 75 increases your car's efficiency by at least 15%. Make sure your tires are properly inflated—consumers waste 500 million gallons of gas annually due to underinflation. Pack light—the less weight, the more miles you get for your gallon!

6. Replace your incandescent bulbs with equally bright compact fluorescent bulbs.

Go to **www.masteringenvironmentalscience.com** for practice quizzes, Pearson eText, videos, current events, and more.

CHAPTER 19

Atmospheric Pollution

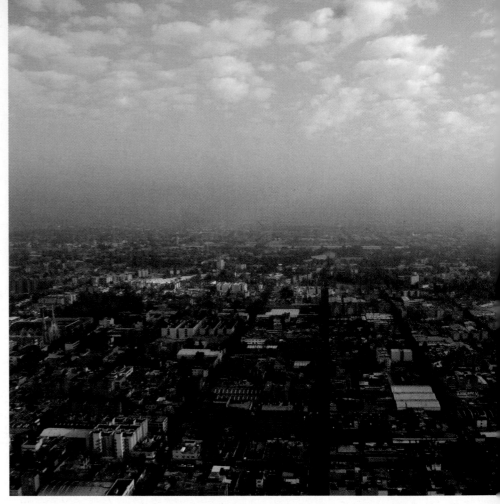

Smog in Mexico City. Because of its geographic location and heavy automobile traffic, Mexico City has had some of the worst air pollution anywhere.

O N TUESDAY MORNING, October 26, 1948, the people of Donora, Pennsylvania (population 13,000), awoke to a dense fog (Fig. 19–1). Donora lies in an industrialized valley along the Monongahela River, south of Pittsburgh. On the outskirts of town, a sooty sign read "DONORA: NEXT TO YOURS, THE BEST TOWN IN THE U.S.A." The town had a large steel mill that used high-sulfur coal and a zinc-reduction plant that roasted ores laden with sulfur. At the time, most homes in the area were heated with coal. At first, the fog did not seem unusual: Most of Donora's fogs lifted by noon, as the Sun warmed the upper atmosphere and then the land. This one, however, didn't lift for five days.

Through Wednesday and Thursday, the air began to smell of sulfur dioxide—an acrid, penetrating odor. By Friday morning, the town's physicians began to get calls from people in trouble. At first, the calls were from elderly citizens and those with asthmatic conditions. They were having difficulty breathing. The calls continued into Friday afternoon. People young and old were complaining of stomach pain, headaches, nausea, and choking. Work at the mills went on, however. The first deaths occurred

Figure 19–1 Donora, Pennsylvania. Donora was the scene of a major air-pollution disaster in 1948.

on Saturday morning. By 10 A.M., one mortician had nine bodies; two other morticians had one each. On Sunday morning, the mills were shut down. Even so, the owners were certain that their plants had nothing to do with the trouble. Mercifully, the rain came on Sunday and cleared the air—but not before more than 6,000 townspeople were stricken and 20 elderly people had died. During the next month, 50 more people died.

Eventually, it was determined that the cause of the deaths was a combination of polluting gases and particles, a thermal inversion in the lower atmosphere, and a stagnant weather system that, together, brought home the deadly potential of air pollution. Today a historical marker in the town commemorates the Donora Smog of 1948, but the lasting legacy of the smog and of those who suffered sickness and death are the state and national laws to control air pollution, culminating in the Clean Air Act of 1970 and its amendments. Donora was a landmark event.

In the previous chapter, we looked at the structure and function of the atmosphere and its links to weather and climate, and we examined the effects of greenhouse gas pollutants on climate change. Here we will consider the impacts of a host of other pollutants on human health and the environment.

19.1 Air-Pollution Essentials

For a long time, we have known that the atmosphere contains numerous *gases*. The major constituents of the atmosphere are nitrogen (N_2), at a level of 78.08%; oxygen (O_2), at 20.95%; argon (Ar), at 0.93%; carbon dioxide (CO_2), at 0.04%; and water vapor, ranging from 0 to 4%. Smaller amounts of at least 40 "trace gases" are normally present as well, including ozone, helium, hydrogen, nitrogen oxides, sulfur dioxide, and neon. In addition, the atmosphere contains *aerosols*—microscopic liquid and solid particles such as dust, carbon particles, pollen, sea salts, and microorganisms, originating primarily from land and water surfaces and carried up into the atmosphere. Outdoor air is a valuable and essential resource, providing gases that sustain life, maintain warmth (Chapter 18), and shield the Earth from harmful radiation. With the advent of the Industrial Revolution, the mixture of gases and particles in our atmosphere began to change, and we learned the hard way about air pollution.

Pollutants and Atmospheric Cleansing

Air pollutants are substances in the atmosphere—certain gases and aerosols—that have harmful effects. Three factors determine the level of air pollution:

- The amount of pollutants entering the air.
- The amount of space into which the pollutants are dispersed, and
- The mechanisms that remove pollutants from the air.

Recall from Chapter 18 that the lower atmosphere, called the *troposphere* (Fig. 19–2), is the site and source of our weather and contains almost all of the water vapor and clouds of the atmosphere. Pollutants that enter the troposphere are usually removed within hours or a few days unless they are carried up into the upper troposphere by towering cumulus clouds. If that happens, they can be carried far and persist for many days before they, too, are removed by cleansing agents.

Some pollutants are resistant to most of the cleansing mechanisms and can be carried up into the next higher layer of the atmosphere, the *stratosphere*. The most notorious of these are the ozone-depleting chemicals (chlorine and bromine compounds), which we will consider in Section 19.5.

Atmospheric Cleansing. Environmental scientists distinguish between natural and *anthropogenic* (human-caused) air pollutants. For millions of years, volcanoes, fires, and dust storms have sent smoke, gases, and particles into the atmosphere. Trees and other plants emit volatile organic compounds into the air around them as they photosynthesize. However, there are mechanisms in the biosphere that remove, assimilate, and recycle these natural pollutants. First, a naturally occurring compound, the **hydroxyl radical (OH)**, oxidizes many gaseous pollutants to products that are harmless or that can be brought down to the ground or water by precipitation (Fig. 19–3). Sea salts—picked up from sea spray as air masses move over the oceans—are a second cleansing agent. These salts, now aerosols, act as excellent nuclei for the formation of raindrops. The rain then brings

Figure 19–2 Layers of the atmosphere closest to Earth's surface. The troposphere extends from the surface to roughly 6 miles above the surface, and the stratosphere is above that. The colored band shows the average temperature of the atmosphere at different altitudes. In the troposphere, temperatures generally decrease with height, while in the stratosphere temperatures increase with height. (*Source:* CCSP 1.1. *Temperature Trends in the Lower Atmosphere: Steps for Understanding and Reconciling Differences.* Thomas R. Karl, Susan J. Hassol, Christopher D. Miller, and William L. Murray, editors, 2006. A Report by the Climate Change Science Program and the Subcommittee on Global Change Research, Washington, D.C.)

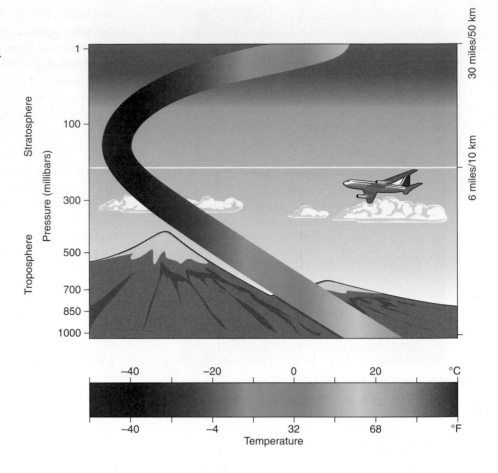

Figure 19–3 The hydroxyl radical. This is a simplified model of atmospheric cleansing by the hydroxyl radical. The first step is the photochemical destruction of ozone, which is the major process leading to ozone breakdown in the troposphere. The second step produces hydroxyl radicals, which react rapidly with many pollutants, converting them to substances that are less harmful or that can be returned to Earth via precipitation.

down many particulate pollutants (other aerosols) from the atmosphere to the ocean, cleansing pollutant-laden air. Third, sunlight breaks organic molecules apart. These three processes hold natural pollutants below toxic levels (except in the immediate area of a source, such as around an erupting volcano).

Many of the pollutants oxidized by the hydroxyl radical are of concern because human activities have raised their concentrations far above normal levels. Recent studies have shown that the hydroxyl radical plays a key role in the removal of anthropogenic pollutants from the atmosphere. Highly reactive hydrocarbons are oxidized within an hour of their appearance in the atmosphere, and nitrogen oxides (NO_x) are converted to nitric acid (HNO_3) within a day. It takes months, however, for less reactive substances, such as carbon monoxide (CO), to be oxidized by hydroxyl. It appears that the atmospheric levels of hydroxyl are in turn determined by the levels of anthropogenic pollutant gases; thus, hydroxyl's cleansing power is "used up" when high concentrations of pollutants are oxidized and the pollutants are subsequently able to build up to damaging levels. However, as Figure 19–3 shows, the photochemical breakdown of tropospheric ozone is the major source of the hydroxyl radical (not shown is a newly discovered additional source of the hydroxyl radical, from the photoexcitation of nitrogen dioxide). As we will see, higher levels of ozone result from higher concentrations of other polluting gases.

Smogs and Brown Clouds

Down through the centuries, the practice of venting the products of combustion and other fumes into the atmosphere remained the natural way to avoid their noxious effects within buildings. With the Industrial Revolution of the 1800s came crowded cities and the use of coal for heating and energy. It was then that air pollution began in earnest.

Industrial Smog. In *Hard Times*, Charles Dickens described a typical English scene: "Coketown lay shrouded in a haze of its own, which appeared impervious to the sun's rays. You only knew the town was there because there could be no such sulky blotch upon the prospect without a town." This shrouding haze became known as **industrial smog** (a combination of *sm*oke and f*og*), an irritating, grayish mixture of soot, sulfurous compounds, and water vapor (Fig. 19–4a). This kind of smog continues to be found wherever industries are concentrated and where coal is the primary energy source; such smog events mostly occur in the winter. The Donora incident is a classic example. Currently, industrial smog can be found in the cities of China, India, Korea, and a number of eastern European countries.

Photochemical Smog. After World War II, the booming economy in the United States led to the creation of vast suburbs and a mushrooming use of cars for commuting to and from work. As other fossil fuels were used in place of coal for heating, industry, and transportation, industrial smog was largely replaced by another kind of smog. Increasingly, Los Angeles and other cities served by huge freeway systems began to be enshrouded daily in a brownish, irritating haze that was different from the more familiar industrial smog (Fig. 19–5). Weather conditions were usually warm and

(a) Industrial smog

(b) Photochemical smog

Figure 19–4 Industrial and photochemical smog. (a) Industrial smog occurs when coal is burned and the atmosphere is humid. (b) Photochemical smog occurs when sunlight acts on vehicle pollutants.

(a)

(b)

Figure 19–5 Los Angeles air. (a) Los Angeles on a good day (b) L.A. under a blanket of photochemical smog.

sunny rather than cool and foggy, and typically, the new haze would arise during the morning commute and begin to dissipate only by the time of the evening commute. This **photochemical smog,** as it came to be called, is produced when several pollutants from automobile exhausts—nitrogen oxides and volatile organic carbon compounds—are acted on by sunlight (Fig. 19–4b).

Inversions.   Certain weather conditions intensify levels of both industrial and photochemical smogs. Under normal conditions, the daytime air temperature is highest near the ground because sunlight strikes Earth and the absorbed heat radiates to the air near the surface. The warm air near the ground rises, carrying pollutants upward and dispersing them at higher altitudes (Fig. 19–6a). At times, however, a warm air layer occurs above a cooler layer. This condition of cooler air below and warmer air above is called a **temperature inversion** (Fig. 19–6b). Inversions often occur at night, when the surface air is cooled by radiative heat loss. Such inversions are usually short lived, as the next morning's sunlight begins the heating process anew and any pollutants that accumulated overnight are carried up and away. During cloudy weather, however, the Sun may not be strong enough to break up the inversion for hours or even days. A mass of high-pressure air may move in and sit above the cool surface air, trapping it. Topography can also intensify smog as cool ocean air flows into valleys and is trapped there by nearby mountain ranges. Los Angeles has both topographical features!

When such long-term temperature inversions occur, pollutants can build up to dangerous levels, prompting local health officials to urge people with breathing problems to stay indoors. For many people, smog causes headaches, nausea, and eye and throat irritation. It may aggravate

preexisting respiratory conditions, such as asthma and emphysema. In some industrial cities (such as Donora), air pollution reached lethal levels under severe temperature inversions. These cases became known as *air-pollution disasters.* London experienced repeated episodes of inversion-related disasters in the mid-1900s. One episode in 1952 resulted in 4,000 pollution-related deaths.

Atmospheric Brown Clouds.   A relative newcomer on the air-pollution scene is the so-called **atmospheric brown cloud (ABC),** a 1- to 3-kilometer-thick blanket of pollution that frequently hovers over south and central Asia. The ABC is in fact a persistent aerosol, similar to the summer aerosol found over industrial regions in the North Temperate Zone (see Chapter 18), but dissimilar in that it persists year-round and has a different makeup. Unique to the ABC is the high proportion of black carbon and soot derived from biomass burning (residential cooking, agricultural clearing) and fossil fuel combustion (coal and diesel exhaust). Ozone and acid rain accompany the ABC.

The impacts of the ABC are serious: major dimming, especially over large cities; reductions in rainfall due to the aerosol particles; solar heating of the atmosphere and decreased reflection by snow and ice (both of which are contributing to a shrinking of the Hindu Kush–Himalaya–Tibetan glaciers, vital water sources for the major river systems of eastern Asia); a weakening of the Indian summer monsoon; reduced crop yields due to the dimming and to ozone levels; and heightened acute and chronic health effects typical of exposure to industrial and urban air pollution.

Industrial smog from coal burning, photochemical smog from vehicular exhaust, and atmospheric brown cloud are all caused by the release of pollutants, and we now turn our attention to those pollutants.

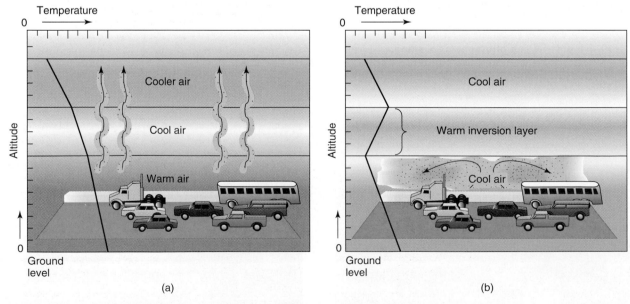

Figure 19–6 Temperature inversion. A temperature inversion may cause episodes of high concentrations of air pollutants. (a) Normally, air temperatures are highest at ground level and decrease at higher elevations. (b) In a temperature inversion, a layer of warmer air overlies cooler air at ground level.

19.2 Major Air Pollutants and Their Sources

In large measure, anthropogenic air pollutants are direct and indirect products of the *combustion* of coal, gasoline, other liquid fuels, and refuse (wastepaper, plaster, and so on). These fuels and wastes are organic compounds. With complete combustion, the products of burning organic compounds are carbon dioxide and water vapor. Unfortunately, combustion is seldom complete, and many complex chemical substances are involved. The combustion of fuels and refuse creates a host of gaseous and particulate products that create unwanted problems before they are cleansed from the air. Other processes producing air pollutants are *evaporation* (of volatile substances) and *strong winds* that pick up dust and other particles.

Primary Pollutants

Table 19–1 presents the names, symbols, major sources, characteristics, and general effects of the major air pollutants. The first seven (**particulates, VOCs, CO, NO$_x$, SO$_2$, lead,** and **air toxics**) are called **primary pollutants** because they are the direct products of combustion and evaporation. Some primary pollutants may undergo further reactions in the atmosphere and produce additional undesirable compounds, called **secondary pollutants (ozone, peroxyacetyl nitrates, sulfuric acid,** and **nitric acid**). Sources of air pollution are summarized in Figure 19–7. Power plants are the major source of sulfur dioxide, industrial processes are the major source of particulates, and transportation accounts for the lion's share of carbon monoxide and nitrogen oxides. This diversity suggests that strategies for controlling air pollutants will be different for the different sources.

TABLE 19–1   Major Anthropogenic Air Pollutants and Their Effects

Name	Symbol	Source	Description and Effects
Primary Pollutants			
Suspended particulate matter	PM	Soot, smoke, metals, and carbon from combustion; dust, salts, metal, and dirt from wind erosion; atmospheric reactions of gases	Complex mixture of solid particles and aerosols; reduces lung function and affects respiration and cardiovascular disease
Volatile organic compounds	VOC	Incomplete combustion of fossil fuels; evaporation of solvents and gasoline; emission from plants	Mixture of compounds, some carcinogenic; major agent of ozone formation
Carbon monoxide	CO	Incomplete combustion of fuels	Invisible, odorless, tasteless gas; poisonous because of ability to bind to hemoglobin and block oxygen delivery to tissues; aggravates cardiovascular disease
Nitrogen oxides	NO$_x$	From nitrogen gas due to high combustion temperatures when burning fuels; also from wood burning	Reddish-brown gas and lung irritant; aggravates respiratory disease; major source of acid rain; contributes to ozone formation
Sulfur dioxide	SO$_2$	Combustion of sulfur-containing fuels, especially coal	Poisonous gas that impairs breathing; major source of acid rain; contributes to particle formation
Lead	Pb	Battery manufacture; lead smelters; combustion of leaded fuels and solid wastes	Toxic at low concentrations; accumulates in body and can lead to brain damage in children
Air toxics	Various	Fuel combustion in vehicles; industrial processes; building materials; solvents	Toxic chemicals, many of which are known human carcinogens (e.g., benzene, asbestos, vinyl chloride); Clean Air Act identifies 187 air toxics
Radon	Rn	Rocks and soil; natural breakdown of radium and uranium	Invisible and odorless radioactive gas can accumulate inside homes; second leading cause of lung cancer in United States
Secondary Pollutants			
Ozone	O$_3$	Photochemical reactions between VOCs and NO$_x$	Toxic to animals and plants and highly reactive in lungs; oxidizes surfaces and rubber tires
Peroxyacetyl nitrates	PAN	Photochemical reactions between VOCs and NO$_x$	Damages plants and forests; irritates mucous membranes of eyes and lungs

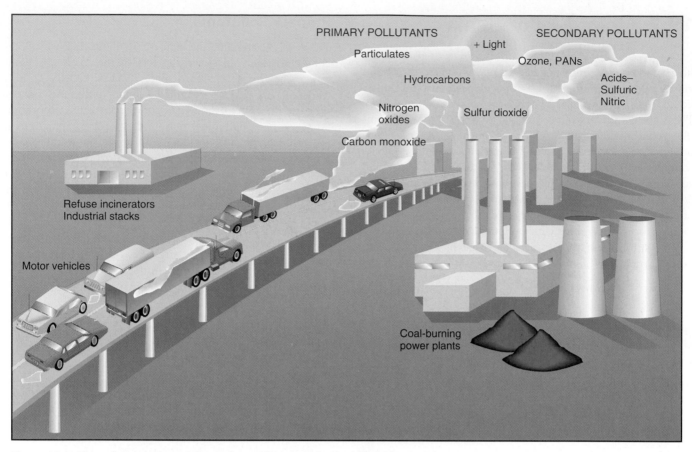

PRIMARY POLLUTANTS

SECONDARY POLLUTANTS

Particulates

+ Light

Ozone, PANs

Hydrocarbons

Acids—
Sulfuric
Nitric

Nitrogen
oxides

Sulfur dioxide

Carbon monoxide

Refuse incinerators
Industrial stacks

Motor vehicles

Coal-burning
power plants

Figure 19–7 The prime sources of the major anthropogenic air pollutants. Industrial processes and transportation generate a toxic brew of primary and secondary air pollutants.

When fuels and wastes are burned, particles consisting mainly of carbon are emitted into the air; these are the particulates we see as soot and smoke (Fig. 19–8). In addition, various unburned fragments of fuel molecules are given off; these are the VOC emissions. Incompletely oxidized carbon is carbon monoxide (CO), in contrast to completely oxidized carbon, which is carbon dioxide (CO_2). Combustion takes place in the air, a mixture of 78% nitrogen and 21% oxygen. At high combustion temperatures, some of the atmospheric nitrogen gas is oxidized to form the gas nitric oxide (NO). In the air, nitric oxide immediately reacts with additional oxygen to form nitrogen dioxide (NO_2) and nitrogen tetroxide (N_2O_4). These compounds are collectively referred to as the **nitrogen oxides**, or **NO_x**. Nitrogen dioxide absorbs light and is largely responsible for the brownish color of photochemical smog.

In addition to organic matter, fuels and refuse contain impurities and additives, and these substances are also emitted into the air during burning. Coal, for example, contains from 0.2% to 5.5% sulfur. In combustion, this sulfur is oxidized, giving rise to several gaseous oxides, the most common being sulfur dioxide (SO_2). Coal also contains heavy-metal impurities such as mercury, and refuse contains an endless array of "impurities."

Tracking Pollutants. The Environmental Protection Agency (EPA) operates the **Clearinghouse for Inventories and Emissions Factors,** which tracks trends in **national emissions**

Figure 19–8 Industrial pollution. Because the stack of this wood-products mill in New Richmond, Quebec Province, Canada, lacks electrostatic precipitators, the plume contains substantial amounts of suspended particulate matter that can be seen from a great distance.

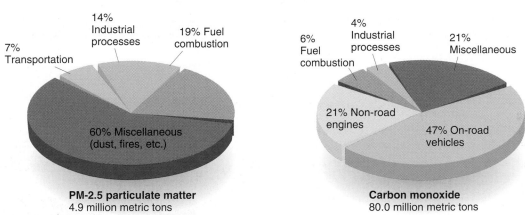

Figure 19–9 U.S. emissions of five primary air pollutants, by source, for 2008. *Fuel combustion* refers to fuels burned for electrical power generation and for space heating. Note especially the different contributions by transportation and fuel combustion/electric utilities, the two major sources of air pollutants. (*Source:* EPA National Emissions Inventory, 2009. http://www.epa.gov/ttn/chief/trends/index.html.)

of the primary pollutants from all sources. The EPA also follows air quality by measuring **ambient concentrations** (where *ambient* means "present in the air") of the pollutants at thousands of monitoring stations across the country. According to EPA data, 2008 emissions in the United States of the first five primary pollutants amounted to 116 million tons. By comparison, in 1970, when the first Clean Air Act became law, these same five pollutants totaled 301 million tons. The relative amounts emitted in the United States in 2008, as well as their major sources, are illustrated in Figure 19–9. The general picture shows significant improvement for emissions over the past 38 years (Fig. 19–10), reflecting the effectiveness of

Clean Air Act regulations. This progressively improving trend over time has occurred in spite of large increases in the gross domestic product, vehicular miles traveled, and population and energy consumption (Fig. 19–11).

Getting the Lead Out. The sixth type of primary pollutant—lead and other heavy metals—is discussed separately because the quantities emitted are far less than the levels for the first five. Before the EPA-directed phaseout in the 1980s and 1990s, lead was added to gasoline as an inexpensive way to prevent engine knock. Emitted with the exhaust from gasoline-burning vehicles, lead remained airborne and traveled great distances before settling as far away as the glaciers

Figure 19–10 Clean Air Act impacts. Comparison of percent reductions in the emission of six criteria air pollutants between 1970 and 2008. $PM_{2.5}$ comparison values are from 1990, when data were first gathered, and 2008. Go to **GRAPHit!** at www .masteringenvironmentalscience.com. (*Source:* EPA Office of Air and Radiation, 2009. http://www.epa.gov/ttn/chief/trends/index.html.)

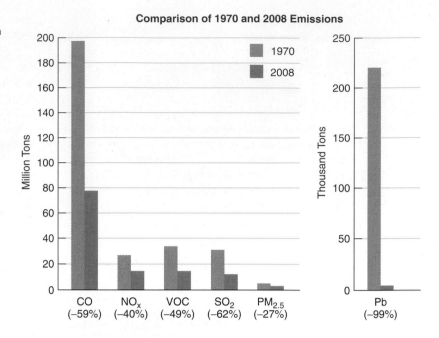

Comparison of 1970 and 2008 Emissions

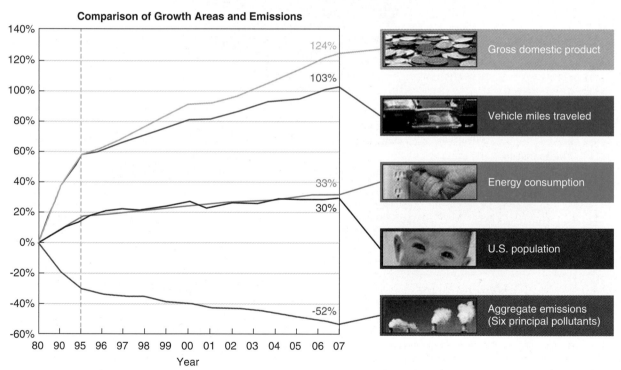

Figure 19–11 Comparison of growth versus emissions. Gross domestic product, vehicle miles traveled, energy consumption, and population are compared with aggregate reductions in emissions of the six principal pollutants from 1970 to 2008. (*Source:* EPA Air Trends, 2009. http://epa.gov/airtrends/aqtrends.html.)

of Greenland. Since the phaseout, concentrations of lead in the air of cities in the United States have declined remarkably. Between 1970 and 2008, lead emissions fell 99% (Fig. 19–10). At the same time, levels of lead in children's blood have dropped 90% (some children still get lead from dust and paint chips). Lead concentrations in Greenland ice have also decreased significantly. All these declines indicate that lead restrictions in the United States and a few other Northern Hemisphere nations have had a global impact. The data indicate that we have reached a steady state in lead emissions and

that any new reductions must target the lead smelters and battery manufacturers, which are now by far the largest sources of lead emissions.

Toxics and Radon. As with lead, the concentrations of toxic chemicals and radon in the air are small compared with those of the other primary pollutants. Some of the air's toxic compounds—benzene, for example—originate with transportation fuels. Most, however, are traceable to industries and small businesses. Radon, by contrast, is produced by the spontaneous decay of fissionable material in rocks and soils. Radon escapes naturally to the surface and seeps into buildings through cracks in foundations and basement floors, sometimes collecting in the structures.

Secondary Pollutants

Ozone and numerous reactive organic compounds are formed as a result of chemical reactions between nitrogen oxides and volatile organic carbons. Because sunlight provides the energy necessary to propel the reactions, these products are also known as **photochemical oxidants**. In preindustrial times, ozone concentrations ranged from 10 to 15 parts per billion (ppb), well below harmful levels. Summer concentrations of ozone in unpolluted air in North America now range from 20 to 50 ppb. Polluted air may contain ozone concentrations of 150 ppb or more, a level considered quite unhealthy if encountered for extended periods. EPA data indicate that ambient ozone levels, monitored across the United States at some 1,100 sites, declined by 20% between 1980 and 1992, leveled off during the 1990s, and then declined only a few percent in the 2000s. Ozone air quality standards are the leaders in nonattainment across the country, and more than 132 million people live in counties that still do not meet ozone standards.

Ozone Formation. A simplified view of the major reactions in the formation of ozone and other photochemical oxidants is shown in Figure 19–12. Nitrogen dioxide absorbs

Major pollutants from vehicles

Figure 19–12 Formation of ozone and other photochemical oxidants. (a) Nitrogen oxides, by themselves, would not cause ozone and other oxidants to reach damaging levels because reactions involving nitrogen oxides are cyclic. (b) When VOCs are also present, however, reactions occur that lead to the accumulation of numerous damaging compounds—most significantly, ozone, the most injurious.

light energy and splits to form nitric oxide and atomic oxygen, which rapidly combines with oxygen gas to form ozone. If other factors are not involved, ozone and nitric oxide then react to form nitrogen dioxide and oxygen gas. A steady-state concentration of ozone results, and there is no appreciable accumulation of the gas (Fig. 19–12a). When VOCs are present, however, the nitric oxide reacts with them instead of with the ozone, causing several serious problems. First, the reaction between nitric oxide and the VOCs leads to highly reactive and damaging compounds known as per-oxyacetyl nitrates, or PANs (Fig. 19–12b). Second, numerous aldehyde and ketone compounds are produced as the VOCs are oxidized by atomic oxygen, and these compounds are also noxious. Finally, with the nitric oxide tied up in this way, the ozone tends to accumulate. Because of the complex air chemistry involved, ozone concentrations may peak 30 to 100 miles (50 to 160 km) downwind of urban centers where the primary pollutants were generated. As a result, ozone concentrations above the air quality standards are sometimes found in rural and wilderness areas.

In a sense, **sulfuric acids** and **nitric acids** can also be considered secondary pollutants, since they are products of sulfur dioxide and nitrogen oxides, respectively, reacting with atmospheric moisture and oxidants such as hydroxyl. Sulfuric and nitric acids are the acids in acid rain (technically referred to as acid deposition). We now turn our attention to this important and widespread problem.

Acid Precipitation and Deposition

Acid precipitation refers to any precipitation—rain, fog, mist, or snow—that is more acidic than usual. Because dry acidic particles are also found in the atmosphere, the combination of precipitation and dry-particle fallout is called **acid deposition**. In the late 1960s, Swedish scientist Svante Odén first documented the acidification of lakes in Scandinavia and traced it to air pollutants originating in other parts of Europe and Great Britain. Since then, careful monitoring has shown that broad areas of North America, as well as most of Europe and other industrialized regions of the world, have been experiencing precipitation that is between 10 and 1,000 times more acidic than usual. This has affected ecosystems in diverse ways, as illustrated in Figure 19–13.

To understand the full extent of the problem, we must first understand some principles about acids and how we measure their concentration.

Acids and Bases.   Acidic properties (for example, a sour taste and corrosiveness) are due to the presence of hydrogen ions (H^+, hydrogen atoms without their electron), which are

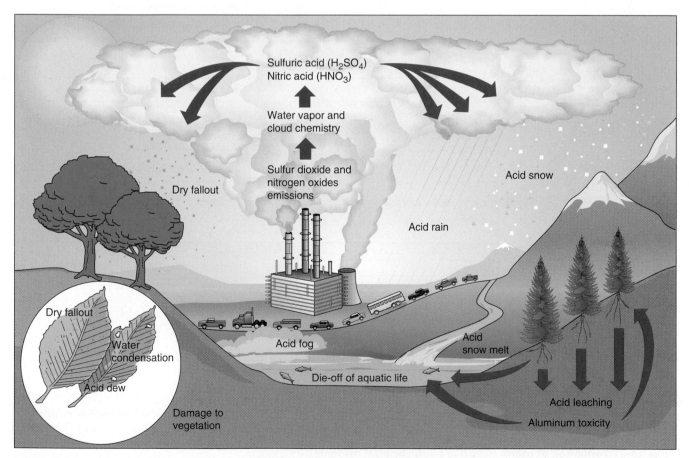

Figure 19–13 Acid deposition. Emissions of sulfur dioxide and nitrogen oxides react with hydroxyl radicals and water vapor in the atmosphere to form their respective acids, which return to the surface either as dry acid deposition or, mixed with water, as acid precipitation. Various effects of acid deposition are noted.

TABLE 19–2 Common Acids and Bases

	Formula	Yields	H⁺ Ion(s)	Plus	Negative Ion
Acid					
Hydrochloric acid	HCl	$\rightarrow$	H^+	$+$	Cl^- Chloride
Sulfuric acid	H_2SO_4	$\rightarrow$	$2H^+$	$+$	SO_4^{2-} Sulfate
Nitric acid	HNO_3	$\rightarrow$	H^+	$+$	NO_3^- Nitrate
Phosphoric acid	H_3PO_4	$\rightarrow$	$3H^+$	$+$	PO_4^{3-} Phosphate
Acetic acid	CH_3COOH	$\rightarrow$	H^+	$+$	CH_3COO^- Acetate
Carbonic acid	H_2CO_3	$\rightarrow$	H^+	$+$	HCO_3^- Bicarbonate
	Formula	**Yields**	**OH⁻ Ion(s)**	**Plus**	**Positive Ion**
Base					
Sodium hydroxide	$NaOH$	$\rightarrow$	OH^-	$+$	Na^+ Sodium ion
Potassium hydroxide	KOH	$\rightarrow$	OH^-	$+$	K^+ Potassium ion
Calcium hydroxide	$Ca(OH)_2$	$\rightarrow$	$2OH^-$	$+$	Ca^{2+} Calcium ion
Ammonium hydroxide	NH_4OH	$\rightarrow$	OH^-	$+$	NH_4^+ Ammonium ion

highly reactive. Therefore, an **acid** is any chemical that releases hydrogen ions when dissolved in water. The chemical formulas of a few common acids are shown in Table 19–2. Note that all of them ionize—that is, their components separate—to create hydrogen ions plus a negative ion. The higher the concentration of hydrogen ions in a solution, the more acidic the solution.

A **base** is any chemical that releases hydroxide ions (OH^-, oxygen-hydrogen groups with an extra electron) when dissolved in water. (See Table 19–2.) The bitter taste and caustic properties of alkaline, or basic, solutions are due to the presence of hydroxide ions.

pH. The concentration of hydrogen ions is expressed as **pH**. The pH scale goes from 0 (highly acidic) through 7 (neutral) to 14 (highly basic) (Fig. 19–14). As the pH scale numbers go from 0 to 7, the concentration of hydrogen ions (H^+) decreases, and the solution becomes less acidic and more neutral. As the numbers go from 7 to 14, the concentration of hydrogen ions remains low, but the concentration of hydroxide ions (OH^-) increases, and the solution becomes more basic. The numbers on the scale (0–14) represent the negative logarithm (power of 10) of the hydrogen ion concentration, expressed in grams per liter (g/L). For example, a solution with a pH of 1 has an H^+ concentration of 10^{-1} g/L, a pH of 2 has a concentration of 10^{-2} g/L, and so on.

Since numbers on the pH scale represent powers of 10, there is a *tenfold difference* between each unit and the next. For example, pH 5 is 10 times as acidic (has 10 times as many H^+ ions) as pH 6, pH 4 is 10 times as acidic as pH 5, and so on.

Extent and Potency of Acid Precipitation. In the absence of any pollution, rainfall is normally somewhat acidic, with a pH of 5.6, because carbon dioxide in the air readily dissolves in, and combines with, water to produce carbonic acid. **Acid precipitation**, then, is any precipitation with a pH less than 5.5.

Figure 19–14 The pH scale. Each unit on the scale represents a tenfold difference in hydrogen ion concentration.

Figure 19–15 Acid deposition in eastern North America.
Five-year average of the pH level in rainfall in the eastern regions of
Canada and the United States. (*Source:* Environment Canada, 2007.)

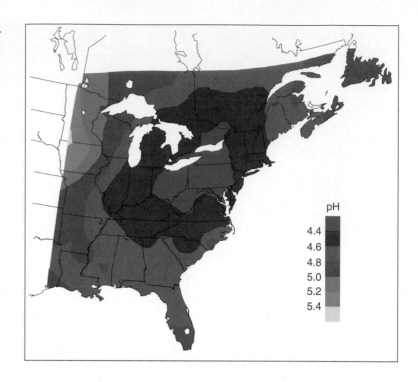

Unfortunately, acid precipitation is now the norm over most of the industrialized world. The pH of rain and snowfall over a large portion of eastern North America is typically below 4.6, reflecting the west-to-east movement of polluted air from the Midwest and industrial Canada (Fig. 19–15). Many areas in this region regularly receive precipitation having a pH of 4.0 and, occasionally, as low as 3.0. Fogs and dews can be even more acidic. In mountain forests east of Los Angeles, scientists found fog water of pH 2.8—almost 1,000 times more acidic than usual—dripping from pine needles. Acid precipitation also has been heavy in Europe, from the British Isles to central Russia. It is now documented in China and Japan as well.

Sources of Acid Deposition.   Chemical analysis of acid precipitation reveals the presence of two acids, **sulfuric acid** (H_2SO_4) and **nitric acid** (HNO_3), in a ratio of about two to one. (In the western United States and in the western provinces of Canada, nitric acid predominates, principally formed from tailpipe emissions.) As we have seen, burning fuels produce sulfur dioxide and nitrogen oxides, so the source of the acid deposition problem is evident. These oxides enter the troposphere in large quantities from both anthropogenic and natural sources. Once in the troposphere, they are oxidized by hydroxyl radicals (Fig. 19–3) to sulfuric and nitric acids, which dissolve readily in water or adsorb to particles and are brought down to Earth in acid deposition. This usually occurs within a week of the oxides' entering the atmosphere.

Natural Versus Anthropogenic Sources. Natural sources contribute substantial quantities of pollutants to the air, including 50 to 70 million tons per year of sulfur dioxide (from volcanoes, sea spray, and microbial processes) and 30 to 40 million tons per year of nitrogen oxides (from lightning, the burning of biomass, and microbial processes). *Anthropogenic sources* are estimated at 100 to 130 million tons per year of sulfur dioxide and 60 to 70 million tons per year of nitrogen oxides. The vital difference between these two sources is that anthropogenic oxides are strongly concentrated in industrialized regions, whereas the emissions from natural sources are spread out and are a part of the global environment. Levels of the anthropogenic oxides have increased at least fourfold since 1900, while levels of the natural emissions have remained fairly constant.

As Figure 19–9 indicates, 11.4 million tons of sulfur dioxide were released into the air in 2008 in the United States; 94% was from fuel combustion (mostly from coal-burning power plants). Some 16.4 million tons of nitrogen oxides were released, 58% of which can be traced to transportation emissions and 41% to fuel combustion at fixed sites. Nitrogen oxides are doubly damaging because the nitrogen contributes to eutrophication in coastal ecosystems (Chapter 20). In the eastern United States, the source of much of the acid deposition was identified as the tall stacks of 50 huge, older coal-burning power plants located primarily in the Midwest (Fig. 19–16). The tall stacks were built to alleviate local sulfur dioxide pollution at ground level. The good news is that many of these same plants, along with more than 200 other large, coal-fired power plants targeted by the EPA, are now reducing their emissions as a result of the Clean Air Act. Acid deposition has decreased by 33–35% over the past 15 years. However, 67% of SO_2 emissions and 22% of NO_x emissions still originate from fossil-fuel-burning electric utility plants.

19.3 Impacts of Air Pollutants

Air pollution is an alphabet soup of gases and particles, mixed with the normal constituents of air. The amount of each pollutant present varies greatly, depending on its proximity to the

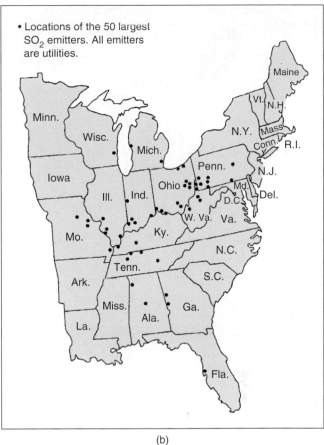

• Locations of the 50 largest SO_2 emitters. All emitters are utilities.

(a)

(b)

Figure 19–16 Midwestern coal-burning power plant. (a) Standard smokestacks of this coal-burning power plant were replaced by new 1,000-foot (330-m) stacks to aid in the dispersion of pollutants into the atmosphere. The taller stacks alleviated local air-pollution problems, but created a more widespread distribution of acid-generating pollutants. (b) Locations of the 50 largest sulfur dioxide emitters, all of which are utilities' coal-burning power plants.

source and various conditions of wind and weather. As a result, we are exposed to a mixture that varies in makeup and concentration from day to day—even from hour to hour—and from place to place. Consequently, the effects we feel or observe are rarely, if ever, the effects of a single pollutant. Instead, they are the combined impact of the whole mixture of pollutants acting over our life span up to that point.

For example, plants may be so stressed by pollution that they become more vulnerable to other environmental factors, such as drought or attack by insects. Humans may develop lung disease due to the combined effects of ozone and NO_x. Given the complexity of this situation, it is often difficult to determine the role of any particular pollutant in causing an observed result. Nevertheless, some significant progress has been made in linking cause and effect.

Human Health

The air-pollution disasters in Donora and London demonstrated that exposure to air pollution can be deadly. Every one of the primary and secondary air pollutants (Table 19–1) is a threat to human health, particularly the health of the respiratory system (Fig. 19–17). *Acute* exposure to some pollutants can be life threatening, but many effects are *chronic*,

acting over a period of years to cause a gradual deterioration of physiological functions and eventual premature mortality. Moreover, some pollutants are *carcinogenic*, adding significantly to the risk of lung cancer as they are breathed into the lungs.

Chronic Effects. Many people living in areas of urban air pollution suffer from chronic effects. Long-term exposure to *sulfur dioxide* can lead to bronchitis (inflammation of the bronchi). Chronic inhalation of *ozone* can cause inflammation and, ultimately, fibrosis of the lungs, a scarring that permanently impairs lung function. *Carbon monoxide* reduces the capacity of the blood to carry oxygen, and extended exposure to carbon monoxide can contribute to heart disease. Chronic exposure to *nitrogen oxides* impairs lung function and is known to affect the immune system, leaving the lungs open to attack by bacteria and viruses. Exposure to airborne *particulate matter* can bring on a broad range of health problems, including respiratory and cardiovascular pathology. Other factors—such as poor diet, lack of exercise, preexisting diseases, and individual genetic makeup—may add to the effects of pollution to bring on adverse health outcomes.

COPD. Any one or combination of these exposures can develop into **chronic obstructive pulmonary disease**

(b)

Figure 19–17 The respiratory system. (a) In the lungs, air passages branch and rebranch and finally end in millions of tiny sacs called alveoli. These sacs are surrounded by capillaries. As blood passes through the capillaries, oxygen from inhaled air diffuses into the blood from the alveoli. Carbon dioxide diffuses in the reverse direction and leaves the body in the exhaled air. (b) On the left is normal lung tissue, and on the right is lung tissue from a person who suffered from emphysema, a chronic lung disease in which some of the structure of the lungs has broken down. Cigarette smoking and heavy air pollution are associated with the development of emphysema and other chronic lung diseases.

(COPD), a slowly progressive lung disease that makes it increasingly hard to breathe. COPD is estimated to affect 18 million people and is the fourth leading cause of death in the United States. Worldwide, COPD afflicts up to 10% of adults aged 40 and older. The leading cause of this disease is smoking, but in many developing countries, the burning of wood and dung in indoor stoves for heating and cooking delivers much the same burden of pollutants as smoking to the women who tend the stoves. COPD involves three separate disease processes: emphysema (destruction of the lung alveoli—see Fig. 19–17), bronchitis (inflammation and obstruction of the airways), and asthma.

Asthma. Those most sensitive to air pollution are small children, asthmatics, people with chronic pulmonary or heart disease, and the elderly. Asthma is an immune system disorder characterized by impaired breathing caused by constricted air passageways. Asthma episodes are triggered by contact with allergens (dust mites, molds, pet dander) and compounds in polluted air (ozone, particulate matter, SO_2). According to the National Center for Health Statistics, almost 500,000 hospitalizations for asthma occur yearly, together with 1.8 million visits to hospital emergency departments. The highest prevalence is among children from birth to 4 years of age. In the last decade, the incidence of asthma in the United States doubled, affecting 23 million people (including 7 million children).

Strong Evidence. Two key studies recently analyzed by the Health Effects Institute provide strong evidence of the harmful effects of fine particles and sulfur pollution. These studies followed thousands of adult subjects living in 154 U.S. cities for as long as 16 years. In the studies, higher concentrations of fine particles were correlated with increased mortality, especially from cardiopulmonary disease and lung cancer. The more polluted the city, the higher the mortality. The studies were used by the EPA to step up its regulatory attention to fine particles, as we will see later.

Fine particles were also cited as a key to heightened costs of air pollution in southern California, along with ozone. Both pollutants exceed federal standards and were judged to cause more deaths in the region than did traffic accidents. Meeting those standards would save $28 billion annually in avoided health care costs, missed work, school absences, and premature deaths, according to a research study from California State University–Fullerton in 2008. Health impacts of the excess pollutants include asthma, chronic bronchitis, and cardiovascular problems.

Lead Again. One of the heavy-metal pollutants, lead, deserves special attention. For several decades, lead poisoning has been recognized as a cause of mental retardation. Researchers once thought that eating paint chips containing lead was the main source of lead contamination in humans. In the early 1980s, however, elevated lead levels in the blood were shown to be much more widespread than previously expected, and they were present in adults as well as children. Learning disabilities in children, as well as high blood pressure in adults, were correlated with levels of lead in the blood. The major source of this widespread contamination was found to be leaded gasoline. (The lead in

exhaust fumes was inhaled directly or settled on food, water, or any number of items that are put into the mouth.) This knowledge led the EPA to mandate the elimination of leaded gasoline by the end of 1996. The result has been a dramatic reduction in lead concentrations in the environment (Fig. 19–10).

Acute Effects.   In severe cases, air pollution reaches levels that cause death, although such deaths usually occur among people already suffering from critical respiratory or heart disease or both. The gases present in air pollution are known to be lethal in high concentrations, but such concentrations occur nowadays only in cases of accidental poisoning. Nevertheless, intense air pollution puts an additional stress on the body, and if a person is already in a weakened condition (as are, for example, the elderly or asthmatics), this additional stress may be fatal. Most of the deaths in air-pollution disasters reflect the acute effects of air pollution. However, recent research shows that even moderate air pollution can cause changes in cardiac rhythms of people with heart disease, triggering fatal heart attacks.

Carcinogenic Effects.   The heavy-metal and organic constituents of air pollution include many chemicals known to be carcinogenic in high doses. According to the industrial reporting required by the EPA, 4.7 million tons of hazardous air pollutants (the air toxics—see Table 19–1) are released annually into the air in the United States. The presence of trace amounts of these chemicals in the air may be responsible for a significant portion of the cancer observed in humans. One major source of such carcinogens is diesel exhaust (Fig. 19–18). The EPA has classified diesel exhaust as a likely human carcinogen and has established the National Clean Diesel Campaign to reduce emissions from the current diesel truck fleet and establish standards for new diesel-powered engines.

In some cases, exposure to a pollutant can be linked directly to cancer and other health problems by way of epidemiological evidence. One pollutant that clearly and indisputably is correlated with cancer and other disorders is **benzene**. This organic chemical is present in motor fuels and is also used as a solvent for fatty substances and in the manufacture of detergents, explosives, and pharmaceuticals. Environmentally, benzene is found in the emissions from fossil fuel combustion that result from operating motor vehicles and burning coal and oil. Benzene is also present in tobacco smoke, which accounts for half of the public's exposure to the chemical. The EPA has classified benzene as a known human carcinogen, linked to leukemia in persons encountering the chemical through occupational exposure. Chronic exposure to benzene can also lead to numerous blood disorders and damage to the immune system.

The Environment

Experiments show that plants are considerably more sensitive to gaseous air pollutants than are humans. Before emissions were controlled, it was common to see wide areas of totally barren land or severely damaged vegetation downwind from

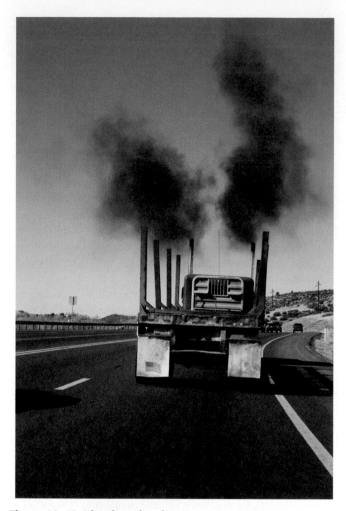

Figure 19–18 Diesel truck exhaust. Diesel truck exhaust is a major source of hazardous air pollutants as well as particulates, and the exhaust from diesel engines has been classified as a probable human carcinogen.

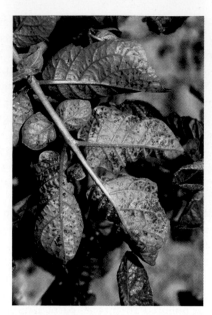

Figure 19–19 Ambient ozone injury. This potato leaf shows the black flecking symptomatic of ozone damage. (*Source:* U.S. Department of Agriculture Agricultural Research Service. Effects of Ozone Air Pollution on Plants. 2009. www.ars.usda.gov/Main/docs.htm?docid=12462.)

smelters and coal-burning power plants. The pollutant responsible was usually sulfur dioxide.

Crop Damage. Damage to crops, orchards, and forests downwind of urban centers is caused mainly by exposure to ozone. The ozone gains access to plants through their stomata, the respiratory pores in all plant leaves. Symptoms of ozone damage include black flecking and yellowing of leaves (Fig. 19–19). Crops vary in their susceptibility to ozone, but damage to many important crops (such as soybeans, corn, and wheat) is observed at common ambient levels of ozone (Fig. 19–20). Unfortunately, ozone levels are highest in summer,

Figure 19–20 Ozone impact on crop yields. Some crop species react more strongly to ozone than others. (*Source:* U.S. Department of Agriculture Agricultural Research Service: Effects of Ozone Air Pollution on Plants. 2009. www.ars.usda.gov/Main/docs.htm?docid=12462.)

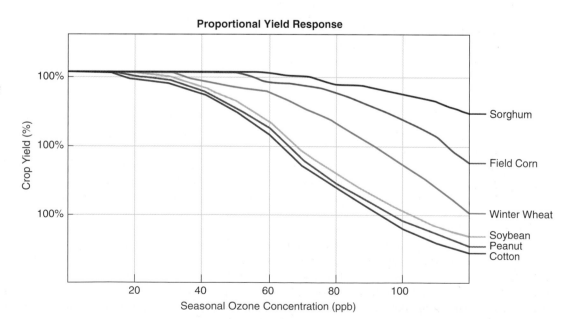

when crops are actively growing, and rural areas often experience high ozone levels as ozone forms downwind of pollution sources. Much of the world's grain production occurs in regions that receive enough ozone pollution to reduce crop yields. India loses $5 billion yearly to ozone crop damage, China loses $2.5 billion, and farmers in Europe lose more than $7.5 billion yearly to ozone.

Forest Damage.   The negative impact of air pollution on wild plants and forest trees may be even greater than on agricultural crops. Significant damage to valuable ponderosa and Jeffrey pines occurred along the entire western foothills of the Sierra Nevada in California. Studies by the U.S. Forest Service concluded that ozone from the nearby Central Valley and San Francisco–Oakland metropolitan region was responsible for this damage. Forests under stress from pollution are more susceptible to damage by insects and other pathogens than are unstressed forests. The deaths of the ponderosa and Jeffrey pines in the Sierra Nevada were attributed to western pine beetles, which invade trees weakened by ozone.

Studies have shown that damage to forest trees begins at about 40 ppb and gets more intense with higher levels. In California and southern Appalachian forests, ozone levels often reached 60 ppb and above, and seasonal losses in tree stem growth in the latter were measured at 30–50% for most species. Studies in European forests found that ozone stress impacts photosynthetic carbon uptake, increases respiration of foliage, amplifies water loss, and generally disrupts carbon and nutrient metabolism. In sum, research has suggested that as temperatures climb and moisture is less predictable under conditions of expected global climate change, ozone levels will have an increasingly serious impact on forest growth.

Effects on Materials and Aesthetics.   Walls, windows, and other exposed surfaces turn gray and dingy as particulates settle on them. Paints and fabrics deteriorate more rapidly, and the sidewalls of tires and other rubber products become hard and checkered with cracks because of oxidation by ozone. Metal corrosion is increased dramatically by sulfur dioxide and acids derived from sulfur and nitrogen oxides, as are weathering and the deterioration of stonework (Fig. 19–21). These and other effects of air pollutants on materials increase the costs for cleaning or replacing them by hundreds of millions of dollars a year. Many of the damaged materials, such as historic statues and carvings, are irreplaceable.

Visibility.   A clear blue sky and good visibility are matters of health, but they also carry significant aesthetic value and can have a deep psychological impact on people. Can a value be put on these benefits? Many of us spend thousands of dollars and hundreds of hours commuting long distances to work so that we can live in a less polluted environment than the one in which we work. Ironically, the resulting traffic and congestion cause much of the very pollution we are trying to escape. Then we travel to national parks and wilderness areas and too often find that the visibility in *these* natural areas is impaired by what has been called "regional haze,"

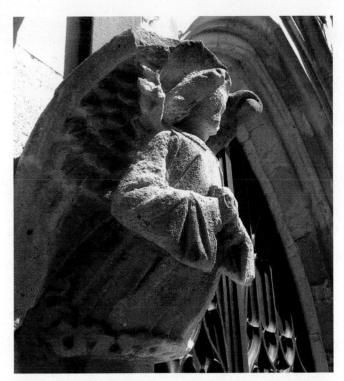

Figure 19–21 Effect of pollution on monuments. The corrosive effects of acids from air pollutants are dissolving away the features of many monuments and statues, as seen in the angel perched on this building buttress.

coming from particulates and gases originating hundreds of miles away. In an encouraging move, the EPA established a Regional Haze Rule in 1999, aimed at improving the visibility at 156 national parks and wilderness areas. The regulations call on all 50 states to establish goals for improving visibility and to develop long-term strategies for reducing the emissions (especially particles) that cause the problem.

Effects of Acid Deposition

Acid deposition has been recognized as a problem in and around industrial centers for over 100 years. Its impact on ecosystems, however, was noted about 45 years ago, when anglers started noticing sharp declines in fish populations in many lakes in Sweden, Ontario, and the Adirondack Mountains of upper New York State. Since that time, as ecological damage continued to spread, studies have revealed many ways in which acid deposition alters and may destroy ecosystems.

Impact on Aquatic Ecosystems.   The pH of an environment is extremely critical because it affects the function of virtually all enzymes, hormones, and other proteins in the bodies of all organisms living in that environment. Ordinarily, organisms are able to regulate their internal pH within the narrow limits necessary to function properly. A consistently low environmental pH, however, often overwhelms the regulatory mechanisms in many life forms, thus weakening or killing them. Most freshwater lakes, ponds, and streams have a natural pH in the range from 6 to 8, and organisms have adapted accordingly. The eggs, sperm, and developing young of these organisms are especially sensitive to changes in pH. Most are severely stressed, and many die, if the environmental pH shifts as little as one unit from the optimum.

As aquatic ecosystems become acidified (pH 5 and below), higher organisms die off, either because the acidified water kills them or because it keeps them from reproducing. Figure 19–13 shows that acid precipitation may leach aluminum and various heavy metals from the soil as the water percolates through it. Normally, the presence of these elements in the soil does not pose a problem because they are bound in insoluble mineral compounds and, therefore, are not absorbed by organisms. As the compounds are dissolved by low-pH water, however, the metals are freed. They may then be absorbed by and are highly toxic to both plants and animals. For example, mercury tends to accumulate in fish as lake waters become more acidic. Indeed, mercury levels are so high in the Great Lakes that many bordering states advise against eating fish caught in those waters.

Acid deposition hit some regions especially hard in the latter part of the 20th century. In Ontario, Canada, approximately 1,200 lakes lost their fish life. In the Adirondacks, a favorite recreational region for New Yorkers, 346 lakes were without fish in 1990. In New England and the eastern Catskills, over 1,000 lakes have suffered from recent acidification. The physical appearance of such lakes is deceiving. From the surface, they are clear and blue, the outward signs of a healthy condition. However, the only life found below the surface is acid-loving mosses growing on the bottom.

Neutralizing Capacity. As Figure 19–15 indicates, wide regions of eastern North America receive substantial amounts of acid precipitation, yet not all areas have acidified lakes. Apparently, many areas remain healthy, whereas others have become acidified to the point of becoming lifeless. How is this possible? The key lies in the system's *acid neutralizing capacity (ANC).* Despite the addition of acid, a system may be protected from changes in pH by a **neutralizer**—a substance that, when present in a solution, has a large capacity to absorb hydrogen ions and thus maintain the pH at a relatively constant value.

Limestone ($CaCO_3$) is a natural acid neutralizer (Fig. 19–22) that protects lakes from the effects of acid

precipitation in many areas of the North American continent, particularly in the Midwest. Lakes and streams receiving their water from rain and melted snow that have percolated through soils derived from limestone will contain dissolved limestone. The regions that are sensitive to acid precipitation are those containing much granitic rock, which does not yield good ANC. The good news is that the sensitive regions (New England, the Adirondacks, the Northern Appalachian Plateau) are recovering their ANC and the percentage of acidic lakes has declined in the past 18 years. The decline is consistent with the decrease in acid deposition in these regions, which has been brought about by the reductions of SO_2 and NO_x emissions mandated by the Clean Air Act.

Impact on Forests. From the Green Mountains of Vermont to the San Bernardino Mountains of California, the die-off of forest trees in the 1980s caused great concern. Red spruce forests were especially vulnerable. In New England, 1.3 million acres of high-elevation forests were devastated. Commonly, the damaged trees lost needles as acidic water drew calcium from them, rendering them more susceptible to winter freezing. Sugar maples, important forest trees in the Northeast, have shown extensive mortality, ranging from 20% to 80% of all trees in some forests.

Much of the damage to forests from acid precipitation is due to chemical interactions within the forest soils. Sustained acid precipitation at first adds nitrogen and sulfur to the soils, which stimulate tree growth. In time, though, these chemicals leach out large quantities of the buffering chemicals (usually calcium and magnesium salts). When these buffering salts no longer neutralize the acid rain, aluminum ions, which are toxic, are dissolved from minerals in the soil. The combination of aluminum and the increasing scarcity of calcium, which is essential to plant growth, leads to reduced tree growth. Research at the Hubbard Brook Experimental Forest in the White Mountains of New Hampshire has shown a marked reduction in calcium and magnesium in the forest soils from the 1960s on, which is reflected in the amount of calcium in tree rings over the same period. The net result of these changes has been a severe decline in forest growth. Experimentally liming of soils in forests showing maple tree dieback was able to restore the trees to health, and aluminum ion concentrations were seen to decrease greatly.

Impact on Human Artifacts. Limestone and marble (which is a form of limestone) are favored materials for the outsides of buildings and for monuments (collectively called **artifacts**). The reaction between acid deposition and limestone is causing these structures to erode at a tremendously accelerated pace. Monuments and buildings that have stood for hundreds or even thousands of years with little change are now dissolving and crumbling away, as seen in Figure 19–21. The corrosion of buildings, monuments, and outdoor equipment by acid precipitation costs billions of dollars for replacement and repair each year in the United States. In an extreme case of corrosion, a 37-foot bronze Buddha in Kamakura, Japan, is slowly dissolving away as precipitation from Korea and China bathes the

Figure 19–22 Acid neutralizing. Acids may be neutralized by certain non-basic compounds. A chemical such as limestone (calcium carbonate) reacts with hydrogen ions as shown. Hence, the pH of a lake or river remains close to neutral despite the additional acid, when it contains adequate acid neutralizing capacity.

statue in rain that is more acidic than that found in the United States.

Although the decay of such artifacts is a tragic loss in itself, it should also stand as a grim reminder of how we are dissolving away the buffering capacity of ecosystems. As the saying goes, "What goes up must come down." The sulfur and nitrogen oxides pumped into the troposphere in the United States gradually increased from some 11.4 million tons in 1900 to their peak of 53 million tons in 1973. Although this number has been reduced by half, the pollutants still come back down as acid deposition, generally to the east of their origin because of the way weather systems flow. The deposits cross national boundaries, too. As a result, Canada receives half of its acid deposition from the United States, and Scandinavia gets it mostly from Great Britain and other western European nations. Japan receives the wind-borne pollution from widespread coal burning in Korea and China. As a result of all the harmful impacts, the problem of acid deposition has been addressed at national and international levels. Accordingly, we turn next to the efforts being made to curb air pollution.

19.4 Bringing Air Pollution Under Control

By the 1960s, it was obvious that pollutants produced by humans were overloading natural cleansing processes in the atmosphere. The unrestricted discharge of pollutants into the atmosphere could no longer be tolerated.

Clean Air Act

Under grassroots pressure from citizens, the U.S. Congress enacted the **Clean Air Act of 1970 (CAA)**. Together with amendments passed in 1977 and 1990, this law, administered by the EPA, represents the foundation of U.S. air-pollution control efforts. The act calls for identifying the most widespread pollutants, setting **ambient standards** (levels that need to be achieved to protect environmental and human health), and establishing control methods and timetables to meet the standards.

NAAQS. The CAA mandated the setting of standards for four of the primary pollutants—particulates, sulfur dioxide, carbon monoxide, and nitrogen oxides—and for the secondary pollutant ozone. At the time, these five pollutants were recognized as the most widespread and objectionable ones. Today, with the addition of lead, they are known as the **criteria pollutants** and are covered by the **National Ambient Air Quality Standards (NAAQS)** (See Chapter 21, Table 21–3). The **primary standard** for each pollutant is based on the presumed highest level that can be tolerated by humans without noticeable ill effects, minus a 10% to 50% margin of safety. For some of the pollutants, long-term and short-term levels are set. The short-term levels are designed to protect against acute effects, while the long-term standards are designed to protect against chronic effects.

TABLE 19–3 National Ambient Air Quality Standards for Criteria Pollutants (NAAQS)

Pollutant	Averaging Time*	Primary Standard
PM_{10} particulates	24 hours	150 $\mu g/m^3$
$PM_{2.5}$ particulates**	1 year	15 $\mu g/m^3$
	24 hours	35 $\mu g/m^3$
Sulfur dioxide	1 year	0.03 ppm
	24 hours	0.14 ppm
Carbon monoxide	8 hours	9 ppm
	1 hour	35 ppm
Nitrogen oxides	1 year	0.053 ppm
Ozone	8 hours	0.075 ppm
Lead	3 months	0.15 $\mu g/m^3$

Source: U.S. EPA, Office of Air Quality Planning and Standards, *National Air Quality: Status and Trends Through 2007* (Washington, D.C., U.S. EPA, November 2008).
*The averaging time is the period over which concentrations are measured and averaged.
**$PM_{2.5}$ is the particulate fraction having a diameter smaller than or equal to 2.5 micrometers.

Table 19–3 reflects two changes announced by the EPA in 2008: a revision of the ozone standard from 0.08 parts per million (ppm) to 0.075 ppm for 8 hours and a revision of the lead standard from 1.5 $\mu g/m^3$ to 0.15 $\mu g/m^3$. The EPA is required to review the pollutants every five years and make adjustments if the science justifies changes. The NAAQS have far-reaching impacts on many regulations, and proposed changes always generate controversies, as we will see.

NESHAPs. In addition, the CAA required the EPA to establish **National Emission Standards for Hazardous Air Pollutants (NESHAPs)** for a number of toxic substances (the air toxics). The Clean Air Act Amendments of 1990 greatly extended this section of the EPA's regulatory work by specifically naming 187 toxic air pollutants for the agency to track and regulate. The EPA tracks both the criteria pollutants and the hazardous air pollutants in a database, the **National Emission Inventory**. This database serves as an essential source of information for the EPA in carrying out its regulatory responsibilities.

Control Strategies

The basic strategy of the 1970 Clean Air Act was to regulate the emissions of air pollutants so that the *ambient* criteria pollutants would remain below the primary standard levels. This approach is called **command and control** because regulations were enacted requiring industry to achieve a set limit on each pollutant through the use of specific control equipment. The assumption was that human and environmental health could be significantly improved by a reduction in the output of pollutants. If a particular region was in violation for a given pollutant, a local government agency would track down the source(s) and order reductions in emissions until the region came into compliance.

Unfortunately, this strategy proved difficult to implement. Most of the regulatory responsibility fell on the states and cities, which were often unable or unwilling to enforce control. In spite of these difficulties, however, total air pollutants were reduced by some 25% during a time when both population and economic activity increased substantially, and even in the noncompliant regions, the severity of air-pollution episodes lessened. Accordingly, some significant changes were made in the law.

1990 Amendments.

The **Clean Air Act Amendments of 1990 (CAAA)** targeted specific pollutants more directly and enforced compliance more aggressively, through such means as the imposition of sanctions. As with the earlier law, the states do much of the work in carrying out the mandates of the 1990 act. Each state must develop a **State Implementation Plan (SIP)** that is required to go through a process of public comment before being submitted to the EPA for approval. The SIP is designed to reduce emissions of every NAAQS pollutant whose control standard (Table 19–3) has not been attained. One major change is the addition of a permit application process (already in place for the release of pollutants into waterways). Polluters must apply for a permit that identifies the kinds of pollutants they release, the quantities of those pollutants, and the steps they are taking to reduce pollution. Permit fees provide funds the states can use to support their air-pollution control activities. The CAAA also affords more flexibility than did the earlier command-and-control approach by allowing polluters to choose the most cost-effective way to accomplish the goals. In addition, the legislation uses a market system to allocate pollution among different utilities.

Under the CAAA, regions of the United States that have failed to attain the required levels must submit *attainment plans* based on **reasonably available control technology (RACT)** measures. Offending regions must convince the EPA that the standards will be reached within a certain time frame.

Reducing Particulates.

Prior to the 1970s, the major sources of particulates were industrial stacks and the open burning of refuse. The CAA mandated the phaseout of open burning of refuse and required that particulates from industrial stacks be reduced to "no visible emissions."

The alternative generally taken to dispose of refuse was landfilling, a solution that has created its own set of environmental problems (discussed in Chapter 21). To reduce stack emissions, many industries were required to install filters, electrostatic precipitators, and other devices. Although these measures have markedly reduced the levels of particulates since the 1970s, particulates continue to be released from steel mills, power plants, cement plants, smelters, construction sites, diesel engines, and so on. Wood-burning stoves and wood and grass fires also contribute to the particulate load, making regulation even more difficult.

PM. The EPA added a new ambient air quality standard for particulates ($PM_{2.5}$) in 1997, on the basis of information indicating that smaller particulate matter (less than 2.5 micrometers, or microns, in diameter) has the greatest effect on health. The coarser material (PM_{10}) is still regulated, but is not considered to play as significant a role in health problems as do the finer particles. There is overwhelming evidence that the finer particles (2.5 microns or less) go right into the lungs when breathed. Responding to this evidence, the EPA announced a revision to the NAAQS in 2006, reducing the 24-hour primary standard for $PM_{2.5}$ from 65 μg/m^3 to 35 μg/m^3. The EPA left the annual standard at 15 μg/m^3, in spite of recommendations from the EPA's Clean Air Scientific Advisory Committee that it be lowered to between 13 μg/m^3 and 14 μg/m^3. As a result, the new standards, although breaking new ground for pollution control, have been heavily criticized by public-health advocates. The EPA points to the lowered daily standard as a major step forward, estimating that it will yield health benefits valued at between $9 billion and $75 billion a year. However, a federal appeals court ruled in February 2009 that the particulate standards were inadequate and contrary to the scientific evidence, sending the issue back to the EPA.

Controlling Air Toxics.

By EPA estimates, the amount of air toxics emitted annually into the air in the United States is around 4.7 million tons. These come from stationary sources (e.g., factories, refineries, power plants), mobile sources (cars, trucks, buses), and indoor sources (solvents, building materials). Cancer and other adverse health effects, environmental contamination, and catastrophic chemical accidents are the major concerns associated with this category of pollutants. Under the CAAA, Congress identified 187 toxic pollutants. It then directed the EPA to identify major sources of these pollutants and to develop **maximum achievable control technology (MACT)** standards. Besides affecting control technologies, the standards include options for substituting nontoxic chemicals, giving industry some flexibility in meeting MACT goals. State and local air-pollution authorities are responsible for seeing that industrial plants achieve the goals. In response to the CAAA, the EPA has mobilized over 300 monitoring sites to track emissions of the affected substances (Fig. 19–23). To reduce the contribution from vehicles, the agency requires cleaner burning fuels (reformulated gasoline) in urban areas. Industrial sources are being addressed by setting emission standards for some 82 stationary-source categories (e.g., paper mills, oil refineries).

Limiting Pollutants from Motor Vehicles.

Cars, trucks, and buses release nearly half of the pollutants that foul our air. Vehicle exhaust sends out VOCs, carbon monoxide, and nitrogen oxides that lead to ground-level ozone and PANs. Additional VOCs come from the evaporation of gasoline and oil vapors from fuel tanks and engine systems. The CAA mandated a 90% reduction in these emissions by 1975. This timing proved to be unrealistic, but enough improvements have been made over the years that a new car in 2007 emitted 86% less pollution than did pre-1970 cars. This improvement is

Figure 19–23 National Air Toxics Trends Station. This site in the Cherokee Heights community, WI, tracks ambient air for emissions of air toxics compounds as well as ozone, particulates, and NO_x.

fortunate because driving in the United States has been increasing much more rapidly than the population has. Between 1970 and 2007, the number of vehicle miles increased from 1 trillion to 3 trillion miles per year, and between 1980 and 2007, the number of vehicles on the road increased more than 63%. It is hard to imagine what the air would be like without the improvements mandated by the CAA.

The reductions in automobile emissions have been achieved with a general reduction in the size of passenger vehicles, along with a considerable array of pollution-control devices, among which is one that affords the computerized control of fuel mixture and ignition timing, allowing more complete combustion of fuel and decreasing VOC emissions. To this day, however, the most significant control device on cars is the **catalytic converter** (Fig. 19–24). As exhaust passes through this device, two different catalysts are at work: a reduction catalyst that converts NO_x emissions to harmless nitrogen gas and an oxidation catalyst that oxidizes unburned hydrocarbons and carbon monoxide to carbon dioxide.

CAAA Changes. The continuing failure to meet standards in many regions of the United States, as well as the aesthetically poor air quality in many cities, made it clear during the 1980s that further legislation was necessary. The following are the highlights of the CAAA related to motor vehicles and fuels:

1. New cars sold in 1994 and thereafter were required to emit 30% less VOCs and 60% less nitrogen oxides than cars sold in 1990. Emission-control equipment had to function properly (as represented in warranties) to 100,000 miles. Buses and trucks were required to meet more-stringent standards. Also, the EPA was given the authority to control emissions from all nonroad engines that contribute to air pollution, including lawn and garden equipment, motorboats, off-road vehicles, and farm equipment.

2. Starting in 1992 in the regions with continuing carbon monoxide problems, oxygen was to be added to gasoline in the form of methyl tertiary butyl ether or alcohols in order to stimulate more-complete combustion and to cut down on carbon monoxide emissions.

3. Before 1990, only a few states and cities required that vehicle inspection stations be capable of measuring emissions accurately. The new amendments required more than 40 metropolitan areas to initiate inspection and maintenance programs, while other cities were to improve their programs.

Two negative factors now affect fuel efficiency and consumption rates. First, the elimination of federal speed limits in 1996 reduced fuel efficiencies because of the higher speeds (and increased the rate of traffic fatalities). Second, the sale of sport-utility vehicles (SUVs), vans, and pickup trucks has surged from 25% of the combined light-duty vehicle sales in 1990 to 50% in 2004 (Fig. 19–25). Automobile sales finally passed those of heavier vehicles in 2008, a response to both the high fuel prices early in the year and the economic recession. The EPA's fuel economy estimates for model year 2008

Gases from engine

To exhaust pipe

Figure 19–24 Cutaway of a catalytic converter. Engine gas is routed through the converter, where catalysts promote chemical reactions that change the harmful hydrocarbons, carbon monoxide, and nitrogen oxides into less harmful gases, such as carbon dioxide, water, and nitrogen.

Figure 19–25 Fuel economy and light-duty vehicles.
Changes in the fuel economy in model years from 1975 to 2008. "Trucks" refers to pickups, family vans, and SUVs. The percentage of "trucks" has risen from below 25% to around 50%, a consequence of SUV popularity. (*Source:* EPA, *Light-Duty Automotive Technology and Fuel Economy Trends: 1975 Through 2008,* EPA420-R-08-015 (Washington, D.C.: EPA, September 2008).)

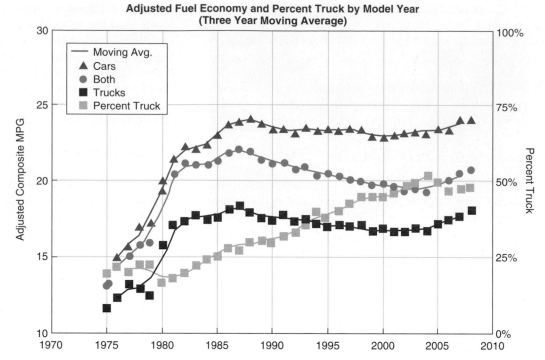

**Adjusted Fuel Economy and Percent Truck by Model Year
(Three Year Moving Average)**

are as follows (these are "real-world estimates," which are some 15% lower than those previously in use): minivans 19.7 miles per gallon (mpg), SUVs 18.7 mpg, and pickups 16.3 mpg, yielding a combined fuel economy for this "truck" class of 18.1 mpg. With the 2008 automobile fuel economy averaging 24.1 mpg, the combined overall value for all passenger vehicles is estimated to be 20.8 mpg, some 5% lower than the peak value achieved in 1987–1988.

CAFE Standards. Under the authority of the **Energy Policy and Conservation Act of 1975** and its amendments, the National Highway Traffic Safety Administration was given the authority to set *corporate average fuel economy* (*CAFE*) standards for motor vehicles. The intention of the law was to conserve oil and promote energy security, but reducing gasoline consumption also means less air pollution. The standard set in 1984 for passenger cars required a fleet average of 27.5 mpg. The standard for light trucks (pickups, SUVs, minivans) was set lower, at 20.7 mpg. Growing concern about the increasing dominance of light trucks in the national fleet led to a new rule raising the light-truck standard to 22.2 mpg, to be phased in by 2007. Actual current vehicle performances in both categories are several mpg lower than the current CAFE standards and have never achieved the CAFE standards (Fig. 19–25). (For comparison though, the fleet fuel economy in 1974 was only 12.9 mpg.)

The **Energy Independence and Security Act**, passed in 2007, mandated a new CAFE fuel economy standard of 35 mpg, a 40% increase in fuel economy, to be achieved by 2020. This standard is to be achieved by the combined fleet of cars and light trucks. Congress passed this legislation as a major step toward a new energy future of reduced reliance on foreign oil and increased renewable energy, with a view toward reducing CO_2 greenhouse gas emissions (Chapter 18). In 2009, the Obama administration set more-stringent

fuel-economy standards, calling for an annual fuel economy increase of 5%, beginning with the 2011 model year. By 2016, the CAFE standard for cars would reach 39 mpg, and for light trucks, the standard would reach 30 mpg. It is calculated that the new standards would mean a 30% reduction in greenhouse gas emissions from new vehicles by 2016.

Getting Around. Raising the CAFE standards helps address (1) our dependence on imported oil (and all that that implies), (2) all the health issues from smog and particulates, and (3) carbon dioxide emissions that are bringing on global climate change. Even though many Americans continue to prefer SUVs and larger automobiles, there are *some* signs of change, particularly in light of the substantial rise in the cost of gasoline in recent years. One is the **hybrid electric vehicle**, which combines a conventional gasoline engine and a battery-powered electric motor to achieve substantial improvements in fuel economy. Such vehicles have been marketed by Honda and Toyota for almost a decade and are the most fuel-efficient cars in the United States. The Toyota Prius averages 46 mpg, while the Honda Insight gets 43 mpg, according to EPA figures.

All of the major automakers are jumping on the hybrid bandwagon, and higher prices of gasoline have boosted hybrid sales in the United States. Sales of hybrids soared in 2007, with 352,000 units sold, but with the downturn in the economy and the decrease in gasoline prices, hybrid sales dropped to 313,000 vehicles in 2008, about 2.5% of the passenger vehicle sales. Most newer hybrids are "green" versions of standard vehicles, including SUVs. For example, the Toyota Camry hybrid employs the same 2.4-liter gasoline engine as the standard four-cylinder Camry, but combines it with a large battery nestled behind the rear seat. Two electric motors work as generators that charge the battery as the car slows down. The car starts out under electric power, and with acceleration, the gasoline engine kicks in. When the car is sitting at a stoplight or coasting down a hill, the gasoline engine shuts off and starts automatically when speeding up. The EPA rates the Camry hybrid at 34 mpg, but users commonly experience mileages up to 40 mpg.

Prices for hybrids are several thousand dollars above those for comparable standard vehicles, but with gasoline ranging as high as $3 or more a gallon, the savings catch up with the extra cost in a few years. As more and more of the new hybrids are sold, prices are expected to drop substantially. For a time, a federal tax credit has been available for hybrid buyers that offsets much of the cost differential. Many states also offer tax incentives for hybrid buyers. For some excellent information on hybrids, consult the Union of Concerned Scientists' special Web site, www.hybridcenter.org.

Another encouraging trend is the increase in mass-transit ridership. The American Public Transportation Association (APTA) reports substantial increases in ridership on buses, subways, and commuter rail lines in most major metropolitan areas, a likely response to the rising price of gasoline. The APTA calculates that if Americans used mass transit at the same rate as Europeans do, our use of imported oil would decline by 40%.

Managing Ozone.   Because ozone is a secondary pollutant, the only way to control ozone levels is to address the compounds that lead to its formation. For a long time, it was assumed that the best way to reduce ozone levels was simply to reduce emissions of VOCs. The steps pertaining to motor vehicles in the CAAA addressed the sources of about half of the VOC emissions. **Point sources** (industries) accounted for another 30% of such emissions, and **area sources** (numerous small emitters, such as dry cleaners, print shops, and users of household products) represented the remaining 20%. RACT measures have already been mandated for many point sources, and much progress has been made through EPA, state, and local regulatory efforts to reduce emissions from those sources. Since passage of the CAAA, VOC emissions have declined by 24%.

However, recent understanding of the complex chemical reactions involving NO_x, VOCs, and oxygen has thrown some uncertainty into this strategy of simply emphasizing a reduction in VOCs. The problem is that *both* NO_x and VOC concentrations are crucial to the generation of ozone. Either one or the other can become the rate-limiting factor in the reaction that forms ozone. Thus, as the ratio of VOCs to NO_x changes, the concentration of NO_x can become the controlling chemical factor, pointing to the need to reduce NO_x emissions as well. This happens more commonly in air-pollution episodes that occur over a period of days than in the daily photochemical smog of the urban city.

New Ozone Standards.   The revised ozone standard announced in 1997 (0.08 ppm instead of 0.12 ppm) met with strong opposition from industry groups, which obtained a court injunction in 1999 prohibiting the EPA from enforcing the new standard. However, the U.S. Supreme Court upheld the EPA in the dispute. This and other delays put off implementation of the revised standard to 2004. Cost-benefit

estimates indicated that the anticipated health benefits far outweighed the costs of compliance. Then the EPA recommended in 2007 that a new ozone standard in the range of 0.06–0.07 ppm be set, based on human health concerns. EPA Administrator Stephen Johnson decided on a level of 0.075 ppm and published the new standard in early 2008—to the dismay of public-health advocates. The new ozone standard is expected to prevent a substantial number of premature deaths, but only a third of those that the lower standard would have prevented. No prevention will happen, however, unless the standard is accompanied by regulations that reduce the source compounds that produce ozone.

Down with NO_x.   In a response to petitions from states in the Northeast that were having trouble achieving ozone and smog ambient standards because of out-of-state emissions, the EPA has implemented regulations to reduce NO_x emissions from *mobile sources*, and from *power plants and large industrial boilers and turbines*. For the mobile sources, which account for 59% of the smog-generating emissions in the eastern United States, the EPA has established the so-called **Tier 2 Standards**, which were phased in gradually between 2004 and 2009. Under these regulations, emissions for all SUVs, pickup trucks, and passenger vans are held to the same standards as those for passenger cars are. For NO_x, this means a gradual reduction to 0.07 grams per mile for the fleet of cars sold by each manufacturer, an 83% reduction over 2003 passenger car standards. These standards apply nationally, and although NO_x emissions are declining, the standards are far from being met; the 2007 average for NO_x exhaust emissions was 0.72 g/mile, due to the large number of older vehicles still in service.

Fixed sources of NO_x in the eastern United States (accounting for 22% of the emissions) are being addressed by the **NO_x Budget Trading Program** (NBP). This program establishes a market-based "cap and trade" system whereby large combustion sources reduce their NO_x emissions during the ozone season (May 1 through September 30). State-by-state trading budgets were set by the EPA, in terms of tons of NO_x emissions, and the states and industries were given flexibility in how they were to attain the target budgets. Some 2,594 industrial entities were affected by the NBP in 2007; the program has achieved a remarkable reduction in emissions (Fig. 19–26), especially in combination with NO_x reductions covered by the Acid Rain Reduction Program. The reductions in emissions have contributed to a significant improvement in eight-hour ozone concentrations in the eastern states, bringing two-thirds of the areas once in violation of the standards back into conformity.

CAIR?   In May 2005, the EPA announced two broad new rules designed to use a cap and trade method to accomplish continued reduction of ozone, particulates, and mercury by mandating new reductions of SO_2, NO_x, and mercury emissions. The **Clean Air Interstate Rule** (CAIR) set new lower caps on SO_2 and NO_x in 28 states, many of which are upwind of eastern states experiencing nonattainment of ozone and particulate matter. CAIR would reduce SO_2 emissions 50% beginning in 2010 and 65% beginning in 2015. NO_x emissions would be reduced by similar percentages. The **Clean Air**

Figure 19–26 NO$_x$ emissions under the NO$_x$ Budget Trading Program. This program, together with the Acid Rain Reduction Program, has achieved a great reduction in the NO$_x$ emissions that lead to ozone formation. (*Source:* NO$_x$ Budget Program 2007 Progress Report, December 2008.)

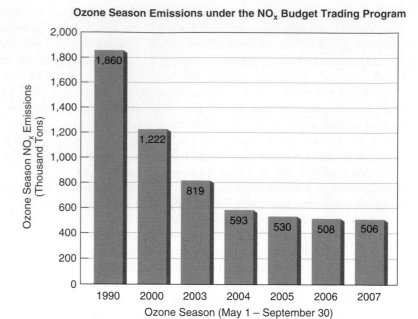

Ozone Season Emissions under the NO$_x$ Budget Trading Program

Mercury Rule (CAMR) would cap and reduce mercury emissions from coal-fired power plants in all 50 states; by 2018, the cap of 15 tons would be in place, representing a reduction of mercury emissions from these plants of almost 70%.

The new rules had a mixed reception. The industries and utilities were disappointed because the rules lacked the curb on lawsuits that a congressional act would provide, but they were pleased with the extended time frame for compliance. Environmental groups and a number of states praised CAIR, but were highly critical of CAMR, pointing out that it should have been possible to achieve greater mercury reductions—and more quickly. Fourteen states and a number of environmental groups filed a lawsuit challenging CAMR, charging that the new rule was much less rigorous than existing CAA regulations. And in early 2008, a federal court ruled that the EPA must redo its mercury program, effectively canceling CAMR. An appeal to the U.S. Supreme Court failed to get a hearing in February 2009, leaving the lower court ruling in place.

CAIR also received a legal setback when a federal appeals court overturned the CAIR regulations in July 2008, stating that the program had "more than several fatal flaws," including not being strict enough. This left the utilities in a state of limbo, uncertain about their expenditures on new pollution-control equipment and the cap and trade purchases they had already made. However, the EPA, numerous states, and the Environmental Defense Fund successfully petitioned the court to reconsider its decision, and in December 2008, the court reversed its ruling and reinstated CAIR. In essence, CAIR will remain in effect until the EPA fixes the rule in accord with CAA requirements.

New Source Review. The Clean Air Act explicitly requires all power plants and other polluting industrial facilities built after 1970 to incorporate "best available" pollution-control technology, such as modern scrubbers. Rather than imposing large and immediate costs on industry and the consumer, the CAA exempted all older facilities from having to install new pollution controls and chose rather to require them to do so only when they upgraded their facilities in any way that would increase pollution. All new facilities and all upgrades must obtain a permit before construction, and the EPA program that grants the permits is the New Source Review Program. Routine maintenance would be permitted, with no new antipollution technology required. For years, however, industries skirted the regulation by making substantial improvements to their facilities and calling the improvements "routine maintenance." The Clinton administration cracked down on this practice and brought enforcement cases against 51 power-plant operators and dozens of refineries.

Prompted by Vice President Cheney's energy task force (see Chapter 14), the EPA announced that it would change the rules for the New Source Review Program. Factories and power plants would no longer be required to update their pollution controls unless the improvements or changes involved more than 20% of the entire facility's value. The EPA's position was that "routine maintenance" is too vague a criterion. Instead, industry needed a more precise definition of when the New Source Review process would be triggered. Many state officials and environmental groups were outraged at the change in the regulations and warned that this would allow some 17,000 facilities (including more than 500 older coal-fired power plants) across the country to increase their pollution emissions simply by making regular repairs of old equipment instead of modernizing pollution controls. The new rule was praised by the National Association of Manufacturers, the American Petroleum Institute, and other groups that lobby for the industries affected by the CAA. However, 15 states mounted a legal challenge to the new rule, and in March 2006, a federal appeals court ruled unanimously that the rule changes violated the CAA as well as simple logic, stating that the plain language of the law required a stricter approach. A Supreme Court ruling in 2007 overturned a lower court ruling that would have allowed Duke Energy to upgrade its older power plants without reducing pollutants, thereby signaling that the EPA rule changes were invalid.

Coping with Acid Deposition

Scientists working on acid-rain issues calculated that a 50% reduction in acid-causing emissions in the United States would effectively prevent further acidification of the environment. This reduction was not expected to correct the already bad situations, but together with natural buffering processes, it was estimated to be capable of preventing further environmental deterioration. Because we know that about 50% of acid-producing emissions come from coal-burning power plants, control strategies focused on these sources.

Political Developments. Although evidence of the link between power-plant emissions and acid deposition was well established by the early 1980s, no legislative action was taken until 1990. The problem was one of different regional interests. Western Pennsylvania and the states of the Ohio River Valley, where older coal-burning power plants produce most of the electrical power for the region, argued that controlling their sulfur dioxide emissions would make electricity unaffordable in the region. Throughout the 1980s, a coalition of politicians from these states, representatives of high-sulfur coal producers, and representatives from the electric power industry effectively blocked all attempts at passing legislation that would take action on acid deposition.

On the other side of the issue were New York and the New England states, as well as most of the environmental and scientific communities, which argued that it was both possible and necessary to address acid deposition and that the best way to do so was to control emissions from power plants. Also, since 70% of Canada's acid-deposition problem came from the United States, diplomatic pressure toward a resolution was applied.

Action. With the passage of Title IV of the CAAA, the outcome of the controversy became history. However, two decades of action were lost because of political delays. In the wake of the passage of the 1990 act, Canada and the United States signed a treaty stipulating that Canada would cut its sulfur dioxide emissions by half and cap them at 3.2 million tons by the year 2000. (The U.S. obligations from the treaty will be discussed shortly.) Canada has done well; its emissions are now about 2 million tons and are expected to stay down. As a result of several treaties in Europe, sulfur dioxide emissions there are down 40% from 1980 levels and should decline another 18% by 2010 if the signatories fulfill their obligations.

The main beneficiaries of these agreements are the remaining healthy aquatic and forested ecosystems, which, if all goes according to plan, will be protected from future acid deposition. In addition, ecosystems already harmed will be able to recover from current damage, and we will be back on the road to a sustainable interaction with the atmosphere.

Title IV of the Clean Air Act Amendments of 1990. Title IV of the CAAA addressed the acid-deposition problem by mandating reductions in both sulfur dioxide and nitrogen oxide levels. The major provisions of Title IV are as follows:

1. By 2010, total SO_2 emissions must be reduced 10 million tons below 1980 levels. This is the 50% reduction called for by scientists, and it involves the setting of a permanent *cap* of 8.9 million tons on such emissions.

2. In a major departure from the command-and-control approach, Title IV promotes a free-market approach to regulation. Each plant is granted **emission allowances** based on formulas in the legislation. The plants are free to choose how they will achieve their allowances.

3. In the future, new utilities will not receive allowances. Instead, they will need to buy into the system by purchasing existing allowances. Thus, there will be a finite number of allowances.

4. Nitrogen oxide emissions from power plants were to be reduced by 2 million tons by 2001. This was to be accomplished by regulating the boilers used by the utilities and by mandating the continuous monitoring of emissions.

Accomplishments of Title IV. The utilities industry has responded to the new law with three actions:

1. Many utilities are switching to low-sulfur coal, available in Appalachia and the western United States.

2. Many older power plants are adding scrubbers—"liquid filters" that put exhaust fumes through a spray of water containing lime. SO_2 reacts with the lime and is precipitated as calcium sulfate ($CaSO_4$). Since the technology for high-efficiency scrubbing is well established, more and more power plants are installing scrubbers, which have been required for all coal-burning power plants built after 1977.

3. Many utilities are trading their emission allowances. The purchase of allowances often represents a less costly way to achieve compliance than by purchasing low-sulfur coal or adding a scrubber. A typical transaction might involve a trade in the rights to emit, say, 10,000 tons of sulfur dioxide at a cost of $200 a ton. The combination of approaches has created so many efficiencies that it has cut compliance costs to 10% of what was expected.

To carry out Title IV of the CAAA, the EPA initiated a two-phase approach. Phase I addressed 398 of the largest coal-fired power plants, aiming at a reduction of 3.5 million tons of SO_2 by 2000. Remarkably, the goal was met and even exceeded—at a cost far below the gloomy industry predictions (Fig. 19–27). Phase II began in 2000 and targets the remaining sources of SO_2 in order to reach the 8.9-million-ton cap by 2010. This goal was met for the first time in 2007, when emissions totaled 8.9 million tons of SO_2.

The 2001 goal for a 2-million-ton reduction in NO_x emissions by fixed sources was met in 2000 and every year since; in fact, emissions reductions of NO_x (4.8 million tons by 2008) are more than double the Title IV reduction objective. The Acid Rain Reduction Program was responsible for a large portion of these reductions, but the NBP has also played a significant role, as we have seen. However, these sources account for less than half of all NO_x emissions, and

Figure 19–27 SO₂ emissions covered under the Acid Rain Reduction Program. Emissions from all power plants were under allowances in Title IV of the CAAA for the first time in 2006, and even more so in 2007. (*Source:* EPA Office of Air and Radiation, United States–Canada Air Quality Agreement: Progress Report 2008.)

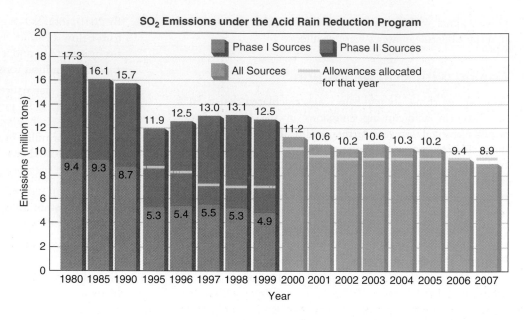

the trend in total NO$_x$ emissions has been a much slower decline than that for SO$_2$.

Field News. The news from the field reflects the trends. Under the Acid Rain Reduction Program, air quality has improved, with significant benefits to human health, reductions in acid deposition, some recovery of freshwater lakes and streams, and improved forest conditions. National ambient SO$_2$ concentrations declined 55% between 1990 and 2007. Concentrations of sulfate in rain and deposition on the land have shown a significant decline (over 30%) over a large part of the eastern United States in the last 18 years. The acidified streams and lakes of the Adirondacks and New England have seen some recovery, in the form of pH increases and increases in acid neutralizing capacity.

One reason for the slow recovery is the continued impact of nitrogen deposition. Apparently, nitrogen plays a much larger role in acid deposition than was once believed. In passing the CAAA, Congress did not set curbs on nitrogen emissions, but simply opted to reduce the emissions from fixed sources. The other reason that slow progress has been made in recovery is the long-term buildup of sulfur deposits in soils. Residual sulfur takes a long time to flush from natural ecosystems. Accordingly, most scientists believe that the recovery of the affected ecosystems will require further reductions in both sulfur and nitrogen emissions. Acid rain still falls on eastern forests, as Figure 19–15 indicates. The 2005 *National Acid Precipitation Assessment Program Report to Congress*, and many other studies, concluded that further reductions in SO$_2$ and NO$_x$ will be needed in order to achieve substantial recovery of acidic lakes and streams and improvement of forest soils. These reductions are in the pipeline, as we have seen.

Costs and Benefits

We have air pollution because we want the goods and services that inevitably generate it. We want to be warm or cool, depending on the weather; we use electricity to power our

homes, institutions, and manufacturing facilities; we crave freedom of movement, so we have our motor vehicles, trains, boats, and planes; and we want the array of goods produced by industries, from food, to DVDs, to computers, to toys—the list is endless. We embrace all of these because we derive some benefit from them. The benefits may be essential or trivial, but we are accustomed to having them, as long as we can pay for them. Satisfying all of these needs and wants is what drives the American economy—and it produces millions of tons of pollutants in the process.

In recent decades, we have been learning that air pollution does not have to get worse—that in fact it can be remedied enough to keep most of us from getting sick and the environment from being degraded. But the remedies come with a price. Without question, measures taken to reduce air pollution carry an economic cost. Some critics have charged that air-pollution controls are not cost effective; that is, the benefits are not nearly as great as the costs. They see lost opportunities for economic growth and tend to disregard the *costs avoided* (from improved health).

The Acid Rain Reduction Program has recently been examined from a benefit-cost perspective.[1] The estimated value of the program's benefits in the year 2010 totals $122 billion, as opposed to the estimated 2010 costs of $3 billion. The costs have turned out to be far below early estimates. The benefit-cost ratio speaks for itself. Benefits extend well beyond the acid deposition issues because the reductions in NO$_x$ and SO$_2$ reduce the levels of PM$_{2.5}$ particulates and ozone. The study points out that many other benefits, such as improvements in forest health and surface waters, were not quantified.

Acid rain is just one of many problems brought on by our involvement in major technologies, and although decades of inaction passed before it was addressed by public policy,

---

[1] Lauraine G. Chestnut and David M. Mill, "A Fresh Look at the Benefits and Costs of the U.S. Acid Rain Program," *Journal of Environmental Management* 77 (2005): 252–266.

TABLE 19–4   Key Findings of the 2006 Scientific Assessment of Ozone Depletion

1. The abundances of anthropogenic ozone-depleting gases in the troposphere and the stratosphere show a downward trend from their peak values in the 1990s.

2. Springtime polar ozone depletion continues to be severe in cold stratospheric winters. Large Antarctic holes continue, and large Arctic ozone depletion occurs during colder Arctic winters. Antarctic ozone levels are projected to return to normal levels around 2060–2075.

3. Ozone abundances over regions outside of the poles no longer continue to decline, in spite of contributions of polar ozone depletion to extrapolar regions. It is likely that global ozone will return to normal levels around the middle of the 21st century.

4. Our basic understanding that anthropogenic ozone-depleting substances have been the principle cause of the ozone depletion over the past decade has been strengthened.

5. The Montreal Protocol is working: There is clear evidence of a decrease in the atmospheric burden of ozone-depleting substances and some early signs of stratospheric ozone recovery.

6. Failure to comply with the Montreal Protocol would delay, or could even prevent, recovery of the ozone layer.

Source: World Meteorological Organization, *Scientific Assessment of Ozone Depletion: 2006,* Global Ozone Research and Monitoring Project report no. 50 (Geneva, Switzerland: World Meteorological Organization, 2007).

the CAAA has made real improvements in the troposphere, ecosystems suffering from acidification, and human health.

We turn now to yet another problem of unintended consequences, where the use of some apparently harmless chemicals led to destruction of the stratospheric ozone that protects Earth from harmful ultraviolet radiation.

19.5 Destruction of the Ozone Layer

The stratospheric ozone layer protects Earth from harmful ultraviolet radiation. The depletion of this layer is another major atmospheric challenge that is the result of human

technology. Environmental scientists have traced the problem to a widely used group of chemicals: the chlorofluorocarbons, or CFCs. As with global warming, there were skeptics who were unconvinced that the problem was serious. However, scientists have achieved a strong consensus on the ozone depletion problem, reflected in the work of the Scientific Assessment Panel of the Montreal Protocol. Table 19–4 summarizes some of the major findings of the panel's 2006 assessment.

Radiation and Importance of the Shield

Solar radiation emits electromagnetic waves with a wide range of energies and wavelengths (Fig. 19–28). Visible light is that part of the electromagnetic spectrum that can be detected by the

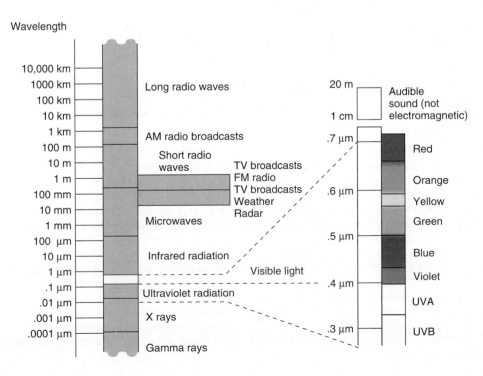

Figure 19–28 The electromagnetic spectrum. Ultraviolet radiation, visible light, infrared radiation, and many other forms of radiation are at different wavelengths of the electromagnetic spectrum.

Figure 19–29 Ozone in the atmosphere. The vertical distribution of ozone in the troposphere (green) and stratosphere (red), showing the location of the stratospheric ozone layer that absorbs 99% of UV radiation. (*Source:* Synthesis and Assessment Product 2.4. Report by the U.S. Climate Change Science Program and the Subcommittee on Global Change Research. November, 2008.)

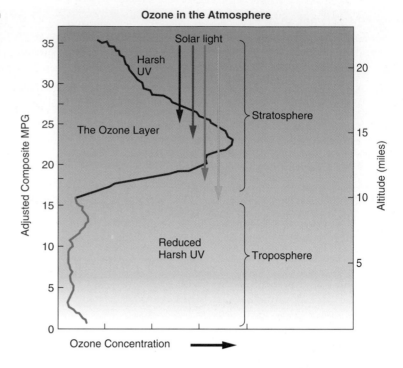

eye's photoreceptors. Ultraviolet (UV) wavelengths are slightly shorter than the wavelengths of violet light, which are the shortest wavelengths visible to the human eye. UVB radiation consists of wavelengths that range from 280 to 315 nanometers (0.28 μm to 0.32 μm), whereas UVA radiation is from 315 to 400 nanometers (0.32 μm to 0.40 μm). Since energy is inversely related to wavelength, UVB is more energetic and, therefore, more dangerous, but UVA can also cause damage.

Upon penetrating the atmosphere and being absorbed by biological tissues, UV radiation damages protein and DNA molecules at the surfaces of all living things. (This damage is what occurs when you get a sunburn.) If the full amount of

UV radiation falling on the stratosphere reached Earth's surface, it is doubtful that any life could survive. We are spared the more damaging effects from UV rays because most UV radiation (more than 99%) is absorbed by ozone in the stratosphere. For that reason, stratospheric ozone is commonly referred to as the **ozone shield** (Fig. 19–29).

However, even the small amount (less than 1%) of UVB radiation that does reach us is responsible for sunburns and more than 700,000 cases of skin cancer and precancerous ailments per year in North America as well as for untold damage to plant crops and other life-forms. (See Sound Science, "Coping with UV Radiation.")

Sound Science

Coping with UV Radiation

People living in Chile, Australia, and New Zealand are no strangers to the effects of the thinner ozone shield. UV alerts are given in those parts of the Southern Hemisphere during their spring months as lobes of ozone-depleted stratospheric air move outward from the Antarctic. At times, the ozone above those countries can be less than half its normal concentration, which means that greater intensities of UV radiation reach Earth's surface. It is now predicted that one out of every three Australians will develop serious and perhaps fatal skin cancer in his or her lifetime.

Even without a thinner ozone layer, UV radiation is a serious problem. Concerned over the rising incidence of skin cancer and cataract surgery in the United States, the EPA, together with the National Oceanic and Atmospheric Administration and the Centers for Disease Control, has initiated a new *UV Index* (see the table at the end of the essay). The index takes the form of daily forecasts of UV exposure, issued by the Weather Service for 58 cities. Satellite measurements of stratospheric ozone are combined with other weather patterns. The goal of the index is to remind people of the dangers of UV

radiation and to prompt them to take appropriate action to avoid cancer, premature aging of the skin, eye damage, cataracts, and blindness. For at least the next two decades, the dangers will be quite noticeable, ranging from a greater incidence of sunburn to a likely epidemic of malignant melanoma.

Less obvious, but no less important, are the chronic effects of exposure to the Sun. The normal aging of skin—coarse wrinkling, yellowing, and the development of irregular patches of heavily pigmented and unpigmented skin—is now known to be largely the result of damage from the Sun. This can happen even in the absence of episodes of sunburn. Further, a large proportion of the million cataract operations performed annually in the United States can be traced to UV exposure, another of the chronic effects of this kind of radiation.

The most serious impact of chronic exposure to the Sun is skin cancer; over 1 million new cases occur each year in the United States, according to the National Cancer Institute. Three types of skin cancer are traced to UV exposure: basal cell carcinoma (BCC), squamous cell carcinoma (SCC), and melanoma. Most of the skin cancers (75%) are BCCs, slow growing and therefore treatable. SCC accounts for about 20% of skin cancers and can also be cured if treated early. However, it metastasizes (spreads away from the source) more readily than BCC and thus is potentially fatal. Melanoma (6%) is the most deadly form of skin cancer, as it metastasizes easily. Melanoma is often traced to occasional sunburns during childhood or adolescence or to sunburns in people who normally stay out of the sun.

What is an appropriate response to this disturbing information? First, know when UV intensity is greatest (three hours on either side of midday), and take precautions accordingly. Protect your eyes with sunglasses and a hat, and apply sunscreen with a sun protection factor (SPF) rating of 15 or higher during such times. Think twice before considering sunbathing or using tanning beds, and never proceed without sunscreen that is strong enough to prevent sunburn. Always protect children with sunscreen. Their years of potential exposure are many, and their skin is easily burned. And if you detect a patch of skin or a mole that is changing in size or color or that is red and fails to heal, consult a dermatologist for treatment.

UV Index (EPA)

Exposure Category	Index Value	Minutes to Burn for "Never Tans" (Most Susceptible)	Minutes to Burn for "Rarely Burns" (Least Susceptible)
Minimal	0–2	35	>120
Low	3–4	20	>120
Moderate	5–6	12	90
High	8–9	8	55
Very high	10–12	6	40

Formation and Breakdown of the Shield

Ozone is formed in the stratosphere when UV radiation acts on oxygen (O_2) molecules. The high-energy UV radiation first causes some molecular oxygen (O_2) to split apart into free oxygen (O) atoms, and these atoms then combine with molecular oxygen to form ozone via the following reactions:

$$O_2 + UVB \rightarrow O + O \qquad (1)$$

$$O + O_2 \rightarrow O_3 \qquad (2)$$

Not all of the molecular oxygen is converted to ozone, however, because free oxygen atoms may also combine with ozone molecules to form two oxygen molecules in the following reaction:

$$O + O_3 \rightarrow O_2 + O_2 \qquad (3)$$

Finally, when ozone absorbs UVB, it is converted back to free oxygen and molecular oxygen:

$$O_3 + UVB \rightarrow O + O_2 \qquad (4)$$

Thus, the amount of ozone in the stratosphere is dynamic. There is an equilibrium due to the continual cycle of reactions of formation (Eqs. 1 and 2) and reactions of destruction (Eqs. 3 and 4). Because of seasonal changes in solar radiation, ozone concentration in the Northern Hemisphere is highest in summer and lowest in winter. Also, in general, ozone concentrations are highest at the equator and diminish as latitude increases—again, a function of higher overall amounts of solar radiation. However, the presence of other chemicals in the stratosphere can upset the normal ozone equilibrium and promote undesirable reactions there.

Halogens in the Atmosphere. Chlorofluorocarbons (**CFCs**) are a type of halogenated hydrocarbon (see Chapter 22). CFCs are nonreactive, nonflammable, nontoxic organic molecules in which both chlorine and fluorine atoms have replaced some hydrogen atoms. At room temperature, CFCs are gases under normal (atmospheric) pressure, but they liquefy under modest pressure, giving off heat in the process and becoming cold. When they revaporize, they reabsorb the heat and become hot. These attributes led to the widespread use of CFCs (over 1 million tons per year in the 1980s) for the following applications:

- In refrigerators, air conditioners, and heat pumps as the heat-transfer fluid;
- In the production of plastic foams;
- By the electronics industry for cleaning computer parts, which must be meticulously purified; and
- As the pressurizing agent in aerosol cans.

Rowland and Molina. All of the preceding uses led to the release of CFCs into the atmosphere, where they mixed with the normal atmospheric gases and eventually reached the stratosphere. In 1974, chemists Sherwood Rowland and Mario Molina published a classic paper[2] (for which they

---

[2] M. J. Molina and F. S. Rowland, "Stratospheric Sink for Chlorofluoro-methanes: Chlorine-atom Catalyzed Distribution of Ozone," *Nature* 249 (1974): 810–812.

were awarded the Nobel Prize in chemistry in 1995) concluding that CFCs could damage the stratospheric ozone layer through the release of chlorine atoms and, as a result, UV radiation would increase and cause more skin cancer. Their story is described in Chapter 2 to illustrate the public-policy life cycle. Rowland and Molina reasoned that, although CFCs would be stable in the troposphere (they have been found to last 70 to 110 years there), in the stratosphere they would be subjected to intense UV radiation, which would break them apart, releasing free chlorine atoms via the following reaction:

$$CFCl_3 + UV \quad Cl + CFCl_2 \tag{5}$$

Ultimately, all of the chlorine of a CFC molecule would be released as a result of further photochemical breakdown. The free chlorine atoms would then attack stratospheric ozone to form chlorine monoxide (ClO) and molecular oxygen:

$$Cl + O_3 \quad ClO + O_2 \tag{6}$$

Furthermore, two molecules of chlorine monoxide may react to release more chlorine and an oxygen molecule:

$$ClO + ClO \quad 2Cl + O_2 \tag{7}$$

Reactions 6 and 7 are called the **chlorine catalytic cycle** because chlorine is continuously regenerated as it reacts with ozone. Thus, chlorine acts as a **catalyst**, a chemical that promotes a chemical reaction without itself being used up in the reaction. Because every chlorine atom in the stratosphere can last from 40 to 100 years, it has the potential to break down 100,000 molecules of ozone. Thus, CFCs are judged to be damaging because they act as transport agents that continuously move chlorine atoms into the stratosphere. The damage can persist because the chlorine atoms are removed from the stratosphere only very slowly. Figure 19–30 shows the basic processes of ozone formation and destruction, including recent refinements to our knowledge of those processes that will be explained shortly.

EPA Action. After studying the evidence, the EPA became convinced that CFCs were a threat and, in 1978, banned their use in aerosol cans in the United States. Manufacturers quickly switched to nondamaging substitutes, such as butane, and things were quiet for several years. CFCs continued to be used in applications other than aerosols, however, and skeptics demanded more convincing evidence of their harmfulness.

Atmospheric scientists reasoned that any substance carrying reactive halogens to the stratosphere had the potential to deplete ozone. These substances include halons, methyl chloroform, carbon tetrafluoride, and methyl bromide. Chemically similar to chlorine, bromine also attacks ozone and forms a monoxide (BrO) in a catalytic cycle. Because of its extensive use as a soil fumigant and pesticide, bromine plays a major role in ozone depletion. (Bromine is 60 times as potent as chlorine in ozone destruction.)

The Ozone "Hole." In the fall of 1985, British atmospheric scientists working in Antarctica reported a gaping "hole" (actually, a serious thinning) in the stratospheric ozone layer over the South Pole. There, in an area the size of the United States, ozone levels were 50% lower than normal. The hole would have been discovered earlier by NASA satellites monitoring ozone levels except that computers were programmed to reject data showing a drop as large as 30% as being due to instrument anomalies. Scientists had assumed that the loss of ozone, if it occurred, would be slow, gradual, and uniform over the whole planet. The ozone hole came as a surprise, and if it had occurred anywhere but over the South Pole, the UV damage would have been extensive. As it is, the limited time and area of ozone depletion there have not apparently brought on any catastrophic ecological events so far.

News of the ozone hole stimulated an enormous scientific research effort. A unique set of conditions was found to be responsible for the hole. In the summer, gases such as nitrogen dioxide and methane react with chlorine monoxide and chlorine to trap the chlorine, forming so-called **chlorine reservoirs** (Fig. 19–30), preventing much ozone depletion.

Polar Vortex. When the Antarctic winter arrives in June, it creates a vortex (like a whirlpool) in the stratosphere, which confines stratospheric gases within a ring of air circulating around the Antarctic. The extremely cold temperatures of the Antarctic winter cause the small amounts of moisture and other chemicals present in the stratosphere to form the south polar stratospheric clouds. During winter, the cloud particles provide surfaces on which chemical reactions release molecular chlorine (Cl_2) from the chlorine reservoirs.

When sunlight returns to the Antarctic in the spring, the Sun's warmth breaks up the clouds. UV light then attacks the molecular chlorine, releasing free chlorine and initiating the chlorine cycle, which rapidly destroys ozone. By November, the beginning of the Antarctic summer, the vortex breaks down, and ozone-rich air returns to the area. However, by that time, ozone-poor air has spread all over the Southern Hemisphere. Shifting patches of ozone-depleted air have caused UV radiation to increase to 20% above normal in Australia. Television stations there now report daily UV readings and warnings for Australians to stay out of the Sun. On the basis of current data, estimates indicate that in Queensland, where the ozone shield is thinnest, two out of three Australians are expected to develop skin cancer by the age of 70. The ozone hole intensified during the 1990s and has leveled off between 22 and 27 million square kilometers, an area as large as North America (Fig. 19–31).

Because severe ozone depletion occurs under polar conditions, observers have been keeping a close watch on the Arctic. To date, no hole has developed. The higher temperatures and weaker vortex formation there are expected to prevent the severe losses that have become routine in the Antarctic. However, ozone depletion does occur in the Arctic—up to a 25% loss in especially cold winters.

Further Ozone Depletion. Ozone losses have not been confined to Earth's polar regions, although they are most spectacular there. A worldwide network of ozone-measuring stations sends data to the World Ozone Data Center in

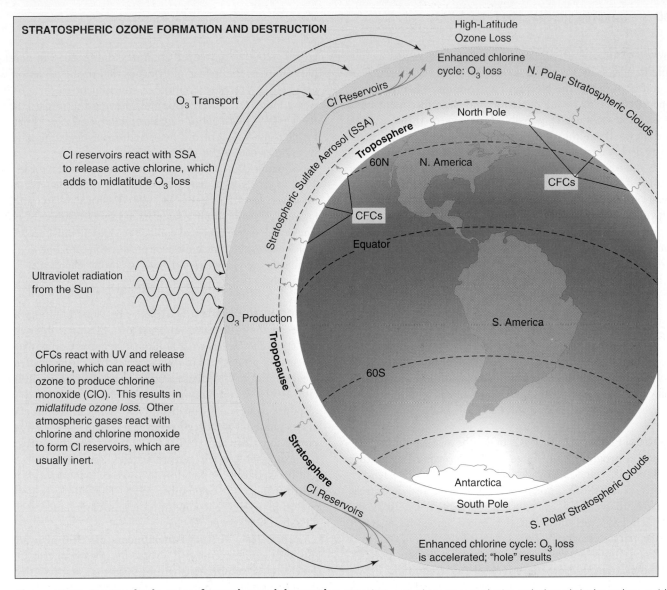

STRATOSPHERIC OZONE FORMATION AND DESTRUCTION

High-Latitude Ozone Loss

Enhanced chlorine cycle: O₃ loss

N. Polar Stratospheric Clouds

O₃ Transport

Cl Reservoirs

North Pole

Cl reservoirs react with SSA to release active chlorine, which adds to midlatitude O₃ loss

Stratospheric Sulfate Aerosol (SSA)

Troposphere

60N   N. America

CFCs

CFCs

Equator

Ultraviolet radiation from the Sun

O₃ Production

S. America

CFCs react with UV and release chlorine, which can react with ozone to produce chlorine monoxide (ClO). This results in *midlatitude ozone loss*. Other atmospheric gases react with chlorine and chlorine monoxide to form Cl reservoirs, which are usually inert.

Tropopause

60S

Stratosphere

Cl Reservoirs

Antarctica

South Pole

S. Polar Stratospheric Clouds

Enhanced chlorine cycle: O₃ loss is accelerated; "hole" results

Figure 19–30 Stratospheric ozone formation and destruction. UV radiation stimulates ozone production at the lower latitudes, and ozone-rich air migrates to high latitudes. At the same time, CFCs and other compounds carry halogens into the stratosphere, where they are broken down by UV radiation, to release chlorine and bromine. Ozone is subject to high-latitude loss during winter, as the chlorine cycle is enhanced by the polar stratospheric clouds. Midlatitude losses occur as chlorine reservoirs are stimulated to release chlorine by reacting with stratospheric sulfate aerosol.

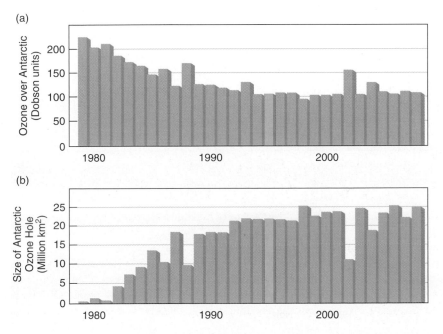

(a)

(b)

Figure 19–31 Ozone loss and extent of ozone hole. (a) The decline in ozone from 1980 to 2008 is shown in Dobson units for September 21 to October 16 each year (the average amount of ozone in the stratosphere is 300 Dobson units). (b) The size of the ozone hole is plotted for the same time frame, showing the origin of the hole and its leveling off in the last few years. No data were acquired in 1995, shown as a gray column in both plots. (*Source:* NASA Ozone Hole Watch.)

Toronto, Canada. Reports from the center reveal ozone depletion levels of 3% and 6% over the period 2002–2007 in the midlatitudes of the Northern and Southern hemispheres, respectively. Polar ozone depletion and the subsequent mixing of air masses are responsible for much of this loss. The good news from the network is that ozone levels are no longer decreasing.

Is the ozone loss significant to our future? The EPA has calculated that the ozone losses of the 1980s will eventually have caused 12 million people in the United States to develop skin cancers over their lifetime and that 93,000 of these cancers will be fatal. Americans are estimated to have developed more than 900,000 new cases of skin cancer a year in the 1990s. The ozone losses of that decade allowed more UVB radiation than ever to reach Earth. Studies have confirmed increased UVB levels, especially at higher latitudes.

Coming to Grips with Ozone Depletion

The dramatic growth of the hole in the ozone layer (Fig. 19–31) has galvanized a response around the world. Scientists and politicians in the United States and other countries have achieved treaties designed to avert a UV disaster.

Montreal Protocol. In 1987, under the auspices of its environmental program, the United Nations convened a meeting in Montreal, Canada, to address ozone depletion. Member nations reached an agreement, known as the **Montreal Protocol,** to scale CFC production back 50% by 2000. To date, 194 countries (including the United States) have signed the original agreement.

The Montreal Protocol was written even before CFCs were so clearly implicated in driving the destruction of ozone and before the threat to Arctic and temperate-zone ozone was recognized. Because ozone losses during the late 1980s were greater than expected, an amendment to the protocol was adopted in June 1990. The amendment required participating nations to phase out the major chemicals destroying the ozone

layer by 2000 in developed countries and by 2010 in developing countries. In the face of evidence that ozone depletion was accelerating even more, another amendment to the protocol was adopted in November 1992, moving the *target date for the complete phaseout of CFCs to January 1, 1996.* Timetables for phasing out all of the suspected ozone-depleting halogens were shortened at the 1992 meeting.

Quantities of CFCs were still being manufactured after 1992 to satisfy legitimate demand in the developing countries. However, according to the Montreal Protocol, all manufacturing of CFCs was to have been halted by December 31, 2005. The results of the protocol can be seen clearly in Figure 19–32, which shows the time course of production of ozone-depleting chemicals and the abundance of those chemicals in the stratosphere.

Action in the United States. The United States was the leader in the production and use of CFCs and other ozone-depleting chemicals, with DuPont Chemical Company being the major producer. Following years of resistance, DuPont pledged in 1988 to phase out CFC production by 2000. In late 1991, a company spokesperson announced that, in response to new data on ozone loss, DuPont would accelerate its phaseout by three to five years. Many of the large corporate users of CFCs (AT&T, IBM, and Northern Telecom, for example) phased out their CFC use by 1994. Also, DuPont spoke in opposition to three bills introduced in the 104th Congress in September 1995. The bills were designed to terminate U.S. participation in, and compliance with, the CFC-banning protocols.

The **Clean Air Act Amendments of 1990** also address this problem, in Title VI, Protecting Stratospheric Ozone. Title VI is a comprehensive program that restricts the production, use, emissions, and disposal of an entire family of chemicals identified as ozone depleting. For example, the program calls for a phaseout schedule for the hydrochlorofluorocarbons (HCFCs), a family of chemicals being used as less damaging substitutes for CFCs until nonchlorine

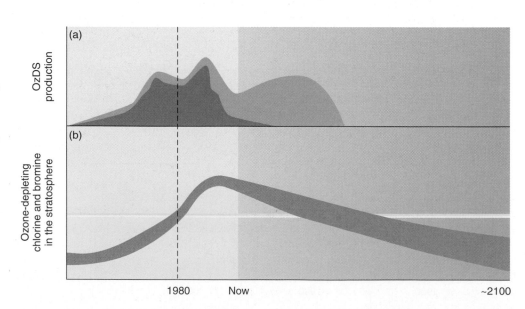

Figure 19–32 Production and presence of ozone-depleting substances in the atmosphere: past, present, and future. (a) Production of ozone-depleting substances (ODS); CFCs are shown in blue, and HCFCs, being used to replace CFCs, are shown in green. (b) Relative abundances of ozone-depleting chlorine and bromine in the stratosphere. The impact of the Montreal Protocol and its amendments is clear. (*Source:* UN Environment Program, Scientific Assessment of Ozone Depletion: 2006.)

substitutes are available. Halons—used in chemical fire extinguishers—were banned in 1994. The CAAA also regulate the servicing of refrigeration and air-conditioning units.

The Montreal Protocol had its 20th anniversary in 2007. As a result of this and subsequent agreements, CFCs are no longer being produced or used. Substitutes for CFCs are readily available and, in some cases, are even less expensive than CFCs. The most commonly used substitutes are HCFCs, which still contain some chlorine and are scheduled for a gradual phaseout. The most *promising* substitutes are hydrofluorocarbons (HFCs), which contain no chlorine and are judged to have no ozone-depleting potential. We saw in Chapter 18 that CFCs are also potent greenhouse gases, equivalent to 15% of the contribution from the major greenhouse gases. Consequently, phasing out CFCs and other ozone-depleting substances is helping the efforts to slow down global climate change.

Methyl bromide continues to be produced and released into the atmosphere; the compound is employed as a soil fumigant to control agricultural pests. Under the Montreal Protocol, it was scheduled to be completely phased out by 2005 in the developed countries and 10 years later in the developing countries. In the meantime, bromine is likely to play a greater role in ozone depletion relative to chlorine. Responding to requests from the agricultural industry, the United States has repeatedly received "critical use" exemptions to the ban for use of some 15 million pounds of the ozone-depleting chemical. Many conscientious growers have found alternatives, although they may be more costly.

Final Thoughts. The ozone story is a remarkable episode in human history. From the first warnings in 1974 that something might be amiss in the stratosphere because of a practically inert and highly useful industrial chemical, through the development of the Montreal Protocol and the final steps of CFC phaseout, the world has shown that it can respond collectively and effectively to a clearly perceived threat. The ozone layer is expected to recover completely by midcentury in the midlatitudes and a couple of decades later in the polar regions. The scientific community has played a crucial role in this episode, first alerting the world and then plunging into intense research programs to ascertain the validity of the threat. Scientists continue to influence the political process that has forged a response to the threat. Although skeptics were stressing the uncertainties in our understanding of ozone loss and its consequences, the strong consensus in the scientific community convinced the world's political leaders that action was clearly needed. This is a most encouraging development as we look forward to the actions that must be taken during the 21st century to prevent catastrophic global climate change.

Revisiting the Themes

Sound Science

Atmospheric science has come a long way in the last half century. Some of its triumphs include understanding causes and effects that are operative in photochemical smog, discovering the cleansing role of the hydroxyl radical, tracing the sources of lead and mercury in humans, working out causes and effects in the acid-rain story, and alerting the world to ozone shield destruction. Sound science is increasingly important in measuring ambient pollutant levels and in devising technologies to prevent more pollution. Many scientists have put their reputations and careers on the line as they have fought for policy changes to correct the human and environmental damage they have uncovered.

seriously, and the ozone shield being destroyed. Less obvious, but just as important, chronic exposure to air pollutants was responsible for poor health and increased mortality for millions. Are we back on a more sustainable track now? The important indicators would be long-term declines in lead in the environment, annual amounts of emissions of pollutants, acid and sulfate deposition in the Northeast, ambient levels of pollutants, asthma hospitalizations, and locations that are noncompliant. Most of these parameters are moving in the right direction, but much remains to be done. The ozone story represents a sustainable pathway. Levels of ozone-destroying chemicals are now declining, and it is very likely that the ozone layer will be "healthy" by midcentury.

Sustainability

Air pollution was on an unsustainable track, with industrial smog producing health disasters, photochemical smog making city living intolerable for many, crops and natural ecosystems deteriorating

Stewardship

Stewardship means doing the right thing when faced with difficult problems and choices. It is fair to say that our society has responded well in caring for children threatened

(*continued*)

(continued)

by high lead levels in their blood and, in general, in taking the costly actions we have documented in this chapter. Further, in addressing acid deposition, we have even shown care for the natural world. The need for stewardship is still very strong, however, and thankfully, there are many in nongovernmental and other organizations, and indeed many in the EPA, that are determined to stay the course and make sure that effective action is taken. Often, though, they have to do battle with industry lobbies and special interests that are concerned with the economic bottom line and that seem unable to embrace the common good represented by effective action to curb pollution.

REVIEW QUESTIONS

1. What naturally occurring cleanser helps to remove pollutants from the atmosphere? Where does it come from? What other mechanisms also act to cleanse the atmosphere of pollutants?

2. Describe the origins of industrial smog, photochemical smog, and atmospheric brown clouds. What are the differences in the cause and appearance of each?

3. What are the major primary air pollutants and their sources?

4. What are secondary pollutants, and how are they formed?

5. Distinguish between emissions and ambient concentrations. How are they measured?

6. What is the difference between an acid and a base? What is the pH scale?

7. What two major acids are involved in acid deposition? Where does each come from?

8. What impacts does air pollution have on human health? Give the three categories of impact and distinguish among them.

9. Describe the negative effects of pollutants on crops, forests, and other materials. Which pollutants are mainly responsible for these effects?

10. How can a shift in environmental pH affect aquatic ecosystems? In what other ecosystems can acid deposition be observed? What are its effects?

11. What are the criteria pollutants and the National Ambient Air Quality Standards, and how are they used?

12. Discuss ways in which the Clean Air Act Amendments of 1990 address the failures of previous legislation.

13. What are the major steps being taken to reduce emissions from passenger vehicles?

14. How do the NO_x Budget Trading Program and CAIR address ozone pollution?

15. What has been happening in the United States to reduce acid deposition, and how successful has it been?

16. How is the protective ozone shield formed? What causes its natural breakdown?

17. How do CFCs affect the concentration of ozone in the stratosphere and contribute to the formation of the ozone hole?

18. Describe the international efforts that are currently in place to protect our ozone shield. What evidence is there that such efforts have been effective?

THINKING ENVIRONMENTALLY

1. Motor vehicles release close to half the pollutants that dirty our air. What alternatives might be introduced to encourage a decrease in our use of automobiles and other vehicles with internal combustion engines?

2. How would our pattern of life be different in the absence of the Clean Air Act and its amendments? Write a short essay describing the possibilities.

3. The wind blows across Country A and into Country B. Country A has fossil-fueling-burning power plants, but a relatively low level of air pollution. Country B has power plants run by moving water (hydroelectric plants), but a higher level of pollution. What's happening? What could the two countries do to correct the situation? Explain how this pertains to international air-pollution laws.

4. A large amount of air pollution is attributed to manufacturing. Some would argue that it is not technologically possible, or economically efficient, to eliminate this pollution. How would you respond to this argument?

5. What arguments might the auto and utility industries present to delay regulatory action on particulates and ozone? How would health officials respond to their arguments?

MAKING A DIFFERENCE

1. At all costs, avoid burning trash in your yard. It's illegal in many places and can lead to birth defects, cancers, and lung diseases.

2. Check your home heating system. Simple wood smoke contains fine particles that are associated with serious negative health effects. The tiny size of these pollutants allows them to be easily inhaled, traveling deep into your lungs where they can cause respiratory and cardiovascular problems, including premature death. Wood stoves, fireplaces, and other wood-burning devices put out hundreds of times more air pollution than other sources of heat such as natural gas, propane, oil, and electricity.

3. Discuss air-pollution issues with your employer. Many employers are realizing the benefits of considering air quality issues. Green travel plans, alternative-powered vehicles, and the provision of cycling facilities are just three ways employers can help to reduce their impacts on air quality, realize a financial saving, and maintain a healthy workforce.

4. If you live in an urban area, take advantage of public transportation. It helps ease urban congestion as well as atmospheric conditions in the area.

5. When you buy a new car, consider purchasing a hybrid electric vehicle.

Go to **www.masteringenvironmentalscience.com** for practice quizzes, Pearson eText, videos, current events, and more.

CHAPTER 20

Water Pollution and Its Prevention

The Mississippi Delta. The Mississippi River Basin consists of 40% of the landmass of the lower 48 states. Runoff from this huge area delivers 500 million tons of sediment and over 1 million tons of nutrients to the Gulf of Mexico yearly. Below: A type of phytoplankton, viewed through a microscope (SEM). (*Source:* Earth Observatory, NASA.)

KEY TOPICS

1. **Perspectives on Water Pollution:** The most important pollutants are disease organisms, organic wastes, chemical pollutants, sediments, and nutrients, all of which have objectionable impacts on human health and the environment.

2. **Wastewater Treatment and Management:** Cities and towns employ a system of sewers and wastewater-treatment plants to deal with human and domestic wastes.

3. **Eutrophication:** Aquatic ecosystems undergo undesirable changes when nutrients are introduced, including heavy algal growth and the creation of anaerobic conditions in deeper lake and bay waters.

4. **Public Policy and Water Pollution:** The Clean Water Act and its amendments give the Environmental Protection Agency jurisdiction over water-pollution issues, though the states are responsible for managing pollution and wastewater programs.

THE MISSISSIPPI RIVER WATERSHED ENCOMPASSES 40% of the land area of the United States; the river collects water from this huge area and eventually delivers it to the Gulf of Mexico. Because much of the watershed is America's agricultural heartland, the Mississippi River is a reflection of what happens on the nation's farms. When U.S. farm production experienced tremendous growth during the latter half of the 20th century, more and more fertilizers were spread on agricultural fields, and more animals were raised on feedlots. At the same time, wetlands bordering river tributaries were drained and no longer intercepted agricultural runoff. The result was a tripling of soluble nitrogen delivered to the river and, eventually, to the Gulf of Mexico.

DEAD ZONE. The impact of this major flow of nitrogen to the Gulf of Mexico was first detected in 1974 by marine scientists who found areas in the gulf where oxygen had disappeared from bottom sediments and much of the water column above them. This absence of oxygen is deadly to the bottom-dwelling animals and any oxygen-breathing creatures in the water column that are not able to escape. At first, the hypoxic (lacking oxygen) area—or

Figure 20–1 The 2008 dead zone in the Gulf of Mexico. Each year oxygen disappears from middle depths to the bottom over thousands of square miles of the Gulf of Mexico. In 2008, the dead zone was 8,000 square miles, an area the size of New Jersey. The black line encloses the hypoxic area, where oxygen is less than 2 mg/L (ppm). (*Source:* N. Rabalais, Louisiana Universities Marine Consortium, 2008.)

dead zone, as it is called—was thought to be similar to those in other bodies of water, such as the lower Chesapeake Bay, that receive high levels of nutrients—a minor disturbance that would disappear seasonally. But then the hypoxic area suddenly doubled in size following the 1993 floods in the Midwest, and it continues to grow (Fig. 20–1). It averaged over 6,700 square miles from 2004 to 2009. The shrimpers and fishers who work these waters have learned to avoid the hypoxic area.

THE CULPRIT. To the scientists working on the problem, the connection between nitrogen and hypoxia was clear. Nitrogen is a common limiting factor in coastal marine waters. Where it is abundant, it promotes the dense growth of phytoplankton—photosynthetic microorganisms that live freely suspended in the water. Zooplankton—microscopic animals in the water—rapidly consume the phytoplankton and multiply. As the abundant phytoplankton and zooplankton (and their fecal pellets) die or sink toward the bottom, they are decomposed by bacteria, a process that consumes dissolved oxygen. Eventually, the oxygen is used up, creating the dead zone. In the Gulf of Mexico, the zone can extend up from the bottom of the sea to within a few meters of the surface. Typically, it stays from May through September, finally being dispersed by the mixing of the water column as colder weather sets in.

Because the gulf's fishery is a $2.8 billion enterprise, the problem eventually got national attention. In 1998, Congress passed the Harmful Algal Bloom and Hypoxia Research and Control Act, establishing an interagency task force of scientists charged with assessing the dead zone in the Gulf of Mexico, as well as harmful algal blooms, another water-pollution problem discussed later in the chapter (see Sound Science, "Harmful Algal Blooms"). The task force released its first report in May 2000, confirming the connection between nitrogen and the dead zone and laying out options for reducing the nitrogen

load coming downriver. Two strategies were key to mitigation of the problem: (1) using less fertilizer, which would greatly reduce the farm runoff and lessen the nitrogen loads of streams and rivers in the Mississippi basin, and (2) restoring and promoting nitrogen retention and denitrification processes in the basin.

The Environmental Protection Agency (EPA) and the National Oceanic and Atmospheric Administration (NOAA) proposed an action plan in 2001 with the goal of reducing the size of the hypoxic area by half by 2015. The plan was voluntary and depended on conservation funding from the U.S. Department of Agriculture. Seven years into the plan, the dead zone had not shown any signs of shrinking; in fact, the 2008 dead zone was the size of the state of New Jersey and only slightly smaller than the record size. The latest action plan recommends a reduction target of 45% of the nitrogen and phosphorus and acknowledges that phosphorus also contributes to the hypoxia. Since there is no specific funding for this work, the task force is coordinating funding from numerous state and federal agencies and conservation programs.

The Gulf of Mexico dead zone turns out to be one more costly lesson about **eutrophication** (enrichment by nutrients): It is unreasonable to allow vital water resources to receive pollutants and then expect those resources to continue to provide us with their usual bounty of goods and services. Since the 1960s, the number of coastal dead zones has doubled every decade; worldwide, over 400 are now known, affecting a total area of over 95,000 square miles.

In this chapter, we will focus on water *quality* (Chapter 10 dealt with water *quantity*—the global water cycle and water resources). Eutrophication and other forms of water pollution are discussed, and we take a close look at wastewater treatment. The following two chapters focus on pollution by solid and hazardous wastes.

20.1 Perspectives on Water Pollution

In the early days of the Industrial Revolution in the United States, little attention was given to the pollution of lakes, rivers, and the coastal ocean. Industries emptied their wastes into the closest waterway, and the general means of disposing of human excrement was the outdoor privy (or behind the nearest bush). Seepage from the privy frequently contaminated drinking water and caused disease, especially in places where privies and wells were located near one another. In the late 1800s, Louis Pasteur and other scientists showed that sewage-borne bacteria were responsible for many infectious diseases. This important discovery led to intensive efforts to rid cities of human excrement as expediently as possible. Cities already had drain systems for storm water, but using these systems for human wastes had been prohibited. With the urgency of the situation, however, minds quickly changed. The flush toilet was introduced, and sewers were tapped into storm drains. Thus, Western civilization adopted the one-way flow of flushing domestic and industrial wastewater into natural waterways. Introducing sewage into a waterway was symptomatic of an approach that believed "the solution to pollution is dilution."

As a result of this practice, receiving waters that had limited capacity for dilution became open cesspools of foul odors, vermin, and filth as the overload of organic matter depleted dissolved oxygen and most aquatic life suffocated. For increasing distances around or downstream from the sewage outfall, the water became unfit for any recreational use because of the sewage contamination. Any further use of the water for human consumption required extensive treatment to remove pathogens.

Industrial wastes only added insult to injury. From the dawn of the Industrial Age to the relatively recent past, it was common practice to flush all waste liquids and contaminated water into sewer systems or directly into natural waterways. Many human health problems followed, but they either were not recognized as being caused by the pollution or were accepted as the "price of progress."

The **Federal Water Pollution Control Act of 1948** was the first federal foray into water pollution, providing technical assistance to state and local governments, but otherwise leaving things up to the states and local municipalities. In the 1950s, as industrial production expanded and synthetic organics came into widespread use in the developed countries, many streams and rivers essentially became open chemical sewers as well as sewers for human waste. These waters not only were devoid of life, but also were themselves hazardous. Finally, in 1969, the Cuyahoga River, which flows through Cleveland, Ohio, was carrying so much flammable material that it actually caught fire and destroyed seven bridges before the fire burned itself out (Fig. 20–2).

Worsening pollution (from both chemicals and sewage) and increasing recognition of the adverse health effects finally created a degree of public outrage that pushed Congress to pass the **Clean Water Act of 1972 (CWA)**. This legislation gave the newly created Environmental Protection Agency (EPA) the responsibility to carry out a very ambitious goal—namely, to restore and maintain the chemical, physical, and biological integrity of the nation's waters. The CWA was to achieve zero discharge of pollutants by 1985, a totally naïve (and impossible) task, given the nature of the pollutants involved. However, imperfect as it is, the Clean Water Act (and its amendments) is still recognized as one of the most far-reaching and effective environmental laws ever enacted.

Figure 20–2 Cuyahoga River on fire. In 1969, the Cuyahoga River in Cleveland, Ohio, caught fire, burning seven bridges. Flammable material from industrial sources was being discharged into the river. This incident captured national attention and contributed to the public outrage over water pollution that led to the Clean Water Act of 1972.

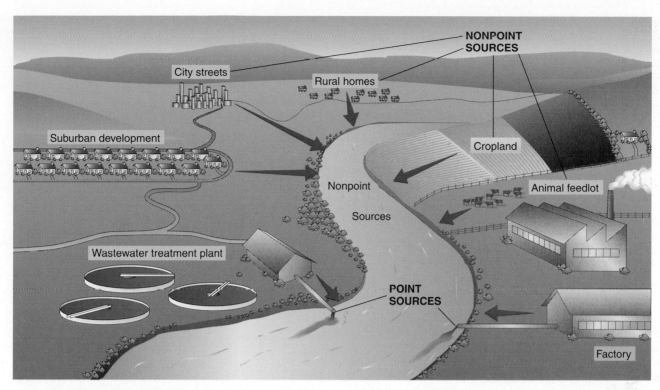

Figure 20–3 Point and nonpoint sources. Point sources are far easier to identify and correct than the diffuse nonpoint sources.

Water Pollution: Sources and Types

Water pollutants originate from a host of human activities and reach surface water or groundwater through an equally diverse host of pathways. For purposes of regulation, it is customary to distinguish between **point sources** and **nonpoint sources** of pollutants (Fig. 20–3). Point sources involve the discharge of substances from factories, sewage systems, power plants, underground coal mines, and oil wells. These sources are relatively easy to identify and, therefore, are easier to monitor and regulate than nonpoint sources, which are poorly defined and scattered over broad areas. Some of the most prominent nonpoint sources of pollution are agricultural runoff (from farm animals and croplands), storm-water drainage (from streets, parking lots, and lawns), and atmospheric deposition (from air pollutants washed to or deposited as dry particles on Earth). Pollution occurs as rainfall and snowmelt move over and through the ground, picking up pollutants as they go.

Two basic strategies are employed in attempting to bring water pollution under control: (1) reduce or remove the sources and (2) treat the water before it is released so as to remove pollutants or convert them to harmless forms. Water treatment is the best option for point sources. Source reduction can be employed for both kinds of sources and is the best option for nonpoint sources.

Pathogens. The most serious and widespread water pollutants are the infectious agents that cause sickness and death. (See Chapter 17.) The excrement from humans and other animals infected with certain **pathogens** (disease-causing

bacteria, viruses, and other parasitic organisms) contains large numbers of these organisms or their eggs. Table 20–1 lists the most common waterborne pathogens. Even after symptoms of disease disappear, an infected person or animal may still harbor low populations of the pathogen, thus continuing to act as a carrier of disease. If wastes from carriers

TABLE 20–1	Pathogens Carried by Sewage
Disease	**Infectious Agent**
Typhoid fever	*Salmonella typhi* (bacterium)
Cholera	*Vibrio cholerae* (bacterium)
Salmonellosis	*Salmonella* species (bacteria)
Diarrhea	*Escherichia coli* (bacterium)
	Campylobacter species (bacteria)
	Cryptosporidium parvum (protozoan)
Infectious hepatitis	Hepatitis A virus
Poliomyelitis	Poliovirus
Dysentery	*Shigella* species (bacteria)
	Entamoeba histolytica (protozoan)
Giardiasis	*Giardia intestinalis* (protozoan)
Numerous parasitic diseases	Roundworms, flatworms

Figure 20–4 The Ganges River in India. In many places in the developing world, the same waterways are used simultaneously for drinking, washing, and disposal of sewage. A high incidence of disease, infant and childhood mortality, and parasites is the result.

contaminate drinking water, food, or water used for swimming or bathing, the pathogens can gain access to, and infect, other individuals (Fig. 20–4).

Public Health. Before the connection between disease and sewage-carried pathogens was recognized in the mid-1800s, disastrous epidemics were common in cities. Epidemics of typhoid fever and cholera killed thousands of people prior to the 20th century. Today public-health measures that prevent this disease cycle have been adopted throughout the developed world and, to some extent, in the developing world. The following measures were far more important than modern medicine in controlling waterborne diseases:

1. Purification and disinfection of public water supplies with chlorine or other agents (Chapter 10),

2. Sanitary collection and treatment of human and animal wastes,

3. Maintenance of sanitary standards in all facilities in which food is processed or prepared for public consumption, and

4. Instruction in personal and domestic hygiene practices.

Standards regarding these measures are set and enforced by government public-health departments. A variety of other measures are enforced as well. For example, if bathing areas are contaminated with raw sewage, health departments close them to swimming. Implicit in all measures is monitoring for sewage contamination. (See Sound Science, "Monitoring for Sewage Contamination.") Our own personal hygiene, sanitation, and health precautions—such as not drinking water from untested sources, washing hands frequently, and making sure that foods like pork, chicken, and hamburger are always well cooked—remain the last and most important line of defense against disease.

Sanitation = Good Medicine. Many people attribute good health in a population to modern medicine, but good health is primarily a result of *the prevention of disease through public-health measures.* According to the 2008 Millennium Development Goals (MDGs) report,[1] almost 1 billion people lack access to safe drinking water, and 2.5 billion live in areas having poor (or no) sewage collection or treatment (Fig. 20–5). Each year more than 2 million deaths (mostly among children under five) are traced to waterborne diseases. Millennium Development Goal 7 addresses these serious public-health needs: "Halve, by 2015, the proportion of people without sustainable access to safe drinking water and basic sanitation" (Table 9–2). A midterm assessment of progress indicates that the world is on track to meet the drinking water target, but is seriously behind in many regions in meeting the sanitation target. Almost one-fourth of the developing world's people have no adequate place to go to the bathroom and generally relieve themselves in streams, backyards, and woods, contaminating the water and the places where children play. When it rains, the drainage ditches and alleys run with human and animal excrement.

Largely because of poor sanitation regarding water and sewage, a significant portion of the world's population is chronically infected with various pathogens. Moreover, populations in areas where there is little or no sewage treatment are extremely vulnerable to deadly epidemics of many diseases spread by way of sewage. This is especially true of cholera. Because of unsanitary conditions, an outbreak of cholera in Peru in 1990 killed several thousand people as it spread through Latin America. The civil unrest and economic collapse in Zimbabwe have brought on the most recent cholera epidemic. Over 4,200 deaths occurred from August 2008 through June 2009, with 98,000 cases reported. The World Health Organization (WHO) reports numerous cholera outbreaks annually, and cases frequently number in the thousands, with hundreds of deaths.

---

[1]www.un.org/millenniumgoals/pdf/The Millennium Development Goals Report 2008.pdf.

Sound Science

Monitoring for Sewage Contamination

An important aspect of public health is the detection of sewage in water supplies and other bodies of water with which humans come in contact. It is virtually impossible to test for each specific pathogen that might be present in a body of water. Therefore, an indirect method called the **fecal coliform test** has been developed. This test is based on the fact that huge populations of a bacterium called *E. coli* (*Escherichia coli*) normally inhabit the lower intestinal tract of humans and other animals and large numbers of the bacterium are excreted with fecal material. In temperate regions at least, *E. coli* does not last long in the outside environment.

Therefore, the presumption is that when *E. coli* is found in natural waters, it is an indication of recent and probably persisting contamination with sewage wastes. In most situations, *E. coli* is not a pathogen itself, but is referred to as an **indicator organism**: Its presence *indicates* that water is contaminated with fecal wastes and that sewage-borne pathogens may be present. Conversely, the absence of *E. coli* is taken to mean that water is free from such pathogens.

The fecal coliform test, one technique of which is shown in the accompanying figure, detects and counts the number of *E. coli* in a sample of water. Thus, the results indicate the relative degree of contamination of the water and the relative risk of pathogens. For example, to be safe for drinking, water may not have any fecal coliform in a 0.4-cup (100-mL) sample. The same sample with as many as 200 *E. coli* is still considered safe for swimming. Beyond that level, a river or beach may be posted as polluted, and swimming and other direct contact should be avoided. By contrast, raw sewage (99.9% water, 0.1% waste) has *E. coli* counts in the millions.

(a) (b) (c) (d)

Testing water for sewage contamination by the Millipore technique. (a) A Millipore filter disk is placed in the filter apparatus. (b) A sample of the water being tested is drawn through the filter, and any bacteria present are entrapped on the disk. (c) The disk is then placed in a petri dish on a special medium that supports the growth of bacteria and that will impart a particular color to fecal *E. coli* bacteria. Next, the dish is incubated for 24 hours at 38°C, during which time the bacteria on the disk will multiply to form a colony visible to the naked eye. (d) *Escherichia coli* bacteria, indicating sewage contamination, are identifiable as the colonies with a metallic green sheen.

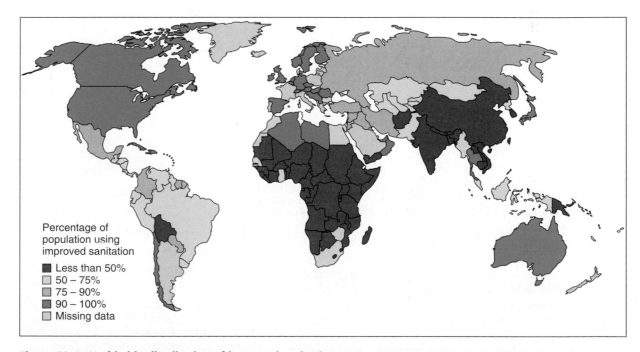

Percentage of population using improved sanitation

- ■ Less than 50%
- ☐ 50 – 75%
- ☐ 75 – 90%
- ■ 90 – 100%
- ☐ Missing data

Figure 20–5 Worldwide distribution of improved sanitation. Less than half the population has adequate sanitation in much of the developing world. The Millennium Development Goal of reducing by half the proportion of people without improved sanitation will not be met without additional effort. (*Source:* World Water Development Report, 2006.)

Organic Wastes.   Along with pathogens, human and animal wastes contain organic matter that creates serious problems if it enters bodies of water untreated. Other kinds of organic matter (leaves, grass clippings, trash, and so forth) can enter bodies of water as a consequence of runoff. With the exception of plastics and some human-made chemicals, these wastes are biodegradable. As mentioned in connection with the Gulf of Mexico dead zone, when bacteria and detritus feeders decompose organic matter in water, they consume oxygen gas dissolved in the water. The amount of oxygen that water can hold in solution is severely limited. In cold water, dissolved oxygen (DO) can reach concentrations up to 10 parts per million (ppm); much less can be held in warm water. Compare this ratio with that of oxygen in air, which is 200,000 ppm (20%), and you can understand why even a moderate amount of organic matter decomposing in water can deplete the water of its DO. Bacteria keep the oxygen supply depleted as long as there is dead organic matter to support their growth and oxygen replenishment is inadequate.

BOD.   Biochemical oxygen demand (BOD) is a measure of the amount of organic material in water, stated in terms of how much oxygen will be required to break it down biologically, chemically, or both. The higher the BOD measure, the greater the likelihood that dissolved oxygen will be depleted in the course of breaking the organic material down. A high BOD causes so much oxygen depletion that animal life is severely limited or precluded, as in the bottom waters of the Gulf of Mexico. Fish and shellfish are killed when the DO drops below 2 or 3 ppm, though some are less tolerant at even higher DO levels. If the system goes anaerobic (has no oxygen), only bacteria can survive, using their abilities to switch to fermentation or anaerobic respiration (metabolic pathways that do not require oxygen). A typical BOD value for raw sewage would be around 250 ppm. Even a moderate amount of sewage, added to natural waters containing at most 10 ppm DO, can deplete the water of its oxygen and produce highly undesirable consequences well beyond those caused by the introduction of pathogens (Fig. 20–6).

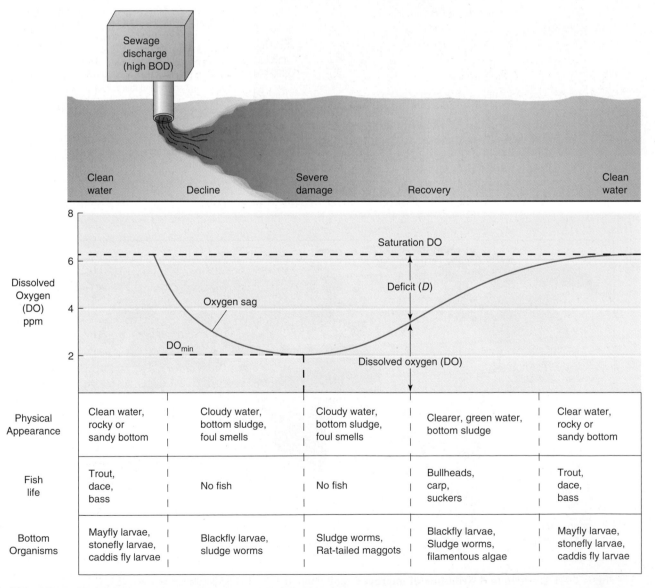

Figure 20–6 The oxygen sag curve. When sewage with a high BOD is discharged into a stream or small river, it creates an oxygen deficit and severely impacts the stream's biology.

Chemical Pollutants. Because water is such an excellent solvent, it is able to hold many chemical substances in solution that have undesirable effects. Water-soluble **inorganic chemicals** constitute an important class of pollutants that include heavy metals (lead, mercury, arsenic, nickel, and so forth), acids from mine drainage (sulfuric acid) and acid precipitation (sulfuric and nitric acids), and road salts employed to melt snow and ice in the colder climates (sodium and calcium chlorides). The **organic chemicals** are another group of substances found in polluted waters. Petroleum products pollute many bodies of water, from the major oil spills in the ocean to the small streams receiving runoff from parking lots. Other organic substances with serious impacts are the pesticides that drift down from aerial spraying or that run off from land areas (Chapter 13) and the various industrial chemicals, such as polychlorinated biphenyls (PCBs), cleaning solvents, and detergents.

Many of these pollutants are toxic even at low concentrations (Chapter 22). Some may become concentrated as they are passed up the food chain in a process called biomagnification (Chapter 13). Even at very low concentrations, they can render water unpalatable to humans and dangerous to aquatic life. At higher concentrations, they can change the properties of bodies of water so as to prevent them from serving any useful purpose except perhaps navigation. Acid mine drainage, for example, pollutes thousands of miles of streams in coal-bearing regions of the United States (Fig. 20–7). Stream bottoms are coated with orange deposits of iron, and the water is so acidic that only a few hardy bacteria and algae can tolerate it.

Sediments. As natural landforms weather, especially during storms, a certain amount of sediment enters streams and rivers. However, erosion from farmlands, deforested slopes, overgrazed rangelands, construction sites, mining sites, stream banks, and roads can greatly increase the load of sediment entering waterways. Frequently, storm drains simply lead to the nearest depression or some natural streambed. Sediments (sand, silt, and clay) have direct and extreme physical impacts on streams and rivers. When erosion is slight, streams and rivers of the watershed run clear and support algae and other

Figure 20–7 Acid mine drainage. Acidic water from mines contains high concentrations of sulfides and iron, which are oxidized when they come into the open, thus producing unsightly and sterile streams.

aquatic plants that attach to rocks or take root in the bottom. These producers, plus miscellaneous detritus from fallen leaves and so on, support a complex food web of bacteria, protozoa, worms, insect larvae, snails, fish, crayfish, and other organisms, which keep themselves from being carried downstream by attaching to rocks or seeking shelter behind or under rocks. Even fish that maintain their position by active swimming occasionally need such shelter to rest (Fig. 20–8a).

Sediment entering waterways in large amounts has an array of impacts. Sand, silt, clay, and organic particles (humus) are quickly separated by the agitation of flowing water and are carried at different rates. Clay and humus are carried in suspension, making the water muddy and reducing the amount of light penetrating the water and, hence, reducing photosynthesis as well. As the material settles, it coats everything and continues to block photosynthesis. It also kills the animals by clogging their gills and feeding structures. The eggs of fish and other aquatic organisms are particularly vulnerable to being smothered by sediment.

Bed Load. Especially destructive is the **bed load** of sand and silt, which is not readily carried in suspension, but is gradually washed along the bottom. As particles roll and tumble along, they scour organisms from the rocks. They also bury and smother the bottom life and fill in the hiding and resting places of fish and crayfish. Aquatic plants and other organisms are prevented from reestablishing themselves because the bottom is a constantly shifting bed of sand (Fig. 20–8b and c). Modern storm-water management is designed to reduce the bed load, usually via storm drains that are periodically emptied of their sediment. Many newer housing developments include a storm-water retention reservoir, which is simply a pond that receives and holds runoff from the area during storms. Water may infiltrate the soil or may create a pocket of natural wetland habitat supporting wildlife (Fig. 20–9, p. 527).

Sediments do not receive the attention the news media give to hazardous wastes and certain other pollution problems, but erosion is so widespread throughout the world that few streams and rivers escape the harsh impact of excessive sediment loads.

Nutrients. Some of the inorganic chemicals carried in solution in all bodies of water are classified as nutrients—essential elements required by plants. The two most important nutrient elements for aquatic plant growth are *phosphorus* and *nitrogen*, and they are often in such low supply in water that they are the limiting factors for phytoplankton or other aquatic plants. More nutrients mean more plant growth, so nutrients become water pollutants when they are added from point or nonpoint sources and stimulate undesirable plant growth in bodies of water. The most obvious point sources of excessive nutrients are sewage outfalls. As we will see in Section 20.2, it is costly to remove nutrients from sewage during the treatment process, so that is not always done. If the sewage is not treated at all, higher levels of nutrients will enter the water, which is what's happening in much of the developing world.

Agricultural runoff is the most notorious nonpoint source of nutrients. The nutrients are applied to agricultural crops as chemical fertilizers, manure, and dissolved in irrigation water, and are present in crop residues. When nutrients

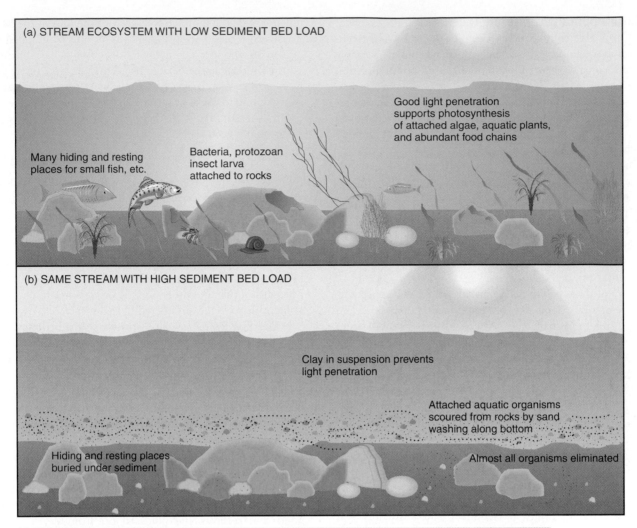

(a) STREAM ECOSYSTEM WITH LOW SEDIMENT BED LOAD

Good light penetration supports photosynthesis of attached algae, aquatic plants, and abundant food chains

Many hiding and resting places for small fish, etc.

Bacteria, protozoan insect larva attached to rocks

(b) SAME STREAM WITH HIGH SEDIMENT BED LOAD

Clay in suspension prevents light penetration

Attached aquatic organisms scoured from rocks by sand washing along bottom

Hiding and resting places buried under sediment

Almost all organisms eliminated

(c)

Figure 20–8 Impact of sediment on streams and rivers. (a) The ecosystem of a stream that is not subjected to a large sediment bed load. (b) The changes that occur with large sediment inputs. (c) Platte River at Lexington, Nebraska. The sandbars seen here constitute the bed load; they shift and move with high water, preventing the reestablishment of aquatic vegetation.

Figure 20–9 Storm-water management. Rather than letting excessive runoff from developed areas cause flooding and other environmental damage, runoff can be funneled into a retention pond, as shown here. Then it can drain away slowly, maintaining natural stream flow, or it can recharge groundwater. The retention pond may be designed to retain a certain amount of water and thus create a pocket of wildlife habitat in an otherwise urban or suburban setting.

are applied in excess of the needs of plants or when runoff from agricultural fields and feedlots is heavy, the nutrients are picked up by water that eventually enters streams, rivers, lakes, and the ocean. Other nonpoint sources of nutrients include lawns and gardens, golf courses, and storm drains.

Water Quality Standards. When is water considered polluted? Many water pollutants, like pesticides, cleaning solvents, and detergents, are substances that are found in water only because of human activities. Others, such as nutrients and sediments, are always found in natural waters, and they are a problem only under certain conditions. In both cases, *pollution* means any quantity that is harmful to human health or the environment, or prevents full use of the environment by humans or natural species. The mere presence of a substance in water does not necessarily pose a problem. Rather, it is the *concentration* of the pollutant that must be of primary concern.

Criteria Pollutants. But what concentration is worrisome? To provide standards for assessing water pollution, the EPA has established the **National Recommended Water Quality Criteria.** On the basis of the latest scientific knowledge, the EPA has listed 167 chemicals and substances as **criteria pollutants.** The majority of these are toxic chemicals, but many are also natural chemicals or conditions that describe the state of water, such as nutrients, hardness (a general measure of dissolved calcium and magnesium salts), and pH (a measure of the acidity of the water). The list identifies the pollutant and then recommends concentrations for fresh water, salt water, and human consumption (usually of fish and shellfish). Values are given for the **criteria maximum concentration** (CMC, the highest single concentration beyond which environmental impacts may be expected) and the **criterion continuous concentration** (CCC, the highest sustained concentration beyond which undesirable impacts may be expected). The criteria are *recommendations* meant to be used by the states, which are given the primary responsibility for upholding water-pollution laws. States and Native American tribes may revise these criteria, but their revisions are subject to EPA approval.

Drinking water standards are stricter. For these, the EPA has established the **Drinking Water Standards and Health Advisories,** a set of tables that are updated periodically. These standards, covering some 94 contaminants, are enforceable under the authority of the **Safe Drinking Water Act (SDWA).** They are presented as **maximum contaminant levels (MCLs).** To see how these two sets of standards work, consider the heavy metal arsenic.

Arsenic. Arsenic is listed as a known human carcinogen by the Department of Health and Human Services; it occurs naturally in groundwater, reaching high concentrations in some parts of the world. The CMC and CCC values for arsenic are 340 μg/L and 150 μg/L (1 μg/L = 1 part per billion) for freshwater bodies and 69 μg/L and 36 μg/L for saltwater bodies. The drinking-water MCL concentration, however, is 10 μg/L, and therein lies a story.

For years, the MCL concentration for arsenic was 50 μg/L, despite warnings by the U.S. Public Health Service and the WHO that this was much too high a level. (Both bodies recommended 10 μg/L.) The problem is the cancer risk. From standard risk assessment processes (see Chapter 17), the cancer mortality risk was assessed at 1 in 100 for people drinking water regularly at the 50 μg/L level. This ratio is more than 100 times higher than the permitted risk for any other contaminant. Before Clinton left office, the EPA determined that the MCL for arsenic should be lowered, but the pending rule was withdrawn by the Bush EPA on the premise that "sound science" did not clearly support the new rule. Many states and rural areas were complaining that attaining the lower arsenic concentration would be too costly. In the end, after a National Academy of Sciences report and urgings from both the House and the Senate, then EPA Administrator Christine Whitman upheld the Clinton era rule and set the arsenic MCL at 10 μg/L.

Other Applications. Two important applications of water quality criteria are the **National Pollution Discharge Elimination System (NPDES)** and **Total Maximum Daily Load (TMDL)** programs. The NPDES program addresses point-source pollution and issues *permits* that regulate discharges from wastewater-treatment plants and industrial sources. The TMDL program, by contrast, evaluates all sources of pollutants entering a body of water, especially nonpoint sources, according to the water body's ability to assimilate the pollutant. For both of these programs, water quality criteria provide an essential means of assessing the condition of the receiving body of water as well as evaluating the levels of the various pollutant discharges. We will examine these two programs in more detail when we consider the problem of eutrophication.

The good news is that, according to the EPA, 92% of the people in the United States have access to drinking water that meets the *drinking water standards*. However, on the basis of data supplied to the EPA by the states, over 42,000 rivers, lakes, and estuaries are not meeting the recommended *water quality standards*, and this excludes over 60% of U.S. waters, which have not been assessed at all. The major problems are pathogens, mercury, nutrients, other heavy metals, sediments, and oxygen depletion. As we proceed, we will address the specifics of public-policy responses to this serious national problem. We turn now to ways of dealing with the biological wastes that are a large part of our pollution problem.

20.2 Wastewater Treatment and Management

To alleviate the problem of sewage-polluted waterways, facilities were designed and constructed to treat the outflow before it entered the receiving waterway. The first wastewater-treatment plants in the United States were built around 1900. However, the combined volumes of sewage and storm water soon proved impossible to handle. During heavy rains, wastewater would overflow the treatment plant and carry raw sewage into the receiving waterway. Gradually, regulations were passed requiring municipalities to install separate systems—**storm drains** to collect and drain runoff from precipitation and **sanitary sewers** to receive all the wastewater from sinks, tubs, and toilets in homes and other buildings. (Note the distinction between the two terms; it is incorrect to speak of storm drains as sewers.)

The ideal modern system, then, is a system in which all wastewater is collected separately from storm water and is fully treated to remove pollutants before the water finally is released. But progress toward this goal has been extremely uneven. Up through the 1970s, even in the United States and other developed countries, thousands of communities still discharged untreated sewage directly into waterways, and for many more, the degree of treatment was minimal. Indeed, the increasing sewage pollution of waterways and beaches was the major impetus behind the passage of the Clean Water Act.

Before addressing methods of treating wastewater, let us clarify which contaminants and pollutants we are talking about.

The Pollutants in Raw Wastewater

Raw wastewater is not only the flushings from toilets, but also the collection of wastewater from all other drains in homes and other buildings. A sewer system brings all tub, sink, and toilet drains from all homes and buildings together into larger and larger sewer pipes, just as the branches of a tree eventually come together into the trunk. This total mixture collected from all drains, which flows from the end of the "trunk" of the collection system, is called **raw sewage** or **raw wastewater**. Because we use such large amounts of water to flush away small amounts of dirt, especially as we stand under the shower or often just run the water, most of what goes down sewer drains is water. Raw wastewater is about 1,000 parts water for every 1 part waste—99.9% water to 0.1% waste.

Given our voluminous use of water, the quantity of raw-wastewater output is on the order of 150–200 gallons (600–800 L) per person per day. That is, a community of 10,000 persons will produce 1.5 to 2.0 million gallons (6–8 million L) of wastewater each day. With the addition of storm water, raw wastewater is diluted still more. Nevertheless, the pollutants are sufficient to make the water brown and foul smelling.

Types of Pollutants. The pollutants in wastewater generally are divided into the following four categories, which correspond to the techniques used to remove them:

1. **Debris and grit:** rags, plastic bags, coarse sand, gravel, and other objects flushed down toilets or washed through storm drains.

2. **Particulate organic material:** fecal matter, food wastes from garbage-disposal units, toilet paper, and other matter that tends to settle in still water.

3. **Colloidal and dissolved organic material:** very fine particles of particulate organic material, bacteria, urine, soaps, detergents, and other cleaning agents.

4. **Dissolved inorganic material:** nitrogen, phosphorus, and other nutrients from excretory wastes and detergents.

In addition to the four categories of pollutants in "standard" raw wastewater, variable amounts of pesticides, heavy metals, and other toxic compounds may be found because people pour unused portions of products containing such materials down sink, tub, or toilet drains. Also, industries may discharge various toxic wastes into sewers.

Removing the Pollutants from Wastewater

The challenge of treating wastewater is more than installing a technology that will do the job; it is also finding one that will do the job *at a reasonable cost*. A diagram of wastewater treatment procedures is shown in Figure 20–10.

Primary Treatment. Because debris and grit will damage or clog pumps and later treatment processes, removing this material is a necessary first step. Usually, removal involves two steps: the screening out of debris and the settling of grit. Debris is removed by letting raw sewage flow through a **bar screen**, a row of bars mounted about 1 inch (2.5 cm) apart. Debris is mechanically raked from the screen and taken to an incinerator. After passing through the screen, the water flows through a **grit chamber**—a swimming-pool-like tank—in which its velocity is slowed just enough to permit the grit to settle. The settled grit is mechanically removed from these tanks and taken to landfills.

After debris and grit are removed, the water then flows very slowly through large tanks called **primary clarifiers**. Because its flow is slow, the water is nearly motionless for several hours. The particulate organic material, about 30% to 50% of the total organic material, settles to the bottom, where it can be removed. At the same time, fatty or oily material floats to the top, where it is skimmed from the surface. All the material that is removed, both particulate

organic material and fatty material, is combined into what is referred to as **raw sludge**, which must be treated separately.

Primary treatment involves nothing more complicated than putting polluted water into a "bucket," letting material settle, and pouring off the water. Nevertheless, such treatment removes much of the particulate organic matter at minimal cost.

Secondary Treatment. **Secondary treatment** is also called **biological treatment** because it uses organisms—natural decomposers and detritus feeders. Basically, an environment

Figure 20–10 A diagram of wastewater treatment. Raw sewage moves from the grit chamber to undergo primary treatment, in which sludge is removed. The clarified water then undergoes secondary treatment (shown here as activated-sludge treatment).

is created that enables these organisms to feed on the colloidal and dissolved organic material and break it down to carbon dioxide, mineral nutrients, and water via their cell respiration. The wastewater from primary treatment is a food- and water-rich medium for the decomposers and detritus feeders. The only thing that needs to be added to the water is oxygen to enhance the organisms' respiration and growth. Either of two systems may be used to add oxygen to the water: a *trickling-filter system* or an *activated-sludge system*. In the **trickling-filter system**, the water exiting from primary treatment is sprinkled onto, and allowed to percolate through, a bed of fist-sized rocks 6–8 feet (2–3 m) deep (Fig. 20–11). The spaces between the rocks provide good aeration. Like a natural stream, this environment supports a complex food web of bacteria, protozoans, rotifers (organisms that consume protozoans), various small worms, and other detritus feeders attached to the rocks. The organic material in the water, including pathogenic organisms, is absorbed and digested by decomposers and detritus feeders as it trickles by.

The **activated-sludge system** (Fig. 20–10, bottom) is the most common secondary-treatment system. Water from primary treatment enters a large tank that is equipped with an air-bubbling system or a rapidly churning system of paddles. A mixture of detritus-feeding organisms, referred to as **activated sludge**, is added to the water as it enters the tank, and the water is vigorously aerated as it moves through the tank. Organisms in this well-aerated environment reduce the biomass of organic material, including pathogens, as they feed. As the organisms feed on each other, they tend to form into clumps, called *floc*, that settle readily when the water is stilled. Then, from the aeration tank, the water is passed into a **secondary clarifier tank**, where the organisms settle out and the water—now with more than 90% of all the organic material removed—flows on. The settled organisms are pumped back into the aeration tank. (They are the activated sludge that is added at the beginning of the process.) Surplus amounts of activated sludge, which occur as populations of organisms grow, are removed and added to the raw sludge. The organic material is oxidized by microbes to form carbon dioxide, water, and mineral nutrients that remain in the water solution.

Biological Nutrient Removal. The original secondary-treatment systems were designed and operated so as to simply maximize biological digestion because the elimination of organic material and its resulting BOD was considered the prime objective. Ending up with a nutrient-rich discharge was not recognized as a problem. Today, with increased awareness of the problem of cultural eutrophication, secondary activated-sludge systems have been added and are being modified and operated in a manner that both removes nutrients and oxidizes detritus in a process known as **biological nutrient removal (BNR)**.

Nitrogen. Recall that, in the natural nitrogen cycle (Chapter 3, Fig. 3–19), various bacteria convert nutrient forms of nitrogen (ammonia and nitrate) back to non-nutritive nitrogen gas in the atmosphere through **denitrification**. For the biological removal of nitrogen, then, the activated-sludge system is partitioned into zones, and the environment in each zone is

(a)

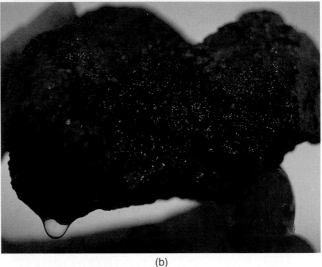

(b)

Figure 20–11 Trickling filters for secondary treatment. (a) The water from primary clarifiers is sprinkled onto, and trickles through, a bed of rocks 6–8 feet deep. (b) Various bacteria and other detritus feeders adhering to the rocks consume and digest organic matter as it trickles by in the water, which is collected at the bottom of the filters and goes on for final clarification and disinfection.

controlled in a manner that promotes the denitrifying process (Fig. 20–12). The resulting nitrogen gas is vented into the air.

Phosphorus. In an environment that is rich in oxygen, but relatively lacking in food (the environment of zone 3 in Figure 20–12), bacteria take up phosphate from the solution and store it in their bodies. Thus, phosphate is removed as the excess organisms are removed from the system. These organisms, together with the phosphate they contain, are added to, and treated with, the raw sludge, ultimately producing a more nutrient-rich treated-sludge product.

Various chemical treatments are often used as an alternative to BNR. One such process is simply to pass the effluent from standard secondary treatment through a filter of lime, which causes the phosphate to precipitate out as insoluble calcium phosphate. Another is to treat the effluent with ferric chloride, which produces insoluble ferric phosphate, or with an organic polymer, which gives rise to a floc. With

Figure 20–12 Biological nutrient removal (BNR). The secondary treatment—the activated-sludge process—may be modified to remove nitrogen and phosphate, while at the same time breaking down organic matter. For this BNR process, the aeration tank is partitioned into three zones, only the third of which is aerated. As seen in the diagram, ammonium (NH_4^+) is converted to nitrate (NO_3^-) in zone 3. Recycled to zone 2, which is without oxygen gas (anoxic), the nitrate supplies the oxygen for cell respiration. In the process, the nitrate is converted to nitrogen gas, which is released into the atmosphere. Phosphate is taken up by bacteria in zone 3 and removed with the excess sludge.

these methods, most of the phosphorus is removed along with the sludge.

Final Cleansing and Disinfection. With or without BNR, the wastewater is subjected to a final clarification and disinfection (Fig. 20–10, bottom). Although few pathogens survive the combined stages of treatment, public-health rigors still demand that the water be disinfected before being discharged into natural waterways. The most widely used disinfecting agent is chlorine gas because it is both effective and relatively inexpensive. But this treatment also introduces chlorine into natural waterways, and even minute levels of chlorine can harm aquatic animals. Also, chlorine gas is dangerous to work with. More commonly, sodium hypochlorite (Clorox®) provides a safer way of adding chlorine to achieve disinfection.

Other disinfecting techniques are coming into use as well. One alternative disinfecting agent is ozone gas, which kills microorganisms and breaks down to oxygen gas, improving water quality in the process. Ozone is unstable and, hence, explosive, however, so it must be generated at the point of use, a step that demands considerable capital investment and energy. Another disinfection technique is to pass the effluent through an array of ultraviolet lights mounted in the water. (Standard fluorescent lights without the white

coating emit ultraviolet light.) The ultraviolet radiation kills microorganisms, but does not otherwise affect the water.

Final Effluent. After all these steps, the wastewater from a modern treatment plant has a lower organic and nutrient content than many bodies of water into which it is discharged. BOD values are commonly 10 to 20 ppm, as opposed to 200 ppm in the incoming sewage. In other words, discharging the wastewater may actually contribute toward improving water quality in the receiving body. In water-short areas, there is every reason to believe that the treated water itself, with little additional treatment, could be recycled into the municipal water-supply system.

Bear in mind that the techniques we have described here represent the state of the art. Many cities, even in the developed world, are still operating with antiquated systems that provide lower-quality treatment. A few coastal cities still persist in discharging sewage that has received only primary treatment directly into the ocean, although in the United States this practice has almost disappeared.

Treatment of Sludge

Recall that the particulate organic matter that settles out or floats to the surface of sewage water in primary treatment forms the bulk of *raw sludge*, although the sewage also contains excesses from activated-sludge and BNR systems. Raw sludge is a gray, foul-smelling, syrupy liquid with a water content of 97% to 98%. Pathogens are certain to be present in raw sludge because it includes material directly from toilets. Indeed, raw sludge is considered to be a biologically hazardous material. However, as nutrient-rich organic material, it has the potential to be used as organic fertilizer if it is suitably treated to kill pathogens and if it does not contain other toxic contaminants.

Three commonly used methods for treating sludge and converting it into organic fertilizer are *anaerobic digestion*, *composting*, and *pasteurization*. Since this is a developing industry, it is unclear which method will prove most cost effective and environmentally acceptable over the long run. Also, none of these methods is capable of removing toxic substances such as heavy metals and nonbiodegradable synthetic organic compounds. The presence of such toxins can preclude the use of sludge as fertilizer.

Anaerobic Digestion.

Anaerobic digestion is a process that allows bacteria to feed on the sludge *in the absence of oxygen*. The raw sludge is put into large airtight tanks called **sludge digesters** (Fig. 20–13). In the absence of oxygen, a consortium of anaerobic bacteria breaks down the organic matter. The end products of this decomposition are carbon dioxide, methane, and water. Thus, a major by-product of anaerobic processes is **biogas**, a gaseous mixture that is about two-thirds *methane*. The other one-third is made up of carbon dioxide and various foul-smelling organic compounds that give sewage its characteristic odor. Because of its methane content, biogas is flammable and can be used for fuel. In fact, it is commonly collected and burned to heat the sludge digesters because the bacteria working on the sludge do best when maintained at about 104°F (38°C).

After four to six weeks, anaerobic digestion is more or less complete, and what remains is called **treated sludge** or **biosolids**, consisting of the remaining organic matter, which is now a relatively stable, nutrient-rich, humuslike material suspended in water. Pathogens have been largely, if not entirely, eliminated, so they no longer present any significant health hazard.

Such treated sludge makes an excellent organic fertilizer that can be applied directly to lawns and agricultural fields in the liquid state in which it comes from the digesters, providing the benefit of both the humus and the nutrient-rich water. Alternatively, the sludge may be dewatered by means of belt presses, whereby the sludge is passed between rollers that squeeze out most of the water (Fig. 20–14a). This leaves the organic material as a semisolid **sludge cake** (Fig. 20–14b), which, after disinfection, is easy to stockpile, distribute, and spread on fields with traditional manure spreaders.

Composting.

Another process sometimes used to treat sewage sludge is **composting**. Raw sludge is mixed with wood chips or some other water-absorbing material to reduce the water content. It is then placed in **windrows**—long, narrow piles that allow air to circulate easily through the material and that can be turned with machinery. Bacteria and other decomposers break down the organic material to rich humuslike material that makes an excellent treatment for poor soil.

Pasteurization.

After the raw sludge is dewatered, the resulting sludge cake may be put through ovens that operate like oversized laundry dryers. In the dryers, the sludge is **pasteurized**—that is, heated sufficiently to kill any pathogens (exactly the same process that makes milk safe to drink). The product is dry, odorless organic pellets. Milwaukee, Wisconsin, which has a particularly rich sludge resulting from the brewing

Figure 20–13 Anaerobic sludge digesters. In these tanks, the bacterial digestion of raw sludge in the absence of oxygen leads to the production of methane gas, which is tapped from the tops of the tanks, and humuslike organic matter, which can be used as a soil conditioner. The egglike shape of the tanks facilitates mixing and digestion.

(a)

(b)

Figure 20–14 Dewatering treated sludge. (a) Sludge (98% water) may be dewatered by means of belt presses, as shown here. The liquid sludge is run between canvas belts going over and between rollers, so that much of the water is pressed out. (b) The resulting "sludge cake" is a semisolid humuslike material that can be used as an organic fertilizer.

industry, has been using this process for more than 60 years. The city bags and sells the pellets throughout the country as an organic fertilizer under the trade name Milorganite®.

Alternative Treatment Systems

Despite the expansion of sewage collection systems, many homes in rural and suburban areas lie outside the reach of a municipal system. For these homes, *on-site* treatment systems are required. Currently, 25% of the U.S. population is served by such systems. The traditional and still most common on-site system is the *septic tank and leaching field* (Fig. 20–15). Wastewater flows into the tank, where particulate organic material settles to the bottom. The tank acts like a primary clarifier in a municipal system. Water containing colloidal and dissolved organic material, as well as dissolved nutrients, flows into the leaching field and gradually percolates into the soil. Organic material that settles in the tank is digested by bacteria, but accumulations still must be pumped out regularly. Soil bacteria decompose the colloidal and dissolved organic material that comes through the leaching field. Some people establish successful vegetable gardens over septic drain fields, thus instantiating the sound principle of recycling the nutrients. Prerequisites for this traditional system are suitable land area for the drain field and subsoil that allows sufficient percolation of water.

Unfortunately, on-site systems frequently fail, resulting in unpleasant sewage backup into homes and pollution of groundwater and surface waters. Homeowners are often unaware of how the systems work, and most local regulatory programs do not require homeowners to be accountable for them. Substantial nutrient and pathogen pollution of groundwater and surface waters is traced to failed on-site systems. For this reason, the EPA has made several resources available to local and state agencies responsible for maintaining water quality: voluntary national guidelines, a handbook for management, and a manual for on-site wastewater systems.

Septic System Primer. The following are some simple suggestions for successful septic system maintenance:

1. Use caution in disposing of materials down the drain. Do not dump coffee grounds, cat litter, diapers, grease, feminine hygiene products, pesticides, paint, gasoline, household chemicals (like chlorine-containing compounds), or other products that could either fill up or

Figure 20–15 Septic tank treatment. Sewage treatment for a private home, using a septic tank and leaching field. Normally, the pipes and the tank are buried underground. They are shown uncovered here only for illustration. (after U.S. Department of Agriculture, Natural Resources Conservation Service. www.or.nrcs.usda.gov/news/factsheets/fs19.pdf.)

clog the septic tank or kill the bacteria that make the system work.

2. Maintain your system by having it inspected regularly and pumped out at least every five years and more frequently if you have a larger family. Keep records of maintenance, and have a map of the location of your septic tank and the leaching field.

3. If you have a garbage disposal unit, get rid of it; the added materials will shorten the life of the system and force it to be pumped out more frequently.

4. Keep heavy equipment and vehicles off your system and leaching field.

Composting Toilet Systems. A viable and relatively inexpensive alternative to septic systems, especially for more remote and occasional use, is the **composting toilet system** (Fig. 20–16). The objective of the system is to provide a sanitary means of treating human wastes that will destroy human pathogens and produce a stable humuslike end product that can be safely disposed of. There are many types of these systems available, but they have some common components:

• A bathroom toilet that connects to a composting reactor. This can have a removable bowl for easy cleaning.

• A composting reactor that receives solid and liquid wastes from the toilet. The reactor can be right under the toilet seat (self-contained) or located in the basement or on the ground outside.

• An exhaust system (may have a fan) to remove the odors and vapors that result from decomposition of the waste matter.

• Some means of ventilation in the reactor to promote aerobic breakdown of waste. This can be active or passive, where *active systems* involve rotating drums, automatic mixers, fans, and the like and *passive systems* simply allow the waste to decompose in cool temperatures.

• A means of managing and draining excess liquid and leachate (liquid that has percolated through the wastes) and an access door for removing the end product.

These systems require active management if they are constantly in use, and the end product must legally be buried or removed by a licensed septage hauler. They are capable of reducing toilet wastes to 10–30% of their original volume. Their most common use is in vacation cottages, roadside rest stops, and parks, but where building codes permit, they can also be employed in urban settings. The Institute of Asian Research in Vancouver, British Columbia, is a 30,000-square-foot office complex that is not connected to the city

(a)

(b)

Figure 20–16 Composting toilet system. (a) A toilet in the bathroom leads to (b) the composting reactor, which receives the wastes and allows them to decompose. These systems are often used in vacation cottages, parks, and roadside rest stops. They reduce human wastes to as little as 10% of their original volume.

sewer system and uses only composting toilets to handle human wastes.

Using Effluents for Irrigation.

The nutrient-rich water coming from the standard secondary-treatment process is beneficial for growing plants. The problem is that we don't want to put that water into waterways, where it will stimulate the growth of undesirable algae. But why not use it for irrigating plants we do want to grow? This is a way of completing the nutrient cycle. Indeed, the concept has been put into practice in a considerable number of locations as an alternative to upgrading the treatment to remove nutrients. Again, effluents must not be contaminated with toxic materials.

To illustrate, the nutrient-rich effluent from standard secondary treatment from St. Petersburg, Florida, was causing cultural eutrophication in Tampa Bay. Now St. Petersburg uses the effluent to irrigate 4,000 acres (1,600 hectares) of urban open space, from parks and residential lawns to a golf course. Revenues from the water sales help offset operating costs. Similarly, Bakersfield, California, receives a $30,000 annual income from a 5,000-acre (2,000-hectares) farm irrigated with its treated effluent. And Clayton County, Georgia, is irrigating 2,500 acres (1,000 hectares) of woodland with partially treated sewage. Hundreds of similar projects are under way around the country.

Growing Crops. A number of developing countries irrigate croplands with raw (untreated) sewage effluents. This practice results in getting the bad with the good. The crops respond well, but parasites and disease organisms can easily be transferred to farm workers and consumers. Therefore, it is important to emphasize that only *treated* effluents should be used for irrigation. In Matimangwe, Mozambique, sewage is treated in dry pits and turned into a rich fertilizer that dramatically increased the food harvest, using a latrine system called EcoSan. The wastes are mixed with soil and ash and left for eight months, while the lack of moisture and long incubation kill off pathogens and turn the wastes into a rich compost.

Reconstructed Wetland Systems.

In treating wastewater, it is also possible to make use of the nutrient-absorbing capacity of wetlands in suitable areas and under suitable climatic conditions. The project may be part of a wetlands recovery program, or artificial wetlands may be constructed.

In the 1960s and 1970s, for example, much of the land around Orlando, Florida, which was originally wetlands, was drained and converted to cattle pasture. At the time, Orlando was discharging 20 million gallons (75 million L) per day of nutrient-rich effluent into the St. Johns River following secondary treatment. Through the Orlando Easterly Wetlands Reclamation Project, 1,200 acres (480 hectares) of pastureland were converted back to wetlands. The project involved scooping soil from pastures and building berms (mounds of earth) around them to create a chain of shallow lakes and ponds. In addition, 2 million wetland plants ranging from bulrushes and cattails to various trees were planted.

The effluent entering the upper end now percolates through the wetland for about 30 days before entering the St. Johns River virtually pure. Thus, the project has recreated a wildlife habitat. Wetland systems can be designed for small as well as large areas and are becoming an increasingly popular alternative for small communities. The key to success for such systems is to ensure that they are kept in balance and not provided more input than they are able to handle.

Many aquatic systems are simply not able to act like wetlands and absorb extra nutrients without a major change in ecosystem function. This response, called eutrophication, is the most widespread water-pollution problem in the United States.

20.3 Eutrophication

Recall from Chapter 5 that the term *trophic* refers to feeding. Literally, *eutrophic* means "well nourished." As we will see, there are some quite undesirable consequences of high levels of "nourishment" in bodies of water. Although eutrophication can be an entirely natural process, the introduction of pollutants into bodies of water has greatly increased the scope and speed of eutrophication.

Different Kinds of Aquatic Plants

To understand eutrophication, you need to be able to distinguish between *benthic* plants and *phytoplankton*.

Benthic Plants. **Benthic plants** (from the Greek *benthos*, meaning "deep") are aquatic plants that grow attached to, or are rooted in, the bottom of a body of water. All common aquarium plants and sea grasses are benthic plants. As shown in Figure 20–17a, benthic plants may be categorized as **submerged aquatic vegetation (SAV)**, which generally grows totally under water, or **emergent vegetation**, which grows with the lower parts in water, but the upper parts emerging from the water.

To thrive, SAV requires water that is clear enough to allow sufficient light to penetrate to allow photosynthesis. As water becomes more turbid (cloudy), light is diminished. In extreme situations, it may be reduced to penetrating to just a few centimeters beneath the water's surface. Thus, increasing turbidity decreases the depth at which SAV can survive.

Another important feature of SAV is that it absorbs its required mineral nutrients from the bottom sediments through the roots, just as land plants do. SAV is not limited by water that is low in nutrients. Indeed, enrichment of the water with nutrients is counterproductive for SAV because it stimulates the growth of phytoplankton.

Phytoplankton. *Phytoplankton* consist of numerous species of photosynthetic algae, protists, and chlorophyll-containing bacteria (cyanobacteria, formerly referred to as blue-green algae) that grow as microscopic single cells or in small groups, or "threads," of cells. Phytoplankton live suspended in the water and are found wherever light and nutrients are available (Fig. 20–17b). In extreme situations, water may become literally pea-soup green (or tea colored, depending on the species involved), and a scum of phytoplankton

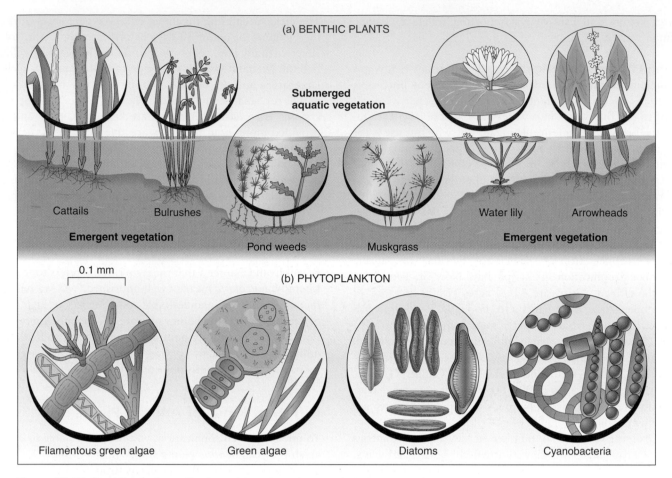

Figure 20–17 Aquatic photosynthesizers. (a) Benthic, or bottom-rooted, plants. These are subdivided into submerged aquatic vegetation (SAV) and emergent vegetation. (b) Phytoplankton, various photosynthetic organisms that are either single cells or small groups or filaments of cells, float freely in the water.

may float on the surface and absorb essentially all the light. However, phytoplankton reach such densities only in nutrient-rich water because, not being connected to the bottom, they must absorb nutrients from the water. A low level of nutrients in the water limits the growth of phytoplankton accordingly.

Considering the different requirements of SAV and phytoplankton, the balance between them is altered when nutrient levels in the water are changed. As long as water remains low in nutrients, populations of phytoplankton are suppressed, the water is clear, and light can penetrate to support the growth of SAV. As nutrient levels increase, phytoplankton can grow more prolifically, making the water turbid and thus shading out the SAV.

The Impacts of Nutrient Enrichment

A lake in which light penetrates deeply—one in which the bottom is visible beyond the immediate shoreline—is **oligotrophic** (low in nutrients). Such a lake is fed by a watershed that holds its nutrients well. A forested watershed, for example, holds nitrogen and phosphorus tightly and allows very little of those elements to enter the water draining into the streams and rivers. If the streams feed into a lake, it will reflect their low nutrient content.

In an oligotrophic lake, the low nutrient levels limit the growth of phytoplankton and allow enough light to penetrate to support the growth of SAV, which draws its nutrients from the bottom sediments. In turn, the benthic plants support the rest of a diverse aquatic ecosystem by providing food, habitats, and dissolved oxygen. The oligotrophic body of water is prized for its aesthetic and recreational qualities as well as for its production of fish.

The Process of Eutrophication.   As the water of an oligotrophic body becomes enriched with nutrients, numerous changes are set in motion. First, the nutrient enrichment allows the rapid growth and multiplication of phytoplankton, increasing the turbidity of the water. The increasing turbidity shades out the SAV that lives in the water. With the die-off of SAV, there is a loss of food, habitats, and dissolved oxygen from their photosynthesis.

Phytoplankton have remarkably high growth and reproduction rates. Under optimal conditions, phytoplankton biomass may double every 24 hours, a capacity far beyond that of benthic plants. Thus, phytoplankton soon reach a maximum population density, and continuing growth and reproduction are balanced by die-off. Dead phytoplankton settle out, resulting in heavy deposits of detritus on the lake or river bottom. In turn, the abundance of detritus supports an

Figure 20–18 Eutrophication. As nutrients are added from sources of pollution, an oligotrophic system rapidly becomes eutrophic and undesirable.

OLIGOTROPHIC

- Low in nutrients
- Phytoplankton limited

- Water clear
- Light penetrates
- Submerged aquatic vegetation (SAV) thrives

NUTRIENT INPUTS

- Nutrient-rich
- Phytoplankton thrive

- Water turbid
- SAV shaded out

EUTROPHIC

- Nutrient-rich
- Rapid turnover of phytoplankton
- Accumulation of detritus of dead algae

- Decomposers feed on detritus
- Depletion of dissolved oxygen
- Fish and shellfish suffocate

abundance of decomposers, mainly bacteria. The explosive growth of bacteria, consuming oxygen via respiration, creates an additional demand for dissolved oxygen. The result is the depletion of dissolved oxygen, creating hypoxic conditions, with the consequent suffocation of fish and shellfish. A dead zone has been created; it can be as limited as a small lake basin or as extensive as the Baltic Sea, and the cause is always nutrient enrichment.

In sum, **eutrophication** refers to this whole sequence of events, starting with nutrient enrichment and proceeding to the growth and die-off of phytoplankton, the accumulation of detritus, the growth of bacteria, and, finally, the depletion of dissolved oxygen and the suffocation of higher organisms (Fig. 20–18). For humans, a eutrophic body of water is unappealing for swimming, boating, and sportfishing. Also, if the lake is a source of drinking water, its value may be greatly impaired because phytoplankton rapidly clog water filters and may cause a foul taste. Even worse,

some species of phytoplankton secrete various toxins into the water that may kill other aquatic life and be injurious to human health as well. (See Sound Science, "Harmful Algal Blooms.")

Shallow Lakes and Ponds. In lakes and ponds whose water depth is 6 feet (2 m) or less, eutrophication takes a somewhat different course. There SAV may grow to a height of a meter or more, reaching the surface. Thus, with nutrient enrichment, the SAV is not shaded out, but grows abundantly, sprawling over and often totally covering the water surface with dense mats of vegetation that make boating, fishing, or swimming impossible (Fig. 20–19). Any vegetation beneath these mats is shaded out. As the mats of vegetation die and sink to the bottom, they create a BOD that often depletes the water of dissolved oxygen, causing the death of aquatic organisms other than bacteria.

Sound Science

Harmful Algal Blooms

Imagine driving along Palos Verdes Drive on the California coast and having a pelican fall out of the sky and crash into your windshield. Well, maybe you'd rather not imagine that, but every spring along the coast of southern California, for the last 10 years, dead marine birds, sea otters, dolphins, and sea lions have been washing up on the beaches, and pelicans have literally dropped out of the sky—victims of an algal toxin, *domoic acid*. The toxin is acquired by the birds and mammals when they ingest marine shellfish and small fish that have passed the toxin along by ingesting smaller planktonic organisms that have been feeding on a diatom occurring in dense "blooms" in the water. Domoic acid doesn't harm the invertebrates or fish, but in mammals and birds, it causes brain damage that incapacitates them, often leading to death. The first reported outbreak occurred in Prince Edward Island, Canada, in 1987, when over 100 people were sickened and three died from eating blue mussels. The spring visit of the toxin on the California coast is a recent phenomenon, and observers note that it seems to be getting worse every year.

In 1988, Dr. Joanne Burkholder, a North Carolina State University plankton biologist, examined some open sores that were found on menhaden fish from a North Carolina estuary. She identified a new organism, a microscopic dinoflagellate (a type of alga) she named *Pfiesteria piscicida*. (Recently, a second toxic species, *Pfiesteria*

shumwayae, has been identified.) Sometimes *Pfiesteria* is an amoeba, sometimes it is a flagellate, and sometimes it is in a cyst. Sometimes it secretes a toxin that stuns the fish, and then it turns into a form that proceeds to feed on the fish, which have been disoriented by the toxin. Sometimes it behaves like a decent, ordinary alga and lives by photosynthesis.

Pfiesteria has been implicated in at least 50 major fish kills in Albemarle and Pamlico sounds, North Carolina, and has been identified in waters from Virginia, Maryland, Delaware, Florida, and even northern Europe. In the summer of 1997, fish kills struck several tributaries of the Chesapeake Bay, particularly the Pocomoke River. The fish kills and catches of fish with ulcerous lesions finally led the state to close the river to all fishing and harvesting of shellfish. When news of the fish kills in the Chesapeake Bay and the impacts on humans became public, there was a predictable panic response, and consumers all around the Chesapeake Bay stopped eating seafood. In Maryland alone, the economic loss was estimated at between $15 and $20 million in sales during 1997.

The domoic acid and *Pfiesteria* episodes join a long list of what are called *harmful algal blooms* (*HABs*). Red tides, consisting of concentrations of different species of toxic dinoflagellates, have been occurring in coastal waters all around North America. They may be responsible for more

than $1 billion in economic damage to coastal systems and fisheries. Scientists believe that nutrients—especially nitrogen and phosphorus—are the basic culprits in feeding the growth of the algae in estuaries and coastal waters. Recently, five federal agencies joined to sponsor the Ecology and Oceanography of Harmful Algal Blooms (ECOHAB), a national research program studying HABs in U.S. coastal waters. Like the dead zones, the blooms of harmful algae are increasing in frequency. These blooms are virtually always associated with nutrient enrichment, and they are found in freshwater lakes as well as coastal marine environments. They are a nasty reminder that our work in cleaning up the nation's waterways still has a long way to go.

Lesions on menhaden fish caused by *Pfiesteria*.

Figure 20–19 Eutrophication in shallow lakes and ponds. In shallow water, sufficient light for photosynthesis continues to reach the submerged aquatic vegetation. The oversupply of nutrients stimulates growth, so that vegetation reaches the surface and forms mats.

Natural Versus Cultural Eutrophication. In nature, apart from human impacts, eutrophication is part of the process of natural succession, discussed in Chapter 5. Over periods of hundreds or thousands of years, bodies of water are subject to gradual enrichment with nutrients. Thus, *natural eutrophication* is a normal process. Wherever nutrients come from sewage-treatment plants, poor farming practices, urban runoff, and certain other human activities, however, humans have inadvertently managed to vastly accelerate the process of nutrient enrichment. The accelerated eutrophication caused by humans is called **cultural eutrophication.**

Combating Eutrophication

There are two approaches to combating the problem of cultural eutrophication. One is to *attack the symptoms*—the growth of vegetation, the lack of dissolved oxygen, or both.

The other is to *get at the root cause*—excessive inputs of nutrients and sediments.

Attacking the Symptoms. Attacking the symptoms is appropriate in certain situations in which immediate remediation is the goal and costs are not prohibitive. Methods of attacking the symptoms of eutrophication include (1) applying chemical treatments (with herbicides), (2) aerating, (3) harvesting aquatic weeds, and (4) drawing water down.

Applying Herbicides. Herbicides are often applied to ponds and lakes to control the growth of nuisance plants. To control phytoplankton growth, copper sulfate and diquat are frequently used. For controlling SAV and emergent vegetation, fluridone, glyphosate, and 2,4-D are employed. However, at the concentrations required to bring the vegetation under control, these compounds are sometimes toxic to fish and aquatic animals. Also, fish are often killed after herbicide is applied because the rotting vegetation depletes the water of dissolved oxygen. In sum, herbicides provide only cosmetic treatment, and as soon as they wear off, the vegetation grows back rapidly.

Aerating. Decomposers deplete the water of dissolved oxygen, thus suffocating other aquatic life, in the final and most destructive stage of eutrophication. *Artificial aeration* of the water can avert this terminal stage. An aeration technique currently gaining in popularity is to lay a network of plastic tubes with microscopic pores on the bottom of the waterway to be treated. High-pressure air pumps force microbubbles from the pores, and the bubbles dissolve directly into the water. The method is proving effective in speeding up the breakdown of accumulated detritus, improving water quality, and enabling the return of more desirable aquatic life. Despite its high cost, the technique is applicable to harbors, marinas, and some water-supply reservoirs, where the demand for better water quality justifies the cost.

Harvesting. In shallow lakes or ponds, where the problem is bottom-rooted vegetation reaching and sprawling over the surface (see Figure 20–19), *harvesting* the aquatic weeds may be an expedient way to improve the water's recreational potential and aesthetics. Commercial mechanical harvesters are used, and nearby residents sometimes get together to remove the vegetation by hand. The harvested vegetation makes good organic fertilizer and mulch. But even harvesting has a limited effect: The vegetation soon grows back because roots are left in the nutrient-rich sediments.

Drawing Water Down. Because many recreational lakes are dammed, another option for shallow-water weed control is to draw the lake down for a period each year. This process kills most of the rooted aquatic plants along the shore, although they grow back in time.

All of these approaches are only temporary fixes for the problem and have to be repeated often, at significant cost, to keep the unwanted plant growth under control.

Getting at the Root Cause. Controlling eutrophication requires long-term strategies for correcting the problem, which ultimately means reducing the inputs of nutrients and sediments—two of the types of pollution discussed at the beginning of this chapter. The first step is to identify the major point and nonpoint sources of nutrients and sediments. Then it is a matter of developing and implementing strategies for correction. Which source or factor is most significant will depend on the human population and the land uses within the particular watershed. Therefore, each watershed must be analyzed as a separate entity, and appropriate measures must be taken to reduce the levels of nutrients and sediments exiting from that watershed. In the sections that follow, we shall discuss major strategies used to control the root causes of eutrophication.

First, recall the concept of limiting factors (Chapter 3), in which the lack of only one nutrient can suppress growth. In natural freshwater systems, **phosphorus** is the most common limiting factor. In marine systems (like the coastal Gulf of Mexico), the limiting nutrient is most often **nitrogen**. Both in the environment and in biological systems, phosphorus (P) is present as phosphate (PO_4^{3-}), and nitrogen is present in a variety of compounds, most commonly as nitrate (NO_3^-) or the ammonium ion (NH_4^+). Therefore, we will focus primarily on phosphate and nitrogen compounds.

Ecoregional Nutrient Criteria. Beginning in 2001, the EPA began publishing water-quality nutrient criteria aimed at preventing and reducing the eutrophication that impacts so many bodies of water. The agency listed its recommended criteria for *causative* factors—*nitrogen* and *phosphorus*—and for *response* factors—*chlorophyll a* as a measure of both phytoplankton density and water clarity. The EPA divided the country into ecoregions and determined criteria levels for each specific region. For example, ecoregion VIII, which comprises the northern Great Lakes and the northeastern states, has the following criteria for lakes and reservoirs: total phosphorus 8.0 μg/L, total nitrogen 0.24 mg/L, chlorophyll *a* 2.43 μg/L, and Secchi depth (a measurement of transparency) 4.93 meters. Other ecoregions often have higher criteria levels, reflecting their differences from the pristine waters around the northern Great Lakes. Like the other water quality criteria, the nutrient criteria are provided as targets for the states as they address their water-pollution (and especially their eutrophication) problems.

NPDES: Control Strategy for Point Sources. In heavily populated areas, discharges from sewage-treatment plants have been major sources of nutrients entering waterways. The levels of phosphate in effluents from these plants were elevated even more in areas where laundry detergents containing phosphate were used. Phosphate contained in detergents goes through the system and out with the discharge unless the system has a nutrient removal component. In regions where eutrophication has been identified as a problem, a key step toward prevention has been to ban the sale of phosphate-based laundry detergents. Because many states established bans on the use of these products, the major laundry detergent manufacturers voluntarily ended their manufacture in 1994. However, the bans do not usually cover dishwashing detergents, some brands of which contain as much as 8.7% phosphate.

Bans on detergents with phosphate and upgrades of sewage-treatment plants have brought about marked improvements in waterways that were heavily damaged by effluents from these plants. In a sense, however, these are the easy measures, as the target for correction (the effluent) is an obvious *point source*, and methods for correcting the problem are clear-cut. The NPDES permitting process is the basic regulatory tool for reducing point-source pollutants. Under the Clean Water Act, "anyone discharging pollutants from any point source into waters of the U.S. [must] obtain an NPDES permit from EPA or an authorized state." More than 400,000 facilities are now regulated by way of NPDES permits. The permit must be drawn up in the context of the total pollutant sources (point and nonpoint) impacting a watershed or body of water.

TMDL: Control Strategy for Nonpoint Sources. Correction becomes more difficult, but no less important, when the source is diffuse, as in the case of farm and urban runoff. Remediation of such *nonpoint sources* will involve thousands—perhaps millions—of individual property owners adopting new practices regarding land management and the use of fertilizer and other chemicals on their properties. Nevertheless, that is the challenge. Section 303(d) of the Clean Water Act requires states to develop management programs to address nonpoint sources of pollution. Thus, there is a legal mandate to address the issues of agricultural and urban runoff.

The EPA has opted to develop regulations via the TMDL program. The concept is straightforward:

- Identify the pollutants responsible for degrading a body of water.

- Estimate the pollution coming from all sources (point and nonpoint).

- Estimate the ability of the body of water to assimilate the pollutants, while retaining water quality.

- Determine the maximum allowable pollution load.

- Within a margin of safety, allocate the allowable level of pollution among the different sources such that the water quality standards are achieved.

As with most attempts at solving environmental problems, the devil is in the details. The states are responsible for administering TMDL pollution abatement, but the EPA maintains oversight and approval of the results. State administrators have a number of options for managing nonpoint-source pollution, including regulatory programs, technical assistance, financial assistance, education and training, and demonstration projects. By 2008, the states had listed over 42,000 water bodies as being impaired. More than 4,100 were on the list because of nutrient overload. The EPA has approved a total of 37,000 TMDLs submitted by the states to correct the impairments and has substantially closed the gap between the state listings and the approved TMDLs.

Best Management Practices. Reducing or eliminating pollution from nonpoint sources will involve different strategies for different sources. For example, where a significant portion of the watershed is devoted to agricultural activities, the major sources of nutrients and sediments are likely to be (1) erosion and leaching of fertilizer from croplands and (2) runoff of animal wastes from barns and feedlots. All the practices that can be used to minimize such erosion, runoff, and leaching are lumped under a single term, **best management practices**, which includes all the methods of soil conservation discussed in Chapter 11. Table 20–2 lists examples of these practices for a variety of nonpoint sources.

Once control measures have been put in place, the polluted body of water must be monitored to determine whether water quality standards are being attained. Several years of data collection may be necessary to detect genuine trends

TABLE 20–2 Best Management Practices for Reducing Pollution from Nonpoint Sources

Agriculture	Forestry
Animal waste management	Ground cover maintenance
Conservation tillage	Limiting disturbed areas
Contour farming	Log removal techniques
Strip-cropping	Pesticide and herbicide management
Cover crops	Proper management of roads
Crop rotation	Removal of debris
Fertilizer management	Riparian zone management
Integrated pest management	**Mining**
Range and pasture management	Underdrains
Terraces	Water diversion
Construction	**Multicategory**
Limiting disturbed areas	Buffer vegetated strips
Nonvegetative soil stabilization	Detention and sedimentation basins
Runoff detention and retention	Devices to encourage infiltration
Urban	Vegetated waterways
Flood storage	Interception and diversion of runoff
Porous pavements	Sediment traps
Runoff detention and retention	Streamside management zones
Street cleaning	Vegetative stabilization and mulching

Source: U.S. Environmental Protection Agency, *Guidance for Water Quality-Based Decisions: The TMDL Process*, EPA440-4-91-001 (Washington, D.C.: USEPA, 1991).

in quality. When water quality standards are reached and sustained, the body of water is removed from the list of impaired waters. To date, a total of 1,145 previously impaired bodies of water have been removed from the list (not a very impressive record). If standards are not met, the TMDL process must be revisited and new pollution allowances allocated.

Recovery. The good news is that cultural eutrophication and other forms of water pollution can be controlled and often reversed, provided that a total watershed management approach is undertaken. For a small lake fed from a modest-sized watershed, the task may be quite straightforward because major sources of nutrients can be identified and addressed. For large bodies of water, such as the Chesapeake Bay—with a watershed of 63,700 square miles (165,000 km²) in six states and the District of Columbia—the task is significantly greater. The Chesapeake Bay has suffered from heavy nutrient pollution, with cultural eutrophication destroying all but a tenth of the 600,000 acres of vital sea grasses that once formed the basis of the bay's rich ecosystem. A dead zone appears every year, involving up to 40% of the main bay for 10 months of the year. The states involved have been cooperating under the Chesapeake Bay Program with the common goal of achieving federal clean water standards by 2010. Unfortunately, new development in the watershed is contributing nutrients and sediments faster than the restoration efforts are reducing them. It is clear that the 2010 goal of the Chesapeake Bay Program will not be met, and the EPA has begun the process of setting a baywide TMDL that will then limit specific nutrient inflows from each of the bay's tributaries. This will give the states clear targets for implementing the restoration activities.

Lake Washington. One remarkable success story is Lake Washington, east of Seattle (Fig. 20–20). This 34-square-mile lake is at the center of a large metropolitan area. During the 1940s and 1950s, 11 sewage-treatment plants were sending state-of-the-art treated water into the lake at a rate of 20 million gallons per day. At the same time, phosphate-based detergents came into wide use. The lake responded to the massive input of nutrients by developing unpleasant "blooms" of noxious blue-green algae. The water lost its clarity, the desirable fish populations declined, and masses of dead algae accumulated on the shores of the lake.

Citizen concern led to the creation of a system that diverted the treatment-plant effluents into nearby Puget Sound, where tidal flushing would mix them with open-ocean water. The diversion was complete by 1968, and the lake responded quickly. The algal blooms diminished, the water regained its clarity, and by 1975, recovery was complete. Careful studies by a group of limnologists from the University of Washington showed that phosphate was the culprit. Before the diversion, the lake was receiving 220 tons of phosphate per year from all sources. Afterward, the total dropped to 40 tons per year, well within the lake's normal capacity to absorb phosphate by depositing it in deep bottom sediments. One clear lesson learned was that sewage-treatment effluent must never be allowed to enter a lake unless nutrients are removed as part of the treatment. The major lesson, however, was that eutrophication can be reversed by paying attention to nutrient inputs and addressing the various sources. Lake management, in other words, is a matter of controlling the phosphate loading into a lake from all sources: human and animal sewage, agricultural and yard fertilizers, street runoff, and failing septic systems. Many lakes are now protected by associations of concerned citizens who see themselves as stewards of their lake.

20.4 Public Policy and Water Pollution

In the United States, the responsibility for overseeing the health of the nation's waters rests with the EPA. However, the EPA can develop regulations only if Congress gives it the authority to do so. Hence, the foundation for public policy must be the laws passed by Congress. Some major legislative milestones in protecting the nation's waters are listed in Table 20–3. The landmark legislation is the **Clean Water Act of 1972 (CWA)**, which gave the EPA jurisdiction over (and for the first time required permits for) all point-source discharges of pollutants. This act and subsequent amendments have provided $78 billion to help cities and towns build treatment plants to meet the federal requirement for secondary treatment of all sewage. As the table shows, the 1987 amendments to the CWA established a revolving loan fund—the **Clean Water State Revolving Fund (SRF)** program—to replace the direct-grants program. To build treatment facilities, local governments borrow at low interest rates, and as they repay the loans, the funds received are used for more loans. The fund may also be used to control nonpoint-source pollution. To date, over $68 billion in SRF loans have been made by the states. However, the EPA estimates that an additional $390 billion will be required over the next 20 years to

Figure 20–20 Lake Washington. A testimony to the fact that bodies of water can recover from cultural eutrophication, Lake Washington is now thriving because sources of nutrients enriching the lake have been removed.

TABLE 20–3 Legislative Milestones in Protecting Natural Waters and Water Supplies

1899—Rivers and Harbors Act: First federal legislation protecting the nation's waters in order to promote commerce.

1948—Water Pollution Control Act: Authorizes the federal government to provide technical assistance and funds to state and local governments to promote efforts to protect water quality.

1972—Clean Water Act: Establishes a comprehensive federal program to achieve the goal of protecting and restoring the physical, chemical, and biological integrity of the nation's waters. Requires permits for any discharge of pollution, strengthens water quality standards, and encourages the use of the best achievable pollution control technology. Provides billions of dollars for construction of sewage-treatment plants.

1972—Marine Protection, Research, and Sanctuaries Act: Prevents unacceptable dumping in oceans.

1974—Safe Drinking Water Act: Authorizes the EPA to regulate the quality and safety of public drinking-water supplies, maintain drinking-water standards for numerous contaminants, and set requirements for the chemical and physical treatment of drinking water.

1977—Clean Water Act amendments: Strengthen controls on toxic pollutants and allow states to assume responsibility for federal programs.

1987—Water Quality Act: Enacts major amendments to the Clean Water Act that (1) create a revolving loan fund to provide ongoing support for the construction of treatment plants, (2) address regional pollution with a watershed approach, and (3) address nonpoint-source pollution, with requirements for states to assess the problem and develop and implement plans for dealing with it. Makes available $400 million in grants to carry out its provisions.

1996—Safe Drinking Water Act amendments: Impose numerous requirements for managing water supplies, establish a revolving loan fund to help municipalities update their water system infrastructure, and provide flexibility for the EPA to base its selection of contaminants on health risk and costs and benefits.

2002—Great Lakes Legacy Act: Enacted amendments to CWA provisions for the Great Lakes and authorized $50 million annually for fiscal years 2004–2008 for the EPA to conduct projects designed to remediate sediment problems in the Great Lakes.

---

meet the funding needs for all eligible municipal wastewater-treatment systems. This amount represents a gap of about $6 billion between current annual expenditures and projected needs.

Reauthorization. Reauthorization of the Clean Water Act is long overdue. Congress has been hung up on debates over whether requirements should be strengthened or weakened, whether federal regulations in the wetlands permit program intrude on private land-use rights, and how regulatory relief should be provided to industries, states, cities, and individuals required to take actions to comply with the regulations. In particular, Congress has been unsure of how to deal with the TMDL program. Its approach for the last two decades has been to reauthorize the provisions of the CWA and its amendments, usually maintaining a status quo in the EPA's regulatory responsibilities and the annual appropriations for CWA programs.

Problems and Progress. The EPA has identified nonpoint-source pollution as the nation's number-one water-pollution problem, with the construction of new wastewater-treatment facilities not far behind. Other significant issues, including storm-water discharges, combined and separate sewer overflows, wetlands protection, and animal feeding operations, are also receiving the EPA's attention and are in Congress's sights for possible action. The continuing high percentage of the nation's water resources that fail to meet

the water quality standards makes it clear that much remains to be accomplished. According to the U.S. Public Interest Research Group, 57% (more than 3,600) of major facilities are polluting waterways above the levels that their NPDES permits allow.

Nevertheless, much progress has been made in the 38 years since the enactment of the Clean Water Act. This act is seen as one of the most successful of our nation's environmental laws. The number of people in the United States served by adequate sewage-treatment plants has risen from 85 million to 208 million. Soil erosion has been reduced by 1 billion tons annually, and two-thirds of the nation's waterways are safe for fishing and swimming, double the number in 1972. Many of the nation's most heavily used rivers, lakes, and bays—including the Androscoggin River in Maine, Boston Harbor, the upper Mississippi River, the South Platte River, the upper Arkansas River, Lake Erie, the Illinois River, and the Delaware River—have been cleaned up and restored. Fish now swim in rivers once so polluted that only bacteria and sludge worms could survive. Levels of toxic chemicals in the Great Lakes have been greatly reduced. A national sense of stewardship has been applied to the rivers, lakes, and bays that are our heritage from a previous generation, and public policy has been enacted and billions of dollars spent to bring our waters back from a polluted condition.

Revisiting the Themes

Sound Science

Solid scientific research and communication were essential in establishing the basic knowledge about pollutants and their impacts on human health and the environment. They continue to be highly important as the EPA carries out its research and regulatory activities, administering the Clean Water Act and the Safe Drinking Water Act. The criteria pollutant standards for natural waters and the maximum contaminant standards for drinking waters must be based on accurate risk assessments as the EPA develops and defends its regulatory activities. Sound science was particularly crucial in the investigation of dead zones and algal blooms and their causes.

Sustainability

Our natural waters represent an enormous bank of ecosystem capital, providing essential goods and services to people everywhere. The sustainability of whole ecosystems is threatened when we fail to handle our wastes in a conscientious manner. The opening story of the dead zone demonstrates how it is possible for nutrient overload to degrade an enormous area of the Gulf of Mexico. Unfortunately, a similar situation exists in the Chesapeake Bay and many other bodies of water. These are just a couple of the more spectacular examples. The more important battlegrounds are the thousands of smaller lakes, rivers, and coastal estuaries that are also suffering from human impacts. More needs to be done, especially at the state level, to improve our national water quality; the Congress can play an important role as it establishes funding for the states to carry out its mandates. Many states are behind in assessing water quality, and the need for updating sewage-treatment systems is enormous.

Stewardship

Stewardship means intentionally caring for people and the natural world. The public-health measures put in place in the last two centuries represent the stewardship of human resources. Also, as we divert public and private resources to cleaning up our waterways, we are acting as good stewards in the best sense of the concept. Lakes and watersheds often have people and associations that have provided leadership in correcting the problems that lead to eutrophication. The Lake Washington story is an excellent example. Stewardship also means caring for our neighbors' needs, and nowhere are those needs more apparent than in the need for clean water and sanitation in so many developing countries. A sense of the global reach of water pollution is difficult to imagine, yet billions of people in the developing countries are held back from achieving their physical and intellectual potential because of the terrible effects of polluted water and lack of sanitation.

REVIEW QUESTIONS

1. What practices and consequences led to passage of the Clean Water Act of 1972?

2. Discuss each of the following categories of water pollutants and the problems they cause: pathogens, organic wastes, chemical pollutants, and sediments.

3. How are water quality standards determined? Distinguish between water quality criteria pollutants and maximum contaminant levels.

4. Name and describe the facility and the process used to remove debris, grit, particulate organic matter, colloidal and dissolved organic matter, and dissolved nutrients from wastewater.

5. Why is secondary treatment also called biological treatment? What is the principle involved? What are the two alternative techniques used?

6. What are the principles involved in removing biological nutrients from waste, and what is accomplished by doing so? Where do nitrogen and phosphate go in the process?

7. Name and describe three methods of treating raw sludge, and give the end product(s) that may be produced from each method.

8. How may sewage from individual homes be handled in the absence of municipal collection systems? What are some problems with these on-site systems?

9. Describe and compare submerged aquatic vegetation and phytoplankton. Where and how does each get nutrients and light?

10. Explain the difference between oligotrophic and eutrophic waters. Describe the sequential process of eutrophication.

11. Distinguish between natural and cultural eutrophication.

12. What is being done to establish nutrient criteria?

13. How does the NPDES program address point-source pollution by nutrients?

14. Describe the TMDL program. How does it address nonpoint-source pollution, and what role do water quality criteria play in the program?

15. What are some of the important public-policy issues relating to water quality?

THINKING ENVIRONMENTALLY

1. A large number of fish are suddenly found floating dead on a lake. You are called in to investigate the problem. You find an abundance of phytoplankton and no evidence of toxic dumping. Suggest a reason for the fish kill.

2. Nitrogen pollution has damaged the San Diego Creek–Newport Bay ecosystem in California by encouraging the heavy growth of intertidal algae. Report on the TMDL for this system. Consult the EPA Web site: www.epa.gov/owow/tmdl/examples/nutrients.html.

3. Arrange a tour to the sewage-treatment plant that serves your community. Compare it with what is described in this chapter. Is the water being purified or handled in a way that will prevent cultural eutrophication? Are sludges being converted to, and used as, fertilizer? What improvements, if any, are in order? How can you help promote such improvements?

4. Suppose a new community of several thousand people is going to be built in Arizona (with a warm, dry climate). You are called in as a consultant to design a complete sewage system, including the collection, treatment, and use or disposal of by-products. Write an essay describing the system you recommend, and give a rationale for the choices involved.

MAKING A DIFFERENCE

1. Check out the lakes and ponds in your area, and determine whether cultural eutrophication is occurring. If you find that it is, follow up by consulting a local environmental group, and get involved in combating the problem.

2. Investigate sewage treatment in your community. To what extent is the sewage treated? Where does the effluent go? The sludge? Are there any problems? Are there alternatives that are more ecologically sound? Consult local environmental groups for help.

3. Bring discarded household chemicals to hazardous waste collection centers; do not pour them down the drain. Pouring chemicals down the drain could upset your septic system or infect treatment-plant sludge.

4. Test your soil before using fertilizers. Excess from overfertilizing can leach into groundwater or contaminate rivers or lakes. Also, avoid using fertilizers and pesticides near surface waters.

Go to **www.masteringenvironmentalscience.com** for practice quizzes, Pearson eText, videos, current events, and more.

E-waste, in the form of circuit boards. Electronic waste is modern technology's unwanted contribution to the household hazardous-waste stream. Worldwide, 50 million tons per year are generated.

CHAPTER 21

Municipal Solid Waste: Disposal and Recovery

M AYOR THOMAS W. DANEHY PARK features four soccer fields, four softball fields, three basketball courts, a baseball diamond, two playgrounds, and a two-mile jogging trail. The 50-acre park, opened in 1990, is adjacent to a heavily populated area of North Cambridge, Massachusetts, and is in almost constant use in good weather (Fig. 21–1). The city of Cambridge sponsors annual family days in September that draw more than 4,000 people to the park, and nearby upscale Walden Park Apartments features Danehy Park as one of its selling points. An unusual feature of the park is a big red light in the public restroom that warns users to vacate the park if the light goes on. The light is tied into an elaborate venting system that prevents methane from building up below ground. The system is necessary because Danehy Park is built on the former city dump. Aside from the presence of the light and a few minor settling problems that have complicated the water drainage on the soccer fields, a newcomer would never know that this was once a blight on the neighborhood—an open, burning dump in the 1950s and then a "sanitary landfill" that was closed in 1972. The park increased Cambridge's open space by 20% when it was created.

Danehy Park is a lesson in land use that is being learned only slowly in the United States—and, for that matter,

Figure 21–1 From landfills to playing fields. Thomas W. Danehy Park in Cambridge, Massachusetts, a former landfill that has been recycled into a recreational park. In addition to playing fields, the park features a half-mile "glassphalt" pathway (built with recycled glass and asphalt), shown in the foreground.

in the rest of the world. The lesson is that, if we could picture more effectively what 50 years can bring in the way of change, we could solve one of the most contentious problems facing local communities: what to do with municipal waste. As old dumps and landfills were closed because of environmental concerns, they created a temporary problem identified as the "solid-waste crisis" in the 1970s and 1980s. Now, many of those dumps and landfills are being converted into parks, golf courses, and nature preserves, and future generations will hardly suspect that their recreational site was once a landfill.

STILL A CRISIS? In some ways, the solid-waste crisis is still with us, as we will see when we examine the problem of

interstate trash movement. It is commonly said that we are running out of space to put our trash and garbage. Perhaps we are, but if so, it is because we are happy to purchase the goods displayed so prominently in our malls and advertised in the media, but we are reluctant to accept the consequences of getting rid of them responsibly.

This chapter is about solid-waste issues. We examine current patterns of disposal—landfills, combustion, and recycling—and look for sustainable solutions to our solid-waste problems. The ideal would be to imitate natural ecosystems and reuse everything. Some solutions do well in conforming to this principle, whereas others do not—and possibly cannot.

21.1 Solid Waste: Landfills and Combustion

The focus of this chapter is **municipal solid waste (MSW)**, defined as the total of all the materials (commonly called trash, refuse, or garbage) thrown away from homes and smaller commercial establishments and collected by local governments. MSW is different from hazardous waste (covered in Chapter 22) and nonhazardous industrial waste. The latter is no small matter: Industrial facilities generate and manage 7.6 billion tons of nonhazardous industrial waste annually. Included in the category are wastes from demolition and construction operations, agricultural and mining residues, combustion ash, sewage-treatment sludge, and wastes generated by industrial processes. The states oversee these wastes because Congress has not delegated any authority to the

Environmental Protection Agency (EPA) to regulate them, as it has for MSW and hazardous waste. The EPA Office of Solid Waste does, however, provide annual reports on municipal solid waste in the United States, and we will make full use of the latest report in this chapter.

Disposal of Municipal Solid Waste

Over the years, the amount of MSW generated in the United States has grown steadily, in part because of a growing population, but also because of changing lifestyles and the increasing use of disposable materials and excessive packaging. In 1960, the nation generated 2.7 pounds (1.2 kg) per person per day. In 2007, we generated a total of 254 million tons (231 million metric tons) of MSW, an average of 4.6 pounds (2 kg) per person per day. In 2007, the U.S. population of

302 million generated enough waste to fill 96,000 garbage trucks each day. The solid-waste problem can be stated simply: *We generate huge amounts of MSW, and it is increasingly expensive to dispose of it in ways that are environmentally responsible and protective of human health.*

The refuse generated by municipalities is a mixture of materials from households and small businesses, in the proportions shown in Figure 21–2. However, the proportions vary greatly, depending on the generator (commercial versus residential), the neighborhood (affluent versus poor), and the time of year (during certain seasons, yard wastes, such as grass clippings and raked leaves, add greatly to the solid-waste burden). Little attention is given to what people throw away in their trash. Even if there are restrictions and prohibitions, they can be bypassed with careful packing of the trash containers. Thus, many environmentally detrimental substances—paint, used motor oil, old TV sets, and so on—are discarded, with the feeling that they are gone forever.

Whose Job? Customarily, local governments have had the responsibility for collecting and disposing of MSW. The local jurisdiction may own trucks and employ workers, or it may contract with a private firm to provide the collection service. Traditionally, the cost of waste pickup is passed along to households via taxes. Alternatively, many municipalities have opted for a pay-as-you-throw (PAYT, as it is called) system, in which households are charged for waste collection on the basis of the amount of trash they throw away. Some municipalities put all trash collection and disposal in the private sector, with the collectors billing each home by volume and weight of trash. The MSW that is collected is then disposed of in a variety of ways, and it is at the point of disposal that state and federal regulations begin to apply.

Past Sins. Until the 1960s, most MSW was burned in open dumps. The waste was burned to reduce its volume and lengthen the life span of the dump site, but refuse does not burn well. Smoldering dumps produced clouds of smoke that could be seen from miles away, smelled bad, and created a

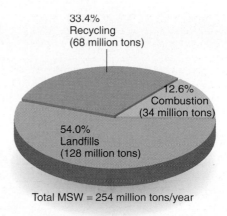

Total MSW = 254 million tons/year

Figure 21–3 U.S. municipal solid-waste disposal. Disposal of solid waste to landfills, combustion, and recycling, 2007. (*Source:* Data from EPA, Office of Solid Waste, *Municipal Solid Waste in the United States: 2007 Facts and Figures* (Washington, D.C.: EPA, November 2008).)

breeding ground for flies and rats. Some cities turned to incinerators (or combustion facilities, as they are called today): huge furnaces in which high temperatures allow the waste to burn more completely than in open dumps. Without controls, however, incinerators were prime sources of air pollution. Public objections and air pollution laws forced the phaseout of open dumps and many incinerators during the 1960s and early 1970s. Open dumps were then converted to landfills.

In the United States in 2007, 54.0% of MSW was disposed of in landfills, 33.4% was recovered for recycling and composting, and the remainder (12.6%) was combusted (Fig. 21–3). Over the last 10 years, the landfill and combustion components have been declining, while recycling has shown a steady increase. The pattern is different in countries where population densities are higher and there is less open space for landfills. High-density Japan, for instance, combusts 93% of its nonrecycled trash. The pattern in western Europe is mixed, ranging from 60% combustion of nonrecycled wastes in the crowded Netherlands (where 50% is recycled) to over 90% of nonrecycled wastes going to the landfills in the United Kingdom (with only 12% recycled).

Landfills

In a **landfill**, the waste is put on or in the ground and is covered with earth. Because there is no burning and because each day's fill is covered with at least six inches of earth, air pollution and populations of vermin are kept down. Unfortunately, aside from those concerns and the minimizing of cost, no other factors were given real consideration when the first landfills were opened. Municipal waste managers generally had little understanding of ecology, the water cycle, or what products decomposing wastes would generate, and they had no regulations to guide them. Therefore, in general, any cheap, conveniently located piece of land on the outskirts of town became the site for a landfill. Frequently, this site was a natural gully or ravine, an abandoned stone quarry, a

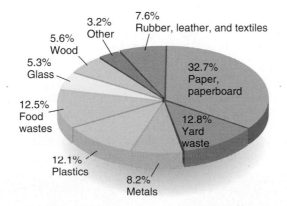

Figure 21–2 U.S. municipal solid-waste composition. The composition of municipal solid waste in the United States in 2007. Go to **GRAPHIT!** at www.masteringenvironmentalscience.com. (*Source:* Data from EPA, Office of Solid Waste, *Municipal Solid Waste in the United States: 2007 Facts and Figures* (Washington, D.C.: EPA, November 2008).)

Figure 21–4 New Orleans dump. This burning dump, sited on wetlands, demonstrates the worst of the MSW disposal practices of the past.

section of wetlands, or a previous dump (Fig. 21–4). Once the municipality acquired the land, dumping commenced without any precautions. After the site was full, it would be covered with earth and ignored. Only recently have abandoned landfills been seen as a valuable open-space resource.

Problems of Landfills. Landfills are subjected to biological and physical factors in the environment and will undergo change over time as a consequence of the operation of those factors on the waste that is deposited. Several of the changes are undesirable because, if they are not dealt with effectively, they present the following problems:

- Leachate generation and groundwater contamination,
- Methane production,
- Incomplete decomposition, and
- Settling.

Leachate Generation and Groundwater Contamination. The most serious problem by far is groundwater contamination. As water percolates through any material, various chemicals in the material may dissolve in the water and get carried along in a process called *leaching* (Chapter 10). The water with various pollutants in it is called *leachate*. As water percolates through MSW, a noxious leachate is generated that consists of residues of decomposing organic matter combined with iron, mercury, lead, zinc, and other metals from rusting cans, discarded batteries, and appliances—all generously "spiced" with paints, pesticides, cleaning fluids, newspaper inks, and other chemicals. The nature of the landfill site and the absence of precautionary measures can funnel this "witches' brew" directly into groundwater aquifers.

All states have some municipal landfills that are, or soon will be, contaminating groundwater, but Florida has some unique problems. Flat and with vast areas of wetlands, much of the state is only a few feet above sea level and rests on water-saturated limestone. No matter where Florida's

landfills were located, they were either in wetlands or just a few feet above the water table. As a result, more than 145 former municipal landfill sites in Florida appear on the Superfund list. (Superfund is the federal program with the responsibility for cleaning up sites that are in imminent danger of jeopardizing human health through groundwater contamination.) All landfills in Florida are now required to have state-of-the-art landfill liners.

Methane Production. Because it is about two-thirds organic material, MSW is subject to natural decomposition. Buried wastes do not have access to oxygen, however, so their decomposition is anaerobic. A major by-product of the process is *biogas*, which is about two-thirds methane and one-third hydrogen and carbon dioxide, a highly flammable mixture. (See Chapter 20.) Produced deep in a landfill, biogas may seep horizontally through the soil and rock, enter basements, and even cause explosions if it accumulates and is ignited. Homes at distances up to 1,000 feet from landfills have been destroyed, and some deaths have occurred as a result of such explosions. Also, gases seeping to the surface kill vegetation by poisoning the roots. Without vegetation, erosion occurs, exposing the unsightly waste.

A number of cities have exploited the problem by installing "gas wells" in old and existing landfills. The wells tap the landfill gas, and the methane is purified and used as fuel. There are now over 450 landfill gas energy projects in the United States. In 2008, commercial landfill gas in the nation produced 11 billion kilowatt-hours of electricity and 77 billion cubic feet of gas for direct-use applications, powering or heating over 1.4 million homes. For example, Waste Management, Inc., of Houston, is now mining the landfill gas from a closed landfill in Westminster, Massachusetts (Fig. 21–5). The methane drives two generators that provide electricity to 3,000 homes. This is one of 22 gas-to-energy facilities in Massachusetts, providing power to an estimated 77,000 homes.

The recovery of landfill gas has significant environmental benefits. It directly reduces greenhouse gas emissions because methane is a powerful greenhouse gas and landfills

Figure 21–5 Westminster landfill. Methane from this closed Massachusetts landfill is now being mined by Waste Management, Inc., and fuels a small power plant that generates 3.2 megawatts of electricity, enough to power 3,000 homes.

are the largest anthropogenic source of methane emissions in the United States. Methane recovery also reduces the use of coal and oil, which are nonrenewable and highly polluting energy resources.

Incomplete Decomposition. The commonly used plastics in MSW resist natural decomposition because of their molecular structure. Chemically, they are polymers of petroleum-based compounds that microbes are unable to digest. Biodegradable plastic polymers have been developed from sources such as cornstarch, cellulose, lactic acid, and soybean protein as well as petroleum. As oil prices rise, plant-based biodegradable plastics are beginning to make economic inroads. Two kinds of bioplastics based on corn sugars—polyhydroxyalkanoate and polylactic acid—are in commercial production; a recent estimate from NatureWorks indicates a coming demand for up to 50 billion pounds of bioplastics a year in the next two to five years. Currently, bioplastics are more expensive than petroleum-based plastics and tend to be used by companies manufacturing natural and certified organic products. They are indeed biodegradable, but some of the corn-based plastics must go to a commercial composting plant to be decomposed.

A team of "garbologists" from the University of Arizona, led by William Rathje, has been carrying out research on old landfills. The research has shown that even materials formerly assumed to be biodegradable—newspapers, wood, and so on—are often degraded only slowly, if at all, in landfills. In one landfill, 30-year-old newspapers were recovered in a readable state, and layers of telephone directories, practically intact, were found marking each year. Since paper materials are 35% of MSW, this is a serious matter. The reason paper and other organic materials decompose so slowly is the absence of suitable amounts of moisture. The more water percolating through a landfill, the better paper materials biodegrade. Unfortunately, the more percolation there is, the greater the amount of toxic leachate produced!

Settling. Finally, waste settles as it compacts and decomposes. Luckily, this eventuality was recognized from the beginning, so buildings have never been put on landfills. Settling does present a problem in landfills that have been converted to playgrounds and golf courses because it creates shallow depressions (and sometimes deep holes) that collect and hold water. This problem can be addressed by continually monitoring the facility and using fill to restore a level surface.

Improving Landfills. Recognizing the foregoing problems, the EPA upgraded siting and construction requirements for new landfills. Under current regulations,

- New landfills are sited on high ground, well above the water table, not in a geologically unstable area, and away from airports (because of bird hazards).

- The floor is contoured so that water will drain into a tile leachate-collection system. The floor and sides are covered with a plastic liner and at least two feet of compacted soil. With such a design, any leachate percolating through the fill will move into the leachate-collection system. Collected leachate can be treated as necessary.

- Layer upon layer of refuse is positioned such that the fill is built up in the shape of a pyramid. Finally, it is capped with at least 18 inches of earthen material and a layer of topsoil and then seeded. The cap and the pyramidal shape help the landfill shed water. In this way, water infiltration into the fill is minimized, and less leachate is formed.

- The entire site is surrounded by a series of groundwater-monitoring wells that are checked periodically, and such checking must go on indefinitely.

These design features are summarized in Figure 21–6. Most landfills currently in operation have the improved technologies, which protect both human health and the environment.

The landfill pyramids may well last as long as the Egyptian pyramids. Fortunately, the creative siting and construction of landfills have the potential to address some highly significant future needs because abandoned landfills can become attractive golf courses, recreational facilities, or wildlife preserves.

Siting New Landfills. Between 1988 and 2007, the number of municipal landfills declined from around 8,000 to 1,754. Because the size of landfills has increased and because recycling is on the rise, the EPA does not believe that landfill capacity is a problem. However, as the agency puts it, "regional dislocations" may be expected. These are due to the one problem associated with landfills that gets more attention than any other: **siting.** It is not that landfills take up enormous amounts of land. One particular landfill occupying 121 acres (Central Landfill) serves the entire state of Rhode Island, and Fresh Kills, the largest landfill in the world, served much of New York City and its environs for many years and was only 2,200 acres. But as old landfills are closed, it has become increasingly difficult to find land for the new ones needed to take their place.

People in residential communities (where the MSW is generated) invariably reject proposals to site landfills anywhere near where they live, and those who already live close to existing landfills are anxious to close them down. Weary of the odor and heavy truck traffic, residents of Staten Island, New York, pressured the city of New York to shut down the huge Fresh Kills landfill, once the recipient of 27,000 tons of garbage per day. The city fulfilled its promise to phase out the landfill in 2001, and Staten Islanders are now anticipating the development of Fresh Kills Park, with soccer fields, mountain bike and jogging trails, natural areas, and possibly a golf course. Gas from the huge landfill is providing the area with a new energy source. Currently, New York City exports its 11,550 tons per day of MSW outside of the city by truck, barge, and rail.

With spreading urbanization, few suburban areas are not already dotted with residential developments. Practically any site selected, then, is met with protests and legal suits. The problem has been repeated in so many parts of the country (and globe!) that it has given rise to some inventive acronyms: **LULU** (locally unwanted land use), **NIMBY** (not in my backyard), **NIMTOO** (not in my term of office), and **BANANA** (build absolutely nothing anywhere near anything)!

Figure 21–6 Features of a modern landfill. The landfill is sited on a high location, well above the water table. The bottom is sealed with compacted soil and a plastic liner, overlaid by a rock or gravel layer, with pipes to drain the leachate. Refuse is built up in layers as the amount generated each day is covered with soil, so that the completed fill has a pyramidal shape that sheds water. The fill is provided with wells for monitoring groundwater.

You can well imagine how these attitudes apply to the landfill-siting problem.

Outsourcing. The siting problem has some undesirable consequences. First, it drives up the costs of waste disposal, as alternatives to local landfills are invariably more expensive. Second, it leads to the inefficient and equally objectionable practice of the long-distance transfer of trash as waste generators look for private landfills whose owners are anxious to receive trash. Often, this transfer occurs across state and even national lines, leading to resentment and opposition on the part of citizens of the recipient state or nation. Led by Pennsylvania, 11 states import more than 1 million tons of MSW a year (Table 21–1), while 13 states export more than 1 million tons per year, led by New York. Not surprisingly, the heaviest importing states are adjacent to the heaviest exporting states.

One positive impact of the siting problem is that it encourages residents to reduce their waste and recycle as much as possible. (This alternative is discussed in Section 21.2.) Another effect is that the problem stimulates the use of combustion as an option for waste disposal.

Combustion: Waste to Energy

Because it has a high organic content, refuse (especially the plastic portion) can be burned. Currently, 89 combustion facilities are operating in the United States, burning about 32 million tons of MSW annually—12.6% of the waste stream. This process is really waste *reduction*, not waste *disposal*, because after incineration the ash must still be disposed of.

Advantages of Combustion. The combustion of MSW has some advantages:

- Combustion can reduce the weight of trash by more than 70% and the volume by 90%, thus greatly extending the life of a landfill (which is still required to receive the ash).

- Toxic or hazardous combustion products are concentrated into two streams of ash, which are easier to handle and control than the original MSW. The *fly ash* (captured from the combustion gases by air-pollution control equipment) contains most of the toxic substances and can be safely put into a landfill. The *bottom ash* (from the bottom of the boiler) can be used as fill in some construction sites and roadbeds. Some combustion facilities process the bottom ash further to recover metals and then convert the remainder into concrete blocks.

- No changes are needed in trash collection procedures or people's behavior. Trash is simply hauled to a combustion facility instead of to the landfill.

- Two-thirds of the combustion facilities are waste-to-energy (WTE) facilities equipped with modern emission-control technology that brings the emissions into compliance with Clean Air Act regulations.

TABLE 21–1   Largest Interstate Transfers of MSW, 2005

Exporting States and Provinces	Exported MSW (millions of tons)	Importing States	Imported MSW (millions of tons)
New York	7.2	Pennsylvania	7.9
New Jersey	5.8	Virginia	5.7
Illinois	4.4	Michigan	5.4
Ontario, Canada	4.0	Indiana	2.4
Missouri	2.4	Wisconsin	2.1
Maryland	2.0	Illinois	2.1
Massachusetts	2.0	Oregon	1.8
Washington	1.7	Georgia	1.7
Minnesota	1.1	New Jersey	1.7
North Carolina	1.1	Ohio	1.7
Indiana	1.1	South Carolina	1.2
District of Columbia	1.1		
Florida	1.0		
Total, all exports	**42.8**	**Total, all imports**	**42.2**

Source: James E. McCarthy, *Interstate Shipment of Municipal Solid Waste: 2007 Update,* CRS Report for Congress RL34043 (Washington, D.C.: Congressional Research Service, June 13, 2007).

- When burned, unsorted MSW releases about 35% as much energy as coal, pound for pound. WTE facilities produce 2,700 megawatts of electricity annually, enough to meet the power needs of 2.3 million homes. Advocates of WTE point to the wasted energy going to landfills—equivalent to 9.4 billion gallons of diesel oil a year.

- To their waste processing, many of these facilities add resource recovery, in which many materials are separated and recovered before (and sometimes after) combustion.

Drawbacks of Combustion.   Combustion has some drawbacks, too:

- Air pollution and offensive odors are two problems that the public associates with combustion facilities. Large reductions in air pollutants have resulted from compliance with stringent air-pollution regulations applied to WTE facilities, making this much less of a problem than it once was. For example, dioxin emissions (dioxin is a highly toxic substance given off when many organic substances are combusted) have been reduced by 99% and mercury emissions by 90%. Odor pollution is best controlled by isolating the plant from residential areas.

- Combustion facilities are expensive to build, and their siting has the same problem as that of landfills: No one wants to live near one. Most combustion facilities are located in industrial areas for this reason.

- Combustion ash is often loaded with metals and other hazardous substances and must be disposed of in secure landfills.

- To justify the cost of its operation, the combustion facility must have a continuing supply of MSW. For that reason, the facility enters into long-term agreements with municipalities, and these agreements can lessen the flexibility of the community's solid-waste management options.

- Even if the combustion facility generates electricity, the process wastes both energy and materials unless it is augmented with recycling and recovery.

An Operating Facility.   Let us look at the operation of a typical modern WTE facility, which might serve a number of communities or a large metropolitan area. Servicing a population of 1 million or more, the plant receives about 3,000 tons of MSW per day. The waste comes in by rail and truck, and the communities pay tipping fees (the costs assessed at the disposal site) that average $65 per ton. Waste processing is efficient because, overall, about 80% of the MSW is burned for energy, 12% is recovered, and 8% is put in landfills. The process, pictured in Figure 21–7, is as follows:

1. Incoming waste is inspected, and obvious recyclable and bulky materials are removed.

2. Waste is then pushed onto conveyers that feed shredders capable of reducing the width of waste particles to 6 inches or less.

Figure 21–7 Waste-to-energy combustion facility. Schematic flow for the separation of materials and combustion in a typical modern WTE combustion facility. Numbers refer to steps in the process as described in the text.

3. Strong magnets remove about two-thirds of ferrous metals for recycling before combustion.

4. The waste is then blown into boilers, where light materials burn in suspension and heavier materials burn on a moving grate.

5. Water circulated through the walls of the boilers produces steam, which drives turbines for generating electricity.

6. After the waste is burned, the bottom ash is conveyed to a processing facility, where further separation of metals may occur in a process that recovers brass, aluminum, gold, copper, and iron. In some facilities, this process nets $1,000 a day in coins alone!

7. Combustion gases are passed through a lime-based spray dryer/absorber to neutralize sulfur dioxide and other noxious gases. Then the gases go through electrostatic precipitators that remove particles. The resulting air emissions are significantly lower in pollutants than those of an energy-equivalent utility based on coal or oil combustion.

8. The fly ash and bottom ash residues are put into landfills.

An appreciation for the impact of such a facility can be gained from looking at the outcome of a year's operation. In one year, 1 million tons of MSW are processed, 40,000 tons of metal are recycled, and 570,000 megawatt-hours of electricity are generated—the equivalent of more than 60 million gallons of fuel oil and enough electricity to power 65,000 homes. All this comes from stuff that people have thrown away!

Because of the drawbacks mentioned previously, WTE facilities are strongly opposed during the siting, permitting, and construction phases. Opponents cite public-health concerns about air pollutants like dioxins and mercury as well as traffic and property value concerns.

Costs of Municipal Solid-Waste Disposal

The costs of disposing of MSW are escalating, and not just because of the new design features of landfills. More and more, they reflect the expenses of acquiring a site and providing transportation. Landfill tipping fees average about $42 a ton (they are lower than those at WTE facilities because of the high capital costs at WTE facilities). The waste collector must recover this cost, as well as transportation costs to the site, and generate a profit (if the facility is privately operated). With the closing of the Fresh Kills landfill, the total cost of disposing of New York City's MSW rose to $263 a ton. At 11,550 tons per day, this amounts to more than $1 billion a year!

Getting rid of *all* trash is becoming more expensive, and one sad consequence of this increasing expense is illegal dumping. Some towns are charging up to $5 a bag for MSW disposal, and it now costs $1 or more to get rid of an automobile tire and $30 or more to dispose of a refrigerator. As a result, tires, refrigerators, yard waste, car parts, and construction waste are appearing all over the landscape. Institutions and apartments that operate with dumpsters often must put padlocks on them in order to prevent their unauthorized use by people trying to avoid disposal costs. Many states have established a corps of environmental police to track down midnight dumpers and bring them to justice.

21.2 Better Solutions: Source Reduction and Recycling

Our domestic wastes and their disposal represent a huge stream of material that flows in one direction: from our resource base to disposal sites. Just as natural ecosystems depend on recycling nutrients, we can move in the direction of sustainability only if we also learn to recycle more of our wastes. There is strong evidence that we are moving in this direction, but the best strategy of all is to reduce waste at its source, called *source reduction*.

Source Reduction

Source reduction is defined by the EPA as "the practice of designing, manufacturing, purchasing, or using materials (such as products and packaging) in ways that reduce the amount or toxicity of trash created." Source reduction accomplishes two goals: It reduces the amount of waste that must be managed, and it conserves resources. It is noteworthy that, after rising rapidly during the latter decades of the 20th century, the amount of waste per capita in the United States peaked at 4.5 pounds in 1990 and has since leveled off. (We still lead the world in this dubious statistic, however.) The leveling off is a signal that some changes in lifestyle may be occurring that have an impact on waste generation.

Source reduction is difficult to measure because it means trying to measure something that no longer exists. The EPA measures source reduction by measuring consumer spending, which reflects the goods and products that ultimately make their way to the trash bin. After 1990, consumer spending continued to grow, but the MSW stream slowed down. If the MSW had grown at the same pace as consumer spending, some 287 million tons would have been generated in 2000 instead of 232 million. Thus, some 55 million tons never made it into the waste stream in 2000, and the EPA considers this to be due to source-reduction activities.

Examples. Source reduction can involve a broad range of activities on the part of homeowners, businesses, communities, manufacturers, and institutions. Consider the following examples:

- Reducing the weight of many items has reduced the amount of materials used in manufacturing. Steel cans are 60% lighter than they used to be; disposable diapers contain 50% less paper pulp, due to absorbent-gel technology; and aluminum cans contain only two-thirds as much aluminum as they did 15 years ago.

- The Information Age may be having an impact on the use of paper. Electronic communication, data transfer, and advertising are increasingly performed on personal computers, now present in most homes in the United States. The growth of the Internet as a medium for information transfer has been phenomenal. These developments have the potential to reduce paper waste. The U.S. Postal Service reports declining mail volume, while Internet advertising is increasing.

- Many durable goods are reusable, and indeed, the United States has a long tradition of resale of furniture, appliances, and carpets. Many goods may be donated to charities and put to good use. The growing popularity of yard sales, flea markets, consignment clothing stores, and other markets for secondhand goods (Fig. 21–8) is an encouraging development. All sorts of used goods are offered for sale on eBay® and Craigslist, two Internet services that link private sellers and buyers. U-Exchange promotes bartering, while free goods can be advertised and acquired on the Freecycle Network™, a concept that has spread to 85 countries.

- Lengthening a product's life can keep that product out of the waste stream. If products are designed to last longer and to be easier to repair, consumers will learn that they are worth the extra cost. (You get what you pay for; cheap products usually take the shortest route to the waste bin.)

- We are all on bulk-mailing lists because these lists are sold and shared widely, so we are guaranteed to receive increasing volumes of advertising. To stay off such lists, simply inform mail-order companies and other organizations involved that you do not want your name and address shared. (You can e-mail Mail Preference Service, Direct Marketing Association, at www.dmachoice.org. The service allows users to control their direct mail as well as email advertising. An interesting alternative method is found at www.stopthejunkmail.com!)

- An increasingly popular way of treating yard waste (currently 12.8% of MSW) is **composting**, especially as more and more states and municipalities are banning yard waste from MSW collections. Composting, the natural biological decomposition (rotting) of organic matter in the presence of air, can be carried out in the backyard by homeowners, or it can be promoted by municipalities as a way of dealing with a large fraction of MSW.

Figure 21–8 Waste reduction by reuse. Yard sales and flea markets are keeping many materials from becoming MSW after a single use.

The Recycling Solution

In addition to reuse, recycling is an obvious solution to the solid-waste problem. More than 75% of MSW is recyclable material. There are two levels of recycling: *primary* and *secondary*. In **primary recycling,** the original waste material is made back into the same material—for example, newspapers recycled to make newsprint. In **secondary recycling,** waste materials are made into different products that may or may not be recyclable—for instance, newspapers recycled to make cardboard.

Why Recycle?   Recycling is a hands-down winner in terms of energy use and pollution:

- **It saves energy and resources.** One ton of recycled steel cans saves 2,500 pounds of iron ore, 1,000 pounds of coal, and more than 5,400 BTUs of energy. One ton of papers recycled saves 17 trees, 6,953 gallons of water, 463 gallons of oil, and 4,000 kilowatt-hours of energy.

- **It decreases pollution.** Making recycled paper requires 64% less energy and generates 74% less air pollution and 35% less water pollution than does using wood from trees. For every ton of waste processed, a thorough recycling program will eliminate 620 pounds of carbon dioxide, 30 pounds of methane, 5 pounds of carbon monoxide, 2.5 pounds of particulate matter, and smaller amounts of other pollutants.

What Gets Recycled?   The primary items from MSW that are currently being heavily recycled are cans (both aluminum and steel), bottles, plastic containers, newspapers, and yard wastes. Yet there are many alternatives for reprocessing various components of refuse, and people are coming up with new ideas and techniques all the time. A few of the major established techniques, together with their current percentages of recovery by recycling, are as follows:

- Paper and paperboard (47% recovery) can be remade into pulp and reprocessed into recycled paper, cardboard, and other paper products; finely ground and sold as cellulose insulation; or shredded and composted.

- Most glass (28% recovery) that is recycled is crushed, remelted, and made into new containers; a smaller amount is used in fiberglass or "glassphalt" for highway construction.

- Some forms of plastic (12% recovery) can be remelted and fabricated into carpet fiber, outdoor wearing apparel, irrigation drainage tiles, building materials (Fig. 21–9), and sheet plastic.

- Metals can be remelted and refabricated. Making aluminum (39% recovery) from scrap aluminum saves up to 90% of the energy required to make aluminum from virgin ore. In addition, aluminum ore is imported and represents part of the mounting U.S. trade deficit. National recycling of aluminum saves energy, creates jobs, and reduces the trade deficit.

- Yard wastes (leaves, grass, and plant trimmings—64% recovery) can be composted to produce a humus soil conditioner.

Figure 21–9 Deck made from Trex® decking and railing. Trex is made primarily with recycled plastic grocery bags, reclaimed pallet wrap, and waste wood. (*Source:* www.trex.com.)

- Textiles (17% recovery) can be shredded and used to strengthen recycled paper products.

- Old tires (35% recovery) can be remelted or shredded and incorporated into highway asphalt. Not included in the 35% figure are over a million tons of tires burned annually in special combustion facilities.

Recycling is both an environmental and an economic issue. Many people are motivated to recycle because of environmental concern, but the use of recyclable and recycled materials is also driven by economic factors. The Global Recycling Network (www.grn.com) is an electronic information exchange that promotes the trade of recyclables from MSW and the marketing of "ecofriendly" products made from recyclable materials.

Municipal Recycling

Recycling is probably the most direct and obvious way that most people can become involved in environmental issues. If you recycle, you save some natural resources from being used

(trees, in the case of paper), and you prevent landfills from becoming "landfulls."

Recycling's popularity is well established: Virtually every state has specific recycling goals, met with varying degrees of success (Fig. 21–10). EPA sources report that only 6.4% of MSW was recycled in 1960 versus 33.4% in 2007 (Fig. 21–11). There is a great diversity of approaches to recycling in municipalities, from recycling centers requiring miles of driving to curbside recycling with sophisticated separation processes. The most successful programs have the following characteristics:

1. There is a strong incentive to recycle where there are PAYT charges for general trash and no charge for recyclable goods.

2. Recycling is not optional; mandatory regulations are in place, with warnings and sanctions for violations.

3. Residential recycling is curbside (Fig. 21–12), with free recycling bins distributed to households. (Curbside recycling rose rapidly and then plateaued. Currently, nearly 60% of the U.S. populace is served via 8,660 curbside programs.)

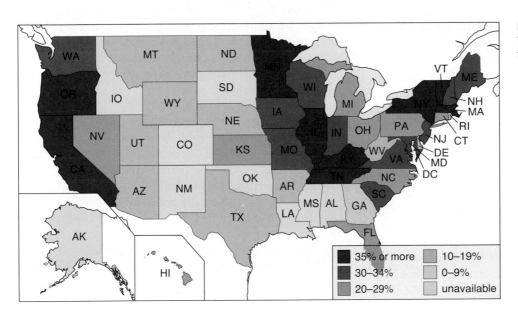

Figure 21–10 State recycling rates. There is great variation in recycling from state to state. (*Source: Biocycle*, December 2008.)

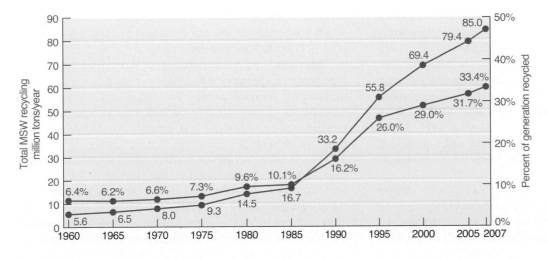

Figure 21–11 MSW recycling in the United States. MSW recycled from 1960 to 2007, as total waste and percentage recycled. The EPA's goal was to reach 35% by 2008. (*Source:* EPA, Office of Solid Waste, *Municipal Solid Waste in the United States: 2007 Facts and Figures* (Washington, D.C.: EPA, November 2008).)

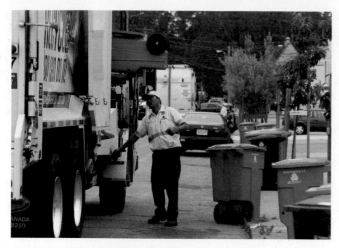

Figure 21–12 Curbside recycling. Recyclable materials are picked up in a San Francisco neighborhood, where residents are issued 3 bins: one for garbage, one for recycling, and one for compostable materials. The city now recycles 72% of its waste.

4. Drop-off sites are provided for bulky goods like sofas, appliances, construction and demolition materials, and yard waste.

5. Recycling goals are ambitious, yet clear and feasible. Some percentage of the waste stream is targeted, and progress is followed and communicated.

6. A concerted effort is made to involve local industries in recycling.

7. The municipality employs an experienced and committed recycling coordinator.

Nevertheless, municipalities experience very different recycling rates. New York City, for example, has made great progress—from recycling only 5% to recycling 16%—by mandating curbside recycling for all of its 3 million households. Boston, by contrast, with only 11% recycling (with voluntary curbside pickup), is holding Massachusetts back in its goal to achieve 56% recycling by 2010. Worcester, farther to the west in the state, has achieved a commendable 41% recycling rate. The prize likely goes to San Francisco, a city that has achieved a recycling rate of 72%.

Economics of Recycling. Recycling has its critics, who base their arguments primarily on economics. If the costs of recycling (from pickup to disposal of recyclable components) are compared with the costs of combusting waste or placing it in landfills, recycling frequently comes out second best. Markets for recyclable materials fluctuate wildly, and residents often end up subsidizing the recycling effort. Competition between landfills and combustion facilities often lowers tipping fees, creating an even greater disincentive to recycle. Thus, critics of current recycling practices argue that, unless recycling pays for itself through the sale of the materials recovered, it should not be done.

However, a true environmental assessment should compare the energy costs of recycling with the energy costs of landfill or combustion disposal as well as comparing the energy costs of making products from recyclable goods with the energy costs of making new products from scratch—called a *life-cycle analysis*. Performed this way, the energy saved by recycling aluminum, for example, is 96%; for newsprint, 45%; for two kinds of plastic, 76–88%; and for glass, 21%.

Unfortunately, all recycling has taken a hit because of the recession of 2008–2009. Demand for recyclable products has plummeted: Early in 2008, recyclers paid Boston $112/ton for paper, but by the end of the year, the city was paying the recyclers $5.50/ton to take the paper. The city of Berkeley, California, received $200/ton for recyclable materials in the fall of 2008, and in 2009, it had to settle for $35/ton. As cities and towns struggle to meet their budgets, the collection of recyclables is often a tempting target for cuts.

Irrespective of these economic considerations, the demonstrated support for recycling programs is strong. Experience has shown that at least two-thirds of households will recycle if presented with a curbside pickup program. The percentage goes up when recycling is combined with a PAYT program, or a program that rewards participants for their participation, like RecycleBank. Dover, New Hampshire—a city of 26,000—was facing escalating costs of MSW collection and disposal after its landfill closed. The city instituted both a PAYT program and curbside recycling in 1991 and saw per household trash plummet from 6 pounds to 2.3 pounds per day in six years. At the same time, the cost per household dropped from $122 to $73 per year, and the recycling rate rose quickly to more than 50%. The city's solid-waste management costs dropped 27% in spite of a population increase.

Paper Recycling. Newspapers are by far the most important item that is recycled because of their predominance in the waste stream. It is a simple matter to tie up or bag household newspapers, and the amount recovered by recycling is increasing dramatically. Currently, far more newspaper is being recycled (78%) than discarded. Since more than 25% of the trees harvested in the United States are used to make paper, recycling paper saves trees: A 1-meter stack of newspapers equals the amount of pulp from one tree.

There is often some confusion about what is meant by "recycled paper." The key is the amount of *postconsumer* recycled paper in a given product. In manufacturing processes, much paper is "wasted"; paper that is routinely recovered and rerouted back into processing is called "recycled paper." Thus, the total recycled content of a paper can be 50%, although the actual postconsumer amount recovered by recycling might be only 10% of the total amount of paper.

Markets. After the wastepaper is incorporated into a final product, the market becomes a critical factor. Is there a demand for recycled paper? The technology for producing high-quality paper from recyclable stock has improved greatly, to the point where it is virtually impossible to distinguish the recycled product from "virgin" paper.

The market for used newspapers has fluctuated greatly over the past several decades. During the late 1980s, the market was saturated, and municipalities often had to pay to get rid of newspapers. In 1995, discarded papers were so valuable—up to $160 a ton—that thieves were stealing them off the sidewalks before the recycling trucks could pick them up.

we pay for them with taxes to clean up litter as well as with our injuries, flat tires, environmental degradation, and so on.

Bottle Laws. In an attempt to reverse these trends, environmental and consumer groups have promoted *bottle laws* that facilitate the recycling or reuse of beverage containers. Such laws generally call for a deposit on all beverage containers, both reusable and throwaway and both glass and plastic. Retailers are required to accept the used containers and pass them along for recycling or reuse. Bottle laws have been proposed in virtually all state legislatures over the last decade. Nevertheless, in every case, the proposals have met with fierce opposition from the beverage and container industries and certain other special-interest groups. The reason for their opposition is economic loss to their operations. The container industry contends that bottle laws will result in the loss of jobs and higher beverage costs for the consumer.

In most cases, the industry's well-financed lobbying efforts have defeated bottle laws. However, some states—11, as of 2009—have adopted bottle laws of varying types despite industry opposition (Table 21–2). The experience of these states has proved the beverage and bottle industry's claims false. In fact, more jobs are gained than lost, costs to the consumer have not risen, a high percentage of bottles is returned, and there is a marked reduction in the can and bottle portion of litter. A final measure of the success of bottle laws is continued public approval: Despite industry efforts to repeal these laws, no state that has one has repealed it.

Repeated attempts have been made to get a national bottle law through Congress—to date, unsuccessfully. Opponents (the same ones that oppose state-level bottle bills) argue that such a law will threaten the continued success of curbside recycling, of which beverage containers represent an important source of revenue. However, since curbside recycling currently reaches only 60% of the U.S. population, a national bottle law would undoubtedly recover a greater proportion of beverage containers. (States with bottle laws report that 75%

Figure 21–13 Wastepaper exports. Bales of wastepaper ready to be loaded for shipment overseas. China imports a third of the wastepaper being sold internationally.

A year later, as more recycling programs came on line, the market collapsed again, and many cities once more paid as much as $25 a ton to have the newspapers hauled away—but this is still less expensive than the tipping fees at landfills. Furthermore, there is a lively international trade in used paper (Fig. 21–13). Forest-poor countries in Europe and Asia purchase wastepaper from the United States and other industrial countries in the Northern Hemisphere, where there continues to be a surplus of such paper.

Glass Recycling and Bottle Laws. The glass in MSW is primarily in the form of containers, most of which held beverages. The average person drinks about a quart of liquid each day. Given that there are 305 million Americans, this daily consumption amounts to some 28 billion gallons of liquid per year for the nation as a whole. Much of this volume is packaged in single-serving containers that are used once and then thrown away. Nonreturnable glass containers constitute 5.3% of the solid-waste stream in the United States and about 50% of the nonburnable portion; they also make up a large proportion of roadside litter. Broken bottles along roads, beaches, and parklands are responsible for innumerable cuts and other injuries, not to mention flat tires. Both the mining of the materials and the process used to manufacture the beverage containers create pollution. All of these factors produce hidden costs that do not appear on the price tag of the item; however,

TABLE 21–2 States with Bottle Laws

State	Year Passed
Connecticut	1972
Oregon	1972
Vermont	1973
Maine	1976
Michigan	1976
Iowa	1978
Massachusetts	1978
Delaware	1982
New York	1983
California	1991
Hawaii	2002

of containers are returned, a significant proportion from the curbside collections from people who don't bother to redeem the containers.) Also, a national bottle law would be labor intensive, employing tens of thousands of new workers.

In 2007, recycling and bottle laws resulted in the recovery of 28% of glass containers, 39% of aluminum containers, 65% of steel containers, and 33% of plastic containers, with an aggregate weight of 6.6 million tons.

Plastics Recycling. Plastics have a bad reputation in the environmental debate for several reasons. First, plastics have many uses that involve rapid throughput—for example, packaging, bottling, the manufacture of disposable diapers, and the incorporation of plastics into a host of cheap consumer goods. Second, plastics are conspicuous in MSW and litter. Ironically, most trash is disposed of in plastic bags manufactured just for that purpose. Finally, plastics do not decompose in the environment because no microbes (or any other organisms) are able to digest them. (Imagine plastics in landfills delighting archaeologists hundreds of years from now.) Therefore, the possibility of recycling at least some of the plastic in products—containers of liquid, for example—interests the environmental community.

Bottled Water. Throughout the world, the number one "new" drink is bottled water. Health consciousness and suspicions about the safety of tap water have made the plastic water bottle the new symbol of healthy living. Some 8.9 billion gallons of bottled water were sold in the United States in 2007, making us the world leader in consumption of bottled water. As a consequence, a new form of litter along our highways and on our beaches is the discarded plastic water bottle. Only two states with bottle laws include bottled water and other noncarbonated beverages. People often pay more for bottled water (which in some brands is only bottled tap water) than for gasoline, and as much as 10,000 times more than for water from the tap. Only one in six plastic water bottles is recycled, and the rest end up in litter, landfills, or combustion facilities. This is a real waste, as there is a market demand for these and other bottles made from polyethylene terephthalate (PETE).

PETE and HDPE. If you look on the bottom of a plastic container, you will see a number (or some letters) inside the little triangle of arrows that has come to represent recycling. The number or the letters are used to differentiate between and sort the many kinds of plastic polymers. The two recyclable plastics in most common use are high-density polyethylene (HDPE; code 2) and polyethylene terephthalate (PETE; code 1). In the recycling process, plastics must be melted down and poured into molds, and some contaminants from the original containers may carry over, making it difficult to reuse some plastics for food containers. While this somewhat restricts uses for recyclable plastic, some interesting uses are appearing, with more in the development stage. PETE, for example, is turned into carpets, jackets, film, strapping, and new PETE bottles; HDPE becomes irrigation drainage tiles, sheet plastic, and, appropriately, recycling bins.

Critics of the recycling of plastics argue that recovering plastics is more costly than starting from scratch—that is, beginning with petroleum derivatives—and that manufacturers are getting involved mainly because environmentally concerned consumers are demanding their involvement. Industry-supported critics also point out that plastics in landfills create no toxic leachate or "dangerous" biogas. Moreover, plastics in combustion facilities burn wonderfully hot and leave almost no ash. However, with the rising costs of petroleum and the many uses of recyclable plastic bottles (including new bottles), closing the loop to recover plastic bottles makes both economic and environmental sense.

Plastic Bags. Hanging from trees and bushes, blowing along highways, clogging sewer pipes, drifting on ocean currents, plastic bags are everywhere on the planet—everywhere, that is, but the trash or recycling barrel. They cause thousands of deaths of marine mammals and turtles each year. Convenient? Indeed they are, as anyone who has shopped in a supermarket knows. A hundred billion of them are distributed in the United States every year, and close to a trillion are used annually worldwide. Made of low-density polyethylene, plastic bags are difficult to recycle and, like all petroleum-based plastics, are virtually indestructible. China, once a user of 100 billion plastic bags a year, has cracked down on them, banning the production of ultrathin bags—bringing China in line with a number of other countries. In the United States, San Francisco initiated the first citywide ban on plastic grocery bags in 2007; many other cities are considering similar bans. Instead of the thin plastic bags, customers are encouraged to carry their own cloth bags or to use paper bags or larger, thicker reusable plastic bags that hold five times as much as the standard flimsy bag. The plastics industry is fighting back with lawsuits and information campaigns touting the benefits of the plastic bag. Perhaps someday soon the "national flower of South Africa," as the bags have been called in that country, will become an endangered species.

Regional Recycling Options

As landfills close down and MSW is transferred out of cities and towns, transfer stations are set up to receive the wastes from collection trucks and transfer them to larger vehicles for long-distance hauling to their final destination. It is an encouraging sign that a number of these transfer stations are being converted into **materials recovery facilities**, referred to in the trade as MRFs, or "murfs." In 2007, there were 567 MRFs operating in the United States, handling over 91,000 tons of MSW a day.

How a MRF Works. Basic sorting takes place when waste is collected, either through curbside collection or by town recycling stations (sites to which residents can bring wastes to be recycled). The waste is then trucked to the MRF, where it is moved through the facility by escalators and conveyor belts, tended by workers who inspect and sort further (Fig. 21–14). The objective of the process is to prepare materials for the recyclable-goods market. Glass is sorted by color, cleaned, crushed into small pebbles, and then shipped to glass companies, where it replaces the raw materials that go into glass manufacture—sand and soda ash—and saves substantially on energy costs. Cans are sorted, flattened, and sent either to detinning plants or to aluminum-processing facilities. Paper is sorted, baled, and sent to reprocessing

Figure 21–14 Materials recovery facility.
A modern MRF in Australia. Its purpose is to sort a mixed stream of waste material (cans, bottles, newspapers, etc.) for eventual recycling.

mills. Plastics are sorted into four categories, depending on their color and type of polymer, and then sold.

The facility's advantages are its economy of scale and its ability to produce a high-quality end product for the recyclable-materials market. Townspeople know where to bring their waste, and they quickly become familiar with the requirements for initial sorting during the collection and transfer of the waste. Some MRFs have gone to high technology, replacing the manual sorting with magnetic pulleys, optical sensors, and air sorters. As the movement grows, this is likely the direction the technology will take to improve the efficiency of operations and cut costs.

Less common than conventional MRFs are mixed-waste processing facilities, which receive unsorted MSW just as if it was going to a landfill or a WTE facility. The waste is loaded on a conveyer and is sorted for recovery of recyclable materials before being landfilled or combusted. A total of 34 such facilities were operating in the United States in 2007, processing about 43,000 tons of MSW a day.

Composting Mixed Waste and Yard Trimmings. Taking advantage of the fact that MSW has a high organic content, a few regional facilities compost the MSW after removing large items and metals. Often, they will mix treated sewage sludge with the MSW (called *co-composting*), which provides a rich source of bacteria and nutrients. These facilities are frequently plagued with odor problems, and some have had serious fires. Sometimes such facilities are combined with a MRF, as in the case of Nantucket Island, Massachusetts. (See Sustainability, "The Nantucket Story.") Much more common are the yard-trimmings composting programs—more than 3,500 in 2007, handling about 57,000 tons per day.

Sustainability

The Nantucket Story

Nantucket Island lies some 30 miles south of Cape Cod and is the year-round home to 10,000 residents, but hosts more than 50,000 during the summer months. Nantucket's MSW was taken to a landfill for many years, but in the late 1980s, the landfill was overflowing and leaching wastes into sensitive surrounding wetlands. In 1994, the state of Massachusetts mandated that the island close its landfill, leaving islanders with few options. The least attractive one was to ship the MSW off the island, which would quadruple the island's waste management costs. In 1996, the island decided to privatize waste-handling operations and signed a 25-year agreement with Waste Options to build an integrated waste management system that encompassed recycling, composting, and cleanup of the existing landfill. After several years of operation, the island's MSW has been reduced to only 14% of what is generated, and it goes to a landfill that is actually shrinking in size.

The process begins with a MRF, which has boosted Nantucket's recycling rate from 17% to 42%. Cardboard, newspaper, magazines, cans, glass, and plastic bottles are sorted, baled, and shipped off the island. A unique feature is a "Take It or Leave It" exchange, where residents

(continued)

(continued)

can drop off used clothing and other unwanted items and others can pick up what they want from the dropped-off items. Other items, such as mattresses, tires, yard wastes, and construction materials, are handled separately and are either shipped to the mainland for further processing or processed on-site.

The remaining MSW is processed together with dewatered sewage sludge from the waste-treatment plant (thus, co-composting) to create a high-grade compost. The heart of this process is the Eweson Digester, a 185-foot-long steel drum that rotates continually and promotes the decomposition of much of the organic matter in the wastes. Air and water are added as needed to optimize biological decomposition. After three days in the digester, the digested material is screened, with the larger particles going to the landfill. The screenings are further processed in a maturation building, where, over several weeks, they are repeatedly turned to reach a stable, high-quality compost that is useful for gardening and landscaping. Air in the building is controlled for odor with a biofilter.

The landfill materials represent only 14% of the original MSW and are added to the island landfill no faster than previously buried waste is taken from the landfill (called landfill mining) to the digester during the winter months, when trash and people are at a low ebb on the island. The price paid by the Nantucket Islanders is $90 a ton, which is quite competitive with MSW costs in other Massachusetts communities. This public-private partnership on Nantucket Island has produced a waste-treatment process that could become a model for the rest of the country.

21.3 Public Policy and Waste Management

The management of MSW used to be entirely under the control of local governments. In recent years, however, state and federal agencies have played an increasingly important role in waste management, partly through regulation and partly through encouragement and facilitation.

The Regulatory Perspective

At the federal level, the following legislation has been passed:

- The first attempt by Congress to address the problem was the **Solid Waste Disposal Act** of 1965. The legislation gave jurisdiction over solid waste to the Bureau of Solid Waste Management, but the agency's mandate was basically financial and technical rather than regulatory.

- The **Resource Recovery Act** of 1970 gave jurisdiction over waste management to the newly created EPA and directed attention to recycling programs and other ways of recovering resources in MSW. The act also encouraged the states to develop some kind of waste management program.

- The passage of the **Resource Conservation and Recovery Act (RCRA)** of 1976 saw a more regulatory (command-and-control) approach to dealing with MSW, as the EPA was given power to close local dumps and set regulations for landfills. Combustion facilities were covered by air-pollution and hazardous-waste regulations, again under the EPA's jurisdiction. RCRA also required the states to develop comprehensive solid-waste management plans.

- The **Superfund Act** of 1980 (described in Chapter 22) addressed abandoned hazardous-waste sites throughout the country, 41% of which are old landfills.

- The **Hazardous and Solid Waste Amendments** of 1984 gave the EPA greater responsibility for setting solid-waste criteria for all hazardous-waste facilities. Since even household waste must be assumed to contain some hazardous substances, this meant that the EPA had to determine and monitor all landfill and combustion criteria more closely.

Largely in response to these federal mandates, the states have been putting pressure on local governments to develop *integrated waste management plans*, with goals for recycling, source reduction, and landfill performance.

Integrated Waste Management

It is not necessary to fasten onto just one method of handling wastes. Source reduction, waste-to-energy combustion, recycling, materials recovery facilities, landfills, and composting all have roles to play in waste management. Different combinations of these options will work in different regions of the country. A system having several processes in operation is called *integrated waste management*. Let us examine the available MSW management options, with a view toward developing recommendations that make environmental sense—that is, ones that push us in a more sustainable direction.

Waste Reduction.   The modern United States leads the world in per capita waste production and per capita energy consumption—we really are a "throwaway society" (Fig. 21–15). How can we turn this situation around? True management of MSW begins in the home (or dorm room). Materialistic lifestyles, affluence, and overconsumption can pack our homes and our trash cans with stuff we simply do not need.

WasteWise.   Several incentives have been put in place on the government level to encourage waste reduction. For example, the EPA sponsors the **WasteWise** program, targeting the reduction of MSW by establishing partnerships with not only local governments, schools, and organizations, but multinational corporations as well. WasteWise is a voluntary partnership that allows the partners to design their own solid-waste reduction programs. Over 2,000 organizations throughout the country have partnered with the EPA to reduce and recycle waste and, in a new focus of the program, to calculate their reductions of waste-related greenhouse gas emissions. The EPA regularly honors partner organizations that have achieved exemplary waste reduction. For example, Arkansas-based Virco Manufacturing Corporation, a furniture manufacturer, has recycled more than 250 million tons of materials since 1991. The company regularly spends $25 million a year on materials with recycled content. Indeed, company employees have been in the forefront of a volunteer effort to help the town of Conway, Arkansas, handle an overload of cardboard, collecting and selling it and giving the profits to the city.

PAYT. The EPA also brings attention to a national trend of unit pricing, or charging households and other customers for the waste they dispose of, with the agency's PAYT program. Instead of using local taxes to pay for trash collection and disposal, which provides no incentive to reduce waste, communities levy curbside charges for all unsorted MSW—say, $1–$5 per container (Fig. 21–16). This is catching on as cities and towns see the advantages of the practice; more than 7,000 cities and towns, serving 25% of the U.S. population, now employ the PAYT method. Following this highly successful strategy, PAYT communities reduce their waste by 25% to 45%, on average, and increase their recycling by 32% to 59%. As with the WasteWise program, the EPA's role is that of a facilitator, not a regulator.

Figure 21–16 Pay-as-you-throw trash pickup. Consumers purchase the empty bags for a set price, and only these bags are picked up by the trash hauler. Next to the bags are three recycling containers, which are emptied free of charge.

EPR. Another means of bringing about waste reduction is to establish a program of *extended product responsibility* (EPR), a concept that involves assigning some responsibility for reducing the environmental impact of a product at each stage of its "life cycle," especially the end. For example, Hewlett-Packard and Xerox make it easy for customers to return spent copier cartridges, and the two companies then recycle the components of the cartridges. EPR originated in western Europe. Germany, the Netherlands, and Sweden have well-developed regulations that require companies to take back many items, that encourage the reuse of products, and that promote the manufacture of more durable products.

Again, the EPA is active in providing information to manufacturers and purchasing departments to help them design and buy more environmentally sound products.

Waste Disposal Issues. No human society can avoid generating solid waste. There will always be MSW, no matter how much we reduce, reuse, and recycle. Most experts see WTE combustion facilities holding their own, a decrease in the percentage of waste going to landfills, and a continued rise in recycling.

It will be important to break the gridlock at local and state levels on the siting of new landfills and WTE combustion facilities. Landfills will still be needed, although they should last longer than in the past. Policy makers have long known about the landfill shortage, but have opted for short-term solutions with the lowest political cost. One result of this choice is the long-distance hauling of MSW by truck and rail described earlier. If regions and municipalities are required to handle their own trash locally, as is the case in some states, they will find places to site landfills, regardless of NIMBY considerations. New landfills must use the best technologies, such as proper lining, leachate collection and treatment, groundwater monitoring, biogas collection, and final capping. If these technologies are employed, a landfill will not be a health hazard. Also, sites should be selected with a view to some future use that is attractive.

Just Say No! With the closing of New York City's huge Fresh Kills landfill, the city is now sending its MSW to other states. Yet most citizens in the other states object to being New York's dumping ground, and there is the deeper stewardship issue of local responsibility for local wastes as well as the inefficiency of long-distance waste hauling. It will take an act of Congress to give states or local jurisdictions the right to ban imports of waste because of the interstate commerce clause in the Constitution. The **Solid Waste Interstate Transportation Act** of 2009 (H.R. 274) was introduced in the 111th Congress, with bipartisan support, especially from states that are currently receiving waste from other states. The legislation would give local and state governments the authority to limit or prohibit the transportation of out-of-state wastes to landfills. The bill is a demand for fairness to states that are working hard to deal responsibly with their own wastes, only to see the unrestricted transport of wastes from other states. Similar bills have been introduced in every Congress since the 100th.

Another policy goal should be to encourage more co-composting of MSW. Nantucket Island was forced to change its ways of managing MSW and has come up with a very attractive waste management program. (See Sustainability, "The Nantucket Story.") As on Nantucket Island, the technologies employed should be the best available, and if these are used, there should be no significant threat to human health. This option seems to be the best way to deal with mixed waste that is nonrecyclable. Recyclable materials are recovered before the composting step, the waste is greatly reduced in volume, and the compost is a highly useful product that can aid in the recovery of damaged lands in need of soil.

Recycling and Reuse. Recycling is certainly the wave of the future. It should not, however, be pursued in lieu of waste reduction and reuse. The move toward more-durable goods is an overlooked and underutilized option. In fact, waste reduction remains the most environmentally sound and least costly goal for MSW management because wastes that are never generated do not need to be managed.

Banning the disposal of recyclable items in landfills and at combustion facilities makes very good sense. Many states have incorporated this regulation into their management programs. For example, Massachusetts phased out yard waste in 1991, metals and glass in 1992, and recyclable papers and plastic in 1994. Landfill and combustion-facility operators are supposed to conduct random truck searches and are authorized to turn back trucks with significant amounts of banned items. Massachusetts banned the disposal of cathode ray tubes (CRTs) in landfills in 2000, becoming the first state to do so. Since then, a number of states have addressed the disposal of CRTs and other forms of e-waste, a growing problem. (See Stewardship, "Between the Cracks: E-Waste.")

Stewardship

Between the Cracks: E-Waste

We all use them—televisions, computers, DVD players, VCRs, cell phones, printers, copiers, video game systems. But what do we do when they are obsolete or broken? The chances are that we put them in the trash and hope they get picked up. Better yet, we take them to a transfer station and pay a small fee to turn in that old television or computer monitor. This is no small matter, however. **E-waste**, as these devices are called once they are discarded, is growing far more rapidly than our ability to deal with it. The EPA estimates that in 2005 alone some 1.9 to 2.2 million tons of electronic devices became obsolete; 80% of this e-waste was disposed of, primarily in landfills. One estimate puts the yearly global generation of e-waste at 50 million tons. In 2009, U.S. residents discarded a record number of televisions as the country made the transition from analog to digital television broadcasts.

E-waste has numerous toxic components and is classified as hazardous waste. Cathode ray tubes (CRTs), which are the vacuum tubes that provide the video display on older televisions and computer monitors, contain an average of four pounds of lead. Many other heavy metals—mercury, barium, chromium, and copper, to name a few—are found in the wiring and circuit boards of electronic devices. In a landfill, these substances can leach into groundwater. If the e-waste is incinerated, they are sent into the air. Federal law regulates the transport, treatment, and disposal of such hazardous wastes under RCRA. However, RCRA does not apply to small businesses and households, so the disposal of this e-waste is left up to the states and municipalities. To date, only 14 states have implemented any kind of e-waste management, and there is broad consensus on the need for federal legislation to manage this waste. The states that do have laws for e-waste are basically trying to make sure this waste is not landfilled or incinerated. The ideal destination for e-waste is a recycling company that will dismantle the devices, recover the useful materials, and dispose of the rest according to RCRA rules (see Chapter 22).

"Recycling" companies may be found to take the products from transfer stations and consumers, but all too often, they are packing the e-waste into large containers and shipping it to China, India, or African countries. An estimated 80% of "recycled" e-waste takes this route, ending up in any of thousands of small shops in developing countries where laborers pull apart the devices, burn wiring to recover copper, fry circuit boards to free the chips, and dip the chips in buckets of acid to extract gold and silver—and they are usually working without protection from the hazardous chemicals. However, there *are* legitimate recycling companies. Recently, the nation's largest electronics recycler, Electronics Recyclers International, joined with the Electronics TakeBack Coalition and the Basel Action Network in the **e-Stewards Initiative**. This initiative certifies e-waste recyclers, assuring consumers and businesses that their e-waste will not be landfilled, incinerated, or exported to developing countries. With a patchwork of state and local regulations for e-waste, the e-Stewards Initiative is a welcome and much-needed measure. However, federal legislation would be better, and given the huge volume of e-waste, it can't come too soon.

Enacting a national bottle-deposit law would be a giant stride toward the reuse and recycling of beverage containers and would also greatly reduce roadside litter in states lacking bottle laws.

Finally, closing the "recycling loop" remains a significant action to be taken by governments to encourage recycling. A number of states have opted for one or more of the following approaches: (1) minimum postconsumer levels of recycled content for newsprint and glass containers; (2) requirements stating that purchases of certain goods must include recycled products even if they are more expensive than "virgin" products; (3) requirements that all packaging be reusable or be made (at least partly) of recycled materials; (4) tax credits or incentives that encourage the use of recycled or recyclable materials in manufacturing; and (5) assistance in the development of recycling markets.

Revisiting the Themes

Sound Science

The role of the EPA and its scientists is crucial. The agency sets the rules for landfill operations, as mandated by Congress, and then takes a proactive role in areas where it has no regulatory authority, encouraging source reduction, recycling, and state MSW management programs. WasteWise and the PAYT program are two excellent examples of initiatives where the EPA plays a facilitating role. The greatest current public-policy needs are a national bottle law, e-waste legislation, and a law dealing with interstate waste transport (like the Solid Waste Interstate Transportation Act currently being considered).

Sustainability

Sustainable natural systems recycle materials, but we create so many products that are difficult to recycle that it becomes necessary to be intentional about source reduction and product responsibility. The greatest departure from sustainable living is the interstate transfer of trash, which wastes energy and allows cities and states to go for the cheapest immediate solution and avoid coming to grips with a responsible approach to waste management. The most positive step toward sustainability is intentional source reduction, which both conserves resources and reduces the amount of waste that must be managed.

Stewardship

Instead of allowing economic considerations to dictate MSW management choices, many municipalities and organizations are consciously choosing options that reflect (and often bear the name of) stewardship. Here is one possible stewardly waste management plan:

1. Emphasize source reduction wherever possible,
2. Employ mandatory curbside recycling and a PAYT collection program,
3. If feasible, establish a MRF for efficient handling of recyclables (and possibly MSW),
4. Make sure all e-waste is delivered to a recycling company certified by e-Stewards,
5. Employ co-composting of the remaining MSW with treated sewage sludge,
6. Deposit residual materials in a local landfill, and
7. Prohibit all interstate transfer of MSW.

This plan features local MSW management and local or regional recycling, composting, and landfilling. Every step of the plan involves decisions on the part of individuals and municipalities; these practices do not happen unless there are people who care enough to promote responsible stewardship of the "stuff" that all of us feel that we need.

REVIEW QUESTIONS

1. List the major components of MSW.

2. Trace the historical development of refuse disposal. What percentages now go to landfills, combustion, and recycling?

3. What are the major problems with placing waste in landfills? How can these problems be managed?

4. Explain the difficulties accompanying landfill siting and outsourcing. How can these processes be handled responsibly?

5. What are the advantages and disadvantages of WTE combustion?

6. What is the evidence for increasing source reduction, and what are some examples of how it is accomplished?

7. What are the environmental advantages of recycling?

8. Describe the materials that are recycled and how recycling is accomplished.

9. Discuss the attributes of successful recycling programs.

10. Distinguish among MRFs, mixed waste processing, and mixed waste co-composting.

11. What laws has the federal government adopted to control solid-waste disposal?

12. What is e-waste, and how is it being managed?

13. What is integrated waste management? What decisions would be part of a stewardly and sustainable solid-waste management plan?

THINKING ENVIRONMENTALLY

1. Compile a list of all the plastic items you used and threw away this week. Was it more or less than you expected? How can you reduce the number of items on the list?

2. How and where does your school dispose of solid waste? E-waste? Is a recycling program in place? How well does it work?

3. Suppose your town planned to build a combustion facility or landfill near your home. Outline your concerns, and explain your decision to be for or against the site.

4. Does your state have a bottle law? If not, what has prevented such a bill from being adopted?

5. Consider the stewardly waste management plan described in the "Stewardship" section of "Revisiting the Themes." How many of these components does your city or town employ?

MAKING A DIFFERENCE

1. Investigate how municipal solid waste is handled in your school or community and what the future plans are for waste management. Promote curbside recycling if it is not happening, and investigate the possibility of adopting a pay-as-you-throw system.

2. Consider your consumption patterns. Live as simply as possible so as to minimize your impact on the environment.

Buy and use durable products, and minimize your use of disposables.

3. When you shop, bring your own reusable bags for your groceries. When buying small items that you can easily carry, ask that they not be put in a bag.

Mastering ENVIRONMENTAL SCIENCE™

Go to **www.masteringenvironmentalscience.com** for practice quizzes, Pearson eText, videos, current events, and more.

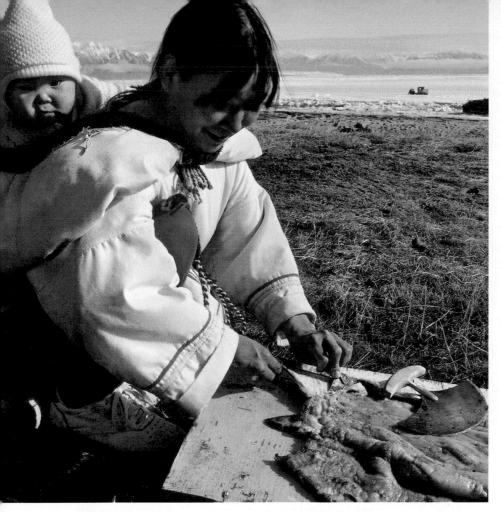

A Nunavut hazard. This Inuit mother is scraping the fat from a sealskin on Baffin Island, Nunavut. Seals and other animals used for food by the Inuit harbor high levels of persistent organic pollutants.

KEY TOPICS

1. **Toxicology and Chemical Hazards:** More than 80,000 chemicals are in use in our society, many of them quite hazardous; the most toxic ones are the heavy metals and the petroleum-based organic compounds.

2. **Hazardous-Waste Disposal:** The unregulated land disposal of hazardous materials in the past has left us with thousands of abandoned and buried waste sites; today land disposal is highly regulated.

3. **Cleaning Up the Mess:** The Environmental Protection Agency administers the Superfund program put in place by Congress to deal with the abandoned toxic sites.

4. **Managing Current Toxic Chemicals and Wastes:** Congress has enacted several laws to regulate the handling and disposal of hazardous wastes and to evaluate new chemicals before they are marketed.

5. **Broader Issues:** The topics of environmental justice and pollution prevention merit special attention in this final section on hazardous chemicals.

There are strange things done in the midnight sun by the men who moil for gold;
The arctic trails have their secret tales that would make your blood run cold.
The Northern Lights have seen queer sights, but the queerest they ever did see
Was that night on the marge of Lake LeBarge
I cremated Sam McGee.

S O BEGINS "THE CREMATION OF SAM MCGEE" by Robert Service.[1] In the poem, Lake LeBarge, in the Canadian Yukon Territory, eventually received the cremated ashes of Sam McGee. Unfortunately, the lake has been receiving a much different burden in recent years, to the point where its fish have become hazardous to eat because of their DDT content. Much to research scientists' surprise, fish, birds, and mammals all over the Arctic are showing elevated body burdens of a number of *persistent organic pollutants* (POPs, as they are called)—in particular, DDT, toxaphene, chlordane, PCBs,

[1]Robert Service, "The Cremation of Sam McGee," in *The Best of Robert Service* (New York: Berkley Publishing Group, 1989).

and dioxins. Moreover, because the Inuit people of the far north depend on these animals for food, they, too, are carrying high loads of POPs—some of the highest in the world.

The real mystery is how these chemicals—all of which are toxic—get to such remote and pristine environments. No one is spraying pesticides there, and there are no industries to produce PCBs and dioxins. The key to the presence of such chemicals can be found in several factors that characterize POPs: persistence, bioaccumulation, and the potential for long-range transport. Added to these factors is the unique climate of the Arctic, where the extreme cold promotes a process called *cold condensation*. POPs are semivolatile chemicals carried in global air patterns to the Arctic, where they condense on the snowpack and are then washed into the water with the spring thaw. Once in the lakes and coastal waters, the chemicals are picked up by plankton and passed up the food chain through bioaccumulation and biomagnification (see Chapter 13, Figure 13–9). The highest concentrations have been found in Arctic ponds located close to nesting sites of seabirds; the birds feed near the top of the food chain and effectively concentrate the contaminants from coastal marine waters. Nutrients from their guano enrich the ponds and surrounding terrestrial areas, but the same guano carries the high levels of contaminants into the Arctic ecosystems.

A Canadian government study found that three-fourths of Inuit women have PCB levels up to five times above those deemed safe. Dioxins in Nunavut territory are transferred to the Inuit via caribou, a major food source, as well as through the aquatic food chain. The caribou get the dioxins from deposition on lichens and mosses. Although some of the POPs are on the decline in parts of the Arctic, some new arrivals—brominated fire retardants (polybrominated diphenyl ethers, or PBDEs)—have begun to accumulate in polar bears, seals, and foxes.

BOON OR BANE? The potential health effects of the chemical load carried by the Arctic residents range from immune-system disorders to disruption of their hormone systems and, over time, to cancer. (All five of the POPs mentioned are on the National Toxicology Program's list of carcinogens.) The latest impact has been a serious imbalance in births, where twice as many girls are born as boys in many Arctic villages. Scientists from the Arctic Monitoring and Assessment Program, an international body that advises the governments of the eight Arctic nations, measured levels of man-made chemicals in women's blood that were high enough to trigger changes in the sex of unborn children during their first few weeks of gestation.

Stories like these, and others that can only be called disasters, have convinced the public that significant dangers are associated with the manufacture, use, and disposal of many chemicals. Very few, however, would advocate giving up the advantages of the innumerable products that derive from our modern chemical industry. Better, instead, to learn to handle and dispose of chemicals in ways that minimize the risks. Thus, over the past 30 years, regulations surrounding the production, transport, use, and disposal of chemicals have mushroomed. Where it once had almost no controls at all, the chemical industry is now heavily regulated, and everyone using or handling chemicals that are deemed hazardous is affected by the regulations in one way or another.

This chapter focuses on hazardous chemicals—their nature and how they are investigated. An overview of laws and regulations designed to protect human and environmental health is presented, and the strengths and weaknesses of those measures are examined.

22.1 Toxicology and Chemical Hazards

Toxicology is the study of the harmful effects of chemicals on human and environmental health. A toxicologist might, for example, investigate the potential for *phenolphthalein* (a common chemical laboratory reagent that has also been used in over-the-counter laxatives) to cause health problems. The toxicologist might study the **acute** toxicity effects that could occur upon ingestion of the chemical or upon its contact with the skin, as well as the effects of **chronic** exposure over a period of years, and, finally, the **carcinogenic** potential of phenolphthalein. Indeed, such studies have shown that phenolphthalein causes tumors in mice, and because of this finding, the Food and Drug Administration (FDA) canceled all laxative uses of phenolphthalein, which is now listed as "reasonably anticipated to be a human carcinogen."

Data on toxic chemicals are made available to health practitioners and the public via a number of sources, including the National Toxicology Program (NTP) Chemical Repository and the National Institute of Environmental Health Sciences (NIEHS), the latter of which provides an annual updated *Report on Carcinogens*, with a complete list of carcinogenic agents. The Environmental Protection Agency (EPA) also makes available data on toxic chemicals in its Integrated Risk Information System (IRIS). All of these sources of information are on the Internet. Environmental Defense, a nonprofit organization linking science, economics, and law to combat environmental problems, uses these and other data sources in its *Scorecard*, an exhaustive Web-based profile of 11,200 chemicals, including information on their manufacture and uses.

Dose Response and Threshold

In our discussion of risk assessment (Chapter 17), we introduced the concepts of *dose-response* and *exposure*. When investigating a suspect chemical, a toxicologist would conduct tests on animals, investigate human involvement with the chemical, and present information linking the **dose** (the level of exposure multiplied by the length of time over which exposure occurs) with the **response** (some acute or chronic effect or the development of tumors). For example, studies of phenolphthalein toxicity indicate that the LDL_0 (lowest dose at which death occurred in animal testing) was 500 milligrams/kilogram. This is a low toxicity, so concern about the chemical would center on *chronic* or *carcinogenic* issues.

Human **exposure** to a hazard is a vital part of its risk characterization, and such exposure can come through the workplace, food, water, or the surrounding environment. For example, the NTP Chemical Repository gives the various uses of phenolphthalein, followed by both precautions to take in handling it and the symptoms of exposure (there are many!). Getting an accurate determination of human exposure is often the most difficult area of risk assessment.

Threshold Level. In the dose-response relationship, there is usually a *threshold*. Organisms are able to deal with certain levels of many substances without suffering ill effects. The level below which no ill effects are observed is called the **threshold level.** Above this level, the effect of a substance depends on both its concentration and the duration of exposure to it. Higher levels may be tolerated if the exposure time is short. Thus, the threshold level is high for short exposures, but gets lower as the exposure time increases (Fig. 22–1). In other words, it is not the absolute amount, but rather the *dose*, that is important.

Where carcinogens are concerned, the EPA generally takes a zero-dose, zero-response approach. That is, there is no evidence of a discrete threshold level for any carcinogenic chemical. However, the lower the dose, the more likely it is that the response cannot be distinguished from the background level of cancers in a population. In such cases, the risk becomes very low and drops below the level for which regulatory action is needed.

These are just some basics of toxicology. The field is well established and is the most important source of sound scientific information supporting the regulatory efforts of the FDA and the EPA. The NTP was established in 1978 to provide this kind of information and has become the world's leader in assessing chemical toxicity and carcinogenicity. We turn our attention now to the chemicals themselves.

The Nature of Chemical Hazards: HAZMATs

A chemical that presents a certain hazard or risk is known as a **hazardous material (HAZMAT).** The EPA categorizes substances on the basis of the following hazardous properties:

- **Ignitability:** substances that catch fire readily (e.g., gasoline and alcohol)
- **Corrosivity:** substances that corrode storage tanks and equipment (e.g., acids)
- **Reactivity:** substances that are chemically unstable and that may explode or create toxic fumes when mixed with water (e.g., explosives, elemental phosphorus [not phosphate], and concentrated sulfuric acid)
- **Toxicity:** substances that are injurious to health when they are ingested or inhaled (e.g., chlorine, ammonia, pesticides, and formaldehyde)

Containers in which HAZMATs are stored and vehicles that carry HAZMATs are required to display placards identifying the hazards (flammability, corrosiveness, the potential for poisonous fumes, and so on) presented by the material inside (Fig. 22–2).

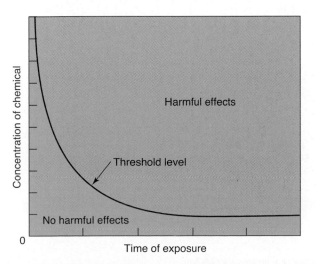

Figure 22–1 The threshold level. The threshold level for harmful effects of toxic pollutants becomes lower as the exposure time increases. It is also different for different chemicals.

Figure 22–2 HAZMAT placards. These are examples of some of the placards that are mandatory on trucks and railcars carrying hazardous materials. Numbers in place of the word on the placard or on an additional orange panel identify the specific material. Placards alert workers, police, and firefighters to the kinds of hazards they face in case an accident occurs.

Radioactive materials, discussed in Chapter 15, are probably the most hazardous of all and are treated as an entirely separate category.

Sources of Chemicals Entering the Environment

To understand how HAZMATs enter our environment, we need to look at how people in our society live and work. First, the materials making up almost everything we use, from the shampoo and toothpaste we apply in the morning to the television we watch in the evening, are products of chemical technology. Our use constitutes only one step in the **total product life cycle,** a term that encompasses all steps, from obtaining raw materials to finally disposing of the product. Implicit in our use of hair spray, for example, is that raw materials were obtained and various chemicals were produced to make both the spray and its container. Chemical wastes and by-products are inevitable in production

processes. In addition, consider the risks of accidents or spills occurring in the manufacturing process and in the transportation of raw materials, the finished product, or wastes. Finally, what are the risks of breathing hair spray? What happens to the container that still holds some of the hair spray when you throw it into the trash because the propellant is used up?

Multiply these steps by all the hundreds of thousands of products used by millions of people in homes, factories, and businesses, and you can appreciate the magnitude of the situation. More than 80,000 chemicals are registered for commercial use within the United States. At every stage—from the mining of raw materials through manufacturing, use, and final disposal—various chemical products and by-products enter the environment, with consequences for both human and environmental health (Fig. 22–3).

The use of many products—pesticides, fertilizers, and road salts, for example—directly introduces them into the environment. Alternatively, the intended use may leave a

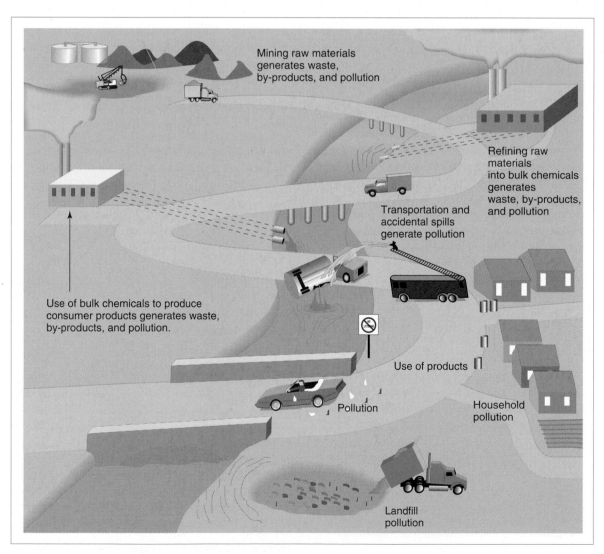

Mining raw materials generates waste, by-products, and pollution

Refining raw materials into bulk chemicals generates waste, by-products, and pollution

Transportation and accidental spills generate pollution

Use of bulk chemicals to produce consumer products generates waste, by-products, and pollution.

Use of products

Pollution

Household pollution

Landfill pollution

Figure 22–3 Total product life cycle. The life cycle of a product begins when the raw materials are obtained and ends when the used-up product is discarded. At each step—or in transport between steps—wastes, by-products, or the product itself may enter the environment, causing pollution and creating various risks to human and environmental health.

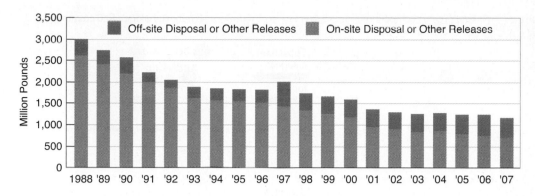

Figure 22–4 Toxics Release Inventory, 1988–2007. The data here follow a consistent set of chemicals and industries from 1988 to 2007, indicating the significant reductions in releases of toxic chemicals to the environment since the program began. (*Source:* EPA, 2007 Toxics Release Inventory, March 2009.)

fraction of the material in the environment—the evaporation of solvents from paints and adhesives, for instance. Then there are the product life cycles of materials that are used tangentially to the desired item. Consider lubricants, solvents, cleaning fluids, cooling fluids, and so on, with whatever contaminants they may contain. Likewise, there are the product life cycles of gasoline, coal, and other fuels that are consumed for energy. In addition to the unavoidable wastes and pollutants produced, in every case there is the potential for accidental releases, ranging from minor leaks from storage tanks to the release of toxic gas from a pesticide plant in Bhopal, India, in 1984 that killed at least 10,500 people.

Toxics Release Inventory. The introduction of chemicals into the environment may occur in every sector, from major industrial plants to small shops and individual homes. Whereas single events involving large amounts of one chemical may constitute a disaster and make headlines, the total amounts entering the environment from millions of homes and businesses are far greater and present much more of a health risk to society. Some idea of the quantities involved can be obtained from the **Toxics Release Inventory (TRI)**. The **Emergency Planning and Community Right-to-Know Act of 1986 (EPCRA)** requires industries to report releases of toxic chemicals to the environment, and the **Pollution Prevention Act of 1990** mandates collection of data on toxic chemicals that are treated on-site, recycled, or combusted for energy. (The TRI does not cover small businesses, such as dry-cleaning establishments and gas stations, or household hazardous waste.) The TRI is disseminated on the Internet and provides an annual record of releases of almost 650 designated chemicals by some 22,000 industrial facilities. For 2007, the TRI reveals the following information about toxic chemicals:

• Total production-related toxic wastes: 24.2 billion pounds

• Releases to the air: 1,311 million pounds

• Releases to the water: 237 million pounds

• Releases to land disposal sites and underground injection: 2,538 million pounds

• Total environmental releases: 4.086 billion pounds (16.9% of all production-related toxic wastes; the remainder are treated on- or off-site or are recycled)

Add to the last figure the estimated 3.2 billion pounds of household hazardous waste released each year and an unknown quantity coming from small businesses (facilities do not have to report releases under 500 pounds), and you have some idea of the dimensions of the problem. The good news is that over the 20 years since the TRI has been in effect, the quantities of virtually all categories of toxic waste have kept going down (Fig. 22–4), and total releases have declined by 61%. (The data in the figure are based on a consistent set of chemicals and industries and so do not agree with the 2007 information presented previously. Many chemicals and industries have been added since the TRI began in 1988.)

The Threat from Toxic Chemicals

All toxic chemicals are, by definition, hazards that pose a risk to humans. Fortunately, a large portion of the chemicals introduced into the environment are gradually broken down and assimilated by natural processes. After that, they pose no long-term human or environmental risk, even though they may be highly toxic in acute doses (high-level, short-term exposures).

Two major classes of chemicals do not readily degrade in the environment, however: (1) *heavy metals and their compounds* and (2) *synthetic organics*. If sufficiently diluted in air or water, most of these compounds do not pose a hazard, although there are some notable exceptions.

Heavy Metals. The most dangerous heavy metals are lead, mercury, arsenic, cadmium, tin, chromium, zinc, and copper. These metals are used widely in industry, particularly in metalworking or metal-plating shops and in such products as batteries and electronics. They are also used in certain pesticides and medicines. In addition, because compounds of heavy metals can have brilliant colors, they once were heavily used in paint pigments, glazes, inks, and dyes. Millions of children have been treated for lead poisoning because of the heavy use of lead-based paint. The paint was banned in the United States in 1978, but many older painted surfaces still hold lead and are a hazard.

Heavy metals are extremely toxic because, as ions or in certain compounds, they are soluble in water and may be readily absorbed into the body, where they tend to combine

Figure 22–5 Halogenated hydrocarbons. These are organic (carbon-based) compounds in which one or more hydrogen atoms have been replaced by halogen atoms (chlorine, fluorine, bromine, or iodine). Such compounds are particularly hazardous to health because they are nonbiodegradable and they tend to bioaccumulate. Shown here are tetrachloroethylene and 1,2-dibromoethane.

Natural organic compound — Ethylene

Substitute chlorine for hydrogen

Synthetic halogenated counterpart — Tetrachloroethylene

Ethane

Substitute bromine for hydrogen

1,2-dibromoethane

with and inhibit the functioning of particular vital enzymes. Even very small amounts can have severe physiological or neurological consequences. The mental retardation caused by lead poisoning and the insanity and crippling birth defects caused by mercury are particularly well-known examples.

Organic Compounds. Petroleum-derived and *synthetic* organic compounds are the chemical basis for all plastics, synthetic fibers, synthetic rubber, modern paintlike coatings, solvents, pesticides, wood preservatives, and hundreds of other products. Because of their chemical structure, many synthetic organics are resistant to biodegradation. Ironically, this resistance is an important part of what makes many such compounds useful: We wouldn't want fungi and bacteria attacking and rotting the tires on our cars, and paints and wood preservatives function only insofar as they are both nonbiodegradable and toxic to decomposer organisms.

These compounds are toxic because they are often readily absorbed into the body, where they interact with particular enzymes, but their nonbiodegradability prevents them from being broken down or processed further. When a person ingests a sufficiently high dose, the effect may be acute poisoning and death. With low doses over extended periods, however, the effects are insidious and can be mutagenic (mutation causing), carcinogenic (cancer causing), or teratogenic (birth defect causing). Synthetic organic compounds also may cause serious liver and kidney dysfunction, sterility, and numerous other physiological and neurological problems.

Dirty Dozen. A particularly troublesome class of synthetic organics is the **halogenated hydrocarbons**, organic compounds in which one or more of the hydrogen atoms have been replaced by atoms of chlorine, bromine, fluorine, or iodine. These four elements are classed as *halogens*—hence, the name (Fig. 22–5). Of the halogenated hydrocarbons, the **chlorinated hydrocarbons** (also called **organic chlorides**) are by far the most common. Organic chlorides are widely used in plastics (polyvinyl chloride), pesticides (DDT, Kepone, and mirex), solvents (carbon tetrachlorophenol and tetrachloroethylene), electrical insulation (polychlorinated biphenyls),

and many other products. Most of the "dirty dozen" POPs (Table 22–1) are halogenated hydrocarbons. All are toxic to varying extents, and most are known animal carcinogens. Many are also suspected endocrine disruptors at very low levels. (See Chapter 13.) These 12 POPs have been banned or highly restricted as a result of the **Stockholm Convention on Persistent Organic Pollutants**, which was signed by 162 countries and entered into force in May 2004.

PERC. Let us consider one of the halogenated hydrocarbon compounds, tetrachloroethylene (Fig. 22–5). This substance, also called perchloroethylene, or PERC, is colorless and nonflammable and is the major substance in dry-cleaning fluid. It is an effective solvent and is used in all kinds of industrial cleaning operations. It can also be found in shoe polish and other household products. PERC evaporates readily when exposed to air, but in the soil, it can enter groundwater easily because it does not bind to soil particles. Human exposure can occur in the workplace, especially in connection with dry-cleaning operations. It can also occur if people use products containing PERC, such as dry-cleaned garments. PERC enters the body most readily when breathed in with contaminated air. Breathing PERC for short periods can bring on dizziness, fatigue, headaches, and unconsciousness. Over longer periods, PERC can cause liver and kidney damage.

Laboratory studies also show PERC to be carcinogenic to rats and mice, and it is listed in NTP's 2009 *Report on Carcinogens*[2] as "reasonably anticipated to be a human carcinogen." Studies have shown a higher risk of cancer, as well as neurological impairment, in dry-cleaning employees. The EPA has issued rules that would phase out PERC (by 2020) in dry-cleaning shops in residential buildings; critics wonder why it should take so long if, as the EPA maintains, PERC is harmful to human health. In response to a court challenge, the Obama EPA has promised to review these rules. PERC is used in large quantities in the United States (370 million

---

[2]U.S. Department of Health and Human Services, Public Health Service, National Toxicology Program, *Report on Carcinogens*, 11th ed. (Washington, D.C.: U.S. Department of Health and Human Services, 2009).

TABLE 22–1 The "Dirty Dozen" Persistent Organic Pollutants

Chemical Substance	Designed Use	Major Concerns
Aldrin	Pesticide to control soil insects and to protect wooden structures from termites	Toxic to humans, may be carcinogenic
Chlordane	Broad-spectrum insecticide to protect crops	Biomagnification in food webs
DDT	Widely used insecticide, malaria control	Biomagnification in food webs
Dieldrin	Insecticide to control termites and crop pests	Toxic, biomagnification in food webs, high persistence
Endrin	Insecticide and rodenticide	Toxic, especially in aquatic systems
Heptachlor	General insecticide	Toxic, carcinogenic
Hexachlorobenzene	Fungicide	Toxic, carcinogenic
Mirex	Insecticide to control ants	High toxicity to aquatic animals, carcinogenic
Toxaphene	General insecticide	High toxicity to aquatic animals, carcinogenic
PCBs	Variety of industrial uses, especially in transformers and capacitors	Toxic, teratogenic, carcinogenic
Dioxins	No known use; by-products of incineration and paper bleaching	Toxic, carcinogenic, reproductive system effects
Furans	No known use; by-products of incineration, PCB production	Toxic, especially in aquatic systems

Source: IPCS Assessment Report: L. Ritter, et al., *Persistent Organic Pollutants: An Assessment Report on DDT, Aldrin, Dieldrin, Endrin, Chlordane, Heptachlor, Hexachlorobenzene, Mirex, Toxaphene, Polychlorinated Biphenyls, Dioxins, and Furans, International Programme on Chemical Safety,* Dec. 1995. Available at http://www.chem.unep.ch/Pops/alts02.html.

pounds in 2007); dry cleaners use about 10% of the PERC, while most of the remainder is employed as a basic raw material in manufacturing hydrofluorocarbons—to replace the ozone-depleting chlorofluorocarbons (see Chapter 19).

Issues with Other Organics. Several synthetic organic chemicals have been in the news recently because of concerns about health effects:

- **Phthalates:** A group of chemicals used to soften plastic toys such as teethers and rubber duckies, phthalates are toxic to rodent reproductive systems in tests and have been implicated as possible reproductive hormone disrupters. A congressionally initiated federal ban on the use of these chemicals in any product intended for the use of children took effect in 2008.

- **Bisphenol A (BPA):** BPA is a starter material for many plastics, including reusable beverage containers like baby bottles. However, BPA exposure in animals contributes to many problems, including obesity, diabetes, infertility, and cancer. The FDA decided in 2008 that BPA does not pose a health hazard when used in food containers, but an FDA scientific advisory panel determined that the FDA ruling was flawed.

- **Perchlorate:** An ingredient of rocket fuel and other flammable products, this compound has been showing up in drinking water in many states and in a wide range of

foods. Most recently, perchlorate was found in every brand of powdered infant formula tested. The chemical has been shown to interfere with thyroid gland function. In late 2008, the EPA refused to set a drinking water standard for perchlorate, in spite of evidence that at least 20 million Americans are exposed to unsafe levels of the chemical.

One common thread to these issues was the tendency of the Bush administration regulatory agencies to choose lax or no rules for regulating the controversial chemicals. Pressured by the Bush White House, the EPA also had weakened the TRI reporting rules. All of these steps are being addressed by the Obama administration, which favors stricter reporting and more science-based regulatory decisions.

Involvement with Food Chains

The trait that makes heavy metals and nonbiodegradable synthetic organics particularly hazardous is their tendency to accumulate in organisms. We discussed the phenomena of bioaccumulation and biomagnification in connection with DDT in Chapter 13 and have just seen these phenomena in operation in Arctic food chains.

A tragic episode in the early 1970s, known as Minamata disease, revealed the potential for biomagnification of mercury and other heavy metals. The disease is named for a small fishing village in Japan where the episode occurred. In the mid-1950s, cats in Minamata began to show spastic movements, followed by partial paralysis, coma, and death. At first, this was thought to be a syndrome peculiar to felines, and little attention was paid to it. However, when the same symptoms began to occur in people, concern escalated quickly. Additional symptoms, such as mental retardation, insanity, and birth defects, were also observed. Scientists and health experts eventually diagnosed the cause as acute mercury poisoning.

A chemical company near the village was discharging wastes containing mercury into a river that flowed into the bay where the Minamata villagers fished. The mercury, which settled with detritus, was first absorbed and bioaccumulated by bacteria and then biomagnified as it passed up the food chain through fish to cats or to humans. Cats had suffered first and most severely because they fed almost exclusively on the remains of fish. By the time the situation was brought under control, some 50 persons had died, and 150 had suffered serious bone and nerve damage. Even now, the tragedy lives on in the crippled bodies and retarded minds of some Minamata victims.

22.2 Hazardous-Waste Disposal

The Clean Air and Clean Water Acts, passed in the early 1970s, put an end to the widespread disposal of hazardous wastes in the atmosphere and waterways (see Chapter 21). However, their passage left an enormous loophole. If you can't vent wastes into the atmosphere or flush them into waterways, what do you do with them? Industry turned to *land disposal*, which was essentially unregulated at the time. Indiscriminate air and water disposal became indiscriminate land disposal. Thus, in retrospect, we see that the Clean Air and Clean Water Acts, for all their benefits in improving air and water quality, also succeeded in transferring pollutants from one part of the environment to another.

Methods of Land Disposal

In the early 1970s, there were three primary land disposal methods: (1) deep-well injection, (2) surface impoundments, and (3) landfills. With the conscientious implementation of safeguards, each of these methods has some merit, and each is still heavily used for hazardous-waste disposal. Without adequate regulations or enforcement, however, contamination of groundwater is virtually inevitable.

Deep-Well Injection. Deep-well injection involves drilling a borehole thousands of feet below groundwater into a porous geological formation. A well consists of concentric pipes and casings that isolate the wastes as they are injected, and the well is sealed at the bottom to prevent wastes from backing up (Fig. 22–6). Over time, the wastes often will undergo reactions with naturally occurring material that will make them less hazardous. This method is presently used for various volatile organic compounds, pesticides, fuels, and explosives. Currently, 121 wells are in operation, mostly in the Gulf Coast region. The EPA's Underground Injection Control Program regulations require that the wells be limited to geologically stable areas so that the hazardous-waste liquids pumped into the well remain isolated indefinitely. The total amount of deep-well injection (203 million pounds in 2007) has declined over the years, but the technique remains useful because it has a greater potential than other methods for keeping toxic wastes from contaminating the hydrologic cycle and the food web.

Surface Impoundments. Surface impoundments are simple excavated depressions (*ponds*) into which liquid wastes are drained and held. They are the least expensive—and, hence, the most widely used—way to dispose of large amounts of water carrying relatively small amounts of chemical wastes. The impoundments continue to be widely used, often to carry out wastewater treatment prior to discharging the wastes. As waste is discharged into the pond, solid wastes settle and accumulate, while water evaporates (Fig. 22–7). If the bottom of the pond is well sealed and if the loss of water through evaporation equals the gain from input, impoundments may receive wastes indefinitely. However, inadequate seals may allow wastes to percolate into groundwater, exceptional storms may cause overflows, and volatile materials can evaporate into the atmosphere, adding to air pollution problems.

A 1996 **Resource Conservation and Recovery Act (RCRA)** amendment prompted the EPA to conduct a study of surface impoundments. The study (*Industrial Surface Impoundments in the United States*[3]), released in 2001, surveyed the scope of impoundments, the characteristics of the wastes they managed, and the potential risks posed by chemical constituents in existing surface impoundments. The 2007 TRI reports 781 million pounds of toxics released to on-site surface impoundments. (*On-site* refers to disposal by manufacturers or institutions on their own facilities; *off-site* means that the wastes have been transferred to a waste-treatment or disposal facility.) Some of the findings in the EPA study (which was limited to the 1990–2000 time frame) are as follows:

- Approximately 18,000 industrial surface impoundments were in use at 7,500 facilities, mostly east of the Mississippi River or along the Pacific Coast.

- At least two-thirds of the impoundments contained hazardous chemicals with carcinogenic and other human health concerns, originating largely from the chemical, paper, concrete, and petroleum industries.

- Most of the impoundments were only a few meters above groundwater, and more than half lacked liner systems that would prevent the movement of chemicals into groundwater or soil.

- More than 20 million people lived within 1.2 miles (2 km) of an industrial impoundment, and some 10% of

---

[3]Office of Solid Waste and Emergency Response, EPA, *Industrial Surface Impoundments in the United States*, EPA530-R-01-005 (Washington, D.C.: EPA, March 2001). Available at www.epa.gov/osw.

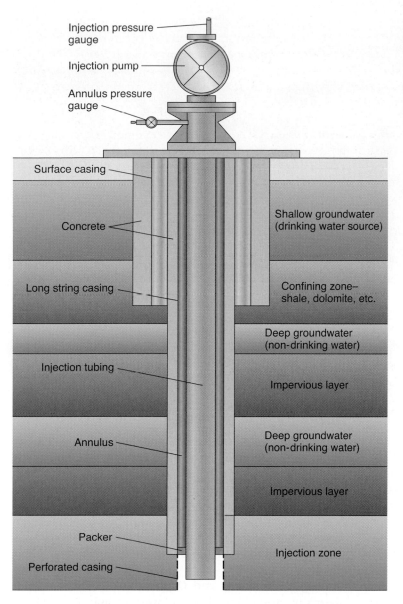

Injection pressure gauge

Injection pump

Annulus pressure gauge

Surface casing

Concrete

Long string casing

Injection tubing

Annulus

Packer

Perforated casing

Shallow groundwater (drinking water source)

Confining zone– shale, dolomite, etc.

Deep groundwater (non-drinking water)

Impervious layer

Deep groundwater (non-drinking water)

Impervious layer

Injection zone

Figure 22–6 Deep-well injection. This technique is used for the disposal of large amounts of liquid wastes. The concept is that toxic wastes may be drained into dry, porous strata below ground, where they may reside harmlessly "forever." The injection tubing is surrounded by a space (the annulus) that is filled with an inert, pressurized fluid. The long string casing is filled with cement from the surface to the injection zone. (Adapted from the EPA Underground Injection Program Web site, http://www.epa.gov/ogwdw000/uic/.)

Figure 22–7 Surface impoundment. Surface impoundments are often used for preliminary treatment of liquid wastes. Shown here is an impoundment at a paper manufacturer in Alabama.

Figure 22–8 Secure landfill. Landfills are a technique that is widely used for disposing of concentrated chemical wastes. Shown is an operating hazardous-waste landfill in Alabama.

impoundments were located within 500 feet (150 m) of a domestic drinking-water well.

- From 2% to 5% of the sites were judged to pose possible risks to human health, whereas 24% were found to pose a risk of release to the environment.

The study concluded that existing state and federal regulations should be adequate to cover most of the impoundment-related problems. However, in view of the existing risks, gaps in regulatory coverage clearly exist and may need future regulations. The EPA has not yet determined what these might be.

Landfills. RCRA sets rigid standards for the disposal of hazardous wastes in landfills. When hazardous wastes are in a concentrated liquid or solid form, they are commonly put into drums, typically collected by a waste management company (for a hefty fee!), and then treated in accordance with their chemical and physical characteristics. *Treatment standards* for wastes are established by the EPA and characterized as **best-demonstrated available technologies (BDATs)**. These standards apply to all solid and liquid hazardous wastes and have the effect of reducing their toxicity and mobility. The BDAT technologies include stabilization and neutralization of sludges, soils, liquids, powders, and slurries; chemical oxidation of organic materials; and various specific techniques applied to metal-bearing wastes that convert them to an insoluble solid material for landfill disposal. Some 403 million pounds of hazardous wastes were delivered to on-site landfills in 2007, and an additional 341 million pounds were deposited in off-site landfills. Only 23 landfills in North America are licensed to receive off-site hazardous wastes, and these are owned by just a few waste management companies (e.g., Safety-Kleen, Waste Management, and Envirosource).

If a landfill is properly lined, supplied with a system to remove leachate, provided with monitoring wells, and appropriately capped, it may be reasonably safe and is referred to as a **secure landfill** (Fig. 22–8). The various barriers are subject to damage and deterioration, however, requiring adequate surveillance and monitoring systems to prevent leakage.

Mismanagement of Hazardous Wastes

Because early land disposal was not regulated, in numerous instances not even the most rudimentary precautions were taken. Indeed, in many cases, deep wells were injecting wastes directly into groundwater, abandoned quarries were sometimes used as landfills with no additional precautions being taken, and surface impoundments frequently had no seals or liners whatsoever. Even worse, considerable amounts of waste failed to get to any disposal facility at all.

Midnight Dumping and Orphan Sites. The need for alternative methods to dispose of waste chemicals created an opportunity for a new enterprise: waste disposal. Many reputable businesses entered the field, but—again in the absence of regulations—there were also disreputable operators. As stacks of drums filled with hazardous wastes "mysteriously" appeared in abandoned warehouses, vacant lots, or municipal landfills, it became clear that some operators simply were pocketing the disposal fee and then unloading the wastes in any available location, frequently under cover of darkness—an activity termed **midnight dumping** (Fig. 22–9). Authorities trying to locate the individuals responsible would learn that they had gone out of business or were nowhere to be found.

Some companies or individuals simply stored wastes on their own properties and then went out of business, abandoning the property and the wastes. These locations became known as **orphan sites**—hazardous-waste sites without a responsible party to clean them up. As drums containing hazardous chemicals corroded and leaked, there was great danger of reactive chemicals combining and causing explosions and fires. One of the most famous abandoned sites is the "Valley of the Drums" (VOD) in Kentucky (Fig. 22–10).

Figure 22–9 Midnight dumping. Hazardous wastes were often left on remote or unoccupied properties by unscrupulous haulers.

Scope of the Mismanagement Problem.

The mounting problem of unregulated land disposal of hazardous wastes was brought vividly to public attention by the events at Love Canal, near Niagara Falls, New York. The area was occupied by a school and a number of houses, all of which were perched on top of a chemical waste dump that had been covered over and developed. The surface of the dump began to collapse, exposing barrels of chemical wastes. Fumes and chemicals began seeping into cellars. Ominously, people began reporting serious health problems, including birth defects and miscarriages. An aroused neighborhood, led by activist Lois Gibbs, began demanding that the state do something about the problem. Following confirmation of the contamination by toxic chemicals, President Carter signed an emergency declaration in 1978 to relocate hundreds of residents. The state and federal governments closed the school and demolished a number of the homes nearest the site (Fig. 22–11). In all, more than 800 families moved out of the area because of the fear of damage to their health from the toxic chemicals.

Occidental? The absence of a public policy for dealing with the disposal of hazardous chemicals contributed to the situation at Love Canal. Hooker Chemical and Plastics Company had purchased an abandoned canal near Niagara Falls in 1942 and proceeded to fill it with an estimated 21,000 tons of hazardous wastes. Hooker covered the canal over with a clay cap and sold it to the school board for a small sum, reportedly after warning the board that there were chemicals buried on the property. Subsequent construction penetrated the clay cap, and rain seeped in and leached chemicals in all directions. Hooker Chemical's parent company, Occidental Petroleum, eventually spent more than $233 million on the cleanup and subsequent lawsuits.

As both government and independent researchers began surveying the extent of the problem, bad disposal practices were found to be rampant. The World Resources Institute estimated that, in the United States in the early 1980s, there existed 75,000 active industrial landfill sites, along with

Figure 22–10 "Valley of the Drums," an orphan waste site. Thousands of drums of waste chemicals, many of them toxic, were unloaded near Louisville, Kentucky, around 1975 and left to rot, seriously threatening the surrounding environment, waterways, and aquifers.

Figure 22–11 Love Canal. These are a few of the nearly 100 homes that the state was forced to buy and later destroy as a result of contamination from Love Canal. The area has now been cleaned up and redeveloped with new homes.

180,000 surface impoundments and 200 other special facilities that were or could be sources of groundwater contamination. In most cases, the contaminated area was relatively small—200 acres (80 hectares) or less—but in total, the problem was immense and affected every state in the country. As studies and tests proceeded, thousands of individual wells and some major municipal wells were closed because they were contaminated with toxic chemicals.

Many cases of private-well contamination caused by careless disposal were discovered only after people experienced "unexplainable" illnesses over prolonged periods. While these incidents did not receive the media attention of Love Canal, they were nevertheless devastating to the people involved. Even more important than the damage already done was the recognition that the problem was ongoing.

The problems concerning toxic chemical wastes can be divided into three areas:

- Cleaning up the "messes" already created, especially where they threaten drinking water supplies;

- Regulating the handling and disposal of wastes currently being produced, so as to protect public and environmental health; and

- Reducing the quantity of hazardous waste produced.

Each of these problems is addressed in the sections that follow.

22.3 Cleaning Up the Mess

A major public-health threat from land-disposed toxic wastes is the contamination of groundwater that is subsequently used for drinking. The first priority is to ensure that people have safe water. The second is to clean up or isolate the source of pollution so that further contamination does not occur.

Ensuring Safe Drinking Water

To protect the public from the risk of toxic chemicals contaminating drinking water supplies, Congress passed the **Safe Drinking Water Act of 1974**. Under this act, the EPA sets national standards to protect the public health, including allowable levels of specific contaminants. If any contaminants are found to exceed **maximum contaminant levels** (MCLs), the water supply should be closed until adequate purification procedures or other alternatives are adopted. The act was amended in 1986, and the EPA now has jurisdiction over groundwater and sets MCLs for 90 contaminants. States and public water agencies are required to monitor drinking water to be sure that it meets the standards (see Chapter 20). The contaminants represent a host of chemicals that are known pollutants and are regarded as dangerous to human health (see the EPA Web site www.epa.gov/safewater/ contaminants/index.html#mcls for a description of the MCLs and why they are being regulated).

Groundwater Remediation. If dumps, leaking storage tanks, or spills of toxic materials have contaminated groundwater and if such groundwater is threatening water supplies, all is not yet lost. **Groundwater remediation** is a developing and growing technology. Techniques involve drilling wells, pumping out the contaminated groundwater, purifying it, and injecting the purified water back into the ground or discharging it into surface waters (Fig. 22–12). Cleaning up the source of the contamination is mandatory. If, however, the contamination is extensive, remediation may not be possible, and the groundwater must be considered unfit for use as drinking water.

Superfund for Toxic Sites

Probably the most monumental task we are facing is the cleanup of the tens of thousands of toxic sites resulting from the years of mismanaged disposal of toxic materials. Wherever

Figure 22–12 Leaking underground storage tank remediation. (a) Typical subsurface contamination from a leaking fuel tank at a gas station. (b) After the leak has been repaired, the vacuum-extraction process causes gasoline and residual hydrocarbons in the soil and on the water table to evaporate and then removes the vapors, preventing further contamination of the groundwater. (c) Contaminated groundwater is pumped out, treated, and returned to the ground. (d) Soil vapor extraction system removing volatile organic compounds in soil at an abandoned gasoline station.

facilities were still operating, pressures were brought to bear on the operators to clean up the sites. Many operators who could not afford to do so, however, simply declared bankruptcy, and the sites joined those already abandoned.

Prompted by the public outrage over the Love Canal fiasco, Congress enacted the **Comprehensive Environmental Response, Compensation, and Liability Act of 1980 (CERCLA)**, popularly known as **Superfund.** Through a tax on chemical raw materials, this legislation provides a trust fund for the identification of abandoned chemical waste sites, protection of groundwater near the sites, remediation of groundwater if it has been contaminated, and cleanup of

the sites. The Superfund program was greatly expanded by the **Superfund Amendments and Reauthorization Act of 1986 (SARA)**. Superfund, one of the EPA's largest ongoing programs, works as described in the next several subsections.

Setting Priorities. Resources are insufficient to clean up all sites at once. Therefore, a system for setting priorities has been developed:

- As sites are identified—note that many abandoned sites had long since been forgotten—their current and potential threats to groundwater supplies are initially assessed by taking samples of the waste, determining its characteristics, and testing groundwater around the site for contamination. If it is determined that no immediate threat exists, nothing more may be done.

- If a threat to human health does exist, the most expedient measures are taken immediately to protect the public. These measures may include digging a deep trench around the site, installing a concrete dike, and recapping it with impervious layers of plastic and clay to prevent infiltration. Thus, the wastes are isolated, at least for the short term. If the situation has gone past the threat stage and contaminated groundwater is or will be reaching wells, remediation procedures are begun immediately.

- The worst sites (those presenting the most immediate and severe threats) are put on the **National Priorities List (NPL)** and scheduled for total cleanup. A site on this list is reevaluated to determine the most cost-effective method of cleaning it up, and finally, cleanup begins with efforts to identify potential responsible parties (PRPs)—namely, industries in which the wastes originated. The industries are "invited" to contribute to the cleanup, and they do so either by contributing financially or by participating in cleanup activities.

Cleanup Technology. If the chemical wastes are contained in drums, the drummed wastes can be picked up, treated, and managed in proper fashion. For the VOD site (Fig. 22–10), the EPA cleanup started with the removal of some 4,200 drums for off-site treatment and disposal. The bigger problem for many sites is the soil (often millions of tons) contaminated by leakage. One procedure is to excavate contaminated soil and run it through an incinerator or kiln assembled on the site to burn off chemicals (Fig. 22–13). (Synthetic organic chemicals can be broken down by incineration, and heavy metals are converted to insoluble, stable

oxides.) Another method is to drill a ring of injection wells around the site and a suction well in the center. Water containing a harmless detergent is injected into the injection wells and drawn into the suction well, cleansing the soil along the way. The withdrawn water is treated to remove the pollutants and is then reused for injection.

The VOD site was situated in a poorly drained shale and limestone deposit. No local use was made of the groundwater. The EPA determined that the most feasible (BDAT) treatment was simply to contain the contaminated soil and groundwater, so the agency installed a clay cap, a perimeter drainage system, monitoring wells, and a security fence. The remedial work was completed in 1989, at a total cost of $2.5 million, and the site was deleted from the NPL in 1996. The third five-year review indicated that the cap was functioning well, the site was still undeveloped, and no human uses were being made of the groundwater.

Bioremediation. Another rapidly developing cleanup technology is *bioremediation*. In many cases, the soil is contaminated with toxic organic compounds that are biodegradable. The problem is that they do not degrade because the soil lacks organisms, oxygen, or both. In **bioremediation**, oxygen and organisms are injected into contaminated zones. The organisms feed on and eliminate the pollutants (as in secondary sewage treatment) and then die when the pollutants are gone. Considerable research is being done to find and develop microbes that will break down certain kinds of wastes more readily. Bioremediation, which may be used in place of detergents to decontaminate soil, is a rapidly expanding and developing technology that is applicable to leaking storage tanks and spills as well as to waste-disposal sites.

Plant Food? When the soil contaminants are heavy metals and nonbiodegradable organic compounds, *phytoremediation* has been employed with some success. **Phytoremediation** uses plants to accomplish a number of desirable cleanup steps: stabilizing the soil, preventing further movement of contaminants by erosion, and extracting the contaminants by direct uptake from the soil. After full growth, the plants are removed and treated as toxic waste products. However, this is a slow process and can be used only at sites where the contaminants and their concentrations are not toxic to plants. Nevertheless, scientists have found sunflowers that will capture uranium, poplar trees that soak up dry-cleaning solvents and mercury, and ferns that thrive on arsenic. Phytoremediation is now used at a level above $100 million a year.

Evaluating Superfund. Of the literally hundreds of thousands of various waste sites, more than 47,000 have been deemed serious enough to be given Superfund status. Over time, the EPA judged that about 33,000 of these sites did not pose a significant public-health or environmental threat. Such sites were accordingly assigned to the category "no further removal action planned" (NFRAP). Still, more than 11,300 sites remain on the master list—a figure that, it is assumed, now includes most (if not all) of the sites in the

Figure 22–13 Mobile incinerator for toxic waste. The EPA is cleaning up some Superfund sites by running contaminated materials through this incinerator, erected at the site.

United States that pose a significant risk. Interestingly, some of the worst that have come to light are on military bases, the result of what many characterize as the totally heedless and unconscionable discarding of toxic materials from military operations (13% of the NPL sites are "federal facilities").

Progress.   As of 2009, 1,264 sites were still on the NPL. Additions are made to the list as sites from the master list are assessed, and subtractions are made when remediation on a site is complete. Since 1980, when CERCLA went into effect, about 1,063 sites have received all the necessary cleanup-related construction, and 332 of these have been deleted from the list (Love Canal was removed in 2004). Cleanup of these sites takes an average of 12 years, at a mean cost of $20 million. Total cleanup often takes that long because groundwater remediation is a slow process. The other NPL sites are in various stages of analysis, remediation, and construction. (An example of a Superfund site and its cleanup is presented in Stewardship, "Woburn's Civil Action.")

Stewardship

Woburn's Civil Action

One of the earliest Superfund sites to appear on the NPL is known as the **Industri-Plex 128** site, in Woburn, Massachusetts. By the time of its official listing by the EPA in 1981, the Industri-Plex site had already become notorious. Starting in the mid–19th century, Woburn Chemical Works built a huge industrial complex of 90 buildings on 400 acres in the city of Woburn. Among the products of the complex were arsenic-based insecticides, sulfuric and acetic acids, glue from raw animal hides, chromium-tanned hides, and organic chemicals such as phenol, benzene, and toluene. The company continued to operate until the late 1960s. When a developer attempted to excavate the site for redevelopment in the 1970s, the uncovered land yielded decaying animal hides and buried chemical waste deposits, all of which gave off offensive odors.

In the 1960s, the city of Woburn had installed two wells slightly downstream of the buried wastes, and these became the subject of a civil action when a cluster of childhood leukemia cases appeared in the city. Lawyer Jan Schlichtmann[4] acted on behalf of eight families who sued W. R. Grace & Co. and Beatrice Corporation, allegedly the parties responsible for polluting the groundwater with toxic wastes. The suit, which eventually ended in an out-of-court settlement, was portrayed in the 1998 film *A Civil Action*, in which John Travolta played the part of Schlichtmann. In spite of many inaccurate depictions in the film, Woburn residents were pleased to see their story told and the chemical companies exposed.

The EPA identified a number of "potential responsible parties" who took a financial role in the cleanup. The cleanup involved removing drums of chemical wastes, treating highly contaminated groundwater, and constructing permanent caps over some 105 acres of contaminated soils and sediments. By 1996, the EPA was entering into agreements with prospective purchasers of parcels on the site, and currently, the area is fully involved in an economic renewal of the city of Woburn. The site now accommodates a regional transportation center, an interstate exchange on I-93, a retail development, an office park, and, next to the former waste site, a new mall and hotel. Woburn received more than $4 million in tax revenue in 2004 from businesses on the site. Although it is still on the NPL list, the Industri-Plex site has been transformed from a noxious waste area to a thriving commercial complex—exactly what the Superfund program is intended to accomplish.

Woburn's "Civil Action." Lawyer Jan Schlichtmann, who represented families who sued W.R. Grace and Co. and Beatrice Corporation. alleged to be responsible for contaminating Woburn's water supply. He was portrayed by John Travolta in the 1998 film A Civil Action. Schlichtmann continues to be active in civil litigation and education involving hazardous wastes.

[4]See Jonathan Harr, *A Civil Action* (New York: Vintage Books, 1996).

Who Pays? CERCLA is based on the principle that "the polluter pays." However, the history of a waste-disposal site may go back 50 years or more, and users may have included everything from schools, hospitals, and small businesses to large corporations, all of which mount legal defenses to disclaim responsibility. When liability is difficult to track down or the responsible parties are unable to pay, the Superfund trust fund kicks in. However, over 70% of the cleanup costs to date have been paid by the polluters (the PRPs).

Over the years, the EPA has gained a great deal of experience in dealing with Superfund sites. The technology has become quite sophisticated, and many hazardous-waste remediation companies have become established. A great deal of progress has been made; however, the American public still rates toxic waste third in a list of eight environmental concerns, according to a 2009 Gallup survey.[5]

Critics. The Superfund program gets its share of criticism. Industries claim that they are unfairly blamed for pollution that reaches back to activities that were legal before the enactment of CERCLA. Many feel that overly stringent standards of cleanup are costing large sums of money without providing any additional benefit to public health. (This problem of finding a suitable balance between costs and benefits is discussed in more depth in Chapter 2.) Congress refused to renew the tax on industry at the end of 1995; the trust fund had a balance of $4 billion at the time, but over the years, this fund has been used up. With the disappearance of the trust fund monies, the public (through taxes) is paying the costs of Superfund cleanups that are not recovered from PRPs. However, the cleanup job is nowhere near complete. The Obama administration has proposed reinstating the tax on oil and chemical producers, but will do so only after the economy recovers from the current recession. Congress, however, has appropriated $600 million for EPA's Superfund program as a component of the **American Recovery and Reinvestment Act of 2009**, aimed at creating jobs and speeding up the cleanup of NPL sites.

Brownfields. One highly successful recent Superfund development is the *brownfields* program. **Brownfields** are "abandoned, idled, or underused industrial and commercial facilities where expansion or redevelopment is complicated by real or perceived environmental contamination."[6] It has been estimated that environmental hazards not serious enough to be put on the Superfund NPL impair the value of $2 trillion worth of real estate in the United States. Bipartisan support led to passage of the **Brownfield Act** in 2002, which provides grants for site assessment and remediation work and authorized $250 million/year for the ensuing five years for the program. Previously authorized under CERCLA, the brownfields program now stands on its own (see Figure 23–1 for an example of a brownfield site

transformation) and continues to be funded by Congress beyond its limit. The legislation limits liability for owners and prospective purchasers of contaminated land, thus clearing the way for more cleanup of the estimated 450,000 sites that would qualify as brownfields. States are also moving on this issue. One of the nation's most successful brownfields programs is in Massachusetts, which has provided a $30 million brownfields redevelopment fund managed by MassDevelopment, the state's economic development agency. Over 300 projects are in the development pipeline under the program.

Many brownfield sites lie in economically disadvantaged communities, and their rehabilitation contributes jobs and exchanges a functional facility for an unsightly blight on the neighborhood. For example, the city of Chicago recently acquired and cleaned up a former bus barn in west Chicago that was becoming an illegal indoor garbage dump. The Illinois EPA cleared the site for reuse, and it was acquired by Scott Peterson Meats, which erected a $5.2 million smokehouse that employs 100 workers. Frequently, the rehabilitation of brownfield sites provides industries and municipalities with centrally located, prime land for facilities that would otherwise have been carved out of suburban or *greenfield* lands (land occupied by natural ecosystems). A further advantage is that the new developments go back on the tax rolls, turning a liability into a community asset that lightens the load of residential taxpayers.

Leaking Underground Storage Tanks (LUSTs). One consequence of our automobile-based society is the millions of underground fuel-storage tanks at service stations and other facilities. Putting such tanks underground greatly diminishes the risk of explosions and fires, but it also hides leaks. Underground storage tanks were traditionally made of bare steel, so they had a life expectancy of about 20 years before they began leaking and contaminating groundwater. By 2009, more than 479,000 failures had been reported, with 7,364 new "releases" in 2008. Without monitoring, small leaks can go undetected until nearby residents begin to smell fuel-tainted water flowing from their faucets.

Underground storage tank (UST) regulations, part of RCRA, now require strict monitoring of fuel supplies, tanks, and piping so that leaks are detected early. When leaks are detected, remediation must begin within 72 hours. Rules now require all USTs to be upgraded with interior lining and cathodic protection (to retard electrolytic corrosion of the steel), and new tanks must be provided with the same protection if they are steel. Many service stations are turning to fiberglass tanks, which do not corrode. A LUST trust fund, financed by a 0.1-cent-per-gallon tax on motor fuel, pays for federal activities involved with oversight and cleanup. The states are required to have UST programs, and they implement the federal regulations, sometimes adding more stringent requirements. By 2009, some 377,000 cleanups were completed, with 103,000 yet to go. The states have reported that LUSTs are the most common source of groundwater contamination.

---

[5]Lydia Saad, *Water Pollution Americans' Top Green Concern* (March 25, 2009). Available at www.gallup.com/poll.

[6]http://epa.gov/enforcement/basics/cleanup.html.

22.4 Managing Current Toxic Chemicals and Wastes

The production of chemical wastes is, and will continue to be, an ongoing phenomenon as long as modern societies persist. Therefore, human and environmental health can be protected only if we have management procedures to handle and dispose of wastes safely so that we will not be creating more Superfund sites. We have already noted that the Clean Water Act and the Clean Air Act limit discharges into water and air, respectively. When problems regarding the disposal of wastes on land became evident in the mid-1970s, Congress passed RCRA in 1976 in order to control land disposal. Thus, a company producing hazardous wastes in the United States today is covered by the regulations of these three major environmental acts.

The Clean Air and Clean Water Acts

The Clean Air Act of 1970, the Clean Water Act of 1972, and their various amendments make up the basic legislation limiting discharges into the air or water. More is said about the Clean Air Act in Chapter 19. Under the Clean Water Act (see Chapter 20), any firm (including facilities such as sewage-treatment plants) discharging more than a certain volume into natural waterways must have a **discharge permit** (specifically, a National Pollution Discharge Elimination System or NPDES permit). Discharge permits are a means of monitoring who is discharging what. Establishments with discharge permits are required to report all discharges of substances covered by the TRI. The renewal of the permits is then made contingent on reducing pollutants to meet certain standards within certain periods. Standards are being made continually stricter as technologies for pollution control improve. Some manufacturing firms discharging wastewater into municipal sewer systems are required to pretreat such water to remove any pollutant that cannot be removed by sewage-treatment plants—that is, nonbiodegradable organics and heavy metals.

Nevertheless, even this restriction does not end all water pollution. Certain amounts of wastes are still legally discharged under permits, and although they may account for a low percentage of total emissions, they still add up to large numbers, as we have seen. Moreover, small firms, homes, and farms are exempt from regulation and contribute an unknown quantity of toxics to the air and water. Further still, a great deal of pollution comes from nonpoint sources such as urban and farm runoff.

The Resource Conservation and Recovery Act (RCRA)

The 1976 RCRA and its subsequent amendments are the cornerstone legislation designed to prevent unsafe or illegal disposal of all solid wastes on land. RCRA has three main features. *First*, it requires that all disposal facilities, such as landfills, be sanctioned by *permit*. The permitting process requires that these facilities have all the safety features mentioned in Section 22.2, including monitoring wells. This requirement caused most old facilities to shut down—many subsequently became Superfund sites—and new high-quality landfills with safety measures to be constructed.

Second, RCRA requires that toxic wastes destined for landfills be pretreated to convert them to forms that will not leach. Such treatment now commonly includes biodegradation or incineration in various kinds of facilities, including cement kilns (Fig. 22–14). Biodegradation involves the use of systems similar to those used for secondary sewage treatment, as discussed in Chapter 20, and perhaps new species of

Figure 22–14 Cement kiln to destroy hazardous wastes. A cement kiln is a huge rotating "pipe," typically 15 feet in diameter and 230 feet long, mounted on an incline. (a) Solid wastes fed in with raw materials are fully incinerated and made to react with cement compounds as they gradually tumble toward the combustion chamber. (b) Flammable liquid wastes added with the fuel and air impart fuel value as they are burned. (c) Waste dust is trapped and recycled into the kiln.

(a)

(b)

(b) Fuel and air input
Preheater
(a)
Clinker cooler
Kiln
(c)
Dust collector

bacteria capable of breaking down synthetic organics. If treatment is thorough, there may be little or nothing to put in a landfill. That is the ultimate objective.

For whatever is still going to disposal facilities, the *third* major feature of RCRA is the requirement of "cradle-to-grave" tracking of all hazardous wastes. The generator (the company that originated the wastes) must fill out a form detailing the exact kinds and amounts of waste generated. Persons transporting the waste, who are also required to be permitted, and those operating the disposal facility must each sign the form, vouching that the amounts of waste transferred are accurate. Copies of their signed forms go to the EPA. All phases are subject to unannounced EPA inspections. The generator remains responsible for any waste "lost" along the way or any inaccuracies in reporting. This provision of RCRA thus ensures that generators will deal only with responsible parties and curtails midnight dumping.

Reduction of Accidents and Accidental Exposures

A significant risk to personal and public health lies in exposures that occur as a result of leaks, accidents, and the misguided use of hazardous chemicals in the home or workplace. A considerable number of laws bear on reducing the probability of accidents and on minimizing the exposure of both workers and the public if accidents should occur.

Department of Transportation Regulations.   Transport is an area that is particularly prone to accidents. As modern society uses increasing amounts and kinds of hazardous materials, the stage is set for accidents to become wide-scale disasters. To reduce this risk, **Department of Transportation Regulations (DOT Regs)** specify the kinds of containers and methods of packing to be used in the transport of various hazardous materials. Such regulations are intended to reduce the risk of spills, fires, and poisonous fumes that could be generated from mixing certain chemicals in case an accident occurs.

In addition, DOT Regs require that every individual container and the outside of a truck or railcar carry a standard placard identifying the hazards (flammability, corrosiveness, the potential for poisonous fumes, and so on) presented by the material inside (Fig. 22–2). Such placards enable police and firefighters to identify the potential hazard and respond appropriately in case of an accident. You may also see highway HAZMAT signs restricting truckers with hazardous materials to using particular routes or lanes or to using a highway only during certain hours.

Worker Protection: OSHA and the "Worker's Right to Know."   In the past, it was not uncommon for industries to require workers to perform jobs that entailed exposure to hazardous materials without informing the workers of the hazards involved. This situation is now addressed by certain amendments to the **Occupational Safety and Health Act of 1970**. These amendments make up the **hazard communication standard** or **"worker's right to know."** Basically, the law

requires businesses, industries, and laboratories to make available both information regarding hazardous materials and suitable protective equipment. One form in which this information is presented is the **material safety data sheet (MSDS)**, which must accompany more than 600 chemicals when they are shipped, stored, and handled. These sheets contain information about the reactivity and toxicity of the chemicals, with precautions to follow when one is using the chemical. Notably, however, the responsibility to read the information and exercise proper precautions remains with the worker.

Community Protection and Emergency Preparedness: SARA, Title III.   In 1984, an accident at a Union Carbide pesticide plant in Bhopal, India, caused the release of some 30 to 40 tons of methyl isocyanate, an extremely toxic gas (Fig. 22–15). An estimated 600,000 people in communities surrounding the plant were exposed to the deadly fumes. The official death toll stands at 10,500, but unofficial estimates range much higher. At least 50,000 people suffered various

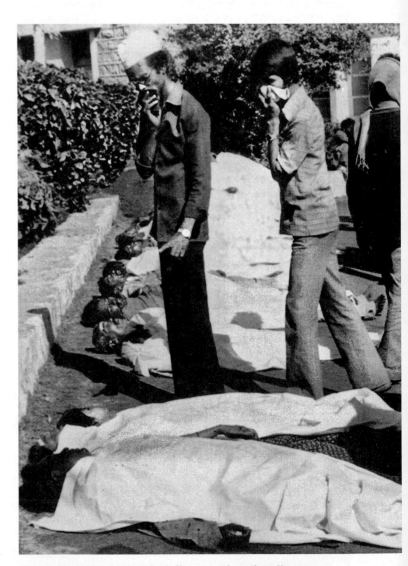

Figure 22–15 Toxic chemical disaster, Bhopal, India. Officials view some of the victims of a deadly gas release at a Union Carbide plant in 1984. At least 10,500 died, and many more suffered injuries.

degrees of visual impairment, respiratory problems, and other injuries from the exposure, and many more have experienced chronic illnesses related to the disaster. An Indian investigation of the accident found that the disaster likely happened because Union Carbide officials had scaled back safety and alarm systems at the plant in order to cut costs. Ironically, most of the deaths and injuries could have been avoided if people had known that methyl isocyanate is highly soluble in water, that a wet towel over the head would have greatly reduced one's exposure, and that showers would have alleviated aftereffects. Unfortunately, neither the people affected nor the medical authorities involved had any idea of what chemical they confronted, much less the way to protect or treat themselves.

After the Bhopal disaster, Congress passed legislation to address the problem of accidents. For expediency, the measure was added as Title III to a bill reauthorizing Superfund: the **Superfund Amendments and Reauthorization Act of 1986 (SARA)**. Title III of SARA is better known as the **Emergency Planning and Community Right-to-know Act (EPCRA)**, which we have already encountered in connection with the TRI.

EPCRA Rules. EPCRA requires companies that handle in excess of 5 tons of any hazardous material to provide a "complete accounting" of storage sites, feed hoppers, and so on. The information goes to a *local emergency planning committee*, which is also required in every government jurisdiction. The committee is made up of officials representing local fire and police departments, hospitals, and any other groups that might be involved in case of an emergency as well as the executive officers of the companies in question.

The task of the committee is to draw up scenarios for accidents involving the chemicals on-site and to have a contingency plan for every case. This means everything from having firefighters trained and properly equipped to fight particular kinds of chemical fires, to having hospitals stocked with medicines to treat exposures to the particular chemicals, to implementing procedures to follow for the different emergencies. Thus, there can be an immediate and appropriate response to any kind of accident. Information required by the act, together with the information in the TRI, enables communities to draw up a chemical profile of their local area and encourages them to initiate pollution-prevention and risk-reduction activities.

The Toxic Substances Control Act. In the past, new synthetic organic compounds were introduced for a specific purpose without any testing of their potential side effects. For example, in the 1960s, a new compound known as TRIS was found to be an effective flame retardant and was widely used in children's sleepwear. It was only later discovered that TRIS was a potent carcinogen. Treated sleepwear was immediately withdrawn from the market. It is not known how many children (if any) developed cancer from TRIS, but this and other such cases revealed a hole in the systems of protection.

Congress responded by passing the **Toxic Substances Control Act of 1976 (TSCA)**, which requires that, before manufacturing a new chemical in bulk, manufacturers submit a "pre-manufacturing notice" to the EPA in which the potential environmental and human health impacts of the substance are assessed (including those that may derive from the ultimate disposal of the chemical). Depending on the results of the assessment, a manufacturer may be required to test the effects of the product on living things. Following the results of testing, the EPA may restrict a product's uses or keep it off the market altogether. The law, however, also tells the EPA to use the "least burdensome" approach to regulation and to compare the costs and benefits of regulation. TSCA also authorizes the EPA to develop an inventory of the chemicals already in use at the time the legislation was passed and to require testing whenever there are indications of potential risks. More than 80,000 chemical substances are currently on the inventory.

There is serious concern over the effectiveness of TSCA. The EPA has required testing for only 200 of the chemicals in commerce and has actually developed regulations for only 5 of the existing chemicals. The Government Accountability Office (the investigative arm of Congress) has actually labeled TSCA as a failed program, largely due to the stringent requirements the act places on the EPA's ability to act.

REACH. A new player in this field comes from the European Union (EU): **REACH** (Registration, Evaluation and Authorization of Chemicals). Recently authorized by the EU member states, the REACH Regulation has resulted in a process of registering an estimated 30,000 chemical substances out of the 100,000 or so in use in European societies. Each chemical must be provided with a chemical safety report (CSR), which involves human health and environmental hazard assessments, and information on how the chemical is used once it is sold. Many chemicals not now tested in the United States will be included. The approach, from screening to regulation, is based on the precautionary principle (see Chapters 12 and 17); the responsibility for managing the risks from chemicals and providing safety information lies with industry rather than the government (as it is in the United States). REACH has left U.S. trade representatives and many industries concerned about the $20 billion in chemical products the United States exports to Europe each year, as they are claiming that the costs of testing would be prohibitive. However, the European chemical industry is facing the same costs, which should make for a level playing field.

Laws applying to hazardous wastes are summarized graphically in Figure 22–16. Note that nongovernmental consumer advocate groups have been a major force behind the passage of these laws as well as of regulations requiring ingredients to be labeled on all products. Citizen action does make a difference!

22.5 Broader Issues

Before we leave the topic of hazardous wastes, we still need to address two important broader issues: the problem of environmental justice and the movement toward pollution prevention.

EPCRA (Emergency Planning and Community Right-to-know Act) Informs public about storage and releases of toxic substances

CAA (Clean Air Act) Limits discharges into the air

DOT (Department of Transportation Regulations) Assures safe transport

RCRA (Resource Conservation and Recovery Act) Assures that wastes get to suitable disposal facilities

HAZARDOUS WASTE

Disposal facility

OSHA (Occupational Safety and Health Act) Protects workers' health and safety

TOXIC WASTE

Superfund: Provides for cleanup of abandoned hazardous waste sites

TSCA (Toxic Substance Control Act) Requires new chemicals to be shown safe for specific uses

CWA (Clean Water Act) Limits discharges into waterways

SDWA (Safe Drinking Water Act) Sets standards for drinking water

Figure 22–16 Major hazardous-waste laws. Summary of the major laws pertaining to the protection of workers, the public, and the environment from hazardous materials.

Environmental Justice and Hazardous Wastes

- The largest commercial hazardous-waste landfill in the United States is located in Emelle, Alabama. African Americans make up 90% of Emelle's population. This landfill receives wastes from Superfund sites and every state in the continental United States.

- A Choctaw reservation in Philadelphia, Mississippi, was targeted to become the home of a 466-acre hazardous-waste landfill. The reservation population is entirely Native American.

- A recent study[7] found that 870,000 U.S. federally subsidized housing units are within a mile of factories that have reported toxic emissions to the EPA. Most of the occupants of these apartments are minorities.

The issue is *environmental justice*, introduced in Chapter 1. The EPA defines **environmental justice** as

the fair treatment and meaningful involvement of all people regardless of race, color, national origin, or income with respect to the development, implementation, and enforcement of environmental laws, regulations, and policies. Fair treatment means that no group

of people, including racial, ethnic, or socioeconomic group[s], should bear a disproportionate share of the negative environmental consequences resulting from industrial, municipal, and commercial operations or the execution of federal, state, local, and tribal programs and policies.[8]

Several recent studies have shown that, all across the United States, waste sites and other hazardous facilities are more likely than not to be located in towns and neighborhoods where most of the residents are non-Caucasian. These same towns and neighborhoods are also less affluent, a further element of environmental injustice. The wastes involved are generated primarily by affluent industries and the affluent majority, but somehow the wastes tend to end up well away from where they were generated and in the backyards of people of color. It seems fair to assume that the siting of hazardous facilities is a matter of political power, and those with the power would like to have the sites well away from their own backyards.

Federal Response. The federal government has taken this problem seriously. In 1994, President Clinton issued Executive Order 12898, focusing federal agency attention on environmental justice. The EPA's response has been to establish

[7]Craig Flournoy and Randy Lee Loftis. "Toxic Neighbors: Residents of Projects find Common Problem: Pollution." *The Dallas Morning News*, October 1, 2000.

[8]Environmental Protection Agency, *Final Guidance for Incorporating Environmental Justice Concerns in EPA's NEPA Compliance Analyses* (April 1998). Available at www.epa.gov.

an Environmental Justice (EJ) Program, put in place early in 1998, to capture the intent of Executive Order 12898 and to further a number of strategies already established by the EPA's Office of Solid Waste and Emergency Response. For example, in connection with the EPA's Brownfields program, communities are putting abandoned properties back into productive use. Many EJ efforts are directed toward addressing justice concerns *before* they become problems. In 1997, the EPA rejected a permit for a plastics plant in Convent, Louisiana, in a predominantly minority community. Over the course of its life, the EPA's EJ Program has awarded over $19 million in small grants to organizations and local governments that confront environmental justice issues.

International EJ.   One form of trade the developing countries can do without is the international trade in toxic wastes. Some countries or local jurisdictions in search of ready cash have found a source in the form of toxic-waste shipments, which invariably come from more developed countries and are usually illegal. This issue arose in Chapter 21 in connection with the overseas shipments of electronic wastes. The **Basel Convention** is an international agreement that bans most international toxic-waste trade, while the *Basel Action Network* is a nongovernmental organization (NGO) that seeks to prevent such trade when it is not in conformity with the convention. The network is vigilant in publicizing and coordinating legal challenges to incidents involving toxic-waste shipments and has made it clear that this trade is a matter of international environmental justice.

Pollution Prevention for a Sustainable Society

Pollution *control* and pollution *prevention* are not the same. **Pollution control** involves adding a filter or some other device at the "end of the pipe" to prevent pollutants from entering the environment. The captured pollutants still have to be disposed of, and this entails more regulation and control. **Pollution prevention**, by contrast, involves changing the production process, the materials used, or both so that harmful pollutants won't be produced in the first place. For example, adding a catalytic converter to the exhaust pipe of your car is pollution control, whereas redesigning the engine so that less pollution is produced (or switching to a hybrid car) is pollution prevention.

Pollution prevention often results in better product or materials management—that is, less wastage. Thus, pollution prevention frequently creates a cost savings. For example, Exxon Chemical Company added simple "floating roofs" to tanks storing its most volatile chemicals, thereby reducing evaporative emissions by 90% and gaining a savings of $200,000 per year. That paid for the cost of the roofs in six months. As another example, United Musical Instruments in Nogales, Arizona, worked with the Arizona Department of Environmental Quality's Pollution Prevention Unit to install a closed-loop system that reduced water use by 500,000 gallons per year. The system also reduced hazardous waste by 58% in three years and saved the company $127,000 in its first year alone. Both are examples of the *minimization* or *elimination* of pollution.

Green Chemistry.   A second angle on pollution avoidance is *substitution*—finding nonhazardous substitutes for hazardous materials. For example, the dry-cleaning industry, which uses large quantities of toxic organic chemicals, is exploring the use of water-based cleaning, or **wet cleaning**. Preliminary results from some entrepreneurs indicate that wet cleaning is quite comparable in cost and performance to dry cleaning. There are now more than 150 wet cleaners in North America, and the number is growing. In another case, Clairol switched from water to foam balls for flushing pipes during the manufacture of hair products, reducing toxin-containing wastewater by 70% and saving $250,000 per year. Water-based inks and paints are being substituted for those containing synthetic organic solvents, and inks and dyes based on biodegradable organic compounds are being substituted for those based on heavy metals.

Reuse.   A third approach is *reuse*—cleaning up and recycling solvents and lubricants. Some military bases, for example, have been able to distill solvents and reuse them, instead of discarding these solvents in a manner that releases them into the environment. This approach has increased greatly during the past decade, as measured by the TRI data. Indeed, all three approaches have been employed increasingly by industries handling hazardous chemicals, and all have contributed to significant reductions in hazardous-waste releases into the environment. (See Figure 22–4) The public disclosure of TRI data is judged to have played a crucial role in these reductions. The EPA holds that the system has been "one of the most effective environmental programs ever legislated by Congress and administered by EPA."

You, the Consumer.   Finally, pollution avoidance can also be applied to the individual consumer. So far as you are able to reduce or avoid the use of products containing harmful chemicals, you are preventing those amounts of chemicals from being released into the environment. You are also reducing the by-products that result from producing those chemicals. The average American home contains as much as 100 pounds of hazardous household waste (HHW)—substances such as paints, stains, varnishes, batteries, pesticides, motor oil, oven cleaners, and more. These materials need to be stored safely, used responsibly, and disposed of properly if they are not to pose a risk of injury or death and a threat to environmental health. (Many communities hold regular HHW collection days; if yours doesn't, you could initiate one!)

Increasingly, companies are gradually beginning to produce and market **green products,** a term used for products that are more environmentally benign than are their traditional counterparts. For example, Method Products and Clorox-based Green Works are now marketing nontoxic cleaners in Target, Costco, and other large retailers. How fast and to what degree these green products replace traditional products will in large part depend on how we behave as consumers. In other words, your buying power can be an extremely potent force.

In conclusion, there are four ways to address the problems of chemical pollution: (1) pollution prevention, (2) recycling, (3) treatment (breaking down or converting the material to harmless products), and (4) safe disposal. The first three of these methods promote a minimum of waste, in harmony with the principles of sustainability. There is every reason to believe that we can have the benefits of modern technology without destroying the sustainability of our environment by polluting it.

Revisiting the Themes

Sound Science

The POPs still plague much of the Arctic region, with toxic substances having accumulated in the tissues of the animals in some of the most remote regions of the Earth. Thankfully, many of these chemicals have now been banned, their effects are diminishing, and many of the species especially affected (e.g., by DDT) have recovered. It took the work of many environmental scientists to uncover these pathways.

Toxicology is a well-established and crucial branch of science that has helped us to develop a thorough inventory of hazardous chemicals and their health effects. Federal agencies and NGOs have made certain that this information is available to the public via the Internet (through the NTP, IRIS, NIEHS, and Scorecard), an excellent wedding of sound science and social needs. The "green chemistry" movement is an outcome of the efforts of chemists (and their sponsors) who are determined to design products that are nontoxic, yet effective.

Sustainability

In our recent past, society was on an unsustainable course in dealing with hazardous chemicals. More and more sites were being contaminated, and we were seeing rising rates of cancer and other health problems. Although the connection between hazardous-waste sites and actual cancer cases has been difficult to prove, the rising tide of hazardous chemicals in the environment was certainly a prime suspect. The legislation and programs discussed in this chapter have turned the situation around, and the move toward pollution prevention and the recycling of hazardous chemicals indicates that we are moving in a sustainable direction in our dealings with toxic chemicals.

Stewardship

One vital task of stewardship is to protect fellow human beings from unnecessary risks. We were not doing well with this responsibility, and a few wake-up calls led to stewardly action on the part of people such as Lois Gibbs, Jan Schlichtmann, and others who put their lives and reputations on the line to bring some sanity to our permissive dealings with toxic-waste dumps. As a result, our society stopped simply accepting contamination as the "price of progress," and after a great deal of public outcry, Congress responded with legislation to address the many problems. We have made real progress in cleaning up the Superfund sites, the leaking underground storage tanks, and the brownfield sites. Industries can no longer release products without testing their potential for doing harm. The TRI, in particular, has provided the incentive for industry to clean up its act. It is not the time, however, to let down our guard. Injustice to minority groups, pressures by industries to ease up on the regulations, and the attitude that hazardous wastes no longer pose a threat indicate that this issue requires constant surveillance and repeated pressure on Congress and the EPA. New chemicals are constantly appearing, and old ones have not yet been thoroughly researched; the Superfund program is still far from completion.

REVIEW QUESTIONS

1. How do toxicologists investigate hazardous chemicals? How is this information disseminated to health practitioners and the public?

2. What four categories are used to define hazardous chemicals?

3. Define *total product life cycle*, and describe the many stages at which pollutants may enter the environment.

4. What are the two classes of chemicals that pose the most serious long-term toxic risk, and how do they affect food chains?

5. What are the "dirty dozen" POPs? Why are they on a list?

6. What two laws pertaining to the disposal of hazardous wastes were passed in the early 1970s? Describe how the passage of the laws shifted pollution from one part of the environment to another.

7. Describe three methods of land disposal that were used in the 1970s. How has their use changed over time?

8. What law was passed to cope with the problem of abandoned hazardous-waste sites? What are the main features of the legislation?

9. What is being done about leaking underground storage tanks and brownfields?

10. What law was passed to ensure the safe land disposal of hazardous wastes? What are the main features of the legislation?

11. What laws exist to protect the public against exposures resulting from hazardous-chemical accidents? What are the main features of the legislation?

12. What role does the Toxic Substances Control Act play in the hazardous-waste arena?

13. Why does the EPA have an Environmental Justice Program?

14. Describe the advantages of pollution-prevention efforts.

THINKING ENVIRONMENTALLY

1. Select a chemical product or drug you suspect could be hazardous and investigate it, using the resources made available on the Internet by federal agencies and Environmental Defense.

2. Before the 1970s, it was not illegal to dispose of hazardous chemicals in unlined pits, and many companies did so. Should they be held responsible today for the contamination those wastes are causing, or should the government (taxpayers) pay for the cleanup? Give a rationale for your position.

3. Use the Toxics Release Inventory on the Internet to investigate the locations and amounts of toxic chemicals released in your state or region.

4. List three toxic items in your house. What do you do to keep them out of the environment? How do you dispose of such chemicals?

MAKING A DIFFERENCE

1. Read labels and become informed about the potentially hazardous materials you use in the home and workplace. Whenever possible, use nontoxic substitutes. If you use toxic substances, make sure that you dispose of them properly.

2. Flammable products, such as gasoline, kerosene, propane gas, and paint thinner, should be stored in approved containers in a garage, shed, or similarly separate, well-ventilated area—never inside the house.

3. Don't dump pesticides on the ground because doing so can contaminate groundwater. Don't put them in the garbage either. Follow the instructions on the container for disposal.

4. Use up your paint—try to buy only the amount of paint that you need, and then finish it. If possible, use only latex-based paint, which can be disposed of in household wastes. DO NOT put oil-based paints in the trash; save them for a household hazardous-waste collection.

Go to **www.masteringenvironmentalscience.com** for practice quizzes, Pearson eText, videos, current events, and more.

PART SIX

Stewardship for a Sustainable Future

Denver, Colorado. As this view from Ferril Lake in City Park shows, Denver's setting, with the snow-capped Rockies to the west, makes it a very attractive place to live. The "mile-high" city is considered one of the most livable cities in America.

H OW WILL WE SHAPE a sustainable future? In previous chapters, we discussed many issues that need to be addressed and resolved if we are indeed to achieve a sustainable relationship with the environment. In Chapter 23, the last one of the book, we explore one of the most difficult problems faced by every nation today: promoting livable and sustainable communities. With half of humanity now living in cities, urbanization is an irresistible trend, particularly in the developing world, where 95% of urban population growth is occurring. There one of the most profound challenges is the slums and shantytowns that receive so many migrants from the countryside. How can the people living under such deplorable conditions be brought into the mainstream of a country's economy and achieve a sustainable well-being?

THERE ARE MANY CHALLENGES in the developed countries as well. How can these countries reverse the urban blight that plagues many older cities, dooming so many to second-class citizenship? How should both developed and developing countries deal with the urban sprawl that spreads out from cities into the countryside, consuming prime agricultural land and wildlife habitat?

IN THIS CHAPTER, we explore answers to these questions. We consider examples of cities moving in a sustainable direction. We also look at some of the work being done by organizations that are grappling with the question of sustainability. Finally, we look at the vital, but difficult, task of adopting lifestyles that embrace some of the stewardship values capable of shaping a sustainable future. The outcome is uncertain, but we know enough now to be able to point in a sustainable and stewardly direction.

CHAPTER 23
Sustainable Communities and Lifestyles

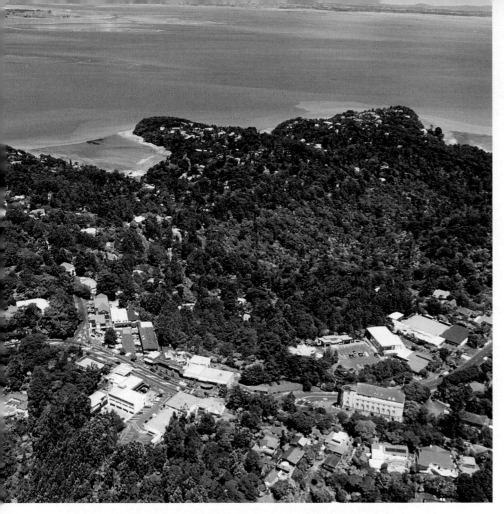

Waitakere City, New Zealand. New Zealand's first ecocity, Waitakere City, covers a large area in the Auckland metropolitan area. Shown here is one of the more developed areas, called Titirangi. Below: Leaves from a eucalyptus tree, one of the many native trees planted yearly in Waitakere City.

CHAPTER 23

Sustainable Communities and Lifestyles

WAITAKERE CITY IS THE SIXTH-LARGEST city in New Zealand, with a population of 186,000 and a land area of 367 square miles (954 km²), of which 7% is designated "living" area. Waitakere City has been experiencing the typical problems of urban growth, such as urban sprawl, traffic congestion, and pressure on natural areas within the city. In addition, the city reflects a diverse ethnic makeup: a distinctly European ethnic majority and minorities of several cultural groups, especially the native Maori. In 1993, this city formally adopted *Agenda 21* (a UN program for sustainable development) and became New Zealand's first ecocity.

GREENPRINT. Adopting the goals of Agenda 21 meant that the city needed to protect its indigenous people and cultures, while meeting other goals of sustainable development. Waitakere City focused first on Maori principles and values and then on New Zealand legislation and Agenda 21 principles to develop the guidelines for the city's efforts. Using the Maori perspective of *tangata whenua*, or "people of the land," which incorporates ecological and stewardship principles into the relationship between people and land, the city developed a strategic

KEY TOPICS

1. **Urban Sprawl:** Urban trends in the United States have been marked by a profound dependency on cars, urban sprawl, and exurban migration; "smart growth" is intended to rein in urban sprawl.

2. **Urban Blight:** Many cities in the United States have lost their vitality as people leave for the suburbs; in the developing world, however, people come to cities for employment, creating vast slums.

3. **Moving Toward Sustainable Communities:** Cities can become more livable and attractive and can be more ecologically sustainable than the suburbs.

4. **Lifestyles and the Common Good:** Many good things are happening to move our society toward greater sustainability, including lifestyle changes on the part of people who want to be responsible stewards.

plan called *Greenprint*.[1] In this plan, the Maori are included in decision making by the city council, and Maori principles serve as guidelines for the city's ongoing growth and development. Greenprint makes specific reference to the principles of *taonga* (a community's treasures, such as forests, rivers, animals, humans, and the health of the environment) and *kaitiaki* (those selected by a tribe to be stewards of the *taonga*).

GREEN NETWORK.

With input from the Maori, the city council developed a Green Network Plan[2] with the objective of "protecting, restoring, and enhancing the natural environment across the whole city, while increasing human enjoyment and appreciation of that environment." This goal places human beings within the natural world, as "stewards of its natural and spiritual health." The whole community is enlisted in "greening the city": restoring native plants, providing corridors for wildlife, and providing safe access to the city's parks and reserves. Because the city's population is growing, the network plan will be continually challenged by issues related to urban sprawl, such as loss of habitat and the impacts of exotic species and pets on native wildlife. Other Greenprint initiatives include improving housing, developing subdivisions, purchasing and retailing goods and services, and fostering local business ventures, all guided by stewardship principles. Since Waitakere City became an ecocity,

- 95% of new homes have been built in the urban area, reducing the pressure to subdivide countryside.
- Over 80,000 native trees and shrubs have been planted each year.
- There has been a 30% drop in the amount of rubbish each resident generates.
- From having some of the worst road-safety statistics in New Zealand, Waitakere now has some of the best, reducing child pedestrian and cycle injuries 50% since 1998.

[1]Greenprint Waitakere (February 1999). Available at www.waitakere.govt.nz/abtcit/ec/pdf/grnprntpt1.pdf.

[2]Available at www.waitakere.govt.nz/Cnlser/pw/greennetwk.

TRENTON, NEW JERSEY.

On the other side of the world, citizens of Trenton, New Jersey, have made remarkable progress in rehabilitating their city, with the aid of an urban-planning organization named Isles, Inc. (from the concept of neighborhood-scale "islands of redevelopment"). Trenton, with a multicultural population of 83,000, is the capital of New Jersey. Isles, Inc., was founded in 1981 by nearby Princeton University personnel, but quickly evolved into a locally owned and controlled development organization.

A thriving industrial city a century ago, Trenton was suffering from the all-too-common problems of a decaying urban center, ethnic segregation, and urban blight. To date, Isles has fostered the following changes: Vacant land has been transformed into 30 community garden sites; hundreds of units of affordable housing have been created; numerous unemployed young people have been given job training in connection with the affordable housing projects; a career center provides academic and job training, counseling, and leadership training to at-risk, inner-city youths; environmental education programs are provided yearly to 3,000 students; a financial self-reliance program helps adults with financial planning and housing counseling; a recent capital campaign raised funding for an urban environmental education center and an endowment; and the city is redeveloping nearly 60 brownfield sites (see Chapter 22). An indication of why Isles has been so successful can be seen in one of its guiding principles: "The only sustainable solutions for urban problems are those which empower people with the knowledge and skills to help themselves."

SHOWCASE COMMUNITY.

Trenton has received more than $2 million in Environmental Protection Agency (EPA) grants to renew vacant and abandoned industrial sites (Fig. 23–1). The city routinely involves local residents and businesses in the redevelopment process. As a consequence, Trenton has been selected as a "Showcase Community" by the Brownfields National Partnership and is a model for cooperative efforts to

(a)

(b)

Figure 23–1 Trenton brownfield site. An abandoned service station with contaminated soil and underground storage tanks (a) was converted to a new fire station (b) in Trenton's West Ward.

accomplish locally based initiatives. Like Waitakere City, Trenton has established a course of action and has found help in its efforts toward renewal.

This tale of two cities is tied together by a dedicated attention to urban living and a desire on the part of city residents to take charge of their own communities and reverse the trends that have been leading to a decline in the livability of their cities. It is significant that the direction they take is toward sustainability. These city residents recognize that things need to change, and they are becoming agents of changes that involve individual lifestyles and institutional redirection. One city is coping with urban sprawl, and the other is tackling urban blight.

Our objective in this chapter is first to show how urban sprawl and urban blight are related and how they can be remedied. Then we will consider the global and national efforts being undertaken to forge sustainable cities and communities. Finally, we will look at some ethical concerns and personal lifestyles as we consider the urgent needs of people and the environment in this new millennium.

23.1 Urban Sprawl

What is urban sprawl? It's been said that, like pornography, you know it when you see it. One thing is certain: Sprawl has been getting a lot of attention from the academic community. A team of researchers from Rutgers University, Cornell University, and Smart Growth America, a nongovernmental organization (NGO), worked for several years on a comprehensive effort to define, measure, and evaluate sprawl and its impacts. Their work, *Measuring Sprawl and Its Impact*,[3] was published on Smart Growth America's Web site. A further study entitled *Measuring the Health Effects of Sprawl*[4] attempts to quantify the health effects of sprawl and was published on the same site. The Federal Transit Administration sponsored a third important study, *Costs of Sprawl—2000*,[5] published by National Academies Press. These studies will inform our analysis of urban sprawl.

Sprawl's signature manifestation is a far-flung urban-suburban network of low-density residential areas, shopping malls, industrial parks, and other facilities loosely laced together by multilane highways. The word *sprawl* is used because the perimeters of the city have simply been extended outward into the countryside, one development after the next, with little plan as to where the expansion is going and no notion as to where it will stop. Almost everywhere we go near urban areas, we are confronted by farms and natural areas giving way to new developments being built, new highways being constructed, and old roadways being upgraded and expanded (Fig. 23–2). In fact, cars and sprawl are codependent; they need each other!

---

[3]Reid Ewing, Rolf Pendall, and Don Chen, *Measuring Sprawl and Its Impact* (Washington, D.C.: Smart Growth America, 2002). Available at www.smart-growthamerica.org/sprawlindex/MeasuringSprawl.PDF.
[4]Barbara A. McCann and Reid Ewing, *Measuring the Health Effects of Sprawl* (Washington, D.C.: Smart Growth America and Surface Transportation Policy Project, September 2003). Available at www.smartgrowthamerica.org/healthreport .html.

[5]Robert W. Burchell, George Lowenstein, William R. Dolphin, and Catherine C. Galley., *The Costs of Sprawl—2000* (Washington, D.C.: National Academies Press, 2002).

Figure 23–2 New-highway construction. The hallmark of development over the past several decades has been the construction of new highways, which, however, have only fed the environmentally damaging car-dependent lifestyle.

Figure 23–3 The integrated city. Before the widespread use of automobiles, cities had an integrated structure. A wide variety of small stores and offices on ground floors, with residences on upper floors, placed everyday needs within walking distance. This scene is from a historic section of Philadelphia.

The Origins of Urban Sprawl

Until the end of World War II, a relatively small percentage of people owned cars. Cities had developed in ways that allowed people to meet their needs by means of the transportation that was available—mainly walking. Every few blocks had a small grocery, a pharmacy, and other stores as well as professional offices integrated with residences. Often, buildings had stores at the street level and residences above (Fig. 23–3). Schools were scattered throughout the city, as were parks for outdoor recreation or relaxation. Walking distances were generally short, and bicycling made going somewhere even more convenient. For more-specialized needs, people boarded public transportation—electric trolleys, cable cars, and buses—that took them from neighborhoods to the "downtown" area, where big department stores, specialty shops, and offices were located. Public transportation did not change the compact structure of cities because people still needed to walk to the transit line. At the outer ends of the transit lines, cities gave way abruptly to farms, which provided most of the food for the city, and natural areas. The small towns and villages surrounding cities—the original suburbs—were compact for the same reasons and mainly served farmers in the immediate area. At the end of World War II, this pattern began to change dramatically.

Despite the many advantages of urban life, many people found cities less-than-pleasant places to live. Especially in industrial cities, poor housing, inadequate sewage systems, inadequate refuse collection, pollution from home furnaces and industry, and generally congested, noisy conditions were common. A decrease in services during World War II aggravated those problems. Hence, many people had the desire—the "American dream"—to live in their own house, on their own piece of land, away from the city.

Suburbs. Because of the massive production of war materiel, consumer goods were in short supply during World War II. As a result, when the war ended, there was a pent-up demand for consumer goods. When mass production of cars resumed at the end of the war, people flocked to buy them. With private cars, people were no longer restricted to living within walking distance of their workplaces or transit lines. They could move out of their cramped city apartments and into homes of their own outside the city. Their cars would allow them to drive back and forth easily to their jobs, shopping, recreation, and so on.

Housing Boom. Returning veterans and the accompanying baby boom created a housing demand that rapidly exhausted the existing inventory. Developers responded quickly to the demand for private homes. They bought farmlands and natural areas outside the cities and put up houses. The government aided this trend by providing low-interest mortgages through the Veterans Administration and the Federal Housing Administration, and interest payments on mortgages and local property taxes were made tax deductible. (Rent was not.) These financial factors meant that, for the first time in history, making monthly payments on one's own home in the suburbs was cheaper than paying rent for equivalent or less living space in the city.

Thus, the mushrooming development around cities did not proceed according to any plan; rather, it happened wherever developers could acquire land (Fig. 23–4). No governing body existed to devise—much less enforce—an overall plan. Cities were usually surrounded by a maze of more or less autonomous local jurisdictions (towns, villages, townships, and counties). Local governments were quickly thrown into the catch-up role of trying to provide schools, sewers, water systems, other public facilities, and, most of all, roads to accommodate the uncontrolled growth. Local zoning laws kept residential and commercial uses separate. It became more difficult to walk to a grocery store or an office.

Highways. The influx of commuters into previously rural areas soon resulted in traffic congestion, creating a need for new and larger roads. To raise money to build and expand highways, Congress passed the Highway Revenue Act of 1956, which created the **Highway Trust Fund**. This legislation placed a tax on gasoline and earmarked the revenues to be used exclusively for building new roads.

Ironically, the Highway Trust Fund perpetuates development. A new highway not only alleviates existing congestion, but also encourages development at farther locations because *time*, not *distance*, is the limiting factor for commuters. (When people describe how far they live from work, it is almost always put in terms of minutes, not distance.) The average person is willing to spend 20 to 40 minutes each way on a daily commute. With cars and expressways, people can live 30 to 50 miles from work and still get there in an acceptable amount of time.

Vicious Cycle. New highways that were intended to reduce congestion actually fostered the development of open land and commuting by more drivers from distant locations (Fig. 23–5). Soon traffic conditions became as congested as ever.

Figure 23–4 Urban sprawl. Prime land is being sacrificed for development around most U.S. cities and towns. The locations of developments are determined more by the availability of land than by any overall urban planning. Here we see sprawl surrounding the fastest-growing major city in the United States, Las Vegas, Nevada.

(a)

Figure 23–5 The new-highway traffic-congestion cycle. (a) Growing traffic congestion creates the need for new and upgraded highways. (b) However, upgraded highways encourage development in more distant locations, thus creating more traffic congestion and, hence, the need for more highways.

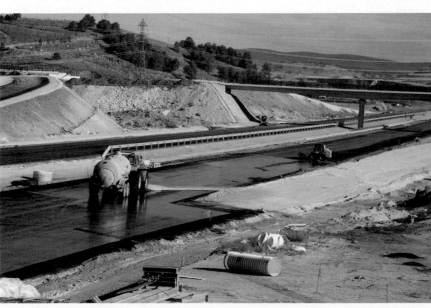

(b)

Figure 23–6 The development cycle spawned by the Highway Trust Fund.

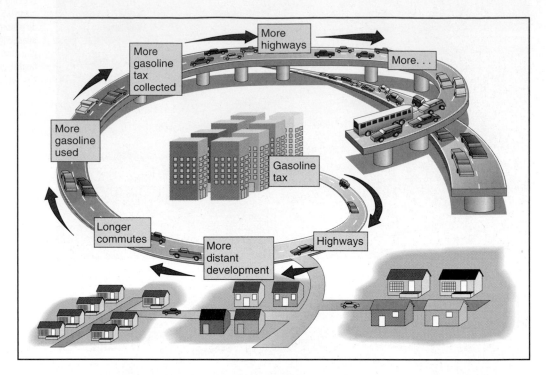

Average commuting distance has doubled since 1960, but average commuting time has remained about the same. The increase in commuting distance, however, requires more fuel, which generates more money for the Highway Trust Fund. Thus, the whole process is repeated in a continuous cycle (Fig. 23–6). As a result, government policy, far from curtailing urban sprawl, helped support and promote it.

Residential developments are inevitably followed (or sometimes led) by shopping malls, industrial parks, big-box stores, and office complexes. These commercial centers are usually situated in such a way that the only possible access to them is by car. This situation in turn has only changed the direction of commuting: Whereas, in the early days of suburban sprawl, the major traffic flow was into and out of cities, it is now between suburban centers. Multilane highways connecting suburban centers are perpetually congested with traffic going in both directions.

Exurbs, Too. In broad perspective, then, urban sprawl is a process of **exurban migration**—that is, a relocation of residences, shopping areas, and workplaces from their traditional spots in the city to outlying areas. The populations of many eastern and midwestern U.S. cities, excluding the suburbs, have been declining over the last 50 years as a result of exurban migration (Table 23–1). At the same time, the suburban population has increased from 35 million in 1950 to 150 million in 2007, half of the U.S. population. Exurban migration is continuing in a leapfrog fashion as people from older suburbs move to **exurbs** (communities farther from cities than suburbs are).

The "love affair" with cars is not just a U.S. phenomenon: Around the world, in both developed and developing countries, people aspire to own cars and adopt the car-dependent lifestyle. Consequently, urban sprawl is occurring around many developing-world cities as people become affluent enough to own cars. Love of the car-dependent lifestyle, however, is insufficient to make it sustainable. In fact, this lifestyle is at the crux of a number of the unsustainable trends described in earlier chapters.

Measuring Sprawl

The Smart Growth America research team labeled *sprawl* as "the process in which the spread of development across the landscape far outpaces population growth."[6] The team members described four dimensions characterizing sprawl and used them to analyze metropolitan regions in order to create a *sprawl index*:

- Population widely dispersed in low-density development; thus, measure *residential density*.

- Sharply separated homes, stores, and workplaces; thus, measure *neighborhood mix* of these three.

- A network of roads marked by poor connections and huge superblocks that concentrate traffic onto a few routes; thus, measure *accessibility of the street networks*.

- A general lack of well-defined downtowns and activity centers; thus, measure *strength of activity centers and downtowns*.

Each of the preceding four dimensions was standardized numerically so as to give 100 points as the average of the factor; higher means less sprawl, and lower means more. Scores for each factor were combined into a total *sprawl score* that

---

[6]Ewing, Pendall, and Chen, *Measuring Sprawl*, op cit.

TABLE 23–1   Decline in Population of American Cities, 1950–2007

City	Population (thousands)						Percent Change, 1950–2007
	1950	1970	1990	2000	2005	2007	
Baltimore, MD	950	905	736	651	640	637	−33
Boston, MA	801	641	574	589	597	599	−25
Buffalo, NY	580	463	328	293	278	273	−53
Cleveland, OH	915	751	505	478	450	438	−52
Detroit, MI	1,850	1,514	1,028	951	921	917	−50
Minneapolis, MN	522	434	368	383	375	377	−28
Philadelphia, PA	2,072	1,949	1,586	1,517	1,460	1,450	−30
St. Louis, MO	857	662	396	348	353	351	−59
Washington, DC	802	757	607	572	582	588	−27
U.S. population (millions)	151	218	249	281	296	302	+100

Source: Data from Population Division, U.S. Bureau of the Census, 2009.

also averaged 100. The team analyzed 83 metropolitan areas, representing almost half of the U.S. population. The results ranged from 14 to 178 (Fig. 23–7), with the Riverside–San Bernardino, California, metropolitan area the most sprawled and New York City the least. The 10 most sprawling metropolitan regions are ranked in Table 23–2. The Smart Growth America authors point out that, even though an area ranks higher on the sprawl score, this does not mean that sprawl is not a problem. Each of the four dimensions and the overall score were used to examine a range of travel- and transportation-related outcomes as the authors examined the impacts of sprawl on daily life.

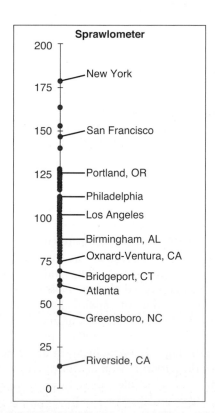

Figure 23–7 Sprawl scores for 83 metropolitan regions. The "sprawlometer" shows rankings of numerous metropolitan areas (each round dot represents a separate area). Low scores mean high sprawl, and vice versa. (*Source: Measuring Sprawl and Its Impact, 2002,* Smart Growth America.)

TABLE 23–2   Ten Most Sprawling Metropolitan Regions

Metropolitan Region	Overall Sprawl Index Score	Rank
Riverside–San Bernardino, CA	14.2	1
Greensboro–Winston-Salem–High Point, NC	46.8	2
Raleigh–Durham, NC	54.2	3
Atlanta, GA	57.7	4
Greenville–Spartanburg, SC	58.6	5
West Palm Beach–Boca Raton–Delray Beach, FL	67.7	6
Bridgeport–Stamford–Danbury, CT	68.4	7
Knoxville, TN	68.7	8
Oxnard–Ventura, CA	75.1	9
Forth Worth–Arlington, TX	77.2	10

Source: Reid Ewing, Rolf Pendall, and Don Chen, *Measuring Sprawl and Its Impact* (Washington, D.C.: Smart Growth America, 2002).

Figure 23–8 Conversion of land to developed uses. The buildup of developed land in the United States is shown. From 1982 to 2003, 35 million acres were converted to developed uses from natural lands and farmlands (2003 data do not provide a breakdown of categories but do give the total of developed land to be 108.1 million acres). (*Source:* 2003 National Resources Inventory, July 2009.)

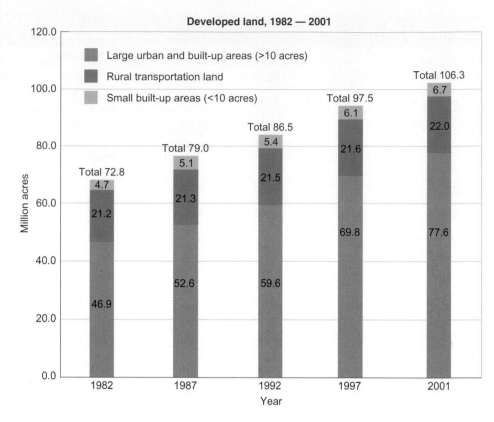

Environmental Impacts of Urban Sprawl

Urban sprawl affects the quality of life and the environment in numerous ways. The *environmental impacts* of urban sprawl are many and serious, as the following discussion indicates.

Energy Resources. Shifting to a car-dependent lifestyle has created an ever-increasing demand for petroleum, with many consequences (Chapter 14). From 1950 to 2007, U.S. oil consumption tripled, while population doubled, a 63% increase in per capita consumption. Transportation accounts for 28% of the U.S. greenhouse gas emissions that are causing climate change.

Air Pollution. Despite improvements in pollution control, many cities still fail to meet desired air quality standards. Vehicles are responsible for an estimated 80% of the air pollution in metropolitan regions. The higher use of gasoline produces greater amounts of carbon dioxide, the most serious anthropogenic greenhouse gas.

Water Resources. All the highways, parking lots, driveways, and other paved areas associated with urban sprawl lead to a substantial increase in runoff, resulting in increased flooding and erosion of stream banks. Also, water quality is degraded by the runoff of fertilizer, pesticides, crankcase oil (the oil that drips from engines), pet droppings, and so on. These non-point-source pollutants are the leading cause of water quality problems, according to the EPA.

Loss of Agricultural Land. Perhaps most serious in the long run is the loss of prime agricultural land. In the United States, the National Resources Inventory shows that cropland is declining at a rate of 1.3 million acres per year. Most of the food for cities used to be locally grown on small, diversified family farms surrounding the city. With so many of these farms turned into housing developments, it is estimated that food now travels an average of 1,000 miles (1,500 km) from where it is produced—mostly on huge commercial farms—to where it is eaten. The loss is not just the locally grown produce, but also the social interactions and ties with the farm community.

Loss of Landscapes and Wildlife. New developments are consuming land at an increasing pace. According to the National Resources Inventory, 1.75 million acres a year were developed from 1997 to 2003, compared with 1.4 million acres a year from 1982 to 1992 (Fig. 23–8). The result is the sacrifice of aesthetic, recreational, and wildlife values in metropolitan areas, the very places where they are most important.

The fragmentation of wildlife habitat due to urban sprawl has led to marked declines in many species, ranging from birds to amphibians. A 2005 report[7] found that the 35 fastest-growing metropolitan areas are home to nearly one-third of the nation's rarest and most endangered species of plants and animals. Unless these local governments act to

---

[7]Reid Ewing and John Kostyack, with Don Chen, Bruce Stein, and Michelle Ernst, *Endangered by Sprawl: How Runaway Development Threatens America's Wildlife* (Washington, D.C.: National Wildlife Federation, Smart Growth America, and NatureServe, January 2005).

protect critical habitats and green infrastructure, the survival of many species is threatened.

Impacts of Sprawl on Quality of Life

Urban sprawl can also have major negative impacts on *quality of life* for people in the surrounding areas.

Higher Vehicle Ownership and Driving Mileages. Cars are driven greater distances per person in high-sprawl areas. In Atlanta (sprawl score of 58), for example, daily use per person was 34 miles each day, while in Portland, Oregon (sprawl score of 126), it was only 24 miles each day. In the top 10 sprawl areas, there were 180 vehicles per 100 households, while in the lowest 10, there were 162. Chauffeuring kids to and from school or after-school activities (such as sports and music lessons) requires more driving in sprawl areas. As these values for miles driven and cars owned aggregate over many metropolitan areas, they add up to millions more miles driven and cars on the road as well as higher gas costs for families. The higher rates of vehicle use also lead to more highway fatalities. In Riverside, California, at the top of the sprawl index, 18 of every 100,000 people die each year in highway accidents. In the eight lowest areas on the index, fewer than 8 of 100,000 residents die annually on the highways.

Less Physical Activity and Greater Health Risks. The percentage of commuter trips by mass transit in the 10 highest-sprawl areas was 2% compared with 7% in the 10 lowest-sprawl areas. Similar trends are found for people walking to their workplace. People living in high-sprawl areas drive more, while those in more compact communities walk and take mass transit more. Walking and other moderate physical activities have many health benefits. The study of sprawl and its health effects found a highly significant correlation between the degree of sprawl, on the one hand, and weight, obesity, and high blood pressure, on the other.

Congestion and Higher Costs. There was no difference in commute times between high- and low-sprawl areas. This indicates that moving out to the suburbs or exurbs to get away from traffic congestion does not work. As more people move to the suburbs, the car traffic increases, and the congestion reaches out farther. Thus, metropolitan regions cannot sprawl their way out of congestion. The developments in outlying regions of metropolitan areas also all have to be serviced with schools, sewers, water, electricity, roads, and other infrastructure elements, often forcing county and town budgets (and, therefore, taxes) to escalate. More-compact forms of development are less expensive to service.

With all of these undesirable effects of sprawl, one might question why people put up with it. The answer is that people perceive that it is better to live in such areas, so they move there. There are benefits to sprawl, and it is important to examine them and compare them with the costs. This type of comparison was made in the *Costs of Sprawl—2000* study referenced previously.

The Benefits of Urban Sprawl

Quality-of-life issues tend to be decisive for most people, and these issues seem to be heavily weighted *in favor of sprawl*. In general, sprawl involves lower-density residential living, larger lot sizes (often 2 acres or more), larger single-family homes, better-quality public schools, lower crime rates, better social services, and greater opportunity for participation in local governments. As people move farther out from an urban center, housing costs often decline, and communities tend to be more homogeneous. Given these actual or perceived benefits, it is no wonder that sprawl is the dominant form of growth occurring in major metropolitan areas.

Net Benefits? As we have seen, there is little or no advantage in commute times or traffic congestion and actually greater distances to drive for those in the outlying suburbs and exurbs. Also, the traffic fatality rate is worse, since speed alone allows people to drive greater distances and achieve an equivalent commuting time. Several of the perceived benefits of sprawl have serious negative side effects for society as a whole. People move to the exurbs to live in neighborhoods with lower crime rates, to have access to better schools, or to live in more homogeneous communities. In one way or another, these choices are often exclusionary in their effects. The result of thousands of people making these choices is to accentuate the concentration of poorer people in high-density, inner-city, or aging suburban communities.

Common Good? The environmental costs of sprawl are very real, but these are seldom perceived as decisive by the people moving to the sprawling suburbs. They are more a matter of the common good, and people tend to make choices based on *personal good* rather than the *common good*. On the whole, most observers who have studied sprawl and its impacts believe that the costs of sprawl outweigh the benefits. They feel it is worth the effort to bring sprawl under control. The key to controlling sprawl is to provide those quality-of-life benefits that attract people, but to do it without incurring the serious environmental and social costs that are clearly associated with sprawl. This is what "smart growth" intends to do.

Reining in Urban Sprawl: Smart Growth

The way to curtail urban sprawl might appear to be simply to pass laws that bring exurban development under control. Indeed, a number of European countries and Japan have such laws. However, laws cannot be passed without the support of the public. In the United States, the strong sentiment favoring the right of a property owner to develop property as he or she sees fit has made such restrictions, with few exceptions, politically impossible to pass.

Yet there are signs that a sea change is occurring in the attitudes of Americans toward uncontrolled growth. For example, the recently emerged concept of *smart growth* is inviting communities and metropolitan areas to address sprawl and purposely choose to develop in more environmentally sustainable ways.

TABLE 23–3   Principles of Smart Growth
Create a range of housing opportunities and choices for people of all income levels.
Create walkable neighborhoods.
Encourage the collaboration of citizens, corporations, businesses, and municipal bodies within communities.
Foster distinctive, attractive communities with a strong sense of place, setting standards for development that respond to community values.
Make development decisions predictable, fair, and cost effective.
Mix land uses to provide a more attractive and healthy place to live.
Preserve open space, farmland, natural beauty, and critical environmental areas.
Provide a variety of transportation choices.
Strengthen and direct development toward existing communities (called *infilling*).
Take advantage of compact building design instead of conventional land-hogging development.

The concept recognizes that growth will occur and focuses on economic, environmental, and community values that together will lead to more sustainable communities. Table 23–3 presents the basic principles of smart growth. Zoning laws are being changed in many localities, and a new generation of architects and developers is beginning to focus on creating integrated communities, as opposed to disassociated facilities. By directing growth to particular places, smart growth provides protection for sensitive lands. Early indications in the marketplace—the final determining factor—are that the new communities are a great success.

Initiatives.   Smart-growth initiatives are appearing on the ballot in many states and municipalities. These initiatives involve several key strategies:

1. **Setting boundaries on urban sprawl.** Oregon led the way in passing a land-use law in the 1970s that requires every city to set a boundary beyond which urban growth is prohibited. The boundary is a line permitting expected future growth—but controlled growth. Portland has had great success with this strategy, and today it is a highly desirable place to live and work.

2. **Saving open space.** The best way to preserve open space is to own it or the development rights to it, and states are enacting programs to acquire crucial remaining open spaces. For example, in 2004, voters approved 162 state and local ballot measures to provide $4.1 billion for the protection of natural areas.

3. **Developing existing urban space.** In 1997, Maryland's General Assembly passed the Smart Growth and Neighborhood Conservation initiative, which includes measures to channel new growth to community sites where key infrastructure is already in place. Abandoned and brownfield sites in metropolitan areas are already well served with roads, electricity, water, and sewerage and can be developed into new homes, workplaces, and shopping areas, as in Trenton, New Jersey.

4. **Creating new towns.** A key to developing sustainable communities is changing zoning laws to allow stores, light industries, professional offices, and higher-density housing to be integrated. Affordable and attractive housing can be created around industrial parks and shopping malls for the people who work and shop in those places. Safe bicycle and pedestrian access can be provided between residential areas and workplaces. Reston, Virginia; Columbia, Maryland; and Valencia, California, were established as self-contained towns along these lines in the 1960s, and they are still highly desirable as business and residential locations. Fairfax County, Virginia, by contrast, next to Washington, D.C., is now almost 100% covered by sprawl-type development. It is estimated that, if Fairfax had pursued smart-growth policies, the county would still be 70% open space and would be far less congested and less polluted.

Raiding the Fund.   In addition to the foregoing key strategies, there have been successful attempts to break the cycle created by the Highway Trust Fund by allowing revenues to be used for purposes other than the construction of more highways. The first significant inroad was made in 1991 with the passage of the **Intermodal Surface Transportation Efficiency Act (ISTEA)**. Under this act, a substantial amount of the money levied by the Highway Trust Fund was eligible to be used for other modes of transportation, including cycling, walking, and mass transit. The objectives of ISTEA were extended with the passage of the **Transportation Equity Act for the Twenty-First Century (TEA–21)** in 1998. Then the **Safe, Accountable, Flexible, Efficient Transportation Equity Act—A Legacy for Users (SAFETEA–LU)** was passed in 2005. This act, funded by the Highway Trust Fund, authorized $244 billion for 2005–2009 for the highway and transit programs, a significant increase over TEA–21 funding. Transit programs received an average of $9 billion a year.

Highway Trust Fund monies are only part of the reason for the resurgence of mass transit; the high gasoline costs of 2008 drove many commuters to buses, trains, and subways, and mass transit ridership remained high even after gasoline prices dropped down again. Mass transit issues fared well in the 2008 election, with voters approving initiatives that would infuse $75 billion into transportation systems. In addition, the economic stimulus package (the American Recovery and Reinvestment Act of 2009) contains $64 billion for roads, bridges, rail, and transit, and it is certain that mass transit will see much of this money, given President Obama's expressed desire to push projects that will counteract urban sprawl.

Another key strategy for reducing sprawl is to revitalize the cities and older suburbs. This means coping with urban blight.

23.2 Urban Blight

In the developed world, the other end of the exurban migration is the city from which people are moving. *Exurban migration* is the major factor underlying an urban decay, or blight, that has occurred over the last 60 years. A completely different set of factors is responsible for urban blight in the developing world. In that case, people are moving *to* the cities at a rate that far exceeds the capacity of the cities to assimilate them. The result is urban slums that surround virtually every city in the poorer developing countries. UN–HABITAT estimates that over 800 million people live in urban slums in the developing world, often under conditions so squalid that they defy description.

We look first at urban blight in the developed countries, with the United States as our focus.

Economic and Ethnic Segregation

Historically, U.S. cities have included people with a wide diversity of economic and ethnic backgrounds. However, moving to the suburbs required some degree of affluence—at least the ability to manage a down payment and mortgage on a home and the ability to buy and drive a car. Therefore, exurban migration often excluded the poor, the elderly, and persons with disabilities. The poor were mostly African Americans and other minorities, largely because a history of racial discrimination had kept them from good education and well-paying jobs. Discriminatory lending practices by banks and dishonest sales practices by real estate agents (called *redlining*) kept minorities out of the suburbs even when money was not a factor. Civil rights laws passed in the 1960s made such practices illegal, but there is evidence that they still persist in various forms. Even without the racial segregation, economic segregation continues to exist and has even intensified.

In short, exurban migration and urban sprawl have led to segregation of the population into groups sharing common economic, social, and cultural backgrounds. Moving into the new suburban and exurban developments (often labeled *white flight*) were the economically advantaged, whereas many areas of the cities and older suburbs were effectively abandoned to the economically depressed, who in large part were ethnic minorities, persons with disabilities, and the elderly (Fig. 23–9).

The Vicious Cycle of Urban Blight

Affluent people moving to the suburbs set into motion a vicious cycle of exurban migration and urban blight that continues today. To understand how this downward cycle occurs, we must note several points about local governments. First, local (usually city or county) governments are responsible for providing public schools, maintaining local roads, furnishing police and fire protection, collecting and disposing of refuse, maintaining public water and sewers, and providing welfare services, libraries, and local parks. Second, a local government's major source of revenue to pay for these services is local property taxes, a yearly tax proportional to the market value of the property and home or other buildings on it. If property values increase or decrease, property taxes are adjusted accordingly. Third, in most cases, the central city is a government jurisdiction that is separate from those of the surrounding suburbs.

Economic Dysfunction. Because of the preceding three characteristics of local government, exurban migration has the following consequences: By the economics of supply and demand, property values in the suburbs escalate with the influx of affluent newcomers. Thus, suburban jurisdictions enjoy an increasing tax revenue with which they can improve and expand local services. At the same time, property values in the city decline because of decreasing demand. The lower property taxes create a powerful disincentive toward maintaining property. Many properties end up being abandoned by the owner for nonpayment of taxes. The city "inherits" such abandoned properties, but they are a liability rather than a source of revenue (Fig. 23–10). The declining tax revenue resulting from falling property values is referred to as an **eroding**, or **declining, tax base**, and it has been a serious handicap for most U.S. cities since the exurban migration started in the late 1940s. Adding to the problem is the fact that many immigrants tend to settle in the inner cities, drawn there by ethnic bonds and the hope of employment. The result is that many inner-city dwellers require public assistance of one sort or another. Thus, besides the declining tax revenues, cities bear a disproportionate burden of welfare obligations.

The eroding tax base forces city governments to cut local services, increase the tax rate, or, generally, both. Hence, the property tax on a home in a city is often two to three times greater than that on a comparably priced home in the suburbs. At the same time, schools, refuse collection, street repair, libraries, parks, and other services and facilities deteriorate from neglect. Increasing taxes and deteriorating services cause yet more people to leave the city and become part of urban sprawl, even if the car-dependent lifestyle is not their primary choice. However, it is still only the relatively more affluent individuals who can make this move; thus, the whole process of exurban migration, a declining tax base, deteriorating real estate, and worsening schools and other services and facilities is perpetuated in a vicious cycle. This downward spiral of conditions is referred to as **urban blight** or **urban decay**.

Economic Exclusion of the Inner City. The exurban migration includes more than just individuals and families: The flight of affluent people from the city removes the purchasing power necessary to support stores, professional establishments, and other enterprises. Faced with declining business, merchants and practitioners of all kinds are forced

Figure 23–9 Segregation by exurban migration. Exurban migration, the driving force behind urban sprawl, has also led to segregation of the population along economic and racial lines. In (a) an area of suburbia and (b) an area of inner-city Baltimore, you see the contrast.

(a)

(b)

Figure 23–10 Abandoned buildings. Many cities' ills are caused by too *little* population, not too much. When the exurban migration causes a city's population to drop below a certain level, businesses are no longer supported. The result is abandonment and urban blight. Shown here is a section of Baltimore only a few blocks from the redeveloped Inner Harbor seen in Figure 23–11.

either to go out of business or to move to new locations in the suburbs. Either way, vacant storefronts are the result, and people remaining in the city lose convenient access to goods, services, and, most important, jobs because each business also represents employment opportunities.

Unemployment rates run at 50% or more in depressed areas of inner cities, and the jobs that do exist are mostly at the minimum wage. By contrast, the new jobs being created are in the shopping malls and industrial parks in the outlying areas, which are largely inaccessible to inner-city residents for lack of public transportation. Therefore, not only are people who remain in the inner city poor from the outset, but also the cycle of exurban migration and urban decay has now led to their economic exclusion from the mainstream. As a result, drug dealing, crime, violence, and other forms of deviant social behavior are widespread in such areas.

To be sure, in the last two decades, many cities have redeveloped core areas. Shops, restaurants, hotels, convention and entertainment centers, office buildings, residences, and places for walking and relaxing are all combining to bring a new spirit of life and hope, as well as the practical assets of taxes and employment, back into cities (Fig. 23–11). This is an encouraging new trend, so far as it goes. However, walk a few blocks in any direction from the sparkling, revitalized core of most U.S. cities, and you will find urban blight continuing unabated.

Urban Blight in Developing Countries

More than half of the 3 million residents of Nairobi, Kenya's capital, live in slums consisting of flimsy shacks made of scrap wood, metal, plastic sheeting, rocks, and mud (Fig. 23–12).

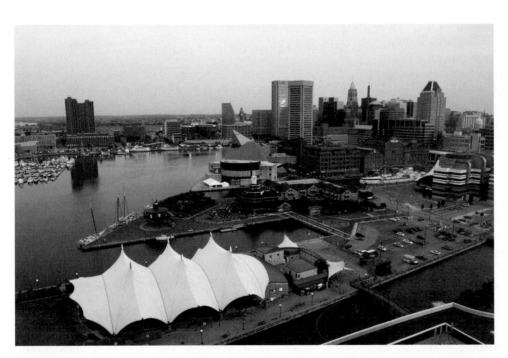

Figure 23–11 Inner Harbor, Baltimore. High population densities help make a city, not destroy it. Redeveloping key areas of cities with heterogeneous mixtures of office buildings, shopping, restaurants, residences, and recreational facilities, all of which attract people, has helped to bring new life back into the cities.

Figure 23–12 Shantytown in Nairobi, Kenya. More than half of Nairobi's residents live in such slums, where conditions defy description.

The slum neighborhoods of Nairobi surround the outskirts of the city. Burning trash and charcoal cooking fires cloud the air, and the lack of sewers adds to the smell. People get their water from private companies, since the city does not provide water or any other public amenities. There is no land ownership because the settlements are built on public land or on private land where landowners exact high rents. Crime and disease are endemic, and AIDS and tuberculosis spread readily in such slums.

Similar slum neighborhoods can be found around most of the developing-world cities; UN–HABITAT estimates that one out of every three people in the developing-world cities lives in slums. Yet in spite of the conditions, the cities are expected to continue growing well into the middle of the century. The continued rapid growth is a function of both natural population growth (which is high in many developing countries) and migration into the city from rural areas. The economies of rural areas, often based only on subsistence farming, simply do not provide the jobs needed by a growing population. So people move to the cities, where they at least have the hope of employment. The city housing supply is overwhelmed by the influx of migrants, who could not afford the rents even if housing was available. The slums actually represent a great deal of entrepreneurial energy because the residents build the dwellings and erect their own infrastructure, usually including food stands, coffee shops, barber shops, and, in some cases, schools for their children.

The Needs. The vast slums surrounding the cities are a tremendous challenge to the institutional structure of developing countries. Because slum areas are unauthorized, cities rarely provide them with electricity, water, sanitation facilities, and other social amenities. Yet the slums often represent a workforce essential to the city, taking the low-paying jobs that keep the city and its inhabitants functioning. A great need in such neighborhoods is *home security*. People live in fear of bulldozers coming at any time and leveling the shantytown. In Brazil, governments are providing legal status to existing slums (*favelas*), granting lawful titles to the land. Peru has undertaken a huge titling program, giving recognition to some 1 million urban land parcels in a four-year period. Providing this level of security to the inhabitants is often the key to mobilizing further assets, such as acquiring access to credit and negotiating with the city government for utilities or additional services. People will improve their living conditions if they are given the assurance that they can stay where they are.

The people in the shantytowns also need more *jobs*, and again, local governments could provide employment at low cost to accomplish many improvements in their own neighborhoods and elsewhere—collecting trash, building sewers, composting organic wastes, and growing food. Providing *cheap transportation* is another need. People need to get to where the jobs are and could often do so if they had bicycles or could afford to ride public transportation. Perhaps the greatest need of people living in the informal neighborhoods of city slums is *government representation*. Their voices are seldom heard because they are among the poorest and most powerless in a society. They are often plagued by government corruption, wherein nothing gets done unless accompanied by bribes (which these people can ill afford).

In some regions, however, community organizations are emerging to deal with the basic issues of the city slums. People are organizing and demanding their rights as citizens. **Shack/Slum Dwellers International (SDI)** now represents these communities to institutions like the World Bank and city governments and even held its first international conference in 2006. Change can happen, and people are learning how to make it happen. As cities devote energy and resources to providing the essential services and infrastructure to the informal slums that surround them, they will only be strengthened. The cities, and even the slums, will then become more livable.

What Makes Cities Livable?

The environmental consequences of urban sprawl and the social consequences of urban blight are two sides of the same coin. A sustainable future will depend on both reining in urban sprawl and revitalizing cities. The urban blight of cities in the developing world requires deliberate policies to address the social needs of the people flocking to the shantytowns. UN–HABITAT reports that half of the world's people now live in cities, with 3 million people added weekly to developing-world cities. Looking to the future, the only possible way to sustain the global population is by having viable, resource-efficient cities, leaving the countryside for agriculture and natural ecosystems. The key word is *viable*, which means *livable*. No one wants to, or should be required to, live in the conditions that have come to typify urban blight and urban slums.

Livability is a general concept based on people's response to the question "Do you like living here, or would you rather live somewhere else?" Crime, pollution, and recreational, cultural, and professional opportunities, as well as many other social and environmental factors, are summed up in the subjective answer to that question. Although many people assume that the social ills of the city are an outcome of high population densities, crime rates and other problems in U.S. cities have climbed, while populations have dwindled as a result of the exurban migration.

Looking at livable cities around the world, we find that the common denominators are

1. Maintaining a high population density;

2. Preserving a heterogeneity of residences, businesses, stores, and shops; and

3. Keeping layouts on a human dimension so that people can meet, visit, or conduct business informally over coffee at a sidewalk café or stroll on a promenade through an open area.

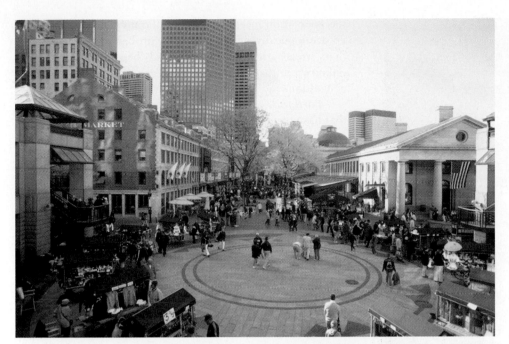

Figure 23–13 Livable cities. The key to having a livable city is to provide a heterogeneity of residences, businesses, and stores and to keep layouts on a human dimension so that people can meet or conduct business incidentally over coffee in a sidewalk café or stroll on a promenade through an open area. In short, the space is designed for, and devoted to, people. Shown here is Quincy Market, Boston.

Figure 23–14 Car-centered cities. In contrast to what is required for livability, city development of the last several decades has focused on moving and parking cars and creating homogeneity. Note the sterility of parking lots surrounding buildings in the foreground of this photograph of Los Angeles.

In short, the space is designed for, and devoted to, people (Fig. 23–13). In contrast, city development of the past 60 years has focused on accommodating automobiles and traffic. Two-thirds of the land in cities that have grown up in the era of the automobile is devoted to moving, parking, or servicing cars, and such space is essentially alien to the human psyche (Fig. 23–14). William Whyte, a well-known city planner, remarked, "It is difficult to design space that will not attract people. What is remarkable is how often this has been accomplished."[8]

A Matter of Design. The world's most livable cities are not those with "perfect" auto access between all points. Instead, they are cities that have taken measures to reduce

---

[8]William H. Whyte, "The Design of Spaces," in *The City Reader*, 3rd ed., ed. Richard T. LeGates (New York: Routledge, 2003), 432.

Figure 23–15 Bicycles in Tokyo. Most people in Tokyo do not own automobiles, but ride bicycles or walk to stations, from which they take fast, efficient, inexpensive subways to reach their destinations.

outward sprawl, diminish automobile traffic, and improve access by foot and bicycle in conjunction with mass transit. For example, Geneva, Switzerland, prohibits automobile parking at workplaces in the city's center, forcing commuters to use the excellent public transportation system. Copenhagen bans all on-street parking in the downtown core. Paris has removed 200,000 parking places in the downtown area. Curitiba, Brazil, with a population of 1.6 million, is cited as the most livable city in all of Latin America. The achievement of Curitiba is due almost entirely to the efforts of Jaime Lerner, who, serving as mayor for many decades, guided development with an emphasis on mass transit rather than cars. The space saved by not building highways and parking lots has been put into parks and shady walkways, causing the amount of green area per inhabitant to increase from 4.5 square feet in 1970 to 450 square feet today.

Charinkos. In Tokyo, millions of people ride *charinkos*, or bicycles (Fig. 23–15), either all the way to work or to subway stations from which they catch fast, efficient trains, including the "bullet train," to their destination. By sharply restricting development outside certain city limits, Japan has maintained population densities within cities and along metropolitan corridors that ensure the viability of commuter trains. Japan's cities have maintained a heterogeneous urban structure that mixes small shops, professional offices, and residences in such a way that a large portion of

the population meets its needs without cars. In maintaining an economically active city, it is probably no coincidence that street crime, vagrancy, and begging are virtually unknown in the vast expanse of Tokyo, despite the seeming congestion.

Portland. Portland, Oregon, is a pioneering U.S. city that has taken giant steps to curtail automobile use. The first step was to encircle the city with an urban growth boundary, a line outside of which new development was prohibited. Thus, compact growth, rather than sprawl, was ensured. Second, an efficient light-rail and bus system was built, which now carries 45% of all commuters to downtown jobs. (In most U.S. cities, only 10% to 25% of commuters ride public transit systems.) By reducing traffic, Portland was able to convert a former expressway and a huge parking lot into the now-renowned Tom McCall Waterfront Park. Portland is now ranked among the world's most livable cities and has been recognized by SustainLane as the most sustainable city in the United States.[9]

The Big Dig. Other U.S. cities are discovering hidden resources in their inner core areas. Expressways, which were built with federal highway funds to speed vehicles through the cities, frequently dissected neighborhoods, separating people from waterfronts and consuming existing open space. Now, with federal help, many cities are removing or burying the intruding expressways. Boston, for example, has recently put the finishing touches on the "Big Dig," a $15 billion project that has moved its central freeway below ground, added parks and open space, and reconnected the city to its waterfront (Fig. 23–16). The waterfronts themselves are often the neglected relics of bygone transportation by water and rail. In recent years, cities such as Cleveland, Chicago, San Francisco, and Baltimore have redeveloped their waterfronts, turning rotting piers and abandoned freight yards into workplaces, high-rise residences, and public and entertainment space.

Livable Equals Sustainable. Even though cities are concentrations of energy use, resource use, and pollution, there are certain environmental benefits to urbanization. Decreased auto traffic and a greater reliance on foot and public transportation reduce energy consumption and pollution. If the cities effectively provide the basic infrastructure needs—clean water, solid-waste and sewage disposal, electricity—the economies of scale greatly reduce per capita resource consumption in comparison with suburbs and rural areas. For example, urban populations use half as much electricity per capita as those in the suburbs. A number of factors—economic opportunities for women, access to health services, and less need for children's household labor—work to reduce fertility rates in urban populations versus rural populations.

Beyond these basics, housing can be retrofitted with passive solar space heating and heating for hot water. Landscaping can provide cooling, as described in Chapter 16.

[9]www.sustainlane.com/us-city-rankings/.

Figure 23–16 Boston, Massachusetts.
The Rose Fitzgerald Kennedy Greenway, on land once covered by an elevated expressway, is a series of four parks made up of gardens, plazas, and tree-lined promenades. The "Big Dig" project has revitalized the core city of Boston by replacing the central freeway with an underground tunnel system that allows traffic to flow to and through the city without disrupting city life above. (*Source:* Rose Fitzgerald Kennedy Greenway Conservancy.)

Figure 23–17 Urban gardens. Urban garden plots on formerly vacant lots in Philadelphia. Urban gardening is becoming recognized as having sociological, economic, environmental, and aesthetic benefits.

A number of cities are developing vacant or cleared areas into garden plots (Fig. 23–17), and rooftop hydroponic gardens are becoming popular. Such gardens do not make cities agriculturally self-sufficient, but they add to urban livability, provide an avenue for recycling compost and nutrients removed from sewage, give a source of fresh vegetables, and have the potential to generate income for many unskilled workers. If urban sprawl is curbed, relatively close-in farms could provide most of the remaining food needs for the city.

To the Point. The preceding discussion of urban issues may seem to have strayed far from nature and environmental science. However, there is a close connection between the two: The decay of our cities is hastening the degradation of our larger environment. As the growing human population spreads outward from the old cities in developed countries, we are co-opting natural lands and prime farmlands and ratcheting up the rate of air pollution and greenhouse gases. Similarly, as increasing numbers in the developing world stream *into* the cities, they are degrading the human and social resources that are vital to all environmental issues. Thus, without the creation of sustainable human communities, there is little chance for the sustainability of the rest of the biosphere.

By making urban areas more appealing and economically stable, we not only improve the lives of those who choose to remain or must remain in cities, but also spare surrounding areas. Our parks, wildernesses, and farms will not be replaced by exurbs and shopping malls, but rather will be saved for future generations. In the last decade, some remarkable developments have begun to accelerate the movement toward building sustainable communities. We turn to these developments now.

23.3 Moving Toward Sustainable Communities

Recall from Chapter 1 that the 1992 UN Conference on Environment and Development (UNCED) created the Commission on Sustainable Development. This new body was given the responsibility of monitoring and reporting on the implementation of the agreements that were made at UNCED—in particular, those related to sustainability. Following Agenda 21, one of the accords signed at UNCED, countries agreed to work on sustainable-development strategies and action plans that would put them on a track toward sustainability. Some 82 countries have now developed national sustainable-development strategies that lay out public-policy priorities. Most of these countries are currently actively engaged in transforming the policies into action plans, and we will investigate two of the outcomes of these activities: the *Sustainable Cities Program* and the *Sustainable Communities movement.*

Sustainable Cities

A joint facility of the UN Environment Program and UN–HABITAT, the Sustainable Cities Program (SCP) is designed to foster the planning and management needed to move cities in the developing countries toward sustainability. The program defines a sustainable city as a city in which "achievements in social, economic and physical development are made to last." Cities are the focus of the program because they are absorbing two-thirds of the population growth in the developing countries and, in the process, are experiencing serious environmental degradation in and around their growing urban centers. Focusing on smaller cities and towns, the Localizing Agenda 21 (LA21) program uses the same basic approaches as the SCP, but emphasizes issues related to the Millennium Development Goals (see Chapter 9).

Bottom Up. The SCP and LA21 programs are viewed as *capacity-building programs*, meaning that they are intended to help cities and countries mobilize resources and capabilities that are mostly available within the cities themselves (building management capacity). The approaches fostered by the programs are thus "bottom up" in nature, calling for the involvement of people at all economic and social levels and reconciling their interests when conflicts are evident. Common principles seen in sustainable-development theory—principles such as social equity, economic efficiency, and environmental planning and management—are put into practice. Working groups specific to unique city issues (e.g., solid-waste management) are the core element of the program. To date, over 120 cities in 33 countries are participating in SCP/LA21. Among the SCP cities are Madras (India), Dar es Salaam (Tanzania), Accra (Ghana), Shenyang (China), and Concepción (Chile) (Fig. 23–18). The process involves a preparatory phase, then moves into demonstration activities at the level of the whole city, and finally extends to replications in other cities of the country. The following are the processes in a city's involvement:

• The preparation of an environmental profile and the identification of priority issues, involving key stakeholders from the public, business, and community sectors.

• A city consultation bringing together key actors from government and the community to deliberate and agree

Figure 23–18 Dar es Salaam, Tanzania. Traffic on Samora Avenue. The city was the first Sustainable Cities Program demonstration city and is trying to cope with deteriorating environmental conditions caused by its rapid growth.

on the issues of prime concern and to establish working groups.

- The creation of appropriate institutions for an environmental planning and management (EPM) process that will continue to serve city needs after the SCP activities end.

Funding for the demonstration phase comes primarily from the city itself, but additional aid often originates from partnerships with other countries. Denmark, Canada, France, the Netherlands, the United Kingdom, and Italy have provided support and expertise. The World Bank and other UN agencies are frequently participants in SCP/LA21 activities. Indeed, the programs' approach is quite conducive to collaboration among NGOs, other countries, and the host country and city. The programs have operated for 18 years and have tackled issues such as providing basic urban services, constructing roads, managing growth, fighting environmental degradation, and providing social services. The programs are working on sustainable development in exactly the places where it is needed most.

SustainLane. In the United States, the 50 most populous cities have been ranked for sustainability by the organization SustainLane. The organization's Web site is a great source of advice, personal accounts, local business and product reviews, and news about sustainability lifestyles. Its most prominent product, however, is a ranking of cities based on 16 criteria, including transit ridership, street congestion, air and water quality, green building, local food and agriculture, energy and climate change, and green economy. At the top of the 2008 ranking was Portland, Oregon, followed by San Francisco, Seattle, Chicago, New York, Boston, Minneapolis, Philadelphia, Oakland, and Baltimore. At the bottom were Las Vegas (#47), Tulsa, Oklahoma City, and Mesa. The goal of the process has been "not to shame, but to inspire, to facilitate the sharing of best practices, and to see all cities continually raise the bar on their efforts," in the words of SustainLane CEO James Elsen.[10]

Sustainable Communities

Revitalizing urban economies and rehabilitating cities require coordinated efforts on the part of all sectors of society. In fact, such efforts are beginning to occur. What is termed a *sustainable communities movement* is taking root in cities around the United States. The **Sustainable Communities Network** and the **Smart Communities Network** provide a cornucopia of information, help, case studies, and linkages, demonstrating that the movement is gaining ground throughout the nation. Chattanooga, Tennessee, is now regarded as a prototype of what can occur with such a movement.

Chattanooga. Forty years ago, Chattanooga, which straddles the Tennessee River, was a decaying industrial city with high levels of pollution. Employment was falling, and residents were fleeing to the suburbs, leaving abandoned properties and increasing crime. In 1969, the EPA presented the city with a special award for "the dirtiest city in America." Then **Chattanooga Venture**, a nonprofit organization founded by community leaders, launched *Vision 2000*, the first step of which was to bring people from all walks of life together to build a consensus about what the city *could* be like. Literally thousands of ideas were gradually distilled into 223 specific projects. Then, with the cooperation of all sectors—including government, businesses, lending institutions, and average persons—and an investment of more than $800 million, work on the projects began, providing employment in construction for more than 7,300 people and

[10]Grist, "Green-City Ranking Group SustainLane Explains Its Methodology," May 12, 2008.

Figure 23–19 Chattanooga, Tennessee. With numerous projects to reduce pollution and traffic and to provide attractions and amenities for people, Chattanooga, Tennessee, has changed its reputation from one of the worst to one of the best places to live in the United States.

permanent employment for 1,380. Among the projects were the following:

- With the support of the Clean Air and Clean Water Acts, local government clamped down hard on industries to control pollution.

- The Chattanooga Neighborhood Enterprise fostered the building or renovation of more than 4,600 units of low- and moderate-income housing.

- A new industry to build pollution-control equipment was spawned, as was another industry to build electric buses, which now serve the city without noise or pollution.

- A recycling center employing adults with mental disabilities to separate materials was built.

- An urban greenway demonstration farm, which schoolchildren may visit, was created.

- A zero-emissions industrial park utilizing pollution-avoidance principles was built.

- The river was cleaned up, and a 22-mile Riverwalk reclaimed the waterfront, with parks, playgrounds, and walkways built along the riverbanks.

- Theaters and museums were renovated, and freshwater and saltwater aquariums were built.

- All facilities and a renovated business district were made pedestrian friendly and accessible.

Revision. With these and numerous other projects, many of them still ongoing, Chattanooga has changed its reputation from one of the worst to one of the best places to live (Fig. 23–19). Chattanooga Venture developed a step-by-step guide to assist community groups in similar efforts

to build sustainable communities, and the city's experience is being modeled in other cities throughout the United States as well as internationally. Recently, Chattanooga sponsored *Revision 2000*, with more than 2,600 participants taking up where Vision 2000 left off. Revision 2000 identified an additional 27 goals and more than 120 recommendations for further improving Chattanooga. The city's experience demonstrates that visioning and change must occur and continue in order to successfully create and maintain sustainable community development. Although Chattanooga Venture is no longer in operation, its past work has built an expectation throughout the city that public projects will involve the public and that the process will produce results. The latest manifestation of this expectation was *Recreate 2008*, in which hundreds of people engaged in a visioning process that created a plan for revitalizing the city's park system.

23.4 Lifestyles and the Common Good

Sound science, stewardship, and sustainability—these are the three strategic themes that have kept our eyes on the basics of how we must live on our planet. As discussed in Chapter 1, sound science is the basis for our understanding of how the world works and how human systems interact with it, stewardship is the actions and programs that manage natural resources and human well-being for the common good, and sustainability is the practical goal that our interactions with the natural world should be working toward. Are we making progress in incorporating these concepts into our society?

Our Dilemma

Our intention in this book has been to avoid dwelling on the bad news and instead to raise the hope—indeed, the certainty—that all environmental problems can be addressed successfully. Throughout the text, we have pointed to policies and possibilities that can help move human affairs in a sustainable direction. There is much at stake. When 2015 arrives, will we be able to say that we have met all the Millennium Development Goals, alleviating much of the developing world's grinding poverty and its consequences? Will we reduce our use of fossil fuels to halt global warming and the rising sea level? Can we feed the billions who will join us by midcentury and simultaneously reduce the malnutrition and hunger that still plague over 1 billion people?

Human Decisions. As the late Nobel Laureate Henry Kendall said, "Environmental problems at root are human, not scientific or technical."[11] Even though our scientific understanding is incomplete and our grasp of sustainable development is still tentative, we know enough to be able to act decisively in most circumstances. Therefore, it is human decisions, at both the personal and the societal levels, that can bring about change, and it is these decisions that define our stewardship relationship with Earth.

However, making stewardly decisions is not a simple task. Decision making is affected by our personal values and needs, and competing values and needs exist at every turn. One group wins, and another loses. If we stop all destruction of the rain forests, the peasant who needs land to farm will not eat. If we force a halt to the harvesting of shrimp because of concerns about the bycatch, thousands of shrimpers will be out of work. Making decisions means considering competing needs and values and reaching the best conclusion in the face of the numbing complexity of demands and circumstances. It is the nature of our many dilemmas that business as usual resolves only a very small fraction of them. Even though public policies at the national and international levels are absolutely essential elements of our success in turning things around, these in the end are the outcome of decisions by very human people; and even if thoughtful decisions are made by those in power, they will fail unless they are supported by the people who are affected by them.

Here are a few examples of significant changes at the level of governments, municipalities, and industries:

- The EPA has established the Green Power Partnership, a program that encourages organizations to buy green power. The number one purchaser is Intel Corp., buying more than 1 billion kilowatt-hours of green power annually.
- Transition Towns, communities that work together to respond to the challenges of peak oil and climate change, are forming in many countries.

[11]Union of Concerned Scientists, *Keeping the Earth: Scientific and Religious Perspectives on the Environment*, videotape transcript (Pittsburgh, Pa.: New Wrinkle, 1996), p. 4.

- The American Recovery and Reinvestment Act of 2009 (the stimulus package) contains more than $50 billion for green energy initiatives.
- The World Business Council for Sustainable Development is a network of more than 200 global companies promoting greater resource productivity and sustainable product manufacture.
- Four hundred cities in the United States have signed the U.S. Mayors' Climate Protection Agreement, which pledges the cities to meet Kyoto Protocol targets for greenhouse gas emission reductions (in the absence of a federal commitment).
- Australia led the way in 2007 to eliminate incandescent lightbulbs, an action that has been followed by the European Union nations and other countries.
- California established the Million Solar Roofs initiative and is on target to meet its 10-year goal of 1,750 megawatts of solar-generating capacity by 2016.

The Common Good. In Chapter 1, we defined the common good in the context of public policy: *to improve human welfare and to protect the natural world*. What compels people to act to promote the common good? We can identify a number of values that can be the basis of stewardship and help us in our decisions involving both public policies and personal lifestyles:

- *Compassion for those less well off*, or those suffering from extreme poverty or other forms of deprivation—making compassion a major element of our foreign policy, which would energize our country's efforts in promoting sustainable development in the countries that are most in need. Compassion can motivate young people to spend two years of their life in the Peace Corps or a hunger relief agency.
- *Concern for justice*—working to make just policies the norm in international economic relations. A concern for justice can move people to protest the placement of hazardous facilities in underprivileged communities.
- *Honesty*, or a concern for the truth—keeping the laws of the land and openly examining issues from different perspectives.
- *Sufficiency*—simplifying lifestyles rather than ever increasing our consumption; using no more than is necessary of Earth's resources.
- *Humility*—instead of demanding our rights, being willing to share and even defer to others.
- *Neighborliness*—being concerned for other members of our communities (even the global community), such that we do not engage in activities that are harmful to them, but instead make positive contributions to their welfare.

To achieve sustainability, we will have to couple values like these with the knowledge that can come from our understanding of how the natural world works and what is happening to it as a result of human activities. For this plan to work, people must be willing to grapple with scientific evidence and basic concepts and to develop a respect for the consensus that scientists have reached in areas of environmental concern. It has been our objective to convey that consensus to you in this text. Recall also that sustainable solutions have to be economically feasible, socially desirable, and ecologically viable (see Figure 1–11). This applies directly to the political decisions that are shaping our future in so many of our environmental concerns.

The good news is that many people are engaging environmental problems—working with governments and industries, bringing relief to people in need, and making exactly the kinds of decisions that are needed for effective stewardship. (See Stewardship, "The Tangier Island Covenant.") Community groups are organizing around sustainable principles, providing models of larger-scale changes that will be essential if we are to achieve a sustainable world. Traditional environmental organizations, as well as people from other sectors of society, are working together on the many issues that we have highlighted in this text, bringing about policy changes as well as changes in personal lifestyles.

Stewardship

The Tangier Island Covenant

The Chesapeake Bay, the largest estuary in the United States, is famous for its blue crabs and the unique "watermen" who fish for the crabs and other marine animals. Watermen are independent fishermen who have lived on the bay for generations and who defend their way of life fiercely. In recent years, the watermen of Tangier, a small island in the Virginia part of the eastern shore of the bay, have been in frequent battles with the Chesapeake Bay Foundation and the state regulators over crab and oyster laws that restrict when and where they can harvest. The watermen skirted restrictions by keeping undersized crabs and dredging clams and oysters out of season. They were also accustomed to throwing their trash overboard, causing the shores of the island to be littered with debris.

In 1997, the island community of Tangier became part of an unusual, but significant, process called *values-based stewardship*, a form of conflict resolution that requires parties to work within the value system of the community to bring about sustained changes in behavior toward the environment. Susan Drake, a doctoral candidate in environmental science at the University of Wisconsin, came to the island to carry out her thesis research on how conflicts are dealt with in homogeneous communities such as Tangier. Originally attracted to the project because of the highly religious community present on the island—the two churches are the center of life and authority there—Drake knew that most of the watermen considered themselves the beneficiaries of a rich bounty in an environment created by God. After working with watermen on their boats and spending time in their churches and with their families, Drake became concerned about the apparent lack of a connection between what the God-fearing watermen believed and what they actually did in their profession.

Convinced that she might be able to intervene in a positive way, Drake requested permission from her thesis committee to engage in "participatory action research," wherein the researcher may enter the conflict and attempt to bring about a resolution. Working within the value system of the community—in this case, a value system based on faith—Drake addressed a joint meeting of the two churches on the island one Sunday morning. Referring to an image of Jesus standing behind a young fisherman at the wheel of a boat in rough seas that almost all Tangier watermen displayed in the cabins of their vessels, Drake addressed the issues of dumping trash overboard and skirting the laws. She showed the picture to the congregation and then put paper blinders over the eyes of Jesus as she listed the ways in which the watermen were disobeying the law. The day of her sermon coincided with a very high tide that gathered the debris from around the shores of the island and washed it up into the streets and people's yards.

The Covenant

The meeting ended with a call by Drake to the watermen to adopt a covenant she had drafted earlier with the help of some of the island's women. Fifty-six watermen went to the altar and promised to abide by the following covenant:

The Watermen's Stewardship Covenant is a covenant among all watermen, regardless of their profession of religious faith. As watermen, we agree to (1) be good stewards of God's creation by setting a high standard of obedience to civil laws (fishery, boat, and pollution laws) and (2) commit to brotherly accountability. If any person who has committed to this covenant is overtaken in any trespass against this covenant, we agree to spiritually restore such a one in a spirit of gentleness. We also agree to fly a red ribbon on the antennas of our boats to signify that we are part of the covenant.

The next day scores of watermen began flying red ribbons, taking trash bags out on their boats, and fishing according to the laws. Trucks began to pick up the trash on the island early in the morning. The two churches proceeded to work with the entire island community to draft a plan for the future sustainability of the island and its way of life. They formed the Tangier Watermen's Stewardship for the Chesapeake in order to implement the plan, which is far reaching and involves the women of the island lobbying the state legislature and opening up lines of communication with scientists and government regulators involved in overseeing the bay's health. In recent years, Tangier Islanders have implemented cleanup, waste reduction, and recycling campaigns; environmental education programs; and a Coastal America Project for shoreline and wetland restoration.

Values-based stewardship is a call to life-changing stewardship, put within the value system of a community. In this case, putting the call within the context of the faith of the watermen resulted in the establishment of structures within the community that show great promise in helping to bring the entire island toward a sustainable future.

Susan Drake Emmerich.

Lifestyle Changes

What are some of the lifestyle changes that are needed, and are they in fact occurring? We should be encouraged and inspired by the literally millions of people in all walks of life who are acutely aware of the problems and who are making outstanding efforts to bring about solutions. Every pathway toward solutions that we have mentioned represents the work of thousands of dedicated professionals and volunteers, ranging from scientists and engineers to businesspeople, lawyers, and public servants. Indeed, we are all involved, whether we recognize it or not. Simply by our existence on the planet, everything we do—the car we drive, the products we use, the wastes we throw away, virtually every choice we make and action we take—has a certain environmental impact and a certain consequence for the future. Therefore, it is not a matter of choosing to have an effect, but a matter of what and how great that effect will be. It is a matter of each of us asking ourselves: Will I be part of the problem or part of the solution? The outcome will depend on how each of us responds to the challenges ahead.

There are a number of levels on which we can participate to work toward a sustainable society:

- Individual lifestyle choices
- Political involvement
- Membership and participation in nongovernmental environmental organizations
- Volunteer work
- Career choices

Lifestyle choices may involve such things as switching to a more fuel-efficient car or walking or using a bicycle for short errands; recycling paper, cans, and bottles; retrofitting your home with solar energy; starting a backyard garden and composting and recycling food and garden wastes into your soil; putting in time at a soup kitchen to help the homeless; choosing low-impact recreation, such as canoeing rather than jet-skiing; and living closer to your workplace.

Political involvement ranges from supporting and voting for particular candidates to expressing your support for particular legislation through letters or phone calls. In effect, you are exercising your citizenship on behalf of the common good.

Membership in nongovernmental environmental organizations can enhance both lifestyle changes and political involvement. As a member of an environmental organization, you will receive, and can help disseminate, information, making you and others more aware of specific environmental problems and things you can do to help. Also, your membership and contribution serve to support the lobbying efforts of the organization. A lobbyist representing a million-member organization that can follow up with thousands of phone calls and letters (and, ultimately, votes) can have a powerful impact.

In cases where enforcement of the existing law has been the weak link, some organizations, such as the Public Interest Research Groups, the Natural Resources Defense Council, and the Environmental Defense Fund, have been highly influential in bringing polluters or the government to court to see that the law is upheld. Again, this can be done only with the support of members.

Another form of involvement is *joining a volunteer organization*. Many effective actions that care for people and the environment are carried out by groups that depend on volunteer labor. Political organizations and virtually all NGOs depend highly on volunteers. Many helping organizations, such as those dedicated to alleviating hunger and to building homes for needy families (Fig. 23–20), are adept at mobilizing volunteers to accomplish some vital tasks that often fill in the gaps left behind by inadequate public policies.

Finally, you may choose to devote your *career* to implementing solutions to environmental problems. Environmental careers go far beyond the traditional occupations of wildlife ranger and park manager. Many lawyers, journalists, teachers, research scientists, engineers, medical personnel, agricultural extension workers, and others are focusing their talents and training on environmental issues or hazards. Business and job opportunities exist in pollution control, recycling, waste management, ecological restoration, city planning, environmental monitoring and analysis, nonchemical pest control, the production and marketing of organically grown produce, the manufacture of solar and wind energy components, and so on. Some developers concentrate on

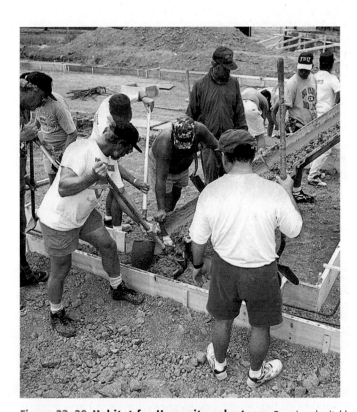

Figure 23–20 Habitat for Humanity volunteers. By using charitable contributions and volunteer labor, Habitat for Humanity creates quality homes—more than 300,000 built throughout the world by 2009—that needy families can afford to buy with no-interest mortgages, enabling the money to be recycled to build more homes. Prospective home buyers participate in the construction and may gain valuable skills in the process.

rehabilitation and the reversal of urban blight, as opposed to contributing more to urban sprawl. Some engineers are working on the development of pollution-free vehicles to help solve the photochemical smog dilemma of our cities. Indeed, it is difficult to think of a vocation that cannot be directed toward promoting solutions to environmental problems.

We began this book with the story of Easter Island. The Easter Islanders failed to grasp the connections between the essential goods and services provided by their island environment and their sustainable existence on the island. Their failure didn't lead to their extinction, but if sustainability means that people should thrive and not just survive, they certainly failed.

Human society is well into a new millennium, and it promises to be an era of rapid changes unprecedented in human history. Will we just survive, or will we thrive and become a sustainable global society? We are engaged in an environmental revolution—a major shift in our worldview and practice from seeing the natural world as resources to be exploited to seeing it as life's supporting structure that needs our stewardship. You are living in the early stages of this revolution, and we invite you to be one of the many who will make it happen. We close with two quotes:

> Imagine a future of relentless storms and floods; islands and heavily inhabited coastal regions inundated by rising sea levels; fertile soils rendered barren by drought and the desert's advance; mass migrations of environmental refugees; and armed conflicts over water and precious natural resources.
>
> Then think again—for one might just as easily conjure a more hopeful picture: of green technologies; livable cities; energy-efficient homes, transport and industry; and rising standards of living for all the people, not just a fortunate minority. The choice between these competing visions is ours to make.[12]
>
> *Kofi Annan, former Secretary General of the United Nations*

> No one could make a greater mistake than he who did nothing because he could do only a little.
>
> *Edmund Burke, Irish orator, philosopher, and politician (1729–1797)*

---

[12]Kofi Annan, *Time*, August 26, 2002.

Revisiting the Themes

Sound Science

City planners and highway engineers have a new set of challenges as our society makes the changes that will move us toward greater sustainability. Creating sustainable cities and towns requires the dedicated work of social scientists, natural scientists, and many average people as they meet to provide a vision of what they would like their future to look like and then carry out that vision. Many scientific careers are great avenues for work on green solutions.

Sustainability

The story of urban sprawl is a story of poor or nonexistent public policy. Land use is traditionally under local control (if it is under any control at all), and absent any public policy dealing with land use, sprawl happens. In fact, public policy fed sprawl, as the Highway Trust Fund taxed gasoline sales in order to build roads and highways. Crafting effective policy to deal with sprawl means recognizing why people choose to live on the suburban fringes in low-density houses and changing the way we organize our communities. It requires changing public policy to do this, but smart-growth initiatives are taking root in state after state. Reining in sprawl is an essential move toward greater sustainability of our urban surroundings.

Urban blight, too, was the outcome of poor or nonexistent public policy. Cities stood by helplessly and watched the flight of wealthier people and businesses out to the suburbs. In a similar way, developing-world cities are simply watching the poor flock in, and those cities often refuse to devote their resources to helping the poor meet their needs. It often takes people with compassion and a vision to bring about change in these situations. Frequently, change can be brought about at the grassroots level, but it helps immensely to have leaders who care about the poor in their midst. Whether in developed or developing countries, addressing urban blight will make cities more livable and, therefore, more sustainable.

Sustainable solutions are becoming more mainstream, in the United States and around the world. Cities and towns are making an effort to respond to global climate change and limited resources as they engage in community efforts to promote sustainability and encourage sustainable lifestyle changes in their people.

Stewardship

The stewardship ethic includes a concern for justice. "White flight" into the suburbs has left minorities trapped in the inner cities, held down by racial and economic discrimination. Equally important is the migration of rural people into the cities of the developing world, where they are forced to live in slums and are unjustly denied ordinary social and environmental services. Both of these developments have energized many individuals and numerous NGOs to make common cause with those who are suffering the injustice and help them mobilize their human and social resources to bring about change. A similar motivation to bring about change led Chattanooga Venture to begin the visioning process that eventually turned that city around. This is exactly what stewards should be doing. It is a fact that every one of us is given many opportunities to work together to create a truly sustainable society, and stewardship is the ethic that is there to guide our actions.

REVIEW QUESTIONS

1. What problems do Trenton and Waitakere City illustrate, and how are they organizing to deal with those problems?

2. How did the structure of cities begin to change after World War II? What factors were responsible for the change?

3. Why are the terms *car dependent* and *urban sprawl* used to describe our current suburban lifestyle and urban layout?

4. What federal laws and policies tend to support urban sprawl?

5. Compare the costs and benefits of urban sprawl, looking at the environmental impacts and the impacts on quality of life.

6. What is *smart growth*? What are four smart-growth strategies that address urban sprawl?

7. What services do local governments provide, and what is their prime source of revenue for these services?

8. What is meant by *erosion of a city's tax base*? Why does it follow from exurban migration? What are the results?

9. How do exurban migration, urban sprawl, and urban decay become a vicious cycle?

10. What is meant by *economic exclusion*? How is it related to the problems of crime and poverty in cities?

11. What are the reasons for urban blight in the cities of many developing countries? What do the people in the shantytowns need most?

12. What are some characteristics of livable cities?

13. How do the UN Sustainable Cities Program and Localizing Agenda 21 work?

14. What is the Sustainable Communities Program? How does Chattanooga illustrate the program?

15. What are some stewardship virtues, and what role could they play in fostering sustainability?

16. State five levels on which people can participate in working toward a sustainable future.

THINKING ENVIRONMENTALLY

1. Interview an older person (say, 75 years of age or older) about what his or her city or community was like before 1950 in terms of meeting people's needs for shopping, participating in recreation, getting to school or work, and so on. Was it necessary to have a car?

2. Do a study of your region. What aspects of urban sprawl and urban blight are evident? What environmental and social problems are evident? Are they still going on? Are efforts being made to correct them?

3. Identify nongovernmental organizations in your region that are working toward reining in urban sprawl or toward preserving or bettering the city. What specific projects are under way in each of these areas? What roles are local governments playing in the process? How can you become involved?

4. Imagine that you are a planner or developer. Design a small community that is people oriented and that integrates the principles of sustainability and self-sufficiency. Consider all the fundamental aspects: food, water, energy, sewage, solid waste, and how people get to school, shopping, work, and recreation.

MAKING A DIFFERENCE

1. Analyze your environmental impacts as a consumer. What modes of transportation do you use, where do you shop, what foods do you consume or throw away, what appliances do you own, how do you heat or cool your home, and how much water do you use? Consider especially your impact on the four leading consumption-related environmental problems: air pollution, global climate change, changes in natural habitats, and water pollution.

2. Investigate land-use planning and policies in your region. Is urban sprawl or urban blight a problem? Become involved in zoning issues, and support zoning and policies that will promote, rather than hinder, ecologically sound land use (*smart growth*).

3. Contact local conservation organizations or land trusts, and offer your help in protecting and preserving open land in your region.

4. Find out if there is a Habitat for Humanity project or group in your area, and if there is, volunteer some time to support its work.

Go to **www.masteringenvironmentalscience.com** for practice quizzes, Pearson eText, videos, current events, and more.

APPENDIX A

Environmental Organizations

This appendix presents a list of selected nongovernmental organizations that are active in environmental matters. Listed are national organizations as well as some smaller, specialized ones. These organizations offer a variety of fact sheets, brochures, newsletters, publications, educational materials, and annual reports, many of which are available on the Internet. Web addresses are provided in the list. Requests for information should be specific and are best made by e-mail; Web sites have e-mail addresses and query options. Some organizations have internship positions available for those wishing to do work for an environmental group. An exhaustive listing of more than 4,200 environmental, governmental, and educational organizations can be found in the *Conservation Directory* maintained by the National Wildlife Federation: **www.nwf.org/conservationdirectory**. The directory is a well-organized Web site with extensive search options.

AMERICAN FARMLAND TRUST. Focuses on the preservation of American farmland. 1200 18th Street, N.W., Suite 800, Washington, DC 20036. **www.farmland.org**

AMERICAN LUNG ASSOCIATION. Research, education, legislation, lobbying, advocacy: indoor and outdoor air pollution effects, smoking. 61 Broadway, 6th Floor, New York, NY 10006. **www.lungusa.org**

AMERICAN RIVERS, INC. Mission is "to preserve and restore America's rivers' systems and to foster a river stewardship ethic." Education, litigation, lobbying: wild and scenic rivers, hydropower relicensing. Information on specific rivers available. 1101 14th Street, N.W., Suite 1400, Washington, DC 20005. **www.amrivers.org**

BREAD FOR THE WORLD. National advocacy group that lobbies for legislation dealing with hunger. 50 F Street, N.W., Suite 500, Washington, DC 20001. **www.bread.org**

CENTER FOR SCIENCE IN THE PUBLIC INTEREST. Research and education: alcohol policies, food safety, health, nutrition, organic agriculture. 1875 Connecticut Avenue, N.W., Suite 300, Washington, DC 20009. **www.cspinet.org**

CHESAPEAKE BAY FOUNDATION. Research, education, and litigation: environmental defense and management of the Chesapeake Bay and surrounding area. Philip Merrill Environmental Center, 6 Herndon Avenue, Annapolis, MD 21403. **www.cbf.org**

CLEAN WATER ACTION PROJECT. Lobbying, education, research: water quality. 4455 Connecticut Avenue, N.W., Suite A300, Washington, DC 20008-2328. **www.cleanwateraction.org**

COMMON CAUSE. Lobbying: government reform, energy reorganization, clean air. 1133 19th Street, N.W., Washington, DC 20036. **www.commoncause.org**

COMMUNITY TRANSPORTATION ASSOCIATION OF AMERICA. Provides technical assistance for rural and specialized transportation systems. 1341 G Street, N.W., 10th Floor, Washington, DC 20005. **www.ctaa.org**

CONGRESS WATCH. Lobbying: consumer health and safety, pesticides (a function of Public Citizen). 215 Pennsylvania Avenue, S.E., Washington, DC 20003. **www.citizen.org/congress**

CONSERVATION INTERNATIONAL. Education and research to preserve and promote awareness about the world's most endangered biodiversity. 1919 M Street, N.W., Suite 600, Washington, DC 20036. **www.conservation.org**

CORAL REEF ALLIANCE. Works with the diving community and others to promote coral reef conservation around the world. 417 Montgomery Street, Suite 205, San Francisco, CA 94104. **www.coral.org**

CRITICAL MASS ENERGY AND ENVIRONMENT PROJECT. Research and education: alternative energy and nuclear power (a function of Public Citizen). 215 Pennsylvania Avenue, S.E., Washington, DC 20003. **www.citizen.org/cmep**

DEFENDERS OF WILDLIFE. Research, education, and lobbying: endangered species. 1130 17th Street, N.W., Washington, DC 20036. **www.defenders.org**

DUCKS UNLIMITED, INC. Hunters working to fulfill the annual life-cycle needs of North American waterfowl by protecting, enhancing, restoring, and managing important wetlands and associated uplands. One Waterfowl Way, Memphis, TN 38120. **www.ducks.org**

EARTHWATCH INSTITUTE. Environmental research is encouraged by organizing teams to make expeditions to various locations all over the world. Team members contribute to the expenses and spend from several weeks to several months on the site. 3 Clock Tower Place, Suite 100, Box 75, Maynard, MA 01754. **www.earthwatch.org**

EMPOWERMENT INSTITUTE. Consulting and training organization specializing in the methodology of empowerment, aimed at behavior change at the community and organizational level. P.O. Box 428, Woodstock, NY 12498. **empowermentinstitute.net**

ENVIRONMENTAL DEFENSE FUND. Research, litigation, and lobbying: cosmetics safety, drinking water, energy, transportation, pesticides, wildlife, air pollution, cancer prevention, radiation. 257 Park Avenue South, New York, NY 10010. **www.edf.org**

ENVIRONMENTAL LAW INSTITUTE. Training, educational workshops, and seminars for environmental professionals, lawyers, and judges on institutional and legal issues affecting the environment. 200 L Street, N.W., Suite 620, Washington, DC 20036. **www.eli.org**

FOREST STEWARDSHIP COUNCIL. Promotes sustainable forest management and a system of certification of sustainable forest products. 1155 30th Street, N.W., Suite 300, Washington, DC 20007. **www.fscus.org**

FREEDOM FROM HUNGER. Develops programs for the elimination of hunger worldwide. 1644 DaVinci Court, Davis, CA 95618. **www.freedomfromhunger.org**

FRIENDS OF THE EARTH. Research, lobbying: all aspects of energy development, preservation, restoration, and rational use of the Earth. 1717 Massachusetts Avenue, N.W., Suite 600, Washington, DC 20036. **www.foe.org**

GREENPEACE, USA, INC. An international environmental organization dedicated to protecting the planet through nonviolent, direct action providing public education, scientific research, and legislative lobbying. 702 H Street, N.W., Washington, DC 20001. **www.greenpeace.org**

HABITAT FOR HUMANITY INTERNATIONAL. Fosters homebuilding and ownership for the poor. Education on housing issues. 121 Habitat Street, Americus, GA 31709-3498. **www.habitat.org**

HEIFER PROJECT INTERNATIONAL. Works throughout the developing world to bring domestic animals to poor families. P.O. Box 8058, Little Rock, AR 72203. **www.heifer.org**

INSTITUTE FOR LOCAL SELF-RELIANCE. Research and education: appropriate technology for community development. 927 15th Street, N.W., 4th Floor, Washington, DC 20005. **www.ilsr.org**

INTERNATIONAL FOOD POLICY RESEARCH INSTITUTE. Mission is to identify and analyze policies for sustainably meeting the food needs of the developing world. Seeks to make its research results available to all those in a position to use them and to strengthen institutions in developing countries that conduct research relevant to its mandate. 2033 K Street, N.W., Washington, DC 20006. **www.ifpri.org**

IZAAK WALTON LEAGUE OF AMERICA, INC. Research, education, endowment grants: conservation, air and water quality, streams. 707 Conservation Lane, Gaithersburg, MD 20878. **www.iwla.org**

KUPENDA FOR THE CHILDREN. Bringing help to the handicapped children of the Malindi region of the Kenyan coast, Kupenda organizes sponsorships to educate and house the children. P.O. Box 473, Hampton, NH 03843. **www.kupenda.org**

LAND TRUST ALLIANCE. Works with local and regional land trusts to enhance their work. 1331 H Street, N.W., Suite 400, Washington, DC 20005. **www.lta.org**

LEAGUE OF CONSERVATION VOTERS. Political arm of the environmental community. Works to elect candidates to the U.S. House and Senate who will vote to protect the nation's environment and holds them accountable by publishing the National Environmental Scorecard each year, available from its Web site. 1920 L Street, N.W., Suite 800, Washington, DC 20036. **www.lcv.org**

LEAGUE OF WOMEN VOTERS OF THE UNITED STATES. Education and lobbying, general environmental issues. Publications on groundwater, agriculture, and farm policy, with additional topics available. 1730 M Street, N.W., Suite 1000, Washington, DC 20036. **www.lwv.org**

MILLENNIUM DEVELOPMENT GOALS. The World Bank site providing information on and monitoring progress toward the Millennium Development Goals. **www.developmentgoals.org**

MILLENNIUM ECOSYSTEM ASSESSMENT (MA). An international work program designed to meet the needs of decision makers and the public for scientific information concerning the consequences of ecosystem change for human well-being and options for responding to those changes. The MA completed its numerous reports in 2005, many of which are referenced in this edition of the text. **www.millenniumassessment.org**

NATIONAL AUDUBON SOCIETY. Research, lobbying, education, litigation, and citizen action: broad-based environmental issues. 700 Broadway, New York, NY 10003. **www.audubon.org**

NATIONAL COUNCIL FOR SCIENCE AND THE ENVIRONMENT. Mission is to improve the scientific basis for making decisions about environmental issues. An umbrella organization, one important affiliate, The National Library for the Environment, provides briefings on environmental issues and the legislation addressing them. 1707 H Street, N.W., Suite 200, Washington, DC 20006. **www.ncseonline.org**

NATIONAL PARK FOUNDATION. Education, land acquisition, management of endowments, grant making. 11 Dupont Circle, N.W., Suite 600, Washington, DC 20036. **www.nationalparks.org**

NATIONAL PARKS CONSERVATION ASSOCIATION. Research and education: parks, wildlife, forestry, general environmental quality. 1300 19th Street, N.W., Suite 300, Washington, DC 20036. **www.npca.org**

NATIONAL RELIGIOUS PARTNERSHIP FOR THE ENVIRONMENT. An association of independent faith groups promoting stewardship of the Creation in the context of the different faiths. 49 South Pleasant Street, Suite 301, Amherst, MA 01002. **www.nrpe.org**

NATIONAL WILDLIFE FEDERATION. Research, education, lobbying: general environmental quality, wilderness, and wildlife. 11100 Wildlife Center Drive, Reston, VA 20190. **www.nwf.org**

NATURAL RESOURCES DEFENSE COUNCIL. Research and litigation: water and air quality, land use, energy, pesticides, toxic waste. 40 West 20th Street, New York, NY 10011. **www.nrdc.org**

NATURE CONSERVANCY. "Preserve plants, animals, and natural communities that represent the diversity of life on Earth by protecting the land and the water they need to survive." 4525 North Fairfax Drive, Suite 100, Arlington, VA 22203. **www.nature.org**

OXFAM AMERICA. Funding agency for projects to benefit the "poorest of the poor" in South America, Africa, India, Central America, the Caribbean, and the Philippines. Provides whatever resources are needed. 226 Causeway Street, 5th Floor, Boston, MA 02114. **www.oxfamamerica.org**

PEW CENTER ON GLOBAL CLIMATE CHANGE. Mission "is to provide credible information, straight answers, and innovative solutions in the effort to address global climate change." 2101 Wilson Boulevard, Suite 550, Arlington, VA 22201. **www.pewclimate.org**

PLANNED PARENTHOOD FEDERATION OF AMERICA. Education, services, and research: fertility control, family planning. 434 West 33rd Street, New York, NY 10001. **www.plannedparenthood.org**

POPULATION CONNECTION. Public education, lobbying, and research toward stable populations. 1400 16th Street, N.W., Suite 320, Washington, DC 20036. **www.populationconnection.org**

THE POPULATION INSTITUTE. Education, research, and speaking engagements: population control. 107 2nd Street, N.E., Washington, DC 20002. **www.populationinstitute.org**

POPULATION REFERENCE BUREAU, INC. Organization engaged in collection and dissemination of objective population information. Excellent publications. 1875 Connecticut Avenue, N.W., Suite 520, Washington, DC 20009. **www.prb.org**

PUBLIC EMPLOYEES FOR ENVIRONMENTAL RESPONSIBILITY. A service organization that helps public employees work as "anonymous activists," enabling them to inform the public and their agencies about environmental abuses. 2000 P Street, N.W., Suite 240, Washington, DC 20036. **www.peer.org**

RACHEL CARSON COUNCIL, INC. Publication and distribution of information on pesticides and toxic substances, educational conferences, and seminars. P.O. Box 10779, Silver Spring, MD 20914. **www.rachelcarsoncouncil.org**

RAIN FOREST ACTION NETWORK. Information and educational resources: world's rain forests. 221 Pine Street, 5th Floor, San Francisco, CA 94104. **www.ran.org**

RAINFOREST ALLIANCE. Education, medicinal plants project, timber project to certify "smart wood," and news bureau in Costa Rica. 665 Broadway, Suite 500, New York, NY 10012. **www.rainforest-alliance.org**

RESOURCES FOR THE FUTURE. Research and education think tank: connects economics and conservation of natural resources, environmental quality. Publishes *Resources*. 1616 P Street, N.W., Washington, DC 20036. **www.rff.org**

SHACK/SLUM DWELLERS INTERNATIONAL. Promoting awareness of and support for the urban poor in developing countries. P.O. Box 14038 Mowbray, 7705 Cape Town, South Africa. **www.sdinet.org**

SIERRA CLUB. Education and lobbying: broad-based environmental issues. 85 Second Street, 2nd Floor, San Francisco, CA 94105. **www.sierraclub.org**

TRUST FOR PUBLIC LAND. Works with citizen groups and government agencies to acquire and preserve open space. 116 New Montgomery, 4th Floor, San Francisco, CA 94105. **www.tpl.org**

UNION OF CONCERNED SCIENTISTS. Connects citizens and scientists; works through programs and lobbying to inform the public and policy makers on science-related issues. 2 Brattle Square, Cambridge, MA 02238. **www.ucsusa.org**

U.S. PUBLIC INTEREST RESEARCH GROUP. Research, education, and lobbying, working through state chapters: alternative energy, consumer protection, utilities regulation, public interest. 44 Winter Street, 4th Floor, Boston, MA 02108. **www.pirg.org**

WATER ENVIRONMENT FEDERATION. Research, education, and lobbying to preserve the global water environment. 601 Wythe Street, Alexandria, VA 22314. **www.wef.org**

WILDERNESS SOCIETY. Research, education, and lobbying: wilderness, public lands. 1615 M Street, N.W., Washington, DC 20036. **www.wilderness.org**

WORLD RESOURCES INSTITUTE. Conducts research and publishes reports for educators, policy makers, and organizations on environmental issues. 10 G Street, N.E., Suite 800, Washington, DC 20002. **www.wri.org**

WORLDWATCH INSTITUTE. Research and education: energy, food, population, health, women's issues, technology, the environment. Publications: *State of the World*, *World Watch* magazine, Worldwatch Papers. 1776 Massachusetts Avenue, N.W., Washington, DC 20036. **www.worldwatch.org**

WORLD WILDLIFE FUND. Preservation of wildlife habitats and protection of endangered species. 1250 24th Street, N.W., P.O. Box 97180, Washington, DC 20037. **www.worldwildlife.org**

APPENDIX B

Units of Measure

DISTANCE	1 centimeter (cm) ? 10 =	1 decimeter (dm) ? 10 =	1 meter (m) ? 1 000 =	1 kilometer (km)
	1 cm = 0.39 in 1 in = 2.54 cm	1 dm = 3.94 in 1 foot = 3.05 dm	1 m = 1.09 yards 1 yard = 0.91 m	1 km = 0.62 miles 1 mile = 1.61 km

AREA

square centimeter (cm²) — ? 10 000 = — square meter (m²) — ? 10 000 = — 1 hectare (ha) — ? 100 = — 1 square kilometer (km²)

1 cm² = 1.55 sq in
1 sq in = 6.45 cm²

1 m² = 10.8 sq ft
= 1.20 sq yard
1 sq yd = .836 m²

1 ha = 2.47 acres
1 acre = 0.405 ha

1 km² = 0.39 sq mi
1 sq mile = 2.6 km²

VOLUME

1 cm — cubic centimeter (cm³) 1 milliliter (mL) — ? 1 000 = — 1 dm — cubic decimeter (dm³) 1 liter (L) — ? 1 000 = — 1 m — 1 cubic meter (m³)

1 mL = 0.203 teaspoons
1 teaspoon = 4.9 mL

1 L = 1.06 qts
1 qt = 0.95 L

1 m³ = 1.31 cubic yards
1 cubic yd = 0.76 m³

MASS (WEIGHT)

1 mL of water at 4°C has a mass of 1 gram (g) — ? 1 000 = — 1 L of water at 4°C has a mass of 1 kilogram (kg) — ? 1 000 = — 1 m³ of water at 4°C has a mass of 1 metric ton (t)

10^6 metric tons = 1 teragram

10^9 metric tons = 1 petagram

1 gram = 0.035 ounces
1 oz = 28.4 g

1 kg = 2.2 pounds
1 lb = 0.45 kg

1 t = 1000 kg
1 English (short) ton =
2 000 lbs = 0.91 t

APPENDIX C

Some Basic Chemical Concepts

Atoms, Elements, and Compounds

All matter, whether gas, liquid, or solid, living or nonliving, or organic or inorganic, is composed of fundamental units called **atoms.** Atoms are extremely tiny. If all the world's people, more than 6 billion of us, were reduced to the size of atoms, there would be room for all of us to dance on the head of a pin. In fact, we would only occupy a tiny fraction (about 1/10,000) of the pin's head. Given the incredibly tiny size of atoms, even the smallest particle that can be seen with the naked eye consists of billions of atoms.

The atoms making up a substance may be all of one kind, or they may be of two or more kinds. If the atoms are all of one kind, the substance is called an **element.** If the atoms are of two or more kinds bonded together, the substance is called a **compound.**

Through countless experiments, chemists have ascertained that there are only 94 distinct kinds of atoms that occur in nature. (An additional 21 have been synthesized.) These natural and most of the synthetic elements are listed in Table C–1, together with their chemical symbols. By scanning the table, you can see that a number of familiar substances, such as aluminum, chlorine, carbon, oxygen, and iron, are elements; that is, they are a single, distinct kind of atom. However, most of the substances with which we interact in everyday life, such as water, stone, wood, protein, and sugar, are not on the list. Their absence from the list is indicative that they are not elements; rather, they are compounds, which means that they are actually composed of two or more different kinds of atoms bonded together.

Atoms, Bonds, and Chemical Reactions

In chemical reactions, atoms are neither created nor destroyed, nor is one kind of atom changed into another. What occurs in chemical reactions, whether mild or explosive, is simply a rearrangement of the ways in which the atoms involved are bonded together. An oxygen atom, for example, may be combined and recombined with different atoms to form any number of different compounds, but a given oxygen atom always has been, and always will be, an oxygen atom. The same can be said for all the other kinds of atoms. To understand how atoms may bond and undergo rearrangement to form different compounds, it is necessary to examine several concepts concerning the structure of atoms.

Structure of Atoms

Every atom consists of a central core called the **nucleus** (not to be confused with a cell nucleus). The nucleus of an atom contains one or more **protons** and, except for hydrogen, one or more **neutrons.** Surrounding the nucleus are particles called **electrons.** Each proton has a positive (+) electric charge, and each electron has an equal and opposite negative (−) electric charge. Thus, in any atom the charge of the protons may be balanced by an equal number of electrons, making the whole atom neutral. Neutrons have no charge.

Atoms of all elements have this same basic structure, consisting of protons, electrons, and neutrons. The distinction between atoms of different elements is in the number of protons. The atoms of each element have a characteristic number of protons that is known as the **atomic number** of the element. (See Table C–1.) The number of electrons characteristic of the atoms of each element also differs, in a manner corresponding to the number of protons. The combined total of protons and neutrons in the nucleus of an element is the **mass number,** or **atomic weight.** The general structure of the atoms of several elements is shown in Figure C–1.

The number of protons and electrons in an atom of an element (i.e., the atomic number of the element) determines the chemical properties of the element. The number of neutrons may vary. For example, most carbon atoms have six neutrons in addition to the six protons, as indicated in Figure C–l. But some carbon atoms have eight neutrons. Atoms of the same element that have different numbers of neutrons are known as **isotopes** of the element. The total number of protons plus neutrons is used to define different isotopes. For example, the usual isotope of carbon is referred to as carbon-12, while the isotope with eight neutrons is referred to as carbon-14. The chemical reactivities of different isotopes of the same element are identical. (However, certain other properties may differ.) Many isotopes of various elements prove to be radioactive, such as carbon-14. The atomic weights in Table C–1 are expressed as the average of the isotopes of an element as they occur in nature. The weights listed in parentheses are those of the most stable known isotope.

Bonding of Atoms

The chemical properties of an element are defined by the ways in which its atoms will react and form bonds with other atoms. By examining how atoms form bonds, we shall see how the number of electrons and protons determines these properties. There are two basic kinds of bonding: (**1**) **covalent bonding** and (**2**) **ionic bonding.**

In both kinds of bonding, it is important to recognize that electrons are not randomly distributed around the atom's nucleus. Rather, there are, in effect, specific spaces in a series of layers, or **orbitals,** around the nucleus. If an orbital is occupied by one or more electrons but is not filled, the atom is unstable; it will then tend to react and form bonds with other atoms to achieve greater stability. A stable state is achieved by having all the spaces in the orbital filled with electrons. It is important, however, to keep the charge neutral (i.e., the total number of electrons should be equal to that of the protons).

Table C–1   The Elements

Element	Symbol	Atomic Number	Atomic Weight	Element	Symbol	Atomic Number	Atomic Weight
Actinium	Ac	89	(227)	Hafnium	Hf	72	178.5
Aluminum	Al	13	27.0	*Hassium	Hs	108	(265)
*Americium	Am	95	(243)	Helium	He	2	4.00
Antimony	Sb	51	121.8	Holmium	Ho	67	164.9
Argon	Ar	18	39.9	Hydrogen	H	1	1.01
Arsenic	As	33	74.9	Indium	In	19	114.8
Astatine	At	85	(210)	Iodine	I	53	126.9
Barium	Ba	56	137.3	Iridium	Ir	77	192.2
*Berkelium	Bk	97	(245)	Iron	Fe	26	55.8
Beryllium	Be	4	9.01	Krypton	Kr	36	83.8
Bismuth	Bi	83	209.0	Lanthanum	La	57	138.9
*Bohrium	Bh	107	(262)	*Lawrencium	Lr	103	(257)
Boron	B	5	10.8	Lead	Pb	82	207.2
Bromine	Br	35	79.9	Lithium	Li	3	6.94
Cadmium	Cd	48	112.4	Lutetium	Lu	71	175.0
Calcium	Ca	20	40.1	Magnesium	Mg	12	24.3
*Californium	Cf	98	(251)	Manganese	Mn	25	54.9
Carbon	C	6	12.0	*Meitnerium	Mt	109	(265)
Cerium	Ce	58	140.1	*Mendelevium	Md	101	(256)
Cesium	Cs	55	132.9	Mercury	Hg	80	200.6
Chlorine	Cl	17	35.5	Molybdenum	Mo	42	95.9
Chromium	Cr	24	52.0	Neodymium	Nd	60	44.2
Cobalt	Co	27	58.9	Neon	Ne	10	20.2
Copper	Cu	29	63.5	*Neptunium	Np	93	237.0
*Curium	Cm	96	(245)	Nickel	Ni	28	58.7
*Dubnium	Db	105	(262)	Niobium	Nb	41	92.9
Dysprosium	Dy	66	162.5	Nitrogen	N	7	14.0
*Einsteinium	Es	99	(254)	*Nobelium	No	102	(254)
Erbium	Er	68	167.3	Osmium	Os	76	190.2
Europium	Eu	63	152.0	Oxygen	O	8	16.0
*Fermium	Fm	100	(254)	Palladium	Pd	46	106.4
Fluorine	F	9	19.0	Phosphorus	P	15	31.0
Francium	Fr	87	(223)	Platinum	Pt	78	195.1
Gadolinium	Gd	64	157.3	Plutonium	Pu	94	(242)
Gallium	Ga	31	69.7	Polonium	Po	84	(210)
Germanium	Ge	32	72.6	Potassium	K	19	39.1
Gold	Au	79	197.0	Praseodymium	Pr	59	140.9

Table C–1   The Elements (*continued*)

Element	Symbol	Atomic Number	Atomic Weight	Element	Symbol	Atomic Number	Atomic Weight
Promethium	Pm	61	(145)	Tantalum	Ta	73	180.9
Protoactinium	Pa	91	231.0	Technetium	Tc	43	98.9
Radium	Ra	88	226.0	Tellurium	Te	52	127.6
Radon	Rn	86	(222)	Terbium	Tb	65	158.9
Rhenium	Re	75	186.2	Thallium	Tl	81	204.4
Rhodium	Rh	45	102.9	Thorium	Th	90	232.0
Ruthenium	Ru	44	101.1	Thulium	Tm	69	168.9
*Rutherfordium	Rf	104	(261)	Tin	Sn	50	118.7
Samarium	Sm	62	150.4	Titanium	Ti	22	47.9
Scandium	Sc	21	45.0	Tungsten	W	74	183.8
*Seaborgium	Sg	106	(263)	Uranium	U	92	238.0
Selenium	Se	34	79.0	Vanadium	V	23	50.9
Silicon	Si	14	28.1	Xenon	Xe	54	131.3
Silver	Ag	47	107.9	Ytterbium	Yb	70	173.0
Sodium	Na	11	23.0	Yttrium	Y	39	88.9
Strontium	Sr	38	87.6	Zinc	Zn	30	65.4
Sulfur	S	16	32.1	Zirconium	Zr	40	91.2

*Elements that do not occur in nature.
Parentheses indicate the most stable isotope of the element.

Covalent Bonding

These two requirements—filling all the spaces and keeping the charge neutral—may be satisfied by adjacent atoms sharing one or more pairs of electrons, as shown in Figure C–2. The sharing of a pair of electrons holds the atoms together in what it is called a **covalent bond.**

Hydrogen   Helium   Lithium

Carbon   Nitrogen   Oxygen

p = proton (+ charge)
n = neutron (no charge)
• = electron (– charge)

Figure C–1 Structure of Atoms. All atoms consist of fundamental particles: protons (p), which have a positive electric charge; neutrons (n), which have no charge; and electrons, which have a negative charge.

Covalent bonding, by satisfying the charge-orbital requirements, leads to discrete units of two or more atoms bonded together. Units of two or more covalently bonded atoms are called **molecules.** A few simple, but important, examples are shown in Figure C–2.

A chemical formula is simply a shorthand description of the number of each kind of atom in a given molecule. The element is given by the chemical symbol, and a subscript following the symbol gives the number of atoms of that element present in the molecule; if there is no subscript, there is only one atom of that chemical. A molecule with two or more different kinds of atoms may be called a compound, but a molecule composed of a single kind of atom—oxygen (O_2), for example—is still defined as an element.

Only a few elements—carbon, hydrogen, oxygen, nitrogen, phosphorus, and sulfur—have configurations of electrons that lend themselves readily to the formation of covalent bonds. Carbon in particular, with its ability to form four covalent bonds, can produce long, straight, or branched chains or rings (Fig. C–3). Thus, an infinite array of molecules can be formed by using covalently bonded carbon atoms as a "backbone" and filling in the sides with atoms of hydrogen or other elements. Accordingly, it is covalent bonding among atoms of carbon and these few other elements that produces all natural organic molecules, those molecules which make up all the tissues of living things, and also synthetic organic compounds such as plastics.

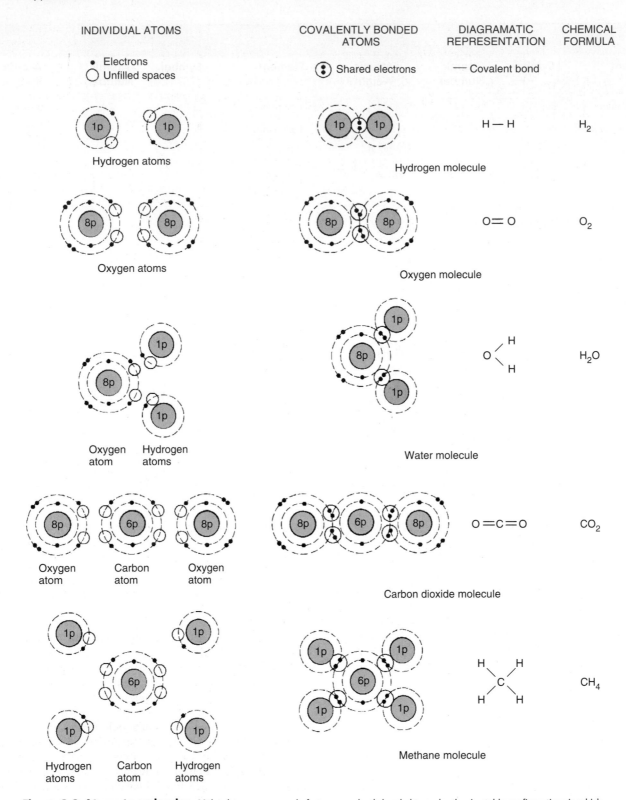

Figure C–2 Atoms to molecules. Molecules are composed of atoms covalently bonded to each other in stable configurations in which electrons are shared.

Ionic Bonding

Another way in which atoms may achieve a stable electron configuration is to gain additional electrons to complete the filling of an orbital or to lose excess electrons in an incompleted orbital. In general, the maximum number of electrons that can be gained or lost by an atom is three. Therefore, an element's atomic number determines whether one or more electrons will be lost or gained. If an atom's outer orbital is one to three electrons short of being filled, it will always tend to gain additional electrons. If an atom has one to three electrons in excess of its last complete orbital, it will always tend to lose those electrons.

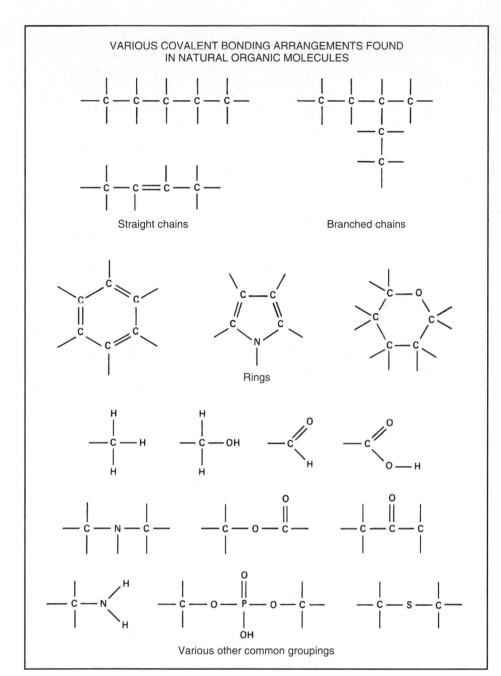

Figure C–3 Covalent bonding and organic molecules. The ability of carbon and a few other elements to readily form covalent bonds leads to an infinite array of complex organic molecules, which constitute all living things. A few major kinds of groupings are shown here. Note that each element forms a characteristic number of bonds: carbon, 4; nitrogen, 3; oxygen, 2; hydrogen, 1; sulfur, 2; and phosphorus, 5. Bonds left hanging (lines without atoms attached) indicate attachments to other atoms or groups of atoms.

Of course, gaining or losing electrons results in the number of electrons in an atom being greater or less than the number of protons; the atom consequently has an electric charge. The charge will be one negative unit for each electron gained and one positive unit for each electron lost (Fig. C–4). A covalently bonded group of atoms may acquire an electric charge in the same way. An atom or group of atoms that has acquired an electric charge in this way is called an **ion**, positive or negative. Ions are designated by a superscript following the chemical symbol, which denotes the number of positive or negative charges. The absence of superscripts indicates that the atom or molecule is neutral. Some important ions are listed in Table C–2.

Since unlike charges attract, positive and negative ions tend to join and pack together in dense clusters in such a way as to neutralize an atom's overall electric charge. This joining together of ions through the attraction of their opposite charges is called **ionic bonding.** The result is the formation of hard, brittle, more or less crystalline substances of which all rocks and minerals are examples (Fig. C–5).

It is significant to note that whereas covalent bonding leads to discrete molecules, ionic bonding does not. Any number and combination of positive and negative ions may enter into an ionic bond to produce crystals of almost any size. The only restriction is that the overall charge of positive ions be balanced by that of negative ions. Thus, substances bonded in an ionic fashion are properly called compounds but not molecules. When chemical formulas are used to describe such compounds, they define the ratio of various elements involved, not specific molecules.

Figure C–4 Formation of ions. Many atoms will tend to gain or lose one or more electrons in order to achieve a state of complete (electron-filled) orbitals. In doing so, these atoms become positively or negatively charged ions, as indicated.

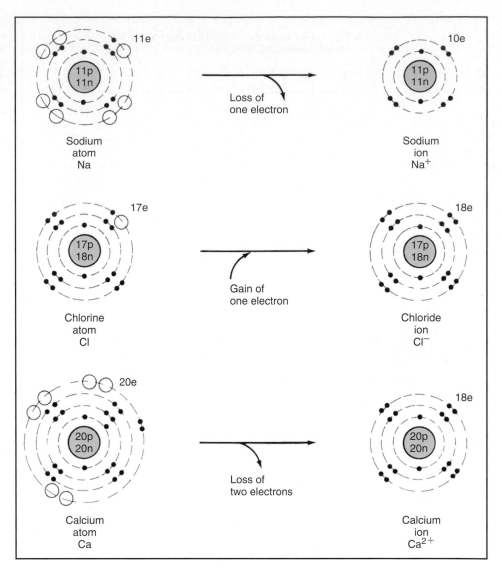

Chemical Reactions

While atoms themselves do not change, the bonds between atoms may be broken and re-formed, with different atoms producing different compounds or molecules. This is essentially

Figure C–5 Crystals. Positive and negative ions bond together by their mutual attraction and form crystals.

Table C–2	Ions of Particular Importance to Biological Systems		
Negative (–) Ions		**Positive (+) Ions**	
Phosphate	PO_4^{3-}	Potassium	K^+
Sulfate	SO_4^{2-}	Calcium	Ca^{2+}
Nitrate	NO_3^-	Magnesium	Mg^{2+}
Hydroxyl	OH^-	Iron	Fe^{2+}, Fe^{3+}
Chloride	Cl^-	Hydrogen	H^+
Bicarbonate	HCO_3^-	Ammonium	NH_4^+
Carbonate	CO_3^{2-}	Sodium	Na^+

what occurs in all chemical reactions. What determines whether a given chemical reaction will occur? Earlier, we noted that atoms form bonds because they achieve a greater stability by doing so. However, some bonding arrangements may provide greater overall stability than others. Consequently, substances with relatively unstable bonding arrangements will tend to react to form one or more different compounds that have more stable bonding arrangements. Common examples are the reaction between hydrogen and oxygen to produce water and the reaction between carbon and oxygen to produce carbon dioxide (Fig. C–6).

Hydrogen + Oxygen   Water
Relatively unstable   Very stable

$$2H_2 + O_2 \quad\longrightarrow\quad 2H_2O$$

Balanced equation

Carbon   Oxygen   Energy   Carbon dioxide
Unstable   Very stable

$$C + O_2 \quad\longrightarrow\quad CO_2$$

Balanced equation

Figure C–6 Stable bonding arrangements. Some bonding arrangements are more stable than others. Chemical reactions will go spontaneously toward more stable arrangements, releasing energy in the process. However, reactions may be driven in the opposite direction with suitable energy inputs.

Energy is always released when the constituents of a reaction gain greater overall stability, as indicated in the figure. Thus, when energy is released in a chemical reaction, the atoms achieve more stable bonding arrangements. Therefore, it may be said that chemical reactions always tend to go in a direction that releases energy, as well as giving greater stability.

However, chemical reactions can be made to go in a reverse direction. With suitable energy inputs and under suitable conditions, stable bonding arrangements may be broken and less stable arrangements formed. As described in Chapter 3 of the text, this is the basis of the photosynthesis that occurs in green plants. Light energy causes the highly stable hydrogen-oxygen bonds of water to split and form less stable carbon-hydrogen bonds, thus creating high-energy organic compounds.

Obviously, much more could be said about the involvement of chemistry in environmental science. We would highly recommend a course (or several courses) in chemistry to the student who wants to gain a deeper understanding of how the subject is involved in environmental concepts and issues.

CREDITS

Photo Credits

Contents

p. vi: Ariana Cubillos/Associated Press; NASA; p. vii: altmarkfoto/Istockphoto; GlobalP/Istockphoto; p. viii: David Grossman/Alamy; toos/Istockphoto; p. ix: Justin Sullivan/Getty Images; Elzbieta Sekowska/Shutterstock; p. xi: Mircea Bezerghean/Shutterstock; PhilAugustavo/Istockphoto; p. xii: Curtis W. Richter/Getty Images; Reinhold Foeger/Shutterstock.

Essays

p. xiii: Sound Science: Casper Wilkins/Istockphoto; Sustainability: globestock/Istockphoto; Stewardship: Sveta/Istockphoto.

About the Authors

p. v: Richard Wright.

Chapter 1

PO-1a: Ariana Cubillos/Associated Press; PO-1b: NASA; CO-1a: Yann Arthus Bertrand/Agence Altitude; CO-1b: oversnap/ Istockphoto; Figure 1–1: oversnap/Istockphoto; Figure 1–2: Diane Cook and Len Jenshel/Getty Images; Figure 1–5: Ford Cristo/Getty Images; Figure 1–5-UN: Tom Stoddart Archive/Getty Images; Figure 1–6: USDA; Figure 1–7a: Norm Dettlaff/Associated Press; Figure 1–7b: Arctic-Images/ Getty Images; Figure 1–7c: Buddy Mays/Alamy; Figure 1–9: Courtesy of Michael Amaranthus/Mycorrhizal Applications; Figure 1–10: Courtesy of JEOL USA; Figure 1–12: Courtesy of the Goldman Environmental Prize; Figure 1–13: Terje Bendiksby/Getty Images.

Chapter 2

CO-2a: Lou Linwei/Alamy; CO-2b: Luisa Fer/Istockphoto; Figure 2–1: Dang Yun/Getty Images; Figure 2–5a: Toklo/Istockphoto; Figure 2–5b: Noah Poritz/MacroWorld/Photo Researchers; Figure 2–5c: David Weintraub/Photo Researchers; Figure 2–8a: Jonathan Larsen/ Shutterstock; Figure 2–8b: VisualField/Istockphoto; Figure 2–8c: Frontpage/Shutterstock; Figure 2–10: Associated Press; Figure 2–12a: John Craig/Bureau of Land Management; Figure 2–12b: Charles Mauzy/ Corbis.

Chapter 3

PO-2a: altmarkfoto/Istockphoto; PO-2b: GlobalP/Istockphoto; CO-3a: miflippo/Istockphoto; CO-3b: Alex Staroseltsev/Shutterstock; Figure 3–2: Jerry and Marcy Monkman/EcoPhotography.com/Alamy; Figure 3–3: Richard T. Wright; Figure 3–20: USDA.

Chapter 4

CO-4a: Dave Pape; CO-4b: Getty Images; Figure 4–5: Steve Maslowski/Visuals Unlimited/Alamy; Figure 4–6a: galam/Fotolia; Figure 4–6b: Sinclair Stammers/Science Photo Library/Photo Researchers; Figure 4–6c: Brian Parker/Tom Stack & Associates; Figure 4–6d: John Watney/Photo Researchers; Figure 4–8: Fran Coleman Trust/Animals Animals/Earth Scenes; Figure 4–9: Jim Zipp/Photo Researchers; Figure 4–10: Johann Schumacher/Peter Arnold; Figure 4–11: Rick Grosberg & David Ayre; Figure 4–13a: Mark Putney/Flickr; Figure 4–13b: Cayambe/Wikimedia Commons; Figure 4–14a: Christopher Badzioch/Istockphoto; Figure 4–14b: Bernard J. Nebel; Figure 4–15: pj malsbury/Istockphoto; Figure 4–16: Bernard J. Nebel; Figure 4–17: Constantinos Petrinos/npl/Minden Pictures; Figure 4–20a: Jeff Lepore/Photo Researchers; Figure 4–20b: Craig K. Lorenz/Photo Researchers; Figure 4-22-UN: David C. Lahti; Figure 4–24: LightScribe/Istockphoto; Figure 4–25a: Ron Peplowski/DTE Energy; Figure 4–25b: Peter Parks/ Oxford Scientific Films/Animals Animals/Earth Scenes; Figure 4-25-UN: John W. Bova/ Photo Researchers.

Chapter 5

CO-5a: Renee Lynn/Corbis; CO-5b: GlobalP/Istockphoto; Figure 5–1a: USDA; Figure 5–1b: Laurie Labar; Table 5–1a: oksana.perkins/Shutterstock; Table 5–1b: steve geer/Istockphoto; Table 5–1c: Dmitry Savinov/Shutterstock; Table 5–1d: Galyna Andrushko/Shutterstock;

Table 5–1e: VeryBigAlex/Shutterstock; Table 5–1f: TTphoto/Shutterstock; Table 5–2a: mikedabell/Istockphoto; Table 5–2b: JohnnyLye/Istockphoto; Table 5–2c: Angelo Garibaldi/Shutterstock; Table 5–2d: Cedric Schmidt/Fotolia; Table 5–2e: pastorscott/Istockphoto; Table 5–2f: Mr.Eckhart/Istockphoto; Figure 5–6: Reproduced from Transmission electron micrograph by Breznak & Pankratz (1977) with permission of the American Society for Microbiology; Figure 5–11: United States Enrichment Corporation; Figure 5–14: Bernard J. Nebel; Figure 5–15b: Bernard J. Nebel; Figure 5–16: Bernard J. Nebel; Figure 5–17: Francis Lepine/Biosphere Photographie/Animals Animals/Earth Scenes; Figure 5–18: Raymond Gehman/CORBIS; Figure 5–19a: szefei/Shutterstock; Figure 5–19b: Marek Slusarczyk/Shutterstock.

Chapter 6

CO-6a: Bill Sholtz; CO-6b: Getty Images; Figure 6–1b: Gary G. Gibson/National Audubon Society/Photo Researchers; Figure 6–2: ECHO: Educational Concerns for Hunger Organization; Figure 6–3: Wildlife GmbH/Alamy; Figure 6–4a: Istockphoto; Figure 6–4b: Flip Nicklin/Minden Pictures; Figure 6–4c: Getty; Figure 6–4d: Punchstock; Figure 6–4e: Wolfgang Kaehler; Figure 6–4f: Punchstock; Figure 6–7: James P. Blair/Getty Images; Figure 6–8: Bob Pool/Getty Images; Figure 6–9: Klaus Uhlenhut/Animals Animals/Earth Scenes; Figure 6–10a & b: Richard N. Mack, Washington State University; Figure 6–11: steve geer/Istockphoto; Figure 6–12: Allen Blake Sheldon/Animals Animals/Earth Scenes; Figure 6–13a: Dinodia Picture Agency; Figure 6–13b: Alain Dragesco-Joffe/Animals Animals/Earth Scenes; Figure 6–14: step2626/Istockphoto; Figure 6–15: Joel Sartore Photography; Figure 6–16: Associated Press; Figure 6–17a: C. Allan Morgan/Peter Arnold; Figure 6–17b: kuma/Shutterstock; Figure 6–17c: Julia Sims/Peter Arnold; Figure 6–17d: A.H. Rider/Photo Researchers; Figure 6–17e: Jeff Lepore/Photo Researchers; Figure 6–17f: Canon Bob/Istockphoto; Figure 6–17g: Tom McHugh/Steinhart Aquarium/Photo Researchers; Figure 6–17h: Mickey Gibson/Animals Animals/Earth Scenes; Figure 6–17i: Nicolas Larento/Fotolia; Figure 6–17-UN: Kennan Ward; Figure 6–18: Richard Langs/Shutterstock; Figure 6–19: Operation Migration Inc.

Chapter 7

CO-7a: Paul A. Souders/Corbis; CO-7b: Dalmatin.o/Fotolia; Figure 7–2: Yann Arthus-Bertrand/Corbis; Figure 7–3: Mark Moffett/Minden Pictures; Figure 7–4: BIOS (A. Compost)/Peter Arnold; Figure 7–5: Jim West/Alamy; Figure 7–6-UN: Mitsuaki Iwago/Minden Pictures; Figure 7–7: Visual&Written SL/Alamy; Figure 7–11: David Austen/Woodfin Camp and Associates; Figure 7–12: T. Hall/Robert Harding Imagery; Figure 7–13: American Association for the Advancement of Science; Figure 7–14-UN: Mark C Burnett/Photo Researchers; Figure 7–17: Associated Press; Figure 7–18: Richard T. Wright; Figure 7–19: Klaus Jost/Peter Arnold.

Chapter 8

PO-3a: David Grossman/Alamy; PO-3b: toos/Istockphoto; CO-8a1: Aubrey Wade/Panos Pictures; CO-8a2 & a3: Getty Images; CO-8b: Istockphoto; Figure 8–2: Anthony Bannister/Animals Animals/Earth Scenes; Figure 8–3: Archiv/Photo Researchers; Figure 8–4: Narinder Nanu/Getty Images; Figure 8–6: SASI Group/University of Sheffield) and Mark Newman (University of Michigan); Figure 8–9: Erik De Castro/Reuters/Landov Media; Figure 8–10: Daniel Beltra/Getty Images; Figure 8–11: Martin Bernetti/Getty Images; Figure 8–12: Robert Nickelsberg/Getty Images.

Chapter 9

CO-9a: Amir Bhargava/Bloomberg News/Landov Media; CO-9b: Getty Images; Figure 9–6: Dario Mitidieri/Getty Images; Figure 9–8-UN: Darrell Plowes; Figure 9–12: Bruno J. Zehner/Peter Arnold; Figure 9–15: John Van Hasselt/Corbis; Figure 9–16: Gordon Sato.

Chapter 10

CO-10a1 & a2: NASA Earth Observing System; CO-10b: IT Photo/Alamy; Figure 10–2a: Elenathewise/Fotolia; Figure 10–2b: Pete Saloutos/Corbis; Figure 10–11: Mark Edwards/Still Pictures/Peter Arnold; Figure 10–13: Buddy Mays/Corbis; Figure 10–14a: Yva Momatiuk & John Eastcott/Minden Pictures; Figure 10–14b: Theo Allofs/Corbis; Figure 10–15: Manoj Shrestha/Associated Press; Figure 10–16b: Stephen Mcsweeny/Shutterstock; Figure 10–16b: Francois Gohier/Photo Researchers; Figure 10–17: Veronica Decker/The Orlando Sentinel,Used with permission of the Orlando Sentinel, copyright (2010); Figure 10–19: Stringer Shanghai/Reuters; Figure 10–20: Courtesy Tampa Bay Water; Figure 10–21: Lowell Georgia/Photo Researchers; Figure 10–22: International Development Enterprises; Figure 10–23: Courtesy of Kohler.

Chapter 11

CO-11a: CIMMYT; CO-11b: fotoIE/Istockphoto; Figure 11–1: Michael Reynolds/Corbis; Figure 11–9: (c) Frank B. Dazzo, Michigan State University; Figure 11–10: Bernard J. Nebel; Figure 11–11: Richard T. Wright; Figure 11–12: N. Allen and G. Barron, University of Guelph, Guelph, Ontario; Figure 11–15: Charles R. Belinky/Photo Researchers; Figure 11–16b: Steve Kaufman/Peter Arnold; Figure 11–17: Ron Haviv/VII/Associated Press; Figure 11–18: Debra Ferguson/AGStockUSA; Figure 11–19a: USDA; Figure 11–19b: Dan McCoy/Rainbow Image Library; Figure 11–20: Betty Derig/Photo Researchers; Figure 11–21: John M. White/Texas Agricultural Extension Service; Figure 11–22: G. Nicolini/Project Keita.

Chapter 12

CO-12a: Denis Wauthg/Getty Images; CO-12b: langdu/Shutterstock; Figure 12–1: Golden Rice Board; Figure 12–2a: Garry D. McMichael/Photo Researchers; Figure 12–2b: Orientaly/Shutterstock; Figure 12–4a & b: Nigel J.H. Smith/Animals Animals/Earth Scenes; Figure 12–6: rest/Istockphoto; Figure 12–7: Roland Seitre/Peter Arnold; Figure 12–8: Robin Smith/Photolibrary.com; Figure 12–9: Heifer Project International; Figure 12–12a: Stephen A. Ferreira, University of Hawaii at Manoa; Figure 12–12b: University of Hawaii; Figure 12–13: Nigel Cattlin/Alamy; Figure 12–15a: Susan Kuklin/Photo Researchers; Figure 12–15b: Ryan Rodrick Beiler/Shutterstock; Figure 12–15c: NASA; Figure 12–17: Robert Patrick/Corbis; Figure 12–19: Scott Peterson/Getty Images.

Chapter 13

CO-13a: Reuters/Corbis; CO-13b: Alamy; Figure 13–2a & b: Richard T. Wright; Figure 13–2c: USDA/Agricultural Research Service Honey Breeding; Figure 13–2d, e: Richard T. Wright; Figure 13–2f: USDA/ARS/Agricultural Research Service; Figure 13–2g: Dan Nickrent/Southern Illinois University; Figure 13–2h: Wally Eberhart/Getty Images; Figure 13–4: dlewis33/Istockphoto; Figure 13–5: Courtesy Dr. Bill Johnson, Purdue University; Figure 13–6: Jef Meul/Minden Pictures; Figure 13–8: Ronnie Howard/Shutterstock; Figure 13–12: Richard T. Wright; Figure 13–12-UNa: Christopher M. Ranger and Arthur M. Hower; Figure 13–12-UNb: Georg Goergen/ITA; Figure 13–13b: ARS/USDA; Figure 13–13c: Harry Rogers/Photo Researchers; Figure 13–14a: USDA; Figure 13–14b: Peggy Greb/USDA; Figure 13–15: Louis Quitt/Photo Researchers; Figure 13–16: Christopher M. Ranger and Arthur M. Hower; Figure 13–17: NatureMark Potatoes, a unit of Monsanto Company; Figure 13–18: Jack Dermid/Photo Researchers; Figure 13–20: Georg Goergen/ITA.

Chapter 14

PO-4a: Justin Sullivan/Getty Images; PO-4b: Elzbieta Sekowska/Shutterstock; CO-14a: Melissa Farlow/National Geographic; CO-14b: Jason Verschoor/Istockphoto; Figure 14–1: Erin Paul Donovan/Alamy; Figure 14–2: Michio Hoshino/Minden Pictures; Figure 14–3: Tony Wells/Mason Dixon Historical Society; Figure 14–6: JensenPhoto/Istockphoto; Figure 14–8: haak78/Shutterstock; Figure 14–14: Associated Press; Figure 14–16: Shell International; Figure 14–17: Tony Wells/Mason Dixon Historical Society; Figure 14–18: Courtesy Alyeska Pipeline Service Company; Figure 14–20: mathieukor/Istockphoto.

Chapter 15

CO-15a: Tennessee Valley Authority/Associated Press; CO-15b: daliborzivotic/Istockphoto; Figure 15–1: Japan Coast Guard/Associated Press; Figure 15–3: Long Island Power Authority; Figure 15–7c: Erich Hartmann/Magnum Photos; Figure 15–12: Nuclear Energy Institute; Figure 15–13: dpa/Landov Media; Figure 15–15a: Maine Yankee; Figure 15–15b: Pat Wellenbach/Associated Press; Figure 15–16: Alexander Tsiaras/Photo Researchers.

Chapter 16

CO-16a: Anke Heuper; CO-16b: Sebastien Burel/Shutterstock; Figure 16–1: Tom McHugh/Photo Researchers; Figure 16–2: MaxFX/ Shutterstock; Figure 16–7a: Bernard J. Nebel; Figure 16–9: Au Sable Institute of Environmental Studies; Figure 16–10: Bernard J. Nebel; Figure 16–12: Kent Bullard/National Renewable Energy Laboratory; Figure 16–13: Nellis Air Force Base; Figure 16–14: Sandia National Laboratories; Figure 16–15: Mario Savoia/Shutterstock; Figure 16–16: iofoto/Shutterstock; Figure 16–17: Alamy; Figure 16–19: Creative Energy Corporation; Figure 16–20: Aroon Thaewchatturat/onasia.com; Figure 16–21: ideeone/Istockphoto; Figure 16–22b: Courtesy of Honda; Figure 16–23: Pacific Gas & Electric Company.

Chapter 17

PO-5a: Mircea Bezerghean/Shutterstock; PO-5b: PhilAugustavo/Istockphoto; CO-17a: Robert Ghement/epa/Corbis Images; CO-17b: Getty Images; Figure 17–4: Associated Press; Figure 17–5: WHO; Figure 17–7a: Kimimasa Mayama/Reuters/Corbis; Figure 17–7b: Smiley N. Pool/Dallas Morning News/Corbis; Figure 17–7c: Mads Nissen/Getty Images; Figure 17–7d: Amber Douthit/Getty Images; Figure 17–8a: Calvin Larsen/Photo Researchers; Figure 17–8b: Gilles Mingasson/Getty Images; Figure 17–8c: Doug Wechsler/Animals Animals/Earth Scenes; Figure 17–8d: How Hwee Young/epa/Corbis; Figure 17–10: Shehzad Noorani/Woodfin Camp and Associates; Figure 17–12: Chris Hondros/Getty Images; Figure 17–14: Copyright WHO/Andy Crump; Figure 17–17: Slide Works/Jackson Laboratory.

Chapter 18

CO-18a: NASA; CO-18b: mostlymozart/Istockphoto; Figure 18–14a & b: Online glacier photograph database, Boulder, CO: National Snow and Ice Data Center/World Data Center for Glaciology, digital media; Figure 18–11-UN: NASA

Chapter 19

CO-19a: Ronaldo Schemidt/Getty Images; CO-19b: Liliboas/Istockphoto; Figure 19–1: Bettmann/Corbis; Figure 19–5a: ekash/Istockphoto; Figure 19–5b: steinphoto/Istockphoto; Figure 19–8: Richard T. Wright; Figure 19–16: Grapes/Michaud/Photo Researchers; Figure 19–17a: Matt Meadows/Peter Arnold; Figure 19–17b: Matt Meadows/Peter Arnold; Figure 19–18: Getty Images; Figure 19–19: AS Heagle/USDA; Figure 19–21: Balliolman/Fotolia; Figure 19–23: Courtesy of the Cherokee Nation Environmental Programs/ITEC.

Chapter 20

CO-20a: NASA; CO-20b: Science Photo Library/Alamy; Figure 20–2: Bettmann/Corbis; Figure 20–4: Robert Harding Picture Library/SuperStock; Figure 20–4-UN: Bob Hudson/George Waclawin/Bernard J. Nebel; Figure 20–7: Thomas R. Fletcher/Alamy;Figure 20–8c: Charles R. Belinky/Photo Researchers; Figure 20–9: Bernard J. Nebel; Figure 20–11a & b: Bernard J. Nebel; Figure 20–13: Bernard J. Nebel; Figure 20–14a & b: Bernard J. Nebel; Figure 20–16a & b: Envirolet®, Sancor Industries Ltd.; Figure 20–18-UN: JoAnn M. Burkholder/Scientific American, August 1999, page 46; Figure 20–19: G. Avila/Alamy; Figure 20–20: Jennifer Weinberg/Alamy.

Chapter 21

CO-21a: © Chris Jordan, All Rights Reserved; CO-21b: imagestock/Istockphoto; Figure 21–1: Photo (c) Cymie Payne, courtesy of Cambridge Arts Council; Figure 21–4: Van Bucher/Photo Researchers; Figure 21–5: Bill Polo/Boston Globe; Figure 21–8: Richard Levine/Alamy; Figure 21–9: Courtesy of Trex; Figure 21–12: Justin Sullivan/Getty Images; Figure 21–13: John Griffin/The Image Bank; Figure 21–14: ropable/Wikapedia Foundation; Figure 21–15: (c) Tribune Media Services, All Rights Reserved. Reprinted with permission; Figure 21–16: Robert J. Fiore, City of Worcester, Massachusetts.

Chapter 22

CO-22a: Gordon Wiltsie/Peter Arnold; CO-22b: Zirafek/istockphoto; Figure 22–7: U.S. Environmental Protection Agency Headquarters;

Figure 22–8: Alon Reininger/Contact Press/Woodfin Camp & Associates; Figure 22–9: Nancy J. Pierce/Photo Researchers; Figure 22–10: Van Bucher/Photo Researchers; Figure 22–11: Joe Traver/Getty Images; Figure 22–12d: Tetra Tech EM Inc; Figure 22–13: TMTS Associates; Figure 22–13-UN: Steven Senne/Associated Press; Figure 22–15: Associated Press.

Chapter 23

PO-6a: Curtis W. Richter/Getty Images; PO-6b: Reinhold Foeger/Shutterstock; CO-23a: David Wall Photography; CO-23b: Robyn Mackenzie/Shutterstock; Figure 23–1a & b: City of Trenton; Figure 23–2: David Pollack/Corbis; Figure 23–3: Esbin-Anderson/The Image Works; Figure 23–4: Brooks Kraft/Corbis; Figure 23–5a: Bart Everett/Shutterstock; Figure 23–5b: asterix0597/Istockphoto; Figure 23–9a: Cameron Davidson/Getty Images; Figure 23–9b: Bernard J. Nebel; Figure 23–10: Bernard J. Nebel; Figure 23–11: Altrendo Travel/Getty Images; Figure 23–12: Sven Torfinn/Panos Pictures; Figure 23–13: Getty Images; Figure 23–14: steinphoto/Istockphoto; Figure 23–15: Kim Newton/Woodfin Camp & Associates; Figure 23–17: 18th & Glenwood Garden, photo courtesy of the Pennsylvania Horticultural Society; Figure 23–18: McPhoto/age footstock; Figure 23–19: Chattanooga Area Convention and Visitors Bureau; Figure 23–19-UN: Jeff Pohorski; Figure 23–20: Karl Gehring/Getty Images.

Text Credits

Chapter 1

Figure 1–4: Courtesy of zFacts.com/p/226.html.

Chapter 2

Figure 2–2: From World Development Report 1992 by The World Bank. Copyright 1992 by The World Bank. Reproduced with permission of The World Bank in the formats Textbook, CD-ROM, and Other Book via Copyright Clearance Center; Figure 2–6: From World Development Report 2003 by The World Bank. Copyright 2003 by The World Bank. Reproduced with permission of The World Bank in the formats Textbook, CD-ROM, and Other Book via Copyright Clearance Center; Figure 2–7: "GDP and GPI per capita, 1950–2004" by John Talberth, Clifford Cobb, and Noah Slattery from *The Genuine Progress Indicator 2006: A Tool for Sustainable Development,* Redefining Progress and Center for Sustainable Economy; Table 2–1: From *Where Is the Wealth of Nations?: Measuring Capital for the 21st Century* by The World Bank. Copyright © 2000 by The World Bank. Reproduced with permission of The World Bank in the format Textbook, Other book, and CD-ROM via Copyright Clearance Center.

Chapter 5

Table 5–3: Reprinted by permission from Macmillan Publishers Ltd: *Nature,* 387 (May 15, 1997): 253, copyright © 1997.

Chapter 7

Figure 7 10: From Global Forest Resources Assessment 2005: Progress towards sustainable forest management, FAO Forestry Paper 147, 2006, Figure 2.6. Used by permission of the Food and Agriculture Organization of the United Nations; Figure 7–13: From Benjamin S. Halpern, et al., "A Global Map of Human Impact on Marine Geosystems" from *Science,* Vol. 319, February 15, 2008: 949. Used by permission from AAAS.

Chapter 8

Figure 8–7: From *World Population Highlights,* September 2008, Vol 63, No 3. Used by permission; Figure 8–8: From *World Population Highlights,* September 2008, Vol 63, No 3. Used by permission; Figure 8–14: From *World Population Prospects: The 2008 Revision.* Copyright © United Nations 2009. Reproduced with permission.

Chapter 9

Figure 9–3: From Poncet, Sandra, 2006, "The Long Term Growth Prospects of the World Economy: Horizon 2050," Working Paper 2006–16. Centre d'Etudes Prospectives et d'Informations Internationales; Figure 9–9: From *Millennium Development Goals Report 2008.* Copyright © United Nations 2009. Reproduced with permission.

Chapter 11

Figure 11–14: From "Global assessment of land degradation and improvement. 1. Identification by remote sensing. Report 2008/01, ISRIC – World Soil Information, Used by permission of the Food and Agriculture Organization of the United Nations.

Chapter 12

Figure 12–5: From World Development Report 2008 (Overview) by The World Bank. Copyright 2007 by The World Bank. Reproduced with permission of The World Bank in the formats Textbook, CD-ROM, and Other Book via Copyright Clearance Center; Figure 12–10: From "Global population and grain and meat production" from FAOSTAT database collections. Used by permission of the Food and Agriculture Organization of the United Nations; Figure 12–11: From "Global Status of Commercialized Biotech/GM Crops: 2008" by Clive James from *ISAAA Briefs,* 37, 2007, p. 3 (International Service for the Acquisition of Agri-Biotech Applications); Figure 12–14: From FAO *Food Outlook,* June 2008, p. 2. Used by permission of the Food and Agriculture Organization of the United Nations; Figure 12–18: From *Millennium Development Goals Report 2008.* Copyright © United Nations 2009. Reproduced with permission; Table 12–1: From FAOSTAT Agricultural Data, 2008. Used by permission of the Food and Agriculture Organization of the United Nations; Table 12–2: From FAO *Food Outlook,* June 2008.Used by permission of the Food and Agriculture Organization of the United Nations.

Chapter 15

Figure 15–14: "Advanced boiling-water reactor" from "Advanced Light-Water Reactors" by M. W. Golay and N.E. Todreas, *Scientific American,* 1990. Used by permission of the illustrator, George Retseck.

Chapter 16

Figure 16–11: REN21. 2008. "Renewables 2007 Global Status Report" (Paris: REN21 Secretariat and Washington, DC:Worldwatch Institute).Copyright © 2008 Deutsche Gesellschaft für Technische Zusammenarbeit (GTZ) GmbH.

Chapter 17

Table 17–5: "Top Domestic Priorities for Obama and Congress" from January 22, 2009 Survey Report, "Economy, Jobs Trump All Other Policy Priorities in 2009," Pew Research Center for the People and the Press, a project of the Pew Research Center. Used by permission.

Chapter 18

Figure 18–21: From "Global Warming: How Urgent Is Climate Change?" by Richard A. Kerr in *Science,* November 23, 2007, 318: 1231. Used by permission from AAAS; Figure 18–25: "Stabilization Wedges: A Concept and Game" used with permission from Carbon Mitigation Initiative of the Princeton Environmental Institute.

Chapter 19

Figure 19–15: Copyright © Her Majesty The Queen in Right of Canada, Environment Canada, 2005. Reproduced with the permission of the Minister of Public Works and Government Services Canada; Table 9–4: "Key Findings of the 2006 Scientific Assessment of Ozone Depletion" from *Scientific Assessment of Ozone Depletion: 2006,* Global Ozone Research and Monitoring Project, report no. 50 (Geneva, Switzerland: World Meteorological Organization, 2007). Used with permission.

Chapter 20

Figure 20–1: Bottom Dissolved Oxygen (mg/L), July 21–27, 2008. N. N. N. Rabalais, Louisiana Universities Marine Consortium, and R. E. Turner, Louisiana State University funded by National Oceanic and Atmospheric Administration, Center for Sponsored Coastal Ocean Research; Figure 20–5: From *World Water Development Report 2006.* Copyright © United Nations 2007. Reproduced with permission.

Chapter 23

Figure 23–7: From *Measuring Sprawl and Its Impact* by Reid Ewing, Rolf Pendall, and Don Chen, 2002, Smart Growth America. Used by permission; Table 23.2: From *Measuring Sprawl and Its Impact* by Reid Ewing, Rolf Pendall, and Don Chen, 2002, Smart Growth America. Used by permission.

GLOSSARY

abiotic Pertaining to factors or things that are separate and independent from living things; nonliving.

abortion The termination of a pregnancy by some form of surgical or medicinal intervention.

acid Any compound that releases hydrogen ions when dissolved in water. Also, a water solution that contains a surplus of hydrogen ions.

acid deposition Any form of acid precipitation and also fallout of dry acid particles. (See **acid precipitation**.)

acid neutralizing capacity (ANC) In a water body, the capacity to neutralize acids due to the presence of **buffer** chemicals in solution.

acid precipitation Includes acid rain, acid fog, acid snow, and any other form of precipitation that is more acidic than normal (i.e., less than pH 5.6). Excess acidity is derived from certain air pollutants—namely, sulfur dioxide and oxides of nitrogen.

activated sludge system A system for removing organic wastes from water. The system uses microorganisms and active aeration to decompose such wastes. The system is used most as a means of secondary sewage treatment following the primary settling of materials.

active safety features Those safety features of nuclear reactors that rely on operator-controlled reactions, external power sources, and other features that are capable of failing. (See **passive safety features**.)

acute A reaction to a disease or chemical pollutant that brings on life-threatening reactions within hours or days.

adaptation An ecological or evolutionary change in structure or function that enables an organism to adjust better to its environment and hence enhances the organism's ability to survive and reproduce.

adiabatic cooling The cooling that occurs when warm air rises and encounters lower atmospheric pressure. **Adiabatic** *warming* is the opposite process, whereby cool air descends and encounters higher pressure.

adsorption The process whereby chemicals (ions or molecules) stick to the surface of other materials.

aeration *Soil*: The exchange within the soil of oxygen and carbon dioxide necessary for the respiration of roots. *Water*: The bubbling of air or oxygen through water to increase the amount of oxygen dissolved in the water.

aerosols Microscopic liquid and solid particles originating from land and water surfaces and carried up into the atmosphere.

Agenda 21 A definitive program to promote sustainable development that was produced by the 1992 Earth Summit in Rio de Janeiro, Brazil, and adopted by the UN General Assembly.

age structure Within a population, the different proportions of people who are old, middle aged, young adults, and children.

AID The U.S. Agency for International Development, the development arm of the U.S. State Department responsible for administering international aid from the United States.

AIDS Acquired immune deficiency syndrome, a fatal disease caused by the human immunodeficiency virus (HIV) and transmitted by sexual contact or the use of nonsterile needles (as in drug addiction).

air pollution disaster A short-term situation in industrial cities in which intense industrial smog brings about a significant increase in human mortality.

air toxics A category of air pollutants that includes radioactive materials and other toxic chemicals that are present at low concentrations but are of concern because they often are carcinogenic.

alga, pl. algae Any of numerous kinds of photosynthetic plants that live and reproduce while entirely immersed in water. Many species (the planktonic forms) exist as single cells or small groups of cells that float freely in the water. Other species (the seaweeds) may be large and attached.

algal bloom A relatively sudden development of a heavy growth of algae, especially planktonic forms. Algal blooms generally result from additions of nutrients, whose scarcity is normally limiting.

alkaline, alkalinity A *basic* substance; chemically, a substance that absorbs hydrogen ions or releases hydroxyl ions; in reference to natural water, a measure of the base content of the water.

Alliance for a Green Revolution An organization promoting a new Green Revolution in sub-Saharan Africa, where the needs of subsistence farmers to grow more food will be addressed.

ambient standards Air-quality standards (set by the EPA) determining certain levels of pollutants that should not be exceeded in order to maintain environmental and human health.

amensalism An interaction between species whereby one species is harmed, while the other is unaffected.

anaerobic Lacking oxygen.

anaerobic digestion The breakdown of organic material by microorganisms in the absence of oxygen. The process results in the release of methane gas as a waste product.

animal testing Procedures used to assess the toxicity of chemical substances using rats, mice, and guinea pigs as surrogates for humans who might be exposed to the substances.

anthropogenic Referring to pollutants and other forms of impact on natural environments that can be traced to human activities.

aquaculture A propagation or rearing of any aquatic (water) organism in a more or less artificial system.

aquifer An underground layer of porous rock, sand, or other material that allows the movement of water between layers of nonporous rock or clay. Aquifers are frequently tapped for wells.

Arctic Climate Impact Assessment (ACIA) An assessment of the serious impacts of global warming on the Arctic region, published in 2004.

artificial selection Plant and animal breeders' practice of selecting individuals with the greatest expression of desired traits to be the parents of the next generation.

asthma A chronic disease of the respiratory system in which the air passages tighten and constrict, causing breathing difficulties; acute attacks may be life threatening.

atmosphere The thin layer of gases surrounding Earth. Nitrogen, oxygen, water vapor, and carbon dioxide are major gases, while many minor gases are also present in trace amounts.

atmosphere-ocean general circulation model A computer-based model for simulating long-term climatic conditions that combines global atmospheric circulation patterns with ocean circulation and cloud-radiation feedback.

atmospheric brown cloud (ABC)   A 1- to 3-km-thick blanket of pollution that frequently sits over south and central Asia, made up of black carbon and soot from biomass and fossil fuel burning.

atom   The fundamental unit of all elements.

autotroph   Any organism that can synthesize all its organic substances from inorganic nutrients, using light or certain inorganic chemicals as a source of energy. Green plants are the principal autotrophs.

Bacillus thuringensis (BT)   A type of bacterium that produces a toxin that can kill the larvae of plant-eating insects; genes from this bacterium have been incorporated into plants using genomic techniques.

background radiation   Radiation that comes from natural sources apart from any human activity. We are all exposed to such radiation.

bacterium, pl. bacteria   Any of numerous kinds of prokaryotic microscopic organisms that exist as simple single cells that multiply by simple division. Along with fungi, bacteria constitute the decomposer component of ecosystems. A few species cause disease.

balance-of-payment deficit   A deficit in money flow resulting from purchasing more from other countries than is sold to other countries.

BANANA   (**"build absolutely nothing anywhere near anything"**) An acronym expressing a common attitude about siting various facilities.

base   Any compound that releases hydroxyl (OH⁻) ions when dissolved in water. A solution that contains a surplus of hydroxyl ions.

Basel Convention   An international agreement that bans most international toxic-waste trade.

base-load power plant   A power plant, usually large and fueled by coal or nuclear energy, that is kept operating continuously to supply electricity.

bed load   The load of coarse sediment, mostly coarse silt and sand, that is gradually moved along the bottom of a riverbed by flowing water rather than being carried in suspension.

benefit-cost analysis   A comparison of the benefits of any particular action or project to its costs. (See **cost-benefit ratio**.)

benthic plants   Plants that grow under water and that are attached to or rooted in the bottom of the body of water. For photosynthesis, these plants depend on light penetrating the water.

benzene   An organic chemical present in crude and refined oil products and tobacco smoke, a known human carcinogen.

best-demonstrated available technologies (BDATs)   Treatment standards for hazardous wastes that have the effect of reducing their toxicity and mobility.

best management practice   Farm management practice that serves best to reduce soil and nutrient runoff and subsequent pollution.

bioaccumulation   The accumulation of higher and higher concentrations of potentially toxic chemicals in organisms. Bioaccumulation occurs in the case of substances that are ingested but that cannot be excreted or broken down (nonbiodegradable substances).

biochemical oxygen demand (BOD)   The amount of oxygen that will be absorbed or "demanded" as wastes are being digested or oxidized in both biological and chemical processes. Potential impacts of wastes are commonly measured in terms of the BOD.

biodegradable   Able to be consumed and broken down to natural substances, such as carbon dioxide and water, by biological organisms, particularly decomposers. Opposite: **nonbiodegradable**.

biodiesel   Diesel fuel made from a mixture of vegetable oil (largely soybean oil) and regular diesel oil.

biodiversity   The diversity of living things found in the natural world. The concept usually refers to the different species but also includes ecosystems and the genetic diversity within a given species.

biodiversity hot spots   Thirty-four locations around the world that are home to some 60% of the world's known biodiversity.

biofuel   Any fuel, but usually liquids, derived from agricultural crops (e.g., ethanol and plant oils).

biogas   The mixture of gases—about two-thirds methane, one-third carbon dioxide, and small portions of foul-smelling compounds—resulting from the anaerobic digestion of organic matter. The methane content enables biogas to be used as a fuel.

biogeochemical cycle   See **nutrient cycle**.

biological control   Control of a pest population by the introduction of predatory, parasitic, or disease-causing organisms.

biological nutrient removal   A process employed in sewage treatment to remove nitrogen and phosphorus from the effluent coming from secondary treatment.

biological wealth   The life-sustaining combination of commercial, scientific, and aesthetic values imparted to a region by its biota.

biomagnification   Bioaccumulation occurring through several levels of a food chain.

biomass   The mass of biological material. Usually the total mass of a particular group or category (e.g., biomass of producers).

biomass energy   Energy produced by burning plant-related materials such as firewood.

biomass fuels   Fuels in the form of plant-related material (e.g., firewood, grasses, organic farm wastes).

biomass pyramid   The structure that is obtained when the respective biomasses of producers, herbivores, and carnivores in an ecosystem are compared. Producers have the largest biomass, followed by herbivores and then carnivores.

biome   A group of ecosystems that are related by having a similar type of vegetation governed by similar climatic conditions. Examples include prairies, deciduous forests, arctic tundra, deserts, and tropical rain forests.

bioremediation   The use of microorganisms for the decontamination of soil or groundwater. Usually involves injecting organisms or oxygen into contaminated zones.

biosphere   The overall ecosystem of Earth. The sum total of all the biomes and smaller ecosystems, which ultimately are all interconnected and interdependent through global processes such as the water cycle and the atmospheric cycle.

biosphere reserves   Natural areas in various parts of the world set aside to protect genetic biodiversity.

biota   The sum total of all living organisms. The term usually is applied to the setting of natural ecosystems.

biotechnology   The use of genetic-engineering techniques, enabling researchers to introduce new genes into food crops and domestic animals and thereby produce valuable products and more nutritious crops.

biotic   Living or derived from living things.

biotic community   All the living organisms (plants, animals, and microorganisms) that live in a particular area.

biotic potential   Reproductive capacity. The potential of a species for increasing its population and/or distribution. The biotic potential of every species is such that, given optimum conditions, its population will increase. (Contrast **environmental resistance**.)

biotic structure   The organization of living organisms in an ecosystem into groups such as producers, consumers, detritus feeders, and decomposers.

birth control Any means, natural or artificial, that may be used to reduce the number of live births.

bisphenol A A starter material for many plastics that is suspected of contributing to obesity, diabetes, infertility, and cancer.

blue water In the hydrologic cycle, precipitation and its further movement in infiltration, runoff, surface water, and groundwater; liquid water wherever it occurs.

BOD See **biochemical oxygen demand**.

bottle law, bottle bill A law that provides for the recycling or reuse of beverage containers, usually by requiring a returnable deposit upon purchase of the item.

bottom-up regulation Basic control of a natural population that occurs as a result of the scarcity of some resource.

breeder reactor A nuclear reactor that, in the course of producing energy, also converts nonfissionable uranium-238 into fissionable plutonium-239, which can be used as fuel. Hence, a reactor that produces at least as much nuclear fuel as it consumes. Also called a **fast-neutron reactor**.

broad-spectrum pesticides Chemical pesticides that kill a wide range of pests. These pesticides also kill a wide range of nonpest and beneficial species; therefore, they may lead to environmental disturbances and resurgences. The opposite of narrow-spectrum pesticides and biorational pesticides.

brownfields Abandoned, idled, or underused industrial and commercial facilities whose further development is inhibited because of real or perceived chemical contamination.

BTU (British thermal unit) A fundamental unit of energy in the English system. The amount of heat required to raise the temperature of 1 pound of water 1 degree Fahrenheit.

buffer A substance that will maintain a constant pH in a solution by absorbing added hydrogen or hydroxyl ions. An example is the bicarbonate ion, the most important buffer in blood.

Bureau of Land Management (BLM) A federal agency that manages public lands other than national forests and national parks. The agency leases land to ranchers for grazing, among other activities.

bush meat Wild game that is harvested for food, especially in developing countries; often involves poaching and overexploitation.

CAFE (corporate average fuel economy) Fuel economy standards for vehicles set by the National Highway Traffic Safety Administration.

calorie A fundamental unit of energy. The amount of heat required to raise the temperature of 1 gram of water 1 degree Celsius. All forms of energy can be converted to heat and measured in calories. Calories used in connection with food are kilocalories, or "big" calories, the amount of heat required to raise the temperature of 1 liter of water 1 degree Celsius.

cancer Any of a number of cellular changes that result in uncontrolled cellular growth.

cap-and-trade system A form of market-based environmental policy that sets a maximum level of pollutant (the cap), distributes pollution permits, and allows industries to trade permits to achieve their allowable pollution.

capillary water Water that clings in small pores, cracks, and spaces against the pull of gravity (e.g., water held in a sponge).

carbon dioxide sinks The oceans and terrestrial environments serve to absorb some of the carbon dioxide added to the atmosphere by human agency; these are called sinks.

carbon monoxide A highly poisonous gas, the molecules of which consist of a carbon atom with one oxygen attached. Not to be confused with carbon dioxide, a natural gas in the atmosphere.

carbon tax A tax levied on all fossil fuels in proportion to the amount of carbon dioxide that is released as they burn.

carcinogen, carcinogenic Having the property of causing cancer, at least in animals and, by implication, in humans.

carnivore An animal that feeds more or less exclusively on other animals.

carrying capacity The maximum population of a given species that an ecosystem can support without being degraded or destroyed in the long run. Represented symbolically by K.

Cartagena Protocol An international agreement governing trade in genetically modified organisms. The Cartagena Protocol was signed in January 2000.

castings The excrements from earthworms that are deposited on the surface of soils, as mixtures of mineral particles and humus compounds.

catalyst A substance that promotes a given chemical reaction without itself being consumed or changed by the reaction. Enzymes are catalysts for biological reactions. Also, catalysts are used in some pollution control devices (e.g., the catalytic converter).

catalytic converter The device used by U.S. automobile manufacturers to reduce the amount of carbon monoxide and hydrocarbons in a car's exhaust. The converter contains a catalyst that oxidizes these compounds to carbon dioxide and water as the exhaust passes through.

catch shares See **individual quota system**.

cell The basic unit of life; the smallest unit that still maintains all the attributes of life. Many microscopic organisms consist of a single cell. Large organisms consist of trillions of specialized cells functioning together.

cell respiration The chemical process that occurs in all living cells whereby organic compounds are broken down to release energy required for life processes. For respiration, higher plants and animals require oxygen and release carbon dioxide and water as waste products, but certain microorganisms do not require oxygen.

cellulose The organic macromolecule that is the prime constituent of plant cell walls and hence the major molecule in wood, wood products, and cotton. Cellulose is composed of glucose molecules, but because it cannot be digested by humans, its dietary value is only as fiber, bulk, or roughage.

center pivot irrigation An irrigation system consisting of a spray arm several hundred meters long, supported by wheels pivoting around a central well from which water is pumped.

centrally planned economy An economic system in which a ruling class makes most of the basic decisions about how the economy will be structured; typical of communist countries.

CFCs See **chlorofluorocarbons**.

chain reaction A nuclear reaction wherein each atom that fissions (splits) causes one or more additional atoms to fission.

character displacement An effect of interspecific competition, whereby a physical change occurs in one or both competing species that lessens competition when the two species co-occur.

chemical barrier A chemical aspect of a plant that makes it resist attacks from certain pests. Used in genetic pest control.

chemical energy The potential energy that is contained in certain chemicals; most importantly, the energy that is contained in organic compounds such as food and fuels and that may be released through respiration or burning.

chemical treatment The use of pesticides and herbicides to control or eradicate agricultural pests.

chemosynthesis The process whereby some microorganisms utilize the chemical energy contained in certain reduced inorganic chemicals (e.g., hydrogen sulfide) to produce organic material.

chlorinated hydrocarbons Synthetic organic molecules in which one or more hydrogen atoms have been replaced by chlorine atoms. Chlorinated hydrocarbons are hazardous compounds because they tend to be nonbiodegradable and therefore to bioaccumulate; many have been shown to be carcinogenic. Also called **organochlorides**.

chlorination   The disinfection process of adding chlorine to drinking water or sewage water in order to kill microorganisms that may cause disease.

chlorine catalytic cycle   In the stratosphere, a cyclical chemical process in which chlorine monoxide breaks down ozone.

chlorofluorocarbons (CFCs)   Synthetic organic molecules that contain one or more of both chlorine and fluorine atoms and that are known to cause ozone destruction.

chlorophyll   The green pigment in plants responsible for absorbing the light energy required for photosynthesis.

chromosome   In a living organism, one of a number of structures on which genes are arranged in a linear fashion.

chronic   A reaction to infection or pollutants that leads to a gradual deterioration of health over a period of years.

chronic obstructive pulmonary disease (COPD)   A progressive lung disease that makes it hard to breathe, involving three syndromes: asthma, bronchitis and emphysema.

CITES (Convention on International Trade in Endangered Species)   An international treaty conveying some protection to endangered and threatened species by restricting trade in those species or their products.

clay   Any particles in soil that are less than 0.004 millimeters in size.

Clean Air Act of 1970   Amended in 1977 and 1990, the act is the foundation of U.S. air pollution control efforts.

Clean Air Interstate Rule (CAIR)   EPA rule published in 2005 establishing cap-and-trade programs for sulfur dioxide and nitrogen oxides in 28 eastern states.

Clean Water Act of 1972   The cornerstone federal legislation addressing water pollution.

climate   A general description of the average temperature and rainfall conditions of a region over the course of a year.

Climate Change Science Program (CCSP)   Now called the **Global Change Research Program (GCRP)**, the U.S. government-sponsored program that addresses a number of issues in climate change with research and reports.

climate sensitivity   The response of climate to greenhouse gas concentrations, usually measured as a rise in temperature as a consequence of a rise in greenhouse gas concentration.

climax ecosystem   The last stage in ecological succession. An ecosystem in which populations of all organisms are in balance with each other and with existing abiotic factors.

cogeneration   Also known as **combined heat and power (CHP)**, the joint production of useful heat and electricity. For example, furnaces may be replaced with gas turbogenerators that produce electricity, while the hot exhaust still serves as a heat source. An important avenue of conservation, cogeneration effectively avoids the waste of heat that normally occurs at centralized power plants.

cold condensation   The process whereby volatile organic pollutants condense on the snowpack in cold climates.

combined heat and power (CHP)   See **cogeneration**.

combustion   The practice of disposing of wastes by incineration in a special facility designed to handle large amounts of waste; modern combustion facilities also capture some of the energy by generating electricity on the site.

command-and-control strategy   The basic strategy behind most public policy having to do with air and water pollution. The strategy involves setting limits on pollutant levels and specifying control technologies that must be used to achieve those limits.

commensalism   A relation between two species in which one is benefited and the other is not affected.

common good   A recognition of the need for protection of the natural world and improvement of human welfare, especially as applied to public policy.

commons, common pool resources   Resources (usually natural ones) owned by many people in common or, as in the case of the air or the open oceans, owned by no one but open to exploitation.

compaction   Packing down. *Soil*: Packing and pressing out air spaces present in the soil. Compaction reduces soil aeration and infiltration and thus diminishes the capacity of the soil to support plants. *Trash*: Packing down trash to reduce the space that it requires.

competitive exclusion principle   The concept that when two species compete very directly for resources, one eventually excludes the other from the area.

composting/compost   The process of letting organic wastes decompose in the presence of air. A nutrient-rich humus, or compost, is the resulting product.

composting toilet   A toilet that does not flush wastes away with water but deposits them in a chamber where they will compost. (See **composting**.)

compound   Any substance (gas, liquid, or solid) that is made up of two or more different kinds of atoms bonded together. (Contrast **element**.)

Comprehensive Environmental Response, Compensation, and Liability Act of 1980 (CERCLA)   See **Superfund**.

Comprehensive Everglades Restoration Plan (CERP)   Multi-billion-dollar plan to restore the Florida Everglades by addressing water flow and storage.

concepts   Valid explanations of data from the natural world that often allow predictions but do not reach the level of validity of natural laws.

condensation   The collecting of molecules from the vapor state to form the liquid state, as, for example, when water vapor condenses on a cold surface and forms droplets. Opposite: **evaporation**.

condenser   A device that turns turbine exhaust steam into water in a power plant.

conditions   Abiotic factors (e.g., temperature) that vary in time and space but are not used up by organisms.

confined animal feeding operation (CAFO)   Animal farming where the animals are kept in crowded pens, producing wastes that are hard to manage and creating other health threats.

confusion technique   A pest control method in which a quantity of sex attractant is applied to an area so that males become confused and are unable to locate females. The actual quantities of pheromones applied are very small because of their extreme potency.

conservation   The management of a resource in such a way as to assure that it will continue to provide maximum benefit to humans over the long run. *Energy*: Saving energy by cutting back on the use of heating, air conditioning, lighting, transportation, and so on and also by increasing the efficiency of energy use.

conservation biology   The branch of science focused on the protection of populations and species.

conservation reserve   An imaginary source of energy that results from policies promoting greater efficiency of energy use, resulting in a reduced energy requirement.

Conservation Reserve Program (CRP)   A federal program whereby farmers are paid to place highly erodible cropland in a reserve. A similar program, the **Conservation Stewardship Program**, encourages farmers to conserve soil and water resources on their lands.

consumers   In an ecosystem, those organisms that derive their energy from feeding on other organisms or their products.

consumptive use   The harvesting of natural resources in order to provide for people's immediate needs for food, shelter, fuel, and clothing.

consumptive water use   The use of water for such things as irrigation, wherein the water does not remain available for potential purification and reuse.

containment building   The reinforced concrete building housing a nuclear reactor. Designed to contain an explosion, should one occur.

continental drift   The theory, first proposed by Alfred Wegener, that the continents were once connected and have since been moving apart; the process is now known as **plate tectonics**.

contour farming   The practice of cultivating land along the contours across, rather than up and down, slopes. In combination with strip cropping, contour farming reduces water erosion.

contraceptive   A device or method employed by couples during sexual intercourse to prevent the conception of a child.

control rods   A part of the core of a nuclear reactor; the rods of neutron-absorbing material that are inserted or removed as necessary to control the rate of nuclear fissioning.

convection/convection currents   The vertical movement of air due to atmospheric heating and cooling, moving generally in a west-to-east pattern.

Convention on Biological Diversity   The biodiversity treaty signed by 158 nations at the Earth Summit in Rio de Janeiro in 1992 calling for various actions and cooperative steps between nations to protect the world's biodiversity.

conversion losses   Unavoidable energy losses in energy production, usually due to the loss of heat energy.

conveyor system   The giant pattern of oceanic currents that moves water masses from the surface to the depths and back again, producing major effects on the climate.

cooling tower   A massive tower designed to dissipate waste heat from a power plant (or other industrial facility) into the atmosphere.

corrosion   In a nuclear power plant, the deterioration of pipes receiving hot, pressurized water in a circulation system that conveys heat from one part of the unit to another.

cosmetic spraying   The spraying of pesticides to control pests that damage only the surface appearance of fruits and vegetables.

cost-benefit analysis   See **benefit-cost analysis**.

cost-benefit ratio, benefit-cost ratio   The value of the benefits to be gained from a project, divided by the costs of the project. If the ratio is greater than unity, the project is economically justified; if the ratio is less than unity, the project is not economically justified.

cost-effectiveness analysis   An alternative to benefit-cost analysis, this approach accepts the merits of a goal and attempts to achieve the goal with the least costly alternative.

covalent bond   A chemical bond between two atoms, formed by sharing a pair of electrons between the atoms. Atoms of all organic compounds are joined by covalent bonds.

credit associations   Groups of poor people who individually lack the collateral to assure loans but who collectively have the wherewithal to assure each other's loans. Associated with microlending.

criteria maximum concentration (CMC)   A water-quality standard; the highest single concentration of a water pollutant beyond which environmental impacts may be expected.

criteria pollutants   Certain pollutants whose levels are used as a gauge for the determination of air (or water) quality.

criterion continuous concentration (CCC)   A water-quality standard; the highest sustained concentration of a water pollutant beyond which undesirable impacts may be expected.

critical habitat   Under the Endangered Species Act, an area provided for a listed species where it can be found or could likely spread as it recovers.

critical level   The level of one or more pollutants above which severe damage begins to occur and below which few, if any, ill effects are noted.

critical number   The minimum number of individuals of a given species that is required to maintain a healthy, viable population of the species. If a population falls below its critical number, it will almost certainly become extinct.

crude birthrate   The number of births per 1,000 individuals per year.

crude death rate   The number of deaths per 1,000 individuals per year.

cryptogamic crust   Made up of primitive plants, called cryptogams, the crust develops on hardened soils and can stabilize soil but also prevents water infiltration and seed germination.

cultural control   A change in the practice of growing, harvesting, storing, handling, or disposing of wastes that reduces their susceptibility or exposure to pests. For example, spraying the house with insecticides to kill flies is a chemical control; putting screens on the windows to keep flies out is a cultural control.

cultural eutrophication   The process of natural eutrophication accelerated by human activities. (See **eutrophication**.)

cycling of matter   See **nutrient cycle**.

DALY (disability-adjusted life year)   In assessing the disease burden of different health risks, 1 DALY equals the loss of one healthy year of a person's life (whether through disability or loss of life).

DDT (dichlorodiphenyltrichloroethane)   The first and most widely used of the synthetic organic pesticides belonging to the chlorinated hydrocarbon class.

dead zone   A region of an aquatic ecosystem (usually a marine coastal area) where oxygen is low or absent from the bottom up well into the water column. Also called a **hypoxic area**.

debt crisis   Brought about by the great debt incurred by many less-developed nations, this crisis involves countries building up debts that can never be paid off.

debt relief   Steps, including cancellation of debts and other means of reducing poverty, taken by the World Bank and donor nations to alleviate the heavy debt of many poor nations.

declining tax base   The loss of tax revenues that occurs when affluent taxpayers and businesses leave an area and property values subsequently decline. Also referred to as an **eroding tax base**.

decommissioning   The taking of nuclear power plants out of service after 25–35 years because the effects of radiation will gradually make them inoperable.

decomposers   Organisms whose feeding action results in decay or rotting of organic material. The primary decomposers are fungi and bacteria.

deep-well injection   A technique used for the disposal of liquid chemical wastes that involves putting them into deep dry wells where they permeate dry strata.

deforestation   The process of removing trees and other vegetation covering the soil and converting the forest to another land use, often leading to erosion and loss of soil fertility.

Delaney clause   A stipulation in the Federal Food, Drug, and Cosmetic Act of 1938 that prohibited the use of any food additive that is found to induce cancer in humans or animals; this has been replaced by the **negligible risk** standard.

demand-side policies   Energy policies that emphasize conserving energy and using more renewable and nuclear energy in the future.

demographic dividend   As a country goes through the demographic transition, there is a stage when the dependency ratio is low, giving the country the opportunity to spend more on poverty alleviation and economic growth.

demographic transition   The transition of a human population from a condition of a high birthrate and a high death rate to a condition of a low birthrate and a low death rate. A demographic transition may result from economic or social development.

demography/demographer   The study of population trends (growth, movement, development, and so on). People who perform such studies and make projections from them.

denitrification   The process of reducing oxidized nitrogen compounds present in soil or water to nitrogen gas to the atmosphere, conducted by certain bacteria and now utilized in the treatment of sewage effluents.

density-dependent   An attribute of population-balancing factors, such as predation, that increase and decrease in intensity in proportion to population density. May lead to population stability.

density-independent   Mortality to a population that occurs regardless of the population density. Tends to be destabilizing to populations.

Department of Transportation regulations (DOT regs)   Regulations intended to reduce the risk of spills, fires, and poisonous fumes by specifying the kinds of containers and methods of packing to be used in transporting hazardous materials.

dependency ratio   In a human population, the ratio of the non-working-age population (under 15 and over 65) to the working-age population.

deregulation   In the electric power industry, a policy that requires utility companies to divest themselves of power-generating facilities and allows the market to determine electrical costs.

desalinization   A process that purifies seawater into high-quality drinking water via distillation or microfiltration.

desertification   The formation and expansion of degraded areas of soil and vegetation cover in arid, semiarid, and seasonally dry areas, caused by climatic variations and human activities.

desertified   Said of land whose productivity has been significantly reduced (25% or more) because of human mismanagement. Erosion is the most common cause of desertified land.

desert pavement   A covering of stones and coarse sand protecting desert soils from further wind erosion. The covering results from the differential erosion of finer material.

detritus   The dead organic matter, such as fallen leaves, twigs, and other plant and animal wastes, that exists in any ecosystem.

detritus feeders   Organisms such as termites, fungi, and bacteria that obtain their nutrients and energy mainly by feeding on dead organic matter.

deuterium (^{2}H)   A stable, naturally occurring isotope of hydrogen that contains one neutron in addition to the single proton normally in the nucleus.

developed countries   The high-income industrialized countries—the United States, Canada, western European nations, Japan, Korea, Australia, and New Zealand, as well as many middle-income countries such as those in Latin America, China, eastern Europe and many Arab states.

developing countries   Countries in which the gross domestic product is less than $936 per capita. The category includes nations of Africa, India, other countries of southern Asia, and some former Soviet republics.

Diclofenac   An anti-inflammatory drug often administered to animals but toxic to birds. Implicated in the die-off of vultures in the Indian subcontinent.

dioxin   A synthetic organic chemical of the chlorinated hydrocarbon class. Dioxin is one of the most toxic compounds known to humans and a widespread environmental pollutant.

discharge permit   (Technically called an **NPDES permit**.) A permit that allows a company to legally discharge certain amounts or levels of pollutants into air or water.

discount rate   In economics, a rate applied to some future benefit or cost in order to calculate its present real value.

dish-engine system   A solar power system consisting of a set of parabolic concentrator dishes that focus sunlight onto a receiver, where the hot fluid is transferred to an engine that generates electricity.

disinfection   The killing (as opposed to removal) of microorganisms in water or other media where they might otherwise pose a health threat. For example, chlorine is commonly used to disinfect water supplies.

dissolved oxygen (DO)   Oxygen gas molecules (O_2) dissolved in water. Fish and other aquatic organisms depend on dissolved oxygen for respiration. Therefore, the concentration of dissolved oxygen is a measure of water quality.

distillation   A process of purifying water or some other liquid by boiling the liquid and recondensing the vapor. Contaminants remain behind in the boiler.

distributive justice   The ethical process of making certain that all parties receive equal rights, regardless of their economic status.

disturbance   A natural or human-induced event or process (e.g., a forest fire) that interrupts ecological succession and creates new conditions on a site.

DNA (deoxyribonucleic acid)   The natural organic macromolecule that carries the genetic or hereditary information for virtually all organisms.

Doha Development Round   Negotiations for trade begun in 2001 with the objective of lowering trade barriers around the world.

donor fatigue   The tendency of donor organizations or countries to withhold their aid because of long-term misuse of aid funds in the recipient countries.

dose   The mathematical product of the concentration of a hazardous material and the length of exposure to it. The effects of any given material or radiation correspond to the dose received.

dose-response assessment   Part of risk assessment, establishing the number of cancers that might develop due to exposure to different doses of a chemical.

doubling time   The time it will take a population to double in size, assuming that the current rate of growth continues.

drip irrigation   A method of supplying irrigation water through tubes that literally drip water onto the soil at the base of each plant.

drought   A local or regional lack of precipitation such that the ability to raise crops and water animals is seriously impaired.

drylands   Ecosystems characterized by low precipitation (25 to 75 cm per year) and often subject to droughts; some 41% of Earth's land is drylands.

earth-sheltered housing   Housing that makes use of the insulating properties of earth materials and that is also oriented for passive solar heating; often involves earth berms built up against the building walls.

easement   In reference to land protection, an arrangement whereby a landowner gives up development rights into the future but retains ownership of the land.

ecdysone   The hormone that promotes molting in insects.

ecological economist   An economist who thoroughly integrates ecological and economic concerns; part of a new breed of economist who disagrees with classical economic theory.

ecological footprint A concept for measuring the demand placed on Earth's resources by individuals from different parts of the world, involving calculations of the natural area required to satisfy human needs.

ecological pest control Natural control of pest populations through understanding various ecological factors and, so far as possible, utilizing those factors, as opposed to using synthetic chemicals.

ecologists Scientists who study ecology.

ecology The study of all processes influencing the distribution and abundance of organisms and the interactions between living things and their environment.

economic exclusion The cutting off of access of certain ethnic or economic groups to jobs, a good education, and other opportunities and thus preventing them from entering the economic mainstream of society—a condition that prevails in poor areas of cities.

economics The social science that deals with the production, distribution, and consumption of goods and services.

economic threshold The level of pest damage that, to be reduced further, would require an application of pesticides that is more costly than the economic damage caused by the pests.

ecosystem An interactive complex of communities and the abiotic environment affecting them within a particular area. Ecosystems have characteristic forms, such as deserts, grasslands, tundra, deciduous forests, and tropical rain forests.

ecosystem capital The sum of goods and services provided by natural and managed ecosystems, both free of charge and essential to human life and well-being.

ecosystem management The management paradigm, adopted by all federal agencies managing public lands, that involves a long-term stewardship approach to maintaining the lands in their natural state.

ecotone A transitional region between two adjacent ecosystems that contains some of the species and characteristics of each one and also certain species of its own.

ecotourism The enterprises involved in promoting tourism of unusual or interesting ecological sites.

edges Breaks between habitats that may expose sensitive species to predators.

electrical power The amount of work done by an electric current over a given time.

electric generator A device that converts mechanical energy into electrical energy by way of wire coils that are rotated in a magnetic field.

electrolysis The use of electrical energy to split water molecules into their constituent hydrogen and oxygen atoms. Hydrogen gas and oxygen gas result.

electrons Fundamental atomic particles that have a negative electrical charge and a very small mass, approximately 9.1×10^{-31} kg. Electrons surround the nuclei of atoms and thus balance the positive charge of protons in the nucleus. A flow of electrons in a wire is identical to an electrical current.

element A substance that is made up of one and only one distinct kind of atom. (Contrast **compound**.)

El Niño A major climatic phenomenon characterized by the movement of unusually warm surface water into the eastern equatorial Pacific Ocean. El Niño results in extensive disruption of weather around the world.

embrittlement Becoming brittle. Pertains especially to the reactor vessel of nuclear power plants gradually becoming prone to breakage or snapping as a result of continuous bombardment by radiation.

Emergency Planning and Community Right-to-know Act (EPCRA) See **SARA**.

emergent vegetation Aquatic plants whose lower parts are under water but whose upper parts emerge from the water.

emissions intensity The ratio of greenhouse gas emissions to economic output, reduction of which is one of the (failed) policies put in place by the George W. Bush administration to address global climate change.

endangered species A species whose total population is declining to relatively low levels such that, if the trend continues, the species will likely become extinct.

Endangered Species Act The federal legislation that mandates the protection of species and their habitats that are determined to be in danger of extinction.

endemic species A species that is found in only one habitat and that is therefore vulnerable.

endocrine disrupters Any of a class of organic compounds, often pesticides, that are suspected of having the capacity to interfere with hormonal activities in animals.

energy The capacity to do work. Common forms of energy are light, heat, electricity, motion, and the chemical bond energy inherent in compounds such as sugar, gasoline, and other fuels.

energy carrier Something (e.g., electricity) that transfers energy from a primary energy source to its point of use.

energy flow The movement of energy through ecosystems, starting with the capture of solar energy by primary producers and ending with the loss of heat energy.

Energy Independence and Security Act of 2007 Legislation establishing energy policy by emphasizing demand-side policies such as conserving energy and using renewable energy sources.

Energy Policy Act of 2005 Legislation establishing U.S. energy policy for years to come, encouraging greater use of fossil fuels and nuclear power, and providing some support for energy conservation and renewable energy.

Energy Star Program The EPA's program focusing on energy conservation in buildings and in appliances. An Energy Star tag signals that the device or building incorporates energy-saving technology.

enrichment With reference to nuclear power, the separation and concentration of uranium-235 so that, in suitable quantities, it will sustain a chain reaction.

entomologist A scientist who studies insects and their life cycles, physiology, behavior, and so on.

entropy A degree of disorder; increasing entropy means increasing disorder.

environment The combination of all things and factors external to the individual or population of organisms in question.

environmental accounting A process of keeping national accounts of economic activity that includes gains and losses of environmental assets.

environmental impact The effect on the natural environment caused by human actions. Includes indirect effects, for example, through pollution, as well as direct effects such as cutting down trees.

environmental impact statement A study of the probable environmental impacts of a development project. The National Environmental Policy Act of 1968 (NEPA) requires such studies prior to proceeding with any project receiving federal funding.

environmentalism An attitude or movement involving concern about pollution, resource depletion, population pressures, loss of biodiversity, and other environmental issues. Usually also implies action to address those concerns.

environmentalist Any person who is concerned about the degradation of the natural world and is willing to act on that concern.

environmental justice   The fair treatment and meaningful involvement of all people, regardless of race, color, national origin, or income, with respect to the development, implementation, and enforcement of environmental laws and regulations.

environmental movement   The upwelling of public awareness and citizen action regarding environmental issues that began during the 1960s.

environmental public policy   All of a society's laws and regulations that deal with that society's interactions with the environment. Its purpose is to promote the common good.

environmental racism   Discrimination against people of color whereby hazardous industries are sited in nonwhite neighborhoods and towns.

environmental resistance   The totality of factors such as adverse weather conditions, shortages of food or water, predators, and diseases that tend to cut back populations and keep them from growing or spreading. (Contrast **biotic potential**.)

Environmental Revolution   In the view of some, a coming change in the adaptation of humans to the rising deterioration of the environment. The Environmental Revolution should bring about sustainable interactions with the environment.

environmental science   The multidisciplinary branch of science concerned with environmental issues.

environmental tobacco smoke (ETS)   Secondhand smoke experienced by those breathing air contaminated with tobacco smoke from others; classified as a known human carcinogen.

enzyme   A protein that promotes the synthesis or breaking of chemical bonds.

EPA (U.S. Environmental Protection Agency)   The federal agency responsible for the control of all forms of pollution and other kinds of environmental degradation.

epidemiologic transition   In human populations, the pattern of change in mortality from high death rates to low death rates; contributes to the demographic transition.

epidemiology, epidemiological study   The study of the causes of disease (e.g., lung cancer) through an examination and comparison of large populations of people living in different locations or following different lifestyles or habits (e.g., smoking versus nonsmoking).

epiphytes   Air plants that are not parasitic but that "perch" on the branches of trees, where they can get adequate light.

equilibrium climate sensitivity   The sustained change in mean global temperature that occurs if atmospheric carbon dioxide is held at double preindustrial values, or 550 parts per million.

equity principle   An ethical principle claiming that those who produce the most pollution should compensate those who produce less, especially since the less fortunate often suffer the most from, say, global climate change.

erosion   The process of soil particles' being carried away by wind or water. Erosion moves the smaller particles first and hence degrades the soil to a coarser, sandier, stonier texture.

estimated reserves   See **reserves**.

estuary   A bay or river system open to the ocean at one end and receiving fresh water at the other. In the estuary, fresh water and saltwater mix, producing brackish water.

ethnobotany   The study of the relationships between plants and people.

euphotic zone   In aquatic systems, the layer or depth of water through which an adequate amount of light penetrates to support photosynthesis.

eutrophic   Characterized by nutrient-rich water supporting an abundant growth of algae or other aquatic plants at the surface.

eutrophication   The process of becoming eutrophic.

evaporation   The process whereby molecules leave the liquid state and enter the vapor or gaseous state as, for example, when water evaporates to form water vapor. Opposite: **condensation**.

evapotranspiration   The combination of evaporation and transpiration that restores water to the atmosphere.

evolution   The theory, attributed to Charles Darwin and Alfred Russell Wallace, that all species now on Earth descended from ancestral species through a process of gradual change brought about by natural selection.

e-waste   Any of a broad array of electronic devices that are discarded and often are hazardous to handle and difficult to recycle.

exotic species   A species introduced into a geographical area to which it is not native; also, species that are rare and subject to illegal trafficking.

experimental group   In an experiment, the group that receives the experimental treatment, in contrast to the control group, used for comparison, which does not receive the treatment. Synonym: **test group**.

exponential increase   The growth produced when a base population increases by a given percentage (as opposed to a given amount) each year. An exponential increase is characterized by doubling again and again, each doubling occurring in the same period of time. It produces a J-shaped curve.

exposure assessment   A part of risk assessment, identifying human groups exposed to a chemical and calculating the doses and length of time of their exposure.

extended product responsibility   Voluntary or government programs that assign some responsibility for reducing the environmental impact of a product at various stages of its life cycle—especially at disposal.

external cost, externality   Any effect of a business process not included in the usual calculations of profit and loss. Pollution of air or water is an example of a *negative* externality—one that imposes a cost on society that is not paid for by the business itself.

extinction   The death of all individuals of a particular species. When this occurs, all the genes of that particular line are lost forever.

extreme poverty   Poverty that is defined as living on less than $1.25 a day. In 2009, some 1.2 billion were in this condition.

exurban migration   The pronounced trend since World War II of relocating homes and businesses from the central city and older suburbs to more outlying suburbs.

exurbs   New developments beyond the traditional suburbs, from which most residents still commute to the associated city for work.

facilitation   A mechanism of succession where earlier species create conditions that are more favorable to newer occupants, and are replaced by them.

family planning   Making contraceptives and other health and reproductive services available to couples in order to enable them to achieve a desired family size.

famine   A severe shortage of food accompanied by a significant increase in the local or regional death rate.

Farm Security and Rural Investment Act of 2002   The farm bill successor to the Federal Agricultural Improvement and Reform Act of 1996 that continued to maintain subsidies to farmers.

fecal coliform test   A test for the presence of *Escherichia coli*, the bacterium that normally inhabits the gut of humans and other mammals. A positive test result indicates sewage contamination and the potential presence of disease-causing microorganisms.

Federal Agricultural Improvement and Reform (FAIR) Act of 1996   Major legislation removing many subsidies and controls from farming.

Federal Food and Drug Administration (FDA) The federal agency with jurisdiction over foods, drugs, and all products that come in contact with the skin or are ingested.

fermentation A form of respiration carried on by yeast cells and many bacteria in the absence of oxygen. Fermentation involves a partial breakdown of glucose (sugar) that yields energy for the yeast and the release of alcohol as a by-product.

fertility rate See **total fertility rate**.

fertility transition The pattern of change in birthrates in a human society from high rates to low; a major component of the demographic transition.

fertilizer Material applied to plants or soil to supply plant nutrients, most commonly nitrogen, phosphorus, and potassium but possibly others. *Organic* fertilizer is natural organic material, such as manure, that releases nutrients as it breaks down. *Inorganic* fertilizer, also called chemical fertilizer, is a mixture of one or more necessary nutrients in inorganic chemical form.

field scouts Persons trained to survey crop fields and determine whether applications of pesticides or other pest-management procedures are actually necessary to avert significant economic loss.

FIFRA (Federal Insecticide, Fungicide, and Rodenticide Act) The key U.S. legislation passed to control pesticides.

fire climax ecosystems Ecosystems that depend on the recurrence of fire to maintain the existing balance.

first-generation pesticides Toxic inorganic chemicals that were the first used to control insects, plant diseases, and other pests. These pesticides included mostly compounds of arsenic and cyanide and various heavy metals, such as mercury and copper.

First Law of Thermodynamics The empirical observation, confirmed innumerable times, that energy is never created or destroyed but may be converted from one form to another (e.g., electricity to light). Also called the **Law of Conservation of Energy**. (See also **Second Law of Thermodynamics**.)

fishery Fish species being exploited, or a limited marine area containing commercially valuable fish.

fish ladder A stepwise series of pools on the side of a dam where the water flows in small falls that fish can negotiate.

fission The splitting of a large atom into two atoms of lighter elements. When large atoms such as uranium or plutonium fission, tremendous amounts of energy are released.

fission products Any and all atoms and subatomic particles resulting from splitting atoms in nuclear reactors or nuclear explosions. Practically all such products are highly radioactive.

fitness The state of an organism whereby it is adapted to survive and reproduce in the environment it inhabits.

flat-plate collector A solar collector that consists of a stationary, flat, black surface oriented perpendicular to the average angle of the sun. Heat absorbed by the surface is removed and transported by air or water (or some other liquid) flowing over or through the surface.

flood irrigation A technique of irrigation in which water is diverted from rivers by means of canals and is flooded through furrows in fields.

food aid Food of various forms that is donated or sold below cost to needy people for humanitarian reasons.

Food and Agriculture Organization of the United Nations (FAO) The UN agency responsible for coordinating international efforts to defeat hunger; also gathers and disseminates information about forests, fisheries, and farming.

food chain The transfer of energy and material through a series of organisms as each one is fed upon by the next.

Food Conservation and Energy Act of 2008 The latest farm bill, maintaining the existing high levels of support and subsidies to farms.

food guide pyramid A graphic presentation of basic food needs arranged in a pyramid to indicate the relative proportions of each type of food needed for good nutrition.

Food Quality Protection Act (FQPA) Passed in 1996, this act amended existing laws governing pesticide use and residues on foods, including the **Delaney clause**.

food security For families, the ability to meet the food needs of everyone in the family, providing freedom from hunger and malnutrition.

Forest Stewardship Council An alliance of organizations directed toward the certification of sustainable wood products.

food web The combination of all the feeding relationships that exist in an ecosystem.

fossil fuels Energy sources—mainly crude oil, coal, and natural gas—that are derived from the prehistoric photosynthetic production of organic matter on Earth.

FQPA (Food Quality Protection Act of 1996) Legislation that removed the Delancy clause and replaced many provisions of FIFRA.

fragmentation The division of a landscape into patches of habitat by road construction, agricultural lands, or residential areas.

Framework Convention on Climate Change A result of the 1992 Earth Summit, this international treaty was a start in negotiating agreements on steps to prevent future catastrophic climate change.

Framework Convention on Tobacco Control Adopted in 2003 and put into effect in 2005, this is a global treaty that aims to reduce the spread of smoking with the use of various strategies.

frankenfood A derogatory term used by opponents to genomics to apply to any food plant that has been developed with the use of genomics.

free-market economy In its purest form, an economy in which the market itself determines what and how goods will be exchanged. The system is wholly in private hands.

fresh water Water that has a salt content of less than 0.1% (1,000 parts per million).

front The boundary where different air masses meet.

fuel assembly In a nuclear reactor, the assembly of many rods containing the nuclear fuel, usually uranium, positioned close together. The chain reaction generated in the fuel assembly is controlled by rods of neutron-absorbing material between the fuel rods.

fuel cell A device that produces electrical energy by combining hydrogen and oxygen, used to power vehicles and other electrical machines.

fuel elements (fuel rods) The pellets of uranium or other fissionable material that are placed in tubes, or fuel rods, and that, together with the control rods, form the core of the nuclear reactor.

fuel load Dead trees and dense understories in many forestlands that tend to burn readily if ignited in a forest fire.

fuelwood The use of wood for cooking and heating; the most common energy source for people in developing countries.

fungus, pl. fungi Any of numerous species of molds, mushrooms, bracket fungi, and other forms of nonphotosynthetic plants. Fungi derive energy and nutrients by consuming other organic material. Along with bacteria, they form the decomposer component of ecosystems.

fusion The joining together of two atoms to form a single atom of a heavier element. When light atoms such as hydrogen are fused, tremendous amounts of energy are released.

gasohol A blend of 90% gasoline and 10% ethanol that can be substituted for straight gasoline. Gasohol serves to stretch gasoline supplies and enable gasoline to burn more cleanly.

gene A segment of DNA that codes for one protein, which in turn determines a particular physical, physiological, or behavioral trait.

gene pool The sum total of all the genes that exist among all the individuals of a species.

gene revolution The widespread development of genetically modified crops.

genetically modified organism (GMO) Any organism that has received genes from a different species through the techniques of genetic engineering; also called a **transgenic organism**.

genetic bank The concept that natural ecosystems with all their species serve as a repository of genes that may be drawn upon to improve domestic plants and animals and to develop new medicines, among other uses.

genetic control The selective breeding of a desired plant or animal to make it resistant to attack by pests. Also, attempting to introduce harmful genes—for example, those that cause sterility—into the pest populations.

genetic engineering The artificial transfer of specific genes from one organism to another.

genomics The science of gene exchange using the techniques of genetic engineering to produce transgenic organisms.

Genuine Progress Indicator (GPI) An alternative measure of economic progress, the GPI calculates positive and negative economic activities to reach a more realistic view of sustainable economic activity. It is an alternative to the **gross domestic product**.

geoengineering Large-scale schemes, proposed to reduce global climate change, involving major intrusions into climatic or ecological systems.

geothermal Refers to the naturally hot interior of Earth, where heat is maintained by naturally occurring nuclear reactions.

geothermal energy Useful energy derived from water heated by the naturally hot interior of Earth.

geothermal heat pump An energy system whereby heat can be obtained or deposited via pipes buried in the ground and an exchange process that allows heated or cooled air to be circulated via ductwork.

GLASOD A map produced by the Global Assessment of Soil Degradation, prepared in the late 20th century and often cited but based on very suspect data.

Global Assessment of Land Degradation and Improvement (GLADA) A new assessment of land degradation based on satellite remote-sensing techniques.

Global Change Research Program (GCRP) See **Climate Change Science Program**.

global climate change The cumulative effects of rising levels of greenhouse gases on Earth's climate. These effects include global warming, weather changes, and a rising sea level.

Global Climate Change Initiative The Bush administration's voluntary response to climate change, involving a reduction in **emissions intensity**.

Global Forest Resources Assessment (FRA) A periodic assessment of forest resources conducted by the UN Food and Agricultural Organization (FAO) and made the basis of regular FAO *State of the World's Forest* reports.

Global Gag Rule The U.S. policy initiated by President Reagan and continued by other Republican presidents whereby U.S. government aid is denied to any agency that provides abortions or abortion counseling.

globalization The accelerating interconnectedness of human activities, ideas, and cultures, especially evident in economic and information exchange.

glucose A simple sugar, the major product of photosynthesis. Glucose serves as the basic building block for cellulose and starches and as the major "fuel" for the release of energy through cell respiration in both plants and animals.

glyphosate (Roundup®) A herbicide that kills most varieties of plants, used to control weeds with crops that have been bioengineered with resistance to the glyphosate.

golden rice New strains of rice produced by genetic engineering that contain carotene (which the body uses to synthesize vitamin A) and iron, intended to compensate for nutritional deficiencies.

goods Products, such as wood and food, that are extracted from natural ecosystems to satisfy human needs.

Grameen Bank A banking network serving the poor with microloans, created by Muhammad Yunus of Bangladesh.

grassroots movement An environmental movement that begins at the level of the populace, often in response to a perceived problem not being addressed by public policy.

gravitational water Water that is not held by capillary action in soil but that percolates downward by the force of gravity.

Gray A unit of measure of radiation used for low-level sources of radiation.

graying The increasing average age in populations in developed countries and in many developing countries that is occurring because of decreasing birthrates and increasing longevity.

gray water Wastewater, as from sinks and tubs, that does not contain human excrement. Such water can be reused without purification for some purposes.

green fee An added cost for a particular service or product that has the effect of internalizing the external costs and that reflects the real environmental cost of the service or product.

greenhouse effect An increase in the atmospheric temperature caused by increasing amounts of carbon dioxide and certain other gases that absorb and trap heat, which normally radiates away from Earth.

greenhouse gases Gases in the atmosphere that absorb infrared energy and contribute to the air temperature. These gases are like a heat blanket and are important in insulating Earth's surface. Among the greenhouse gases are carbon dioxide, water vapor, methane, nitrous oxide, chlorofluorocarbons, and other halocarbons.

green products Manufactured products that are more environmentally benign than their traditional counterparts.

Green Revolution The development and introduction of new varieties of (mainly) wheat and rice that increased yields per acre dramatically in many countries since the 1960s.

green water In the hydrologic cycle, water that is evaporated or transpired and returned as water vapor to the atmosphere; water in vapor form.

grit chamber A part of preliminary treatment in wastewater-treatment plants; a swimming pool–like tank in which the velocity of the water is slowed enough to let sand and other gritty material settle.

gross domestic product The total value of all goods and services exchanged in a year within a country.

groundwater Water that has accumulated in the ground, completely filling and saturating all pores and spaces in rock or soil. Free to move more or less readily, it is the reservoir for springs and wells and is replenished by infiltration of surface water.

groundwater remediation The repurification of contaminated groundwater by any of a number of techniques.

gully erosion Soil erosion produced by running water and resulting in the formation of gullies.

habitat The specific environment (woods, desert, swamp) in which an organism lives.

habitat conservation plan Under the Endangered Species Act, a plan that may be drafted by the U.S. Fish and Wildlife Service in working with landowners to help mitigate conflicts resulting from an application of the act.

Hadley cell A system of vertical and horizontal air circulation predominating in tropical and subtropical regions and creating major weather patterns.

half-life The length of time it takes for half of an unstable isotope to decay. The length of time is the same regardless of the starting amount. Also refers to the amount of time it takes compounds to break down in the environment.

halogenated hydrocarbon A synthetic organic compound containing one or more atoms of the halogen group, which includes chlorine, fluorine, and bromine.

hard water Water that contains relatively large amounts of calcium or certain other minerals that cause soap to precipitate. (Contrast **soft water**.)

hazard Anything that can cause (1) injury, disease, or death to humans; (2) damage to property; or (3) degradation of the environment. *Cultural* hazards include factors that are often a matter of choice, such as smoking or sunbathing. *Biological* hazards are pathogens and parasites that infect humans. *Physical* hazards are natural disasters like earthquakes and tornadoes. *Chemical* hazards refer to the chemicals in use in different technologies and household products.

hazard assessment The process of examining evidence linking a particular hazard to its harmful effects.

hazardous material (HAZMAT) Any material having one or more of the following attributes: ignitability, corrosivity, reactivity, and toxicity.

health According to the World Health Organization, "a state of complete mental, physical and social well-being and not merely the absence of disease or infirmity."

heavy metal Any of the high-atomic-weight metals, such as lead, mercury, cadmium, and zinc. All may be serious pollutants in water or soil because they are toxic in relatively low concentrations and they tend to bioaccumulate.

heavy water Deuterium oxide (D_2O), a form of water incorporating a heavy isotope of hydrogen.

herbicide A chemical used to kill or inhibit the growth of undesired plants.

herbivore An organism such as a rabbit or deer that feeds primarily on green plants or plant products such as seeds or nuts. Such an organism is said to be herbivorous. (Synonym: **primary consumer**.)

herbivory The feeding on plants that occurs in an ecosystem. The total feeding of all plant-eating organisms.

heterotroph Any organism that consumes organic matter as a source of energy. Such an organism is said to be heterotrophic.

Highway Trust Fund The monies collected from the gasoline tax and designated for the construction of new highways.

HIPPO An acronym for the major threats to biodiversity: Habitat destruction, Invasive species, Pollution, Population, and Overexploitation.

hormones Natural chemical substances that control the development, physiology, and behavior of an organism. Hormones are produced internally and affect only the individual organism. Hormones are coming into use in pest control. (See also **pheromones**.)

host In feeding relationships, particularly parasitism, refers to the organism that is being fed upon (i.e., the organism that is supporting the feeder).

host-parasite relationship The relationship between a parasite and the organism upon which it feeds.

host-specific An attribute whereby insects, fungal diseases, and other parasites are unable to attack species other than their particular host.

household hazardous wastes An important component of the hazardous-waste problem in the United States.

Hubbert Peak The concept proposed by M. King Hubbert that peak oil production will follow a bell-shaped curve, when half of available oil would have been withdrawn.

human capital See **intangible capital**.

Human Poverty Index, Human Development Index Based on life expectancy, literacy, and living standards, these indices are used by the United Nations Development Program to measure progress in alleviating poverty.

human system The entire system that humans have created for their own support, consisting of agriculture, industry, transportation, communication networks, and so on.

humidity The amount of water vapor in the air. (See also **relative humidity**.)

humus A dark brown or black, soft, spongy residue of organic matter that remains after the bulk of dead leaves, wood, or other organic matter has decomposed. Humus does oxidize but relatively slowly. It is extremely valuable in enhancing the physical and chemical properties of soil.

hunger A condition wherein the basic food required for meeting nutritional and energy needs is lacking and the individual is unable to lead a normal, healthy life.

hunter-gatherers Humans surviving by hunting wild game and gathering seeds, nuts, berries, and other edible things from the natural environment.

hybrid A plant or animal resulting from a cross between two closely related species that do not normally cross.

hybrid electric vehicle An automobile combining a gasoline motor and a battery-powered electric motor that produces less pollution and gets higher gasoline mileages than do conventional gasoline-powered vehicles.

hybridization Cross-mating between two more or less closely related species.

hydrocarbons *Chemistry*: Natural or synthetic organic substances that are composed mainly of carbon and hydrogen. Crude oil, fuels from crude oil, coal, animal fats, and vegetable oils are examples. *Pollution*: A wide variety of relatively small carbon-hydrogen molecules resulting from incomplete burning of fuel that are emitted into the atmosphere. (See **volatile organic compounds**.)

hydroelectric dam A dam and an associated reservoir used to produce electrical power by letting the high-pressure water behind the dam flow through and drive a turbogenerator.

hydroelectric power Electric power that is produced from hydroelectric dams or, in some cases, natural waterfalls.

hydrogen bonding A weak attractive force that occurs between a hydrogen atom of one molecule and, usually, an oxygen atom of another molecule. Hydrogen bonding is responsible for holding water molecules together to produce the liquid and solid states.

hydrogen economy A future economy where renewable energy via hydrogen as a major energy carrier prevails.

hydrogen ions Hydrogen atoms that have lost their electrons (chemical symbol, H^+).

hydrologic cycle The movement of water from points of evaporation, through the atmosphere, through precipitation, and through or over the ground, returning to points of evaporation.

hydrosphere The water on Earth, in all of its liquid and solid compartments: oceans, rivers, lakes, ice, and groundwater.

hydroxyl radical The hydroxyl group (OH), missing the electron. The hydroxyl radical is a natural cleansing agent of the atmosphere. It is highly reactive, readily oxidizes many pollutants upon contact, and thus contributes to their removal from the air.

hypothesis A tentative guess concerning the cause of an observed phenomenon that is then subjected to experiment to test its logical or empirical consequences.

hypoxia See dead zone.

ICPD (International Conference on Population and Development) A UN-sponsored conference held in 1994 in Cairo, Egypt, that produced a program of action that was drafted and later adopted by the United Nations.

immigration The movement of people from countries of their birth to other countries, usually with the intent to reside there permanently.

indicator organism An organism, the presence or absence of which indicates certain conditions. For example, the presence of *Escherichia coli* indicates that water is contaminated with fecal wastes and that pathogens may be present.

individual quota system A system of fishery management wherein a quota is set and individual fishers are given or sold the right to harvest some proportion of the quota. Also called **catch shares.**

industrialized agriculture The use of fertilizer, irrigation, pesticides, and energy from fossil fuels to produce large quantities of crops and large numbers of livestock with minimal labor for domestic and foreign sale.

Industrial Revolution During the 19th century, the development of manufacturing processes using fossil fuels and based on applications of scientific knowledge.

industrial smog The grayish mixture of moisture, soot, and sulfurous compounds that occurs in local areas in which industries are concentrated and coal is the primary energy source.

infant mortality The number of babies that die before one year, per 1,000 babies born.

infiltration The process in which water soaks into soil as opposed to running off the surface of the soil.

infiltration-runoff ratio The ratio of the amount of water soaking into the soil to that running off the surface. The ratio is obtained by dividing the first amount by the second.

infrared radiation Radiation of somewhat longer wavelengths than red light, which comprises the longest wavelengths of the visible spectrum. Such radiation manifests itself as heat.

infrastructure The sewer and water systems, roadways, bridges, and other facilities that underlie the functioning of a city or a society.

inherently safe reactor In theory, a nuclear reactor that is designed in such a way that any accident would be automatically corrected with no radioactivity released.

inorganic compounds/molecules *Classical definition*: All things such as air, water, minerals, and metals that are neither living organisms nor products uniquely produced by living things. *Chemical definition*: All chemical compounds or molecules that do not contain carbon atoms as an integral part of their molecular structure. (Contrast **organic compounds/molecules.**)

inorganic fertilizer See **fertilizer.**

insecticide Any chemical used to kill insects.

instrumental value The value that living organisms or species have by virtue of their benefit to people; the degree to which they benefit humans. (Contrast **intrinsic value.**)

insurance spraying The spraying of pesticides when it is not really needed, in the belief that it will ensure against loss due to pests.

intangible capital One component of the wealth of nations. Comprises the *human capital*, or the population and its attributes; the *social capital*, or the social and political environment that people have created in a society; and the *knowledge assets*, or the human fund of knowledge.

integrated gasification combined cycle (IGCC) plant A coal-fired technology where coal is mixed with water and oxygen and heated under pressure to produce a synthetic gas.

integrated pest management (IPM) A program consisting of two or more methods of pest control carefully integrated and designed to avoid economic loss from pests. The objective of IPM is to minimize the use of environmentally hazardous synthetic chemicals.

integrated waste management The approach to municipal solid waste that provides for several options for dealing with wastes, including recycling, composting, waste reduction, and landfilling as well as incineration where unavoidable.

intergenerational equity Meeting the needs of the present without jeopardizing the ability of future generations to meet their needs in an equitable way.

Intergovernmental Panel on Climate Change (IPCC) The UN-sponsored organization charged with continually assessing the science of global climate change, its potential impacts, and the means of responding to the threat.

International Whaling Commission (IWC) The international body that regulates the harvesting of whales; the IWC placed a ban on all whaling in 1986.

interspecific competition Competition for resources between members of two or more species.

intragenerational equity The attempt to achieve an equitable distribution of assets within a given generation or across existing people-groups.

intraspecific competition Competition for resources between members of the same species.

intrinsic value The value that living organisms or species have in their own right; in other words, organisms and species do not have to be useful to have value. (Contrast **instrumental value.**)

intrusion The alteration of habitats by placing lights, towers, or glass surfaces that interfere with the normal behavior of species and often lead to their death.

invasive species An introduced species that spreads out and often has harmful ecological effects on other species or ecosystems.

inversion See **temperature inversion.**

inverter In a photovoltaic (PV) system, the device that acts as an interface between the solar PV modules and the electric grid or batteries.

ion An atom (or a group of atoms) that has lost or gained one or more electrons and, consequently, has acquired a positive or negative charge. Ions are designated by "+" or "−" superscripts following the chemical symbol for the element(s) involved.

ion-exchange capacity See **nutrient-holding capacity.**

ionic bond The bond formed by the attraction between a positive and a negative ion.

ionizing radiation Radiation that displaces ions in tissues as it penetrates them, leaving behind charged particles (ions).

IPAT formula A conceptual formula relating environmental impact (I) to population (P), affluence (A), and technology (T).

irrigation Any method of artificially adding water to crops.

isotope A form of an element in which the atoms have more (or less) than the usual number of neutrons. Isotopes of a given element have identical chemical properties but differ in mass (weight) as a result of the superfluity (or deficiency) of neutrons. Many isotopes are unstable and radioactive. (See **radioactive decay, radioactive emissions,** and **radioactive materials.**)

ISTEA (Intermodal Surface Transportation Efficiency Act) Legislation that provides for the funding of alternative transportation (e.g., mass transit or bicycle paths) using money from the Highway Trust Fund. Succeeded by the Transportation Equity Act for the Twenty-first Century (TEA–21) and then the Safe, Accountable, Flexible, Efficient Transportation Equity Act (SAFETEA).

IUCN (International Union for Conservations of Nature) Formerly the **World Conservation Union.** An international organization that maintains a Red List of threatened species worldwide.

jet stream   A river of air, generated by Earth's rotation and air-pressure gradients, high in the troposphere that flows eastward at high speeds but meanders considerably.

junk science   Information presented as valid science but unsupported by peer-reviewed research. Often, politically motivated and biased results are selected to promote a particular point of view.

juvenile hormone   The insect hormone that, at sufficient levels, preserves the larval state. Pupation requires diminished levels of juvenile hormone; hence, artificial applications of the hormone may block pupal development.

keystone species   A species whose role is essential for the survival of many other species in an ecosystem.

kinetic energy   The energy inherent in motion or movement, including molecular movement (heat) and the movement of waves (hence, radiation and therefore light).

knowledge assets   See **intangible capital**.

K-strategist   A reproductive strategy for a species whereby there is a low reproductive rate but good survival of young due to care and protection by adults. Also called **equilibrium species**.

Kyoto Protocol   An international agreement among the developed nations to curb greenhouse gas emissions. The Kyoto Protocol was forged in December 1997 and became binding in 2004.

Lacey Act   Passed in 1900, the first national act that gave protection to wildlife by forbidding interstate commerce in illegally killed animals.

land degradation   The processes where land is rendered less capable of supporting plant growth, often through soil erosion.

land ethic   First described by Aldo Leopold, this is an ethic that promotes protection of natural areas based on their integrity, stability, and beauty.

landfill   A site where municipal, industrial, or chemical wastes are disposed of by burying them in the ground or placing them on the ground and covering them with earth.

landscape   A group of interacting ecosystems occupying adjacent geographical areas.

land subsidence   The gradual sinking of land. The condition may result from the removal of groundwater or oil, which is frequently instrumental in supporting the overlying rock and soil.

land trust   Land that is purchased and held by various private organizations specifically for the purpose of protecting the region's natural environment and the biota that inhabit it.

La Niña   A complete reversal of El Niño conditions in the tropical Pacific Ocean. Easterly trade winds are intense, and global weather patterns are often the reverse of those brought on by an El Niño event.

larva, pl. larvae   A free-living immature form that occurs in the life cycle of many organisms and that is structurally distinct from the adult. For example, caterpillars are the larval stage of moths and butterflies.

Law of Conservation of Energy   See **First Law of Thermodynamics**.

Law of Conservation of Matter   The law stating that, in chemical reactions, atoms are neither created nor changed nor destroyed; they are only rearranged.

law of limiting factors   The law stating that a system may be limited by the absence or minimum amount (in terms of that needed) of any required factor. Also known as **Liebig's law of minimums**. (See **limiting factor**.)

leachate   The mixture of water and materials that are leaching.

leaching   The process in which materials in or on the soil gradually dissolve and are carried by water seeping through the soil. Leaching may eventually remove valuable nutrients from the soil, or it may carry buried wastes into groundwater, thereby contaminating it.

legumes   The group of pod-bearing land plants that is virtually alone in its ability to fix nitrogen; legumes include such common plants as peas, beans, clovers, alfalfa, and locust trees but no major cereal grains. (See **nitrogen fixation**.)

Liebig's law of minimums   See **law of limiting factors**.

life expectancy   The number of years a newborn can expect to live under mortality rates current for its country of residence.

light-water reactor (LWR)   A reactor that uses extra-pure ordinary water as a moderator. Nuclear plants in the United States are all LWRs.

limiting factor   A factor primarily responsible for determining the growth or reproduction of an organism or a population. The limiting factor in a given environment may be a physical factor such as temperature or light, a chemical factor such as a particular nutrient, or a biological factor such as a competing species.

limits of tolerance   The extremes of any factor (e.g., temperature) that an organism or a population can tolerate and still survive and reproduce.

lipids   A class of natural organic molecules that includes animal fats, vegetable oils, and phospholipids, the last being an integral part of cellular membranes.

lithosphere   The Earth's crust, made up of rocks and minerals.

litter   In an ecosystem, the natural cover of dead leaves, twigs, and other dead plant material. This natural litter is subject to rapid decomposition and recycling in the ecosystem, whereas human litter, such as bottles, cans, and plastics, is not.

livability   A subjective index of how enjoyable a city is to live in.

loam   A soil texture consisting of a mixture of about 40% sand, 40% silt, and 20% clay; loam is usually good soil.

LOCA (loss of cooling accident)   A situation in a nuclear reactor where an accident causes the loss of the water from around the reactor, resulting in overheating and a possible meltdown of the nuclear core.

logistic growth   A pattern of growth of a population that results in an S-shaped curve plotted over time, such that the population levels off at the carrying capacity (K).

longevity   The maximum life span of individuals of a given species. The known record for humans is 122 years.

low-till farming   Similar to no-till farming, except that instead of using herbicides, farmers plow just once over residues of a previous crop and plant their new crop.

LULU ("locally unwanted land use")   An acronym expressing the difficulty in siting a facility that is necessary but that no one wants in his or her immediate locality.

macromolecules   Very large organic molecules, such as proteins and nucleic acids, that constitute the structural and functional parts of cells.

Magnuson Act   An act passed in 1976 that extended the limits of jurisdiction over coastal waters and fisheries of the United States to 200 miles offshore. Reauthorized in 2006 as the **Magnuson-Stevens Fishery Conservation and Management Reauthorization Act**.

malnutrition   The lack of essential nutrients such as vitamins, minerals, and amino acids. Malnutrition ranges from mild to severe and life threatening.

marine protected area   An area of the coast or open ocean that is closed to commercial fishing and mineral mining.

marker-assisted breeding   The science of locating genes with desirable traits using DNA sequencing and then using crossbreeding with standard crop lines, thus avoiding the use of transgenics.

market-based policy   An approach to environmental regulation that allows the use of markets to achieve a desired outcome (e.g., a **cap-and-trade system** for reducing emissions of a pollutant).

mass number The number that accompanies the chemical name or symbol of an element or isotope. The mass number represents the number of neutrons and protons in the nucleus of the atom.

material safety data sheets (MSDS) Documents that contain information on the reactivity and toxicity of over 600 chemicals and that must accompany the shipping, storage, and handling of these chemicals.

materials recovery facility (MRF) A processing plant in which regionalized recycling is carried out. Recyclable municipal solid waste, usually presorted, is prepared in bulk for the recycling market.

matter Any gas, liquid, or solid that occupies space and has mass. (Contrast **energy**.)

maximum achievable control technology (MACT) The best technology available for reducing the output of especially toxic industrial pollutants.

maximum contaminant level (MCL) A drinking water standard; the highest allowable concentration of a pollutant in a drinking water source.

maximum sustainable yield The maximum amount of a renewable resource that can be taken year after year without depleting the resource. The maximum sustainable yield is the maximum rate of use or harvest that will be balanced by the regenerative capacity of the system.

MDG Support Team The successor to the Millennium Project, this team works with developing countries to achieve the targets of the Millennium Development Goals (MDGs).

Medical Revolution Medical advances and public sanitation led to spectacular reductions in mortality, beginning in the late 1800s and extending to the present.

Megatons to Megawatts Program An exchange program between the United States and Russia whereby weapons-grade uranium from Russia is diluted to produce power-plant uranium that is used in U.S. power plants.

meltdown The event of a nuclear reactor's getting out of control or losing its cooling water so that it melts from its own production of heat. The melted reactor would continue to produce heat and could melt its way out of its containment vessel and eventually down into groundwater, where it would cause a violent eruption of steam that could spread radioactive materials over a wide area. (See **LOCA**.)

metabolism The sum of all the chemical reactions that occur in an organism.

meteorology The scientific study of the atmosphere, involving both weather and climate.

methane A gas, CH_4. Methane is the primary constituent of natural gas and is also a product of fermentation by microbes. Methane is one of the greenhouse gases.

methyl bromide A soil fumigant to control agricultural pests; its use releases bromine to the atmosphere, and it destroys ozone when it gets up into the stratosphere.

microbe Any microscopic organism, although primarily a bacterium, a virus, or a protozoan.

microclimate The actual conditions experienced by an organism in its particular location. Owing to numerous factors, such as shading, drainage, and sheltering, the microclimate may be quite distinct from the overall climate.

microfiltration A technology for desalination, based on forcing seawater through very small pores in a membrane filter that removes the salt.

microlending The process of providing very small loans (usually $50–$500) to poor people to facilitate their starting a small enterprise and becoming economically self-sufficient.

microorganism See **microbe**.

midnight dumping The illicit dumping of materials, particularly hazardous wastes, frequently under the cover of darkness.

Milankovitch cycle A cycle of major oscillations in the Earth's orbit, taking place over frequencies of thousands of years and known to influence the distribution of solar radiation and therefore global weather patterns.

Millennium Development Goals (MDGs) A comprehensive set of goals aimed at addressing the most important needs of people in the developing countries to improve their well-being, adopted by the United Nations at the UN Millennium Summit in 2000.

Millennium Ecosystem Assessment (MA) A four-year effort by over 1,360 scientists to produce reports on the state of Earth's ecosystems. The reports are aimed at policy makers and the public and are all available on the Internet.

Millennium Project A UN-commissioned project given a mandate to develop an action plan to coordinate the efforts of various agencies and organizations to achieve the Millennium Development Goals. Succeeded by the **MDG Support Team**.

Minamata disease A disease named for a fishing village in Japan where an "epidemic" was first observed. Symptoms, which included spastic movements, mental retardation, coma, death, and crippling birth defects in the next generation, were found to be the result of mercury poisoning.

mineral Any hard, brittle, stonelike material that occurs naturally in Earth's crust. All minerals consist of various combinations of positive and negative ions held together by ionic bonds. A pure mineral, or crystal, is one specific combination of elements. Common rocks are composed of mixtures of two or more minerals.

mineralization The process of gradual oxidation of the organic matter (humus) present in soil that leaves just the gritty mineral component of the soil.

mitigation Reducing greenhouse gas emissions by a variety of strategies and policies.

mixture A combination of elements in which there is no chemical bonding between the molecules. For example, air contains (is a mixture of) oxygen, nitrogen, and carbon dioxide.

moderator In a nuclear reactor, any material that slows down neutrons from fission reactions so that they are traveling at the right speed to trigger another fission. Water and graphite are two types of moderators.

molecule The smallest unit of two or more atoms forming a compound. A molecule has all the characteristics of the compound of which it is a unit.

monsoon The seasonal airflow created by major differences in cooling and heating between oceans and continents, usually bringing extensive rain.

Montreal Protocol An agreement made in 1987 by a large group of nations to cut back the production of chlorofluorocarbons in order to protect the ozone shield. A 1990 amendment called for the complete phaseout of these chemicals by 2000 in developed nations and by 2010 in less developed nations.

morbidity The incidence of disease in a population.

mortality The incidence of death in a population.

mountaintop removal mining A form of **strip mining** for coal that takes off whole tops of mountains to get at the seams of coal that are buried as deeply as 1,000 feet below the surface.

municipal solid waste (MSW) The entirety of refuse or trash generated by a residential and business community. Distinct from agricultural and industrial wastes, municipal solid waste is the refuse that a municipality is responsible for collecting and disposing of.

mutagenic Causing mutations.

mutation A random change in one or more genes of an organism. Mutations may occur spontaneously in nature, but their number and degree are vastly increased by exposure to radiation or certain chemicals. Mutations generally result in a physical deformity or metabolic malfunction.

mutualism A close relationship between two organisms from which both derive a benefit.

mycelia The threadlike feeding filaments of fungi.

mycorrhiza, pl mycorrhizae The mycelia of certain fungi that grow symbiotically with the roots of some plants and provide for additional nutrient uptake.

NASA National Aeronautics and Space Administration.

National Ambient Air Quality Standards (NAAQS) The allowable levels of ambient criteria air pollutants set by EPA regulation.

National Emission Standards for Hazardous Air Pollutants (NESHAP) The standards for allowable emissions of certain toxic substances.

National Energy Policy Report The 2001 report by a task force chaired by Vice President Cheney that emphasized dependence on **supply-side policies** to generate the energy needed by the United States into the future.

national forests Public forests and woodlands administered by the National Forest Service for multiple uses, such as logging, mineral exploitation, livestock grazing, and recreation.

national parks Lands and coastal areas of great scenic, ecological, or historical importance administered by the National Park Service with the dual goals of protecting them and providing public access.

National Pollution Discharge Elimination System (NPDES) An EPA-administered program that addresses point-source water pollution through the issuance of permits that regulate pollution discharge.

National Priorities List (NPL) A list of the chemical waste sites presenting the most immediate and severe threats. Such sites are scheduled for cleanup ahead of other sites.

national wildlife refuges Administered by the U.S. Fish and Wildlife Service, these lands are maintained for the protection and enhancement of wildlife and for the provision of public access.

natural Produced as a normal part of nature, apart from any activity or intervention of humans. Opposite of artificial, synthetic, human made, or caused by humans.

natural capital The natural assets and the services they perform. One form of the wealth of a nation is its complement of natural capital.

natural chemical control The use of one or more natural chemicals, such as hormones or pheromones, to control a pest.

natural control methods Any of many techniques of controlling a pest population without resorting to the use of synthetic organic or inorganic chemicals. (See **biological control, cultural control, genetic control, hormones,** and **pheromones.**)

natural enemies All the predators and parasites that may feed on a given organism. Organisms used to control a specific pest through predation or parasitism.

natural goods The food, fuel, wood, fibers, oils, alcohols, and the like derived from the natural world, on which the world economy and human well-being depend.

natural increase The number of births minus the number of deaths in a given population. The natural increase does not take into account immigration and emigration and is the percent of growth (or decline) of a given population during a year. It is found by subtracting the crude death rate from the crude birthrate and changing the result to a percent.

natural laws Generalizations derived from our observations of matter, energy, and other phenomena. Though not absolute, natural laws have been empirically confirmed to a high degree and are often derivable from higher-level theory.

Natural Resources Conservation Service (NRCS) Formerly the Soil Conservation Service, a nationwide network of offices that provide extension services to farmers and encourage soil and water conservation throughout the United States.

natural resources Features of natural ecosystems and species that are of economic value and that may be exploited. Also, features of particular segments of ecosystems, such as air, water, soil, and minerals.

natural selection The evolutionary process whereby the natural factors of environmental resistance tend to eliminate those members of a population that are least well adapted to cope with their environment and thus, in effect, tend to select those best adapted for survival and reproduction.

natural services Functions performed free of charge by natural ecosystems, such as control of runoff and erosion, absorption of nutrients, and assimilation of air pollutants.

negligible risk standard A new safety standard based on risk analysis, established by the **FQPA** of 1996, that replaces the **Delaney clause** in dealing with pesticide residues on food products.

Neolithic Revolution The development of agriculture begun by human societies around 12,000 years ago, leading to more permanent settlement and population increases.

net metering The practice of allowing individuals on the power grid to subtract the cost of the energy they get from photovoltaics or wind turbines from their electric bills.

neutron A fundamental atomic particle found in the nuclei of atoms (except hydrogen) and having one unit of atomic mass but no electrical charge.

new forestry Now part of the U.S. Forest Service's management practice, a forestry management strategy that places priority on protecting the ecological health and diversity of forests rather than maximizing the harvest of logs.

niche (ecological) The total of all the relationships that bear on how an organism copes with the biotic and abiotic factors it faces.

NIMBY ("not in my backyard") A common attitude regarding undesirable facilities such as incinerators, nuclear facilities, and hazardous-waste treatment plants whereby people do everything possible to prevent such facilities from being located near their residences.

NIMTOO ("not in my term of office") An attitude expressing the reluctance of officeholders to make unpopular decisions on matters such as siting waste facilities.

nitric acid A secondary air pollutant and a major component of acid deposition, nitric acid is produced when NO_x in the air reacts with the hydroxyl radical and water.

nitrogen fixation The process of chemically converting nitrogen gas (N_2) from the air into compounds such as nitrates (NO_3) or ammonia (NH_3) that can be used by plants in building amino acids and other nitrogen-containing organic molecules.

nitrogen oxides (NO_X) A group of nitrogen-oxygen compounds formed when some of the nitrogen gas in the air combines with oxygen during high-temperature combustion. Nitrogen oxides are a major category of air pollutants and, along with hydrocarbons, are a primary factor in the production of ozone and other photochemical oxidants that are the most harmful components of photochemical smog. Nitrogen oxides also contribute to acid precipitation.

nitrous oxide A gas (N_2O). Nitrous oxide, which comes from biomass burning, fossil fuel burning, and the use of chemical fertilizers, is of concern because it is a greenhouse gas in the troposphere and it contributes to ozone destruction in the stratosphere.

NOAA   National Oceanic and Atmospheric Administration.

nonbiodegradable   Not able to be consumed or broken down by biological organisms. Nonbiodegradable substances include plastics, aluminum, and many chemicals used in industry and agriculture. Particularly dangerous are synthetic chemicals that are also toxic and tend to accumulate in organisms. (See **biodegradable** and **bioaccumulation**.)

nonconsumptive water use   The use of water for such purposes as washing and rinsing, wherein the water, albeit polluted, remains available for further uses. With suitable purification, such water may be recycled indefinitely.

nongovernmental organization (NGO)   Any of a number of private organizations involved in studying or advocating action on different environmental issues.

nonpersistent   Said of chemicals that break down readily to harmless compounds, as, for example, natural organic compounds that break down to carbon dioxide and water; applied to pesticides, any pesticide that breaks down within days or a few weeks into nontoxic products.

nonpoint sources   Sources of pollution such as the general runoff of sediments, fertilizers, pesticides, and other materials from farms and urban areas, as opposed to specific points of discharge such as factories. Also called **diffuse sources**. (Contrast **point sources**.)

nonreactive nitrogen   The main reservoir of nitrogen as nitrogen gas in the atmosphere.

nonrenewable resources   Resources, such as ores of various metals, oil, and coal that exist as finite deposits in Earth's crust and that are not replenished by natural processes as they are mined. (Contrast **renewable resources**.)

no-till agriculture   The farming practice in which weeds are killed with chemicals (or other means) and seeds are planted and grown without resorting to plowing or cultivation. The practice is highly effective in reducing soil erosion.

NRCS (Natural Resources Conservation Service)   Formerly the U.S. Soil Conservation Service). The NRCS provides information to landowners regarding soil and water conservation practices.

nuclear power   Electrical power generated by using a nuclear reactor to boil water and produce steam that in turn drives a turbogenerator.

Nuclear Regulatory Commission (NRC)   The agency within the U.S. Department of Energy that sets and enforces safety standards for the operation and maintenance of nuclear power plants.

nucleic acids   The class of natural organic macromolecules that function in the storage and transfer of genetic information.

nucleus   *Biology:* The large body residing in most living cells that contains the genes or hereditary material (DNA). *Physics:* The central core of atoms, which is made up of neutrons and protons. Electrons surround the nucleus.

nutrient   *Animal:* Material such as protein, vitamins, and minerals required for growth, maintenance, and repair of the body and material such as carbohydrates required for energy. *Plant:* An essential element (C, P, N, Ca) in a particular ion or molecule that can be absorbed and used by the plant.

nutrient cycle   The repeated pathway of particular nutrients or elements from the environment through one or more organisms and back to the environment. Nutrient cycles include the carbon cycle, the nitrogen cycle, the phosphorus cycle, and so on.

nutrient-holding capacity   The capacity of a soil to bind and hold nutrients (fertilizer) against their tendency to be leached from the soil.

observations   Things or phenomena that are perceived through one or more of the basic five senses in their normal state. In addition, to be accepted as factual, observations must be verifiable by others.

Occupational Safety and Health Act of 1970 (OSHA)   The act protecting the rights of workers to know about hazards in their workplace.

ocean acidification   An outcome of the rise in atmospheric carbon dioxide; as the oceans take up more and more of the CO_2, the carbonate ion concentration is reduced, making it more difficult for coral animals to build their calcium carbonate skeletons.

ocean thermal-energy conversion (OTEC)   The concept of harnessing the difference in temperature between surface water heated by the sun and colder deep water to produce power.

Official Development Assistance (ODA)   Financial assistance provided to developing countries from the developed countries and coordinated by United Nations agencies.

oil field   The underground area in which exploitable oil is found.

oil sand   Sedimentary material containing bitumen, a tarlike hydrocarbon that is subject to exploitation under favorable economic conditions.

oil shale   A natural sedimentary rock that contains kerogen, a material that can be extracted and refined into oil and oil products.

oligotrophic   Nutrient poor and hence unable to support much phytoplankton (said of a body of water).

omnivore   An animal that feeds on both plant material and other animals.

OPEC (Organization of Petroleum Exporting Countries)   A cartel of oil-producing nations that attempts to control the market price of oil.

optimal population   The population of a harvested biological resource that yields the greatest harvest for exploitation; according to maximum-sustained-yield equations, the optimal population is half the carrying capacity.

optimal range   With respect to any particular factor or combination of factors, the maximum variation that still supports optimal or near-optimal growth of the species in question.

optimum   The condition or amount of any factor or combination of factors that will produce the best result. For example, the amount of heat, light, moisture, nutrients, and so on that will produce the best plant growth.

organically grown   Grown without the use of hard chemical pesticides or inorganic fertilizer. Any produce grown according to USDA standards for organic foods is organically grown.

organic compounds/molecules   *Classical definition:* All living things and products that are uniquely produced by living things, such as wood, leather, and sugar. *Chemical definition:* All chemical compounds or molecules, natural or synthetic, that contain carbon atoms as an integral part of their molecular structure. The structure of organic compounds is based on bonded carbon atoms with hydrogen atoms attached. (Contrast **inorganic compounds**.)

organic fertilizer   See **fertilizer**.

organic food/gardening/farming   Gardening or farming without the use of inorganic fertilizers, synthetic pesticides, or other human-made materials; food raised with organic farming techniques.

organic phosphate   Phosphate (PO_4^{-3}) bonded to an organic molecule.

organism   Any living thing—plant, animal, or microbe.

orphan site   The location of a hazardous waste site abandoned by former owners.

OSHA (Occupational Safety and Health Administration) The federal agency that promulgates regulations to protect workers.

osmosis The phenomenon whereby water diffuses through a semipermeable membrane toward an area where there is more material in solution (i.e., where there is a relatively lower concentration of water). Osmosis has particular application in the salinization of those soils in which plants are unable to grow because of osmotic water loss.

outbreak A population explosion of a particular pest. Often caused by an application of pesticides that destroys the pest's natural enemies.

overcultivation The practice of repeatedly cultivating and growing crops more rapidly than the soil can regenerate, leading to a decline in soil quality and productivity.

overexploitation The overharvesting of a species or ecosystem that leads to its decline.

overgrazing The phenomenon of grazing animals in greater numbers than the land can support in the long term.

overnourishment The condition of obesity and overweight that results from overeating.

oxidation A chemical reaction that generally involves a breakdown of some substance through its combining with oxygen. Burning and cellular respiration are examples of oxidation. In both cases, organic matter is combined with oxygen and broken down to carbon dioxide and water.

ozone A gas, O_3, that is a secondary air pollutant in the lower atmosphere but that is necessary to screen out ultraviolet radiation in the upper atmosphere. May also be used for disinfecting water.

ozone hole First discovered over the Antarctic, a region of stratospheric air that is severely depleted of its normal levels of ozone during the Antarctic spring because of CFCs from anthropogenic (human-made) sources.

ozone shield The layer of ozone gas (O_3) in the upper atmosphere that screens out harmful ultraviolet radiation from the Sun.

parasites Organisms (plant, animal, or microbial) that attach themselves to another organism, the host, and feed on it over a period of time without killing it immediately, although usually doing harm to it. Parasites are commonly divided into ectoparasites—those that attach to the outside of their hosts—and endoparasites—those that live inside their hosts.

parasitoid Any parasitic insect, usually a wasp, that attacks other insects and can be responsible for bringing pest insects under control.

parent material Rock material whose weathering and gradual breakdown are the source of the mineral portion of soil.

particulates See $PM_{2.5}$ and PM_{10} and **suspended particulate matter.**

parts per million (ppm) A frequently used expression of concentration. The number of units of one substance present in a million units of another. Equivalent to milligrams per liter.

passive safety features Those safety features of nuclear facilities that involve processes that are not vulnerable to operator intrusion or electrical power failures. Passive safety features enhance the degree of safety of nuclear reactors. (See **active safety features.**)

passive solar-heating system A solar-heating system that does not use pumps or blowers to transfer heated air or water. Instead, natural convection currents are used, or the interior of the building itself acts as the solar collector.

pasteurization The process of applying enough heat to milk or some other substance to kill pathogens and sufficient other bacteria to extend the shelf life of the product.

pastoralist One involved in animal husbandry, usually in subsistence agriculture.

pathogen An organism, usually a microbe, that is capable of causing disease. Such an organism is said to be pathogenic.

pathogen pollution A situation where human wastes spread diseases to wild species.

Payments for Ecosystem Services (PES) A market-based system that pays owners who hold properties that provide valuable services.

PAYT (pay as you throw) A system for collecting municipal wastes that charges residents by the container for nonrecycled wastes.

percolation The process of water seeping downward through cracks and pores in soil or rock.

permafrost The ground of arctic regions that remains permanently frozen. Defines tundra, since only small herbaceous plants can be sustained on the thin layer of soil that thaws each summer.

peroxyacetyl nitrates (PANs) A group of compounds present in photochemical smog that are extremely toxic to plants and irritating to the eyes, nose, and throat membranes of humans.

persistent organic pollutants (POPs) Any members of a class of organic pollutants that are resistant to biodegradation and that are often toxic; for example, DDT, PCBs, and dioxin are persistent organic pollutants. Such chemicals may remain present in the environment for periods of years.

Personal Responsibility and Work Opportunity Reconciliation Act of 1996 (PRWOR) Legislation intended to move welfare recipients from dependency on public assistance to independence.

pest Any organism that is noxious, destructive, or troublesome; usually an agricultural pest.

pesticide A chemical used to kill pests. Pesticides are further categorized according to the pests they are designed to kill—for example, herbicides kill plants, insecticides kill insects, fungicides kill fungi, and so on.

pesticide treadmill The idea that use of chemical pesticides simply creates a vicious cycle of "needing more pesticides" to overcome developing resistance and secondary outbreaks caused by the pesticide applications.

pest-loss insurance Insurance that a grower can buy that will pay in the event of loss of a crop due to pests.

pH The scale used to designate the acidity or basicity (alkalinity) of solutions or soil, expressed as the logarithm of the concentration of hydrogen ions (H^+). pH 7 is neutral; values decreasing from 7 indicate increasing acidity, and values increasing from 7 indicate increasing basicity. Each unit from 7 indicates a 10-fold increase over the preceding unit.

pharma crops Crop plants that produce pharmaceutical products with the use of genomics.

pheromones Chemical substances secreted externally by certain members of a species that affect the behavior of other members of the same species. The most common examples are sex attractants, which female insects secrete to attract males. Pheromones are coming into use in pest control. (See also **hormones.**)

phosphorus, phosphate An ion composed of a phosphorus atom with four oxygen atoms attached. Denoted PO_4^{-3}, phosphate is an important plant nutrient. In natural waters, it is frequently the limiting factor; therefore, additions of phosphate to natural water are often responsible for algal blooms.

photochemical oxidants A major category of secondary air pollutants, including ozone, that are highly toxic and damaging, especially to plants and forests. Formed as a result of interactions between nitrogen oxides and hydrocarbons driven by sunlight.

photochemical smog The brownish haze that frequently forms on otherwise clear, sunny days over large cities with significant amounts of automobile traffic. Photochemical smog results largely from sunlight-driven chemical reactions among nitrogen oxides and hydrocarbons, both of which come primarily from auto exhausts.

photosynthesis The chemical process carried on by green plants through which light energy is used to produce glucose from carbon dioxide and water. Oxygen is released as a by-product.

photovoltaic cell (PV cell) A device that converts light energy into an electrical current.

phthalates A group of chemicals used to soften plastic toys and suspected of being hormone disrupters.

physical barrier A genetic feature on a plant, such as sticky hairs, that physically blocks attack by pests.

phytoplankton Any of the many species of photosynthetic microorganisms that consist of single cells or small groups of cells that live and grow freely suspended near the surface in bodies of water.

phytoremediation The use of plants to accomplish the cleanup of some hazardous chemical wastes.

planetary albedo The reflection of solar radiation back into space due to cloud cover, contributing to the cooling of the atmosphere.

plankton Any and all living things that are found freely suspended in the water and that are carried by currents, as opposed to being able to swim against currents. Plankton includes both plant (phytoplankton) and animal (zooplankton) forms.

plant community The array of plant species, including their numbers, ages, and distribution, that occupy a given area.

plate tectonics The theory that postulates the movement of sections of the Earth's crust, creating earthquakes, volcanic activity, and the buildup of continental masses over long periods of time.

PM_{25} and PM_{10} Standard criterion pollutants for suspended particulate matter. PM_{10} refers to particles smaller than 10 micrometers in diameter; $PM_{2.5}$ refers to particles smaller than 2.5 micrometers in diameter. The smaller particles are readily inhaled directly into the lungs.

point sources Specific points of origin of pollutants, such as factory drains or outlets from sewage-treatment plants. (Contrast **nonpoint sources.**)

policy life cycle The typical course of events occurring over time in the recognition of, formulation of an approach to, implementation of a solution to, and control of an environmental problem requiring public policy action.

pollutant A substance that contaminates air, water, or soil.

polluter-pays principle An ethical principle whereby those who pollute the environment should pay for the damage their pollution causes.

pollution According to the EPA, "the presence of a substance in the environment that, because of its chemical composition or quantity, prevents the functioning of natural processes and produces undesirable environmental and health effects."

polychlorinated biphenyls (PCBs) A group of widely used industrial chemicals of the chlorinated hydrocarbon class. PCBs have become serious and widespread pollutants because they are resistant to breakdown and are subject to bioaccumulation. They are also carcinogenic.

population A group within a single species whose individuals can and do freely interbreed.

population density The number of individuals per unit of area.

population equilibrium A state of balance between births and deaths in a population.

population explosion The exponential increase observed to occur in a population when conditions are such that a large percentage of the offspring are able to survive and reproduce

in turn. A population explosion frequently leads to overexploitation and eventual collapse of the ecosystem.

population growth rate The change in population numbers divided by the time over which that change occurs.

population momentum A property whereby a rapidly growing human population may be expected to grow for 50–60 years after replacement fertility (2.1 live births per female) is reached. Momentum is sustained because of increasing numbers entering reproductive age.

population profile A bar graph that shows the number of individuals at each age or in each five-year age group, starting with the youngest ages at the bottom of the profile.

population structure See **age structure.**

positive feedback A phenomenon where some process leads to even more intensification of that process; an example is the evaporative rise in water vapor with global warming, adding water, a greenhouse gas, to the atmosphere.

potential energy The ability to do work that is stored in some chemical or physical state. For example, gasoline is a form of potential energy because the ability to do work is stored in the chemical state and is released as the fuel is burned in an engine.

poverty trap In a developing country, the situation where people are so poor that they are unable to earn or work their way out of their poverty and their condition deteriorates.

power grid The combination of central power plants, high-voltage transmission lines, poles, wires, and transformers that makes up an electrical power system.

power tower A solar energy converter that collects solar energy with an array of sun-tracking mirrors and focuses the energy onto a receiver on a tower in the center of the array where the heat energy drives a conventional turbogenerator.

precautionary principle The principle that says that where there are threats of serious or irreversible damage, the absence of scientific certainty shall not be used as a reason for postponing cost-effective measures to prevent environmental degradation.

precipitation Any form of moisture condensing in the air and depositing on the ground.

predator An animal that feeds on another living organism, either plant or animal.

predator-prey relationship A feeding relationship existing between two kinds of organisms. The predator is the animal feeding on the prey. Such relationships are frequently instrumental in controlling populations of herbivores.

preliminary treatment The removal of debris and grit from wastewater by passing the water through a coarse screen and a grit-settling chamber.

preservation In protecting natural areas, the objective of preservation is to ensure the continuity of species and ecosystems, regardless of their potential utility.

prey An organism that is fed on by a predator.

primary air pollutants Pollutants released directly into the atmosphere mainly as a result of burning fuels and wastes, as opposed to **secondary air pollutants.**

primary clarifier In a sewage-treatment system, a large tank that allows particulate matter to settle to the bottom before the water moves on to **secondary treatment.**

primary consumer An organism, such as a rabbit or deer, that feeds more or less exclusively on green plants or their products, such as seeds and nuts. (Synonym: **herbivore.**)

primary energy sources Fossil fuels, radioactive material, and solar, wind, water, and other energy sources that exist as natural resources subject to exploitation.

primary producers/primary production Photosynthetic organisms. The activities of these organisms in creating new organic matter in ecosystems.

primary recovery In an oil well or oil field, the oil that can be removed by conventional pumping. (See **secondary recovery**.)

primary succession The gradual establishment, through a series of stages, of a climax ecosystem in an area that has not been occupied before (e.g., a rock face).

primary treatment The process that follows preliminary sewage treatment and that consists of passing the water very slowly through a large tank so that the particulate organic material can settle out. The settled material is **raw sludge**.

prior informed consent (PIC) A UN-approved procedure whereby countries exporting pesticides must inform importing countries of actions the exporting countries have taken to restrict the use of the pesticides.

produced capital The stock of buildings, machinery, vehicles, and other elements of a country's infrastructure that are essential to the production of economic goods and services. One component of the wealth of nations.

producers In an ecosystem, those organisms (mostly green plants) that use light energy to construct their organic constituents from inorganic compounds.

productive use The exploitation of ecosystem resources for economic gain.

property taxes Taxes that the local government levies on privately owned properties, proportional to the value of the property. Property taxes are the major source of revenue for local governments.

protein Organic macromolecules made up of amino acids; the major structural component of all animal tissues as well as enzymes in both plants and animals.

proton A fundamental atomic particle with a positive charge, found in the nuclei of atoms. The number of protons present equals the atomic number and is distinct for each element.

protozoon, pl. protozoa Any of a large group of eukaryotic microscopic organisms that consist of a single, relatively large complex cell or, in some cases, small groups of cells. All have some means of movement. Amoebae and paramecia are examples.

proven reserves See **reserves**.

proxies These are measurable records, like tree rings, ice cores, and marine sediments, that can provide extensions back in time of climatic factors like temperatures and greenhouse gases.

Public Trust Doctrine The legal arrangement whereby wildlife resources are considered to be public resources, held in trust for all the people and protected or managed by state or federal agencies.

pumped-storage power plant A hydroelectric plant component that pumps water up to an elevated reservoir when power demand is low and allows it to return to a lower reservoir and generate electricity when demand is higher.

purification For water, purification occurs whenever water is separated from the solutes and particles it contains.

radiation sickness The outcome of a high dose of radiation, often leading to death.

radiative forcing Refers to the internal or external factors of the climate system that can influence the climate, usually to promote warming or cooling of the atmosphere as they affect the Earth's energy balance.

radioactive decay The reduction in radioactivity that occurs as an unstable isotope (radioactive substance) gives off radiation, ultimately becoming stable.

radioactive emissions Any of various forms of radiation or particles that may be given off by unstable isotopes. Many such emissions have very high energy and can destroy biological tissues or cause mutations leading to cancer or birth defects.

radioactive materials Substances that are or that contain unstable isotopes and that consequently give off radioactive emissions. (See **isotope** and **radioactive emissions**.)

radioactive wastes Waste materials that contain, are contaminated with, or are radioactive substances. Many materials used in the nuclear industry become radioactive wastes because of such contamination.

radioisotope An isotope of an element that is unstable and that gives off radioactive emissions. (See **isotope** and **radioactive decay**.)

radon A radioactive gas produced by natural processes in Earth that is known to seep into buildings. Radon can be a major hazard within homes and is a known carcinogen.

rain shadow The low-rainfall region that exists on the leeward (downwind) side of a mountain range. This rain shadow is the result of the mountain range causing precipitation on the windward side.

range of tolerance The range of conditions within which an organism or population can survive and reproduce—for example, the range from the highest to the lowest temperature that can be tolerated. Within the range of tolerance is the optimum, or best, condition.

raw sewage The water from a sewer system as it reaches a sewage-treatment plant.

raw sludge The untreated organic matter that is removed from sewage water by letting it settle. Raw sludge consists of organic particles from feces, garbage, paper, and bacteria.

REACH (Registration, Evaluation and Authorization of Chemicals) The European Union's new, rigorous approach to screening hazardous chemicals for toxicity.

reactive nitrogen All forms of nitrogen in ecosystems that are usable by organisms, as opposed to the nonreactive nitrogen in the form of nitrogen gas.

reactor vessel A steel-walled vessel that contains a nuclear reactor.

reasonably available control technology (RACT) Applied to the goals of the Clean Air Act, EPA-approved forms of technology that will reduce the output of industrial air pollutants. (See **maximum achievable control technology**.)

recharge area The area over which groundwater will infiltrate and resupply an aquifer.

recruitment The maturation and successful entry of young into an adult breeding population.

recycling The recovery of materials that would otherwise be buried in landfills or be combusted. Primary recycling involves remaking the same material from the waste; in secondary recycling, waste materials are made into different products.

redlining The practice of discriminatory lending and real estate showing designed to keep minorities out of white suburbs.

Red List A list of threatened and endangered species maintained by the International Union for Conservation of Nature (IUCN) and accessible on the Internet.

reformulated gasoline Gasoline with oxygen-containing additives (MTBE, ethanol) to enhance clean burning.

relative humidity The percentage of moisture in the air, compared with how much the air can hold at the given temperature.

rem An older unit of measurement of the ability of radioactive emissions to penetrate biological tissue. (See **sievert**.)

remediation A return to an original uncontaminated state. (See **bioremediation** and **groundwater remediation**.)

remittances Money sent back to the country of birth by guest workers and immigrants to help family and friends meet their financial needs.

renewable energy Energy sources—namely, solar, wind, and geothermal—that will not be depleted by use.

renewable fuel standard (RFS) A policy that requires a minimum volume of renewable fuel in gasoline.

renewable portfolio standard A policy that requires a percentage of power from renewable energy sources.

renewable resources   Biological resources, such as trees, that may be renewed by reproduction and regrowth. Conservation to prevent overcutting and to protect the environment is still required, however. (Contrast **nonrenewable resources.**)

replacement fertility rate/level   The fertility rate (level) that will just sustain a stable population. It is 2.1 for developed countries.

reprocessing   The recovery of radioactive materials from nuclear reactors and the conversion of those materials into useful nuclear fuel.

reproductive health   Health care directed primarily toward the needs of women and infants, strongly emphasized in the 1994 Cairo population conference.

reproductive isolation   One of the processes of speciation, involving anything that keeps individuals or subpopulations from interbreeding.

reproductive strategy   A life history property of a species that involves a balance between reproduction and death. (See *r*-**strategists** and *K*-**strategists.**)

reserves   The amount of a mineral resource (including oil, coal, and natural gas) remaining in Earth that can be exploited using current technologies and at current prices. Usually given as *proven* reserves (those that have been positively identified), *estimated* reserves (those that are presumed to exist based on geological exploration), and *undiscovered* reserves (reserves from fields that have not yet been found).

resilience   The tendency of ecosystems to recover from disturbances through a number of processes known as resilience mechanisms (e.g., succession after a fire).

resistance   The development of more hardy pests through the effects of the repeated use of pesticides in selecting against sensitive individuals in a pest population.

resource partitioning   The outcome of competition by a group of species where natural selection favors the division of a resource in time or space by specialization of the different species.

resources   Biotic and abiotic factors that are consumed by organisms.

Resources Conservation and Recovery Act of 1976 (RCRA)   The cornerstone legislation to control indiscriminate land disposal of hazardous wastes.

respiration   See **cell respiration.**

restoration ecology   The branch of ecology devoted to restoring degraded and altered ecosystems to their natural state.

resurgence   The rapid comeback of a population (especially of pests after a severe die-off, usually caused by pesticides) and the return to even higher levels than before the treatment.

reverse osmosis   See **microfiltration.**

riparian woodlands   The strip of woody plants that grow along natural watercourses.

risk   The probability of suffering injury, disease, death, or some other loss as a result of exposure to a hazard.

risk assessment/analysis   The process of evaluating the risks associated with a particular hazard before taking some action.

risk characterization   The process of determining the level of a risk and its accompanying uncertainties after hazard assessment, dose-response assessment, and exposure assessment have been accomplished.

risk management   The task of regulators whose job is to review risk data and make regulatory decisions based on the data. The process often is influenced by considerations of costs and benefits as well as by public perception.

risk perception   Nonexperts' intuitive judgments about risks, which often are not in agreement with the level of risk as judged by experts.

roadless rule   A moratorium placed on building new roads in the national forests, put in place by the Clinton administration and contested by subsequent administrations.

rooftop garden   A form of aboveground garden on the top of a flat residential roof, using shallow soil beds and containers such as old automobile tires or plastic wading pools.

rotavirus   A little-known virus that causes many childhood deaths, responsible for most cases of severe diarrhea.

Roundup®   See **glyphosate.**

r-strategist   A reproductive strategy of a species that involves producing large numbers of young (a high reproductive rate) and survival of smaller numbers over time. Also called **opportunistic species.**

runoff   The portion of precipitation that runs off the surface, as opposed to soaking into the ground.

Safe, Accountable, Flexible, Efficient Transportation Equity Act of 2005 (SAFETEA)   Successor to TEA–21 and ISTEA, this act authorizes funding for highway and transit programs.

Safe Drinking Water Act of 1974   Legislation to protect the public from the risk that toxic chemicals will contaminate drinking water supplies. The act mandates regular testing of municipal water supplies.

safety net   In a society, the availability of food and other necessities extended to people who are unable to meet their own needs for a variety of reasons.

salinization   The process whereby soil becomes saltier and saltier until, finally, the salt prevents the growth of plants. Salinization is caused by irrigation because salts brought in with the water remain in the soil as the water evaporates.

saltwater intrusion, saltwater encroachment   The phenomenon of seawater's moving back into aquifers or estuaries. Such intrusion occurs when the normal outflow of freshwater is diverted or removed for use.

salvage logging   Established by the Bush administration, this refers to harvesting timber from recent burns in national forestlands; it is controversial because it is suspected of preventing natural regeneration of trees on the burned land.

sand   Mineral particles 0.063 to 2.0 mm in diameter.

sanitary sewer   A separate drainage system used to receive all the wastewater from sinks, tubs, and toilets.

SARA (Title III)   A section of the Superfund Amendments and Reauthorization Act that addresses hazardous waste accidents and promulgates community right-to-know requirements. Also known as the **Emergency Planning and Community Right-to-Know Act (EPCRA) of 1986.** (See **Toxics Release Inventory.**)

savanna   A type of grassland typical of subtropical regions, particularly in Africa, usually dotted with trees and supported by wet and dry seasons and frequent natural fires.

scale insect   Any of a group of minute insect pests that suck the juices from plant cells.

scientific method   The process of making observations and logically integrating those observations into a model of how the world works. Often involves forming hypotheses, experimenting, and conducting further testing for confirmation.

secondary air pollutants   Air pollutants resulting from reactions of primary air pollutants resident in the atmosphere. Secondary air pollutants include ozone, other reactive organic compounds, and sulfuric and nitric acids. (See **ozone, peroxyacetyl nitrates,** and **photochemical oxidants.**)

secondary consumer   An organism such as a fox or coyote that feeds more or less exclusively on other animals that feed on plants.

secondary energy source   A form of energy such as electricity that must be produced from a primary energy source such as coal or radioactive material.

secondary pest outbreak   The phenomenon of a small and therefore harmless population of a plant-eating insect suddenly exploding to become a serious pest problem. Often caused by the elimination of competitors through the use of pesticides.

secondary recovery In an oil well or oil field, the oil that can be removed by manipulating pressure in the oil reservoir by injecting brine or other substances; more costly than **primary recovery.**

secondary succession The reestablishment, through a series of stages, of a climax ecosystem in an area from which it was previously cleared.

secondary treatment Also called **biological treatment.** A sewage-treatment process that follows primary treatment. Any of a variety of systems that remove most of the remaining organic matter by enabling organisms to feed on it and oxidize it through their respiration. Trickling filters and activated-sludge systems are the most commonly used secondary-treatment methods.

second-generation pesticides Synthetic organic compounds used to kill insects and other pests. Started with the use of DDT in the 1940s.

Second Law of Thermodynamics The empirical observation, confirmed innumerable times, that in every energy conversion (e.g., from electricity to light), some of the energy is converted to heat and some heat always escapes from the system because it always moves toward a cooler place. Therefore, in every energy conversion, a portion of energy is lost, and since, by the First Law of Thermodynamics, energy cannot be created, the functioning of any system requires an energy input.

secure landfill A landfill with suitable barriers, leachate drainage, and monitoring systems such that it is deemed secure against contaminating groundwater with hazardous wastes.

sediment Soil particles—namely, sand, silt, and clay—carried by flowing water. The same material after it has been deposited. Because of different rates of settling, deposits generally are pure sand, pure silt, or pure clay.

sedimentation The filling in of lakes, reservoirs, stream channels, and so on with soil particles, mainly sand and silt. The soil particles come from erosion. Also called **siltation.**

seep An area where groundwater seeps from the ground. Contrast with a spring, which is a single point from which groundwater exits.

selective pressure An environmental factor that causes individuals with certain traits that are not the norm for the population to survive and reproduce more than the rest of the population. The result is a shift in the genetic makeup of the population. Selective pressure is a fundamental mechanism of evolution.

septic system An onsite method of treating sewage that is widely used in suburban and rural areas. Requires a suitable amount of land and porous soil.

services Ecosystem functions that are essential to human life and economic well-being, such as waste breakdown, climate regulation, and erosion control. These can be further categorized as regulating, supporting, and provisioning services.

sex attractant A natural chemical substance (see **pheromones**) secreted by the female of many insect species that serves to attract males for the function of mating. Sex attractants may be used by humans in traps or to otherwise confuse insect pests in order to control them.

shadow pricing In cost-benefit analysis, a technique used to estimate benefits when normal economic analysis is ineffective. For example, people could be asked how much they might be willing to pay monthly to achieve some improvement in their environment.

sheet erosion The loss of a more or less even layer of soil from the land surface due to the impact of rain and runoff from a rainstorm.

shelterbelts Rows of trees around cultivated fields for the purpose of reducing wind erosion.

sievert A unit of measurement of the ability of radioactive emissions to penetrate biological tissues. 1 sievert = 100 rem.

silt Soil particles between the size of sand particles and that of clay particles—namely, particles 0.004 to 0.063 mm in diameter.

siltation See **sedimentation.**

simplification The human use of habitats that removes natural objects, such as maintaining a forest to produce one kind of tree or removing fallen logs.

sinkhole A large hole in the ground resulting from the collapse of an underground cavern.

slash-and-burn agriculture The practice, commonly found in tropical regions, of cutting and burning vegetation to make room for agriculture. The process destroys soil humus and may lead to rapid degradation of the soil.

sludge cake Treated sewage sludge that has been dewatered to make a moist solid.

sludge digesters Large tanks in which raw sludge (removed from sewage) is treated through anaerobic digestion by bacteria.

slum A concentrated habitation where squalor and disease reign, usually consisting of flimsy shacks and characterized by an almost complete lack of public amenities.

Smart Growth A movement that addresses urban sprawl by protecting sensitive lands and directing growth to limited areas.

smog See **industrial smog** and **photochemical smog.**

social attraction A method of reestablishing species that involves simulating the social conditions conducive to colonization by the introduced individuals.

social capital See **intangible capital.**

social modernization A process of sustainable development that leads developing countries through the demographic transition by promoting education, family planning, and health improvements rather than simply economic growth.

soft water Water with little or no calcium, magnesium, or other ions in solution that will cause soap to precipitate. (The soap would otherwise form a curd that makes a "ring" around the bathtub.) (Contrast **hard water.**)

soil A dynamic terrestrial system involving three components: mineral particles, detritus, and soil organisms feeding on the detritus.

soil classes Taxonomically arranged major groupings of soil that define the properties of the soils of different regions.

soil erosion The loss of soil caused by particles being carried away by wind or water.

soil fertility A soil's ability to support plant growth; often refers specifically to the presence of proper amounts of nutrients. The soil's ability to fulfill all the other needs of plants is also involved.

soil horizons Distinct layers within a soil that convey different properties to the soil and that derive from natural processes of soil formation.

soil profile A description of the different naturally formed layers, called horizons, within a soil.

soil structure The composition of soil in terms of particles (sand, silt, and clay) stuck together to form clumps and aggregates, generally with considerable air spaces in between. Soil structure affects infiltration and aeration and develops as organisms feed on organic matter in and on the soil.

soil texture The relative size of the mineral particles that make up the soil. Generally defined in terms of the soil's sand, silt, and clay content.

solar cells See **photovoltaic cells.**

solar constant The radiant energy reaching the top of Earth's atmosphere from the Sun, measured at 1,366 watts per square meter.

solar energy Energy derived from the Sun; includes direct solar energy (the use of sunlight directly for heating or the production of electricity) and indirect solar energy (the use of wind, which results from the solar heating of the atmosphere, and biological materials such as wood, which result from photosynthesis).

solar-trough collectors   Reflectors, in the shape of a parabolic trough, that reflect sunlight onto a tube of oil at the focal point of the trough. Thus heated, the oil is used to boil water to drive a steam turbine.

solar variability   One proposed mechanism for global warming involving increases in solar output; this explanation has been discredited because of lack of convincing data.

solid waste   The total of materials discarded as "trash" and handled as solids, as opposed to those that are flushed down sewers and handled as liquids.

solubility   The degree to which a substance will dissolve and enter into solution.

solution   A mixture of molecules (or ions) of one material in another; most commonly, molecules of air or ions of various minerals in water. For example, seawater contains salt in solution.

sound science   *The basis for our understanding of how the world works and how human systems interact with it.* It stems from scientific work based on peer-reviewed research and is one of the unifying themes of this text.

source reduction   According to the EPA, "the practice of designing, manufacturing, purchasing, or using materials in ways that reduce the amount or toxicity of trash created."

special-interest group   In politics, people organized around a particular issue who lobby for their cause and exert their influence on the political process to bring about changes they favor.

specialization   In evolution, the phenomenon whereby species become increasingly adapted to exploit one particular niche but thereby are less able to exploit other niches.

speciation   The evolutionary process whereby populations of a single species separate and, through being exposed to different forces of natural selection, gradually develop into distinct species.

species   All the organisms (plant, animal, or microbe) of a single kind. The "single kind" is determined by similarity of appearance or by the fact that members do or can mate and produce fertile offspring. Physical, chemical, or behavioral differences block breeding between species.

splash erosion   The compaction of soil that results when rainfall hits bare soil.

springs   Natural exits of groundwater to the surface.

stakeholders   Groups of people who have an interest in the management of a given environmental resource, from those dependent on the resource to those managing or studying the resource.

standing crop biomass   The biomass of primary producers in an ecosystem at any given time.

starvation   The failure to get enough calories to meet energy needs over a prolonged period of time. Starvation results in a wasting away of body tissues until death occurs.

Statement on Forest Principles   One of the treaties signed at the 1992 Earth Summit whereby the signatory countries agreed to a number of nonbinding principles stressing sustainable management.

sterile-male technique   Saturating an infested area with males of a pest species that have been artificially reared and sterilized by radiation. Matings between normal females and sterile males render the eggs infertile.

steward/stewardship   A steward is one to whom a trust has been given. As one of the unifying themes of this text, stewardship is *the actions and programs that manage natural resources and human well-being for the common good.*

Stockholm Convention on Persistent Organic Pollutants   An international treaty that works toward the banning of most uses of the persistent organic pollutants.

storm drains   Separate drainage systems used for collecting and draining runoff from precipitation in urban areas.

stratosphere   The layer of Earth's atmosphere between 10 and 40 miles above the surface that contains the ozone shield. This layer mixes only slowly; pollutants that enter it may remain for long periods of time. (See **troposphere**.)

strip mining   A mining procedure in which all the earth covering a desired material, such as coal, is stripped away with huge power shovels in order to facilitate removal of the material.

submerged aquatic vegetation (SAV)   Aquatic plants rooted in bottom sediments and growing underwater. Submerged aquatic vegetation depends on the penetration of light through the water for photosynthesis.

subsidy   A payment or other support of agricultural products that helps farmers to maintain their production and earn income above what the market would normally provide.

subsistence farming   Farming that meets the food needs of farmers and their families but little more. Subsistence farming involves hand labor and is practiced extensively in the developing world.

subsoil   In a natural situation, the soil beneath topsoil. In contrast to topsoil, subsoil is compacted and has little or no humus or other organic material, living or dead. In many areas, topsoil has been lost or destroyed as a result of erosion or development, and subsoil is at the surface.

succession   The gradual or sometimes rapid change in the species that occupy a given area, with some species invading and becoming more numerous, while others decline in population and disappear. Succession is caused by a change in one or more abiotic or biotic factors that benefit some species at the expense of others.

sulfur dioxide (SO_2)   A major air pollutant and toxic gas formed as a result of burning sulfur. The major sources are burning coal (in coal-burning power plants) that contains some sulfur and refining metal ores (in smelters) that contain sulfur.

sulfuric acid (H_2SO_4)   The major constituent of acid precipitation. Formed when sulfur dioxide emissions react with water vapor in the atmosphere. (See **sulfur dioxide**.)

Superfund   The popular name for the **Comprehensive Environmental Response, Compensation, and Liability Act of 1980**. This act is the cornerstone legislation that provides the mechanism and funding for the cleanup of potentially dangerous hazardous-waste sites and the protection of groundwater.

supply-side policies   Those energy policies that emphasize providing greater amounts of energy through more development of oil, coal and gas resources.

surface impoundments   Closed ponds used to collect and hold liquid chemical wastes.

surface water   All bodies of water, lakes, rivers, ponds, and so on that are on Earth's surface. Contrast with **groundwater**, which lies below the surface.

suspended particulate matter (SPM)   A category of major air pollutants consisting of solid and liquid particles suspended in the air. (See **$PM_{2.5}$** and **PM_{10}**.)

sustainability   A property whereby a process can be continued indefinitely without depleting the energy or material resources on which it depends. As one of the unifying themes of the text, *sustainability is the practical goal toward which our interactions with the natural world should be working.*

sustainable agriculture   Agriculture that maintains the integrity of soil and water resources such that it can be continued indefinitely.

Sustainable Cities Program (SCP)   A UN program designed to foster the planning and management needed to move cities in the developing countries toward sustainability.

Sustainable Communities Network   A grassroots organization that provides information, help, case studies, and linkages to encourage communities in the developed countries to move toward sustainability.

sustainable development   Development that provides people with a better life without sacrificing or depleting resources or causing environmental impacts that will undercut the ability of future generations to meet their needs.

Sustainable Fisheries Act   The 1996 reauthorization of the Magnuson Act for managing the coastal fisheries of the United States.

sustainable forest management   The management of forests as ecosystems wherein the primary objective is to maintain the biodiversity and function of the ecosystem.

sustainable society   A society that functions in a way so as not to deplete the energy or material resources on which it depends. Such a society interacts with the natural world in ways that maintain existing species and ecosystems.

sustainable yield   The taking of a biological resource (e.g., fish or forests) that does not exceed the capacity of the resource to reproduce and replace itself.

symbiosis   The intimate living together or association of two kinds of organisms.

synergism   The phenomenon whereby two factors acting together have a greater effect than would be indicated by the sum of their effects separately—as, for example, the sometimes fatal mixture of modest doses of certain drugs in combination with modest doses of alcohol.

synfuels, synthetic fuels   Fuels similar or identical to those that come from crude oil or natural gas. Synfuels are produced from coal, oil shale, or tar sands.

synthetic organics   Any of a large group of organic compounds that may be synthesized in chemical laboratories and are known by the difficulty with which they degrade in the environment.

tariff   Any tax placed on imported goods that raises the price of those goods in the recipient country.

tar sands   Sedimentary, bitumen-containing material that can be "melted out" using heat and then refined in the same way as crude oil.

taxonomy   The science of identifying and classifying organisms according to their presumed natural relationships.

temperature inversion   The weather phenomenon in which a layer of warm air overlies cooler air near the ground and prevents the rising and dispersion of air pollutants.

tenure   Property rights over land and water resources that involve different patterns of use of those resources.

teratogenic   Causing birth defects.

terracing   The practice of grading sloping farmland into a series of steps and cultivating only the level portions in order to reduce erosion.

territoriality   The behavioral characteristic exhibited by many animal species, especially birds and mammalian carnivores, to mark and defend a given territory against other members of the same species. A form of intraspecific competition.

TESRA (Threatened and Endangered Species Recovery Act)   One of a number of acts proposed in Congress in an attempt to weaken the Endangered Species Act.

theory   A conceptual formulation that provides a rational explanation or framework for numerous related observations.

thermal pollution   The addition of abnormal and undesirable amounts of heat to air or water. Thermal pollution is most significant with respect to the discharging of waste heat from electric generating plants—especially nuclear power plants—into bodies of water.

third world   A dated term for **developing countries**.

threatened species   A species whose population is declining precipitously because of direct or indirect human impacts.

threshold level   The maximum degree of exposure to a pollutant, drug, or some other factor that can be tolerated with no ill effect. The threshold level varies, depending on the species, the sensitivity of the individual, the length of exposure, and the presence of other factors that may produce synergistic effects.

tidal barrage   A form of renewable energy where barriers are placed across the mouth of a coastal bay and turbines are able to convert tidal water flow to power.

tidal wetlands   Areas of marsh grasses and reeds along coasts and estuaries where the ground is covered by high tides but drained at low tide.

tilth   In farming, the ability of a soil to support crop growth.

tipping point   A situation in a human-impacted ecosystem where a small action or reaction is the catalyst for a major change in the state of the system.

top-down regulation   Basic control of a population (or species) occurs as a result of predation.

topsoil   The surface layer of soil, which is rich in humus and other organic material, both living and dead. As a result of the activity of organisms living in the topsoil, it generally has a loose, crumbly structure, as opposed to being a compact mass.

total allowable catch (TAC)   In fisheries management, a yearly quota set for the harvest of a species by managers of fisheries.

total fertility rate   The average number of children that will be born alive to each woman during her total reproductive years if her fertility is average at each age.

Total Maximum Daily Load (TMDL) program   An EPA-administered program to address nonpoint-source water pollution that sets pollution limits according to the ability of a body of water to assimilate different pollutants.

total product life cycle   The sum of all steps in the manufacture of a product, from obtaining the raw materials through producing, using, and, finally, disposing of the product. By-products and the pollution resulting from each step are taken into account in a consideration of the total product life cycle.

toxicity   The condition of being harmful, deadly, or poisonous.

toxicology   The study of the impacts of toxic substances on human health and the pathways by which such substances reach humans.

Toxics Release Inventory   An annual record of releases of toxic chemicals to the environment and the locations and quantities of toxic chemicals stored at all U.S. sites. Required by the **Emergency Planning and Community Right-to-Know Act of 1986 (EPCRA)**. See **SARA**.

Toxic Substances Control Act of 1976 (TSCA)   A law requiring the assessment of the potential hazards of a chemical before the chemical is put on the market.

trace elements   Those essential elements, like copper or iron, that are needed in only very small amounts.

trade winds   The more or less constant winds blowing in horizontal directions over the surface as part of Hadley cells.

traditional agriculture   Farming methods as they were practiced before the advent of modern industrialized agriculture.

TRAFFIC network   An acronym for the Trade Records Analysis for Flora and Fauna in International Commerce network, monitoring trade in wildlife and wildlife parts.

tragedy of the commons   The overuse or overharvesting and consequent depletion or destruction of a renewable resource that tends to occur when the resource is treated as a commons—that is, when it is open to be used or harvested by any and all with the means to do so.

trait   Any physical or behavioral characteristic or talent that an individual is born with.

transgenic organism   Any organism with genes introduced from other species via biotechnology to convey new characteristics.

transpiration   The loss of water vapor from plants. Water evaporates from cells within the leaves and exits through the stomata.

trapping technique   The use of sex attractants to lure male insects into traps.

treadle pump   A pump worked like a step exercise machine that pulls water up from groundwater for irrigation.

treated sludge   Solid organic material that has been removed from sewage and treated so that it is nonhazardous.

trickling filter system   A system in which wastewater trickles over rocks or a framework coated with actively feeding microorganisms. The feeding action of the organisms in a well-aerated environment results in the decomposition of organic matter. Used in secondary or biological treatment of sewage.

tritium (^{3}H)   An unstable isotope of hydrogen that contains two neutrons in addition to the usual single proton in the nucleus. Tritium does not occur in significant amounts in nature.

trophic level   A feeding level defined with respect to the primary source of energy. Green plants are at the first trophic level, primary consumers at the second, secondary consumers at the third, and so on.

trophic structure   The major feeding relationships between organisms within ecosystems, organized into trophic levels.

troposphere/tropopause   The layer of Earth's atmosphere from the surface to about 10 miles in altitude. The tropopause is the boundary between the troposphere and the stratosphere above. The troposphere is well mixed and is the site and source of our weather as well as the primary recipient of air pollutants. (See also **stratosphere**.)

tsunami   A series of ocean waves (often called tidal waves) originating with underwater or coastal earthquakes and moving rapidly across ocean expanses, often resulting in major inundations of coastal areas.

turbid   Cloudy due to particles present. (Said of water and water purity.)

turbine   A rotary engine driven at a very high speed by steam, water, or exhaust gases from combustion, employed in generating electrical energy.

turbogenerator   A turbine coupled to and driving an electric generator. Virtually all commercial electricity is produced by such devices.

ultraviolet radiation   Radiation similar to light but with slightly shorter wavelengths and with more energy than violet light. Its greater energy causes ultraviolet light to severely burn and otherwise damage biological tissues.

underground storage tanks   Tanks used to store petroleum products at service stations, now subject to mandated protection against leakage. (See **UST legislation**.)

undernourishment   A form of hunger in which the individual lacks adequate food energy, as measured in calories. Starvation is the most severe form of undernourishment.

underweight   The world's number one health risk factor, this refers to the effects of undernourishment on children that prevent their normal growth.

United Nations Convention to Combat Desertification (UNCCD)   A UN agency with the mission of helping drylands countries to take steps to reduce soil degradation.

United Nations Development Program (UNDP)   A UN agency with offices in 131 countries, with the mission of helping developing countries to attract and use development aid effectively and, in particular, to reach the Millennium Development Goals.

United Nations Environment Program (UNEP)   A UN agency with the mission of providing leadership and encouraging partnership in caring for the environment by inspiring, informing, and enabling nations and peoples to improve their quality of life without compromising that of future generations.

unmet need   A situation where women who would like to postpone or prevent childbearing are not currently using contraception, often because it is not available to them.

upwelling   In oceanic systems, the upward vertical movement of water masses, caused by diverging currents and offshore winds, bringing nutrient-rich water to the surface.

urban blight/decay   The general deterioration of structures and facilities such as buildings and roadways, in addition to the decline in quality of services such as education, that has occurred in inner-city areas.

urban sprawl   The widespread expansion of metropolitan areas through building houses and shopping centers farther and farther from urban centers and lacing them together with more and more major highways, usually without planning.

USDA Organic   A food that is certified by the U.S. Department of Agriculture to have been produced without the use of chemical fertilizers, pesticides, or antibiotics.

UST legislation   Amendments to the Resources Conservation and Recovery Act of 1976 passed in 1984 to address the mounting problem of leaking underground storage tanks (USTs).

value   In ethical considerations, a property attributed to objects, species, or individuals that implies a moral duty to the one assigning the value; intrinsic value and instrumental value are two types of value.

vector/vector control   An agent, such as an insect or tick, that carries a parasite from one host to another; methods of controlling a parasite or disease by attacking the vector.

virtual water   In place of shipping water to water-stressed regions, food itself may be exported to those regions, thus preventing the direct need for water and replacing it with "virtual" water.

vitamin   A specific organic molecule that is required by the body in small amounts but that cannot be made by the body and therefore must be present in the diet.

volatile organic compounds (VOCs)   A category of major air pollutants present in the air in the vapor state. The category includes fragments of hydrocarbon fuels from incomplete combustion and evaporated organic compounds such as paint solvents, gasoline, and cleaning solutions. Volatile organic compounds are major factors in the formation of photochemical smog.

vulnerability   A measure of the likelihood of exposure to a hazard.

Waste Isolation Pilot Plant (WIPP)   A facility built by the U.S. Department of Energy in New Mexico to receive defense-related nuclear wastes.

waste-to-energy   A process of combustion of solid wastes that also generates electrical energy.

WasteWise   An EPA-sponsored program that targets the reduction of municipal solid waste by a number of partnerships.

water cycle   See **hydrologic cycle**.

water-holding capacity   The ability of a soil to hold water so that it will be available to plants.

waterlogging   The total saturation of soil with water. Waterlogging results in plant roots not being able to get air and dying as a result.

watershed   The total land area that drains directly or indirectly into a particular stream or river. The watershed is generally named from the stream or river into which it drains.

water table   The upper surface of groundwater, rising and falling with the amount of groundwater.

water vapor   Water molecules in the gaseous state.

weather   The day-to-day variations in temperature, air pressure, wind, humidity, and precipitation mediated by the atmosphere in a given region.

weathering   The gradual breakdown of rock into smaller and smaller particles, caused by natural chemical, physical, and biological factors.

weed   Any plant that competes with agricultural crops, forests, lawns, or forage grasses for light and nutrients.

wet cleaning   A water-based alternative to dry cleaning that avoids the use of hazardous chemicals.

wetlands   Areas that are constantly wet and are flooded at more or less regular intervals—especially marshy areas along coasts that are regularly flooded by tides.

wetland system   A biological aquatic system (usually a restored wetland) used to remove nutrients from treated sewage wastewater and return it virtually pure to a river or stream. Wetland systems are sometimes used when the application of treated wastewater for irrigation is not feasible.

Wild and Scenic Rivers Act of 1968   Legislation that protects rivers designated as "wild and scenic," preventing them from being dammed or otherwise harmfully impacted.

Wilderness Act of 1964   Federal legislation that provides for the permanent protection of undeveloped and unexploited areas so that natural ecological processes can operate freely in them.

Wildlife Services   An agency within the U.S. Department of Agriculture that removes "nuisance" animals for landowners; the agency dispatches more than 2 million animals and birds yearly.

wind farms   Arrays of numerous, modestly sized wind turbines for the purpose of producing electrical power.

windrows   Piles of organic material extended into long rows to facilitate turning and aeration in order to enhance composting.

wind turbines   Large "windmills" designed for the purpose of producing electrical power.

work   The change associated with the motion or state of matter. Any such change requires the expenditure of energy.

workability   The relative ease with which a soil can be cultivated.

World Bank   A branch of the United Nations that acts as a conduit for handling loans to developing countries.

World Food Program (WFP). A UN administration that receives donations of money and food from donor countries and distributes food aid to needy regions and countries.

World Trade Organization (WTO)   A body that sets the rules for international trade agreements, made up of representatives from many countries and often dominated by the rich nations.

worldview   A set of assumptions that a person holds regarding the world and how it works.

xeriscaping   Landscaping with drought-resistant plants that need no watering.

yard wastes   Grass clippings and other organic wastes from lawn and garden maintenance.

yellowcake   So-called because of its yellowish color, uranium ore that has been partially purified and is ready for further purification and enrichment.

Younger Dryas event   A rapid change in global climate between 10,000 and 12,000 years ago that brought on 1,500 years of colder weather.

Yucca Mountain   A remote location in Nevada chosen to be the nation's commercial nuclear waste repository; although well along in development, its future is uncertain.

zones of stress   Regions where a species finds conditions tolerable but suboptimal. The species survives but is under stress.

zooplankton   Any of a number of animal groups, including protozoa, crustaceans, worms, molluscs, and cnidarians, that live suspended in the water column and that feed on phytoplankton and other zooplankton.

zooxanthellae   Photosynthetic algae that live within the tissues of coral species and enable them to grow productively in warm, shallow coastal areas.

INDEX

Note: **Boldface** pages locate definitions.